Mechanical Engineers' Handbook

Mechanical Engineers' Handbook
Third Edition

Energy and Power

Edited by
Myer Kutz

WILEY

JOHN WILEY & SONS, INC.

Copyright © 2006 by John Wiley & Sons, Inc. All rights reserved.

Published by John Wiley & Sons, Inc., Hoboken, New Jersey.
Published simultaneously in Canada.

For general information on our other products and services, please contact our Customer Care Department within the United States at (800) 762-2974, outside the United States at (317) 572-3993 or fax (317) 572-4002.

Wiley also publishes its books in a variety of electronic formats. Some content that appears in print may not be available in electronic books. For more information about Wiley products, visit our web site at www.wiley.com.

Library of Congress Cataloging-in-Publication Data:
Mechanical engineers' handbook/edited by Myer Kutz.—3rd ed.
        p. cm.
    Includes bibliographical references and index.
    ISBN-13 978-0-471-44990-4
    ISBN-10 0-471-44990-3 (cloth)
  1. Mechanical engineering—Handbooks, manuals, etc.   I. Kutz, Myer.
    TJ151.M395 2005
    621—dc22

                                                                          2005008603

Printed in the United States of America.

10 9 8 7 6 5 4 3 2 1

To Arthur and Bess, Tony and Mary-Ann, for all the good times

Contents

PART 2   POWER   573

Preface

The fourth volume of the Third Edition of the *Mechanical Engineers' Handbook* comprises 32 chapters. The volume begins with a chapter on thermophysical properties of fluids, then covers fundamentals of fluid mechanics, thermodynamics (including a chapter on exergy and entropy generation minimization), heat transfer, combustion, and furnaces. Additional heat transfer topics in this volume include heat exchangers, heat pipes, air heating, and electronic equipment cooling. This volume includes chapters on both conventional energy sources—gaseous and liquid fuels, coal, and nuclear—and alternative energy sources—solar, geothermal, and fuel cells (in a chapter entirely new to the handbook). There are six chapters on power machinery: one on fans, blowers, compressors, and pumps; one each on gas, wind (in a chapter entirely new to the handbook), and steam turbines; one on internal combustion engines; and one on fluid power. Refrigeration and cryogenics are covered in two chapters. Four chapters deal with environmental issues: energy auditing, indoor environmental control, and air and water pollution control technologies. A chapter on thermal systems optimization rounds out this volume of the handbook.

The contributors to this volume include engineers working in industry in the United States and Canada, as well as in U.S. government agencies, and business owners, consultants, and academics from all around the United States. Three contributors, Reuben Olsen, Carroll Cone, and Leonard Wenzel, whose chapters first appeared in previous editions, are deceased. Their distinguished work survives.

Vision for the Third Edition

Basic engineering disciplines are not static, no matter how old and well established they are. The field of mechanical engineering is no exception. Movement within this broadly based discipline is multidimensional. Even the classic subjects on which the discipline was founded, such as mechanics of materials and heat transfer, continue to evolve. Mechanical engineers continue to be heavily involved with disciplines allied to mechanical engineering, such as industrial and manufacturing engineering, which are also constantly evolving. Advances in other major disciplines, such as electrical and electronics engineering, have significant impact on the work of mechanical engineers. New subject areas, such as neural networks, suddenly become all the rage.

In response to this exciting, dynamic atmosphere, the *Mechanical Engineers' Handbook* is expanding dramatically, from one volume to four volumes. The third edition not only is incorporating updates and revisions to chapters in the second edition, which was published in 1998, but also is adding 24 chapters on entirely new subjects as well, incorporating updates and revisions to chapters in the *Handbook of Materials Selection,* which was published in 2002, as well as to chapters in *Instrumentation and Control,* edited by Chester Nachtigal and published in 1990.

The four volumes of the third edition are arranged as follows:

Volume I: *Materials and Mechanical Design*—36 chapters
    Part 1. Materials—14 chapters
    Part 2. Mechanical Design—22 chapters

Volume II: *Instrumentation, Systems, Controls, and MEMS*—21 chapters
    Part 1. Instrumentation—8 chapters
    Part 2. Systems, Controls, and MEMS—13 chapters

Volume III: *Manufacturing and Management*—24 chapters
    Part 1. Manufacturing—12 chapters
    Part 2. Management, Finance, Quality, Law, and Research—12 chapters

Volume IV: *Energy and Power*—31 chapters
    Part 1: Energy—15 chapters
    Part 2: Power—16 chapters

The mechanical engineering literature is extensive and has been so for a considerable period of time. Many textbooks, reference works, and manuals as well as a substantial number of journals exist. Numerous commercial publishers and professional societies, particularly in the United States and Europe, distribute these materials. The literature grows continuously, as applied mechanical engineering research finds new ways of designing, controlling, measuring, making and maintaining things, and monitoring and evaluating technologies, infrastructures, and systems.

Most professional-level mechanical engineering publications tend to be specialized, directed to the specific needs of particular groups of practitioners. Overall, however, the mechanical engineering audience is broad and multidisciplinary. Practitioners work in a variety of organizations, including institutions of higher learning, design, manufacturing, and consulting firms as well as federal, state, and local government agencies. A rationale for an expanded general mechanical engineering handbook is that every practitioner, researcher, and bureaucrat cannot be an expert on every topic, especially in so broad and multidisciplinary a field, and may need an authoritative professional summary of a subject with which he or she is not intimately familiar.

Starting with the first edition, which was published in 1986, our intention has always been that the *Mechanical Engineers' Handbook* stand at the intersection of textbooks, research papers, and design manuals. For example, we want the handbook to help young engineers move from the college classroom to the professional office and laboratory where they may have to deal with issues and problems in areas they have not studied extensively in school.

With this expanded third edition, we have produced a practical reference for the mechanical engineer who is seeking to answer a question, solve a problem, reduce a cost, or improve a system or facility. The handbook is not a research monograph. The chapters offer design techniques, illustrate successful applications, or provide guidelines to improving the performance, the life expectancy, the effectiveness, or the usefulness of parts, assemblies, and systems. The purpose is to show readers what options are available in a particular situation and which option they might choose to solve problems at hand.

The aim of this expanded handbook is to serve as a source of practical advice to readers. We hope that the handbook will be the first information resource a practicing engineer consults when faced with a new problem or opportunity—even before turning to other print sources, even officially sanctioned ones, or to sites on the Internet. (The second edition has been available online on knovel.com.) In each chapter, the reader should feel that he or she is in the hands of an experienced consultant who is providing sensible advice that can lead to beneficial action and results.

Can a single handbook, even spread out over four volumes, cover this broad, interdisciplinary field? We have designed the third edition of the *Mechanical Engineers' Handbook* as if it were serving as a core for an Internet-based information source. Many chapters in the handbook point readers to information sources on the Web dealing with the subjects addressed. Furthermore, where appropriate, enough analytical techniques and data are provided to allow the reader to employ a preliminary approach to solving problems.

The contributors have written, to the extent their backgrounds and capabilities make possible, in a style that reflects practical discussion informed by real-world experience. We would like readers to feel that they are in the presence of experienced teachers and consultants who know about the multiplicity of technical issues that impinge on any topic within mechanical engineering. At the same time, the level is such that students and recent graduates can find the handbook as accessible as experienced engineers.

Contributors

Andrew Alleyne
University of Illinois, Urbana-Champaign
Urbana, Illinois

Avram Bar-Cohen
University of Maryland
College Park, Maryland

Adrian Bejan
Duke University
Durham, North Carolina

Peter D. Blair
National Academy of Sciences
Washington, DC

Carl Blumstein
University of California
Berkeley, California

Carl A. Brunner
U.S. Environmental Protection Agency

Carroll Cone (deceased)
Toledo, Ohio

Eric G. Eddings
University of Utah
Salt Lake City, Utah

James G. Keppeler
Progress Materials, Inc.
St. Petersburg, Florida

William Kerr
University of Michigan
Ann Arbor, Michigan

Allan Kraus
Beachwood, Ohio

Jan F. Kreider
Kreider and Associates, LLC
and
University of Colorado
Boulder, Colorado

J. F. Kreissl
Villa Hills, Kentucky

Peter Kuhn
Kuhn and Kuhn
Sausalito, California

Peter E. Liley
Purdue University
West Lafayette, Indiana

Hogbin Ma
University of Missouri
Columbia, Missouri

Keith Marchildon
Keith Marchildon Chemical Process
Design, Inc.
Kingston, Ontario, Canada

Ronald Douglas Matthews
The University of Texas at Austin
Austin, Texas

Matthew M. Mench
The Pennsylvania State University
University Park, Pennsylvania

C. A. Miller
U.S. Environmental Protection Agency
Research Triangle Park, North Carolina

Harold E. Miller
GE Energy
Schenectady, New York

David Mody
Fluor Canada
Kingston, Ontario, Canada

Todd S. Nemec
GE Energy
Schenectady, New York

Reuben M. Olson (deceased)
Ohio University
Athens, Ohio

Dennis L. O'Neal
Texas A&M University
College Station, Texas

Joseph W. Palen
Eugene, Oregon

G. P. Peterson
Rensselaer Polytechnic Institute
Troy, New York

Reinhard Radermacher
University of Maryland
College Park, Maryland

Richard J. Reed
North American Manufacturing Company
Cleveland, Ohio

Jelena Srebric
The Pennsylvania State University
University Park, Pennsylvania

William G. Steltz
Turboflow International, Inc.
Palm City, Florida

William Updegrove
Tucson, Arizona

Leonard A. Wenzel (deceased)
Lehigh University
Bethlehem, Pennsylvania

Mechanical Engineers' Handbook

PART 1
ENERGY

CHAPTER 1

THERMOPHYSICAL PROPERTIES OF FLUIDS

Peter E. Liley
School of Mechanical Engineering
Purdue University
West Lafayette, Indiana

In this chapter, information is usually presented in the System International des Unités, called in English the International System of Units and abbreviated SI. Various tables of conversion factors from other unit systems into the SI system and vice versa are available. The following table is intended to enable rapid conversion to be made with moderate, that is, five significant figure, accuracy, usually acceptable in most engineering calculations. The references listed should be consulted for more exact conversions and definitions.

Table 1 Conversion Factors

Density: 1 kg/m³ = 0.06243 lb_m/ft³ = 0.01002 lb_m/U.K. gallon = 8.3454 × 10⁻³ lb_m/U.S. gallon = 1.9403 × 10⁻³ slug/ft³ = 10⁻³ g/cm³

Energy: 1 kJ = 737.56 ft·lb_f = 239.01 cal_{th} = 0.94783 Btu = 3.7251 × 10⁻⁴ hp hr = 2.7778 × 10⁻⁴ kWhr

Specific energy: 1 kJ/kg = 334.54 ft·lb_f/lb_m = 0.4299 Btu/lb_m = 0.2388 cal/g

Specific energy per degree: 1 kJ/kg·K = 0.23901 Btu_{th}/lb·°F = 0.23901 cal_{th}/g·°C

Mass: 1 kg = 2.20462 lb_m = 0.06852 slug = 1.1023 × 10⁻³ U.S. ton = 10⁻³ tonne = 9.8421 × 10⁻⁴ U.K. ton

Pressure: 1 bar = 10⁵ N/m² = 10⁵ Pa = 750.06 mm Hg at 0°C = 401.47 in. H_2O at 32°F = 29.530 in. Hg at 0°C = 14.504 lb/in.² = 14.504 psia = 1.01972 kg/cm² = 0.98692 atm = 0.1 MPa

Temperature: T(K) = T(°C) + 273.15 = [T(°F) + 459.69]/1.8 = T(°R)/1.8

Temperature difference: ΔT(K) = ΔT(°C) = ΔT(°F)/1.8 = ΔT(°R)/1.8

Thermal conductivity: 1 W/m·K = 0.8604 kcal/m·hr·°C = 0.5782 Btu/ft·hr·°F = 0.01 W/cm·K = 2.390 × 10⁻³ cal/cm·s·°C

Thermal diffusivity: 1 m²/s = 38750 ft²/hr = 3600 m²/hr = 10.764 ft²/sec

Viscosity, dynamic: 1 N·s/m² = 1 Pa·s = 10⁷ μP = 2419.1 lb_m/ft·hr = 10³ cP = 75.188 slug/ft·hr = 10 P = 0.6720 lb_m/ft·sec = 0.02089 lb_f·sec/ft²

Viscosity, kinematic (*see* thermal diffusivity)

Source: E. Lange, L. F. Sokol, and V. Antoine, *Information on the Metric System and Related Fields,* 6th ed., G. C. Marshall Space Flight Center, AL (exhaustive bibliography); B. N. Taylor, *The International System of Units,* NBS S.P. 330, Washington, D.C., 2001; E. A. Mechtly, *The International System of Units. Physical Constants and Conversion Factors,* NASA S.P. 9012, 1973. Numerous revisions periodically appear: see, for example, *Pure Appl. Chem.,* **51,** 1–41 (1979) and later issues.

Table 2 Phase Transition Data for the Elements[a]

Name	Symbol	Formula Weight	T_m (K)	Δh_{fus} (kJ/kg)	T_b (K)	T_c (K)
Actinium	Ac	227.028	1323	63	3475	
Aluminum	Al	26.9815	933.5	398	2750	7850
Antimony	Sb	121.75	903.9	163	1905	5700
Argon	Ar	39.948	83	30	87.2	151
Arsenic	As	74.9216	885			2100
Barium	Ba	137.33	1002	55.8		4450
Beryllium	Be	9.01218	1560	1355	2750	6200
Bismuth	Bi	208.980	544.6	54.0	1838	4450
Boron	B	10.81	2320	1933	4000	3300
Bromine	Br	159.808	266	66.0	332	584
Cadmium	Cd	112.41	594	55.1	1040	2690
Calcium	Ca	40.08	1112	213.1	1763	4300
Carbon	C	12.011	3810		4275	7200
Cerium	Ce	140.12	1072	390		9750
Cesium	Cs	132.905	301.8	16.4	951	2015
Chlorine	Cl_2	70.906	172	180.7	239	417

Table 2   (*Continued*)

Name	Symbol	Formula Weight	T_m (K)	Δh_{fus} (kJ/kg)	T_b (K)	T_c (K)
Chromium	Cr	51.996	2133	325.6	2950	5500
Cobalt	Co	58.9332	1766	274.7	3185	6300
Copper	Cu	63.546	1357	206.8	2845	8280
Dysprosium	Dy	162.50	1670	68.1	2855	6925
Erbium	Er	167.26	1795	119.1	3135	7250
Europium	Eu	151.96	1092	60.6	1850	4350
Fluorine	F$_2$	37.997	53.5	13.4	85.0	144
Gadolinum	Gd	157.25	1585	63.8	3540	8670
Gallium	Ga	69.72	303	80.1	2500	7125
Germanium	Ge	72.59	1211	508.9	3110	8900
Gold	Au	196.967	1337	62.8	3130	7250
Hafnium	Hf	178.49	2485	134.8	4885	10400
Helium	He	4.00260	3.5	2.1	4.22	5.2
Holmium	Ho	164.930	1744	73.8	2968	7575
Hydrogen	H$_2$	2.0159	14.0		20.4	
Indium	In	114.82	430	28.5	2346	6150
Iodine	I$_2$	253.809	387	125.0	457	785
Iridium	Ir	192.22	2718	13.7	4740	7800
Iron	Fe	55.847	1811	247.3	3136	8500
Krypton	Kr	83.80	115.8	19.6	119.8	209.4
Lanthanum	La	138.906	1194	44.6	3715	10500
Lead	Pb	207.2	601	23.2	2025	5500
Lithium	Li	6.941	454	432.2	1607	3700
Lutetium	Lu	174.967	1937	106.6	3668	
Magnesium	Mg	24.305	922	368.4	1364	3850
Manganese	Mn	54.9380	1518	219.3	2334	4325
Mercury	Hg	200.59	234.6	11.4	630	1720
Molybdenum	Mo	95.94	2892	290.0	4900	1450
Neodymium	Nd	144.24	1290	49.6	3341	7900
Neon	Ne	20.179	24.5	16.4	27.1	44.5
Neptunium	Np	237.048	910		4160	12000
Nickel	Ni	58.70	1728	297.6	3190	8000
Niobium	Nb	92.9064	2740	283.7	5020	12500
Nitrogen	N$_2$	28.013	63.2	25.7	77.3	126.2
Osmium	Os	190.2	3310	150.0	5300	12700
Oxygen	O$_2$	31.9988	54.4	13.8	90.2	154.8
Palladium	Pd	106.4	1826	165.0	3240	7700
Phosphorus	P	30.9738	317		553	995
Platinum	Pt	195.09	2045	101	4100	10700
Plutonium	Pu	244	913	11.7	3505	10500
Potassium	K	39.0983	336.4	60.1	1032	2210
Praseodymium	Pr	140.908	1205	49	3785	8900
Promethium	Pm	145	1353		2730	

Table 2   (*Continued*)

Name	Symbol	Formula Weight	T_m (K)	Δh_{fus} (kJ/kg)	T_b (K)	T_c (K)
Protactinium	Pa	231	1500	64.8	4300	
Radium	Ra	226.025	973		1900	
Radon	Rn	222	202	12.3	211	377
Rhenium	Re	186.207	3453	177.8	5920	18900
Rhodium	Rh	102.906	2236	209.4	3980	7000
Rubidium	Rb	85.4678	312.6	26.4	964	2070
Ruthenium	Ru	101.07	2525	256.3	4430	9600
Samarium	Sm	150.4	1345	57.3	2064	5050
Scandium	Sc	44.9559	1813	313.6	3550	6410
Selenium	Se	78.96	494	66.2	958	1810
Silicon	Si	28.0855	1684	1802	3540	5160
Silver	Ag	107.868	1234	104.8	2435	6400
Sodium	Na	22.9898	371	113.1	1155	2500
Strontium	Sr	87.62	1043	1042	1650	4275
Sulfur	S	32.06	388	53.4	718	1210
Tantalum	Ta	180.948	3252	173.5	5640	16500
Technetium	Tc	98	2447	232	4550	11500
Tellurium	Te	127.60	723	137.1	1261	2330
Terbium	Tb	158.925	1631	67.9	3500	8470
Thallium	Tl	204.37	577	20.1	1745	4550
Thorium	Th	232.038	2028	69.4	5067	14400
Thulium	Tm	168.934	1819	99.6	2220	6450
Tin	Sn	118.69	505	58.9	2890	7700
Titanium	Ti	47.90	1943	323.6	3565	5850
Tungsten	W	183.85	3660	192.5	5890	15500
Uranium	U	238.029	1406	35.8	4422	12500
Vanadium	V	50.9415	2191	410.7	3680	11300
Xenon	Xe	131.30	161.3	17.5	164.9	290
Ytterbium	Yb	173.04	1098	44.2	1467	4080
Yttrium	Y	88.9059	1775	128.2	3610	8950
Zinc	Zn	65.38	692.7	113.0	1182	
Zirconium	Zr	91.22	2125	185.3	4681	10500

$^a T_m$ = normal melting point; Δh_{fus} = enthalpy of fusion; T_b = normal boiling point; T_c = critical temperature.

Table 3   Phase Transition Data for Compounds[a]

Substance	T_m (K)	Δh_m (kJ/kg)	T_b (K)	Δh_v (kJ/kg)	T_c (K)	P_c (bar)
Acetaldehyde	149.7	73.2	293.4	584	461	55.4
Acetic acid	289.9	195.3	391.7	405	594	57.9
Acetone	178.6	98	329.5	501	508	47
Acetylene		96.4	189.2	687	309	61.3
Air	60				133	37.7
Ammonia	195.4	331.9	239.7	1368	405.6	112.8
Aniline	267.2	113.3	457.6	485	699	53.1
Benzene	267.7	125.9	353.3	394	562	49
n-Butane	134.8	80.2	261.5	366	425.2	38
Butanol	188	125.2	391.2	593	563	44.1
Carbon dioxide	216.6	184	194.7	573	304.2	73.8
Carbon disulfide	161	57.7	319.6	352	552	79
Carbon monoxide	68.1	29.8	81.6	215	133	35
Carbon tetrachloride	250.3	173.9	349.9	195	556	45.6
Carbon tetrafluoride	89.5		145.2	138	227.9	37.4
Chlorobenzene	228		405	325	632.4	45.2
Chloroform	210	77.1	334.4	249	536.6	54.7
m-Cresol	285.1		475.9	421	705	45.5
Cyclohexane	279.6	31.7	356	357	554.2	40.7
Cyclopropane	145.5	129.4	240.3	477	397.8	54.9
n-Decane	243.2	202.1	447.3	276	617	21
Ethane	89.9	94.3	184.6	488	305.4	48.8
Ethanol	158.6	109	351.5	840	516	63.8
Ethyl acetate	190.8	119	350.2	366	523.3	38.3
Ethylene	104	119.5	169.5	480	283.1	51.2
Ethylene oxide	161.5	117.6	283.9	580	469	71.9
Formic acid	281.4	246.4	373.9	502	576	34.6
Heptane	182.6	140.2	371.6	316	540	27.4
Hexane	177.8	151.2	341.9	335	507	29.7
Hydrazine	274.7	395	386.7	1207	653	147
Hydrogen peroxide	271.2	310	431	1263		
Isobutane	113.6	78.1	272.7	386	408.1	36.5
Methane	90.7	58.7	111.5	512	191.1	46.4
Methanol	175.4	99.2	337.7	1104	513.2	79.5
Methyl acetate	174.8		330.2	410	507	46.9
Methyl bromide	180	62.9	277.7	252	464	71.2
Methyl chloride	178.5	127.4	249.3	429	416.3	66.8
Methyl formate	173.4	125.5	305	481	487.2	60
Methylene chloride	176.4	54.4	312.7	328	510.2	60.8
Naphthalene	353.2	148.1	491	341	747	39.5
Nitric oxide	111	76.6	121.4	460	180.3	65.5
Octane	216.4	180.6	398.9	303	569.4	25
Pentane	143.7	116.6	309.2	357	469.8	33.7

Table 3 (*Continued*)

Substance	T_m (K)	Δh_m (kJ/kg)	T_b (K)	Δh_v (kJ/kg)	T_c (K)	P_c (bar)
Propane	86	80	231.1	426	370	42.5
Propanol	147	86.5	370.4	696	537	51.7
Propylene	87.9	71.4	225.5	438	365	46.2
Refrigerant 12	115	34.3	243.4	165	385	41.2
Refrigerant 13	92		191.8	148	302.1	38.7
Refrigerant 13B1	105.4		215.4	119	340	39.6
Refrigerant 21	138		282.1	242	451.7	51.7
Refrigerant 22	113	47.6	232.4	234	369	49.8
Steam/water	273.2	334	373.2	2257	647.3	221.2
Sulfuric acid	283.7	100.7	v	v	v	v
Sulfur dioxide	197.7	115.5	268.4	386	430.7	78.8
Toluene	178.2		383.8	339	594	41

$^a v$ = variable; T_m = normal melting point; Δh_m = enthalpy of fusion; T_b = normal boiling point; Δh_v = enthalpy of vaporization; T_c = critical temperature; P_c = critical pressure.

Table 4 Thermodynamic Properties of Liquid and Saturated Vapor Air^a

T (K)	P_f (MPa)	P_g (MPa)	v_f (m³/kg)	v_g (m³/kg)	h_f (kJ/kg)	h_g (kJ/kg)	s_f (kJ/kg·K)	s_g (kJ/kg·K)
60	0.0066	0.0025	0.001027	6.876	−144.9	59.7	2.726	6.315
65	0.0159	0.0077	0.001078	2.415	−144.8	64.5	2.727	6.070
70	0.0340	0.0195	0.001103	1.021	−144.0	69.1	2.798	5.875
75	0.0658	0.0424	0.001130	0.4966	−132.8	73.5	2.896	5.714
80	0.1172	0.0826	0.001160	0.2685	−124.4	77.5	3.004	5.580
85	0.1954	0.1469	0.001193	0.1574	−115.2	81.0	3.114	5.464
90	0.3079	0.2434	0.001229	0.0983	−105.5	84.1	3.223	5.363
95	0.4629	0.3805	0.001269	0.0644	−95.4	86.5	3.331	5.272
100	0.6687	0.5675	0.001315	0.0439	−84.8	88.2	3.436	5.189
105	0.9334	0.8139	0.001368	0.0308	−73.8	89.1	3.540	5.110
110	1.2651	1.1293	0.001432	0.0220	−62.2	89.1	3.644	5.033
115	1.6714	1.5238	0.001512	0.0160	−49.7	87.7	3.750	4.955
120	2.1596	2.0081	0.001619	0.0116	−35.9	84.5	3.861	4.872
125	2.743	2.614	0.001760	0.0081	−19.7	78.0	3.985	4.770
132.5^b	3.770	3.770	0.00309	0.0031	33.6	33.6	4.38	4.38

$^a v$ = specific volume; h = specific enthalpy; s = specific entropy; f = saturated liquid; g = saturated vapor. 1 MPa = 10 bar.
^{b}Approximate critical point. Air is a multicomponent mixture.

Table 5   Ideal Gas Thermophysical Properties of Air[a]

T (K)	v (m³/kg)	h (kJ/kg)	s (kJ/kg·K)	c_p (kJ/kg·K)	γ	$\bar{v}_s$ (m/s)	η (N·s/m²)	λ (W/m·K)	Pr
200	0.5666	−103.0	6.4591	1.008	1.398	283.3	1.33.−5[b]	0.0183	0.734
210	0.5949	−92.9	6.5082	1.007	1.399	290.4	1.39.−5	0.0191	0.732
220	0.6232	−82.8	6.5550	1.006	1.399	297.3	1.44.−5	0.0199	0.730
230	0.6516	−72.8	6.5998	1.006	1.400	304.0	1.50.−5	0.0207	0.728
240	0.6799	−62.7	6.6425	1.005	1.400	310.5	1.55.−5	0.0215	0.726
250	0.7082	−52.7	6.6836	1.005	1.400	317.0	1.60.−5	0.0222	0.725
260	0.7366	−42.6	6.7230	1.005	1.400	323.3	1.65.−5	0.0230	0.723
270	0.7649	−32.6	6.7609	1.004	1.400	329.4	1.70.−5	0.0237	0.722
280	0.7932	−22.5	6.7974	1.004	1.400	335.5	1.75.−5	0.0245	0.721
290	0.8216	−12.5	6.8326	1.005	1.400	341.4	1.80.−5	0.0252	0.720
300	0.8499	−2.4	6.8667	1.005	1.400	347.2	1.85.−5	0.0259	0.719
310	0.8782	7.6	6.8997	1.005	1.400	352.9	1.90.−5	0.0265	0.719
320	0.9065	17.7	6.9316	1.006	1.399	358.5	1.94.−5	0.0272	0.719
330	0.9348	27.7	6.9625	1.006	1.399	364.0	1.99.−5	0.0279	0.719
340	0.9632	37.8	6.9926	1.007	1.399	369.5	2.04.−5	0.0285	0.719
350	0.9916	47.9	7.0218	1.008	1.398	374.8	2.08.−5	0.0292	0.719
360	1.0199	57.9	7.0502	1.009	1.398	380.0	2.12.−5	0.0298	0.719
370	1.0482	68.0	7.0778	1.010	1.397	385.2	2.17.−5	0.0304	0.719
380	1.0765	78.1	7.1048	1.011	1.397	390.3	2.21.−5	0.0311	0.719
390	1.1049	88.3	7.1311	1.012	1.396	395.3	2.25.−5	0.0317	0.719
400	1.1332	98.4	7.1567	1.013	1.395	400.3	2.29.−5	0.0323	0.719
410	1.1615	108.5	7.1817	1.015	1.395	405.1	2.34.−5	0.0330	0.719
420	1.1898	118.7	7.2062	1.016	1.394	409.9	2.38.−5	0.0336	0.719
430	1.2181	128.8	7.2301	1.018	1.393	414.6	2.42.−5	0.0342	0.718
440	1.2465	139.0	7.2535	1.019	1.392	419.3	2.46.−5	0.0348	0.718
450	1.2748	149.2	7.2765	1.021	1.391	423.9	2.50.−5	0.0355	0.718
460	1.3032	159.4	7.2989	1.022	1.390	428.3	2.53.−5	0.0361	0.718
470	1.3315	169.7	7.3209	1.024	1.389	433.0	2.57.−5	0.0367	0.718
480	1.3598	179.9	7.3425	1.026	1.389	437.4	2.61.−5	0.0373	0.718
490	1.3882	190.2	7.3637	1.028	1.388	441.8	2.65.−5	0.0379	0.718
500	1.4165	200.5	7.3845	1.030	1.387	446.1	2.69.−5	0.0385	0.718
520	1.473	221.1	7.4249	1.034	1.385	454.6	2.76.−5	0.0398	0.718
540	1.530	241.8	7.4640	1.038	1.382	462.9	2.83.−5	0.0410	0.718
560	1.586	262.6	7.5018	1.042	1.380	471.0	2.91.−5	0.0422	0.718
580	1.643	283.5	7.5385	1.047	1.378	479.0	2.98.−5	0.0434	0.718
600	1.700	304.5	7.5740	1.051	1.376	486.8	3.04.−5	0.0446	0.718
620	1.756	325.6	7.6086	1.056	1.374	494.4	3.11.−5	0.0458	0.718
640	1.813	346.7	7.6422	1.060	1.371	501.9	3.18.−5	0.0470	0.718
660	1.870	368.0	7.6749	1.065	1.369	509.3	3.25.−5	0.0482	0.717
680	1.926	389.3	7.7067	1.070	1.367	516.5	3.32.−5	0.0495	0.717
700	1.983	410.8	7.7378	1.075	1.364	523.6	3.38.−5	0.0507	0.717
720	2.040	432.3	7.7682	1.080	1.362	530.6	3.45.−5	0.0519	0.716
740	2.096	453.9	7.7978	1.084	1.360	537.5	3.51.−5	0.0531	0.716
760	2.153	475.7	7.8268	1.089	1.358	544.3	3.57.−5	0.0544	0.716
780	2.210	497.5	7.8551	1.094	1.356	551.0	3.64.−5	0.0556	0.716
800	2.266	519.4	7.8829	1.099	1.354	557.6	3.70.−5	0.0568	0.716

Table 5 (*Continued*)

T (K)	v (m³/kg)	h (kJ/kg)	s (kJ/kg·K)	c_p (kJ/kg·K)	γ	$\bar{v}_s$ (m/s)	η (N·s/m²)	λ (W/m·K)	Pr
820	2.323	541.5	7.9101	1.103	1.352	564.1	3.76.−5	0.0580	0.715
840	2.380	563.6	7.9367	1.108	1.350	570.6	3.82.−5	0.0592	0.715
860	2.436	585.8	7.9628	1.112	1.348	576.8	3.88.−5	0.0603	0.715
880	2.493	608.1	7.9885	1.117	1.346	583.1	3.94.−5	0.0615	0.715
900	2.550	630.4	8.0136	1.121	1.344	589.3	4.00.−5	0.0627	0.715
920	2.606	652.9	8.0383	1.125	1.342	595.4	4.05.−5	0.0639	0.715
940	2.663	675.5	8.0625	1.129	1.341	601.5	4.11.−5	0.0650	0.714
960	2.720	698.1	8.0864	1.133	1.339	607.5	4.17.−5	0.0662	0.714
980	2.776	720.8	8.1098	1.137	1.338	613.4	4.23.−5	0.0673	0.714
1000	2.833	743.6	8.1328	1.141	1.336	619.3	4.28.−5	0.0684	0.714
1050	2.975	800.8	8.1887	1.150	1.333	633.8	4.42.−5	0.0711	0.714
1100	3.116	858.5	8.2423	1.158	1.330	648.0	4.55.−5	0.0738	0.715
1150	3.258	916.6	8.2939	1.165	1.327	661.8	4.68.−5	0.0764	0.715
1200	3.400	975.0	8.3437	1.173	1.324	675.4	4.81.−5	0.0789	0.715
1250	3.541	1033.8	8.3917	1.180	1.322	688.6	4.94.−5	0.0814	0.716
1300	3.683	1093.0	8.4381	1.186	1.319	701.6	5.06.−5	0.0839	0.716
1350	3.825	1152.3	8.4830	1.193	1.317	714.4	5.19.−5	0.0863	0.717
1400	3.966	1212.2	8.5265	1.199	1.315	726.9	5.31.−5	0.0887	0.717
1450	4.108	1272.3	8.5686	1.204	1.313	739.2	5.42.−5	0.0911	0.717
1500	4.249	1332.7	8.6096	1.210	1.311	751.3	5.54.−5	0.0934	0.718
1550	4.391	1393.3	8.6493	1.215	1.309	763.2	5.66.−5	0.0958	0.718
1600	4.533	1454.2	8.6880	1.220	1.308	775.0	5.77.−5	0.0981	0.717
1650	4.674	1515.3	8.7256	1.225	1.306	786.5	5.88.−5	0.1004	0.717
1700	4.816	1576.7	8.7622	1.229	1.305	797.9	5.99.−5	0.1027	0.717
1750	4.958	1638.2	8.7979	1.233	1.303	809.1	6.10.−5	0.1050	0.717
1800	5.099	1700.0	8.8327	1.237	1.302	820.2	6.21.−5	0.1072	0.717
1850	5.241	1762.0	8.8667	1.241	1.301	831.1	6.32.−5	0.1094	0.717
1900	5.383	1824.1	8.8998	1.245	1.300	841.9	6.43.−5	0.1116	0.717
1950	5.524	1886.4	8.9322	1.248	1.299	852.6	6.53.−5	0.1138	0.717
2000	5.666	1948.9	8.9638	1.252	1.298	863.1	6.64.−5	0.1159	0.717
2050	5.808	2011.6	8.9948	1.255	1.297	873.5	6.74.−5	0.1180	0.717
2100	5.949	2074.4	9.0251	1.258	1.296	883.8	6.84.−5	0.1200	0.717
2150	6.091	2137.3	9.0547	1.260	1.295	894.0	6.95.−5	0.1220	0.717
2200	6.232	2200.4	9.0837	1.263	1.294	904.0	7.05.−5	0.1240	0.718
2250	6.374	2263.6	9.1121	1.265	1.293	914.0	7.15.−5	0.1260	0.718
2300	6.516	2327.0	9.1399	1.268	1.293	923.8	7.25.−5	0.1279	0.718
2350	6.657	2390.5	9.1672	1.270	1.292	933.5	7.35.−5	0.1298	0.719
2400	6.800	2454.0	9.1940	1.273	1.291	943.2	7.44.−5	0.1317	0.719
2450	6.940	2517.7	9.2203	1.275	1.291	952.7	7.54.−5	0.1336	0.720
2500	7.082	2581.5	9.2460	1.277	1.290	962.2	7.64.−5	0.1354	0.720

[a]v = specific volume; h = specific enthalpy; s = specific entropy; c_p = specific heat at constant pressure; γ = specific heat ratio, c_p/c_v (dimensionless); $\bar{v}_s$ = velocity of sound; η = dynamic viscosity; λ = thermal conductivity; Pr = Prandtl number (dimensionless). Condensed from S. Gordon, *Thermodynamic and Transport Combustion Properties of Hydrocarbons with Air*, NASA Technical Paper 1906, 1982, Vol. 1. These properties are based on constant gaseous composition. The reader is reminded that at the higher temperatures the influence of pressure can affect the composition and the thermodynamic properties.

[b]The notation 1.33.−5 signifies 1.33×10^{-5}.

Table 6   Thermophysical Properties of the U.S. Standard Atmosphere[a]

Z (m)	H (m)	T (K)	P (bar)	ρ (kg/m^3)	g (m/s^2)	$\bar{v}_s$ (m/s)
0	0	288.15	1.0133	1.2250	9.8067	340.3
1000	1000	281.65	0.8988	1.1117	9.8036	336.4
2000	1999	275.15	0.7950	1.0066	9.8005	332.5
3000	2999	268.66	0.7012	0.9093	9.7974	328.6
4000	3997	262.17	0.6166	0.8194	9.7943	324.6
5000	4996	255.68	0.5405	0.7364	9.7912	320.6
6000	5994	249.19	0.4722	0.6601	9.7882	316.5
7000	6992	242.70	0.4111	0.5900	9.7851	312.3
8000	7990	236.22	0.3565	0.5258	9.7820	308.1
9000	8987	229.73	0.3080	0.4671	9.7789	303.9
10000	9984	223.25	0.2650	0.4135	9.7759	299.5
11000	10981	216.77	0.2270	0.3648	9.7728	295.2
12000	11977	216.65	0.1940	0.3119	9.7697	295.1
13000	12973	216.65	0.1658	0.2667	9.7667	295.1
14000	13969	216.65	0.1417	0.2279	9.7636	295.1
15000	14965	216.65	0.1211	0.1948	9.7605	295.1
16000	15960	216.65	0.1035	0.1665	9.7575	295.1
17000	16954	216.65	0.0885	0.1423	9.7544	295.1
18000	17949	216.65	0.0756	0.1217	9.7513	295.1
19000	18943	216.65	0.0647	0.1040	9.7483	295.1
20000	19937	216.65	0.0553	0.0889	9.7452	295.1
22000	21924	218.57	0.0405	0.0645	9.7391	296.4
24000	23910	220.56	0.0297	0.0469	9.7330	297.7
26000	25894	222.54	0.0219	0.0343	9.7269	299.1
28000	27877	224.53	0.0162	0.0251	9.7208	300.4
30000	29859	226.51	0.0120	0.0184	9.7147	301.7
32000	31840	228.49	0.00889	0.01356	9.7087	303.0
34000	33819	233.74	0.00663	0.00989	9.7026	306.5
36000	35797	239.28	0.00499	0.00726	9.6965	310.1
38000	37774	244.82	0.00377	0.00537	9.6904	313.7
40000	39750	250.35	0.00287	0.00400	9.6844	317.2
42000	41724	255.88	0.00220	0.00299	9.6783	320.7
44000	43698	261.40	0.00169	0.00259	9.6723	324.1
46000	45669	266.93	0.00131	0.00171	9.6662	327.5
48000	47640	270.65	0.00102	0.00132	9.6602	329.8
50000	49610	270.65	0.00080	0.00103	9.6542	329.8

[a]Z = geometric attitude; H = geopotential attitude; ρ = density; g = acceleration of gravity; $\bar{v}_s$ = velocity of sound. Condensed and in some cases converted from *U.S. Standard Atmosphere 1976*, National Oceanic and Atmospheric Administration and National Aeronautics and Space Administration, Washington, DC. Also available as NOAA-S/T 76-1562 and Government Printing Office Stock No. 003-017-00323-0.

Table 7 Thermophysical Properties of Condensed and Saturated Vapor Carbon Dioxide from 200 K to the Critical Point[a]

T (K)	P (bar)	Specific Volume		Specific Enthalpy		Specific Entropy		Specific Heat (c_p)		Thermal Conductivity		Viscosity		Prandtl Number	
		Condensed[b]	Vapor	Condensed[b]	Vapor	Condensed[b]	Vapor	Condensed[b]	Vapor	Liquid	Vapor	Liquid	Vapor	Liquid	Vapor
200	1.544	0.000644	0.2362	164.8	728.3	1.620	4.439								
205	2.277	0.000649	0.1622	171.5	730.0	1.652	4.379								
210	3.280	0.000654	0.1135	178.2	730.9	1.682	4.319								
215	4.658	0.000659	0.0804	185.0	731.3	1.721	4.264								
216.6	5.180	0.000661	0.0718	187.2	731.5	1.736	4.250								
216.6	5.180	0.000848	0.0718	386.3	731.5	2.656	4.250	1.707	0.958	0.182	0.011	2.10	0.116	1.96	0.96
220	5.996	0.000857	0.0624	392.6	733.1	2.684	4.232	1.761	0.985	0.178	0.012	1.86	0.118	1.93	0.97
225	7.357	0.000871	0.0515	401.8	735.1	2.723	4.204	1.820	1.02	0.171	0.012	1.75	0.120	1.87	0.98
230	8.935	0.000886	0.0428	411.1	736.7	2.763	4.178	1.879	1.06	0.164	0.013	1.64	0.122	1.84	0.99
235	10.75	0.000901	0.0357	420.5	737.9	2.802	4.152	1.906	1.10	0.160	0.013	1.54	0.125	1.82	1.01
240	12.83	0.000918	0.0300	430.2	738.9	2.842	4.128	1.933	1.15	0.156	0.014	1.45	0.128	1.80	1.02
245	15.19	0.000936	0.0253	440.1	739.4	2.882	4.103	1.959	1.20	0.148	0.015	1.36	0.131	1.80	1.04
250	17.86	0.000955	0.0214	450.3	739.6	2.923	4.079	1.992	1.26	0.140	0.016	1.28	0.134	1.82	1.06
255	20.85	0.000977	0.0182	460.8	739.4	2.964	4.056	2.038	1.34	0.134	0.017	1.21	0.137	1.84	1.08
260	24.19	0.001000	0.0155	471.6	738.7	3.005	4.032	2.125	1.43	0.128	0.018	1.14	0.140	1.89	1.12
265	27.89	0.001026	0.0132	482.8	737.4	3.047	4.007	2.237	1.54	0.122	0.019	1.08	0.144	1.98	1.17
270	32.03	0.001056	0.0113	494.4	735.6	3.089	3.981	2.410	1.66	0.116	0.020	1.02	0.150	2.12	1.23
275	36.59	0.001091	0.0097	506.5	732.8	3.132	3.954	2.634	1.81	0.109	0.022	0.96	0.157	2.32	1.32
280	41.60	0.001130	0.0082	519.2	729.1	3.176	3.925	2.887	2.06	0.102	0.024	0.91	0.167	2.57	1.44
285	47.10	0.001176	0.0070	532.7	723.5	3.220	3.891	3.203	2.40	0.095	0.028	0.86	0.178	2.90	1.56
290	53.15	0.001241	0.0058	547.6	716.9	3.271	3.854	3.724	2.90	0.088	0.033	0.79	0.191	3.35	1.68
295	59.83	0.001322	0.0047	562.9	706.3	3.317	3.803	4.68		0.081	0.042	0.71	0.207	4.1	1.8
300	67.10	0.001470	0.0037	585.4	690.2	3.393	3.742			0.074	0.065	0.60	0.226		
304.2[c]	73.83	0.002145	0.0021	636.6	636.6	3.558	3.558								

[a]Specific volume, m³/kg; specific enthalpy, kJ/kg; specific entropy, kJ/kg·K; specific heat at constant pressure, kJ/kg·K; thermal conductivity, W/m·K; viscosity, 10^{-4} Pa·s. Thus, at 250 K the viscosity of the saturated liquid is 1.28×10^{-4} N·s/m² = 0.000128 N·s/m² = 0.000128 Pa·s. The Prandtl number is dimensionless.
[b]Above the solid line the condensed phase is solid; below the line, it is liquid.
[c]Critical point.

13

Table 8 Thermophysical Properties of Gaseous Carbon Dioxide at 1 Bar Pressure[a]

								T (K)							
	300	350	400	450	500	550	600	650	700	750	800	850	900	950	1000
v (m³/kg)	0.5639	0.6595	0.7543	0.8494	0.9439	1.039	1.133	1.228	1.332	1.417	1.512	1.606	1.701	1.795	1.889
h (kJ/kg)	809.3	853.1	899.1	947.1	997.0	1049	1102	1156	1212	1269	1327	1386	1445	1506	1567
s (kJ/kg·K)	4.860	4.996	5.118	5.231	5.337	5.435	5.527	5.615	5.697	5.775	5.850	5.922	5.990	6.055	6.120
c_p (kJ/kg·K)	0.852	0.898	0.941	0.980	1.014	1.046	1.075	1.102	1.126	1.148	1.168	1.187	1.205	1.220	1.234
λ (W/m·K)	0.0166	0.0204	0.0243	0.0283	0.0325	0.0364	0.0407	0.0445	0.0481	0.0517	0.0551	0.0585	0.0618	0.0650	0.0682
μ (10^{-4} Pa·s)	0.151	0.175	0.198	0.220	0.242	0.261	0.281	0.299	0.317	0.334	0.350	0.366	0.381	0.396	0.410
Pr	0.778	0.770	0.767	0.762	0.755	0.750	0.742	0.742	0.742	0.742	0.742	0.742	0.742	0.743	0.743

[a]v = specific volume; h = enthalpy; s = entropy; c_p = specific heat at constant pressure; λ = thermal conductivity; η = viscosity (at 300 K the gas viscosity is 0.0000151 N·s/m² = 0.0000151 Pa·s); Pr = Prandtl number.

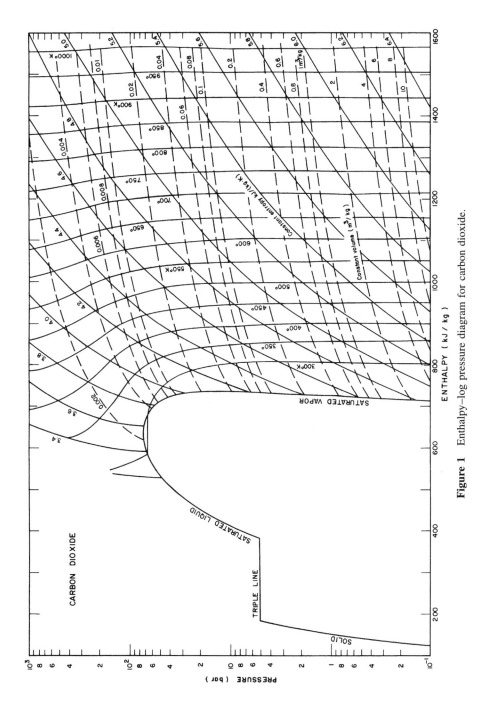

Figure 1 Enthalpy–log pressure diagram for carbon dioxide.

15

Table 9   Thermodynamic Properties of Saturated Mercury[a]

T (K)	P (bar)	v (m³/kg)	h (kJ/kg)	s (kJ/kg·K)	c_p (kJ/kg·K)
0		6.873.−5[b]	0	0	0
20		6.875.−5	0.466	0.0380	0.0513
40		6.884.−5	1.918	0.0868	0.0894
60		6.897.−5	3.897	0.1267	0.1067
80		6.911.−5	6.129	0.1588	0.1156
100		6.926.−5	8.497	0.1852	0.1209
120		6.942.−5	10.956	0.2076	0.1248
140		6.958.−5	13.482	0.2270	0.1278
160		6.975.−5	16.063	0.2443	0.1304
180		6.993.−5	18.697	0.2598	0.1330
200		7.013.−5	21.386	0.2739	0.1360
220		7.034.−5	24.139	0.2870	0.1394
234.3	7.330.−10	7.050.−5	26.148	0.2959	0.1420
234.3	7.330.−10	7.304.−5	37.585	0.3447	0.1422
240	1.668.−9	7.311.−5	38.395	0.3481	0.1420
260	6.925.−8	7.339.−5	41.224	0.3595	0.1409
280	5.296.−7	7.365.−5	44.034	0.3699	0.1401
300	3.075.−6	7.387.−5	46.829	0.3795	0.1393
320	1.428.−5	7.413.−5	49.609	0.3885	0.1386
340	5.516.−5	7.439.−5	52.375	0.3969	0.1380
360	1.829.−4	7.472.−5	55.130	0.4048	0.1375
380	5.289.−4	7.499.−5	57.874	0.4122	0.1370
400	1.394.−3	7.526.−5	60.609	0.4192	0.1366
450	0.01053	7.595.−5	67.414	0.4352	0.1357
500	0.05261	7.664.−5	74.188	0.4495	0.1353
550	0.1949	7.735.−5	80.949	0.4624	0.1352
600	0.5776	7.807.−5	87.716	0.4742	0.1356
650	1.4425	7.881.−5	94.508	0.4850	0.1360
700	3.153	7.957.−5	101.343	0.4951	0.1372
750	6.197	8.036.−5	108.242	0.5046	0.1382
800	11.181	8.118.−5	115.23	0.5136	0.1398
850	18.816	8.203.−5	122.31	0.5221	0.1416
900	29.88	8.292.−5	129.53	0.5302	0.1439
950	45.23	8.385.−5	137.16	0.5381	0.1464
1000	65.74	8.482.−5	144.41	0.5456	0.1492

[a]v = specific volume; h = specific enthalpy; s = specific entropy, c_p = specific heat at constant pressure. Properties above the solid line are for the solid; below they are for the liquid. Condensed, converted, and interpolated from the tables of M. P. Vukalovich, A. I. Ivanov, L. R. Fokin, and A. T. Yakovlev, *Thermophysical Properties of Mercury,* Standartov, Moscow, USSR, 1971.

[b]The notation 6.873.−5 signifies 6.873×10^{-5}.

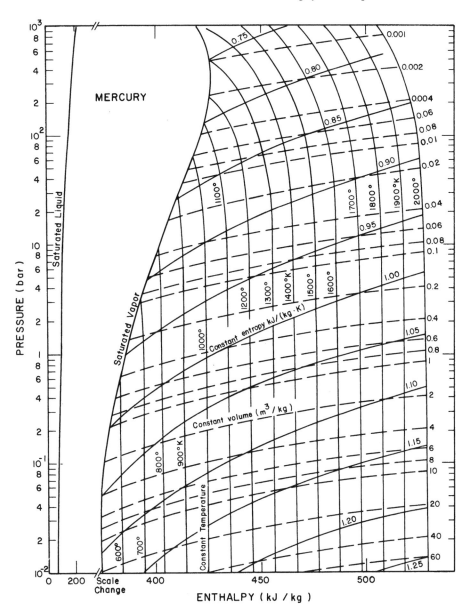

Figure 2 Enthalpy–log pressure diagram for mercury.

Table 10   Thermodynamic Properties of Saturated Methane[a]

T (K)	P (bar)	v_f (m³/kg)	v_g (m³/kg)	h_f (kJ/kg)	h_g (kJ/kg)	s_f (kJ/kg·K)	s_g (kJ/kg·K)	c_{p_f} (kJ/kg·K)	$\bar{v}_s$ (m/s)
90.68	0.117	2.215.−3[b]	3.976	216.4	759.9	4.231	10.225	3.288	1576
92	0.139	2.226.−3	3.410	220.6	762.4	4.279	10.168	3.294	1564
96	0.223	2.250.−3	2.203	233.2	769.5	4.419	10.006	3.326	1523
100	0.345	2.278.−3	1.479	246.3	776.9	4.556	9.862	3.369	1480
104	0.515	2.307.−3	1.026	259.6	784.0	4.689	9.731	3.415	1437
108	0.743	2.337.−3	0.732	273.2	791.0	4.818	9.612	3.458	1393
112	1.044	2.369.−3	0.536	287.0	797.7	4.944	9.504	3.497	1351
116	1.431	2.403.−3	0.401	301.1	804.2	5.068	9.405	3.534	1308
120	1.919	2.438.−3	0.306	315.3	810.8	5.187	9.313	3.570	1266
124	2.523	2.475.−3	0.238	329.7	816.2	5.305	9.228	3.609	1224
128	3.258	2.515.−3	0.187	344.3	821.6	5.419	9.148	3.654	1181
132	4.142	2.558.−3	0.150	359.1	826.5	5.531	9.072	3.708	1138
136	5.191	2.603.−3	0.121	374.2	831.0	5.642	9.001	3.772	1093
140	6.422	2.652.−3	0.0984	389.5	834.8	5.751	8.931	3.849	1047
144	7.853	2.704.−3	0.0809	405.2	838.0	5.858	8.864	3.939	999
148	9.502	2.761.−3	0.0670	421.3	840.6	5.965	8.798	4.044	951
152	11.387	2.824.−3	0.0558	437.7	842.2	6.072	8.733	4.164	902
156	13.526	2.893.−3	0.0467	454.7	843.2	6.177	8.667	4.303	852
160	15.939	2.971.−3	0.0392	472.1	843.0	6.283	8.601	4.470	802
164	18.647	3.059.−3	0.0326	490.1	841.6	6.390	8.533	4.684	749
168	21.671	3.160.−3	0.0278	508.9	839.0	6.497	8.462	4.968	695
172	25.034	3.281.−3	0.0234	528.6	834.6	6.606	8.385	5.390	637
176	28.761	3.428.−3	0.0196	549.7	827.9	6.720	8.301	6.091	570
180	32.863	3.619.−3	0.0162	572.9	818.1	6.843	8.205	7.275	500
184	37.435	3.890.−3	0.0131	599.7	802.9	6.980	8.084	9.831	421
188	42.471	4.361.−3	0.0101	634.0	776.4	7.154	7.912	19.66	327
190.56	45.988	6.233.−3	0.0062	704.4	704.4	7.516	7.516		

[a]v = specific volume; h = specific enthalpy; s = specific entropy; c_p = specific heat at constant pressure; $\bar{v}_s$ = velocity of sound; f = saturated liquid; g = saturated vapor. Condensed and converted from R. D. Goodwin, N.B.S. Technical Note 653, 1974.
[b]The notation 2.215.−3 signifies 2.215×10^{-3}.

Table 11 Thermophysical Properties of Methane at Atmospheric Pressure[a]

	Temperature (K)									
	250	300	350	400	450	500	550	600	650	700
v	1.275	1.532	1.789	2.045	2.301	2.557	2.813	3.068	3.324	3.580
h	1090	1200	1315	1437	1569	1709	1857	2016	2183	2359
s	11.22	11.62	11.98	12.30	12.61	12.91	13.19	13.46	13.73	14.00
c_p	2.04	2.13	2.26	2.43	2.60	2.78	2.96	3.16	3.35	3.51
Z	0.997	0.998	0.999	1.000	1.000	1.000	1.000	1.000	1.000	1.000
$\bar{v}_s$	413	450	482	511	537	562	585	607	629	650
λ	0.0276	0.0342	0.0417	0.0486	0.0571	0.0675	0.0768	0.0863	0.0956	0.1052
η	0.095	0.112	0.126	0.141	0.154	0.168	0.180	0.192	0.202	0.214
Pr	0.701	0.696	0.683	0.687	0.690	0.693	0.696	0.700	0.706	0.714

[a] v = specific volume (m³/kg); h = specific enthalpy (kJ/kg); s = specific entropy (kJ/kg·K); c_p = specific heat at constant pressure (kJ/kg ·K); Z = compressibility factor = Pv/RT; $\bar{v}_s$ = velocity of sound (m/s); λ = thermal conductivity (W/m·K); η = viscosity 10^{-4} N·s/m² (thus, at 250 K the viscosity is 0.095×10^{-4} N·s/m² = 0.0000095 Pa·s); Pr = Prandtl number.

Table 12 Thermophysical Properties of Saturated Refrigerant 22[a]

T (K)	P (bar)	v (m³/kg) Liquid	v (m³/kg) Vapor	h (kJ/kg) Liquid	h (kJ/kg) Vapor	s (kJ/kg·K) Liquid	s (kJ/kg·K) Vapor	c_p (Liquid)	η (Liquid)	λ (Liquid)	τ (Liquid)
150	0.0017	6.209.-4^b	83.40	268.2	547.3	3.355	5.215	1.059		0.161	
160	0.0054	6.293.-4	28.20	278.2	552.1	3.430	5.141	1.058		0.156	
170	0.0150	6.381.-4	10.85	288.3	557.0	3.494	5.075	1.057	0.770	0.151	
180	0.0369	6.474.-4	4.673	298.7	561.9	3.551	5.013	1.058	0.647	0.146	
190	0.0821	6.573.-4	2.225	308.6	566.8	3.605	4.963	1.060	0.554	0.141	
200	0.1662	6.680.-4	1.145	318.8	571.6	3.657	4.921	1.065	0.481	0.136	0.024
210	0.3116	6.794.-4	0.6370	329.1	576.5	3.707	4.885	1.071	0.424	0.131	0.022
220	0.5470	6.917.-4	0.3772	339.7	581.2	3.756	4.854	1.080	0.378	0.126	0.021
230	0.9076	7.050.-4	0.2352	350.6	585.9	3.804	4.828	1.091	0.340	0.121	0.019
240	1.4346	7.195.-4	0.1532	361.7	590.5	3.852	4.805	1.105	0.309	0.117	0.0172
250	2.174	7.351.-4	0.1037	373.0	594.9	3.898	4.785	1.122	0.282	0.112	0.0155
260	3.177	7.523.-4	0.07237	384.5	599.0	3.942	4.768	1.143	0.260	0.107	0.0138
270	4.497	7.733.-4	0.05187	396.3	603.0	3.986	4.752	1.169	0.241	0.102	0.0121
280	6.192	7.923.-4	0.03803	408.2	606.6	4.029	4.738	1.193	0.225	0.097	0.0104
290	8.324	8.158.-4	0.02838	420.4	610.0	4.071	4.725	1.220	0.211	0.092	0.0087
300	10.956	8.426.-4	0.02148	432.7	612.8	4.113	4.713	1.257	0.198	0.087	0.0071
310	14.17	8.734.-4	0.01643	445.5	615.1	4.153	4.701	1.305	0.186	0.082	0.0055

Table 12 (*Continued*)

T (K)	P (bar)	v (m³/kg) Liquid	v (m³/kg) Vapor	h (kJ/kg) Liquid	h (kJ/kg) Vapor	s (kJ/kg·K) Liquid	s (kJ/kg·K) Vapor	c_p (Liquid)	η (Liquid)	λ (Liquid)	τ (Liquid)
320	18.02	9.096.−4	0.01265	458.6	616.7	4.194	4.688	1.372	0.176	0.077	0.0040
330	22.61	9.535.−4	9.753.−3	472.4	617.3	4.235	4.674	1.460	0.167	0.072	0.0026
340	28.03	1.010.−3	7.479.−3	487.2	616.5	4.278	4.658	1.573	0.151	0.067	0.0014
350	34.41	1.086.−3	5.613.−3	503.7	613.3	4.324	4.637	1.718	0.130	0.062	0.0008
360	41.86	1.212.−3	4.036.−3	523.7	605.5	4.378	4.605	1.897	0.106		
369.3	49.89	2.015.−3	2.015.−3	570.0	570.0	4.501	4.501	∝	—	—	0

[a]c_p in units of kJ/kg·K; η = viscosity (10^{-4} Pa·s); λ = thermal conductivity (W/m·K); τ = surface tension (N/m). *Sources: P, v, T, h, s* interpolated and extrapolated from I. I. Perelshteyn, *Tables and Diagrams of the Thermodynamic Properties of Freons 12, 13, 22,* Moscow, USSR, 1971. c_p, η, λ interpolated and converted from *Thermophysical Properties of Refrigerants,* ASHRAE, New York, 1976. τ calculated from V. A. Gruzdev et al., *Fluid Mech. Sov. Res.,* **3**, 172 (1974).

[b]The notation 6.209.−4 signifies 6.209×10^{-4}.

Table 13 Thermophysical Properties of Refrigerant 22 at Atmospheric Pressure[a]

	Temperature (K)					
	250	300	350	400	450	500
v	0.2315	0.2802	0.3289	0.3773	0.4252	0.4723
h	597.8	630.0	664.5	702.5	740.8	782.3
s	4.8671	4.9840	5.0905	5.1892	5.2782	5.3562
c_p	0.587	0.647	0.704	0.757	0.806	0.848
Z	0.976	0.984	0.990	0.994	0.995	0.996
$\bar{v}_s$	166.4	182.2	196.2	209.4	220.0	233.6
λ	0.0080	0.0110	0.0140	0.0170	0.0200	0.0230
η	0.109	0.130	0.151	0.171	0.190	0.209
Pr	0.800	0.765	0.759	0.761	0.766	0.771

[a]v = specific volume (m³/kg); h = specific enthalpy (kJ/kg); s = specific entropy (kJ/kg·K); c_p = specific heat at constant pressure (kJ/kg·K); Z = compressibility factor = Pv/RT; $\bar{v}_s$ = velocity of sound (m/s); λ = thermal conductivity (W/m·K); η = viscosity 10^{-4} N·s/m² (thus, at 250 K the viscosity is 0.109×10^{-4} N·s/m² = 0.0000109 Pa·s); Pr = Prandtl number.

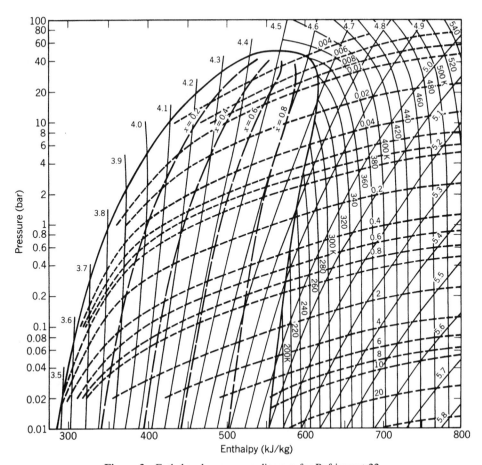

Figure 3   Enthalpy–log pressure diagram for Refrigerant 22.

Table 14   Thermodynamic Properties of Saturated Refrigerant 134a[a]

P (bar)	t (°C)	v_f (m³/kg)	v_g (m³/kg)	h_f (kJ/kg)	h_g (kJ/kg)	s_f (kJ/kg·K)	s_g (kJ/kg·K)	c_{pf} (kJ/kg·K)	c_{pg} (kJ/kg·K)
0.5	−40.69	0.000 707	0.3690	−0.9	225.5	−0.0036	0.9698	1.2538	0.7476
0.6	−36.94	0.000 712	0.3109	3.3	227.5	0.0165	0.9645	1.2600	0.7584
0.7	−33.93	0.000 716	0.2692	7.7	229.4	0.0324	0.9607	1.2654	0.7680
0.8	−31.12	0.000 720	0.2375	11.2	231.2	0.0472	0.9572	1.2707	0.7771
0.9	−28.65	0.000 724	0.2126	14.4	233.0	0.0601	0.9545	1.2755	0.7854
1	−26.52	0.000 729	0.1926	16.8	234.2	0.0720	0.9519	1.2799	0.7931
1.5	−17.26	0.000 742	0.1314	28.6	240.0	0.1194	0.9455	1.2992	0.8264
2	−10.18	0.000 754	0.1001	37.8	243.6	0.1547	0.9379	1.3153	0.8540
2.5	−4.38	0.000 764	0.0809	45.5	247.3	0.1834	0.9340	1.3297	0.8782
3	0.59	0.000 774	0.0679	52.1	250.1	0.2077	0.9312	1.3428	0.9002
4	8.86	0.000 791	0.0514	63.3	254.9	0.2478	0.9270	1.3670	0.9400
5	15.68	0.000 806	0.0413	72.8	258.8	0.2804	0.9241	1.3894	0.9761
6	21.54	0.000 820	0.0344	81.0	262.1	0.3082	0.9219	1.4108	0.9914
8	31.35	0.000 846	0.0257	95.0	267.1	0.3542	0.9185	1.4526	1.0750
10	39.41	0.000 871	0.0204	106.9	270.9	0.3921	0.9157	1.4948	1.1391
12	46.36	0.000 894	0.01672	117.5	273.9	0.4246	0.9132	1.539	1.205
14	52.48	0.000 918	0.01411	127.0	276.2	0.4533	0.9107	1.586	1.276
16	57.96	0.000 941	0.01258	135.7	277.9	0.4794	0.9081	1.637	1.353
18	62.94	0.000 965	0.01056	143.9	279.3	0.5031	0.9052	1.695	1.439
20	67.52	0.000 990	0.00929	151.4	280.1	0.5254	0.9021	1.761	1.539
22.5	72.74	0.001 023	0.00800	160.9	280.8	0.5512	0.8976	1.859	1.800
25	77.63	0.001 058	0.00694	169.8	280.8	0.5756	0.8925	1.983	1.883
27.5	82.04	0.001 096	0.00605	178.2	280.3	0.5989	0.8865	2.151	2.149
30	86.20	0.001 141	0.00528	186.5	279.4	0.6216	0.8812	2.388	2.527
35	93.72	0.001 263	0.00397	203.7	274.1	0.6671	0.8589	3.484	4.292
40	100.34	0.001 580	0.00256	227.4	257.2	0.7292	0.8090	26.33	37.63
40.59	101.06	0.001 953	0.00195	241.5	241.5	0.7665	0.7665		

[a]Converted and reproduced from R. Tillner-Roth and H. D. Baehr, *J. Phys. Chem. Ref. Data*, **23** (5), 657–730 (1994). $h_f = s_f = 0$ at 233.15 K = −40°C.

Table 15   Thermophysical Properties of Refrigerant 134a

t (°C)	Property	1	2.5	5	7.5	10	12.5	15	Sat. Vapor
					P (bar)				
0	c_p (kJ/kg·K)	0.8197	0.8740						0.8975
	μ (10^{-6} Pa·s)	11.00	10.95						10.94
	k (W/m·K)	0.0119	0.0120						0.0120
	Pr	0.763	0.798						0.809
10	c_p (kJ/kg·K)	0.8324	0.8815						0.9408
	μ (10^{-6} Pa·s)	11.38	11.42						11.42
	k (W/m·K)	0.0126	0.0127						0.0129
	Pr	0.753	0.786						0.821
20	c_p (kJ/kg·K)	0.8458	0.8726	0.9642					0.9864
	μ (10^{-6} Pa·s)	11.78	11.83	11.91					11.93
	k (W/m·K)	0.0134	0.0135	0.0138					0.0139
	Pr	0.747	0.774	0.830					0.838
30	c_p (kJ/kg·K)	0.8602	0.8900	0.9587	1.044				1.048
	μ (10^{-6} Pa·s)	12.27	12.28	12.29	12.36				12.37
	k (W/m·K)	0.0141	0.0143	0.0145	0.0150				0.0150
	Pr	0.743	0.764	0.805	0.857				0.859
40	c_p (kJ/kg·K)	0.8747	0.8998	0.9547	1.027	1.134			1.145
	μ (10^{-6} Pa·s)	12.57	12.61	12.66	12.75	12.88			12.89
	k (W/m·K)	0.0148	0.0150	0.0153	0.0156	0.0161			0.0161
	Pr	0.740	0.757	0.789	0.839	0.907			0.916
50	c_p (kJ/kg·K)	0.8891	0.9112	0.9555	1.017	1.120	1.213		1.246
	μ (10^{-6} Pa·s)	12.96	13.00	13.05	13.14	13.23	13.33		13.47
	k (W/m·K)	0.0156	0.0158	0.0160	0.0163	0.0167	0.0171		0.0173
	Pr	0.739	0.752	0.778	0.820	0.887	0.946		0.960
60	c_p (kJ/kg·K)	0.9045	0.9230	0.9589	1.003	1.060	1.151	1.248	1.387
	μ (10^{-6} Pa·s)	13.35	13.39	13.44	13.51	13.60	13.75	13.96	14.16
	k (W/m·K)	0.0164	0.0165	0.0167	0.0170	0.0173	0.0178	0.0184	0.0185
	Pr	0.739	0.750	0.772	0.801	0.829	0.889	0.935	1.059
70	c_p (kJ/kg·K)	0.9201	0.9359	0.9652	0.9972	1.046	1.100	1.175	1.606
	μ (10^{-6} Pa·s)	13.74	13.77	13.82	13.89	13.97	14.10	14.27	15.04
	k (W/m·K)	0.0171	0.0172	0.0174	0.0176	0.0179	0.0183	0.0189	0.0197
	Pr	0.739	0.750	0.759	0.787	0.813	0.848	0.886	1.226
80	c_p (kJ/kg·K)	0.9359	0.9520	0.9715	0.9992	1.038	1.222	1.313	2.026
	μ (10^{-6} Pa·s)	14.11	14.14	14.19	14.25	14.33	14.43	14.59	16.31
	k (W/m·K)	0.0178	0.0179	0.0180	0.0182	0.0185	0.0188	0.0193	0.0205
	Pr	0.741	0.752	0.757	0.782	0.804	0.938	0.993	1.612
Sat. vapor	c_p (kJ/kg·K)	0.7931	0.8782	0.9761	1.059	1.139	1.223	1.314	—
	μ (10^{-6} Pa·s)			11.72	12.34	12.86	13.34	13.81	—
	k (W/m·K)			0.0136	0.0149	0.0161	0.0173	0.0184	—
	Pr			0.841	0.877	0.910	0.943	0.986	—

Note. At 0°C, 1 bar the viscosity is 11×10^{-6} Pa·s.; Pr = Prandtl number.

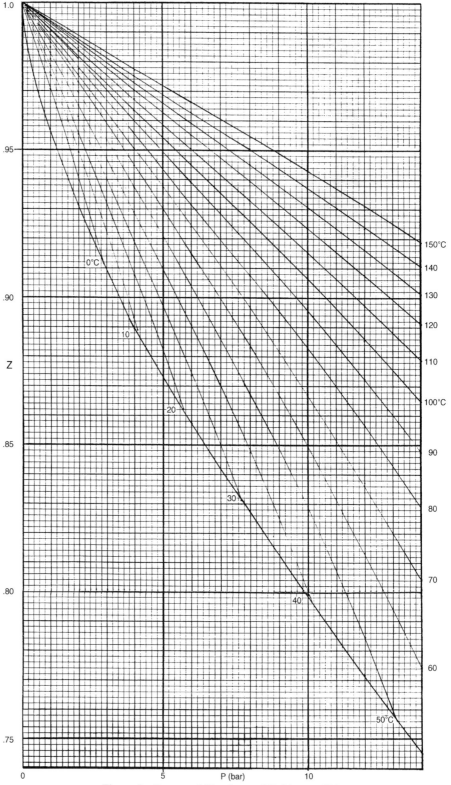

Figure 4 Compressibility factor of Refrigerant 134a.

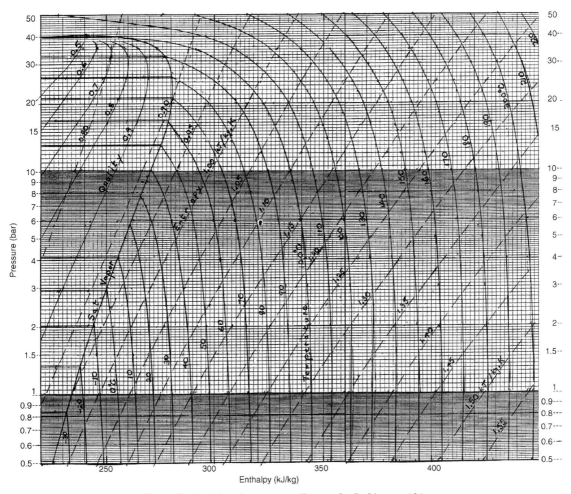

Figure 5 Enthalpy–log pressure diagram for Refrigerant 134a.

Table 16 Thermodynamic Properties of Saturated Sodium[a]

T (K)	P (bar)	v_f	v_g	h_f	h_g	s_f	s_g	c_{pf}	c_{pg}
380	2.55.−10[b]	1.081.−3	5.277.+9	0.219	4.723	2.853	14.705	1.384	0.988
400	1.36.−9	1.086.−3	2.173.+9	0.247	4.740	2.924	14.158	1.374	1.023
420	6.16.−9	1.092.−3	2.410.+8	0.274	4.757	2.991	13.665	1.364	1.066
440	2.43.−8	1.097.−3	6.398.+7	0.301	4.773	3.054	13.219	1.355	1.117
460	8.49.−8	1.103.−3	1.912.+7	0.328	4.790	3.114	12.814	1.346	1.176
480	2.67.−7	1.109.−3	6.341.+6	0.355	4.806	3.171	12.443	1.338	1.243
500	7.64.−7	1.114.−3	2.304.+6	0.382	4.820	3.226	12.104	1.330	1.317
550	7.54.−6	1.129.−3	2.558.+5	0.448	4.856	3.352	11.367	1.313	1.523
600	5.05.−5	1.145.−3	41511	0.513	4.887	3.465	10.756	1.299	1.745
650	2.51.−4	1.160.−3	9001	0.578	4.915	3.569	10.241	1.287	1.963
700	9.87−4	1.177.−3	2449	0.642	4.939	3.664	9.802	1.278	2.160
750	0.00322	1.194.−3	794	0.705	4.959	3.752	9.424	1.270	2.325
800	0.00904	1.211.−3	301	0.769	4.978	3.834	9.095	1.264	2.452
850	0.02241	1.229.−3	128.1	0.832	4.995	3.910	8.808	1.260	2.542
900	0.05010	1.247.−3	60.17	0.895	5.011	3.982	8.556	1.258	2.597
1000	0.1955	1.289.−3	16.84	1.021	5.043	4.115	8.137	1.259	2.624
1200	1.482	1.372.−3	2.571	1.274	5.109	4.346	7.542	1.281	2.515
1400	6.203	1.469.−3	0.688	1.535	5.175	4.546	7.146	1.330	2.391
1600	17.98	1.581.−3	0.258	1.809	5.225	4.728	6.863	1.406	2.301
1800	40.87	1.709.−3	0.120	2.102	5.255	4.898	6.649	1.516	2.261
2000	78.51	1.864.−3	0.0634	2.422	5.256	5.064	6.480	1.702	2.482
2200	133.5	2.076.−3	0.0362	2.794	5.207	5.235	6.332	2.101	3.307
2400	207.6	2.480.−3	0.0196	3.299	5.025	5.447	6.166	3.686	8.476
2500	251.9	3.323.−3	0.0100	3.850	4.633	5.666	5.980		

[a]v = specific volume (m³/kg); h = specific enthalpy (MJ/kg); s = specific entropy (kJ/kg·K); c_p = specific heat at constant pressure (kJ/kg·K); f = saturated liquid; g = saturated vapor. Converted from the tables of J. K. Fink, Argonne Nat. Lab. rept. ANL-CEN-RSD-82-4, 1982.

[b]The notation 2.55.−10 signifies 2.55×10^{-10}.

Table 17 Thermodynamic Properties of Ice/Water[a]

T (K)	P (bar)	v (m³/kg)	h (kJ/kg)	s (kJ/kg·K)	c_p (kJ/kg·K)
150	6.30.−11[b]	0.001073	94.7	1.328	1.224
160	7.72.−10	0.001074	107.3	1.409	1.291
170	7.29.−9	0.001076	120.6	1.489	1.357
180	5.38.−8	0.001077	134.5	1.569	1.426
190	3.23.−7	0.001078	149.1	1.648	1.495
200	1.62.−6	0.001079	164.4	1.726	1.566
210	7.01.−6	0.001081	180.4	1.805	1.638
220	2.65.−5	0.001082	197.1	1.882	1.711
230	8.91.−5	0.001084	214.6	1.960	1.785
240	2.73.−4	0.001085	232.8	2.038	1.860
250	7.59.−4	0.001087	251.8	2.115	1.936
260	0.00196	0.001088	271.5	2.192	2.013
270	0.00469	0.001090	292.0	2.270	2.091
273.15	0.00611	0.001091	298.7	2.294	2.116
273.15	0.00611	0.001000	632.2	3.515	4.217
280	0.00990	0.001000	661.0	3.619	4.198
290	0.01917	0.001001	702.9	3.766	4.184
300	0.03531	0.001003	744.7	3.908	4.179

[a] v = specific volume; h = specific enthalpy; s = specific entropy; c_p = specific heat at constant pressure. Properties above the solid line are for the solid; below they are for the liquid. Ice values ($T \leq 273.15$ K) converted and rounded off from S. Gordon, NASA Tech. Paper 1906, 1982.

[b] The notation 6.30.−11 signifies 6.30×10^{-11}.

Table 18   Thermophysical Properties of Saturated Steam/Water[a]

P (bar)	T (K)	v_f	v_g	h_f	h_g	η_g	λ_f	λ_g	Pr_f	Pr_g
1.0	372.78	1.0434.−3[b]	1.6937	417.5	2675.4	0.1202	0.6805	0.0244	1.735	1.009
1.5	384.52	1.0530.−3	1.1590	467.1	2693.4	0.1247	0.6847	0.0259	1.538	1.000
2.0	393.38	1.0608.−3	0.8854	504.7	2706.3	0.1280	0.6866	0.0268	1.419	1.013
2.5	400.58	1.0676.−3	0.7184	535.3	2716.4	0.1307	0.6876	0.0275	1.335	1.027
3.0	406.69	1.0735.−3	0.6056	561.4	2724.7	0.1329	0.6879	0.0281	1.273	1.040
3.5	412.02	1.0789.−3	0.5240	584.3	2731.6	0.1349	0.6878	0.0287	1.224	1.050
4.0	416.77	1.0839.−3	0.4622	604.7	2737.6	0.1367	0.6875	0.0293	1.185	1.057
4.5	421.07	1.0885.−3	0.4138	623.2	2742.9	0.1382	0.6869	0.0298	1.152	1.066
5	424.99	1.0928.−3	0.3747	640.1	2747.5	0.1396	0.6863	0.0303	1.124	1.073
6	432.00	1.1009.−3	0.3155	670.4	2755.5	0.1421	0.6847	0.0311	1.079	1.091
7	438.11	1.1082.−3	0.2727	697.1	2762.0	0.1443	0.6828	0.0319	1.044	1.105
8	445.57	1.1150.−3	0.2403	720.9	2767.5	0.1462	0.6809	0.0327	1.016	1.115
9	448.51	1.1214.−3	0.2148	742.6	2772.1	0.1479	0.6788	0.0334	0.992	1.127
10	453.03	1.1274.−3	0.1943	762.6	2776.1	0.1495	0.6767	0.0341	0.973	1.137
12	461.11	1.1386.−3	0.1632	798.4	2782.7	0.1523	0.6723	0.0354	0.943	1.156
14	468.19	1.1489.−3	0.1407	830.1	2787.8	0.1548	0.6680	0.0366	0.920	1.175
16	474.52	1.1586.−3	0.1237	858.6	2791.8	0.1569	0.6636	0.0377	0.902	1.191
18	480.26	1.1678.−3	0.1103	884.6	2794.8	0.1589	0.6593	0.0388	0.889	1.206
20	485.53	1.1766.−3	0.0995	908.6	2797.2	0.1608	0.6550	0.0399	0.877	1.229
25	497.09	1.1972.−3	0.0799	962.0	2800.9	0.1648	0.6447	0.0424	0.859	1.251
30	506.99	1.2163.−3	0.0666	1008.4	2802.3	0.1684	0.6347	0.0449	0.849	1.278
35	515.69	1.2345.−3	0.0570	1049.8	2802.0	0.1716	0.6250	0.0472	0.845	1.306
40	523.48	1.2521.−3	0.0497	1087.4	2800.3	0.1746	0.6158	0.0496	0.845	1.331
45	530.56	1.2691.−3	0.0440	1122.1	2797.7	0.1775	0.6068	0.0519	0.849	1.358
50	537.06	1.2858.−3	0.0394	1154.5	2794.2	0.1802	0.5981	0.0542	0.855	1.386
60	548.70	1.3187.−3	0.0324	1213.7	2785.0	0.1854	0.5813	0.0589	0.874	1.442
70	558.94	1.3515.−3	0.0274	1267.4	2773.5	0.1904	0.5653	0.0638	0.901	1.503
80	568.12	1.3843.−3	0.0235	1317.1	2759.9	0.1954	0.5499	0.0688	0.936	1.573
90	576.46	1.4179.−3	0.0205	1363.7	2744.6	0.2005	0.5352	0.0741	0.978	1.651
100	584.11	1.4526.−3	0.0180	1408.0	2727.7	0.2057	0.5209	0.0798	1.029	1.737
110	591.20	1.4887.−3	0.0160	1450.6	2709.3	0.2110	0.5071	0.0859	1.090	1.837
120	597.80	1.5268.−3	0.0143	1491.8	2689.2	0.2166	0.4936	0.0925	1.163	1.963
130	603.98	1.5672.−3	0.0128	1532.0	2667.0	0.2224	0.4806	0.0998	1.252	2.126
140	609.79	1.6106.−3	0.0115	1571.6	2642.4	0.2286	0.4678	0.1080	1.362	2.343
150	615.28	1.6579.−3	0.0103	1611.0	2615.0	0.2373	0.4554	0.1172	1.502	2.571
160	620.48	1.7103.−3	0.0093	1650.5	2584.9	0.2497	0.4433	0.1280	1.688	3.041
170	625.41	1.7696.−3	0.0084	1691.7	2551.6	0.2627	0.4315	0.1404	2.098	3.344
180	630.11	1.8399.−3	0.0075	1734.8	2513.9	0.2766	0.4200	0.1557	2.360	3.807
190	634.58	1.9260.−3	0.0067	1778.7	2470.6	0.2920	0.4087	0.1749	2.951	8.021
200	638.85	2.0370.−3	0.0059	1826.5	2410.4	0.3094	0.3976	0.2007	4.202	12.16

Table 18 (*Continued*)

P (bar)	s_f	s_g	c_{pf}	c_{pg}	η_f	γ'_f	γ_g	$\bar{v}'_{sf}$	$\bar{v}_{sg}$	τ
1.0	1.3027	7.3598	4.222	2.048	2.801	1.136	1.321	438.74	472.98	0.0589
1.5	1.4336	7.2234	4.231	2.077	2.490	1.139	1.318	445.05	478.73	0.0566
2.0	1.5301	7.1268	4.245	2.121	2.295	1.141	1.316	449.51	482.78	0.0548
2.5	1.6071	7.0520	4.258	2.161	2.156	1.142	1.314	452.92	485.88	0.0534
3.0	1.6716	6.9909	4.271	2.198	2.051	1.143	1.313	455.65	488.36	0.0521
3.5	1.7273	6.9392	4.282	2.233	1.966	1.143	1.311	457.91	490.43	0.0510
4.0	1.7764	6.8943	4.294	2.266	1.897	1.144	1.310	459.82	492.18	0.0500
4.5	1.8204	6.8547	4.305	2.298	1.838	1.144	1.309	461.46	493.69	0.0491
5	1.8604	6.8192	4.315	2.329	1.787	1.144	1.308	462.88	495.01	0.0483
6	1.9308	6.7575	4.335	2.387	1.704	1.144	1.306	465.23	497.22	0.0468
7	1.9918	6.7052	4.354	2.442	1.637	1.143	1.304	467.08	498.99	0.0455
8	2.0457	6.6596	4.372	2.495	1.581	1.142	1.303	468.57	500.55	0.0444
9	2.0941	6.6192	4.390	2.546	1.534	1.142	1.302	469.78	501.64	0.0433
10	2.1382	6.5821	4.407	2.594	1.494	1.141	1.300	470.76	502.64	0.0423
12	2.2161	6.5194	4.440	2.688	1.427	1.139	1.298	472.23	504.21	0.0405
14	2.2837	6.4651	4.472	2.777	1.373	1.137	1.296	473.18	505.33	0.0389
16	2.3436	6.4175	4.504	2.862	1.329	1.134	1.294	473.78	506.12	0.0375
18	2.3976	6.3751	4.534	2.944	1.291	1.132	1.293	474.09	506.65	0.0362
20	2.4469	6.3367	4.564	3.025	1.259	1.129	1.291	474.18	506.98	0.0350
25	2.5543	6.2536	4.640	3.219	1.193	1.123	1.288	473.71	507.16	0.0323
30	2.6455	6.1837	4.716	3.407	1.143	1.117	1.284	472.51	506.65	0.0300
35	2.7253	6.1229	4.792	3.593	1.102	1.111	1.281	470.80	505.66	0.0280
40	2.7965	6.0685	4.870	3.781	1.069	1.104	1.278	468.72	504.29	0.0261
45	2.8612	6.0191	4.951	3.972	1.040	1.097	1.275	466.31	502.68	0.0244
50	2.9206	5.9735	5.034	4.168	1.016	1.091	1.272	463.67	500.73	0.0229
60	3.0273	5.8908	5.211	4.582	0.975	1.077	1.266	457.77	496.33	0.0201
70	3.1219	5.8162	5.405	5.035	0.942	1.063	1.260	451.21	491.31	0.0177
80	3.2076	5.7471	5.621	5.588	0.915	1.048	1.254	444.12	485.80	0.0156
90	3.2867	5.6820	5.865	6.100	0.892	1.033	1.249	436.50	479.90	0.0136
100	3.3606	5.6198	6.142	6.738	0.872	1.016	1.244	428.24	473.67	0.0119
110	3.4304	5.5595	6.463	7.480	0.855	0.998	1.239	419.20	467.13	0.0103
120	3.4972	5.5002	6.838	8.384	0.840	0.978	1.236	409.38	460.25	0.0089
130	3.5616	5.4408	7.286	9.539	0.826	0.956	1.234	398.90	453.00	0.0076
140	3.6243	5.3803	7.834	11.07	0.813	0.935	1.232	388.00	445.34	0.0064
150	3.6859	5.3178	8.529	13.06	0.802	0.916	1.233	377.00	437.29	0.0053
160	3.7471	5.2531	9.456	15.59	0.792	0.901	1.235	366.24	428.89	0.0043
170	3.8197	5.1855	11.30	17.87	0.782	0.867	1.240	351.19	420.07	0.0034
180	3.8765	5.1128	12.82	21.43	0.773	0.838	1.248	336.35	410.39	0.0026
190	3.9429	5.0332	15.76	27.47	0.765	0.808	1.260	320.20	399.87	0.0018
200	4.0149	4.9412	22.05	39.31	0.758	0.756	1.280	298.10	387.81	0.0011

$^a v$ = specific volume (m^3/kg); h = specific enthalpy (kJ/kg); s = specific entropy (kJ/kg·K); c_p = specific heat at constant pressure (kJ/kg ·K); η = viscosity (10^{-4} Pa·s); λ = thermal conductivity (W/m·K); Pr = Prandtl number; γ = c_p/c_v ratio; $\bar{v}_s$ = velocity of sound (m/s); τ = surface tension (N/m); f' = wet saturated vapor; g = saturated vapor. Rounded off from values of C. M. Tseng, T. A. Hamp, and E. O. Moeck, Atomic Energy of Canada Report AECL-5910, 1977.

^{b} The notation 1.0434.−3 signifies 1.0434 × 10^{-3}.

Table 19   Thermophysical Properties of Miscellaneous Substances at Atmospheric Pressure[a]

	Temperature (K)					
	250	300	350	400	450	500
n-Butane						
v	0.0016	0.411	0.485	0.558	0.630	0.701
h	236.6	718.9	810.7	913.1	1026	1149
s	3.564	5.334	5.616	5.889	6.155	6.414
c_p	2.21	1.73	1.94	2.15	2.36	2.56
Z	0.005	0.969	0.982	0.988	0.992	0.993
$\bar{v}_s$	1161	211	229	245	259	273
λ	0.0979	0.0161	0.0220	0.0270	0.0327	0.0390
η	2.545	0.076	0.088	0.101	0.111	0.124
Pr	5.75	0.84	0.82	0.81	0.80	0.80
Ethane						
v	0.672	0.812	0.950	1.088	1.225	1.362
h	948.7	1068	1162	1265	1380	1505
s	7.330	7.634	7.854	8.198	8.467	8.730
c_p	1.58	1.76	1.97	2.18	2.39	2.60
Z	0.986	0.992	0.995	0.997	0.998	0.998
$\bar{v}_s$	287	312	334	355	374	392
λ	0.0103	0.0157	0.0219	0.0288	0.0361	0.0438
η	0.079	0.094	0.109	0.123	0.135	0.148
Pr	1.214	1.056	0.978	0.932	0.900	0.878
Ethylene						
v	0.734	0.884	1.034	1.183	1.332	1.482
h	966.8	1039	1122	1215	1316	1415
s	7.556	7.832	8.085	8.331	8.568	8.800
c_p	1.40	1.57	1.75	1.93	2.10	2.26
Z	0.991	0.994	0.997	0.998	0.999	1.000
$\bar{v}_s$	306	330	353	374	394	403
λ	0.0149	0.0206	0.0271	0.0344	0.0425	0.0506
η	0.087	0.103	0.119	0.134	0.148	0.162
Pr	0.816	0.785	0.767	0.751	0.735	0.721
n-Hydrogen						
v	10.183	12.218	14.253	16.289	18.324	20.359
h	3517	4227	4945	5669	6393	7118
s	67.98	70.58	72.79	74.72	76.43	77.96
c_p	14.04	14.31	14.43	14.48	14.50	14.51
Z	1.000	1.000	1.000	1.000	1.000	1.000
$\bar{v}_s$	1209	1319	1423	1520	1611	1698
λ	0.162	0.187	0.210	0.230	0.250	0.269
η	0.079	0.089	0.099	0.109	0.118	0.127
Pr	0.685	0.685	0.685	0.684	0.684	0.684

Table 19 (*Continued*)

	Temperature (K)					
	250	300	350	400	450	500
Nitrogen						
v	0.7317	0.8786	1.025	1.171	1.319	1.465
h	259.1	311.2	363.3	415.4	467.8	520.4
s	6.650	6.840	7.001	7.140	7.263	7.374
c_p	1.042	1.041	1.042	1.045	1.050	1.056
Z	0.9992	0.9998	0.9998	0.9999	1.0000	1.0002
$\bar{v}_s$	322	353	382	407	432	455
λ	0.0223	0.0259	0.0292	0.0324	0.0366	0.0386
η	0.155	0.178	0.200	0.220	0.240	0.258
Pr	0.724	0.715	0.713	0.710	0.708	0.706
Oxygen						
v	0.6402	0.7688	0.9790	1.025	1.154	1.282
h	226.9	272.7	318.9	365.7	413.1	461.3
s	6.247	6.414	6.557	6.682	6.793	6.895
c_p	0.915	0.920	0.929	0.942	0.956	0.972
Z	0.9987	0.9994	0.9996	0.9998	1.0000	1.0000
$\bar{V}_s$	301	330	356	379	401	421
λ	0.0226	0.0266	0.0305	0.0343	0.0380	0.0416
η	0.179	0.207	0.234	0.258	0.281	0.303
Pr	0.725	0.716	0.713	0.710	0.708	0.707
Propane						
v	0.451	0.548	0.644	0.738	0.832	0.926
h	877.2	957.0	1048	1149	1261	1384
s	5.840	6.131	6.409	6.680	6.944	7.202
c_p	1.50	1.70	1.96	2.14	2.35	2.55
Z	0.970	0.982	0.988	0.992	0.994	0.996
$\bar{v}_s$	227	248	268	285	302	317
λ	0.0128	0.0182	0.0247	0.0296	0.0362	0.0423
η	0.070	0.082	0.096	0.108	0.119	0.131
Pr	0.820	0.772	0.761	0.765	0.773	0.793
Propylene						
v	0.482	0.585	0.686	0.786	0.884	0.972
h	891.8	957.6	1040	1131	1235	1338
s	6.074	6.354	6.606	6.851	7.095	7.338
c_p	1.44	1.55	1.73	1.91	2.09	2.27
Z	0.976	0.987	0.992	0.994	0.995	0.996
$\bar{v}_s$	247	257	278	298	315	333
λ	0.0127	0.0177	0.0233	0.0296	0.0363	0.0438
η	0.072	0.087	0.101	0.115	0.128	0.141
Pr	0.814	0.769	0.754	0.742	0.731	0.728

[a] v = specific volume (m^3/kg); h = specific enthalpy (kJ/kg); s = specific entropy (kJ/kg·K); c_p = specific heat at constant pressure (kJ/kg·K); Z = compressibility factor = Pv/RT; $\bar{v}_s$ = velocity of sound (m/s); λ = thermal conductivity (W/m·K); η = viscosity (10^{-4})(N·s/m²) (thus, at 250 K for *n*-butane the viscosity is 2.545×10^{-4} N·s/m² = 0.0002545 Pa·s); Pr = Prandtl number.

Table 20   Physical Properties of Numbered Refrigerants[a]

Number	Formula, Composition, Synonym	Molecular Weight	n.b.p. (°C)	Crit. P (bar)	Crit. T (°C)
4	R-32/125/134a/143a (10/33/21/36)	94.50	−49.4	40.1	77.5
10	CCl_4 (carbon tetrachloride)	153.8	76.8	45.6	283.2
CFC-11	CCl_3F	137.37	23.8	44.1	198.0
11B1	$CBrCl_2F$	181.82	52		
11B2	CBr_2ClF	226.27	80		
11B3	CBr_3F	270.72	107		
CFC-12	CCl_2F_2	120.91	−29.8	41.1	112.0
12B1	$CBrClF_2$	165.36	−2.5	42.5	153.0
12B2	CF_2Br_2	209.81	24.5	40.7	204.9
CFC-13	$CClF_3$	104.46	−81.4	38.7	28.8
BFC-13B1	$CBrF_3$ (Halon 1301)	148.91	−57.8	39.6	67.0
FC-14	CF_4 (carbon tetrafluoride)	88.00	−127.9	37.5	−45.7
20	$CHCl_3$ (chloroform)	119.38	61.2	54.5	263.4
21	$CHCl_2F$	102.92	8.9	51.7	178.5
HCFC-22	$CHClF_2$	86.47	−40.8	49.9	96.2
HFC-23	CHF_3	70.01	−82.1	48.7	26.3
HCC-30	CH_2Cl_2 (methylene chloride)	84.93	40.2	60.8	237.0
31	CH_2ClF	68.47	−9.1	56.2	153.8
HFC-32	CH_2F_2	52.02	−51.7	58.0	78.2
33	R-22/124/152a (40/43/17)	96.62	−28.8		
40	CH_3Cl (methyl chloride)	50.49	−12.4	66.7	143.1
FX-40	R-32/125/143a (10/45/45)	90.70	−48.4	40.5	72.0
HFC-41	CH_3F (methyl fluoride)	34.03	−78.4	58.8	44.3
50	CH_4 (methane)	16.04	−161.5	46.4	−82.5
FX-57	R-22/124/142b (65/25/10)	96.70	−35.2	47.0	105.0
110	CCl_3CCl_3	236.8	185	33.4	401.8
111	CCl_3CCl_2F	220.2	137		
112	CCl_2FCCl_2F	203.8	92.8	33.4	278
CFC-113	$CClF_2CCl_2F$	187.38	47.6	34.4	214.1
113a	CCl_3CF_3	187.36	47.5		
CFC-114	$CClF_2CClF_2$	170.92	3.8	32.5	145.7
114a	CF_3CCl_2F	170.92	3.0	33.0	145.5
CFC-115	$CClF_2CF_3$	154.47	−39.1	31.5	79.9
FC-116	CF_3CF_3 (perfluoroethane)	138.01	−78.2	30.4	19.9
120	$CHCl_2CCl_3$	202.3	162	34.8	373
121	$CHCl_2CCl_2F$	185.84	116.6		
122	$CClF_2CHCl_2$	131.39	72.0		
HCFC-123	$CHCl_2CF_3$	152.93	27.9	36.7	183.7
HCFC-123a	$CHClFCClF_2$	152.93	28.0	44.7	188.5
HCFC-124	$CHClFCF_3$	136.48	−12.0	36.4	122.5
E-125	CHF_2OCF_2	136.02	−41.9	33.3	80.4

Table 20   (*Continued*)

Number	Formula, Composition, Synonym	Molecular Weight	n.b.p. (°C)	Crit. P (bar)	Crit. T (°C)
HFC-125	CHF_2CF_3	120.02	−48.1	36.3	66.3
131	$CHCl_2CHClF$	151.4	102.5		
132	$CHClFCHClF$	134.93	58.5		
133	$CHClFCHF_2$	118.5	17.0		
E-134	CHF_2OCHF_2	118.03	6.2	42.3	153.5
HFC-134	CHF_2CHF_2	102.03	−23.0	46.2	118.7
HFC-134a	CH_2FCF_3	102.03	−26.1	40.6	101.1
141	$CH_2ClCHClF$	116.95	76		
141a	$CHCl_2CH_2F$	116.95			
HCFC-141b	CH_3CCl_2F	116.95	32.2	42.5	204.4
142	CHF_2CH_2Cl	100.49	35.1		
142a	$CHClFCH_2F$	100.49			
HCFC-142b	CH_3CClF_2	100.50	−9.8	41.2	137.2
143	CHF_2CH_2F	84.04	5.		
E-143a	CH_3OCF_3	100.04	−24.1	35.9	104.9
HFC-143a	CH_3CF_3	84.04	−47.4	38.3	73.6
151	CH_2FCH_2Cl	82.50	53.2		
152	CH_2FCH_2F	66.05	10.5	43.4	171.8
HFC-152a	CH_3CHF_2	66.05	−24.0	45.2	113.3
HCC-160	CH_3CH_2Cl	64.51	12.4	52.4	186.6
HFC-161	CH_3CH_2F	48.06	−37.1	47.0	102.2
E-170	CH_3OCH_3 (dimethyl ether)	46.07	−24.8	53.2	128.8
HC-170	CH_3CH_3 (ethane)	30.07	−88.8	48.9	32.2
216	$C_3Cl_2F_6$	220.93	35.7	27.5	180.0
FC-218	$CF_3CF_2CF_3$	188.02	−36.7	26.8	71.9
HFC-227ca	$CHF_2CF_2CF_3$	170.03	−17.0	28.7	106.3
HFC-227ea	CF_3CHFCF_3	170.03	−18.3	29.5	103.5
234da	$CF_3CHClCHClF$	114.03	70.1		
235ca	$CF_3CF_2CH_2Cl$	156.46	28.1		
HFC-236ca	$CHF_2CF_2CHF_2$	152.04	5.1	34.1	153.2
HFC-236cb	$CH_2FCF_2CF_3$	152.04	−1.4	31.2	130.2
HFC-236ea	CHF_2CHFCF_3	152.04	6.6	35.3	141.2
HFC-235fa	$CF_3CH_2CF_3$	152.04	−1.1	31.8	130.7
HFC-245ca	$CH_2FCF_2CHF_2$	134.05	25.5	38.6	178.5
E-245cb	$CHF_2OCH_2CF_3$	150.05	34.0		185.2
HFC-245cb	$CH_3CF_2CF_3$	134.05	−18.3	32.6	108.5
E-245fa	$CHF_2OCH_2CF_3$	150.05	29.2	37.3	170.9
HFC-245fa	$CHF_2CH_2CF_3$	134.05	15.3	36.4	157.6
HFC-254cb	$CH_3CF_2CHF_2$	116.06	−0.8	37.5	146.2
HC-290	$CH_3CH_2CH_3$ (propane)	44.10	−42.1	42.5	96.8
RC-318	cyclo-$CF_2CF_2CF_2CF_2$	200.04	−5.8	27.8	115.4
400	R-12/114				

Table 20   (*Continued*)

Number	Formula, Composition, Synonym	Molecular Weight	n.b.p. (°C)	Crit. *P* (bar)	Crit. *T* (°C)
R-401a	R-22/124/152a (53/34/13)	94.44	−33.1	46.0	108.0
R-401b	R-22/124/152a (61/28/11)	92.84	−34.7	46.8	106.1
R-401c	R-22/124/152a (33/52/15)	101.04	−28.4	43.7	112.7
R-402a	R-22/125/290 (38/60/2)	101.55	−49.2	41.3	75.5
R-402b	R-22/125/290 (60/38/2)	94.71	−47.4	44.5	82.6
R-403a	R-22/218/290 (75/20/5)	91.06	−50.0	50.8	93.3
R-403b	R-22/218/290 (56/39/5)	102.06	−49.5	50.9	90.0
R-404a	R-125/134a/143a (44/4/52)	97.60	−46.5	37.3	72.1
R-405a	R-22/142b/152a/C318 (45/5/7/43)	116.00	−27.3	42.6	106.1
R-406a	R-22/142b/600a (55/41/4)	89.85	−30.0	47.4	123.0
R-407a	R-32/125/134a (20/40/40)	90.10	−45.5	45.4	82.8
R-407b	R-32/125/134a (10/70/20)	102.94	−47.3	41.6	75.8
R-407c	R-32/125/134a (23/25/52)	86.20	−43.6	46.2	86.7
R-408a	R-22/125/143a (47/7/46)	87.02	−43.5	43.4	83.5
R-409a	R-22/124/142b (60/25/15)	97.45	−34.2	45.0	107.0
R-410a	R-32/125 (50/50)	72.56	−50.5	49.6	72.5
R-410b	R-32/125 (45/55)				
R-411a	R-22/152a/1270 (88/11/2)				
R-411b	R-22/152a/1270 (94/3/3)				
R-412a	R-22/142b/218 (70/25/5)				
R-500	R-12/152a (74/26)	99.31	−33.5	44.2	105.5
R-501	R-12/22 (25/75)				
R-502	R-22/115 (49/51)	111.64	−45.4	40.8	82.2
R-503	R-13/23 (60/40)	87.28	−87.8	43.6	19.5
R-504	R-32/115 (48/52)	79.2	−57.2	47.6	66.4
R-505	R-12/31 (78/22)	103.5	−30		
R-506	R-31/114 (55/45)	93.7	−12		
R-507	R-125/143a (50/50)	98.90	−46.7	37.9	70.9
R-508	R-23/116 (39/61)	100.10	−85.7		23.1
R-509	R-22/218 (44/56)	124.0	−47		
R-600	$CH_3CH_2CH_2CH_3$ (butane)	58.13	−0.4	38.0	152.0
R-600a	$CH_3CH_3CH_3CH$ (isobutane)	58.13	−11.7	36.5	135.0
R-610	$C_4H_{10}O$ (ethyl ether)	74.12	−116.3	36.0	194.0
R-611	$C_2H_4O_2$ (methyl formate)	60.05	31.8	59.9	204
630	CH_3NH (methyl amine)	31.06	−6.7	74.6	156.9
631	$C_2H_5NH_2$ (ethyl amine)	45.08	16.6	56.2	183.0
702	H_2 (hydrogen)	2.016	−252.8	13.2	−239.9
702p	Parahydrogen	2.016	−252.9	12.9	−240.2
704	He (helium)	4.003	−268.9	2.3	−267.9
717	NH_3 (ammonia)	17.03	−33.3	114.2	133.0
718	H_2O (water)	18.02	100.0	221.0	374.2

Table 20   (*Continued*)

Number	Formula, Composition, Synonym	Molecular Weight	n.b.p. (°C)	Crit. P (bar)	Crit. T (°C)
720	Ne (neon)	20.18	−246.1	34.0	−228.7
728	N_2 (nitrogen)	28.01	−198.8	34.0	−146.9
728a	CO (carbon monoxide)	28.01	−191.6	35.0	−140.3
729	- (air)	28.97	−194.3	37.6	−140.6
732	O_2 (oxygen)	32.00	−182.9	50.8	−118.4
740	A (argon)	39.95	−185.9	49.0	−122.3
744	CO_2 (carbon dioxide)	44.01	−78.4	73.7	31.1
744a	N_2O (nitrous oxide)	44.02	−89.5	72.2	36.5
R-764	SO_2 (sulfur dioxide)	64.07	−10.0	78.8	157.5
1113	C_2ClF_3	116.47	−27.9	40.5	106
1114	C_2F_4	100.02	−76.0	39.4	33.3
1120	$CHClCCl_2$	131.39	87.2	50.2	271.1
1130	CHClCHCl	96.95	47.8	54.8	243.3
1132a	$C_2H_2F_2$	64.03	−85.7	44.6	29.7
1141	C_2H_3F (vinyl fluoride)	46.04	−72.2	52.4	54.7
1150	C_2H_4 (ethylene)	28.05	−103.7	51.1	9.3
1270	C_3H_6 (propylene)	42.09	−185	46.2	91.8
R-7146	SF_6 (sulfur hexafluoride)	146.05	−63.8	37.6	45.6

[a]Refrigerant numbers in some cases are tentative and subject to revision. Compositions rounded to nearest weight percent. Based on data supplied by M. O. McLinden, NIST, Boulder, CO, PCR Chemicals, Gainesville, FL, G. H. Thomson, DIPPR, Bartlesville, OK, and literature sources.

Table 21 Specific Heat (kJ/kg·K) at Constant Pressure of Saturated Liquids

Substance	Temperature (K)															
	250	260	270	280	290	300	310	320	330	340	350	360	370	380	390	400
Acetic acid	—[a]	—	—	—	2.03	2.06	2.09	2.12	2.16	2.19	2.23	2.26	2.29	2.33	2.36	2.39
Acetone	2.05	2.07	2.10	2.13	2.16	2.19	2.22	2.26	2.30	2.35	2.40					
Ammonia	4.48	4.54	4.60	4.66	4.73	4.82	4.91	5.02	5.17	5.37	5.64	6.04	6.68	7.80	10.3	21
Aniline	—	—	2.03	2.04	2.05	2.07	2.10	2.13	2.16	2.19	2.22	2.26	2.31	2.38	2.47	2.58
Benzene	—	—	—	1.69	1.71	1.73	1.75	1.78	1.81	1.84	1.87	1.91	1.94	1.98	2.03	2.09
n-Butane	2.19	2.23	2.27	2.32	2.37	2.43	2.50	2.58	2.67	2.76	2.86	2.97	3.08			
Butanol	2.13	2.17	2.22	2.27	2.33	2.38	2.44	2.51	2.58	2.65	2.73	2.82	2.93	3.06	3.20	3.36
Carbon tetrachloride	0.833	0.838	0.843	0.848	0.853	0.858	0.864	0.870	0.879	0.891	0.912	0.941	0.975			
Chlorobenzene	1.29	1.31	1.32	1.32	1.33	1.33	1.34	1.36	1.38	1.40	1.42	1.44	1.46	1.47	1.49	1.51
m-Cresol	—	—	—	—	2.04	2.07	2.11	2.14	2.18	2.21	2.24	2.27	2.30	2.32	2.35	2.38
Ethane	2.97	3.20	3.50	4.00	5.09	9.92										
Ethanol	—	2.24	2.28	2.33	2.38	2.45	2.54	2.64	2.75	2.86	2.99	3.12	3.26	3.41	3.56	3.72
Ethyl acetate	—	—	—	1.89	1.92	1.94	1.97	2.00	2.03	2.06	2.09	2.13	2.16	2.20	2.24	2.28
Ethyl sulfide	1.96	1.97	1.97	1.98	2.00	2.01	2.02	2.03								
Ethylene	3.25	3.78	5.0													
Formic acid	—	—	2.13	—	2.15	2.16	2.17	2.18	2.20	2.22	2.24	2.26	2.28	2.30	2.33	2.36
Heptane	2.08	2.10	2.13	2.17	2.20	2.24	2.28	2.32	2.36	2.41	2.45	2.49	2.54	2.59	2.64	2.70
Hexane	2.09	2.12	2.15	2.19	2.22	2.26	2.31	2.36	2.41	2.46	2.51	2.56	2.62	2.69	2.76	2.83
Methanol	2.31	2.34	2.37	2.41	2.46	2.52	2.58	2.65	2.73	2.82	2.91	3.01	3.12	3.24	3.36	3.49
Methyl formate					2.16	2.16										
Octane	2.07	2.10	2.13	2.16	2.19	2.22	2.26	2.31	2.35	2.39	2.43	2.47	2.52	2.57	2.62	2.69
Oil, linseed	1.58	1.61	1.65	1.69	1.73	1.78	1.82	1.87	1.91	1.95	1.99	2.03	2.08	2.13	2.17	2.21
Oil, olive	1.90	1.92	1.95	1.98	2.01	2.05	2.09	2.13	2.16	2.20	2.24	2.28	2.32	2.37	2.41	2.46
Pentane	1.96	2.02	2.08	2.14	2.21	2.28	2.35	2.42	2.49	2.56	2.63	2.70	2.77	2.84	2.91	2.98
Propane	2.35	2.41	2.48	2.56	2.65	2.76	2.89	3.06	3.28	3.62	4.23	5.98				
Propanol	2.04	2.10	2.16	2.24	2.32	2.41	2.51	2.62	2.74	2.86	2.99	3.12	3.26	3.40	3.55	3.71
Propylene	2.22	2.27	2.34	2.43	2.55	2.69	2.87	3.12	3.44	3.92	4.75	6.75				
Sulfuric acid	—	—	—	—	1.39	1.41	1.43	1.45	1.46	1.48	1.50	1.51	1.53	1.54	1.56	1.57
Sulfur dioxide	1.31	1.32	1.33	1.34	1.36	1.39	1.42	1.46	1.50	1.55	1.61	1.68	1.76	1.85	1.99	2.14
Turpentine	1.60	1.65	1.70	1.75	1.80	1.85	1.90	1.95	2.00	2.05	2.10	2.15	2.20	2.25	2.30	2.35

[a]Dashes indicate inaccessible states.

36

Table 22 Ratio of Principal Specific Heats, c_p/c_v, for Liquids and Gases at Atmospheric Pressure

Substance	200	220	240	260	280	300	320	340	360	380	400	420	440	460	480	500
Acetylene	1.313	1.294	1.277	1.261	1.247	1.234	1.222	1.211	1.200							
Air	1.399	1.399	1.399	1.399	1.399	1.399	1.399	1.398	1.397	1.396	1.395	1.394	1.392	1.390	1.388	1.386
Ammonia							1.307	1.299	1.291	1.284	1.278	1.271	1.265	1.260	1.255	1.249
Argon	1.663	1.665	1.666	1.666	1.666	1.666	1.666	1.666	1.666	1.666	1.666	1.666	1.666	1.666	1.666	1.666
n-Butane	1.418	1.413	1.409	1.406	1.112	1.103	1.096	1.089	1.084	1.079	1.075	1.072	1.069	1.066	1.063	1.061
Carbon dioxide		1.350	1.332	1.317	1.303	1.290	1.282	1.273	1.265	1.258	1.253	1.247	1.242	1.238	1.233	1.229
Carbon monoxide	1.405	1.404	1.403	1.402	1.402	1.401	1.401	1.400	1.399	1.398	1.396	1.395	1.393	1.391	1.389	1.387
Ethane		1.250	1.233	1.219	1.205	1.193	1.182	1.172	1.163	1.155	1.148	1.141	1.135	1.130	1.125	1.120
Ethylene				1.268	1.251	1.236	1.223	1.212	1.201	1.192	1.183	1.175	1.168	1.162	1.156	1.151
Fluorine	1.393	1.387	1.380	1.374	1.368	1.362										
Helium	1.667	1.667	1.667	1.667	1.667	1.667	1.667	1.667	1.667	1.667	1.667	1.667	1.667	1.667	1.667	1.667
n-Hydrogen	1.439	1.428	1.418	1.413	1.409	1.406	1.403	1.402	1.401	1.400	1.399	1.398	1.398	1.398	1.397	1.397
Isobutane	1.357	1.356	1.357	1.362	1.113	1.103	1.096	1.089	1.084	1.079	1.075	1.071	1.068	1.065	1.063	1.060
Krypton	1.649	1.654	1.657	1.659	1.661	1.662	1.662	1.662	1.662	1.662	1.662	1.663	1.663	1.664	1.666	1.667
Methane	1.337	1.333	1.328	1.322	1.314	1.306	1.296	1.287	1.278	1.268	1.258	1.249	1.241	1.233	1.226	1.219
Neon	1.667	1.667	1.667	1.667	1.667	1.667	1.667	1.667	1.667	1.667	1.667	1.667	1.667	1.667	1.667	1.667
Nitrogen	1.399	1.399	1.399	1.399	1.399	1.399	1.399	1.399	1.398	1.398	1.397	1.396	1.395	1.393	1.392	1.391
Oxygen	1.398	1.397	1.397	1.396	1.395	1.394	1.392	1.389	1.387	1.384	1.381	1.378	1.375	1.371	1.368	1.365
Propane	1.513	1.506	1.173	1.158	1.145	1.135	1.126	1.118	1.111	1.105	1.100	1.095	1.091	1.087	1.084	1.081
Propylene			1.133	1.122	1.111	1.101	1.091	1.082	1.072	1.063	1.055	1.046	1.038	1.030	1.023	1.017
R12					1.148	1.139	1.132	1.127	1.122	1.118	1.115	1.112	1.110	1.108	1.106	1.104
R21			1.171	1.159		1.179	1.165	1.156	1.148	1.142	1.137	1.133	1.129	1.126	1.123	1.120
R22					1.204	1.190	1.174	1.167	1.160	1.153	1.148	1.144	1.140	1.136	1.132	1.129
Steam									1.323	1.321	1.319	1.317	1.315	1.313	1.311	1.309
Xenon	1.623	1.632	1.639	1.643	1.651	1.655	1.658	1.661	1.662	1.662	1.662	1.662	1.662	1.662	1.662	1.662

Temperature (K)

Table 23 Surface Tension (N/m) of Liquids

Substance	Temperature (K)															
	250	260	270	280	290	300	310	320	330	340	350	360	370	380	390	400
Acetone	0.0291	0.0279	0.0266	0.0253	0.0240	0.0228	0.0214	0.0201	0.0187	0.0174	0.0162	0.0150	0.0139	0.0128	0.0117	0.0106
Ammonia	0.0317	0.0294	0.0271	0.0248	0.0226	0.0203	0.0181	0.0159	0.0138	0.0117	0.0099	0.0080	0.0059	0.0040	0.0021	0.0003
Benzene			0.0320	0.0306	0.0292	0.0278	0.0265	0.0252	0.0239	0.0227	0.0215	0.0203	0.0191	0.0179	0.0167	0.0155
Butane	0.0177	0.0165	0.0153	0.0141	0.0129	0.0116	0.0104	0.0092	0.0080	0.0069	0.0059	0.0049	0.0040	0.0031	0.0023	0.0016
CO_2	0.0092	0.0071	0.0051	0.0032	0.0016	0.0003	—	—	—	—	—	—	—	—	—	—
Chlorine	0.0244	0.0228	0.0213	0.0198	0.0183	0.0168	0.0153	0.0138	0.0123	0.0108	0.0094	0.0080	0.0066	0.0052	0.0044	0.0037
Ethane	0.0059	0.0047	0.0035	0.0024	0.0013	0.0005	—	—	—	—	—	—	—	—	—	—
Ethanol	0.0271	0.0261	0.0251	0.0242	0.0232	0.0223	0.0214	0.0205	0.0196	0.0186	0.0177	0.0167	0.0158	0.0148	0.0137	0.0126
Ethylene	0.0032	0.0019	0.0009	0.0002	—	—	—	—	—	—	—	—	—	—	—	—
Heptane	0.0244	0.0234	0.0224	0.0214	0.0205	0.0195	0.0185	0.0176	0.0166	0.0156	0.0147	0.0137	0.0127	0.0118	0.0109	0.0099
Hexane	0.0229	0.0218	0.0207	0.0197	0.0186	0.0175	0.0165	0.0154	0.0145	0.0135	0.0125	0.0115	0.0106	0.0096	0.0086	0.0076
Mercury	0.474	0.472	0.470	0.468	0.466	0.464	0.462	0.460	0.458	0.456	0.454	0.452	0.450	0.448	0.446	0.444
Methanol						0.0223	0.0214	0.0205	0.0196	0.0187	0.0178	0.0168	0.0159	0.0149	0.0139	0.0128
Octane	0.0251	0.0243	0.0234	0.0225	0.0216	0.0207	0.0197	0.0188	0.0179	0.0170	0.0161	0.0152	0.0143	0.0135	0.0127	0.0120
Propane	0.0132	0.0118	0.0104	0.0091	0.0079	0.0067	0.0056	0.0046	0.0037	0.0028	0.0018	0.0009	—	—	—	—
Propylene	0.0133	0.0120	0.0106	0.0091	0.0078	0.0065	0.0053	0.0042	0.0032	0.0023	0.0014	0.0006	—	—	—	—
R-12	0.0148	0.0135	0.0122	0.0109	0.0096	0.0083	0.0070	0.0058	0.0047	0.0036	0.0027	0.0018	0.0010	0.0003	—	—
R-13	0.0057	0.0042	0.0029	0.0018	0.0009	0.0003	—	—	—	—	—	—	—	—	—	—
Toluene	0.0342	0.0327	0.0312	0.0298	0.0285	0.0272	0.0260	0.0249	0.0236	0.0225	0.0214	0.0203	0.0193	0.0183	0.0173	0.0163
Water		—	—	0.0746	0.0732	0.0716	0.0699	0.0684	0.0666	0.0650	0.0636	0.0614	0.0595	0.0575	0.0555	0.0535

Table 24 Thermal Conductivity (W/m·K) of Saturated Liquids

Substance	Temperature (K)															
	250	260	270	280	290	300	310	320	330	340	350	360	370	380	390	400
Acetic acid	—	—	—	—	0.165	0.164	0.162	0.161	0.160	0.158	0.157	0.156	0.154	0.153	0.152	0.150
Acetone	0.179	0.175	0.171	0.167	0.163	0.160	0.156	0.152	0.149	0.145	0.141	0.137	0.133	0.130	0.126	0.122
Ammonia	0.562	0.541	0.522	0.503	0.484	0.466	0.445	0.424	0.403	0.382	0.359	0.335	0.309	0.284	0.252	0.227
Aniline	—	—	0.175	0.175	0.174	0.173	0.172	0.172	0.171	0.170	0.169	0.168	0.167	0.166	0.165	0.165
Benzene	—	—	—	0.151	0.149	0.145	0.141	0.138	0.135	0.132	0.129	0.126	0.123	0.120	0.117	0.114
Butane	0.130	0.126	0.122	0.118	0.115	0.111	0.108	0.104	0.101	0.097	0.094	0.090	0.086	0.082	0.078	0.075
Butanol	0.161	0.159	0.157	0.155	0.153	0.151	0.149	0.147	0.145	0.143	0.141	0.139	0.138	0.136	0.135	0.133
Carbon tetrachloride	0.114	0.112	0.110	0.107	0.105	0.103	0.100	0.098	0.096	0.094	0.092	0.089	0.087	0.085	0.083	0.080
Chlorobenzene	0.137	0.135	0.133	0.131	0.129	0.127	0.125	0.123	0.121	0.119	0.117	0.115	0.113	0.111	0.109	0.107
m-Cresol	—	—	—	—	0.150	0.149	0.148	0.148	0.147	0.146	0.146	—	—	—	—	—
Ethane	0.105	0.098	0.091	0.084	0.074	0.065	—	—	—	—	—	—	—	—	—	—
Ethanol	—	0.178	0.176	0.173	0.169	0.166	0.163	0.160	0.158	0.155	0.153	0.151	0.149	0.147	0.145	0.143
Ethyl acetate	0.158	0.155	0.152	0.149	0.146	0.143	0.140	0.137	0.134	0.131	0.128	0.124	0.121	0.118	0.115	0.112
Ethyl sulfide	0.144	0.142	0.140	0.138	0.136	0.134	—	—	—	—	—	—	—	—	—	—
Ethylene	0.100	0.090	0.078	0.064	—	—	—	—	—	—	—	—	—	—	—	—
Formic acid	—	—	—	—	0.293	0.279	0.255	0.241	—	—	—	—	—	—	—	—
Heptane	0.138	0.136	0.133	0.130	0.127	0.125	0.122	0.119	0.117	0.114	0.111	0.108	0.106	0.104	0.102	0.100
Hexane	0.138	0.135	0.132	0.129	0.125	0.122	0.119	0.116	0.112	0.110	0.107	0.104	0.101	0.099	0.096	0.093
Methanol	0.218	0.215	0.211	0.208	0.205	0.202	0.199	0.196	0.194	0.191	0.188	0.186	0.183	0.180	0.178	0.176
Methyl formate	0.202	0.199	0.196	0.193	0.190	0.187	0.184	0.181	0.178	0.175	0.172	—	—	—	—	—
Octane	0.141	0.138	0.136	0.133	0.130	0.128	0.125	0.123	0.120	0.118	0.115	0.112	0.110	0.107	0.105	0.103
Oil, linseed	0.175	0.173	0.171	0.169	0.167	0.165	0.163	0.161	0.159	0.157	0.154	0.152	0.150	0.148	0.146	0.144
Oil, olive	0.170	0.170	0.169	0.169	0.168	0.168	0.167	0.167	0.166	0.165	0.165	0.164	0.163	0.163	0.162	0.161
Pentane	0.133	0.128	0.124	0.120	0.116	0.112	0.108	0.106	0.102	0.098	0.094	0.090	0.086	0.082	0.078	0.074
Propane	0.120	0.116	0.112	0.107	0.102	0.097	0.093	0.088	0.084	0.079	0.075	0.070	—	—	—	—
Propanol	0.166	0.163	0.161	0.158	0.156	0.154	0.152	0.150	0.148	0.146	0.144	0.142	0.141	0.139	0.138	0.136
Propylene	0.137	0.131	0.127	0.122	0.118	0.111	0.107	0.101	0.096	0.090	0.083	0.077	—	—	—	—
Sulfuric acid	—	—	—	—	0.324	0.329	0.332	0.336	0.339	0.342	0.345	0.349	0.352	0.355	0.358	0.361
Sulfur dioxide	0.227	0.221	0.214	0.208	0.202	0.196	0.190	0.184	0.172	0.158						
Turpentine				0.128	0.127	0.127	0.126									

Table 25 Viscosity (10^{-4} Pa·s) of Saturated Liquids

Substance	Temperature (K)															
	250	260	270	280	290	300	310	320	330	340	350	360	370	380	390	400
Acetic acid	—	—	—	—	13.1	11.3	9.77	8.49	7.40	6.48	5.70	5.03	4.45	3.95	3.52	3.15
Acetone	5.27	4.63	4.11	3.68	3.32	3.01	2.73	2.49	2.28	2.10	1.94					
Ammonia	2.20	1.94	1.74	1.58	1.44	1.31	1.20	1.08	0.98	0.88	0.78	0.70	0.64	0.58	0.53	0.48
Aniline	—	—	96.9	71.6	53.5	40.3	30.7	23.6	18.2	14.2	11.2	8.84	7.03	5.63	4.54	3.68
Benzene	—	—	—	8.08	6.80	5.84	5.13	4.52	4.05	3.60	3.29	2.98	2.70	2.45	2.24	2.04
Butane	2.63	2.37	2.15	1.95	1.77	1.61	1.56	1.33	1.21	1.10	1.00	0.90	0.81	0.73	0.66	0.60
Butanol	94.2	70.9	53.9	41.4	32.1	25.1	19.8	15.7	12.6	10.1	8.22	6.71	5.50	4.53	3.75	3.13
Carbon tetrachloride	—	—	14.0	11.9	10.2	8.78	7.70	6.81	6.06	5.43	4.89	4.41	4.00	3.65	3.34	3.07
Chlorobenzene	13.9	12.2	10.7	9.50	8.44	7.53	6.74	6.05	5.46	4.93	4.47	4.07	3.70	3.39	3.10	2.85
m-Cresol	—	—	—	—	261	136	72.0	39.0	21.5	12.1	6.90					
Ethane	0.81	0.70	0.60	0.51	0.42	0.35										
Ethanol	—	23.3	18.9	15.5	12.8	10.6	8.81	7.39	6.24	5.29	4.51	3.86	3.31	2.86	2.48	2.15
Ethyl acetate	7.21	6.53	5.86	5.28	4.78	4.33	3.95	3.60	3.30	3.03	2.79	2.57	2.38	2.20	2.04	1.90
Ethyl sulfide	7.69	6.68	5.86	5.19	4.64	4.18	3.78	3.45	3.16	2.92	2.70	2.51	2.34	2.20	2.06	1.95
Ethylene	0.582	0.505	0.418	0.31	—	—										
Formic acid	48.3	37.4	29.5	23.6	19.2	15.9	13.3	11.2	9.56	8.24	7.16	6.27	5.53	4.91	4.39	3.94
Heptane	7.25	6.25	5.46	4.82	4.28	3.83	3.45	3.14	2.87	2.62	2.39	2.20	2.03	1.87	1.72	1.59
Hexane	5.03	4.46	3.97	3.57	3.22	2.93	2.68	2.45	2.24	2.07	1.92	1.78	1.63	1.50	1.38	1.26
Methanol	12.3	10.1	8.43	7.15	6.14	5.32	4.60	4.09	3.60	3.19	2.84	2.53	2.28	2.06	1.85	1.65
Methyl formate	5.71	5.04	4.50	4.04	3.66	3.34	3.06	2.82	2.61	2.43	2.27	2.13	2.01	1.89	1.79	1.70
Octane	10.5	8.80	7.49	6.45	5.66	5.01	4.48	4.03	3.65	3.33	3.05	2.80	2.58	2.38	2.18	2.02
Oil, linseed	2300	1500	1000	700	490	356	263	198	152	118	93.5	74.8	60.6	49.6	41.1	34.3
Oil, olive	8350	4600	2600	1600	1000	630	410	278	193	136	98.3	72.2	54.0	40.9	31.0	24.6
Pentane	3.50	3.16	2.87	2.62	2.39	2.20	2.03	1.87	1.73	1.59	1.46	1.34	1.22	1.11	1.01	0.92
Propane	1.71	1.53	1.37	1.23	1.10	0.98	0.87	0.76	0.66	0.58	0.51	0.45	—	—	—	—
Propanol	67.8	50.0	37.7	29.0	22.7	18.0	14.6	11.9	9.87	8.27	6.99	5.97	5.14	4.46	3.90	3.44
Propylene	1.41	1.27	1.16	1.05	0.94	0.84	0.75	0.67	0.60	0.53	0.47	0.42	—	—	—	—
Sulfuric acid				363	259	189	141	107	82.6	64.7	51.4	41.4	33.7	27.7	23.1	19.4
Sulfur dioxide	5.32	4.53	3.90	3.39	2.85	2.42	2.09	1.83	1.62	1.45	1.30	1.16	1.02	0.89	0.77	0.66
Turpentine	36.3	28.8	23.3	19.1	15.9	13.4	11.4	9.80	8.50	7.44	6.56	5.83	5.21	4.68	4.23	3.85

40

Table 26 Thermochemical Properties at 1.013 Bar, 298.15 K

Substance	Formula	ΔH_f° (kJ/kg·mol)	ΔG_f° (kJ/kg·mol)	S° (kJ/kg·mol·K)
Acetaldehyde	$C_2H_4O(g)$	$-166,000$	$-132,900$	265.2
Acetic acid	$C_2H_4O_2(g)$	$-436,200$	$+315,500$	282.5
Acetone	$C_3H_6O(l)$	$-248,000$	$-155,300$	200.2
Acetylene	$C_2H_2(g)$	$+266,740$	$+209,190$	200.8
Ammonia	$NH_3(g)$	$-46,190$	$-16,590$	192.6
Aniline	$C_6H_7N(l)$	$+35,300$	$+153,200$	191.6
Benzene	$C_6H_6(l)$	$+49,030$	$+117,000$	172.8
Butanol	$C_4H_{10}O(l)$	$-332,400$	$-168,300$	227.6
n-Butane	$C_4H_{10}(l)$	$-105,900$		
n-Butane	$C_4H_{10}(g)$	$-126,200$	$-17,100$	310.1
Carbon dioxide	$CO_2(g)$	$-393,510$	$-394,390$	213.7
Carbon disulfide	$CS_2(g)$	$-109,200$		237.8
Carbon monoxide	$CO(g)$	$-110,520$	$-137,160$	197.6
Carbon tetrachloride	$CCl_4(g)$	$-103,000$	$-66,100$	311.3
Carbon tetrafluoride	$CF_4(g)$	$-921,300$	$-878,200$	261.5
Chloroform	$CHCl_3(g)$	$-104,000$	$-70,500$	295.6
Cyclohexane	$C_6H_{12}(g)$	$-123,100$	$+31,800$	298.2
Cyclopropane	$C_3H_6(g)$	$+53,300$	$+104,300$	237.7
n-Decane	$C_{10}H_{22}(l)$	$-332,600$	$-17,500$	425.5
Diphenyl	$C_{12}H_{10}(g)$	$-172,800$	$-283,900$	348.5
Ethane	$C_2H_6(g)$	$-84,670$	$-32,900$	229.5
Ethanol	$C_2H_6O(g)$	$-235,200$	$-168,700$	282.7
Ethanol	$C_2H_6O(l)$	$-277,600$	$-174,600$	160.7
Ethyl acetate	$C_4H_8O_2(g)$	$-432,700$	$-325,800$	376.8
Ethyl chloride	$C_2H_5Cl(g)$	$-107,600$	$-55,500$	274.8
Ethyl ether	$C_4H_{10}O(g)$	$-250,800$	$-118,400$	352.5
Ethylene	$C_2H_4(g)$	$+52,280$	$+68,130$	219.4
Ethylene oxide	$C_2H_4O(g)$	$-38,500$	$-11,600$	242.9
Heptane	$C_7H_{16}(g)$	$-187,800$	$-427,800$	166.0
Hexane	$C_6H_{18}(g)$	$-208,400$	$16,500$	270.7
Hydrazine	$N_2H_4(l)$	$-50,600$	$-149,200$	121.2
Hydrazine	$N_2H_4(g)$	$+95,400$	$+159,300$	238.4
Hydrogen peroxide	$H_2O_2(l)$	$-187,500$	$-120,400$	109.6
Isobutane	$C_4H_{10}(g)$	$-134,500$	$-20,900$	294.6
Isopentane	$C_5H_{12}(g)$	$-154,500$	$-14,600$	343.6
Methane	$CH_4(g)$	$-74,840$	$-50,790$	186.2
Methanol	$CH_4O(l)$	$-238,600$	$-126,800$	81.8
Methanol	$CH_4O(g)$	$-200,900$	$-162,100$	247.9
Methyl acetate	$C_3H_6O_2(l)$	$-444,300$		
Methyl bromide	$CH_3Br(g)$	$-36,200$	$-25,900$	246.1
Methyl chloride	$CH_3Cl(g)$	$-86,300$	$-63,000$	234.2

Table 26   (*Continued*)

Substance	Formula	ΔH_f° (kJ/kg·mol)	ΔG_f° (kJ/kg·mol)	S° (kJ/kg·mol·K)
Methyl formate	$C_2H_4O_2(g)$	−335,100	−301,000	292.8
Methylene chloride	$CH_2Cl_2(g)$	−94,000	−67,000	270.2
Naphthalene	$C_{10}H_8(g)$	−151,500	−224,200	336.5
Nitric oxide	$NO(g)$	+90,300	86,600	210.6
Nitrogen peroxide	$NO_2(g)$	+33,300		240.0
Nitrous oxide	$N_2O(g)$	+82,000	+104,000	219.9
Octane	$C_8H_{18}(l)$	−250,000	6,610	360.8
Octane	$C_8H_{18}(g)$	−208,400	16,500	466.7
n-Pentane	$C_5H_{12}(g)$	−146,400	−8,370	348.9
Propane	$C_3H_8O(g)$	−103,800	−107,200	269.9
Propanol	$C_3H_8(g)$	−258,800	−164,100	322.6
Propylene	$C_3H_6(g)$	+20,400	+62,700	267.0
R11	$CFCl_3(g)$	−284,500	−238,000	309.8
R12	$CCl_2F_2(g)$	−468,600	−439,300	300.9
R13	$CClF_3(g)$	−715,500	−674,900	285.6
R13B1	$CF_3Br(g)$	−642,700	−616,300	297.6
R23	$CHF_3(g)$	−682,000	−654,900	259.6
Sulfur dioxide	$SO_2(g)$	−296,900	−300,200	248.1
Sulfur hexafluoride	$SF_6(g)$	−1,207,900	−1,105,000	291.7
Toluene	$C_7H_8(g)$	−50,000	−122,300	320.2
Water	$H_2O(l)$	−285,830	−237,210	70.0
Water	$H_2O(g)$	−241,820	−228,600	188.7

Table 27   Ideal Gas Sensible Enthalpies (kJ/kg·mol) of Common Products of Combustion[a,b]

T (K)	Substance							
	CO	CO_2	$H_2O(g)$	NO	NO_2	N_2	O_2	SO_2
200	−2858	−3414	−3280	−2950	−3494	−2855	−2860	−3736
220	−2276	−2757	−2613	−2345	−2803	−2270	−2279	−3008
240	−1692	−2080	−1945	−1749	−2102	−1685	−1698	−2266
260	−1110	−1384	−1277	−1148	−1390	−1105	−1113	−1508
280	−529	−667	−608	−547	−668	−525	−529	−719
300	54	67	63	54	67	55	60	75
320	638	822	736	652	813	636	650	882
340	1221	1594	1411	1248	1573	1217	1238	1702
360	1805	2383	2088	1847	2344	1800	1832	2541
380	2389	3187	2768	2444	3130	2383	2429	3387
400	2975	4008	3452	3042	3929	2967	3029	4251
420	3562	4841	4138	3641	4739	3554	3633	5123
440	4152	5689	4828	4247	5560	4147	4241	6014
460	4642	6550	5522	4849	6395	4737	4853	6923
480	5334	7424	6221	5054	7243	5329	5468	7827
500	5929	8314	6920	6058	8100	5921	6088	8749
550	7427	10580	8695	7589	10295	7401	7655	11180
600	8941	12915	10500	9146	12555	8902	9247	13545
650	10475	15310	12325	10720	14875	10415	10865	16015
700	12020	17760	14185	12310	17250	11945	12500	18550
750	13590	20270	16075	13890	19675	13490	14160	21150
800	15175	22820	17990	15550	22140	15050	15840	23720
850	16780	25410	19945	17200	24640	16630	17535	26390
900	18400	28040	21920	18860	27180	18220	19245	29020
950	20030	30700	23940	20540	29740	19835	20970	31700
1000	21690	33410	25980	22230	32340	21460	22710	34430
1050	23350	36140	28060	23940	34950	23100	24460	37180
1100	25030	38890	30170	25650	37610	24750	26180	39920
1150	26720	41680	32310	27380	40260	26430	27990	42690
1200	28430	44480	34480	29120	42950	28110	29770	45460
1250	30140	47130	36680	30870	45650	29810	31560	48270
1300	31870	50160	38900	32630	48350	31510	33350	51070
1350	33600	53030	41130	34400	51090	33220	35160	53900
1400	35340	55910	43450	36170	53810	34940	36970	56720
1450	37090	58810	45770	37950	56560	36670	38790	59560
1500	38850	61710	48100	39730	59310	38410	40610	62400
1550	40610	64800	50460	41520	62070	40160	42440	65260
1600	42380	67580	52840	43320	64850	41910	44280	68120
1650	44156	70530	55240	45120	67640	43670	46120	71000
1700	45940	73490	57680	46930	70420	45440	47970	73870
1750	47727	76460	60130	48740	73220	47210	49830	76760
1800	49520	79440	62610	50560	76010	49000	51690	79640
1850	51320	82430	65100	52380	78810	50780	53560	82540
1900	53120	85430	67610	54200	81630	52570	55440	85440
1950	54930	88440	70140	56020	84450	54370	57310	88350

Table 27   (*Continued*)

T (K)	Substance							
	CO	CO_2	$H_2O(g)$	NO	NO_2	N_2	O_2	SO_2
2000	56740	91450	72690	57860	87260	56160	59200	91250
2100	60380	97500	77830	61530	92910	59760	62990	97080
2200	64020	103570	83040	65220	98580	63380	66800	102930
2300	67680	109670	88290	68910	104260	67010	70630	108790
2400	71350	115790	93600	72610	109950	70660	74490	114670
2500	75020	121930	98960	76320	115650	74320	78370	120560
2600	78710	128080	104370	80040	121360	77990	82270	126460
2700	82410	134260	109810	83760	127080	81660	86200	132380
2800	86120	140440	115290	87490	132800	85360	90140	138300
2900	89830	146650	120810	91230	138540	89050	94110	144240
3000	93540	152860	126360	94980	144270	92750	98100	150180

[a]Converted and usually rounded off from *JANAF Thermochemical Tables,* NSRDS-NBS-37, 1971.

[b]To illustrate the term *sensible enthalpy,* which is the difference between the actual enthalpy and the enthalpy at the reference temperature, 298.15 K (= 25°C = 77°F = 537°R), the magnitude of the heat transfer, in kJ/kg·mol fuel and in kJ/kg fuel, will be calculated for the steady-state combustion of acetylene in excess oxygen, the reactants entering at 298.15 K and the products leaving at 2000 K. All substances are in the gaseous phase.

The basic equation is

$$Q + W = \sum_P n_i\,(\Delta h_f^\circ + \Delta h_s)_i - \sum_R n_i(\Delta h_f^\circ + \Delta h_s)_i$$

where P signifies products and R reactants, s signifies sensible enthalpy, and the Δh_s are looked up in the table for the appropriate temperatures.

If the actual reaction was

$$C_2H_2 + 1\tfrac{1}{2}O_2 \rightarrow 2CO_2 + H_2O + 3O_2$$

then $W = 0$ and $Q = 2(-393{,}510 + 91{,}450) + 1(-241{,}810 + 72{,}690) + 3(0 + 59{,}200) - (226{,}740 + 0) - 1\tfrac{1}{2}(0 + 0) = -604{,}120 + (-169{,}120) + 177{,}600 - 226{,}740 = -822{,}380$ kJ/mg mol. $C_2H_2 = -31{,}584$ kJ/kg C_2H_2. Had the fuel been burnt in air one would write the equation with an additional 3.76(5.5) N_2 on each side of the equation. In the above, the enthalpy of formation of the stable elements at 298.15 K has been set equal to zero. For further information, most undergraduate engineering thermodynamics texts may be consulted.

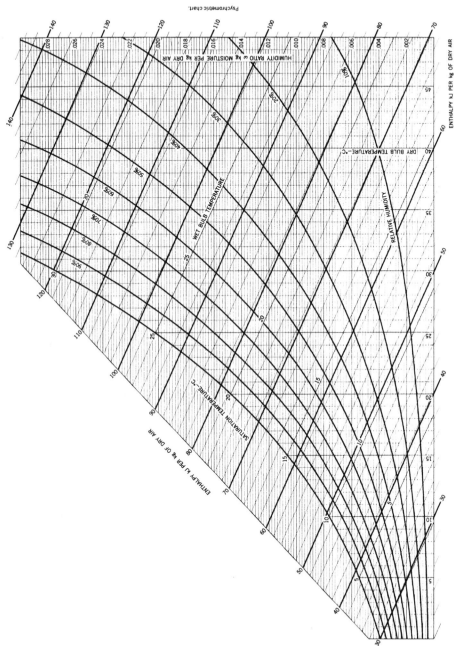

Figure 6 Psychrometric chart. (ASHRAE chart No. 1, 1981. Prepared at the Center for Applied Thermodynamic Studies, University of Idaho, Moscow, ID. Copyright 1981 by the American Society of Heating, Refrigerating and Air Conditioning Engineers, Inc. and reproduced by permission of the copyright owner.)

CHAPTER 2

FLUID MECHANICS

Reuben M. Olson
College of Engineering and Technology
Ohio University
Athens, Ohio

1   DEFINITION OF A FLUID

A solid generally has a definite shape; a fluid has a shape determined by its container. Fluids include liquids, gases, and vapors, or mixtures of these. A fluid continuously deforms when shear stresses are present; it cannot sustain shear stresses at rest. This is characteristic of all real fluids, which are viscous. Ideal fluids are nonviscous (and nonexistent), but have been studied in great detail because in many instances viscous effects in real fluids are very small and the fluid acts essentially as a nonviscous fluid. Shear stresses are set up as a result of relative motion between a fluid and its boundaries or between adjacent layers of fluid.

2   IMPORTANT FLUID PROPERTIES

Density ρ and surface tension σ are the most important fluid properties for liquids at rest. Density and viscosity μ are significant for all fluids in motion; surface tension and vapor pressure are significant for cavitating liquids; and bulk elastic modulus K is significant for compressible gases at high subsonic, sonic, and supersonic speeds.

Sonic speed in fluids is $c = \sqrt{K/\rho}$. Thus, for water at 15°C, $c = \sqrt{2.18 \times 10^9/999} = 1480$ m/sec. For a mixture of a liquid and gas bubbles at nonresonant frequencies, $c_m = \sqrt{K_m/\rho_m}$, where m refers to the mixture. This becomes

$$c_m = \sqrt{\frac{p_g K_l}{[xK_l + (1 - x)p_g][x\rho_g + (1 - x)\rho_l]}}$$

where the subscript l is for the liquid phase and g is for the gas phase. Thus, for water at 20°C containing 0.1% gas nuclei by volume at atmospheric pressure, $c_m = 312$ m/sec. For a gas or a mixture of gases (such as air), $c = \sqrt{kRT}$, where $k = c_p/c_v$, R is the gas constant, and T is the absolute temperature. For air at 15°C, $c = \sqrt{(1.4)(287.1)(288)} = 340$ m/sec. This sonic property is thus a combination of two properties, density and elastic modulus.

Kinematic viscosity is the ratio of dynamic viscosity and density. In a Newtonian fluid, simple laminar flow in a direction x at a speed of u, the shearing stress parallel to x is $\tau_L = \mu(du/dy) = \rho\nu(du/dy)$, the product of dynamic viscosity and velocity gradient. In the more general case, $\tau_L = \mu(\partial u/\partial y + \partial v/\partial x)$ when there is also a y component of velocity v. In turbulent flows the shear stress resulting from lateral mixing is $\tau_T = -\rho\overline{u'v'}$, a Reynolds stress, where u' and v' are instantaneous and simultaneous departures from mean values $\overline{u}$ and $\overline{v}$. This is also written as $\tau_T = \rho\epsilon(du/dy)$, where ϵ is called the turbulent eddy viscosity or diffusivity, an indirectly measurable flow parameter and not a fluid property. The eddy viscosity may be orders of magnitude larger than the kinematic viscosity. The total shear stress in a turbulent flow is the sum of that from laminar and from turbulent motion: $\tau = \tau_L + \tau_T = \rho(\nu + \epsilon)du/dy$ after Boussinesq.

3   FLUID STATICS

The differential equation relating pressure changes dp with elevation changes dz (positive upward parallel to gravity) is $dp = -\rho g\, dz$. For a constant-density liquid, this integrates to $p_2 - p_1 = -\rho g\,(z_2 - z_1)$ or $\Delta p = \gamma h$, where γ is in N/m^3 and h is in m. Also $(p_1/\gamma) + z_1 = (p_2/\gamma) + z_2$; a constant piezometric head exists in a homogeneous liquid at rest, and since $p_1/\gamma - p_2/\gamma = z_2 - z_1$, a change in pressure head equals the change in potential head. Thus, horizontal planes are at constant pressure when body forces due to gravity act. If body forces are due to uniform linear accelerations or to centrifugal effects in rigid-body rotations, points equidistant below the free liquid surface are all at the same pressure. Dashed lines in Figs. 1 and 2 are lines of constant pressure.

Pressure differences are the same whether all pressures are expressed as gage pressure or as absolute pressure.

3.1 Manometers

Pressure differences measured by barometers and manometers may be determined from the relation $\Delta p = \gamma h$. In a barometer, Fig. 3, $h_b = (p_a - p_v)/\gamma_b$ m.

An open manometer, Fig. 4, indicates the inlet pressure for a pump by $p_{\text{inlet}} = -\gamma_m h_m - \gamma y$ Pa gauge. A differential manometer, Fig. 5, indicates the pressure drop across an orifice, for example, by $p_1 - p_2 = h_m(\gamma_m - \gamma_0)$ Pa.

Manometers shown in Figs. 3 and 4 are a type used to measure medium or large pressure differences with relatively small manometer deflections. Micromanometers can be designed to produce relatively large manometer deflections for very small pressure differences. The relation $\Delta p = \gamma \Delta h$ may be applied to the many commercial instruments available to obtain pressure differences from the manometer deflections.

3.2 Liquid Forces on Submerged Surfaces

The liquid force on any flat surface submerged in the liquid equals the product of the gage pressure at the centroid of the surface and the surface area, or $F = \bar{p}A$. The force F is not applied at the centroid for an inclined surface, but is always below it by an amount that diminishes with depth. Measured parallel to the inclined surface, $\bar{y}$ is the distance from 0 in Fig. 6 to the centroid and $y_F = \bar{y} + I_{CG}/A\bar{y}$, where I_{CG} is the moment of inertia of the flat surface with respect to its centroid. Values for some surfaces are listed in Table 1.

For curved surfaces, the horizontal component of the force is equal in magnitude and point of application to the force on a projection of the curved surface on a vertical plane, determined as above. The vertical component of force equals the weight of liquid above the curved surface and is applied at the centroid of this liquid, as in Fig. 7. The liquid forces on opposite sides of a submerged surface are equal in magnitude but opposite in direction. These statements for curved surfaces are also valid for flat surfaces.

Buoyancy is the resultant of the surface forces on a submerged body and equals the weight of fluid (liquid or gas) displaced.

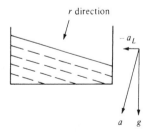

Figure 1 Constant linear acceleration.

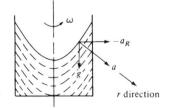

Figure 2 Constant centrifugal acceleration.

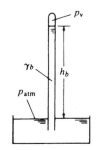

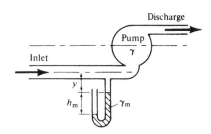

Figure 3 Barometer.

Figure 4 Open manometer.

3.3 Aerostatics

The U.S. standard atmosphere is considered to be dry air and to be a perfect gas. It is defined in terms of the temperature variation with altitude (Fig. 8), and consists of isothermal regions and polytropic regions in which the polytropic exponent n depends on the lapse rate (temperature gradient).

Conditions at an upper altitude z_2 and at a lower one z_1 in an isothermal atmosphere are obtained by integrating the expression $dp = -\rho g\, dz$ to get

$$\frac{p_2}{p_1} = \exp \frac{-g(z_2 - z_1)}{RT}$$

In a polytropic atmosphere where $p/p_1 = (\rho/\rho_1)^n$,

$$\frac{p_2}{p_1} = \left(1 - g\,\frac{n-1}{n}\,\frac{z_2 - z_1}{RT_1}\right)^{n/(n-1)}$$

from which the lapse rate is $(T_2 - T_1)/(z_2 - z_1) = -g(n-1)/nR$ and thus n is obtained from $1/n = 1 + (R/g)(dt/dz)$. Defining properties of the U.S. standard atmosphere are listed in Table 2.

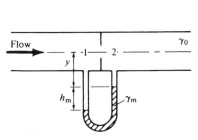

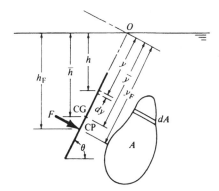

Figure 5 Differential manometer.

Figure 6 Flat inclined surface submerged in a liquid.

Table 1 Moments of Inertia for Various Plane Surfaces about Their Center of Gravity

Surface		I_{CG}
Rectangle or square		$\dfrac{1}{12}Ah^2$
Triangle		$\dfrac{1}{18}Ah^2$
Quadrant of circle (or semicircle)		$\left(\dfrac{1}{4}-\dfrac{16}{9\pi^2}\right)Ar^2 = 0.0699\,Ar^2$
Quadrant of ellipse (or semiellipse)		$\left(\dfrac{1}{4}-\dfrac{16}{9\pi^2}\right)Aa^2 = 0.0699\,Aa^2$
Parabola		$\left(\dfrac{3}{7}-\dfrac{9}{25}\right)Ah^2 = 0.0686\,Ah^2$
Circle		$\dfrac{1}{16}Ad^2$
Ellipse		$\dfrac{1}{16}Ah^2$

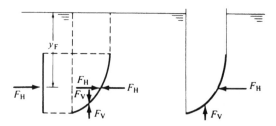

Figure 7 Curved surfaces submerged in a liquid.

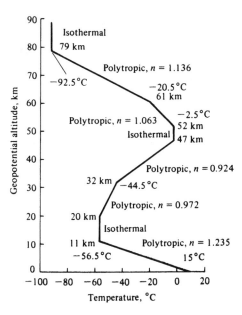

Figure 8   U.S. standard atmosphere.

Table 2   Defining Properties of the U.S. Standard Atmosphere

Altitude (m)	Temperature (°C)	Type of Atmosphere	Lapse Rate (°C/km)	$\bar{g}$ (m/s²)	n	Pressure, p (Pa)	Density, ρ (kg/m³)
0	15.0					1.013×10^5	1.225
		Polytropic	−6.5	9.790	1.235		
11,000	−56.5					2.263×10^4	3.639×10^{-1}
		Isothermal	0.0	9.759			
20,000	−56.5					5.475×10^3	8.804×10^{-2}
		Polytropic	+1.0	9.727	0.972		
32,000	−44.5					8.680×10^2	1.323×10^{-2}
		Polytropic	+2.8	9.685	0.924		
47,000	−2.5					1.109×10^2	1.427×10^{-3}
		Isothermal	0.0	9.654			
52,000	−2.5					5.900×10^1	7.594×10^{-4}
		Polytropic	−2.0	9.633	1.063		
61,000	−20.5					1.821×10^1	2.511×10^{-4}
		Polytropic	−4.0	9.592	1.136		
79,000	−92.5					1.038	2.001×10^{-5}
		Isothermal	0.0	9.549			
88,743	−92.5					1.644×10^{-1}	3.170×10^{-6}

The U.S. standard atmosphere is used in measuring altitudes with altimeters (pressure gauges) and, because the altimeters themselves do not account for variations in the air temperature beneath an aircraft, they read too high in cold weather and too low in warm weather.

3.4   Static Stability

For the *atmosphere* at rest, if an air mass moves very slowly vertically and remains there, the atmosphere is neutral. If vertical motion continues, it is unstable; if the air mass moves to return to its initial position, it is stable. It can be shown that atmospheric stability may be defined in terms of the polytropic exponent. If $n < k$, the atmosphere is stable (see Table 2); if $n = k$, it is neutral (adiabatic); and if $n > k$, it is unstable.

The stability of a body *submerged* in a fluid at rest depends on its response to forces which tend to tip it. If it returns to its original position, it is stable; if it continues to tip, it is unstable; and if it remains at rest in its tipped position, it is neutral. In Fig. 9 G is the center of gravity and B is the center of buoyancy. If the body in (a) is tipped to the position in (b), a couple Wd restores the body toward position (a) and thus the body is stable. If B were below G and the body displaced, it would move until B becomes above G. Thus stability requires that G is below B.

Floating bodies may be stable even though the center of buoyancy B is below the center of gravity G. The center of buoyancy generally changes position when a floating body tips because of the changing shape of the displaced liquid. The floating body is in equilibrium in Fig. 10a. In Fig. 10b the center of buoyancy is at B_1, and the restoring couple rotates the body toward its initial position in Fig. 10a. The intersection of BG is extended and a vertical line through B_1 is at M, the metacenter, and GM is the metacentric height. The body is stable if M is above G. Thus, the position of B relative to G determines stability of a submerged body, and the position of M relative to G determines the stability of floating bodies.

4   FLUID KINEMATICS

Fluid flows are classified in many ways. Flow is *steady* if conditions at a point do not vary with time, or for turbulent flow, if mean flow parameters do not vary with time. Otherwise the flow is *unsteady.* Flow is considered *one dimensional* if flow parameters are considered constant throughout a cross section, and variations occur only in the flow direction. *Two-dimensional* flow is the same in parallel planes and is not one dimensional. In *three-dimensional* flow gradients of flow parameters exist in three mutually perpendicular directions (x, y, and z). Flow may be *rotational* or *irrotational,* depending on whether the

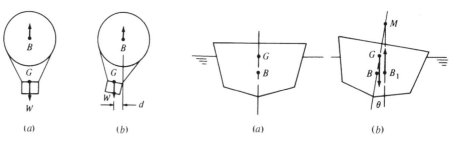

| (a) | (b) | | (a) | (b) |

Figure 9   Stability of a submerged body.   **Figure 10**   Floating body.

fluid particles rotate about their own centers or not. Flow is *uniform* if the velocity does not change in the direction of flow. If it does, the flow is *nonuniform. Laminar* flow exists when there are no lateral motions superimposed on the mean flow. When there are, the flow is *turbulent.* Flow may be intermittently laminar and turbulent; this is called flow in *transition.* Flow is considered *incompressible* if the density is constant, or in the case of gas flows, if the density variation is below a specified amount throughout the flow, 2–3%, for example. Low-speed gas flows may be considered essentially incompressible. Gas flows may be considered as *subsonic, transonic, sonic, supersonic,* or *hypersonic* depending on the gas speed compared with the speed of sound in the gas. Open-channel water flows may be designated as *subcritical, critical,* or *supercritical* depending on whether the flow is less than, equal to, or greater than the speed of an elementary surface wave.

4.1 Velocity and Acceleration

In Cartesian coordinates, velocity components are u, v, and w in the x, y, and z directions, respectively. These may vary with position and time, such that, for example, $u = dx/dt = u(x, y, z, t)$. Then

$$du = \frac{\partial u}{\partial x} dx + \frac{\partial u}{\partial y} dy + \frac{\partial u}{\partial z} dz + \frac{\partial u}{\partial t} dt$$

and

$$a_x = \frac{du}{dt} = \frac{\partial u}{\partial x}\frac{dx}{dt} + \frac{\partial u}{\partial y}\frac{dy}{dt} + \frac{\partial u}{\partial z}\frac{dz}{dt} + \frac{\partial u}{\partial t}$$

$$= \frac{Du}{Dt} = u\frac{\partial u}{\partial x} + v\frac{\partial u}{\partial y} + w\frac{\partial u}{\partial z} + \frac{\partial u}{\partial t}$$

The first three terms on the right hand side are the *convective* acceleration, which is zero for uniform flow, and the last term is the *local* acceleration, which is zero for steady flow.

In natural coordinates (streamline direction s, normal direction n, and meridional direction m normal to the plane of s and n), the velocity V is always in the streamline direction. Thus, $V = V(s, t)$ and

$$dV = \frac{\partial V}{\partial s} ds + \frac{\partial V}{\partial t} dt$$

$$a_s = \frac{dV}{dt} = V\frac{\partial V}{\partial s} + \frac{\partial V}{\partial t}$$

where the first term on the right-hand side is the *convective* acceleration and the last is the *local* acceleration. Thus, if the fluid velocity changes as the fluid moves throughout space, there is a convective acceleration, and if the velocity at a point changes with time, there is a local acceleration.

4.2 Streamlines

A *streamline* is a line to which, at each instant, velocity vectors are tangent. A *pathline* is the path of a particle as it moves in the fluid, and for steady flow it coincides with a streamline.

The equations of streamlines are described by stream functions ψ, from which the velocity components in two-dimensional flow are $u = -\partial\psi/\partial y$ and $v = +\partial\psi/\partial x$. Streamlines are lines of constant stream function. In polar coordinates

$$v_r = -\frac{1}{r}\frac{\partial\psi}{\partial\theta} \quad \text{and} \quad v_\theta = +\frac{\partial\psi}{\partial r}$$

Some streamline patterns are shown in Figs. 11, 12, and 13. The lines at right angles to the streamlines are potential lines.

4.3 Deformation of a Fluid Element

Four types of deformation or movement may occur as a result of spatial variations of velocity: translation, linear deformation, angular deformation, and rotation. These may occur singly or in combination. Motion of the face (in the x-y plane) of an elemental cube of sides δx, δy, and δz in a time dt is shown in Fig. 14. Both translation and rotation involve motion or deformation without a change in shape of the fluid element. Linear and angular deformations, however, do involve a change in shape of the fluid element. Only through these linear and angular deformations are heat generated and mechanical energy dissipated as a result of viscous action in a fluid.

For linear deformation the relative change in volume is at a rate of

$$(\mathcal{V}_{dt} - \mathcal{V}_0)/\mathcal{V}_0 = \frac{\partial u}{\partial x} + \frac{\partial v}{\partial y} + \frac{\partial w}{\partial z} = \text{div V}$$

which is zero for an incompressible fluid, and thus is an expression for the continuity equation. Rotation of the face of the cube shown in Fig. 14d is the average of the rotations of the bottom and left edges, which is

$$\frac{1}{2}\left(\frac{\partial v}{\partial x} - \frac{\partial u}{\partial y}\right) dt$$

The rate of rotation is the angular velocity and is

$$\omega_z = \frac{1}{2}\left(\frac{\partial v}{\partial x} - \frac{\partial u}{\partial y}\right) \quad \text{about the } z \text{ axis in the } x\text{-}y \text{ plane}$$

$$\omega_x = \frac{1}{2}\left(\frac{\partial w}{\partial y} - \frac{\partial v}{\partial z}\right) \quad \text{about the } x \text{ axis in the } y\text{-}z \text{ plane}$$

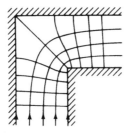

Figure 11   Flow around a corner in a duct.

Figure 12   Flow around a corner into a duct.

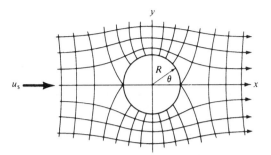

Figure 13 Inviscid flow past a cylinder.

and

$$\omega_y = \frac{1}{2}\left(\frac{\partial u}{\partial z} - \frac{\partial w}{\partial x}\right) \quad \text{about the } y \text{ axis in the } x\text{-}z \text{ plane}$$

These are the components of the angular velocity vector Ω,

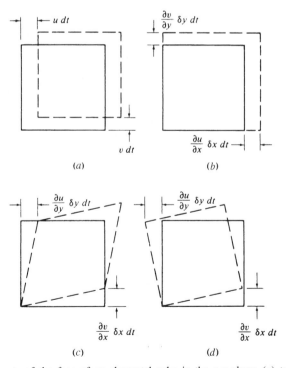

Figure 14 Movements of the face of an elemental cube in the x-y plane: (a) translation; (b) linear deformation; (c) angular deformation; (d) rotation.

$$\Omega = \frac{1}{2} \text{ curl } \mathbf{V} = \frac{1}{2} \begin{vmatrix} \mathbf{i} & \mathbf{j} & \mathbf{k} \\ \frac{\partial}{\partial x} & \frac{\partial}{\partial y} & \frac{\partial}{\partial z} \\ u & v & w \end{vmatrix} = \omega_x \mathbf{i} + \omega_y \mathbf{j} + \omega_z \mathbf{k}$$

If the flow is irrotational, these quantities are zero.

4.4 Vorticity and Circulation

Vorticity is defined as twice the angular velocity, and thus is also zero for irrotational flow. Circulation is defined as the line integral of the velocity component along a closed curve and equals the total strength of all vertex filaments that pass through the curve. Thus, the vorticity at a point within the curve is the circulation per unit area enclosed by the curve. These statements are expressed by

$$\Gamma = \oint \mathbf{V} \cdot d\mathbf{l} = \oint (u \, dx + v \, dy + w \, dz) \qquad \text{and} \qquad \zeta_A = \lim_{A \to 0} \frac{\Gamma}{A}$$

Circulation—the product of vorticity and area—is the counterpart of volumetric flow rate as the product of velocity and area. These are shown in Fig. 15.

Physically, fluid rotation at a point in a fluid is the instantaneous average rotation of two mutually perpendicular infinitesimal line segments. In Fig. 16 the line δx rotates positively and δy rotates negatively. Then $\omega_x = (\partial v/\partial x - \partial u/\partial y)/2$. In natural coordinates (the n direction is opposite to the radius of curvature r) the angular velocity in the s-n plane is

$$\omega = \frac{1}{2} \frac{\Gamma}{\delta A} = \frac{1}{2} \left(\frac{V}{r} - \frac{\partial V}{\partial n} \right) = \frac{1}{2} \left(\frac{V}{r} + \frac{\partial V}{\partial r} \right)$$

This shows that for irrotational motion $V/r = \partial V/\partial n$ and thus the peripheral velocity V increases toward the center of curvature of streamlines. In an irrotational vortex, $Vr = C$ and in a solid-body-type or rotational vortex, $V = \omega r$.

A combined vortex has a solid-body-type rotation at the core and an irrotational vortex beyond it. This is typical of a tornado (which has an inward sink flow superimposed on the vortex motion) and eddies in turbulent motion.

4.5 Continuity Equations

Conservation of mass for a fluid requires that in a *material* volume, the mass remains constant. In a *control* volume the net rate of influx of mass into the control volume is equal to

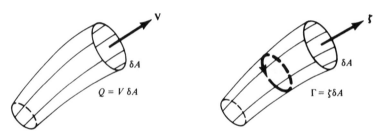

Figure 15 Similarity between a stream filament and a vortex filament.

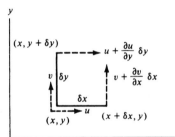

Figure 16 Rotation of two line segments in a fluid.

the rate of change of mass in the control volume. Fluid may flow into a control volume either through the control surface or from internal sources. Likewise, fluid may flow out through the control surface or into internal sinks. The various forms of the continuity equations listed in Table 3 do not include sources and sinks; if they exist, they must be included.

The most commonly used forms for duct flow are $\dot{m} = VA\rho$ in kg/sec, where V is the average flow velocity in m/sec, A is the duct area in m³, and ρ is the fluid density in kg/

Table 3 Continuity Equations

General	$\dfrac{\partial \rho}{\partial t} + \nabla \cdot \rho \mathbf{V} = 0$ or $\dfrac{D\rho}{Dt} + \rho \nabla \cdot \mathbf{V} = 0$	Vector
Unsteady, compressible	$\dfrac{\partial \rho}{\partial t} + \dfrac{\partial(\rho u)}{\partial x} + \dfrac{\partial(\rho v)}{\partial y} + \dfrac{\partial(\rho w)}{\partial z} = 0$	Cartesian
	$\dfrac{\partial \rho}{\partial t} + \dfrac{\partial(\rho v_r)}{\partial r} + \dfrac{1}{r}\dfrac{\partial(\rho v_\theta)}{\partial \theta} + \dfrac{\partial(\rho v_z)}{\partial z} + \dfrac{\rho v_r}{r} = 0$	Cylindrical
	$\dfrac{\partial(\rho A)}{\partial t} + \dfrac{\partial}{\partial s}(\rho \mathbf{V} \cdot \mathbf{A}) = 0$	Duct
Steady, compressible	$\nabla \cdot \rho \mathbf{V} = 0$	Vector
	$\dfrac{\partial(\rho u)}{\partial x} + \dfrac{\partial(\rho v)}{\partial y} + \dfrac{\partial(\rho w)}{\partial z} = 0$	Cartesian
	$\dfrac{\partial(\rho v_r)}{\partial r} + \dfrac{1}{r}\dfrac{\partial(\rho v_\theta)}{\partial \theta} + \dfrac{\partial(\rho v_z)}{\partial z} + \dfrac{\rho v_r}{r} = 0$	Cylindrical
	$\rho \mathbf{V} \cdot \mathbf{A} = \dot{m}$	
Incompressible,	$\nabla \cdot \mathbf{V} = 0$	Vector
steady or unsteady	$\dfrac{\partial u}{\partial x} + \dfrac{\partial v}{\partial y} + \dfrac{\partial w}{\partial z} = 0$	Cartesian
	$\dfrac{\partial v_r}{\partial r} + \dfrac{1}{r}\dfrac{\partial v_\theta}{\partial \theta} + \dfrac{\partial v_z}{\partial z} + \dfrac{v_r}{r} = 0$	Cylindrical
	$\mathbf{V} \cdot \mathbf{A} = Q$	Duct

m³. In differential form this is $dV/V + dA/A + d\rho/\rho = 0$, which indicates that all three quantities may not increase nor all decrease in the direction of flow. For incompressible duct flow $Q = VA$ m³/sec, where V and A are as above. When the velocity varies throughout a cross section, the average velocity is

$$V = \frac{1}{A} \int u \, dA = \frac{1}{n} \sum_{i=1}^{n} u_i$$

where u is a velocity at a point, and u_i are point velocities measured at the centroid of n equal areas. For example, if the velocity is u at a distance y from the wall of a pipe of radius R and the centerline velocity is u_m, $u = u_m(y/R)^{1/7}$ and the average velocity is $V = {}^{49}\!/_{60}\, u_m$.

5 FLUID MOMENTUM

The momentum theorem states that the net external force acting on the fluid within a control volume equals the time rate of change of momentum of the fluid plus the net rate of momentum flux or transport out of the control volume through its surface. This is one form of the Reynolds transport theorem, which expresses the conservation laws of physics for fixed mass systems to expressions for a control volume:

$$\Sigma \mathbf{F} = \frac{D}{Dt} \int_{\substack{\text{material} \\ \text{volume}}} \rho \mathbf{V} \, d\forall$$

$$= \frac{\partial}{\partial t} \int_{\substack{\text{control} \\ \text{volume}}} \rho \mathbf{V} \, d\forall + \int_{\substack{\text{control} \\ \text{surface}}} \rho \mathbf{V}(\mathbf{V} \cdot d\mathbf{s})$$

5.1 The Momentum Theorem

For steady flow the first term on the right-hand side of the preceding equation is zero. Forces include normal forces due to pressure and tangential forces due to viscous shear over the surface S of the control volume, and body forces due to gravity and centrifugal effects, for example. In scalar form the net force equals the total momentum flux leaving the control volume minus the total momentum flux entering the control volume. In the x direction

$$\Sigma F_x = (\dot{m}V_x)_{\text{leaving } S} - (\dot{m}V_x)_{\text{entering } S}$$

or when the same fluid enters and leaves,

$$\Sigma F_x = \dot{m}(V_{x \text{ leaving } S} - V_{x \text{ entering } S})$$

with similar expressions for the y and z directions.

For one-dimensional flow $\dot{m}V_x$ represents momentum flux passing a section and V_x is the average velocity. If the velocity varies across a duct section, the true momentum flux is $\int_A (u\rho dA)u$, and the ratio of this value to that based upon average velocity is the momentum correction factor β,

$$\beta = \frac{\int_A u^2 \, dA}{V^2 A} \geq 1$$

$$\approx \frac{1}{V^2 n} \sum_{i=1}^{n} u_i^2$$

For laminar flow in a circular tube, $\beta = 4/3$; for laminar flow between parallel plates, $\beta = 1.20$; and for turbulent flow in a circular tube, β is about 1.02–1.03.

5.2 Equations of Motion

For steady irrotational flow of an incompressible nonviscous fluid, Newton's second law gives the Euler equation of motion. Along a streamline it is

$$V \frac{\partial V}{\partial s} + \frac{1}{\rho} \frac{\partial p}{\partial s} + g \frac{\partial z}{\partial s} = 0$$

and normal to a streamline it is

$$\frac{V^2}{r} + \frac{1}{\rho} \frac{\partial p}{\partial n} + g \frac{\partial z}{\partial n} = 0$$

When integrated, these show that the sum of the kinetic, displacement, and potential energies is a constant along streamlines as well as across streamlines. The result is known as the Bernoulli equation:

$$\frac{V^2}{2} + \frac{p}{\rho} + gz = \text{constant energy per unit mass}$$

$$\frac{\rho V_1^2}{2} + p_1 + \rho g z_1 = \frac{\rho V_2^2}{2} + p_2 + \rho g z_2 = \text{constant total pressure}$$

and

$$\frac{V_1^2}{2g} + \frac{p_1}{g\rho} + z_1 = \frac{V_2^2}{2g} + \frac{p_2}{g\rho} + z_2 = \text{constant total head}$$

For a reversible adiabatic compressible gas flow with no external work, the Euler equation integrates to

$$\frac{V_1^2}{2} + \frac{k}{k-1} \left(\frac{p_1}{\rho_1} \right) + g z_1 = \frac{V_2^2}{2} + \frac{k}{k-1} \left(\frac{p_2}{\rho_2} \right) + g z_2$$

which is valid whether the flow is reversible or not, and corresponds to the steady-flow energy equation for adiabatic no-work gas flow.

Newton's second law written normal to streamlines shows that in horizontal planes $dp/dr = \rho V^2/r$, and thus dp/dr is positive for both rotational and irrotational flow. The pressure increases away from the center of curvature and decreases toward the center of curvature of curvilinear streamlines. The radius of curvature r of straight lines is infinite, and thus no pressure gradient occurs across these.

For a liquid rotating as a solid body

$$-\frac{V_1^2}{2g} + \frac{p_1}{\rho g} + z_1 = -\frac{V_2^2}{2g} + \frac{p_2}{\rho g} + z_2$$

The negative sign balances the increase in velocity and pressure with radius.

The differential equations of motion for a viscous fluid are known as the Navier–Stokes equations. For incompressible flow the x-component equation is

$$\frac{\partial u}{\partial t} + u\frac{\partial u}{\partial x} + v\frac{\partial u}{\partial y} + w\frac{\partial u}{\partial z} = X - \frac{1}{\rho}\frac{\partial p}{\partial x} + v\left(\frac{\partial^2 u}{\partial x^2} + \frac{\partial^2 u}{\partial y^2} + \frac{\partial^2 u}{\partial z^2}\right)$$

with similar expressions for the y and z directions. X is the body force per unit mass. Reynolds developed a modified form of these equations for turbulent flow by expressing each velocity as an average value plus a fluctuating component ($u = \bar{u} + u'$ and so on). These modified equations indicate shear stresses from turbulence ($\tau_T = -\rho u'v'$, for example) known as the Reynolds stresses, which have been useful in the study of turbulent flow.

6 FLUID ENERGY

The Reynolds transport theorem for fluid passing through a control volume states that the heat added to the fluid less any work done by the fluid increases the energy content of the fluid in the control volume or changes the energy content of the fluid as it passes through the control surface. This is

$$Q - Wk_{\text{done}} = \frac{\partial}{\partial t}\int_{\substack{\text{control}\\\text{volume}}} (e\rho)\,d\mathcal{V} + \int_{\substack{\text{control}\\\text{surface}}} e\rho(\mathbf{V}\cdot d\mathbf{S})$$

and represents the first law of thermodynamics for control volume. The energy content includes kinetic, internal, potential, and displacement energies. Thus, mechanical and thermal energies are included, and there are no restrictions on the direction of interchange from one form to the other implied in the first law. The second law of thermodynamics governs this.

6.1 Energy Equations

With reference to Fig. 17, the steady flow energy equation is

$$\alpha_1\frac{V_1^2}{2} + p_1v_1 + gz_1 + u_1 + q - w = \alpha_2\frac{V_2^2}{2} + p_2v_2 + gz_2 + u_2$$

in terms of energy per unit mass, and where α is the kinetic energy correction factor:

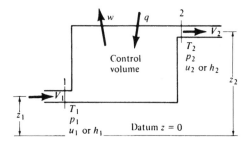

Figure 17 Control volume for steady-flow energy equation.

$$\alpha = \frac{\int_A u^3 \, dA}{V^3 A} \approx \frac{1}{V^3 n} \sum_{i=1}^{n} u_i^3 \geq 1$$

For laminar flow in a pipe, $\alpha = 2$; for turbulent flow in a pipe, $\alpha = 1.05$–1.06; and if one-dimensional flow is assumed, $\alpha = 1$.

For one-dimensional flow of compressible gases, the general expression is

$$\frac{V_1^2}{2} + h_1 + gz_1 + q - w = \frac{V_2^2}{2} + h_2 + gz_2$$

For adiabatic flow, $q = 0$; for no external work, $w = 0$; and in most instances changes in elevation z are very small compared with changes in other parameters and can be neglected. Then the equation becomes

$$\frac{V_1^2}{2} + h_1 = \frac{V_2^2}{2} + h_2 = h_0$$

where h_0 is the stagnation enthalpy. The stagnation temperature is then $T_0 = T_1 + V_1^2/2c_p$ in terms of the temperature and velocity at some point 1. The gas velocity in terms of the stagnation and static temperatures, respectively, is $V_1 = \sqrt{2c_p(T_0 - T_1)}$. An increase in velocity is accompanied by a decrease in temperature, and vice versa.

For one-dimensional flow of liquids and constant-density (low-velocity) gases, the energy equation generally is written in terms of energy per unit weight as

$$\frac{V_1^2}{2g} + \frac{p_1}{\gamma} + z_1 - w = \frac{V_2^2}{2g} + \frac{p_2}{\gamma} + z_2 + h_L$$

where the first three terms are velocity, pressure, and potential heads, respectively. The head loss $h_L = (u_2 - u_1 - q)/g$ and represents the mechanical energy dissipated into thermal energy irreversibly (the heat transfer q is assumed zero here). It is a positive quantity and increases in the direction of flow.

Irreversibility in compressible gas flows results in an entropy increase. In Fig. 18 reversible flow between pressures p' and p is from a to b or from b to a. Irreversible flow

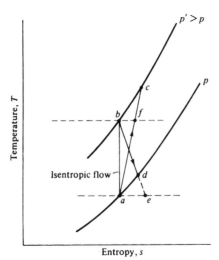

Figure 18 Reversible and irreversible adiabatic flows.

from p' to p is from b to d, and from p to p' it is from a to c. Thus, frictional duct flow from one pressure to another results in a higher final temperature, and a lower final velocity, in both instances. For frictional flow between given temperatures (T_a and T_b, for example), the resulting pressures are lower than for frictionless flow (p_c is lower than p_a and p_f is lower than p_b).

6.2   Work and Power

Power is the rate at which work is done, and is the work done per unit mass times the mass flow rate, or the work done per unit weight times the weight flow rate.

Power represented by the work term in the energy equation is $P = w(VA\gamma) = w(VA\rho)$ W.

Power in a jet at a velocity V is $P = (V^2/2)(VA\rho) = (V^2/2g)(VA\gamma)$ W.

Power loss resulting from head loss is $P = h_L(VA\gamma)$ W.

Power to overcome a drag force is $P = FV$ W.

Power available in a hydroelectric power plant when water flows from a headwater elevation z_1 to a tailwater elevation z_2 is $P = (z_1 - z_2)(Q\gamma)$ W, where Q is the volumetric flow rate.

6.3   Viscous Dissipation

Dissipation effects resulting from viscosity account for entropy increases in adiabatic gas flows and the heat loss term for flows of liquids. They can be expressed in terms of the rate at which work is done—the product of the viscous shear force on the surface of an elemental fluid volume and the corresponding component of velocity parallel to the force. Results for a cube of sides dx, dy, and dz give the dissipation function Φ:

$$\Phi = 2\mu \left[\left(\frac{\partial u}{\partial x} \right)^2 + \left(\frac{\partial v}{\partial y} \right)^2 + \left(\frac{\partial w}{\partial z} \right)^2 \right]$$

$$+ \mu \left[\left(\frac{\partial v}{\partial x} + \frac{\partial u}{\partial y} \right)^2 + \left(\frac{\partial w}{\partial y} + \frac{\partial v}{\partial z} \right)^2 + \left(\frac{\partial u}{\partial z} + \frac{\partial w}{\partial x} \right)^2 \right]$$

$$- \frac{2}{3} \mu \left(\frac{\partial u}{\partial x} + \frac{\partial v}{\partial y} + \frac{\partial w}{\partial z} \right)^2$$

The last term is zero for an incompressible fluid. The first term in brackets is the linear deformation, and the second term in brackets is the angular deformation and in only these two forms of deformation is there heat generated as a result of viscous shear within the fluid. The second law of thermodynamics precludes the recovery of this heat to increase the mechanical energy of the fluid.

7   CONTRACTION COEFFICIENTS FROM POTENTIAL FLOW THEORY

Useful engineering results of a conformal mapping technique were obtained by von Mises for the contraction coefficients of two-dimensional jets for nonviscous incompressible fluids in the absence of gravity. The ratio of the resulting cross-sectional area of the jet to the area of the boundary opening is called the *coefficient of contraction, C_c*. For flow geometries shown in Fig. 19, von Mises calculated the values of C_c listed in Table 4. The values agree well with measurements for low-viscosity liquids. The results tabulated for two-dimensional flow may be used for axisymmetric jets if C_c is defined by $C_c = b_{jet}/b = (d_{jet}/d)^2$ and if d and D are diameters equivalent to widths b and B, respectively. Thus, if a small round hole

Potential flow

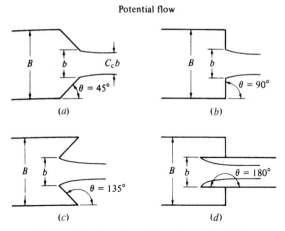

Figure 19 Geometry of two-dimensional jets.

of diameter d in a large tank ($d/D \approx 0$), the jet diameter would be $(0.611)^{1/2} = 0.782$ times the hole diameter, since $\theta = 90°$.

8 DIMENSIONLESS NUMBERS AND DYNAMIC SIMILARITY

Dimensionless numbers are commonly used to plot experimental data to make the results more universal. Some are also used in designing experiments to ensure dynamic similarity between the flow of interest and the flow being studied in the laboratory.

8.1 Dimensionless Numbers

Dimensionless numbers or groups may be obtained from force ratios, by a dimensional analysis using the Buckingham Pi theorem, for example, or by writing the differential equations of motion and energy in dimensionless form. Dynamic similarity between two geo-

Table 4 Coefficients of Contraction for Two-Dimensional Jets

b/B	C_c $\theta = 45°$	C_c $\theta = 90°$	C_c $\theta = 135°$	C_c $\theta = 180°$
0.0	0.746	0.611	0.537	0.500
0.1	0.747	0.612	0.546	0.513
0.2	0.747	0.616	0.555	0.528
0.3	0.748	0.622	0.566	0.544
0.4	0.749	0.631	0.580	0.564
0.5	0.752	0.644	0.599	0.586
0.6	0.758	0.662	0.620	0.613
0.7	0.768	0.687	0.652	0.646
0.8	0.789	0.722	0.698	0.691
0.9	0.829	0.781	0.761	0.760
1.0	1.000	1.000	1.000	1.000

metrically similar systems exists when the appropriate dimensionless groups are the same for the two systems. This is the basis on which model studies are made, and results measured for one flow may be applied to similar flows.

The dimensions of some parameters used in fluid mechanics are listed in Table 5. The mass–length–time (MLT) and the force–length–time (FLT) systems are related by $F = Ma = ML/T^2$ and $M = FT^2/L$.

Force ratios are expressed as

$$\frac{\text{Inertia force}}{\text{Viscous force}} = \frac{\rho L^2 V^2}{\mu VL} = \frac{\rho LV}{\mu}, \quad \text{the Reynolds number Re}$$

$$\frac{\text{Inertia force}}{\text{Gravity force}} = \frac{\rho L^2 V^2}{\rho L^3 g} = \frac{V^2}{Lg} \quad \text{or} \quad \frac{V}{\sqrt{Lg}}, \quad \text{the Froude number Fr}$$

$$\frac{\text{Pressure force}}{\text{Inertia force}} = \frac{\Delta p L^2}{\rho L^2 V^2} = \frac{\Delta p}{\rho V^2} \quad \text{or} \quad \frac{\Delta p}{\rho V^2/2}, \quad \text{the pressure coefficient } C_p$$

Table 5   Dimensions of Fluid and Flow Parameters

	FLT	MLT
Geometrical characteristics		
Length (diameter, height, breadth, chord, span, etc.)	L	L
Angle	None	None
Area	L^2	L^2
Volume	L^3	L^3
Fluid properties[a]		
Mass	FT^2/L	M
Density (ρ)	FT^2/L^4	M/L^3
Specific weight (γ)	F/L^3	M/L^2T^2
Kinematic viscosity (v)	L^2/T	L^2/T
Dynamic viscosity (μ)	FT/L^2	M/LT
Elastic modulus (K)	F/L^2	M/LT^2
Surface tension (σ)	F/L	M/T^2
Flow characteristics		
Velocity (V)	L/T	L/T
Angular velocity (ω)	$1/T$	$1/T$
Acceleration (a)	L/T^2	L/T^2
Pressure (Δp)	F/L^2	M/LT^2
Force (drag, lift, shear)	F	ML/T^2
Shear stress (τ)	F/L^2	M/LT^2
Pressure gradient ($\Delta p/L$)	F/L^3	M/L^2T^2
Flow rate (Q)	L^3/T	L^3/T
Mass flow rate ($\dot{m}$)	FT/L	M/T
Work or energy	FL	ML^2/T^2
Work or energy per unit weight	L	L
Torque and moment	FL	ML^2/T^2
Work or energy per unit mass	L^2/T^2	L^2/T^2

[a]Density, viscosity, elastic modulus, and surface tension depend on temperature, and therefore temperature will not be considered a property in the sense used here.

$$\frac{\text{Inertia force}}{\text{Surface tension force}} = \frac{\rho L^2 V^2}{\sigma L} = \frac{V^2}{\sigma/\rho L} \quad \text{or} \quad \frac{V}{\sqrt{\sigma/\rho L}}, \quad \text{the Weber number We}$$

$$\frac{\text{Inertia force}}{\text{Compressibility force}} = \frac{\rho L^2 V^2}{K L^2} = \frac{V^2}{K/\rho} \quad \text{or} \quad \frac{V}{\sqrt{K/\rho}}, \quad \text{the Mach number M}$$

If a system includes n quantities with m dimensions, there will be at least $n - m$ independent dimensionless groups, each containing m repeating variables. Repeating variables (1) must include all the m dimensions, (2) should include a geometrical characteristic, a fluid property, and a flow characteristic and (3) should not include the dependent variable.

Thus, if the pressure gradient $\Delta p/L$ for flow in a pipe is judged to depend on the pipe diameter D and roughness k, the average flow velocity V, and the fluid density ρ, the fluid viscosity μ, and compressibility K (for gas flow), then $\Delta p/L = f(D, k, V, \rho, \mu, K)$ or in dimensions, $F/L^3 = f(L, L, L/T, FT^2/L^4, FT/L^2, F/L^2)$, where $n = 7$ and $m = 3$. Then there are $n - m = 4$ independent groups to be sought. If D, ρ, and V are the repeating variables, the results are

$$\frac{\Delta p}{\rho V^2/2} = f\left(\frac{DV\rho}{\mu}, \frac{k}{D}, \frac{V}{\sqrt{K/\rho}}\right)$$

or that the friction factor will depend on the Reynolds number of the flow, the relative roughness, and the Mach number. The actual relationship between them is determined experimentally. Results may be determined analytically for laminar flow. The seven original variables are thus expressed as four dimensionless variables, and the Moody diagram of Fig. 32 shows the result of analysis and experiment. Experiments show that the pressure gradient does depend on the Mach number, but the friction factor does not.

The Navier–Stokes equations are made dimensionless by dividing each length by a characteristic length L and each velocity by a characteristic velocity U. For a body force X due to gravity, $X = g_x = g(\partial z/\partial x)$. Then $x' = x/L$, etc., $t' = t(LU)$, $u' = u/U$, etc., and $p' = p/\rho U^2$. Then the Navier–Stokes equation (x component) is

$$u' \frac{\partial u'}{\partial x'} + v' \frac{\partial u'}{\partial y'} + w' \frac{\partial u'}{\partial z'} + \frac{\partial u'}{\partial t'}$$

$$= \frac{gL}{U^2} - \frac{\partial p'}{\partial x'} + \frac{\mu}{\rho UL}\left(\frac{\partial^2 u'}{\partial x'^2} + \frac{\partial^2 u'}{\partial y'^2} + \frac{\partial^2 u'}{\partial z'^2}\right)$$

$$= \frac{1}{\text{Fr}^2} - \frac{\partial p'}{\partial x'} + \frac{1}{\text{Re}}\left(\frac{\partial^2 u'}{\partial x'^2} + \frac{\partial^2 u'}{\partial y'^2} + \frac{\partial^2 u'}{\partial z'^2}\right)$$

Thus for incompressible flow, similarity of flow in similar situations exists when the Reynolds and the Froude numbers are the same.

For compressible flow, normalizing the differential energy equation in terms of temperatures, pressure, and velocities gives the Reynolds, Mach, and Prandtl numbers as the governing parameters.

8.2 Dynamic Similitude

Flow systems are considered to be dynamically similar if the appropriate dimensionless numbers are the same. Model tests of aircraft, missiles, rivers, harbors, breakwaters, pumps, turbines, and so forth are made on this basis. Many practical problems exist, however, and it is not always possible to achieve complete dynamic similarity. When viscous forces govern the flow, the Reynolds number should be the same for model and prototype, the length in the Reynolds number being some characteristic length. When gravity forces govern the flow, the Froude number should be the same. When surface tension forces are significant, the Weber number is used. For compressible gas flow, the Mach number is used; different gases may be used for the model and prototype. The pressure coefficient $C_p = \Delta p/(\rho V^2/2)$, the drag coefficient $C_D = \mathrm{drag}/(\rho V^2/2)A$, and the lift coefficient $C_L = \mathrm{lift}/(\rho V^2/2)A$ will be the same for model and prototype when the appropriate Reynolds, Froude, or Mach number is the same. A cavitation number is used in cavitation studies, $\sigma_v = (p - p_v)/(\rho V^2/2)$ if vapor pressure p_v is the reference pressure or $\sigma_c = (p - p_c)/(\rho V^2/2)$ if a cavity pressure is the reference pressure.

Modeling ratios for conducting tests are listed in Table 6. Distorted models are often used for rivers in which the vertical scale ratio might be 1/40 and the horizontal scale ratio 1/100, for example, to avoid surface tension effects and laminar flow in models too shallow.

Incomplete similarity often exists in Froude–Reynolds models since both contain a length parameter. Ship models are tested with the Froude number parameter, and viscous effects are calculated for both model and prototype.

Table 6 Modeling Ratios[a]

Ratio	Reynolds Number	Froude Number, Undistorted Model[b]	Froude Number, Distorted Model[b]	Mach Number, Same Gas[d]	Mach Number, Different Gas[d]
Velocity $\dfrac{V_m}{V_p}$	$\dfrac{L_p}{L_m}\dfrac{\rho_p}{\rho_m}\dfrac{\mu_m}{\mu_p}$	$\left(\dfrac{L_m}{L_p}\right)^{1/2}$	$\left(\dfrac{L_m}{L_p}\right)^{1/2}_V$	$\left(\dfrac{\theta_m}{\theta_p}\right)^{1/2}$	$\left(\dfrac{k_m R_m \theta_m}{k_p R_p \theta_p}\right)^{1/2}$
Angular velocity $\dfrac{\omega_m}{\omega_p}$	$\left(\dfrac{L_p}{L_m}\right)^2\dfrac{\rho_p}{\rho_m}\dfrac{\mu_m}{\mu_p}$	$\left(\dfrac{L_p}{L_m}\right)^{1/2}$	—[c]	$\left(\dfrac{\theta_m}{\theta_p}\right)^{1/2}\dfrac{L_p}{L_m}$	$\left(\dfrac{k_m R_m \theta_m}{k_p R_p \theta_p}\right)^{1/2}\dfrac{L_p}{L_m}$
Volumetric flow rate $\dfrac{Q_m}{Q_p}$	$\dfrac{L_m}{L_p}\dfrac{\rho_p}{\rho_m}\dfrac{\mu_m}{\mu_p}$	$\left(\dfrac{L_m}{L_p}\right)^{5/2}$	$\left(\dfrac{L_m}{L_p}\right)^{3/2}_V\left(\dfrac{L_m}{L_p}\right)_H$	—[c]	—[c]
Time $\dfrac{t_m}{t_p}$	$\left(\dfrac{L_m}{L_p}\right)^2\dfrac{\rho_m}{\rho_p}\dfrac{\mu_p}{\mu_m}$	$\left(\dfrac{L_m}{L_p}\right)^{1/2}\left(\dfrac{g_p}{g_m}\right)^{1/2}$	$\left(\dfrac{L_m}{L_p}\right)_H\left(\dfrac{L_p}{L_m}\right)^{1/2}_V\left(\dfrac{g_p}{g_m}\right)^{1/2}$	$\left(\dfrac{\theta_p}{\theta_m}\right)^{1/2}\dfrac{L_m}{L_p}$	$\left(\dfrac{k_p R_p \theta_p}{k_m R_m \theta_m}\right)^{1/2}\dfrac{L_m}{L_p}$
Force $\dfrac{F_m}{F_p}$	$\left(\dfrac{\mu_m}{\mu_p}\right)^2\dfrac{\rho_p}{\rho_m}$	$\left(\dfrac{L_m}{L_p}\right)^3\dfrac{\rho_m}{\rho_p}$	$\dfrac{\rho_m}{\rho_p}\left(\dfrac{L_m}{L_p}\right)_H\left(\dfrac{L_m}{L_p}\right)^2_V$	$\dfrac{\rho_m}{\rho_p}\dfrac{\theta_m}{\theta_p}\left(\dfrac{L_m}{L_p}\right)^2$	$\dfrac{K_m}{K_p}\left(\dfrac{L_m}{L_p}\right)^2$

[a]Subscript m indicates model, subscript p indicates prototype.
[b]For the same value of gravitational acceleration for model and prototype.
[c]Of little importance.
[d]Here θ refers to temperature.

The specific speed of pumps and turbines results from combining groups in a dimensional analysis of rotary systems. That for pumps is $N_{s\,(\text{pump})} = N\sqrt{Q}/e^{3/4}$ and for turbines it is $N_{s\,(\text{turbines})} = N\sqrt{\text{power}}/\rho^{1/2}e^{5/4}$, where N is the rotational speed in rad/sec, Q is the volumetric flow rate in m³/sec, and e is the energy in J/kg. North American practice uses N in rpm, Q in gal/min, e as energy per unit weight (head in ft), power as brake horsepower rather than watts, and omits the density term in the specific speed for turbines. The numerical value of specific speed indicates the type of pump or turbine for a given installation. These are shown for pumps in North America in Fig. 20. Typical values for North American turbines are about 5 for impulse turbines, about 20–100 for Francis turbines, and 100–200 for propeller turbines. Slight corrections in performance for higher efficiency of large pumps and turbines are made when testing small laboratory units.

9 VISCOUS FLOW AND INCOMPRESSIBLE BOUNDARY LAYERS

In viscous flows, adjacent layers of fluid transmit both normal forces and tangential shear forces, as a result of relative motion between the layers. There is no relative motion, however, between the fluid and a solid boundary along which it flows. The fluid velocity varies from zero at the boundary to a maximum or free stream value some distance away from it. This region of retarded flow is called the boundary layer.

9.1 Laminar and Turbulent Flow

Viscous fluids flow in a laminar or in a turbulent state. There are, however, transition regimes between them where the flow is intermittently laminar and turbulent. Laminar flow is smooth, quiet flow without lateral motions. Turbulent flow has lateral motions as a result of eddies superimposed on the main flow, which results in random or irregular fluctuations of velocity, pressure, and, possibly, temperature. Smoke rising from a cigarette held at rest in still air has a straight threadlike appearance for a few centimeters; this indicates a laminar flow. Above that the smoke is wavy and finally irregular lateral motions indicate a turbulent flow. Low velocities and high viscous forces are associated with laminar flow and low Reynolds

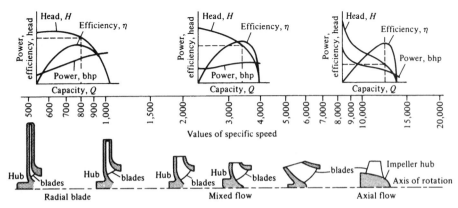

Figure 20 Pump characteristics and specific speed for pump impellers. (Courtesy Worthington Corporation)

numbers. High speeds and low viscous forces are associated with turbulent flow and high Reynolds numbers. Turbulence is a characteristic of flows, not of fluids. Typical fluctuations of velocity in a turbulent flow are shown in Fig. 21.

The axes of eddies in turbulent flow are generally distributed in all directions. In *isotropic* turbulence they are distributed equally. In flows of low turbulence, the fluctuations are small; in highly turbulent flows, they are large. The turbulence level may be defined as (as a percentage)

$$T = \frac{\sqrt{(\overline{u'^2} + \overline{v'^2} + \overline{w'^2})/3}}{\overline{u}} \times 100$$

where u', v', and w' are instantaneous fluctuations from mean values and $\overline{u}$ is the average velocity in the main flow direction (x, in this instance).

Shear stresses in turbulent flows are much greater than in laminar flows for the same velocity gradient and fluid.

9.2 Boundary Layers

The growth of a boundary layer along a flat plate in a uniform external flow is shown in Fig. 22. The region of retarded flow, δ, thickens in the direction of flow, and thus the velocity changes from zero at the plate surface to the free stream value u_s in an increasingly larger distance δ normal to the plate. Thus, the velocity gradient at the boundary, and hence the shear stress as well, decreases as the flow progresses downstream, as shown. As the laminar boundary thickens, instabilities set in and the boundary layer becomes turbulent. The transition from the laminar boundary layer to a turbulent boundary layer does not occur at a well-defined location; the flow is intermittently laminar and turbulent with a larger portion of the flow being turbulent as the flow passes downstream. Finally, the flow is completely turbulent, and the boundary layer is much thicker and the boundary shear greater in the turbulent region than if the flow were to continue laminar. A viscous sublayer exists within the turbulent boundary layer along the boundary surface. The shape of the velocity profile also changes when the boundary layer becomes turbulent, as shown in Fig. 22. Boundary surface roughness, high turbulence level in the outer flow, or a decelerating free stream causes transition to occur nearer the leading edge of the plate. A surface is considered rough if the roughness elements have an effect outside the viscous sublayer, and smooth if they do not. Whether a surface is rough or smooth depends not only on the surface itself but also on the character of the flow passing it.

A boundary layer will separate from a continuous boundary if the fluid within it is caused to slow down such that the velocity gradient du/dy becomes zero at the boundary. An adverse pressure gradient will cause this.

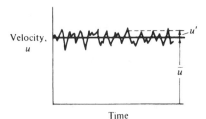

Velocity, u

Time

Figure 21   Velocity at a point in steady turbulent flow.

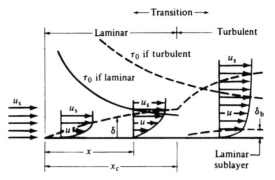

Figure 22   Boundary layer development along a flat plate.

One parameter of interest is the boundary layer thickness δ, the distance from the boundary in which the flow is retarded, or the distance to the point where the velocity is 99% of the free stream velocity (Fig. 23). The displacement thickness is the distance the boundary is displaced such that the boundary layer flow is the same as one-dimensional flow past the displaced boundary. It is given by (see Fig. 23)

$$\delta_1 = \frac{1}{u_s} \int_0^\delta (u_s - u)\, dy = \int_0^\delta \left(1 - \frac{u}{u_s}\right) dy$$

A momentum thickness is the distance from the boundary such that the momentum flux of the free stream within this distance is the deficit of momentum of the boundary layer flow. It is given by (see Fig. 23)

$$\delta_2 = \int_0^\delta \left(1 - \frac{u}{u_s}\right)\frac{u}{u_s}\, dy$$

Also of interest is the viscous shear drag $D = C_f(\rho u_s^2/2)A$, where C_f is the average skin friction drag coefficient and A is the area sheared.

These parameters are listed in Table 7 as functions of the Reynolds number $\mathrm{Re}_x = u_s\rho x/\mu$, where x is based on the distance from the leading edge. For Reynolds numbers between 1.8×10^5 and 4.5×10^7, $C_f = 0.045/\mathrm{Re}_x^{1/6}$, and for Re_x between 2.9×10^7 and 5×10^8, $C_f = 0.0305/\mathrm{Re}_x^{1/7}$. These results for turbulent boundary layers are obtained from pipe flow friction measurements for smooth pipes, by assuming the pipe radius equivalent to the bound-

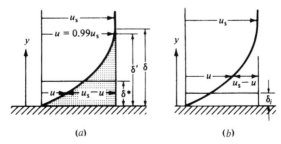

Figure 23   Definition of boundary layer thickness: (*a*) displacement thickness; (*b*) momentum thickness.

Table 7 Boundary Layer Parameters

Parameter	Laminar Boundary Layer	Turbulent Boundary Layer
$\dfrac{\delta}{x}$	$\dfrac{4.91}{\text{Re}_x^{1/2}}$	$\dfrac{0.382}{\text{Re}_x^{1/5}}$
$\dfrac{\delta_1}{x}$	$\dfrac{1.73}{\text{Re}_x^{1/2}}$	$\dfrac{0.048}{\text{Re}_x^{1/5}}$
$\dfrac{\delta_2}{x}$	$\dfrac{0.664}{\text{Re}_x^{1/2}}$	$\dfrac{0.037}{\text{Re}_x^{1/5}}$
C_f	$\dfrac{1.328}{\text{Re}_x^{1/2}}$	$\dfrac{0.074}{\text{Re}_x^{1/5}}$
Re_x range	Generally not over 10^6	Less than 10^7

ary layer thickness, the centerline pipe velocity equivalent to the free stream boundary layer flow, and appropriate velocity profiles. Results agree with measurements.

When a turbulent boundary layer is preceded by a laminar boundary layer, the drag coefficient is given by the Prandtl–Schlichting equation:

$$C_f = \frac{0.455}{(\log \text{Re}_x)^{2.58}} - \frac{A}{\text{Re}_x}$$

where A depends on the Reynolds number Re_c at which transition occurs. Values of A for various values of $\text{Re}_c = u_s x_c / v$ are

| Re_c | 3×10^5 | 5×10^5 | 9×10^5 | 1.5×10^6 |
| A | 1035 | 1700 | 3000 | 4880 |

Some results are shown in Fig. 24 for transition at these Reynolds numbers for completely laminar boundary layers, for completely turbulent boundary layers, and for a typical ship hull. (The other curves are applicable for smooth model ship hulls.) Drag coefficients for flat plates may be used for other shapes that approximate flat plates.

The thickness of the viscous sublayer δ_b in terms of the boundary layer thickness is approximately

$$\frac{\delta_b}{\delta} = \frac{80}{(\text{Re}_x)^{7/10}}$$

At $\text{Re}_x = 10^6$, $\delta_b/\delta = 0.0050$ and when $\text{Re}_x = 10^7$, $\delta_b/\delta = 0.001$, and thus the viscous sublayer is very thin.

Experiments show that the boundary layer thickness and local drag coefficient for a turbulent boundary layer preceded by a laminar boundary layer at a given location are the same as though the boundary layer were turbulent from the beginning of the plate or surface along which the boundary layer grows.

10 GAS DYNAMICS

In gas flows where density variations are appreciable, large variations in velocity and temperature may also occur and then thermodynamic effects are important.

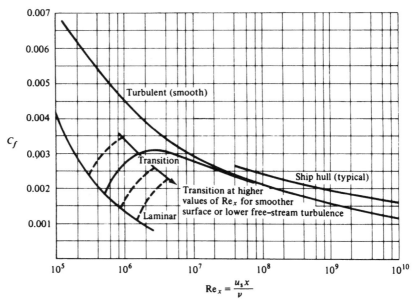

Figure 24 Drag coefficients for smooth plane surfaces parallel to flow.

10.1 Adiabatic and Isentropic Flow

In adiabatic flow of a gas with no external work and with changes in elevation negligible, the steady-flow energy equation is

$$\frac{V_1^2}{2} + h_1 = \frac{V_2^2}{2} + h_2 = h_0 = \text{constant}$$

for flow from point 1 to point 2, where V is velocity and h is enthalpy. Subscript 0 refers to a stagnation condition where the velocity is zero.

The speed of sound is $c = \sqrt{(\partial p/\partial s)_{\text{isentropic}}} \sqrt{K/\rho} = \sqrt{kp/\rho} = \sqrt{kRT}$. For air, $c = 20.04\sqrt{T}$ m/sec, where T is in degrees kelvin. A local Mach number is then $M = V/c = V/\sqrt{kRT}$.

A gas at rest may be accelerated adiabatically to any speed, including sonic ($M = 1$) and theoretically to its maximum speed when the temperature reduces to absolute zero. Then,

$$c_p T_0 = c_p T + \frac{V^2}{2} = c_p T^* + \frac{V^{*2}}{2} = \frac{V_{\text{max}}^2}{2}$$

where the asterisk (*) refers to a sonic state where the Mach number is unity.

The stagnation temperature T_0 is $T_0 = T + V^2/2c_p$, or in terms of the Mach number $[c_p = Rk/(k-1)]$

$$\frac{T_0}{T} = 1 + \frac{k-1}{2} M^2 = 1 + 0.2M^2 \text{ for air}$$

The stagnation temperature is reached adiabatically from any velocity V where the Mach number is M and the temperature T. The temperature T^* in terms of the stagnation temperature T_0 is $T^*/T_0 = 2/(k+1) = \frac{5}{6}$ for air.

The stagnation pressure is reached reversibly and is thus the isentropic stagnation pressure. It is also called the reservoir pressure, since for any flow a reservoir (stagnation) pressure may be imagined from which the flow proceeds isentropically to a pressure p at a Mach number M. The stagnation pressure p_0 is a constant in isentropic flow; if nonisentropic, but adiabatic, p_0 decreases:

$$\frac{p_0}{p} = \left(\frac{T_0}{T}\right)^{k/(k-1)} = \left(1 + \frac{k-1}{2} M^2\right)^{k/(k-1)} = (1 + 0.2M^2)^{3.5} \text{ for air}$$

Expansion of this expression gives

$$p_0 = p + \frac{\rho V^2}{2}\left[1 + \frac{1}{4}M^2 + \frac{2-k}{24}M^4 + \frac{(2-k)(3-2k)}{192}M^6 + \cdots\right]$$

where the term in brackets is the compressibility factor. It ranges from 1 at very low Mach numbers to a maximum of 1.27 at $M = 1$, and shows the effect of increasing gas density as it is brought to a stagnation condition at increasingly higher initial Mach numbers. The equations are valid to or from a stagnation state for subsonic flow, and from a stagnation state for supersonic flow at M^2 less than $2/(k-1)$, or M less than $\sqrt{5}$ for air.

10.2 Duct Flow

Adiabatic flow in short ducts may be considered reversible, and thus the relation between velocity and area changes is $dA/dV = (A/V)(M^2 - 1)$. For subsonic flow, dA/dV is negative and velocity changes relate to area changes in the same way as for incompressible flow. At supersonic speed, dA/dV is positive and an expanding area is accompanied by an increasing velocity; a contracting area is accompanied by a decreasing velocity, the opposite of incompressible flow behavior. Sonic flow in a duct (at $M = 1$) can exist only when the duct area is constant $(dA/dV = 0)$, in the throat of a nozzle or in a pipe. It can also be shown that velocity and Mach numbers always increase or decrease together, that temperature and Mach numbers change in opposite directions, and that pressure and Mach numbers also change in opposite directions.

Isentropic gas flow tables give pressure ratios p/p_0, temperature ratios T/T_0, density ratios ρ/ρ_0, area ratios A/A^*, and velocity ratios V/V^* as functions of the upstream Mach number M_x and the specific heat ratio k for gases.

The mass flow rate through a converging nozzle from a reservoir with the gas at a pressure p_0 and temperature T_0 is calculated in terms of the pressure at the nozzle exit from the equation $\dot{m} = (VA\rho)_{\text{exit}}$, where $\rho_e = p_e/RT_e$ and the exit temperature is $T_e = T_0(p_e/p_0)^{(k-1)/k}$ and the exit velocity is

$$V_e = \sqrt{2c_p T_0 \left[1 - \left(\frac{p_e}{p_0}\right)^{(k-1)/k}\right]}$$

The mass flow rate is maximum when the exit velocity is sonic. This requires the exit pressure to be critical, and the receiver pressure to be critical or below. Supersonic flow in the nozzle is impossible. If the receiver pressure is below critical, flow is not affected in the nozzle, and the exit flow remains sonic. For air at this condition, the maximum flow rate is $\dot{m} = 0.0404 A_1 p_0/\sqrt{T_0}$ kg/sec.

Flow through a converging–diverging nozzle (Fig. 25) is subsonic throughout if the throat pressure is above critical (dashed lines in Fig. 25). When the receiver pressure is at

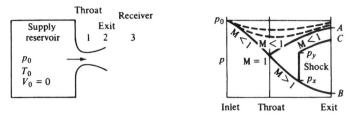

Figure 25   Gas flow through converging–diverging nozzle.

A, the exit pressure is also, and sonic flow exists at the throat, but is subsonic elsewhere. Only at B is there sonic flow in the throat with isentropic expansion in the diverging part of the nozzle. The flow rate is the same whether the exit pressure is at A or B. Receiver pressures below B do not affect the flow in the nozzle. Below A (at C, for example) a shock forms as shown and then the flow is isentropic to the shock, and beyond it, but not through it. When the throat flow is sonic, the mass flow rate is given by the same equation as for a converging nozzle with sonic exit flow. The pressures at A and B in terms of the reservoir pressure p_0 are given in isentropic flow tables as a function of the ratio of exit area to throat area, A_c/A^*.

10.3   Normal Shocks

The plane of a normal shock is at right angles to the flow streamlines. These shocks may occur in the diverging part of a nozzle, the diffuser of a supersonic wind tunnel, in pipes and forward of blunt-nosed bodies. In all instances the flow is supersonic upstream and subsonic downstream of the shock. Flow through a shock is not isentropic, although nearly so for very weak shocks. The abrupt changes in gas density across a shock allow for optical detection. The interferometer responds to density changes, the Schlieren method to density gradients, and the spark shadowgraph to the rate of change of density gradient. Density ratios across normal shocks in air are 2 at $M = 1.58$, 3 at $M = 2.24$, and 4 at $M = 3.16$ to a maximum value of 6.

Changes in fluid and flow parameters across normal shocks are obtained from the continuity, energy, and momentum equations for adiabatic flow. They are expressed in terms of upstream Mach numbers with upstream conditions designated with subscript x and downstream with subscript y. Mach numbers M_x and M_y are related by

$$\frac{1 + kM_x^2}{M_x\left(1 + \dfrac{k-1}{2}M_x^2\right)^{1/2}} = \frac{1 + kM_y^2}{M_y\left(1 + \dfrac{k-1}{2}M_y^2\right)^{1/2}} = f(M,k)$$

which is plotted in Fig. 26. The requirement for an entropy increase through the shock indicates M_x to be greater than M_y. Thus, the higher the upstream Mach number, the lower the downstream Mach number, and vice versa. For normal shocks, values of downstream Mach number M_y; temperature ratios T_y/T_x; pressure ratios p_y/p_x, p_{0y}/p_x, and p_{0y}/p_{0x}; and density ratios ρ_y/ρ_x depend only on the upstream Mach number M_x and the specific heat ratio k of the gas. These values are tabulated in books on gas dynamics and in books of gas tables.

The density ratio across the shock is given by the Rankine–Hugoniot equation

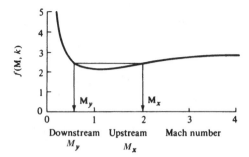

Downstream Upstream Mach number
M_y M_x

Figure 26 Mach numbers across a normal shock, $k = 1.4$.

$$\frac{\rho_y}{\rho_x} = \left[\left(\frac{k+1}{k-1}\right)\frac{p_y}{p_x} + 1\right]\bigg/\left[\frac{p_y}{p_x} + \left(\frac{k+1}{k-1}\right)\right]$$

and is plotted in Fig. 27, which shows that weak shocks are nearly isentropic, and that the density ratio approaches a limit of 6 for gases with $k = 1.4$.

Gas tables show that at an upstream Mach number of 2 for air, $M_y = 0.577$, the pressure ratio is $p_y/p_x = 4.50$, the density ratio is $\rho_y/\rho_x = 2.66$, the temperature ratio is $T_y/T_x = 1.68$, and the stagnation pressure ratio is $p_{0y}/p_{0x} = 0.72$, which indicates an entropy increase of $s_y - s_x = -R \ln(p_{0y}/p_{0x}) = 94$ J/kg.

10.4 Oblique Shocks

Oblique shocks are inclined from a direction normal to the approaching streamlines. Figure 28 shows that the normal velocity components are related by the normal shock relations. From a momentum analysis, the tangential velocity components are unchanged through

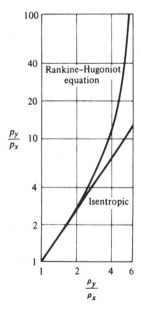

Figure 27 Rankine–Hugoniot curve, $k = 1.4$.

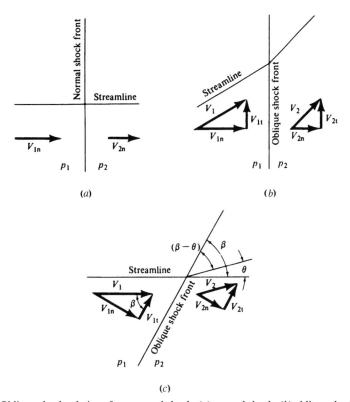

Figure 28   Oblique shock relations from normal shock; (*a*) normal shock; (*b*) oblique shock; (*c*) oblique shock angles.

oblique shocks. The upstream Mach number M_1 is given in terms of the deflection angle θ, the shock angle β, and the specific heat ratio k for the gas as

$$\frac{1}{M_1^2} = \sin^2 \beta - \frac{k+1}{2} \frac{\sin \beta \sin \theta}{\cos(\beta - \theta)}$$

The geometry is shown in Fig. 29, and the variables in this equation are illustrated in Fig. 30. For each M_1 there is the possibility of two wave angles β for a given deflection angle θ. The larger wave angle is for strong shocks, with subsonic downstream flow. The smaller

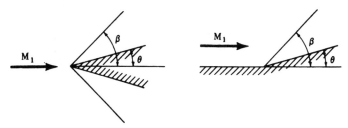

Figure 29   Supersonic flow past a wedge and an inside corner.

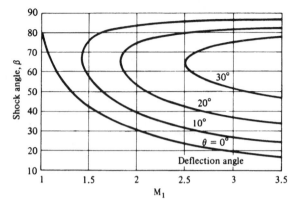

Figure 30 Oblique shock relations, $k = 1.4$.

wave angle is for weak shocks, generally with supersonic downstream flow at a Mach number less than M_1.

Normal shock tables are used for oblique shocks if M_x is used for $M_1 \sin \beta$. Then $M_y = M_2 \sin(\beta - \theta)$ and other ratios of property values (pressure, temperature, and density) are the same as for normal shocks.

11 VISCOUS FLUID FLOW IN DUCTS

The development of flow in the entrance of a pipe with the development of the boundary layer is shown in Fig. 31. Wall shear stress is very large at the entrance, and generally decreases in the flow direction to a constant value, as does the pressure gradient dp/dx. The

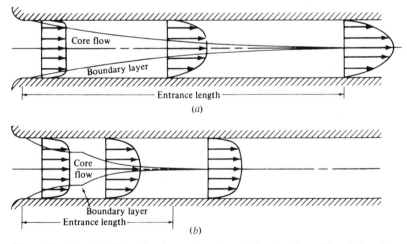

Figure 31 Growth of boundary layers in a pipe: (*a*) laminar flow; (*b*) turbulent flow.

velocity profile also changes and becomes adjusted to a fixed shape. When these have reached constant conditions, the flow is called *fully developed* flow.

The momentum equation for a pipe of diameter D gives the pressure gradient as

$$-\frac{dp}{dx} = \frac{4}{D}\tau_0 + \rho V^2 \frac{d\beta}{dx} + \beta\rho V \frac{dV}{dx}$$

which shows that a pressure gradient overcomes wall shear and increases momentum of the fluid either as a result of changing the shape of the velocity profile ($d\beta/dx$) or by changing the mean velocity along the pipe (dV/dx is not zero for gas flows).

For fully developed incompressible flow

$$-\frac{dp}{dx} = \frac{\Delta p}{L} = \frac{4\tau_0}{D}$$

and a pressure drop simply overcomes wall shear.

For developing flow in the entrance, $\beta = 1$ initially and increases to a constant value downstream. Thus, the pressure gradient overcomes wall shear and also increases the flow momentum according to

$$-\frac{dp}{dx} = \frac{4\tau_0}{D} + \rho V^2 \frac{d\beta}{dx}$$

For fully developed flow, $\beta = \frac{4}{3}$ for laminar flow and $\beta \approx 1.03$ for turbulent flow in round pipes.

For compressible gas flow beyond the entrance, the velocity profile becomes essentially fixed in shape, but the velocity changes because of thermodynamic effects that change the density. Thus, the pressure gradient is

$$-\frac{dp}{dx} = \frac{4\tau_0}{D} + \beta\rho V \frac{dV}{dx}$$

Here β is essentially constant but dV/dx may be significant.

11.1 Fully Developed Incompressible Flow

The pressure drop is $\Delta p = (fL/D)(\rho V^2/2)$ Pa, where f is the Darcy friction factor. The Fanning friction factor $f' = f/4$ and then $\Delta p = (4f'/D)(\rho V^2/2)$, and the head loss from pipe friction is

$$h_f = \frac{\Delta p}{\gamma} = f\left(\frac{L}{D}\right)\frac{V^2}{2g} = (4f')\left(\frac{L}{D}\right)\frac{V^2}{2g} \quad \text{m}$$

The shear stress varies linearly with radial position, $\tau = (\Delta p/L)(r/2)$, so that the wall shear is $\tau_0 = (\Delta p/L)(D/4)$, which may then be written $\tau_0 = f\rho V^2/8 = f'\rho V^2/2$.

A shear velocity is defined as $v\cdot = \sqrt{\tau_0/\rho} = V\sqrt{f/8} = V\sqrt{f'/2}$ and is used as a normalizing parameter.

For noncircular ducts the diameter D is replaced by the hydraulic or equivalent diameter $D_h = 4A/P$, where A is the flow cross section and P is the wetted perimeter. Thus, an annulus between pipes of diameter D_1 and D_2, D_1 being larger, the hydraulic diameter is $D_2 - D_1$.

11.2 Fully Developed Laminar Flow in Ducts

The velocity profile in circular tubes is that of a parabola, and the centerline velocity is

$$u_{max} = \frac{\Delta p}{L}\left(\frac{R^2}{4\mu}\right)$$

and the velocity profile is

$$\frac{u}{u_{max}} = 1 - \left(\frac{r}{R}\right)^2$$

where r is the radial location in a pipe of radius R. The average velocity is one-half the maximum velocity, $V = u_{max}/2$.

The pressure gradient is

$$\frac{\Delta p}{L} = \frac{128\mu Q}{\pi D^4}$$

which indicates a linear increase with increasing velocity or flow rate. The friction factor for circular ducts is $f = 64/\mathrm{Re}_D$ or $f' = 16/\mathrm{Re}_D$ and applies to both smooth as well as rough pipes, for Reynolds numbers up to about 2000.

For *noncircular* ducts the value of the friction factor is $f = C/\mathrm{Re}$ and depends on the duct geometry. Values of $f\,\mathrm{Re} = C$ are listed in Table 8.

11.3 Fully Developed Turbulent Flow in Ducts

Knowledge of turbulent flow in ducts is based on physical models and experiments. Physical models describe lateral transport of fluid as a result of mixing due to eddies. Prandtl and von Kármán both derived expressions for shear stresses in turbulent flow based on the Reynolds stress ($\tau = -\rho\overline{u'v'}$) and obtained velocity defect equations for pipe flow. Prandtl's equation is

$$\frac{u_{max} - u}{\sqrt{\tau_0/\rho}} = \frac{u_{max} - u}{v\cdot} = 2.5 \ln \frac{R}{y}$$

where u_{max} is the centerline velocity and u is the velocity a distance y from the pipe wall. von Kármán's equation is

$$\frac{u_{max} - u}{\sqrt{\tau_0/\rho}} = \frac{u_{max} - u}{v\cdot}$$

$$= -\frac{1}{\kappa}\left[\ln\left(1 - \sqrt{1 - \frac{y}{R}}\right) + \sqrt{1 - \frac{y}{R}}\right]$$

In both, κ is an experimentally determined constant equal to 0.4 (some experiments show better agreement when $\kappa = 0.36$). Similar expressions apply to external boundary layer flow when the pipe radius R is replaced by the boundary layer thickness δ. Friction factors for smooth pipes have been developed from these results. One is the Blasius equation for $\mathrm{Re}_D = 10^5$ and is $f = 0.316/\mathrm{Re}_D^{1/4}$ obtained by using a power-law velocity profile $u/u_{max} = (y/R)^{1/7}$. The value 7 here increases to 10 at higher Reynolds numbers. The use of a logarithmic form of velocity profile gives the Prandtl law of pipe friction for smooth pipes:

Table 8 Friction Factors for Laminar Flow

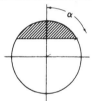

r_1/r_2	f Re		a/b	f Re		x	f Re
0.0001	71.78		0	96.00		0	62.2
0.001	74.68		1/20	89.91		10	62.2
0.01	80.11		1/10	84.68		20	62.3
0.05	86.27		1/8	82.34		30	62.4
0.10	89.37		1/6	78.81		40	62.5
0.20	92.35		1/4	72.93		60	62.8
0.40	94.71		2/5	65.47		90	63.1
0.60	95.59		1/2	62.19		120	63.3
0.80	95.92		3/4	57.89		150	63.7
1.00	96.00		1	56.91		180	64.0

	Circular Sector	Isosceles Triangle	Right Triangle
α	f Re	f Re	f Re
0	48.0	48.0	48.0
10	51.8	51.6	49.9
20	54.5	52.9	51.2
30	56.7	53.3	52.0
40	58.4	52.9	52.4
50	59.7	52.0	52.4
60	60.8	51.1	52.0
70	61.7	49.5	51.2
80	62.5	48.3	49.9
90	63.1	48.0	48.0

$$\frac{1}{\sqrt{f}} = 2 \log(\mathrm{Re}_D \sqrt{f}) - 0.8$$

which agrees well with experimental values. A more explicit formula by Colebrook is $1/\sqrt{f} = 1.8 \log(\mathrm{Re}_D/6.9)$, which is within 1% of the Prandtl equation over the entire range of turbulent Reynolds numbers.

The logarithmic velocity defect profiles apply for *rough* pipes as well as for smooth pipes, since the velocity defect $(u_{\max} - u)$ decreases linearly with the shear velocity $v\cdot$, keeping the ratio of the two constant. A relation between the centerline velocity and the average velocity is $u_{max}/V = 1 + 133\sqrt{f}$, which may be used to estimate the average velocity from a single centerline measurement.

The Colebrook–White equation encompasses all turbulent flow regimes, for both smooth and rough pipes:

$$\frac{1}{\sqrt{f}} = 1.74 - 2 \log \left(\frac{2k}{D} + \frac{18.7}{Re_D\sqrt{f}} \right)$$

and this is plotted in Fig. 32, where k is the equivalent sand-grain roughness. A simpler equation by Haaland is

$$\frac{1}{\sqrt{f}} = -1.8 \log \left[\frac{6.9}{Re_D} + \left(\frac{k}{3.7D} \right)^{1.11} \right]$$

which is explicit in f and is within 1.5% of the Colebrook–White equation in the range $4000 \leq Re_D \leq 10^8$ and $0 \leq k/D \leq 0.05$.

Three types of problems may be solved:

1. *The pressure drop or head loss.*   The Reynolds number and relative roughness are determined and calculations are made directly.

2. *The flow rate for given fluid and pressure drops or head loss.*   Assume a friction factor, based on a high Re_D for a rough pipe, and determine the velocity from the Darcy equation. Calculate a Re_D, get a better f, and repeat until successive velocities are the same. A second method is to assume a flow rate and calculate the pressure drop or head loss. Repeat until results agree with the given pressure drop or head loss. A plot of Q versus h_L, for example, for a few trials may be used.

3. *A pipe size.*   Assume a pipe size and calculate the pressure drop or head loss. Compare with given values: Repeat until agreement is reached. A plot of D versus h_L, for example, for a few trials may be used. A second method is to assume a reasonable friction factor and get a first estimate of the diameter from

$$D = \left(\frac{8fLQ^2}{\pi^2 g h_f} \right)^{1/5}$$

From the first estimate of D, calculate the Re_D and k/D to get a better value of f. Repeat until successive values of D agree. This is a rapid method.

Results for circular pipes may be applied to noncircular ducts if the hydraulic diameter is used in place of the diameter of a circular pipe. Then the relative roughness is k/D_h and the Reynolds number is $Re = VD_h/\nu$. Results are reasonably good for square ducts, rectangular ducts of aspect ratio up to about 8, equilateral ducts, hexagonal ducts, and concentric annular ducts of diameter ratio to about 0.75. In eccentric annular ducts where the pipes touch or nearly touch, and in tall narrow triangular ducts, both laminar and turbulent flow may exist at a section. Analyses mentioned here do not apply to these geometries.

11.4   Steady Incompressible Flow in Entrances of Ducts

The increased pressure drop in the entrance region of ducts as compared with that for the same length of fully developed flow is generally included in a correction term called a loss coefficient, k_L. Then,

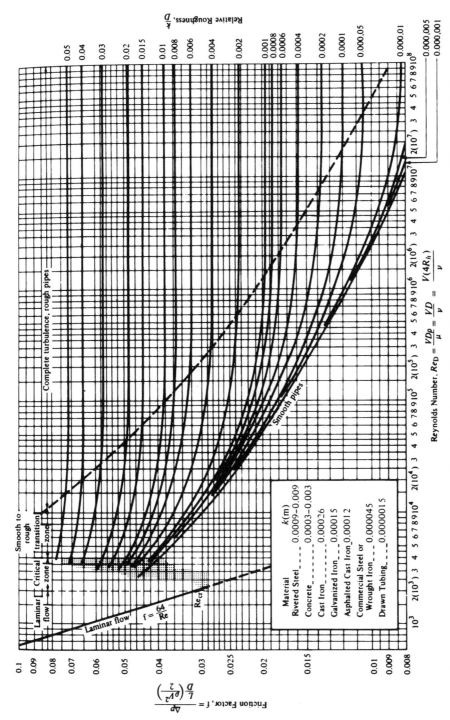

Figure 32 Friction factors for commercial pipe. [From L. F. Moody, "Friction Factors for Pipe Flow," *Trans. ASME*, 66 (1944). Courtesy of The American Society of Mechanical Engineers.]

$$\frac{p_1 - p}{\rho V^2/2} = \frac{fL}{D_h} + k_L$$

where p_1 is the pressure at the duct inlet and p is the pressure a distance L from the inlet. The value of k_L depends on L but becomes a constant in the fully developed region, and this constant value is of greatest interest.

For *laminar* flow the pressure drop in the entrance length L_e is obtained from the Bernoulli equation written along the duct axis where there is no shear in the core flow. This is

$$p_1 - p_e = \frac{\rho u_{\max}^2}{2} - \frac{\rho V^2}{2} = \left[\left(\frac{u_{\max}}{V}\right)^2 - 1\right]\frac{\rho V^2}{2}$$

for any duct for which $u_{\max}/V$ is known. When both friction factor and k_L are known, the entrance length is

$$\frac{L_e}{D_h} = \frac{1}{f}\left[\left(\frac{u_{\max}}{V}\right)^2 - 1 - k_L\right]$$

For a circular duct, experiments and analyses indicate that $k_L \approx 1.30$. Thus, for a circular duct, $L_e/D = (\mathrm{Re}_D/64)(2^2 - 1 - 1.30) = 0.027\mathrm{Re}_D$. The pressure drop for fully developed flow in a length L_e is $\Delta p = 1.70\rho V^2/2$ and thus the pressure drop in the entrance is $3/1.70 = 1.76$ times that in an equal length for fully developed flow. Entrance effects are important for short ducts.

Some values of k_L and $(L_e/D_h)\mathrm{Re}$ for laminar flow in various ducts are listed in Table 9.

Table 9 Entrance Effects, Laminar Flow (See Table 8 for Symbols)

r_1/r_2	k_L		a/b	k_L	$L_c D_h \, \mathrm{Re}$		x	k_L
0.0001	1.13		0	0.69	0.0059		0	1.74
0.001	1.07		1/8	0.88	0.0094		10	1.73
0.01	0.97		1/5	1.00	0.0123		20	1.72
0.05	0.86		1/4	1.08	0.0146		30	1.69
0.10	0.81		1/2	1.38	0.0254		40	1.65
0.20	0.75		3/4	1.52	0.0311		60	1.57
0.40	0.71		1	1.55	0.0324		90	1.46
0.60	0.69						120	1.39
0.80	0.69						150	1.34
1.00	0.69						180	1.33

α	Circular Sector k_L	Isosceles Triangle k_L	Right Triangle k_L
0	2.97	2.97	2.97
10	2.06	2.14	2.40
20	1.71	1.85	2.09
30	1.58	1.79	1.94
40	1.53	1.83	1.88
50	1.50	1.95	1.88
60	1.49	2.14	1.94
70	1.48	2.38	2.09
80	1.47	2.72	2.40
90	1.46	2.97	2.97

For turbulent flow, loss coefficients are determined experimentally. Results are shown in Fig. 33. Flow separation accounts for the high loss coefficients for the square and reentrant shapes for circular tubes and concentric annuli. For a rounded entrance, a radius of curvature of $D/7$ or more precludes separation. The boundary layer starts laminar then changes to turbulent, and the pressure drop does not significantly exceed the corresponding value for fully developed flow in the same length. (It may even be less with the laminar boundary layer—a trip or slight roughness may force a turbulent boundary layer to exist at the entrance.)

Entrance lengths for circular ducts and concentric annuli are defined as the distance required for the pressure gradient to become within a specified percentage of the fully developed value (5%, for example). On this basis L_e/D_h is about 30 or less.

11.5 Local Losses in Contractions, Expansions, and Pipe Fittings; Turbulent Flow

Calculations of local head losses generally are approximate at best unless experimental data for given fittings are provided by the manufacturer.

Losses in *contractions* are given by $h_L = k_L V^2/2g$. Loss coefficients for a sudden contraction are shown in Fig. 34. For gradually contracting sections k_L may be as low as 0.03 for D_2/D_1 of 0.5 or less.

Losses in *expansions* are given by $h_L = k_L(V_1 - V_2)^2/2g$, section 1 being upstream. For a sudden expansion, $k_L = 1$, and for gradually expanding sections with divergence angles of 7° or 8°, k_L may be as low as 0.14 or even 0.06 for diffusers for low-speed wind tunnels or cavitation-testing water tunnels with curved inlets to avoid separation.

Losses in pipe fittings are given in the form $h_L = k_L V^2/2g$ or in terms of an equivalent pipe length by pipe-fitting manufacturers. Typical values for various fittings are given in Table 10.

11.6 Flow of Compressible Gases in Pipes with Friction

Subsonic gas flow in pipes involves a decrease in gas density and an increase in gas velocity in the direction of flow. The momentum equation for this flow may be written as

$$\frac{dp}{\rho V^2/2} + f\,\frac{dx}{D} + 2\,\frac{dV}{V} = 0$$

For *isothermal* flow the first term is $(2/\rho_1 V_1^2 p_1)p\,dp$, where the subscript 1 refers to an upstream section where all conditions are known. For $L = x_2 - x_1$, integration gives

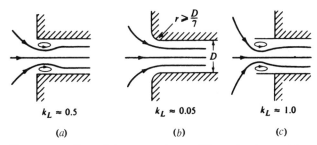

$$k_L \approx 0.5 \qquad\qquad k_L \approx 0.05 \qquad\qquad k_L \approx 1.0$$

(a) (b) (c)

Figure 33 Pipe entrance flows: (*a*) square entrance; (*b*) round entrance; (*c*) reentrant inlet.

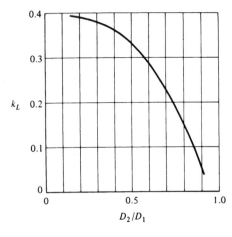

Figure 34 Loss coefficients for abrupt contract in pipes.

Table 10 Typical Loss Coefficients for Valves and Fittings

Valve or Fitting	Nominal Diameter, cm					
	2.5	5	10	15	20	25
Globe valve, wide open						
Screwed	9	7	5.5			
Flanged	12	9	6	6	5.5	5.5
Gate valve, wide open						
Screwed	0.24	0.18	0.13			
Flanged		0.35	0.16	0.11	0.08	0.06
Foot valve, wide open			0.80 for all sizes			
Swing check valve, wide open						
Screwed	3.0	2.3	2.1			
Flanged			2.0 for all sizes			
Angle valve, wide open						
Screwed	4.5	2.1	1.0			
Flanged		2.4	2.1	2.1	2.1	2.1
Regular elbow, 90°						
Screwed	1.5	1.0	0.65			
Flanged	0.42	0.37	0.31	0.28	0.26	0.25
Long-radius elbow, 90°						
Screwed	0.75	0.4	0.25			
Flanged		0.3	0.22	0.18	0.15	0.14

Note: The k_L values listed may be expressed in terms of an equivalent pipe length for a given installation and flow by equating $k_L = fL_c/D$ so that $L_e = k_L D/f$.

Source: Reproduced, with permission, from *Engineering Data Book: Pipe Friction Manual*, Hydraulic Institute, Cleveland, 1979.

$$p_1^2 - p_2^2 = \rho_1 V_1^2 p_1 \left(f \frac{L}{D} - 2 \ln \frac{p_2}{p_1} \right)$$

or, in terms of the initial Mach number,

$$p_1^2 - p_2^2 = k M_1^2 p_1^2 \left(f \frac{L}{D} - 2 \ln \frac{p_2}{p_1} \right)$$

The downstream pressure p_2 at a distance L from section 1 may be obtained by trial by neglecting the term $2 \ln(p_2/p_1)$ initially to get a p_2, then including it for an improved value. The distance L is a section where the pressure is p_2 is obtained from

$$f \frac{L}{D} = \frac{1}{kM_1^2} \left[1 - \left(\frac{p_2}{p_1} \right)^2 \right] - 2 \ln \frac{p_1}{p_2}$$

A limiting condition (designated by an asterisk) at a length L^* is obtained from an expression dp/dx to get

$$\frac{dp}{dx} = \frac{pf/2D}{1 - p/\rho V^2} = \frac{(f/D)(\rho V^2/2)}{kM^2 - 1}$$

For a low subsonic flow at an upstream section (as from a compressor discharge) the pressure gradient increases in the flow direction with an infinite value when $M^* = 1/\sqrt{k} = 0.845$ for $k = 1.4$ (air, for example). For M approaching zero, this equation is the Darcy equation for incompressible flow. The limiting pressure is $p^* = p_1 M_1 \sqrt{k}$, and the limiting length is given by

$$\frac{fL^*}{D} = \frac{1}{kM_1^2} - 1 - \ln \frac{1}{kM_1^2}$$

Since the gas at any two locations 1 and 2 in a long pipe has the same limiting condition, the distance L between them is

$$\frac{fL}{D} = \left(\frac{fL^*}{D} \right)_{M_1} - \left(\frac{fL^*}{D} \right)_{M_2}$$

Conditions along a pipe for various initial Mach numbers are shown in Fig. 35.

For *adiabatic* flow the limiting Mach number is $M^* = 1$. This is from an expression for dp/dx for adiabatic flow:

$$\frac{dp}{dx} = -\frac{fkp}{2D} M^2 \left[\frac{1 + (k - 1)M^2}{1 - M^2} \right] = -\frac{f}{D} \frac{\rho V^2}{2} \left[\frac{1 + (k - 1)M^2}{1 - M^2} \right]$$

The limiting pressure is

$$\frac{p^*}{p_1} = M_1 \sqrt{\frac{2[1 + \frac{1}{2}(k - 1)M_1^2]}{k + 1}}$$

and the limiting length is

$$\frac{\bar{f}L^*}{D} = \frac{1 - M_1^2}{kM_1^2} + \frac{k + 1}{2k} \ln \frac{(k + 1)M_1^2}{2[1 + \frac{1}{2}(k - 1)M_1^2]}$$

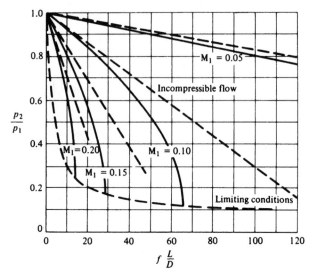

Figure 35 Isothermal gas flow in a pipe for various initial Mach numbers, $k = 1.4$.

Except for subsonic flow at high Mach numbers, isothermal and adiabatic flow do not differ appreciably. Thus, since flow near the limiting condition is not recommended in gas transmission pipelines because of the excessive pressure drop, and since purely isothermal or purely adiabatic flow is unlikely, either adiabatic or isothermal flow may be assumed in making engineering calculations. For example, for methane from a compressor at 2000 kPa absolute pressure, 60°C temperature and 15 m/sec velocity ($M_1 = 0.032$) in a 30-cm commercial steel pipe, the limiting pressure is 72 kPa absolute at $L^* = 16.9$ km for isothermal flow, and 59 kPa at $L^* = 17.0$ km for adiabatic flow. A pressure of 500 kPa absolute would exist at 16.0 km for either type of flow.

12 DYNAMIC DRAG AND LIFT

Two types of forces act on a body past which a fluid flows: a pressure force normal to any infinitesimal area of the body and a shear force tangential to this area. The components of these two forces integrate over the entire body in a direction parallel to the approach flow is the *drag* force, and in a direction normal to it is the *lift* force. *Induced* drag is associated with a lift force on finite airfoils or blank elements as a result of downwash from tip vortices. Surface waves set up by ships or hydrofoils, and compression waves in gases such as Mach cones are the source of *wave* drag.

12.1 Drag

A drag force is $D = C (\rho u_s^2/2)A$, where C is the drag coefficient, $\rho u_s^2/2$ is the dynamic pressure of the free stream, and A is an appropriate area. For pure viscous shear drag C is C_f, the skin friction drag coefficient of Section 9.2 and A is the area sheared. In general, C is designated C_D, the drag coefficient for drag other than that from viscous shear only, and A is the chord area for lifting vanes or the projected frontal area for other shapes.

The drag coefficient for incompressible flow with pure pressure drag (a flat plate normal to a flow, for example) or for combined skin friction and pressure drag, which is called *profile* drag, depends on the body shape, the Reynolds number, and, usually, the location of boundary layer transition.

Drag coefficients for spheres and for flow normal to infinite circular cylinders are shown in Fig. 36. For spheres at $Re_D < 0.1$, $C_D = 24/Re_D$ and for $Re_D < 100$, $C_D = (24/Re_D)(1 + 3 Re_D/16)^{1/2}$. The boundary layer for both shapes up to and including the flat portion of the curves before the rather abrupt drop in the neighborhood of $Re_D = 10^5$ is laminar. This is called the *subcritical* region; beyond that is the *supercritical* region. Table 11 lists typical drag coefficients for two-dimensional shapes, and Table 12 lists them for three-dimensional shapes.

The drag of spheres, circular cylinders, and streamlined shapes is affected by boundary layer separation, which, in turn, depends on surface roughness, the Reynolds number, and free stream turbulence. These factors contribute to uncertainties in the value of the drag coefficient.

12.2 Lift

Lift in a nonviscous fluid may be produced by prescribing a circulation around a cylinder or lifting vane. In a viscous fluid this may be produced by spinning a ping-pong ball, a golf ball, or a baseball, for example, Circulation around a lifting vane in a viscous fluid results from the bound vortex or countercirculation that is equal and opposite to the starting vortex, which peels off the trailing edge of the vane. The lift is calculated from $L = C_L(\rho u_s^2/2)A$, where C_L is the lift coefficient, $\rho u_s^2/2$ is the dynamic pressure of the free stream, and A is the chord area of the lifting vane. Typical values of C_L as well as C_D are shown in Fig. 37. The induced drag and the profile drag are shown. The profile drag is the difference between the dashed and solid curves. The induced drag is zero at zero lift.

13 FLOW MEASUREMENTS

Fluid flow measurements generally involve determining static pressures, local and average velocities, and volumetric or mass flow rates.

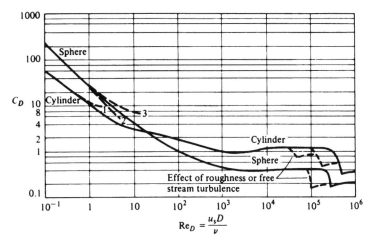

Figure 36 Drag coefficients for infinite circular cylinders and spheres: (1) Lamb's solution for cylinder; (2) Stokes' solution for sphere; (3) Oseen's solution for sphere.

Table 11 Drag Coefficients for Two-Dimensional Shapes at Re $= 10^5$ Based on Frontal Projected Area (Flow Is from Left to Right)

Shape		C_D	Shape	C_D	
Plate	|	2.0	Rectangle		
			1:1	1.18	
Open tube	(	1.2	5:1	1.2	
	)	2.3	10:1	1.3	
			20:1	1.5	
Half cylinder	⪗	1.16	Elliptical	Below	Above
	⪘	1.7	cylinder	Re$_c$	Re$_c$
			2:1	0.6	0.20
Square cylinder	□	2.05	4:1	0.36	0.10
	◇	1.55	8:1	0.26	0.10
Equilateral	▷	2.0			
triangle	◁	1.6			

13.1 Pressure Measurements

Static pressures are measured by means of a small hole in a boundary surface connected to a sensor—a manometer, a mechanical pressure gage, or an electrical transducer. The surface may be a duct wall or the outer surface of a tube, such as those shown in Fig. 38. In any case, the surface past which the fluid flows must be smooth, and the tapped holes must be at right angles to the surface.

Total or stagnation pressures are easily measured accurately with an open-ended tube facing into the flow, as shown in Fig. 38.

Table 12 Drag Coefficients for Three-Dimensional Shapes Re between 10^4 and 10^6 (Flow Is from Left to Right)

Shape		C_D
Disk	|	1.17
Open hemisphere	(	0.38
	)	1.42
Solid hemisphere	⪗	0.42
	⪘	1.17
Cube	▭	1.05[a]
	◇	0.80[a]
Cone, 60°	◁	0.50

[a]Mounted on a boundary wall.

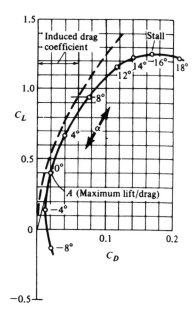

Figure 37 Typical polar diagram showing lift–drag characteristics for an airfoil of finite span.

13.2 Velocity Measurements

A combined pitot tube (Fig. 38) measures or detects the difference between the total or stagnation pressure p_0 and the static pressure p. For an incompressible fluid the velocity being measured is $V = \sqrt{2(p_0 - p)/\rho}$. For subsonic gas flow the velocity of a stream at a temperature T and pressure p in

$$V = \sqrt{\frac{2kRT}{k-1}\left[\left(\frac{p_0}{p}\right)^{(k-1)/k} - 1\right]}$$

and the corresponding Mach number is

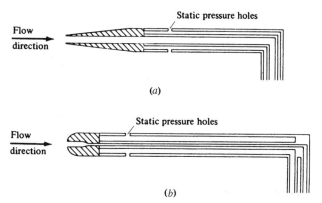

Figure 38 Combined pitot tubes: (a) Brabbee's design; (b) Prandtl's design—accurate over a greater range of yaw angles.

$$M = \sqrt{\frac{2}{k-1}\left[\left(\frac{p_0}{p}\right)^{(k-1)/k} - 1\right]}$$

For supersonic flow the stagnation pressure p_{0y} is downstream of a shock, which is detached and ahead of the open stagnation tube, and the static pressure p_x is upstream of the shock. In a wind tunnel the static pressure could be measured with a pressure tap in the tunnel wall. The Mach number M of the flow is

$$\frac{p_{0y}}{p} = \left(\frac{k+1}{2}M^2\right)^{k/(k-1)}\left(\frac{2k}{k+1}M^2 - \frac{k-1}{k+1}\right)^{1/(1-k)}$$

which is tabulated in gas tables.

In a mixture of gas bubbles and a liquid for gas concentrations C no more than 0.6 by volume, the velocity of the mixture with the pitot tube and manometer free of bubbles is

$$V_{mixture} = \sqrt{\frac{2(p_0 - p_1)}{(1-C)\rho_{liquid}}} = \sqrt{\frac{2gh_m}{(1-C)}\left(\frac{\gamma_m}{\gamma_{liquid}} - 1\right)}$$

where h_m is the manometer deflection in meters for a manometer liquid of specific weight γ_m. The error in this equation from neglecting compressible effects for the gas bubbles is shown in Fig. 39. A more correct equation based on the gas–liquid mixture reaching a stagnation pressure isentropically is

$$\frac{V_1^2}{2} = \frac{p_0 - p_1}{\rho_u(1-C)} + \frac{C}{1-C}\left(\frac{p_1}{\rho_u}\right)\left[\frac{k}{k-1}\left(\frac{p_0}{p_1}\right)^{(k-1)/k} - \frac{1}{k-1} - \left(\frac{p_0}{p_1}\right)\right]$$

but is cumbersome to use. As indicated in Fig. 39 the error in using the first equation is very small for high concentrations of gas bubbles at low speeds and for low concentrations at high speeds.

If n velocity readings are taken at the centroid of n subareas in a duct, the average velocity V from the point velocity readings u_i is

$$V = \frac{1}{n}\sum_{i=1}^{n} u_i$$

In a circular duct, readings should be taken at $(r/R)^2 = 0.055, 0.15, 0.25, \ldots, 0.95$. Velocities measured at other radial positions may be plotted versus $(r/R)^2$, and the area under the curve may be integrated numerically to obtain the average velocity.

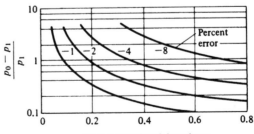

Figure 39 Error in neglecting compressibility of air in measuring velocity of air–water mixture with a combined pitot tube.

Other methods of measuring fluid velocities include length–time measurements with floats or neutral-buoyancy particles, rotating instruments such as anemometers and current meters, hot-wire and hot-film anemometers, and laser-doppler anemometers.

13.3   Volumetric and Mass Flow Fluid Measurements

Liquid flow rates in pipes are commonly measured with commercial water meters; with rotameters; and with venturi, nozzle, and orifice meters. These latter types provide an obstruction in the flow and make use of the resulting pressure change to indicate the flow rate.

The continuity and Bernoulli equations for liquid flow applied between sections 1 and 2 in Fig. 40 give the ideal volumetric flow rate as

$$Q_{\text{ideal}} = \frac{A_2 \sqrt{2g\,\Delta h}}{\sqrt{1 - (A_2/A_1)^2}}$$

where Δh is the change in piezometric head. A form of this equation generally used is

$$Q = K \left(\frac{\pi d^2}{4}\right) \sqrt{2g\,\Delta h}$$

where K is the flow coefficient, which depends on the type of meter, the diameter ratio d/D, and the viscous effects given in terms of the Reynolds number. This is based on the length parameter d and the velocity V through the hole of diameter d. Approximate flow coefficients are given in Fig. 41. The relation between the flow coefficient K and this Reynolds number is

$$\text{Re}_d = \frac{Vd}{v} = \frac{Qd}{\frac{1}{4}\pi d^2 v} = K\frac{d\sqrt{2g\,\Delta h}}{v}$$

The dimensionless parameter $d\sqrt{2g\,\Delta h}/v$ can be calculated, and the intersection of the appropriate line for this parameter and the appropriate meter curve gives an approximation to the flow coefficient K. The lower values of K for the orifice result from the contraction of the jet beyond the orifice where pressure taps may be located. Meter throat pressures

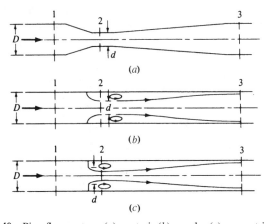

Figure 40   Pipe flow meters: (*a*) venturi; (*b*) nozzle; (*c*) concentric orifice.

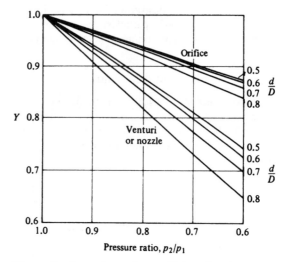

Figure 41 Approximate flow coefficients for pipe meters.

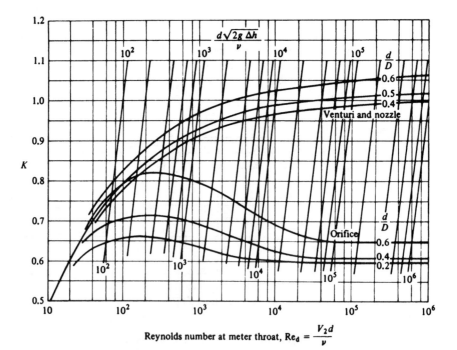

Figure 42 Expansion factors for pipe meters, $k = 1.4$.

should not be so low as to create cavitation. Meters should be calibrated in place or purchased from a manufacturer and installed according to instructions.

Elbow meters may be calibrated in place to serve as metering devices, by measuring the difference in pressure between the inner and outer radii of the elbow as a function of flow rate.

For compressible gas flows, isentropic flow is assumed for flow between sections 1 and 2 in Fig. 40. The mass flow rate is $\dot{m} = KYA_2\sqrt{2\rho_1(p_1 - p_2)}$, where K is as shown in Fig. 41 and $Y = Y(k, p_2/p_1, d/D)$ and is the expansion factor shown in Fig. 42. For nozzles and venturi tubes

$$Y = \sqrt{\frac{\left(\dfrac{k}{k-1}\right)\left(\dfrac{p_2}{p_1}\right)^{2/k}\left[1 - \left(\dfrac{p_2}{p_1}\right)^{(k-1)/k}\right]\left[1 - \left(\dfrac{d}{D}\right)^4\right]}{\left[1 - \left(\dfrac{p_2}{p_1}\right)\right]\left[1 - \left(\dfrac{d}{D}\right)^4\left(\dfrac{p_2}{p_1}\right)^{2/k}\right]}}$$

and for orifice meters

$$Y = 1 - \frac{1}{k}\left[0.41 + 0.35\left(\frac{d}{D}\right)^4\right]\left(1 - \frac{p_2}{p_1}\right)$$

These are the basic principles of fluid flow measurements. Utmost care must be taken when accurate measurements are necessary, and reference to meter manufacturers' pamphlets or measurements handbooks should be made.

BIBLIOGRAPHY

General
Olson, R. M., *Essentials of Engineering Fluid Mechanics,* 4th ed., Harper & Row, New York, 1980.
Streeter, V. L. (ed.), *Handbook of Fluid Dynamics,* McGraw-Hill, New York, 1961.
Streeter, V. L., and E. B. Wylie, *Fluid Mechanics,* McGraw-Hill, New York, 1979.

Section 9
Schlichting, H., *Boundary Layer Theory* (translated by J. Kestin), 7th ed., McGraw-Hill, New York, 1979.

Section 10
Shapiro, A. H., *The Dynamics and Thermodynamics of Compressible Fluid Flow,* Ronald Press, New York, 1953, Vol. I.

Section 12
Hoerner, S. F., *Fluid-Dynamic Drag,* S. F. Hoerner, Midland Park, NJ, 1958.

Section 13
Miller, R. W., *Flow Measurement Engineering Handbook,* McGraw-Hill, New York, 1983.
Ower, E., and R. C. Pankhurst, *Measurement of Air Flow,* Pergamon Press, Elmsford, NY, 1977.

CHAPTER 3

THERMODYNAMICS FUNDAMENTALS

Adrian Bejan
Department of Mechanical Engineering and Materials Science
Duke University
Durham, North Carolina

1 INTRODUCTION

Thermodynamics describes the relationship between mechanical work and other forms of energy. There are two facets of contemporary thermodynamics that must be stressed in a review such as this. The first is the equivalence of *work* and *heat* as two possible forms of energy exchange. This facet is expressed by the first law of thermodynamics. The second aspect is the one-way character, or irreversibility, of all flows that occur in nature. As expressed by the second law of thermodynamics, irreversibility or entropy generation is what prevents us from extracting the most possible work from various sources; it is also what prevents us from doing the most with the work that is already at our disposal. The objective of this chapter is to review the first and second laws of thermodynamics and their implications in mechanical engineering, particularly with respect to such issues as energy conversion and conservation. The analytical aspects (the formulas) of engineering thermodynamics are reviewed primarily in terms of the behavior of a pure substance, as would be the case of the working fluid in a heat engine or in a refrigeration machine. In the next chapter we review in greater detail the newer field of entropy generation minimization (thermodynamic optimization) and the generation of system configuration (constructal theory).

SYMBOLS AND UNITS

c	specific heat of incompressible substance, J/(kg·K)
c_P	specific heat at constant pressure, J/(kg·K)
c_T	constant temperature coefficient, m^3/kg
c_v	specific heat at constant volume, J/(kg·K)
COP	coefficient of performance
E	energy, J

f	specific Helmholtz free energy $(u - Ts)$, J/kg
$\mathbf{F}$	force vector, N
g	gravitational acceleration, m/s^2
g	specific Gibbs free energy $(h - Ts)$, J/kg
h	specific enthalpy $(u + Pv)$, J/kg
K	isothermal compressibility, m^2/N
m	mass of closed system, kg
$\dot{m}$	mass flow rate, kg/s
m_i	mass of component in a mixture, kg
M	mass inventory of control volume, kg
M	molar mass, g/mol or kg/kmol
n	number of moles, mol
N_0	Avogadro's constant
P	pressure
δQ	infinitesimal heat interaction, J
$\dot{Q}$	heat transfer rate, W
$\mathbf{r}$	position vector, m
R	ideal gas constant, J/(kg·K)
s	specific entropy, J/(kg·K)
S	entropy, J/K
S_{gen}	entropy generation, J/K
$\dot{S}_{\text{gen}}$	entropy generation rate, W/K
T	absolute temperature, K
u	specific internal energy, J/kg
U	internal energy, J
v	specific volume, m^3/kg
$\bar{v}$	specific volume of incompressible substance, m^3/kg
V	volume, m^3
V	velocity, m/s
δW	infinitesimal work interaction, J
$\dot{W}_{\text{lost}}$	rate of lost available work, W
$\dot{W}_{\text{sh}}$	rate of shaft (shear) work transfer, W
x	linear coordinate, m
x	quality of liquid and vapor mixture
Z	vertical coordinate, m
β	coefficient of thermal expansion, 1/K
γ	ratio of specific heats, c_P/c_v
η	"efficiency" ratio
η_{I}	first-law efficiency
η_{II}	second-law efficiency
θ	relative temperature, °C

SUBSCRIPTS

$(\ )_f$	saturated liquid state (f = "fluid")
$(\ )_g$	saturated vapor state (g = "gas")
$(\ )_s$	saturated solid state (s = "solid")
$(\ )_{\text{in}}$	inlet port
$(\ )_{\text{out}}$	outlet port
$(\ )_{\text{rev}}$	reversible path
$(\ )_H$	high-temperature reservoir

$(\;)_L$ low-temperature reservoir
$(\;)_{max}$ maximum
$(\;)_T$ turbine
$(\;)_C$ compressor
$(\;)_N$ nozzle
$(\;)_D$ diffuser
$(\;)_0$ reference state
$(\;)_1$ initial state
$(\;)_2$ final state
$(\;)_*$ moderately compressed liquid state
$(\;)_+$ slightly superheated vapor state

Definitions

Boundary: The real or imaginary surface delineating the thermodynamic system. The boundary separates the system from its environment. The boundary is an unambiguously defined surface. The boundary has zero thickness and zero volume.

Closed System: A thermodynamic system whose boundary is not crossed by mass flow.

Cycle: The special process in which the final state coincides with the initial state.

Environment: The thermodynamic system external to the thermodynamic system.

Extensive Properties: Properties whose values depend on the size of the system (e.g., mass, volume, energy, enthalpy, entropy).

Intensive Properties: Properties whose values do not depend on the system size (e.g., pressure, temperature). The collection of all intensive properties constitutes the *intensive state*.

Open System: A thermodynamic system whose boundary is permeable to mass flow. Open systems (flow systems) have their own nomenclature: the thermodynamic system is usually referred to as the *control volume,* the boundary of the open system is the *control surface,* and the particular regions of the boundary that are crossed by mass flows are the *inlet* and *outlet ports*.

Phase: The collection of all system elements that have the same intensive state (e.g., the liquid droplets dispersed in a liquid–vapor mixture have the same intensive state, that is, the same pressure, temperature, specific volume, specific entropy, etc.).

Process: The change of state from one initial state to a final state. In addition to the end states, knowledge of the process implies knowledge of the *interactions* experienced by the system while in communication with its environment (e.g., work transfer, heat transfer, mass transfer, and entropy transfer). To know the process also means to know the *path* (the history, or the succession of states) followed by the system from the initial to the final state.

State: The condition (the being) of a thermodynamic system at a particular point in time, as described by an ensemble of quantities called *thermodynamic properties* (e.g., pressure, volume, temperature, energy, enthalpy, entropy). Thermodynamic properties are only those quantities that do not depend on the "history" of the system between two different states. Quantities that depend on the system evolution (path) between states are not thermodynamic properties (examples of nonproperties are the work, heat, and mass transfer; the entropy transfer; the entropy generation; and the destroyed exergy—see also the definition of *process*).

Thermodynamic System: The region or the collection of matter in space selected for analysis.

2   THE FIRST LAW OF THERMODYNAMICS FOR CLOSED SYSTEMS

The first law of thermodynamics is a statement that brings together three concepts in thermodynamics: work transfer, heat transfer, and energy change. Of these concepts, only energy change or, simply, energy, is a thermodynamic property. We begin with a review[1] of the concepts of work transfer, heat transfer, and energy change.

Consider the force F_x experienced by a system at a point on its boundary. The infinitesimal *work transfer* between system and environment is

$$\delta W = -F_x \, dx$$

where the boundary displacement dx is defined as positive in the direction of the force F_x. When the force $\mathbf{F}$ and the displacement of its point of application $d\mathbf{r}$ are not collinear, the general definition of infinitesimal work transfer is

$$\delta W = -\mathbf{F} \cdot d\mathbf{r}$$

The work-transfer interaction is considered positive when the system does work on its environment—in other words, when $\mathbf{F}$ and $d\mathbf{r}$ are oriented in opposite directions. This sign convention has its origin in heat engine engineering, because the purpose of heat engines as thermodynamic systems is to deliver work while receiving heat.

For a system to experience work transfer, two things must occur: (1) a force must be present on the boundary, and (2) the point of application of this force (hence, the boundary) must move. The mere presence of forces on the boundary, without the displacement or the deformation of the boundary, does not mean work transfer. Likewise, the mere presence of boundary displacement without a force opposing or driving this motion does not mean work transfer. For example, in the free expansion of a gas into an evacuated space, the gas system does not experience work transfer because throughout the expansion the pressure at the imaginary system–environment interface is zero.

If a closed system can interact with its environment only via work transfer (i.e., in the absence of heat transfer δQ discussed later), then measurements show that the work transfer during a change of state from state 1 to state 2 is the same for all processes linking states 1 and 2,

$$-\left(\int_1^2 \delta W \right)_{\delta Q=0} = E_2 - E_1$$

In this special case the work-transfer interaction $(W_{1-2})_{\delta Q=0}$ is a property of the system, because its value depends solely on the end states. This thermodynamic property is the *energy change* of the system, $E_2 - E_1$. The statement that preceded the last equation is the first law of thermodynamics for closed systems that do not experience heat transfer.

Heat transfer is, like work transfer, an energy interaction that can take place between a system and its environment. The distinction between δQ and δW is made by the second law of thermodynamics discussed in the next section: Heat transfer is the energy interaction accompanied by entropy transfer, whereas work transfer is the energy interaction taking place in the absence of entropy transfer. The transfer of heat is driven by the *temperature difference* established between the system and its environment.[2] The system temperature is measured by placing the system in thermal communication with a test system called *thermometer*. The result of this measurement is the *relative temperature* θ expressed in degrees Celsius, θ (°C), or Fahrenheit, θ (°F); these alternative temperature readings are related through the conversion formulas

$$\theta(°C) = \tfrac{5}{9}[\theta(°F) - 32]$$

$$\theta(°F) = \tfrac{5}{9}\theta(°C) + 32$$

$$1°F = \tfrac{5}{9}°C$$

The boundary that prevents the transfer of heat, regardless of the magnitude of the system–environment temperature difference, is termed *adiabatic*. Conversely, the boundary that is crossed by heat even in the limit of a vanishingly small system–environment temperature difference is termed *diathermal*.

Measurements also show that a closed system undergoing a change of state $1 \rightarrow 2$ in the absence of work transfer experiences a heat interaction whose magnitude depends solely on the end states:

$$\left(\int_1^2 \delta Q\right)_{\delta W=0} = E_2 - E_1$$

In the special case of zero work transfer, the heat-transfer interaction is a thermodynamic property of the system, which is by definition equal to the energy change experienced by the system in going from state 1 to state 2. The last equation is the first law of thermodynamics for closed systems incapable of experiencing work transfer. Note that, unlike work transfer, the heat transfer is considered positive when it increases the energy of the system.

Most thermodynamic systems do not manifest purely mechanical ($\delta Q = 0$) or purely thermal ($\delta W = 0$) behavior. Most systems manifest a *coupled* mechanical and thermal behavior. The preceding first-law statements can be used to show that the first law of thermodynamics for a process executed by a closed system experiencing both work transfer and heat transfer is

$$\underbrace{\underbrace{\int_1^2 \delta Q}_{\substack{\text{heat} \\ \text{transfer}}} - \underbrace{\int_1^2 \delta W}_{\substack{\text{work} \\ \text{transfer}}}}_{\substack{\text{energy interactions} \\ \text{(nonproperties)}}} = \underbrace{E_2 - E_1}_{\substack{\text{energy} \\ \text{change} \\ \text{(property)}}}$$

The first law means that the net heat transfer into the system equals the work done by the system on the environment, plus the increase in the energy of the system. The first law of thermodynamics for a cycle or for an integral number of cycles executed by a closed system is

$$\oint \delta Q = \oint \delta W = 0$$

Note that the net change in the thermodynamic property energy is zero during a cycle or an integral number of cycles.

The energy change term $E_2 - E_1$ appearing on the right-hand side of the first law can be replaced by a more general notation that distinguishes between macroscopically identifiable forms of energy storage (kinetic, gravitational) and energy stored internally,

$$\underbrace{E_2 - E_1}_{\substack{\text{energy} \\ \text{change}}} = \underbrace{U_2 - U_1}_{\substack{\text{internal} \\ \text{energy} \\ \text{change}}} + \underbrace{\frac{mV_2^2}{2} - \frac{mV_1^2}{2}}_{\substack{\text{kinetic} \\ \text{energy} \\ \text{change}}} + \underbrace{mgZ_2 - mgZ_1}_{\substack{\text{gravitaional} \\ \text{energy} \\ \text{change}}}$$

If the closed system expands or contracts *quasistatically* (i.e., slowly enough, in mechanical equilibrium internally and with the environment) so that at every point in time the pressure P is uniform throughout the system, then the work-transfer term can be calculated as being equal to the work done by all the boundary pressure forces as they move with their respective points of application,

$$\int_1^2 \delta W = \int_1^2 P \, dV$$

The work-transfer integral can be evaluated provided the path of the quasistatic process, $P(V)$, is known; this is another reminder that the work transfer is path-dependent (i.e., not a thermodynamic property).

3   THE SECOND LAW OF THERMODYNAMICS FOR CLOSED SYSTEMS

A *temperature reservoir* is a thermodynamic system that experiences only heat transfer and whose temperature remains constant during such interactions. Consider first a closed system executing a cycle or an integral number of cycles *while in thermal communication with no more than one temperature reservoir*. To state the second law for this case is to observe that the net work transfer during each cycle cannot be positive,

$$\oint \delta W = 0$$

In other words, a closed system cannot deliver work during one cycle, while in communication with one temperature reservoir or with no temperature reservoir at all. Examples of such cyclic operation are the vibration of a spring–mass system, or a ball bouncing on the pavement: for these systems to return to their respective initial heights, that is, for them to execute cycles, the environment (e.g., humans) must perform work on them. The limiting case of frictionless cyclic operation is termed *reversible,* because in this limit the system returns to its initial state without intervention (work transfer) from the environment. Therefore, the distinction between reversible and irreversible cycles executed by closed systems in communication with no more than one temperature reservoir is

$$\oint \delta W = 0 \qquad \text{(reversible)}$$

$$\oint \delta W < 0 \qquad \text{(irreversible)}$$

To summarize, the first and second laws for closed systems operating cyclically in contact with no more than one temperature reservoir are (Fig. 1)

$$\oint \delta W = \oint \delta Q \leq 0$$

This statement of the second law can be used to show[1] that in the case of a closed system executing one or an integral number of cycles *while in communication with two temperature reservoirs,* the following inequality holds (Fig. 1)

$$\frac{Q_H}{T_H} + \frac{Q_L}{T_L} \leq 0$$

where H and L denote the high-temperature and the low-temperature reservoirs, respectively. Symbols Q_H and Q_L stand for the value of the cyclic integral $\oint \delta Q$, where δQ is in one case

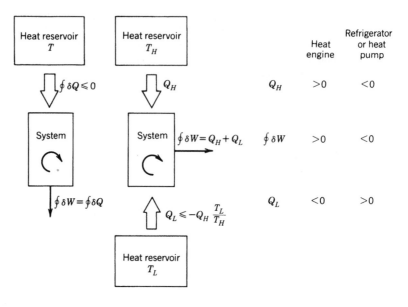

One heat reservoir   Two heat reservoirs

Figure 1 The first and second laws of thermodynamics for a closed system operating cyclically while in communication with one or two heat reservoirs.

exchanged only with the H reservoir, and in the other with the L reservoir. In the reversible limit, the second law reduces to $T_H/T_L = -Q_H/Q_L$, which serves as definition for the absolute *thermodynamic temperature* scale denoted by symbol T. Absolute temperatures are expressed either in kelvins, T (K), or in degrees Rankine, T (°R); the relationships between absolute and relative temperatures are

$$T \text{ (K)} = \theta \text{ (°C)} + 273.15 \text{ K} \qquad T \text{ (°R)} = \theta \text{ (°F)} + 459.67 \text{ °R}$$

$$1 \text{ K} = 1°\text{C} \qquad\qquad 1 \text{ °R} = 1°\text{F}$$

A *heat engine* is a special case of a closed system operating cyclically while in thermal communication with two temperature reservoirs, a system that during each cycle receives heat and delivers work:

$$\oint \delta W = \oint \delta Q = Q_H + Q_L > 0$$

The goodness of the heat engine can be described in terms of the heat engine efficiency or the first-law efficiency

$$\eta = \frac{\oint \delta W}{Q_H} \leq 1 - \frac{T_L}{T_H}$$

Alternatively, the second-law efficiency of the heat engine is defined as[1,3,4]

$$\eta_{\mathrm{II}} = \frac{\oint \delta W}{\left(\oint \delta W\right)_{\mathrm{maximum\ (reversible\ case)}}} = \frac{\eta_1}{1 - T_L/T_H}$$

A *refrigerating machine* or a *heat pump* operates cyclically between two temperature reservoirs in such a way that during each cycle it receives work and delivers net heat to the environment,

$$\oint \delta W = \oint \delta Q = Q_H + Q_L < 0$$

The goodness of such machines can be expressed in terms of a coefficient of performance (COP)

$$\mathrm{COP}_{\mathrm{refrigerator}} = \frac{Q_L}{-\oint \delta W} \le \frac{1}{T_H/T_L - 1}$$

$$\mathrm{COP}_{\mathrm{heat\ pump}} = \frac{-Q_H}{-\oint \delta W} \le \frac{1}{1 - T_L/T_H}$$

Generalizing the second law for closed systems operating cyclically, one can show that if during each cycle the system experiences any number of heat interactions Q_i with any number of temperature reservoirs whose respective absolute temperatures are T_i, then

$$\sum_i \frac{Q_i}{T_i} \le 0$$

Note that T_i is the absolute temperature of the boundary region crossed by Q_i. Another way to write the second law in this case is

$$\oint \frac{\delta Q}{T} \le 0$$

where, again, T is the temperature of the boundary pierced by δQ. Of special interest is the reversible cycle limit, in which the second law states ($\oint \delta Q/T)_{\mathrm{rev}} = 0$). According to the definition of thermodynamic property, the second law implies that during a reversible process the quantity $\delta Q/T$ is the infinitesimal change in a property of the system: by definition, that property is the *entropy change*

$$dS = \left(\frac{\delta Q}{T}\right)_{\mathrm{rev}} \quad \mathrm{or} \quad S_2 - S_1 = \left(\int_1^2 \frac{\delta Q}{T}\right)_{\mathrm{rev}}$$

Combining this definition with the second law for a cycle, $\oint \delta Q/T \le 0$, yields the second law of thermodynamics for *any process* executed by a closed system,

$$\underbrace{S_2 - S_1}_{\substack{\text{entropy} \\ \text{change} \\ \text{(property)}}} - \underbrace{\int_1^2 \frac{\delta Q}{T}}_{\substack{\text{entropy} \\ \text{transfer} \\ \text{(nonproperty)}}} \ge 0$$

The entire left-hand side in this inequality is by definition the *entropy generated* by the process,

$$S_{gen} = S_2 - S_1 - \int_1^2 \frac{\delta Q}{T}$$

The entropy generation is a measure of the inequality sign in the second law and hence a measure of the irreversibility of the process. As shown in the next chapter, the entropy generation is proportional to the useful work destroyed during the process.[1,3,4] Note again that any heat interaction (δQ) is accompanied by entropy transfer ($\delta Q/T$), whereas the work transfer δW is not.

4   ENERGY MINIMUM PRINCIPLE

Consider now a closed system that executes an infinitesimally small change of state, which means that its state changes from $(U, S, ...)$ to $(U + dU, S + dS, ...)$. The first law and the second law statements are

$$\delta Q - \delta W = dU$$

$$dS - \frac{\delta Q}{T} \geq 0$$

If the system is *isolated* from its environment, then $\delta W = 0$ and $\delta Q = 0$, and the two laws dictate that during any such process the energy inventory stays constant ($dU = 0$), and the entropy inventory cannot decrease,

$$dS \geq 0$$

Isolated systems undergo processes when they experience internal changes that do not require intervention from the outside, e.g., the removal of one or more of the *internal constraints* plotted qualitatively in the vertical direction in Fig. 2. When all the constraints are removed, changes cease, and, according to $dS \geq 0$, the entropy inventory reaches its highest possible level. This *entropy maximum principle* is a consequence of the first and second laws. When all the internal constraints have disappeared, the system has reached the *unconstrained equilibrium state*.

Alternatively, if changes occur in the absence of work transfer and at constant S, the first law and the second law require, respectively, $dU = \delta Q$ and $\delta Q \leq 0$, hence

$$dU \leq 0$$

The energy inventory cannot increase, and when the unconstrained equilibrium state is reached the system energy inventory is minimum. This *energy minimum principle* is also a consequence of the first and second laws for closed systems.

The interest in this classical formulation of the laws (e.g., Fig. 2) has been renewed by the emergence of an analogous principle of performance increase (the constructal law) in the search for optimal configurations in the design of open (flow) systems.[5] This analogy is based on the *constructal law* of maximization of flow access,[1,6] and is summarized in the next chapter.

5   THE LAWS OF THERMODYNAMICS FOR OPEN SYSTEMS

If $\dot{m}$ represents the mass flow rate through a port in the control surface, the principle of *mass conservation* in the control volume is

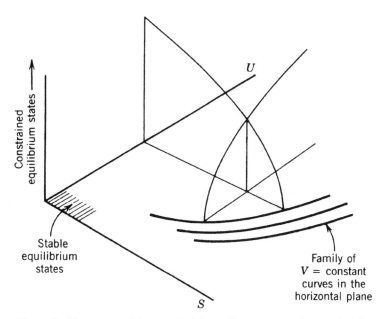

Figure 2 The energy minimum principle, or the entropy maximum principle.

$$\sum_{\text{in}} \dot{m} - \sum_{\text{out}} \dot{m} = \frac{\partial M}{\partial t}$$

$$\underbrace{\sum_{\text{in}} \dot{m} - \sum_{\text{out}} \dot{m}}_{\text{mass transfer}} = \underbrace{\frac{\partial M}{\partial t}}_{\text{mass change}}$$

Subscripts in and out refer to summation over all the inlet and outlet ports, respectively, while M stands for the instantaneous mass inventory of the control volume.

The first law of thermodynamics is more general than the statement encountered earlier for closed systems, because this time we must account for the flow of energy associated with the $\dot{m}$ streams:

$$\underbrace{\sum_{\text{in}} \dot{m}\left(h + \frac{V^2}{2} + gZ\right) - \sum_{\text{out}} \dot{m}\left(h + \frac{V^2}{2} + gZ\right) + \sum_i \dot{Q}_i - \dot{W}}_{\text{energy transfer}} = \underbrace{\frac{\partial E}{\partial t}}_{\text{energy change}}$$

On the left-hand side we have the energy interactions: heat, work, and the energy transfer associated with mass flow across the control surface. The specific enthalpy h, fluid velocity V, and height Z are evaluated right at the boundary. On the right-hand side, E is the instantaneous system energy integrated over the control volume.

The second law of thermodynamics for an open system assumes the form

$$\underbrace{\sum_{\text{in}} \dot{m}s - \sum_{\text{out}} \dot{m}s + \sum_i \frac{\dot{Q}_i}{T_i}}_{\text{entropy transfer}} \le \underbrace{\frac{\partial S}{\partial t}}_{\text{entropy change}}$$

The specific entropy s is representative of the thermodynamic state of each stream right at the system boundary. The *entropy generation rate* is defined by

$$\dot{S}_{\text{gen}} = \frac{\partial S}{\partial t} + \sum_{\text{out}} \dot{m}s - \sum_{\text{in}} \dot{m}s - \sum_i \frac{\dot{Q}_i}{T_i}$$

and is a measure of the irreversibility of open system operation. The engineering importance of $\dot{S}_{\text{gen}}$ stems from its proportionality to the rate of destruction of available work. If the following parameters are fixed—all the mass flows ($\dot{m}$), the peripheral conditions (h, s, V, Z), and the heat interactions (Q_i, T_i) except (Q_0, T_0)—then one can use the first law and the second law to show that the work-transfer rate cannot exceed a theoretical maximum.[1,3,4]

$$\dot{W} \leq \sum_{\text{in}} \dot{m} \left(h + \frac{V^2}{2} + gZ - T_0 s \right) - \sum_{\text{out}} \dot{m} \left(h + \frac{V^2}{2} + gZ - T_0 s \right) - \frac{\partial}{\partial t} (E - T_0 s)$$

The right-hand side in this inequality is the maximum work transfer rate $\dot{W}_{\text{sh,max}}$, which would exist only in the ideal limit of reversible operation. The rate of *lost work*, or the rate of exergy (availability) destruction, is defined as

$$\dot{W}_{\text{lost}} = \dot{W}_{\text{max}} - \dot{W}$$

Again, using both laws, one can show that lost work is directly proportional to entropy generation,

$$\dot{W}_{\text{lost}} = T_0 \dot{S}_{\text{gen}}$$

This result is known as the Gouy-Stodola theorem.[1,3,4] Conservation of useful work (exergy) in thermodynamic systems can only be achieved based on the systematic minimization of entropy generation in all the components of the system. Engineering applications of entropy generation minimization as a design optimization philosophy may be found in Refs. 1, 3, and 4, and in the next chapter.

6 RELATIONS AMONG THERMODYNAMIC PROPERTIES

The analytical forms of the first and second laws of thermodynamics contain properties such as internal energy, enthalpy, and entropy, which cannot be measured directly. The values of these properties are derived from measurements that can be carried out in the laboratory (e.g., pressure, volume, temperature, specific heat); the formulas connecting the derived properties to the measurable properties are reviewed in this section. Consider an infinitesimal change of state experienced by a closed system. If kinetic and gravitational energy changes can be neglected, the first law reads

$$\delta Q_{\text{any path}} - \delta W_{\text{any path}} = dU$$

which emphasizes that dU is path-independent. In particular, for a reversible path (rev), the same dU is given by

$$\delta Q_{\text{rev}} - \delta W_{\text{rev}} = dU$$

Note that from the second law for closed systems we have $\delta Q_{\text{rev}} = T\,dS$. Reversibility (or zero entropy generation) also requires internal mechanical equilibrium at every stage during the process; hence, $\delta W_{\text{rev}} = P\,dV$, as for a quasistatic change in volume. The infinitesimal change experienced by U is therefore

$$T\,dS - P\,dV = dU$$

Note that this formula holds for an infinitesimal change of state along any path (because dU is path-independent); however, $T\,dS$ matches δQ and $P\,dV$ matches δW only if the path is reversible. In general, $\delta Q < T\,dS$ and $\delta W < P\,dV$. The formula derived above for dU can be written for a unit mass: $T\,ds - P\,dv = du$. Additional identities implied by this relation are

$$T = \left(\frac{\partial u}{\partial s}\right)_v \qquad -P = \left(\frac{\partial u}{\partial v}\right)_s$$

$$\frac{\partial^2 u}{\partial s\,\partial v} = \left(\frac{\partial T}{\partial v}\right)_s = -\left(\frac{\partial P}{\partial s}\right)_v$$

where the subscript indicates which variable is held constant during partial differentiation. Similar relations and partial derivative identities exist in conjunction with other derived functions such as enthalpy, Gibbs free energy, and Helmholtz free energy:

- Enthalpy (defined as $h = u + Pv$)

$$dh = T\,ds + v\,dP$$

$$T = \left(\frac{\partial h}{\partial s}\right)_P \qquad v = \left(\frac{\partial h}{\partial P}\right)_s$$

$$\frac{\partial^2 h}{\partial s\,\partial P} = \left(\frac{\partial T}{\partial P}\right)_s = \left(\frac{\partial v}{\partial s}\right)_P$$

- Gibbs free energy (defined as $g = h - Ts$)

$$dg = -s\,dT + v\,dP$$

$$-s = \left(\frac{\partial g}{\partial T}\right)_P \qquad v = \left(\frac{\partial g}{\partial P}\right)_T$$

$$\frac{\partial^2 g}{\partial T\,\partial P} = -\left(\frac{\partial s}{\partial P}\right)_T = \left(\frac{\partial v}{\partial T}\right)_P$$

- Helmholtz free energy (defined as $f = u - Ts$)

$$df = -s\,dT - P\,dv$$

$$-s = \left(\frac{\partial f}{\partial T}\right)_v \qquad -P = \left(\frac{\partial f}{\partial v}\right)_T$$

$$\frac{\partial^2 f}{\partial T\,\partial v} = -\left(\frac{\partial s}{\partial v}\right)_T = -\left(\frac{\partial P}{\partial T}\right)_v$$

In addition to the (P, v, T) surface, which can be determined based on measurements (Fig. 3), the following partial derivatives are furnished by special experiments[1]:

- The specific heat at constant volume, $c_v = (\partial u/\partial T)_v$, follows directly from the constant volume $(\partial W = 0)$ heating of a unit mass of pure substance.
- The specific heat at constant pressure, $c_P = (\partial h/\partial T)_P$, is determined during the constant-pressure heating of a unit mass of pure substance.

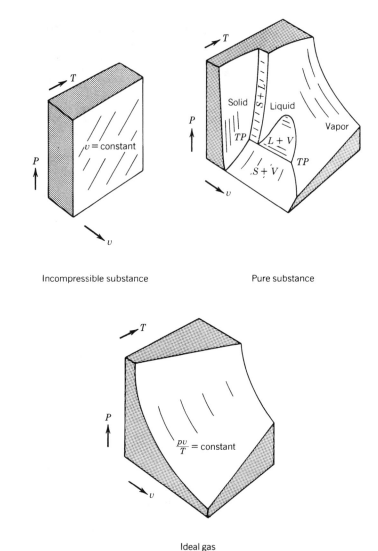

Incompressible substance   Pure substance

Ideal gas

Figure 3   The (P, v, T) surface for a pure substance that contracts upon freezing, showing regions of ideal gas and incompressible fluid behavior. In this figure, S = solid, V = vapor, L = liquid, TP = triple point.

- The Joule-Thompson coefficient, $\mu = (\partial T / \partial P)_h$, is measured during a throttling process, that is, during the flow of a stream through an adiabatic duct with friction (see the first law for an open system in the steady state).
- The coefficient of thermal expansion, $\beta = (1/v)(\partial v / \partial T)_P$.
- The isothermal compressibility, $K = (-1/v)(\partial v / \partial P)_T$.
- The constant temperature coefficient, $c_T = (\partial h / \partial P)_T$.

Two noteworthy relationships between some of the partial-derivative measurements are

$$c_P - c_v = \frac{Tv\beta^2}{K}$$

$$\mu = \frac{1}{c_P}\left[T\left(\frac{\partial v}{\partial T}\right)_P - v\right]$$

The general equations relating the derived properties (u, h, s) to measurable quantities are

$$du = c_v\,dT + \left[T\left(\frac{\partial P}{\partial T}\right)_v - P\right]dv$$

$$dh = c_P\,dT + \left[-T\left(\frac{\partial v}{\partial T}\right)_P + v\right]dP$$

$$ds = \frac{c_v}{T}\,dT + \left(\frac{\partial v}{\partial T}\right)_v dv \quad \text{or} \quad ds = \frac{c_P}{T}\,dT - \left(\frac{\partial v}{\partial T}\right)_P dP$$

These relations also suggest the following identities:

$$\left(\frac{\partial u}{\partial T}\right)_v = T\left(\frac{\partial s}{\partial T}\right)_v = c_v \qquad \left(\frac{\partial h}{\partial T}\right)_P = T\left(\frac{\partial s}{\partial T}\right)_P = c_P$$

The relationships between thermodynamic properties and the analyses associated with applying the laws of thermodynamics are simplified considerably in cases where the pure substance exhibits *ideal gas* behavior. As shown in Fig. 3, this behavior sets in at sufficiently high temperatures and low pressures; in this limit, the (P, v, T) surface is fitted closely by the simple expression

$$\frac{Pv}{T} = R \text{ (constant)}$$

where R is the ideal gas constant of the substance of interest (Table 1). The formulas for internal energy, enthalpy, and entropy, which concluded the preceding section, assume the following form in the ideal-gas limit:

$$du = c_v\,dT \qquad c_v = c_v(T)$$

$$dh = c_P\,dT \qquad c_P = c_P(T) = c_v + R$$

$$ds = \frac{c_v}{T}\,dT + \frac{R}{v}\,dv \quad \text{or} \quad ds = \frac{c_P}{T}\,dT - \frac{R}{P}\,dP \quad \text{or} \quad ds = \frac{c_v}{P}\,dP + \frac{c_P}{v}\,dv$$

If the coefficients c_v and c_P are constant in the temperature domain of interest, then the *changes* in specific internal energy, enthalpy, and entropy relative to a reference state ( )$_0$ are given by the formulas

$$u - u_0 = c_v\,(T - T_0)$$

$$h - h_0 = c_P\,(T - T_0) \qquad \text{(where } h_0 = u_0 + RT_0\text{)}$$

$$s - s_0 = c_v\ln\frac{T}{T_0} + R\ln\frac{v}{v_0}$$

$$s - s_0 = c_P\ln\frac{T}{T_0} - R\ln\frac{P}{P_0}$$

Table 1   Values of Ideal-Gas Constant and Specific Heat at Constant Volume for Gases Encountered in Mechanical Engineering[1]

Ideal Gas	R $\left(\dfrac{J}{kg \cdot K}\right)$	c_P $\left(\dfrac{J}{kg \cdot K}\right)$
Air	286.8	715.9
Argon, Ar	208.1	316.5
Butane, C_4H_{10}	143.2	1595.2
Carbon dioxide, CO_2	188.8	661.5
Carbon monoxide, CO	296.8	745.3
Ethane, C_2H_6	276.3	1511.4
Ethylene, C_2H_4	296.4	1423.5
Helium, He_2	2076.7	3152.7
Hydrogen, H	4123.6	10216.0
Methane, CH_4	518.3	1687.3
Neon, Ne	412.0	618.4
Nitrogen, N_2	296.8	741.1
Octane, C_8H_{18}	72.85	1641.2
Oxygen, O_2	259.6	657.3
Propane, C_3H_8	188.4	1515.6
Steam, H_2O	461.4	1402.6

$$s - s_0 = c_v \ln \frac{P}{P_0} + c_p \ln \frac{v}{v_0}$$

The ideal-gas model rests on two empirical constants, c_v and c_P, or c_v and R, or c_P and R. The ideal-gas limit is also characterized by

$$\mu = 0 \qquad \beta = \frac{1}{P} \qquad K = \frac{1}{P} \qquad c_T = 0$$

The extent to which a thermodynamic system destroys available work is intimately tied to the system's entropy generation, that is, to the system's departure from the theoretical limit of reversible operation. Idealized processes that can be modeled as reversible occupy a central role in engineering thermodynamics, because they can serve as standard in assessing the goodness of real processes. Two benchmark reversible processes executed by closed ideal-gas systems are particularly simple and useful. A *quasistatic adiabatic process* $1 \rightarrow 2$ executed by a closed ideal-gas system has the following characteristics:

$$\int_1^2 \delta Q = 0$$

$$\int_1^2 \delta W = \frac{P_2 V_2}{\gamma - 1}\left[\left(\frac{V_2}{V_1}\right)^{\gamma - 1} - 1\right]$$

where $\gamma = c_P/c_v$

• Path

$$PV^\gamma = P_1 V_1^\gamma = P_2 V_2^\gamma \qquad \text{(constant)}$$

- Entropy change

$$S_2 - S_1 = 0$$

hence the name *isoentropic* or *isentropic* for this process

- Entropy generation

$$S_{\text{gen}_{1 \to 2}} = S_2 - S_1 - \int_1^2 \frac{\delta Q}{T} = 0 \qquad \text{(reversible)}$$

A *quasistatic isothermal process* $1 \to 2$ executed by a closed ideal-gas system in communication with a single temperature reservoir T is characterized by

- Energy interactions

$$\int_1^2 \delta Q = \int_1^2 \delta W = m\,RT \ln \frac{V_2}{V_1}$$

- Path

$$T = T_1 = T_2 \quad \text{(constant)} \quad \text{or} \quad PV = P_1 V_1 = P_2 V_2 \quad \text{(constant)}$$

- Entropy change

$$S_2 - S_1 = m\,R \ln \frac{V_2}{V_1}$$

- Entropy generation

$$S_{\text{gen}_{1 \to 2}} = S_2 - S_1 - \int_1^2 \frac{\delta Q}{T} = 0 \qquad \text{(reversible)}$$

Mixtures of ideal gases also behave as ideal gases in the high-temperature, low-pressure limit. If a certain mixture of mass m contains ideal gases mixed in mass proportions m_i, and if the ideal-gas constants of each component are (c_{v_i}, c_{P_i}, R_i), then the equivalent ideal gas constants of the mixture are

$$c_v = \frac{1}{m} \sum_i m_i c_{v_i}$$

$$c_p = \frac{1}{m} \sum_i m_i c_{P_i}$$

$$R = \frac{1}{m} \sum_i m_i R_i$$

where $m = \sum_i m_i$.

One mole is the amount of substance of a system that contains as many elementary entities (e.g., molecules) as there are in 12 g of carbon 12; the number of such entities is Avogadro's constant, $N_0 \cong 6.022 \times 10^{23}$. The mole is not a mass unit, because the mass of 1 mole is not the same for all substances. The *molar mass* M of a given molecular species is the mass of 1 mole of that species, so that the total mass m is equal to M times the number of moles n,

$$m = nM$$

Thus, the ideal-gas equation of state can be written as

$$PV = nMRT$$

where the product MR is the *universal gas constant*

$$\overline{R} = MR = 8.314 \; \frac{J}{mol \cdot K}$$

The equivalent molar mass of a mixture of ideal gases with individual molar masses M_i is

$$M = \frac{1}{n} \sum n_i M_i$$

where $n = \sum n_i$. The molar mass of air, as a mixture of nitrogen, oxygen, and traces of other gases, is 28.966 g/mol (or 28.966 kg/kmol). A more useful model of the air gas mixture relies on only nitrogen and oxygen as constituents, in the proportion 3.76 moles of nitrogen to every mole of oxygen; this simple model is used frequently in the field of combustion.[1]

At the opposite end of the spectrum is the *incompressible substance* model. At sufficiently high pressures and low temperatures in Fig. 3, solids and liquids behave so that their density or specific volume is practically constant. In this limit the (P, v, T) surface is adequately represented by the equation

$$v = \overline{v} \qquad \text{(constant)}$$

The formulas for calculating changes in internal energy, enthalpy, and entropy become (see the end of the section on relations among thermodynamic properties)

$$du = c \, dT$$

$$dh = c \, dT + \overline{v} \, dP$$

$$ds = \frac{c}{T} \, dT$$

where c is the sole specific heat of the incompressible substance,

$$c = c_v = c_P$$

The specific heat c is a function of temperature only. In a sufficiently narrow temperature range where c can be regarded as constant, the finite changes in internal energy, enthalpy, and entropy relative to a reference state denoted by $(\;)_0$ are

$$u - u_0 = c \, (T - T_0)$$

$$h - h_0 = c \, (T - T_0) + \overline{v} \, (P - P_0) \qquad \text{(where } h_0 = u_0 + P_0 \overline{v})$$

$$s - s_0 = c \ln \frac{T}{T_0}$$

The incompressible substance model rests on two empirical constants, c and $\overline{v}$.

As shown in Fig. 3, the domains in which the pure substance behaves either as an ideal gas or as an incompressible substance intersect over regions where the substance exists as a mixture of two phases, liquid and vapor, solid and liquid, or solid and vapor. The two-phase regions themselves intersect along the *triple point* line labeled TP-TP on the middle sketch of Fig. 3. In engineering cycle calculations, the projections of the (P, v, T) surface on the P-v plane or, through the relations reviewed earlier, on the T-s plane are useful. The terminology associated with two-phase equilibrium states is defined on the P-v diagram of Fig. 4a, where we imagine the isothermal compression of a unit mass of substance (a closed

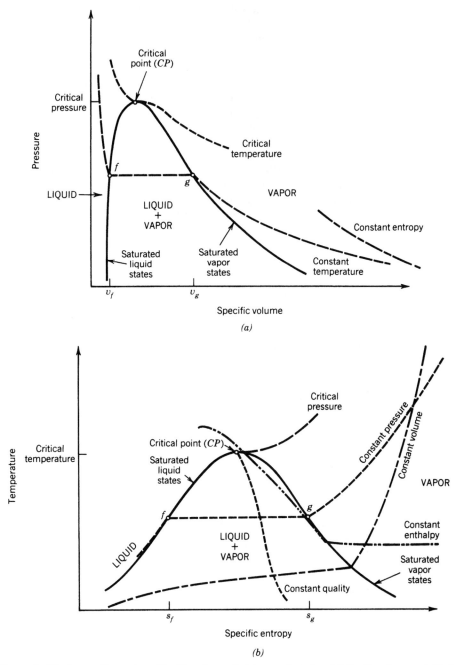

Figure 4 The locus of two-phase (liquid and vapor) states, projected on (a) the P-v plane, and (b) the T-s plane.

system). As the specific volume v decreases, the substance ceases to be a pure vapor at state g, where the first droplets of liquid are formed. State g is a *saturated vapor state*. It is observed that isothermal compression beyond g proceeds at constant pressure up to state f, where the last bubble (immersed in liquid) is suppressed. State f is a *saturated liquid state*. Isothermal compression beyond f is accompanied by a steep rise in pressure, depending on the compressibility of the liquid phase. The *critical state* is the intersection of the locus of saturated vapor states with the locus of saturated liquid states (Fig. 4a). The temperature and pressure corresponding to the critical state are the *critical temperature* and *critical pressure*. Table 2 contains a compilation of critical-state properties of some of the more common substances.

Figure 4b shows the projection of the liquid and vapor domain on the T-s plane. On the same drawing is shown the relative positioning (the relative slopes) of the traces of various constant-property cuts through the three-dimensional surface on which all the equilibrium states are positioned. In the two-phase region, the temperature is a unique function of pressure. This one-to-one relationship is indicated also by the *Clapeyron* relation

$$\left(\frac{dP}{dT}\right)_{sat} = \frac{h_g - h_f}{T(v_g - v_f)} = \frac{s_g - s_f}{v_g - v_f}$$

where the subscript sat is a reminder that the relation holds for saturated states (such as g and f) and for mixtures of two saturated phases. Subscripts g and f indicate properties

Table 2   Critical-State Properties[1]

Fluid	Critical Temperature [K (°C)]	Critical Pressure [MPa (atm)]		Critical Specific Volume (cm³/g)
Air	133.2 (−140)	3.77	(37.2)	2.9
Alcohol (methyl)	513.2 (240)	7.98	(78.7)	3.7
Alcohol (ethyl)	516.5 (243.3)	6.39	(63.1)	3.6
Ammonia	405.4 (132.2)	11.3	(111.6)	4.25
Argon	150.9 (−122.2)	4.86	(48)	1.88
Butane	425.9 (152.8)	3.65	(36)	4.4
Carbon dioxide	304.3 (31.1)	7.4	(73)	2.2
Carbon monoxide	134.3 (−138.9)	3.54	(35)	3.2
Carbon tetrachloride	555.9 (282.8)	4.56	(45)	1.81
Chlorine	417 (143.9)	7.72	(76.14)	1.75
Ethane	305.4 (32.2)	4.94	(48.8)	4.75
Ethylene	282.6 (9.4)	5.85	(57.7)	4.6
Helium	5.2 (−268)	0.228	(2.25)	14.4
Hexane	508.2 (235)	2.99	(29.5)	4.25
Hydrogen	33.2 (−240)	1.30	(12.79)	32.3
Methane	190.9 (−82.2)	4.64	(45.8)	6.2
Methyl chloride	416.5 (143.3)	6.67	(65.8)	2.7
Neon	44.2 (−288.9)	2.7	(26.6)	2.1
Nitric oxide	179.2 (−93.9)	6.58	(65)	1.94
Nitrogen	125.9 (−147.2)	3.39	(33.5)	3.25
Octane	569.3 (296.1)	2.5	(24.63)	4.25
Oxygen	154.3 (−118.9)	5.03	(49.7)	2.3
Propane	368.7 (95.6)	4.36	(43)	4.4
Sulfur dioxide	430.4 (157.2)	7.87	(77.7)	1.94
Water	647 (373.9)	22.1	(218.2)	3.1

corresponding to the saturated vapor and liquid states found at temperature T_{sat} (and pressure P_{sat}). Built into the last equation is the identity

$$h_g - h_f = T(s_g - s_f)$$

which is equivalent to the statement that the Gibbs free energy is the same for the saturated states and their mixtures found at the same temperature, $g_g = g_f$.

The properties of a two-phase mixture depend on the proportion in which saturated vapor, m_g, and saturated liquid, m_f, enter the mixture. The composition of the mixture is described by the property called *quality,*

$$x = \frac{m_g}{m_f + m_g}$$

The quality varies between 0 at state f and 1 at state g. Other properties of the mixture can be calculated in terms of the properties of the saturated states found at the same temperature,

$$u = u_f + x u_{fg} \qquad s = s_f + x s_{fg}$$
$$h = h_f + x h_{fg} \qquad v = v_f + x v_{fg}$$

with the notation $( )_{fg} = ( )_g - ( )_f$. Similar relations can be used to calculate the properties of two-phase states other than liquid and vapor, namely, solid and vapor or solid and liquid. For example, the enthalpy of a solid and liquid mixture is given by $h = h_s + x h_{sf}$, where subscript s stands for the *saturated solid state* found at the same temperature as for the two-phase state, and where h_{sf} is the latent heat of melting or solidification.

In general, the states situated immediately outside the two-phase dome sketched in Figs. 3 and 4 do not follow very well the limiting models discussed earlier in this section (ideal gas, incompressible substance). Because the properties of closely neighboring states are usually not available in tabular form, the following approximate calculation proves useful. For a *moderately compressed liquid state,* which is indicated by the subscript $( )_*$, that is, for a state situated close to the left of the dome in Fig. 4, the properties may be calculated as slight deviations from those of the saturated liquid state found at the same temperature as the compressed liquid state of interest,

$$h_* \cong (h_f)_{T*} + (v_f)_{T*}[P_* - (P_f)_{T*}]$$
$$s \cong (s_f)_{T*}$$

For a *slightly superheated vapor state,* that is, a state situated close to the right of the dome in Fig. 4, the properties may be estimated in terms of those of the saturated vapor state found at the same temperature:

$$h_+ \cong (h_g)_{T+}$$
$$s_+ \cong (s_g)_{T+} + \left(\frac{P_g v_g}{T_g}\right)_{T+} \ln \frac{(P_g)_{T+}}{P_+}$$

In these expressions, subscript $( )_+$ indicates the properties of the slightly superheated vapor state.

7   ANALYSIS OF ENGINEERING SYSTEM COMPONENTS

This section contains a summary[1] of the equations obtained by applying the first and second laws of thermodynamics to the components encountered in most engineering systems, such as power plants and refrigeration plants. It is assumed that each component operates in *steady flow*.

- *Valve* (throttle) or adiabatic duct with friction (Fig. 5a):

First law $h_1 = h_2$

Second law $\dot{S}_{gen} = \dot{m}(s_2 - s_1) > 0$

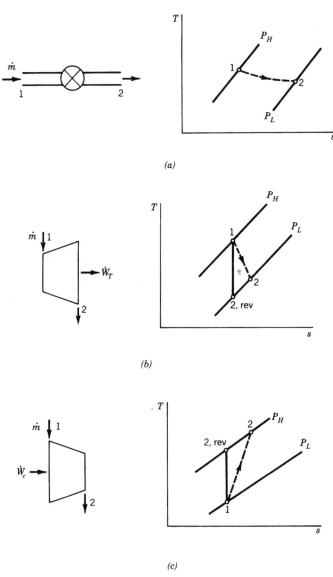

(a)

(b)

(c)

Figure 5 Engineering system components, and their inlet and outlet states on the *T-s* plane, P_H = high pressure; P_L = low pressure.

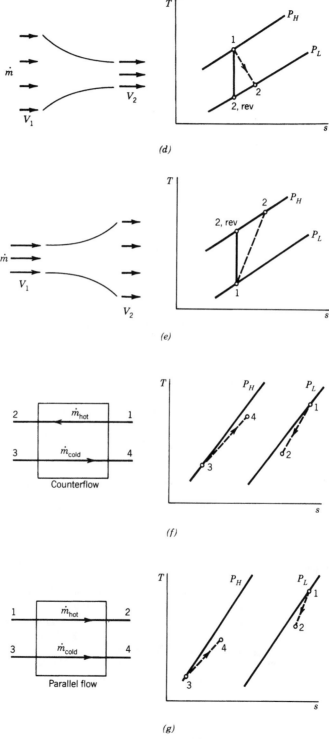

Figure 5 (*Continued*)

- *Expander* or *turbine* with negligible heat transfer to the ambient (Fig. 5b)

  First law $\dot{W}_T = \dot{m}\,(h_1 - h_2)$

  Second law $\dot{S}_{gen} = \dot{m}\,(s_2 - s_1) \geq 0$

  Efficiency $\eta_T = \dfrac{h_1 - h_2}{h_1 - h_{2,\text{rev}}} \leq 1$

- *Compressor* or *pump* with negligible heat transfer to the ambient (Fig. 5c):

  First law $\dot{W}_C = \dot{m}(h_2 - h_1)$

  Second law $\dot{S}_{gen} = \dot{m}(s_2 - s_1) \geq 0$

  Efficiency $\eta_C = \dfrac{h_{2,\text{rev}} - h_1}{h_2 - h_1} \leq 1$

- *Nozzle* with negligible heat transfer to the ambient (Fig. 5d):

  First law $\frac{1}{2}\,(V_2^2 - V_1^2) = h_1 - h_2$

  Second law $\dot{S}_{gen} = \dot{m}(s_2 - s_1) \geq 0$

  Efficiency $\eta_N = \dfrac{V_2^2 - V_1^2}{V_{2,\text{rev}}^2 - V_1^2} \leq 1$

- *Diffuser* with negligible heat transfer to the ambient (Fig. 5e):

  First law $h_2 - h_1 = \frac{1}{2}\,(V_1^2 - V_2^2)$

  Second law $\dot{S}_{gen} = \dot{m}(s_2 - s_1) \geq 0$

  Efficiency $\eta_D = \dfrac{h_{2,\text{rev}} - h_1}{h_2 - h_1} \leq 1$

- *Heat exchangers* with negligible heat transfer to the ambient (Figs. 5f and g)

  First law $\dot{m}_{\text{hot}}(h_1 - h_2) = \dot{m}_{\text{cold}}(h_4 - h_3)$

  Second law $\dot{S}_{gen} - \dot{m}_{\text{hot}}(s_2 - s_1) + \dot{m}_{\text{cold}}(s_4 - s_3) \geq 0$

Figures 5f and g show that a pressure drop always occurs in the direction of flow, in any heat exchanger flow passage.

REFERENCES

1. A. Bejan, *Advanced Engineering Thermodynamics,* 2nd ed., Wiley, New York, 1997.

2. A. Bejan, *Heat Transfer,* Wiley, New York, 1993.

3. A. Bejan, *Entropy Generation through Heat and Fluid Flow,* Wiley, New York, 1982.

4. A. Bejan, *Entropy Generation Minimization,* CRC Press, Boca Raton, FL, 1996.

5. A. Bejan and S. Lorente, "The Constructal Law and the Thermodynamics of Flow Systems with Configuration," *Int. J. Heat Mass Transfer* **47**, 3203–3214 (2004).

6. A. Bejan, *Shape and Structure, from Engineering to Nature,* Cambridge University Press, Cambridge, UK, 2000.

CHAPTER 4

EXERGY ANALYSIS, ENTROPY GENERATION MINIMIZATION, AND CONSTRUCTAL THEORY

Adrian Bejan
Department of Mechanical Engineering and Materials Science
Duke University
Durham, North Carolina

1 INTRODUCTION

In this chapter, we review three methods that account for much of the newer work in engineering thermodynamics and thermal design and optimization. The method of *exergy analysis* rests on thermodynamics alone. The first law, the second law, and the environment are used simultaneously to determine (1) the theoretical operating conditions of the system in the reversible limit and (2) the entropy generated (or exergy destroyed) by the actual system, that is, the departure from the reversible limit. The focus is on *analysis*. Applied to the system components individually, exergy analysis shows us quantitatively how much each component contributes to the overall irreversibility of the system.[1,2]

Entropy generation minimization (EGM) is a method of *modeling* and *optimization*. The entropy generated by the system is first developed as a function of the physical characteristics of the system (dimensions, materials, shapes, constraints). An important preliminary step is the construction of a system model that incorporates not only the traditional building blocks of engineering thermodynamics (systems, laws, cycles, processes, interactions), but also the fundamental principles of fluid mechanics, heat transfer, mass transfer and other transport phenomena. This combination makes the model "realistic" by accounting for the inherent irreversibility of the actual device. Finally, the minimum entropy generation design ($S_{\text{gen,min}}$) is determined for the model, and the approach of any other design (S_{gen}) to the limit of realistic ideality represented by $S_{\text{gen,min}}$ is monitored in terms of the entropy generation number $N_S = S_{\text{gen}}/S_{\text{gen,min}} > 1$.

To calculate S_{gen} and minimize it, the analyst does not need to rely on the concept of exergy. The EGM method represents an important step beyond thermodynamics. It is a self-standing method[3] that combines thermodynamics, heat transfer, and fluid mechanics into a powerful technique for modeling and optimizing real systems and processes. The use of the EGM method has expanded greatly.[4]

The most recent development in thermodynamics is the focus on the generation of flow system *configuration* (architecture, shape, structure).[1,5,6] Flow systems are thermodynamically imperfect because of resistances for fluid, heat, electricity, etc. Resistances cannot be eliminated. At best, they can be distributed (balanced) such that their combined effect is minimum. Distribution means configuration, drawing, design. The search for design is being pursued on the basis of principle—the *constructal* law, which is the statement that as configurations change on the designer's table the ones that survive are those that offer greater access (less resistance) to currents. The numerous observations that the flow configurations generated by the constructal law (e.g., tree networks) also occur in nature have led to *constructal theory,*[1,5] which is the thought that natural flow structures can be predicted based on the same principle. The constructal law bridges the gap between engineered and natural flow systems, and elevates "design" to the rank of scientific method. Moreover, the evolutionary search for the best design can be summarized in terms analogous to the classical energy minimum principle, as shown in the concluding section of this chapter.

SYMBOLS AND UNITS

a	specific nonflow availability, J/kg
A	nonflow availability, J
A	area, m^2
b	specific flow availability, J/kg
B	flow availability, J
B	duty parameter for plate and cylinder
B_s	duty parameter for sphere
B_0	duty parameter for tube
Be	Bejan number, $\dot{S}'''_{\mathrm{gen},\Delta T}/(\dot{S}'''_{\mathrm{gen},\Delta T} + \dot{S}'''_{\mathrm{gen},\Delta P})$
c_P	specific heat at constant pressure, J/(kg·K)
C	specific heat of incompressible substance, J/(kg·K)
C	heat leak thermal conductance, W/K
C^*	time constraint constant, s/kg
D	diameter, m
e	specific energy, J/kg
E	energy, J
$\bar{e}_{\mathrm{ch}}$	specific flow chemical exergy, J/kmol
$\bar{e}_t$	specific total flow exergy, J/kmol
e_x	specific flow exergy, J/kg
$\bar{e}_x$	specific flow exergy, J/kmol
E_Q	exergy transfer via heat transfer, J
$\dot{E}_W$	exergy transfer rate, W
E_x	flow exergy, J
EGM	the method of entropy generation minimization
f	friction factor
F_D	drag force, N
g	gravitational acceleration, m/s^2
G	mass velocity, kg/(s·m^2)
h	specific enthalpy, J/kg
h	heat transfer coefficient, W/(m^{2}K)
h°	total specific enthalpy, J/kg
H°	total enthalpy, J

k	thermal conductivity, W/(m K)
L	length, m
m	mass, kg
$\dot{m}$	mass flow rate, kg/s
M	mass, kg
N	mole number, kmol
$\dot{N}$	molal flow rate, kmol/s
N_S	entropy generation number, $S_{gen}/S_{gen,min}$
Nu	Nusselt number
N_{tu}	number of heat transfer units
P	pressure, N/m^2
Pr	Prandtl number
q'	heat transfer rate per unit length, W/m
Q	heat transfer, J
$\dot{Q}$	heat transfer rate, W
r	dimensionless insulation resistance
R	ratio of thermal conductances
Re_D	Reynolds number
s	specific entropy, J/(kg·K)
S	entropy, J/K
S_{gen}	entropy generation, J/K
$\dot{S}_{gen}$	entropy generation rate, W/K
$\dot{S}'_{gen}$	entropy generation rate per unit length, W/(m·K)
$\dot{S}'''_{gen}$	entropy generation rate per unit volume, W/(m^3 K)
t	time, s
t_c	time constraint, s
T	temperature, K
U	overall heat transfer coefficient, W/(m^2 K)
U_∞	free stream velocity, m/s
v	specific volume, m^3/kg
V	volume, m^3
V	velocity, m/s
$\dot{W}$	power, W
x	longitudinal coordinate, m
z	elevation, m
ΔP	pressure drop, N/m^2
ΔT	temperature difference, K
η	first law efficiency
η_{II}	second law efficiency
θ	dimensionless time
μ	viscosity, kg/(s·m)
μ_i^*	chemical potentials at the restricted dead state, J/kmol
$\mu_{0,i}$	chemical potentials at the dead state, J/kmol
ν	kinematic viscosity, m^2/s
ξ	specific nonflow exergy, J/kg
Ξ	nonflow exergy, J
Ξ_{ch}	nonflow chemical exergy, J
Ξ_t	nonflow total exergy, J
$\prime$	density, kg/m^3

SUBSCRIPTS

$(\ )_B$ base
$(\ )_C$ collector
$(\ )_C$ Carnot
$(\ )_H$ high
$(\ )_L$ low
$(\ )_m$ melting
$(\ )_{max}$ maximum
$(\ )_{min}$ minimum
$(\ )_{opt}$ optimal
$(\ )_p$ pump
$(\ )_{rev}$ reversible
$(\ )_t$ turbine
$(\ )_0$ environment
$(\ )_\infty$ free stream

2   EXERGY ANALYSIS

Figure 1 shows the general features of an open thermodynamic system that can interact thermally $(\dot{Q}_0)$ and mechanically $(P_0\, dV/dt)$ with the atmospheric temperature and pressure reservoir (T_0, P_0). The system may have any number of inlet and outlet ports, even though only two such ports are illustrated. At a certain point in time, the system may be in communication with any number of additional temperature reservoirs $(T_1, \ldots, T_n)$, experiencing the instantaneous heat interactions, $\dot{Q}_1, \ldots, \dot{Q}_n$. The work transfer rate $\dot{W}$ represents all the possible modes of work transfer: the work done on the atmosphere $(P_0\, dV/dt)$ and the remaining (useful, deliverable) portions such as $P\, dV/dt$, shaft work, shear work, electrical work, and magnetic work. The useful part is known as available work (or simply exergy) or, on a unit time basis,

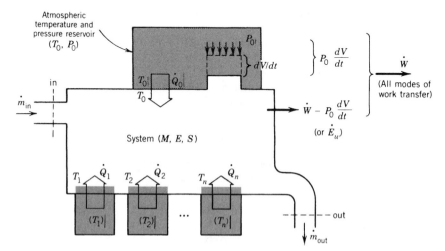

Figure 1   Open system in thermal and mechanical communication with the ambient. (From A. Bejan, *Advanced Engineering Thermodynamics.* © 1997 John Wiley & Sons, Inc. Reprinted by permission.)

$$\dot{E}_W = \dot{W} - P_0 \frac{dV}{dt}$$

The first and second laws of thermodynamics can be combined to show that the available work transfer rate from the system of Fig. 1 is given by the equation[1]

$$\dot{E}_W = -\frac{d}{dt}(E - T_0 S + P_0 V) + \sum_{i=1}^{n} \left(1 - \frac{T_0}{T_i}\right)\dot{Q}_i$$

$\underbrace{\hspace{5cm}}$ $\underbrace{\hspace{3cm}}$
Accumulation Exergy transfer
of nonflow exergy via heat transfer

$$+ \sum_{in} \dot{m}(h° - T_0 s) - \sum_{out} \dot{m}(h° - T_0 s) - T_0 \dot{S}_{gen}$$

$\underbrace{\hspace{4cm}}$ $\underbrace{\hspace{3cm}}$ $\underbrace{\hspace{2cm}}$
Intake of Release of Destruction
flow exergy via flow exergy via of exergy
mass flow mass flow

where E, V, and S are the instantaneous energy, volume, and entropy of the system, and $h°$ is shorthand for the specific enthalpy plus the kinetic and potential energies of each stream, $h° = h + \frac{1}{2}V^2 + gz$. The first four terms on the right-hand side of the $\dot{E}_W$ equation represent the exergy rate delivered as useful power (to an external user) in the limit of reversible operation ($\dot{E}_{W,rev}$, $\dot{S}_{gen} = 0$). It is worth noting that the $\dot{E}_W$ equation is a restatement of the Gouy-Stodola theorem (see Section 5), which is the proportionality between the rate of exergy destruction and the rate of entropy generation

$$\dot{E}_{W,rev} - \dot{E}_W = T_0 \dot{S}_{gen}$$

A special exergy nomenclature has been devised for the terms formed on the right side of the $\dot{E}_W$ equation. The exergy content associated with a heat transfer interaction (Q_i, T_i) and the environment (T_0) is the *exergy of heat transfer,*

$$E_{Qi} = Q_i \left(1 - \frac{T_0}{T_i}\right)$$

This means that the heat transfer with the environment (Q_0, T_0) carries zero exergy.

Associated with the system extensive properties (E, S, V) and the two specified intensive properties of the environment (T_0, P_0) is a new extensive property: the thermomechanical or physical *nonflow availability,*

$$A = E - T_0 S + P_0 V$$

$$a = e - T_0 s + P_0 v$$

Let A_0 represent the nonflow availability when the system is at the *restricted dead state* (T_0, P_0), that is, in thermal and mechanical equilibrium with the environment, $A_0 = E_0 - T_0 S_0 + P_0 V_0$. The difference between the nonflow availability of the system in a given state and its nonflow availability in the restricted dead state is the thermomechanical or physical *nonflow exergy,*

$$\Xi = A - A_0 = E - E_0 - T_0(S - S_0) + P_0(V - V_0)$$

$$\xi = a - a_0 = e - e_0 - T_0(s - s_0) + P_0(v - v_0)$$

The nonflow exergy represents the most work that would become available if the system were to reach its restricted dead state reversibly, while communicating thermally only with the environment. In other words, the nonflow exergy represents the exergy content of a given closed system relative to the environment.

Associated with each of the streams entering or exiting an open system is the thermomechanical or physical *flow availability,*

$$B = H° - T_0 S$$

$$b = h° - T_0 s$$

At the restricted dead state, the nonflow availability of the stream is $B_0 = H_0° - T_0 S_0$. The difference $B - B_0$ is known as the thermomechanical or physical *flow exergy* of the stream

$$E_x = B - B_0 = H° - H_0° - T_0 (S - S_0)$$

$$e_x = b - b_0 = h° - h_0° - T_0 (s - s_0)$$

The flow exergy represents the available work content of the stream relative to the restricted dead state (T_0, P_0). This work could be extracted in principle from a system that operates reversibly in thermal communication only with the environment (T_0), while receiving the given stream $(\dot{m}, h°, s)$ and discharging the same stream at the environmental pressure and temperature $(\dot{m}, h_0°, s_0)$.

In summary, *exergy analysis* means that the $\dot{E}_W$ equation can be rewritten more simply as

$$\dot{E}_W = -\frac{d\Xi}{dt} + \sum_{i=1}^{n} \dot{E}_{Q_i} + \sum_{in} \dot{m} e_x - \sum_{out} \dot{m} e_x - T_0 \dot{S}_{gen}$$

Examples of how these exergy concepts are used in the course of analyzing component by component the performance of complex systems can be found in Refs. 1–4. Figure 2 shows one such example.[1] The upper part of the drawing shows the traditional description of the four components of a simple Rankine cycle. The lower part shows the exergy streams that enter and exit each component, with the important feature that the heater, the turbine, and the cooler destroy significant portions (shaded, fading away) of the entering exergy streams. The numerical application of the $\dot{E}_W$ equation to each component tells the analyst the exact widths of the exergy streams to be drawn in Fig. 2. In graphical or numerical terms, the "exergy wheel" diagram[1] shows not only *how much* exergy is being destroyed but also *where.* It tells the designer how to rank order the components as candidates for optimization according to the method of *entropy generation minimization* (Sections 3–8).

To complement the traditional (first-law) energy conversion efficiency, $\eta = (\dot{W}_t - \dot{W}_p)/\dot{Q}_H$ in Fig. 2, exergy analysis recommends as figure of merit the *second-law efficiency,*

$$\eta_{II} = \frac{\dot{W}_t - \dot{W}_p}{\dot{E}_{Q_H}}$$

where $\dot{W}_t - \dot{W}_p$ is the net power output (i.e., $\dot{E}_W$ earlier in this section). The second-law efficiency can have values between 0 and 1, where 1 corresponds to the reversible limit. Because of this limit, η_{II} describes very well the fundamental difference between the method of exergy analysis and the method of entropy generation minimization (EGM), because in EGM the system always operates *irreversibly.* The question in EGM is how to change the system such that its $\dot{S}_{gen}$ value (always finite) approaches the minimum $\dot{S}_{gen}$ allowed by the system constraints.

Consider next a nonflow system that can experience heat, work, and mass transfer in communication with the environment. The environment is represented by T_0, P_0, and the n

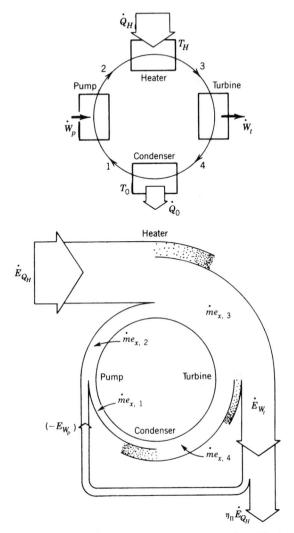

Figure 2 The exergy wheel diagram of a simple Rankine cycle. *Top:* The traditional notation and energy interactions. *Bottom:* The exergy flows and the definition of the second law efficiency. (From A. Bejan, *Advanced Engineering Thermodynamics.* © 1997 John Wiley & Sons, Inc. Reprinted by permission.)

chemical potentials $\mu_{0,i}$ of the environmental constituents that are also present in the system. Taken together, the $n + 2$ intensive properties of the environment (T_0, P_0, $\mu_{0,i}$) are known as the *dead state*.

Reading Fig. 3 from left to right, we see the system in its initial state represented by E, S, V, and its composition (mole numbers $N_1, \ldots, N_n$), and by its $n + 2$ intensities (T, P, μ_i). The system can reach its dead state in two steps. In the first, it reaches only thermal and mechanical equilibrium with the environment (T_0, P_0), and delivers the nonflow exergy Ξ defined in the preceding section. At the end of this first step, the chemical potentials of the constituents have changed to μ_i^* ($i = 1, \ldots, n$). During the second step, mass transfer

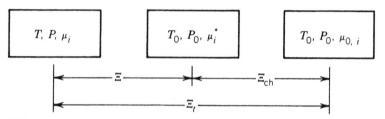

Figure 3 The relationship between the nonflow total (Ξ_t), physical (Ξ), and chemical (Ξ_{ch}) exergies. (From A. Bejan, *Advanced Engineering Thermodynamics*. © 1997 John Wiley & Sons, Inc. Reprinted by permission.)

occurs (in addition to heat and work transfer) and, in the end, the system reaches chemical equilibrium with the environment, in addition to thermal and mechanical equilibrium. The work made available during this second step is known as *chemical exergy,*[1]

$$\Xi_{ch} = \sum_{i=1}^{n} (\mu_i^* - \mu_{0,i})N_i$$

The total exergy content of the original nonflow system (E, S, V, N_i) relative to the environmental dead state (T_0, P_0, $\mu_{0,i}$) represents the *total nonflow exergy,*

$$\Xi_t = \Xi + \Xi_{ch}$$

The *total flow exergy* of a mixture stream of total molal flow rate $\dot{N}$ (composed of n species, with flow rates $\dot{N}_i$) and intensities T, P, and μ_i ($i = 1, \ldots, n$) is, on a mole of mixture basis,

$$\bar{e}_t = \bar{e}_x + \bar{e}_{ch}$$

where the physical flow exergy $\bar{e}_x$ was defined above, and $\bar{e}_{ch}$ is the *chemical exergy* per mole of mixture,

$$\bar{e}_{ch} = \sum_{i=1}^{n} (\mu_i^* - \mu_{0,i}) \frac{\dot{N}_i}{\dot{N}}$$

In the $\bar{e}_{ch}$ expression μ_i^* ($i = 1, \ldots, n$) are the chemical potentials of the stream constituents at the restricted dead state (T_0, P_0). The chemical exergy is the additional work that could be extracted (reversibly) as the stream evolves from the restricted dead state to the dead state (T_0, P_0, $\mu_{0,i}$) while in thermal, mechanical, and chemical communication with the environment.

3 ENTROPY GENERATION MINIMIZATION

The EGM method[3,4] is distinct from exergy analysis, because in exergy analysis the analyst needs only the first law, the second law, and a convention regarding the values of the intensive properties of the environment. The critically new aspects of the EGM method are the modeling of the system, the development of S_{gen} as a function of the physical parameters of the model, and the *minimization* of the calculated entropy generation rate. To minimize the irreversibility of a proposed design, the engineer must use the relations between temperature differences and heat transfer rates, and between pressure differences and mass flow rates.

The engineer must relate the degree of thermodynamic nonideality of the design to the physical characteristics of the system, namely, to finite dimensions, shapes, materials, finite speeds, and finite-time intervals of operation. For this, the engineer must rely on heat transfer and fluid mechanics principles, in addition to thermodynamics. Only by varying one or more of the physical characteristics of the system can the engineer bring the design closer to the operation characterized by minimum entropy generation subject to finite-size and finite-time constraints.

The modeling and optimization progress made in EGM is illustrated by some of the simplest and most fundamental results of the method, which are reviewed in the following sections. The structure of the EGM field is summarized in Fig. 4 by showing on the vertical the expanding list of applications. On the horizontal, we see the two modeling approaches that are being used. One approach is to focus from the start on the total system, to divide the system into compartments that account for one or more of the irreversibility mechanisms, and to declare the rest of the system irreversibility-free. In this approach, success depends on the modeler's intuition, because the assumed compartments do not always correspond to the pieces of hardware of the real system.

In the alternative approach (from the right in Fig. 4), modeling begins with dividing the system into its real components, and recognizing that each component may contain large numbers of one or more elemental features. The approach is to minimize S_{gen} in a fundamental way at each level, starting from the simple and proceeding toward the complex. Important to note is that when a component or elemental feature is imagined separately from the larger system, the quantities assumed specified at the points of separation act as con-

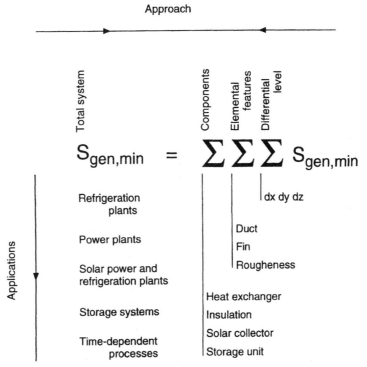

Figure 4 Approaches and applications of the method of entropy generation minimization (EGM).

straints on the optimization of the smaller system. The principle of thermodynamic isolation (Ref. 4, p. 125) must be kept in mind during the later stages of the optimization procedure, when the optimized elements and components are integrated into the total system, which itself is optimized for *minimum cost* in the final stage.[2]

4   CRYOGENICS

The field of low-temperature refrigeration was the first where EGM became an established method of modeling and optimization. Consider a path for heat leak ($\dot{Q}$) from room temperature (T_H) to the cold end (T_L) of a low-temperature refrigerator or liquefier. Examples of such paths are mechanical supports, insulation layers without or with radiation shields, counterflow heat exchangers, and electrical cables. The total rate of entropy generation associated with the heat leak path is

$$\dot{S}_{\text{gen}} = \int_{T_L}^{T_H} \frac{\dot{Q}}{T^2} \, dT$$

where $\dot{Q}$ is in general a function of the local temperature T. The proportionality between the heat leak and the local temperature gradient along its path, $\dot{Q} = kA(dT/dx)$, and the finite size of the path [length L, cross section A, material thermal conductivity $k(T)$] are accounted for by the integral constraint

$$\int_{T_L}^{T_H} \frac{k(T)}{\dot{Q}(T)} \, dT = \frac{L}{A} \quad \text{(constant)}$$

The optimal heat leak distribution that minimizes $\dot{S}_{\text{gen}}$ subject to the finite-size constraint is[3,4]

$$\dot{Q}_{\text{opt}}(T) = \left(\frac{A}{L} \int_{T_L}^{T_H} \frac{k^{1/2}}{T} \, dT \right) k^{1/2} T$$

$$\dot{S}_{\text{gen,min}} = \frac{A}{L} \left(\int_{T_L}^{T_H} \frac{k^{1/2}}{T} \, dT \right)^2$$

The technological applications of the variable heat leak optimization principle are numerous and important. In the case of a mechanical support, the optimal design is approximated in practice by placing a stream of cold helium gas in counterflow (and in thermal contact) with the conduction path. The heat leak varies as $d\dot{Q}/dT = \dot{m}c_P$, where $\dot{m}c_P$ is the capacity flow rate of the stream. The practical value of the EGM method is that it pinpoints the optimal flow rate for minimum entropy generation. To illustrate, if the support conductivity is temperature-independent, then the optimal flow rate is $\dot{m}_{\text{opt}} = (Ak/Lc_P) \ln (T_H/T_L)$. In reality, the conductivity of cryogenic structural materials varies strongly with the temperature, and the single-stream intermediate cooling technique can approach $\dot{S}_{\text{gen,min}}$ only approximately.

Other applications include the optimal cooling (e.g., optimal flow rate of boil-off helium) for cryogenic current leads, and the optimal temperatures of cryogenic radiation shields. The main counterflow heat exchanger of a low-temperature refrigeration machine is another important path for heat leak in the end-to-end direction ($T_H \rightarrow T_L$). In this case, the optimal variable heat leak principle translates into[3,4]

$$\left(\frac{\Delta T}{T} \right)_{\text{opt}} = \frac{\dot{m}c_P}{UA} \ln \frac{T_H}{T_L}$$

where ΔT is the local stream-to-stream temperature difference of the counterflow, $\dot{m}c_P$ is the capacity flow rate through one branch of the counterflow, and UA is the fixed size (total thermal conductance) of the heat exchanger.

5   HEAT TRANSFER

The field of heat transfer adopted the techniques developed in cryogenic engineering and applied them to a vast selection of devices for promoting heat transfer. The EGM method was applied to complete components (e.g., heat exchangers) and elemental features (e.g., ducts, fins). For example, consider the flow of a single-phase stream ($\dot{m}$) through a heat exchanger tube of internal diameter D. The heat transfer rate per unit of tube length q' is given. The entropy generation rate per unit of tube length is

$$\dot{S}'_{gen} = \frac{q'^2}{\pi k T^2 \mathrm{Nu}} + \frac{32\dot{m}^3 f}{\pi^2 \rho^2 T D^5}$$

where Nu and f are the Nusselt number and the friction factor, $\mathrm{Nu} = hD/k$ and $f = (-dP/dx)\rho D/(2G^2)$ with $G = \dot{m}/(\pi D^2/4)$. The $\dot{S}'_{gen}$ expression has two terms: the irreversibility contributions made by heat transfer and fluid friction. These terms compete against one another such that there is an optimal tube diameter for minimum entropy generation rate,[3,4]

$$\mathrm{Re}_{D,opt} \cong 2B_0^{0.36}\,\mathrm{Pr}^{-0.07}$$

$$B_0 = \frac{q'\dot{m}\rho}{(kT)^{1/2}\mu^{5/2}}$$

where $\mathrm{Re}_D = VD/\nu$ and $V = \dot{m}/(\rho\pi D^2/4)$. This result is valid in the range $2500 < \mathrm{Re}_D < 10^6$ and $\mathrm{Pr} > 0.5$. The corresponding entropy generation number is

$$N_S = \frac{\dot{S}'_{gen}}{\dot{S}'_{gen,min}} = 0.856\left(\frac{\mathrm{Re}_D}{\mathrm{Re}_{D,opt}}\right)^{-0.8} + 0.144\left(\frac{\mathrm{Re}_D}{\mathrm{Re}_{D,opt}}\right)^{4.8}$$

where $\mathrm{Re}_D/\mathrm{Re}_{D,opt} = D_{opt}/D$ because the mass flow rate is fixed. The N_S criterion was used extensively in the literature to monitor the approach of actual designs to the optimal irreversible designs conceived subject to the same constraints. This work is reviewed in Refs. 3 and 4.

The EGM of elemental features was extended to the optimization of augmentation techniques such as extended surfaces (fins), roughened walls, spiral tubes, twisted tape inserts, and full-size heat exchangers that have such features. For example, the entropy generation rate of a body with heat transfer and drag in an external stream (U_∞, T_∞) is

$$\dot{S}_{gen} = \frac{\dot{Q}_B(T_B - T_\infty)}{T_B T_\infty} + \frac{F_D U_\infty}{T_\infty}$$

where $\dot{Q}_B$, T_B, and F_D are the heat transfer rate, body temperature, and drag force. The relation between $\dot{Q}_B$ and temperature difference ($T_B - T_\infty$) depends on body shape and external fluid and flow, and is provided by the field of convective heat transfer.[7] The relation between F_D, U_∞, geometry and fluid type comes from fluid mechanics.[7] The $\dot{S}_{gen}$ expression has the expected two-term structure, which leads to an optimal body size for minimum entropy generation rate.

The simplest example is the selection of the swept length L of a plate immersed in a parallel stream (Fig. 5 inset). The results for $\mathrm{Re}_{L,opt} = U_\infty L_{opt}/\nu$ are shown in Fig. 5, where B is a constraint (duty parameter)

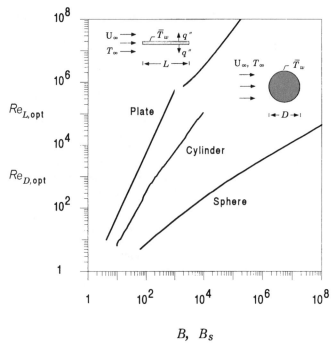

Figure 5   The optimal size of a plate, cylinder and sphere for minimum entropy generation. (From A. Bejan, G. Tsatsaronis, and M. Moran, *Thermal Design and Optimization.* © 1996 John Wiley & Sons, Inc. Reprinted by permission.)

$$B = \frac{\dot{Q}_B / W}{U_\infty (k\mu T_\infty \mathrm{Pr}^{1/3})^{1/2}}$$

and W is the plate dimension perpendicular to the figure. The same figure shows the corresponding results for the optimal diameter of a cylinder in cross-flow, where $\mathrm{Re}_{D,\mathrm{opt}} = U_\infty D_\mathrm{opt} / \nu$, and B is given by the same expression as for the plate. The optimal diameter of the sphere is referenced to the sphere duty parameter defined by

$$B_s = \frac{\dot{Q}_B}{\nu (k\mu T_\infty \mathrm{Pr}^{1/3})^{1/2}}$$

The fins built on the surfaces of heat exchanges act as bodies with heat transfer in external flow. The size of a fin of given shape can be optimized by accounting for the internal heat-transfer characteristics (longitudinal conduction) of the fin, in addition to the two terms (convective heat and fluid flow) shown in the last $\dot{S}_\mathrm{gen}$ formula. The EGM method has also been applied to complete heat exchangers and heat exchanger networks. This vast literature is reviewed in Ref. 4. One technological benefit of EGM is that it shows how to select certain dimensions of a device so that the device destroys minimum power while performing its assigned heat and fluid flow duty.

Several computational heat and fluid flow studies recommended that future commercial CFD packages have the capability of displaying entropy generation rate fields (maps) for both laminar and turbulent flows. For example, Paoletti et al.[8] recommend the plotting of contour lines for constant values of the dimensionless group Be $= \dot{S}'''_{\mathrm{gen},\Delta T} / (\dot{S}'''_{\mathrm{gen},\Delta T} +$

$\dot{S}'''_{gen,\Delta P}$), which they termed the Bejan number, where $\dot{S}'''_{gen}$ means local (volumetric) entropy generation rate, and ΔT and ΔP refer to the heat transfer and fluid flow irreversibilities, respectively. This dimensionless group should not be confused with the dimensionless pressure drop that is useful in the optimization of spacings in heat-transfer assemblies with forced convection, cf. Ref. 7, pp. 136–141.

6   STORAGE SYSTEMS

In the optimization of time-dependent heating or cooling processes the search is for optimal histories, that is, optimal ways of executing the processes. Consider as a first example the sensible heating of an amount of incompressible substance (mass M, specific heat C), by circulating through it a stream of hot ideal gas ($\dot{m}$, c_P, T_∞) (Fig. 6). Initially, the storage material is at the ambient temperature T_0. The total thermal conductance of the heat exchanger placed between the storage material and the gas stream is UA and the pressure drop is negligible. After flowing through the heat exchanger, the gas stream is discharged into the atmosphere. The entropy generated from $t = 0$ until a time t reaches a minimum when t is of the order of $MC/(\dot{m}c_P)$. Charts for calculating the optimal heating (storage) time interval are available in Refs. 3 and 4. For example, when ($T_\infty - T_0$), the optimal heating time is given by $\theta_{opt} = 1.256/[1 - \exp(-N_{tu})]$, where $\theta_{opt} = t_{opt} \dot{m}c_P/(MC)$ and $N_{tu} = UA/(\dot{m}c_P)$.

Another example is the optimization of a sensible-heat cooling process subject to an overall time constraint. Consider the cooling of an amount of incompressible substance (M, C) from a given initial temperature to a given final temperature, during a prescribed time interval t_c. The coolant is a stream of cold ideal gas with flow rate $\dot{m}$ and specific heat $c_P(T)$. The thermal conductance of the heat exchanger is UA; however, the overall heat transfer coefficient generally depends on the instantaneous temperature, $U(T)$. The cooling process requires a minimum amount of coolant m (or minimum refrigerator work for producing the cryogen m),

$$m = \int_0^{t_c} \dot{m}(t)\, dt$$

when the gas flow rate has the optimal history[3,4]

$$\dot{m}_{opt}(t) = \left[\frac{U(T)A}{C^* c_P(T)} \right]^{1/2}$$

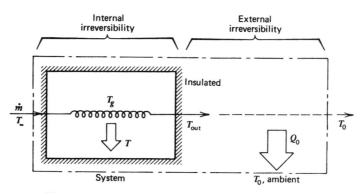

Figure 6   Entropy generation during sensible-heat storage.[3]

In this expression, $T(t)$ is the corresponding optimal temperature history of the object that is being cooled, and C^* is a constant that can be evaluated based on the time constraint, as shown in Refs. 3 and 4. The optimal flow rate history result ($\dot{m}_{opt}$) tells the operator that at temperatures where U is small the flow rate should be decreased. Furthermore, because during cooldown the gas c_p increases, the flow rate should decrease as the end of the process nears.

In the case of energy storage by melting there is an optimal melting temperature (i.e., optimal type of storage material) for minimum entropy generation during storage. If T_∞ and T_0 are the temperatures of the heat source and the ambient, the optimal melting temperature of the storage material has the value $T_{m,opt} = (T_\infty T_0)^{1/2}$.

7   SOLAR ENERGY CONVERSION

The generation of power and refrigeration based on energy from the sun has been the subject of some of the oldest EGM studies, which cover a vast territory. A characteristic of these EGM models is that they account for the irreversibility due to heat transfer in the two temperature gaps (sun-collector and collector-ambient) and that they reveal an optimal *coupling* between the collector and the rest of the plant.

Consider, for example, the steady operation of a power plant driven by a solar collector with convective heat leak to the ambient, $\dot{Q}_0 = (UA)_c(T_c - T_0)$, where $(UA)_c$ is the collector-ambient thermal conductance and T_c is the collector temperature (Fig. 7). Similarly, there is a finite size heat exchanger $(UA)_i$ between the collector and the hot end of the power cycle (T), such that the heat input provided by the collector is $\dot{Q} = (UA)_i(T_c - T)$. The power cycle is assumed reversible. The power output $\dot{W} = \dot{Q}(1 - T_0/T)$ is maximum, or the total entropy generation rate is minimum, when the collector has the optimal temperature.[3,4]

$$\frac{T_{c,opt}}{T_0} = \frac{\theta_{max}^{1/2} + R\theta_{max}}{1 + R}$$

where $R = (UA)_c/(UA)_i$, $\theta_{max} = T_{c,max}/T_0$ and $T_{c,max}$ is the maximum (stagnation) temperature of the collector.

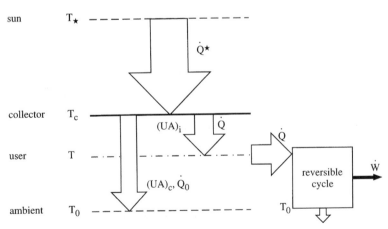

Figure 7   Solar power plant model with collector-ambient heat loss and collector-engine heat exchanger.[3]

Another type of optimum is discovered when the overall size of the installation is fixed. For example, in an extraterrestrial power plant with collector area A_H and radiator area A_L, if the total area is constrained[1]

$$A_H + A_L = A \qquad \text{(constant)}$$

then the optimal way to allocate the area is $A_{H,\text{opt}} = 0.35A$ and $A_{L,\text{opt}} = 0.65A$. Other examples of optimal allocation of hardware between various components subject to overall size constraints are given in Ref. 4.

8 POWER PLANTS

There are several EGM models and optima of power plants that have fundamental implications. The loss of heat from the hot end of a power plant can be modeled by using a thermal resistance in parallel with an irreversibility-free compartment that accounts for the power output $\dot{W}$ of the actual power plant (Fig. 8). The hot-end temperature of the working fluid cycle T_H can vary. The heat input $\dot{Q}_H$ is fixed. The bypass heat leak is proportional to the temperature difference, $\dot{Q}_C = C(T_H - T_L)$, where C is the thermal conductance of the power plant insulation. The power output is maximum (and $\dot{S}_{\text{gen}}$ is minimum) when the hot-end temperature reaches the optimal level[3,9]

$$T_{H,\text{opt}} = T_L \left(1 + \frac{\dot{Q}_H}{CT_L}\right)^{1/2}$$

The corresponding efficiency ($\dot{W}_{\text{max}}/\dot{Q}_H$) is

$$\eta = \frac{(1 + r)^{1/2} - 1}{(1 + r)^{1/2} + 1}$$

where $r = \dot{Q}_H/(CT_L)$ is a dimensionless way of expressing the size (thermal resistance) of the power plant. An optimal T_H value exists because when $T_H < T_{H,\text{opt}}$, the Carnot efficiency

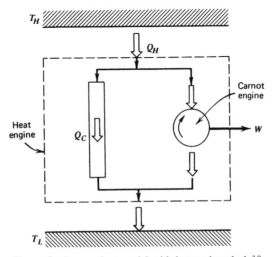

Figure 8 Power plant model with bypass heat leak.[3,9]

of the power-producing compartment is too low, while when $T_H > T_{H,\text{opt}}$, too much of the unit heat input $\dot{Q}_H$ bypasses the power compartment.

Another optimal hot-end temperature is revealed by the power plant model shown in Fig. 9. The power plant is driven by a stream of hot single-phase fluid of inlet temperature T_H and constant specific heat c_P. The model has two compartments. The one sandwiched between the heat exchanger surface (T_{HC}) and the ambient (T_L) operates reversibly. The other is a heat exchanger: for simplicity, the area of the T_{HC} surface is assumed sufficiently large that the stream outlet temperature is equal to T_{HC}. The stream is discharged into the ambient. The optimal hot-end temperature for maximum $\dot{W}$ (or minimum $\dot{S}_{\text{gen}}$) is[1,4] $T_{HC,\text{opt}} = (T_H T_L)^{1/2}$. The corresponding first-law efficiency, $\eta = \dot{W}_{\text{max}}/\dot{Q}_H$, is $\eta = 1 - (T_L/T_H)^{1/2}$.

The optimal allocation of a finite heat exchanger inventory between the hot end and the cold end of a power plant is illustrated by the model with two heat exchangers proposed in Ref. 3 and Fig. 10. The heat transfer rates are proportional to the respective temperature differences, $\dot{Q}_H = (UA)_H \Delta T_H$ and $\dot{Q}_L = (UA)_L \Delta T_L$, where the thermal conductances $(UA)_H$ and $(UA)_L$ account for the sizes of the heat exchangers. The heat input $\dot{Q}_H$ is *fixed* (e.g., the optimization is carried out for one unit of fuel burnt). The role of overall heat exchanger inventory constraint is played by

$$(UA)_H + (UA)_L = UA \qquad \text{(constant)}$$

where UA is the total thermal conductance available. The power output is maximized, and the entropy generation rate is minimized, when UA is allocated according to the rule

$$(UA)_{H,\text{opt}} = (UA)_{L,\text{opt}} = \tfrac{1}{2} UA$$

The corresponding maximum efficiency is, as expected, lower than the Carnot efficiency,

$$\eta = 1 - \frac{T_L}{T_H}\left(1 - \frac{4\dot{Q}_H}{T_H UA}\right)^{-1}$$

The EGM modeling and optimization progress on power plants is extensive, and is reviewed in Ref. 4. Similar models have also been used in the field of refrigeration, as we saw already in Section 4. For example, in a steady-state refrigeration plant with two heat

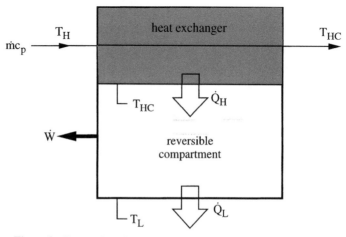

Figure 9 Power plant driven by a stream of hot single-phase fluid.[1,4]

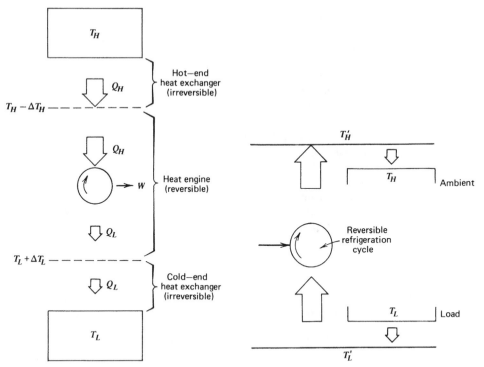

Figure 10 Power plant with two finite-size heat exchangers.[3]

Figure 11 Refrigerator model with two finite-size heat exchangers.[3]

exchangers (Fig. 11) subjected to the total UA constraint listed above, the refrigerator power input is minimum when UA is divided equally among the two heat exchangers, $(UA)_{H,\text{opt}} = \frac{1}{2} UA = (UA)_{L,\text{opt}}$.

9 CONSTRUCTAL THEORY

Flow systems are imperfect thermodynamically because of the resistances that their flows must overcome. Depending on system purpose and complexity, the currents may carry fluids, heat, electricity, and chemical species. The resistances are an integral and unavoidable presence because of the finite-size constraints that define the flow system. For example, the resistance to the flow of heat between two streams in a balanced counterflow heat exchanger can be made vanishingly small if the heat transfer surface can be made infinitely large. In reality, the surface size is fixed, and this means that the heat current is destined to encounter a thermal resistance. The current flows irreversibly, and this feature has a negative effect on global thermodynamic performance. The flow system is destined to be imperfect.

When the flow system is complex, the currents and resistances are many and diverse. The route to higher global performance consists of balancing each resistance against the rest. The distributing and redistributing of imperfection through the complex flow system is accomplished by making changes in the flow architecture. A prerequisite then is for the flow system to be free to change its configuration—free to morph. The morphing of structure is the result of the collision between the global objective and the global constraints. The generation of flow architecture is the means by which the flow system achieves its global objective under the constraints.

In recent years, this activity of thermodynamic optimization through the selection of flow configuration has become more focused on the end result, which is the generation of the architecture of the flow system.[1,5] This is particularly evident in modern computational heat and fluid flow, where large numbers of flow configurations can be simulated, compared, and optimized. The generation of flow architecture is a phenomenon at work everywhere, not only in engineered flow systems but also in natural flow systems (animate and inanimate). The universality of this observation was expressed in a compact statement (the constructal law) that proclaims a natural tendency in time: the maximization of access for the currents that flow through a morphing flow system. The thought that this principle can be used to rationalize the occurrence of optimized flow structures in nature (e.g., tree networks, round tubes) was named constructal theory.[1,5,10–12]

In this section the constructal law is formulated in analytical and graphical terms that are analogous to terms employed in thermodynamics.[6] This formulation makes the universality of the constructal law more evident.

A flow system, or nonequilibrium thermodynamic system, is characterized by "properties" (constraints), such as total volume and total volume occupied by all the ducts. A flow system is also characterized by "performance" (function, objective) and "flow structure" (configuration, layout, geometry, architecture). Unlike the black box of classical thermodynamics, which represents a system at equilibrium, a flow system has performance and especially configuration. Each flow system has a *drawing*.

Consider one of the simplest examples of how the collision between global objective and global constraints generates the complete architecture of the flow system: the flow between two points (Fig. 12), where "simple" are only the optimal and near-optimal architectures. This makes the example easy to present graphically. The rest of the design process is conceptually as vast and complicated as in any other example. When the flow architecture is free to morph, the design space is infinite. There is an infinity of flow architectures that can be chosen to guide a fluid stream ($\dot{m}$) from one point to another point.

Constructal theory begins with the global objective(s) and the global constraint(s) of the flow system, and the fact that in the beginning geometry is the unknown. In Fig. 12 the objective is to force the single-phase fluid stream $\dot{m}$ to flow from one point to another, while using minimal pumping power. When $\dot{m}$ is fixed, this objective is the same as seeking flow architectures with minimal pressure overall difference (ΔP), minimal overall flow resistance (R), or minimal rate of entropy generation by fluid friction.

There are two global constraints, one external and the other internal. The external constraint is the "system size," which is represented by the distance between the two points, L. The internal global constraint is the "amount" invested in making the flow architecture. In Fig. 12 that amount is the total volume (V) of all the ducts of the flow structure. Without such an investment there is no flow—not even a drawing that would show the flow. A flow must be guided. Flow means direction, geometry, and architecture, in addition to flow rate.

There are many reasons why there is an infinity of eligible flow architectures that meet the global objective and global constraints recognized above, i.e., many thoughts in the direction of which the number of possible architectures increases without bounds: (1) the flow pattern may be two dimensional (in the plane of Fig. 12), or three dimensional; (2) any number of ducts may be connected in parallel between the two points; (3) a duct may have any number of branches or tributaries at any location between the two points; (4) a single duct may have any length; (5) the cross-sectional shape may vary along the duct; (6) a duct may have any cross-sectional shape.

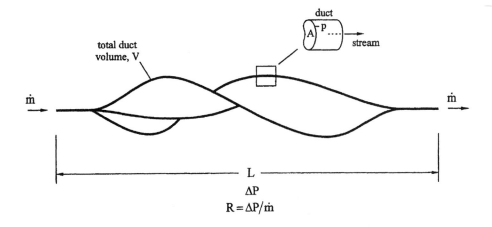

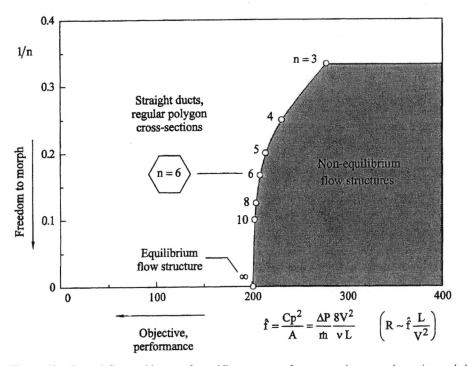

Figure 12 General flow architecture for guiding a stream from one point to another point, and the approach to the minimal global flow resistance when the number of sides of the regular-polygon cross-section (n) increases (data from Table 1).[6]

How do we identify the geometric features that bring a flow architecture to the highest level of global performance? There are many lessons of this type throughout engineering, and, if remembered, they constitute *strategy*—they shorten dramatically the search for the geometry in which all the features are "useful" in serving the global objective. Constructal theory is about strategy, about compact lessons of optimal shape and structure, which are fundamental and universally applicable. They are geometric relatives of truths such as the universal observation that all things flow naturally from high to low (the second law of thermodynamics).

Here are the classical lessons that abbreviate the search through the broad categories listed as (1)–(6). Again, for simplicity assume that ducts are slender, and the flows are slow so that in each cross section the regime is laminar and fully developed. Each lesson is identified by the symbol of the geometric feature that it addresses: (1)–(3) a single duct with large cross section offers a smaller flow resistance than two ducts with smaller cross sections connected in parallel; (4) the lowest resistance belongs to the shortest duct, in this case the straight duct between the two points; (5) the duct with cross-sectional geometry that does not vary longitudinally has a lower resistance than the duct with variable cross section.

Summing up, out of the infinity of designs represented by (1)–(5) we have selected a single straight duct with a cross-sectional shape that does not vary from one end of the duct to the other. According to (6), however, there is still an infinite number of possible cross-section shapes: symmetric vs. asymmetric, smooth vs. polygonal, etc. Which impedes the flow the least? The answer becomes visible if we assume cross sections with polygonal shapes. Start with an arbitrary cross-section shaped as a triangle. The area of the cross section A is fixed because the total duct volume V and the duct length L are fixed, namely $A = V/L$. Triangular cross sections constrict the flow when one of the angles is much smaller than the other two.

The least resistance is offered by the most "open" triangular cross section, which is shaped as an equilateral triangle. Once again, if one very small angle and two larger ones represent a nonuniform distribution of geometric features of imperfection (i.e., features that impede the flow), then the equilateral triangle represents the constructal architecture, i.e., the one with "optimal distribution of imperfection."

The same holds for any other polygonal shape. The least resistance is offered by a cross section shaped as a regular polygon. In conclusion, out of the infinity of flow architectures recognized in class (6) we have selected an infinite number of candidates. They are ordered according to the number of sides (n) of the regular polygon, from the equilateral triangle ($n = 3$) to the circle ($n = \infty$). The flow resistance for Hagen-Poiseuille flow through a straight duct with polygonal cross section can be written as (Ref. 5, pp. 127–128)

$$\frac{\Delta P}{\dot{m}} = \frac{\nu L}{8V^2} \frac{Cp^2}{A}$$

where p is the perimeter of the cross section. As shown in Table 1, the dimensionless perimeter $p/A^{1/2}$ is only a function of n. The same is true about C, which appears in the solution for friction factor in Hagen-Poiseuille flow,

$$f = \frac{C}{\text{Re}}$$

where $\text{Re} = \overline{U}D_h/\nu$, $D_h = 4A/p$, and $\overline{U} = \dot{m}/(\rho A)$. In conclusion, the group Cp^2/A depends only on n, and accounts for how this last geometric degree of freedom influences global performance. The group Cp^2/A is the dimensionless global flow resistance of the flow system. The smallest Cp^2/A value is the best, and the best is the round cross section.

Table 1 The Laminar Flow Resistances of Straight Ducts with Regular Polygonal Cross Sections with n Sides[6]

n	C	$p/A^{1/2}$	Cp^2/A
3	40/3	4.559	277.1
4	14.23	4	227.6
5	14.74	3.812	214.1
6	15.054	3.722	208.6
8	15.412	3.641	204.3
10	15.60	3.605	202.7
∞	16	$2\pi^{1/2}$	201.1

Figure 12 shows a plot of the flow-resistance data of Table 1. The flow structure with minimal global resistance is approached gradually (with diminishing decrements) as n increases. The polygonal cross section with $n = 10$ performs nearly as well as the round cross section ($n = \infty$). The "evolution" of the cross-sectional shape stops when the number of features (n) has become infinite, i.e., when the structure has become the most free. This configuration where changes in global performance have stopped is the *equilibrium flow architecture*.[6]

The curve plotted in Fig. 12 was generated by calculations for regular-polygon cross sections. The curve is in reality a sequence of discrete points, one point for each n value. We drew a continuous line through these points to stress an additional idea. Regardless of n, the regular polygon and straight duct with constant cross section is already the "winner" from an infinitely larger group of competing architectures. This means that the global flow resistances of all the designs that are not covered by Table 1 fall to the right of the curve plotted in Fig. 12.

In sum, the immensely large world of possible designs occupies only a portion of the two-dimensional domain illustrated in Fig. 12. This domain can be described qualitatively as "performance versus freedom," when global properties such as L and V are specified. The boundary of the domain is formed by a collection of the better flow structures. The best is achieved by putting more freedom in the geometry of the flow structure (e.g., a larger n). The best performance belongs to the structure that was most free to morph—the equilibrium configuration. In its immediate vicinity, however, we find many configurations that are different (they have finite n values), but have practically the same global performance level. These are *near-equilibrium* flow structures.[6]

The evolution of flow configuration illustrated in Figs. 1 and 2 for point-to-point flows is a universal phenomenon, which manifests itself during any search for optimal flow architectures. Additional examples are given in Ref. 6. Some of the more complex architectures that have been optimized recently are the flow structures that connect one point (source, or sink) with an infinity of points (line, area, or volume). According to constructal theory, the best flow path that makes such a connection is shaped like a tree.[1,5] The tree is for point–area flows what the straight duct is for point–point flows.

All the possible configurations inhabit the hyperspace suggested in Fig. 13a. All the constant-L flow configurations that are possible inhabit the volume visualized by the constant-V and constant-R cuts. The bottom figure shows the view of all the possible flow structures, projected on the base plane. Plotted on the R axis is the global resistance of the flow system, namely $R = \Delta P/\dot{m}$ in the preceding examples. The abscissa accounts for the total volume occupied by the ducts (V): this is a global measure of how "porous" or "per-

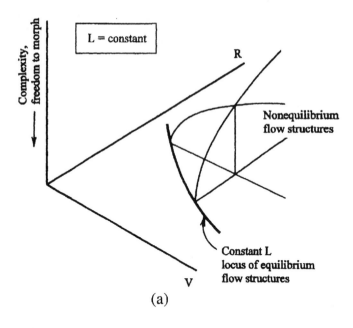

(a)

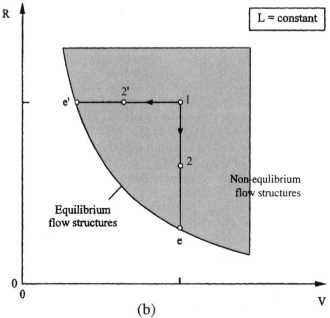

(b)

Figure 13   The space occupied by all the flow architectures when the global external size (L) is fixed.[6]

meable" the flow system is. The constant-V plane that cuts through Fig. 13a is the same as the plane of Fig. 12.

The constructal law is the statement that summarizes the common observation that flow structures that survive are those that morph in one direction in time: toward configurations that make it easier for currents to flow. This holds for natural and engineered flow structures. The first such statement was[1,5]

> For a finite-size system to persist in time (to live), it must evolve in such a way that it provides easier access to the imposed currents that flow through it.

If the flow structures are free to change (free to approach the base plane in Fig. 13a), they will move at constant L and constant V in the direction of progressively smaller R. If the initial configuration is represented by point 1 in Fig.13b, then a more recent configuration is represented by point 2. The relation between the two configurations is $R_2 \leq R_1$ (constant L, V). If freedom to morph persists, then the flow structure will continue toward smaller R values. Any such change is characterized by

$$dR \leq 0 \qquad \text{(constant } L, V)$$

The end of this migration is the equilibrium flow structure (point e), where the geometry of the flow enjoys total freedom. Equilibrium is characterized by minimal R at constant L and V. In the vicinity of the equilibrium point we have

$$dR = 0 \quad \text{and} \quad d^2 R > 0 \qquad \text{(constant } L, V)$$

The $R(V)$ curve shown in Fig. 13b is the edge of the cloud of possible flow architectures with the same global size L. The curve has negative slope because of the physics of flow: the flow resistance always decreases when the flow channels open up, $(\partial R / \partial V)_L < 0$.

The constant-R cut through the configuration space shows another way of expressing the constructal law. If free to morph, the flow system will evolve from point 1 to point 2' at constant L and R. In the limit of total freedom, the geometry will reach another equilibrium configuration, which is represented by point e'. The alternative analytical statement of the constructal law is

$$dV \leq 0 \qquad \text{(constant } L, R)$$

For changes in structure in the immediate vicinity of the equilibrium structure, we note

$$dV = 0 \quad \text{and} \quad d^2V > 0 \qquad \text{(constant } L, R)$$

Paraphrasing the original statement of the constructal law, we may describe processes of type 1–2'–e' as follows:

> For a system with fixed global size and global performance to persist in time (to live), it must evolve in such a way that its flow structure occupies a smaller fraction of the available space.

The constant-V alternative to Fig. 13 is shown in Fig. 14. The lower drawing is the projection of the space of possible flow architectures on the base plane R-L. The continuous line is the locus of equilibrium flow structures at constant V, namely the curve $R(V)$, where $(\partial R / \partial L)_V > 0$. The fact that the slope is positive is flow physics: the flow resistance always increases as the distance that must be overcome by the flow increases.

The constructal law statement can be read off Fig. 14b in two ways. One is the original statement[1,5]: at constant V and L, the evolution is from a suboptimal structure (point 1) to one that has a lower global resistance (point 2). If the flow geometry continues to morph freely, the structure approaches the equilibrium configuration (point e).

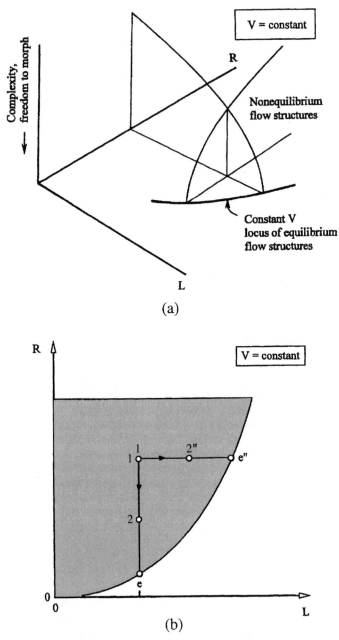

Figure 14   The space occupied by all the flow architectures when the global internal size (total duct volume V) is fixed.[6]

The alternative is when structural changes are made such that R remains constant while V is also fixed. Then the evolution in Fig. 14b is from point 1 to point 2″. Such changes mean that

$$dL \geq 0 \quad (\text{constant } R, V)$$

and that the constructal law statement becomes

> For a flow system with fixed global resistance (R) and internal size (V) to persist in time, the architecture must evolve in such a way that it covers a progressively larger territory.

Equilibrium is reached at point e''. The changes in flow structures in the immediate vicinity of the equilibrium structure are such that the global external dimension at equilibrium is maximal,

$$dL = 0 \quad d^2L < 0 \quad (\text{constant } R, V)$$

Accordingly, the constructal law states that the ultimate flow structure with specified global resistance (R) and internal size (V) is the largest. A flow architecture with specified R and V has a maximum size, and this global size belongs to the equilibrium architecture. A flow structure larger than this does not exist. This formulation of the constructal law has implications in natural design, e.g., the spreading of species and river deltas without access to the sea.

The original statement of the constructal law was about the maximization of flow access under global size constraints (external L, internal V). This behavior is illustrated by the structural changes 1–2–e in Figs. 13b and 14b. This means survival by increasing efficiency—survival of the fittest. This is the physics principle behind Darwin's observations, the principle that rules not only the animate natural flow systems, but also the inanimate natural flow systems and the engineered flow systems. Engineered systems are diverse species of "man + machine" beings.

The alternative shown by the changes 1–2″–e'' in Fig. 14b is survival by spreading: growth as the mechanism for being able to persist in time. The limit to growth is set by the specified constraints, in this case the fixed global flow resistance R and the global internal size V. A given living species (river delta, animal population) will spread over a certain, maximal territory.

An equivalent interpretation of the constructal principle is based on processes of type 1–2′–e', Fig. 13b. Flow architectures with the same performance (R) and size (L) evolve toward compactness—smaller volumes dedicated to the internal ducts, i.e., larger volumes dedicated to the working volume elements, which are the interstices. This is survival based on the maximization of the use of the available space.

In summary, changes in performance (R) can be achieved through changes of three types:

1. Flow configuration
2. Global external size, or covered territory, L.
3. Global internal size, or duct volume, V.

The examples discussed so far showed that changes may occur in one category, or simultaneously in two or three. The simplest illustration is possible for the case of equilibrium flow architectures. For them the solid curves shown in Figs. 13b and 14b proclaim the existence of the fundamental relation $R(L, V)$, the differential of which is

$$dR = Y_L \, dL + Y_V \, dV \quad (\text{equilibrium})$$

Physics requires that the first partial derivatives of R have opposite signs, $Y_L > 0$ and $Y_V < 0$, as noted earlier in this section. In general, when the flow architecture has not reached equilibrium, R can be decreased by means 1, 2, and 3. Then the general version of the last equation is

$$dR \leq Y_L \, dL + Y_V \, dV$$

where the inequality sign refers to the time arrow of structural changes in a flow configuration that, at least initially, was not of the equilibrium type. This inequality is a concise statement of the three analytical formulations of the constructal law that we discussed so far:

   R minimum at constant L and V

   V minimum at constant R and L

   L maximum at constant V and R

Another way to summarize the analytical formulation that we have just constructed is by recognizing the analogy between the analytical constructal law and the analytical formulation of classical thermodynamics (cf. the preceding chapter in this handbook). The analogy is presented in Table 2. It is stressed further by Fig. 2 in Chapter 3, which is from present-day thermodynamics.[1] Figure 2 in Chapter 3 expresses the energy minimum principle, which states that as the internal constraints of a closed system are removed at constant volume and entropy, the energy approaches a minimal value. Figure 2 in Chapter 3 is analogous to Fig. 14a in this Chapter.

The analytical formulation of the constructal law presented in this section expresses a universal phenomenon: figures such as Fig. 12 characterize the evolution toward equilibrium configuration in any flow system with global objective, global constraints, and freedom to morph. In Ref. 6, this was demonstrated through examples from three wide classes of flow architectures: flow between two points, flow between a circle and its center, and flow between one point and an area. Many other examples can be contemplated, and they will all reveal the image of Fig. 12 on the road to equilibrium flow architectures.

At equilibrium the flow configuration achieves the most that its freedom to morph has to offer. Equilibrium does not mean that the flow architecture (structure, geometry, configuration) stops changing. On the contrary, it is here at equilibrium that the flow geometry enjoys most freedom to change. Equilibrium means that the global performance does not change when changes occur in the flow architecture.

Table 2   The Concepts and Principles of Classical Thermodynamics and Constructal Theory[6]

Thermodynamics	Constructal Theory
State	Flow architecture (geometry, configuration, structure)
Process, removal of internal constraints	Morphing, change in flow configuration
Properties (U, S, Vol, . . .)	Global objective and global constraints (R, L, V, . . .)
Equilibrium state	Equilibrium flow architecture
Fundamental relation, $U(S, \text{Vol}, \ldots)$	Fundamental relation, $R(L, V, \ldots)$
Constrained equilibrium states	Nonequilibrium flow architectures
Removal of constraints	Increased freedom to morph
Energy minimum principle:	Constructal principle:
$\quad U$ minimum at constant S and Vol	$\quad R$ minimum at constant L and V
$\quad$Vol minimum at constant F and T	$\quad V$ minimum at constant R and L
$\quad S$ maximum at constant U and Vol	$\quad L$ maximum at constant V and R

REFERENCES

1. A. Bejan, *Advanced Engineering Thermodynamics,* 2nd ed., Wiley, New York, 1997.
2. A. Bejan, G. Tsatsaronis, and M. Moran, *Thermal Design and Optimization,* Wiley, New York, 1996.
3. A. Bejan, *Entropy Generation Through Heat and Fluid Flow,* Wiley, New York, 1982.
4. A. Bejan, *Entropy Generation Minimization,* CRC Press, Boca Raton, FL, 1996.
5. A. Bejan, *Shape and Structure, from Engineering to Nature,* Cambridge University Press, Cambridge, UK, 2000.
6. A. Bejan and S. Lorente, "The Constructal Law and the Thermodynamics of Flow Systems with Configuration," *Int. J. Heat Mass Transfer,* **47,** 3203–3214 (2004).
7. A. Bejan, *Convection Heat Transfer,* 3rd ed., Wiley, New York, 2004.
8. S. Paoletti, F. Rispoli, and E. Sciubba, "Calculation of Exergetic Losses in Compact Heat Exchanger Passages," *ASME AES,* **10**(2), 21–29 (1989).
9. A. Bejan, *Solved Problems in Thermodynamics,* Massachusetts Institute of Technology, Department of Mechanical Engineering, 1976, Problem VII-D.
10. H. Poirier, "A theory explains the intelligence of nature, *Science & Vie,*" **1034,** 44–63 (2003).
11. A. Bejan, I. Dincer, S. Lorente, A. F. Miguel, and A. H. Reis, *Porous and Complex Flow Structures in Modern Technologies,* Springer-Verlag, New York, 2004.
12. R. N. Rosa, A. H. Reis, and A. F. Miguel, *Bejan's Constructal Theory of Shape and Structure,* Évora Geophysics Center, University of Évora, Portugal, 2004.

CHAPTER 5

HEAT-TRANSFER FUNDAMENTALS

G. P. Peterson
Rensselaer Polytechnic Institute
Troy, New York

SYMBOLS AND UNITS

A	area of heat transfer
Bi	Biot Number, hL/k, dimensionless
C	circumference, m, constant defined in text
C_p	specific heat under constant pressure, J/kg·K
D	diameter, m
e	emissive power, W/m^2
f	drag coefficient, dimensionless
F	cross flow correction factor, dimensionless
$F_{i\text{-}j}$	configuration factor from surface i to surface j, dimensionless
Fo	Fourier Number, $\alpha t A^2/V^2$, dimensionless
$F_{o\text{-}\lambda T}$	radiation function, dimensionless
G	irradiation, W/m^2; mass velocity, kg/m^2·s
g	local gravitational acceleration, 9.8 m/s^2
g_c	proportionality constant, 1 kg·m/N·s^2
Gr	Grashof number, $gL^3\beta\Delta T/v^2$ dimensionless
h	convection heat transfer coefficient, equals $q/A\Delta T$, W/m^2·K
h_{fg}	heat of vaporization, J/kg
J	radiosity, W/m^2
k	thermal conductivity, W/m·K
K	wick permeability, m^2
L	length, m
Ma	Mach number, dimensionless
N	screen mesh number, m^{-1}

$\underline{\text{Nu}}$	Nusselt number, $\text{Nu}_L = hL/k$, $\text{Nu}_D = hD/k$, dimensionless
$\overline{\text{Nu}}$	Nusselt number averaged over length, dimensionless
P	pressure, N/m^2, perimeter, m
Pe	Peclet number, RePr, dimensionless
Pr	Prandtl number, $C_p\mu/k$, dimensionless
q	rate of heat transfer, W
q''	rate of heat transfer per unit area, W/m^2
R	distance, m; thermal resistance, K/W
r	radial coordinate, m; recovery factor, dimensionless
Ra	Rayleigh number, GrPr; $\text{Ra}_L = \text{Gr}_L\text{Pr}$, dimensionless
Re	Reynolds Number, $\text{Re}_L = \rho VL/\mu$, $\text{Re}_D = \rho VD/\mu$, dimensionless
S	conduction shape factor, m
T	temperature, K or °C
t	time, sec
T_{as}	adiabatic surface temperature, K
T_{sat}	saturation temperature, K
T_b	fluid bulk temperature or base temperature of fins, K
T_e	excessive temperature, $T_s - T_{\text{sat}}$, K or °C
T_f	film temperature, $(T_\infty + T_s)/2$, K
T_i	initial temperature; at $t = 0$, K
T_0	stagnation temperature, K
T_s	surface temperature, K
T_∞	free stream fluid temperature, K
U	overall heat transfer coefficient, $W/m^2\cdot K$
V	fluid velocity, m/s; volume, m^3
w	groove width, m; or wire spacing, m
We	Weber number, dimensionless
x	one of the axes of Cartesian reference frame, m

GREEK SYMBOLS

α	thermal diffusivity, $k/\rho C_p$, m^2/s; absorptivity, dimensionless
β	coefficient of volume expansion, 1/K
Γ	mass flow rate of condensate per unit width, kg/m·s
γ	specific heat ratio, dimensionless
ΔT	temperature difference, K
δ	thickness of cavity space, groove depth, m
$\in$	emissivity, dimensionless
ε	wick porosity, dimensionless
λ	wavelength, μm
η_f	fin efficiency, dimensionless
μ	viscosity, kg/m·s
ν	kinematic viscosity, m^2/s
ρ	reflectivity, dimensionless; density, kg/m^3
σ	surface tension, N/m; Stefan-Boltzmann constant, 5.729×10^{-8} $W/m^2\cdot K^4$
τ	transmissivity, dimensionless, shear stress, N/m^2
Ψ	angle of inclination, degrees or radians

SUBSCRIPTS

a	adiabatic section, air
b	boiling, black body
c	convection, capillary, capillary limitation, condenser

e	entrainment, evaporator section
eff	effective
f	fin
i	inner
l	liquid
m	mean, maximum
n	nucleation
o	outer
0	stagnation condition
p	pipe
r	radiation
s	surface, sonic or sphere
w	wire spacing, wick
v	vapor
λ	spectral
∞	free stream
$-$	axial hydrostatic pressure
$+$	normal hydrostatic pressure

Transport phenomena represents the overall field of study and encompasses a number of subfields. One of these is heat transfer, which focuses primarily on the energy transfer occurring as a result of an energy gradient that manifests itself as a temperature difference. This form of energy transfer can occur as a result of a number of different mechanisms, including *conduction,* which focuses on the transfer of energy through the direct impact of molecules; *convection,* which results from the energy transferred through the motion of a fluid; and *radiation,* which focuses on the transmission of energy through electromagnetic waves. In the following review, as is the case with most texts on heat transfer, *phase change heat transfer,* i.e., *boiling* and *condensation,* will be treated as a subset of convection heat transfer.

1   CONDUCTION HEAT TRANSFER

The exchange of energy or heat resulting from the kinetic energy transferred through the direct impact of molecules is referred to as *conduction,* and takes place from a region of high energy (or temperature) to a region of lower energy (or temperature). The fundamental relationship that governs this form of heat transfer is *Fourier's law of heat conduction,* which states that in a one-dimensional system with no fluid motion, the rate of heat flow in a given direction is proportional to the product of the temperature gradient in that direction and the area normal to the direction of heat flow. For conduction heat transfer in the x direction this expression takes the form

$$q_x = -kA \frac{\partial T}{\partial x}$$

where q_x is the heat transfer in the x direction, A is the area normal to the heat flow, $\partial T/\partial x$ is the temperature gradient, and k is the thermal conductivity of the substance.

Writing an energy balance for a three-dimensional body and utilizing Fourier's law of heat conduction yields an expression for the transient diffusion occurring within a body or substance:

$$\frac{\partial}{\partial x}\left(k\,\frac{\partial T}{\partial x}\right) + \frac{\partial}{\partial y}\left(k\,\frac{\partial T}{\partial y}\right) + \frac{\partial}{\partial z}\left(k\,\frac{\partial T}{\partial z}\right) + \dot{q} = \rho c_p\,\frac{\partial}{\partial x}\frac{\partial T}{\partial t}$$

This expression, usually referred to as the *heat diffusion equation* or heat equation, provides a basis for most types heat conduction analyses. Specialized cases of this equation can be used to solve many steady-state or transient problems. Some of these specialized cases are

Thermal conductivity is a constant

$$\frac{\partial^2 T}{\partial x^2} + \frac{\partial^2 T}{\partial y^2} + \frac{\partial^2 T}{\partial z^2} + \frac{\dot{q}}{k} = \frac{\rho c_p}{k}\frac{\partial T}{\partial t}$$

Steady-state with heat generation

$$\frac{\partial}{\partial x}\left(k\,\frac{\partial T}{\partial x}\right) + \frac{\partial}{\partial y}\left(k\,\frac{\partial T}{\partial y}\right) + \frac{\partial}{\partial z}\left(k\,\frac{\partial T}{\partial z}\right) + \dot{q} = 0$$

Steady-state , one-dimensional heat transfer with no heat sink (i.e., a fin)

$$\frac{\partial}{\partial x}\left(\frac{\partial T}{\partial x}\right) + \frac{\dot{q}}{k} = 0$$

One-dimensional heat transfer with no internal heat generation

$$\frac{\partial}{\partial x}\left(\frac{\partial T}{\partial x}\right) = \frac{\rho c_p}{k}\frac{\partial T}{\partial t}$$

In the following sections, the heat diffusion equation will be utilized for several specific cases. However, in general, for a three-dimensional body of constant thermal properties without heat generation under steady-state heat conduction the temperature field satisfies the expression

$$\nabla^2 T = 0$$

1.1 Thermal Conductivity

The ability of a substance to transfer heat through conduction can be represented by the constant of proportionality, *k,* referred to as the thermal conductivity. Figure 1 illustrates the characteristics of the thermal conductivity as a function of temperature for several solids, liquids, and gases. As shown, the thermal conductivity of solids is higher than liquids, and liquids higher than gases. Metals typically have higher thermal conductivities than nonmetals, with pure metals having thermal conductivities that decrease with increasing temperature, while the thermal conductivity of nonmetallic solids generally increases with increasing temperature and density. The addition of other metals to create alloys, or the presence of impurities, usually decreases the thermal conductivity of a pure metal.

In general, the thermal conductivity of liquids decreases with increasing temperature. Alternatively, the thermal conductivity of gases and vapors, while lower, increases with increasing temperature and decreases with increasing molecular weight. The thermal conductivities of a number of commonly used metals and nonmetals are tabulated in Tables 1

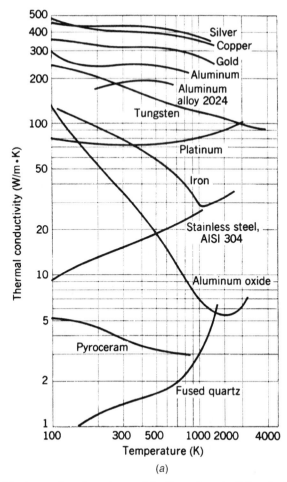

Figure 1a   The temperature dependence of the thermal conductivity of selected solids.[1]

and 2, respectively. Insulating materials, which are used to prevent or reduce the transfer of heat between two substances or a substance and the surroundings, are listed in Tables 3 and 4, along with the thermal properties. The thermal conductivities for liquids, molten metals, and gasses are given in Tables 5, 6 and 7, respectively.

1.2   One-Dimensional Steady-State Heat Conduction

The steady-state rate of heat transfer resulting from heat conduction through a homogeneous material can be expressed in terms of the rate of heat transfer, q, or $q = \Delta T/R$, where ΔT is the temperature difference and R is the *thermal resistance*. This thermal resistance is the reciprocal of the *thermal conductance* ($C = 1/R$) and is related to the thermal conductivity by the cross-sectional area. Expressions for the thermal resistance, the temperature distribution, and the rate of heat transfer are given in Table 8 for a plane wall, a cylinder, and a sphere. For a plane wall, the heat transfer is typically assumed to be one dimensional (i.e.,

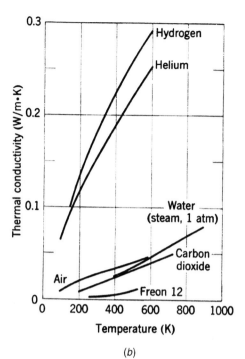

(b)

Figure 1b The temperature dependence of the thermal conductivity of selected nonmetallic liquids under saturated conditions.[1]

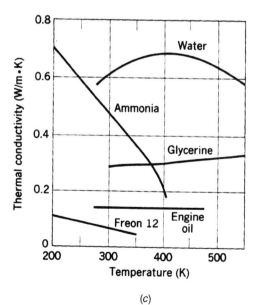

(c)

Figure 1c The temperature dependence of the thermal conductivity of selected gases at normal pressures.[1]

Table 1 Thermal Properties of Metallic Solids[a]

Composition	Melting Point (K)	Properties at 300 K				Properties at Various Temperatures (K) k (W/m·K); C_p (J/kg·K)		
		ρ (kg/m³)	C_p (J/kg·K)	k (W/m·K)	$\alpha \times 10^6$ (m²/s)	100	600	1200
Aluminum	933	2702	903	237	97.1	302; 482	231; 1033	
Copper	1358	8933	385	401	117	482; 252	379; 417	339; 480
Gold	1336	19300	129	317	127	327; 109	298; 135	255; 155
Iron	1810	7870	447	80.2	23.1	134; 216	54.7; 574	28.3; 609
Lead	601	11340	129	35.3	24.1	39.7; 118	31.4; 142	
Magnesium	923	1740	1024	156	87.6	169; 649	149; 1170	
Molybdenum	2894	10240	251	138	53.7	179; 141	126; 275	105; 308
Nickel	1728	8900	444	90.7	23.0	164; 232	65.6; 592	76.2; 594
Platinum	2045	21450	133	71.6	25.1	77.5; 100	73.2; 141	82.6; 157
Silicon	1685	2330	712	148	89.2	884; 259	61.9; 867	25.7; 967
Silver	1235	10500	235	429	174	444; 187	412; 250	361; 292
Tin	505	7310	227	66.6	40.1	85.2; 188		
Titanium	1953	4500	522	21.9	9.32	30.5; 300	19.4; 591	22.0; 620
Tungsten	3660	19300	132	174	68.3	208; 87	137; 142	113; 152
Zinc	693	7140	389	116	41.8	117; 297	103; 436	

[a]Adapted from Ref. 1.

heat is conducted in only the x direction) and for a cylinder and sphere, only in the radial direction.

Aside from the heat transfer in these simple geometric configurations, other common problems encountered in practical applications is that of heat transfer through layers or composite walls consisting of N layers, where the thickness of each layer is represented by

Table 2 Thermal Properties of Nonmetals

Description/Composition	Temperature (K)	Density, ρ (kg/m³)	Thermal Conductivity, k (W/m·K)	Specific Heat, C_p (J/kg·K)	$\alpha \times 10^6$ (m²/s)
Bakelite	300	1300	0.232	1465	0.122
Brick, refractory					
Carborundum	872	—	18.5	—	—
Chrome-brick	473	3010	2.32	835	0.915
Fire clay brick	478	2645	1.0	960	0.394
Clay	300	1460	1.3	880	1.01
Coal, anthracite	300	1350	0.26	1260	0.153
Concrete (stone mix)	300	2300	1.4	880	0.692
Cotton	300	80	0.059	1300	0.567
Glass, window	300	2700	0.78	840	0.344
Rock, limestone	300	2320	2.15	810	1.14
Rubber, hard	300	1190	0.160	—	—
Soil, dry	300	2050	0.52	1840	0.138
Teflon	300	2200	0.35	—	—
	400	—	0.45	—	—

Table 3   Thermal Properties of Building and Insulating Materials (at 300 K)[a]

Description/Composition	Density ρ (kg/m^3)	Thermal Conductivity, k (W/m·K)	Specific Heat, C_p (J/kg·K)	$\alpha \times 10^6$ (m^2/s)
Building boards				
Plywood	545	0.12	1215	0.181
Acoustic tile	290	0.058	1340	0.149
Hardboard, siding	640	0.094	1170	0.126
Woods				
Hardwoods (oak, maple)	720	0.16	1255	0.177
Softwoods (fir, pine)	510	0.12	1380	0.171
Masonry materials				
Cement mortar	1860	0.72	780	0.496
Brick, common	1920	0.72	835	0.449
Plastering materials				
Cement plaster, sand aggregate	1860	0.72	—	—
Gypsum plaster, sand aggregate	1680	0.22	1085	0.121
Blanket and batt				
Glass fiber, paper faced	16	0.046	—	—
Glass fiber, coated; duct liner	32	0.038	835	1.422
Board and slab				
Cellular glass	145	0.058	1000	0.400
Wood, shredded/cemented	350	0.087	1590	0.156
Cork	120	0.039	1800	0.181
Loose fill				
Glass fiber, poured or blown	16	0.043	835	3.219
Vermiculite, flakes	80	0.068	835	1.018

[a] Adapted from Ref. 1.

Δx_n and the thermal conductivity by k_n for $n = 1, 2, \ldots, N$. Assuming that the interfacial resistance is negligible, (i.e., there is no thermal resistance at the contacting surfaces), the overall thermal resistance can be expressed as

$$R = \sum_{n=1}^{N} \frac{\Delta x_n}{k_n A}$$

Similarly, for conduction heat transfer in the radial direction through a number of N concentric cylinders with negligible interfacial resistance, the overall thermal resistance can be expressed as

$$R = \sum_{n=1}^{N} \frac{\ln(r_{n+1}/r_n)}{2\pi k_n L}$$

where r_1 = inner radius, r_{N+1} = outer radius.

For N *concentric spheres* with negligible interfacial resistance, the thermal resistance can be expressed as

$$R = \sum_{n=1}^{N} \left(\frac{1}{r_n} - \frac{1}{r_{n+1}} \right) \Big/ 4\pi k$$

where r_1 = inner radius, r_{N+1} = outer radius.

Table 4 Thermal Conductivities for Some Industrial Insulating Materials[a]

Description/Composition	Maximum Service Temperature (K)	Typical Density (kg/m³)	Typical Thermal Conductivity, k(W/m·K), at Various Temperatures (K)			
			200	300	420	645
Blankets						
Blanket, mineral fiber, glass; fine fiber organic bonded	450	10		0.048		
		48		0.033		
Blanket, alumina-silica fiber	1530	48				0.105
Felt, semirigid; organic bonded	480	50–125		0.038	0.063	
Felt, laminated; no binder	920	120			0.051	0.087
Blocks, boards, and pipe insulations						
Asbestos paper, laminated and corruagated, 4-ply	420	190		0.078		
Calcium silicate	920	190			0.063	0.089
Polystyrene, rigid						
Extruded (R-12)	350	56	0.023	0.027		
Molded beads	350	16	0.026	0.040		
Rubber, rigid foamed	340	70		0.032		
Insulating cement						
Mineral fiber (rock, slag, or glass)						
With clay binder	1255	430			0.088	0.123
With hydraulic setting binder	922	560			0.123	
Loose fill						
Cellulose, wood, or paper pulp	—	45		0.039		
Perlite, expanded	—	105	0.036	0.053		
Vermiculite, expanded	—	122		0.068		

[a] Adapted from Ref. 1.

Table 5 Thermal Properties of Saturated Liquids[a]

Liquid	T (K)	ρ (kg/m³)	C_p (kJ/kg·K)	$v \times 10^6$ (m²/s)	$k \times 10^3$ (W/m·K)	$\alpha \times 10^7$ (m²/s)	Pr	$\beta \times 10^3$ (K⁻¹)
Ammonia, NH_3	223	703.7	4.463	0.435	547	1.742	2.60	2.45
	323	564.3	5.116	0.330	476	1.654	1.99	2.45
Carbon dioxide, CO_2	223	1156.3	1.84	0.119	85.5	0.402	2.96	14.0
	303	597.8	36.4	0.080	70.3	0.028	28.7	14.0
Engine oil (unused)	273	899.1	1.796	4280	147	0.910	47,000	0.70
	430	806.5	2.471	5.83	132	0.662	88	0.70
Ethylene glycol, $C_2H_4(OH)_2$	273	1130.8	2.294	57.6	242	0.933	617.0	0.65
	373	1058.5	2.742	2.03	263	0.906	22.4	0.65
Clycerin, $C_3H_5(OH)_3$	273	1276.0	2.261	8310	282	0.977	85,000	0.47
	320	1247.2	2.564	168	287	0.897	1,870	0.50
Freon (Refrigerant-12), CCl_2F_2	230	1528.4	0.8816	0.299	68	0.505	5.9	1.85
	320	1228.6	1.0155	0.190	68	0.545	3.5	3.50

[a] Adapted from Ref. 2. See Table 22 for H_2O.

Table 6   Thermal Properties of Liquid Metals[a]

Composition	Melting Point (K)	T (K)	ρ (kg/m³)	C_p (kJ/kg·K)	$v \times 10^7$ (m²/s)	k (W/m·K)	$\alpha \times 10^5$ (m²/s)	Pr
Bismuth	544	589	10,011	0.1444	1.617	16.4	0.138	0.0142
		1033	9,467	0.1645	0.8343	15.6	1.001	0.0083
Lead	600	644	10,540	0.159	2.276	16.1	1.084	0.024
		755	10,412	0.155	1.849	15.6	1.223	0.017
Mercury	234	273	13,595	0.140	1.240	8.180	0.429	0.0290
		600	12,809	0.136	0.711	11.95	0.688	0.0103
Potassium	337	422	807.3	0.80	4.608	45.0	6.99	0.0066
		977	674.4	0.75	1.905	33.1	6.55	0.0029
Sodium	371	366	929.1	1.38	7.516	86.2	6.71	0.011
		977	778.5	1.26	2.285	59.7	6.12	0.0037
NaK (56%/44%)	292	366	887.4	1.130	6.522	25.6	2.55	0.026
		977	740.1	1.043	2.174	28.9	3.74	0.0058
PbBi (44.5%/55.5%)	398	422	10,524	0.147	—	9.05	0.586	—
		644	10,236	0.147	1.496	11.86	0.790	0.189

[a]Adapted from *Liquid Metals Handbook,* The Atomic Energy Commission, Department of the Navy, Washington, DC, 1952.

Table 7   Thermal Properties of Gases at Atmospheric Pressure[a]

Gas	T (K)	ρ (kg/m³)	C_p (kJ/kg·K)	$v \times 10^6$ (m²/s)	k (W/m·K)	$\alpha \times 10^4$ (m²/s)	Pr
Air	100	3.6010	1.0266	1.923	0.009246	0.0250	0.768
	300	1.1774	1.0057	16.84	0.02624	0.2216	0.708
	2500	0.1394	1.688	543.0	0.175	7.437	0.730
Ammonia, NH_3	220	0.3828	2.198	19.0	0.0171	0.2054	0.93
	473	0.4405	2.395	37.4	0.0467	0.4421	0.84
Carbon dioxide	220	2.4733	0.783	4.490	0.01081	0.0592	0.818
	600	0.8938	1.076	30.02	0.04311	0.4483	0.668
Carbon monoxide	220	1.5536	1.0429	8.903	0.01906	0.1176	0.758
	600	0.5685	1.0877	52.06	0.04446	0.7190	0.724
Helium	33	1.4657	5.200	3.42	0.0353	0.04625	0.74
	900	0.05286	5.200	781.3	0.298	10.834	0.72
Hydrogen	30	0.8472	10.840	1.895	0.0228	0.02493	0.759
	300	0.0819	14.314	109.5	0.182	1.554	0.706
	1000	0.0819	14.314	109.5	0.182	1.554	0.706
Nitrogen	100	3.4808	1.0722	1.971	0.009450	0.02531	0.786
	300	1.1421	1.0408	15.63	0.0262	0.204	0.713
	1200	0.2851	1.2037	156.1	0.07184	2.0932	0.748
Oxygen	100	3.9918	0.9479	1.946	0.00903	0.02388	0.815
	300	1.3007	0.9203	15.86	0.02676	0.2235	0.709
	600	0.6504	1.0044	52.15	0.04832	0.7399	0.704
Steam (H_2O vapor)	380	0.5863	2.060	21.6	0.0246	0.2036	1.060
	850	0.2579	2.186	115.2	0.0637	1.130	1.019

[a]Adapted from Ref. 2.

Table 8 One-Dimensional Heat Conduction

Geometry	Heat-Transfer Rate and Temperature Distribution	Heat-Transfer Rate and Overall Heat-Transfer Coefficient with Convection at the Boundaries
Plane wall	$$q = \frac{T_1 - T_2}{(x_2 - x_1)/kA}$$ $$T = T_1 + \frac{T_2 - T_1}{x_x - x_1}(x - x_1)$$ $$R = (x_x - x_1)/kA$$	$$q = UA(T_{\infty,1} - T_{\infty,2})$$ $$U = \frac{1}{\dfrac{1}{h_1} + \dfrac{x_2 - x_1}{k} + \dfrac{1}{h_2}}$$
Hollow cylinder	$$q = \frac{T_1 - T_2}{[\ln(r_2/r_1)]/2\pi kL}$$ $$T = T_1 + \frac{T_2 - T_1}{\ln(r_2/r_1)}\ln\frac{r}{r_1}$$ $$R = \frac{\ln(r_2/r_1)}{2\pi kL}$$	$$q = 2\pi r_1 L U_1(T_{\infty,1} - T_{\infty,2})$$ $$= 2\pi r_1 L U_2(T_{\infty,1} - T_{\infty,2})$$ $$U_1 = \frac{1}{\dfrac{1}{h_1} + \dfrac{r_1\ln(r_2/r_1)}{k} + \dfrac{r_1}{r_2}\dfrac{1}{h_2}}$$ $$U_2 = \frac{1}{\left(\dfrac{r_2}{r_1}\right)\dfrac{1}{h_1} + \dfrac{r_2\ln(r_2/r_1)}{k} + \dfrac{1}{h_2}}$$
Hollow sphere	$$q = \frac{T_1 - T_2}{\left(\dfrac{1}{r_1} - \dfrac{1}{r_2}\right)\Big/4\pi k}$$ $$T = \frac{1}{\left(1 - \dfrac{r_1}{r_2}\right)}\left[\frac{r_1}{r}(T_1 - T_2) + \left(T_2 - T_1\frac{r_1}{r_2}\right)\right]$$ $$R = \left(\frac{1}{r_1} - \frac{1}{r_2}\right)\Big/4\pi k$$	$$q = 4\pi r_1^2 U_1(T_{\infty,1} - T_{\infty,2})$$ $$= 4\pi r_2^2 U_2(T_{\infty,1} - T_{\infty,2})$$ $$U_1 = \frac{1}{\dfrac{1}{h_1} + r_1^2\left(\dfrac{1}{r_1} - \dfrac{1}{r_2}\right)\Big/k + \left(\dfrac{r_1}{r_2}\right)^2\dfrac{1}{h_2}}$$ $$U_2 = \frac{1}{\left(\dfrac{r_1}{r_2}\right)^2\dfrac{1}{h_1} + r_2^2\left(\dfrac{1}{r_1} - \dfrac{1}{r_2}\right)\Big/k + \dfrac{1}{h_2}}$$

1.3 Two-Dimensional Steady-State Heat Conduction

Two-dimensional heat transfer in an isotropic, homogeneous material with no internal heat generation requires solution of the heat-diffusion equation of the form $\partial^2 T / \partial X^2 + \partial T / \partial y^2 = 0$, referred to as the *Laplace equation*. For certain geometries and a limited number of fairly simple combinations of boundary conditions, exact solutions can be obtained analytically. However, for anything but simple geometries or for simple geometries with complicated boundary conditions, development of an appropriate analytical solution can be difficult and other methods are usually employed. Among these are solution procedures involving the use of *graphical* or *numerical* approaches. In the first of these, the rate of heat transfer between two isotherms, T_1 and T_2, is expressed in terms of the conduction shape factor, defined by

$$q = kS(T_1 - T_2)$$

Table 9 illustrates the shape factor for a number of common geometric configurations. By combining these shape factors, the heat-transfer characteristics for a wide variety of geometric configurations can be obtained.

Prior to the development of high-speed digital computers, shape factor and analytical methods were the most prevalent methods utilized for evaluating steady-state and transient conduction problems. However, more recently, solution procedures for problems involving complicated geometries or boundary conditions utilize the finite difference method (FDM). Using this approach, the solid object is divided into a number of distinct or discrete regions, referred to as *nodes,* each with a specified boundary condition. An energy balance is then written for each nodal region and these equations are solved simultaneously. For interior nodes in a two-dimensional system with no internal heat generation, the energy equation takes the form of the Laplace equation discussed earlier. However, because the system is characterized in terms of a nodal network, a finite difference approximation must be used. This approximation is derived by substituting the following equation for the x-direction rate of change expression

$$\left. \frac{\partial^2 T}{\partial x^2} \right|_{m,n} \approx \frac{T_{m+1,n} + T_{m-1,n} - 2T_{m,n}}{(\Delta x)^2}$$

and for the y-direction rate of change expression

$$\left. \frac{\partial^2 T}{\partial y^2} \right|_{m,n} \quad \frac{T_{m,n+1} + T_{m,n-1} + T_{m,n}}{(\Delta y)^2}$$

Assuming $\Delta x = \Delta y$ and substituting into the Laplace equation and results in the following expression

$$T_{m,n+1} + T_{m,n-1} + T_{m+1,n} + T_{m-1,n} - 4T_{m,n} = 0$$

which reduces the exact difference to an approximate algebraic expression.

Combining this temperature difference with Fourier's law yields an expression for each internal node

$$T_{m,n+1} + T_{m,n+1} + T_{m-1,n} + T_{m-1,n} + \frac{\dot{q}\Delta x \cdot \Delta y \cdot 1}{k} - 4T_{m,n} = 0$$

Similar equations for other geometries (i.e., corners) and boundary conditions (i.e., convection) and combinations of the two are listed in Table 10. These equations must then be solved using some form of matrix inversion technique, Gauss-Seidel iteration method or other method for solving large numbers of simultaneous equations.

Table 9   Conduction Shape Factors

System	Schematic	Restrictions	Shape Factor
Isothermal sphere buried in a semiinfinite medium having isothermal surface		$z > D/2$	$\dfrac{2\pi D}{1 - D/4z}$
Horizontal isothermal cylinder of length L buried in a semiinfinite medium having isothermal surface		$\left.\begin{array}{l} L \gg D \\ L \gg D \\ z > 3D/2 \end{array}\right\}$	$\dfrac{2\pi L}{\cosh^{-1}(2z/D)}$ $\dfrac{2\pi L}{\ln(4z/D)}$
The cylinder of length L with eccentric bore		$L \gg D_1, D_2$	$\dfrac{2\pi L}{\cosh^{-1}\left(\dfrac{D_1^2 + D_2^2 - 4\varepsilon^2}{2D_1 D_2}\right)}$
Conduction between two cylinders of length L in infinite medium		$L \gg D_1, D_2$	$\dfrac{2\pi L}{\cosh^{-1}\left(\dfrac{4W^2 - D_1^2 - D_2^2}{2D_1 D_2}\right)}$
Circular cylinder of length L in a square solid		$L \gg W$ $w > D$	$\dfrac{2\pi L}{\ln(1.08\, w/D)}$
Conduction through the edge of adjoining walls		$D > L/5$	$0.54\,D$
Conduction through corner of three walls with inside and outside temperature, respectively, at T_1 and T_2		$L \ll$ length and width of wall	$0.15L$

1.4   Heat Conduction with Convection Heat Transfer on the Boundaries

In physical situations where a solid is immersed in a fluid, or a portion of the surface is exposed to a liquid or gas, heat transfer will occur by convection, (or when there is a large temperature difference, through some combination of convection and/or radiation). In these situations, the heat transfer is governed by *Newton's law of cooling*, which is expressed as

$$q = hA\,\Delta T$$

Table 10 Summary of Nodal Finite-Difference Equations

Configuration	Finite-Difference Equation for $\Delta x = \Delta y$
Case 1. Interior node 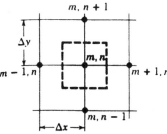	$T_{m,n+1} + T_{m,n-1} + T_{m-1,n} - 4T_{m,n} = 0$
Case 2. Node at an internal corner with convection 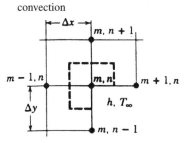	$2(T_{m-1,n} + T_{m,n+1}) + (T_{m+1,n} + T_{m,n-1})$ $+ 2\dfrac{h\,\Delta x}{k}T_\infty - 2\left(3 + \dfrac{h\,\Delta x}{k}\right)T_{m,n} = 0$
Case 3. Node at a plane surface with convection 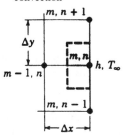	$(2T_{m-1,n} + T_{m,n+1} + T_{m,n-1}) + \dfrac{2h\,\Delta x}{k}T_\infty$ $- 2\left(\dfrac{h\,\Delta x}{k} + 2\right)T_{m,n} = 0$
Case 4. Node at an external corner with convection 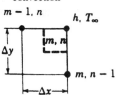	$(T_{m,n-1} + T_{m-1,n}) + 2\dfrac{h\,\Delta x}{k}T_\infty$ $- 2\left(\dfrac{h\,\Delta x}{k} + 1\right)T_{m,n} = 0$

Table 10 (*Continued*)

Configuration	Finite-Difference Equation for $\Delta x = \Delta y$
Case 5. Node near a curved surface maintained at a nonuniform temperature	$\dfrac{2}{a+1} T_{m+1,n} + \dfrac{2}{b+1} T_{m,n-1}$ $+ \dfrac{2}{a(a+1)} T_1 + \dfrac{2}{b(b+1)} T_2$ $- \left(\dfrac{2}{a} + \dfrac{2}{b} \right) T_{m,n} = 0$

where h is the *convection heat-transfer coefficient* (Section 2), ΔT is the temperature difference between the solid surface and the fluid, and A is the surface area in contact with the fluid. The resistance occurring at the surface abounding the solid and fluid is referred to as the *thermal resistance* and is given by $1/hA$, i.e., the *convection resistance*. Combining this resistance term with the appropriate conduction resistance yields an *overall heat-transfer coefficient U*. Usage of this term allows the overall heat transfer to be defined as $q = UA \Delta T$.

Table 8 shows the overall heat-transfer coefficients for some simple geometries. Note that U may be based either on the inner surface (U_1) or on the outer surface (U_2) for the cylinders and spheres.

Critical Radius of Insulation for Cylinders

A large number of practical applications involve the use of insulation materials to reduce the transfer of heat into or out of cylindrical surfaces. This is particularly true of steam or hot water pipes where concentric cylinders of insulation are typically added to the outside of the pipes to reduce the heat loss. Beyond a certain thickness, however, the continued addition of insulation may not result in continued reductions in the heat loss. To optimize the thickness of insulation required for these types of applications, a value typically referred to as the critical radius, defined as $r_{cr} = k/h$, is used. If the outer radius of the object to be insulated is less than r_{cr} then the addition of insulation will increase the heat loss, while for cases where the outer radii is greater than r_{cr} any additional increases in insulation thickness will result in a decrease in heat loss.

Extended Surfaces

In examining Newton's law of cooling, it is clear that the rate of heat transfer between a solid and the surrounding ambient fluid may be increased by increasing the surface area of the solid that is exposed to the fluid. This is typically done through the addition of extended surfaces or fins to the primary surface. Numerous examples often exist, including the cooling fins on air-cooled engines, i.e., motorcycles or lawn mowers or the fins attached to automobile radiators.

Figure 2 illustrates a common uniform cross-section extended surface, fin, with a constant base temperature, T_b, a constant cross-sectional area, A, a circumference of $C = 2W + 2t$, and a length, L, which is much larger than the thickness, t. For these conditions, the temperature distribution in the fin must satisfy the following expression:

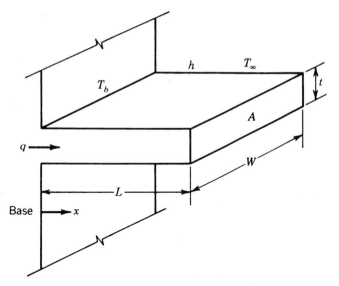

Figure 2   Heat transfer by extended surfaces.

$$\frac{d^2T}{dx^2} - \frac{hC}{kA}(T - T_\infty) = 0$$

The solution of this equation depends on the boundary conditions existing at the tip, i.e., at $x = L$. Table 11 shows the temperature distribution and heat-transfer rate for fins of uniform cross section subjected to a number of different tip conditions, assuming a constant value for the heat-transfer coefficient, h.

Two terms are used to evaluate fins and their usefulness. The first of these is the *fin effectiveness,* defined as the ratio of the heat-transfer rate with the fin to the heat-transfer rate that would exist if the fin were not used. For most practical applications, the use of a fin is justified only when the fin effectiveness is significantly greater than 2. A second term used to evaluate the usefulness of a fin is the *fin efficiency, η_f,* This term represents the ratio

Table 11   Temperature Distribution and Heat-Transfer Rate at the Fin Base ($m = \sqrt{hc/kA}$)

Condition at $x = L$	$\dfrac{T - T_\infty}{T_b - T_\infty}$	Heat-Transfer Rate $q/mkA\,(T_b - T_\infty)$
$h(T_{x=L} - T_\infty) = -k\left(\dfrac{dT}{dx}\right)_{x=L}$ (convection)	$\dfrac{\cosh m(L-x) + \dfrac{h}{mk}\sinh m(L-x)}{\cosh mL + \dfrac{h}{mk}\sinh mL}$	$\dfrac{\sinh mL + \dfrac{h}{mk}\cosh mL}{\cosh mL + \dfrac{h}{mk}\sinh mL}$
$\left(\dfrac{dT}{dx}\right)_{x=L} = 0$ (insulated)	$\dfrac{\cosh m(L-x)}{\cosh mL}$	$\tanh mL$
$T_{x=L} = T_L$ (prescribed temperature)	$\dfrac{(T_L - T_\infty)/(T_b - T_\infty)\sinh mx + \sinh m(L-x)}{\sinh ml}$	$\dfrac{\cosh mL - (T_L - T_\infty)/(T_b - T_\infty)}{\sinh ml}$
$T_{x=L} = T_\infty$ (infinitely long fin, $L \to \infty$)	e^{-mx}	1

of actual the heat-transfer rate from a fin to the heat-transfer rate that would occur if the entire fin surface could be maintained at a uniform temperature equal to the temperature of the base of the fin. For this case, Newton's law of cooling can be written as

$$q = \eta_f h A_f (T_b - T_\infty)$$

where A_f is the total surface area of the fin and T_b is the temperature of the fin at the base. The application of fins for heat removal can be applied to either forced or natural convection of gases, and while some advantages can be gained in terms of increasing the liquid–solid or solid–vapor surface area, fins as such are not normally utilized for situations involving phase change heat transfer, such as boiling or condensation.

1.5 Transient Heat Conduction

Given a solid body, at a uniform temperature, $T_{\infty i}$, immersed in a fluid of different temperature T_∞, the surface of the solid body will be subject to heat losses (or gains) through convection from the surface to the fluid. In this situation, the heat lost (or gained) at the surface results from the conduction of heat from inside the body. To determine the significance of these two heat-transfer modes, a dimensionless parameter referred to as the *Biot number* is used. This dimensionless number is defined as $Bi = hL/k$, where $L = V/A$ or the ratio of the volume of the solid to the surface area of the solid, and really represents a comparative relationship of the importance of convections from the outer surface to the conduction occurring inside. When this value is less than 0.1, the temperature of the solid may be assumed uniform and dependent on time alone. When this value is greater than 0.1, there is some spatial temperature variation that will affect the solution procedure.

For the first case, $Bi < 0.1$, an approximation referred to as the *lumped heat-capacity* method may be used. In this method, the temperature of the solid is given by

$$\frac{T - T_\infty}{T_i - T_\infty} = \exp\left(\frac{-t}{\tau_t}\right) = \exp(-BiFo)$$

where τ_t is the *time constant* and is equal to $\rho C_p V / hA$. Increasing the value of the time constant, τ_t, will result in a decrease in the thermal response of the solid to the environment and hence, will increase the time required for it to reach thermal equilibrium (i.e., $T = T_\infty$). In this expression, Fo represents the dimensionless time and is called the *Fourier number*, the value of which is equal to $\alpha t A^2 / V^2$. The Fourier number, along with the Biot number, can be used to characterize transient heat conduction problems. The total heat flow through the surface of the solid over the time interval from $t = 0$ to time t can be expressed as

$$Q = \rho V C_p (T_i - T_\infty)[1 - \exp(-t/\tau_t)]$$

Transient Heat Transfer for Infinite Plate, Infinite Cylinder, and Sphere Subjected to Surface Convection
Generalized analytical solutions to transient heat-transfer problems involving infinite plates, cylinders, and finite diameter spheres subjected to surface convection have been developed. These solutions can be presented in graphical form through the use of the *Heisler charts,*[3] illustrated in Figs. 3–11 for plane walls, cylinders, and spheres. In this procedure, the solid is assumed to be at a uniform temperature, T_i, at time $t = 0$ and then is suddenly subjected to or immersed in a fluid at a uniform temperature T_∞. The convection heat-transfer coefficient, h, is assumed to be constant, as is the temperature of the fluid. Combining Figs. 3 and 4 for plane walls, Figs. 6 and 7 for cylinders, and Figs. 9 and 10 for spheres allows the resulting time-dependent temperature of any point within the solid to be found. The total

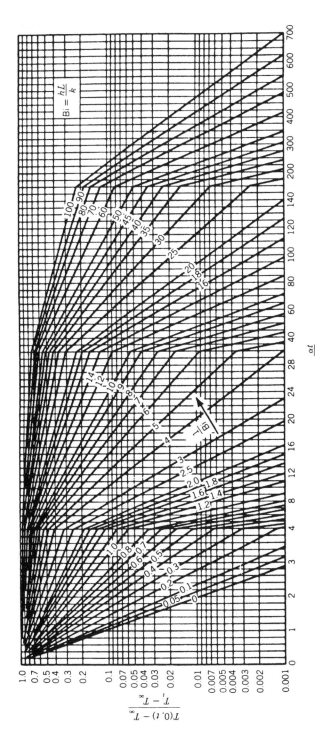

Figure 3 Midplane temperature as a function of time for a plane wall of thickness $2L$. (Adapted from Heisler.[3])

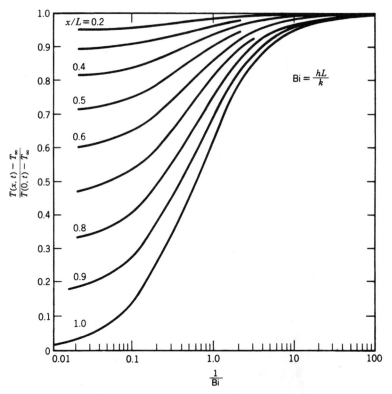

Figure 4 Temperature distribution in a plane wall of thickness $2L$. (Adapted from Heisler.[3])

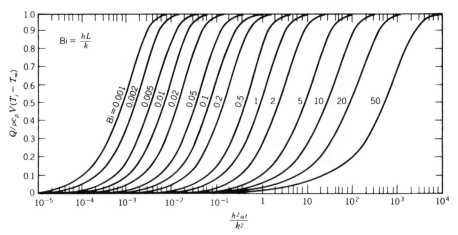

Figure 5 Internal energy change as a function of time for a plane wall of thickness $2L$.[4] (Used with the permission of McGraw-Hill Book Company.)

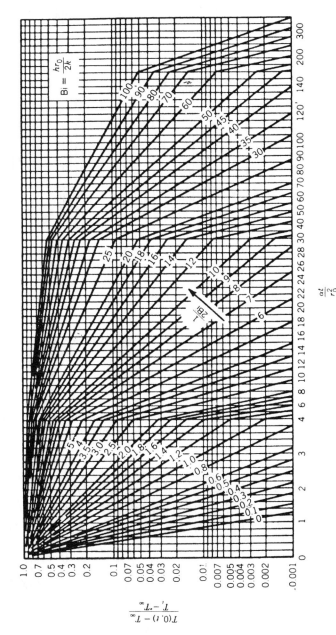

Figure 6 Centerline temperature as a function of time for an infinite cylinder of radius r_o. (Adapted from Heisler.[3])

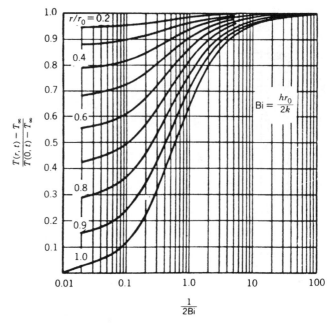

Figure 7   Temperature distribution in an infinite cylinder of radius r_o. (Adapted from Heisler.[3])

amount of energy, Q, transferred to or from the solid surface from time $t = 0$ to time t can be found from Figs. 5, 8, and 11.

1.6   Conduction at the Microscale

The mean free path of electrons and the size of the volume involved has long been recognized as having a pronounced effect on electron-transport phenomena. This is particularly true in

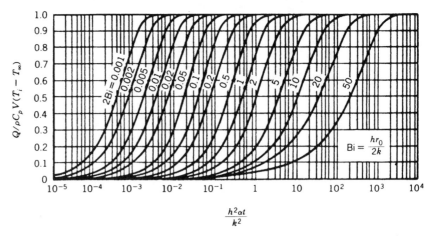

Figure 8   Internal energy change as a function of time for an infinite cylinder of radius r_o.[4] (Used with the permission of McGraw-Hill Book Company.)

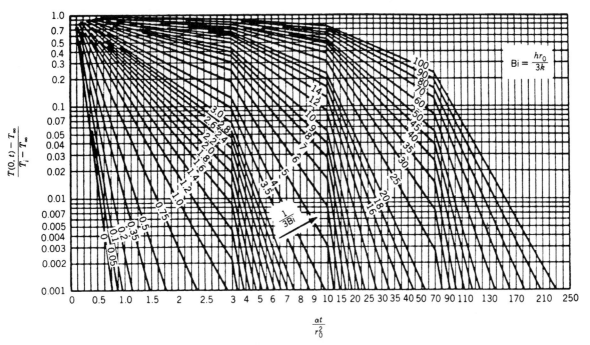

Figure 9 Center temperature as a function of time in a sphere of radius r_o. (Adapted from Heisler.[3])

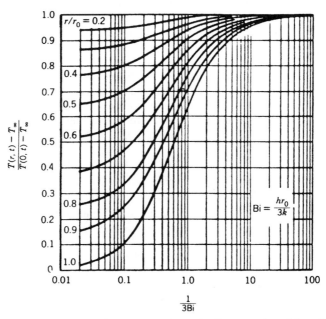

Figure 10 Temperature distribution in a sphere of radius r_o. (Adapted from Heisler.[3])

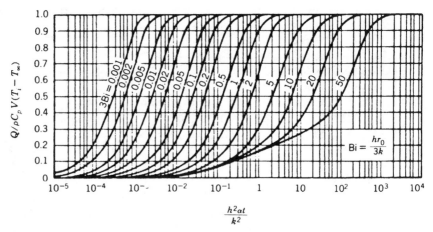

Figure 11 Internal energy change as a function of time for a sphere of radius r_o.[4] (Used with the permission of McGraw-Hill Book Company.)

applications involving thin metallic films or wires where the characteristic length may be close to the same order of magnitude as the scattering mean free path of the electrons (Duncan and Peterson, 1994). The first notable work in this area was performed by Tien et al. (1968), where the thermal conductivity of thin metallic films and wires were calculated at cryogenic temperatures. Because the length of the mean free path in these types of applications is shortened near the surface, due to termination at the boundary, a reduction in transport coefficients, such as electrical and thermal conductivities, was observed. Tests at cryogenic temperatures were first performed, because the electron mean free path increases as temperature decreases, and the size effects were expected to become especially significant in this range. The primary purpose of this investigation was to outline in a systematic manner a method by which the thermal conductivity of such films and wires at cryogenic temperatures could be determined. The results indicated that, particularly in the case of thin metallic films, size effects may become an increasingly important part of the design and analysis required for application. Due to the increased use of thin films in optical components and solid-state devices and systems, there has been an increasing interest in the effect of decreasing size on the transport properties of thin solid films and wires.

The most common method for calculating the thermal conductivities in thin films and wires consists of three essential steps:

1. Identifying the appropriate expression for the electrical conductivity size effect
2. Determining the mean free path for electrical conductivity, which is essential in calculations of all electron-transport properties
3. Applying the electrical–thermal transport analogy for calculating the thermal conductivity size effect (Duncan and Peterson, 1994)

For domain thicknesses on the order of the carrier mean free path, jump boundary conditions significantly affect the solution of the conduction problem. This problem can be resolved through the solution of the hyperbolic heat equation-based analysis, which is generally justifiable engineering applications (Bai and Lavine, 1993).

2 CONVECTION HEAT TRANSFER

As discussed earlier, convection heat transfer is the mode of energy transport in which the energy is transferred by means of fluid motion. This transfer can be the result of the random molecular motion or bulk motion of the fluid. If the fluid motion is caused by external forces, the energy transfer is called *forced convection*. If the fluid motion arises from a buoyancy effect caused by density differences, the energy transfer is called *free convection* or *natural convection*. For either case, the heat-transfer rate, q, can be expressed in terms of the surface area, A, and the temperature difference, ΔT, by Newton's law of cooling:

$$q = hA\ \Delta T$$

In this expression, h is referred to as the convection heat-transfer coefficient or film coefficient and a function of the velocity and physical properties of the fluid, and the shape and nature of the surface. The nondimensional heat-transfer coefficient $Nu = hL/k$ is called the *Nusselt number,* where L is a characteristic length and k is the thermal conductivity of the fluid.

2.1 Forced Convection—Internal Flow

For internal flow in a tube or pipe, the convection heat-transfer coefficient is typically defined as a function of the temperature difference existing between the temperature at the surface of the tube and the *bulk* or *mixing-cup temperature, T_b,* i.e., $\Delta T = T_s - T_b$ can be defined as

$$T_b = \frac{\int C_p T\ d\dot{m},}{\int C_p\ d\dot{m}}$$

where $\dot{m}$ is the axial flow rate. Using this value, heat transfer between the tube and the fluid can be written as $q = hA(T_s - T_b)$.

In the entrance region of a tube or pipe, the flow is quite different from that occurring downstream from the entrance. The rate of heat transfer differs significantly, depending on whether the flow is *laminar* or *turbulent*. From fluid mechanics, the flow is considered to be turbulent when $Re_D = V_m D/v > 2300$ for a smooth tube. This transition from laminar to turbulent, however, also depends on the roughness of tube wall and other factors. The generally accepted range for transition is $200 < Re_D < 4000$.

Laminar Fully Developed Flow
For situations where both the thermal and velocity profiles are fully developed, the Nusselt number is constant and depends only on the thermal boundary conditions. For *circular tubes* with $Pr \geq 0.6$, and $x/DRe_D Pr > 0.05$, the Nusselt numbers have been shown to be $Nu_D = 3.66$ and 4.36, for constant temperature and constant heat flux conditions, respectively. Here, the fluid properties are based on the mean bulk temperature.

For *noncircular tubes,* the hydraulic diameter, $D_h = 4 \times$ the flow cross-sectional area/wetted perimeter, is used to define the Nusselt number Nu_D and the Reynolds number Re_D. Table 12 shows the Nusselt numbers based on hydraulic diameter for various cross-sectional shapes.

Laminar Flow for Short Tubes
At the entrance of a tube, the Nusselt number is infinite, and decreases asymptotically to the value for fully developed flow as the flow progresses down the tube. The Sieder-Tate

Table 12 Nusselt Numbers for Fully Developed Laminar Flow for Tubes of Various Cross Sections[a]

Geometry $(L/DH > 100)$	Nu_{H1}	Nu_{H2}	Nu_r
$2b$ ☐ $\dfrac{2b}{2a} = 1$ $2a$	3.608	3.091	2.976
$2b$ ▭ $\dfrac{2b}{2a} = \dfrac{1}{2}$ $2a$	4.123	3.017	3.391
$2b$ ▭ $\dfrac{2b}{2a} = \dfrac{1}{4}$ $2a$	5.099	4.35	3.66
$2b$ ▭ $\dfrac{2b}{2a} = \dfrac{1}{8}$ $2a$	6.490	2.904	5.597
▬▬ $\dfrac{2b}{2a} = 0$	8.235	8.235	7.541
▬▬ $\dfrac{b}{a} = 0$	5.385	—	4.861
◯	4.364	4.364	3.657

[a] Nu_{H1} = average Nusselt number for uniform heat flux in flow direction and uniform wall temperature at particular flow cross section.
Nu_{H2} = average Nusselt number for uniform heat flux both in flow direction and around periphery.
Nu_{Hrr} = average Nusselt number for uniform wall temperature.

equation[5] gives good correlation for the combined entry length, i.e., that region where the thermal and velocity profiles are both developing or for short tubes:

$$\overline{Nu_D} = \frac{\overline{h}D}{k} = 1.86(Re\ D\ Pr)^{1/3}\left(\frac{D}{L}\right)^{1/3}\left(\frac{\mu}{\mu_s}\right)^{0.14}$$

for T_s = constant, $0.48 < Pr < 16,700$, $0.0044 < \mu/\mu_s < 9.75$, and $(Re_D\ Pr\ D/L)^{1/3}$ $(\mu/\mu_s)^{0.14} > 2$.

In this expression, all of the fluid properties are evaluated at the mean bulk temperature except for μ_s, which is evaluated at the wall surface temperature. The average convection heat-transfer coefficient $\overline{h}$ is based on the arithmetic average of the inlet and outlet temperature differences.

Turbulent Flow in Circular Tubes

In turbulent flow, the velocity and thermal entry lengths are much shorter than for a laminar flow. As a result, with the exception of short tubes, the fully developed flow values of the Nusselt number are frequently used directly in the calculation of the heat transfer. In general, the Nusselt number obtained for the constant heat flux case is greater than the Nusselt number obtained for the constant temperature case. The one exception to this is the case of liquid metals, where the difference is smaller than for laminar flow and becomes negligible for $Pr > 1.0$. The Dittus-Boelter equation[6] is typically used if the difference between the pipe surface temperature and the bulk fluid temperature is less than 6°C (10°F) for liquids or 56°C (100°F) for gases:

$$\text{Nu}_D = 0.023 \, \text{Re}_D^{0.8} \, \text{Pr}^n$$

for $0.7 \leq \text{Pr} \leq 160$, $\text{Re}_D \geq 10{,}000$, and $L/D \geq 60$, where

$$n = 0.4 \text{ for heating, } T_s > T_b$$

$$= 0.3 \text{ for cooling, } T_s < T_b$$

For temperature differences greater than specified above, use[5]

$$\text{Nu}_D = 0.027 \, \text{Re}_D^{0.8} \text{Pr}^{1/3} \left(\frac{\mu}{\mu_s} \right)^{0.14}$$

for $0.7 \leq \text{Pr} \leq 16{,}700$, $\text{Re}_D \geq 10{,}000$, and $L/D \geq 60$. In this expression, the properties are all evaluated at the mean bulk fluid temperature with the exception of μ_s, which is again evaluated at the tube surface temperature.

For *concentric tube annuli*, the hydraulic diameter $D_h = D_o - D_i$ (outer diameter − inner diameter) must be used for Nu_D and Re_D, and the coefficient h at either surface of the annulus must be evaluated from the Dittus-Boelter equation. Here, it should be noted that the foregoing equations apply for smooth surfaces and that the heat-transfer rate will be larger for rough surfaces, and are not applicable to liquid metals.

Fully Developed Turbulent Flow of Liquid Metals in Circular Tubes

Because the Prandtl number for liquid metals, is on the order of 0.01, the Nusselt number is primarily dependent on a dimensionless parameter number referred to as the *Peclet number,* which in general is defined as $\text{Pe} = \text{RePr}$:

$$\text{Nu}_D = 5.0 + 0.025 \text{Pe}_D^{0.8}$$

which is valid for situations where T_s = a constant and $\text{Pe}_D > 100$ and $L/D > 60$.

For $q'' = $ constant and $3.6 \times 10^3 < \text{Re}_D < 9.05 \times 10^5$, $10^2 < \text{Pe}_D < 10^4$, and $L/D > 60$, the Nusselt number can be expressed as

$$\text{Nu}_D = 4.8 + 0.0185 \text{Pe}_D^{0.827}$$

2.2 Forced Convection—External Flow

In forced convection heat transfer, the heat-transfer coefficient, h, is based on the temperature difference between the wall surface temperature and the fluid temperature in the free stream outside the thermal boundary layer. The total heat-transfer rate from the wall to the fluid is given by $q = hA \, (T_s - T_\infty)$. The Reynolds numbers are based on the free stream velocity. The fluid properties are evaluated either at the free stream temperature T_∞ or at the film temperature $T_f = (T_s + T_\infty)/2$.

Laminar Flow on a Flat Plate

When the flow velocity along a constant temperature semi-infinite plate is uniform, the boundary layer originates from the leading edge and is laminar and the flow remains laminar until the local Reynolds number $\text{Re}_x = U_\infty x/\nu$ reaches the *critical Reynolds number,* Re_c. When the surface is smooth, the Reynolds number is generally assumed to be $\text{Re}_c = 5 \times 10^5$, but the value will depend on several parameters, including the surface roughness.

For a given distance x from the leading edge, the *local Nusselt number* and the *average Nusselt number* between $x = 0$ and $x = L$ are given below (Re_x and $\text{Re}_L \leq 5 \times 10^5$):

$$\left. \begin{aligned} \mathrm{Nu}_x &= hx/k = 0.332\mathrm{Re}_x^{0.5}\mathrm{Pr}^{1/3} \\ \overline{\mathrm{Nu}}_L &= \overline{h}L/k = 0.664\mathrm{Re}_L^{0.5}\mathrm{Pr}^{1/3} \end{aligned} \right\} \quad \text{for Pr} \geq 0.6$$

$$\left. \begin{aligned} \mathrm{Nu}_x &= 0.565(\mathrm{Re}_x\ \mathrm{Pr})^{0.5} \\ \overline{\mathrm{Nu}}_L &= 1.13(\mathrm{Re}_L\ \mathrm{Pr})^{0.5} \end{aligned} \right\} \quad \text{for Pr} \leq 0.6$$

Here, all of the fluid properties are evaluated at the mean or average film temperature.

Turbulent Flow on a Flat Plate

When the flow over a flat plate is turbulent from the leading edge, expressions for the *local Nusselt number* can be written as

$$\mathrm{Nu}_x = 0.0292\mathrm{Re}_x^{0.8}\mathrm{Pr}^{1/3}$$

$$\overline{\mathrm{Nu}}_L = 0.036\mathrm{Re}_L^{0.8}\mathrm{Pr}^{1/3}.$$

where the fluid properties are all based on the mean film temperature and $5 \times 10^5 \leq \mathrm{Re}_x$ and $\mathrm{Re}_L \leq 10^8$ and $0.6 \leq \mathrm{Pr} \leq 60$.

The Average Nusselt Number between x = 0 and x = L with Transition

For situations where transition occurs immediately once the critical Reynolds number Re_c has been reached[7]

$$\overline{\mathrm{Nu}}_L = 0.036\mathrm{Pr}^{1/3}[\mathrm{Re}_L^{0.8} - \mathrm{Re}_c^{0.8} + 18.44\mathrm{Re}_c^{0.5}]$$

provided that $5 \times 10^5 \leq \mathrm{Re}_L \leq 10^8$ and $0.6 \leq \mathrm{Pr} \leq 60$. Specialized cases exist for this situation i.e.,

$$\overline{\mathrm{Nu}}_L = 0.036\mathrm{Pr}^{1/3}(\mathrm{Re}_L^{0.8} - 18{,}700)$$

for $\mathrm{Re}_c = 4 \times 10^5$, or

$$\overline{\mathrm{Nu}}_L = 0.036\mathrm{Pr}^{1/3}(\mathrm{Re}_L^{0.8} - 23{,}000)$$

for $\mathrm{Re}_c = 5 \times 10^5$. Again, all fluid properties are evaluated at the mean film temperature.

Circular Cylinders in Cross-Flow

For circular cylinders in cross-flow, the Nusselt number is based upon the diameter and can be expressed as

$$\overline{\mathrm{Nu}}_D = (0.4\mathrm{Re}_D^{0.5} + 0.06\mathrm{Re}^{2/3})\mathrm{Pr}^{0.4}(\mu_\infty/\mu_s)^{0.25}$$

for $0.67 < \mathrm{Pr} < 300$, $10 < \mathrm{Re}_D < 10^5$, and $0.25 < 5.2$. Here, the fluid properties are evaluated at the free stream temperature except μ_s, which is evaluated at the surface temperature.[8]

Cylinders of Noncircular Cross Section in Cross-Flow of Gases

For noncircular cylinders in cross-flow, the Nusselt number is again based on the diameter, but is expressed as

$$\overline{\mathrm{Nu}}_D = C(\mathrm{Re}_D)^m\mathrm{Pr}^{1/3}$$

where C and m are listed in Table 13, and the fluid properties are evaluated at the mean film temperature.[9]

Table 13 Constants and m for Noncircular Cylinders in Cross-Flow

Geometry	Re_D	C	m
Square			
$V \rightarrow \diamondsuit \updownarrow D$	5×10^3–10^5	0.246	0.588
	5×10^3–10^5	0.102	0.675
$V \rightarrow \square \; D\uparrow$			
Hexagon			
$V \rightarrow \;\; \updownarrow D$	5×10^3–1.95×10^4	0.160	0.538
	1.95×10^4–10^5	0.0385	0.782
$V \rightarrow \;\; D\uparrow$			
	5×10^3–10^5	0.153	0.638
Vertical plate			
$V \rightarrow \square \; \updownarrow D$	4×10^3–1.5×10^4	0.228	0.731

Flow past a Sphere

For flow over a sphere, the Nusselt number is based on the sphere diameter and can be expressed as

$$\overline{Nu}_D = 2 + (0.4Re_D^{0.5} + 0.06Re_D^{2/3})Pr^{0.4}(\mu_\infty/\mu_s)^{0.25}$$

for the case of $3.5 < Re_D < 8 \times 10^4$, $0.7 < Pr < 380$, and $1.0 < \mu_\infty/\mu_s < 3.2$. The fluid properties are calculated at the free stream temperature except μ_s, which is evaluated at the surface temperature.[8]

Flow across Banks of Tubes

For banks of tubes, the tube arrangement may be either *staggered* or *aligned* (Fig. 12), and the heat-transfer coefficient for the first row is approximately equal to that for a single tube.

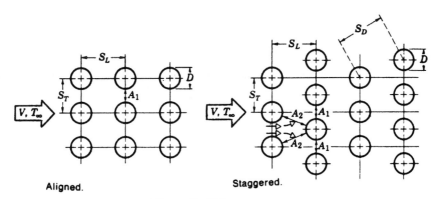

Aligned. Staggered.

Figure 12 Tube arrangement.

In turbulent flow, the heat-transfer coefficient for tubes in the first row is smaller than that of the subsequent rows. However, beyond the fourth or fifth row, the heat-transfer coefficient becomes approximately constant. For tube banks with more than twenty rows, $0.7 < \text{Pr} < 500$, and $1000 < \text{Re}_{D,\text{max}} < 2 \times 10^6$, the average Nusselt number for the entire tube bundle can be expressed as[10]

$$\overline{\text{Nu}}_D = C(\text{Re}_{D,\text{max}})^m \text{Pr}^{0.36}(\text{Pr}_\infty/\text{Pr}_s)^{0.25}$$

where all fluid properties are evaluated at T_∞ except Pr_s, which is evaluated at the surface temperature. The constants C and m used in this expression are listed in Table 14, and the Reynolds number is based on the maximum fluid velocity occurring at the minimum free flow area available for the fluid. Using the nomenclature shown in Fig. 12, the maximum fluid velocity can be determined by

$$V_{\max} = \frac{S_T}{S_T - D} V$$

for the aligned or staggered configuration provided

$$\sqrt{S_L^2 + (S_T/2)^2} > (S_T + D)/2$$

or as

$$V_{\max} = \frac{S_T}{\sqrt[2]{S_L^2 + (S_T/2)^2}} V$$

for staggered if

$$\sqrt{S_L^2 + (S_T/2)^2} < (S_T + D)/2$$

Liquid Metals in Cross-Flow over Banks of Tubes

The average Nusselt number for tubes in the inner rows can be expressed as

$$\overline{\text{Nu}}_D = 4.03 + 0.228(\text{Re}_{D,\text{max}}\text{Pr})^{0.67}$$

which is valid for $2 \times 10^4 < \text{Re}_{D,\text{max}} < 8 \times 10^4$ and $\text{Pr} < 0.03$ and the fluid properties are evaluated at the mean film temperature.[11]

High-Speed Flow over a Flat Plate

When the free stream velocity is very high, the effects of viscous dissipation and fluid compressibility must be considered in the determination of the convection heat transfer. For

Table 14   Constants C and m of Heat-Transfer Coefficient for the Banks in Cross-Flow

Configuration	$\text{Re}_{D,\text{max}}$	C	m
Aligned	10^3–2×10^5	0.27	0.63
Staggered $(S_T/S_L < 2)$	10^3–2×10^5	$0.35(S_T/S_L)^{1/5}$	0.60
Staggered $(S_G/S_L > 2)$	10^3–2×10^5	0.40	0.60
Aligned	2×10^5–2×10^6	0.21	0.84
Staggered	2×10^5–2×10^6	0.022	0.84

these types of situations, the convection heat transfer can be described as $q = hA (T_s - T_{as})$, where T_{as} is the *adiabatic surface temperature* or *recovery temperature,* and is related to the *recovery factor* by $r = (T_{as} - T_\infty)/(T_0 - T_\infty)$. The value of the *stagnation temperature, T_0,* is related to the free stream static temperature, T_∞, by the expression

$$\frac{T_0}{T_\infty} = 1 + \frac{\gamma - 1}{2} M_\infty^2$$

where γ is the specific heat ratio of the fluid and M_∞ is the ratio of the free stream velocity and the acoustic velocity. For the case where $0.6 < Pr < 15$,

$$r = Pr^{1/2} \quad \text{for laminar flow } (Re_x < 5 \times 10^5)$$

$$r = Pr^{1/3} \quad \text{for turbulent flow } (Re_x > 5 \times 10^5)$$

Here, all of the fluid properties are evaluated at the reference temperature $T_{ref} = T_\infty + 0.5(T_s - T_\infty) + 0.22(T_{as} - T_\infty)$. Expressions for the local heat-transfer coefficients at a given distance x from the leading edge are given as[2]

$$Nu_x = 0.332Re_x^{0.5} Pr^{1/3} \quad \text{for } Re_x < 5 \times 10^5$$

$$Nu_x = 0.0292Re_x^{0.8} Pr^{1/3} \quad \text{for } 5 \times 10^5 < Re_x < 10^7$$

$$Nu_x = 0.185Re_x(logRe_x)^{-2.584} \quad \text{for } 10^7 < Re_x < 10^9$$

In the case of gaseous fluids flowing at very high free stream velocities, dissociation of the gas may occur, and will cause large variations in the properties within the boundary layer. For these cases, the heat-transfer coefficient must be defined in terms of the enthalpy difference, i.e., $q = hA(i_s - i_{as})$, and the recovery factor will be given by $r = (i_s - i_{as})/(i_0 - i_\infty)$, where i_{as} represents the enthalpy at the adiabatic wall conditions. Similar expressions to those shown above for Nu_x can be used by substituting the properties evaluated at a reference enthalpy defined as $i_{ref} = i_\infty + 0.5(i_s - i_\infty) + 0.22(i_{as} - i_\infty)$.

High-Speed Gas Flow past Cones
For the case of high-speed gaseous flows over conical shaped objects the following expressions can be used:

$$Nu_x = 0.575Re_x^{0.5} Pr^{1/3} \quad \text{for } Re_x < 10^5$$

$$Nu_x = 0.0292Re_x^{0.8} Pr^{1/3} \quad \text{for } Re_x > 10^5$$

where the fluid properties are evaluated at T_{ref} as in the plate.[12]

Stagnation Point Heating for Gases
When the conditions are such that the flow can be assumed to behave as *incompressible,* the Reynolds number is based on the free stream velocity and $\bar{h}$ is defined as $q = \bar{h}A(T_s - T_\infty)$[13]. Estimations of the Nusselt can be made using the following relationship

$$Nu_D = CRe_D^{0.5} Pr^{0.4}$$

where $C = 1.14$ for cylinders and 1.32 for spheres, and the fluid properties are evaluated at the mean film temperature. When the flow becomes *supersonic,* a bow shock wave will occur just off the front of the body. In this situation, the fluid properties must be evaluated at the stagnation state occurring behind the bow shock and the Nusselt number can be written as

$$\overline{\mathrm{Nu}}_D = C\mathrm{Re}_D^{0.5}\,\mathrm{Pr}^{0.4}(\rho_\infty/\rho_0)^{0.25}$$

where $C = 0.95$ for cylinders and 1.28 for spheres; ρ_∞ is the free stream gas density and ρ_0 is the stagnation density of stream behind the bow shock. The heat-transfer rate for this case, is given by $q = \overline{h}A(T_s - T_0)$.

2.3   Free Convection

In free convection the fluid motion is caused by the buoyant force resulting from the density difference near the body surface, which is at a temperature different from that of the free fluid far removed from the surface where velocity is zero. In all free convection correlations, except for the enclosed cavities, the fluid properties are usually evaluated at the mean film temperature $T_f = (T_1 + T_\infty)/2$. The thermal expansion coefficient β, however, is evaluated at the free fluid temperature T_∞. The convection heat-transfer coefficient h is based on the temperature difference between the surface and the free fluid.

Free Convection from Flat Plates and Cylinders

For free convection from flat plates and cylinders, the average Nusselt number $\overline{\mathrm{Nu}}_L$ can be expressed as[4]

$$\overline{\mathrm{Nu}}_L = C(\mathrm{Gr}_L\,\mathrm{Pr})^m$$

where the constants C and m are given as shown in Table 15. The *Grashof Prandtl number* product, $(\mathrm{Gr}_L\mathrm{Pr})$ is called the *Rayleigh number* (Ra_L) and for certain ranges of this value, Figs. 13 and 14 are used instead of the above equation. Reasonable approximations for other types of *three-dimensional shapes,* such as short cylinders and blocks, can be made for $10^4 < \mathrm{Ra}_L < 10^9$, by using this expression and $C = 0.6$, $m = \frac{1}{4}$, provided that the characteristic length, L, is determined from $1/L = 1/L_{\mathrm{hor}} + 1/L_{\mathrm{ver}}$, where L_{ver} is the height and L_{hor} is the horizontal dimension of the object in question.

For *unsymmetrical horizontal* square, rectangular, or circular surfaces, the characteristic length L can be calculated from the expression $L = A/P$, where A is the area and P is the wetted perimeter of the surface.

Table 15   Constants for Free Convection from Flat Plates and Cylinders

Geometry	$\mathrm{Gr}_K\mathrm{Pr}$	C	m	L
Vertical flat plates and cylinders	10^{-1}–10^4	Use Fig. 12	Use Fig. 12	Height of plates and cylinders; restricted to $D/L \geq 35/\mathrm{Gr}_L^{1/4}$ for cylinders
	10^4–10^9	0.59	$\frac{1}{4}$	
	10^9–10^{13}	0.10	$\frac{1}{3}$	
Horizontal cylinders	0–10^{-5}	0.4	0	Diameter D
	10^{-5}–10^4	Use Fig. 13	Use Fig. 13	
	10^4–10^9	0.53	$\frac{1}{4}$	
	10^9–10^{13}	0.13	$\frac{1}{3}$	
Upper surface of heated plates or lower surface of cooled plates	2×10^4–8×10^6	0.54	$\frac{1}{4}$	Length of a side for square plates, the average length of the two sides for rectangular plates
	8×10^6–10^{11}	0.15	$\frac{1}{3}$	
Lower surface of heated plates or upper surface of cooled plates	10^5–10^{11}	0.58	$\frac{1}{5}$	0.9D for circular disks

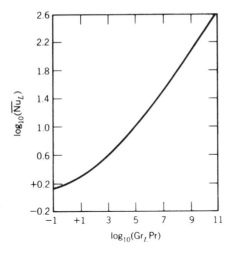

Figure 13 Free convection heat-transfer correlation for heated vertical plates and cylinders. (Adapted from Ref. 14. Used with permission of McGraw-Hill Book Company.)

Free Convection from Spheres

For free convection from spheres, the following correlation has been developed:

$$\overline{Nu}_D = 2 + 0.43(Gr_D\, Pr)^{0.25} \qquad \text{for } 1 < Gr_D < 10^5$$

Although this expression was designed primarily for gases, $Pr \approx 1$, it may be used to approximate the values for liquids as well.[15]

Free Convection in Enclosed Spaces

Heat transfer in an enclosure occurs in a number of different situations and with a variety of configurations. Then a temperature difference is imposed on two opposing walls that enclose a space filled with a fluid, convective heat transfer will occur. For small values of the Rayleigh number, the heat transfer may be dominated by conduction, but as the Rayleigh number increases, the contribution made by free convection will increase. Following are a

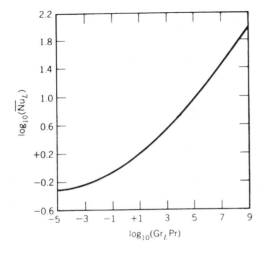

Figure 14 Free convection heat-transfer correlation from heated horizontal cylinders. (Adapted from Ref. 14. Used with permission of McGraw-Hill Book Company.)

number of correlations, each designed for a specific geometry. For all of these, the fluid properties are evaluated at the average temperature of the two walls.

Cavities between Two Horizontal Walls at Temperatures T_1 and T_2 Separated by Distance δ (T_1 for Lower Wall, $T_1 > T_2$)

$$q'' = \overline{h}(T_1 - T_2)$$

$$\overline{Nu}_\delta = 0.069 Ra_\delta^{1/3} Pr^{0.074} \qquad \text{for } 3 \times 10^5 < Ra_\delta < 7 \times 10^9$$

$$= 1.0 \qquad \text{for } Ra_\delta < 1700$$

where $Ra_\delta = g\beta (T_1 - T_2) \delta^3 / \alpha\nu$; δ is the thickness of the space.[16]

Cavities between Two Vertical Walls of Height H at Temperatures by Distance T_1 and T_2 Separated by Distance δ[17,18]

$$q'' = \overline{h}(T_1 - T_2)$$

$$\overline{Nu}_\delta = 0.22\left(\frac{Pr}{0.2 + Pr} Ra_\delta\right)^{0.28} \left(\frac{\delta}{H}\right)^{0.25}$$

for $2 < H/\delta < 10$, $Pr < 10^5$ $Ra_\delta < 10^{10}$;

$$\overline{Nu}_\delta = 0.18\left(\frac{Pr}{0.2 + Pr} Ra_\delta\right)^{0.29}$$

for $1 < H/\delta < 2$, $10^3 < Pr < 10^5$, and $10^3 < Ra_\delta Pr/(0.2 + Pr)$; and

$$\overline{Nu}_\delta = 0.42 Ra_\delta^{0.25} Pr^{0.012}(\delta/H)^{0.3}$$

for $10 < H/\delta < 40$, $1 < Pr < 2 \times 10^4$, and $10^4 < Ra_\delta < 10^7$.

2.4 The Log-Mean Temperature Difference

The simplest and most common type of heat exchanger is the *double-pipe heat exchanger* illustrated in Fig. 15. For this type of heat exchanger, the heat transfer between the two fluids can be found by assuming a constant overall heat transfer coefficient found from Table 8 and a constant fluid specific heat. For this type, the heat transfer is given by

$$q = UA \, \Delta T_m$$

where

$$\Delta T_m = \frac{\Delta T_2 - \Delta T_1}{\ln(\Delta T_2/\Delta T_1)}$$

In this expression, the temperature difference, ΔT_m, is referred to as the *log-mean temperature difference* (LMTD); ΔT_1 represents the temperature difference between the two fluids at one end and ΔT_2 at the other end. For the case where the ratio $\Delta T_2/\Delta T_1$ is less than two, the *arithmetic mean temperature difference*, $(\Delta T_2 + \Delta T_1)/2$, may be used to calculate heat-transfer rate without introducing any significant error. As shown in Fig. 15,

$$\Delta T_1 = T_{h,i} - T_{c,i} \qquad \Delta T_2 = T_{h,o} - T_{c,o} \qquad \text{for parallel flow}$$

$$\Delta T_1 = T_{h,i} - T_{c,o} \qquad \Delta T_2 = T_{h,o} - T_{c,i} \qquad \text{for counterflow}$$

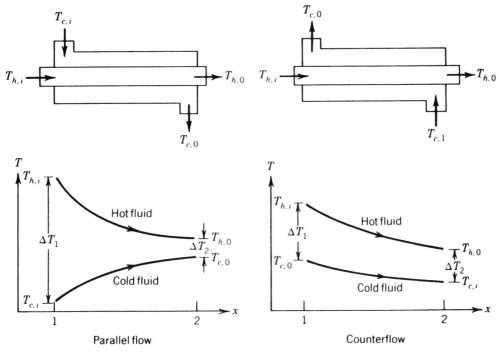

Figure 15 Temperature profiles for parallel flow and counterflow in double-pipe heat exchanger.

Cross-Flow Coefficient

In other types of heat exchangers, where the values of the overall heat transfer coefficient, U, may vary over the area of the surface, the LMTD may not be representative of the actual average temperature difference. In these cases, it is necessary to utilize a correction factor such that the heat transfer, q, can be determined by

$$q = UAF \, \Delta T_m$$

Here the value of ΔT_m is computed assuming counterflow conditions, i.e., $\Delta T_1 = T_{h,i} - T_{c,i}$ and $\Delta T_2 = T_{h,o} - T_{c,o}$. Figures 16 and 17 illustrate some examples of the *correction factor* F for various multiple-pass heat exchangers.

3 RADIATION HEAT TRANSFER

Heat transfer can occur in the absence of a participating medium through the transmission of energy by electromagnetic waves, characterized by a wavelength, λ, and frequency, v, which are related by $c = \lambda v$. The parameter c represents the velocity of light, which in a vacuum is $c_o = 2.9979 \times 10^8$ m/sec. Energy transmitted in this fashion is referred to as *radiant energy* and the heat-transfer process that occurs is called radiation heat transfer or simply *radiation*. In this mode of heat transfer, the energy is transferred through electromagnetic waves or through photons, with the energy of a photon being given by hv, where h represents Planck's constant.

In nature, every substance has a characteristic wave velocity that is smaller than that occurring in a vacuum. These velocities can be related to c_o by $c = c_o/n$, where n indicates

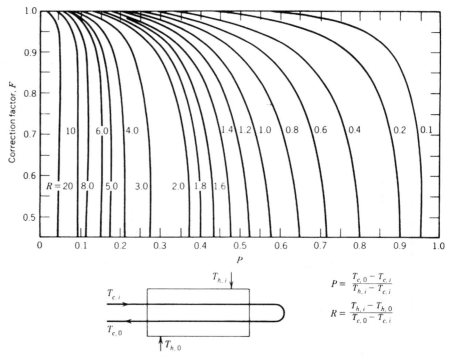

Figure 16 Correction factor for a shell-and-tube heat exchanger with one shell and any multiple of two tube passes (two, four, etc., tube passes).

the refractive index. The value of the refractive index, n, for air is approximately equal to 1. The wavelength of the energy given or for the radiation that comes from a surface depends on the nature of the source and various wavelengths sensed in different ways. For example, as shown in Fig. 18, the electromagnetic spectrum consists of a number of different types of radiation. Radiation in the visible spectrum occurs in the range $\lambda = 0.4$–0.74 μm, while radiation in the wavelength range 0.1–100 μm is classified as *thermal radiation* and is sensed as heat. For radiant energy in this range, the amount of energy given off is governed by the temperature of the emitting body.

3.1 Blackbody Radiation

All objects in space are continuously being bombarded by radiant energy of one form or another and all of this energy is either absorbed, reflected, or transmitted. An ideal body that absorbs all the radiant energy falling upon it, regardless of the wavelength and direction, is referred to as a *blackbody*. Such a body emits maximum energy for a prescribed temperature and wavelength. Radiation from a blackbody is independent of direction and is referred to as a *diffuse emitter*.

The Stefan-Boltzmann Law
The Stefan-Boltzmann law describes the rate at which energy is radiated from a blackbody and states that this radiation is proportional to the fourth power of the absolute temperature of the body,

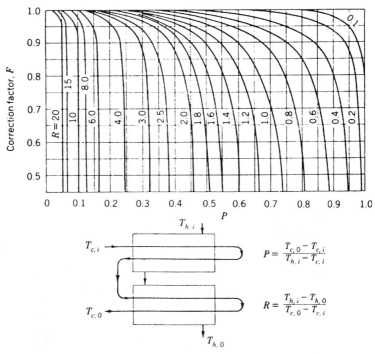

Figure 17 Correction factor for a shell-and-tube heat exchanger with two shell passes and any multiple of four tubes passes (four, eight, etc., tube passes).

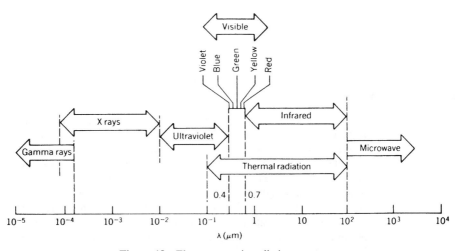

Figure 18 Electromagnetic radiation spectrum.

$$e_b = \sigma T^4$$

where e_b is the *total emissive power* and σ is the Stefan-Boltzmann constant, which has the value 5.729×10^{-8} W/m$^2 \cdot$K^4 (0.173×10^{-8} Btu/hr$\cdot$ft$^2 \cdot$°R^4).

Planck's Distribution Law

The temperature amount of energy leaving a blackbody is described as the *spectral emissive power*, $e_{\lambda b}$, and is a function of wavelength. This function, which was derived from quantum theory by Planck, is

$$e_{\lambda b} = \frac{2\pi C_1}{\lambda^5 [\exp(C_2/\lambda T) - 1]}$$

where $e_{\lambda b}$ has a unit W/m$^2 \cdot \mu$m (Btu/hr$\cdot$ft$^2 \cdot \mu$m).

Values of the constants C_1 and C_2 are 0.59544×10^{-16} W$\cdot$m^2 (0.18892×10^8 Btu$\cdot \mu$m^4/hr ft^2) and $14,388 \ \mu$m$\cdot$K ($25,898 \ \mu$m$\cdot$°R), respectively. The distribution of the spectral emissive power from a blackbody at various temperatures is shown in Fig. 19, which shows that the energy emitted at all wavelengths increases as the temperature increases. The maximum or peak values of the constant temperature curves illustrated in Fig. 20 shift to the left for shorter wavelengths as the temperatures increase.

The fraction of the emissive power of a blackbody at a given temperature and in the wavelength interval between λ_1 and λ_2 can be described by

$$F_{\lambda_1 T - \lambda_2 T} = \frac{1}{\sigma T^4} \left(\int_0^{\lambda_1} e_{\lambda b} \, d\lambda - \int_0^{\lambda_2} e_{\lambda b} \, d\lambda \right) = F_{o - \lambda_1 T} - F_{o - \lambda_2 T}$$

where the function $F_{o - \lambda T} = (1/\sigma T^4) \int_o^{\lambda} e_{\lambda b} \, d\lambda$ is given in Table 16. This function is useful for the evaluation of total properties involving integration on the wavelength in which the spectral properties are piecewise constant.

Wien's Displacement Law

The relationship between these peak or maximum temperatures can be described by *Wien's displacement law*,

$$\lambda_{\max} T = 2897.8 \ \mu\text{m} \cdot \text{K}$$

or

$$\lambda_{\max} T = 5216.0 \ \mu\text{m} \cdot \text{°R}$$

3.2   Radiation Properties

While to some degree, all surfaces follow the general trends described by the Stefan-Boltzmann and Planck laws, the behavior of real surfaces deviates somewhat from these. In fact, because blackbodies are ideal, all real surfaces emit and absorb less radiant energy than a blackbody. The amount of energy a body emits can be described in terms of the emissivity and is, in general, a function of the type of material, the temperature, and the surface conditions, such as roughness, oxide layer thickness, and chemical contamination. The emissivity is, in fact, a measure of how well a real body radiates energy as compared with a blackbody of the same temperature. The radiant energy emitted into the entire hemispherical space above a real surface element, including all wavelengths is given by $e = \varepsilon \sigma T^4$, where ε is less than 1.0 and is called the *hemispherical emissivity* (or *total hemispherical emissivity* to

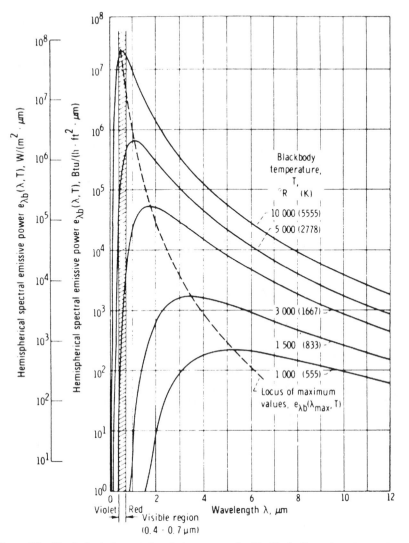

Figure 19 Hemispherical spectral emissive power of a blackbody for various temperatures.

indicate averaging over the total wavelength spectrum). For a given wavelength the *spectral hemispherical emissivity* ε_λ of a real surface is defined as

$$\varepsilon_\lambda = e_\lambda / e_{\lambda b}$$

where e_λ is the hemispherical emissive power of the real surface and $e_{\lambda b}$ is that of a blackbody at the same temperature.

  Spectral irradiation, G_λ, (W/m$^2 \cdot \mu$m), is defined as the rate at which radiation is incident upon a surface per unit area of the surface, per unit wavelength about the wavelength λ, and encompasses the incident radiation from all directions.

  Spectral hemispherical reflectivity, ρ_λ is defined as the radiant energy reflected per unit time, per unit area of the surface, per unit wavelength/G_λ.

$$dA_i \, dF_{di-dj} = dA_j \, dF_{dj-di}$$

$$dA_i F_{di-j} = A_j \, dF_{j-di}$$

$$A_i F_{i-j} = A_j F_{j-i}$$

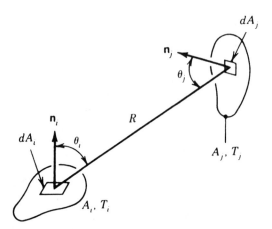

Figure 20   Configuration factor for radiation exchange between surfaces of area dA_i and dA_j.

Spectral hemispherical absorptivity, α_λ, is defined as the radiant energy absorbed per unit area of the surface, per unit wavelength about the wavelength $/G_\lambda$.

Spectral hemispherical transmissivity is defined as the radiant energy transmitted per unit area of the surface, per unit wavelength about the wavelength $/G_\lambda$.

For any surface, the sum of the reflectivity, absorptivity, and transmissivity must equal unity, i.e.,

$$\alpha_\lambda + \rho_\lambda \tau_\lambda = 1$$

When these values are averaged over the entire wavelength from $\lambda = 0$ to ∞ they are referred to as *total* values. Hence, the *total hemispherical reflectivity, total hemispherical absorptivity,* and *total hemispherical transmissivity* can be written as

$$\rho = \int_0^\infty \rho_\lambda G_\lambda \, d\lambda / G$$

$$\alpha = \int_0^\infty \alpha_\lambda G_\lambda \, d\lambda / G$$

and

$$\tau = \int_0^\infty \tau_\lambda G_\lambda \, d\lambda / G$$

respectively, where

Table 16 Radiation Function $F_{o-\lambda T}$

λT			λT			λT		
$\mu m \cdot K$	$\mu m \cdot °R$	$F_{o-\lambda T}$	$\mu m \cdot K$	$\mu m \cdot °R$	$F_{o-\lambda T}$	$\mu m \cdot K$	$\mu m \cdot °R$	$F_{o-\lambda T}$
400	720	0.1864×10^{-11}	3400	6120	0.3617	6400	11,520	0.7692
500	900	0.1298×10^{-8}	3500	6300	0.3829	6500	11,700	0.7763
600	1080	0.9290×10^{-7}	3600	6480	0.4036	6600	11,880	0.7832
700	1260	0.1838×10^{-5}	3700	6660	0.4238	6800	12,240	0.7961
800	1440	0.1643×10^{-4}	3800	6840	0.4434	7000	12,600	0.8081
900	1620	0.8701×10^{-4}	3900	7020	0.4624	7200	12,960	0.8192
1000	1800	0.3207×10^{-3}	4000	7200	0.4809	7400	13,320	0.8295
1100	1980	0.9111×10^{-3}	4100	7380	0.4987	7600	13,680	0.8391
1200	2160	0.2134×10^{-2}	4200	7560	0.5160	7800	14,040	0.8480
1300	2340	0.4316×10^{-2}	4300	7740	0.5327	8000	14,400	0.8562
1400	2520	0.7789×10^{-2}	4400	7920	0.5488	8200	14,760	0.8640
1500	2700	0.1285×10^{-1}	4500	8100	0.5643	8400	15,120	0.8712
1600	2880	0.1972×10^{-1}	4600	8280	0.5793	8600	15,480	0.8779
1700	3060	0.2853×10^{-1}	4700	8460	0.5937	8800	15,840	0.8841
1800	3240	0.3934×10^{-1}	4800	8640	0.6075	9000	16,200	0.8900
1900	3420	0.5210×10^{-1}	4900	8820	0.6209	10,000	18,000	0.9142
2000	3600	0.6673×10^{-1}	5000	9000	0.6337	11,000	19,800	0.9318
2100	3780	0.8305×10^{-1}	5100	9180	0.6461	12,000	21,600	0.9451
2200	3960	0.1009	5200	9360	0.6579	13,000	23,400	0.9551
2300	4140	0.1200	5300	9540	0.6694	14,000	25,200	0.9628
2400	4320	0.1402	5400	9720	0.6803	15,000	27,000	0.9689
2500	4500	0.1613	5500	9900	0.6909	20,000	36,000	0.9856
2600	4680	0.1831	5600	10,080	0.7010	25,000	45,000	0.9922
2700	4860	0.2053	5700	10,260	0.7108	30,000	54,000	0.9953
2800	5040	0.2279	5800	10,440	0.7201	35,000	63,000	0.9970
2900	5220	0.2505	5900	10,620	0.7291	40,000	72,000	0.9979
3000	5400	0.2732	6000	10,800	0.7378	45,000	81,000	0.9985
3100	5580	0.2958	6100	10,980	0.7461	50,000	90,000	0.9989
3200	5760	0.3181	6200	11,160	0.7541	55,000	99,000	0.9992
3300	5940	0.3401	6300	11,340	0.7618	60,000	108,000	0.9994

$$G = \int_0^\infty G_\lambda \, d\lambda$$

As was the case for the wavelength-dependent parameters, the sum of the total reflectivity, total absorptivity, and total transmissivity must be equal to unity, i.e.,

$$\alpha + \rho + \tau = 1$$

It is important to note that while the emissivity is a function of the material, temperature, and surface conditions, the absorptivity and reflectivity depend on both the surface characteristics and the nature of the incident radiation.

The terms *reflectance, absorptance,* and *transmittance* are used by some authors for the real surfaces and the terms reflectivity, absorptivity, and transmissivity are reserved for the properties of the ideal surfaces (i.e., those optically smooth and pure substances perfectly uncontaminated). Surfaces that allow no radiation to pass through are referred to as *opaque,* i.e., $\tau_\lambda = 0$, and all of the incident energy will be either reflected or absorbed. For such a surface,

$$\alpha_\lambda + \rho_\lambda = 1$$

and

$$\alpha + \rho = 1$$

Light rays reflected from a surface can be reflected in such a manner that the incident and reflected rays are symmetric with respect to the surface normal at the point of incidence. This type of radiation is referred to as *specular*. The radiation is referred to as *diffuse* if the intensity of the reflected radiation is uniform over all angles of reflection and is independent of the incident direction, and the surface is called a *diffuse surface* if the radiation properties are independent of the direction. If they are independent of the wavelength, the surface is called a *gray surface,* and a *diffuse-gray surface* absorbs a fixed fraction of incident radiation from any direction and at any wavelength, and $\alpha_\lambda = \varepsilon_\lambda = \alpha = \varepsilon$.

Kirchhoff's Law of Radiation

The directional characteristics can be specified by the addition of a prime to the value; for example, the spectral emissivity for radiation in a particular direction would be denoted by α_λ'. For radiation in a particular direction, the spectral emissivity is equal to the directional spectral absorptivity for the surface irradiated by a blackbody at the same temperature. The most general form of this expression states that $\alpha_\lambda' = \varepsilon_\lambda'$. If the incident radiation is independent of angle or if the surface is diffuse, then $\alpha_\lambda = \varepsilon_\lambda$ for the hemispherical properties. This relationship can have various conditions imposed on it, depending on whether spectral, total, directional, or hemispherical quantities are being considered.[19]

Emissivity of Metallic Surfaces

The properties of pure smooth metallic surfaces are often characterized by low emissivity and absorptivity values and high values of reflectivity. The spectral emissivity of metals tends to increase with decreasing wavelength, and exhibits a peak near the visible region. At wavelengths $\lambda > {\sim}5\ \mu m$ the spectral emissivity increases with increasing temperature, but this trend reverses at shorter wavelengths ($\lambda < {\sim}1.27\ \mu m$). Surface roughness has a pronounced effect on both the hemispherical emissivity and absorptivity, and large *optical roughnesses,* defined as the mean square roughness of the surface divided by the wavelength, will increase the hemispherical emissivity. For cases where the optical roughness is small, the directional properties will approach the values obtained for smooth surfaces. The presence of impurities, such as oxides or other nonmetallic contaminants, will change the properties significantly and increase the emissivity of an otherwise pure metallic body. A summary of the normal total emissivities for metals are given in Table 17. It should be noted that the hemispherical emissivity for metals is typically 10–30% higher than the values normally encountered for normal emissivity.

Emissivity of Nonmetallic Materials

Large values of total hemispherical emissivity and absorptivity are typical for nonmetallic surfaces at moderate temperatures and, as shown in Table 18, which lists the normal total emissivity of some nonmetals, the temperature dependence is small.

Absorptivity for Solar Incident Radiation

The spectral distribution of solar radiation can be approximated by blackbody radiation at a temperature of approximately 5800 K (10,000°R) and yields an average solar irradiation at

Table 17 Normal Total Emissivity of Metals[a]

Materials	Surface Temperature (K)	Normal Total Emissivity
Aluminum		
Highly polished plate	480–870	0.038–0.06
Polished plate	373	0.095
Heavily oxidized	370–810	0.20–0.33
Bismuth, bright	350	0.34
Chromium, polished	310–1370	0.08–0.40
Copper		
Highly polished	310	0.02
Slightly polished	310	0.15
Black oxidized	310	0.78
Gold, highly polished	370–870	0.018–0.035
Iron		
Highly polished, electrolytic	310–530	0.05–0.07
Polished	700–760	0.14–0.38
Wrought iron, polished	310–530	0.28
Cast iron, rough, strongly oxidized	310–530	0.95
Lead		
Polished	310–530	0.06–0.08
Rough unoxidized	310	0.43
Mercury, unoxidized	280–370	0.09–0.12
Molybdenum, polished	310–3030	0.05–0.29
Nickel		
Electrolytic	310–530	0.04–0.06
Electroplated on iron, not polished	293	0.11
Nickel oxide	920–1530	0.59–0.86
Platinum, electrolytic	530–810	0.06–0.10
Silver, polished	310–810	0.01–0.03
Steel		
Polished sheet	90–420	0.07–0.14
Mild steel, polished	530–920	0.27–0.31
Sheet with rough oxide layer	295	0.81
Tin, polished sheet	310	0.05
Tungsten, clean	310–810	0.03–0.08
Zinc		
Polished	310–810	0.02–0.05
Gray oxidized	295	0.23–0.28

[a] Adapted from Ref. 19.

the outer limit of the atmosphere of approximately 1353 W/m^2 (429 Btu/ft^2·hr). This solar irradiation is called the *solar constant* and is greater than the solar irradiation received at the surface of the earth, due to the radiation scattering by air molecules, water vapor, and dust, and the absorption by O_3, H_2O, and CO_2 in the atmosphere. The absorptivity of a substance depends not only on the surface properties but also on the sources of incident radiation. Since solar radiation is concentrated at a shorter wavelength, due to the high source temperature, the absorptivity for certain materials when exposed to solar radiation may be quite different from that which occurs for low temperature radiation, where the radiation is

Table 18 Normal Total Emissivity of Nonmetals[a]

Materials	Surface Temperature (K)	Normal Total Emissivity
Asbestos, board	310	0.96
Brick		
White refractory	1370	0.29
Rough Red	310	0.93
Carbon, lampsoot	310	0.95
Concrete, rough	310	0.94
Ice, smooth	273	0.966
Magnesium oxide, refractory	420–760	0.69–0.55
Paint		
Oil, all colors	373	0.92–0.96
Lacquer, flat black	310–370	0.96–0.98
Paper, white	310	0.95
Plaster	310	0.91
Porcelain, glazed	295	0.92
Rubber, hard	293	0.92
Sandstone	310–530	0.83–0.90
Silicon carbide	420–920	0.83–0.96
Snow	270	0.82
Water, deep	273–373	0.96
Wood, sawdust	310	0.75

[a] Adapted from Ref. 19.

concentrated in the longer-wavelength range. A comparison of absorptivities for a number of different materials is given in Table 19 for both solar and low-temperature radiation.

3.3 Configuration Factor

The magnitude of the radiant energy exchanged between any two given surfaces is a function of the emisssivity, absorptivity, and transmissivity. In addition, the energy exchange is a strong function of how one surface is viewed from the other. This aspect can be defined in terms of the *configuration factor* (sometimes called the *radiation shape factor,, view factor, angle factor, or interception factor*). As shown in Fig. 20, the configuration factor, F_{i-j}, is defined as that fraction of the radiation leaving a black surface, i, that is intercepted by a black or gray surface, j, and is based on the relative geometry, position, and shape of the two surfaces. The configuration factor can also be expressed in terms of the differential fraction of the energy or dF_{i-dj}, which indicates the differential fraction of energy from a finite area A_i that is intercepted by an infinitesimal area dA_j. Expressions for a number of different cases are given below for several common geometries.

Infinitesimal area dA_j to infinitesimal area dA_j

$$dF_{di-dj} = \frac{\cos \theta_i \cos \theta_j}{\pi R^2} dA_j$$

Infinitesimal area dA_j to finite area A_j

Table 19 Comparison of Absorptivities of Various Surfaces to Solar and Low-Temperature Thermal Radiation[a]

Surface	For Solar Radiation	For Low-Temperature Radiation ($\sim$300 K)
		Absorptivity
Aluminum, highly polished	0.15	0.04
Copper, highly polished	0.18	0.03
Tarnished	0.65	0.75
Cast iron	0.94	0.21
Stainless steel, No. 301, polished	0.37	0.60
White marble	0.46	0.95
Asphalt	0.90	0.90
Brick, red	0.75	0.93
Gravel	0.29	0.85
Flat black lacquer	0.96	0.95
White paints, various types of pigments	0.12–0.16	0.90–0.95

[a] Adapted from Ref. 20 after J. P. Holman, *Heat Transfer,* McGraw-Hill, New York, 1981.

$$F_{di-j} = \int_{Aj} \frac{\cos \theta_i \cos \theta_j}{\pi R^2} \, dA_j$$

Finite area A_i to finite area A_j

$$F_{i-j} = \frac{1}{A_i} \int_{Aj} \int_{Aj} \frac{\cos \theta_i \cos \theta_j}{\pi R^2} \, dA_i \, dA_j$$

Analytical expressions of other configuration factors have been found for a wide variety of simple geometries and a number of these are presented in Figs. 21–24 for surfaces that emit and reflect diffusely.

Reciprocity Relations

The configuration factors can be combined and manipulated using algebraic rules referred to as configuration factor geometry. These expressions take several forms, one of which is the reciprocal properties between different configuration factors, which allow one configuration factor to be determined from knowledge of the others:

$$dA_i \, dF_{di-dj} = dA_j \, dF_{dj-di}$$

$$dA_i \, dF_{di-j} = A_j \, dF_{j-di}$$

$$A_i F_{i-j} = A_j F_{j-i}$$

These relationships can be combined with other basic rules to allow the determination of the configuration of an infinite number of complex shapes and geometries form a few select, known geometries. These are summarized in the following sections.

The Additive Property

For a surface A_i subdivided into N parts ($A_{i_1}, A_{i_2}, \ldots, A_{i_N}$) and a surface A_j subdivided into M parts ($A_{j_1}, A_{j_2}, \ldots, A_{j_M}$),

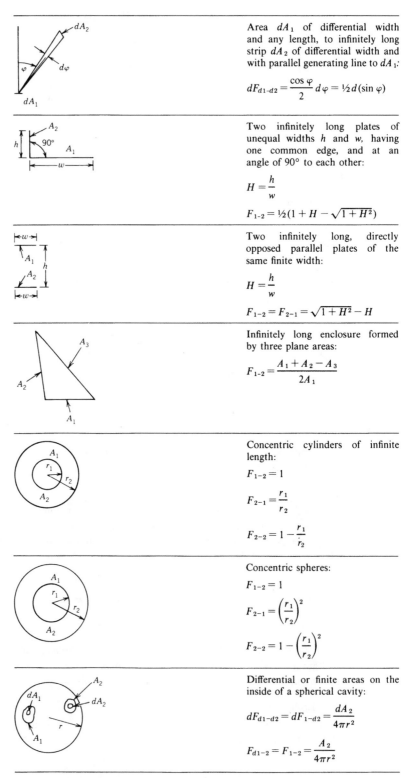

Area dA_1 of differential width and any length, to infinitely long strip dA_2 of differential width and with parallel generating line to dA_1:

$$dF_{d1\text{-}d2} = \frac{\cos \varphi}{2} d\varphi = \tfrac{1}{2} d(\sin \varphi)$$

Two infinitely long plates of unequal widths h and w, having one common edge, and at an angle of 90° to each other:

$$H = \frac{h}{w}$$

$$F_{1\text{-}2} = \tfrac{1}{2}(1 + H - \sqrt{1+H^2})$$

Two infinitely long, directly opposed parallel plates of the same finite width:

$$H = \frac{h}{w}$$

$$F_{1\text{-}2} = F_{2\text{-}1} = \sqrt{1+H^2} - H$$

Infinitely long enclosure formed by three plane areas:

$$F_{1\text{-}2} = \frac{A_1 + A_2 - A_3}{2A_1}$$

Concentric cylinders of infinite length:

$$F_{1\text{-}2} = 1$$

$$F_{2\text{-}1} = \frac{r_1}{r_2}$$

$$F_{2\text{-}2} = 1 - \frac{r_1}{r_2}$$

Concentric spheres:

$$F_{1\text{-}2} = 1$$

$$F_{2\text{-}1} = \left(\frac{r_1}{r_2}\right)^2$$

$$F_{2\text{-}2} = 1 - \left(\frac{r_1}{r_2}\right)^2$$

Differential or finite areas on the inside of a spherical cavity:

$$dF_{d1\text{-}d2} = dF_{1\text{-}d2} = \frac{dA_2}{4\pi r^2}$$

$$F_{d1\text{-}2} = F_{1\text{-}2} = \frac{A_2}{4\pi r^2}$$

Figure 21 Configuration factors for some simple geometries.[19]

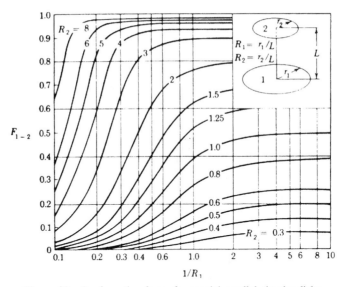

Figure 22 Configuration factor for coaxial parallel circular disks.

$$A_i F_{i-j} = \sum_{n=1}^{N} \sum_{m=1}^{M} A_{i_n} F_{i_n - j_m}$$

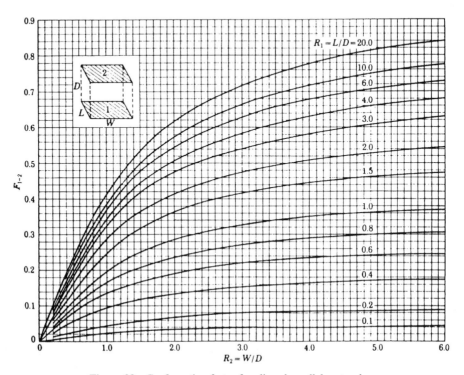

Figure 23 Configuration factor for aligned parallel rectangles.

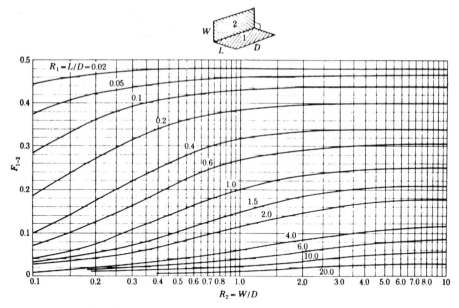

Figure 24   Configuration factor for rectangles with common edge.

$$A_i F_{i-j} = \sum_{n=1}^{N} \sum_{m=1}^{M} A_{i_n} F_{i_n - j_m}$$

Relation in an Enclosure

When a surface is completely enclosed, the surface can be subdivided into N parts having areas $A_1, A_2, \ldots, A_N$, respectively, and

$$\sum_{j=1}^{N} F_{i-j} = 1$$

Blackbody Radiation Exchange

For black surfaces A_i, and A_j at temperatures T_i and T_j, respectively, the net radiative exchange, q_{ij}, can be expressed as

$$q_{ij} = A_i F_{i-j} \sigma (T_i^4 - T_j^4)$$

and for a surface completely enclosed and subdivided into N surfaces maintained at temperatures $T_1, T_2, \ldots, T_N$, the net radiative heat transfer, q_i, to surface area A_i is

$$q_i = \sum_{j=1}^{N} A_i F_{i-j} \sigma (T_i^4 - T_j^4) = \sum_{j=1}^{N} q_{ij}$$

3.4   Radiative Exchange among Diffuse-Gray Surfaces in an Enclosure

One method for solving for the radiation exchange between a number of surfaces or bodies is through the use of the *radiocity, J,* defined as the total radiation that leaves a surface per unit time and per unit area. For an opaque surface, this term is defined as

$$J = \varepsilon \sigma T^4 + (1 - \varepsilon)G$$

For an enclosure consisting of N surfaces, the irradiation on a given surface i can be expressed as

$$G_i = \sum_{j=1}^{N} J_j F_{i-j}$$

and the net radiative heat-transfer rate at given surface i is

$$q_i = A_i(J_i - G_i) = \frac{\varepsilon_i A_i}{1 - \varepsilon_i}(\sigma T_i^4 - J_i)$$

For every surface in the enclosure, a uniform temperature or a constant heat transfer rate can be specified. If the surface temperature is given, the heat-transfer rate can be determined for that surface and vice versa. Shown below, are several specific cases that are commonly encountered.

Case I. The temperatures of the surfaces, T_i ($i = 1, 2, \ldots, N$), are known for each of the N surfaces and the values of the radiocity, J_i, are solved from the expression

$$\sum_{j=1}^{N} \{\delta_{ij} - (1 - \varepsilon_i)F_{i-j}\}J_i = \varepsilon_i \sigma T_i^4, \qquad 1 \le i \le N$$

The net heat-transfer rate to surface i can then be determined from the fundamental relationship

$$q_i = A_i \frac{\varepsilon_i}{1 - \varepsilon_i}(\sigma T_i^4 - J_i), \qquad 1 \le i \le N$$

where $\delta_{ij} = 0$ for $i \ne j$ and $\delta_{ij} = 1$ for $i = j$.

Case II. The heat-transfer rates, q_i ($i = 1, 2, \ldots, N$), to each of the N surfaces are known and the values of the radiocity, J_i, are determined from

$$\sum_{j=1}^{N} \{\delta_{ij} - F_{i-j}\}J_j = q_i/A_i \qquad 1 \le i \le N$$

The surface temperature can then be determined from

$$T_i = \left[\frac{1}{\sigma}\left(\frac{1 - \varepsilon_i}{\varepsilon_i}\frac{q_i}{A_i} + J_i\right)\right]^{1/4} \qquad 1 \le i \le N$$

Case III. The temperatures, T_i ($i = 1, \ldots, N_1$), for N_i surfaces and heat-transfer rates q_i ($i = N_1 + 1, \ldots, N$) for ($N - N_i$) surfaces are known and the radiocities are determined by

$$\sum_{j=1}^{N} \{\delta_{ij} - (1 - \varepsilon_i)F_{i-j}\}J_j = \varepsilon_i \alpha T_i^4 \qquad 1 \le i \le N_1$$

$$\sum_{j=1}^{N} \{\delta_{ij} - F_{i-j}\}J_j = \frac{q_i}{A_i} \qquad N_1 + 1 \le i \le N$$

The net heat-transfer rates and temperatures van be found as

$$q_i = A_i \frac{\varepsilon_i}{1 - \varepsilon_i} (\sigma T_i^4 - J_i) \qquad 1 \le i \le N_1$$

$$T_i = \left[\frac{1}{\sigma} \left(\frac{1 - \varepsilon_i}{\varepsilon_i} \frac{q_i}{A_i} + J_i \right) \right]^{1/4} \qquad N_1 + 1 \le i \le N$$

Two Diffuse-Gray Surfaces Forming an Enclosure

The net radiative exchange, q_{12}, for two diffuse-gray surfaces forming an enclosure are shown in Table 20 for several simple geometries.

Radiation Shields

Often in practice, it is desirable to reduce the radiation heat transfer between two surfaces. This can be accomplished by placing a highly reflective surface between the two surfaces. For this configuration, the ratio of the net radiative exchange with the shield to that without the shield can be expressed by the relationship

$$\frac{q_{12 \text{ with shield}}}{q_{12 \text{ without shield}}} = \frac{1}{1 + \chi}$$

Values for this ratio, χ, for shields between parallel plates, concentric cylinders, and concentric spheres are summarized in Table 21. For the special case of parallel plates involving more than one or N shields, where all of the emissivities are equal, the value of χ equals N.

Radiation Heat-Transfer Coefficient

The rate at which radiation heat transfer occurs can be expressed in a form similar to Fourier's law or Newton's law of cooling, by expressing it in terms of the temperature difference $T_1 - T_2$, or as

Table 20   Net Radiative Exchange between Two Surfaces Forming an Enclosure

Large (infinite) parallel planes $\underline{A_1, T_1, \varepsilon_1}$ A_2, T_2, ε_2	$A_1 = A_2 = A$	$q_{12} = \dfrac{A\sigma(T_1^4 - T_2^4)}{\dfrac{1}{\varepsilon_1} + \dfrac{1}{\varepsilon_2} - 1}$
Long (infinite) concentric cylinders 	$\dfrac{A_1}{A_2} = \dfrac{r_1}{r_2}$	$q_{12} = \dfrac{\sigma A_1(T_1^4 - T_2^4)}{\dfrac{1}{\varepsilon_1} + \dfrac{1 - \varepsilon_2}{\varepsilon_2} \left(\dfrac{r_1}{r_2} \right)}$
Concentric sphere 	$\dfrac{A_1}{A_2} = \dfrac{r_1^2}{r_2^2}$	$q_{12} = \dfrac{\sigma A_1(T_1^4 - T_2^4)}{\dfrac{1}{\varepsilon_1} + \dfrac{1 - \varepsilon_2}{\varepsilon_2} \left(\dfrac{r_1}{r_2} \right)^2}$
Small convex object in a large cavity 	$\dfrac{A_1}{A_2} \approx 0$	$q_{12} = \sigma A_1 \varepsilon_1 (T_1^4 - T_2^4)$

Table 21 Values of X for Radiative Shields

Geometry		X
Shield	$$\dfrac{\dfrac{1}{\varepsilon_{s1}} + \dfrac{1}{\varepsilon_{s2}} - 1}{\dfrac{1}{\varepsilon_1} + \dfrac{1}{\varepsilon_2} - 1}$$	Infinitely long parallel plates
	$$\dfrac{\left(\dfrac{r_1}{r_2}\right)^2 \left(\dfrac{1}{\varepsilon_{s1}} + \dfrac{1}{\varepsilon_{s2}} - 1\right)}{\dfrac{1}{\varepsilon_1} + \left(\dfrac{1}{\varepsilon_2} - 1\right)\left(\dfrac{r_1}{r_2}\right)^2}$$	$n = 1$ for infinitely long concentric cylinders $n = 2$ for concentric spheres

$$q = h_r A (T_1 - T_2)$$

where h_r is the radiation heat-transfer coefficient or *radiation film coefficient*. For the case of radiation between two large parallel plates with emissivities, respectively, of ε_1 and ε_2,

$$h_r \frac{\sigma(T_1^4 - T_2^4)}{T_1 - T_2 \left(\dfrac{1}{\varepsilon_1} + \dfrac{1}{\varepsilon_2} - 1\right)}$$

3.5 Thermal Radiation Properties of Gases

All of the previous expressions assumed that the medium present between the surfaces did not affect the radiation exchange. In reality, gases such as air, oxygen (O_2), hydrogen (H_2), and nitrogen (N_2) have a symmetrical molecular structure and neither emit nor absorb radiation at low to moderate temperatures. Hence, for most engineering applications, such *nonparticipating gases* can be ignored. However, polyatomic gases such as water vapor (H_2O), carbon dioxide (CO_2), carbon monoxide (CO), sulfur dioxide (SO_2), and various hydrocarbons, emit and absorb significant amounts of radiation. These *participating gases* absorb and emit radiation in limited spectral ranges, referred to as spectral *bands*. In calculating the emitted or absorbed radiation for a gas layer, its thickness, shape, surface area, pressure, and temperature distribution must be considered. Although a precise method for calculating the effect of these participating media is quite complex, an approximate method developed by Hottel[21] will yield results that are reasonably accurate.

The effective total emissivities of carbon dioxide and water vapor are a function of the temperature and the product of the partial pressure and the mean beam length of the substance as indicated in Figs. 25 and 26, respectively. The *mean beam length, L_e,* is the characteristic length that corresponds to the radius of a hemisphere of gas, such that the energy flux radiated to the center of the base is equal to the average flux radiated to the area of interest by the actual gas volume. Table 22 lists the mean beam lengths of several simple

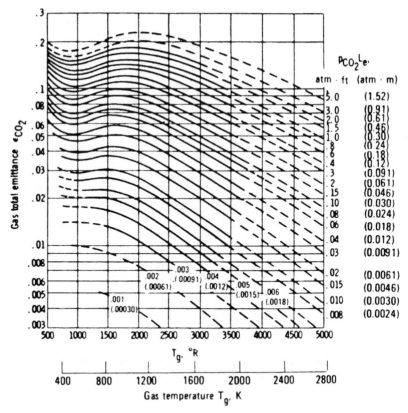

Figure 25 Total emissivity of CO_2 in a mixture having a total pressure of 1 atm. (From Ref. 21. Used with the permission of McGraw-Hill Book Company.)

shapes. For a geometry for which L_e has not been determined, it is generally approximated by $L_e = 3.6V/A$ for an entire gas volume V radiating to its entire boundary surface A. The data in Figs. 25 and 26 were obtained for a total pressure of 1 atm and zero partial pressure of the water vapor. For other total and partial pressures the emissivities are corrected by multiplying C_{CO_2} (Fig. 27) and C_{H_2O} (Fig. 28), respectively, to ε_{CO_2} and ε_{H_2O} which are found from Figs. 25 and 26.

These results can be applied when water vapor or carbon dioxide appear separately or in a mixture with other nonparticipating gases. For mixtures of CO_2 and water vapor in a nonparticipating gas, the total emissivity of the mixture, ε_g, can be estimated from the expression

$$\varepsilon_g = C_{CO_2}\varepsilon_{CO_2} + C_{H_2O}\varepsilon_{H_2O} - \Delta\varepsilon$$

where $\Delta\varepsilon$ is a correction factor given in Fig. 29.

Radiative Exchange between Gas Volume and Black Enclosure of Uniform Temperature
When radiative energy is exchanged between a gas volume and a black enclosure, the exchange per unit area, q'', for a gas volume at uniform temperature, T_g, and a uniform wall temperature, T_w, is given by

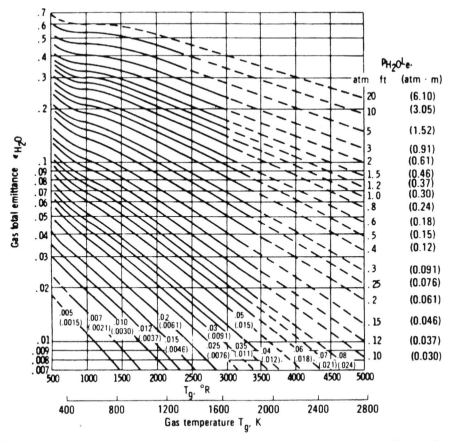

Figure 26 Total emissivity of H_2O at 1 atm total pressure and zero partial pressure (From Ref. 21. Used with the permission of McGraw-Hill Book Company.)

$$q'' = \varepsilon_g(T_g)\sigma T_g^4 - \alpha_g(T_w)\sigma T_w^4$$

where $\varepsilon_g(T_g)$ is the gas emissivity at a temperature T_g and $\alpha_g(T_w)$ is the absorptivity of gas for the radiation from the black enclosure at T_w. As a result of the nature of the band structure of the gas, the absorptivity, α_g, for black radiation at a temperature T_w is different from the emissivity, ε_g, at a gas temperature of T_g. When a mixture of carbon dioxide and water vapor is present, the empirical expression for α_g is

$$\alpha_g = \alpha_{CO_2} + \alpha_{H_2O} - \Delta\alpha$$

where

$$\alpha_{CO_2} = C_{CO_2}\varepsilon'_{CO_2}\left(\frac{T_g}{T_w}\right)^{0.65}$$

$$\alpha_{H_2O} = C_{H_2O}\varepsilon'_{H_2O}\left(\frac{T_g}{T_w}\right)^{0.45}$$

where $\Delta\alpha = \Delta\varepsilon$ and all properties are evaluated at T_w.

Table 22 Mean Beam Length[a]

Geometry of Gas Volume	Characteristic Length	L_e
Hemisphere radiating to element at center of base	Radius R	R
Sphere radiating to its surface	Diameter D	$0.65D$
Circular cylinder of infinite height radiating to concave bounding surface	Diameter D	$0.95D$
Circular cylinder of semi-infinite height radiating to:		
Element at center of base	Diameter D	$0.90D$
Entire base	Diameter D	$0.65D$
Circular cylinder of height equal to diameter radiating to:		
Element at center of base	Diameter D	$0.71D$
Entire surface	Diameter D	$0.60D$
Circular cylinder of height equal to two diameters radiating to:		
Plane end	Diameter D	$0.60D$
Concave surface	Diameter D	$0.76D$
Entire surface	Diameter D	$0.73D$
Infinite slab of gas radiating to:		
Element on one face	Slab thickness D	$1.8D$
Both bounding planes	Slab thickness D	$1.8D$
Cube radiating to a face	Edge X	$0.6X$
Gas volume surrounding an infinite tube bundle and radiating to a single tube:		
Equilateral triangular array:		
$S = 2D$	Tube diameter D and	$3.0(S - D)$
$S = 3D$	spacing between	$3.8(S - D)$
	tube centers, S	
Square array:		$3.5(S - D)$
$S = 2D$		

[a] Adapted from Ref. 19.

In this expression, the values of ε'_{CO_2} and ε'_{H_2O} can be found from Figs. 25 and 26 using an abscissa of T_w, but substituting the parameters $p_{CO_2}L_eT_w/T_g$ and $p_{H_2O}L_eT_w/T_g$ for $p_{CO_2}L_e$ and $p_{H_2O}L_e$, respectively.

Radiative Exchange between a Gray Enclosure and a Gas Volume

When the emissivity of the enclosure, ε_w, is larger than 0.8, the rate of heat transfer may be approximated by

$$q_{gray} = \left(\frac{\varepsilon_w + 1}{2} \right) q_{black}$$

where q_{gray} is the heat-transfer rate for gray enclosure and q_{black} is that for black enclosure. For values of $\varepsilon_w < 0.8$, the band structures of the participating gas must be taken into account for heat-transfer calculations.

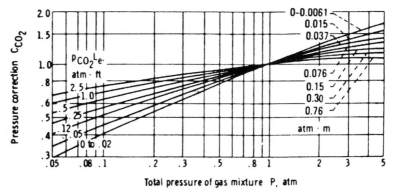

Figure 27 Pressure correction for CO_2 total emissivity for values of P other than 1 atm. (Adapted from Ref. 21. Used with the permission of McGraw-Hill Book Company.)

4 BOILING AND CONDENSATION HEAT TRANSFER

Boiling and condensation are both forms of convection in which the fluid medium is undergoing a change of phase. When a liquid comes into contact with a solid surface maintained at a temperature above the saturation temperature of the liquid, the liquid may vaporize, resulting in boiling. This process is always accompanied by a change of phase, from the liquid to the vapor state, and results in large rates of heat transfer from the solid surface, due to the latent heat of vaporization of the liquid. The process of condensation is usually accomplished by allowing the vapor to come into contact with a surface at a temperature below the saturation temperature of the vapor, in which case the liquid undergoes a change in state from the vapor state to the liquid state, giving up the latent heat of vaporization.

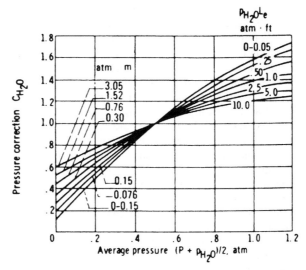

Figure 28 Pressure correction for water vapor total emissivity for values of P_{H_2O} and P other than 0 and 1 atm. (Adapted from Ref. 21. Used with the permission of McGraw-Hill Book Company.)

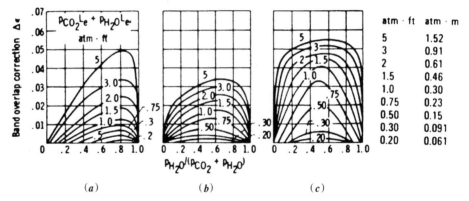

Figure 29   Correction on total emissivity for band overlap when both CO_2 and water vapor are present: (a) gas temperature $T_g = 400$ K (720°R); (b) gas temperature $T_g = 810$ K (1460°R); (c) gas temperature $T_g = 1200$ K (2160°R). (Adapted from Ref. 21. Used with the permission of McGraw-Hill Book Company.)

The heat-transfer coefficients for condensation and boiling are generally larger than that for convection without phase change, sometimes by as much as several orders of magnitude. Application of boiling and condensation heat transfer may be seen in a closed-loop power cycle or in a device referred to as a *heat pipe* which will be discussed in the following section. In power cycles, the liquid is vaporized in a boiler at high pressure and temperature. After producing work by means of expansion through a turbine, the vapor is condensed to the liquid state in a condenser, and then returned to the boiler where the cycle is repeated.

4.1   Boiling

The formation of vapor bubbles on a hot surface in contact with a quiescent liquid without external agitation, it is called *pool boiling*. This differs from *forced-convection boiling,* in which forced convection occurs simultaneously with boiling. When the temperature of the liquid is below the saturation temperature, the process is referred to as *subcooled boiling*. When the liquid temperature is maintained or exceeds the saturation temperature, the process is referred to as *saturated or saturation boiling*. Figure 30 depicts the surface heat flux, q'', as a function of the excess temperature, $\Delta T_e = T_s - T_{\text{sat}}$, for typical pool boiling of water using an electrically heated wire. In the region $0 < \Delta T_e < \Delta T_{e,A}$ bubbles occur only on selected spots of the heating surface, and the heat transfer occurs primarily through free convection. This process is called *free convection boiling*. When $\Delta T_{e,A} < \Delta T_e < \Delta T_{e,C}$, the heated surface is densely populated with bubbles, and the bubble separation and eventual rise due to buoyancy induce a considerable stirring action in the fluid near the surface. This stirring action substantially increases the heat transfer from the solid surface. This process or region of the curve is referred to as *nucleate boiling*. When the excess temperature is raised to $\Delta T_{e,C}$, the heat flux reaches a maximum value, and further increases in the temperature will result in a decrease in the heat flux. The point at which the heat flux is at a maximum value, is called the *critical heat flux*.

Film boiling occurs in the region where $\Delta T_e > \Delta T_{e,D}$, and the entire heating surface is covered by a vapor film. In this region the heat transfer to the liquid is caused by conduction and radiation through the vapor. Between points C and D, the heat flux decreases with

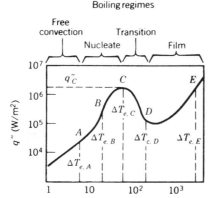

Figure 30   Typical boiling curve for a wire in a pool of water at atmospheric pressure.

increasing ΔT_e. In this region, part of the surface is covered by bubbles and part by a film. the vaporization in this region is called *transition boiling* or *partial film boiling*. The point of maximum heat flux, point C, is called the *burnout point* or the *Lindenfrost point*. Although it is desirable to operate the vapor generators at heat fluxes close to q_c'', to permit the maximum use of the surface area, in most engineering applications it is necessary to control the heat flux and great care is taken to avoid reaching this point. The primary reason for this is that, as illustrated, when the heat flux is increased gradually, the temperature rises steadily until point C is reached. Any increase of heat flux beyond the value of q_c'', however, will dramatically change the surface temperature to $T_s = T_{\text{sat}} + T_{e,E}$, typically exceeding the solid melting point and leading to failure of the material in which the liquid is held or from which the heater is fabricated.

Nucleate Pool Boiling
The heat flux data are best correlated by[26]

$$q'' = \mu_l h_{fg} \left[\frac{g(\rho_l - \rho_v)}{g_c \sigma} \right]^{1/2} \left(\frac{c_{p,l}\Delta T_e}{Ch_{fg}\,\mathrm{Pr}_l^{1.7}} \right)^3$$

where the subscripts l and v denote saturated liquid and vapor, respectively. The surface tension of the liquid is σ (N/m). The quantity g_c is the proportionality constant equal to 1 kg·m/N·s². The quantity g is the local gravitational acceleration in m/sec². The values of C are given in Table 23. The above equation may be applied to different geometries, such as plates, wire, or cylinders.

Table 23   Values of the Constant C for Various Liquid–Surface Combinations

Fluid–Heating Surface Combinations	C
Water with polished copper, platinum, or mechanically polished stainless steel	0.0130
Water with brass or nickel	0.006
Water with ground and polished stainless steel	0.008
Water with Teflon-plated stainless steel	0.008

The *critical heat flux* (point C of Fig. 30) is given by[28]

$$q_c'' = \frac{\pi}{24} h_{fg}\rho_v \left[\frac{\sigma g g_c(\rho_l - \rho_v)}{\rho_v^2}\right]^{0.25}\left(1 + \frac{\rho_v}{\rho_l}\right)^{0.5}$$

For a water–steel combination, $q_c'' \approx 1290$ kW/m^2 and $\Delta T_{e,c} \approx 30°$C. For water–chrome-plated copper, $q_c'' \approx 940$–1260 KW/m^2 and $\Delta T_{e,c} \approx 23$–$28°$C.

Film Pool Boiling

The heat transfer from the surface to the liquid is due to both convection and radiation. A total heat-transfer coefficient is defined by the combination of convection and radiation heat-transfer coefficients of the following form[29] for the outside surfaces of horizontal tubes:

$$h^{4/3} = h_c^{4/3} + h_r h^{1/3}$$

where

$$h_c = 0.62\left[\frac{k_v^3 \rho_v(\rho_l - \rho_v)g(h_{fg} + 0.4c_{p,v}\Delta T_e)}{\mu_v D\,\Delta T_e}\right]^{1/4}$$

and

$$h_r = \frac{5.73 \times 10^{-8}\varepsilon(T_s^4 - T_{sat}^r)}{T_s - T_{sat}}$$

The vapor properties are evaluated at the film temperature $T_f = (T_s + T_{sat})/2$. The temperatures T_s and T_{sat} are in kelvins for the evaluation of h_r. The emissivity of the metallic solids can be found from Table 17. Note that $q = hA(T_s - T_{sat})$.

Nucleate Boiling in Forced Convection

The total heat-transfer rate can be obtained by simply superimposing the heat transfer due to nucleate boiling and forced convection:

$$q'' = q''_{boiling} + q''_{forced\ convection}$$

For forced convection, it is recommended that the coefficient 0.023 be replaced by 0.014 in the Dittus-Boelter equation (Section 2.1). The above equation is generally applicable to forced convection where the bulk liquid temperature is subcooled (*local forced convection boiling*).

Simplified Relations for Boiling in Water

For *nucleate boiling*,[30]

$$h = C(\Delta T_e)^n\left(\frac{p}{p_a}\right)^{0.4}$$

where p and p_a are, respectively, the system pressure and standard atmospheric pressure. The constants C and n are listed in Table 24.

For *local forced convection boiling inside vertical tubes,* valid over a pressure range of 5–170 atm,[31]

$$h = 2.54(\Delta T_e)^3 e^{p/1.551}$$

where h has the unit W/m^2·°C, ΔT_e is in °C, and p is the pressure in 10^6 N/m^3.

Table 24   Values of C and n for Simplified Relations for Boiling in Water

Surface	q'' (kW/m^2)	C	n
Horizontal	$q'' < 16$	1042	1/3
	$16 < q'' < 240$	5.56	3
Vertical	$q'' < 3$	5.7	1/7
	$3 < q'' < 63$	7.96	3

4.2   Condensation

Depending on the surface conditions, the condensation may be a *film condensation* or a *dropwise condensation*. Film condensation usually occurs when a vapor, relatively free of impurities, is allowed to condense on a clean, uncontaminated surface. Dropwise condensation occurs on highly polished surfaces or on surfaces coated with substances that inhibit wetting. The condensate provides a resistance to heat transfer between the vapor and the surface. Therefore, it is desirable to use short vertical surfaces or horizontal cylinders to prevent the condensate from growing too thick. The heat-transfer rate for dropwise condensation is usually an order of magnitude larger than that for film condensation under similar conditions. Silicones, Teflon, and certain fatty acids can be used to coat the surfaces to promote dropwise condensation. However, such coatings may lose their effectiveness owing to oxidation or outright removal. Thus, except under carefully controlled conditions, film condensation may be expected to occur in most instances, and the condenser design calculations are often based on the assumption of film condensation.

For condensation on surface at temperature T_s the total heat-transfer rate to the surface is given by $q = \bar{h}_L A\,(T_{sat} - T_s)$, where T_{sat} is the saturation temperature of the vapor. The mass flow rate is determined by $\dot{m} = q/h_{fg}$; h_{fg} is the latent heat of vaporization of the fluid (see Table 25 for saturated water). Correlations are based on the evaluation of liquid properties at $T_f = (T_s + T_{sat})/2$, except h_{fg}, which is to be taken at T_{sat}.

Film Condensation on a Vertical Plate
The Reynolds number for *condensate flow* is defined by $\mathrm{Re}_\Gamma = \rho_l V_m D_h / \mu_l$, where ρ_l and μ_l are the density and viscosity of the liquid, V_m is the average velocity of condensate, and D_h is the hydraulic diameter defined by $D_h = 4 \times$ condensate film cross-sectional area/ wetted perimeter. For the condensation on a vertical plate $\mathrm{Re}_\Gamma = 4\Gamma/\mu_l$, where Γ is the mass flow rate of condensate per unit width evaluated at the lowest point on the condensing surface. The condensate flow is generally considered to be laminar for $\mathrm{Re}_\Gamma < 1800$, and turbulent for $\mathrm{Re}_\Gamma > 1800$. The average Nusselt number is given by[22]

$$\overline{\mathrm{Nu}}_L = 1.13\left[\frac{g\rho_l(\rho_l - \rho_v)h_{fg}L^3}{\mu_l k_l(T_{sat} - T_s)}\right]^{0.25} \qquad \text{for } \mathrm{Re}_\Gamma < 1800$$

$$\overline{\mathrm{Nu}}_L = 0.0077\left[\frac{g\rho_l(\rho_l - \rho_v)L^3}{\mu_l^2}\right]^{1/3}\mathrm{Re}_\Gamma^{0.4} \qquad \text{for } \mathrm{Re}_\Gamma > 1800$$

Film Condensation on the Outside of Horizontal Tubes and Tube Banks

$$\overline{\mathrm{Nu}}_D = 0.725\left[\frac{g\rho_l(\rho_l - \rho_v)h_{fg}D^3}{N\mu_l k_l(T_{sat} - T_s)}\right]^{0.25}$$

where N is the number of horizontal tubes placed one above the other; $N = 1$ for a single tube.[23]

Table 25 Thermophysical Properties of Saturated Water

Temperature, T (K)	Pressure, P (bar)^a	Specific Volume (m^3/kg)		Heat of Vaporization, h_{fg} (kJ/kg)	Specific Heat (kJ/kg·K)		Viscosity (N·sec/m^2)		Thermal Conductivity (W/m·K)		Prandtl Number		Surface Tension $\sigma_l \times 10^3$ (N/m)	Expansion Coefficient, $\beta l \times 10^6$ (K^{-1})
		$vf \times 10^3$	v_u		$C_{p,l}$	$C_{p,u}$	$\mu_l \times 10^6$	$\mu_v \times 10^3$	$k_l \times 10^3$	$k_v \times 10^3$	Pr_l	Pr_v		
273.15	0.00611	1.000	206.3	2502	4.217	1.854	1750	8.02	659	18.2	12.99	0.815	75.5	−68.05
300	0.03531	1.003	39.13	2438	4.179	1.872	855	9.09	613	19.6	5.83	0.857	71.7	276.1
320	0.1053	1.011	13.98	2390	4.180	1.895	577	9.89	640	21.0	3.77	0.894	68.3	436.7
340	0.2713	1.021	5.74	2342	4.188	1.930	420	10.69	660	22.3	2.66	0.925	64.9	566.0
360	0.6209	1.034	2.645	2291	4.203	1.983	324	11.49	674	23.7	2.02	0.960	61.4	697.9
380	1.2869	1.049	1.337	2239	4.226	2.057	260	12.29	683	25.4	1.61	0.999	57.6	788
400	2.455	1.067	0.731	2183	4.256	2.158	217	13.05	688	27.2	1.34	1.033	63.6	896
450	9.319	1.123	0.208	2024	4.40	2.56	152	14.85	678	33.1	0.99	1.14	42.9	
500	26.40	1.203	0.0766	1825	4.66	3.27	118	16.59	642	42.3	0.86	1.28	31.6	
550	61.19	1.323	0.0317	1564	5.24	4.64	97	18.6	580	58.3	0.87	1.47	19.7	
600	123.5	1.541	0.0137	1176	7.00	8.75	81	22.7	497	92.9	1.14	2.15	8.4	
647.3	221.2	3.170	0.0032	0	∞	∞	45	45	238	238	∞	∞	0.0	

202

Film Condensation Inside Horizontal Tubes

For low vapor velocities such that Re_D based on the vapor velocities at the pipe inlet is less than 3500[24]

$$\overline{Nu}_D = 0.555 \left[\frac{g\rho_l(\rho_l - \rho_v)h'_{fg}D^3}{\mu_l k_l(T_{sat} - T_s)} \right]^{0.25}$$

where $h'_{fg} + \tfrac{3}{8}C_{p,l}(T_{sat} - T_s)$. For higher flow rate,[25] $Re_G > 5 \times 10^4$,

$$\overline{Nu}_D = 0.0265\, Re_G^{0.8}\, Pr^{1/3}$$

where the Reynolds number $Re_G = GD/\mu_l$ is based on the equivalent mass velocity $G = G_l + G_v\,(\rho_l/\rho_v)^{0.5}$. The mass velocity for the liquid G_l and that for vapor G_v are calculated as if each occupied the entire flow area.

The Effect of Noncondensable Gases

If noncondensable gas such as air is present in a vapor, even in a small amount, the heat-transfer coefficient for condensation may be greatly reduced. It has been found that the presence of a few percent of air by volume in steam reduces the coefficient by 50% or more. Therefore, it is desirable in the condenser design to vent the noncondensable gases as much as possible.

4.3 Heat Pipes

Heat pipes are two-phase heat-transfer devices that operate on a closed two-phase cycle[32] and come in a wide variety of sizes and shapes.[33,34] As shown in Fig. 31, they typically consist of three distinct regions, the evaporator or heat addition region, the condenser or heat

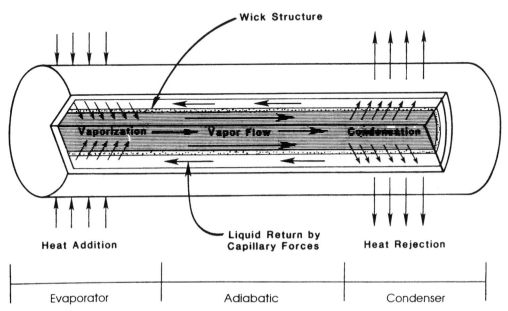

Figure 31 Typical heat pipe construction and operation.[35]

rejection region, and the adiabatic or isothermal region. Heat added to the evaporator region of the container causes the working fluid in the evaporator wicking structure to be vaporized. The high temperature and corresponding high pressure in this region result in flow of the vapor to the other, cooler end of the container where the vapor condenses, giving up its latent heat of vaporization. The capillary forces existing in the wicking structure then pump the liquid back to the evaporator section. Other similar devices, referred to as two-phase thermosyphons have no wick, and utilize gravitational forces to provide the liquid return. Thus, the heat pipe functions as a nearly isothermal device, adjusting the evaporation rate to accommodate a wide range of power inputs, while maintaining a relatively constant source temperature.

Transport Limitations

The transport capacity of a heat pipe is limited by several important mechanisms. Among these are the capillary wicking limit, viscous limit, sonic limit, entrainment, and boiling limits. The capillary wicking limit and viscous limits deal with the pressure drops occurring in the liquid and vapor phases, respectively. The sonic limit results from the occurrence of choked flow in the vapor passage, while the entrainment limit is due to the high liquid vapor shear forces developed when the vapor passes in counterflow over the liquid saturated wick. The boiling limit is reached when the heat flux applied in the evaporator portion is high enough that nucleate boiling occurs in the evaporator wick, creating vapor bubbles that partially block the return of fluid.

To function properly, the net capillary pressure difference between the condenser and the evaporator in a heat pipe must be greater than the pressure losses throughout the liquid and vapor flow paths. This relationship can be expressed as

$$\Delta P_c \geq \Delta P_+ + \Delta P_- + \Delta P_l + \Delta P_v$$

where ΔP_c = net capillary pressure difference
ΔP_+ = normal hydrostatic pressure drop
ΔP_- = axial hydrostatic pressure drop
ΔP_l = viscous pressure drop occurring in the liquid phase
ΔP_v = viscous pressure drop occurring in the vapor phase.

If these conditions are not met, the heat pipe is said to have reached the *capillary limitation*.

Expressions for each of these terms have been developed for steady-state operation, and are summarized below.

$$\textit{Capillary pressure} \qquad \Delta P_{c,m} = \left(\frac{2\sigma}{r_{c,e}}\right)$$

Values for the effective capillary radius, r_c, can be found theoretically for simple geometries or experimentally for pores or structures of more complex geometry. Table 26 gives values for some common wicking structures.

$$\textit{Normal and axial hydrostatic pressure drop} \qquad \Delta P_+ + \rho_l g d_v \cos \psi$$

$$\Delta P_- = \rho_l g L \sin \psi$$

In a gravitational environment, the axial hydrostatic pressure term may either assist or hinder the capillary pumping process, depending on whether the tilt of the heat pipe promotes or hinders the flow of liquid back to the evaporator (i.e., the evaporator lies either below or above the condenser). In a zero-*g* environment, both this term and the normal hydrostatic pressure drop term can be neglected because of the absence of body forces.

Table 26 Expressions for the Effective Capillary Radius for Several Wick Structures

Structure	r_c	Data
Circular cylinder (artery or tunnel wick)	r	r = radius of liquid flow passage
Rectangular groove	ω	ω = groove width
Triangular groove	$\omega/\cos\beta$	ω = groove width
		β = half-included angle
Parallel wires	ω	ω = wire spacing
Wire screens	$(\omega + d_\omega)/2 = 1/2N$	d = wire diameter
		N = screen mesh number
		ω = wire spacing
Packed spheres	$0.41r_s$	r_s = sphere radius

$$\textit{Liquid pressure drop}\qquad \Delta P_l = \left(\frac{\mu_l}{KA_w h_{fg}\rho_l}\right)L_{\text{eff}}\,q$$

where L_{eff} is the effective heat pipe length defined as

$$L_{\text{eff}} = 0.5L_e + L_a + 0.5L_c$$

and K is the liquid permeability as shown in Table 27.

$$\textit{Vapor pressure drop}\qquad \Delta P_v = \left(\frac{C(f_v \text{Re}_v)\mu_v}{2(r_{h,v})^2\,A_v\rho_v h_{fg}}\right)L_{\text{eff}}\,q$$

Although during steady-state operation the liquid flow regime is always laminar, the vapor flow may be either laminar or turbulent. It is therefore necessary to determine the vapor flow regime as a function of the heat flux. This can be accomplished by evaluating the local axial Reynolds and Mach numbers and substituting the values as shown below:

$$\text{Re}_v < 2300,\quad \text{Ma}_v < 0.2$$

$$(f_v \text{Re}_v) = 16$$

Table 27 Wick Permeability for Several Wick Structures

Structure	K	Data
Circular cylinder (artery or tunnel wick)	$r^2/8$	r = radius of liquid flow passage
Open rectangular grooves	$2\varepsilon(r_{h,l})^2/(f_l\text{Re}_l) = \omega/s$	ε = wick porosity
		ω = groove width
		s = groove pitch
		δ = groove depth
		$(r_{h,l}) = 2\omega\delta/(\omega+2\delta)$
Circular annular wick	$2(r_{h,l})^2/(f_l\text{Re}_l)$	$(r_{h,l}) = r_1 - r_2$
Wrapped screen wick	$1/122\,d_\omega^2\varepsilon^3/(1-\varepsilon)^2$	d_ω = wire diameter
		$\varepsilon = 1 - (1.05\pi N d\omega/4)$
		N = mesh number
Packed sphere	$1/37.5r_s^2\varepsilon^3/(1-\varepsilon)^2$	r_s = sphere radius
		ε = porosity (dependent on packing mode)

$$C = 1.00$$

$$\text{Re}_v < 2300, \quad \text{Ma}_v > 0.2$$

$$(f_v \text{Re}_v) = 16$$

$$C = \left[1 + \left(\frac{\gamma_v - 1}{2} \right) \text{Ma}_v^2 \right]^{1/2}$$

$$\text{Re}_v > 2300, \quad \text{Ma}_v < 0.2$$

$$(f_v \text{Re}_v) = 0.038 \left(\frac{2(r_{h,v})q}{A_v \mu_v h_{fg}} \right)^{3/4}$$

$$C = 1.00$$

$$\text{Re}_v > 2300, \quad \text{Ma}_v > 0.2$$

$$(f_v \text{Re}_v) = 0.038 \left(\frac{2(r_{h,v}q}{A_v \mu_v h_{fg}} \right)^{3/4}$$

$$C = \left[1 + \left(\frac{\gamma_v - 1}{2} \right) \text{Ma}_v^2 \right]^{-1/2}$$

Since the equations used to evaluate both the Reynolds number and the Mach number are functions of the heat transport capacity, it is necessary to first assume the conditions of the vapor flow. Using these assumptions, the maximum heat capacity, $q_{c,m}$, can be determined by substituting the values of the individual pressure drops into Eq. (1) and solving for $q_{c,m}$. Once the value of $q_{c,m}$ is known, it can then be substituted into the expressions for the vapor Reynolds number and Mach number to determine the accuracy of the original assumption. Using this iterative approach, accurate values for the capillary limitation as a function of the operating temperature can be determined in units of W-m or watts for $(qL)_{c,m}$ and $q_{c,m}$, respectively.

The *viscous limitation* in heat pipes occurs when the viscous forces within the vapor region are dominant and limit the heat pipe operation:

$$\frac{\Delta P_v}{P_v} < 0.1$$

for determining when this limit might be of a concern. Due to the operating temperature range, this limitation will normally be of little consequence in the design of heat pipes for use in the thermal control of electronic components and devices.

The *sonic limitation* in heat pipes is analogous to the sonic limitation in a converging–diverging nozzle and can be determined from

$$q_{s,m} = A_v \rho_v h_{fg} \left[\frac{\gamma_v R_v T_v}{2(\gamma_v + 1)} \right]^{1/2}$$

where T_v is the mean vapor temperature within the heat pipe.

Since the liquid and vapor flow in opposite directions in a heat pipe, at high enough vapor velocities, liquid droplets may be picked up or entrained in the vapor flow. This entrainment results in excess liquid accumulation in the condenser and, hence, dryout of the evaporator wick. Using the Weber number, We, defined as the ratio of the viscous shear force to the force resulting from the liquid surface tension, an expression for the *entrainment limit* can be found as

$$q_{e,m} = A_v h_{fg} \left[\frac{\sigma \rho_v}{2(r_{h,w})} \right]^{1/2}$$

where $(r_{h,w})$ is the hydraulic radius of the wick structure, defined as twice the area of the wick pore at the wick–vapor interface divided by the wetted perimeter at the wick–vapor interface.

The *boiling limit* occurs when the input heat flux is so high that nucleate boiling occurs in the wicking structure and bubbles may become trapped in the wick, blocking the liquid return and resulting in evaporator dryout. This phenomenon, referred to as the boiling limit, differs from the other limitations previously discussed in that it depends on the evaporator heat flux as opposed to the axial heat flux. This expression, which is a function of the fluid properties, can be written as

$$q_{b,m} = \left(\frac{2\pi L_{\text{eff}} k_{\text{eff}} T_v}{h_{fg} \rho_v \ln(r_i/r_v)} \right) \left(\frac{2\sigma}{r_n} - \Delta P_{c,m} \right)$$

where k_{eff} is the effective thermal conductivity of the liquid–wick combination, given in Table 28, r_i is the inner radius of the heat pipe wall, and r_n is the nucleation site radius. After the power level associated with each of the four limitations is established, determination of the maximum heat transport capacity is only a matter of selecting the lowest limitation for any given operating temperature.

Heat Pipe Thermal Resistance

The *heat pipe thermal resistance* can be found using an analogous electrothermal network. Figure 32 illustrates the electrothermal analog for the heat pipe illustrated in Fig. 31. As shown, the overall thermal resistance is composed of nine different resistances arranged in a series/parallel combination, which can be summarized as follows:

R_{pe} The radial resistance of the pipe wall at the evaporator
R_{we} The resistance of the liquid–wick combination at the evaporator
R_{ie} The resistance of the liquid–vapor interface at the evaporator
R_{ya} The resistance of the adiabatic vapor section
R_{pa} The axial resistance of the pipe wall
R_{wa} The axial resistance of the liquid–wick combination
R_{ic} The resistance of the liquid–vapor interface at the condenser

Table 28   Effective Thermal Conductivity for Liquid-Saturated Wick Structures

Wick Structures	k_{eff}
Wick and liquid in series	$\dfrac{k_l k_w}{\varepsilon k_w + k_l(1 - \varepsilon)}$
Wick and liquid in parallel	$\varepsilon k_l + k_w(1 - \varepsilon)$
Wrapped screen	$\dfrac{k_l[(k_l + k_w) - (1 - \varepsilon)(k_l - k_w)]}{(k_l + k_w) + (1 - \varepsilon)(k_l - k_w)]}$
Packed spheres	$\dfrac{k_l[(2k_l + k_w) - 2(1 - \varepsilon)(k_l - k_w)]}{(2k_l + k_w) + (1 - \varepsilon)(k_l - k_w)}$
Rectangular grooves	$\dfrac{w_f k_l k_w \delta) + w k_l(0.185 w_f k_w + \delta k_l}{(w + w_f)(0.185 w_f k_f + \delta k_l)}$

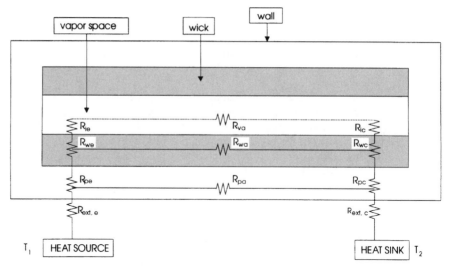

Figure 32   Equivalent thermal resistance of a heat pipe.

R_{wc} The resistance of the liquid–wick combination at the condenser
R_{pc} The radial resistance of the pipe wall at the condenser

Because of the comparative magnitudes of the resistance of the vapor space and the axial resistances of the pipe wall and liquid–wick combinations, the axial resistance of both the pipe wall and the liquid–wick combination may be treated as open circuits and neglected. Also, because of the comparative resistances, the liquid–vapor interface resistances and the axial vapor resistance can, in most situations, be assumed to be negligible. This leaves only the pipe wall radial resistances and the liquid–wick resistances at both the evaporator and condenser. The radial resistances at the pipe wall can be computed from Fourier's law as

$$R_{pe} = \frac{\delta}{k_p A_e}$$

for flat plates, where δ is the plate thickness and A_e is the evaporator area, or

$$R_{pe} = \frac{\ln(D_o/D_i)}{2\pi L_e k_p}$$

for cylindrical pipes, where L_e is the evaporator length. An expression for the equivalent thermal resistance of the liquid–wick combination in circular pipes is

$$R_{we} = \frac{\ln(D_o/D_i)}{2\pi L_e k_{\text{eff}}}$$

where values for the effective conductivity, k_{eff}, can be found in Table 28. The adiabatic vapor resistance, although usually negligible, can be found as

$$R_{va} = \frac{T_v(P_{v,e} - P_{v,c})}{\rho_v h_{fg} q}$$

where $P_{v,e}$ and $P_{v,c}$ are the vapor pressures at the evaporator and condenser. Combining these individual resistances provides a mechanism by which the overall thermal resistance can be computed and hence the temperature drop associated with various axial heat fluxes can be computed.

REFERENCES

1. F. P. Incropera and D. P. Dewitt, *Fundamentals of Heat Transfer,* Wiley, New York, 1981.
2. E. R. G. Eckert and R. M. Drake, Jr., *Analysis of Heat and Mass Transfer,* McGraw-Hill, New York, 1972.
3. M. P. Heisler, "Temperature Charts for Induction and Constant Temperature Heating," *Trans. ASME,* **69,** 227 (1947).
4. H. Grober and S. Erk, *Fundamentals of Heat Transfer,* McGraw-Hill, New York, 1961.
5. E. N. Sieder and C. E. Tate, "Heat Transfer and Pressure Drop of Liquids in Tubes," *Ind. Eng. Chem.,* **28,** 1429 (1936).
6. F. W. Dittus and L. M. K. Baelter, *Univ. Calif., Berkeley, Pub. Eng.,* **2,** 443 (1930).
7. A. J. Chapman, *Heat Transfer,* Macmillan, New York, 1974.
8. S. Whitaker, *AICHE J.,* **18,** 361(1972).
9. M. Jakob, *Heat Transfer,* Wiley, New York, 1949, Vol. 1.
10. A. Zhukauska, "Heat Transfer from Tubes in Cross Flow," in *Advances in Heat Transfer,* J. P. Hartnett and T. F. Irvine, Jr. (eds.), Academic, New York, 1972, Vol. 8.
11. F. Kreith, *Principles of Heat Transfer,* Harper & Row, New York, 1973.
12. H. A. Johnson and M. W. Rubesin, "Aerodynamic Heating and Convective Heat Transfer," *Trans. ASME,* **71,** 447 (1949).
13. C. C. Lin (Ed.), *Turbulent Flows and Heat Transfer, High Speed Aerodynamics and Jet Propulsion,* Princeton University Press, Princeton, NJ, 1959, Vol. V.
14. W. H. McAdams, *Heat Transmission,* McGraw-Hill, New York, 1954.
15. T. Yuge, "Experiments on Heat Transfer from Spheres Including Combined Natural and Forced Convection," *J. Heat Transfer,* **82,** 214 (1960).
16. S. Globe and D. Dropkin, "Natural Convection Heat Transfer in Liquids Confined between Two Horizontal Plates," *J. Heat Transfer,* **81C,** 24 (1959).
17. I. Catton, "Natural Convection in Enclosures," *Proc. 6th International Heat Transfer Conference,* 6, Toronto, Canada, 1978.
18. R. K. MacGregor and A. P. Emery, "Free Convection through Vertical Plane Layers: Moderate and High Prandtl Number Fluids," *J. Heat Transfer,* **91,** 391(1969).
19. R. Siegel and J. R. Howell, *Thermal Radiation Heat Transfer,* McGraw-Hill, New York, 1981.
20. G. G. Gubareff, J. E. Janssen, and R. H. Torborg, *Thermal Radiation Properties Survey,* 2nd ed., Minneapolis Honeywell Regulator Co., Minneapolis, MN, 1960.
21. H. C. Hottel, *Heat Transmission,* W. C. McAdams (ed.), McGraw-Hill, New York, 1954, Chap. 2.
22. W. H. McAdams, *Heat Transmission,* 3rd ed., McGraw-Hill, New York, 1954.
23. W. M. Rohsenow, "Film Condensation" in W. M. Rohsenow and J. P. Hartnett (eds.), *Handbook of Heat Transfer,* McGraw-Hill, New York, 1973.
24. J. C. Chato, "Laminar Condensation Inside Horizontal and Inclined Tubes," *ASHRAE J.,* **4,** 52 (1962).
25. W. W. Akers, H. A. Deans, and O. K. Crosser, "Condensing Heat Transfer Within Horizontal Tubes," *Chem. Eng. Prog., Sym. Ser.,* **55** (29), 171 (1958).
26. W. M. Rohsenow, "A Method of Correlating Heat Transfer Data for Surface Boiling Liquids," *Trans. ASME,* **74,** 969 (1952).
27. J. P. Holman, *Heat Transfer,* McGraw-Hill, New York, 1981.
28. N. Zuber, "On the Stability of Boiling Heat Transfer," *Trans. ASME,* **80,** 711 (1958).
29. L. A. Bromley, "Heat Transfer in Stable Film Boiling," *Chem. Eng. Prog.,* **46,** 221 (1950).

30. M. Jacob and G. A. Hawkins, *Elements of Heat Transfer,* Wiley, New York, 1957.

31. M. Jacob, *Heat Transfer,* Wiley, New York, 1957, Vol. 2, p. 584.

32. G. P. Peterson, *An Introduction to Heat Pipes: Modeling, Testing and Applications,* Wiley, New York, 1994.

33. G. P. Peterson, A. B. Duncan and M. H. Weichold, "Experimental Investigation of Micro Heat Pipes Fabricated in Silicon Wafers," *ASME J. Heat Transfer,* **115**, 3, 751 (1993).

34. G. P. Peterson, "Capillary Priming Characteristics of a High Capacity Dual Passage Heat Pipe," *Chem. Eng. Commun.,* **27**, 1, 119 (1984).

35. G. P. Peterson and L. S. Fletcher, "Effective Thermal Conductivity of Sintered Heat Pipe Wicks," *AIAA J. Thermophys. Heat Transfer,* **1** (3), 36 (1987).

BIBLIOGRAPHY

American Society of Heating, Refrigerating and Air Conditioning Engineering, *ASHRAE Handbook of Fundamentals,* 1972.

Arpaci, V. S., *Conduction Heat Transfer,* Addison-Wesley, Reading, MA, 1966.

Bai, C., and A. S. Lavine, "Thermal Boundary Conditions for Hyperbolic Heat Conduction," *ASME HTD,* **253**, 37–44 (1993).

Carslaw, H. S., and J. C. Jager, *Conduction of Heat in Solid,* Oxford University Press, London, 1959.

Chi, S. W., *Heat Pipe Theory and Practice,* McGraw-Hill, New York, 1976.

Duffie, J. A., and W. A. Beckman, *Solar Engineering of Thermal Process,* Wiley, New York, 1980.

Duncan, A. B., and G. P. Peterson, "A Review of Microscale Heat Transfer," Invited Review Article, *Applied Mechanics Review,* **47** (9), 397–428 (1994).

Dunn, P. D., and D. A. Reay, *Heat Pipes,* 3rd ed., Pergamon Press, New York, 1983.

Gebhart, B., *Heat Transfer,* McGraw-Hill, New York, 1971.

Hottel, H. C., and A. F. Saroffin, *Radiative Transfer,* McGraw-Hill, New York, 1967.

Kays, W. M., *Convective Heat and Mass Transfer,* McGraw-Hill, New York, 1966.

Knudsen, J. G., and D. L. Katz, *Fluid Dynamics and Heat Transfer,* McGraw-Hill, New York, 1958.

Ozisik, M. N., *Radiative Transfer and Interaction with Conduction and Convection,* Wiley, New York, 1973.

Ozisik, M. N., *Heat Conduction,* Wiley, New York, 1980.

Peterson, G. P., *An Introduction to Heat Pipes: Modeling, Testing and Applications,* Wiley, New York, 1994.

Planck, M., *The Theory of Heat Radiation,* Dover, New York, 1959.

Rohsenow, W. M., and H. Y. Choi, *Heat, Mass, and Momentum Transfer,* Prentice-Hall, Englewood Cliffs, NJ, 1961.

Rohsenow, W. M., and J. P. Hartnett, *Handbook of Heat Transfer,* McGraw-Hill, New York, 1973.

Schlichting, H., *Boundary-Layer Theory,* McGraw-Hill, New York, 1979.

Schneider, P. J., *Conduction Heat Transfer,* Addison-Wesley, Reading, MA, 1955.

Sparrow, E. M., and R. D. Cess, *Radiation Heat Transfer,* Wadsworth, Belmont, CA, 1966.

Tien, C. L., B. F. Armaly, and P. S. Jagannathan, P. S., "Thermal Conductivity of Thin Metallic Films," *Proc. 8th Conference on Thermal Conductivity,* October 7–10, 1968.

Tien, C. L., "Fluid Mechanics of Heat Pipes," *Annu. Rev. Fluid Mechanics,* 167 (1975).

Turner, W. C., and J. F. Malloy, *Thermal Insulation Handbook,* McGraw-Hill, New York, 1981.

Vargafik, N. B., *Table of Thermophysical Properties of Liquids and Gases,* Hemisphere, Washington, DC, 1975.

Wiebelt, J. A., *Engineering Radiation Heat Transfer,* Holt, Rinehart & Winston, New York, 1966.

CHAPTER 6
FURNACES

Carroll Cone
Toledo, Ohio

1 SCOPE AND INTENT

This chapter has been prepared for the use of engineers with access to an electronic calculator and to standard engineering reference books, but not necessarily to a computer terminal. The intent is to provide information needed for the solution of furnace engineering problems in areas of design, performance analysis, construction and operating cost estimates, and improvement programs.

In selecting charts and formulas for problem solutions, some allowance has been made for probable error, where errors in calculations will be minor compared with errors in the assumptions on which calculations are based. Conscientious engineers are inclined to carry calculations to a far greater degree of accuracy than can be justified by probable errors in data assumed. Approximations have accordingly been allowed to save time and effort without adding to probable margins for error. The symbols and abbreviations used in this chapter are given in Table 1.

2 STANDARD CONDITIONS

Assuming that the user will be using English rather than metric units, calculations have been based on pounds, feet, Btus, and degrees Fahrenheit, with conversion to metric units provided in the following text (see Table 2).

Assumed standard conditions include ambient temperature for initial temperature of loads for heat losses from furnace walls or open cooling of furnace loads—70°F. Condition of air entering system for combustion or convection cooling: temperature, 70°F; absolute pressure, 14.7 psia; relative humidity, 60% at 70°F, for a water vapor content of about 1.4% by volume.

2.1 Probable Errors

Conscientious furnace engineers are inclined to carry calculations to a far greater degree of accuracy than can be justified by uncertainties in basic assumptions such as thermal properties of materials, system temperatures and pressures, radiation view factors and convection coefficients. Calculation procedures recommended in this chapter will, accordingly, include some approximations, identified in the text, that will result in probable errors much smaller than those introduced by basic assumptions, where such approximations will expedite problem solutions.

3 FURNACE TYPES

Furnaces may be grouped into two general types:

1. As a source of energy to be used elsewhere, as in firing steam boilers to supply process steam, or steam for electric power generation, or for space heating of buildings or open space

2. As a source of energy for industrial processes, other than for electric power

The primary concern of this chapter is the design, operation, and economics of industrial furnaces, which may be classified in several ways:

Table 1   Symbols and Abbreviations

A	area in ft²
a	absorptivity for radiation, as fraction of black body factor for receiver temperature:
	a_g combustion gases
	a_w furnace walls
	a_s load surface
	a_m combined emissivity–absorptivity factor for source and receiver
C	specific heat in Btu/lb·°F or cal/g·°C
cfm	cubic feet per minute
D	diameter in ft or thermal diffusivity (k/dC)
d	density in lb/ft³
e	emissivity for radiation as fraction of blackbody factor for source temperature, with subscripts as for a above
F	factor in equations as defined in text
fpm	velocity in ft/min
G	mass velocity in lb/ft²·hr
g	acceleration by gravity (32.16 ft/sec²)
H	heat-transfer coefficient (Btu/hr·ft²·°F)
	H_r for radiation
	H_c for convection
	H_t for combined $H_r + H_c$
HHV	higher heating value of fuel
h	pressure head in units as defined
k	thermal conductivity (Btu/hr·ft·°F)
L	length in ft, as in effective beam length for radiation, decimal rather than feet and inches
LHV	lower heating value of fuel
ln	logarithm to base e
MTD	log mean temperature difference
N	a constant as defined in text
psi	pressure in lb/in²
	psig, pressure above atmospheric
	psia, absolute pressure
Pr	Prandtl number $(\mu C/k)$
Q	heat flux in Btu/hr
R	thermal resistance (r/k) or ratio of external to internal thermal resistance (k/rH)
Re	Reynolds number (DG/μ)
r	radius or depth of heat penetration in ft
T	temperature in °F, except for radiation calculations where °S = (°F + 460)/100
	T_g, combustion gas temperature
	T_w, furnace wall temperature
	T_s, heated load surface
	T_c, core or unheated surface of load
t	time in hr
μ	viscosity in lb/hr·ft
wc	inches of water column as a measure of pressure
V	volume in ft³
v	velocity in ft/sec
W	weight in lb
X	time factor for nonsteady heat transfer (tD/r^2)
x	horizontal coordinate
y	vertical coordinate
z	coordinate perpendicular to plane xy

Table 2 Conversion of Metric to English Units

Length	1 m = 3.281 ft
	1 cm = 0.394 in
Area	1 m² = 10.765 ft²
Volume	1 m³ = 35.32 ft³
Weight	1 kg = 2.205 lb
Density	1 g/cm³ = 62.43 lb/ft²
Pressure	1 g/cm² = 2.048 lb/ft² = 0.0142 psi
Heat	1 kcal = 3.968 Btu
	1 kwh = 3413 Btu
Heat content	1 cal/g = 1.8 Btu/lb
	1 kcal/m² = 0.1123 Btu/ft³
Heat flux	1 W/cm² = 3170 Btu/hr·ft²
Thermal conductivity	$\dfrac{1\ \text{cal}}{\text{s·cm·°C}} = \dfrac{242\ \text{Btu}}{\text{hr·ft·°F}}$
Heat transfer	$\dfrac{1\ \text{cal}}{\text{s·cm}^2\text{·°C}} = \dfrac{7373\ \text{Btu}}{\text{hr·ft}^2\text{·°F}}$
Thermal diffusivity	$\dfrac{1\ \text{cal/s·cm·°C}}{\text{C·g/cm}^3} = \dfrac{3.874\ \text{Btu/hr·ft·°F}}{\text{C·lb/ft}^3}$

By function:

Heating for forming in solid state (rolling, forging)

Melting metals or glass

Heat treatment to improve physical properties

Preheating for high-temperature coating processes, galvanizing, vitreous enameling, other coatings

Smelting for reduction of metallic ores

Firing of ceramic materials

Incineration

By method of load handling:

Batch furnaces for cyclic heating, including forge furnaces arranged to heat one end of a bar or billet inserted through a wall opening, side door, stationary-hearth-type car bottom designs

Continuous furnaces with loads pushed through or carried by a conveyor

Tilting-type furnace

To avoid the problem of door warpage or leakage in large batch-type furnaces, the furnace can be a refractory-lined box with an associated firing system, mounted above a stationary hearth, and arranged to be tilted around one edge of the hearth for loading and unloading by manual handling, forklift trucks, or overhead crane manipulators.

For handling heavy loads by overhead crane, without door problems, the furnace can be a portable cover unit with integral firing and temperature control. Consider a cover-type furnace for annealing steel strip coils in a controlled atmosphere. The load is a stack of coils with a common vertical axis, surrounded by a protective inner cover and an external heating cover. To improve heat transfer parallel to coil laminations, they are loaded with open coil separators between them, with heat transferred from the inner cover to coil ends by a recirculating fan. To start the cooling cycle, the heating cover is removed by an overhead crane, while atmosphere circulation by the base fan continues. Cooling may be enhanced by air-blast cooling of the inner cover surface.

For heating heavy loads of other types, such as weldments, castings, or forgings, car bottom furnaces may be used with some associated door maintenance problems. The furnace hearth is a movable car, to allow load handling by an overhead traveling crane. In one type of furnace, the door is suspended from a lifting mechanism. To avoid interference with an overhead crane, and to achieve some economy in construction, the door may be mounted on one end of the car and opened as the car is withdrawn. This arrangement may impose some handicaps in access for loading and unloading.

Loads such as steel ingots can be heated in pit-type furnaces, preferably with units of load separated to allow radiating heating from all sides except the bottom. Such a furnace would have a cover displaced by a mechanical carriage and would have a compound metal and refractory recuperator arrangement. Loads are handled by overhead crane equipped with suitable gripping tongs.

Continuous-Type Furnaces

The simplest type of continuous furnace is the hearth-type pusher furnace. Pieces of rectangular cross section are loaded side by side on a charge table and pushed through the furnace by an external mechanism. In the design shown, the furnace is fired from one end, counterflow to load travel, and is discharged through a side door by an auxiliary pusher lined up by the operator.

Furnace length is limited by thickness of the load and alignment of abutting edges, to avoid buckling up from the hearth. A more complex design would provide multiple zone firing above and below the hearth, with recuperative air preheating.

Long loads can be conveyed in the direction of their length in a roller-hearth-type furnace. Loads can be bars, tubes, or plates of limited width, heated by direct firing, by radiant tubes, or by electric-resistor-controlled atmosphere, and conveyed at uniform speed or at alternating high and low speeds for quenching in line.

Sequential heat treatment can be accomplished with a series of chain or belt conveyors. Small parts can be loaded through an atmosphere seal, heated in a controlled atmosphere on a chain belt conveyor, discharged into an oil quench, and conveyed through a washer and tempering furnace by a series of mesh belts without intermediate handling.

Except for pusher-type furnaces, continuous furnaces can be self-emptying. To secure the same advantage in heating slabs or billets for rolling and to avoid scale loss during interrupted operation, loads can be conveyed by a walking-beam mechanism. Such a walking-beam-type slab heating furnace would have loads supported on water-cooled rails for over- and underfiring, and would have an overhead recuperator.

Thin strip materials, joined in continuous strand form, can be conveyed horizontally or the strands can be conveyed in a series of vertical passes by driven support rolls. Furnaces of this type can be incorporated in continuous galvanizing lines.

Unit loads can be individually suspended from an overhead conveyor, through a slot in the furnace roof, and can be quenched in line by lowering a section of the conveyor.

Small parts or bulk materials can be conveyed by a moving hearth, as in the rotary-hearth-type or tunnel kiln furnace. For roasting or incineration of bulk materials, the shaft-type furnace provides a simple and efficient system. Loads are charged through the open top of the shaft and descend by gravity to a discharge feeder at the bottom. Combustion air can be introduced at the bottom of the furnace and preheated by contact with the descending load before entering the combustion zone, where fuel is introduced through sidewalls. Combustion gases are then cooled by contact with the descending load, above the combustion zone, to preheat the charge and reduce flue gas temperature.

With loads that tend to agglomerate under heat and pressure, as in some ore-roasting operations, the rotary kiln may be preferable to the shaft-type furnace. The load is advanced by rolling inside an inclined cylinder. Rotary kilns are in general use for sintering ceramic materials.

Classification by Source of Heat

The classification of furnaces by source of heat is as follows:

Direct-firing with gas or oil fuels

Combustion of material in process, as by incineration with or without supplemental fuel

Internal heating by electrical resistance or induction in conductors, or dielectric heating of nonconductors

Radiation from electric resistors or radiant tubes, in controlled atmospheres or under vacuum

4 FURNACE CONSTRUCTION

The modern industrial furnace design has evolved from a rectangular or cylindrical enclosure, built up of refractory shapes and held together by a structural steel binding. Combustion air was drawn in through wall openings by furnace draft, and fuel was introduced through the same openings without control of fuel/air ratios except by the judgment of the furnace operator. Flue gases were exhausted through an adjacent stack to provide the required furnace draft.

To reduce air infiltration or outward leakage of combustion gases, steel plate casings have been added. Fuel economy has been improved by burner designs providing some control of fuel/air ratios, and automatic controls have been added for furnace temperature and furnace pressure. Completely sealed furnace enclosures may be required for controlled atmosphere operation, or where outward leakage of carbon monoxide could be an operating hazard.

With the steadily increasing costs of heat energy, wall structures are being improved to reduce heat losses or heat demands for cyclic heating. The selection of furnace designs and materials should be aimed at a minimum overall cost of construction, maintenance, and fuel or power over a projected service life. Heat losses in existing furnaces can be reduced by adding external insulation or rebuilding walls with materials of lower thermal conductivity. To reduce losses from intermittent operation, the existing wall structure can be lined with a material of low heat storage and low conductivity, to substantially reduce mean wall temperatures for steady operation and cooling rates after interrupted firing.

Thermal expansion of furnace structures must be considered in design. Furnace walls have been traditionally built up of prefired refractory shapes with bonded mortar joints. Except for small furnaces, expansion joints will be required to accommodate thermal expansion. In sprung arches, lateral expansion can be accommodated by vertical displacement, with longitudinal expansion taken care of by lateral slots at intervals in the length of the furnace. Where expansion slots in furnace floors could be filled by scale, slag, or other debris, they can be packed with a ceramic fiber that will remain resilient after repeated heating.

Differential expansion of hotter and colder wall surfaces can cause an inward-bulging effect. For stability in self-supporting walls, thickness must not be less than a critical fraction of height.

Because of these and economic factors, cast or rammed refractories are replacing pre-fired shapes for lining many types of large, high-temperature furnaces. Walls can be retained by spaced refractory shapes anchored to the furnace casing, permitting reduced thickness as compared to brick construction. Furnace roofs can be suspended by hanger tile at closer spacing, allowing unlimited widths.

Cast or rammed refractories, fired in place, will develop discontinuities during initial shrinkage that can provide for expansion from subsequent heating, to eliminate the need for expansion joints.

As an alternative to cast or rammed construction, insulating refractory linings can be gunned in place by jets of compressed air and retained by spaced metal anchors, a construction increasingly popular for stacks and flues.

Thermal expansion of steel furnace casings and bindings must also be considered. Where the furnace casing is constructed in sections, with overlapping expansion joints, individual sections can be separately anchored to building floors or foundations. For gas-tight casings, as required for controlled atmosphere heating, the steel structure can be anchored at one point and left free to expand elsewhere. In a continuous galvanizing line, for example, the atmosphere furnace and cooling zone can be anchored to the foundation near the casting pot, and allowed to expand toward the charge end.

5 FUELS AND COMBUSTION

Heat is supplied to industrial furnaces by combustion of fuels or by electrical power. Fuels now used are principally fuel oil and fuel gas. Because possible savings through improved design and operation are much greater for these fuels than for electric heating or solid fuel firing, they are given primary consideration in this section.

Heat supply and demand may be expressed in units of *Btu* or *kcal* or as gallons or barrels of fuel oil, tons of coal or *kWh* of electric power. For the large quantities considered for national or world energy loads, a preferred unit is the "quad," one quadrillion or 10^{15} Btu. Conversion factors are

$$
\begin{aligned}
1 \text{ quad} &= 10^{15} \text{ Btu} \\
&= 172 \times 10^6 \text{ barrels of fuel oil} \\
&= 44.34 \times 10^6 \text{ tons of coal} \\
&= 10^{12} \text{ cubic feet of natural gas} \\
&= 2.93 \times 10^{11} \text{ kWh electric power}
\end{aligned}
$$

At 30% generating efficiency, the fuel required to produce 1 quad of electrical energy is 3.33 quads. One quad fuel is accordingly equivalent to 0.879×10^{11} kWh net power.

Fuel demand, in the United States during recent years, has been about 75 quads per year from the following sources:

Coal	15 quads
Fuel oil	
Domestic	18 quads
Imported	16 quads
Natural gas	23 quads
Other, including nuclear	3 quads

Hydroelectric power contributes about 1 quad net additional. Combustion of waste products has not been included, but will be an increasing fraction of the total in the future.

Distribution of fuel demand by use is estimated at

Power generation	20 quads
Space heating	11 quads
Transportation	16 quads
Industrial, other than power	25 quads
Other	4 quads

Net demand for industrial furnace heating has been about 6%, or 4.56 quads, primarily from gas and oil fuels.

The rate at which we are consuming our fossil fuel assets may be calculated as (annual demand)/(estimated reserves). This rate is presently highest for natural gas, because, besides being available at wellhead for immediate use, it can be transported readily by pipeline and burned with the simplest type of combustion system and without air pollution problems. It has also been delivered at bargain prices, under federal rate controls.

As reserves of natural gas and fuel oil decrease, with a corresponding increase in market prices, there will be an increasing demand for alternative fuels such as synthetic fuel gas and fuel oil, waste materials, lignite, and coal.

Synthetic fuel gas and fuel oil are now available from operating pilot plants, but at costs not yet competitive.

As an industrial fuel, coal is primarily used for electric power generation. In the form of metallurgical coke, it is the source of heat and the reductant in the blast furnace process for iron ore reduction, and as fuel for cupola furnaces used to melt foundry iron. Powdered coal is also being used as fuel and reductant in some new processes for solid-state reduction of iron ore pellets to make synthetic scrap for steel production.

Since the estimated life of coal reserves, particularly in North America, is so much greater than for other fossil fuels, processes for conversion of coal to fuel gas and fuel oil have been developed almost to the commercial cost level, and will be available whenever they become economical. Processes for coal gasification, now being tried in pilot plants, include

1. *Producer gas.* Bituminous coal has been commercially converted to fuel gas of low heating value, around 110 Btu/scf LHV, by reacting with insufficient air for combustion and steam as a source of hydrogen. Old producers delivered a gas containing sulfur, tar volatiles, and suspended ash, and have been replaced by cheap natural gas. By reacting coal with a mixture of oxygen and steam, and removing excess carbon dioxide, sulfur gases, and tar, a clean fuel gas of about 300 Btu/scf LHV can be supplied. Burned with air preheated to 1000°F and with a flue gas temperature of 2000°F, the available heat is about 0.69 HHV, about the same as for natural gas.

2. *Synthetic natural gas.* As a supplement to dwindling natural gas supplies, a synthetic fuel gas of similar burning characteristics can be manufactured by adding a fraction of hydrogen to the product of the steam–oxygen gas producer and reacting with carbon monoxide at high temperature and pressure to produce methane. Several processes are operating successfully on a pilot plant scale, but with a product costing much more than market prices for natural gas. The process may yet be practical for extending available natural gas supplies by a fraction, to maintain present market demands. For gas mixtures or synthetic gas supplies to be interchangeable with present gas fuels, without readjustment of fuel/air ratio controls, they must fit the Wobbe index:

$$\frac{\text{HHV Btu/scf}}{(\text{specific gravity})^{0.5}}$$

The fuel gas industry was originally developed to supply fuel gas for municipal and commercial lighting systems. Steam was passed through incandescent coal or coke, and fuel oil vapors were added to provide a luminous flame. The product had a heating value of around 500 HHV, and a high carbon monoxide content, and was replaced as natural gas or coke oven gas became available. Coke oven gas is a by-product of the manufacture of metallurgical coke that can be treated to remove sulfur compounds and volatile tar compounds to provide a fuel suitable for pipeline distribution. Blast furnace gas can be used as an industrial or steam-generating fuel, usually after enrichment with coke oven gas. Gas will be made from replaceable sources such as agricultural and municipal wastes, cereal grains, and wood, as market economics for such products improve.

Heating values for fuels containing hydrogen can be calculated in two ways:

1. Higher heating value (HHV) is the total heat developed by burning with standard air in a ratio to supply 110% of net combustion air, cooling products to ambient temperature, and condensing all water vapor from the combustion of hydrogen.

2. Lower heating value (LHV) is equal to HHV less heat from the condensation of water vapor. It provides a more realistic comparison between different fuels, since flue gases leave most industrial processes well above condensation temperatures.

HHV factors are in more general use in the United States, while LHV values are more popular in most foreign countries. For example, the HHV value for hydrogen as fuel is 319.4 Btu/scf, compared to a LHV of 270.2.

The combustion characteristics for common fuels are tabulated in Table 3, for combustion with 110% standard air. Weights in pounds per 10^6 Btu HHV are shown, rather than corresponding volumes, to expedite calculations based on mass flow. Corrections for flue gas and air temperatures other than ambient are given in charts to follow.

Table 3 Combustion Characteristics of Common Fuels

Fuel	Btu/scf	Weight in lb/10^6 Btu		
		Fuel	Air	Flue Gas
Natural gas (SW U.S.)	1073	42	795	837
Coke oven gas	539	57	740	707
Blast furnace gas	92	821	625	1446
Mixed blast furnace and coke oven gas:				
Ratio CO/BF 1/1	316	439	683	1122
1/3	204	630	654	1284
1/10	133	752	635	1387
Hydrogen	319	16	626	642
	Btu/lb			
No. 2 fuel oil	19,500	51	810	861
No. 6 fuel oil	18,300	55	814	
With air atomization				869
With steam atomization at 3 lb/gal				889
Carbon	14,107	71	910	981

The heat released in a combustion reaction is

Total heats of formation of combustion products − Total heats of formation of reactants

Heats of formation can be conveniently expressed in terms of Btu per pound mol, with the pound mol for any substance equal to a weight in pounds equal to its molecular weight. The heat of formation for elemental materials is zero. For compounds involved in common combustion reactions, values are shown in Table 4.

Data in Table 4 can be used to calculate the higher and lower heating values of fuels. For methane:

$$CH_4 + 2O_2 = CO_2 + 2H_2O$$

HHV

$$169,290 + (2 \times 122,976) - 32,200 = 383,042 \text{ Btu/lb·mol}$$
$$383,042/385 = 995 \text{ Btu/scf}$$

LHV

$$169,290 + (2 \times 104,040) - 32,200 = 345,170 \text{ Btu/lb·mol}$$
$$345,170/385 = 897 \text{ Btu/scf}$$

Available heats from combustion of fuels, as a function of flue gas and preheated air temperatures, can be calculated as a fraction of the HHV. The net ratio is one plus the fraction added by preheated air less the fraction lost as sensible heat and latent heat of water vapor, from combustion of hydrogen, in flue gas leaving the system.

Available heats can be shown in chart form, as in the following figures for common fuels. On each chart, the curve on the right is the fraction of HHV available for combustion with 110% cold air, while the curve on the left is the fraction added by preheated air, as functions of air or flue gas temperatures. For example, the available heat fraction for methane burned with 110% air preheated to 1000°F, and with flue gas out at 2000°F, is shown in Fig. 1: 0.41 + 0.18 − 0.59 HHV.

Values for other fuels are shown in charts that follow:

Fig. 2, fuel oils with air or steam atomization

Fig. 3, by-product coke oven gas

Fig. 4, blast furnace gas

Fig. 5, methane

Table 4   Heats of Formation

Material	Formula	Molecular Weight	Heats of Formation (Btu/lb·mol[a])
Methane	CH_4	16	32,200
Ethane	C_2H_6	30	36,425
Propane	C_3H_8	44	44,676
Butane	C_4H_{10}	58	53,662
Carbon monoxide	CO	28	47,556
Carbon dioxide	CO_2	44	169,290
Water vapor	H_2O	18	104,040
Liquid water			122,976

[a]The volume of 1 lb mol, for any gas, is 385 scf.

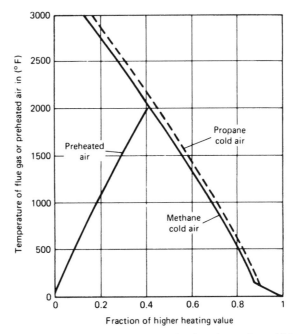

Figure 1 Available heat for methane and propane combustion. Approximate high and low limits for commercial natural gas.[1]

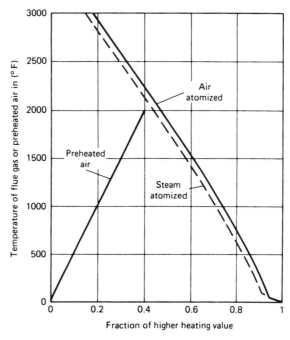

Figure 2 Available heat ratios for fuel oils with air or steam atomization.[1]

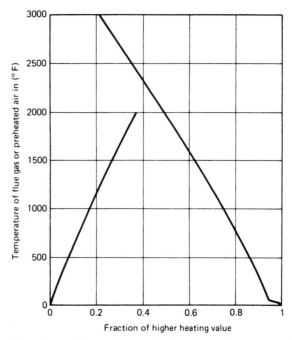

Figure 3 Available heat ratios for by-product coke oven gas.[1]

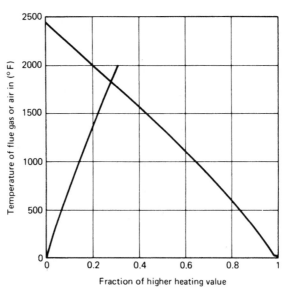

Figure 4 Available heat ratios for blast furnace gas.[1]

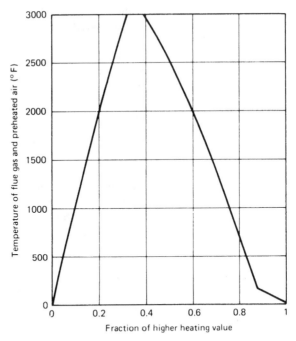

Figure 5   Available heat ratios for combustion of methane with 110% air containing 35% O_2.[1]

For combustion with other than 110% of net air demand, the corrected available heat can be calculated as follows. For methane with preheated air at 1000°F and flue gas out at 2000°F and 150% net air supply:

$$
\begin{array}{lr}
\text{Available heat from Fig. 1 in Chapter 20} & 0.59 \\
\text{Add excess air} + 0.18\,(1.5 - 1.1) \quad = & 0.072 \\
- 0.41\,(1.5 - 1.1) = & \underline{-0.164} \\
\text{Net total at 150\%} & 0.498
\end{array}
$$

Available heats for fuel gas mixtures can be calculated by adding the fractions for either fuel and dividing by the combined volume. For example, a mixture of one-quarter coke oven gas and three-quarters blast furnace gas is burned with 110% combustion air preheated to 1000°F, and with flue gas out at 2000°F. Using data from Table 3 and Figs. 3 and 4,

$$
\begin{array}{lr}
\text{CO } (539 \times 0.25 = 134.75)\,(0.49 + 0.17) \quad = & 88.93 \\
\text{BF } (92 \times 0.75 = \underline{\;69.00})\,(0.21 + 0.144) = & \underline{24.43} \\
\text{HHV } 203.75 \qquad \text{Available} = & 113.36
\end{array}
$$
$$
\text{Net: } 113.36/203.75 = 0.556 \text{ combined HHV}
$$

6   OXYGEN ENRICHMENT OF COMBUSTION AIR

The available heats of furnace fuels can be improved by adding oxygen to combustion air. Some studies have been based on a total oxygen content of 35%, which can be obtained by adding 21.5 scf pure oxygen or 25.45 scf of 90% oxygen per 100 scf of dry air. The available heat ratios are shown in the chart in Fig. 5.

At present market prices, the power needed to concentrate pure oxygen for enrichment to 35% will cost more than the fuel saved, even with metallurgical oxygen from an in-plant source. As plants are developed for economical concentration of oxygen to around 90%, the cost balance may become favorable for very-high-temperature furnaces.

In addition to fuel savings by improvement of available heat ratios, there will be additional savings in recuperative furnaces by increasing preheated air temperature at the same net heat demand, depending on the ratio of heat transfer by convection to that by gas radiation in the furnace and recuperator.

7   THERMAL PROPERTIES OF MATERIALS

The heat content of some materials heated in furnaces or used in furnace construction is shown in the chart in Fig. 6, in units of Btu/lb. Vertical lines in curves represent latent heats of melting or other phase transformations. The latent heat of evaporation for water in flue gas has been omitted from the chart. The specific heat of liquid water is, of course, about 1.

Thermal conductivities in English units are given in reference publications as: (Btu/(ft^2·hr))/(°F/in.) or as (Btu/(ft^2·hr))/(°F/ft). To keep dimensions consistent, the latter term, abbreviated to k = Btu/ft·hr·°F will be used here. Values will be $\frac{1}{12}$th of those in terms of °F/in.

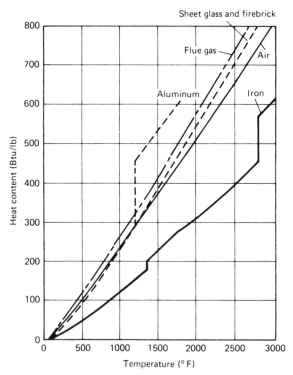

Figure 6   Heat content of materials at temperature.[1]

Thermal conductivities vary with temperature, usually inversely for iron, steel, and some alloys, and conversely for common refractories. At usual temperatures of use, average values of k in Btu/(ft·hr·°F) are in Table 5.

To expedite calculations for nonsteady conduction of heat, it is convenient to use the factor for "thermal diffusivity," defined as

$$D = \frac{k}{dC} = \frac{\text{Thermal conductivity}}{\text{Density} \times \text{Specific heat}}$$

in consistent units. Values for common furnace loads over the usual range of temperatures for heating are:

Carbon steels, 70–1650°F	0.32
70–2300°F	0.25
Low-alloy steels, 70–2000°F	0.23
Stainless steels, 70–2000°F	
300 type	0.15
400 type	0.20
Aluminum, 70–1000°F	3.00
Brass, 70/30, 70–1500°F	1.20

In calculating heat losses through furnace walls with multiple layers of materials with different thermal conductivities, it is convenient to add thermal resistance $R = r/k$, where r is thickness in ft. For example,

	r	k	r/k
9-in. firebrick	0.75	0.9	0.833
4½-in. insulating firebrick	0.375	0.20	1.875
2¼-in. block insulation	0.208	0.15	1.387
	Total R for wall materials		4.095

Overall thermal resistance will include the factor for combined radiation and convection from the outside of the furnace wall to ambient temperature. Wall losses as a function of wall surface temperature, for vertical surfaces in still air, are shown in Fig. 7, and are included in the overall heat loss data for furnace walls shown in the chart in Fig. 8.

Table 5 Average Values of k (Btu·ft·hr·°F)

	Mean Temperature (°F)				
	100	1000	1500	2000	2500
Steel, SAE 1010	33	23	17	17	
Type HH HRA	8	11	14	16	
Aluminum	127	133			
Copper	220	207	200		
Brass, 70/30	61	70			
Firebrick	0.81	0.82	0.85	0.89	0.93
Silicon carbide	11	10	9	8	6
Insulating firebrick	0.12	0.17	0.20	0.24	

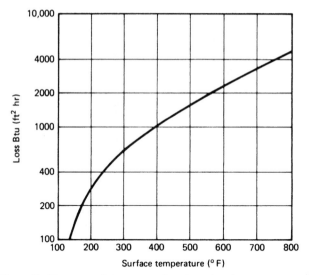

Figure 7 Furnace wall losses as a function of surface temperature.[1]

The chart in Fig. 9 shows the thermodynamic properties of air and flue gas, over the usual range of temperatures, for use in heat-transfer and fluid flow problems. Data for other gases, in formula form, are available in standard references.

Linear coefficients of thermal expansion are the fractional changes in length per °F change in temperature. Coefficients in terms of $10^6 \times$ net values are listed below for materials used in furnace construction and for the usual range of temperatures:

Carbon steel	9
Cast HRA	10.5
Aluminum	15.6
Brass	11.5
Firebrick, silicon carbide	3.4
Silica brick	3.4

Coefficients for cubical expansion of solids are about $3 \times$ linear coefficients. The cubical coefficient for liquid water is about 185×10^{-6}.

8 HEAT TRANSFER

Heat may be transmitted in industrial furnaces by radiation—gas radiation from combustion gases to furnace walls or direct to load, and solid-state radiation from walls, radiant tubes, or electric heating elements to load—or by convection—from combustion gases to walls or load. Heat may be generated inside the load by electrical resistance to an externally applied voltage or by induction, with the load serving as the secondary circuit in an alternating current transformer. Nonconducting materials may be heated by dielectric heating from a high-frequency source.

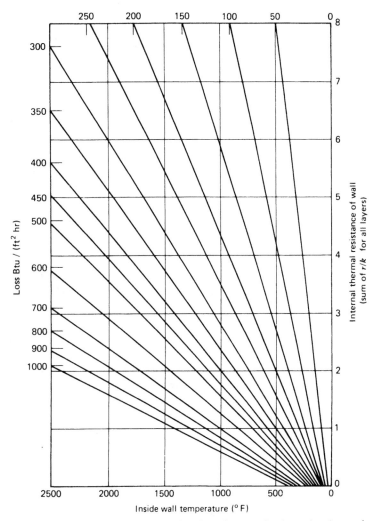

Figure 8   Furnace wall losses as a function of composite thermal resistance.[1]

Heat transfer in the furnace structure or in solid furnace loads will be by conduction. If the temperature profile is constant with time, the process is defined as "steady-state conduction." If temperatures change during a heating cycle, it is termed "non-steady-state conduction."

Heat flow is a function of temperature differentials, usually expressed as the "log-mean temperature difference" with the symbol MTD. MTD is a function of maximum and minimum temperature differences that can vary with position or time. Three cases encountered in furnace design are illustrated in Fig. 10. If the maximum differential, in any system of units, is designated as A and the minimum is designated by B:

$$\text{MTD} = \frac{A - B}{\ln(A/B)}$$

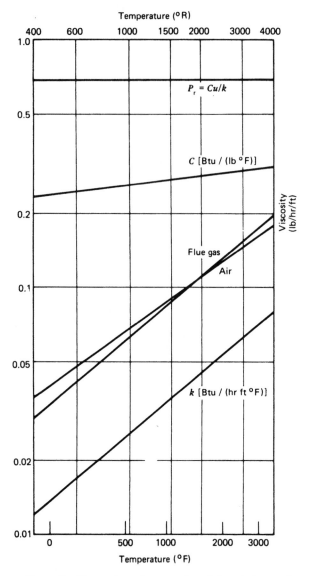

Figure 9 Thermodynamic properties of air and flue gas.[1]

8.1 Solid-State Radiation

"Blackbody" surfaces are those that absorb all radiation received, with zero reflection, and exist only as limits approached by actual sources or receivers of solid radiation. Radiation between black bodies is expressed by the Stefan-Boltzmann equation:

$$Q/A = N(T^4 - T_0^4) \quad \text{Btu/hr·ft}^2$$

where N is the Stefan-Boltzmann constant, now set at about 0.1713×10^{-8} for T and T_0, source and receiver temperatures, in °R. Because the fourth powers of numbers representing

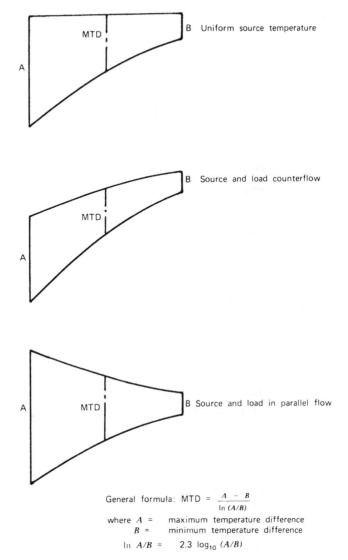

General formula: $\text{MTD} = \dfrac{A - B}{\ln(A/B)}$

where A = maximum temperature difference
B = minimum temperature difference

$\ln A/B$ = $2.3 \log_{10}(A/B)$

Figure 10 Diagrams of log-mean temperature difference (MTD).[1]

temperatures in °R are large and unwieldy, it is more convenient to express temperatures in °S, equivalent to (°F + 460)/100. The constant N is then reduced to 0.1713.

With source and receiver temperatures identified as T_s and T_r in °S, and with allowance for emissivity and view factors, the complete equation becomes

$$Q/A = 0.1713 \times \text{em} \times \text{Fr}(T_s^4 - T_r^4) \quad \text{Btu/hr·ft}^2$$

at the receiving surface,

where em = combined emissivity and absorptivity factors for source and receiving surfaces
Fr = net radiation view factor for receiving surface
T_s and T_r = source and receiving temperature in °S

The factor *em* will be somewhat less than *e* for the source or *a* for the receiving surface, and can be calculated:

$$em = 1 \left/ \sqrt{\frac{1}{a} + \frac{A_r}{A_s}\left(\frac{1}{e} - 1\right)} \right.$$

where a = receiver absorptivity at T_r
A_r/A_s = area ratio, receiver/source
e = source emissivity at T_s

8.2 Emissivity–Absorptivity

While emissivity and absorptivity values for solid materials vary with temperatures, values for materials commonly used as furnace walls or loads, in the usual range of temperatures, are:

Refractory walls	0.80–0.90
Heavily oxidized steel	0.85–0.95
Bright steel strip	0.25–0.35
Brass cake	0.55–0.60
Bright aluminum strip	0.05–0.10
Hot-rolled aluminum plate	0.10–0.20
Cast heat-resisting alloy	0.75–0.85

For materials such as sheet glass, transparent in the visible light range, radiation is reflected at both surfaces at about 4% of incident value, with the balance absorbed or transmitted. Absorptivity decreases with temperature, as shown in Fig. 11. The absorptivity of liquid water is about 0.96.

8.3 Radiation Charts

For convenience in preliminary calculations, black-body radiation, as a function of temperature in °F, is given in chart form in Fig. 12. The value for the receiver surface is subtracted from that of the source to find net interchange for blackbody conditions, and the result is corrected for emissivity and view factors. Where heat is transmitted by a combination of solid-state radiation and convection, a blackbody coefficient, in Btu/hr·°F, is shown in the chart in Fig. 13. This can be added to the convection coefficient for the same temperature interval, after correcting for emissivity and view factor, to provide an overall coefficient (H) for use in the formula

$$Q/A = H(T - T_r)$$

8.4 View Factors for Solid-State Radiation

For a receiving surface completely enclosed by the source of radiation, or for a flat surface under a hemispherical radiating surface, the view factor is unity. Factors for a wide range of geometrical configurations are given in available references. For cases commonly involved in furnace heat-transfer calculations, factors are shown by the following charts.

For two parallel planes, with edges in alignment as shown in Fig. 14*a*, view factors are given in Fig. 15 in terms of ratios of *x*, *y*, and *z*. For two surfaces intersecting at angle of

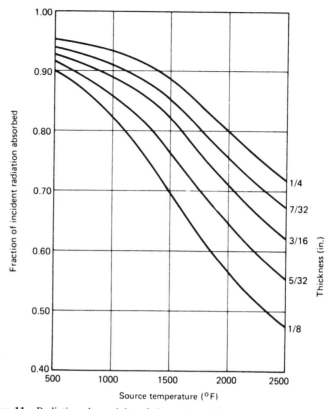

Figure 11   Radiation absorptivity of sheet glass with surface reflection deducted.[1]

90° at a common edge, the view factor is shown in Fig. 16. If surfaces do not extend to a common intersection, the view factor for the missing areas can be calculated and deducted from that with surfaces extended as in the figure, to find the net value for the remaining areas.

For spaced cylinders parallel to a furnace wall, as shown in Fig. 17, the view factor is shown in terms of diameter and spacing, including wall reradiation. For tubes exposed on both sides to source or receiver radiation, as in some vertical strip furnaces, the following factors apply if sidewall reradiation is neglected:

Ratio C/D	1.0	1.5	2.0	2.5	3.0
Factor	0.67	0.793	0.839	0.872	0.894

For ribbon-type electric heating elements, mounted on a back-up wall as shown in Fig. 18, exposure factors for projected wall area and for total element surface area are shown as a function of the (element spacing)/(element width) ratio. Wall reradiation is included, but heat loss through the backup wall is not considered. The emission rate from resistor surface will be $W/in.^2 = Q/491A$, where

$$\frac{Q}{A} = \frac{Btu/hr}{ft^2}$$

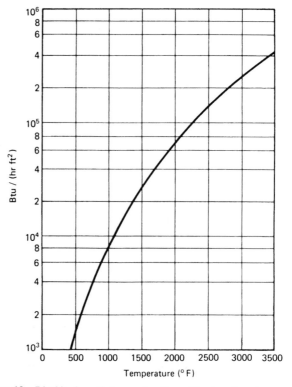

Figure 12   Blackbody radiation as function of load surface temperature.

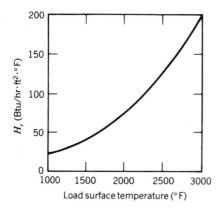

Figure 13   Blackbody radiation coefficient for source temperature uniform at 50–105° above final load surface temperature.

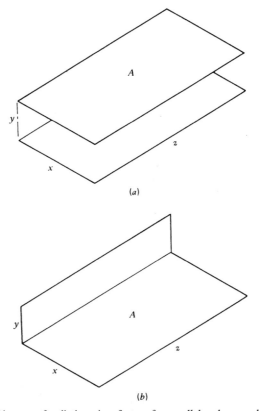

Figure 14 Diagram of radiation view factors for parallel and perpendicular planes.[1]

For parallel planes of equal area, as shown in Fig. 14, connected by reradiating walls on four sides, the exposure factor is increased as shown in Fig. 19. Only two curves, for $z/x = 1$ and $z/x = 10$ have been plotted for comparison with Fig. 13.

8.5 Gas Radiation

Radiation from combustion gases to walls and load can be from luminous flames or from nonluminous products of combustion. Flame luminosity results from suspended solids in combustion gases, either incandescent carbon particles or ash residues, and the resulting radiation is in a continuous spectrum corresponding to that from solid-state radiation at the same source temperature. Radiation from nonluminous gases is in characteristic bands of wavelengths, with intensity depending on depth and density of the radiating gas layer, its chemical composition, and its temperature.

For combustion of hydrocarbon gases, flame luminosity is from carbon particles formed by cracking of unburned fuel during partial combustion, and is increased by delayed mixing of fuel and air in the combustion chamber. With fuel and air thoroughly premixed before ignition, products of combustion will be nonluminous in the range of visible light, but can radiate strongly in other wavelength bands for some products of combustion including carbon

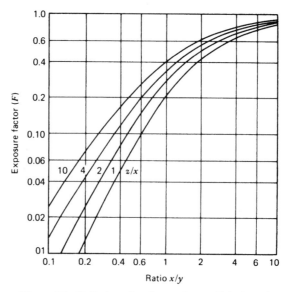

Figure 15 Radiation view factors for parallel planes.[1]

dioxide and water vapor. Published data on emissivities of these gases show intensity of radiation as a function of temperature, partial pressure, and beam length. The combined emissivity for mixtures of carbon dioxide and water vapor requires a correction factor for mutual absorption. To expedite calculations, a chart has been prepared for the overall emissivity of some typical flue gases, including these correction factors. The chart in Fig. 20 has been calculated for products of combustion of methane with 110% of net air demand, and is approximately correct for other hydrocarbon fuels of high heating value, including

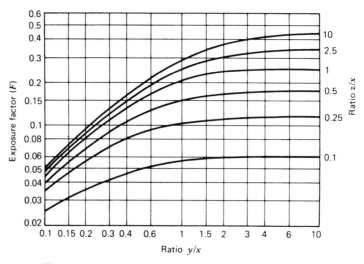

Figure 16 Radiation view factors for perpendicular planes.[1]

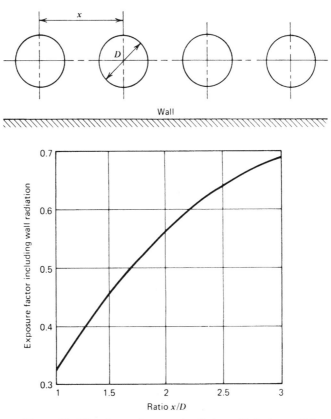

Figure 17 View factors for spaced cylinders with backup wall.[1]

coke oven gas and fuel oils. Emissivities for producer gas and blast furnace gas will be lower, because of dilution of radiating gases by nitrogen.

The emissivity of a layer of combustion gases does not increase directly with thickness or density, because of partial absorption during transmission through the depth of the layer. The chart provides several curves for a range of values of L, the effective beam length in feet, at a total pressure of 1 atm. For other pressures, the effective beam length will vary directly with gas density.

Beam lengths for average gas densities will be somewhat less than for very low density because of partial absorption. For some geometrical configurations, average beam lengths are:

Between two large parallel planes, $1.8 \times$ spacing

Inside long cylinder, about $0.85 \times$ diameter in feet

For rectangular combustion chambers, $3.4V/A$ where V is volume in cubic feet and A is total wall area in square feet

Transverse radiation to tube banks, with tubes of D outside diameter spaced at x centers: L/D ranges from 1.48 for staggered tubes at $x/D = 1.5$ to 10.46 for tubes in line and $x/D = 3$ in both directions

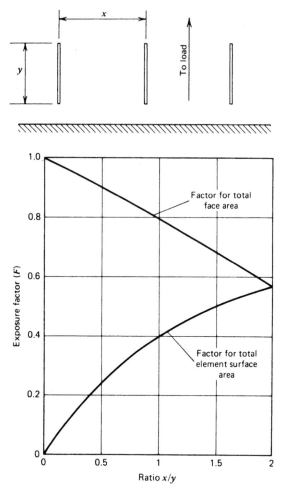

Figure 18   View factors for ribbon-type electric heating elements mounted on backup wall.[1]

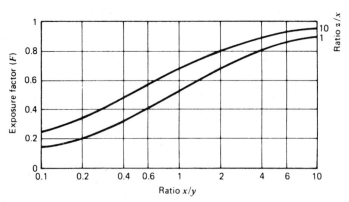

Figure 19   View factors for parallel planes connected by reradiating sidewalls.[1]

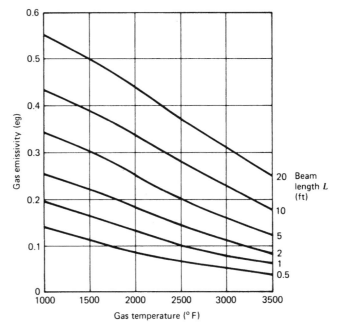

Figure 20 Gas-emissivity for products of combustion of methane burned with 110% air. Approximate for fuel oils and coke oven gas.[1]

8.6 Evaluation of Mean Emissivity–Absorptivity

For a gas with emissivity e_g radiating to a solid surface at a temperature of T_s °F, the absorptivity a_g will be less than e_g at T_s because the density of the gas is still determined by T_g. The effective PL becomes $T_s/T_g \times PL$ at T_s. Accurate calculation of the combined absorptivity for carbon dioxide and water vapor requires a determination of a_g for either gas and a correction factor for the total. For the range of temperatures and PL factors encountered in industrial heat transfer, the net heat transfer can be approximated by using a factor e_{gm} somewhat less than e_g at T_g in the formula:

$$Q/A = 0.1713 e_{gm} F(T_g^4 - T_s^4)$$

where T_g is an average of gas temperatures in various parts of the combustion chamber; the effective emissivity will be about $e_{gm} = 0.9 e_g$ at T_g and can be used with the chart in Fig. 20 to approximate net values.

8.7 Combined Radiation Factors

For a complete calculation of heat transfer from combustion gases to furnace loads, the following factors will need to be evaluated in terms of the equivalent fraction of blackbody radiation per unit area of the exposed receiving surface:

F_{gs} = Coefficient for gas direct to load, plus radiation reflected from walls to load.

F_{gw} = Coefficient for gas radiation absorbed by walls.

F_{ws} = Coefficient for solid-state radiation from walls to load.

Convection heat transfer from gases to walls and load is also involved, but can be eliminated from calculations by assuming that gas to wall convection is balanced by wall losses, and that gas to load convection is equivalent to a slight increase in load surface absorptivity. Mean effective gas temperature is usually difficult to measure, but can be calculated if other factors are known. For example, carbon steel slabs are being heated to rolling temperature in a fuel-fired continuous furnace. At any point in the furnace, neglecting convection,

$$F_{gw} (T_g^4 - T_w^4) = F_{ws} (T_w^4 - T_s^4)$$

where T_g, T_w, and T_s are gas, wall, and load surface temperatures in °S.

For a ratio of 2.5 for exposed wall and load surfaces, and a value of 0.17 for gas-to-wall emissivity, $F_{gw} = 2.5 \times 0.17 = 0.425$. With wall to load emissivity equal to $F_{ws} = 0.89$, wall temperature constant at 2350°F (28.1°S), and load temperature increasing from 70 to 2300°F at the heated surface ($T_s = 5.3-27.6$°S), the mean value of gas temperature (T_g) can be determined:

$$\text{MTD, walls to load} = \frac{2280 - 50}{\ln(2280/50)} = 584°F$$

Mean load surface temperature $T_{sm} = 2350 - 584 = 1766°F$ (22.26°S)

Q/A per unit of load surface, for reradiation:

$$0.425 \times 0.1713(T_g^4 - 28.1^2) = 0.89 \times 0.1713(28.1^4 - 22.26^4) = 57,622 \text{ Btu/hr·ft}^2$$

$$T_g = 34.49°S \ (2989°F)$$

With a net wall emissivity of 0.85, 15% of gas radiation will be reflected to the load, with the balance being absorbed and reradiated. Direct radiation from gas to load is then

$$1.15 \times 0.17 \times 0.1713(34.49^4 - 22.26^4) = 47,389 \text{ Btu/hr·ft}^2$$

$$\text{Total radiation: } 57,622 + 47,389 = 105,011 \text{ Btu/hr·ft}^2$$

For comparison, blackbody radiation from walls to load, without gas radiation, would be 64,743 Btu/hr·ft² or 62% of the combined total.

With practical furnace temperature profiles, in a counterflow, direct-fired continuous furnace, gas and wall temperatures will be depressed at the load entry end to reduce flue gas temperature and stack loss. The resulting net heating rates will be considered in Section 8.12.

Overall heat-transfer coefficients have been calculated for constant wall temperature, in the upper chart in Fig. 21, or for constant gas temperature in the lower chart. Coefficients vary with mean gas emissivity and with A_w/A_s, the ratio of exposed surface for walls and load, and are always less than one for overall radiation from gas to load, or greater than one for wall to load radiation. Curves can be used to find gas, wall, or mean load temperatures when the other two are known.

8.8   Steady-State Conduction

Heat transfer through opaque solids and motionless layers of liquids or gases is by conduction. For constant temperature conditions, heat flow is by "steady-state" conduction and does not vary with time. For objects being heated or cooled, with a continuous change in internal temperature gradients, conduction is termed "non-steady-state."

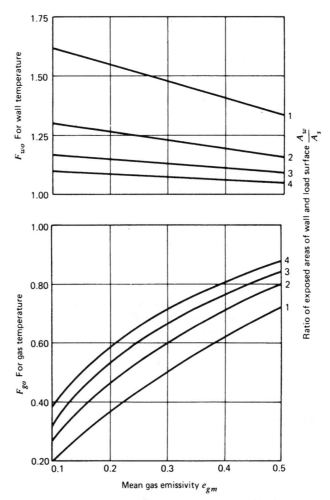

Figure 21 Overall heat-transfer coefficients for gas and solid radiation, as function of gas emissivity and wall-to-load area ratio, for uniform gas or wall temperature, compared to blackbody radiation.[1]

Thermal conduction in some solid materials is a combination of heat flow through the material, radiation across internal space resulting from porosity, and convection within individual pores or through the thickness of porous layers.

Conductivities of refractory and insulating materials tend to increase with temperature, because of porosity effects. Values for most metals decrease with temperature, partly because of reduced density. Conductivity coefficients for some materials used in furnace construction or heated in furnaces are listed in Table 5.

A familiar problem in steady-state conduction is the calculation of heat losses through furnace walls made up of multiple layers of materials of different thermal conductivities. A convenient method of finding overall conductance is to find the thermal resistance (r/k = thickness/conductivity in consistent units) and add the total for all layers. Because conductivities vary with temperature, mean temperatures for each layer can be estimated from a preliminary temperature profile for the composite wall. Overall resistance will include the effects of radiation and conduction between the outer wall surface and its surroundings.

A chart showing heat loss from walls to ambient surrounding at 70°F, combining radiation and convection for vertical walls, is shown in Fig. 7. The corresponding thermal resistance is included in the overall heat-transfer coefficient shown in Fig. 8 as a function of net thermal resistance of the wall structure and inside face temperature.

As an example of application, assume a furnace wall constructed as follows:

Material	r	k	r/k
9 in. firebrick	0.75	0.83	0.90
4½ in. 2000°F insulation	0.375	0.13	2.88
2½ in. ceramic fiber block	0.208	0.067	3.10
Total R for solid wall			6.88

With an inside surface temperature of 2000°F, the heat loss from Fig. 7 is about 265 Btu/ft·hr². The corresponding surface temperature from Fig. 8 is about 200°F, assuming an ambient temperature of 70°F.

Although not a factor affecting wall heat transfer, the possibility of vapor condensation in the wall structure must be considered by the furnace designer, particularly if the furnace is fired with a sulfur-bearing fuel. As the sulfur dioxide content of fuel gases is increased, condensation temperatures increase to what may exceed the temperature of the steel furnace casing in normal operation. Resulting condensation at the outer wall can result in rapid corrosion of the steel structure.

Condensation problems can be avoided by providing a continuous membrane of aluminum or stainless steel between layers of the wall structure, at a point where operating temperatures will always exceed condensation temperatures.

8.9 Non-Steady-State Conduction

Heat transfer in furnace loads during heating or cooling is by transient or non-steady-state conduction, with temperature profiles within loads varying with time. With loads of low internal thermal resistance, heating time can be calculated for the desired load surface temperature and a selected time–temperature profile for furnace temperature. With loads of appreciable thermal resistance from surface to center, or from hot to colder sides, heating time will usually be determined by a specified final load temperature differential, and a selected furnace temperature profile for the heating cycle.

For the case of a slab-type load being heated on a furnace hearth, with only one side exposed, and with the load entering the furnace at ambient temperature, the initial gradient from the heated to the unheated surface will be zero. The heated surface will heat more rapidly until the opposite surface starts to heat, after which the temperature differential between surfaces will taper off with time until the desired final differential is achieved.

In Fig. 22 the temperatures of heated and unheated surface or core temperature are shown as a function of time. In the lower chart temperatures are plotted directly as a function of time. In the upper chart the logarithm of the temperature ratio (Y = load temperature/source temperature) is plotted as a function of time for a constant source temperature. After a short initial heating time, during which the unheated surface or core temperature reaches its maximum rate of increase, the two curves in the upper diagram become parallel straight lines.

Factors considered in non-steady-state conduction and their identifying symbols are listed in Table 6.

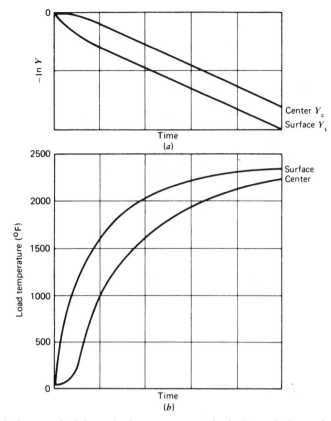

Figure 22   Maximum and minimum load temperatures, and $-\ln Y_s$ or $-\ln Y_c$ as a function of heating time with constant source temperature.[1]

Table 6   Non-Steady-State Conduction Factors and Symbols

T_f = Furnace temperature, gas or wall as defined
T_s = Load surface temperature
T_c = Temperature at core or unheated side of load
T_0 = Initial load temperature with all temperatures in units of (°F − 460)/100 or °S

$$Y_s = \frac{T_f - T_s}{T_f - T_0}$$

$$Y_c = \frac{T_f - T_c}{T_f - T_0}$$

R = External/internal thermal resistance ratio = k/rH
X = Time factor = tD/r^2
D = Diffusivity as defined in Section 45.7
r = Depth of heat penetration in feet
k = Thermal conductivity of load (Btu/ft·hr·°F)
H = External heat transfer coefficient (Btu/ft²·hr·°F)

Charts have been prepared by Gurney-Lurie, Heisler, Hottel, and others showing values for Y_s and Y_c for various R factors as a function of X. Separate charts are provided for Y_s and Y_c, with a series of curves representing a series of values of R. These curves are straight lines for most of their length, curving to intersect at $Y = 1$ and $X = 0$. If straight lines are extended to $Y = 1$, the curves for Y_c at all values of R converge at a point near $X = 0.1$ on the line for $Y_c = 1$. It is accordingly possible to prepare a single line chart for $-\ln Y_c/(X - 0.1)$ to fit selected geometrical shapes. This has been done in Fig. 23 for slabs, long cylinders, and spheres. Values of Y_c determined with this chart correspond closely with those from conventional charts for $X - 0.1$ greater than 0.2.

Because the ratio Y_s/Y_c remains constant as a function of R after initial heating, it can be shown in chart form, as in Fig. 24, to allow Y_s to be determined after Y_c has been found.

By way of illustration, a carbon steel slab 8 in. thick is being heated from cold to $T_s = 2350°F$ in a furnace with a constant wall temperature of 2400°F, with a view factor of 1 and a mean emissivity–absorptivity factor of 0.80. The desired final temperature of the unheated surface is 2300°F, making the Y_c factor

$$Y_c = \frac{2400 - 2300}{2400 - 70} = 0.0429$$

From Fig. 23 $H_r = 114 \times 0.80 = 91$; $r = 8/12 = 0.67$; R is assumed at 17. The required heating time is determined from Fig. 24:

$$R = \frac{17}{0.67 \times 91} = 0.279$$

$$\frac{-\ln Y_c}{X - 0.1} = 1.7$$

and

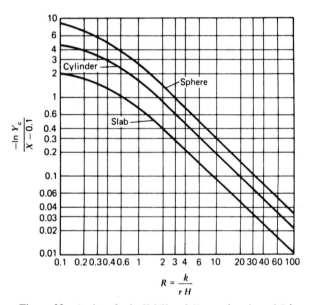

Figure 23 A plot of $-\ln Y_c/(X - 0.1)$ as a function of R.[1]

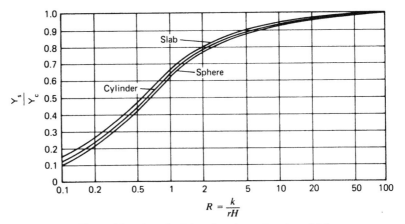

Figure 24 The ratio Y_s/Y_c plotted as a function of R.[1]

$$X = \frac{-\ln 0.0429}{1.7} + 0.1 = 1.95 = tD/r^2$$

With $D = 0.25$, from Section 7,

$$t = \frac{Xr^2}{D} = \frac{1.95 \times 0.67^2}{0.25} = 3.50 \text{ hr}$$

Slabs or plates heated from two sides are usually supported in the furnace in a horizontal position on spaced conveyor rolls or rails. Support members may be uncooled, in which case radiation to the bottom surface will be reduced by the net view factor. If supports are water cooled, the additional heat input needed to balance heat loss from load to supports can be balanced by a higher furnace temperature on the bottom side. In either case, heating times will be greater than for a uniform input from both sides.

Furnace temperatures are normally limited to a fraction above final load temperatures, to avoid local overheating during operating delays. Without losses to water cooling, top and bottom furnace temperature will accordingly be about equal.

8.10   Heat Transfer with Negligible Load Thermal Resistance

When heating thin plates or small-diameter rods, with internal thermal resistance low enough to allow heating rates unlimited by specified final temperature differential, the non-steady-state-conduction limits on heating rates can be neglected. Heating time then becomes

$$t = \frac{W \times C \times (T_s - T_0)}{A \times H \times \text{MTD}}$$

The heat-transfer coefficient for radiation heating can be approximated from the chart in Fig. 13 or calculated as follows:

$$H_r = \frac{0.1713 e_m F_s [T_f^4 - (T_f - \text{MTD})^4]}{\text{MTD} \times A_s}$$

As an illustration, find the time required to heat a steel plate to 2350°F in a furnace at a uniform temperature of 2400°F. The plate is 0.25 in. thick with a unit weight of 10.2 lb/ft² and is to be heated from one side. Overall emissivity–absorptivity is $e_m = 0.80$. Specific heat is 0.165. The view factor is $F_s = 1$. MTD is

$$\frac{(2400 - 70) - (2400 - 2350)}{\ln(2400 - 70)/(2400 - 2350)} = 588°F$$

$$H_r = \frac{0.1713 \times 0.80 \times 1[28.6^4 - (28.6 - 5.88)^4]}{588} = 93.8$$

$$t = \frac{10.2 \times 0.165(2350 - 70)}{1 \times 93.8 \times 588} = 0.069 \text{ hr}$$

8.11 Newman Method

For loads heated from two or more perpendicular sides, final maximum temperatures will be at exposed corners, with minimum temperatures at the center of mass for heating from all sides, or at the center of the face in contact with the hearth for hearth-supported loads heated equally from the remaining sides. For surfaces not fully exposed to radiation, the corrected H factor must be used.

The Newman method can be used to determine final load temperatures with a given heating time t. To find time required to reach specified maximum and minimum final load temperatures, trial calculations with several values of t will be needed.

For a selected heating time t, the factors Y_s and Y_c can be found from charts in Figs. 23 and 24 for the appropriate values of the other variables—T_s, T_c, H, k, and r—for each of the heat flow paths involved—r_x, r_y, and r_z. If one of these paths is much longer than the others, it can be omitted from calculations:

$$Y_c = Y_{cx} \times Y_{cy} \times Y_{cz}$$
$$Y_s = Y_{sx} \times Y_{sy} \times Y_{sz}$$

For two opposite sides with equal exposure only one is considered. With T_c known, T_s and T_f (furnace temperature, T_g or T_w) can be calculated.

As an example, consider a carbon steel ingot, with dimensions 2 ft × 4 ft × 6 ft, being heated in a direct-fired furnace. The load is supported with one 2 ft × 4 ft face in contact with the refractory hearth and other faces fully exposed to gas and wall radiation. Maximum final temperature will be at an upper corner, with minimum temperature at the center of the 2 ft × 4 ft bottom surface. Assuming that the load is a somewhat brittle steel alloy, the initial heating rate should be suppressed and heating with a constant gas temperature will be assumed. Heat-transfer factors are then

Flow paths $r_s = 1$ ft and $r_y = 2$ ft, the contribution of vertical heat flow, on axis r_z, will be small enough to be neglected.

Desired final temperatures: $T_c = 2250°F$ and T_s (to be found) about 2300°F, with trial factor $t = 9$ hr.

H from gas to load = 50

k mean value for load = 20 and $D = 0.25$

Radial heat flow path	r_x	r_y
r	1	2
$X = tD/r^2$	2.25	0.5625
$R = k/H_r$	0.4	0.2
$-\ln Y_c/(X - 0.1)$ from Fig. 23	1.3	1.7
Y_s/Y_c from Fig. 24	0.41	0.26
Y_c	0.0611	0.455
Y_s	0.025	0.119

Combined factors:

$$Y_c = 0.0611 \times 0.455 = 0.0278 = \frac{T_g - T_c}{T_g - 70}$$

$$Y_s = 0.025 \times 0.119 = 0.003 = \frac{T_g - T_s}{T_g - 70}$$

For $T_c = 2250°F$, $\quad T_g = 2316°F$
$$T_s = 2309°F$$

This is close enough to the desired $T_s = 2300°F$.

The time required to heat steel slabs to rolling temperature, as a function of the thickness heated from one side and the final load temperature differential, is shown in Fig. 25. Relative heating times for various hearth loading arrangements, for square billets, are shown in Fig. 26. These have been calculated by the Newman method, which can also be used to evaluate other loading patterns and cross sections.

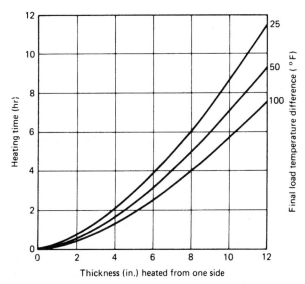

Figure 25 Relative heating time for square billets as a function of loading pattern.[1]

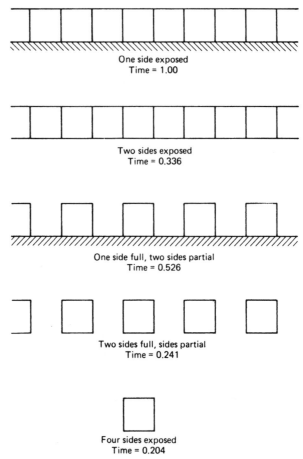

One side exposed
Time = 1.00

Two sides exposed
Time = 0.336

One side full, two sides partial
Time = 0.526

Two sides full, sides partial
Time = 0.241

Four sides exposed
Time = 0.204

Figure 26 Heating time for carbon steel slabs to final surface temperature of 2300°F, as a function of thickness and final load temperature differential.[1]

8.12 Furnace Temperature Profiles

To predict heating rates and final load temperatures in either batch or continuous furnaces, it is convenient to assume that source temperatures, gas (T_g) or furnace wall (T_w), will be constant in time. Neither condition is achieved with contemporary furnace and control system designs. With constant gas temperature, effective heating rates are unnecessarily limited, and the furnace temperature control system is dependent on measurement and control of gas temperatures, a difficult requirement. With uniform wall temperatures, the discharge temperature of flue gases at the beginning of the heating cycle will be higher than desirable. Three types of furnace temperature profiles, constant T_g, constant T_w, and an arbitrary pattern with both variables, are shown in Fig. 27.

Contemporary designs of continuous furnaces provide for furnace temperature profiles of the third type illustrated, to secure improved capacity without sacrificing fuel efficiency. The firing system comprises three zones of length: a preheat zone that can be operated to maintain minimum flue gas temperatures in a counterflow firing arrangement, a firing zone

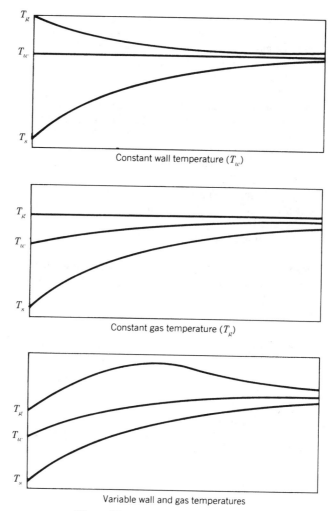

Figure 27 Furnace temperature profiles.

with a maximum temperature and firing rate consistent with furnace maintenance requirements and limits imposed by the need to avoid overheating of the load during operating delays, and a final or soak zone to balance furnace temperature with maximum and minimum load temperature specifications. In some designs, the preheat zone is unheated except by flue gases from the firing zone, with the resulting loss of furnace capacity offset by operating the firing zone at the maximum practical limit.

8.13 Equivalent Furnace Temperature Profiles

Furnace heating capacities are readily calculated on the assumption that furnace temperature, either combustion gases or radiating walls, is constant as a function of position or time. Neither condition is realized in practice; and to secure improved capacity with reduced fuel demand in a continuous furnace, contemporary designs are based on operation with a variable temperature profile from end to end, with furnace wall temperature reduced at the load charge and flue gas discharge end, to improve available heat of fuel, and at the load discharge end, to balance the desired maximum and minimum load temperatures. Any loss in capacity can be recovered by operating the intermediate firing zones at a somewhat elevated temperature.

Consider a furnace designed to heat carbon steel slabs, 6 in. thick, from the top only to final temperatures of 2300°F at top and 2250°F at the bottom. To hold exit flue gas temperature to about 2000°F, wall temperature at the charge end will be about 1400°F. The furnace will be fired in four zones of length, each 25 ft long for an effective total length of 100 ft. The preheat zone will be unfired, with a wall temperature tapering up to 2400°F at the load discharge end. That temperature will be held through the next two firing zones and dropped to 2333°F to balance final load temperatures in the fourth or soak zone. With overall heating capacity equal to the integral of units of length times their absolute temperatures, effective heat input will be about 87% of that for a uniform temperature of 2400°F for the entire length.

Heat transfer from combustion gases to load will be by direct radiation from gas to load, including reflection of incident radiation from walls, and by radiation from gas to walls, absorbed and reradiated from walls to load. Assuming that wall losses will be balanced by convection heat transfer from gases, gas radiation to walls will equal solid-state radiation from walls to load:

$$A_w/A_s \times 0.1713 \times e_{gm}(T_g^4 - T_w^4) = e_{ws} \times 0.1713(T_w^4 - T_s^4)$$

where A_w/A_s = exposed area ratio for walls and load

e_{gm} = emissivity–absorptivity, gas to walls

e_{ws} = emissivity–absorptivity, walls to load

At the midpoint in the heating cycle, MTD = 708°F and mean load surface temperature = T_{sm} = 1698°F.

With a_s = 0.85 for refractory walls, 15% of gas radiation will be reflected to load, and total gas to load radiation will be:

$$1.15 \times e_{gm} \times 0.1713(T_g^4 - T_s^4)$$

For A_w/A_s = 2.5, e_{gm} = 0.17, and e_{ws} = 0.89 from walls to load, the mean gas temperature = T_g = 3108°F, net radiation, gas to load = 47,042 Btu/hr·ft² and gas to walls = walls to load = 69,305 Btu/hr·ft² for a total of 116,347 Btu/hr·ft². This illustrates the relation shown in Fig. 21, since blackbody radiation from walls to load, without gas radiation, would be 77,871 Btu/hr·ft². Assuming black-body radiation with a uniform wall temperature from end to end, compared to combined radiation with the assumed wall temperature, overall heat transfer ratio will be

$$(0.87 \times 116,347)/77,871 = 1.30$$

As shown in Fig. 26, this ratio will vary with gas emissivity and wall to load areas exposed. For the range of possible values for these factors, and for preliminary estimates of heating times, the chart in Fig. 26 can be used to indicate a conservative heating time as a function of final load temperature differential and depth of heat penetration, for a furnace temperature profile depressed at either end.

Radiation factors will determine the mean coefficient of wall to load radiation, and the corresponding non-steady-state conduction values. For black-body radiation alone, H_r is about 77,871/708 = 110. For combined gas and solid-state radiation, in the above example, it becomes $0.87 \times 116{,}347/708 = 143$. Values of R for use with Figs. 23 and 24, will vary correspondingly ($R = k/4H$).

8.14 Convection Heat Transfer

Heat transferred between a moving layer of gas and a solid surface is identified by "convection." Natural convection occurs when movement of the gas layer results from differentials in gas density of the boundary layer resulting from temperature differences and will vary with the position of the boundary surface: horizontal upward, horizontal downward, or vertical. A commonly used formula is

$$H_c = 0.27(T_g - T_s)^{0.25}$$

where H_c = Btu/hr·ft²·°F

$T_g - T_s$ = temperature difference between gas and surface, in °F

Natural convection is a significant factor in estimating heat loss from the outer surface of furnace walls or from uninsulated pipe surfaces.

"Forced convection" is heat transfer between gas and a solid surface, with gas velocity resulting from energy input from some external source, such as a recirculating fan.

Natural convection can be increased by ambient conditions such as building drafts and gas density. Forced convection coefficients will depend on surface geometry, thermal properties of the gas, and Reynolds number for gas flow. For flow inside tubes, the following formula is useful:

$$H_c = 0.023 \frac{k}{D} \, \mathrm{Re}^{0.8} \mathrm{Pr}^{0.4} \mathrm{Btu/hr \cdot ft^2 \cdot °F}$$

where k = thermal conductivity of gas
D = inside diameter of tube in ft
Re = Reynolds number
Pr = Prandtl number

Forced convection coefficients are given in chart form in Fig. 28 for a Prandtl number assumed at 0.70.

For forced convection over plane surfaces, it can be assumed that the preceding formula will apply for a rectangular duct of infinitely large cross section, but only for a length sufficient to establish uniform velocity over the cross section and a velocity high enough to reach the Re value needed to promote turbulent flow.

In most industrial applications, the rate of heat transfer by forced convection as a function of power demand will be better for perpendicular jet impingement from spaced nozzles than for parallel flow. For a range of dimensions common in furnace design, the heat-transfer coefficient for jet impingement of air or flue gas is shown in Fig. 29, calculated for impingement from slots 0.375 in. wide spaced at 18–24 in. centers and with a gap of 8 in. from nozzle to load.

Forced convection factors for gas flow through banks of circular tubes are shown in the chart in Fig. 30 and for tubes spaced as follows:

A: staggered tubes with lateral spacing equal to diagonal spacing.
B: tubes in line, with equal spacing across and parallel to direction of flow.
C: tubes in line with lateral spacing less than half longitudinal spacing.
D: tubes in line with lateral spacing over twice longitudinal spacing.

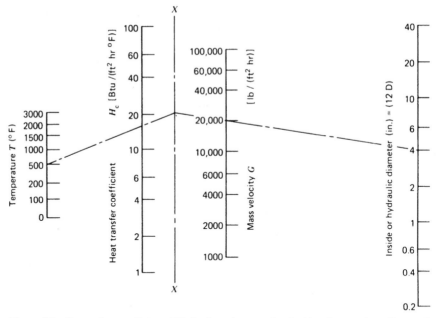

Figure 28 Convection coefficient (H_c) for forced convection inside tubes to air or flue gas.[1]

With F the configuration factor from Fig. 30, heat-transfer coefficients are

$$H_c = Fk\mathrm{Re}^{0.6}/D$$

Convection coefficients from this formula are approximately valid for 10 rows of tubes or more, but are progressively reduced to a factor of 0.65 for a single row.

For gas to gas convection in a cross-flow tubular heat exchanger, overall resistance will be the sum of factors for gas to the outer diameter of tubes, tube wall conduction, and inside diameter of tubes to gas. Factors for the outer diameter of tubes may include gas radiation as calculated in Section 7.5.

8.15 Fluidized-Bed Heat Transfer

For gas flowing upward through a particular bed, there is a critical velocity when pressure drop equals the weight of bed material per unit area. Above that velocity, bed material will be suspended in the gas stream in a turbulent flow condition. With the total surface area of suspended particles on the order of a million times the inside surface area of the container, convection heat transfer from gas to bed material is correspondingly large. Heat transfer from suspended particles to load is by conduction during repeated impact. The combination can provide overall coefficients upward of 10 times those available with open convection, permitting the heating of thick and thin load sections to nearly uniform temperatures by allowing a low gas to load thermal head.

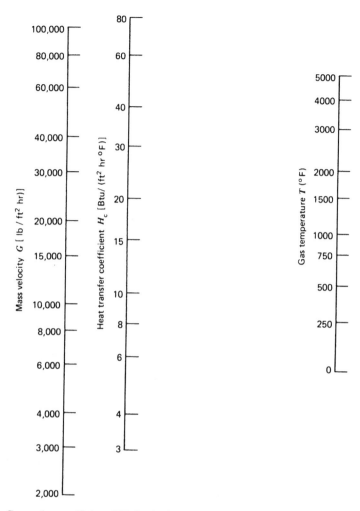

Figure 29 Convection coefficient (H_c) for jet impingement of air or flue gas on plane surfaces, for spaced slots, 0.375 in. wide at 18–24 in. centers, 8 in. from load.[1]

8.16 Combined Heat-Transfer Coefficients

Many furnace heat-transfer problems will combine two or more methods of heat transfer, with thermal resistances in series or in parallel. In a combustion chamber, the resistance to radiation from gas to load will be parallel to the resistance from gas to walls to load, which is two resistances in series. Heat flow through furnace walls combines a series of resistances in series, combustion gases to inside wall surface, consecutive layers of the wall structure, and outside wall surface to surroundings, the last a combination of radiation and convection in parallel.

As an example, consider an insulated, water-cooled tube inside a furnace enclosure. With a tube outside diameter of 0.5 ft and a cylindrical insulation enclosure with an outside

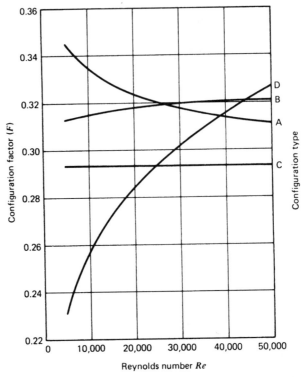

Figure 30 Configuration factors for convection heat transfer, air or flue gas through tube banks.[1]

diameter of 0.75 ft, the net thickness will be 0.125 ft. The mean area at midthickness is $\pi(0.5 + 0.75)/2$, or 1.964 ft^2 per ft of length. Outer surface area of insulation is 0.75π, or 2.36 ft^2 per linear foot. Conductivity of insulation is $k = 0.20$. The effective radiation factor from gas to surface is assumed at 0.5 including reradiation from walls. For the two resistances in series,

$$0.1713 \times 0.5 \times 2.36(29.6^4 - T_s^4) = 1.964(T_s - 150) \times \frac{0.20}{0.125}$$

By trial, the receiver surface temperature is found to be about 2465°F. Heat transfer is about 7250 Btu/hr·linear ft or 9063 Btu/hr·ft^2 water-cooled tube surface.

If the insulated tube in the preceding example is heated primarily by convection, a similar treatment can be used to find receiver surface temperature and overall heat transfer.

For radiation through furnace wall openings, heat transfer in Btu/hr·ft^2·°F is reduced by wall thickness, and the result can be calculated similarly to the problem of two parallel planes of equal size connected by reradiating walls, as shown in Fig. 19.

Heat transfer in internally fired combustion tubes ("radiant tubes") is a combination of convection and gas radiation from combustion gases to tube wall. External heat transfer from tubes to load will be direct radiation and reradiation from furnace walls, as illustrated in Fig. 19. The overall factor for internal heat transfer can be estimated from Fig. 31, calculated for 6 in. and 8 in. inside diameter tubes. The convection coefficient increases with firing rate

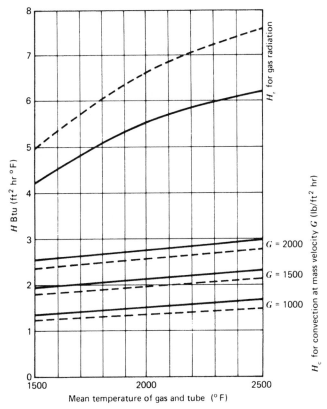

Figure 31 Gas radiation (H_r) and convection (H_c) coefficients for flue gas inside radiant tubes.[1]

and to some extent with temperature. The gas radiation factor depends on temperature and inside diameter. The effect of flame luminosity has not been considered.

9 FLUID FLOW

Fluid flow problems of interest to the furnace engineer include the resistance to flow of air or flue gas, over a range of temperatures and densities through furnace ductwork, stacks and flues, or recuperators and regenerators. Flow of combustion air and fuel gas through distribution piping and burners will also be considered. Liquid flow, of water and fuel oil, must also be evaluated in some furnace designs but will not be treated in this chapter.

To avoid errors resulting from gas density at temperature, velocities will be expressed as mass velocities in units of $G = $ lb/hr·ft^2. Because the low pressure differentials in systems for flow of air or flue gas are usually measured with a manometer, in units of inches of water column (in. H$_2$O), that will be the unit used in the following discussion.

The relation of velocity head h_v in in. H$_2$O to mass velocity G is shown for a range of temperatures in Fig. 32. Pressure drops as multiples of h_v are shown, for some configurations used in furnace design, in Figs. 33 and 34. The loss for flow across tube banks, in multiples of the velocity head, is shown in Fig. 35 as a function of the Reynolds number.

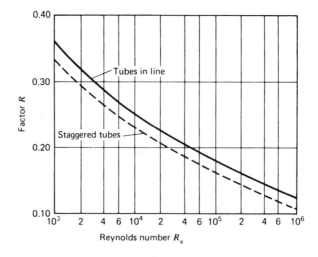

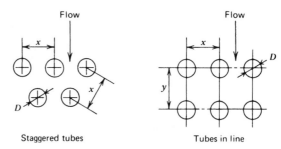

Figure 32 Heat loss for flow of air or flue gas across tube banks at atmospheric pressure (velocity head) $\times F \times R$.

The Reynolds number Re is a dimensionless factor in fluid flow defined as Re = DG/μ, where D is inside diameter or equivalent dimension in feet, G is mass velocity as defined above, and μ is viscosity as shown in Fig. 9. Values for Re for air or flue gas, in the range of interest, are shown in Fig. 36. Pressure drop for flow through long tubes is shown in Fig. 37 for a range of Reynolds numbers and equivalent diameters.

9.1 Preferred Velocities

Mass velocities used in contemporary furnace design are intended to provide an optimum balance between construction costs and operating costs for power and fuel; some values are listed on the next page:

Medium	Mass Velocity G	Velocity Head (in. H_2O)
Cold air	15,000	0.7
800°F air	10,000	0.3
2200°F flue gas	1,750	0.05
1500°F flue gas	2,000	0.05

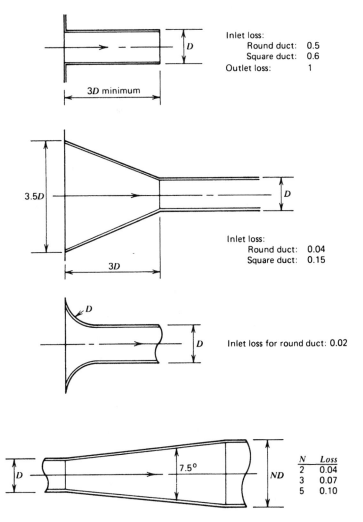

Figure 33 Pressure drop in velocity heads for flow of air or flue gas through entrance configurations or expansion sections.[1]

The use of these factors will not necessarily provide an optimum cost balance. Consider a furnace stack of self-supporting steel construction, lined with 6 in. of gunned insulation. For $G = 2000$ and $h_v = 0.05$ at 1500°F, an inside diameter of 12 ft will provide a flow of 226,195 lb/hr. To provide a net draft of 1 in. H_2O with stack losses of about $1.75h_v$ or 0.0875 in., the effective height from Fig. 38 is about 102 ft. By doubling the velocity head to 0.10 in. H_2O, G at 1500°F becomes 3000. For the same mass flow, the inside diameter is reduced to 9.8 ft. The pressure drop through the stack increases to about 0.175 in., and the height required to provide a net draft of 1 in. increases to about 110 ft. The outside diameter area of the stack is reduced from 4166 ft² to $11 \times 3.1416 \times 110 = 3801$ ft². If the cost per square foot of outside surface is the same for both cases, the use of a higher stack velocity will save construction costs. It is accordingly recommended that specific furnace designs receive a more careful analysis before selecting optimum mass velocities.

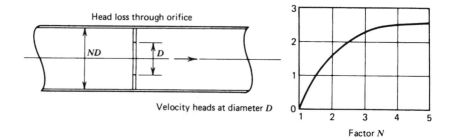

Head loss through orifice

Velocity heads at diameter D

Factor N

Head loss in pipe or duct elbows

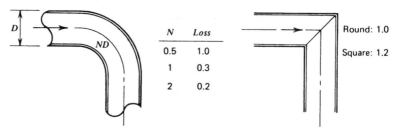

N	$Loss$
0.5	1.0
1	0.3
2	0.2

Round: 1.0

Square: 1.2

Proportioning Piping for uniform distribution

Total pressure = static pressure + velocity head

Area at D should exceed 2.5 × combined areas of A, B, and C

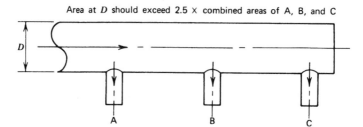

A　　　B　　　C

Figure 34 Pressure drop in velocity heads for flow of air or flue gas through orifices, elbows, and lateral outlets.[1]

Staggered Tubes		Tubes in Line		Factor F for x/D		
x/D	Factor F	y/D	1.5	2	3	4
1.5	2.00	1.25	1.184	0.576	0.334	0.268
2	1.47	1.5	1.266	0.656	0.387	0.307
3	1.22	2	1.452	0.816	0.497	0.390
4	1.14	3	1.855	1.136	0.725	0.572
		4	2.273	1.456	0.957	0.761

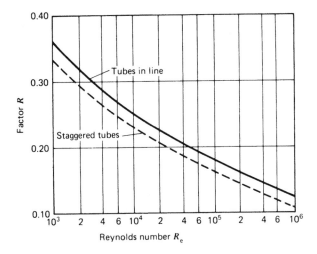

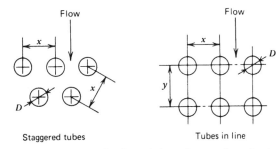

Figure 35 Pressure drop factors for flow of air or flue gas through tube banks.[1]

Staggered Tubes		Tubes in Line		Factor F for x/D		
x/D	Factor F	y/D	1.5	2	3	4
1.5	2.00	1.25	1.184	0.576	0.334	0.268
2	1.47	1.5	1.266	0.656	0.387	0.307
3	1.22	2	1.452	0.816	0.497	0.390
4	1.14	3	1.855	1.136	0.725	0.572
		4	2.273	1.456	0.957	0.761

Stack draft, at ambient atmospheric temperature of 70°F, is shown in Fig. 38 as a function of flue gas temperature. Where greater drafts are desirable with a limited height of stack, a jet-type stack can be used to convert the momentum of a cold air jet into stack draft. Performance data are available from manufacturers.

9.2 Centrifugal Fan Characteristics

Performance characteristics for three types of centrifugal fans are shown in Fig. 39. More exact data are available from fan manufacturers. Note that the backward curved blade has

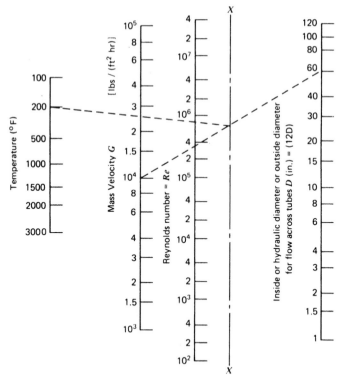

Figure 36 Reynolds number (Re) for flow of air or flue gas through tubes or across tube banks.[1]

the advantage of limited horsepower demand with reduced back pressure and increasing volume, and can be used where system resistance is unpredictable. The operating point on the pressure–volume curve is determined by the increase of duct resistance with flow, matched against the reduced outlet pressure, as shown in the upper curve.

9.3 Laminar and Turbulent Flows

The laminar flow of a fluid over a boundary surface is a shearing process, with velocity varying from zero at the wall to a maximum at the center of cross section or the center of the top surface for liquids in an open channel. Above a critical Reynolds number, between 2000 and 3000 in most cases, flow becomes a rolling action with a uniform velocity extending almost to the walls of the duct, and is identified as turbulent flow.

With turbulent flow the pressure drop is proportional to D; the flow in a large duct can be converted from turbulent to laminar by dividing the cross-sectional area into a number of parallel channels. If flow extends beyond the termination of these channels, the conversion from laminar to turbulent flow will occur over some distance in the direction of flow.

Radial mixing with laminar flow is by the process of diffusion, which is the mixing effect that occurs in a chamber filled with two different gases separated by a partition after the partition is removed. Delayed mixing and high luminosity in the combustion of hydrocarbon gases can be accomplished by "diffusion combustion," in which air and fuel enter the combustion chamber in parallel streams at equal and low velocity.

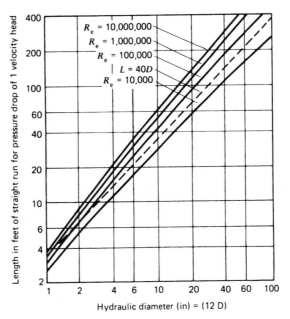

Figure 37 Length in feet for pressure drop of one velocity head, for flow of air or flue gas, as a function of Re and D.[1]

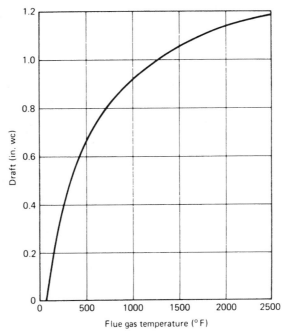

Figure 38 Stack draft for ambient $T_g = 70°F$ and psia $= 14.7$ lb/in.[2,1]

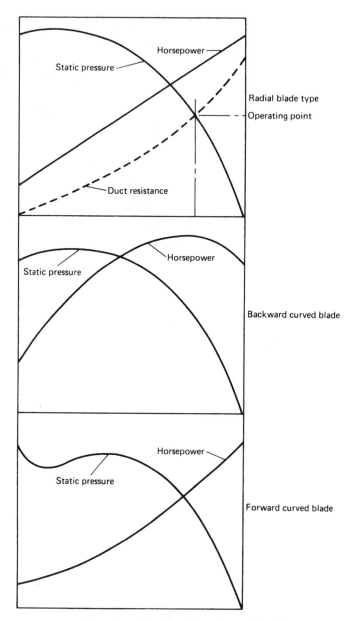

Figure 39   Centrifugal fan characteristics.[1]

10   BURNER AND CONTROL EQUIPMENT

With increasing costs of fuel and power, the fraction of furnace construction and maintenance costs represented by burner and control equipment can be correspondingly increased. Burner designs should be selected for better control of flame pattern over a wider range of turndown and for complete combustion with a minimum excess air ratio over that range.

Furnace functions to be controlled, manually or automatically, include temperature, internal pressure, fuel/air ratio, and adjustment of firing rate to anticipated load changes. For intermittent operation, or for a wide variation in required heating capacity, computer control may be justified to anticipate required changes in temperature setting and firing rates, particularly in consecutive zones of continuous furnaces.

10.1   Burner Types

Burners for gas fuels will be selected for the desired degree of premixing of air and fuel, to control flame pattern, and for the type of flame pattern, compact and directional, diffuse or flat flame coverage of adjacent wall area. Burners for oil fuels, in addition, will need provision for atomization of fuel oil over the desired range of firing rates.

The simplest type of gas burner comprises an opening in a furnace wall, through which combustion air is drawn by furnace draft, and a pipe nozzle to introduce fuel gas through that opening. Flame pattern will be controlled by gas velocity at the nozzle and by excess air ratio. Fuel/air ratio will be manually controlled for flame appearance by the judgment of the operator, possibly supplemented by continuous or periodic flue gas analysis. In regenerative furnaces, with firing ports serving alternately as exhaust flues, the open pipe burner may be the only practical arrangement.

For one-way fired furnaces, with burner port areas and combustion air velocities subject to control, fuel/air ratio control can be made automatic over a limited range of turndown with several systems, including:

Mixing in venturi tube, with energy supplied by gas supply inducing atmospheric air. Allows simplest piping system with gas available at high pressure, as from some natural gas supplies.

Venturi mixer with energy from combustion air at intermediate pressure. Requires air supply piping and distribution piping from mixing to burners.

With both combustion air and fuel gas available at intermediate pressures, pressure drops through adjustable orifices can be matched or proportioned to hold desired flow ratios. For more accurate control, operation of flow control valves can be by an external source of energy.

Proportioning in venturi mixers depends on the conservation of momentum—the product of flow rate and velocity or of orifice area and pressure drop. With increased back pressure in the combustion chamber, fuel/air ratio will be increased for the high pressure gas inspirator, or decreased with air pressure as the source of energy, unless the pressure of the induced fluid is adjusted to the pressure in the combustion chamber.

The arrangement of a high-pressure gas inspirator system is illustrated in Fig. 40. Gas enters the throat of the venturi mixer through a jet on the axis of the opening. Air is induced through the surrounding area of the opening, and ratio control can be adjusted by varying the air inlet opening by a movable shutter disk. A single inspirator can supply a number of burners in one firing zone, or a single burner.

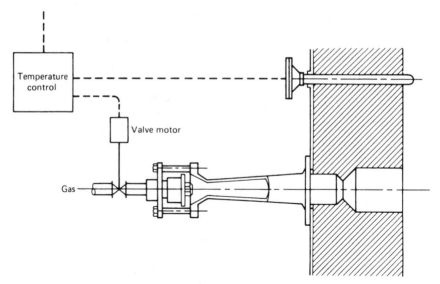

Figure 40 Air/gas ratio control by high-pressure gas inspirator.[1]

For the air primary mixing system, a representative arrangement is shown in Fig. 41. The gas supply is regulated to atmospheric, or to furnace gas pressure, by a diaphragm-controlled valve. Ratio control is by adjustment of an orifice in the gas supply line. With air flow the only source of energy, errors in proportioning can be introduced by friction in the gas-pressure control valve. Each mixer can supply one or more burners, representing a control zone.

With more than one burner per zone, the supply manifold will contain a combustible mixture that can be ignited below a critical port velocity to produce a backfire that can extinguish burners and possibly damage the combustion system. This hazard has made the single burner per mixer combination desirable, and many contemporary designs combine mixer and burner in a single structure.

With complete premixing of fuel and air, the flame will be of minimum luminosity, with combustion complete near the burner port. With delayed mixing, secured by introducing fuel and air in separate streams, through adjacent openings in the burner, or by providing a partial premix of fuel with a fraction of combustion air, flame luminosity can be controlled to increase flame radiation.

In a burner providing no premix ahead of the combustion chamber, flame pattern is determined by velocity differentials between air and fuel streams, and by the subdivision of air flow into several parallel streams. This type of burner is popular for firing with preheated combustion air, and can be insulated for that application.

Partial premix can be secured by dividing the air flow between a mixing venturi tube and a parallel open passage.

With the uncertainty of availability of contemporary fuel supplies, dual fuel burners, optionally fired with fuel gas or fuel oil, can be used. Figure 42 illustrates the design of a large burner for firing gas or oil fuel with preheated air. For oil firing, an oil-atomizing nozzle is inserted through the gas tube. To avoid carbon buildup in the oil tube from cracking of residual oil during gas firing, the oil tube assembly is removable.

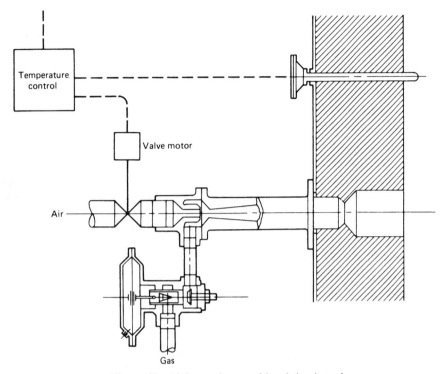

Figure 41   Air/gas ratio control by air inspirator.[1]

Oil should be atomized before combustion in order to provide a compact flame pattern. Flame length will depend on burner port velocity and degree of atomization. Atomization can be accomplished by delivery of oil at high pressure through a suitable nozzle; by intermediate pressure air, part or all of the combustion air supply, mixing with oil at the discharge nozzle; or by high-pressure air or steam. For firing heavy fuel oils of relatively high viscosity, preheating in the storage tank, delivery to the burner through heated pipes, and atomization by high-pressure air or steam will be needed. If steam is available, it can be used for both

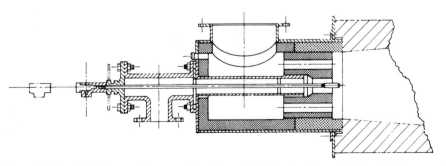

Figure 42   Dual fuel burner with removable oil nozzle.[1] (Courtesy Bloom Engineering Company.)

tank and pipe heating and for atomization. Otherwise, the tank and supply line can be electrically heated, with atomization by high-pressure air.

10.2   Burner Ports

A major function of fuel burners is to maintain ignition over a wide range of demand and in spite of lateral drafts at the burner opening. Ignition can be maintained at low velocities by recirculation of hot products of combustion at the burner nozzle, as in the bunsen burner, but stability of ignition is limited to low port velocities for both the entering fuel/air mixture and for lateral drafts at the point of ignition. Combustion of a fuel/air mixture can be catalyzed by contact with a hot refractory surface. A primary function of burner ports is to supply that source of ignition. Where combustion of a completely mixed source of fuel and air is substantially completed in the burner port, the process is identified as "surface combustion." Ignition by contact with hot refractory is also effective in flat flame burners, where the combustion air supply enters the furnace with a spinning motion and maintains contact with the surrounding wall.

Burner port velocities for various types of gas burners can vary from 3000 to 13,000 lb/hr·ft^2, depending on the desired flame pattern and luminosity. Some smaller sizes of burners are preassembled with refractory port blocks.

10.3   Combustion Control Equipment

Furnace temperature can be measured by a bimetallic thermocouple inserted through the wall or by an optical sensing of radiation from furnace walls and products of combustion. In either case, an electrical impulse is translated into a temperature measurement by a suitable instrument and the result indicated by a visible signal and optionally recorded on a moving chart. For automatic temperature control, the instrument reading is compared to a preset target temperature, and the fuel and air supply adjusted to match through a power-operated valve system.

Control may be on–off, between high and low limits; three position, with high, normal, and off valve openings; or proportional with input varying with demand over the full range of control. The complexity and cost of the system will, in general, vary in the same sequence. Because combustion systems have a lower limit of input for proper burner operation or fuel/air ratio control, the proportioning temperature control system may cut off fuel input when it drops to that limit.

Fuel/air ratios may be controlled at individual burners by venturi mixers or in multiple burner firing zones by similar mixing stations. To avoid back firing in burner manifolds, the pressures of air and gas supplies can be proportioned to provide the proper ratio of fuel and air delivered to individual burners through separate piping. Even though the desired fuel/air ratio can be maintained for the total input to a multiple burner firing zone, errors in distribution can result in excess air or fuel being supplied to individual burners. The design of distribution piping, downstream from ratio control valves, will control delayed combustion of excess fuel and air from individual burners.

In batch-type furnaces for interrupted heating cycles, it may be advantageous to transfer temperature control from furnace temperature to load temperature as load temperature approaches the desired level, in order to take advantage of higher furnace temperatures in the earlier part of the heating cycle. An example is a furnace for annealing steel strip coils. Because heat flow through coil laminations is a fraction of that parallel to the axis of the coil, coils may be stacked vertically with open coil separators between them, to provide for heat transfer from recirculated furnace atmosphere to the end surfaces of coils. For bright annealing, the furnace atmosphere will be nonoxidizing, and the load will be enclosed in an inner cover during heating and cooling, with the atmosphere recirculated by a centrifugal fan in the load support base, to transfer heat from the inner cover to end faces of coils. There will also be some radiation heat transfer from the inner cover to the cylindrical surface of the coil stack.

Inner covers are usually constructed of heat-resisting alloy, with permissible operating temperatures well above the desired final load temperature. A preferred design provides for initial control of furnace inside wall temperature from a thermocouple inserted through the furnace wall, with control switched to a couple in the support base, in control with the bottom of the coil stack, after load temperature reaches a present level below the desired final temperature.

To avoid leakage of combustion gases outward through furnace walls, with possible overheating of the steel enclosure, or infiltration of cold air that could cause nonuniform wall temperatures, control of internal furnace pressure to slightly above ambient is desirable. This can be accomplished by an automatic damper in the outlet flue, adjusted to hold the desired pressure at the selected point in the furnace enclosure. In furnaces with door openings at either end, the point of measurement should be close to hearth level near the discharge end. A practical furnace pressure will be 0.01–0.05 in. H_2O.

With recuperative or regenerative firing systems, the preferred location of the control damper will be between the waste-heat recovery system and the stack, to operate at minimum temperature. In high-temperature furnaces without waste-heat recovery, a water-cooled damper may be needed.

With combustion air preheated before distribution to several firing zones, the ratio control system for each zone will need adjustment to entering air temperature. However, if each firing zone has a separate waste-heat recovery system, the zone air supply can be measured before preheating to maintain the balance with fuel input.

The diagram of a combustion control system in Fig. 43 shows how these control functions can be interlocked with the required instrumentation.

For automatic furnace pressure control to be effective, it should be used in combination with proportioning-type temperature control. With on–off control, for example, the control of furnace pressure at zero firing rate cannot be accomplished by damper adjustment, and with a continuous variation in firing rate between maximum and minimum limits, or between maximum and off, the adjustment of damper position to sudden changes in firing rate will involve a time-lag factor that can make control ineffective.

An important function of a furnace control system is to guard against safety hazards, such as explosions, fires, and personal injury. Requirements have been well defined in codes issued by industrial insurers, and include provision for continuous ignition of burners in low-temperature furnaces, purging of atmosphere furnaces and combustion of hydrogen or carbon monoxide in effluent atmospheres, and protection of operating personnel from injury by burning, mechanical contact, electrical shock, poisoning by inhalation of toxic gases, or asphyxiation. Plants with extensive furnace operation should have a safety engineering staff to supervise selection, installation, and maintenance of safety hazard controls and to coordinate the instruction of operating personnel in their use.

10.4 Air Pollution Control

A new and increasing responsibility of furnace designers and operators is to provide controls for toxic, combustible, or particulate materials in furnace flue gases, to meet federal or local

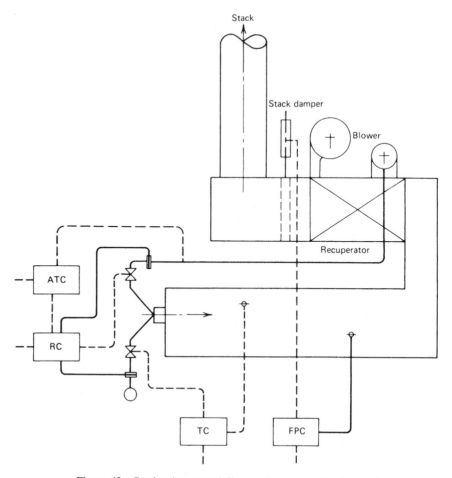

Figure 43 Combustion control diagram for recuperative furnace.[1]

standards for air quality. Designs for furnaces to be built in the immediate future should anticipate a probable increase in restrictions of air pollution in coming years.

Toxic contaminants include sulfur and chlorine compounds, nitrogen oxides, carbon monoxide, and radioactive wastes. The epidermic of "acid rain" in areas downwind from large coal-burning facilities is an example.

Combustible contaminants include unburned fuel, soot, and organic particulates from incinerators, and the visible constituents of smoke, except for steam. Other particulates include suspended ash and suspended solids from calcination processes.

Types of control equipment include

1. Bag filters or ceramic fiber filters to remove suspended solids. Filters require periodic cleaning or replacement, and add to the pressure drop in flue gases leaving the system.

2. Electrostatic filters, in which suspended particles pass through a grid to be electrically charged, and are collected on another grid or on spaced plates with the opposite

potential. Smaller units are cleaned periodically by removal and washing. Large industrial units are cleaned in place. A possible objection to their use is a slight increase in the ozone content of treated air.

3. Wet scrubbers are particularly effective for removing water-soluble contaminants such as sulfur and chlorine compounds. They can be used in place of filters for handling heavy loads of solid particulates such as from foundry cupola furnaces, metal-refining processes, and lime kilns. Waste material is collected as a mud or slurry, requiring proper disposal to avoid solid-waste problems.

4. Combustible wastes, such as the solvent vapors from organic coating ovens, may be burned in incinerator units by adding combustion air and additional fuel as required. Fuel economy may be improved by using waste heat from combustion to preheat incoming gases through a recuperator. The same system may be used for combustible solid particulates suspended in flue gases.

5. Radioactive wastes from nuclear power plants will usually be in the form of suspended solids that can be treated accordingly if suitable facilities for disposal of collected material are available, or as radioactive cooling water for which a suitable dumping area will be needed.

11 WASTE HEAT RECOVERY SYSTEMS

In fuel-fired furnaces, a fraction of the energy from combustion leaves the combustion chamber as sensible heat in waste gases, and the latent heat of evaporation for any water vapor content resulting from the combustion of hydrogen. Losses increase with flue gas temperature and excess air, and can reach 100% of input when furnace temperatures equal theoretical flame temperatures.

Waste heat can be recovered in several ways:

1. Preheating incoming loads in a separate enclosure ahead of the furnace.

2. Generating process steam, or steam for electric power generation. Standby facilities will be needed for continuous demand, to cover interruptions of furnace operation.

3. Preheating combustion air, or low-Btu fuels, with regenerative or recuperative firing systems.

11.1 Regenerative Air Preheating

For the high flue gas temperatures associated with glass- and metal-melting processes, for which metallic recuperators are impractical, air may be preheated by periodical reversal of the direction of firing, with air passing consecutively through a hot refractory bed or checker chamber, the furnace combustion chamber, and another heat-storage chamber in the waste-gas flue. The necessary use of the furnace firing port as an exhaust port after reversal limits the degree of control of flame patterns and the accuracy of fuel/air control in multiple port furnaces. Regenerative firing is still preferred, however, for open hearth furnaces used to convert blast furnace iron to steel, for large glass-melting furnaces, and for some forging operations.

A functional diagram of a regenerative furnace is shown in Fig. 44. The direction of flow of combustion air and flue gas is reversed by a valve arrangement, connecting the low-temperature end of the regenerator chamber to either the combustion air supply or the exhaust

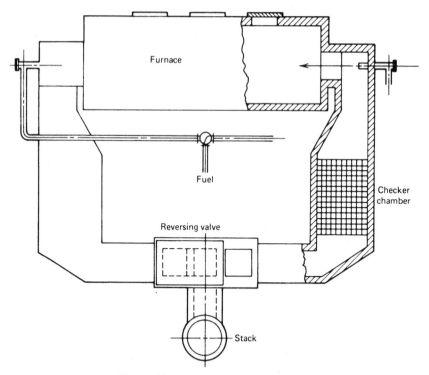

Figure 44 Regenerative furnace diagram.[1]

stack. Fuel input is reversed simultaneously, usually by an interlocked control. Reversal can be in cycles of from 10 to 30 min duration, depending primarily on furnace size.

11.2 Recuperator Systems

Recuperative furnaces are equipped with a heat exchanger arranged to transfer heat continuously from outgoing flue gas to incoming combustion air. Ceramic heat exchangers, built up of refractory tubes or refractory block units arranged for cross flow of air and flue gas, have the advantage of higher temperature limits for incoming flue gas, and the disadvantage of leakage of air or flue gas between passages, with leakage usually increasing with service life and pressure differentials. With the improvement in heat-resistant alloys to provide useful life at higher temperatures, and with better control of incoming flue gas temperatures, metallic recuperators are steadily replacing ceramic types.

Metal recuperators can be successfully used with very high flue gas temperatures if entering temperatures are reduced by air dilution or by passing through a high-temperature waste-heat boiler.

Familiar types of recuperators are shown in the accompanying figures:

Figure 45: radiation or stack type. Flue gases pass through an open cylinder, usually upward, with heat transfer primarily by gas radiation to the surrounding wall. An annular passage is provided between inner and outer cylinders, in which heat is transferred to air at

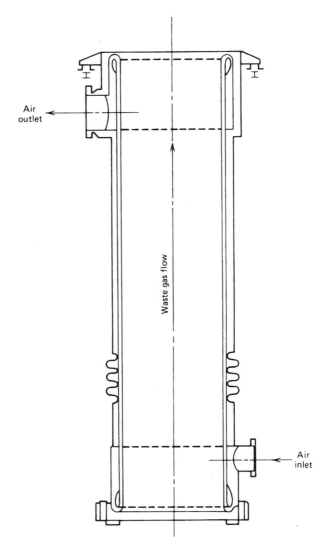

Figure 45   Stack-type recuperator.[1] (Courtesy Morgan Engineering Company.)

high velocity by gas radiation and convection, or by solid-state radiation from inner to outer cylinders and convection. The radiation recuperator has the advantage of acting as a portion of an exhaust stack, usually with flue gas and air counterflow. Disadvantages are distortion and resulting uneven distribution of air flow, resulting from differential thermal expansion of the inner tube, and the liability of damage from secondary combustion in the inner chamber.

Figure 46: cross-flow tubular type. By passing air through a series of parallel passes, as in a tube assembly, with flue gas flowing across tubes, relatively high heat-transfer rates can be achieved. It will ordinarily be more practical to use higher velocities on the air side, and use an open structure on the flue gas side to take some advantage of gas radiation. Figure

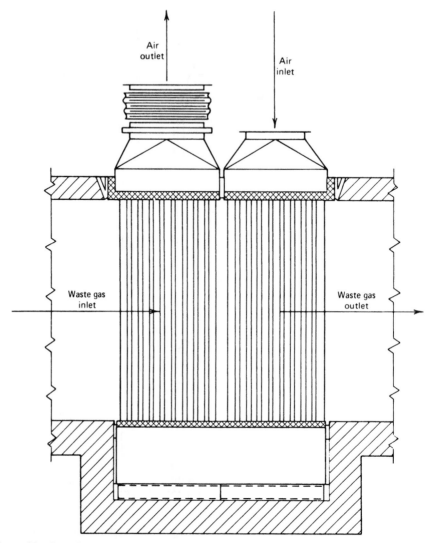

Figure 46 Cross-flow-type recuperator in waste-gas flue.[1] (Courtesy Thermal Transfer Corporation.)

46 shows a basic arrangement, with air tubes in parallel between hot and cold air chambers at either end. Some problems may be introduced by differential thermal expansion of parallel tubes, and tubes may be curved to accommodate variations in length by lateral distortion.

A popular design avoids the problems of thermal expansion by providing heat-exchange tubes with concentric passages and with connections to inlet and outlet manifolds at the same end. Heat transfer from flue gas to air is by gas radiation and convection to the outer tube surface, by convection from the inner surface to high-velocity air, and by solid-state radiation between outer and inner tubes in series with convection from inner tubes to air. Concentric tube recuperators are usually designed for replacement of individual tube units without a complete shutdown and cooling of the enclosing flue. The design is illustrated in Fig. 47.

11.3 Recuperator Combinations

To provide preheated air at pressure required for efficient combustion, without excessive air leakage from the air to the flue gas side in refractory recuperators, the air pressure can be increased between the recuperator and burner by a booster fan. Top air temperatures will be limited by fan materials. As an alternative, air temperatures can be boosted by a jet pump with tolerance for much higher temperatures.

In a popular design for recuperator firing of soaking pits, flue gases pass through the refractory recuperator at low pressure, with air flowing counterflow at almost the same pressure. Air flow is induced by a jet pump, and, to increase the jet pump efficiency, the jet air can be preheated in a metal recuperator between the refractory recuperator and the stack. Because the metal recuperator can handle air preheated to the limit of the metal structure, power demand can be lowered substantially below that for a cold air jet.

Radiant tubes can be equipped with individual recuperators, as shown in Fig. 48. Some direct-firing burners are available with integral recuperators.

12 FURNACE COMPONENTS IN COMPLEX THERMAL PROCESSES

An industrial furnace, with its auxiliaries, may be the principal component in a thermal process with functions other than heating and cooling. For example, special atmosphere

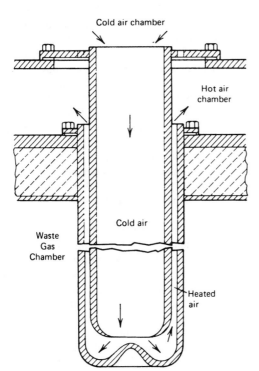

Figure 47 Concentric tube recuperator, Hazen type.[1] (Courtesy C-E Air Preheater Division, Combustion Engineering, Inc.)

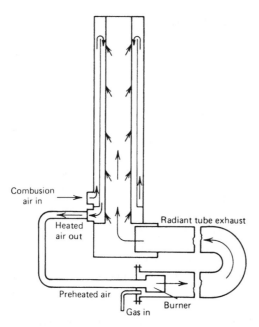

Combusion air in

Heated air out

Radiant tube exhaust

Preheated air

Gas in   Burner

Figure 48   Radiant tube recuperator.[1] (Courtesy Holcroft Division, Thermo-Electron Corp.)

treatment of load surfaces, to increase or decrease carbon content of ferrous alloys, can be accomplished in a furnace heated by radiant tubes or electrical heating elements or by electric induction. A source of the required controlled atmosphere is usually part of the furnace process equipment, designed and supplied by the furnace manufacturer.

Continuous heat treatment of strip or wire, to normalize or anneal ferrous materials, followed by coating in molten metal, such as zinc or aluminum, or electroplating can be accomplished by one of two arrangements for furnace coating lines. One arrangement has a sequence of horizontal passes, with a final cooling zone to regulate strip temperature to the approximate temperature of the coating bath, and an integral molten-metal container. Strip is heat treated in a controlled atmosphere to avoid oxidation, with the same atmosphere maintained to the point of immersion in molten metal. The second arrangement is for higher velocities and longer strands in heating and cooling passes. In this arrangement, strip may be processed in a series of vertical strands, supported by conveyor rolls.

Furnace lines designed for either galvanizing or aluminum coating may be designed with two molten-metal pots, with the entry strand arranged to be diverted to either one, and with the cooling zone adjustable to discharge the strand to either pot at the required temperature.

Thermal processing lines may include furnace equipment for heating the load to the temperature required for annealing, normalizing, or hardening, a quench tank for oil or water cooling to develop hardness, a cleaning station to remove quench oil residues, and a separate tempering furnace to develop the desired combination of hardness and toughness. Loads may be in continuous strand form, or in units carried by trays or fixtures that go through the entire process or carried on a series of conveyors. The required atmosphere generator will be part of the system.

Where exposure to hydrogen or nitrogen in furnace atmospheres may be undesirable, as in heat treatment of some ferrous alloys, heating and cooling can be done in a partial vacuum, usually with heat supplied by electrical resistors. Quenching can be done in a separate chamber with a controlled atmosphere suitable for brief exposure.

Systems for collecting operating data from one or more furnaces, and transmitting the data to a central recording or controlling station, may also be part of the responsibility of the furnace supplier.

13 FURNACE CAPACITY

Factors limiting the heating capacity of industrial furnaces include building space limitations, available fuel supplies, limited temperature of heat sources such as electric resistors or metal radiant tubes, and limits on final load temperature differentials. Other factors under more direct control by furnace designers are the choice between batch and continuous heating cycles; time–temperature cycles to reach specified final load temperatures; fuel firing arrangements; and control systems for furnace temperature, furnace pressure, and fuel/air ratios. In addition, the skills and motivation of furnace operating personnel, as the result of training, experience, and incentive policies, will directly affect furnace efficiency.

14 FURNACE TEMPERATURE PROFILES

Time–temperature patterns can be classified as uniform wall temperature (T_w), uniform combustion gas temperature (T_g), or variable T_w and T_g designed to secure the best combination of heating capacity and fuel efficiency.

In a batch-type furnace with fairly massive loads, the temperature control system can be arranged to allow firing at the maximum burner capacity until a preset wall temperature limit is reached, adjusting firing rate to hold that wall temperature, until load temperature approaches the limit for the heated surface, and reducing the wall temperature setting to hold maximum load temperature T_s while the minimum T_c reaches the desired level.

In continuous furnaces, control systems have evolved from a single firing zone, usually fired from the discharge end with flue gas vented from the load charge end, to two or three zone firing arranged for counterflow relation between furnace loads and heating gases.

Progress from single to multiple zone firing has improved heating rates, by raising furnace temperatures near the charge end, while increasing fuel demand by allowing higher temperatures in flue gas leaving the preheat zone. Load temperature control has been improved by allowing lower control temperatures in the final zone at the discharge end.

With multiple zone firing, the control system can be adjusted to approach the constant-gas-temperature model, constant wall temperature, or a modified system in which both T_g and T_w vary with time and position. Because gas temperatures are difficult to measure directly, the constant-gas-temperature pattern can be simulated by an equivalent wall temperature profile. With increasing fuel costs, temperature settings in a three-zone furnace can be arranged to discharge flue gases at or below the final load temperature, increasing the temperature setting in the main firing zone to a level to provide an equilibrium wall and load temperature, close to the desired final load temperature, during operating delays, and setting a temperature in the final or soak zone slightly above the desired final load surface temperature.

15 REPRESENTATIVE HEATING RATES

Heating times for various furnace loads, loading patterns, and time–temperature cycles can be calculated from data on radiation and non-steady-state conduction. For preliminary estimates, heating times for steel slabs to rolling temperatures, with a furnace temperature profile depressed at the entry end, have been estimated on a conservative basis as a function of thickness heated from one side and final load temperature differential and are shown in Fig. 26. The ratios for heating time required for square steel billets, in various loading patterns, are shown in Fig. 25. For other rectangular cross sections and loading patterns, heating times can be calculated by the Newman method.

Examples of heating times required to reach final load temperatures of $T_s = 2300°F$ and $T_c = 2350°F$, with constant furnace wall temperatures, are

1. 12-in.-thick carbon steel slab on refractory hearth with open firing: 9 hr at 54.4 lb/hr·ft².

2. 4-in.-thick slab, same conditions as 1: 1.5 hr at 109 lb/hr·ft².

3. 4 in. square carbon steel billets loaded at 8 in. centers on a refractory hearth: 0.79 hr at 103 lb/hr·ft².

4. 4 in. square billets loaded as in 3, but heated to $T_s = 1650°F$ and $T_c = 1600°F$ for normalizing: 0.875 hr at 93 lb/hr·ft².

5. Thin steel strip, heated from both sides to 1350°F by radiant tubes with a wall temperature of 1700°F, total heating rate for both sides: 70.4 lb/hr·ft².

6. Long aluminum billets, 6 in. diameter, are to be heated to 1050°F. Billets will be loaded in multiple layers separated by spacer bars, with wind flow parallel to their length. With billets in lateral contact and with wind at a mean temperature of 1500°F, estimated heating time is 0.55 hr.

7. Small aluminum castings are to be heated to 1000°F on a conveyor belt, by jet impingement of heated air. Assuming that the load will have thick and thin sections, wind temperature will be limited to 1100°F to avoid overheating thinner sections. With suitable nozzle spacing and wind velocity, the convection heat-transfer coefficient can be $H_c = 15$ Btu/hr·ft² and the heating rate 27 lb/hr·ft².

16 SELECTING NUMBER OF FURNACE MODULES

For a given heating capacity and with no limits on furnace size, one large furnace will cost less to build and operate than a number of smaller units with the same total hearth area. However, furnace economy may be better with multiple units. For example, where reheating furnaces are an integral part of a continuous hot strip mill, the time required for furnace repairs can reduce mill capacity unless normal heating loads can be handled with one of several furnaces down for repairs. For contemporary hot strip mills, the minimum number of furnaces is usually three, with any two capable of supplying normal mill demand.

Rolling mills designed for operation 24 hr per day may be supplied by batch-type furnaces. For example, soaking-pit-type furnaces are used to heat steel ingots for rolling into slabs. The mill rolling rate is 10 slabs/hr. Heating time for ingots with residual heat from casting averages 4 hr, and the time allowed for reloading an empty pit is 2 hr, requiring an average turnover time of 6 hr. The required number of ingots in pits and spaces for loading is accordingly 60, requiring six holes loaded 10 ingots per hole.

If ingots are poured after a continuous steelmaking process, such as open hearth furnaces or oxygen retorts, and are rolled on a schedule of 18 turns per week, it may be economical at present fuel costs to provide pit capacity for hot storage of ingots cast over weekends, rather than reheating them from cold during the following week.

With over- and underfired slab reheating furnaces, with slabs carried on insulated, water-cooled supports, normal practice has been to repair pipe insulation during the annual shutdown for furnace maintenance, by which time some 50% of insulation may have been lost. By more frequent repair, for example, after 10% loss of insulation, the added cost of lost furnace time, material, and labor may be more than offset by fuel savings, even though total furnace capacity may be increased to offset idle time.

17   FURNACE ECONOMICS

The furnace engineer may be called on to make decisions, or submit recommendations for the design of new furnace equipment or the improvement of existing furnaces. New furnaces may be required for new plant capacity or addition to existing capacity, in which case the return on investment will not determine the decision to proceed. Projected furnace efficiency will, however, influence the choice of design.

If new furnace equipment is being considered to replace obsolete facilities, or if the improvement of existing furnaces is being considered to save fuel or power, or to reduce maintenance costs, return on investment will be the determining factor. Estimating that return will require evaluation of these factors:

Projected service life of equipment to be improved

Future costs of fuel, power, labor for maintenance, or operating supervision and repairs, for the period assumed

Cost of production lost during operating interruptions for furnace improvement or strikes by construction trades

Cost of money during the improvement program and interest available from alternative investments

Cost of retraining operating personnel to take full advantage of furnace improvements

17.1   Operating Schedule

For a planned annual capacity, furnace size will depend on the planned hours per year of operation, and fuel demand will increase with the ratio of idle time to operating time, particularly in furnaces with water-cooled load supports. If furnace operation will require only a two- or three-man crew, and if furnace operation need not be coordinated with other manufacturing functions, operating costs may be reduced by operating a smaller furnace two or three turns per day, with the cost of overtime labor offset by fuel savings.

On the other hand, where furnace treatment is an integral part of a continuous manufacturing process, the provision of standby furnace capacity to avoid plant shutdown for furnace maintenance or repairs may be indicated.

If furnace efficiency deteriorates rapidly between repairs, as with loss of insulation from water-cooled load supports, the provision of enough standby capacity to allow more frequent repairs may reduce overall costs.

17.2   Investment in Fuel-Saving Improvements

At present and projected future costs of gas and oil fuels, the added cost of building more efficient furnaces or modifying existing furnaces to improve efficiency can usually be justified. Possible improvements include better insulation of the furnace structure, modified firing arrangements to reduce flue gas temperatures or provide better control of fuel/air ratios, programmed temperature control to anticipate load changes, more durable insulation of water-cooled load supports and better maintenance of insulation, proportioning temperature control rather than the two position type, and higher preheated air temperatures. For intermittent furnace operation, the use of a low-density insulation to line furnace walls and roofs can result in substantial savings in fuel demand for reheating to operating temperature after idle periods.

The relative costs and availability of gas and oil fuels may make a switch from one fuel to another desirable at any future time, preferably without interrupting operations. Burner equipment and control systems are available, at some additional cost, to allow such changeovers.

The replacement of existing furnaces with more fuel-efficient designs, or the improvement of existing furnaces to save fuel, need not be justified in all cases by direct return on investment. Where present plant capacity may be reduced by future fuel shortages, or where provision should be made for increasing capacity with fuel supplies limited to present levels, cost savings by better fuel efficiency may be incidental.

Government policies on investment tax credits or other incentives to invest in fuel-saving improvements can influence the return on investment for future operation.

REFERENCE

1. C. Cone, *Energy Management for Industrial Furnaces,* Wiley, New York, 1980.

CHAPTER 7
ENERGY AUDITING

Carl Blumstein
Universitywide Energy Research Group
University of California
Berkeley, California

Peter Kuhn
Kuhn and Kuhn
Industrial Energy Consultants
Golden Gate Energy Center
Sausalito, California

1 ENERGY MANAGEMENT AND THE ENERGY AUDIT

Energy auditing is the practice of surveying a facility to identify opportunities for increasing the efficiency of energy use. A facility may be a residence, a commercial building, an industrial plant, or other installation where energy is consumed for any purpose. Energy management is the practice of organizing financial and technical resources and personnel to increase the efficiency with which energy is used in a facility. Energy management typically involves the keeping of records on energy consumption and equipment performance, optimization of operating practices, regular adjustment of equipment, and replacement or modification of inefficient equipment and systems.

Energy auditing is a part of an energy management program. The auditor, usually someone not regularly associated with the facility, reviews operating practices and evaluates energy using equipment in the facility in order to develop recommendations for improvement. An energy audit can be, and often is, undertaken when no formal energy management program exists. In simple facilities, particularly residences, a formal program is impractical and informal procedures are sufficient to alter operating practices and make simple improvements such as the addition of insulation. In more complex facilities, the absence of a formal energy management program is usually a serious deficiency. In such cases a major recommendation of the energy audit will be to establish an energy management program.

There can be great variation in the degree of thoroughness with which an audit is conducted, but the basic procedure is universal. The first step is to collect data with which

277

to determine the facility's major energy uses. These data always include utility bills, name-plate data from the largest energy-using equipment, and operating schedules. The auditor then makes a survey of the facility. Based on the results of this survey, he or she chooses a set of energy conservation measures that could be applied in the facility and estimates their installed cost and the net annual savings that they would provide. Finally, the auditor presents his or her results to the facility's management or operators. The audit process can be as simple as a walkthrough visit followed by a verbal report or as complex as a complete analysis of all of a facility's energy using equipment that is documented by a lengthy written report.

The success of an energy audit is ultimately judged by the resulting net financial return (value of energy saved less costs of energy saving measures). Since the auditor is rarely in a position to exercise direct control over operating and maintenance practices or investment decisions, his or her work can come to naught because of the actions or inaction of others. Often the auditor's skills in communication and interpersonal relations are as critical to obtaining a successful outcome from an energy audit as his or her engineering skills. The auditor should stress from the outset of his or her work that energy management requires a sustained effort and that in complex facilities a formal energy management program is usu-ally needed to obtain the best results. Most of the auditor's visits to a facility will be spent in the company of maintenance personnel. These personnel are usually conscientious and can frequently provide much useful information about the workings of a facility. They will also be critical to the success of energy conservation measures that involve changes in operating and maintenance practices. The auditor should treat maintenance personnel with respect and consideration and should avoid the appearance of "knowing it all." The auditor must also often deal with nontechnical managers. These managers are frequently involved in the decision to establish a formal energy management program and in the allocation of capital for energy saving investments. The auditor should make an effort to provide clear explanations of his or her work and recommendations to nontechnical managers and should be careful to avoid the use of engineering jargon when communicating with them.

While the success of an energy audit may depend in some measure on factors outside the auditor's control, a good audit can lead to significant energy savings. Table 1 shows the percentage of energy saved as a result of implementing energy audit recommendations in 172 nonresidential buildings. The average savings is more than 20%. The results are espe-cially impressive in light of the fact that most of the energy-saving measures undertaken in these buildings were relatively inexpensive. The median value for the payback on energy-saving investments was in the 1- to 2-year range (i.e., the value of the energy savings exceeded the costs in 1–2 years). An auditor can feel confident in stating that an energy saving of 20% or more is usually possible in facilities where systematic efforts to conserve energy have not been undertaken.

2 PERFORMING AN ENERGY AUDIT—ANALYZING ENERGY USE

A systematic approach to energy auditing requires that an analysis of existing energy-using systems and operating practices be undertaken before efforts are made to identify opportu-nities for saving energy. In practice, the auditor may shift back and forth from the analysis of existing energy-use patterns to the identification of energy-saving opportunities several times in the course of an audit—first doing the most simple analysis and identifying the most obvious energy-saving opportunities, then performing more complex analyses, and so on. This strategy may be particularly useful if the audit is to be conducted over a period of

Table 1 The Percentage of Energy Saved as a Result of Implementing Energy Audit Recommendations in 172 Nonresidential Buildings[a,4]

Building Category	Site		Source	
	Savings (%)	Sample Size	Savings (%)	Sample Size
Elementary school	24	72	21	72
Secondary school	30	38	28	37
Large office	23	37	21	24
Hospital	21	13	17	10
Community center	56	3	23	18
Hotel	25	4	24	4
Corrections	7	4	5	4
Small office	33	1	30	1
Shopping center	11	1	11	1
Multifamily apartment	44	1	43	1

[a]Electricity is counted at 3413 Btu/kWhr for site energy and 11,500 Btu/kWhr for source energy (i.e., including generation and transmission losses).

time that is long enough for some of the early audit recommendations to be implemented. The resultant savings can greatly increase the auditor's credibility with the facility's operators and management, so that he or she will receive more assistance in completing his or her work and his or her later recommendations will be attended to more carefully.

The amount of time devoted to analyzing energy use will vary, but, even in a walk-through audit, the auditor will want to examine records of past energy consumption. These records can be used to compare the performance of a facility with the performance of similar facilities. Examination of the seasonal variation in energy consumption can give an indication of the fractions of a facility's use that are due to space heating and cooling. Records of energy consumption are also useful in determining the efficacy of past efforts to conserve energy.

In a surprising number of facilities the records of energy consumption are incomplete. Often records will be maintained on the costs of energy consumed but not on the quantities. In periods of rapidly escalating prices, it is difficult to evaluate energy performance with such records. Before visiting a facility to make an audit, the auditor should ask that complete records be assembled and, if the records are not on hand, suggest that they be obtained from the facility's suppliers. Good record keeping is an essential part of an energy management program. The records are especially important if changes in operation and maintenance are to be made, since these changes are easily reversed and often require careful monitoring to prevent backsliding.

In analyzing the energy use of a facility, the auditor will want to focus his or her attention on the systems that use the most energy. In industrial facilities these will typically involve production processes such as drying, distillation, or forging. Performing a good audit in an industrial facility requires considerable knowledge about the processes being used. Although some general principles apply across plant types, industrial energy auditing is generally quite specialized. Residential energy auditing is at the other extreme of specialization. Because a single residence uses relatively little energy, highly standardized auditing procedures must be used to keep the cost of performing an audit below the value of potential energy savings. Standardized procedures make it possible for audits to be performed quickly by technicians with relatively limited training.

Commercial buildings lie between these extremes of specialization. The term "commercial building" as used here refers to those nonresidential buildings that are not used for the production of goods and includes office buildings, schools, hospitals, and retail stores. The largest energy-using systems in commercial buildings are usually lighting and HVAC (heating, ventilating, and air conditioning). Refrigeration consumes a large share of the energy used in some facilities (e.g., food stores) and other loads may be important in particular cases (e.g., research equipment in laboratory buildings). Table 2 shows the results of a calculation of the amount of energy consumed in a relatively energy-efficient office building for lighting and HVAC in different climates. Office buildings (and other commercial buildings) are quite variable in their design and use. So, while the proportions of energy devoted to various uses shown in Table 2 are not unusual, it would be unwise to treat them (or any other proportions) as "typical." Because of the variety and complexity of energy-using systems in commercial buildings and because commercial buildings frequently use quite substantial amounts of energy in their operation, an energy audit in a commercial building often warrants the effort of a highly trained professional. In the remainder of this section commercial buildings will be used to illustrate energy auditing practice.

Lighting systems are often a good starting point for an analysis of energy in commercial buildings. They are the most obvious energy consumers, are usually easily accessible, and can provide good opportunities for energy saving. As a first step the auditor should determine the hours of operation of the lighting systems and the watts per square foot of floorspace that they use. These data, together with the building area, are sufficient to compute the energy consumption for lighting and can be used to compare the building's systems with efficient lighting practice. Next, lighting system maintenance practices should be examined. As shown in Fig. 1, the accumulation of dirt on lighting fixtures can significantly reduce light output. Fixtures should be examined for cleanliness and the auditor should determine whether or not a regular cleaning schedule is maintained. As lamps near the end of their rated life, they lose efficiency. Efficiency can be maintained by replacing lamps in groups before they reach the end of their rated life. This practice also reduces the higher maintenance costs associated with spot relamping. Fixtures should be checked for lamps that are burned out or show signs of excessive wear, and the auditor should determine whether or not a group-relamping program is in effect.

After investigating lighting operation and maintenance practices, the auditor should measure the levels of illumination being provided by the lighting systems. These measurements can be made with a relatively inexpensive photometer. Table 3 gives recommended levels of

Table 2   Results of a Calculation of the Amount of Energy Consumed in a Relatively Energy-Efficient Office Building for Lighting and HVAC[5]

	Energy Use (kBtu/ft^2/yr)			
	Miami	Los Angeles	Washington	Chicago
Lights	34.0	34.0	34.0	34.0
HVAC auxiliaries	8.5	7.7	8.8	8.8
Cooling	24.4	9.3	10.2	7.6
Heating	0.2	2.9	17.7	28.4
Total	67.1	53.9	70.7	78.8

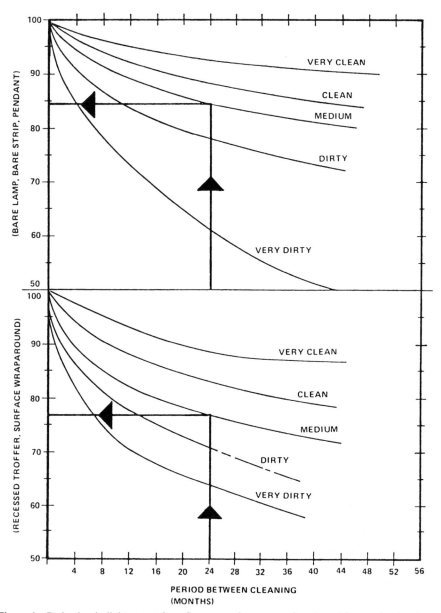

Figure 1   Reduction in light output from fluorescent fixtures as a function of fixture cleaning frequency and the cleanliness of the fixture's surroundings.[3]

illumination for a variety of activities. A level much in excess of these guidelines usually indicates an opportunity for saving energy. However, the auditor should recognize that good seeing also depends on other factors such as glare and contrast and that the esthetic aspects of lighting systems (i.e., their appearance and the effect they create) can also be important. More information about the design of lighting systems can be found in Ref. 1.

Table 3 Range of Illuminances Appropriate for Various Types of Activities and Weighting Factors for Choosing the Footcandle Level[a] within a Range of Illuminance[6]

Category	Range of Illuminances (footcandles)	Type of Activity
A	2–3–5	Public areas with dark surroundings
B	5–7.5–10	Simple orientation for short temporary visits
C	10–15–20	Working spaces where visual tasks are only occasionally performed
D	20–30–50	Performance of visual tasks of high contrast or large size: for example, reading printed material, typed originals, handwriting in ink and good xerography; rough bench and machine work; ordinary inspection; rough assembly
E	50–75–100	Performance of visual tasks of medium contrast or small size: for example, reading medium-pencil handwriting, poorly printed or reproduced material; medium bench and machine work; difficult inspection; medium assembly
F	100–150–200	Performance of visual tasks of low contrast or very small size: for example, reading handwriting in hard pencil or very poorly reproduced material; very difficult inspection
G	200–300–500	Performance of visual tasks of low contrast and very small size over a prolonged period: for example, fine assembly; very difficult inspection; fine bench and machine work
H	500–750–1000	Performance of very prolonged and exacting visual tasks: for example, the most difficult inspection; extra-fine bench and machine work; extrafine assembly
I	1000–1500–2000	Performance of very special visual tasks of extremely low contrast and small size: for example, surgical procedures

Weighting Factors			
Worker or task charactristics	−1	0	+1
Worker's age	Under 40	40–65	Over 65
Speed and/or accuracy	Not important	Important	Critical
Reflectance of task background	Greater than 70%	30–70%	Less than 30%

[a]To determine a footcandle level within a range of illuminance, find the weighting factor for each worker or task characteristic and sum the weighting factors to obtain a score. If the score is −3 or −2, use the lowest footcandle level; if −1, 0, or 1, use the middle footcandle level; if 2 or 3, use the highest level.

Analysis of HVAC systems in a commercial building is generally more complicated and requires more time and effort than lighting systems. However, the approach is similar in that the auditor will usually begin by examining operating and maintenance practices and then proceed to measure system performance.

Determining the fraction of a building's energy consumption that is devoted to the operation of its HVAC systems can be difficult. The approaches to this problem can be classified as either deterministic or statistical. In the deterministic approaches an effort is made to calculate HVAC energy consumption from engineering principles and data. First, the building's heating and cooling loads are calculated. These depend on the operating schedule and thermostat settings, the climate, heat gains and losses from radiation and conduction, the rate of air exchange, and heat gains from internal sources. Then energy use is calculated by taking account of the efficiency with which the HVAC systems meet these loads. The efficiency of the HVAC systems depends on the efficiency of equipment such as boilers and chillers and losses in distribution through pipes and ducts; equipment efficiency and distribution losses are usually dependent on load. In all but the simplest buildings, the calculation of HVAC energy consumption is sufficiently complex to require the use of computer programs; a number of such programs are available (see, for example, Ref. 2). The auditor will usually make some investigation of all of the factors necessary to calculate HVAC energy consumption. However, the effort involved in obtaining data that are sufficiently accurate and preparing them in suitable form for input to a computer program is quite considerable. For this reason, the deterministic approach is not recommended for energy auditing unless the calculation of savings from energy conservation measures requires detailed information on building heating and cooling loads.

Statistical approaches to the calculation of HVAC energy consumption involve the analysis of records of past energy consumption. In one common statistical method, energy consumption is analyzed as a function of climate. Regression analysis with energy consumption as the dependent variable and some function of outdoor temperature as the independent variable is used to separate "climate-dependent" energy consumption from "base" consumption. The climate-dependent fraction is considered to be the energy consumption for heating and cooling, and the remainder is assumed to be due to other uses. This method can work well in residences and in some small commercial buildings where heating and cooling loads are due primarily to the climate. It does not work as well in large commercial buildings because much of the cooling load in these buildings is due to internal heat gains and because a significant part of the heating load may be for reheat (i.e., air that is precooled to the temperature required for the warmest space in the building may have to be reheated in other spaces). The easiest statistical method to apply, and the one that should probably be attempted first, is to calculate the energy consumption for all other end uses (lighting, domestic hot water, office equipment, etc.) and subtract this from the total consumption; the remainder will be HVAC energy consumption. If different fuel types are used for heating and cooling, it will be easy to separate consumption for these uses; if not, some further analysis of the climate dependence of consumption will be required. Energy consumption for ventilation can be calculated easily if the operating hours and power requirements for the supply and exhaust fans are known.

Whatever approach is to be taken in determining the fraction of energy consumption that is used for HVAC systems, the auditor should begin his or her work on these systems by determining their operating hours and control settings. These can often be changed to save energy with no adverse effects on a building's occupants. Next, maintenance practices should be examined. This examination will usually be initiated by determining whether or not a preventive maintenance (PM) program is being conducted. If there is a PM program, much can be learned about the adequacy of maintenance practices by examining the PM records. Often only a few spot checks of the HVAC systems will be required to verify that the records are consistent with actual practice. If there is no PM program, the auditor will usually find that the HVAC systems are in poor condition and should be prepared to make extensive checks for energy-wasting maintenance problems. Establishment of a PM program as part of the energy management program is a frequent recommendation from an energy audit.

Areas for HVAC maintenance that are important to check include heat exchanger surfaces, fuel-air mixture controls in combustors, steam traps, and temperature controllers. Scale on the water side of boiler tubes and chiller condenser tubes reduces the efficiency of heat transfer. Losses of efficiency can also be caused by the buildup of dirt on finned-tube air-cooled condensers. Improper control of fuel-air mixtures can cause significant losses in combustors. Leaky steam traps are a common cause of energy losses. Figure 2 shows the annual rate of heat loss through a leaky trap as a function of the size of the trap orifice and steam pressure. Poorly maintained room thermostats and other controls such as temperature reset

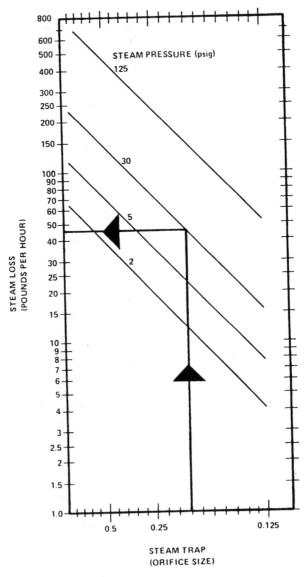

Figure 2   Steam loss through leaking steam traps as a function of stem pressure and trap orifice size.[3]

controllers can also cause energy waste. While major failures of thermostats can usually be detected as a result of occupant complaints or behavior (e.g., leaving windows open on cold days), drifts in these controls that are too small to cause complaints can still lead to substantial waste. Other controls, especially reset controls, can sometimes fail completely and cause an increase in energy consumption without affecting occupant comfort.

After investigating HVAC operation and maintenance practices, the auditor should make measurements of system performance. Typical measurements will include air temperature in rooms and ducts, water temperatures, air flow rates, pressure drops in air ducts, excess air in stack gases, and current drawn by electric motors operating fans and pumps. Instruments required include a thermometer, a pitot tube or anemometer, a manometer, a strobe light, a combustion test kit, and an ammeter. The importance of making measurements instead of relying on design data cannot be emphasized too strongly. Many, if not most, buildings operate far from their design points. Measurements may point to needed adjustments in temperature settings or air flow rates. Table 4 gives recommended air flow rates for various applications. Detailed analysis of the measured data requires a knowledge of HVAC system principles.

After measuring HVAC system performance, the auditor should make rough calculations of the relative importance of the different sources of HVAC system loads. These are primarily radiative and conductive heat gains and losses through the building's exterior surfaces, gains and losses from air exchange, and gains from internal heat sources. Rough calculations are usually sufficient to guide the auditor in selecting conservation measures for consideration. More detailed analyses can await the selection of specific measures.

Table 4 Recommended Rates of Outside-Air Flow for Various Applications[3]

1. Office Buildings	
Work space	5 cfm/person
Heavy smoking areas	15 cfm/person
Lounges	5 cfm/person
Cafeteria	5 cfm/person
Conference rooms	15 cfm/person
Doctors' offices	5 cfm/person
Toilet rooms	10 air changes/hr
Lobbies	0
Unoccupied spaces	0
2. Retail Stores	
Trade areas	6 cfm/customer
Street level with heavy use (less than 5000 ft.² with single or double outside door)	0
Unoccupied spaces	0
3. Religious Buildings	
Halls of worship	5 cfm/person
Meeting rooms	10 cfm/person
Unoccupied spaces	0

While lighting and HVAC systems will usually occupy most of the auditor's time in a commercial building, other systems such as domestic hot water may warrant attention. The approach of first investigating operation and maintenance practices and then measuring system performance is usually appropriate for these systems.

3 PERFORMING AN ENERGY AUDIT—IDENTIFYING OPPORTUNITIES FOR SAVING ENERGY

In almost every facility one can discover a surprisingly large number of opportunities to save energy. These opportunities range from the obvious such as use of light switches to exotic approaches involving advanced energy conversion technologies. Identification of ways to save energy requires imagination and resourcefulness as well as a sound knowledge of engineering principles.

The auditor's job is to find ways to *eliminate unnecessary energy-using tasks* and ways to *minimize the work required to perform necessary tasks*. Some strategies that can be used to eliminate unnecessary tasks are improved controls, "leak plugging," and various system modifications. Taking space conditioning as an example, it is necessary to provide a comfortable interior climate for building occupants, but it is usually not necessary to condition a building when it is unoccupied, it is not necessary to heat and cool the outdoors, and it is not necessary to cool air from inside the building if air outside the building is colder. Controls such as time clocks can turn space-conditioning equipment off when a building is unoccupied, heat leaks into or out of a building can be plugged using insulation, and modification of the HVAC system to add an air-conditioner economizer can eliminate the need to cool inside air when outside air is colder.

Chapter 4 discusses methods of analyzing the minimum amount of work required to perform tasks. While the theoretical minimum cannot be achieved in practice, analysis from this perspective can reveal inefficient operations and indicate where there may be opportunities for large improvements. Strategies for minimizing the work required to perform necessary tasks include heat recovery, improved efficiency of energy conversion, and various system modifications. Heat recovery strategies range from complex systems to cogenerate electrical and thermal energy to simple heat exchangers that can be used to heat water with waste heat from equipment. Examples of improved conversion efficiency are more efficient motors for converting electrical energy to mechanical work and more efficient light sources for converting electrical energy to light. Some system modifications that can reduce the work required to perform tasks are the replacement of resistance heaters with heat pumps and the replacement of dual duct HVAC systems with variable air volume systems.

There is no certain method for discovering all of the energy-saving opportunities in a facility. The most common approach is to review lists of energy conservation measures that have been applied elsewhere to see if they are applicable at the facility being audited. A number of such lists have been compiled (see, for example, Ref. 3). However, while lists of measures are useful, they cannot substitute for intelligent and creative engineering. The energy auditor's recommendations need to be tailored to the facility, and the best energy conservation measures often involve novel elements.

In the process of identifying energy saving opportunities, the auditor should concentrate first on low-cost conservation measures. The savings potential of these measures should be estimated before more expensive measures are evaluated. Estimates of the savings potential of the more expensive measures can then be made from the reduced level of energy consumption that would result from implementing the low-cost measures. While this seems obvious, there have been numerous occasions on which costly measures have been used but simpler, less expensive alternatives have been ignored.

3.1 Low-Cost Conservation

Low-cost conservation measures include turning off energy-using equipment when it is not needed, reducing lighting and HVAC services to recommended levels, rescheduling of electricity-intensive operations to off-peak hours, proper adjustment of equipment controls, and regular equipment maintenance. These measures can be initiated quickly, but their benefits usually depend on a sustained effort. An energy management program that assigns responsibility for maintaining these low-cost measures and monitors their performance is necessary to ensure good results.

In commercial buildings it is often possible to achieve very large energy savings simply by shutting down lighting and HVAC systems during nonworking hours. This can be done manually or, for HVAC systems, by inexpensive time clocks. If time clocks are already installed, they should be maintained in good working order and set properly. During working hours lights should be turned off in unoccupied areas. Frequent switching of lamps does cause some decrease in lamp life, but this decrease is generally not significant in comparison to energy savings. As a rule of thumb, lights should be turned out in a space that will be unoccupied for more than 5 min.

Measurements of light levels, temperatures, and air flow rates taken during the auditor's survey will indicate if lighting or HVAC services exceed recommended levels. Light levels can be decreased by relamping with lower-wattage lamps or by removing lamps from fixtures. In fluorescent fixtures, except for instant-start lamps, ballasts should also be disconnected because they use some energy when the power is on even when the lamps are removed.

If the supply of outside air is found to be excessive, reducing the supply can save heating and cooling energy (but see below on air-conditioner economizers). If possible, the reduction in air supply should be accomplished by reducing fan speed rather than by restricting air flow by the use of dampers, since the former procedure is more energy efficient. Also, too much air flow restriction can cause unstable operation in some fans.

Because most utilities charge more for electricity during their peak demand periods, rescheduling the operation of some equipment can save considerable amounts of money. It is not always easy to reschedule activities to suit the utility's peak demand schedule, since the peak demand occurs when most facilities are engaging in activities requiring electricity. However, a careful examination of major electrical equipment will frequently reveal some opportunities for rescheduling. Examples of activities that have been rescheduled to save electricity costs are firing of electric ceramic kilns, operation of swimming pool pumps, finish grinding at cement plants, and pumping of water from wells to storage tanks.

Proper adjustment of temperature and pressure controls in HVAC distribution systems can cut losses in these systems significantly. Correct temperature settings in air supply ducts can greatly reduce the energy required for reheat. Temperature settings in hot water distribution systems can usually be adjusted to reduce heat loss from the pipes. Temperatures are often set higher than necessary to provide enough heating during the coldest periods; during milder weather, the distribution temperature can be reduced to a lower setting. This can be done manually or automatically using a reset control. Reset controls are generally to be preferred, since they can adjust the temperature continuously. In steam distribution systems, lowering the distribution pressure will reduce heat loss from the flashing of condensate (unless the condensate return system is unvented) and also reduce losses from the surface of the pipes. Figure 3 shows the percentage of the heat in steam that is lost due to condensate flashing at various pressures. Raising temperatures in chilled-water distribution systems also saves energy in two ways. Heat gain through pipe surfaces is reduced, and the chiller's efficiency increases due to the higher suction head on the compressor (see Fig. 4).

A PM program is needed to ensure that energy-using systems are operating efficiently. Among the activities that should be conducted regularly in such a program are cleaning of heat exchange surfaces, surveillance of steam traps so that leaky traps can be found and repaired, combustion efficiency testing, and cleaning of light fixtures. Control equipment such as thermostats, time clocks, and reset controllers need special attention. This equipment should be checked and adjusted frequently.

3.2   Capital-Intensive Energy Conservation Measures

Major additions, modifications, or replacement of energy-using equipment usually require significant amounts of capital. These measures consequently undergo a more detailed scrutiny before a facility's management will decide to proceed with them. While the fundamental approach of eliminating unnecessary tasks and minimizing the work required for necessary tasks is unchanged, the auditor must pay much more attention to the tasks of estimating costs and savings when considering capital-intensive conservation measures.

This subsection will describe only a few of the many possible capital-intensive measures. These measures have been chosen because they illustrate some of the more common approaches to energy saving. However, they are not appropriate in all facilities and they will not encompass the majority of savings in many facilities.

Energy Management Systems

An energy management system (EMS) is a centralized computer control system for building services, especially HVAC. Depending on the complexity of the EMS, it can function as a simple time clock to turn on equipment when necessary, it can automatically cycle the operation of large electrical equipment to reduce peak demand, and it can program HVAC system operation in response to outdoor and indoor temperature trends so that, for example, the "warm-up" heating time before a building is occupied in the morning is minimized. While such a system can be a valuable component of complex building energy service systems, the energy auditor should recognize that the functions of an EMS often duplicate the services of less costly equipment such as time clocks, temperature controls, and manual switches.

Air-Conditioner Economizers

In many areas, outdoor temperatures are lower than return air temperatures during a large part of the cooling season. An air-conditioner economizer uses outside air for cooling during

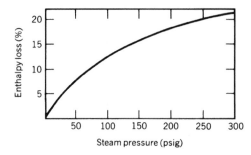

Figure 3   Percentage of heat that is lost due to condensate flashing at various pressures.

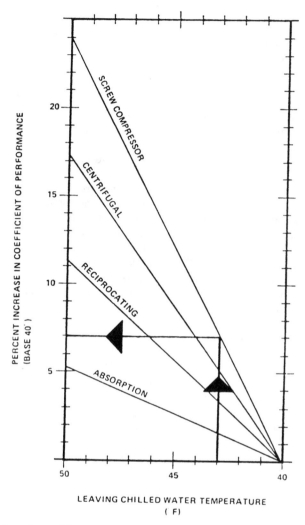

Figure 4 Adjusting air-conditioner controls to provide higher chilled-water temperatures improves chiller efficiency.[3]

these periods so that the load on the compressor is reduced or eliminated. The economizer is a system of automatic dampers on the return air duct that are controlled by return-air and outside-air temperature sensors. When the outside air is cooler than the return air, the dampers divert the return air to the outdoors and let in enough fresh outside air to supply all the air to the building. In humid climates, economizers must be fitted with "enthalpy" sensors that measure wet-bulb as well as dry-bulb temperature so that the economizer will not let in outside air when it is too humid for use in the building.

Building Exhaust-Air Heat Recovery Units
Exhaust-air heat recovery can be practical for facilities with large outside-air flow rates in relatively extreme climates. Hospitals and other facilities that are required to have once-through ventilation are especially good candidates. Exhaust-air heat recovery units reduce

the energy loss in exhaust air by transferring heat between the exhaust air and the fresh air intake.

The common types of units available are heat wheels, surface heat exchangers, and heat-transfer-fluid loops. Heat wheels are revolving arrays of corrugated steel plates or other media. In the heating season, the plates absorb heat in the exhaust air duct, rotate to the intake air duct, and reject heat to the incoming fresh air. Surface heat exchangers are air-to-air heat exchangers. Some of these units are equipped with water sprays on the exhaust air side of the heat exchanger for indirect evaporative cooling. When a facility's exhaust- and fresh-air intakes are physically separated by large distances, heat-transfer-fluid loops (sometimes called run-around systems) are the only practical approach to exhaust-air heat recovery. With the fluid loop, heat exchangers are installed in both the exhaust and intake ducts and the fluid is circulated between the exchangers.

A key factor in estimating savings from exhaust air heat recovery is the unit's effectiveness, expressed as the percentage of the theoretically possible heat transfer that the unit actually achieves. With a 40°F temperature difference between the exhaust and intake air in the heating mode, a 60% effective unit will raise the intake air temperature by 24°F. In units with indirect evaporative cooling, the effectiveness indicates the extent to which the unit can reduce the difference between the intake air dry-bulb temperature and the exhaust air wet-bulb temperature. The effectiveness of commercially available exhaust air heat recovery units ranges from 50 to 80%; greater effectiveness is usually obtained at a higher price per unit of heat recovery capacity.

Refrigeration Heat Recovery

Heat recovery from refrigerators and air conditioners can replace fuel that would otherwise be consumed for low-temperature heating needs. Heat recovery units that generate hot water consist of water storage tanks with an integral refrigerant condenser that supplements or replaces the existing condenser on the refrigerator or air conditioner. These units reduce the facility's fuel or electricity consumption for water heating, and also increase the refrigeration or air conditioning system's efficiency due to the resulting cooler operating temperature of the condenser.

The most efficient condensing temperature will vary, depending on the compressor design and refrigerant, but in most cases it will be below 100°F. In facilities requiring water at higher temperatures, the refrigeration heat recovery unit can preheat water for the existing water heater, which will then heat the water to the final temperature.

Boiler Heat Recovery Devices

Part of the energy conversion losses in a boiler room can be reduced by installing a boiler economizer, air preheater, or a blowdown heat recovery unit. Both the economizer and the air preheater recover heat from the stack gases. The economizer preheats boiler feedwater and the air preheater heats combustion air. The energy savings from these devices are typically 5–10% of the boiler's fuel consumption. The savings depend primarily on the boiler's stack gas temperature. Blowdown heat recovery units are used with continuous blowdown systems and can either supply low-pressure steam to the deaerator or preheat makeup water for the boiler. Their energy savings are typically 1–2% of boiler fuel consumption. The actual savings will depend on the flow rate of the boiler blowdown and the boiler's steam pressure or hot-water temperature.

More Efficient Electric Motors

Replacement of integral-horsepower conventional electric motors with high-efficiency motors will typically yield an efficiency improvement of 2–5% at full load (see Table 5). While this

Table 5 Comparative Efficiencies and Power Factors (%) for U-Frame, T-Frame, and Energy-Efficient Motors[7]

For Smaller Motors

Horsepower Range:	3–30 hp 3600 rpm			3–30 hp 1800 rpm			1.5–20 hp 1200 rpm		
Speed: Type:	U	T	EEM	U	T	EEM	U	T	EEM
Efficiency									
4/4 load	84.0	84.7	86.9	86.0	86.2	89.2	84.1	82.9	86.1
3/4 load	82.6	84.0	87.4	85.3	85.8	91.1	83.5	82.3	86.1
1/2 load	79.5	81.4	85.9	82.8	83.3	83.3	81.0	79.6	83.7
Power factor									
4/4 load	90.8	90.3	86.6	85.3	83.5	85.8	78.1	77.0	73.7
3/4 load	88.7	87.8	84.1	81.5	79.2	81.9	72.9	70.6	67.3
1/2 load	83.5	81.8	77.3	72.8	70.1	73.7	60.7	59.6	56.7

For Larger Motors

Horsepower Range:	40–100 hp 3600 rpm			40–100 hp 1800 rpm			25–75 hp 1200 rpm		
Speed: Type:	U	T	EEM	U	T	EEM	U	T	EEM
Efficiency									
4/4 load	89.7	89.6	91.6	90.8	90.9	92.9	90.4	90.1	92.1
3/4 load	88.6	89.0	92.1	90.2	90.7	93.2	90.3	90.3	92.8
1/2 load	85.9	87.2	91.3	88.1	89.2	92.5	89.2	89.3	92.7
Power factor									
4/4 load	91.7	91.5	89.1	88.7	87.4	87.6	88.3	88.5	86.0
3/4 load	89.9	89.8	88.8	87.1	85.4	86.3	86.6	86.4	83.8
1/2 load	84.7	85.0	85.2	82.0	79.2	81.1	80.9	80.3	77.8

saving is relatively small, replacement of fully loaded motors can still be economical for motors that operate continuously in areas where electricity costs are high. Motors that are seriously underloaded are better candidates for replacement. The efficiency of conventional motors begins to fall sharply at less than 50% load, and replacement with a smaller high-efficiency motor can yield a quick return. Motors that must run at part load for a significant part of their operating cycle are also good candidates for replacement, since high-efficiency motors typically have better part-load performance than conventional motors.

High-efficiency motors typically run faster than conventional motors with the same speed rating because high-efficiency motors operate with less slip. The installation of a high-efficiency motor to drive a fan or pump may actually increase energy consumption due to the increase in speed, since power consumption for fans and pumps increases as the cube of the speed. The sheaves in the fan or pump drive should be adjusted or changed to avoid this problem.

More Efficient Lighting Systems

Conversion of lighting fixtures to more efficient light sources is often practical when the lights are used for a significant portion of the year. Table 6 lists some of the more common conversions and the difference in power consumption. Installation of energy-saving ballasts in fluorescent lights provides a small (5–12%) percentage reduction in fixture power consumption, but the cost can be justified by energy cost savings if the lights are on most of

Table 6 Some Common Lighting Conversions

Present Fixture				Replacement Fixture			
Type	Power Consumption (W)	Light Output (lumens)	Lifetime (hr)	Type	Power Consumption (W)	Light Output (lumens)	Lifetime (hr)
100-W incandescent	100	1,740	750	8-in. 22-W circline adapter in same fixture	34	980	7,500
500-W incandescent	500	10,600	1,000	175-W metal halide fixture	205	12,000	7,500
Four 40-W rapid-start warm-white 48-in. fluorescent tubes in two-ballast fixture	184	12,800	18,000	Four 34-W energy saver rapid-start tubes in same fixture	160	11,600	18,000
Two 75-W slimline warm-white 96-in. fluorescent tubes in one-ballast fixture	172	12,800	12,000	Two 60-W energy saver slimline tubes in same fixture	142	10,680	12,000
150-W incandescent reflector floodlight (R-40)	150	1,200 (beam candlepower)	5,000	75-W incandescent projector floodlight (PAR-38)	75	1,430 (beam candlepower)	5,000
250-W mercury vapor streetlamp	285	10,400	24,000	150-W high-pressure sodium streetlamp	188	14,400	24,000

the time. Additional lighting controls such as automatic dimmers can reduce energy consumption by making better use of daylight. Attention should also be given to the efficiency of the luminaire and (for indoor lighting) interior wall surfaces in directing light to the areas where it is needed. Reference 1 provides data for estimating savings from more efficient luminaires and more reflective wall and ceiling surfaces.

4 EVALUATING ENERGY CONSERVATION OPPORTUNITIES

The auditor's evaluation of energy conservation opportunities should begin with a careful consideration of the possible effects of energy conservation measures on safety, health, comfort, and productivity within a facility. A conscientious effort should be made to solicit information from knowledgeable personnel and those who have experience with conservation measures in similar facilities. For energy conservation measures that do not interfere with the main business of a facility and the health and safety of its occupants, the determinant of action is the financial merit of a given measure.

Most decisions regarding the implementation of an energy conservation measure are based on the auditor's evaluation of the annual dollar savings and the initial capital cost, if any, associated with the measure. Estimation of the cost and savings from energy conservation measures is thus a critically important part of the analytical work involved in energy auditing.

When an energy conservation opportunity is first identified, the auditor should make a rough estimate of costs and savings in order to assess the value of further investigation. A rough estimate of the installed cost of a measure can often be obtained by consulting a local contractor or vendor who has experience with the type of equipment that the measure would involve. For commercial building energy conservation measures, a good guide to costs can be obtained from one of the annually published building construction cost estimating guides. The most valuable guides provide costs for individual mechanical, electrical, and structural components in a range of sizes or capacities. Rough estimates of the annual dollar savings from a measure can use simplified approaches to estimating energy savings such as assuming that a motor operates at its full nameplate rating for a specified percentage of the time.

If further analysis of a measure is warranted, a more accurate estimate of installed cost can be developed by preparing a clear and complete specification for the measure and obtaining quotations from experienced contractors or vendors. In estimating savings, one should be careful to calculate the measure's effect on energy use using accurate data for operating schedules, temperatures, flow rates, and other parameters. One should also give careful consideration to the measure's effect on maintenance requirements and equipment lifetimes, and include a dollar figure for the change in labor or depreciation costs in the savings estimate.

5 PRESENTING THE RESULTS OF AN ENERGY AUDIT

Effective presentation of the energy audit's results is crucial to achieving energy savings. The presentation may be an informal conversation with maintenance personnel, or it may be a formal presentation to management with a detailed financial analysis. In some cases the auditor may also need to make a written application to an outside funding source such as a government agency.

The basic topics that should be covered in most presentations are the following:

1. The facility's historical energy use, in physical and dollar amounts broken down by end use.

2. A review of the existing energy management program (if any) and recommendations for improvement.

3. A description of the energy conservation measures being proposed and the means by which they will save energy.

4. The cost of undertaking the measures and the net benefits the facility will receive each year.

5. Any other effects the measure will have on the facility's operation, such as changes in maintenance requirements or comfort levels.

The auditor should be prepared to address these topics with clear explanations geared to the interests and expertise of the audience. A financial officer, for example, may want considerable detail on cash flow analysis. A maintenance foreman, however, will want information on the equipment's record for reliability under conditions similar to those in his or her facility. Charts, graphs, and pictures may help to explain some topics, but they should be used sparingly to avoid inundating the audience with information that is of secondary importance.

The financial analysis will be the most important part of a presentation that involves recommendations of measures requiring capital expenditures. The complexity of the analysis will vary, depending on the type of presentation, from a simple estimate of the installed cost and annual savings to an internal rate of return or discounted cash flow calculation.

The more complex types of calculations involve assumptions regarding future fuel and electricity price increases, interest rates, and other factors. Because these assumptions are judgmental and may critically affect the results of the analysis, the more complex analyses should not be used in presentations to the exclusion of simpler indices such as simple payback time or after-tax return on investment. These methods do not involve numerous projections about the future.

REFERENCES

1. J. E. Kaufman (ed.), *IES Lighting Handbook,* Illuminating Engineers Society of North America, New York, 1981.

2. M. Lokmanhekim et al., *DOE-2: A New State-of-the-Art Computer Program for the Energy Utilization Analysis of Buildings,* Lawrence Berkeley Laboratory Report, LBL-8974, Berkeley, CA, 1979.

3. U.S. Department of Energy, *Architects and Engineers Guide to Energy Conservation in Existing Buildings,* Federal Energy Management Program Manual, U.S. Department of Energy, Federal Programs Office, Conservation and Solar Energy, NTIS Report DOE/CS-1302, February 1, 1980.

4. L. W. Wall and J. Flaherty, *A Summary Review of Building Energy Use Compilation and Analysis (BECA) Part C: Conservation Progress in Retrofitted Commercial Buildings,* Lawrence Berkeley Laboratory Report, LBL-15375, Berkeley, CA, 1982.

5. F. C. Winkelmann and M. Lokmanhekim, *Life-Cycle Cost and Energy-Use Analysis of Sun Control and Daylighting Options in a High-Rise Office Building,* Lawrence Berkeley Laboratory Report, LBL-12298, Berkeley, CA, 1981.

6. California Energy Commission, *Institutional Conservation Program Energy Audit Report: Minimum Energy Audit Guidelines,* California Energy Commission, Publication No. P400-82-022, Sacramento, CA, 1982.

7. W. C. Turner (ed.), *Energy Management Handbook,* Wiley-Interscience, New York, 1982.

CHAPTER **8**

HEAT EXCHANGERS, VAPORIZERS, CONDENSERS

Joseph W. Palen
Consultant
Eugene, Oregon

1 HEAT EXCHANGER TYPES AND CONSTRUCTION

Heat exchangers permit exchange of energy from one fluid to another, usually without permitting physical contact between the fluids. The following configurations are commonly used in the power and process industries.

1.1 Shell and Tube Heat Exchangers

Shell and tube heat exchangers normally consist of a bundle of tubes fastened into holes, drilled in metal plates called tubesheets. The tubes may be rolled into grooves in the tubesheet, welded to the tubesheet, or both to ensure against leakage. When possible, U-tubes are used, requiring only one tubesheet. The tube bundle is placed inside a large pipe called a shell, see Fig. 1. Heat is exchanged between a fluid flowing inside the tubes and a fluid flowing outside the tubes in the shell.

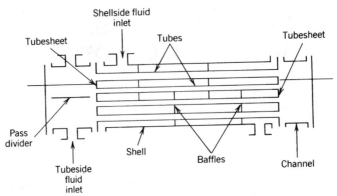

Figure 1 Schematic illustration of shell and tube heat exchanger construction.

When the tubeside heat-transfer coefficient is as high as three times the shellside heat-transfer coefficient, it may be advantageous to use low integral finned tubes. These tubes can have outside heat-transfer coefficients as high as plain tubes, or even higher, but increase the outside heat-transfer area by a factor of about 2.5–4. For design methods using finned tubes, see Ref. 11 for single-phase heat exchangers and Ref. 14 for condensers. Details of construction practices are described by Saunders.[58]

The Tubular Exchanger Manufacturers Association (TEMA) provides a manual of standards for construction of shell and tube heat exchangers,[1] which contains designations for various types of shell and tube heat exchanger configurations. The most common types are summarized below.

E-Type
The E-type shell and tube heat exchanger, illustrated in Fig. 2, is the workhorse of the process industries, providing economical rugged construction and a wide range of capabilities. Baffles support the tubes and increase shellside velocity to improve heat transfer. More than one pass is usually provided for tubeside flow to increase the velocity, Fig. 2a. However, for some cases, notably vertical thermosiphon vaporizers, a single tubepass is used, as shown in Fig. 2b.

The E-type shell is usually the first choice of shell types because of lowest cost, but sometimes requires more than the allowable pressure drop, or produces a temperature "pinch" (see Section 4.4), so other, more complicated types are used.

F-Type Shell
If the exit temperature of the cold fluid is greater than the exit temperature of the hot fluid, a temperature cross is said to exist. A slight temperature cross can be tolerated in a multitubepass E-type shell (see below), but if the cross is appreciable, either units in series or complete countercurrent flow is required. A solution sometimes used is the F-type or two-pass shell, as shown in Fig. 3.

The F-type shell has a number of potential disadvantages, such as thermal and fluid leakage around the longitudinal baffle and high pressure drop, but it can be effective in some cases if well designed.

J-Type
When an E-type shell cannot be used because of high pressure drop, a J-type or divided flow exchanger, shown in Fig. 4, is considered. Since the flow is divided and the flow length

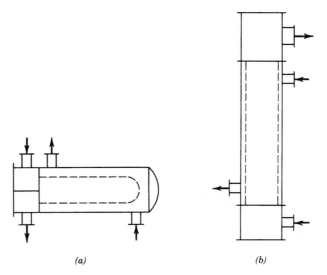

Figure 2 TEMA E-type shell: (*a*) horizontal multitubepass; (*b*) vertical single tubepass.

is also cut in half, the shellside pressure drop is only about one-eighth to one-fifth that of an E-type shell of the same dimensions.

X-Type

When a J-type shell would still produce too high a pressure drop, an X-type shell, shown in Fig. 5, may be used. This type is especially applicable for vacuum condensers, and can be equipped with integral finned tubes to counteract the effect of low shellside velocity on heat transfer. It is usually necessary to provide a flow distribution device under the inlet nozzle. When the ratio of tube length to shell diameter is greater than about 4.5, a second set of shellside nozzles should be added.

G-Type

This shell type, shown in Fig. 6, is sometimes used for horizontal thermosiphon shellside vaporizers. The horizontal baffle is used especially for boiling range mixtures and provides better flow distribution than would be the case with the X-type shell. The G-type shell also permits a larger temperature cross than the E-type shell with about the same pressure drop.

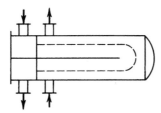

Figure 3 TEMA F-type shell.

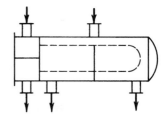

Figure 4 TEMA J-type shell.

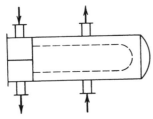

Figure 5   TEMA X-type shell.

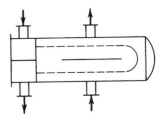

Figure 6   TEMA G-type shell.

H-Type

If a G-type is being considered but pressure drop would be too high, an H-type may be used. This configuration is essentially just two G-types in parallel, as shown in Fig. 7.

K-Type

This type is used exclusively for kettle reboilers and vaporizers, and is characterized by the oversized shell intended to separate vapor and liquid phases, Fig. 8. Shell-sizing relationships are given in Ref. 25. Usually, the shell diameter is about 1.6–2.0 times the bundle diameter. Design should consider amount of acceptable entrainment, height required for flow over the weir, and minimum clearance in case of foaming.

Baffle Types

Baffles are used to increase velocity of the fluid flowing outside the tubes ("shellside" fluid) and to support the tubes. Higher velocities have the advantage of increasing heat transfer and decreasing fouling (material deposit on the tubes), but have the disadvantage of increasing pressure drop (more energy consumption per unit of fluid flow). The amount of pressure drop on the shellside is a function of baffle spacing, baffle cut, baffle type, and tube pitch.

Baffle types commonly used are shown in Fig. 9, with pressure drop decreasing from Fig. 9a to Fig. 9c. The helical baffle (Section 6) has several advantages.

Baffle spacing is increased when it is necessary to decrease pressure drop. A limit must be imposed to prevent tube sagging or flow-induced tube vibration. Recommendations for maximum baffle spacing are given in Ref. 1. Tube vibration is discussed in more detail in Section 4.2. When the maximum spacing still produces too much pressure drop, a baffle type is considered that produces less cross flow and more longitudinal flow, for example,

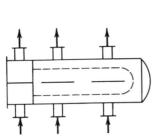

Figure 7   TEMA H-type shell.

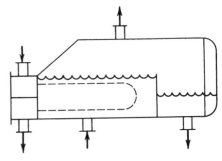

Figure 8   TEMA K-type shell.

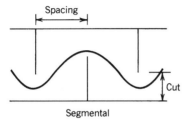

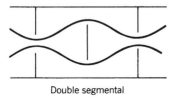

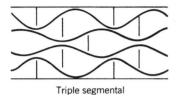

Figure 9 Baffle types.

double segmental instead of segmental. Minimum pressure drop is obtained if baffles are replaced by rod-type tube supports.[52]

1.2 Plate-Type Heat Exchangers

Composed of a series of corrugated or embossed plates clamped between a stationary and a movable support plate, these exchangers were originally used in the food-processing industry. They have the advantages of low fouling rates, easy cleaning, and generally high heat-transfer coefficients, and are becoming more frequently used in the chemical process and power industries. They have the disadvantage that available gaskets for the plates are not compatible with all combinations of pressure, temperature, and chemical composition. Suitability for specific applications must be checked. The maximum operating pressure is usually considered to be about 1.5 MPa (220 psia).[3] However, welded plate versions are now available for much higher pressures. A typical plate heat exchanger is shown in Fig. 10. Welded plate exchangers and other compact types are discussed in Section 6.

1.3 Spiral Plate Heat Exchangers

These exchangers are also becoming more widely used, despite limitations on maximum size and maximum operating pressure. They are made by wrapping two parallel metal plates, separated by spacers, into a spiral to form two concentric spiral passages. A schematic example is shown in Fig. 11.

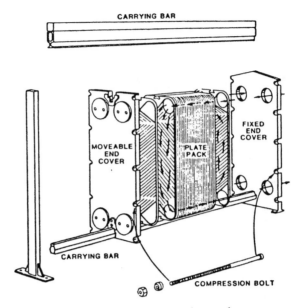

Figure 10 Typical plate-type heat exchanger.

Spiral plate heat exchangers can provide completely countercurrent flow, permitting temperature crosses and close approaches, while maintaining high velocity and high heat-transfer coefficients. Since all flow for each fluid is in a single channel, the channel tends to be flushed of particles by the flow, and the exchanger can handle sludges and slurries more effectively than can shell and tube heat exchangers. The most common uses are for difficult-to-handle fluids with no phase change. However, the low-pressure-drop characteristics are beginning to promote some use in two-phase flow as condensers and reboilers. For this purpose the two-phase fluid normally flows axially in a single pass rather than spirally.

1.4 Air-Cooled Heat Exchangers

It is sometimes economical to condense or cool hot streams inside tubes by blowing air across the tubes rather than using water or other cooling liquid. They usually consist of a

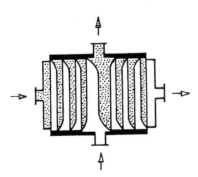

Figure 11 Spiral plate heat exchanger.

horizontal bank of finned tubes with a fan at the bottom (forced draft) or top (induced draft) of the bank, as illustrated schematically in Fig. 12.

Tubes in air-cooled heat exchangers (Fig. 12) are often 1 in. (25.4 mm) in outside diameter with ⅝ in. (15.9 mm) high annular fins, 0.4–0.5 mm thick. The fins are usually aluminum and may be attached in a number of ways, ranging from tension wrapped to integrally extruded (requiring a steel or alloy insert), depending on the severity of service. Tension wrapped fins have an upper temperature limit (~300°F) above which the fin may no longer be in good contact with the tube, greatly decreasing the heat-transfer effectiveness. Various types of fins and attachments are illustrated in Fig. 13.

A more detailed description of air-cooled heat exchanger geometries is given Refs. 2 and 3.

1.5 Compact Heat Exchangers

The term compact heat exchanger normally refers to one of the many types of plate fin exchangers used extensively in the aerospace and cryogenics industries. The fluids flow alternately between parallel plates separated by corrugated metal strips that act as fins and that may be perforated or interrupted to increase turbulence. Although relatively expensive to construct, these units pack a very large amount of heat-transfer surface into a small volume, and are therefore used when exchanger volume or weight must be minimized. A detailed description with design methods is given in Ref. 4. Also see Section 6.

1.6 Boiler Feedwater Heaters

Exchangers to preheat feedwater to power plant boilers are essentially of the shell and tube type but have some special features, as described in Ref. 5. The steam that is used for preheating the feedwater enters the exchanger superheated, is condensed, and leaves as sub-cooled condensate. More effective heat transfer is achieved by providing three zones on the shellside: desuperheating, condensing, and subcooling. A description of the design requirements of this type of exchanger is given in Ref. 5.

1.7 Recuperators and Regenerators

These heat exchangers are used typically to conserve heat from furnace off-gas by exchanging it against the inlet air to the furnace. A recuperator does this in the same manner as any

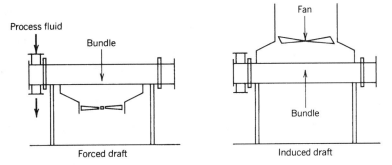

Figure 12 Air-cooled heat exchangers.

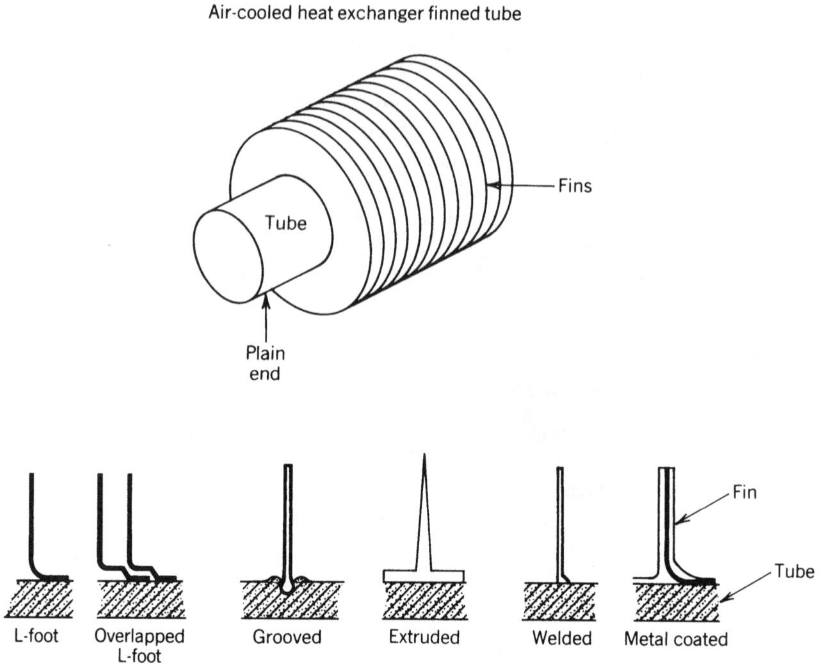

Air-cooled heat exchanger finned tube

Figure 13 Typical finned tube and attachments.

other heat exchanger except the construction may be different to comply with requirements for low pressure drop and handling of the high-temperature, often dirty, off-gas stream. Heat pipes (Chapter 9) with fins are now sometimes used.

The regenerator is a transient batch-type exchanger in which packed beds are alternately switched from the hot stream to the cold stream. A description of the operating characteristics and design of recuperators and regenerators is given in Refs. 6 and 59.

2 ESTIMATION OF SIZE AND COST

In determining the overall cost of a proposed process plant or power plant, the cost of heat exchangers is of significant importance. Since cost is roughly proportional to the amount of heat-transfer surface required, some method of obtaining an estimate of performance is necessary, which can then be translated into required surface. The term "surface" refers to the total area across which the heat is transferred. For example, with shell and tube heat exchangers "surface" is the tube outside circumference times the tube length times the total number of tubes. Well-known basic equations taken from Newton's law of cooling relate the required surface to the available temperature difference and the required heat duty.

2.1 Basic Equations for Required Surface

The following well-known equation is used (equation terms are defined in the Nomenclature):

$$A_o = \frac{Q}{U_o \times \text{MTD}} \tag{1}$$

The required duty (Q) is related to the energy change of the fluids:

(a) Sensible Heat Transfer

$$Q = W_1 C_{p1}(T_2 - T_1) \qquad (2a)$$

$$= W_2 C_{p2}(t_1 - t_2) \qquad (2b)$$

(b) Latent Heat Transfer

$$Q = W\lambda \qquad (3)$$

where W = flow rate of boiling or condensing fluid
λ = latent heat of respective fluid

The mean temperature difference (MTD) and the overall heat transfer coefficient (U_o) in Eq. (1) are discussed in Sections 2.2 and 2.3, respectively. Once the required surface, or area, (A_o) is obtained, heat exchanger cost can be estimated. A comprehensive discussion on cost estimation for several types of exchangers is given in Ref. 7. Cost charts for small- to medium-sized shell and tube exchangers, developed in 1982, are given in Ref. 8.

2.2 Mean Temperature Difference

The mean temperature difference (MTD) in Eq. (1) is given by the equation

$$\text{MTD} = \frac{F(T_A - T_B)}{\ln(T_A/T_B)} \qquad (4)$$

where

$$T_A = T_1 - t_2 \qquad (5)$$

$$T_B = T_2 - t_1 \qquad (6)$$

The temperatures (T_1, T_2, t_1, t_2) are illustrated for the base case of countercurrent flow in Fig. 14.

The factor F in Eq. (4) is the multitubepass correction factor. It accounts for the fact that heat exchangers with more than one tubepass can have some portions in concurrent flow or cross flow, which produce less effective heat transfer than countercurrent flow. Therefore, the factor F is less than 1.0 for multitubepass exchangers, except for the special case of isothermal boiling or condensing streams for which F is always 1.0. Charts for calculating F are available in most heat-transfer textbooks. A comprehensive compilation for various types of exchangers is given by Taborek.[9]

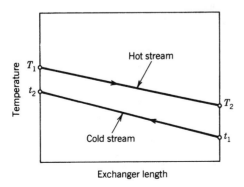

Figure 14 Temperature profiles illustrated for countercurrent flow.

In a properly designed heat exchanger, it is unusual for F to be less than 0.7, and if there is no temperature cross $(T_2 > t_2)$, F will be 0.8 or greater. As a first approximation for preliminary sizing and cost estimation, F may be taken as 0.85 for multitubepass exchangers with temperature change of both streams and 1.0 for other cases.

2.3   Overall Heat-Transfer Coefficient

The factor (U_o) in Eq. (1) is the overall heat-transfer coefficient. It may be calculated by procedures described in Section 3, and is the reciprocal of the sum of all heat-transfer resistances, as shown in the equation

$$U_o = 1/(R_{h_o} + R_{f_o} + R_w + R_{h_i} + R_{f_i}) \tag{7}$$

where

$$R_{h_o} = 1/h_o \tag{8}$$

$$R_{h_i} = (A_o/A_i h_i) \tag{9}$$

$$R_w = \frac{A_o x_w}{A_m k_w} \tag{10}$$

Calculation of the heat-transfer coefficients h_o and h_i can be time consuming, since they depend on the fluid velocities, which, in turn, depend on the exchanger geometry. This is usually done now by computer programs that guess correct exchanger size, calculate heat-transfer coefficients, check size, adjust, and reiterate until satisfactory agreement between guessed and calculated size is obtained. For first estimates by hand before size is known, values of h_o and h_i, as well as values of the fouling resistances, R_{f_o} and R_{f_i}, are recommended by Bell for shell and tube heat exchangers.[10]

Very rough, first approximation values for the overall heat-transfer coefficient are given in Table 1.

2.4   Pressure Drop

In addition to calculation of the heat-transfer surface required, it is usually necessary to consider the pressure drop consumed by the heat exchanger, since this enters into the overall

Table 1   Approximate Values for Overall Heat Transfer Coefficient of Shell and Tube Heat Exchangers (Including Allowance for Fouling)

Fluids	U_o Btu/hr·ft²·°F	W/m²·K
Water–water	250	1400
Oil–water	75	425
Oil–oil	45	250
Gas–oil	15	85
Gas–water	20	115
Gas–gas	10	60

cost picture. Pressure drop is roughly related to the individual heat-transfer coefficients by an equation of the form,

$$\Delta P = Ch^m + EX \qquad (11)$$

where ΔP = shellside or tubeside pressure drop

$\quad h$ = heat-transfer coefficient

$\quad C$ = coefficient depending on geometry

$\quad m$ = exponent depending on geometry—always greater than 1.0, and usually about 3.0

$\quad EX$ = extra pressure drop from inlet, exit, and pass turnaround momentum losses

See Section 3 for actual pressure drop calculations.

Pressure drop is sensitive to the type of exchanger selected. In the final design it is attempted, where possible, to define the exchanger geometry so as to use all available pressure drop and thus maximize the heat-transfer coefficient. This procedure is subject to some constraints, however, as follows. The product of density times velocity squared ρv^2 is limited to minimize the possibility of erosion or tube vibration. A limit often used is $\rho v^2 < 4000$ lbm/ft·sec^2. This results in a velocity for liquids in the range of 7–10 ft/sec. For flow entering the shellside of an exchanger and impacting the tubes, an impingement plate is recommended to prevent erosion if $\rho v^2 > 1500$. Other useful design recommendations may be found in Ref. 1.

For condensing vapors, pressure drop should be limited to a fraction of the operating pressure for cases with close temperature approach to prevent severe decrease of the MTD owing to lowered equilibrium condensing temperature. As a safe "rule of thumb," the pressure drop for condensing is limited to about 10% of the operating pressure. For other cases, "reasonable" design pressure drops for heat exchangers roughly range from about 5 psi for gases and boiling liquids to as high as 20 psi for pumped nonboiling liquids.

3 RATING METHODS

After the size and basic geometry of a heat exchanger has been proposed, the individual heat-transfer coefficients h_o and h_i may be calculated based on actual velocities, and the required surface may be checked, based on these updated values. The pressure drops are also checked at this stage. Any inadequacies are adjusted and the exchanger is rechecked. This process is known as "rating." Different rating methods are used depending on exchanger geometry and process type, as covered in the following sections.

3.1 Shell and Tube Single-Phase Exchangers

Before the individual heat-transfer coefficients can be calculated, the heat exchanger tube geometry, shell diameter, shell type, baffle type, baffle spacing, baffle cut, and number of tubepasses must be decided. As stated above, lacking other insight, the simplest exchanger—E-type with segmental baffles—is tried first.

Tube Length and Shell Diameter
For shell and tube exchangers the tube length is normally about 5–8 times the shell diameter. Tube lengths are usually 8–20 ft long in increments of 2 ft. However, very large size exchangers with tube lengths up to 40 ft are more frequently used as economics dictate smaller

MTD and larger plants. A reasonable trial tube length is chosen and the number of tubes (NT) required for surface A_o, Section 2, is calculated as follows:

$$NT = \frac{A_o}{a_o L} \tag{12}$$

where a_o = the surface/unit length of tube.
For plain tubes (as opposed to finned tubes),

$$a_o = \pi D_o \tag{13}$$

where D_o = the tube outside diameter
L = the tube length

The tube bundle diameter (D_b) can be determined from the number of tubes, but also depends on the number of tubepasses, tube layout, and bundle construction. Tube count tables providing this information are available from several sources. Accurate estimation equations are given by Taborek.[11] A simple basic equation that gives reasonable first approximation results for typical geometries is the following:

$$D_b = P_t \left(\frac{NT}{\pi/4}\right)^{0.5} \tag{14}$$

where P_t = tube pitch (spacing between tube diameters). Normally, P_t/D_o = 1.25, 1.33, or 1.5.

The shell diameter D_s is larger than the bundle diameter D_b by the amount of clearance necessary for the type of bundle construction. Roughly, this clearance ranges from about 0.5 in. for U-tube or fixed tubesheet construction to 3–4 in. for pull-through floating heads, depending on the design pressure and bundle diameter. (For large clearances, sealing strips are used to prevent flow bypassing the bundles.) After the bundle diameter is calculated, the ratio of length to diameter is checked to see if it is in an acceptable range, and the length is adjusted if necessary.

Baffle Spacing and Cut
Baffle spacing L_{bc} and cut B_c (see Fig. 9) cannot be decided exactly until pressure drop is evaluated. However, a reasonable first guess ratio of baffle spacing to shell diameter (L_{bc}/D_s) is about 0.45. The baffle cut (B_c, a percentage of D_s) required to give good shellside distribution may be estimated by the following equation:

$$B_c = 16.25 + 18.75 \left(\frac{L_{bc}}{D_s}\right) \tag{15}$$

For more detail, see the recommendations of Taborek.[11]

Cross-Sectional Flow Areas and Flow Velocities
The cross-sectional flow areas for tubeside flow S_t and for shellside flow S_s are calculated as follows:

$$S_t = \left(\frac{\pi}{4} D_i^2\right)\left(\frac{NT}{NP}\right) \tag{16}$$

$$S_s = 0.785(D_b)(L_{bc})(P_t - D_o)/P_t \tag{17}$$

where L_{bc} = baffle spacing.

Equation (17) is approximate in that it neglects pass partition gaps in the tube field, it approximates the bundle average chord, and it assumes an equilateral triangular layout. For more accurate equations see Ref. 11.

The tubeside velocity V_t and the shellside velocity V_s are calculated as follows:

$$V_t = \frac{W_t}{S_t \, \rho_t} \tag{18}$$

$$V_s = \frac{W_s}{S_s \, \rho_s} \tag{19}$$

Heat-Transfer Coefficients

The individual heat-transfer coefficients, h_o and h_i, in Eq. (1) can be calculated with reasonably good accuracy (± 20–30%) by semiempirical equations found in several design-oriented textbooks.[11,12] Simplified approximate equations are the following:

(a) *Tubeside Flow*

$$\mathrm{Re} = \frac{D_o V_t \, \rho_t}{\mu_t} \tag{20}$$

where μ_t = tubeside fluid viscosity.

If Re < 2000, laminar flow,

$$h_i = 1.86 \left(\frac{k_f}{D_i} \right) \left(\mathrm{Re} \, \mathrm{Pr} \, \frac{D_i}{L} \right)^{0.33} \left(\frac{\mu_f}{\mu_w} \right)^{0.14} \tag{21}$$

If Re > 10,000, turbulent flow,

$$h_i = 0.024 \left(\frac{k_f}{D_i} \right) \mathrm{Re}^{0.8} \, \mathrm{Pr}^{0.4} \left(\frac{\mu_f}{\mu_w} \right)^{0.14} \tag{22}$$

If 2000 < Re < 10,000, prorate linearly.

(b) *Shellside Flow*

$$\mathrm{Re} = \frac{D_o V_s \, \rho_s}{\mu_s} \tag{23}$$

where μ_s = shellside fluid viscosity.

If Re < 500, see Refs. 11 and 12.

If Re > 500,

$$h_o = 0.38 \, C_b^{0.6} \left(\frac{k_f}{D_o} \right) \mathrm{Re}^{0.6} \, \mathrm{Pr}^{0.33} \left(\frac{\mu_f}{\mu_w} \right)^{0.14} \tag{24}$$

The term Pr is the Prandtl number and is calculated as $C_p \, \mu/k$.

The constant (C_b) in Eq. (24) depends on the amount of bypassing or leakage around the tube bundle.[13] As a first approximation, the values in Table 2 may be used.

Pressure Drop

Pressure drop is much more sensitive to exchanger geometry, and, therefore, more difficult to accurately estimate than heat transfer, especially for the shellside. The so-called Bell–Delaware method[11] is considered the most accurate method in open literature, which can be

Table 2 Approximate Bypass Coefficient, C_b

Bundle Type	C_b
Fixed tubesheet or U-tube	0.70
Split-ring floating head, seal strips	0.65
Pull-through floating head, seal strips	0.55

calculated by hand. The following very simplified equations are provided for a rough idea of the range of pressure drop, in order to minimize preliminary specification of unrealistic geometries.

(a) Tubeside (contains about 30% excess for nozzles)

$$\Delta P_t = \left[\frac{0.025(L)(NP)}{D_i} + 2(NP - 1) \right] \frac{\rho_t V_t^2}{g_c} \left(\frac{\mu_w}{\mu_f} \right)^{0.14} \tag{25}$$

where NP = number of tubepasses.

(b) Shellside (contains about 30% excess for nozzles)

$$\Delta P_s = \frac{0.24(L)(D_b)(\rho_s)(C_b V_s)^2}{g_c L_{bc} P_t} \left(\frac{\mu_w}{\mu_f} \right)^{0.14} \tag{26}$$

where g_c = gravitational constant (4.17×10^8 for velocity in ft/hr and density in lb/ft^3).

3.2 Shell and Tube Condensers

The condensing vapor can be on either the shellside or tubeside depending on process constraints. The "cold" fluid is often cooling tower water, but can also be another process fluid, which is sensibly heated or boiled. In this section, the condensing-side heat-transfer coefficient and pressure drop are discussed. Single-phase coolants are handled, as explained in the previous section. Boiling fluids will be discussed in a later section.

Selection of Condenser Type

The first task in designing a condenser, before rating can proceed, is to select the condenser configuration. Mueller[14] presents detailed charts for selection based on the criteria of system pressure, pressure drop, temperature, fouling tendency of the coolant, fouling tendency of the vapor, corrosiveness of the vapor, and freezing potential of the vapor. Table 3 is an abstract of the recommendations of Mueller.

The suggestions in Table 3 may, of course, be ambiguous in case of more than one important criterion, for example, corrosive vapor together with a fouling coolant. In these cases, the most critical constraint must be respected, as determined by experience and engineering judgment. Corrosive vapors are usually put on the tubeside, and chemical cleaning used for the shellside coolant, if necessary. Since most process vapors are relatively clean (not always the case!), the coolant is usually the dirtier of the two fluids and the tendency is to put it on the tubeside for easier cleaning. Therefore, the most common shell and tube condenser is the shellside condenser using TEMA types E, J, or X, depending on allowable pressure drop; see Section 1. An F-type shell is sometimes specified if there is a large condensing range and a temperature cross (see below), but, owing to problems with the F-type, E-type units in series are often preferred in this case.

Table 3 Condenser Selection Chart

Process Condition	Suggested Condenser Type[a]
Potential coolant fouling	HS/E, J, X
High condensing pressure	VT/E
Low condensing pressure drop	HS/J, X
Corrosive or very-high-temperature vapors	VT/E
Potential condensate freezing	HS/E
Boiling coolant	VS/E or HT/K, G, H

[a] V, vertical; H, horizontal; S, shellside condensation; T, tubeside condensation; /E, J, H, K, X, TEMA shell styles.

In addition to the above condenser types the vertical E-type tubeside condenser is sometimes used in a "reflux" configuration with vapor flowing up and condensate flowing back down inside the tubes. This configuration may be useful in special cases, such as when it is required to strip out condensable components from a vent gas that is to be rejected to the atmosphere. The disadvantage of this type of condenser is that the vapor velocity must be very low to prevent carryover of the condensate (flooding), so the heat-transfer coefficient is correspondingly low, and the condenser rather inefficient. Methods used to predict the limiting vapor velocity are given in Refs. 14 and 64.

Temperature Profiles

For a condensing pure component, if the pressure drop is less than about 10% of the operating pressure, the condensing temperature is essentially constant and the LMTD applied ($F = 1.0$) for the condensing section. If there are desuperheating and subcooling sections,[5] the MTD and surface for these sections must be calculated separately. For a condensing mixture, with or without noncondensables, the temperature profile of the condensing fluid with respect to fraction condensed should be calculated according to vapor–liquid equilibrium (VLE) relationships.[15] A number of computer programs are available to solve VLE relationships; a version suitable for programmable calculator is given in Ref. 16.

Calculations of the condensing temperature profile may be performed either integrally, which assumes vapor and liquid phases are well mixed throughout the condenser, or differentially, which assumes separation of the liquid phase from the vapor phase. In most actual condensers the phases are mixed near the entrance where the vapor velocity is high and separated near the exit where the vapor velocity is lower. The "differential" curve produces a lower MTD than the "integral" curve and is safer to use where separation is expected.

For most accuracy, condensers are rated incrementally by stepwise procedures such as those explained by Mueller.[14] These calculations are usually performed by computers.[17] As a first approximation, to get an initial size, a straight-line temperature profile is often assumed for the condensing section (not including desuperheating or subcooling sections!). As illustrated in Fig. 15, the true condensing curve is usually more like curve I, which gives a larger MTD than the straight line, curve II, making the straight-line approximation conservative. However, a curve such as curve III is certainly possible, especially with immiscible condensates, for which the VLE should always be calculated. For the straight-line approximation, the condensing heat-transfer coefficient is calculated at average conditions, as shown below.

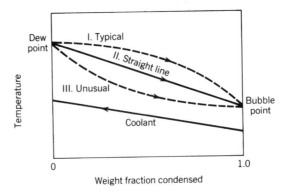

Figure 15   Condensation profiles illustrated.

Heat-Transfer Coefficients, Pure Components

For condensers, it is particularly important to be able to estimate the two-phase flow regime in order to predict the heat-transfer coefficient accurately. This is because completely different types of correlations are required for the two major flow regimes.

Shear-Controlled Flow. The vapor shear force on the condensate is much greater than the gravity force. This condition can be expected, according to Ref. 18, when

$$J_g > 1.5 \tag{27}$$

where

$$J_g = \left[\frac{(Gy)^2}{g D_j \rho_v (\rho_l - \rho_v)} \right]^{0.5} \tag{28}$$

For shear-controlled flow, the condensate film heat-transfer coefficient (h_{cf}) is a function of the convective heat-transfer coefficient for liquid flowing alone and the two-phase pressure drop[18]:

$$h_{cf} = h_l (\phi_l^2)^{0.45} \tag{29}$$

$$h_l = h_i (1 - y)^{0.8} \tag{30}$$

or

$$h_l = h_o (1 - y)^{0.6} \tag{31}$$

$$\phi_l^2 = 1 + \frac{C}{X_{tt}} + \frac{1}{X_{tt}^2} \tag{32}$$

$$C = 20 \text{ (tubeside flow)}, \qquad C = 9 \text{ (shellside flow)}$$

$$X_{tt} = \left[\frac{1 - y}{y} \right]^{0.9} \left[\frac{\rho_v}{\rho_l} \right]^{0.5} \left[\frac{\mu_l}{\mu_v} \right]^{0.1} \tag{33}$$

$$\mu_l = \text{liquid viscosity}, \qquad \mu_v = \text{vapor viscosity}$$

Gravity-Controlled Flow. The vapor shear force on the condensate is small compared to the gravity force, so condensate drains by gravity. This condition can be expected, according to Ref. 18, when $J_g < 0.5$. Under gravity-controlled conditions, the condensate film heat-transfer coefficient is calculated as follows:

$$h_{cf} = F_g h_N \tag{34}$$

The term h_N is the heat-transfer coefficient from the well-known Nusselt derivation, given in Ref. 14 as

Horizontal Tubes

$$h_N = 0.725 \left[\frac{k_l^3 \rho_l (\rho_l - \rho_v) g \lambda}{\mu_l (T_s - T_w) D} \right]^{0.25} \tag{35}$$

where λ = latent heat.

Vertical Tubes

$$h_N = 1.1 k_l \left[\frac{\rho_l (\rho_l - \rho_v) g}{\mu_l^2 \, \text{Re}_c} \right]^{0.33} \tag{36}$$

$$\text{Re}_c = \frac{4 W_c}{\pi D \mu_l} \tag{37}$$

The term F_g in Eq. (34) is a correction for condensate loading, and depends on the exchanger geometry.[14]

On horizontal X-type tube bundles

$$F_g = N_{rv}^{-1/6} \tag{38}$$

(Ref. 12), where N_{rv} = number of tubes in a vertical row.

On baffled tube bundles (owing to turbulence)

$$F_g = 1.0 \qquad \text{(frequent practice)} \tag{39}$$

In horizontal tubes

$$F_g = \left[\frac{1}{1 + (1/y - 1)(\rho_v/\rho_l)^{0.667}} \right]^{0.75} \qquad \text{(from Ref. 14)} \tag{40}$$

or

$$F_g = 0.8 \qquad \text{(from Ref. 18)} \tag{41}$$

Inside or outside vertical tubes

$$F_g = 0.73 \, \text{Re}_c^{0.11} \qquad \text{(rippled film region)} \tag{42}$$

or

$$F_g = 0.021 \, \text{Re}_c^{0.58} \, \text{Pr}^{0.33} \qquad \text{(turbulent film region)} \tag{43}$$

Use higher value of Eq. (42) or (43).

For quick hand calculations, the gravity-controlled flow equations may be used for h_{cf}, and will usually give conservative results.

Correction for Mixture Effects

The above heat-transfer coefficients apply only to the condensate film. For mixtures with a significant difference between the dew-point and bubble-point temperatures (condensing range), the vapor-phase heat-transfer coefficient must also be considered as follows:

$$h_c = \frac{1}{(1/h_{cf} + 1/h_v)} \qquad (44)$$

The vapor-phase heat-transfer rate depends on mass diffusion rates in the vapor. The well-known Colburn–Hougen method and other more recent approaches are summarized by Butterworth.[19] Methods for mixtures forming immiscible condensates are discussed in Ref. 20.

Diffusion-type methods require physical properties not usually available to the designer except for simple systems. Therefore, the vapor-phase heat-transfer coefficient is often estimated in practice by a "resistance-proration"-type method such as the Bell–Ghaly method.[21] In these methods the vapor-phase resistance is prorated with respect to the relative amount of duty required for sensible cooling of the vapor, resulting in the following expression:

$$h_v = \left(\frac{q_t}{q_{sv}}\right)^n h_{sv} \qquad (44a)$$

The exponent n can range from about 0.7 to 1.0 depending on the amount of mixing of the light and heavy components. Use $n = 1.0$ for a well-mixed (high-velocity) vapor and decrease n for low-velocity systems with large molecular weight range.

For more detail in application of the resistance proration method for mixtures, see Ref. 14 or 21.

Pressure Drop

For the condensing vapor, pressure drop is composed of three components—friction, momentum, and static head—as covered in Ref. 14. An approximate estimate on the conservative side can be obtained in terms of the friction component, using the Martinelli separated flow approach:

$$\Delta P_f = \Delta P_l \, \phi_l^2 \qquad (45)$$

where ΔP_f = two-phase friction pressure drop
ΔP_l = friction loss for liquid phase alone

The Martinelli factor ϕ_l^2 may be calculated as shown in Eq. (32). Alternative methods for shellside pressure drop are presented by Diehl[22] and by Grant and Chisholm.[23] These methods were reviewed by Ishihara[24] and found reasonably representative of the available data. However, Eq. (32), also evaluated in Ref. 24 for shellside flow, should give about equivalent results.

3.3   Shell and Tube Reboilers and Vaporizers

Heat exchangers are used to boil liquids in both the process and power industries. In the process industry they are often used to supply vapors to distillation columns and are called reboilers. The same types of exchangers are used in many applications in the power industry, for example, to generate vapors for turbines. For simplicity these exchangers will all be called "reboilers" in this section. Often the heating medium is steam, but it can also be any hot process fluid from which heat is to be recovered, ranging from chemical reactor effluent to geothermal hot brine.

Selection of Reboiler Type

A number of different shell and tube configurations are in common use, and the first step in design of a reboiler is to select a configuration appropriate to the required job. Basically, the type of reboiler should depend on expected amount of fouling, operating pressure, mean temperature difference (MTD), and difference between temperatures of the bubble point and the dew point (boiling range).

The main considerations are as follows: (1) fouling fluids should be boiled on the tubeside at high velocity; (2) boiling either under deep vacuum or near the critical pressure should be in a kettle to minimize hydrodynamic problems unless means are available for very careful design; (3) at low MTD, especially at low pressure, the amount of static head must be minimized; (4) for wide boiling range mixtures, it is important to maximize both the amount of mixing and the amount of countercurrent flow. Often, fairly clean wide boiling mixtures are wrongly assigned a high fouling factor as a "safety" factor. This should not be done, and these fluids normally should be boiled on the shellside because of greater mixing.

These and other criteria are discussed in more detail in Ref. 25, and summarized in a selection guide, which is abstracted in Table 4.

In addition to the above types covered in Ref. 25, falling film evaporators[26] may be preferred in cases with very low MTD, viscous liquids, or very deep vacuum for which even a kettle provides too much static head.

Temperature Profiles

For pure components or narrow boiling mixtures, the boiling temperature is nearly constant and the LMTD applies with $F = 1.0$. Temperature profiles for boiling range mixtures are very complicated, and although the LMTD is often used, it is not a recommended practice, and may result in underdesigned reboilers unless compensated by excessive design fouling factors. Contrary to the case for condensers, using a straight-line profile approximation always tends to give too high MTD for reboilers, and can be tolerated only if the temperature rise across the reboiler is kept low through a high circulation rate.

Table 5 gives suggested procedures to determine an approximate MTD to use for initial size estimation, based on temperature profiles illustrated in Fig. 16. It should be noted that

Table 4 Reboiler Selection Guide

Process Conditions	Suggested Reboiler Type[a]
Moderate pressure, MTD, and fouling	VT/E
Very high pressure, near critical	HS/K or (F)HT/E
Deep vacuum	HS/K
High or very low MTD	HS/K, G, H
Moderate to heavy fouling[b]	VT/E
Very heavy fouling[b]	(F)HT/E
Wide boiling range mixture	HS/G or /H
Very wide boiling range, viscous liquid	(F)HT/E

[a] V, vertical; H, horizontal; S, shellside boiling; T, tubeside boiling; (F), forced flow, else natural convection; /E, G, H, K, TEMA shell styles.
[b] True fouling demonstrated. Not just high fouling factor (which is often assigned for "safety").

Table 5   Reboiler MTD Estimation

Reboiler Type[a]	T_A	T_B	MTD
HS/K	$T_1 - t_2$	$T_2 - t_2$	Eq. (4), $F = 1$
HS/X, G, H	$T_1 - t_1$	$T_2 - t_2$	Eq. (4), $F = 0.9$
VT/E	$T_1 - t_2$	$T_2 - t_1$	Eq. (4), $F = 1$
(F)HT/E or (F)HS/E	$T_1 - t_2$	$T_2 - t_1$	Eq. (4), $F = 0.9$
All types	Isothermal	$T_A = T_B$	T_A

[a] V, vertical; H, horizontal; S, shellside boiling; T, tubeside boiling; (F), forced flow, else natural convection; /E, G, H, K, TEMA shell styles.

the MTD values in Table 5 are intended to be on the safe side and that excessive fouling factors are not necessary as additional safety factors if these values are used. See Section 4.1 for suggested fouling factor ranges.

Heat-Transfer Coefficients

The two basic types of boiling mechanisms that must be taken into account in determining boiling heat-transfer coefficients are nucleate boiling and convective boiling. A detailed description of both types is given by Collier.[27] For all reboilers, the nucleate and convective boiling contributions are additive, as follows:

$$h_b = \alpha h_{nb} + h_{cb} \tag{46a}$$

or

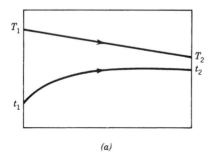

(a)

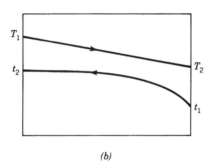

(b)

Figure 16   Reboiler temperature profiles illustrated: (*a*) use for kettle and horizontal thermosiphon; (*b*) use for tubeside boiling vertical thermosiphon.

$$h_b = [h_{nb}^2 + h_{cb}^2]^{0.5} \tag{46b}$$

Equation (46a) includes a nucleate boiling suppression factor, α, that originally was correlated by Chen.[60]

Equation (46b) is a simple asymptotic proration that was found to work well by Steiner and Taborek.[61]

The convective boiling coefficient h_{cb} depends on the liquid-phase convective heat-transfer coefficient h_l, according to the same relationship, Eq. (29), given for shear-controlled condensation. For all reboiler types, except forced flow, the flow velocities required to calculate h_l depend on complex pressure balances for which computers are necessary for practical solution. Therefore, the convective component is sometimes approximated as a multiplier to the nucleate boiling component for quick estimations,[25] as in the following equation:

$$h_b = h_{nb}F_b \tag{47}$$

$$F_b = \frac{h_{nb} + h_{cb}}{h_{nb}} \tag{48}$$

where F_b is approximated as follows:

For tubeside reboilers (VT/E thermosiphon)

$$F_b = 1.5 \tag{49}$$

For shellside reboilers (HS/X, G, H, K)

$$F_b = 2.0 \tag{50}$$

Equations (49) and (50) are intended to give conservative results for first approximations. For more detailed calculations see Refs. 28–30.

The nucleate boiling heat-transfer coefficient (h_{nb}) is dependent not only on physical properties, but also on the temperature profile at the wall and the microscopic topography of the surface. For a practical design, many simplifications must be made, and the approximate nature of the resulting coefficients should be recognized. A reasonable design value is given by the following simple equation[25]:

$$h_{nb} = 0.025F_c P_c^{0.69}q^{0.70}(P/P_c)^{0.17} \tag{51}$$

The term F_c is a correction for the effect of mixture composition on the boiling heat-transfer coefficient. The heat-transfer coefficient for boiling mixtures is lower than that of any of the pure components if boiled alone, as summarized in Ref. 27. This effect can be explained in terms of the change in temperature profile at the wall caused by the composition gradient at the wall, as illustrated in Ref. 31. Since the liquid-phase diffusional methods necessary to predict this effect theoretically are still under development and require data not usually available to the designer, an empirical relationship in terms of mixture boiling range (BR) is recommended in Ref. 25:

$$F_c = [1 + 0.018q^{0.15}BR^{0.75}]^{-1} \tag{52}$$

(BR = difference between dew-point and bubble-point temperatures, °F.)

Maximum Heat Flux

Above a certain heat flux, the boiling heat-transfer coefficient can decrease severely, owing to vapor blanketing, or the boiling process can become very unstable, as described in Refs.

27, 31, and 32. Therefore, the design heat flux must be limited to a practical maximum value. For many years the limit used by industry was in the range of 10,000–20,000 Btu/hr·ft² for hydrocarbons and about 30,000 Btu/hr·ft² for water. These rules of thumb are still considered reasonable at moderate pressures, although the limits, especially for water, are considerably conservative for good designs. However, at both very high and very low pressures the maximum heat fluxes can be severely decreased. Also, the maximum heat fluxes must be a function of geometry to be realistic. Empirical equations are presented in Ref. 25; the equations give much more accurate estimates over wide ranges of pressure and reboiler geometry.

(a) For kettle (HS/K) and horizontal thermosiphon (HS/X, G, H)

$$q_{max} = 803P_c \left(\frac{P}{P_c}\right)^{0.35} \left(1 - \frac{P}{P_c}\right)^{0.9} \phi_b \tag{53}$$

$$\phi_b = 3.1\left(\frac{\pi D_b L}{A_o}\right) \tag{54}$$

In the limit, for $\phi_b > 1.0$, let $\phi_b = 1.0$. For $\phi_b < 0.1$, consider larger tube pitch or vapor relief channels.[25] Design heat flux should be limited to less than 0.7 q_{max}.

(b) For vertical thermosiphon (VT/E)

$$q_{max} = 16,080 \left(\frac{D_i^2}{L}\right)^{0.35} P_c^{0.61} \left(\frac{P}{P_c}\right)^{0.25} \left(1 - \frac{P}{P_c}\right) \tag{55}$$

(c) For tubeside forced circulation boiling

Recent research by Heat Transfer Research, Inc. has provided new proprietary data. For horizontal flow the phenomena are very complicated, involving preferential vapor flow at the top of the tube. The best published predictive methods are by Katto[62] and Kattan et al.[63]

In addition to the preceding check, the vertical tubeside thermosiphon should be checked to insure against mist flow (dryout). The method by Fair[28] was further confirmed in Ref. 33 for hydrocarbons. For water, extensive data and empirical correlations are available as described by Collier.[27] To determine the flow regime by these methods it is necessary to determine the flow rate, as described, for example, in Ref. 28. However, for preliminary specification, it may be assumed that the exit vapor weight fraction will be limited to less than 0.35 for hydrocarbons and less than 0.10 for aqueous solutions and that under these conditions dryout is unlikely.

For some applications, such as LNG vaporization, it is required to fully vaporize and superheat the gas. For tubeside vaporization, this is very difficult due to droplet entrainment. Twisted tape inserts, such as supplied by Brown Fin Tube, solve the problem.

3.4   Air-Cooled Heat Exchangers

Detailed rating of air-cooled heat exchangers requires selection of numerous geometrical parameters, such as tube type, number of tube rows, length, width, number and size of fans, etc., all of which involve economic and experience considerations beyond the scope of this chapter. Air-cooled heat exchangers are still designed primarily by the manufacturers using proprietary methods. However, recommendations for initial specifications and rating are given by Paikert[2] and by Mueller.[3] A preliminary rating method proposed by Brown[34] is also sometimes used for first estimates owing to its simplicity.

Heat-Transfer Coefficients

For a first approximation of the surface required, the bare-surface-based overall heat-transfer coefficients recommended by Smith[35] may be used. A list of these values from Ref. 3 is abstracted in Table 6. The values in Table 6 were based on performance of finned tubes, having a 1 in. outside diameter base tube on $2\frac{3}{8}$-in. triangular pitch, $\frac{5}{8}$-in.-high aluminum fins ($\frac{1}{8}$-in. spacing between fin tips), with eight fins per inch. However, the values may be used as first approximations for other finned types.

As stated by Mueller, air-cooled heat exchanger tubes have had approximately the preceding dimensions in the past, but fin densities have tended to increase and now more typically range from 10 to 12 fins/in. For a more detailed estimate of the overall heat-transfer coefficient, the tubeside coefficients are calculated by methods given in the preceding sections and the airside coefficients are obtained as functions of fin geometry and air velocity from empirical relationships such as given by Gnielinski et al.[36] Rating at this level of sophistication is now done mostly by computer.

Temperature Difference

Air-cooled heat exchangers are normally "cross-flow" arrangements with respect to the type of temperature profile calculation. Charts for determination of the F-factor for such arrangements are presented by Taborek.[9] Charts for a number of arrangements are also given by Paikert[2] based on the "NTU method." According to Paikert, optimum design normally requires NTU to be in the range of 0.8–1.5, where

$$\text{NTU} = \frac{t_2 - t_1}{\text{MTD}} \tag{56}$$

For first approximations, a reasonable air-temperature rise ($t_2 - t_1$) may be assumed, MTD calculated from Eq. (4) using $F = 0.9$–1.0, and NTU checked from Eq. (56). It is assumed that if the air-temperature rise is adjusted so that NTU is about 1, the resulting preliminary size estimation will be reasonable. Another design criterion often used is that the face velocity V_f should be in the range of 300–700 ft/min (1.5–3.5 m/sec):

Table 6 Typical Overall Heat-Transfer Coefficients (U_o), Based on Bare Tube Surface, for Air-Cooled Heat Exchangers

Service	U_o	
	Btu/hr·ft²·°F	W/m²·K
Sensible Cooling		
Process water	105–120	600–680
Light hydrocarbons	75–95	425–540
Fuel oil	20–30	114–170
Flue gas, 10 psig	10	57
Condensation		
Steam, 0–20 psig	130–140	740–795
Ammonia	100–200	570–680
Light hydrocarbons	80–95	455–540
Refrigerant 12	60–80	340–455
Mixed hydrocarbons, steam, and noncondensables	60–70	340–397

$$V_f = \frac{W_a}{L \, W_d \, \rho_v} \tag{57}$$

where W_a = air rate, lb/min
$\quad\quad L$ = tube length, ft
$\quad\quad W_d$ = bundle width, ft
$\quad\quad \rho_v$ = air density, lb/ft^3

Fan Power Requirement

One or more fans may be used per bundle. Good practice requires that not less than 40–50% of the bundle face area be covered by the fan diameter. The bundle aspect ratio per fan should approach 1 for best performance. Fan diameters range from about 4 to 12 ft (1.2 to 3.7 m), with tip speeds usually limited to less than 12,000 ft/min (60 m/sec) to minimize noise. Pressure drops that can be handled are in the range of only 1–2 in. water (0.035–0.07 psi, 250–500 Pa). However, for typical bundle designs and typical air rates, actual bundle pressure drops may be in the range of only ¼–1 in. water.

Paikert[2] gives the expression for fan power as follows:

$$P_f = \frac{V(\Delta p_s + \Delta p_d)}{E_f} \tag{58}$$

where V = volumetric air rate, m^3/sec
$\quad \Delta p_s$ = static pressure drop, Pa
$\quad \Delta p_d$ = dynamic pressure loss, often 40–60 Pa
$\quad E_f$ = fan efficiency, often 0.6–0.7
$\quad P_f$ = fan power, W

3.5 Other Exchangers

For spiral, plate, and compact heat exchangers the heat-transfer coefficients and friction factors are sensitive to specific proprietary designs and such units are best sized by the manufacturer. However, preliminary correlations have been published. For spiral heat exchangers, see Mueller[3] and Minton.[37] For plate-type heat exchangers, Figs. 9 and 10, recommendations are given by Cooper[38] and Marriott.[39] For plate-fin and other compact heat exchangers, a comprehensive treatment is given by Webb.[4] For recuperators and regenerators the methods of Hausen are recommended.[6] Heat pipes are extensively covered by Chisholm.[40] Design methods for furnaces and combustion chambers are presented by Truelove.[41] Heat transfer in agitated vessels is discussed by Penney.[42] Double-pipe heat exchangers are described by Guy.[43]

4   COMMON OPERATIONAL PROBLEMS

When heat exchangers fail to operate properly in practice, the entire process is often affected, and sometimes must be shut down. Usually, the losses incurred by an unplanned shutdown are many times more costly than the heat exchanger at fault. Poor heat-exchanger performance is usually due to factors having nothing to do with the heat-transfer coefficient. More often the designer has overlooked the seriousness of some peripheral condition not even addressed in most texts on heat-exchanger design. Although only long experience, and numerous "experiences," can come close to uncovering all possible problems waiting to plague the heat-exchanger designer, the following subsections relating the more obvious problems are included to help make the learning curve less eventful.

4.1 Fouling

The deposit of solid insulating material from process streams on the heat-transfer surface is known as fouling, and has been called "the major unresolved problem in heat transfer."[44] Although this problem is recognized to be important (see Ref. 45) and is even being seriously researched,[45,46] the nature of the fouling process makes it almost impossible to generalize. As discussed by Mueller,[3] fouling can be caused by (1) precipitation of dissolved substances, (2) deposit of particulate matter, (3) solidification of material through chemical reaction, (4) corrosion of the surface, (5) attachment and growth of biological organisms, and (6) solidification by freezing. The most important variables affecting fouling (besides concentration of the fouling material) are velocity, which affects types 1, 2, and 5, and surface temperature, which affects types 3–6. For boiling fluids, fouling is also affected by the fraction vaporized. As stated in Ref. 25, it is usually impossible to know ahead of time what fouling mechanism will be most important in a particular case. Fouling is sometimes catalyzed by trace elements unknown to the designer. However, most types of fouling are retarded if the flow velocity is as high as possible, the surface temperature is as low as possible (exception is biological fouling[48]), the amount of vaporization is as low as possible, and the flow distribution is as uniform as possible.

The expected occurrence of fouling is usually accounted for in practice by assignment of fouling factors, which are additional heat-transfer resistances, Eq. (7). The fouling factors are assigned for the purpose of oversizing the heat exchanger sufficiently to permit adequate on-stream time before cleaning is necessary. Often in the past the fouling factor has also served as a general purpose "safety factor" expected to make up for other uncertainties in the design. However, assignment of overly large fouling factors can produce poor operation caused by excessive overdesign.[49,50]

For shell and tube heat exchangers it has been common practice to rely on the fouling factors suggested by TEMA.[1] Fouling in plate heat exchangers is usually less, and is discussed in Ref. 38. The TEMA fouling factors have been used for over 30 years and, as Mueller states, must represent some practical validity or else complaints would have forced their revision. A joint committee of TEMA and HTRI members has reviewed the TEMA fouling recommendations and slightly updated for the latest edition. In addition to TEMA, fouling resistances are presented by Bell[10] and values recommended for reboiler design are given in Ref. 25. For preliminary estimation, the minimum value commonly used for design is $0.0005°F \cdot hr \cdot ft^2/Btu$ for condensing steam or light hydrocarbons. Typical conservative estimates for process streams or treated cooling water are around $0.001–0.002°F \cdot hr \cdot ft^2/Btu$, and for heavily fouling streams values in the range of $0.003–0.01°F \cdot hr \cdot ft^2/Btu$ are used. For reboilers (which have been properly designed) a design value of $0.001°F \cdot hr \cdot ft^2/Btu$ is usually adequate, although for wide boiling mixtures other effects in addition to fouling tend to limit performance. These commonly used estimates can contain large built-in safety factors, and should not necessarily be accepted for modern computerized designs. A more realistic approach for most fluids is proposed in Section 5.6 under Fouling.

On the other hand, heavily fouling fluids such as crude oils may require even larger fouling factors for reasonable on-stream times than those given in TEMA. In this case, detailed physical characteristics of the fluid must be determined by experiment before realistic design fouling allowances can be assigned.

Heat Transfer Research, Inc., in cooperation with a task force of industry experts, has an ongoing research program to measure fouling rates over a range of process conditions and compare with fluid characteristics.

4.2 Vibration

A problem with shell and tube heat exchangers that is becoming more frequent as heat exchangers tend to become larger and design velocities tend to become higher is tube failure due to flow-induced tube vibration. Summaries including recommended methods of analysis are given by Chenoweth[51] and by Mueller.[3] In general, tube vibration problems tend to occur when the distance between baffles or tube-support plates is too great. Maximum baffle spacings recommended by TEMA were based on the maximum unsupported length of tube that will not sag significantly. Experience has shown that flow-induced vibration can still occur at TEMA maximum baffle spacing, but for less than about 0.7 times this spacing most vibration can be eliminated at normal design velocities (see Section 2.4). Taborek[11] gives the following equations for TEMA maximum unsupported tube lengths (L_{su}), inches.

Steel and Steel Alloy Tubes

$$\text{For } D_o = \tfrac{3}{4}\text{–}2 \text{ in.,} \tag{59}$$

$$L_{su} = 52D_o + 21$$

$$\text{For } D_o = \tfrac{1}{4}\text{–}\tfrac{3}{4} \text{ in.,} \tag{60}$$

$$L_{su} = 68D_o + 9$$

Aluminum and Copper Alloy Tubes

$$\text{For } D_o = \tfrac{3}{4}\text{–}2 \text{ in.,} \tag{61}$$

$$L_{su} = 46D_o + 17$$

$$\text{For } D_o = \tfrac{1}{4}\text{–}\tfrac{3}{4} \text{ in.,} \tag{62}$$

$$L_{su} = 60D_o + 7$$

For segmental baffles with tubes in the windows, Fig. 9, the maximum baffle spacing is one-half the maximum unsupported tube length.

For very large bundle diameters, segmental or even double segmental baffles may not be suitable, since the spacing required to prevent vibration may produce too high pressure drops. (In addition, flow distribution considerations require that the ratio of baffle spacing to shell diameter not be less than about 0.2.) In such cases, one commonly used solution is to eliminate tubes in the baffle windows so that intermediate support plates can be used and baffle spacing can be increased; see Fig. 17. Another solution, with many advantages is the rod-type tube support in which the flow is essentially longitudinal and the tubes are supported

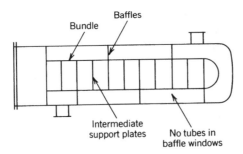

Figure 17 Segmental baffles with no tubes in window.

by a cage of rods. A proprietary design of this type exchanger (RODbaffle) is licensed by Phillips Petroleum Co. Calculation methods are published in Ref. 52.

4.3 Flow Maldistribution

Several types of problems can occur when the flow velocities or fluid phases become distributed in a way not anticipated by the designer. This occurs in all types of exchangers, but the following discussion is limited to shell and tube and air-cooled exchangers, in which maldistribution can occur on either shellside or tubeside.

Shellside Flow

Single-phase flow can be maldistributed on the shellside owing to bypassing around the tube bundle and leakage between tubes and baffle and between baffle and shell. Even for typical well-designed heat exchangers, these ineffective streams can comprise as much as 40% of the flow in the turbulent regime and as much as 60% of the flow in the laminar regime. It is especially important for laminar flow to minimize these bypass and leakage streams, which cause both lower heat-transfer coefficients and lower effective MTD.[13] This can, of course, be done by minimizing clearances, but economics dictate that more practical methods include use of bypass sealing strips, increasing tube pitch, increasing baffle spacing, and using an optimum baffle cut to provide more bundle penetration. Methods for calculating the effects of these parameters are described by Taborek.[11] One method to minimize leakage and bypass inefficiencies is to use helical baffles, which cause flow to proceed through the exchanger along a spiral path. Elimination of sharp flow reversals provides a much more uniform shellside distribution. A proprietary version of the helical baffle option is provided by the ABB Lummus Company.

Another type of shellside maldistribution occurs in gas–liquid two-phase flow in horizontal shells when the flow velocity is low enough that the vapor and liquid phases separate, with the liquid flowing along the bottom of the shell. For condensers this is expected and taken into account. However, for some other types of exchangers, such as vapor–liquid contactors or two-phase reactor feed-effluent exchangers, separation may cause unacceptable performance. For such cases, if it is important to keep the phases mixed, a vertical heat exchanger is recommended. Improvement in mixing is obtained for horizontal exchangers if horizontal rather than vertical baffle cut is used.

Tubeside Flow

Several types of tubeside maldistribution have been experienced. For single-phase flow with axial nozzles into a single-tubepass exchanger, the dynamic head of the entering fluid can cause higher flow in the central tubes, sometimes even producing backflow in the peripheral tubes. This effect can be prevented by using an impingement plate on the centerline of the axial nozzle.

Another type of tubeside maldistribution occurs in cooling viscous liquids. Cooler tubes in parallel flow will tend to completely plug up in this situation, unless a certain minimum pressure drop is obtained, as explained by Mueller.[53]

For air-cooled single pass condensers, a backflow can occur owing to the difference in temperature driving force between bottom and top tube rows, as described by Berg and Berg.[54] This can cause an accumulation of noncondensables in air-cooled condensers, which can significantly affect performance, as described by Breber et al.[55] In fact, in severe cases, this effect can promote freezeup of tubes, or even destruction of tubes by water hammer. Backflow effects are eliminated if a small amount of excess vapor is taken through the main condenser to a backup condenser or if the number of fins per inch on bottom rows is less than on top rows to counteract the difference in temperature driving force.

For multipass tubeside condensers, or tubeside condensers in series, the vapor and liquid tend to separate in the headers with liquid running in the lower tubes. The fraction of tubes filled with liquid tends to be greater at higher pressures. In most cases the effect of this separation on the overall condenser heat-transfer coefficient is not serious. However, for multicomponent mixtures the effect on the temperature profile will be such as to decrease the MTD. For such cases, the temperature profile should be calculated by the differential flash procedure, Section 3.2. In general, because of unpredictable effects, entering a pass header with two phases should be avoided when possible.

4.4 Temperature Pinch

When the hot and cold streams reach approximately the same temperature in a heat exchanger, heat transfer stops. This condition is referred to as a temperature pinch. For shellside single-phase flow, unexpected temperature pinches can be the result of excessive bypassing and leakage combined with a low MTD and possibly a temperature cross. An additional factor, "temperature profile distortion factor," is needed as a correction to the normal F factor to account for this effect.[11,13] However, if good design practices are followed with respect to shellside geometry, this effect normally can be avoided.

In condensation of multicomponent mixtures, unexpected temperature pinches can occur in cases where the condensation curve is not properly calculated, especially when the true curve happens to be of type III in Fig. 15. This can happen when separation of liquid containing heavy components occurs, as mentioned above, and also when the condensing mixture has immiscible liquid phases with more than one dew point.[20] In addition, condensing mixtures with large desuperheating and subcooling zones can produce temperature pinches and must be carefully analyzed. In critical cases it is safer and may even be more effective to do desuperheating, condensing, and subcooling in separate heat exchangers. This is especially true of subcooling.[3]

Reboilers can also suffer from temperature-pinch problems in cases of wide boiling mixtures and inadequate liquid recirculation. Especially for poorly designed thermosiphon reboilers, with the circulation rate is less than expected, the temperature rise across the reboiler will be too high and a temperature pinch may result. This happens most often when the reboiler exit piping is too small and consumes an unexpectedly large amount of pressure drop. This problem normally can be avoided if the friction and momentum pressure drop in the exit piping is limited to less than 30% of the total driving head and the exit vapor fraction is limited to less than 0.25 for wide boiling range mixtures. For other recommendations, see Ref. 25.

4.5 Critical Heat Flux in Vaporizers

Owing to a general tendency to use lower temperature differences for energy conservation, critical heat flux problems are not now frequently seen in the process industries. However, for waste heat boilers, where the heating medium is usually a very hot fluid, surpassing the critical heat flux is a major cause of tube failure. The critical heat flux is that flux (Q/A_o) above which the boiling process departs from the nucleate or convective boiling regimes and a vapor film begins to blanket the surface, causing a severe rise in surface temperature, approaching the temperature of the heating medium. This effect can be caused by either of two mechanisms: (1) flow of liquid to the hot surface is impeded and is insufficient to supply the vaporization process or (2) the local temperature exceeds that for which a liquid phase can exist.[32] Methods of estimating the maximum design heat flux are given in Section 3.3, and the subject of critical heat flux is covered in great detail in Ref. 27. However, in most cases where failures have occurred, especially for shellside vaporizers, the problem has been caused by local liquid deficiency, owing to lack of attention to flow distribution considerations.

4.6 Instability

The instability referred to here is the massive large-scale type in which the fluid surging is of such violence as to at least disrupt operations, if not to cause actual physical damage. One version is the boiling instability seen in vertical tubeside thermosiphon reboilers at low operating pressure and high heat flux. This effect is discussed and analyzed by Blumenkrantz and Taborek.[56] It is caused when the vapor acceleration loss exceeds the driving head, producing temporary flow stoppage or backflow, followed by surging in a periodic cycle. This type of instability can always be eliminated by using more frictional resistance, a valve or orifice, in the reboiler feed line. As described in Ref. 32, instability normally only occurs at low reduced pressures, and normally will not occur if design heat flux is less than the maximum value calculated from Eq. (55).

Another type of massive instability is seen for oversized horizontal tubeside pure component condensers. When more surface is available than needed, condensate begins to subcool and accumulate in the downstream end of the tubes until so much heat-transfer surface has been blanketed by condensate that there is not enough remaining to condense the incoming vapor. At this point the condensate is blown out of the tube by the increasing pressure and the process is repeated. This effect does not occur in vertical condensers since the condensate can drain out of the tubes by gravity. This problem can sometimes be controlled by plugging tubes or injecting inert gas, and can always be eliminated by taking a small amount of excess vapor out of the main condenser to a small vertical backup condenser.

4.7 Inadequate Venting, Drainage, or Blowdown

For proper operation of condensers it is always necessary to provide for venting of noncondensables. Even so-called pure components will contain trace amounts of noncondensables that will eventually build up sufficiently to severely limit performance unless vented. Vents should always be in the vapor space near the condensate exit nozzle. If the noncondensable vent is on the accumulator after the condenser, it is important to ensure that the condensate nozzle and piping are large enough to provide unrestricted flow of noncondensables to the accumulator. In general, it is safer to provide vent nozzles directly on the condenser.

If condensate nozzles are too small, condensate can accumulate in the condenser. It is recommended that these nozzles be large enough to permit weir-type drainage (with a gas core in the center of the pipe) rather than to have a full pipe of liquid. Standard weir formulas[57] can be used to size the condensate nozzle. A rule of thumb used in industry is that the liquid velocity in the condensate piping, based on total pipe cross section, should not exceed 3 ft/sec (0.9 m/sec).

The problem of inadequate blowdown in vaporizers is similar to the problem of inadequate venting for condensers. Especially with kettle-type units, trace amounts of heavy, high-boiling, or nonboiling components can accumulate, not only promoting fouling but also increasing the effective boiling range of the mixture, thereby decreasing the MTD as well as the effective heat-transfer coefficient. Therefore, means of continuous or at least periodic removal of liquid from the reboiler (blowdown) should be provided to ensure good operation. Even for thermosiphon reboilers, if designed for low heat fluxes (below about 2000 BTU/hr/ft², 6300 W/m²), the circulation through the reboiler may not be high enough to prevent heavy components from building up, and some provision for blowdown may be advisable in the bottom header.

5 USE OF COMPUTERS IN THERMAL DESIGN OF PROCESS HEAT EXCHANGERS

5.1 Introduction

The approximate methods for heat transfer coefficient and pressure drop given in the preceding sections will be used mostly for orientation. For an actual heat exchanger design, it only makes sense to use a computer. Standard programs can be obtained for most geometries in practical use. These allow reiterations and incrementation to an extent impossible by hand and also supply physical properties for a wide range of industrial fluids. However, computer programs by no means solve the whole problem of producing a workable efficient heat exchanger. Many experience-guided decisions must be made both in selection of the input data and in interpreting the output data before even the thermal design can be considered final. We will first review why a computer program is effective. This has to do with (1) incrementation and (2) convergence loops.

5.2 Incrementation

The method described in Section 2.1 for calculation of required surface can only be applied accurately to the entire exchanger if the overall heat transfer coefficient is constant and the temperature profiles for both streams are linear. This often is not a good approximation for typical process heat exchangers because of variation in physical properties and/or vapor fraction along the exchanger length. The rigorous expression for Eq. (1) is as follows:

$$A_o = \int \frac{dQ}{U_o \, \text{MTD}}$$

Practical solution of this integral equation requires dividing the heat transfer process into finite increments of ΔQ that are small enough so that U_o may be considered constant and the temperature profiles may be considered linear. The incremental area, ΔA_o, is then calculated for each increment and summed to obtain the total required area. An analogous procedure is followed for the pressure drop. This procedure requires determining a full set of fluid physical properties for all phases of both fluids in each increment and the tedious calculations can be performed much more efficiently by computer. Furthermore, in each increment several trial and error convergence loops may be required, as discussed next.

5.3 Main Convergence Loops

Within each of the increments discussed above, a number of implicit equations must be solved, requiring convergence loops. The two main types of loops found in any heat exchanger calculation are as follows.

Intermediate Temperature Loops

These convergence loops normally are used to determine either wall temperature or, less commonly, interface temperature. The discussion here will be limited to the simpler case of wall temperature. Because of the variation of physical properties between the wall and the bulk of the fluid, heat transfer coefficients depend on the wall temperature. Likewise, the wall temperature depends on the relative values of the heat transfer coefficients of each fluid. Wall temperatures on each side of the surface can be estimated by the following equations:

$$T_{w,\,\text{hot}} = T_{\text{hot}} - \frac{U_o}{h_{\text{hot}}}\,(T_{\text{hot}} - T_{\text{cold}})$$

$$T_{w,\,\text{cold}} = T_{\text{cold}} + \frac{U_o}{h_{\text{cold}}}\,(T_{\text{hot}} - T_{\text{cold}})$$

It is assumed in the above equations that the heat transfer coefficient on the inside surface is corrected to the outside area. Convergence on the true wall temperature can be done in several ways. Figure 18 shows a possible convergence scheme.

Pressure Balance Loops

These convergence loops are needed whenever the equations to be solved are implicit with respect to velocity. The two most frequent cases encountered in heat exchanger design are (1) flow distribution and (2) natural circulation. The first case, flow distribution, is the heart of the shell and tube heat exchanger shellside flow calculations, and involves solution for the fraction of flow across the tube bundle, as opposed to the fraction of flow leaking around baffles and bypassing the bundle. Since the resistance coefficients of each stream are functions of the stream velocity, the calculation is reiterative. The second case, natural circulation, is encountered in thermosiphon and kettle reboilers where the flow rate past the heat transfer surface is a function of the pressure balance between the two-phase flow in the bundle, or tubes, and the liquid static head outside the bundle. In this case the heat transfer coefficients that determine the vaporization rate are functions of the flow velocity, which is in turn a function of the amount of vaporization. Figure 19 shows a flow velocity convergence loop applicable to the flow distribution case.

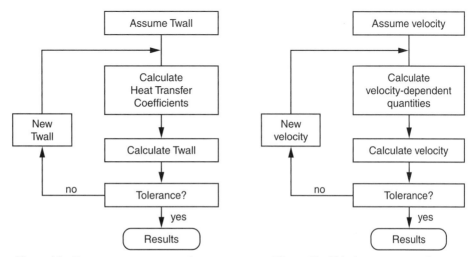

Figure 18 Temperature convergence loop.   **Figure 19** Velocity convergence loop.

5.4 Rating, Design, or Simulation

Several types of solutions are possible by computer. The better standard programs allow the user to choose. It is important to understand what the program is doing in order to properly interpret the results. The above three types of calculations are described as follows.

Rating

This is the normal mode for checking a vendor's bid. All geometry and all process conditions are specified. The program calculates the required heat transfer area and pressure drop and compares with the specified values. Normally this is done including the specified fouling factor. This means that on startup the amount of excess surface will be greater, sometimes excessively greater, causing severe operating adjustments. It is therefore advisable to review clean conditions also.

Design

This mode is used by the process engineer to obtain a size based on process heat transfer requirements. In this case, most of the geometry specifications still are not determined by the program and must be determined by the designer based on experience. Required, but unknown, specifications, in addition to the process requirements of temperatures, flow rates, and pressure drops, include

- Exchanger type (shell and tube, plate-and-frame, plate-fin, air-cooled, etc.)

If shell and tube

- TEMA shell type (E, F, J, G, H, X, K)
- TEMA front and rear head types (flat, dished, fixed tube sheet, split ring, pull-through)
- Baffle type (segmental, double segmental, triple segmental, rod, etc.)
- Tube type (plain, low-finned, enhanced surface, etc.)
- Tube length (usually standard lengths of 8, 12, 16, 20 ft)
- Tube diameter (usually ⅝, ¾, 1, 1¼ in. or 1.25 in.)
- Tube pitch (pitch ratios 1.25, 1.3, 1.5)
- Tube layout (30, 45, 60, 90°)
- Tube material (carbon steel, stainless steel, copper alloys, titanium, etc.)
- Exchanger orientation (horizontal, vertical)

As shown, even with a good computer program, an overwhelming number of combinations of geometry parameters is possible and presently the engineer is required to select the best combination based on mechanical considerations, process considerations, fouling tendencies, and allowable pressure drop. Some general guidelines are given in Section 5.6. Once the above parameters are specified to the computer program, it can proceed to calculate the number of tubes required and the baffle spacing and number of tube passes consistent with the required pressure drops for both streams.

Simulation

This mode of calculation is used most to predict the performance of a field heat exchanger under different operating conditions. Usually the engineer "zeros" the program first by adjusting fouling factors and friction factor multipliers to match existing field performance. Then the adjusted process conditions are imposed and the computer program predicts the heat transfer rates and pressure drops under the new conditions. This mode of calculation can also be used to monitor apparent fouling resistance increase on operating units in order to better schedule maintenance.

5.5 Program Quality and Selection

All heat exchanger programs are not created equal. Heat exchange is not yet an exact science and all of the heat transfer coefficients and friction factors used in calculation are from correlations with empirically determined constants. Therefore, the data base used for correlation development is important.

Methods Source
The methods used for the program should be available and documented in a readable form. Good methods will be based on theoretically derived equation forms that either are limited in range or automatically achieve theoretically justified limits. "Black box" methods, for which this may not be true, should be avoided.

Data Base
Good programs are also backed by a sizable data bank covering the range of conditions of interest as well as demonstrated successes in predicting field performance. No non-tested methods, including so-called rigorous incremental methods, should be accepted without some data-based support.

Suitability
Completely general programs that apply to all geometries and process conditions and fulfill the above data base requirements probably will not exist for sometime. The program manual should list recommended ranges of applicability. When in doubt, consult the supplier.

5.6 Determining and Organizing Input Data

As of this writing, available programs still require a large number of input data decisions to be made by the user. The quality of the answers obtained is crucially dependent on the quality of these input decisions.

Process Data
The basis for the calculation is the heat duty, which usually comes from the process flow sheet. There must, of course, be a heat balance between the hot and cold sides of the exchanger. The temperature profiles are much more significant to a good design than are the heat transfer coefficients. Only in rare cases are these straight lines. For multicomponent phase-change cases, the condensing or vaporization curves should be calculated by a good process simulator program containing state-of-the-art vapor–liquid equilibria methods. Most good heat exchanger programs will accept these curves as the basis for the heat-transfer calculations.

It is important to specify realistic pressure drop limitations, since the heat-transfer coefficient and the fouling rate are functions of the velocity, which is a function of the available pressure drop. For phase change, too much pressure drop can mean a significant loss in available temperature difference and one rule of thumb suggests a limit of 10% of the operating pressure. For liquid flow, erosion velocity often is the limiting factor, and this is usually taken to be in the range of 7–10 ft/sec tubeside or 3–5 ft/sec shellside. Velocities also are sometimes limited to a value corresponding to ρv^2 less than 4000, where ρ is in lb/ft^3 and v is in ft/sec.

Geometry Data

It is necessary for the program user to make a large number of geometry decisions, starting with the type of exchanger, which decides the type of program to be used. Only a brief list of suggestions can be accommodated in this chapter, so recommendations will be limited to some of the main shell-and-tube geometries mentioned in Section 5.4.

TEMA Shell Style. The types E, J, and X are selected based on available pressure drop, highest E, lowest X, and intermediate J. Types G and H are used mostly for horizontal thermosiphon reboilers, although they also obtain a slightly better MTD correction factor than the E-type shell and are sometimes used even for single phase for that purpose. Pressure drop for G and E shells are about the same. For horizontal thermosiphon reboilers, the longitudinal baffle above the inlet nozzle prevents the light vaporizing component to shortcut directly to the exit nozzle. If pressure drop for the less expensive G-shell is too high, H-shell (two G's in parallel) is used. Type F is used when it is required to have a combination of countercurrent flow and two tube passes in a single shell. This type has the disadvantage of leakage around the longitudinal baffle, which severely decreases performance. A welded baffle prevents this but prevents bundle removal. Type K is used only for kettle reboilers.

TEMA Front and Rear Head Types. These are selected based on pressure and/or maintenance considerations. TEMA Standards should be consulted. With respect to maintenance, rear heads permitting bundle removal should be specified for shellside fouling fluids. These are the split ring and pull-through types.

Baffle Types. These are selected based on a combination of pressure drop and vibration considerations. In general, the less expensive, higher-velocity segmental baffle is tried first, going to the double segmental and possibly the triple segmental types if necessary to lower pressure drop. Allowable pressure drop is a very important design parameter and should not be allocated arbitrarily. In the absence of other process limits, the allowable pressure drop should be about 10% of the operating pressure or the ρv^2 should be less than about 4000 (lb/ft^3)(ft/sec)^2, whichever gives the lower velocity. However, vibration limits override these limits. Good thermal design programs also check for tube vibration and warn the user if vibration problems are likely due to high velocity or insufficient tube support. In case of potential vibration problems, it is necessary to decrease velocity or provide more tube support, the latter being preferable. The two best ways of eliminating vibration problems within allowable pressure drop limitations are 1) no-tube-in-window baffles, or 2) RoDbaffles, as discussed in Section 4.2. As mentioned in Section 4.3, the ABB Lummus Company offers software, based on Heat Transfer Research, Inc. technology, containing a helical baffle option. Helical baffles can both decrease vibration tendencies and improve shellside flow distribution.

Tube Types. For low temperature differences and low heat-transfer coefficients, low-finned or enhanced tubes should be investigated. In proper applications these can decrease the size of the exchanger dramatically. Previously, enhanced tubes were considered only for very clean streams. However, recent research is beginning to indicate that finned tubes fare as well in fouling services as plain tubes, and sometimes much better, providing longer on-stream time and often even easier cleaning. In addition, the trend in the future will be to stop assigning arbitrary fouling factors, but rather to design for conditions minimizing fouling. A relatively new option available from the Brown Fin Tube Company is the twisted tube. This tube provides spiral corrugations through which fluids flow in spiral counterflow on the shellside and tubeside. No baffles are needed.

Tube Length. This is usually limited by plant requirements. In general, longer exchangers are economically preferable within pressure drop restrictions, except possibly for vertical thermosiphon reboilers.

Tube Diameter. Small diameters are more economical in the absence of restrictions. Cleaning restrictions normally limit outside diameters to not less than ⅝ or ¾ in. However, some manufacturers now offer microchannel exchangers, which are very effective for some fluids, such as clean gases. Pressure drop restrictions, especially in vacuum, may require larger sizes. Vacuum vertical thermosiphon reboilers often require 1¼-in. tubes, and vacuum falling film evaporators frequently use as large as 2-in. tubes. Excessive pressure drop can be quickly decreased by going to the next standard tube diameter, since pressure drop is inversely proportional to the fifth power of the inside diameter.

Tube Pitch. Tube pitch for shellside flow is analogous to tube diameter for tubeside flow. Small pitches are more economical and also can cause pressure drop or cleaning problems. In laminar flow, here too-small tube pitch can prevent bundle penetration and force more bypassing and leakage. A pitch-to-tube diameter ratio of 1.25 or 1.33 is often used in absence of other restrictions depending on allowable pressure drop. For shellside reboilers operating at high heat flux, a ratio of as much as 1.5 is often required. Equation (54) shows that the maximum heat flux for kettle reboilers increases with increasing tube pitch.

Tube Layout. Performance is not critically affected by tube layout, although some minor differences in pressure drop and vibration characteristics are seen. In general, either 30 or 60° layouts are used for clean fluids, while 45 or 90° layouts are more frequently seen for fluids requiring shellside fouling maintenance.

Tube Material. The old standby for noncorrosive moderate-temperature hydrocarbons is the less expensive and sturdy carbon steel. Corrosive or very high-temperature fluids require stainless steel or other alloys. Titanium and hastelloy are becoming more frequently used for corrosion or high temperature despite the high cost, as a favorable economic balance is seen in comparison with severe problems of tube failure.

Exchanger Orientation. Exchangers normally are horizontal except for tubeside thermosiphons, falling film evaporators, and tubeside condensers requiring very low pressure drop or extensive subcooling. However, it is becoming more frequent practice to specify vertical orientation for two-phase feed-effluent exchangers to prevent phase separation, as mentioned in Section 4.3.

Fouling

All programs require the user to specify a fouling factor, which is the heat-transfer resistance across the deposit of solid material left on the inside and/or outside of the tube surface due to decomposition of the fluid being heated or cooled. Considerations involved in the determination of this resistance are discussed in Section 4.1. Since there are presently no thermal design programs available that can make this determination, the specification of a fouling resistance, or fouling factor, for each side is left up to the user. Unfortunately, this input is probably more responsible than any other for causing inefficient designs and poor operation.

The major problem is that there is very little relationship between actual fouling and the fouling factor specified. Typically, the fouling factor contains a safety factor that has evolved from practice, lived a charmed life as it is passed from one handbook to another, and may no longer be necessary if modern accurate design programs are used. An example is the frequent use of a fouling factor of 0.001 hr ft^2 °F/Btu for clean overhead condenser vapors. This may have evolved as a safety or correction from the failure of early methods to account for mass transfer effects and is completely unnecessary with modern calculation methods. Presently, the practice is to use fouling factors from TEMA Standards. However, these often result in heat exchangers that are oversized by as much as 50% on startup, causing operating problems that actually tend to enhance fouling tendencies. Hopefully, with ongoing research on fouling threshold conditions, it will be possible to design exchangers to essentially clean conditions. In the meantime, the user of computer programs should use common sense in assigning fouling factors only to actual fouling conditions. Startup conditions should also be checked as an alternative case.

Industrial experience has shown for a long time that arbitrary fouling factors may actually contribute to fouling by greatly oversizing exchangers and lowering velocities. Gilmour[65] presented evidence of this years ago. In general, crude oils may need fouling factors, as may polymerizing fluids, but light hydrocarbons may not. We now recommend designing with no fouling factor, then adding about 20% surface, as length, and rechecking pressure drop.

NOMENCLATURE

Note: Dimensional equations should use U.S. units only.

	Description	U.S. Units	S.I. Units
A_i	Inside surface area	ft^2	m^2
A_m	Mean surface area	ft^2	m^2
A_o	Outside surface area	ft^2	m^2
a_o	Outside surface per unit length	ft	m
B_c	Baffle cut % of shell diameter	%	%
BR	Boiling range (dew–bubble points)	°F	(U.S. only)
C	Two-phase pressure drop constant	—	—
C_b	Bundle bypass constant	—	—
C_{p1}	Heat capacity, hot fluid	Btu/lb·°F	J/kg·K
C_{p2}	Heat capacity, cold fluid	Btu/lb·°F	J/kg·K
D	Tube diameter, general	ft	m
D_b	Bundle diameter	ft	m
D_i	Tube diameter, inside	ft	m
D_o	Tube diameter, outside	ft or in.	m or U.S. only
D_s	Shell diameter	ft	m
D_f	Effective length: = D_i for tubeside = $P_t - D_o$ for shellside	ft	m
E_f	Fan efficiency (0.6–0.7, typical)	—	—
F	MTD correction factor	—	—
F_b	Bundle convection factor	—	—
F_c	Mixture correction factor	—	—
F_g	Gravity condensation factor	—	—
g	Acceleration of gravity	ft/hr^2	m/sec^2
G	Total mass velocity	lb/hr·ft^2	kg/sec·m^2

g_c	Gravitational constant	4.17×10^8 lb$_f$·ft/lb·hr^2	1.0
h_{hot}	Heat transfer coeff., hot fluid	Btu/hr·ft^2·°F	W/m^2·K
h_{cold}	Heat transfer coeff., cold fluid	Btu/hr·ft^2·°F	W/m^2·K
h_b	Heat transfer coeff., boiling	Btu/hr·ft^2·°F	W/m^2·K
h_c	Heat transfer coeff., condensing	Btu/hr·ft^2·°F	W/m^2·K
h_{cb}	Heat transfer coeff., conv. boiling	Btu/hr·ft^2·°F	W/m^2·K
h_{cf}	Heat transfer coeff., cond. film	Btu/hr·ft^2·°F	W/m^2·K
h_i	Heat transfer coeff., inside	Btu/hr·ft^2·°F	W/m^2·K
h_l	Heat transfer coeff., liq. film	Btu/hr·ft^2·°F	W/m^2·K
h_N	Heat transfer coeff., Nusselt	Btu/hr·ft^2·°F	W/m^2·K
h_{nb}	Heat transfer coeff., nucleate boiling	Btu/hr·ft^2·°F	W/m^2·K
h_o	Heat transfer coeff., outside	Btu/hr·ft^2·°F	W/m^2·K
h_{sv}	Heat transfer coeff., sens. vapor	Btu/hr·ft^2·°F	W/m^2·K
h_v	Heat transfer coeff., vapor phase	Btu/hr·ft^2·°F	W/m^2·K
J_g	Wallis dimensionless gas velocity	—	—
k_f	Thermal conductivity, fluid	Btu/hr·ft·°F	W/m·K
k_l	Thermal conductivity, liquid	Btu/hr·ft·°F	W/m·K
k_w	Thermal conductivity, wall	Btu/hr·ft·°F	W/m·K
L	Tube length	ft	m
L_{bc}	Baffle spacing	ft	m
L_{su}	Maximum unsupported length	in.	use U.S. only
MTD	Mean temperature difference	°F	K
NP	Number of tube passes	—	—
NT	Number of tubes	—	—
NTU	Number of transfer units	—	—
P	Pressure	psia	use U.S. only
P_c	Critical pressure	psia	use U.S. only
P_f	Fan power	use S.I. only	W
Pr	Prandtl number	—	—
P_t	Tube pitch	ft	m
q_{max}	Maximum allowable heat flux	Btu/hr·ft^2	use U.S. only
q	Heat flux	Btu/hr ft^2	use U.S. only
Q	Heat duty	Btu/hr	W
q_{sv}	Sensible vapor heat flux	Btu/hr ft^2	W/m^2
q_t	Total heat flux	Btu/hr ft^2	W/m^2
Re	Reynolds number	—	—
Re$_c$	Reynolds number, condensate	—	—
R_{f_i}	Fouling resistance, inside	°F ft^2 hr/Btu	K m^2/W
R_{f_o}	Fouling resistance, outside	°F ft^2 hr/Btu	K m^2/W
R_{h_i}	Heat transfer resistance, inside	°F ft^2 hr/Btu	K m^2/W
R_{h_o}	Heat transfer resistance, outside	°F ft^2 hr/Btu	K m^2/W
R_w	Heat transfer resistance, wall	°F ft^2 hr/Btu	K m^2/W
S_s	Crossflow area, shellside	ft^2	m^2
S_t	Crossflow area, tubeside	ft^2	m^2
t_1	Temperature, cold fluid inlet	°F	°C
T_1	Temperature, hot fluid inlet	°F	°C
t_2	Temperature, cold fluid outlet	°F	°C
T_2	Temperature, hot fluid outlet	°F	°C
T_A	Hot inlet—cold outlet temperature	°F	°C
T_B	Hot outlet—cold inlet temperature	°F	°C
T_{hot}	Temperature, hot fluid	°F	°C
T_{cold}	Temperature, cold fluid	°F	°C
T_s	Saturation temperature	°F	°C
T_w	Wall temperature	°F	°C
$T_{w,\,hot}$	Wall temperature, hot fluid side	°F	°C

$T_{w,\,cold}$	Wall temperature, cold fluid side	°F	°C
U_o	Overall heat transfer coefficient	Btu/hr·ft²·°F	W/m²·K
V	Volumetric flow rate	use S.I. only	m³/s
V_f	Face velocity	ft/min	use S.I. only
V_s	Shellside velocity	ft/hr	m/hr
V_t	Tubeside velocity	ft/hr	m/hr
W_a	Air flow rate	lb/min	use U.S. only
W_1	Flow rate, hot fluid	lb/hr	kg/hr
W_2	Flow rate, cold fluid	lb/hr	kg/hr
W_c	Flow rate, condensate	lb/hr	kg/hr
W_d	Air-cooled bundle width	ft	use U.S. only
W_s	Flow rate, shellside	lb/hr	kg/hr
W_t	Flow rate, tubeside	lb/hr	kg/hr
X_{tt}	Martinelli parameter	—	—
x_w	Wall thickness	ft	m
y	Weight fraction vapor	—	—
α	Nucleate boiling suppression factor	—	—
Δp_d	Dynamic pressure loss (typically 40–60 Pa)	use S.I.	Pa
ΔP_f	Two-phase friction pressure drop	psi	kPa
ΔP_l	Liquid phase friction pressure drop	psi	kPa
Δp_s	Static pressure drop, air cooler	use S.I. only	Pa
ΔP_s	Shellside pressure drop	lb/ft²	use U.S. only
ΔP_t	Tubeside pressure drop	lb/ft²	use U.S. only
λ	Latent heat	Btu/lb	J/kg
μ	Viscosity, general	lb/ft·hr	Pa
μ_f	Viscosity, bulk fluid	lb/ft·hr	Pa
μ_w	Viscosity, at wall	lb/ft·hr	Pa
ρ_l	Density, liquid	lb/ft³	kg/m³
ρ_s	Density, shellside fluid	lb/ft³	kg/m³
ρ_t	Density, tubeside fluid	lb/ft³	kg/m³
ρ_v	Density, vapor	lb/ft³	kg/m³
ϕ_b	Bundle vapor blanketing correction	—	—
ϕ_l	Two-phase pressure drop correction	—	—

REFERENCES

Note: Many of the following references are taken from the *Heat Exchanger Design Handbook* (HEDH), Hemisphere, Washington, DC, 1982, which will be referred to for simplicity as HEDH.

1. *Standards of Tubular Heat Exchanger Manufacturers Association,* 6th ed., TEMA, New York, 1978.
2. P. Paikert, "Air-Cooled Heat Exchangers," Section 3.8, HEDH.
3. A. C. Mueller, in *Handbook of Heat Transfer,* Rohsenow and Hartnet (eds.), McGraw-Hill, New York, 1983, Chap. 18.
4. R. L. Webb, "Compact Heat Exchangers," Section 3.9, HEDH.
5. F. L. Rubin, "Multizone Condensers, Desuperheating, Condensing, Subcooling," *Heat Transfer Eng.* **3**(1), 49–59 (1981).
6. H. Hausen, *Heat Transfer in Counterflow, Parallel Flow, and Crossflow,* McGraw-Hill, New York, 1983.
7. D. Chisholm et al., "Costing of Heat Exchangers," Section 4.8, HEDH.
8. R. S. Hall, J. Matley, and K. J. McNaughton, "Current Costs of Process Equipment," *Chem. Eng.* **89**(7), 80–116 (Apr. 5, 1982).
9. J. Taborek, "Charts for Mean Temperature Difference in Industrial Heat Exchanger Configurations," Section 1.5, HEDH.

10. K. J. Bell, "Approximate Sizing of Shell-and-Tube Heat Exchangers," Section 3.1.4, HEDH.

11. J. Taborek, "Shell and Tube Heat Exchangers, Single-Phase Flow," Section 3.3, HEDH.

12. D. Q. Kern, *Process Heat Transfer,* McGraw-Hill, New York, 1950.

13. J. W. Palen and J. Taborek, "Solution of Shellside Heat Transfer and Pressure Drop by Stream Analysis Method," *Chem. Eng. Prog. Symp. Series* **65**(92) (1969).

14. A. C. Mueller, "Condensers," Section 3.4, HEDH.

15. B. D. Smith, *Design of Equilibrium Stage Processes,* McGraw-Hill, New York, 1963.

16. V. L. Rice, "Program Performs Vapor-Liquid Equilibrium Calculations," *Chem. Eng.,* 77–86 (June 28, 1982).

17. R. S. Kistler and A. E. Kassem, "Stepwise Rating of Condensers," *Chem. Eng. Prog.* **77**(7), 55–59 (1981).

18. G. Breber, J. Palen, and J. Taborek, "Prediction of Horizontal Tubeside Condensation of Pure Components Using Flow Regime Criteria," *Heat Transfer Eng.* **1**(2), 72–79 (1979).

19. D. Butterworth, "Condensation of Vapor Mixtures," Section 2.6.3, HEDH.

20. R. G. Sardesai, "Condensation of Mixtures Forming Immiscible Liquids," Section 2.5.4, HEDH.

21. K. J. Bell and A. M. Ghaly, "An Approximate Generalized Design Method for Multicomponent/Partial Condensers," *AIChE Symp. Ser.,* No. 131, 72–79 (1972).

22. J. E. Diehl, "Calculate Condenser Pressure Drop," *Pet. Refiner* **36**(10), 147–153 (1957).

23. I. D. R. Grant and D. Chisholm, "Two-Phase Flow on the Shell-side of a Segmentally Baffled Shell-and-Tube Heat Exchanger," *Trans. ASME J. Heat Transfer* **101**(1), 38–42 (1979).

24. K. Ishihara, J. W. Palen, and J. Taborek, "Critical Review of Correlations for Predicting Two-Phase Flow Pressure Drops Across Tube Banks," *Heat Transfer Eng.* **1**(3) (1979).

25. J. W. Palen, "Shell and Tube Reboilers," Section 3.6, HEDH.

26. R. A. Smith, "Evaporaters," Section 3.5, HEDH.

27. J. G. Collier, "Boiling and Evaporation," Section 2.7, HEDH.

28. J. R. Fair, "What You Need to Design Thermosiphon Reboilers," *Pet. Refiner* **39**(2), 105 (1960).

29. J. R. Fair and A. M. Klip, "Thermal Design of Horizontal Type Reboilers," *Chem. Eng. Prog.* **79**(3) (1983).

30. J. W. Palen and C. C. Yang, "Circulation Boiling Model of Kettle and Internal Reboiler Performance," Paper presented at the 21st National Heat Transfer Conference, Seattle, WA, 1983.

31. J. W. Palen, A. Yarden, and J. Taborek, "Characteristics of Boiling Outside Large Scale Multitube Bundles," *Chem. Eng. Prog. Symp. Ser.* **68**(118), 50–61 (1972).

32. J. W. Palen, C. C. Shih, and J. Taborek, "Performance Limitations in a Large Scale Thermosiphon Reboiler," *Proceedings of the 5th International Heat Transfer Conference,* Tokyo, 1974, Vol. 5, pp. 204–208.

33. J. W. Palen, C. C. Shih, and J. Taborek, "Mist Flow in Thermosiphon Reboilers," *Chem. Eng. Prog.* **78**(7), 59–61 (1982).

34. R. Brown, "A Procedure for Preliminary Estimate of Air-Cooled Heat Exchangers," *Chem. Eng.* **85**(8), 108–111 (Mar. 27, 1978).

35. E. C. Smith, "Air-Cooled Heat Exchangers," *Chem. Eng.* (Nov. 17, 1958).

36. V. Gnielinski, A. Zukauskas, and A. Skrinska, "Banks of Plain and Finned Tubes," Section 2.5.3, HEDH.

37. P. Minton, "Designing Spiral-Plate Heat Exchangers," *Chem. Eng.* **77**(9) (May 4, 1970).

38. A. Cooper and J. D. Usher, "Plate Heat Exchangers," Section 3.7, HEDH.

39. J. Marriott, "Performance of an Alfaflex Plate Heat Exchanger," *Chem. Eng. Prog.* **73**(2), 73–78 (1977).

40. D. Chisholm, "Heat Pipes," Section 3.10, HEDH.

41. J. S. Truelove, "Furnaces and Combustion Chambers," Section 3.11, HEDH.

42. W. R. Penney, "Agitated Vessels," Section 3.14, HEDH.

43. A. R. Guy, "Double-Pipe Heat Exchangers," Section 3.2, HEDH.

44. J. Taborek et al., "Fouling—The Major Unresolved Problem in Heat Transfer," *Chem. Eng. Prog.* **65**(92), 53–67 (1972).

45. *Proceedings of the Conference on Progress in the Prevention of Fouling in Process Plants,* sponsored by the Institute of Corrosion Science Technology and the Institute of Chemical Engineers, London, 1981.

46. J. W. Suitor, W. J. Marner, and R. B. Ritter, "The History and Status of Research in Fouling of Heat Exchangers in Cooling Water Service," *Canad. J. Chem. Eng.* **55** (Aug., 1977).

47. A. Cooper, J. W. Suitor, and J. D. Usher, "Cooling Water Fouling in Plate Exchangers," *Heat Transfer Eng.* **1**(3) (1979).

48. R. B. Ritter and J. W. Suitor, "Seawater Fouling of Heat Exchanger Tubes," in *Proceedings of the 2nd National Conference on Complete Water Reuse,* Chicago, 1975.

49. C. H. Gilmour, "No Fooling–No Fouling," *Chem. Eng. Prog.* **61**(7), 49–54 (1965).

50. J. V. Smith, "Improving the Performance of Vertical Thermosiphon Reboilers," *Chem. Eng. Prog.* **70**(7), 68–70 (1974).

51. J. C. Chenoweth, "Flow-Induced Vibration," Section 4.6, HEDH.

52. C. C. Gentry, R. K. Young, and W. M. Small, "RODbaffle Heat Exchanger Thermal-Hydraulic Predictive Methods," in *Proceedings of the 7th International Heat Transfer Conference,* Munich, 1982.

53. A. C. Mueller, "Criteria for Maldistribution in Viscous Flow Coolers," in *Proceedings of the 5th International Heat Transfer Conference,* HE 1.4, Tokyo, Vol. 5, pp. 170–174.

54. W. F. Berg and J. L. Berg, "Flow Patterns for Isothermal Condensation in One-Pass Air-Cooled Heat Exchangers," *Heat Transfer Eng.* **1**(4), 21–31 (1980).

55. G. Breber, J. W. Palen, and J. Taborek, "Study on Non-Condensable Vapor Accumulation in Air-Cooled Condensers," in *Proceedings of the 7th International Heat Transfer Conference,* Munich, 1982.

56. A. Blumenkrantz and J. Taborek, "Application of Stability Analysis for Design of Natural Circulation Boiling Systems and Comparison with Experimental Data," *AIChE Symp. Ser.* **68**(118) (1971).

57. V. L. Streeter, *Fluid Mechanics,* McGraw-Hill, New York, 1958.

58. E. A. D. Saunders, "Shell and Tube Heat Exchangers, Elements of Construction," Section 4.2, HEDH.

59. F. W. Schmidt, "Thermal Energy Storage and Regeneration," in *Heat Exchangers Theory and Practice,* J. Taborek et al. (eds.), Hemisphere, McGraw-Hill, New York.

60. J. C. Chen, "Correlation for Boiling Heat Transfer to Saturated Fluids in Convective Flow," *Ind. Eng. Chem. Proc. Design and Dev.* **5**(3), 322–339 (1966).

61. D. Steiner and J. Taborek, "Flow Boiling Heat Transfer in Vertical Tubes Correlated by an Asymptotic Method," *Heat Transfer Engineering* **13**(3), 43 (1992).

62. Y. Katto, "Generalized Correlation of Critical Heat Flux for Forced Convection Boiling in Vertical Uniformly Heated Round Tubes," *International Journal of Heat Mass Transfer* **21**(12), 1527–1542 (1978).

63. N. Kattan, J. R. Thome, and D. Farrat, "Flow Boiling in Horizontal Tubes: Part 3—Development of a New Heat Transfer Model based on Plow Pattern," *Journal of Heat Transfer* **120**(1), 156–164 (1998).

64. K. W. McQuillan and P. B. Whalley, "A Comparison Between Flooding Correlations and Experimental Flooding Data for Gas–Liquid Flow in Vertical Circular Tubes," *Chemical Engineering Science* **40**(8), 1425–1440 (1988).

65. C. H. Gilmour, "No Fooling, No Fouling," *Chemical Engineering Progress,* **61**(7), 49–54 (1965).

CHAPTER 9

HEAT PIPES

Hongbin Ma
Department of Mechanical and Aerospace Engineering
University of Missouri
Columbia, Missouri

1 INTRODUCTION

The heat pipe is a device that utilizes the evaporation heat transfer in the evaporator and condensation heat transfer in the condenser, in which the vapor flow from the evaporator to the condenser is caused by the vapor pressure difference and the liquid flow from the condenser to the evaporator is produced by the capillary force, gravitational force, electrostatic force, or other forces directly acting on it. The first heat-pipe concept can be traced to the Perkins tube.[1,2] Based on the structure, a heat pipe typically consists of a sealed container charged with a working fluid. Heat pipes operate on a closed two-phase cycle and only pure liquid and vapor are present in the cycle. The working fluid remains at saturation conditions as long as the operating temperature is between the triple point and the critical state. As illustrated in Fig. 1, a typical heat pipe consists of three sections: an evaporator or heat addition section, an adiabatic section, and a condenser or heat rejection section. When heat is added to the evaporator section of the heat pipe, the heat is transferred through the shell and reaches the liquid. When the liquid in the evaporator section receives enough thermal energy, the liquid vaporizes. The vapor carries the thermal energy through the adiabatic section to the condenser section, where the vapor is condensed into the liquid and releases the latent heat of vaporization. The condensate is pumped back from the condenser to the evaporator by the driving force acting on the liquid.

For a heat pipe to be functional, the liquid in the evaporator must be sufficient to be vaporized. There are a number of limitations to affect the return of the working fluid. When

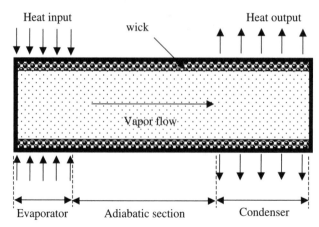

Figure 1 Schematic of a heat pipe.

the pumping pressure produced by the surface tension cannot overcome the summation of the total pressures, the heat transport occurring in the heat pipe reaches a limit known as the capillary limit. There are several other limitations disconnecting the return of the working fluid from the evaporator to the condenser or from the condenser to the evaporator. Among these are the boiling limit, sonic limit, entrainment limit, and viscous limit. When the heat flux added to the evaporator is sufficiently high, nucleate boiling occurs. The bubble formed in the wick significantly increases the thermal resistance, causing the heat-transfer perform-ance to be significantly reduced. More importantly, when the heat flux is so high, the bubbles block the return of the working fluid and lead to a dryout of the evaporator. The boiling limit plays a key role in a high heat flux heat pipe. When the vapor velocity is high and the cross-sectional area variation of the vapor space in a heat pipe cannot meet the flow con-dition, chocked flow occurs and the vapor flow rate will not respond with the amount of heat added in the evaporator. This will lead to a sonic limit. The entrainment limit is due to the frictional shear stresses caused by the vapor flow at the vapor–liquid interface. The viscous limit occurs in a low heat flux heat pipe, where the vapor pressure difference in the vapor phase cannot overcome the vapor pressure drop in the vapor phase.

From a thermodynamics point of view, the thermal energy added to the evaporator in a functional heat pipe produces the mechanical work to pump the working fluid. No external power is needed for a typical heat pipe. The phase-change heat transfer occurs almost in the quasi-equilibrium state. The heat pipe has a very high efficiency to transfer the thermal energy from a higher-temperature heat source to a lower-temperature heat source. An oper-ational heat pipe can provide an extra-high effective thermal conductivity and reach a higher level of temperature uniformity. The working fluid medium in a heat pipe can be selected from a variety of fluids, depending on the operating temperature and compatibility with the shell material. The heat pipe can be operated from a temperature lower than 4 K to a high temperature up to 3000 K. Because the evaporator and condenser of a heat pipe function independently, the heat pipe can be made into any shape, depending on the design require-ment. Due to these unique features, the heat pipe has been widely used in a wide range of applications.

2 FUNDAMENTALS

2.1 Surface Tension

Surface tension is a force that operates on a surface and acts perpendicular and inward from the boundaries of the surface, tending to decrease the area of the interface. As a result, a liquid will tend to take up a shape having minimum area. In the case of zero gravity in vacuum this liquid drop will be a perfect sphere. Surface tension can be viewed as a consequence of attractive and repulsive forces among molecules near the interface. From a thermodynamic point of view, it may be interpreted in terms of energy stored in the molecules near the interface. Surface tension consists of the dispersion force and other specific forces such as metallic or hydrogen bonding, i.e., $\sigma = \sigma_d + \sigma_m$. Surface tension in nonpolar liquids is entirely caused by dispersion forces. In hydrogen-bonded liquids, both dispersion forces and hydrogen bonding have contributions resulting in relatively larger values of surface tension. In liquid metals, the metallic force combining with the dispersion force results in higher values of surface tension. And it is easy to understand that the surface tensions of liquid metals are higher than those of hydrogen-bonded liquid such as water, which in turn are higher than those of nonpolar liquids such as pure hydrocarbons. The surface tension significantly depends on the temperature. As temperature increases, the surface tension decreases. The surface tension of water, for example, decreases almost linearly with temperature, $\sigma = 75.83 - 0.1477T$ (mN/m), where T is temperature (°C).

2.2 Contact Angle

A physical property that is closely related to the surface tension is the contact angle. The contact angle α is defined as the angle (measured in the liquid) formed between the liquid–vapor interface and the solid–liquid interface as shown in Fig. 2, which may be expressed as

$$\cos \alpha = \frac{\sigma_{sv} - \sigma_{sl}}{\sigma} \tag{1}$$

where σ_{sv} is the surface tension between the solid and vapor, and σ_{sl} is the surface tension between the solid and liquid. The surface tension between the liquid and vapor is a function of temperature and decreases as the temperature increases. For a given solid surface and liquid, when the surface tension between the solid and liquid is a fixed constant, the contact angle will decrease as the temperature increases. When the temperature increases, the wetting characteristic of a liquid on a given solid surface becomes better.

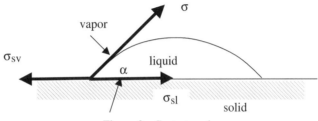

Figure 2   Contact angle.

2.3 Laplace-Young Equation

For a liquid–vapor curved surface, there exists a pressure difference across the interface, which can be calculated by

$$p_{\mathrm{I}} - p_{\mathrm{II}} = \sigma\left(\frac{1}{r_1} + \frac{1}{r_2}\right) \tag{2}$$

where r_1 is the meniscus radius along the x direction and r_2 is the meniscus radius along the y direction, as shown in Fig. 3. This expression is known as the Laplace-Young equation and was derived in 1805. Equation (2) can be used easily to find the pressure differences across the liquid–vapor interfaces for a number of structures shown in Table 1.

3 HEAT TRANSPORT LIMITATIONS

3.1 Capillary Limit

For a heat pipe to function, the capillary pressure difference occurring in the heat pipe must always be greater than the summation of all the pressure losses occurring throughout the liquid and vapor flow paths. When the heat-transfer rate increases, the pressure losses increase, which will be overcome by the increase of the capillary pressure difference. The continuous increase of the heat-transfer rate in a heat pipe will significantly increase the pressure losses, and at one heat-transfer rate the total capillary pressure difference is no longer equal or greater than the total pressure losses. This relationship, referred to as the capillary limit, can be expressed mathematically as

$$\Delta p_{c,\mathrm{max}} \geq \Delta p_l + \Delta p_v + \Delta p_g \tag{3}$$

where $\Delta p_{c,\mathrm{max}}$ = maximum capillary pressure difference generated within capillary wicking structure

Δp_l = sum of inertial and viscous pressure drops occurring in liquid phase
Δp_v = sum of inertial and viscous pressure drops occurring in vapor phase
Δp_g = hydrostatic pressure drop

When the maximum capillary pressure difference is equal to or greater than the summation of these pressure drops, the capillary structure is capable of returning an adequate amount of working fluid to prevent dryout of the evaporator wicking structure. This condition varies

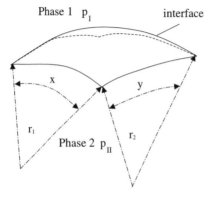

Figure 3 Curved interface between two phases.

Table 1 Pressure Differences Across the Liquid–Vapor Curved Surfaces

Names/Structures	Pressure Difference Across the Liquid–Vapor Interface	Notes
Vapor bubble in liquid phase	$$p_v - p_l = \sigma\left(\frac{1}{r_1} + \frac{1}{r_2}\right) = \frac{2\sigma}{r}$$	$r = r_1 = r_2$
Liquid drop in vapor phase	$$p_l - p_v = \sigma\left(\frac{1}{r_1} + \frac{1}{r_2}\right) = \frac{2\sigma}{r}$$	$r = r_1 = r_2$
Liquid in a micro triangular groove	$$p_v - p_l = \sigma\left(\frac{1}{r_1} + \frac{1}{\not{r_2}}\right) = \frac{\sigma}{r}$$	$r_2 = \infty$ $r = $ constant.
Capillary wicking in a capillary tube	$$p_v - p_l = \sigma\left(\frac{1}{r_1} + \frac{1}{r_2}\right) = \frac{2\sigma}{r} = \frac{2\sigma}{\dfrac{r_b}{\cos\alpha}} = \rho g h$$	$r = r_1 = r_2$

according to the wicking structure, working fluid, evaporator heat flux, vapor flow channel, and operating temperature.

Capillary Pressure

When the meniscus radius exists at a liquid–vapor interface, there is the pressure difference across the interface, which can be determined by the Laplace-Young equation shown in Eq. (2). During steady-state operation, it is generally defined that the maximum capillary pressure exists when the capillary radius in the condenser approaches infinity and the capillary radius in the evaporator reaches the smallest one. To generalize the application, the maximum capillary pressure can be expressed as a function of only the effective capillary radius of the evaporator wick, i.e.,

$$\Delta p_{c,\max} = \frac{2\sigma}{r_{c,e}} \tag{4}$$

where $r_{c,e}$ is the effective capillary radius depending on the wick structures. Table 2 lists effective capillary radii for four typical wicks. For other kinds of wicks, the effective capillary radii can be readily determined using the Laplace-Young equation.

Liquid Pressure Drop

The liquid pressure drop is the result of the combined effect of both viscous and inertial forces. If the flow rate in the wick is very small, the effect of inertial force can be neglected, and the pressure difference in the liquid phase is caused only by the frictional forces at the liquid–solid interface and the liquid–vapor interface due to the vapor flow effect. The total liquid pressure drop can be determined by integrating the pressure gradient over the length of the flow passage, or

$$\Delta p_l(x) = -\int_0^x \frac{dp_l}{dx} \, dx \tag{5}$$

where the limits of integration are from the evaporator end ($x = 0$) to the condenser end ($x = L$), and dp_l/dx is the gradient of the liquid pressure resulting from the frictional forces from the solid wick, wall, and vapor flow at the liquid–vapor interface, which can be written as

$$\frac{dp_l}{dx} = \left(\frac{\mu_l}{KA_w\rho_l}\right)\dot{m}_l \tag{6}$$

where $\dot{m}_l$ is the local mass flow rate in the wick, and K, the permeability, can be expressed as

$$K = \frac{2\varepsilon r_{h,l}^2}{f_l \mathrm{Re}_l} \tag{7}$$

The wick porosity, ε, in Eq. (7) is defined as the ratio of the pore volume V_{por} to the total volume V_{tot} of wick structure. The hydraulic radius, $r_{h,l}$, in Eq. (7) is defined as twice the cross-sectional area divided by the wetted perimeter, or $r_{h,l} = 2A_c/P$. It should be noted that Eq. (7) is valid for both the circular and noncircular channels/grooves. If the Reynolds number of the working fluid flowing through the wick structure is small and less than the critical value, then laminar flow is assumed. The product of the friction factor and Reynolds number, $f_l \mathrm{Re}_l$, for laminar flow is constant and depends only on the passage shape. It should be noticed that when the liquid–vapor interface is affected by the vapor flow, the friction factor Reynolds number product, $f_l \mathrm{Re}_l$, depends on the vapor flow in addition to the contact angle and channel angle.[4]

Table 2   Effective Capillary Radii[3]

Structures	$r_{c,e}$	Note
Rectangular groove	$\dfrac{w}{\cos \alpha}$	w = groove width α = contact angle
Triangular groove	$\dfrac{3 \cos \theta w}{4 \cos \alpha}$	w = groove width α = contact angle θ = half groove angle
Wire screen	$\dfrac{w + d_w}{2 \cos \alpha}$	d_w = wire diameter w = mesh spacing α = contact angle
Packed or sintered particles	$0.41 r_s$	r_s = particle radius

For uniform heat addition and rejection, Eq. (5) can be expressed as

$$\Delta p_l = \left(\frac{\mu_l}{K A_w h_{fg} \rho_l} \right) L_{\text{eff}} q \tag{8}$$

where $q = \dot{m}_l h_{fg}$ and the effective heat pipe length can be found as

$$L_{\text{eff}} = 0.5 L_e + L_a + 0.5 L_c \tag{9}$$

In many cases, an analytical expression for the permeability, K, shown in Eq. (8), is not available. In such a case, semiempirical correlations based on experimental data are usually employed. For example, Marcus[5] has described a method for calculating the permeability of wrapped, screened wicks. This expression, which is a modified form of Blake-Kozeny equation, can be expressed as

$$K = \frac{d^2 \varepsilon^3}{122 (1 - \varepsilon)^2} \tag{10}$$

In this expression, d is the wire diameter and ε is the wick porosity, which can be determined as

$$\varepsilon = 1 - \tfrac{1}{4} \pi S N d \tag{11}$$

where N is the mesh number per unit length and S is the crimping factor (approximately 1.05).[6] For the sintered particles, this equation takes the form

$$K = \frac{d_s^2 \varepsilon^3}{37.5 (1 - \varepsilon)^2} \tag{12}$$

where d_s is the average diameter of the sintered particles.

Vapor Pressure Drop
If the heat pipe is charged with an appropriate amount of working fluid and the wetting point occurs at the cap end of condenser, the vapor pressure drop can be calculated by the approach recommended by Peterson,[6] Chi,[7] and Dunn and Reay.[8] Based on the one-dimensional vapor flow approximation, the vapor pressure drop can be determined by

$$\Delta p_v = \left[\frac{C (f_v \text{Re}_v) \mu_v}{2 r_{h,v}^2 A_v \rho_v h_{fg}} \right] L_{\text{eff}} q \tag{13}$$

where C is the constant that depends on the Mach number defined by

$$Ma_v = \frac{q}{A_v \rho_v h_{fg} (R_v T_v \gamma_v)^{1/2}} \tag{14}$$

The ratio of specific heats, γ_v, in Eq. (14) depends on the molecule types, which is equal to 1.67, 1.4, and 1.33 for monatomic, diatomic, and polyatomic molecules, respectively. Previous investigation summarized by Peterson[6] have demonstrated that the friction factor Reynolds number product, $f_v \text{Re}_v$, and the constant, C, shown in Eq. (13) can be determined by

$$\text{Re}_v < 2300 \quad \text{and} \quad Ma_v < 0.2$$
$$f_v \text{Re}_v = \text{constant}, \ C = 1.0 \tag{15}$$

$$\text{Re}_v < 2300 \quad \text{and} \quad Ma_v > 0.2$$
$$f_v \text{Re}_v = \text{constant}, \ C = \left[1 + \left(\frac{\gamma_v - 1}{2} \right) Ma_v^2 \right]^{-1/2} \tag{16}$$

$$\text{Re}_v > 2300 \quad \text{and} \quad Ma_v < 0.2$$

$$f_v \text{Re}_v = 0.038 \left(\frac{2r_{h,v}q}{A_v \mu_v h_{fg}} \right)^{3/4}, \quad C = 1.0 \tag{17}$$

It should be noted that Eq. (17) was determined based on a round channel. Because the equations used to evaluate both the Reynolds number and the Mach number are functions of the heat-transport capacity, it is first necessary to assume the conditions of the vapor flow, and an iterative procedure must be used to determine the vapor pressure.

If the heat pipe is overcharged and/or the heat pipe is operating at a high cooling rate, the location of the wet point, where the pressures in the vapor and the liquid are equal, should be close to the beginning of the condensing section. In this case, only the total pressure drop in the evaporating and condensing sections is needed in the calculation of the capillary limitation.[9] Equation (13) becomes

$$\Delta p_v = \left[\frac{C(f_v \text{Re}_v)\mu_v}{2r_{h,v}^2 A_v \rho_v h_{fg}} \right] q[0.5L_e(1 + F\,\text{Re}_r) + L_a] \tag{18}$$

where the correction factor, F, in Eq. (18) can be determined by

$$F = \frac{7}{9} - \frac{1.7\text{Re}_r}{36 + 10\,\text{Re}_r} \exp\left(-\frac{7.51L_a}{\text{Re}_r\,L_e} \right) \tag{19}$$

The radial Reynolds number, Re_r, in Eqs. (18) and (19) is defined by

$$\text{Re}_r = \frac{\rho_v v_{lv} r_v}{\mu_v} \tag{20}$$

where v_{lv} is the interfacial velocity. For evaporation, $v_{lv} > 0$; for condensation, $v_{lv} < 0$.

3.2 Boiling Limit

When boiling occurs near the evaporating wall in the wick, two consequences result. First, the amount of thin film evaporation at the solid–liquid–vapor interface dramatically decreases as the boiling condition dominates the phase change behavior of the system. Second, the vapor forming at the base of the wick structure forms a blanket of vapor, preventing reentry of the working fluid. Since the vapor conductivity of working fluid is much lower than the fluid conductivity, the overall conductivity of the wick structure will experience a significant decrease. Obviously, boiling heat transfer in the wick should be avoided as this condition could lead to early dryout of the heat pipe.

If the wick is constructed such that the temperature difference between the wall temperature of the evaporator and the saturation temperature, $T_w - T_s$, remains less than the boiling superheat for a given pressure, bubble formation will not occur near the wall in the wick. When the local working fluid temperature inside the wick exceeds the saturation temperature corresponding to the local pressure and nucleation occurs, the bubbles that are formed will collapse and boiling will be avoided if the superheat is not sufficiently large. The equilibrium state for the bubbles, or the state at which the bubbles no longer collapse, is that thermodynamic state for which the Gibbs free energy between the liquid and vapor phases is minimized. Using the Clausius-Clapeyron relation, the superheat can be found as

$$T_l - T_s(p_l) = \frac{2\sigma T_s(p_l)}{h_{fg} r_e} \left(\frac{1}{\rho_v} - \frac{1}{\rho_l} \right) \tag{21}$$

If the vapor density is much smaller than the liquid density, Eq. (21) may be reduced to

$$T_l - T_s(p_l) = \frac{2\sigma T_s(p_l)}{\rho_v h_{fg} r_e} \tag{22}$$

where r_e is the meniscus radius of the vapor bubble formed in the wick structure of the heat pipe, which is directly related to the pore size of the wick structure. According to the theory presented by Hsu[10] and the derivation presented by Carey,[11] an embryo bubble will grow and a cavity will become an active nucleation site if the equilibrium superheat becomes equaled or exceeded around the perimeter of the embryo. To avoid boiling near the base of the wick structure, the temperature difference between the wall and the saturation temperature must be less than the superheat required for bubble formation. Using the superheat obtained above, the critical heat flux related to the boiling limit can be found as

$$q'' = k_{\text{eff}}[T_l - T_s(p_l)] = \frac{k_{\text{eff}} 2\sigma T_s(p_l)}{\rho_v h_{fg} r_e} \tag{23}$$

As can be seen, the boiling limit is sensitive to the effective thermal conductivity, k_{eff}, and the meniscus radius of vapor bubble, r_e, at the wick–wall interface.

3.3 Entrainment Limit

In an operating heat pipe, when the vapor flow direction is opposite to the liquid flow direction, the frictional shear stress occurring at the liquid–vapor interface may slow down the return of liquid to the evaporator. As the vapor velocity increases, the vapor flow effect on the liquid–vapor interface increases, depending on surface tension, viscosities, and densities of both the vapor and liquid phases. When the influence caused by the frictional shear stress acting on the liquid–vapor interface by the frictional vapor flow is large enough, the liquid flow cannot flow back to the evaporator. When this occurs, the liquid in the evaporator dries out. At this point the heat pipe reaches a heat-transport limit, which is known as the entrainment limit. Based on a Weber number equal to one, i.e., $We = F_{lv}/F_\sigma = 1$, where F_{lv} is the shear stress at the liquid–vapor interface and F_σ is the surface tension force, Cotter[12] developed an approximation of the entrainment limit as follows:

$$q_{ent} = A_v h_{fg} \left(\frac{\sigma \rho_v}{2 r_{h,w}} \right)^{0.5} \tag{24}$$

where $r_{h,w}$ is the hydraulic radius of the wick surface pore.

3.4 Viscous Limit

When the vapor pressure from the evaporator to the condenser cannot overcome the vapor pressure drop caused by the viscous forces, the heat pipe reaches a heat-transport limit, which is called the viscous limit. In particular, the viscous limit is reached when the vapor pressure in the condenser is equal to zero. Using assumptions of laminar flow, ideal gas, and zero pressure in the cap end of the condenser, the viscous limit can be determined by

$$q_{vis} = \frac{4 r_v^2 h_{fg} \rho_{v,e} p_{v,e} A_v}{f_v \text{Re}_v \mu_v L_{\text{eff}}} \tag{25}$$

where $\rho_{v,e}$, $p_{v,e}$ are the vapor density and vapor pressure in the cap end of the evaporator, respectively.

3.5 Sonic Limit

When the vapor velocity at the exit of the evaporating section reaches the local sound speed, the vapor flow is choked. As the chocked flow occurs, the vapor flow rate will not respond with the amount of heat added in the evaporator. The heat pipe has reached the maximum heat transport, which is called the sonic limit. If the vapor flow in the heat pipe can be approximated as one-dimensional flow with the assumptions of negligible frictional force and ideal gas, the sonic limit can be readily derived from the conservation of energy and momentum equations as follows:

$$q_s = A_v \rho_v h_{fg} \left[\frac{\gamma_v R_v T_0}{2(\gamma_v + 1)} \right]^{1/2} \tag{26}$$

where T_0 is the stagnation temperature, A_v and ρ_v are the cross-sectional area and vapor density, respectively, at the location where the local sound speed is reached. It should be noted that this might occur at the exit to the evaporator or any location in the condenser section if the cross-sectional area of vapor flow path changes.

3.6 Effective Thermal Conductivity

While those operating limitations described above dominate the design of a heat pipe, the effective thermal conductivity provided by a heat pipe is a key factor for designing a highly efficient heat-pipe cooling device. The effective thermal conductivity, k_{eff}, is related to the total temperature drop, ΔT_{total}, as

$$k_{\text{eff}} = \frac{q L_{\text{eff}}}{A_h \Delta T_{\text{total}}} \tag{27}$$

where A_h is the total cross-sectional area of heat pipe. The total temperature drop, ΔT_{total}, across the heat pipe is the sum of the temperature drop across the evaporator shell, $\Delta T_{e,\text{shell}}$; the temperature drop across the wick structure in the evaporator, $\Delta T_{e,\text{wick}}$; the temperature drop through the evaporating thin film, $\Delta T_{e,\text{film}}$; the temperature drop in the vapor flow, ΔT_v; the temperature drop across the condensate film, $\Delta T_{c,\text{film}}$; the temperature drop across the wick structure at the condenser, $\Delta T_{c,\text{wick}}$; and the temperature drop across the condenser shell, $\Delta T_{c,\text{shell}}$, i.e.,

$$\Delta t_{\text{total}} = \Delta T_{e,\text{shell}} + \Delta T_{e,\text{wick}} + \Delta T_{e,\text{film}} + \Delta T_v + \Delta T_{c,\text{film}} + \Delta T_{c,\text{wick}} + \Delta T_{c,\text{shell}} \tag{28}$$

Temperature Drops across Shell and Wick
The temperature drop through the case shell material at both the evaporator and the condenser can be calculated by

$$\Delta T_{e,\text{shell}} = \frac{q_e'' \delta_{e,\text{shell}}}{k_{\text{shell}}} \tag{29}$$

and

$$\Delta T_{c,\text{shell}} = \frac{q_c'' \delta_{c,\text{shell}}}{k_{\text{shell}}} \tag{30}$$

respectively. After heat is traveled through the wall, the heat reaches the working fluid in the wick. Provided that the wick is saturated with the working fluid and no boiling occurs in the wick, the heat will transfer through the wick and evaporation will only occur at the liquid–vapor–solid interface. The heat transfer across the wick is dominated by the heat conduction. The temperature drop across the wick in the evaporator can be determined by

$$\Delta T_{e,\text{wick}} = \frac{q_c'' \delta_{\text{wick}}}{k_{\text{eff}}} \tag{31}$$

As shown in Eq. (31), the effective thermal conductivity of a wick plays a key role in the temperature drop in the wick. Since the effective thermal conductivity depends on the working fluid, porosity, and geometric configuration, it is very hard to find the exact effective thermal conductivity theoretically. Nevertheless, several approximations for the effective thermal conductivity have been developed. A list of common expressions for determining the effective thermal conductivity of the wick is presented in Table 3. It is clear from each of the listed expressions that the effective thermal conductivity of the wick is a function of the solid conductivity, k_s, the working fluid conductivity, k_l, and the porosity. In each of the relations, $\lim_{\varepsilon \to 0} k_{\text{eff}} = k_s$ and $\lim_{\varepsilon \to 1} k_{\text{eff}} = k_l$; however, the manner by which the effective conductivity varies between the limiting cases is drastically different depending on the type of arrangement of the wick structure. Comparing the various effective conductivity relations, as illustrated in Table 3 and Fig. 4, it becomes clear that sintering the metallic particles dramatically enhances the overall thermal conductivity.

Similar to the temperature drop calculation for the evaporator, the temperature drop across the wick in the condenser can be found as

$$\Delta T_{c,\text{wick}} = \frac{q_c'' \delta_{\text{wick}}}{k_{\text{eff}}} \tag{32}$$

Temperature Drop across the Liquid–Vapor Interface
If heat is added to the wick structure, the heat is transferred through the wick filled with the working fluid, reaching the surface where the liquid–vapor–solid interface exists. There, by utilizing the thin-film evaporation, the heat is removed. The temperature drop across the evaporating thin film, as shown in Fig. 5, can be calculated from the thin-film thickness by solving the equations governing the heat transfer and fluid flow in the thin-film region.[13] The film thickness profile for a flat surface can be described by

Table 3 Expressions for Wick Effective Thermal Conductivities for Various Geometries

Wick Condition	Expression for Effective Thermal Conductivity
Sintered	$k_{\text{Sintered}} = \dfrac{k_s[2k_s + k_1 - 2\varepsilon(k_s - k_1)]}{2k_s + k_1 + \varepsilon(k_s - k_1)}$
Packed spheres	$k_{\text{PackedSpheres}} = \dfrac{k_1[(2k_1 + k_s) - 2(1 - \varepsilon)(k_1 - k_s)]}{(2k_1 + k_s) + (1 - \varepsilon)(k_1 - k_s)}$
Wick and liquid in series	$k_{\text{Series}} = \dfrac{k_1 k_s}{k_s \varepsilon + k_1(1 - \varepsilon)}$
Wick and fluid in parallel	$k_{\text{Parallel}} = k_1 \varepsilon + k_s(1 - \varepsilon)$
Wrapped screens	$k_{\text{WrappedScreen}} = \dfrac{k_1[(k_1 + k_s) - (1 - \varepsilon)(k_1 - k_s)]}{(k_1 + k_s) + (1 - \varepsilon)(k_1 - k_s)}$

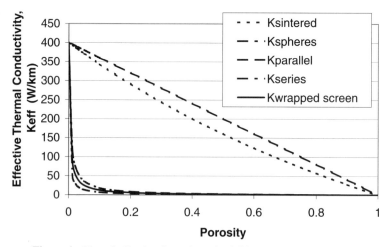

Figure 4   Plot of effective thermal conductivities presented in Table 3.

$$\sigma \frac{dK}{ds} - \frac{dp_d}{ds} = -\frac{f_l^+ \mathrm{Re}_s \mu_l \int_0^s \dfrac{q''(s)}{h_{fg}}\, ds}{2\delta^3(s)\rho_l} \tag{33}$$

for the steady-state evaporating process of a thin film, where the meniscus curvature, K, can be found by

$$K = \frac{d^2\delta/ds^2}{[1 + (d\delta/ds)^2]^{3/2}} \tag{34}$$

Using the following boundary conditions

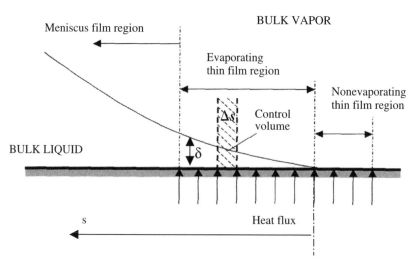

Figure 5   Thin-film evaporation.

$$\delta = \delta_0 \qquad K = 0 \qquad \frac{d\delta}{ds} = 0 \qquad \text{at } s = 0 \tag{35}$$

with Eq. (33), the evaporating thin-film profile is obtained and the temperature drop across the evaporating thin film can be determined. As shown in Fig. 6, the heat-flux level through the thin-film region can reach up to 1400 W/cm^2 with a superheat of $1.0°\text{C}$. The optimization of thin-film evaporation in a high-heat-flux heat-pipe design plays a key role.

Temperature Drop in Vapor Flow

To find the vapor velocity distribution and vapor pressure drop in a heat pipe, a three-dimensional model should be developed, in particularly, when the vapor space shape is irregular and evaporation occurs near the interline region. To obtain an effective tool, a simplified model can be used, wherein the pressure drop at a given z location is found using a two-dimensional model, i.e.,

$$\frac{\partial^2 u_v}{\partial x^2} + \frac{\partial^2 u_v}{\partial y^2} = \frac{1}{\mu_v} \frac{dp_v}{dz} \tag{36}$$

The friction factor can be obtained based on the vapor channel cross section[4] and the vapor flow along the z direction can be expressed as a one-dimensional momentum equation shown as

$$\frac{dp_v}{dz} + \rho_v g \sin \psi + \rho_v \bar{u}_v \frac{d\bar{u}_v}{dz} = -f_v \frac{2\rho_v \bar{u}_v^2}{d_{h,v}} \tag{37}$$

The vapor pressure varies from the evaporator section to the condenser section, due to frictional vapor flow, resulting in a temperature variation, which can be predicted by the Clapeyron equation, i.e.,

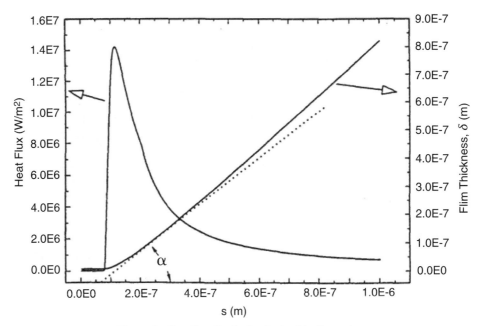

Figure 6   Heat-flux distribution in the thin-film region.

$$T_v = \frac{T_{v,e}}{1 - (RT_{v,e}/h_{fg}) \ln (p_e/p_{v,e})} \qquad (38)$$

where $T_{v,e}$ is the vapor temperature in the evaporator and $p_{v,e}$ is the vapor pressure in the evaporator. Vapor flows through the adiabatic section to the condenser region. At the condensing section, vapor condenses due to the phase-change driving force. The condensed fluid then is drawn into the wicking structure by a capillary force. Assuming that the heat pipe is charged with the proper amount of working fluid, the wick surface can be assumed to directly contact the saturated vapor. Therefore, the film thickness on the wicking structure is neglected while the temperature drop across the liquid–vapor interface in the condenser is equal to zero.

4 HEAT-PIPE FABRICATION PROCESSES

4.1 Wicks

The wicks in a heat pipe are used to pump the condensate from the condenser to the evaporator. Figure 7 lists several of wick structures currently being used in heat pipes. Among the wick structures shown in Fig. 7, the grooved, sintered metal, and wrapped screens are the most common wicks. The grooved wicks have been widely used in the laptop computers for a relatively low heat load. The groove dimensions significantly affect the capillary limitation and effective thermal conductivity. When the groove configuration is different, as shown in Fig. 8, the heat-transfer performance changes. For a given application, there exists an optimized groove configuration for the best heat transfer performance.

Another wick having higher effective thermal conductivity is the sintered metal wick. The sintered wick can be produced at a temperature 50–200°C below the melting point of the sintering material. The porosity of the sintered wick depends on the surrounding pressure, sintering temperature, and time. Also, the environmental gas can significantly help the sintering process. As hydrogen is used as the environmental gas, for example, it helps sinter the copper material. The heat-transfer performance in a heat pipe with sintered wicks largely depends on the wick thickness, particle size, and porosity. For a given heat flux, there exists an optimum design for the evaporating heat transfer in a heat pipe with sintered wicks. As shown in Fig. 9, when the particle radius decreases, the optimum thickness increases.[14] And it is concluded that, if it is possible to decrease the particle radius while maintaining a constant porosity, the particle radius should be as small as possible. The impact of these results is that thicker sintered wicks, which are more readily manufactured and assembled into heat pipes, can provide heat removal capabilities equivalent to the thinner wicks.

The screen wick has been widely used in the conventional heat pipe. Stainless steel, copper, nickel, and aluminum meshes are commercially available for the screen wicks. Although the smallest pore size for the copper meshes is about 100 pores per inch, the smallest pore size for the stainless steel meshes can reach less than 5 μm in the hydraulic diameter.

4.2 Working Fluid Selections

Because a heat pipe cannot function below the freezing point or above the critical temperature of its working fluid, the selected working fluid must be within this range. Table 4 lists some working fluids that can be used in the heat pipe for a given operating temperature. In addition, the vapor pressure, surface tension, contact angle, and viscosity in the heat pipe

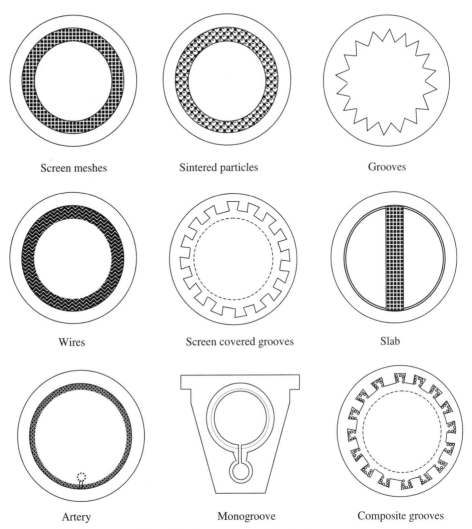

Screen meshes   Sintered particles   Grooves

Wires   Screen covered grooves   Slab

Artery   Monogroove   Composite grooves

Figure 7   Some common wick structures.

must be considered in the selection of working fluid. For example, the vapor pressure in the heat pipe should not be less than 0.1 atm or higher than 10 atm. The detailed information of physical properties of all available working fluids can be found in the heat-pipe design software developed by the ISoTherM Research Consortium at the University of Missouri. Typical working fluids for cryogenic heat pipes include helium, argon, oxygen, and krypton. For most common low-temperature heat pipes ranging from 200–550 K, ammonia, acetone, and water are commonly employed. The typical working fluids being used in high-temperature heat pipes are sodium, lithium, silver, and potassium. The more important factor in the selection of working fluid is its compatibility with the case and wick materials. Table 5 lists results of compatibility tests of materials with some working fluids.

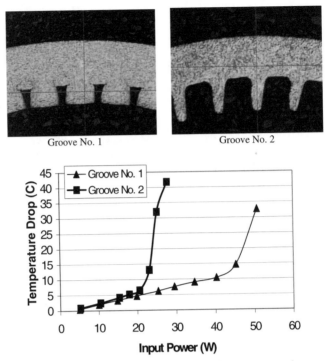

Groove No. 1   Groove No. 2

Figure 8   Groove configuration effect on the temperature drops.

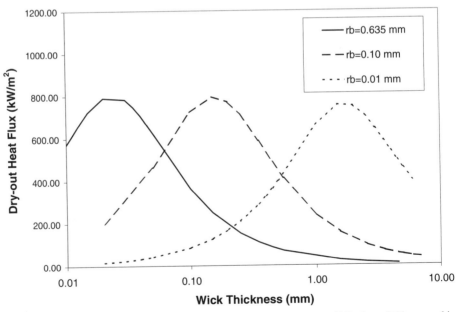

Figure 9   Particle size effect on the dryout heat flux (L = 0.254 m; ε = 43%; L_H = 0.01 m; working fluid = water).

Table 4   Working Fluids and Temperature Ranges[9]

Working Fluids	Melting Point, K at 1 atm	Boiling Point, K at 1 atm	Useful Range, K
Helium	1.0	4.21	2–4
Hydrogen	13.8	20.38	14–31
Neon	24.4	27.09	27–37
Nitrogen	63.1	77.35	70–103
Argon	83.9	87.29	84–116
Oxygen	54.7	90.18	73–119
Methane	90.6	111.4	91–150
Krypton	115.8	119.7	116–160
Ethane	89.9	184.6	150–240
Freon 22	113.1	232.2	193–297
Ammonia	195.5	239.9	213–373
Freon 21	138.1	282.0	233–360
Freon 11	162.1	296.8	233–393
Pentane	143.1	309.2	253–393
Freon 113	236.5	320.8	263–373
Acetone	180.0	329.4	273–393
Methanol	175.1	337.8	283–403
Flutec PP2	223.1	349.1	283–433
Ethanol	158.7	351.5	273–403
Heptane	182.5	371.5	273–423
Water	273.1	373.1	303–473
Toluene	178.1	383.7	323–473
Flutec PP9	203.1	433.1	273–498
Naphthalene	353.4	490	408–478
Dowtherm	285.1	527	423–668
Mercury	234.2	630.1	523–923
Sulphur	385.9	717.8	530–947
Cesium	301.6	943.0	723–1173
Rubidium	312.7	959.2	800–1275
Potassium	336.4	1032	773–1273
Sodium	371	1151	873–1473
Lithium	453.7	1615	1273–2073
Calcium	1112	1762	1400–2100
Lead	600.6	2013	1670–2200
Indium	429.7	2353	2000–3000
silver	1234	2485	2073–2573

4.3   Cleaning and Charging

All of the materials used in a heat pipe must be clean. Cleanliness achieves two important objectives: it ensures that the working fluid will wet the materials, and that no foreign matter is present that could hinder capillary action or create incompatibilities. The presence of contaminants either in solid, liquid, or gaseous state may be detrimental to heat-pipe performance. If the interior of a heat pipe is not clean, degradation of the performance can result over time. Solid particles can physically block the wick structure, decreasing the liquid flow rate and increasing the likelihood of encountering the capillary limit. Oils from machining or from the human hand can decrease the wettability of the wick. Oxides formed on the wall and wick can also decrease the ability for the liquid to wet the surface. Therefore, proper cleaning of all of the parts in contact with the interior of the heat pipe (pipe, end

Table 5   Experimental Compatibility Tests

	Aluminum	Brass	Copper	Inconel	Iron	Nickel	Niobium	Silica	Stainless steel	Tantalum	Titanium	Tungsten
Acetone	C	C	C						C			
Ammonia	C		I		C	C			C			
Cesium							C				C	
Dowtherm			C					C	C			
Freon-11	C											
Heptane	C											
Lead				I		I	I		I	C	I	C
Lithium				I		I	C		I	C	I	C
Mercury					I	I	I		C	I	I	
Methanol		C	C		C			C	C			
Silver				I		I	I		I	C	I	C
Sodium				C		C	C		C		I	
Water	I		C	I		C		C	C		C	

Note. C, compatible; I, incompatible.

caps, wick, and working fluid) is necessary for maximum reliability and performance. Several steps are needed to properly clean the heat-pipe container and wick structure, such as solvent cleaning, vapor degreasing, alkaline cleaning, acid cleaning, passivation, pickling, ultrasonic cleaning, and vacuum baking. Many of these steps are used in a single cleaning operation.[6,8]

It is necessary to treat the working fluid used in a heat pipe with the same care as that given to the wick and container. The working fluid should be the most highly pure available, and further purification may be necessary following purchase. This may be carried out by distillation. The case of low-temperature working fluids, such as acetone, methanol, and ammonia, in the presence of water can lead to incompatibilities, and the minimum possible water content should be achieved. The amount of working fluid required for a heat pipe can be approximately calculated by estimating the volume occupied by the working fluid in the wicking structure, including the volume of any arteries, grooves, cornered regions, and so on. The amount of working fluid charged to a heat pipe significantly affects the heat-transfer performance of the heat pipe. For example, the heat-transfer performance of a grooved heat pipe currently being used in the laptop computer is very sensitive to the charging amount of working fluid.

Once the amount of working fluid required is determined, the working fluid can be introduced into the heat pipe by an evacuation and backfilling technique, a liquid fill and vapor generation technique, a solid fill and sublimation technique, or a supercritical vapor technique[6]. The most common among those for the low- or moderate-temperature working fluids currently being used for the electronics cooling is the evacuation and backfill techniques. All these charging methods are to prevent noncondensable gases from entering the heat pipe during the charging process. To charge the working fluid in, a suitable evacuation/ filling rig must be applied. The rig must be able to evacuate the container to 10^{-4} torr or less. The filling rig is used to evacuate the pipe and charge it with the proper amount of working fluid. Details of the charging process depend on the state of the working fluid at the ambient temperature. The material of construction is generally glass, stainless steel, or plastic materials. Glass has advantages when handling liquids in that the presence of liquid droplets within the ductwork can be observed and their vaporization under vacuum noted. Stainless steel has obvious strength benefits. If plastic materials are chosen, the working fluid used in the charging process would have to not react with the plastic materials.

4.4 Testing

A test facility needs to be established to test the heat-transfer performance and heat-transport limitations. Both transient and steady-state tests should be conducted for a heat pipe. For low-temperature heat pipes, however, the steady-state test is the most important. A typical experimental system for low-temperature heat pipes similar to the one shown in Fig. 10 would normally be used. The test facility shown in Fig. 10 consists of the heat pipe, a heat power supply and measuring unit, a cooling unit, and a data-acquisition unit for the temperature measurements. The operating temperature of heat pipe can be controlled by a cooling block connected to a cooling bath, where the temperature of the coolant is maintained at a constant temperature of the designated operating temperature. The heat source is directly connected to the evaporator. Power input can be supplied by an ac or dc power supply and recorded by multimeters with signals sent directly to a personal computer, which can be used to control the entire system. The heat source should be well insulated to reduce convective losses. A number of temperature sensors are attached to the heat-pipe surfaces to measure the temperature distribution on the heat pipe and temperature variation with the power input. All of the measured data are sent to the data-acquisition system controlled by a personal computer. Prior to the start of the experiment, the system is allowed to equilibrate and reach steady-state such that the temperatures of the cooling media and the heat pipe are constant. When the desired steady-state condition has been obtained, the input power is increased in small increments. Previous tests indicate that a time of approximately 5–30 minutes is necessary to reach steady state. To obtain the data for the next successive power level, the power is incremented every 5–30 minutes. During the tests, the power input and the temperature data are simultaneously recorded using a data-acquisition system controlled by a personal computer.

5 OTHER TYPES OF HEAT PIPES

5.1 Thermosyphon

One of the simplest heat pipes is a thermosyphon. As shown in Fig. 11, a thermosyphon is a vertically oriented, wickless heat pipe with a liquid pool at the bottom. A typical thermosyphon consists of three sections, similar to the convention heat pipe. As heat is added on the evaporator section where a liquid pool exists, the liquid vaporizes into vapor. The vapor rises and passes through the adiabatic section to the condenser section, where the vapor gives up its latent heat and condenses into liquid. The condensate is then pumped from the condenser to the evaporator by the gravitational force. In the conventional thermosyphon, the evaporator must be located below the condenser for satisfactory operation because the device has to rely on gravity for condensate return. Therefore, thermosyphons are ineffective in zero gravity or microgravity. The thermosyphon has been widely used in the preservation of permafrost, electronics cooling, and heat-pipe exchangers due to its highly efficient heat-transfer performance, high level of temperature distribution, simplicity/reliability, and cost-effectiveness. Heat-transfer limitations, such as the dryout limit, flooding and entrainment limit, and boiling limit, should be considered during the design of thermosyphons. The dryout limitation might occur when the amount of working fluid in the evaporator is not sufficient and the heat added on the evaproator cannot be removed which would result

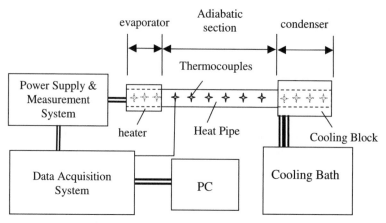

Figure 10 A heat-pipe test apparatus.

in an significant temeprature increases. This is similar to the critical heat flux in pool boiling. Flooding and entrainment limits involve the interfacial shear stress between the countercurrent vapor and liquid flows. Boiling limitation can be found in a thermosyphon with large fill volumes and high radial heat flux in the evaporator.

5.2 Loop Heat Pipes/Capillary Pumped Loop

The loop heat pipe (LHP) and capillary pumped loop (CPL) utilize the capillary pressure developed in a fine-pore wick to circulate the working fluid in a closed loop system. Figure 12 shows the schematic of the CPL. The liquid phase flows through the liquid line from the condenser to the evaporator and the vapor flows from the evaporator to the condenser through the vapor line. The LHP or CPL can significantly reduce or eliminate the liquid pressure drop and vapor flow effect on the liquid flow, resulting in a significant increase in the capillary pumping ability. Because the evaporator in a LHP or CPL is used not only as the heat sink to remove the heat, but also as the source to provide the total capillary pumping pressure, a highly efficient evaporator is a key to a LHP or CPL. Compared with the con-

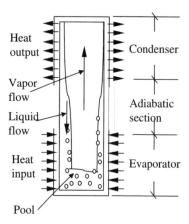

Figure 11 Schematic of a thermosyphon.

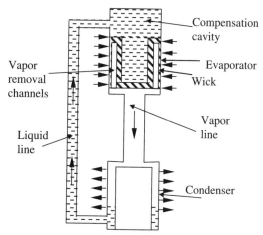

Figure 12 Schematic of a capillary pumped loop/loop heat pipe.

ventional heat pipe, the LHP or CPL systems have the potential to transport large amounts of heat over long distances at various orientations with minimal temperature drops and no external pumping power. Due to this unique feature, the LHP or CPL is especially suitable to the space station program, advanced communication satellite, high-powered spacecraft, and electronics cooling, which require large heat dissipation. It is anticipated that the CPL or LHP will play an important role in the thermal management of space and terrestrial systems in the future.

5.3 Pulsating Heat Pipes

Over the last several years since Akachi[15] invented the pulsating heat pipe (PHP) in 1990, extensive investigations have been conducted. As shown in Fig. 13, a typical PHP consists of an evaporating section, adiabatic section, and a condensing section. As heat is added on the evaporating section, vaporization in the evaporating section causes vapor volume expansion, and heat removal in the condensing section causes vapor volume contraction. The expansion and contraction of vapor volume plus the thermal energy added on the system

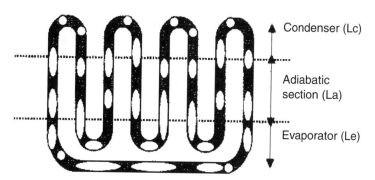

Figure 13 Schematic of a pulsating heat pipe.

generates the pulsating motion of liquid plugs and vapor bubbles in the system. In addition to the pulsating motion observed in the system, the oscillating motions of liquid plugs and vapor bubbles exist at the same time. For this reason, the PHP is sometimes called an oscillating heat pipe (OHP). The oscillating motions in the PHP depend on the dimensions, working fluids, operating temperature, surface conditions, heat flux and total heat load, orientation, turns, and, most importantly, the filled ratio, V_l/V_t, where V_l is the liquid volume occupied by the liquid in the system, and V_t is the total volume.[16,17] Utilizing phase-change heat transfer and forced convection, the heat is transferred from the evaporating section to the condensing section. Compared with a conventional heat pipe, the successful pulsating heat pipe (PHP) has the following features: (1) there is a low-pressure drop in the working fluid because most or all of working fluid does not flow through the wick structure; (2) the PHP is very simple and the manufacturing cost be very low because no wick structures are needed in most or all sections of the PHP; (3) the liquid pressure drop caused by the frictional vapor flow can be significantly reduced because the vapor flow direction is the same as liquid flow; (4) the thermally driven, pulsating flow inside the capillary tube will effectively produce some "blank" surfaces that produce thin-film regions and significantly enhance evaporating and condensing heat transfer; and (5) the heat added on the evaporating area can be distributed by the forced convection in addition to the phase-change heat transfer due to the oscillating motion in the capillary tube. Clearly, the PHP creates a potential to remove an extra-high level of heat flux. On the other hand, the diameter of the pulsating heat pipe must be small enough so that vapor plugs can be formed by the capillary action.

5.4 Micro Heat Pipes

In 1984, Cotter[18] first introduced the concept of very small "micro" heat pipes, which was incorporated into semiconductor devices to promote more uniform temperature distributions and improve thermal control. The micro heat pipe was defined as a heat pipe in which the mean curvature of the liquid–vapor interface is comparable in magnitude to the reciprocal of the hydraulic radius of the total flow channel. Based on this definition, the hydraulic diameter of a typical micro heat pipe ranges from 10 to 500 μm. The fundamental operating principle of micro heat pipes is essentially the same as those occurring in relatively large conventional heat pipes. A typical micro heat pipe shown in Fig. 14 is using the cornered region to pump the condensate from the condenser to the evaporator. As heat is added on the evaporating section, the liquid vaporizes and the vapor brings the heat through the adiabatic section to the condensing section, where the vapor condenses into the liquid and releases the latent heat. The heat addition on the evaporating section causes the liquid to recede into the cornered region and directly reduces the meniscus radius at the liquid–vapor interface in the evaporator. This vaporization and condensation process causes the liquid–vapor interface in the liquid arteries to change continually along the pipe and results in a capillary pressure difference between the evaporator and condenser regions. This capillary pressure difference promotes the flow of the working fluid from the condenser back to the evaporator. As the size of the heat pipe decreases, however, the micro heat pipe may encounter the vapor continuum limitation. This limitation may prevent the micro heat pipe from working under lower temperature. In addition to the vapor continuum limitation, the micro heat pipe is also subject to the operating limits occurring in the conventional heat pipe. Of those operating limits, the capillary limitation remains the most important for the micro heat pipe. Micro heat pipes have been widely used in the electronics cooling.

5.5 Variable-Conductance Heat Pipes

For a typical conventional heat pipe, the operating temperature can be determined by the heat-removal rate from the condenser. When the heat load increases, the temperature drop

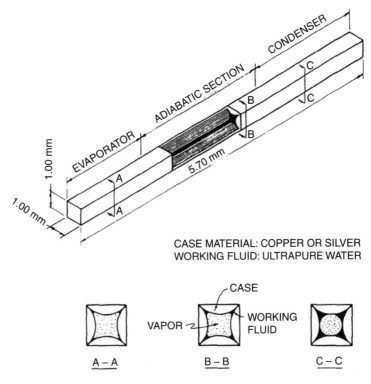

CASE MATERIAL: COPPER OR SILVER
WORKING FLUID: ULTRAPURE WATER

Figure 14 Schematic of a micro heat pipe.

from the evaporator to the condenser increases if the condensing area is constant. However, there are some applications where the evaporator or condenser temperature needs to be kept constant with a varying heat input. The variable-conductance heat pipe is suitable for this function design and its unique feature is the ability to maintain a device mounted at the evaporator at a near constant temperature, independent of the amount of power being generated by the device. To keep the operating temperature independent of the heat input, the whole conductance of heat pipe should be varied with the heat load. Figure 15 illustrates a typical variable conductance heat pipe (VCHP). Comparing the conventional heat pipe, this VCHP as shown in Fig. 15 includes a gas reservoir containing noncondensable gas. When

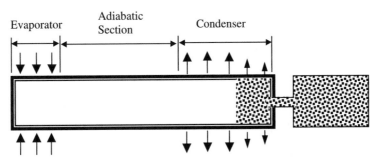

Figure 15 Schematic of a variable-conductance heat pipe.

the input power is low, the vapor pressure inside the heat pipe is low. The volume of noncondensable gas expands and reduces the condensing area. When the input power is high, the vapor pressure increases, resulting in the contraction of vapor volume and directly increasing the condensing area. As a result, the temperature drop from the evaporator to the condenser can be maintained fairly constant. As shown in Fig. 15, the noncondensable gas is used to moderate the conductance as the input power is varied, giving the device the name: gas-loaded variable heat pipe. Based on the same principle, several other VCHPs have been developed to moderate the conductance change in the heat pipe, such as vapor flow-modulated heat pipes, excess-liquid heat pipes and liquid-flow-modulated heat pipes (thermal diodes).[6]

5.6 Rotating Heat Pipes

The rotating heat pipe consists of a sealed hollow shaft, which contains a fixed amount of working fluid. The rotating heat pipe can be divided into two types: those with an internal taper and those without an internal taper. Comparing with the conventional heat pipe, the rotating heat pipe has the following features: (1) the condensate in the rotating heat pipe is returned to the evaporator by the centrifugal force; (2) the rotational speed plays the most important role for its heat-transfer performance; (3) the heat-transfer performance is enhanced for a rotating heat pipe with an internal taper because the removal of the condensate from the cooled liquid surface by the centrifugal action helps the condensate back toward the evaporator; and (4) the sonic, entrainment, boiling, and condensing limits are the primary factors limiting the heat-transfer capacity in the rotating heat pipe. For the detailed information related to the rotating heat pipes, please see the books written by Peterson[6] and Faghri.[9]

5.7 High-Temperature Heat Pipes (Metal Heat Pipes)

Because most working fluids in high-temperature heat pipes are metal, the high-temperature heat pipe is also called the metal heat pipe. Due to the higher surface tension, higher latent heat of vaporization, and higher thermal conductivity, the high-temperature heat pipes can transport large heat load and reach a very high level of temperature uniformity, and have been employed for the advanced energy system such as advanced thermophotovoltaic, advanced gas turbine engines, and nuclear reactors. Table 4 lists some common used working fluids and their temperature ranges. However, comparing with those heat pipes operating at other temperature ranges, within the high-temperature heat pipe exists some obstructions such as corrosion/reliability concern, high chemical reactivity, start-up control, and severe work conditions due to too high an operating temperature.

5.8 Cryogenic Heat Pipes

Cryogenic fluids are used in the cryogenic heat pipe. These working fluids are either a chemically pure material such as helium, argon, krypton, nitrogen, or oxygen or a chemical compound such as methane, ethane, or Freon. The operating temperature ranges for these fluids are listed in Table 4. For cryogenic fluids at a low temperature, the surface tension, thermal conductivity, and latent heat of vaporization are relatively low and the liquid viscosity is much higher. As a result, the heat pipe optimized for zero-g operation would not properly prime in a one-g environment due to the low surface tension. The very high vapor pressure in the heat pipe during storage and the low operating temperature, where the cooling methods to remove heat from the condenser are limited, are also of concern for cryogenic heat pipes. The capillary limit, sonic limit, entrainment limit, or boiling limit governs the heat transport limitation. Of those operating limits, the capillary limit from the low surface tension is the primary factor affecting the heat-transfer performance in the cryogenic heat pipes. For this reason, considerable effort has been conducted to develop wicking structures that further increase the capillary limitation.

NOMENCLATURE

A	cross-sectional area, m^2
C	constant
c	specific heat, J/kg-K
d	diameter, m
f	friction factor
F	correction factor
g	gravitational acceleration, m/s^2
h_{fg}	latent heat, J/kg
k	thermal conductivity, W/m-K
K	curvature, m^{-1}; permeability, m^{-2}
L	length, m
Ma	Mach number
$\dot{m}$	mass flow rate, kg/s
N	mesh number
r	radius, m
R	universal constant, J/kg-K
p	pressure, N/m^2
P	perimeter, m
q	heat transfer, W
q''	heat flux, W/m^2
Re	Reynolds number
s	coordinate, m
S	crimping factor
T	temperature, °C
u	vertical velocity, m/s
$\bar{u}$	average velocity, m/s
We	Weber number
x	coordinate, m
y	coordinate, m
z	coordinate, m

Greek Symbols

α	contact angle, degree
δ	film thickness, m
ε	porosity
γ	ratio of specific heats
ψ	tilt angle, degree
μ	dynamic viscosity, $N\text{-}s/m^2$
ρ	mass density, kg/m^3
σ	surface tension, N/m
τ	shear stress, N/m^2

Subscripts

$+$	normal direction
a	adiabatic section
c	capillary, condensation, condenser
d	disjoining
e	evaporation, evaporator
eff	effective
ent	entrainment
film	film
g	gravity
h	hydraulic
l	liquid
lv	liquid-vapor
m	meniscus
max	maximum
r	radial
s	sintered particles, saturation, solid, sonic
shell	shell
sl	solid-liquid
sv	solid-vapor
total	total
v	vapor
vis	viscous
w	wire, wick

REFERENCES

1. L. P. Perkins and W. E. Buck, "Improvement in Devices for the Diffusion or Transference of Heat," UK Patent 22,272, London, England, 1892.

2. C. R. King, "Perkins Hermetic Tube Boiler," *The Engineer,* **152,** 405–406 (1931).

3. H. B. Ma and G. P. Peterson, "The Minimum Meniscus Radius and Capillary Heat Transport Limit in Micro Heat Pipes," *ASME Journal of Heat Transfer,* **120**(1), 227–233 (1998).

4. H. B. Ma, G. P. Peterson, and X. J. Lu, "The Influence of Vapor–Liquid Interactions on the Liquid Pressure Drop in Triangular Microgrooves," *International Journal of Heat and Mass Transfer,* **37**(15), 2211–2219 (1994).

5. B. D. Marcus, "Theory and Design of Variable Conductance Heat Pipes," Report No. NASA CR, 2018, NASA, Washington, DC, April 1972.

6. G. P. Peterson, *An Introduction to Heat Pipes,* Wiley, New York, 1994.

7. S. W. Chi, *Heat Pipe Theory and Practice,* McGraw-Hill, New York, 1976.

8. P. D. Dunn and D. A. Reay, *Heat Pipes,* Pergamon, New York, 1995.

9. A. Faghri, *Heat Pipe Science and Technology,* Taylor & Francis, New York, 1995.

10. Y. Y. Hsu, "On the Size Range of Active Nucleation Cavities on a Heating Surface," *ASME Journal of Heat Transfer,* **84,** 207–213 (1962).

11. V. P. Carey, *Liquid–Vapor Phase-Change Phenomena,* Taylor & Francis, New York, 1992.

12. T. P. Cotter, "Heat Pipe Startup Dynamics," in *Proc. SAE Thermionic Conversion Specialist Conference,* Palo Alto, CA, 1967.

13. H. B. Ma and G. P. Peterson, "Temperature Variation and Heat Transfer in Triangular Grooves with an Evaporating Film," *AIAA Journal of Thermophysics and Heat Transfer,* **11**(1), 90–97 (1997).

14. M. A. Hanlon and H. B. Ma, "Evaporation Heat Transfer in Sintered Porous Media," *ASME Journal of Heat Transfer,* **125,** 644–653 (2003).

15. H. Akachi, "Structure of a Heat Pipe," U.S. Patent #4,921,041 (1990).

16. H. B. Ma, M. R. Maschmann, and S. B. Liang, "Heat Transport Capability in Pulsating Heat Pipes," *Proceedings of the 8th AIAA/ASME Joint Thermophysica and Heat Transfer Conference,* 24–27 June 2002, St. Louis, MO.

17. H. B. Ma, M. A. Hanlon, and C. L. Chen, "Investigation of the Oscillating Motion in a Pulsating Heat Pipe," *Proceedings of NHTC01, 35th National Heat Transfer Conference,* NHTC2001, 20149, 10–12 June 2001, Anaheim, CA.

18. T. P. Cotter, "Principles and Prospects of Micro Heat Pipes," *Proc. 5th International Heat Pipe Conference,* Tsukuba, Japan, 1984, pp. 328–335.

CHAPTER 10

AIR HEATING

Richard J. Reed
North American Manufacturing Company
Cleveland, Ohio

1 AIR-HEATING PROCESSES

Air can be heated by burning fuel or by recovering waste heat from another process. In either case, the heat can be transferred to air directly or indirectly. *Indirect air heaters* are heat exchangers wherein the products of combustion never contact or mix with the air to be heated. In waste heat recovery, the heat exchanger is termed a *recuperator.*

Direct air heaters or *direct-fired air heaters* heat the air by intentionally mixing the products or combustion of waste gas with the air to be heated. They are most commonly used for ovens and dryers. It may be impractical to use them for space heating or for preheating combustion air because of lack of oxygen in the resulting mixture ("vitiated air"). In some cases, direct-fired air heating may be limited by codes and/or by presence of harmful matter of undesirable odors from the heating stream. Direct-fired air heaters have lower first cost and lower operating (fuel) cost than indirect air heaters.

Heat requirements for direct-fired air heating. Table 1 lists the gross Btu of fuel input required to heat one standard cubic foot of air from a given inlet temperature to a given outlet temperature. It is based on natural gas at 60°F, having 1000 gross Btu/ft³, 910 net Btu/ft³, and stoichiometric air/gas ratio of 9.4:1. The oxygen for combustion is supplied by the air that is being heated. The hot outlet "air" includes combustion products obtained from burning sufficient natural gas to raise the air to the indicated outlet temperature.

Recovered waste heat from another nearby heating process can be used for process heating, space heating, or for preheating combustion air (Ref. 4). If the waste stream is largely nitrogen, and if the temperatures of both streams are between 0 and 800°F, where specific heats are about 0.24, a simplified heat balance can be used to evaluate the mixing conditions:

Heat content of the waste stream + Heat content of the fresh air = Heat content of the mixture or

$$W_w T_w + W_f T_f = W_m T_m = (W_w + W_f) T_m$$

where W = weight and T = temperature of waste gas, fresh air, and mixture (subscripts w, f, and m).

Example 1
If a 600°F waste gas stream flowing at 100 lb/hr is available to mix with 10°F fresh air and fuel, how many pounds per hour of 110°F makeup air can be produced?

362

Table 1 Heat Requirements for Direct-Fired Air Heating, Gross Btu of Fuel Input per scf of Outlet "Air"

Inlet Air Temperature, °F	Outlet Air Temperature, °F														
	100	200	300	400	500	600	700	800	900	1000	1100	1200	1300	1400	1500
−20	2.39	4.43	6.51	8.63	10.8	13.0	15.2	17.5	19.9	22.2	24.7	27.1	29.7	32.2	34.9
0	2.00	4.04	6.11	8.23	10.4	12.6	14.8	17.1	19.5	21.8	24.3	26.7	29.3	31.8	34.4
+20	1.60	3.64	5.71	7.83	9.99	12.2	14.4	16.7	19.0	21.4	23.8	26.3	28.8	31.4	34.0
40	1.20	3.24	5.31	7.43	9.58	11.8	14.0	16.3	18.6	21.0	23.4	25.9	28.4	31.0	33.6
60	0.802	2.84	4.91	7.02	9.18	11.4	13.6	15.9	18.2	20.6	23.0	25.5	28.0	30.6	33.2
80	0.402	2.43	4.51	6.62	8.77	11.0	13.2	15.5	17.8	20.2	22.6	25.1	27.6	30.1	32.7
100		2.03	4.10	6.21	8.36	10.6	12.8	15.1	17.4	19.8	22.2	24.6	27.2	29.7	32.3
200			2.06	4.17	6.31	8.50	10.7	13.0	15.3	17.7	20.1	22.5	25.0	27.6	30.2
300				2.10	4.23	6.41	8.63	10.9	13.2	15.5	17.9	20.4	22.9	25.4	28.0
400					2.13	4.30	6.51	8.76	11.1	13.4	15.8	18.2	20.7	23.2	25.8
500						2.16	4.36	6.61	8.90	11.2	13.6	16.0	18.5	21.0	23.6
600							2.19	4.43	6.71	9.03	11.4	13.8	16.3	18.8	21.3
700								2.23	4.50	6.81	9.16	11.6	14.0	16.5	19.0
800									2.26	4.56	6.91	9.30	11.7	14.2	16.7
900										2.29	4.63	7.01	9.43	11.9	14.4
1000											2.32	4.69	7.11	9.57	12.1

Example: Find the amount of natural gas required to heat 1000 scfm of air from 400°F to 1400°F.

Solution: From the table, read 23.2 gross Btu/scf air. Then $\left(\dfrac{23.2 \text{ gross Btu}}{\text{scf air}} \times \dfrac{1000 \text{ scf air}}{\text{min}} \times \dfrac{60 \text{ min}}{1 \text{ hr}} \right) \div \dfrac{1000 \text{ gross Btu}}{\text{ft}^3 \text{ gas}} = 1392 \text{ cfh gas.}$

The conventional formula derived from the specific heat equation is: $Q = wc\Delta T$; so Btu/hr = weight/hr × specific heat × temp rise = $\dfrac{\text{scf}}{\text{min}} \times \dfrac{60 \text{ min}}{\text{hr}} \times \dfrac{0.076 \text{ lb}}{\text{ft}^3} \times \dfrac{0.24 \text{ Btu}}{\text{lb }^\circ\text{F}} \times {}^\circ\text{rise} =$ scfm × 1.1 × °rise.

The table above incorporates many refinements not considered in the conventional formulas: (a) % available heat which corrects for heat loss to dry flue gases and the heat loss due to heat of vaporization in the water formed by combustion, (b) the specific heats of the products of combustion (N_2, CO_2, and H_2O) are not the same as that of air, and (c) the specific heats of the combustion products change at higher temperatures.

For the example above, the rule of thumb would give 1000 scfm × 1.1 × (1400 − 400) = 1 100 000 gross Btu/hr: whereas the example finds 1392 × 1000 = 1,392,000 gross Btu/hr required. *Reminder:* The fuel being burned adds volume and weight to the stream being heated.

Solution:

$$(100 \times 600) + 10W_f = (100 + W_f) \times (110)$$

Solving, we find $W_f = 490$ lb/hr of fresh air can be heated to 110°F, but the 100 lb/hr of waste gas will be mixed with it; so the delivered stream, W_m will be $100 + 490 = 590$ lb/hr.

If "indirect" air heating is necessary, a heat exchanger (recuperator or regenerator) must be used. These may take many forms such as plate-type heat exchangers, shell and tube heat exchangers, double-pipe heat exchangers, heat-pipe exchangers, heat wheels, pebble heater recuperators, and refractory checkerworks. The supplier of the heat exchanger should be able to predict the air preheat temperature and the final waste gas temperature. The amount of heat recovered Q is then $Q = W c_p (T_2 - T_1)$, where W is the weight of air heated, c_p is the specific heat of air (0.24 when below 800°F), T_2 is the delivered hot air temperature, and T_1 is the cold air temperature entering the heat exchanger. Tables and graphs later in this chapter permit estimation of fuel savings and efficiencies for cases involving preheating of combustion air.

If a waste gas stream is only a few hundred degrees Fahrenheit hotter than the air stream temperature required for heating space, an oven, or a dryer, such uses of recovered heat are highly desirable. For higher waste gas stream temperatures, however, the second law of thermodynamics would say that we can make better use of the energy by stepping it down in smaller temperature increments, and preheating combustion air usually makes more sense. This also simplifies accounting, since it returns the recovered heat to the process that generated the hot waste stream.

Preheating combustion air is a very logical method for recycling waste energy from flue gases in direct-fired industrial heating processes such as melting, forming, ceramic firing, heat treating, chemical and petroprocess heaters, and boilers. (It is always wise, however, to check the economics of using flue gases to preheat the load or to make steam in a waste heat boiler.)

2   COSTS

In addition to the cost of the heat exchanger for preheating the combustion air, there are many other costs that have to be weighed. Retrofit or add-on recuperators or regenerators may have to be installed overhead to keep the length of heat-losing duct and pipe to a minimum; therefore, extra foundations and structural work may be needed. If the waste gas or air is hotter than about 800°F, carbon steel pipe and duct should be insulated on the inside. For small pipes or ducts where this would be impractical, it is necessary to use an alloy with strength and oxidation resistance at the higher temperature, and to insulate on the outside.

High-temperature air is much less dense; therefore, the flow passages of burners, valves, and pipe must be greater for the same input rate and pressure drop. Burners, valves, and piping must be constructed of better materials to withstand the hot air stream. The front face of the burner is exposed to more intense radiation because of the higher flame temperature resulting from preheated combustion air.

If the system is to be operated at a variety of firing rates, the output air temperature will vary; so temperature-compensating fuel/air ratio controls are essential to avoid wasting fuel. Also, to protect the investment in the heat exchanger, it is only logical that it be protected with high-limit temperature controls.

3 WARNINGS

Changing temperatures from end to end of high-temperature heat exchangers and from time to time during high-temperature furnace cycles cause great thermal stress, often resulting in leaks and shortened heat-exchanger life. Heat-transfer surfaces fixed at both ends (welded or rolled in) can force something to be overstressed. Recent developments in the form of high-temperature slip seal methods, combined with sensible location of such seals in cool air entrance sections, are opening a whole new era in recuperator reliability.

Corrosion, fouling, and condensation problems continue to limit the applications of heat-recovery equipment of all kinds. Heat-transfer surfaces in air heaters are never as well cooled as those in water heaters and waste heat boilers; therefore, they must exist in a more hostile environment. However, they may experience fewer problems from acid-dew-point condensation. If corrosives, particulates, or condensables are emitted by the heating process at limited times, perhaps some temporary bypassing arrangement can be instituted. High waste gas design velocities may be used to keep particulates and condensed droplets in suspension until they reach an area where they can be safely dropped out.

Figure 1 shows recommended minimum temperatures to avoid "acid rain" in the heat exchanger.[2] Although a low final waste gas temperature is desirable from an efficiency standpoint, the shortened equipment life seldom warrants it. Acid forms from combination of water vapor with SO_3, SO_2, or CO_2 in the flue gases.

4 BENEFITS

Despite all the costs and warnings listed above, combustion air preheating systems *do* pay. As fuel costs rise, the payback is more rewarding, even for small installations. Figure 2 shows percent available heat[3] (best possible efficiency) with various amounts of air preheat and a variety of furnace exit (flue) temperatures. All curves for hot air are based on 10%

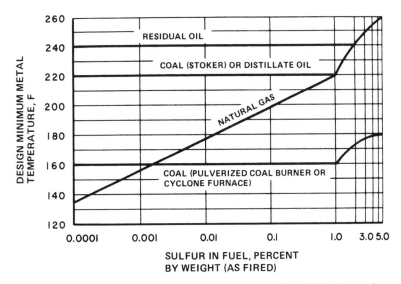

Figure 1 Recommended minimum temperatures to avoid "acid rain" in heat exchangers.

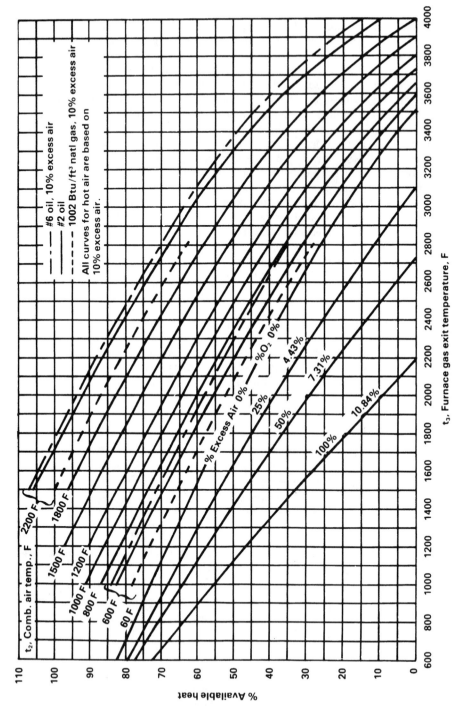

Figure 2 Available heat with preheated combustion air at 10% excess air. Applicable only if there is no unburned fuel in the products of combustion. Corrected for dissociation. (Reproduced with permission from *Combustion Handbook.*[3]) See also Figs. 3 and 4 in Chapter 16.

Table 2 Fuel Savings (%) Resulting from Use of Preheated Air with Natural Gas and 10% Excess Air[a]

t_3, Furnace Gas Exit Temperature (°F)	t_2, Combustion Air Temperature (°F)													
	600	700	800	900	1000	1100	1200	1300	1400	1500	1600	1800	2000	2200
1000	13.4	15.5	17.6	19.6	—	—	—	—	—	—	—	—	—	—
1100	13.8	16.0	18.2	20.2	22.2	—	—	—	—	—	—	—	—	—
1200	14.3	16.6	18.7	20.9	22.9	24.8	—	—	—	—	—	—	—	—
1300	14.8	17.1	19.4	21.5	23.6	25.6	27.5	—	—	—	—	—	—	—
1400	15.3	17.8	20.1	22.3	24.4	26.4	28.4	30.2	—	—	—	—	—	—
1500	16.0	18.5	20.8	23.1	25.3	27.3	29.3	31.2	33.0	—	—	—	—	—
1600	16.6	19.2	21.6	24.0	26.2	28.3	30.3	32.2	34.1	35.8	—	—	—	—
1700	17.4	20.2	22.5	24.9	27.2	29.4	31.4	33.4	35.3	37.0	38.7	—	—	—
1800	18.2	20.9	23.5	26.0	28.3	30.6	32.7	34.6	36.5	38.3	40.1	—	—	—
1900	19.1	21.9	24.6	27.1	29.6	31.8	34.0	36.0	37.9	39.7	41.5	44.7	—	—
2000	20.1	23.0	25.8	28.4	30.9	33.2	35.4	37.5	39.4	41.3	43.0	46.3	51.0	—
2100	21.2	24.3	27.2	29.9	32.4	34.8	37.0	39.1	41.1	43.0	44.7	48.0	52.8	—
2200	22.5	25.7	28.7	31.5	34.1	36.5	38.8	40.9	42.9	44.8	46.6	49.9	54.9	57.5
2300	24.0	27.3	30.4	33.3	36.0	38.5	40.8	42.9	45.0	46.9	48.7	52.0	57.1	59.7
2400	25.7	29.2	32.4	35.3	38.1	40.6	43.0	45.2	47.2	49.2	51.0	54.2	59.6	62.2
2500	27.7	31.3	34.7	37.7	40.5	43.1	45.5	47.7	49.8	51.7	53.5	56.8	62.4	64.9
2600	30.1	33.9	37.3	40.5	43.4	46.0	48.4	50.6	52.7	54.6	56.4	59.6	65.5	67.9
2700	33.0	37.0	40.6	43.8	46.7	49.4	51.8	54.0	56.1	58.0	59.7	62.8	69.1	71.3
2800	36.7	40.8	44.5	47.8	50.8	53.4	55.8	58.0	60.0	61.9	63.5	66.5	73.2	75.2
2900	41.4	45.7	49.5	52.8	55.7	58.4	60.7	62.8	64.7	66.4	68.0	70.8	78.0	79.8
3000	47.9	52.3	56.0	59.3	62.1	64.6	66.7	68.7	70.4	72.0	73.5	75.9	83.8	85.2
3100	57.3	61.5	65.0	68.0	70.5	72.7	74.6	76.2	77.7	79.0	80.2	82.2	90.9	91.8
3200	72.2	75.6	78.3	80.4	82.2	83.7	85.0	86.1	87.1	87.9	88.7	89.9		

[a]These figures are for evaluating a proposed change to preheated air—not for determining system capacity.

Source: Reproduced with permission from *Combustion Handbook*, Vol. I, North American Manufacturing Co.

Table 3 Effect of Combustion Air Preheat on Flame Temperature

Excess Air (%)	Preheated Combustion Air Temperature (°F)	Adiabatic Flame Temperatured (°F)		
		With 1000 Btu/scf Natural Gas	With 137,010 Btu/gal Distillate Fuel Oil	With 153,120 Btu/gal Residual Fuel Oil
0	60	3468	3532	3627
10	60	3314	3374	3475
10	600	3542	3604	3690
10	700	3581	3643	3727
10	800	3619	3681	3763
10	900	3656	3718	3798
10	1000	3692	3754	3831
10	1100	3727	3789	3864
10	1200	3761	3823	3896
10	1300	3794	3855	3927
10	1400	3826	3887	3957
10	1500	3857	3918	3986
10	1600	3887	3948	4014
10	1700	3917	3978	4042
10	1800	3945	4006	4069
10	1900	3973	4034	4095
10	2000	4000	4060	4121
0	2000	4051	4112	4171

excess air.* The percentage of fuel saved by addition of combustion air preheating equipment can be calculated by the formula

$$\% \text{ fuel saved} = 100 \times \left(1 - \frac{\% \text{ available heat before}}{\% \text{ available heat after}}\right)$$

Table 2 lists fuel savings calculated by this method.[4]

Preheating combustion air raises the flame temperature and thereby enhances radiation heat transfer in the furnace, which should lower the exit gas temperature and further improve fuel efficiency. Table 3 and the x intercepts of Fig. 2 show adiabatic flame temperatures when operating with 10% excess air,† but it is difficult to quantify the resultant saving from this effect.

---

*It is advisable to tune a combustion system for closer to stoichiometric air/fuel ratio *before* attempting to preheat combustion air. This is not only a quicker and less costly fuel conservation measure, but it then allows use of smaller heat-exchange equipment.

†Although 0% excess air (stoichiometric air/fuel ratio) is ideal, practical considerations usually dictate operation with 5–10% excess air. During changes in firing rate, time lag in valve operation may result in smoke formation if some excess air is not available prior to the change. Heat exchangers made of 300 series stainless steels may be damaged by alternate oxidation and reduction (particularly in the presence of sulfur). For these reasons, it is wise to have an accurate air/fuel ratio controller with very limited time-delay deviation from air/fuel ratio setpoint.

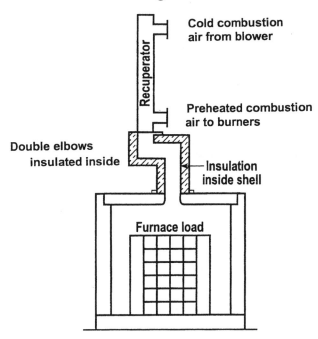

Cooled flue gas exit

Cold combustion air from blower

Recuperator

Preheated combustion air to burners

Double elbows insulated inside

Insulation inside shell

Furnace load

Figure 3 A recuperator using heat from waste flue gas to preheat combustion air to be fed to the burners. The double elbow (insulated inside) in the flue uptake to the recuperator (1) prevents the recuperator from causing a cool spot on top of the furnace load, and (2) prevents the hot furnace load and interior walls from possibly radiating damaging overheat into the recuperator. (Reproduced with permission from *Industrial Furnaces,* 6th ed. by Trinks, Mawhinney, Shannon, Reed, and Garvey, 2004.)

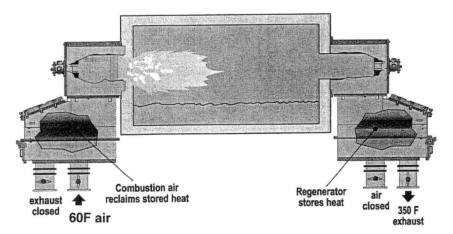

Combustion air reclaims stored heat

exhaust closed

60F air

Regenerator stores heat

air closed

350 F exhaust

Figure 4 A melting furnace with a pair of compact regenerative burners. The regenerator on the right is storing waste heat in its bed of refractory nuggets. After 20 seconds firing, as shown, the systems switch from firing from the left and exhausting through the right to firing from the right and exhausting through the left. At the moment shown, the regenerator on the right is storing waste heat, and the burner on the left is receiving reclaimed stored heat in the form of preheated combustion air. (Reproduced with permission from *Industrial Furnaces,* 6th ed. by Trinks, Mawhinney, Shannon, Reed, and Garvey, 2004.)

Preheating combustion air has some lesser benefits. Flame stability is enhanced by the faster flame velocity and broader flammability limits. If downstream pollution control equipment is required (scrubber, baghouse), such equipment can be smaller and of less costly materials because the heat exchanger will have cooled the waste gas stream before it reaches such equipment.

Somewhat related to air heating (other than heating air for a subsequent process) is preheating combustion air for burners to make any heating process more efficient by recirculating waste heat through a heat exchange device (often built into a burner). This can be either a regenerator (steady-state heat exchanger, Fig. 3) or a recuperator (alternating flow heat exchanger, Fig. 4). Both use hot waste (flue) gases as the heat source to preheat combustion air being fed to the burners on almost any kind of furnace. Preheating air not only improves the thermal efficiency by recycling waste flue gas energy, but also raises the flame temperature (Table 3), thereby increasing the heat transfer rate to the furnace loads.

REFERENCES

1. *Heat Requirements for Direct-Fired Air Heating,* North American Mfg. Co., Cleveland, OH 44105, 1981.
2. *Steam—Its Generation and Use,* Babcock & Wilcox, New York, 1978.
3. R. J. Reed, *Combustion Handbook,* 3rd ed., Vol. 1, North American Manufacturing Co., Cleveland, OH 44105, 1986.
4. R. J. Reed, *Combustion Handbook,* 4th ed., Vol. 2, North American Manufacturing. Co., Cleveland, OH 44105, 1997.
5. W. Trinks, M. H. Mawhinney, R. A. Shannon, R. J. Reed, and J. R. Garvey, *Industrial Furnaces,* 6th ed., Wiley, Hoboken, NJ, 2003.

CHAPTER **11**

COOLING ELECTRONIC EQUIPMENT

Allan Kraus
Beachwood, Ohio

Avram Bar-Cohen
Department of Mechanical Engineering
University of Maryland
College Park, Maryland

Abhay A. Wative
Intel Corp
Chandler, Arizona

1 THERMAL MODELING

1.1 Introduction

To determine the temperature differences encountered in the flow of heat within electronic systems, it is necessary to recognize the relevant heat transfer mechanisms and their governing relations. In a typical system, heat removal from the active regions of the microcircuit(s) or chip(s) may require the use of several mechanisms, some operating in series and others in parallel, to transport the generated heat to the coolant or ultimate heat sink. Practitioners of the thermal arts and sciences generally deal with four basic thermal transport modes: conduction, convection, phase change, and radiation.

1.2 Conduction Heat Transfer

One-Dimensional Conduction
Steady thermal transport through solids is governed by the Fourier equation, which, in one-dimensional form, is expressible as

$$q = -kA \frac{dT}{dx} \quad \text{(W)} \tag{1}$$

where q is the heat flow, k is the thermal conductivity of the medium, A is the cross-sectional area for the heat flow, and dT/dx is the temperature gradient. Here, heat flow produced by a negative temperature gradient is considered positive. This convention requires the insertion of the minus sign in Eq. (1) to assure a positive heat flow, q. The temperature difference resulting from the steady-state diffusion of heat is thus related to the thermal conductivity of the material, the cross-sectional area and the path length, L, according to

$$(T_1 - T_2)_{cd} = q \frac{L}{kA} \quad \text{(K)} \tag{2}$$

The form of Eq. (2) suggests that, by analogy to Ohm's Law governing electrical current flow through a resistance, it is possible to define a thermal resistance for conduction, $R_{cd,}$ as

$$R_{cd} \equiv \frac{T_1 - T_2}{q} = \frac{L}{kA} \tag{3}$$

One-Dimensional Conduction with Internal Heat Generation

Situations in which a solid experiences internal heat generation, such as that produced by the flow of an electric current, give rise to more complex governing equations and require greater care in obtaining the appropriate temperature differences. The axial temperature variation in a slim, internally heated conductor whose edges (ends) are held at a temperature T_o is found to equal

$$T = T_o + q_g \frac{L^2}{2k} \left[\left(\frac{x}{L} \right) - \left(\frac{x}{L} \right)^2 \right]$$

When the volumetic heat generation rate, q_g, in W/m³ is uniform throughout, the peak temperature is developed at the center of the solid and is given by

$$T_{max} = T_o + q_g \frac{L^2}{8k} \quad \text{(K)} \tag{4}$$

Alternatively, because q_g is the volumetric heat generation, $q_g = q/LW\delta$, the center–edge temperature difference can be expressed as

$$T_{max} - T_o = q \frac{L^2}{8kLW\delta} = q \frac{L}{8kA} \tag{5}$$

where the cross-sectional area, A, is the product of the width, W, and the thickness, δ. An examination of Eq. (5) reveals that the thermal resistance of a conductor with a distributed heat input is only one quarter that of a structure in which all of the heat is generated at the center.

Spreading Resistance

In chip packages that provide for lateral spreading of the heat generated in the chip, the increasing cross-sectional area for heat flow at successive "layers" below the chip reduces the internal thermal resistance. Unfortunately, however, there is an additional resistance associated with this lateral flow of heat. This, of course, must be taken into account in the determination of the overall chip package temperature difference.

For the circular and square geometries common in microelectronic applications, an engineering approximation for the spreading resistance for a small heat source on a thick substrate or heat spreader (required to be 3 to 5 times thicker than the square root of the heat source area) can be expressed as[1]

$$R_{sp} = \frac{0.475 - 0.62\epsilon + 0.13\epsilon^2}{k\sqrt{A_c}} \quad (K/W) \qquad (6)$$

where ϵ is the ratio of the heat source area to the substrate area, k is the thermal conductivity of the substrate, and A_c is the area of the heat source.

For relatively thin layers on thicker substrates, such as encountered in the use of thin lead-frames, or heat spreaders interposed between the chip and substrate, Eq. (6) cannot provide an acceptable prediction of R_{sp}. Instead, use can be made of the numerical results plotted in Fig. 1 to obtain the requisite value of the spreading resistance.

Interface/Contact Resistance

Heat transfer across the interface between two solids is generally accompanied by a measurable temperature difference, which can be ascribed to a contact or interface thermal resistance. For perfectly adhering solids, geometrical differences in the crystal structure (lattice mismatch) can impede the flow of phonons and electrons across the interface, but this resistance is generally negligible in engineering design. However, when dealing with real interfaces, the asperities present on each of the surfaces, as shown in an artist's conception in Fig. 2, limit actual contact between the two solids to a very small fraction of the apparent interface area. The flow of heat across the gap between two solids in nominal contact is thus seen to involve solid conduction in the areas of actual contact and fluid conduction across the "open" spaces. Radiation across the gap can be important in a vacuum environment or when the surface temperatures are high.

The heat transferred across an interface can be found by adding the effects of the solid–to–solid conduction and the conduction through the fluid and recognizing that the solid–to–solid conduction, in the contact zones, involves heat flowing sequentially through the two solids. With the total contact conductance, h_{co}, taken as the sum of the solid–to–solid conductance, h_c, and the gap conductance, h_g

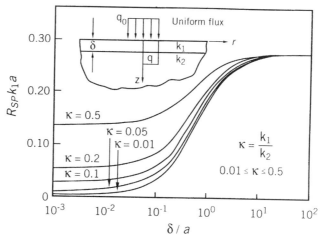

Figure 1 The thermal spreading resistance for a circular heat source on a two layer substrate (from Ref. 2).

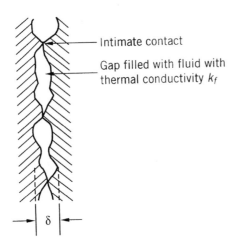

Intimate contact

Gap filled with fluid with
thermal conductivity k_f

δ

Figure 2 Physical contact between two nonideal surfaces.

$$h_{\text{co}} = h_c + h_g \qquad (\text{W}/\text{m}^2\cdot\text{K}) \tag{7a}$$

the contact resistance based on the apparent contact area, A_a, may be defined as

$$R_{\text{co}} \equiv \frac{1}{h_{\text{co}}A_a} \qquad (\text{K}/\text{W}) \tag{7b}$$

In Eq. (7a), h_c is given by

$$h_c = 54.25k_s \left(\frac{m}{\sigma}\right) \left(\frac{P}{H}\right)^{0.95} \tag{8a}$$

where k_s is the harmonic mean thermal conductivity for the two solids with thermal conductivities, k_1 and k_2,

$$k_s = \frac{2k_1 k_2}{k_1 + k_2} \qquad (\text{W}/\text{m}\cdot\text{K})$$

σ is the effective rms surface roughness developed from the surface roughnesses of the two materials, σ_1 and σ_2,

$$\sigma = \sqrt{\sigma_1^2 + \sigma_2^2} \qquad (\mu\cdot\text{m})$$

and m is the effective absolute surface slope composed of the individual slopes of the two materials, m_1 and m_2,

$$m = \sqrt{m_1^2 + m_2^2}$$

where P is the contact pressure and H is the microhardness of the softer material, both in N/m^2. In the absence of detailed information, the σ/m ratio can be taken equal to 5–9 μm for relatively smooth surfaces.[1,2]

In Eq. (7a), h_g is given by

$$h_g = \frac{k_g}{Y + M} \tag{8b}$$

where k_g is the thermal conductivity of the gap fluid, Y is the distance between the mean planes (Fig. 2) given by

$$\frac{Y}{\sigma} = 54.185 \left[-\ln\left(3.132\, \frac{P}{H} \right) \right]^{0.547}$$

and M is a gas parameter used to account for rarefied gas effects

$$M = \alpha\beta\Lambda$$

where α is an accommodation parameter (approximately equal to 2.4 for air and clean metals), Λ is the mean free path of the molecules (equal to approximately 0.06 μm for air at atmospheric pressure and 15°C), and β is a fluid property parameter (equal to approximately 54.7 for air and other diatomic gases).

Equations (8a) and (8b) can be added and, in accordance with Eq. (7b), the contact resistance becomes

$$R_{co} \equiv \left\{ \left[1.25 k_s \left(\frac{m}{\sigma} \right) \left(\frac{P}{H} \right)^{0.95} + \frac{k_g}{Y + M} \right] A_a \right\}^{-1} \tag{9}$$

1.3   Convective Heat Transfer

The Heat-Transfer Coefficient

Convective thermal transport from a surface to a fluid in motion can be related to the heat transfer coefficient, h, the surface–to–fluid temperature difference, and the "wetted" surface area, S, in the form

$$q = hS(T_s - T_{fl}) \qquad \text{(W)} \tag{10}$$

The differences between convection to a rapidly moving fluid, a slowly flowing or stagnant fluid, as well as variations in the convective heat-transfer rate among various fluids, are reflected in the values of h. For a particular geometry and flow regime, h may be found from available empirical correlations and/or theoretical relations. Use of Eq. (10) makes it possible to define the convective thermal resistance as

$$R_{cv} \equiv \frac{1}{hS} \qquad \text{(K/W)} \tag{11}$$

Dimensionless Parameters

Common dimensionless quantities that are used in the correlation of heat-transfer data are the *Nusselt number,* Nu, which relates the convective heat-transfer coefficient to the conduction in the fluid where the subscript, fl, pertains to a fluid property,

$$\text{Nu} \equiv \frac{h}{k_{fl}/L} = \frac{hL}{k_{fl}}$$

the *Prandtl number,* Pr, which is a fluid property parameter relating the diffusion of momentum to the conduction of heat,

$$\text{Pr} \equiv \frac{c_p \mu}{k}$$

the *Grashof number,* Gr, which accounts for the bouyancy effect produced by the volumetric expansion of the fluid,

$$\text{Gr} \equiv \frac{\rho^2 \beta g L^3 \Delta T}{\mu^2}$$

and the *Reynolds number,* Re, which relates the momentum in the flow to the viscous dissipation,

$$\text{Re} \equiv \frac{\rho V L}{\mu}$$

Natural Convection

In natural convection, fluid motion is induced by density differences resulting from temperature gradients in the fluid. The heat-transfer coefficient for this regime can be related to the buoyancy and the thermal properties of the fluid through the *Rayleigh number,* which is the product of the Grashof and Prandtl numbers,

$$\text{Ra} = \frac{\rho^2 \beta g c_p}{\mu k} L^3 \Delta T$$

where the fluid properties, ρ, β, c_p, μ, and k, are evaluated at the fluid bulk temperature and ΔT is the temperature difference between the surface and the fluid.

Empirical correlations for the natural convection heat-transfer coefficient generally take the form

$$h = C \left(\frac{k_\text{fl}}{L}\right) (\text{Ra})^n \qquad (\text{W/m}^2 \cdot \text{K}) \tag{12}$$

where n is found to be approximately 0.25 for $10^3 < \text{Ra} < 10^9$, representing laminar flow, 0.33 for $10^9 < \text{Ra} < 10^{12}$, the region associated with the transition to turbulent flow, and 0.4 for $\text{Ra} > 10^{12}$, when strong turbulent flow prevails. The precise value of the correlating coefficient, C, depends on the fluid, the geometry of the surface, and the Rayleigh number range. Nevertheless, for common plate, cylinder, and sphere configurations, it has been found to vary in the relatively narrow range of 0.45–0.65 for laminar flow and 0.11–0.15 for turbulent flow past the heated surface.[42]

Natural convection in vertical channels such as those formed by arrays of longitudinal fins is of major significance in the analysis and design of heat sinks and experiments for this configuration have been conducted and confirmed.[4,5]

These studies have revealed that the value of the Nusselt number lies between two extremes associated with the separation between the plates or the channel width. For wide spacing, the plates appear to have little influence upon one another and the Nusselt number in this case achieves its *isolated plate limit.* On the other hand, for closely spaced plates or for relatively long channels, the fluid velocity attains its *fully developed* value and the Nusselt number reaches its *fully developed limit.* Intermediate values of the Nusselt number can be obtained from a form of a correlating expression for smoothly varying processes and have been verified by detailed experimental and numerical studies.[19,20]

Thus, the correlation for the average value of h along isothermal vertical placed separated by a spacing, z

$$h = \frac{k_\text{fl}}{z} \left[\frac{576}{(\text{El})^2} + \frac{2.873}{(\text{El})^{1/2}} \right]^{1/2} \tag{13}$$

where El is the *Elenbaas number*

$$\text{El} \equiv \frac{\rho^2 \beta g c_p z^4 \Delta T}{\mu k_\text{fl} L}$$

and $\Delta T = T_s - T_\text{fl}$.

Several correlations for the coefficient of heat transfer in natural convection for various configurations are provided in Section 2.1.

Forced Convection

For forced flow in long, or very narrow, parallel-plate channels, the heat-transfer coefficient attains an asymptotic value (a fully developed limit), which for symmetrically heated channel surfaces is equal approximately to

$$h = \frac{4k_{fl}}{d_e} \qquad (\text{W/m}^2\cdot\text{K}) \tag{14}$$

where d_e is the *hydraulic diameter* defined in terms of the flow area, A, and the wetted perimeter of the channel, P_w

$$d_e \equiv \frac{4A}{P_w}$$

Several correlations for the coefficient of heat transfer in forced convection for various configurations are provided in Section 2.2.

Phase Change Heat Transfer

Boiling heat transfer displays a complex dependence on the temperature difference between the heated surface and the saturation temperature (boiling point) of the liquid. In nucleate boiling, the primary region of interest, the ebullient heat-transfer rate can be approximated by a relation of the form

$$q_\phi = C_{sf}A(T_s - T_{sat})^3 \qquad (\text{W}) \tag{15}$$

where C_{sf} is a function of the surface/fluid combination and various fluid properties. For comparison purposes, it is possible to define a boiling heat-transfer coefficient, h_ϕ,

$$h_\phi = C_{sf}(T_s - T_{sat})^2 \qquad [\text{W/m}^2\cdot\text{K}]$$

which, however, will vary strongly with surface temperature.

Finned Surfaces

A simplified discussion of finned surfaces is germane here and what now follows is not inconsistent with the subject matter contained Section 3.1. In the thermal design of electronic equipment, frequent use is made of finned or "extended" surfaces in the form of *heat sinks* or *coolers*. While such finning can substantially increase the surface area in contact with the coolant, resistance to heat flow in the fin reduces the average temperature of the exposed surface relative to the fin base. In the analysis of such finned surfaces, it is common to define a fin efficiency, η, equal to the ratio of the actual heat dissipated by the fin to the heat that would be dissipated if the fin possessed an infinite thermal conductivity. Using this approach, heat transferred from a fin or a fin structure can be expressed in the form

$$q_f = hS_f\eta(T_b - T_s) \qquad (\text{W}) \tag{16}$$

where T_b is the temperature at the base of the fin and where T_s is the surrounding temperature and q_f is the heat entering the base of the fin, which, in the steady state, is equal to the heat dissipated by the fin.

The thermal resistance of a finned surface is given by

$$R_f \equiv \frac{1}{hS_f\eta} \tag{17}$$

where η, the fin efficiency, is 0.627 for a thermally optimum rectangular cross-section fin,[11]

Flow Resistance

The transfer of heat to a flowing gas or liquid that is not undergoing a phase change results in an increase in the coolant temperature from an inlet temperature of T_{in} to an outlet temperature of T_{out}, according to

$$q = \dot{m}c_p(T_{out} - T_{in}) \quad \text{(W)} \tag{18}$$

Based on this relation, it is possible to define an effective flow resistance, R_{fl}, as

$$R_{fl} \equiv \frac{1}{\dot{m}c_p} \quad \text{(K/W)} \tag{19}$$

where $\dot{m}$ is in kg/s.

1.4 Radiative Heat Transfer

Unlike conduction and convection, radiative heat transfer between two surfaces or between a surface and its surroundings is not linearly dependent on the temperature difference and is expressed instead as

$$q = \sigma S \mathscr{F}(T_1^4 - T_2^4) \quad \text{(W)} \tag{20}$$

where $\mathscr{F}$ includes the effects of surface properties and geometry and σ is the Stefan–Boltzman constant, $\sigma = 5.67 \times 10^{-8}$ W/m^2·K^4. For modest temperature differences, this equation can be linearized to the form

$$q = h_r S(T_1 - T_2) \quad \text{(W)} \tag{21}$$

where h_r is the effective "radiation" heat-transfer coefficient

$$h_r = \sigma \mathscr{F}(T_1^2 + T_2^2)(T_1 + T_2) \quad \text{(W/m}^2\text{·K)} \tag{22a}$$

and, for small $\Delta T = T_1 - T_2$, h_r is approximately equal to

$$h_r = 4\sigma \mathscr{F}(T_1 T_2)^{3/2} \quad \text{(W/m}^2\text{·K)} \tag{22b}$$

It is of interest to note that for temperature differences of the order of 10 K, the radiative heat-transfer coefficient, h_r, for an ideal (or "black") surface in an absorbing environment is approximately equal to the heat-transfer coefficient in natural convection of air.

Noting the form of Eq. (21), the radiation thermal resistance, analogous to the convective resistance, is seen to equal

$$R_r \equiv \frac{1}{h_r S} \quad \text{(K/W)} \tag{23}$$

1.5 Chip Module Thermal Resistances

Thermal Resistance Network

The expression of the governing heat-transfer relations in the form of thermal resistances greatly simplifies the first-order thermal analysis of electronic systems. Following the established rules for resistance networks, thermal resistances that occur sequentially along a thermal path can be simply summed to establish the overall thermal resistance for that path. In similar fashion, the reciprocal of the effective overall resistance of several parallel heat-transfer paths can be found by summing the reciprocals of the individual resistances. In refining the thermal design of an electronic system, prime attention should be devoted to reducing the largest resistances along a specified thermal path and/or providing parallel paths for heat removal from a critical area.

While the thermal resistances associated with various paths and thermal transport mechanisms constitute the "building blocks" in performing a detailed thermal analysis, they have also found widespread application as "figures-of-merit" in evaluating and comparing the thermal efficacy of various packaging techniques and thermal management strategies.

Definition

The thermal performance of alternative chip and packaging techniques is commonly compared on the basis of the overall (junction-to-coolant) thermal resistance, R_T. This packaging figure-of-merit is generally defined in a purely empirical fashion,

$$R_T \equiv \frac{T_j - T_{fl}}{q_c} \quad \text{(K/W)} \tag{24}$$

where T_j and T_{fl} are the junction and coolant (fluid) temperatures, respectively, and q_c is the chip heat dissipation.

Unfortunately, however, most measurement techniques are incapable of detecting the actual junction temperature, that is, the temperature of the small volume at the interface of p-type and n-type semiconductors. Hence, this term generally refers to the average temperature or a representative temperature on the chip. To lower chip temperature at a specified power dissipation, it is clearly necessary to select and/or design a chip package with the lowest thermal resistance.

Examination of various packaging techniques reveals that the junction-to-coolant thermal resistance is, in fact, composed of an internal, largely conductive, resistance and an external, primarily convective, resistance. As shown in Fig. 3, the internal resistance, R_{jc}, is

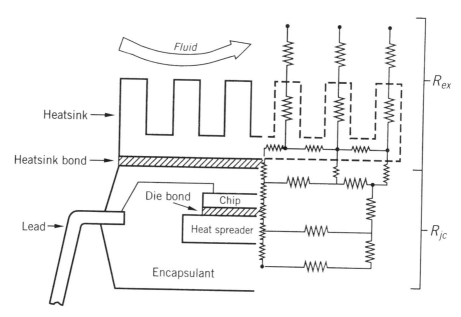

Figure 3 Primary thermal resistances in a single chip package.

encountered in the flow of dissipated heat from the active chip surface through the materials used to support and bond the chip and on to the case of the integrated circuit package. The flow of heat from the case directly to the coolant, or indirectly through a fin structure and then to the coolant, must overcome the external resistance, R_{ex}.

The thermal design of single-chip packages, including the selection of die-bond, heat spreader, substrate, and encapsulant materials, as well as the quality of the bonding and encapsulating processes, can be characterized by the internal, or so-called junction–to–case, resistance. The convective heat removal techniques applied to the external surfaces of the package, including the effect of finned heat sinks and other thermal enhancements, can be compared on the basis of the external thermal resistance. The complexity of heat flow and coolant flow paths in a multichip module generally requires that the thermal capability of these packaging configurations be examined on the basis of overall, or chip-to-coolant, thermal resistance.

Internal Thermal Resistance

As discussed in Section 1.2, conductive thermal transport is governed by the Fourier equation, which can be used to define a conduction thermal resistance, as in Eq. (3). In flowing from the chip to the package surface or case, the heat encounters a series of resistances associated with individual layers of materials such as silicon, solder, copper, alumina, and epoxy, as well as the contact resistances that occur at the interfaces between pairs of materials. Although the actual heat flow paths within a chip package are rather complex and may shift to accommodate varying external cooling situations, it is possible to obtain a first-order estimate of the internal resistance by assuming that power is dissipated uniformly across the chip surface and that heat flow is largely one-dimensional. To the accuracy of these assumptions,

$$R_{jc} = \frac{T_j - T_c}{q_c} = \sum \frac{x}{kA} \quad \text{(K/W)} \qquad (25)$$

can be used to determine the internal chip module resistance where the summed terms represent the conduction thermal resistances posed by the individual layers, each with thickness x. As the thickness of each layer decreases and/or the thermal conductivity and cross-sectional area increase, the resistance of the individual layers decreases. Values of R_{cd} for packaging materials with typical dimensions can be found via Eq. (25) or Fig 4, to range from 2 K/W for a 1000-mm^2 by 1-mm-thick layer of epoxy encapsulant to 0.0006 K/W for a 100-mm^2 by 25-μm (1-mil)-thick layer of copper. Similarly, the values of conduction resistance for typical "soft" bonding materials are found to lie in the range of approximately 0.1 K/W for solders and 1–3 K/W for epoxies and thermal pastes for typical x/A ratios of 0.25 to 1.0.

Commercial fabrication practice in the late 1990s yields internal chip package thermal resistances varying from approximately 80 K/W for a plastic package with no heat spreader to 15–20 K/W for a plastic package with heat spreader, and to 5–10 K/W for a ceramic package or an especially designed plastic chip package. Large and/or carefully designed chip packages can attain even lower values of R_{jc}, down perhaps to 2 K/W.

Comparison of theoretical and experimental values of R_{jc} reveals that the resistances associated with compliant, low-thermal-conductivity bonding materials and the spreading resistances, as well as the contact resistances at the lightly loaded interfaces within the package, often dominate the internal thermal resistance of the chip package. It is thus not only necessary to determine the bond resistance correctly but also to add the values of R_{sp}, obtained from Eq. (6) and/or Fig. 1, and R_{co} from Eq. (7b) or (9) to the junction–to–case resistance calculated from Eq. (25). Unfortunately, the absence of detailed information on

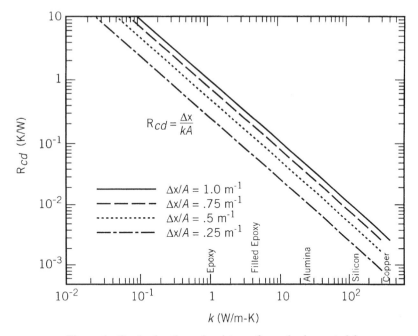

Figure 4   Conductive thermal resistance for packaging materials.

the voidage in the die-bonding and heat-sink attach layers and the present inability to determine, with precision, the contact pressure at the relevant interfaces, conspire to limit the accuracy of this calculation.

Substrate or PCB Conduction
In the design of airborne electronic systems and equipment to be operated in a corrosive or damaging environment, it is often necessary to conduct the heat dissipated by the components down into the substrate or printed circuit board and, as shown in Fig. 5, across the substrate/PCB to a cold plate or sealed heat exchanger. For a symmetrically cooled substrate/PCB with approximately uniform heat dissipation on the surface, a first estimate of the peak temperature, at the center of the board, can be obtained by use of Eq. (5).

Setting the heat generation rate equal to the heat dissipated by all the components and using the volume of the board in the denominator, the temperature difference between the center at T_{ctr} and the edge of the substrate/PCB at T_o is given by

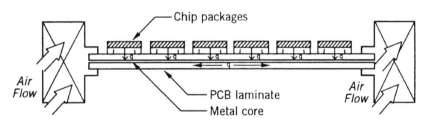

Figure 5   Edge-cooled printed circuit board populated with components.

$$T_{ctr} - T_o = \left(\frac{Q}{LW\delta}\right)\left(\frac{L^2}{8k_e}\right) = \frac{QL}{8W\delta k_e} \tag{26}$$

where Q is the total heat dissipation, W, L, and δ are the width, length, and thickness, respectively, and k_e is the effective thermal conductivity of the board.

This relation can be used effectively in the determination of the temperatures experienced by conductively cooled substrates and conventional printed circuit boards, as well as PCBs with copper lattices on the surface, metal cores, or heat-sink plates in the center. In each case it is necessary to evaluate or obtain the effective thermal conductivity of the conducting layer. As an example, consider an alumina substrate 0.20 m long, 0.15 m wide and 0.005 m thick with a thermal conductivity of 20 W/m·K, whose edges are cooled to 35°C by a cold plate. Assuming that the substrate is populated by 30 components, each dissipating 1 W, use of Eq. (26) reveals that the substrate center temperature will equal 85°C.

External Resistance

To determine the resistance to thermal transport from the surface of a component to a fluid in motion, that is, the convective resistance as in Eq. (11), it is necessary to quantify the heat transfer coefficient, h. In the natural convection air cooling of printed circuit board arrays, isolated boards, and individual components, it has been found possible to use smooth-plate correlations, such as

$$h = C\left(\frac{k_{fl}}{L}\right)Ra^n \tag{27}$$

and

$$h = \frac{k_{fl}}{b}\left[\frac{576}{(El^1)^2} + \frac{2.073}{(El^1)^{0.5}}\right]^{-1/2} \tag{28}$$

to obtain a first estimate of the peak temperature likely to be encountered on the populated board. Examination of such correlations suggests that an increase in the component/board temperature and a reduction in its length will serve to modestly increase the convective heat-transfer coefficient and thus to modestly decrease the resistance associated with natural convection. To achieve a more dramatic reduction in this resistance, it is necessary to select a high-density coolant with a large thermal expansion coefficient—typically a pressurized gas or a liquid.

When components are cooled by forced convection, the laminar heat-transfer coefficient, given in Eq. (54) (page 395), is found to be directly proportional, to the square root of fluid velocity and inversely proportional to the square root of the characteristic dimension. Increases in the thermal conductivity of the fluid and in Pr, as are encountered in replacing air with a liquid coolant, will also result in higher heat-transfer coefficients. In studies of low-velocity convective air cooling of simulated integrated circuit packages, the heat-transfer coefficient, h, has been found to depend somewhat more strongly on Re (using channel height as the characteristic length) than suggested in Eq. (54), and to display a Reynolds number exponent of 0.54 to 0.72.[8–10] When the fluid velocity and the Reynolds number increase, turbulent flow results in higher heat-transfer coefficients, which, following Eq. (56) (page 395), vary directly with the velocity to the 0.8 power and inversely with the characteristic dimension to the 0.2 power. The dependence on fluid conductivity and Pr remains unchanged.

An application of Eq. (27) or (28) to the transfer of heat from the case of a chip module to the coolant shows that the external resistance, $R_{ex} = 1/hS$, is inversely proportional to the wetted surface area and to the coolant velocity to the 0.5 to 0.8 power and directly proportional to the length scale in the flow direction to the 0.5 to 0.2 power. It may thus be observed that the external resistance can be strongly influenced by the fluid velocity and package dimensions and that these factors must be addressed in any meaningful evaluation of the external thermal resistances offered by various packaging technologies.

Values of the external resistance, for a variety of coolants and heat transfer mechanisms are shown in Fig. 6 for a typical component wetted area of 10 cm^2 and a velocity range of 2–8 m/s. They are seen to vary from a nominal 100 K/W for natural convection in air, to 33 K/W for forced convection in air, to 1 K/W in fluorocarbon liquid forced convection, and to less than 0.5 K/W for boiling in fluorocarbon liquids. Clearly, larger chip packages will experience proportionately lower external resistances than the displayed values. Moreover, conduction of heat through the leads and package base into the printed circuit board or substrate will serve to further reduce the effective thermal resistance.

In the event that the direct cooling of the package surface is inadequate to maintain the desired chip temperature, it is common to attach finned heat sinks, or compact heat exchangers, to the chip package. These heat sinks can considerably increase the wetted surface area, but may act to reduce the convective heat transfer coefficient by obstructing the flow channel. Similarly, the attachment of a heat sink to the package can be expected to introduce additional conductive resistances, in the adhesive used to bond the heat sink and in the body of the heat sink. Typical air-cooled heat sinks can reduce the external resistance to approximately 15 K/W in natural convection and to as low as 5 K/W for moderate forced convection velocities.

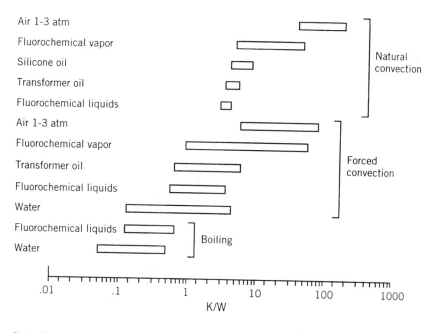

Note: For wetted area = 10 cm^2

Figure 6 Typical external (convective) thermal resistances for various coolants and cooling nodes.

When a heat sink or compact heat exchanger is attached to the package, the external resistance accounting for the bond-layer conduction and the total resistance of the heat sink, R_{sk}, can be expressed as

$$R_{ex} = \frac{T_c - T_{fl}}{q_c} = \sum \left(\frac{x}{kA}\right)_b + R_{sk} \qquad (K/W) \qquad (29)$$

where R_{sk}

$$R_{sk} = \left(\frac{1}{nhS_f\eta} + \frac{1}{h_bS_b}\right)^{-1}$$

is the the parallel combination of the resistance of the n fins

$$R_f = \frac{1}{nhS_f\eta}$$

and the *bare* or base surface not occupied by the fins

$$R_b = \frac{1}{h_bS_b}$$

Here, the base surface is $S_b = S - S_f$ and the heat-transfer coefficient, h_b, is used because the heat transfer coefficient that is applied to the base surfaces is not necessarily equal to that applied to the fins.

An alternative expression for R_{sk} involves and *overall surface efficiency*, η_o, defined by

$$\eta_o = 1 - \frac{nS_f}{S}(1 - \eta)$$

where S is the total surface composed of the base surface and the finned surfaces of n fins

$$S = S_b + nS_f$$

In this case, it is presumed that $h_b = h$ so that

$$R_{sk} = \frac{1}{h\eta_o S}$$

In an optimally designed fin structure, η can be expected to fall in the range of 0.50 to 0.70.[11] Relatively thick fins in a low-velocity flow of gas are likely to yield fin efficiencies approaching unity. This same unity value would be appropriate, as well, for an unfinned surface and, thus, serve to generalize the use of Eq. (29) to all package configurations.

Flow Resistance
In convectively cooled systems, determination of the component temperature requires knowledge of the fluid temperature adjacent to the component. The rise in fluid temperature relative to the inlet value can be expressed in a flow thermal resistance, as done in Eq. (19). When the coolant flow path traverses many individual components, care must be taken to use R_{fl} with the total heat absorbed by the coolant along its path, rather than the heat dissipated by an individual component. For system-level calculations, aimed at determining the average component temperature, it is common to base the flow resistance on the average rise in fluid temperature, that is, one-half the value indicated by Eq. (19).

Total Resistance—Single-Chip Packages

To the accuracy of the assumptions employed in the preceding development, the overall single-chip package resistance, relating the chip temperature to the inlet temperature of the coolant, can be found by summing the internal, external, and flow resistances to yield

$$R_T = R_{jc} + R_{ex} + R_{fl} = \sum \frac{x}{kA} + R_{int} + R_{sp} + R_{sk} + \left(\frac{Q}{q}\right)\left(\frac{1}{2\rho Q c_p}\right) \quad \text{(K/W)} \quad (30)$$

In evaluating the thermal resistance by this relationship, care must be taken to determine the effective cross-sectional area for heat flow at each layer in the module and to consider possible voidage in any solder and adhesive layers.

As previously noted in the development of the relationships for the external and internal resistances, Eq. (30) shows R_T to be a strong function of the convective heat-transfer coefficient, the flowing heat capacity of the coolant, and geometric parameters (thickness and cross-sectional area of each layer). Thus, the introduction of a superior coolant, use of thermal enhancement techniques that increase the local heat transfer coefficient, or selection of a heat-transfer mode with inherently high heat-transfer coefficients (boiling, for example) will all be reflected in appropriately lower external and total thermal resistances. Similarly, improvements in the thermal conductivity and reduction in the thickness of the relatively low-conductivity bonding materials (such as soft solder, epoxy or silicone) would act to reduce the internal and total thermal resistances.

Frequently, however, even more dramatic reductions in the total resistance can be achieved simply by increasing the cross-sectional area for heat flow within the chip module (such as chip, substrate, and heat spreader) as well as along the wetted, exterior surface. The implementation of this approach to reducing the internal resistance generally results in a larger package footprint or volume but is rewarded with a lower thermal resistance. The use of heat sinks is, of course, the embodiment of this approach to the reduction of the external resistance.

2   HEAT-TRANSFER CORRELATIONS FOR ELECTRONIC EQUIPMENT COOLING

The reader should use the material in this section that pertains to heat-transfer correlations in geometries peculiar to electronic equipment.

2.1   Natural Convection in Confined Spaces

For natural convection in confined horizontal spaces the recommended correlations for air are[12]

$$\text{Nu} = 0.195(\text{Gr})^{1/4}, \quad 10^4 < \text{Gr} < 4 \times 10^5 \tag{31}$$

$$\text{Nu} = 0.068(\text{Gr})^{1/3}, \quad \quad \text{Gr} > 10^5$$

where Gr is the Grashof number,

$$\text{Gr} = \frac{g\rho^2 \beta L^2 \Delta T}{\mu^2} \tag{32}$$

and where, in this case, the significant dimension L is the gap spacing in both the Nusselt and Grashof numbers.

For liquids[13]

$$Nu = 0.069(Gr)^{1/3}Pr^{0.407}, \quad 3 \times 10^5 < Ra < 7 \times 10^9 \tag{33a}$$

where Ra is the Rayleigh number,

$$Ra = GrPr \tag{33b}$$

For horizontal gaps with $Gr < 1700$, the conduction mode predominates and

$$h = \frac{k}{b} \tag{34}$$

where b is the gap spacing. For $1700 < Gr < 10,000$, use may be made of the Nusselt-Grashof relationship given in Fig. 7.[14,15]

The historical work of Elenbaas[4] provides the foundation for much of the effort dealing with natural convection in such smooth, isothermal, parallel-plate channels. Many studies showing that the value of the convective heat-transfer coefficient lies between two extremes associated with the separation distance between the plates or the channel width have been reported in literature.[19–21] Adjacent plates appear to have little influence on one another when the spacing between them is large, and the heat-transfer coefficient in this case achieves its isolated plate limit. When the plates are closely spaced or if the adjacent plates form relatively long channels, the fluid attains the fully developed velocity profile and the heat-transfer rate reaches its fully developed value. Intermediate values of the heat-transfer coefficient can be obtained from a judicious superposition of these two limiting phenomena, as presented in the composite expressions proposed by Bar-Cohen and Rohsenow.[7] Composite correlation for other situations such as symmetrically heated isothermal or isoflux surfaces are available in the literature.[42]

Table 1 shows a compilation of these natural convection heat transfer correlations for an array of vertically heated channels. The Elenbaas number, El, used in these correlations is defined as

$$El = \frac{C_p \rho^2 g \beta (T_w - T_0) b^4}{\mu k_f L} \tag{35}$$

where b is the channel spacing, L is the channel length, and $(T_w - T_0)$ is the temperature difference between the channel wall and the ambient, or channel inlet. The equations for the uniform heat flux boundary condition are defined in terms of the modified Elenbaas number, El', which is defined as

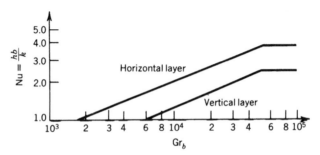

Figure 7   Heat transfer through enclosed air layers.[14,15]

Table 1 Natural Convection Heat-Transfer Correlations for an Array of Heated Vertical Channels

Condition	Fully Developed Limit	Composite Correlation
Symmetric isothermal plates	$Nu_0 = \dfrac{El}{24}$	$Nu_0 = \left(\dfrac{576}{El^2} + \dfrac{2.873}{\sqrt{El}}\right)^{(-1/2)}$
Asymmetric isothermal plates	$Nu_0 = \dfrac{El}{12}$	$Nu_0 = \left(\dfrac{144}{El^2} + \dfrac{2.873}{\sqrt{El}}\right)^{(-1/2)}$
Symmetric isoflux plates	$Nu_0 = \sqrt{\dfrac{El'}{48}}$	
Asymmetric isoflux plates	$Nu_0 = \sqrt{\dfrac{El'}{24}}$	
Symmetric isoflux plates based on mid-height temperature	$Nu_0 = \sqrt{\dfrac{El'}{12}}$	$Nu_0 = \left(\dfrac{12}{El'} + \dfrac{1.88}{(El')^{2/5}}\right)^{(-1/2)}$
Asymmetric isoflux plates based on mid-height temperature	$Nu_0 = \sqrt{\dfrac{El'}{6}}$	$Nu_0 = \left(\dfrac{6}{El'} + \dfrac{1.88}{(El')^{2/5}}\right)^{(-1/2)}$

$$El' = \frac{C_p \rho^2\, g\beta q''b^5}{\mu k_f^2 L} \tag{36}$$

where q'' is the heat flux leaving the channel wall(s).

Asymmetry can also occur if adjacent channel walls are isothermal but at different temperatures or isoflux but dissipating different heat fluxes. Aung[21] defined an asymmetry parameter for the case where the walls are isothermal but at different wall temperatures T_{w1} and T_{w2} as

$$r_T = \frac{T_{w1} - T_0}{T_{w2} - T_0} \tag{37}$$

in which T_0 is the air temperature at the channel inlet. Then the heat transfer could be calculated using the parameters listed in Table 2.

In the case of symmetric isoflux plates, if the heat flux on the adjacent walls is not identical, the equations in Table 1 can be used with an average value of the heat flux on the two walls. The composite relations listed in Table 1 can be used to optimize the spacing between PCBs in a PCB card array. For isothermal arrays, the optimum spacing maximizes the total heat transfer from a given base area or the volume assigned to an array of PCBs.

Table 2 Nusselt Number for Symmetric Isothermal Walls at Different Temperatures[21]

r_T	$\overline{Nu_0}/El$
1.0	1/24
0.5	17/405
0.1	79/1815
0.0	2/45

In the case of isoflux parallel-plate arrays, the power dissipation may be maximized by increasing the number of plates indefinitely. Thus, it is more practical to define the optimum channel spacing for an array of isoflux plates as the spacing, which will yield the maximum volumetric heat dissipation rate per unit temperature difference. Despite this distinction, the optimum spacing is found in the same manner. The optimal spacing for different conditions is listed in Table 3.[42]

The parameter b_{opt} in Table 3 represents the optimal spacing, and the plate to air parameter P is given as

$$P = \frac{C_p(\rho_f)^2 g\beta \, \Delta T_0}{\mu_f k_f L} \tag{38}$$

where ΔT_0 is the temperature difference between the printed circuit board and the ambient temperature at the inlet of the channel. The parameter R in Table 3 is given by

$$R = \frac{C_p(\rho_f)^2 g\beta q''}{\mu_f k^2 L} \tag{39}$$

These smooth-plate relations have proven useful in a wide variety of applications and have been shown to yield very good agreement with measured empirical results for heat transfer from arrays of PCBs. However, when applied to closely spaced printed circuit boards these equations tend to underpredict heat transfer in the channel due to the presence of between-package "wall flow" and the nonsmooth nature of the channel surfaces.

2.2 Natural Convection Heat Sinks

Despite the decades-long rise in component heat dissipation, the inherent simplicity and reliability of buoyantly driven flow continues to make the use of natural convection heat sinks, the cooling technology of choice for a large number of electronic applications. An understanding of natural convection heat transfer from isothermal, parallel-plate channels provides the theoretical underpinning for the conceptual design of natural convection cooled plate–fin heat sinks. However, detailed design and optimization of such fin structures requires an appreciation for the distinct characteristics of such phenomena as buoyancy-induced fluid flow in the interfin channels and conductive heat flow in the plate fins.

The presence of the heat-sink "base," or prime surface area, along one edge of the parallel-plate channel, contrasting with the open edge at the "tip" of the fins, introduces an inherent asymmetry in the flow field. The resulting three-dimensional flow pattern generally involves some inflow from (and possibly outflow through) the open edge. For relatively small

Table 3 Optimum Spacing for Natural Convection Cooled Arrays of Vertical Plates or Printed Circuit Boards

Condition	Optimum Spacing
Symmetric isothermal plates	$b_{opt} = \dfrac{2.714}{P^{1/4}}$
Asymmetric isothermal plates	$b_{opt} = \dfrac{2.154}{P^{1/4}}$
Symmetric isoflux plates	$b_{opt} = 1.472R^{-0.2}$
Asymmetric isoflux plates	$b_{opt} = 1.169R^{-0.2}$

fin spacings with long and low fins, this edge flow may result in a significant decrease in the air temperature between the fins and dramatically alter the performance of such heat sinks. For larger fin spacings, and especially with wide, thick fins, the edge flow may well be negligible.

Heat flow in extended surfaces must result in a temperature gradient at the fin base. When heat flow is from the base to the ambient, the temperature decreases along the fin, and the average fin surface excess temperature (i.e., $T_{fin} - T_{air}$) is typically between 50 and 90% of the base excess temperature. As a consequence of the temperature distribution on the fin surface, exact analytic determination of the heat-sink capability requires a combined (or conjugate) solution of the fluid flow in the channel and heat flow in the fin. Due to the complexity of such a conjugate analysis, especially in the presence of three-dimensional flow effects, the thermal performance of heat sinks is frequently based on empirical results. In recent years, extensive use has also been made of detailed numerical solutions to quantify heat-sink performance. Alternatively, a satisfactory estimate of heat-sink capability can generally be obtained by decoupling the flow and temperature fields and using an average heat-transfer coefficient, along with an average fin surface temperature, to calculate the thermal transport from the fins to the ambient air.

Starner and McManus[55] investigated the thermal performance of natural convection heat sinks, as a function of the geometry (spacing and height) and angle of base-plate orientation (vertical, horizontal and 45°), in detail. Their configuration, with the present terminology, is shown in Figure 8. They found that the measured heat-transfer coefficients for the vertical orientation were generally lower than the values expected for parallel-plate channels. The inclined orientation (45°) resulted in an additional 5–20% reduction in the heat-transfer coefficient. Results for the horizontal orientation showed a strong contribution from three-dimensional flow.

Welling and Wooldridge[56] performed an extensive study of heat transfer from vertical arrays of 2- to 3-mm-thick fins attached to an identical 203 × 66.3-mm base. Their results revealed that, in the range of $0.6 < El < 100$, associated with 4.8- to 19-mm spacings and fin heights from 6.3 to 19 mm, the heat-transfer coefficients along the total wetted surface were lower than attained by an isolated, flat plate but generally above those associated with parallel-plate flow. This behavior was explained in terms of the competing effects of channel flow, serving to preheat the air, and inflow from the open edge, serving to mix the heated

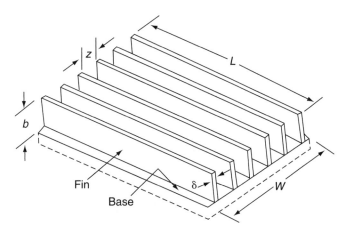

Figure 8   Geometric parameters for Starner and McManus[55] fin arrays.

air with the cooler ambient fluid. In this study it was observed, for the first time, that for any given interfin spacing there is an optimum fin height, b, beyond which thermal performance, per unit surface area, deteriorates.

In 1986 Bilitzky[57] completed a comprehensive investigation of natural convection heat transfer from multiple heat-sink geometries that differed, primarily, in fin height and spacing. The heat sinks were operated at different heat dissipations as well as different angles of inclination and orientation. Twelve distinct heat sinks and a flat plate were tested in a room within which extraneous convection had been suppressed. The range of angles is indicated in Figure 9. The base was first kept vertical, while the fins were rotated through four different positions (90°, 60°, 30°, and 0°). Then, the base was tilted backward toward the horizontal orientation through four different positions (90°, 60°, 30°, and 0°). Six of the heat sinks used 144-mm-long by 115-mm-wide bases to support plate fins, nominally 2 mm in thickness and 6–13.8 mm apart, ranging in height from 8.6 to 25.5 mm. Six additional heat sinks, with identical fin geometries, were supported on 280-mm-long and 115-mm-wide bases. The geometric parameters of the 12 heat sinks were selected to span the base and fin dimensions encountered in electronics cooling applications and are summarized in Table 4.

In all 12 heat sinks studied, the vertical–vertical orientation, that is, a vertically oriented base with vertical fins and channels, yielded the highest heat-transfer coefficients most often. However, in a relatively large number of situations, the thermal performance of the vertical–vertical arrays was indistinguishable from that attained by a vertical base plate with fins rotated 30° from the axis, a horizontal base plate, and a base plate inclined 60° from the horizontal with unrotated fins. On the other hand, the vertical–horizontal orientation, that is, the base plate vertical and the fins rotated 90° from the axis, led to the lowest heat-dissipation rates. For the unrotated fins, the lowest heat-transfer coefficients were almost always found to occur at a base-plate angle of 30° from the horizontal. The use of smoke revealed a

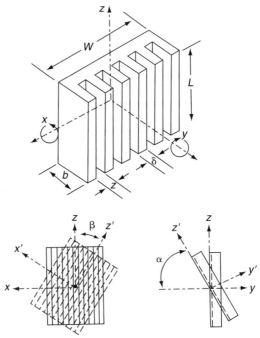

Figure 9 Geometry and orientation of Bilitzky[57] fin arrays.

Table 4 Geometric Parameters for Bilitzky[57] Fin Arrays

Array	L in.	L mm	W in.	W mm	b in.	b mm	z in.	z mm	δ in.	δ mm
1	5.67	144	4.53	115	1.004	25.5	0.236	6.0	0.075	1.9
2	5.67	144	4.53	115	0.677	17.2	0.232	5.9	0.079	2.0
3	5.67	144	4.53	115	0.339	8.6	0.228	5.8	0.083	2.1
4	5.67	144	4.53	115	1.004	25.5	0.547	13.9	0.075	1.9
5	5.67	144	4.53	115	0.677	17.2	0.543	13.8	0.079	2.0
6	5.67	144	4.53	115	0.339	8.6	0.539	13.7	0.083	2.1
7	11.02	280	4.53	115	1.004	25.5	0.236	6.0	0.075	1.9
8	11.02	280	4.53	115	0.669	17.0	0.232	5.9	0.079	2.0
9	11.02	280	4.53	115	0.335	8.5	0.228	5.8	0.083	2.1
10	11.02	280	4.53	115	1.004	25.5	0.547	13.9	0.075	1.9
11	11.02	280	4.53	115	0.669	17.0	0.543	13.8	0.079	2.0
12	11.02	280	4.53	115	0.335	8.5	0.539	13.7	0.083	2.1

relatively complex three-dimensional flow pattern around the heat sinks, with very substantial inflow from the direction of the fin tips when the base plate was strongly inclined and when the heat sinks were in the vertical base–horizontal fins orientation.

The influence of the spacing, z, between the fins, for short and long base plates, was examined by comparing pairs of heat sinks that differed only in geometric parameters, z (fin arrays 1 and 4, 2 and 5, 3 and 6, 7 and 10, 8 and 11, and 9 and 12 in Table 4). Bilitzky[57] observed that, in nearly all of the configurations and operating conditions examined, the highest heat-transfer coefficients were attained with the larger fin spacing. However, the improvement in the heat-transfer coefficient was not always sufficient to compensate for the loss of wetted fin surface area. Moreover, for the horizontal base-plate configuration, as well as for the vertical base with horizontal fins, the total array dissipation appeared not to depend on this parameter.

Analysis of data from the vertically oriented heat sinks led Bilitzky[57] to recognize that closely spaced fins, typical of actual heat sinks used for electronics cooling, display substantially higher heat-transfer coefficients than predicted by the fully developed channel flow equations listed in Table 1. Bilitzky[57] proposed the following modification to the fully developed Nusselt number for symmetric isothermal plates:

$$Nu_0 = \frac{El}{24\psi} \tag{40}$$

where the correction factor ψ was given as

$$\psi = \frac{A_1}{[(1 + (a/2)(1 + A_2 A_3)]^2} \tag{41}$$

The parameter a in Eq. (41) was the ratio of the fin pitch and the fin height and the parameters A_1, A_2, and A_3 were given as follows:

$$A_1 = 1 - 0.483 e^{-0.17/a} \tag{42}$$

$$A_2 = (1 - e^{-0.83/a}) \tag{43}$$

$$A_3 = 9.14 a^{1/2} e^{-1.25(1+a/2)} - 0.61 \tag{44}$$

Use of Eq. (40) was reported by Bilitzky[57] to yield agreement within 5% for his data.

2.3 Thermal Interface Resistance

Heat transfer across a solid interface is accompanied by a temperature difference, caused by imperfect contact between the two solids. Even when perfect adhesion is achieved between the solids, the transfer of heat is impeded by the acoustic mismatch in the properties of the phonons on either side of the interface. Traditionally the thermal resistance arising due to imperfect contact has been called the "thermal contact" resistance. The resistance due to the mismatch in the acoustic properties is usually termed the "thermal boundary" resistance. The thermal contact resistance is a macroscopic phenomenon, whereas thermal boundary resistance is a microscopic phenomenon. This section primarily focuses on thermal contact resistance and methods to reduce the contact resistance.

When two surfaces are joined, as shown in Figure 10, asperities on each of the surfaces limit the actual contact between the two solids to a very small fraction, perhaps just 1–2% for lightly loaded interfaces, of the apparent area. As a consequence, the flow of heat across such an interface involves solid-to-solid conduction in the area of actual contact, A_{co}, and conduction through the fluid occupying the noncontact area, A_{nc}, of the interface. At elevated temperatures or in vacuum, radiation heat transfer across the open spaces may also play an important role.

The pressure imposed across the interface, along with the microhardness of the softer surface and the surface roughness characteristics of both solids, determines the interfacial gap, δ, and the contact area, A_{co}. Assuming plastic deformation of the asperities and a Gaussian distribution of the asperities over the apparent area, Cooper et al.[58] proposed the following relation for the contact resistance R_{co}:

$$R_{co} = \left(1.45 \frac{k_s (P/H)^{0.985}}{\sigma} \overline{|\tan \theta|}\right)^{-1} \tag{45}$$

where k_s is the harmonic mean thermal conductivity defined as $k_s = 2k_1 k_2/(k_1 + k_2)$, P is the apparent contact pressure, H is the hardness of the softer material, and σ is the rms roughness given by

$$\sigma_1 = \sqrt{\sigma_1^2 + \sigma_2^2} \tag{46}$$

where σ_1 and σ_2 are the roughness of surfaces 1 and 2, respectively. The term $\overline{|\tan \theta|}$ in Eq. (45) is the average asperity angle:

$$\overline{|\tan \theta|}^2 = |\tan \theta_1|^2 + |\tan \theta_2|^2 \tag{47}$$

This relation neglects the heat-transfer contribution of any trapped fluid in the interfacial gap.

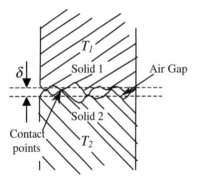

Figure 10 Contact and heat flow at a solid/solid interface.

In the pursuit of a more rigorous determination of the contact resistance, Yovanovich and Antonetti[59] found it possible to predict the area-weighted interfacial gap, Y, in the following form:

$$Y = 1.185\sigma \left[-\ln\left(\frac{3.132P}{H}\right) \right]^{0.547} \tag{48}$$

where σ is the effective root mean square (rms) as given by Eq. (46), P is the contact pressure (Pa), and H is the surface microhardness (Pa) of the softer material, to a depth of the order of the penetration of the harder material. Using Y as the characteristic gap dimension and incorporating the solid–solid and fluid gap parallel heat flow paths, they derived the following equation for the total interfacial thermal resistance.

$$R_{co} = \left[1.25k_s\left(\frac{|\tan\theta|}{\sigma}\right)\left(\frac{P}{H}\right)^{0.95} + \frac{k_g}{Y} \right]^{-1} \tag{49}$$

where k_g is the interstitial fluid thermal conductivity. In the absence of detailed information, $\sigma/|\tan\theta|$ can be expected to range from 5 to 9 μm for relatively smooth surfaces.

In describing heat flow across an interface, Eq. (49) assumed the existence of a fluid gap, which provides a parallel heat flow path to that of the solid–solid contact. Because the noncontact area may occupy in excess of 90% of the projected area, heat flow through the interstitial spaces can be of great importance. Consequently, the use of high thermal conductivity interstitial materials, such as soft metallic foils and fiber disks, conductive epoxies, thermal greases, and polymeric "phase-change" materials, can substantially reduce the contact resistance. The enhanced thermal capability of many of the high-performance epoxies, thermal greases, and "phase-change" materials, commonly in use in the electronic industry, is achieved through the use of large concentrations of thermally conductive particles. Successful design and development of thermal packaging strategies, thus, requires the determination of the effective thermal conductivity of such particle-laden interstitial materials and their effect on the overall interfacial thermal resistance.

Comprehensive reviews of the general role of interstitial materials in controlling contact resistance have been published by several authors including Sauer.[60] When interstitial materials are used for control of the contact resistance, it is desirable to have some means of comparing their effectiveness. Fletcher[61] proposed two parameters for this purpose. The first of these parameters is simply the ratio of the logarithms of the conductances, which is the inverse of the contact resistance, with and without the filler:

$$\chi = \frac{\ln(\kappa_{cm})}{\ln(\kappa_{bj})} \tag{50}$$

in which κ is the contact conductance, cm and bj refer to control material and bare junctions, respectively. The second parameter takes the thickness of the filler material into account and is defined as

$$\eta' = \frac{(\kappa\delta_{filler})_{cm}}{(\kappa\delta_{gap})_{bj}} \tag{51}$$

in which δ is the equivalent thickness and η' is the effectiveness of the interstitial material.

The performance of an interstitial interface material as decided by the parameter defined by Fletcher,[61] in Eqs. (50) and (51) includes the bulk as well as the contact resistance contribution. It is because of this reason that in certain cases that the thermal resistance of these thermal interface materials is higher than that for a bare metallic contact because the bulk resistance is the dominant factor in the thermal resistance. To make a clear comparison of only the contact resistance arising from the interface of the substrate and various thermal interface materials, it is important to measure it exclusively. Separation of the contact resistance and bulk resistance will also help researchers to model the contact resistance and the bulk resistance separately.

Equations (50) and (51) by Fletcher,[61] show that the thermal resistance of any interface material depends on both the bond line thickness and thermal conductivity of the material. As a consequence, for materials with relatively low bulk conductivity, the resistance of the added interstitial layer may dominate the thermal behavior of the interface and may result in an overall interfacial thermal resistance that is higher than that of the bare solid–solid contact. Thus, both the conductivity and the achievable thickness of the interstitial layer must be considered in the selection of an interfacial material. Indeed, while the popular "phase-change" materials have a lower bulk thermal conductivity (at a typical value of 0.7W/mK) than the silicone-based greases (with a typical value of 3.1 W/mK), due to thinner "phase-change" interstitial layers, the thermal resistance of these two categories of interface materials is comparable.

To understand the thermal behavior of such interface materials, it is useful to separate the contribution of the bulk conductivity from the interfacial resistance, which occurs where the interstitial material contacts one of the mating solids. Following Prasher,[62] who studied the contact resistance of phase-change materials (PCM) and silicone-based thermal greases, the thermal resistance associated with the addition of an interfacial material, R_{TIM} can be expressed as

$$R_{TIM} = R_{bulk} + R_{co_1} + R_{co_2} \tag{52}$$

where R_{bulk} is the bulk resistance of the thermal interface material, and R_{co} is the contact resistance with the substrate and subscripts 1 and 2 refer to substrate 1 and 2. Prasher[62] rewrote Eq. (52) as

$$R_{TIM} = \frac{\delta}{k_{TIM}} + \frac{\sigma_1}{2k_{TIM}}\left(\frac{A_{nom}}{A_{real}}\right) + \frac{\sigma_2}{2k_{TIM}}\left(\frac{A_{nom}}{A_{real}}\right) \tag{53}$$

where R_{TIM} is the total thermal resistance of the thermal interface material, δ the bond-line thickness, k_{TIM} the thermal conductivity of the interface material, σ_1 and σ_2 are the roughness of surfaces 1 and 2, respectively, A_{nom} is the nominal area, and A_{real} is the real area of contact of the interface material with the two surfaces. Equation (53) assumes that the thermal conductivity of the substrate is much higher compared to that of the thermal interface material. The first term on right hand side of Eq. (53) is the bulk resistance and other terms are the contact resistances.

Figure 11 shows the temperature variation at the interface between two solids, in the presence of a thermal interface material, associated with Eq. (53). Unlike the situation with the more conventional interface materials, the actual contact area between a polymeric material and a solid is determined by capillary forces, rather than the surface hardness, and an alternative approach is required to determine, A_{real}, in Eq. (53). Modeling each of the relevant surfaces as a series of notches, and including the effects of surface roughness, the slope of the asperities, the contact angle of the polymer with each the substrates, the surface energy of the polymer, and the externally applied pressure, a surface chemistry model was found to match very well with the experimental data for PCM and greases at low pressures.[62] Unfortunately, it has not been possible, as yet, to determine the contact area with a closed form expression. It is also to be noted that Eq. (53) underpredicts the interface thermal resistance data at high pressures.

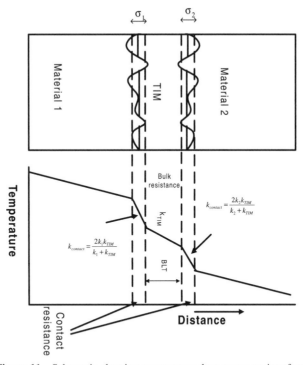

Figure 11 Schematic showing temperature drop across an interface.

2.4 Forced Convection

External Flow on a Plane Surface

For an unheated starting length of the plane surface, x_0, in laminar flow, the local Nusselt number can be expressed by

$$\mathrm{Nu}_x = \frac{0.332\mathrm{Re}^{1/2}\mathrm{Pr}^{1/3}}{[1 - (x_0/x)^{3/4}]^{1/3}} \tag{54}$$

Where Re is the Reynolds number, Pr is the Prandtl number, and Nu is the Nusselt number.

For flow in the inlet zones of parallel plate channels and along isolated plates, the heat-transfer coefficient varies with L, the distance from the leading edge.[3] in the range Re $\leq 3 \times 10^5$,

$$h = 0.664 \left(\frac{k_{fl}}{L}\right) \mathrm{Re}^{0.5}\mathrm{Pr}^{0.33} \tag{55}$$

and for Re $> 3 \times 10^5$

$$h = 0.036 \left(\frac{k_{fl}}{L}\right) \mathrm{Re}^{0.8}\mathrm{Pr}^{0.33} \tag{56}$$

Cylinders in Cross-flow

For airflow around single cylinders at all but very low Reynolds numbers, Hilpert[23] has proposed

$$\text{Nu} = \frac{hd}{k_f} = B \left(\frac{\rho V_\infty d}{\mu_f} \right)^n \tag{57}$$

where V_∞ is the free stream velocity and where the constants B and n depend on the Reynolds number as indicated in Table 5.

It has been pointed out[12] that Eq. (57) assumes a natural turbulence level in the oncoming air stream and that the presence of augmentative devices can increase n by as much as 50%. The modifications to B and n due to some of these devices are displayed in Table 6.

Equation (57) can be extended to other fluids[24] spanning a range of $1 < \text{Re} < 10^5$ and $0.67 < \text{Pr} < 300$:

$$\text{Nu} = \frac{hd}{k} = (0.4\text{Re}^{0.5} + 0.06\text{Re}^{0.67})\text{Pr}^{0.4} \left(\frac{\mu}{\mu_w} \right)^{0.25} \tag{58}$$

where all fluid properties are evaluated at the free stream temperature except μ_w, which is the fluid viscosity at the wall temperature.

Noncircular Cylinders in Cross-flow
It has been found[12] that Eq. (57) may be used for noncircular geometries in cross-flow provided that the characteristic dimension in the Nusselt and Reynolds numbers is the diameter of a cylinder having the same wetted surface equal to that of the geometry of interest and that the values of B and n are taken from Table 7.

Flow across Spheres
For airflow across a single sphere, it is recommended that the average Nusselt number when $17 < \text{Re} < 7 \times 10^4$ be determined from[22]

$$\text{Nu} = \frac{hd}{k_f} = 0.37 \left(\frac{\rho V_\infty d}{\mu_f} \right)^{0.6} \tag{59}$$

and for $1 < \text{Re} < 25$[25],

$$\text{Nu} = \frac{hd}{k} = 2.2\text{Pr} + 0.48\text{Pr}(\text{Re})^{0.5} \tag{60}$$

For both gases and liquids in the range $3.5 < \text{Re} < 7.6 \times 10^4$ and $0.7 < \text{Pr} < 380$[24]

$$\text{Nu} = \frac{hd}{k} = 2 + (4.0\text{Re}^{0.5} + 0.06\text{Re}^{0.67})\text{Pr}^{0.4} \left(\frac{\mu}{\mu_w} \right)^{0.25} \tag{61}$$

Table 5   Constants for Eq. (11)

Reynolds Number Range	B	n
1–4	0.891	0.330
4–40	0.821	0.385
40–4000	0.615	0.466
4000–40,000	0.174	0.618
40,000–400,000	0.0239	0.805

Table 6 Flow Disturbance Effects on B and n in Eq. (57)

Disturbance	Re Range	B	n
1. Longitudinal fin, $0.1d$ thick on front of tube	1000–4000	0.248	0.603
2. 12 longitudinal grooves, $0.7d$ wide	3500–7000	0.082	0.747
3. Same as 2 with burrs	3000–6000	0.368	0.86

Flow across Tube Banks

For the flow of fluids flowing normal to banks of tubes,[26]

$$\text{Nu} = \frac{hd}{k_f} = C \left(\frac{\rho V_\infty d}{\mu_f}\right)^{0.6} \left(\frac{c_p \mu}{k}\right)_f^{0.33} \phi \tag{62}$$

which is valid in the range $2000 < \text{Re} < 32{,}000$.

For in-line tubes, $C = 0.26$, whereas for staggered tubes, $C = 0.33$. The factor ϕ is a correction factor for sparse tube banks, and values of ϕ are provided in Table 8.

For air in the range where Pr is nearly constant ($\text{Pr} \approx 0.7$ over the range 25–200°C), Eq. (62) can be reduced to

$$\text{Nu} = \frac{hd}{k_f} = C' \left(\frac{\rho V_\infty d}{\mu_f}\right)^{n'} \tag{63}$$

where C' and n' may be determined from values listed in Table 9. This equation is valid in the range $2000 < \text{Re} < 40{,}000$ and the ratios x_L and x_T denote the ratio of centerline diameter to tube spacing in the longitudinal and transverse directions, respectively.

For fluids other than air, the curve shown in Fig. 12 should be used for staggered tubes.[22] For in-line tubes, the values of

$$j = \left(\frac{hd_0}{k}\right) \left(\frac{c_p \mu}{k}\right)^{-1/3} \left(\frac{\mu}{\mu_w}\right)^{-0.14}$$

should be reduced by 10%.

Table 7 Values of B and n for Eq. (57)[a]

Flow Geometry	B	n	Range of Reynolds Number
	0.224	0.612	2,500–15,000
	0.085	0.804	3,000–15,000
◇	0.261	0.624	2,500–7,500
◇	0.222	0.588	5,000–100,000
□	0.160	0.699	2,500–8,000
□	0.092	0.675	5,000–100,000
○	0.138	0.638	5,000–100,000
○	0.144	0.638	5,000–19,500
○	0.035	0.782	19,500–100,000
	0.205	0.731	4,000–15,000

[a]From Ref. 12.

Table 8 Correlation Factor ϕ for Sparse Tube Banks

Number of Rows, N	In Line	Staggered
1	0.64	0.68
2	0.80	0.75
3	0.87	0.83
4	0.90	0.89
5	0.92	0.92
6	0.94	0.95
7	0.96	0.97
8	0.98	0.98
9	0.99	0.99
10	1.00	1.00

Flow across Arrays of Pin Fins
For air flowing normal to banks of staggered cylindrical pin fins or spines,[28]

$$\text{Nu} = \frac{hd}{k} = 1.40 \left(\frac{\rho v_\infty d}{\mu}\right)^{0.8} \left(\frac{c_p \mu}{k}\right)^{1/3} \tag{64}$$

Flow of Air over Electronic Components
For single prismatic electronic components, either normal or parallel to the sides of the component in a duct,[29] for $2.5 \times 10^3 < \text{Re} < 8 \times 10^3$,

$$\text{Nu} = 0.446 \left[\frac{\text{Re}}{(1/6) + (5A_n/6A_0)}\right]^{0.57} \tag{65}$$

where the Nusselt and Reynolds numbers are based on the prism side dimension and where A_0 and A_n are the gross and net flow areas, respectively.

Table 9 Values of the Constants C' and n' in Eq. (63)

$x_L = \dfrac{S_L}{d_0}$	$x_T = \dfrac{S_T}{d_0} = 1.25$		$x_T = \dfrac{S_T}{d_0} = 1.50$		$x_T = \dfrac{S_T}{d_0} = 2.00$		$x_T = \dfrac{S_T}{d_0} = 3.00$	
	C'	n'	C'	n'	C'	n'	C'	n'
Staggered								
0.600							0.213	0.636
0.900					0.446	0.571	0.401	0.581
1.000			0.497	0.558				
1.125					0.478	0.565	0.518	0.560
1.250	0.518	0.556	0.505	0.554	0.519	0.556	0.522	0.562
1.500	0.451	0.568	0.460	0.562	0.452	0.568	0.488	0.568
2.000	0.404	0.572	0.416	0.568	0.482	0.556	0.449	0.570
3.000	0.310	0.592	0.356	0.580	0.440	0.562	0.421	0.574
In Line								
1.250	0.348	0.592	0.275	0.608	0.100	0.704	0.0633	0.752
1.500	0.367	0.586	0.250	0.620	0.101	0.702	0.0678	0.744
2.000	0.418	0.570	0.299	0.602	0.229	0.632	0.198	0.648
3.000	0.290	0.601	0.357	0.584	0.374	0.581	0.286	0.608

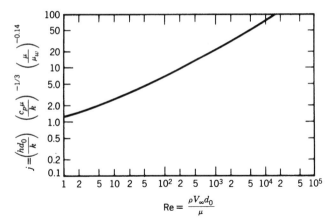

Figure 12 Recommended curve for estimation of heat-transfer coefficient for fluids flowing normal to staggered tubes 10 rows deep (from Ref. 22).

For staggered prismatic components, Eq. (65) may be modified to[29]

$$\text{Nu} = 0.446 \left[\frac{\text{Re}}{(1/6) + (5A_n/A_0)} \right]^{0.57} \left[1 + 0.639 \left(\frac{S_T}{S_{T,\max}} \right) \left(\frac{d}{S_L} \right)^{0.172} \right] \tag{66}$$

where d is the prism side dimension, S_L is the longitudinal separation, S_T is the transverse separation, and $S_{T,\max}$ is the maximum transverse spacing if different spacings exist.

When cylindrical heat sources are encountered in electronic equipment, a modification of Eq. (57) has been proposed[30]:

$$\text{Nu} = \frac{hd}{k_f} = FB \left(\frac{\rho V_\infty d}{\mu} \right)^n \tag{67}$$

where F is an arrangement factor depending on the cylinder geometry (see Table 10) and where the constants B and n are given in Table 11.

Table 10 Values of F to Be Used in Eq. (67)[a]

Single cylinder in free stream: $F = 1.0$
Single cylinder in duct: $F = 1 + d/w$
In-line cylinders in duct:

$$F = \left(1 + \sqrt{\frac{1}{S_T}} \right) \left\{ 1 + \left(\frac{1}{S_L} - \frac{0.872}{S_L^2} \right) \left(\frac{1.81}{S_T^2} - \frac{1.46}{S_T} + 0.318 \right) [\text{Re}^{0.526 - (0.354/S_T)}] \right\}$$

Staggered cylinders in duct:

$$F = \left(1 + \sqrt{\frac{1}{S_T}} \right) \left\{ 1 + \left[\frac{1}{S_L} \left(\frac{15.50}{S_T^2} - \frac{16.80}{S_T} + 4.15 \right) - \frac{1}{S_L} \left(\frac{14.15}{S_T^2} - \frac{15.33}{S_T} + 3.69 \right) \right] \text{Re}^{0.13} \right\}$$

[a]Re to be evaluated at film temperature. S_L = ratio of longitudinal spacing to cylinder diameter. S_T = ratio of transverse spacing to cylinder diameter.

Table 11   Values of B and n for Use in Eq. (67)

Reynolds Number Range	B	n
1000–6000	0.409	0.531
6000–30,000	0.212	0.606
30,000–100,000	0.139	0.806

Forced Convection in Tubes, Pipes, Ducts, and Annuli

For heat transfer in tubes, pipes, ducts, and annuli, use is made of the equivalent diameter

$$d_e = \frac{4A}{WP} \tag{68}$$

in the Reynolds and Nusselt numbers unless the cross section is circular, in which case d_e and $d_i = d$.

In the laminar regime[31] where Re < 2100,

$$\text{Nu} = hd_e/k = 1.86[\text{RePr}(d_e/L)]^{1/3}(\mu/\mu_w)^{0.14} \tag{69}$$

with all fluid properties except μ_w evaluated at the bulk temperature of the fluid.

For Reynolds numbers above transition, Re > 2100,

$$\text{Nu} = 0.023(\text{Re})^{0.8}(\text{Pr})^{1/3}(\mu/\mu_w)^{0.14} \tag{70}$$

and in the transition region, 2100 < Re < 10,000,[32]

$$\text{Nu} = 0.116[(\text{Re})^{2/3} - 125](\text{Pr})^{1/3}(\mu/\mu_w)^{0.14}[1 + (d_e/L)^{2/3}] \tag{71}$$

London[33] has proposed a correlation for the flow of air in rectangular ducts. It is shown in Fig. 13. This correlation may be used for air flowing between longitudinal fins.

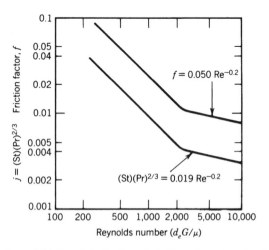

Figure 13   Heat-transfer and friction data for forced air through rectangular ducts. St is the Stanton number, St = hG/c_p.

3   THERMAL CONTROL TECHNIQUES

3.1   Extended Surface and Heat Sinks

The heat flux from a surface, q/A, can be reduced if the surface area A is increased. The use of extended surface or fins in a common method of achieving this reduction. Another way of looking at this is through the use of Newton's law of cooling:

$$q = hA\ \Delta T \tag{72}$$

and considering that ΔT can be reduced for a given heat flow q by increasing h, which is difficult for a specified coolant, or by increasing the surface area A.

The common extended surface shapes are the longitudinal fin of rectangular profile, the radial fin of rectangular profile, and the cylindrical spine shown, respectively, in Figs. 14a, e, and g.

Assumptions in Extended Surface Analysis
The analysis of extended surface is subject to the following simplifying assumptions:[34,35]

1. The heat flow is steady; that is, the temperature at any point does not vary with time.
2. The fin material is homogeneous, and the thermal conductivity is constant and uniform.
3. The coefficient of heat transfer is constant and uniform over the entire face surface of the fin.
4. The temperature of the surrounding fluid is constant and uniform.
5. There are no temperature gradients within fin other than along the fin height.

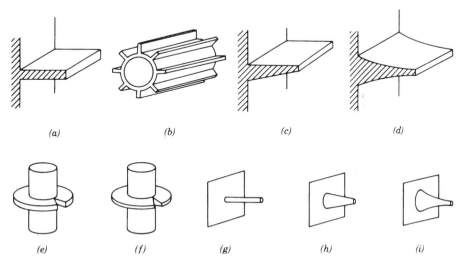

(a) (b) (c) (d)

(e) (f) (g) (h) (i)

Figure 14   Some typical examples of extended surfaces: (a) longitudinal fin of rectangular profile: (b) cylindrical tube equipped with longitudinal fins; (c) longitudinal fin of trapezoidal profile; (d) longitudinal fin of truncated concave parabolic profile; (e) cylindrical tube equipped with radial fin of rectangular profile; (f) cylindrical tube equipped with radial fin of truncated triangular profile; (g) cylindrical spine; (h) truncated conical spine; (i) truncated concave parabolic spine.

6. There is no bond resistance to the flow of heat at the base of the fin.

7. The temperature at the base of the fin is uniform and constant.

8. There are no heat sources within the fin itself.

9. There is a negligible flow of heat from the tip and sides of the fin.

10. The heat flow from the fin is proportioned to the temperature difference or temperature excess, $\theta(x) = T(x) - T_s$, at any point on the face of the fin.

The Fin Efficiency

Because a temperature gradient always exists along the height of a fin when heat is being transferred to the surrounding environment by the fin, there is a question regarding the temperature to be used in Eq. (72). If the base temperature T_b (and the base temperature excess, $\theta_b = T_b - T_s$) is to be used, then the surface area of the fin must be modified by the computational artifice known as the fin efficiency, defined as the ratio of the heat actually transferred by the fin to the ideal heat transferred if the fin were operating over its entirety at the base temperature excess. In this case, the surface area A in Eq. (54) becomes

$$A = A_b + \eta_f A_f \tag{73}$$

The Longitudinal Fin of Rectangular Profile

With the origin of the height coordinate x taken at the fin tip, which is presumed to be adiabatic, the temperature excess at any point on the fin is

$$\theta(x) = \theta_b \frac{\cosh mx}{\cosh mb} \tag{74}$$

where

$$m = \left(\frac{2h}{k\delta}\right)^{1/2} \tag{75}$$

The heat dissipated by the fin is

$$q_b = Y_0 \theta_b \tanh mb \tag{76}$$

where Y_0 is called the characteristic admittance

$$Y_0 = (2hk\delta)^{1/2}L \tag{77}$$

and the fin efficiency is

$$\eta_f = \frac{\tanh mb}{mb} \tag{78}$$

The heat-transfer coefficient in natural convection may be determined from the symmetric isothermal case pertaining to vertical plates in Section 2.1. For forced convection, the London correlation described in Section 2.2 applies.

The Radial Fin of Rectangular Profile

With the origin of the radial height coordinate taken at the center of curvature and with the fin tip at $r = r_a$ presumed to be adiabatic, the temperature excess at any point on the fin is

$$\theta(r) = \theta_b \left[\frac{K_1(mr_a)I_0(mr) + I_1(mr_a)K_0(mr)}{I_0(mr_b)K_1(mr_a) + I_1(mr_a)K_0(mr_b)} \right] \tag{79}$$

where m is given by Eq. (75). The heat dissipated by the fin is

$$q_b = 2\pi r_b km\theta_b \left[\frac{I_1(mr_a)K_1(mr_b) - K_1(mr_a)I_1(mr_b)}{I_0(mr_b)K_1(mr_a) + I_1(mr_a)K_0(mr_b)} \right] \tag{80}$$

and the fin efficiency is

$$\eta_f = \frac{2r_b}{m(r_a^2 - r_b^2)} \left[\frac{I_1(mr_a)K_1(mr_b) - K_1(mr_a)I_1(mr_b)}{I_0(mr_b)K_1(mr_a) + I_1(mr_a)K_0(mr_b)} \right] \tag{81}$$

Tables of the fin efficiency are available,[36] and they are organized in terms of two parameters, the radius ratio

$$\rho = \frac{r_b}{r_a} \tag{82a}$$

and a parameter ϕ

$$\phi = (r_a - r_b) \left(\frac{2h}{kA_p} \right)^{1/2} \tag{82b}$$

where A_p is the profile area of the fin:

$$A_p = \delta(r_a - r_b) \tag{82c}$$

For air under forced convection conditions, the correlation for the heat-transfer coefficient developed by Briggs and Young[37] is applicable:

$$\frac{h}{2r_b k} = \left(\frac{2\rho V r_b}{\mu} \right)^{0.681} \left(\frac{c_p \mu}{k} \right)^{1/3} \left(\frac{s}{r_a - r_b} \right)^{0.200} \left(\frac{s}{\delta} \right)^{0.1134} \tag{83}$$

where all thermal properties are evaluated at the bulk air temperature, s is the space between the fins, and r_a and r_b pertain to the fins.

The Cylindrical Spine
With the origin of the height coordinate x taken at the spine tip, which is presumed to be adiabatic, the temperature excess at any point on the spine is given by Eq. (72), but for the cylindrical spine

$$m = \left(\frac{4h}{kd} \right)^{1/2} \tag{84}$$

where d is the spine diameter. The heat dissipated by the spine is given by Eq. (76), but in this case

$$Y_0 = (\pi^2 hkd^3)^{1/2}/2 \tag{85}$$

and the spine efficiency is given by Eq. (78).

Algorithms for Combining Single Fins into Arrays
The differential equation for temperature excess that can be developed for any fin shape can be solved to yield a particular solution, based on prescribed initial conditions of fin base temperature excess and fin base heat flow, that can be written in matrix form[38,39] as

$$\begin{bmatrix} \theta_a \\ q_a \end{bmatrix} = [\Gamma] \begin{bmatrix} \theta_b \\ q_b \end{bmatrix} = \begin{bmatrix} \gamma_{11} & \gamma_{12} \\ \gamma_{21} & \gamma_{22} \end{bmatrix} \begin{bmatrix} \theta_b \\ q_b \end{bmatrix} \tag{86}$$

The matrix $[\Gamma]$ is called the thermal transmission matrix and provides a linear transformation from tip to base conditions. It has been cataloged for all of the common fin shapes.[38-40] For the longitudinal fin of rectangular profile

$$[\Gamma] = \begin{bmatrix} \cosh mb & -\dfrac{1}{Y_0} \sinh mb \\ -Y_0 \sinh mb & \cosh mb \end{bmatrix} \tag{87}$$

and this matrix possesses an inverse called the inverse thermal transmission matrix

$$[\Lambda] = [\Gamma]^{-1} = \begin{bmatrix} \cosh mb & \dfrac{1}{Y_0} \sinh mb \\ Y_0 \sinh mb & \cosh mb \end{bmatrix} \tag{88}$$

The assembly of fins into an array may require the use of any or all of three algorithms.[40-42] The objective is to determine the input admittance of the entire array

$$Y_{\text{in}} = \left. \frac{q_b}{\theta_b} \right|_A \tag{89}$$

which can be related to the array (fin) efficiency by

$$\eta_f = \frac{Y_{\text{in}}}{hA_f} \tag{90}$$

The determination of Y_{in} can involve as many as three algorithms for the combination of individual fins into an array.

The Cascade Algorithm. For n fins in cascade as shown in Fig. 15a, an equivalent inverse thermal transmission matrix can be obtained by a simple matrix multiplication, with the individual fins closest to the base of the array acting as permultipliers:

$$\{\Lambda\}_e = \{\Lambda\}_n \{\Lambda\}_{n-1} \{\Lambda\}_{n-2} \cdots \{\Lambda\}_2 \{\Lambda\}_1 \tag{91}$$

For the case of the tip of the most remote fin adiabatic, the array input admittance will be

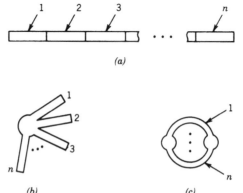

Figure 15 (a) n fins in cascade, (b) n fins in cluster, and (c) n fins in parallel.

$$Y_{in} = \frac{\lambda_{21.e}}{\lambda_{11.e}} \tag{92}$$

If the tip of the most remote fin is not adiabatic, the heat flow to temperature excess ratio at the tip, which is designated as μ,

$$\mu = \frac{q_a}{\theta_a} \tag{93}$$

will be known. For example, for a fin dissipating to the environment through its tip designated by the subscript a:

$$\mu = h A_a \tag{94}$$

In this case, Y_{in} may be obtained through successive use of what is termed the reflection relationship (actually a bilinear transformation):

$$Y_{in,k-1} = \frac{\lambda_{21,k-1} + \lambda_{22,k-1}(q_a/\theta_a)}{\lambda_{11,k-1} + \lambda_{12,k-1}(q_a/\theta_a)} \tag{95}$$

The Cluster Algorithm. For n fins in cluster, as shown in Fig. 15b, the equivalent thermal transmission ratio will be the sum of the individual fin input admittances:

$$\mu_e = \sum_{k=1}^{n} Y_{in,k} = \sum_{k=1}^{n} \frac{q_b}{\theta_b}\bigg|_k \tag{96}$$

Here, $Y_{in,k}$ can be determined for each individual fin via Eq. (93) if the fin has an adiabatic tip or via Eq. (95) if the tip is not adiabatic. It is obvious that this holds if subarrays containing more than one fin are in cluster.

The Parallel Algorithm. For n fins in parallel, as shown in Fig. 15c, an equivalent thermal admittance matrix $[Y]_e$ can be obtained from the sum of the individual thermal admittance matrices:

$$[Y]_e = \sum_{k=1}^{n} [Y]_k \tag{97}$$

where the individual thermal admittance matrices can be obtained from

$$[Y] = \begin{bmatrix} y_{11} & y_{12} \\ y_{21} & y_{22} \end{bmatrix} = \begin{bmatrix} -\dfrac{\gamma_{11}}{\gamma_{12}} & \dfrac{1}{\gamma_{12}} \\ -\dfrac{1}{\gamma_{12}} & \dfrac{\gamma_{22}}{\gamma_{12}} \end{bmatrix} = \begin{bmatrix} \dfrac{\lambda_{22}}{\lambda_{12}} & -\dfrac{1}{\lambda_{12}} \\ \dfrac{1}{\lambda_{12}} & -\dfrac{\lambda_{21}}{\lambda_{22}} \end{bmatrix} \tag{98}$$

If necessary, $[\Lambda]$ may be obtained from $[Y]$ using

$$[\Lambda] = \begin{bmatrix} \lambda_{11} & \lambda_{12} \\ \lambda_{21} & \lambda_{22} \end{bmatrix} = \begin{bmatrix} -\dfrac{y_{22}}{y_{21}} & \dfrac{1}{y_{21}} \\ -\dfrac{\Delta\gamma}{y_{21}} & \dfrac{y_{11}}{y_{21}} \end{bmatrix} \tag{99}$$

where $\Delta_Y = y_{11}y_{22} - y_{12}y_{21}$

Singular Fans. There will be occasions when a singular fin, one whose tip comes to a point, will be used as the most remote fin in an array. In this case the [Γ] and [Λ] matrices do not exist and the fin is characterized by its input admittance.[38–40] Such a fin is the longitudinal fin of triangular profile where

$$Y_{in} = \frac{q_b}{\theta_b} = \frac{2hI_1(2mb)}{mI_0(2mb)} \tag{100}$$

where

$$m = \left(\frac{2h}{k\delta_b}\right)^{1/2} \tag{101}$$

3.2   The Cold Plate

The cold-plate heat exchanger or forced cooled electronic chassis is used to provide a "cold wall" to which individual components and, for that matter, entire packages of equipment may be mounted. Its design and performance evaluation follows a certain detailed procedure that depends on the type of heat loading and whether the heat loading is on one or two sides of the cold plate. These configurations are displayed in Fig. 16.

The design procedure is based on matching the available heat-transfer effectiveness ϵ to the required effectiveness ϵ determined from the design specifications. These effectivenesses are for the isothermal case in Fig. 16a

$$\epsilon = \frac{t_2 - t_1}{T_s - t_1} = e^{-NTU} \tag{102}$$

and for the isoflux case in Fig. 16b

$$\epsilon = \frac{t_2 - t_1}{T_2 - t_1} \tag{103}$$

where the "number of transfer units" is

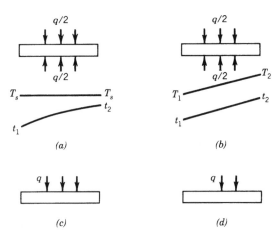

Figure 16   (*a*) Double-sided, evenly loaded cold plate—isothermal case; (*b*) double-sided, evenly loaded cold plate—isoflux case; (*c*) single-sided, evenly loaded cold plate—isothermal case; and (*d*) single-sided, evenly loaded cold plate—isoflux case.

$$\text{NTU} = \frac{h\eta_0 A}{Wc_p} \tag{104}$$

and the overall passage efficiency is

$$\eta_0 = 1 - \frac{A_f}{A}(1 - \eta_f) \tag{105}$$

The surfaces to be used in the cold plate are those described by Kays and London[41] where physical, heat-transfer, and friction data are provided.

The detailed design procedure for the double-side-loaded isothermal case is as follows:

1. Design specification
   (a) Heat load, q, W
   (b) Inlet air temperature, t_1, °C
   (c) Airflow, W, kg/sec
   (d) Allowable pressure loss, cm H_2O
   (e) Overall envelope, H, W, D
   (f) Cold-plate material thermal conductivity, k_m, W/m·°C
   (g) Allowable surface temperature, T_s, °C

2. Select surface[41]
   (a) Type
   (b) Plate spacing, b, m
   (c) Fins per meter, fpm
   (d) Hydraulic diameter, d_e, m
   (e) Fin thickness, δ, m
   (f) Heat-transfer area/volume, β, m²/m³
   (g) Fin surface area/total surface area, A_f/A, m²/m³

3. Plot of j and f data[41]

$$j = (\text{St})(\text{Pr})^{2/3} = f_1(\text{Re}) = f_1\left(\frac{d_e G}{\mu}\right)$$

where St is the Stanton number

$$\text{St} = \frac{hG}{c_p} \tag{106}$$

and f is the friction factor

$$f = f_2(\text{Re}) = f_2\left(\frac{d_e G}{\mu}\right)$$

4. Establish physical data
   (a) $a = (b/H)\beta$, m²/m³
   (b) $r_h = d_e/4$, m
   (c) $\sigma = ar_h$
   (d) $A_{fr} = WH$, m² (frontal area)
   (e) $A_c = \sigma A_{fr}$, m² (flow areas)

(f) $V = DWH$ (volume)

(g) $A = aV$, m² (total surface)

5. Heat balance

(a) Assume average fluid specific heat, c_p, J/kg·°C.

(b) $\Delta t = t_2 - t_1 = q/Wc_p$, °C

(c) $t_2 = t_1 + \Delta t$, °C

(d) $t_{av} = \frac{1}{2}(t_1 + t_2)$

(e) Check assumed value of c_p. Make another assumption if necessary.

6. Fluid properties at t_{av}

(a) c_p (already known), J/kg·°C

(b) μ, N/sec·m²

(c) k, W/m·°C

(d) $(\mathrm{Pr})^{2/3} = (c_p \mu / k)^{2/3}$

7. Heat-transfer coefficient

(a) $G = W/A_c$, kg/sec·m²

(b) $\mathrm{Re} = d_e G / \mu$

(c) Obtain j from curve (see item 3).

(d) Obtain f from curve (see item 3).

(e) $h = jGc_p/(\mathrm{Pr})^{2/3}$, W/m²·°C

8. Fin efficiency

(a) $m = (2h/k\delta)^{1/2}$, m⁻¹

(b) $mb/2$ is a computation

(c) $\eta_f = (\tanh mb/2)/mb/2$

9. Overall passage efficiency

(a) Use Eq. (105).

10. Effectiveness

(a) Required $\epsilon = (t_2 - t_1)/(T_s - T_1)$

(b) Form NTU from Eq. (104).

(c) Actual available $\epsilon = 1 - e^{-\mathrm{NTU}}$

(d) Compare required ϵ and actual ϵ and begin again with step 1 if the comparison fails. If comparison is satisfactory go on to pressure loss calculation.

11. Pressure loss

(a) Establish v_1 (specific volume), m³/k.

(b) Establish v_2, m³/kg.

(c) $v_m = \frac{1}{2}(v_1 + v_2)$, m³/kg

(d) Form v_m/v_1.

(e) Form v_2/v_1.

(f) Obtain K_c and K_e[41]

(g) Determine ΔP, cm.

$$\Delta P = 0.489 \frac{G^2 v_1}{2g} \left[(1 + K_c - \sigma^2) + f \frac{A}{A_c} \frac{v_m}{v_1} + 2 \left(\frac{v_2}{v_1} - 1 \right) - (1 - \sigma^2 - K_c) \frac{v_2}{v_1} \right]$$

(107)

(h) Compare ΔP with specified ΔP. If comparison fails select a different surface or adjust the dimensions and begin again with step 1.

If the cold plate is loaded on one side only, an identical procedure is followed except in steps 8 and 9. For single-side loading and for double and triple stacks, use must be made of the cascade and cluster algorithms for the combination of fins described in Section 3.1. Detailed examples of both of the foregoing cases may be found in Kraus and Bar-Cohen.[11]

3.3 Thermoelectric Coolers

Two thermoelectric effects are traditionally considered in the design and performance evaluation of a thermoelectric cooler:

The Seebeck effect concerns the net conversion of thermal energy into electrical energy under zero current conditions when two dissimilar materials are brought into contact. When the junction temperature differs from a reference temperature, the effect is measured as a voltage called the Seebeck voltage E_s.

The Peltier effect concerns the reversible evolution or absorption of heat that occurs when an electric current traverses the junction between two dissimilar materials. The Peltier heat absorbed or rejected depends on and is proportional to the current flow. There is an additional thermoelectric effect known as the Thomson effect, which concerns the reversible evolution or absorption of heat that occurs when an electric current traverses a single homogeneous material in the presence of a temperature gradient. This effect, however, is a negligible one and is neglected in considerations of thermoelectric coolers operating over moderate temperature differentials.

Equations for Thermoelectric Effects

Given a pair of thermoelectric materials, A and B, with each having a thermoelectric power α_A and α_B,[42] the Seebeck coefficient is

$$\alpha = |\alpha_A| + |\alpha_B|$$

(108)

The Seebeck coefficient is the proportionality constant between the Seebeck voltage and the junction temperature with respect to some reference temperature

$$dE_s = \pm \, \alpha \, dT$$

and it is seen that

$$\alpha = \frac{dE_s}{dt}$$

The Peltier heat is proportional to the current flow and the proportionality constant is Π, the Peltier voltage

$$q_p = \pm \, \pi I$$

(109)

The Thomson heat is proportional to a temperature difference dT and the proportionality constant is σ, the Thomson coefficient. With $dq_T = \pm \sigma I\, dT$, it is observed that $\sigma\, dT$ is a voltage and the Thomson voltage is defined by

$$E_T = \pm \int_{T_1}^{T_2} \sigma\, dT$$

Considerations of the second laws of thermodynamics and the Kirchhoff voltage laws show that the Peltier voltage is related to the Seebeck coefficient[42]

$$\pi = \alpha T \tag{110}$$

and if the Seebeck coefficient is represented as a polynomial[42]

$$\alpha = a + bT + \cdots$$

then

$$\pi = aT + bT^2 + \cdots$$

Design Equations

In Fig. 17, which shows a pair of materials arranged as a thermoelectric cooler, there is a cold junction at T_c and a hot junction at T_h. The materials possess a thermal conductivity k and an electrical resistivity ρ. A voltage is provided so that a current I flows through the cold junction from B to A and through the hot junction from A to B. This current direction is selected to guarantee that $T_c < T_h$.

The net heat absorbed at the cold junction is the Peltier heat

$$q_p = \pi T_c = \alpha I T_c \tag{111a}$$

minus one-half of the I^2R loss (known as the Joule heat or Joule effect)

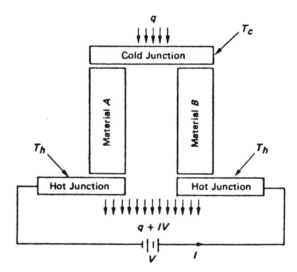

Figure 17   Thermoelectric cooler.

$$q_j = \tfrac{1}{2}I^2R \tag{111b}$$

and minus the heat regained at the cold junction (known as the Fourier heat or Fourier effect) due to the temperature difference $\Delta T = T_h - T_c$

$$q_F = K\,\Delta T = K(T_h - T_c) \tag{111c}$$

Thus, the net heat absorbed at the cold junction is

$$q = \alpha I T_c - \tfrac{1}{2}I^2R - K\,\Delta T \tag{112}$$

where the total resistance of the couple is the series resistance of material A and material B having areas A_A and A_B, respectively (both have length L),

$$R = \left(\frac{\rho_A}{A_A} + \frac{\rho_B}{A_B}\right)L \tag{113}$$

and where the overall conductance K is the parallel conductance of the elements A and B:

$$K = \frac{1}{L}(k_A A_A + k_B A_B) \tag{114}$$

To power the device, a voltage equal to the sum of the Seebeck voltages at the hot and cold junctions plus the voltage necessary to overcome the resistance drop must be provided:

$$V = \alpha T_h - \alpha T_c + RI = \alpha\,\Delta T + RI$$

and the power is

$$P = VI = (\alpha\,\Delta T + RI)I \tag{115}$$

The coefficient of performance (COP) is the ratio of the net cooling effect to the power provided:

$$\text{COP} = \frac{q}{P} = \frac{\alpha T_c I - \tfrac{1}{2}I^2R - K\,\Delta T}{\alpha\Delta T I + I^2R} \tag{116}$$

Optimizations
The maximum possible temperature differential $\Delta T = T_h - T_c$ will occur when there is no net heat absorbed at the cold junction:

$$\Delta T_m = \tfrac{1}{2}zT_c^2 \tag{117}$$

where z is the figure of merit of the material

$$z = \frac{\alpha^2}{KR} \tag{118}$$

The current that yields the maximum amount of heat absorbed at the cold junction can be shown to be[42]

$$I = I_m = \frac{\alpha T_c}{R} \tag{119}$$

and the coefficient of performance in this case will be

$$\text{COP}_m = \frac{1 - \Delta T / \Delta T_m}{2(1 + \Delta T / T_c)} \tag{120}$$

The current that optimizes or maximizes the coefficient of performance can be shown to be

$$I_0 = \frac{\alpha \, \Delta T}{R[(1 + z T_a)^{1/2} - 1]} \tag{121}$$

where $T_a = \frac{1}{2}(T_h + T_c)$. In this case, the optimum coefficient of performance will be

$$\text{COP}_0 = \frac{T_c}{\Delta T} \left[\frac{\gamma - (T_h/T_c)}{\gamma + 1} \right] \tag{122}$$

where

$$\gamma = [1 + \frac{1}{2} z \, (T_h + T_c)]^{1/2} \tag{123}$$

Analysis of Thermoelectric Coolers

In the event that a manufactured thermoelectric cooling module is being considered for a particular application, the designer will need to specify the number of junctions required. A detailed procedure for the selection of the number of junctions is as follows:

1. Design specifications
   - (a) Total cooling load, q_T, W
   - (b) Cold-side temperature, T_c, K
   - (c) Hot-side temperature, T_h, K
   - (d) Cooler specifications
       - i. Materials A and B
       - ii. α_A and α_B, V/°C
       - iii. ρ_A and ρ_B, ohm·cm
       - iv. k_A and k_B, W/cm·°C
       - v. A_A and A_B, cm^2
       - vi. L, cm

2. Cooler calculations
   - (a) Establish $\alpha = |\alpha_A| + |\alpha_B|$.
   - (b) Calculate R from Eq. (113).
   - (c) Calculate K from Eq. (114).
   - (d) Form $\Delta T = T_h - T_c$, K or °C.
   - (e) Obtain z from Eq. (118), 1/°C.

3. For maximum heat pumping per couple
   - (a) Calculate I_m from Eq. (119), A.
   - (b) Calculate the heat absorbed by each couple q, from Eq. (112), W.
   - (c) Calculate ΔT_m from Eq. (117), K or °C.
   - (d) Determine COP$_m$ from Eq. (120).
   - (e) The power required per couple will be $p = q/\text{COP}_m$, W.

(**f**) The heat rejected per couple will be $p + q$, W.

(**g**) The required number of couples will be $n = q_T/q$.

(**h**) The total power required will be $p_T = nP$, W.

(**i**) The total heat rejected will be $q_{RT} = nq_R$, W.

3A. For optimum coefficient of performance

(**a**) Determine $T_a = \frac{1}{2}(T_h + T_c)$, K.

(**b**) Calculate I_0 from Eq. (121), A.

(**c**) Calculate the heat absorbed by each couple, q, from Eq. (112), W.

(**d**) Determine γ from Eq. (123).

(**e**) Determine COP_0 from Eq. (122).

(**f**) The power required per couple will be $P = q/\text{COP}_0$, W.

(**g**) The heat rejected per couple will be $q_R = P + q$, W.

(**h**) The required number of couples will be $n = q_T/q$.

(**i**) The total power required will be $P_T = nP$, W.

(**j**) The total heat rejected will be $q_{RT} = nq_R$, W.

3.4 Spray Cooling

The use of impinging fluid jets for the thermal management of electronic components has received extensive attention in recent years. The high heat-transfer coefficients that can be attained in this cooling mode, the ability to vary and control the heat transfer rates across a large substrate or printed circuit board with an appropriately configured distribution plate or nozzle array, and the freedom to tailor the jet flow to the local cooling requirements, have made spray cooling one of the most promising alternatives for the cooling of high heat-flux components.

Spray cooling may involve a single jet or multiple jets directed at a single component or an array of electronic components. The jets may be formed using circular or slot-shaped orifices, or nozzles of various cross sections. The axis of the impinging jet may be perpendicular or inclined to the surface of the component. Moreover, in the application of liquid jets, a distinction can be made between "free jets," which are surrounded by ambient air, and "submerged jets," for which the volume surrounding the jet is filled with the working liquid. While heat transfer associated with gas jets has been the subject of active research since the mid-1950s, spray cooling cooling with dielectric liquids is a far more recent development. Several reviews of the many pioneering and more recent studies on spray cooling can be found in literature.[43–48] This discussion focuses primarily on single-phase, forced convection.

Despite the complex behavior of the local heat-transfer coefficient resulting from parametric variations in the impinging jet flow, it has been found possible to correlate the average heat transfer coefficient with a single expression, for both individual jets and arrays of jets impinging on isothermal surfaces. Martin[49] proposed a relation of the form shown below to capture the effects of jet Reynolds number, nondimensional distance of separation (H/D), impinging area ratio (f), Prandtl number (Pr), and fluid thermal conductivity, on the jet Nusselt number

$$Nu_D = \left[1 + \left(\frac{H/D}{0.6/\sqrt{f}} \right)^3 \right]^{-0.05} \left[\sqrt{f} \, \frac{1 - 2.2\sqrt{f}}{1 + 0.2(H/D - 6)\sqrt{f}} \right] \text{Re}_D^{0.667} \, \text{Pr}^{0.42} \quad (124)$$

The range of validity[49] for this correlation, developed from extensive gas jet data, as well as some data for water and other, higher Prandtl number liquids, and including some high Schmidt number mass-transfer data, is $2 \times 10^3 \le Re_D \le 10^5$, $0.6 < Pr(Sc) < 7(900)$, $0.004 \le f \le 0.04$ and $2 \le H/D \le 12$. This correlation has a predictive accuracy of 10–20% over the stated parametric range. The average Nu was also found to be nearly unaffected by the angle of inclination of the jet. It is to be noted that for jets produced by sharp-edged orifices, jet contraction immediately after the orifice exit must be taken into consideration in calculating the average velocity, jet diameter, and nozzle area ratio, f. In applying this correlation to the cooling of electronic components, constituting discrete heat sources on a large surface, it is necessary to alter the definition of the jet area ratio, f. Recognizing that, in this application, the impingement area is usually equal to the component area, f can be expressed as

$$f = \frac{nA_{jet}}{A} = \frac{0.785D^2n}{A} \tag{125}$$

Figure 18 shows a comparison correlations by Womac,[50] Brdlik and Savin,[51] and Sitharamayya and Raju[52] with the predictions from Eq. (43) for $f = 0.008$, $H/D = 3$, and $Pr = 13.1$. The Maximum deviation of the data from predictions made using Eq. (24) was 32% in Figure 18. The Martin[49] correlation (Eq. (124)) is recommended over these alternatives primarily because of the broad range of parameters in the database from which it was developed and because it often falls below the other correlations, making it a conservative choice.

The variation of the Nusselt number with each of the three primary factors influencing spray cooling in the range of the Martin[49] correlation, Eq. (124), is shown in Figs 19–21. These figures show that the Nusselt number increases steadily with Reynolds number (Fig. 19), and decreases with the ratio of jet distance to jet diameter (Fig. 20). More surprisingly, the curves shown in Fig. 21 indicate that Nusselt number reaches an asymptote in its dependence on the ratio of jet area to component area.

Returning to Eq. (124), it may be observed that, in the range of interest, the first term on the right side can be approximated as

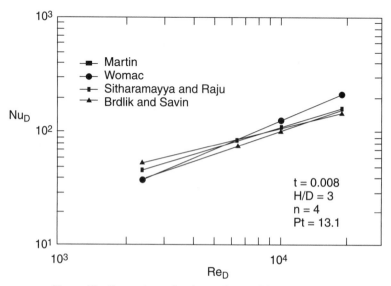

Figure 18 Comparison of various submerged jet correlations.

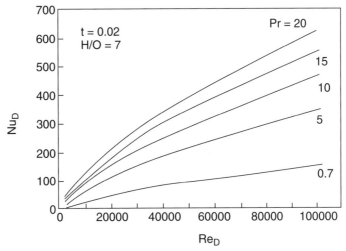

Figure 19   Effect of Reynolds and Prandtl numbers on heat transfer.

$$\left\{ 1 + \left[\frac{H/D}{0.6/f^{0.5}} \right]^6 \right\}^{-0.05} \cong \left\{ \frac{H/D}{0.6/f^{0.5}} \right\}^{-0.3} \tag{126}$$

and that the second term is not far different from $0.6f^{0.5}$. Reexpressing Eq. (126) with these simplifications, the average heat-transfer coefficient, h, is found to approximately equal

$$Nu \cong 0.5 \left(\frac{H}{D} \right)^{-0.3} f^{0.35} \, \text{Re}_D^{0.667} \, \text{Pr}^{0.42} \tag{127}$$

This approximation falls within 30% of Eq. (124) throughout the parametric range indicated, but is within 10% for $H/D < 3$.

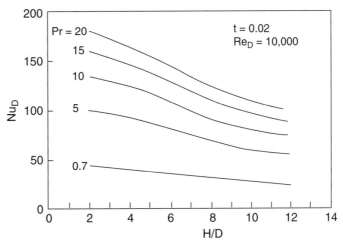

Figure 20   Effect of jet aspect ratio on heat transfer.

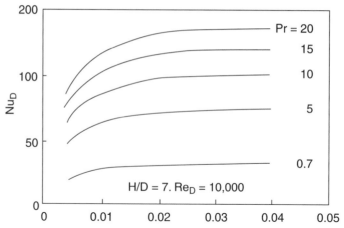

Figure 21   Effect of jet area to heater area ratio on heat transfer.

Recalling the definition of the jet Nusselt number, $Nu = hD/k$, and substituting for the area ratio, f, from Eq. (125), the heat-transfer coefficient produced by impinging liquid jet(s) is found to be proportional to

$$h \propto kH^{-0.3} \left(\frac{n}{A}\right)^{0.35} \mathrm{Re}_D^{0.67} \, \mathrm{Pr}^{0.42} \tag{128}$$

Or, expanding the Reynolds and Prandtl numbers,

$$h \propto [k^{0.58} \, \rho^{0.67} \, \mu^{-0.25}] \left[\left(\frac{n}{A}\right)^{0.35} D^{0.67}\right] \left[\frac{V^{0.67}}{H^{0.3}}\right] \tag{129}$$

The first bracketed term in Eq. (129) represents a fluid figure-of-merit for submerged-jet heat transfer, the second term constitutes a thermal figure-of-merit for the jet plate, and the third the operating conditions of an impingement cooling system. Clearly, to maximize the jet heat-transfer rate, it is desirable to choose a fluid with high thermal conductivity and density but relatively low viscosity. Within the accuracy of the approximations used to derive Eq. (129) (and especially in the low f range), the thermally preferred jet plate would contain many large-diameter nozzles per component. Due to the strong dependence of the heat-transfer rate on the jet Reynolds number, maximization of the heat-transfer coefficient also requires increasing the fluid velocity at the nozzle and decreasing the distance of separation between the nozzle and the component. Alternatively, if a fluid has been selected and if the jet Reynolds number is to remain constant, a higher heat-transfer coefficient can be obtained only by increasing n/A or decreasing H.

Although the thermal relations discussed in the previous section can be used to establish the gross feasibility of submerged spray cooling cooling for high-power chips, successful implementation of this thermal management technique requires consideration of system-level issues and design trade-offs.[53] The minimization of life-cycle costs is a crucial element in electronic systems and, consequently, attention must be devoted to the "consumed" fluid flow rate, pressure drop, and pumping power as well as to the limitations imposed by manufacturing tolerances and costs. The gross impact of these considerations on the design of impinging jet cooling systems can be seen with the aid of Eq. (129).

The nozzle pressure drop and, hence, plenum pressure required to achieve a specified jet velocity have direct bearing on the choice and cost of the coolant circulation system and the structural design, as well as cost, of the jet plate. Since Eqs. (124) and (129) show the jet heat-transfer coefficient to increase with velocity to the 0.67 power, while nozzle pressure losses generally depend on the square of the velocity, concern about the plenum operating pressure would lead the designer to choose the lowest possible jet velocity.

Examination of the approximate relation for the jet heat-transfer coefficient, Eq. (129), suggests that to maintain high heat-transfer rates at low jet velocities would necessitate increasing the number of nozzles (n/A), increasing the diameter of each nozzle (D), or decreasing the spacing between the nozzle exit and the component (H). The minimum value of H is likely to be determined by the precision of assembly and deflection under pressure of the jet plate and, thus, it will benefit from reduced operating pressure. Since the maximum heat-transfer rates are approached asymptotically as the total jet area increases to approximately 4% of the component area (see Figure 21), there is coupling between the number of jets and the jet diameter. The heat-transfer rate can, thus, be improved by increasing both jet diameter and the number of jets up to this value, but if operating near the maximum rate, the jet diameter is inversely related to the square root of n/A.

In the use of liquid jets, the operating costs are often dominated by the pumping power, or product of total volumetric flow rate and pressure drop, needed to provide a specified heat-removal rate. The pumping power can be easily shown to vary with $D^2 V^3 n/A$. Examining this dependence in light of the approximate heat-transfer coefficient relation, Eq. (48), it is again clear that reduced costs are associated with low liquid velocity and a relatively large total nozzle area.

These results suggest that optimum performance, based on system-level as well as thermal considerations, and as represented by the average heat-transfer coefficient, would be achieved by designing spray cooling systems to provide approximately 4% jet-to-component area ratios and operate at relatively low jet velocities. Improved surface coverage, more uniform heat removal capability, and decreased vulnerability to blockage of a single (or a few) nozzles would appear to be favored by the use of a relatively large number of jets per component, allowing reduction in the diameter of individual jets. Alternately, the cost of manufacturing and the probability of nozzle blockage can be expected to increase for small diameter nozzles and, thus, place a lower practical limit on this parameter. Given the approximate nature of Eq. (129), these relationships must be viewed as indicative, rather than definitive and the complete Martin[49] correlation Eq. (124) should be used for any detailed exploration of these trends.

Heat-transfer correlations for impingement-cooled, pin-fin heat sinks have also been reported in literature.[54] Results were presented in terms of Nu_{base}, based on the heat-sink footprint area, and on Nu_{HS}, based on the total heat-sink surface area, as

$$Nu_{base} = 3.361 Re^{0.724} \, Pr^{0.4} \left(\frac{D}{d}\right)^{-0.689} \left(\frac{S}{d}\right)^{-0.210} \tag{130}$$

$$Nu_{HS} = 1.92 \, Re^{0.716} \, Pr^{0.4} \left(\frac{A_{HS}}{A_d}\right)^{-0.689} \left(\frac{D}{d}\right)^{0.678} \left(\frac{S}{d}\right)^{-0.181} \tag{131}$$

These equations are valid for $2000 \leq Re \leq 23{,}000$, $S/d = 2$ and 3. The term D in the above equations refers to the jet diameter, whereas d refers to the pin diameter. The term S refers to the jet to jet pitch in the case of multiple jets.

REFERENCES

1. K. J. Negus, R. W. Franklin, and M. M. Yovanovich, "Thermal Modeling and Experimental Techniques for Microwave Bi-Polar Devices," *Proceedings of the Seventh Thermal and Temperature Symposium,*" San Diego, CA, 1989, pp. 63–72.

2. M. M. Yovanovich and V. W. Antonetti, "Application of Thermal Contact Resistance Theory to Electronic Packages," in *Advances in Thermal Modeling of Electronic Components and Systems,* A. Bar-Cohen and A. D. Kraus (eds.), Hemisphere, New York, 1988, pp. 79–128.

3. *Handbook of Chemistry and Physics (CRC),* Chemical Rubber Co., Cleveland, OH, 1954.

4. W. Elenbaas, "Heat Dissipation of Parallel Plates by Free Convection," *Physica* **9**(1), 665–671 (1942).

5. J. R. Bodoia and J. F. Osterle, "The Development of Free Convection Between Heated Vertical Plates," *J. Heat Transfer* **84,** 40–44 (1964).

6. A. Bar-Cohen, "Fin Thickness for an Optimized Natural Convection Array of Rectangular Fins," *J. Heat Transfer* **101,** 564–566.

7. A. Bar-Cohen and W. M. Rohsenow, "Thermally Optimum Arrays of Cards and Fins in Natural Convection," *Trans. IEEE Chart, CHMT-6,* 154–158.

8. E. M. Sparrow, J. E. Niethhammer, and A. Chaboki, "Heat Transfer and Pressure Drop Characteristics of Arrays of Rectangular Modules Encountered in Electronic Equipment," *Int. J. Heat Mass Transfer* **25**(7), 961–973 (1982).

9. R. A. Wirtz and P. Dykshoorn, "Heat Transfer from Arrays of Flatpacks in Channel Flow," *Proceedings of the Fourth Int. Electronic Packaging Society Conference,* New York, 1984, pp. 318–326.

10. S. B. Godsell, R. J. Dischler, and S. M. Westbrook, "Implementing a Packaging Strategy for High Performance Computers," *High Performance Systems,* 28–31 (January 1990).

11. A. D. Kraus and A. Bar-Cohen, *Design and Analysis of Heat Sinks,* Wiley, New York, 1995.

12. M. Jakob, *Heat Transfer,* Wiley, New York, 1949.

13. S. Globe and D. Dropkin, "Natural Convection Heat Transfer in Liquids Confined by Two Horizontal Plates and Heated from Below," *J. Heat Transfer, Series C* **81,** 24–28 (1959).

14. W. Mull and H. Rieher, "Der Warmeschutz von Luftschichten," *Gesundh-Ing. Beihefte* **28** (1930).

15. J. G. A. DeGraaf and E. F. M. von der Held, "The Relation Between the Heat Transfer and the Convection Phenomena in Enclosed Plane Air Layers," *Appl. Sci. Res., Sec. A* **3,** 393–410 (1953).

16. A. Bar-Cohen and W. M. Rohsenow, "Thermally Optimum Spacing of Vertical, Natural Convection Cooled, Parallel Plates," *J. Heat Transfer* **106,** 116–123 (1984).

17. S. W. Churchill and R. A. Usagi, "A General Expression for the Correlation of Rates of Heat Transfer and Other Phenomena," *AIChE J.* **18**(6), 1121–1138 (1972).

18. N. Sobel, F. Landis, and W. K. Mueller, "Natural Convection Heat Transfer in Short Vertical Channels Including the Effect of Stagger," *Proceedings of the Third International Heat Transfer Conference,* Vol. 2, Chicago, IL, 1966, pp. 121–125.

19. W. Aung, L. S. Fletcher, and V. Sernas, "Developing Laminar Free Convection Between Vertical Flat Plates with Asymmetric Heating," *Int. J. Heat Mass Transfer* **15,** 2293–2308 (1972).

20. O. Miyatake, T. Fujii, M. Fujii, and H. Tanaka, "Natural Convection Heat Transfer Between Vertical Parallel Plates—One Plate with a Uniform Heat Flux and the Other Thermally Insulated," *Heat Transfer Japan Research* **4,** 25–33 (1973).

21. W. Aung, "Fully Developed Laminar Free Convection Between Vertical Flat Plates Heated Asymmetrically," *Int. J. Heat Mass Transfer* **15,** 1577–1580 (1972).

22. W. H. McAdams, *Heat Transmission,* 3rd ed., McGraw-Hill, New York, 1954.

23. R. Hilpert, Warmeabgue von Geheizten Drähten and Rohren in Lufstrom, *Forsch, Ing-Wes* **4,** 215–224 (1933).

24. S. Whitaker, "Forced Convection Heat Transfer Correlations for Flow in Pipes, Past Flat Plates, Single Cylinders, Single Spheres and for Flow in Packed Beds and Tube Bundles," *AIChE Journal* **18,** 361–371 (1972).

25. F. Kreith, *Principles of Heat Transfer,* International Textbook Co., Scranton, PA, 1959.

26. A. P. Colburn, "A Method of Correlating Forced Convection Heat Transfer Data and a Comparison of Fluid Friction," *Trans AIChE* **29,** 174–210 (1933).

27. "Standards of the Tubular Exchanger Manufacturer's Association," New York, 1949.

28. W. Drexel, "Convection Cooling," *Sperry Engineering Review* **14,** 25–30 (December 1961).

29. W. Robinson and C. D. Jones, *The Design of Arrangements of Prismatic Components for Crossflow Forced Air Cooling,* Ohio State University Research Foundation Report No. 47, Columbus, OH, 1955.

30. W. Robinson, L. S. Han, R. H. Essig, and C. F. Heddleson, *Heat Transfer and Pressure Drop Data for Circular Cylinders in Ducts and Various Arrangements,* Ohio State University Research Foundation Report No. 41, Columbus, OH, 1951.

31. E. N. Sieder and G. E. Tate, "Heat Transfer and Pressure Drop of Liquids in Tubes," *Ind. Eng. Chem.* **28,** 1429–1436 (1936).

32. H. Hausen, *Z VDI, Beih. Verfahrenstech.* **4,** 91–98 (1943).

33. A. L. London, "Air Coolers for High Power Vacuum Tubes," *Trans. IRE* **ED-1,** 9–26 (April, 1954).

34. K. A. Gardner, "Efficiency of Extended Surfaces," *Trans. ASME* **67,** 621–631 (1945).

35. W. M. Murray, "Heat Transfer Through an Annular Disc or Fin of Uniform Thickness," *J. Appl. Mech.* **5,** A78–A80 (1938).

36. D. Q. Kern and A. D. Kraus, *Extended Surface Heat Transfer,* McGraw-Hill, New York, 1972.

37. D. E. Briggs and E. H. Young, "Convection Heat Transfer and Pressure Drop of Air Flowing across Triangular Pitch Banks of Finned Tubes," *Chem. Eng. Prog. Symp. Ser.* **41**(59), 1–10 (1963).

38. A. D. Kraus, A. D. Snider, and L. F. Doty, "An Efficient Algorithm for Evaluating Arrays of Extended Surface," *J. Heat Transfer* **100,** 288–293 (1978).

39. A. D. Kraus, *Analysis and Evaluation of Extended Surface Thermal Systems,* Hemisphere, New York, 1982.

40. A. D. Kraus and A. D. Snider, "New Parametrizations for Heat Transfer in Fins and Spines," *J. Heat Transfer* **102,** 415–419 (1980).

41. W. M. Kays and A. L. London, *Compact Heat Exchangers,* 3rd ed., McGraw-Hill, New York, 1984.

42. A. D. Kraus and A. Bar-Cohen, *Thermal Analysis and Control of Electronic Equipment,* Hemisphere, New York, 1983.

43. A. E. Bergles and A. Bar-Cohen, "Direct Liquid Cooling of Microelectronic Components," in *Advances in Thermal Modeling of Electronic Components and Systems,* Vol. 2, A. Bar-Cohen and A. D. Kraus (eds.), ASME Press, New York, 1990, pp. 233–342.

44. J. Stevens and B. W. Webb, "Local Heat Transfer Coefficients Under an Axisymmetric, Single-phase Liquid Jet," National Heat Transfer Conference, Philadelphia, Pennsylvania, 1989, pp. 113–119.

45. T. Nonn, Z. Dagan, and L. M. Jiji, "Boiling Jet Impingement Cooling of Simulated Microelectronic Heat Sources," ASME Paper 88-WA/EEP-3, ASME, New York, 1988.

46. X. S. Wang, Z. Dagan, and L. M. Jiji, "Heat Transfer Between a Laminar Free-surface Impinging Jet and a Composite Disk," *Proceedings of the Ninth International Heat Transfer Conference,* Vol. 4, Hemisphere, New York, 1990, p. 137.

47. D. J. Womac, G. Aharoni, S. Ramadhyani, and F. P. Incropera, "Single Phase Liquid Jet Impingement Cooling of Small Heat Sources," *Heat Transfer 1990* (Proceedings of the Ninth International Heat Transfer Conference), Vol. 4, Hemisphere, New York, 1990, pp. 149–154.

48. D. C. Wadsworth and L. Mudawar, "Cooling of a Multichip Electronic Module by Means of Confined Two Dimensional Jets of Dielectric Liquid," *J. Heat Transfer* **112,** 891–898 (1990).

49. H. Martin, "Heat and Mass Transfer Between Impinging Gas Jets and Solid Surfaces," in *Advanced in Heated Transfer,* Vol. 13, J. P. Hartnett and T. F. Irvine, Jr. (eds.), Academic Press, New York, 1977, pp. 1–60.

50. D. H. Womac, *Single-phase Axisymmetric Liquid Jet Impingement Cooling of Discrete Heat Sources,* Thesis, Department of Mechanical Engineering, Purdue University, Lafayette, IN, 1989.

51. P. M. Brdlik and V. K. Savin, "Heat Transfer Between an Axisymmetric Jet and a Plate Normal to the Flow," *J. Eng. Phys.* **8,** 91–98 (1965).

52. S. Sitharamayya and K. S. Raju, "Heat Transfer Between an Axisymmetric Jet and a Plate Hold Normal to the Flow," *Can. J. Chem. Eng.* **47,** 365–368 (1969).

53. D. E. Maddox and A. Bar-Cohen, "Thermofluid Design of Submerged-jet Impingement Cooling for Electronic Components," Proceedings, ASME/AICHE National Heat Transfer Conference, Minneapolis, Minnesota, 1991.

54. H. A. El Sheikh and S. V. Garimella, "Heat Transfer from Pin-Fin Heat Sinks Under Multiple Impinging Jets," *IEEE Transactions on Advanced Packaging* **23**(1), 113–120 (2000).

55. K. E. Starner and H. N. McManus, "An Experimental Investigation of Free Convection Heat Transfer from Rectangular Fin Arrays," *J. Heat Mass Transfer* **85,** 273–278 (1963).

56. J. R. Welling and C. B. Wooldridge, "Free Convection Heat Transfer Coefficients from Rectangular Vertical Fins," *J. Heat Transfer* **87,** 439–444 (1965).

57. A. Bilitzky, *The Effect of Geometry on Heat Transfer by Free Convection from a Fin Array,* Thesis, Department of Mechanical Engineering, Ben-Gurion University of the Negev, Beer Sheva, Israel, 1986.

58. M. G. Cooper, B. B. Mikic, and M. M. Yovanovich, "Thermal Contact Resistance," *Int. J. Heat Mass Transfer* **12,** 279–300 (1969).

59. M. M. Yovanovich and V. W. Antonetti, "Application of Thermal Contact Resistance Theory to Electronic Packages," in A. Bar-Cohen and A. D. Kraus, (eds.), *Advances in Thermal Modeling of Electronic Components and Systems,* Vol. 1, Hemisphere, New York, 1988.

60. H. J. Sauer, Jr., "Comparative Enhancement of Thermal Contact Conductance of Various Classes of Interstitial Materials, NSF/DITAC Workshop, Melbourne, Monash University, Victoria, Australia, 1992, pp. 103–115.

61. L. S. Fletcher, "A Review of Thermal Control Materials for Metallic Junctions," *J. Spacecrafts Rockets* 849–850 (1972).

62. R. S. Prasher, "Surface Chemistry and Characteristics Based Model for the Contact Resistance of Polymeric Interstitial Thermal Interface Materials," *J. Heat Transfer* **123** (2001).

CHAPTER 12
REFRIGERATION

Dennis L. O'Neal
Department of Mechanical Engineering
Texas A&M University
College Station, Texas

1   INTRODUCTION

Refrigeration is the use of mechanical or heat-activated machinery for cooling purposes. The use of refrigeration equipment to produce temperatures below −150°C is known as cryogenics.[1] When refrigeration equipment is used to provide human comfort, it is called air conditioning. This chapter focuses primarily on refrigeration applications, which cover such diverse uses as food processing and storage, supermarket display cases, skating rinks, ice manufacture, and biomedical applications such as blood and tissue storage or hypothermia used in surgery.

The first patent on a mechanically driven refrigeration system was issued to Jacob Perkins in 1834 in London.[2] The system used ether as the refrigerant. The first viable commercial system was produced in 1857 by James Harrison and D. E. Siebe and used ethyl ether as the refrigerant.[2]

Refrigeration is used in installations covering a broad range of cooling capacities and temperatures. While the variety of applications results in a diversity of mechanical specifications and equipment requirements, the methods for producing refrigeration are well standardized.

2   BASIC PRINCIPLES

Most refrigeration systems utilize the vapor compression cycle to produce the desired refrigeration effect. A less common method used to produce refrigeration is the absorption cycle, which is described later in this chapter. With the vapor compression cycle, a working fluid, called the refrigerant, evaporates and condenses at suitable pressures for practical equipment designs. The ideal (no pressure or frictional losses) vapor compression refrigeration cycle is illustrated in Fig. 1 on a pressure–enthalpy diagram. This cycle has no pressure loss in the evaporator or condenser, and no heat or frictional losses in the compressor.

There are four basic components in every vapor compression refrigeration system: (1) compressor, (2) condenser, (3) expansion device, and (4) evaporator. The compressor raises the pressure of the refrigerant vapor so that the refrigerant saturation temperature is slightly above the temperature of the cooling medium used in the condenser. The condenser is a heat exchanger used to reject heat from the refrigerant to a cooling medium. The refrigerant enters the condenser and usually leaves as a subcooled liquid. Typical cooling mediums used in condensers are air and water. After leaving the condenser, the liquid refrigerant expands to

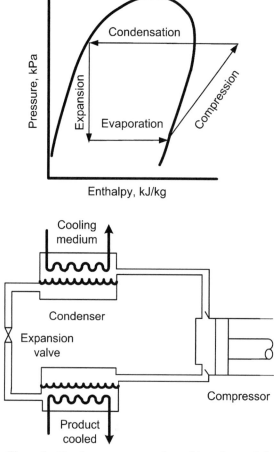

Figure 1   Simple vapor compression refrigeration cycle.[3]

a lower pressure in the expansion valve. The expansion valve can be a passive device, such as a capillary tube or short tube orifice, or an active device, such as a thermal expansion valve or electronic expansion valve. At the exit of the expansion valve, the refrigerant is at a temperature below that of the product to be cooled. As the refrigerant travels through the evaporator, it absorbs energy and is converted from a low-quality, two-phase fluid to a superheated vapor under normal operating conditions. The vapor formed must be removed by the compressor at a sufficient rate to maintain the low pressure in the evaporator and keep the cycle operating.

Pumped recirculation of a liquid secondary refrigerant rather than direct evaporation of a refrigerant is often used to service remotely located or specially designed heat exchangers. This technique is called indirect refrigeration and provides the user with wide flexibility in applying refrigeration to complex processes and greatly simplifies operation. Secondary refrigerants or brines are also commonly used for simple control and operation. Direct application of ice and brine storage tanks may be used to level off batch cooling loads and reduce equipment size. This approach provides stored refrigeration where temperature control is vital as a safety consideration to prevent runaway reactions or pressure buildup.

All mechanical cooling results in the production of a greater amount of heat energy than cooling energy. In many instances, this heat energy is rejected to the environment directly to the air in the condenser or indirectly to water where it is rejected in a cooling tower. Under some specialized applications, it may be possible to utilize this heat energy in another process at the refrigeration facility. This may require special modifications to the condenser. Recovery of this waste heat at temperatures up to 65°C can be used to achieve improved operating economy.

Historically, in the United States, capacities of mechanical refrigeration systems have been stated in tons of refrigeration, which is a unit of measure related to the ability of an ice plant to freeze one short ton (907 kg) of ice in a 24-hr period. A ton is equal to 3.51 kW (12,000 Btu/hr).

3 REFRIGERATION CYCLES AND SYSTEM OVERVIEW

Refrigeration can be accomplished in either closed-cycle or open-cycle systems. In a closed cycle, the refrigerant fluid is confined within the system and recirculates through the components (compressor, heat exchangers, and expansion valve). The system shown at the bottom of Fig. 1 is a closed cycle. In an open cycle, the fluid used as the refrigerant passes through the system once on its way to be used as a product or feedstock outside the refrigeration process. An example is the cooling of natural gas to separate and condense heavier components.

In addition to the distinction between open- and closed-cycle systems, refrigeration processes are also described as simple cycles, compound cycles, or cascade cycles. Simple cycles employ one set of components (compressor, condenser, evaporator, and expansion valve) in a single refrigeration cycle as shown in Fig. 1. Compound and cascade cycles use multiple sets of components and two or more refrigeration cycles. The cycles interact to accomplish cooling at several temperatures or to allow a greater span between the lowest and highest temperatures in the system than can be achieved with the simple cycle.

3.1 Closed-Cycle Operation

For a simple cycle, the lowest evaporator temperature that is practical in a closed-cycle system (Fig. 1) is set by the pressure-ratio capability of the compressor and by the properties of the refrigerant. Most high-speed reciprocating compressors are limited to a pressure ratio of 9:1, so that the simple cycle is used for evaporator temperatures from 2 to −50°C. Below these temperatures, the application limits of a single-stage compressor are reached. Beyond that limit, there is a risk of excessively high temperatures at the end of compression, which may produce lubricant breakdown, high bearing loads, excessive oil foaming at startup, and inefficient operation because of reduced volumetric efficiency in the compressor.

Centrifugal compressors with multiple stages can generate a pressure ratio up to 18:1, but their high discharge temperatures limit the efficiency of the simple cycle at these high pressure ratios. As a result, they operate with evaporator temperatures in the same range as reciprocating compressors.

The compound cycle (Fig. 2) can achieve temperatures of approximately −100°C by using two or three compressors in series and a common refrigerant. This keeps the individual compressors within their application limits. A refrigerant gas cooler (also called a flash intercooler) is normally used between compressors to keep the final discharge temperature from the compressor at a satisfactory level. A common practice is to operate the gas cooler at about the geometric mean of the evaporating and condensing pressures. This provides nearly identical pressure ratios for the two compressors. Besides producing very low temperatures, the compound cycle can also be used in applications where multiple evaporators are needed to produce different temperatures.

Below −100°C, most refrigerants with suitable evaporator pressures have excessively high condensing pressures. For some refrigerants, the specific volume of refrigerant at low temperatures may be so great as to require compressors and other equipment of uneconomical size. With other refrigerants, the specific volume of refrigerant may be satisfactory at low temperature, but the specific volume may become too small at the condensing condition. In some circumstances, although none of the above limitations is encountered and a single refrigerant is practical, the compound cycle is not used because of oil-return problems or difficulties of operation.

To satisfy these conditions, the cascade cycle is used (Fig. 3). This consists of two or more separate refrigerants, each in its own closed cycle. The cascade condenser–evaporator

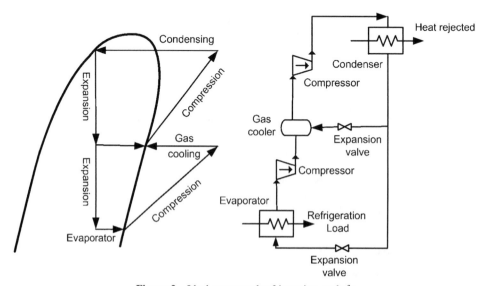

Figure 2   Ideal compound refrigeration cycle.[3]

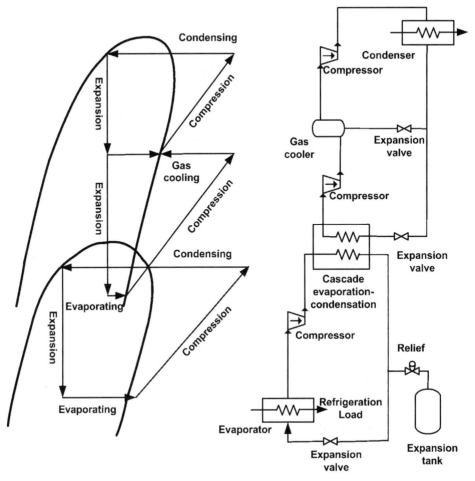

Figure 3 Ideal cascade refrigeration cycle.[3]

rejects heat to the evaporator of the high-temperature cycle, which condenses the refrigerant of the low-temperature cycle. Refrigerants are selected for each cycle with pressure–temperature characteristics that are well suited for application at either the higher or lower portion of the cycle. For extremely low temperatures, more than two refrigerants may be cascaded to produce evaporator temperatures at cryogenic conditions (below −150°C). Expansion tanks, sized to handle the low-temperature refrigerant as a gas at ambient temperatures, are used during standby to hold pressure at levels suitable for economical equipment design.

Compound cycles using reciprocating compressors or any cycle using a multistage centrifugal compressor allows the use of economizers or intercoolers between compression stages. Economizers reduce the discharge gas temperature from the preceding stage by mixing relatively cool gas with discharge gas before entering the subsequent stage. Either flash-type economizers, which cool the refrigerant by reducing its pressure to the intermediate level, or surface-type economizers, which subcool refrigerant at the condensing pressure, may be used to provide the cooler gas for mixing. This keeps the final discharge gas temperature low enough to avoid overheating of the compressor and improves compression efficiency.

Compound compression with economizers also affords the opportunity to provide refrigeration at an intermediate temperature. This provides a further thermodynamic efficiency gain because some of the refrigeration is accomplished at a higher temperature, and less refrigerant must be handled by the lower-temperature stages. This reduces the power consumption and the size of the lower stages of compression.

Figure 4 shows a typical system schematic with flash-type economizers. Process loads at several different temperature levels can be handled by taking suction to an intermediate compression stage as shown. The pressure–enthalpy diagram illustrates the thermodynamic cycle.

Flooded refrigeration systems are a version of the closed cycle that may reduce design problems in some applications. In flooded systems, the refrigerant is circulated to heat exchangers or evaporators by a pump. Figure 5 shows a liquid recirculator, which can use any of the simple or compound closed-refrigeration cycles.

The refrigerant recirculating pump pressurizes the refrigerant liquid and moves it to one or more evaporators or heat exchangers, which may be remote from the receiver. The low-pressure refrigerant may be used as a single-phase heat-transfer fluid as in (A) of Fig. 5, which eliminates the extra heat-exchange step and increased temperature difference encountered in a conventional system that uses a secondary refrigerant or brine. This approach may simplify the design of process heat exchangers, where the large specific volumes of evaporating refrigerant vapor would be troublesome. Alternatively, the pumped refrigerant in the flooded system may be routed through conventional evaporators as in (B) and (C), or special heat exchangers as in (D).

The flooded refrigeration system is helpful when special heat exchangers are necessary for process reasons, or where multiple or remote exchangers are required.

3.2   Open-Cycle Operation

In many chemical processes, the product to be cooled can itself be used as the refrigerating liquid. An example of this is in a natural gas gathering plant. Gas from the wells is cooled, usually after compression and after some of the heavier components are removed as liquid. This liquid may be expanded in a refrigeration cycle to further cool the compressed gas, which causes more of the heavier components to condense. Excess liquid not used for refrigeration is drawn off as product. In the transportation of liquefied petroleum gas (LPG) and of ammonia in ships and barges, the LPG or ammonia is compressed, cooled, and expanded. The liquid portion after expansion is passed on as product until the ship is loaded.

Open-cycle operations are similar to closed-cycle operations, except that one or more parts of the closed cycle may be omitted. For example, the compressor suction may be taken directly from gas wells, rather than from an evaporator. A condenser may be used, and the liquefied gas may be drained to storage tanks.

Compressors may be installed in series or parallel for operating flexibility or for partial standby protection. With multiple reciprocating compressors, or with a centrifugal compressor, gas streams may be picked up or discharged at several pressures if there is refrigerating duty to be performed at intermediate temperatures. It always is more economical to refrigerate at the highest temperature possible.

Principal concerns in the open cycle involve dirt and contaminants, wet gas, compatibility of materials and lubrication circuits, and piping to and from the compressor. The possibility of gas condensing under various ambient temperatures either during operation or

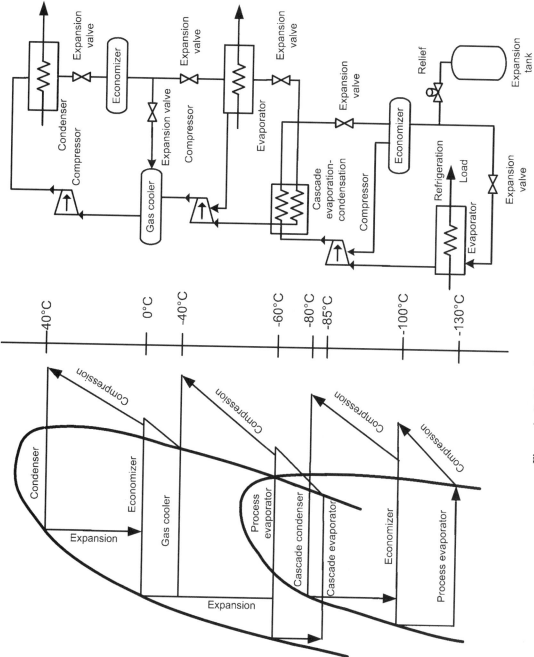

Figure 4 Refrigeration cycle with flash economizers.[3]

427

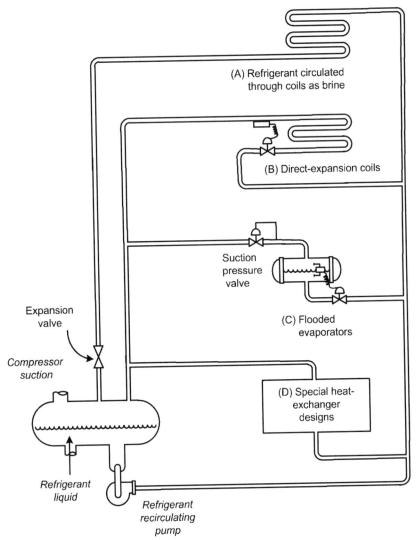

Figure 5 Liquid recirculator.

during standby must be considered. Beyond these considerations, the open-cycle design and its operation are governed primarily by the process requirements. The open system can use standard refrigeration hardware.

3.3 Losses in Refrigeration Cycles

The ideal cycles portrayed in Figs. 1–3 do not include frictional, heat transfer, or pressure losses found in real cycles. These reduce the overall performance of the cycle, both in terms of capacity and efficiency. For example, friction in the compressor produces higher discharge temperatures in the refrigerant than the ideal cycle. These higher temperatures require a larger condenser to reject heat energy from the cycle. Pressure losses in the evaporator and condenser require the compressor to work harder to circulate refrigerant in the system. Predicting the performance of a real system must include these effects, or the system capacity and efficiency will be overestimated.

4 REFRIGERANTS

No one refrigerant is capable of providing cost-effective cooling over the wide range of temperatures and the multitude of applications found in modern refrigeration systems. Ammonia is a popular refrigerant for many industrial and large warehouse applications.[4] Both chlorofluorocarbons (CFCs) and hydrochlorofluorocarbon (HCFC) refrigerants have historically been used in many refrigeration applications, ranging from domestic refrigerators to supermarket and food storage applications. Most of these refrigerants are generally nontoxic and nonflammable. Recent U.S. federal and international regulations[5–7] have placed restrictions on the production and use of CFCs and HCFCs. Hydrofluorocarbons (HFCs) and HFC mixtures are now being used in some applications where CFCs and HCFCs have been used. Regulations affecting refrigerants are discussed in the next section.

The chemical industry uses low-cost fluids such as propane and butane whenever they are available in the process. These hydrocarbon refrigerants, often thought of as too hazardous because of flammability, are suitable for use in modern compressors, and frequently add no more hazard than already exists in an oil refinery or petrochemical plant. These low-cost refrigerants are used in simple, compound, and cascade systems, depending on operating temperatures.

A standard numbering system, shown in Table 1, has been devised to identify refrigerants without the use of their cumbersome chemical names. There are many popular refrigerants in the methane and ethane series. These are called halocarbons or halogenated hydrocarbons because of the presence of halogen elements such as fluorine or chlorine.[8] Halocarbons include CFCs, HCFCs, and HFCs.

Numbers assigned to the hydrocarbons and halohydrocarbons of the methane, ethane, propane, and cyclobutane series are such that the number uniquely specifies the refrigerant compound. The American National Standards Institute (ANSI) and American Society of Heating, Refrigerating and Air-Conditioning Engineers (ASHRAE) Standard 34-2001 describes the method of coding.[9]

Zeotropes and azeotropes are mixtures of two or more different refrigerants. A zeotropic mixture changes saturation temperatures as it evaporates (or condenses) at constant pressure. This phenomena is called temperature glide. For example, R-407C has a boiling (bubble) point of $-44°C$ and a condensation (dew) point of $-37°C$, which means it has a temperature glide of 7°C. An azeotropic mixture behaves much like a single-component refrigerant in that it does not change saturation temperatures appreciably as it evaporates or condenses at constant pressure. Some zeotropic mixtures, such as R-410A, actually have a small enough temperature glide (less than 5.5°C) that they are considered a near-azeotropic refrigerant mixture.

Because the bubble-point and dew-point temperatures are not the same for a given pressure, some zeotropic mixtures have been used to help control the temperature differences in low-temperature evaporators. These mixture have been used in the lowest stage of some liquified natural gas (LNG) plants.[10]

Refrigerants are grouped by their toxicity and flammability (Table 2).[9,11] Group A1 is nonflammable and least toxic, while group B3 is flammable and most toxic. Toxicity is quantified by the threshold limit value–time weighted average (TLV-TWA), which is the

Table 1 Refrigerant Numbering System (ANSI/ASHRAE 34-2001)[a]

Refrigerant Number Designation	Chemical Name	Chemical Formula	Molecular Mass	Normal Boiling Point, °C	Safety Group
Methane Series					
10	Tetrachloramethane	CCl_4	153.8	77	B1
11	Trichlorofluoromethane	CCl_3F	137.4	24	A1
12	Dichlorodifluoromethane	CCl_2F_2	120.9	−30	A1
13	Chlorotrifluoromethane	$CClF_3$	104.5	−81	A1
22	Chlorodifluoromethane	$CHClF_2$	86.5	−41	A1
32	Difluoromethane	CH_2F_2	52.0	−52	A2
50	Methane	CH_4	16.0	−161	A3
Ethane Series					
113	1,1,2-Trichlorotrifluoro-ethane	CCl_2FCClF_2	187.4	48	A1
114	1,2-Dichlorotetrafluoro-ethane	$CClF_2CClF_2$	170.9	4	A1
123	2,2-Dichloro-1,1,1-trifluoroethane	$CHCL2CF3$	152.9	27	B1
125	Pentafluoroethane	CHF_2CF_3	120.0	−49	A1
134a	1,1,1,2-Tetrafluoroethane	CH_2FCF_3	102.0	−26	A1
170	Ethane	CH_3CH_3	30.0	−89	A3
Propane Series					
290	Propane	$CH_3CH_2CH_3$	44.0	−42	A3
Zeotropes Composition					
407C	R-32/R-125/R-134a (23/25/52 wt %)		95.0	−44	A1
410A	R-32/R-125 (50/50 wt%)		72.6	−53	A1
Azeotropes Composition					
500	R-12/152a (73.8/26.2 wt %)		99.31	−33	A1
502	R-22/115 (48.8/51.2 wt %)		112.0	−45	A1
Hydrocarbons					
600	Butane	$CH_3CH_2CH_2CH_3$	58.1		
600a	Isobutane	$CH(CH_3)_3$	58.1	−12	A3
Inorganic Compounds					
717	Ammonia	NH_3	17.0	−33	B2
728	Nitrogen	N_2	28.0	−196	A1
744	Carbon dioxide	CO_2	44.0	−78	A1
764	Sulfur dioxide	SO_2	64.1	−10	B1
Unsaturated Organic Compounds					
1140	Vinyl chloride	CH_2—$CHCl$	62.5	−14	B3
1150	Ethylene	CH_2—CH_2	28.1	−104	A3
1270	Propylene	CH_3CH—CH_2	42.1	−48	A3

[a]Reference 9, reprinted by permission from *ASHRAE Standard 34-2001*. ©American Society of Heating, Refrigerating and Air-Conditioning Engineers, Inc., *www.ashrae.org*.

upper safety limit for airborne exposure to the refrigerant. If the refrigerant is nontoxic in quantities less than 400 parts per million, then it is a class A refrigerant. If exposure to less than 400 parts per million is toxic, then the substance is given the class B designation. The numerical designation refers to the flammability of the refrigerant. The last column of Table 1 shows the toxicity and flammability rating of many of the common refrigerants.

The A1 group of refrigerants generally fulfills the basic requirements for an ideal refrigerant, with considerable flexibility as to refrigeration capacity. Many are used for comfort air conditioning since they are nontoxic and nonflammable. These refrigerants are also used

Table 2 ANSI/ASHRAE Toxicity and Flammability Rating System[9]

Flammability	Group	Group
High	A3	B3
Moderate	A2	B2
Non	A1	B1
Threshold limit value (parts per million)	<400	>400

extensively in refrigeration applications. Many CFCs are in the A1 group. With regulations banning the production and restricting the sale of all CFCs, the CFCs will eventually cease to be used. Common refrigerants in the A1 group include R-11, R-12, R-13, R-22, R-114, R-134a, and R-502.

Refrigerant 11, trichlorofluoromethane, is a CFC. It has a low-pressure–high-volume characteristic suitable for use in close-coupled centrifugal compressor systems for water or brine cooling. Its temperature range extends no lower than −7°C.

Refrigerant 12, dichlorodifluoromethane, is a CFC. It was the most widely known and used refrigerant for U.S. domestic refrigeration and automotive air-conditioning applications until the early 1990s. It is ideal for close-coupled or remote systems ranging from small reciprocating to large centrifugal units. It has been used for temperatures as low as −90°C, although −85°C is a more practical lower limit because of the high gas volumes necessary for attaining these temperatures. It is suited for single-stage or compound cycles using reciprocating and centrifugal compressors.

Refrigerant 13, chlorotrifluoromethane, is a CFC. It is used in low-temperature applications to approximately −126°C. Because of its low volume, high condensing pressure, or both, and because of its low critical pressure and temperature, R-13 is usually cascaded with other refrigerants at a discharge pressure corresponding to a condensing temperature in the range of −56 to −23°C.

Refrigerant 22, chlorodifluoromethane, is an HCFC. It is used in many of the same applications as R-12, but its lower boiling point and higher latent heat permit the use of smaller compressors and refrigerant lines than R-12. The higher pressure characteristics also extend its use to lower temperatures in the range of −100°C.

Refrigerant 114, dichlorotetrafluoroethane, is a CFC. It is similar to R-11, but its slightly higher pressure and lower volume characteristic than R-11 extend its use to −17°C and higher capacities.

Refrigerant 123, dichlorotrifluoroethane, is an HCFC. It is a replacement refrigerant for R-11 in low-pressure centrifugal chillers. New centrifugal equipment designed for R-123 can provide exceptionally high energy efficiency. In retrofits of older existing centrifugal chillers, modifications are often needed to increase capacity or avoid material incompatibility (especially elastomers).

Refrigerant 125, pentafluoroethane, is an HFC. It is used in some refrigerant mixtures, including R-407C and R-410A.

Refrigerant 134a, 1,1,1,2-tetrafluoroethane, is an HFC. It is a replacement refrigerant for R-12 in both refrigeration and air-conditioning applications. It has operating characteristics similar to R-12. R-134a is commonly used in domestic refrigeration applications in the United States.

Refrigerants 407C and *410A* are both mixtures of HFCs. R-407C can be used in some retrofit applications for R-22. Because of its much higher operating pressures, R-410A cannot be used as a replacement refrigerant in an R-22 system. However, manufacturers have begun designing new systems that use R-410A, and these systems can be applied in situations where R-22 systems were used. The higher operating pressures as well as lubricant incompatibility with mineral oils has required manufacturers to completely redesign systems with R-410A.

Refrigerant 502 is an azeotropic mixture of R-22 and R-115. Its pressure characteristics are similar to those of R-22, but it is has a lower discharge temperature.

The B1 refrigerants are nonflammable, but have lower toxicity limits than those in the A1 group. *Refrigerant 123,* an HCFC, is used in many new low-pressure centrifugal chiller applications. Industry standards, such as ANSI/ASHRAE Standard 15-1994, provide detailed guidelines for safety precautions when using R-123 or any other refrigerant that is toxic or flammable.[11]

One of the most widely used refrigerants is *ammonia,* even though it is moderately flammable and has a class B toxicity rating. Ammonia liquid has a high specific heat, an acceptable density and viscosity, and high conductivity. Its enthalpy of vaporization is typically 6–8 times higher than that of the commonly used halocarbons. These properties make it an ideal heat-transfer fluid with reasonable pumping costs, pressure drop, and flow rates. As a refrigerant, ammonia provides high heat transfer, except when affected by oil at temperatures below approximately −29°C, where oil films become viscous. To limit the ammonia-discharge-gas temperature to safe values, its normal maximum condensing temperature is 38°C. Generally, ammonia is used with reciprocating compressors; although relatively large centrifugal compressors (≥3.5 MW), with 8–12 impeller stages required by its low molecular weights, are in use today. Systems using ammonia should contain no copper (with the exception of Monel metal).

The flammable refrigerants (groups A3 and B3) are generally applicable where a flammability or explosion hazard is already present and their use does not add to the hazard. These refrigerants have the advantage of low cost. Although they have fairly low molecular weight, they are suitable for centrifugal compressors of larger sizes. Because of the high acoustic velocity in these refrigerants, centrifugal compressors may be operated at high impeller tip speeds, which partly compensates for the higher head requirements than some of the nonflammable refrigerants.

Flammable refrigerants should be used at pressures greater than atmospheric to avoid increasing the explosion hazard by the admission of air in case of leaks. In designing the system, it also must be recognized that these refrigerants are likely to be impure in refrigerant applications. For example, commercial propane liquid may contain about 2% (by mass) ethane, which in the vapor phase might represent as much as 16–20% (by volume) ethane. Thus, ethane may appear as a noncondensable. Either this gas must be purged or the compressor displacement must be increased about 20% if it is recycled from the condenser; otherwise, the condensing pressure will be higher than required for pure propane and the power requirement will be increased.

Refrigerant 290, propane, is the most commonly used flammable refrigerant. It is well suited for use with reciprocating and centrifugal compressors in close-coupled or remote systems. Its operating temperature range extends to −40°C.

Refrigerant 600, butane, occasionally is used for close-coupled systems in the medium temperature range of 2°C. It has a low-pressure and high-volume characteristic suitable for centrifugal compressors, where the capacity is too small for propane and the temperature is within range.

Refrigerant 170, ethane, normally is used for close-coupled or remote systems at −87 to −7°C. It must be used in a cascade cycle because of its high-pressure characteristics.

Refrigerant 1150, ethylene, is similar to ethane but has a slightly higher-pressure, lower-volume characteristic, which extends its use to −104 to −29°C. Like ethane, it must be used in the cascade cycle.

Refrigerant 50, methane, is used in an ultralow range of −160 to −110°C. It is limited to cascade cycles. Methane condensed by ethylene, which is in turn condensed by propane, is a cascade cycle commonly employed to liquefy natural gas.

Refrigerant 744, carbon dioxide, is currently receiving attention as a possible refrigerant for use in cooling and refrigeration applications. It has the appeal of being a natural substance. Systems can be designed with R-744, but must operate at elevated pressures. Solid carbon dioxide (dry ice) is commonly used in the food industry for chilling and freezing applications.

Table 3 shows the comparative performance of different refrigerants at conditions more typical of some freezer applications. The data show the relatively large refrigerating effect that can be obtained with ammonia. Note also that for these conditions, both R-11 and R-123 would operate with evaporator pressures below atmospheric pressure.

4.1 Regulations on the Production and Use of Refrigerants

In 1974, Molina and Rowland published a paper where they put forth the hypothesis that CFCs destroyed the ozone layer.[13] By the late 1970s, the United States and Canada had banned the use of CFCs in aerosols. In 1985, Farmer noted a depletion in the ozone layer of approximately 40% over what had been measured in earlier years.[4] This depletion in the ozone layer became known as the ozone hole. In September 1987, 43 countries signed an agreement called the Montreal Protocol[7] in which the participants agreed to freeze CFC production levels by 1990, then to decrease production by 20% by 1994 and 50% by 1999. The protocol was ratified by the United States in 1988 and, for the first time, subjected the refrigeration industry to major CFC restrictions.

Recent regulations have imposed restrictions on the production and use of refrigerants.[4,6,14] Production of CFCs in the United States was prohibited after January 1, 1996.[14] A schedule was also imposed that started a gradual phase-out of the production of HCFCs in 2004 and will end complete production by 2030. Refrigerants are divided into two classes:

Table 3 Comparative Refrigeration Performance of Different Refrigerants at −23°C Evaporating Temperature and +37°C Condensing Temperature[a]

Refrigerant Number	Refrigerant Name	Evaporator Pressure (MPa)	Condenser Pressure (MPa)	Net Refrigerating Effect (kJ/kg)	Refrigerant Circulated (kg/h)	Compressor Displacement (L/s)	Power Input (kW)
11	Trichlorofluoromethane	0.013	0.159	145.8	24.7	7.65	0.297
12	Dichlorodifluoromethane	0.134	0.891	105.8	34.0	1.15	0.330
22	Chlorodifluoromethane	0.218	1.390	150.1	24.0	0.69	0.326
123	Dichlorotrifluoroethane	0.010	0.139	130.4	27.6	10.16	0.306
125	Pentafluoroethane	0.301	1.867	73.7	48.9	0.71	0.444
134a	Tetrafluoroethane	0.116	0.933	135.5	26.6	1.25	0.345
502	R-22/R-115 azeotrope	0.260	1.563	91.9	39.2	0.72	0.391
717	Ammonia	0.166	1.426	1057.4	3.42	0.67	0.310

[a]Reference 12, reprinted by permission from 2001 *ASHRAE Handbook of Fundamentals.* ©American Society of Heating, Refrigerating and Air-Conditioning Engineers, Inc., *www.ashrae.org.*

class I, including CFCs, halons, and other major ozone-depleting chemicals; and class II, HCFCs.

Two ratings are used to classify the harmful effects of a refrigerant on the environment.[15] The first, the ozone depletion potential (ODP), quantifies the potential damage that the refrigerant molecule has in destroying ozone in the stratosphere. When a CFC molecule is struck by ultraviolet light in the stratosphere, a chlorine atom breaks off and reacts with ozone to form oxygen and a chlorine/oxygen molecule. This molecule can then react with a free oxygen atom to form an oxygen molecule and a free chlorine. The chlorine can then react with another ozone molecule to repeat the process. The estimated atmospheric life of a given CFC or HCFC is an important factor in determining the value of the ODP. The ODP for CFC-11 is 1.0. All other ODP values for substances are normalized to that of CFC-11.

The second rating is known as the global warming potential (GWP), which represents how much a given mass of a chemical contributes to global warming over a given time period compared to the same mass of carbon dioxide.[16] Carbon dioxide's GWP is defined as 1.0. The GWP of all other substances is normalized to that of carbon dioxide. Refrigerants such as CFCs, HCFCs, and HFCs can block energy from the earth from radiating back into space. One molecule of R-12 can absorb as much energy as 10,000 molecules of CO_2.

Table 4 shows the ODP and GWP for a variety of refrigerants. As a class of refrigerants, the CFCs have the highest ODP and GWP. Because HCFCs tend to be more unstable compounds and, therefore, have much shorter atmospheric lifetimes, their ODP and GWP values are much smaller than those of the CFCs. All HFCs and their mixtures have zero ODP because fluorine does not react with ozone. However, some of the HFCs, such as R-125, R-134a, and R-143a, do have GWP values that are as large or larger than some of the HCFCs. From the standpoint of ozone depletion and global warming, hydrocarbons provide zero ODP and GWP. However, hydrocarbons are flammable, which makes them unsuitable in many applications.

4.2 Refrigerant Selection for the Closed Cycle

In any closed cycle, the choice of the operating fluid is based on the refrigerant with properties best suited to the operating conditions. The choice depends on a variety of factors, some of which may not be directly related to the refrigerant's ability to remove heat. For example, flammability, toxicity, density, viscosity, availability, and similar characteristics are often deciding factors. The suitability of a refrigerant also depends on factors such as the kind of compressor to be used (i.e., centrifugal, rotary, or reciprocating), safety in application, heat-exchanger design, application of codes, size of the job, and temperature ranges. The factors below should be taken into account when selecting a refrigerant.

Discharge (condensing) pressure should be low enough to suit the design pressure of commercially available pressure vessels, compressor casings, etc. However, discharge pressure, that is, condenser liquid pressure, should be high enough to feed liquid refrigerant to all the parts of the system that require it.

Suction (evaporating) pressure should be above approximately 3.45 kPa (0.5 psia) for a practical compressor selection. When possible, it is preferable to have the suction pressure above atmospheric to prevent leakage of air and moisture into the system. Positive pressure normally is considered a necessity when dealing with hydrocarbons, because of the explosion hazard presented by any air leakage into the system.

Standby pressure (saturation at ambient temperature) should be low enough to suit equipment design pressure, unless there are other provisions in the system for handling the refrigerant during shutdown—for example, inclusion of expansion tanks.

Table 4 Ozone Depletion Potential and Halocarbon Global Warming Potential of Popular Refrigerants and Mixtures[a]

Refrigerant Number	Chemical Formula	Ozone Depletion Potential (ODP)	100-yr Global Warming Potential (GWP)
Chlorofluorocarbons			
11	CCl_3F	1.0	4,600
12	CCl_2F_2	1.0	10,600
113	CCl_2FCClF_2	0.80	14,000
114	$CClF_2CClF_2$	1.0	9,800
115	$CClF_2CF_3$	0.6	7,200
Hydrochlorofluorocarbons			
22	$CHClF_2$	0.055	1,700
123	$CHCl_2CF_3$	0.020	120
124	$CHClFCF_3$	0.020	620
141b	CH_3CCl_2F	0.11	700
142b	CH_3CClF_2	0.065	2,400
Hydrofluorocarbons			
32	CH_2F_2	0	550
125	CHF_2CF_3	0	3,400
134a	CH_2FCF_3	0	1,100
143a	CH_3CF_3	0	750
152a	CH_3CHF_2	0	43
Hydrocarbons			
50	CH_4	0	0
290	$CH_3CH_2CH_3$	0	0
Zeotropes			
407C	R-32/125/134a (23/25/52%wt)	0	1,700
410A	R-32/125 (50/50%wt)	0	2,000
Azeotropes			
500	R-12/152a (73.8/26.2 wt%)	0.74	6,310
502	R-22/115 (48.8/51.2 wt%)	0.31	5,494

[a]Compiled from Refs. 4, 15, and 16.

Critical temperature and pressure should be well above the operating level. As the critical pressure is approached, less heat is rejected as latent heat compared to the sensible heat from desuperheating the compressor discharge gas, and cycle efficiency is reduced. Methane (R-50) and chlorotrifluoromethane (R-13) are usually cascaded with other refrigerants because of their low critical points.

Suction volume sets the size of the compressor. High suction volumes require centrifugal or screw compressors, and low suction volumes dictate the use of reciprocating compressors. Suction volumes also may influence evaporator design, particularly at low temperatures, since they must include adequate space for gas-liquid separation.

Freezing point should be lower than minimum operating temperature. This generally is no problem unless the refrigerant is used as a brine.

Theoretical power required for adiabatic compression of the gas is slightly less with some refrigerants than others. However, this is usually a secondary consideration offset by the effects of particular equipment selections, for example, line-pressure drops, etc., on system power consumption.

Vapor density (or molecular weight) is an important characteristic when the compressor is centrifugal because the lighter gases require more impellers for a given pressure rise, that is, head, or temperature lift. On the other hand, centrifugal compressors have a limitation connected with the acoustic velocity in the gas, and this velocity decreases with the increasing molecular weight. Low vapor densities are desirable to minimize pressure drop in long suction and discharge lines.

Liquid density should be taken into account. Liquid velocities are comparatively low, so that pressure drop is usually no problem. However, static head may affect evaporator temperatures, and should be considered when liquid must be fed to elevated parts of the system.

Latent heat should be high because it reduces the quantity of refrigerant that needs to be circulated. However, large flow quantities are more easily controlled because they allow use of larger, less sensitive throttling devices and apertures.

Refrigerant cost depends on the size of the installation and must be considered both from the standpoint of initial charge, and of composition owing to losses during service. Although a domestic refrigerator contains only a few dollars worth of refrigerant, the refrigerant in a cooling system for a typical chemical plant may cost thousands of dollars.

Other desirable properties. Refrigerants should be stable and noncorrosive. For heat-transfer considerations, a refrigerant should have low viscosity, high thermal conductivity, and high specific heat. For safety to life or property, a refrigerant should be nontoxic and nonflammable, should not contaminate products in case of a leak, and should have a low-leakage tendency through normal materials of construction.

With a flammable refrigerant, extra precautions have to be taken in the engineering design if it is required to meet the explosion-proof classification. It may be more economical to use a higher cost, but nonflammable, refrigerant.

4.3 Refrigerant Selection for the Open Cycle

Process gases used in the open cycle include chlorine, ammonia, and mixed hydrocarbons. These gases create a wide variety of operating conditions and corrosion problems. Gas characteristics affect both heat exchangers and compressors, but their impact is far more critical on compressor operation. All gas properties and conditions should be clearly specified to obtain the most economical and reliable compressor design. If the installation is greatly overspecified, design features result that not only add significant cost but also complicate the operation of the system and are difficult to maintain. Specifications should consider the following:

Composition. Molecular weight, enthalpy–entropy relationship, compressibility factor, and operating pressures and temperatures influence the selection and performance of compressors. If process streams are subject to periodic or gradual changes in composition, the range of variations must be indicated.

Corrosion. Special materials of construction and types of shaft seals may be necessary for some gases. Gases that are not compatible with lubricating oils or that must remain oil-free may necessitate reciprocating compressors designed with carbon rings or otherwise made oilless, or the use of centrifugal compressors designed with isolation seals. However, these features are unnecessary on most installations. Standard designs usually can be used to provide savings in initial cost, simpler operation, and reduced maintenance.

Dirt and Liquid Carryover. Generally, the carryover of dirt and liquids can be controlled more effectively by suction scrubbers than by costly compressor design features. Where this is not possible, all anticipated operating conditions should be stated clearly so that suitable materials and shaft seals can be provided.

Polymerization. Gases that tend to polymerize may require cooling to keep the gas temperature low throughout compression. This can be handled by liquid injection or by providing external cooling between stages of compression. Provision may be necessary for internal cleaning with steam.

These factors are typical of those encountered in open-cycle gas compression. Each job should be thoroughly reviewed to avoid unnecessary cost, and to obtain the simplest possible compressor design for ease of operation and maintenance. Direct coordination between the design engineer and manufacturer during final stages of system design is strongly recommended.

5 ABSORPTION SYSTEMS

Ferdinand Carré patented the first absorption machine in 1859.[2] He employed an ammonia/water solution. His design was soon produced in France, England, and Germany. By 1876, over 600 absorption systems had been sold in the United States. One of the primary uses for these machines was in the production of ice. During the late 1800s and early 1900s, different combinations of fluids were tested in absorption machines. These included such diverse combinations as ammonia with copper sulfate, camphor and naphthol with SO_2, and water with lithium chloride. The modern solution of lithium bromide and water was not used industrially until 1940.[2]

Absorption systems offer three distinct advantages over conventional vapor compression refrigeration. First, they do not use CFC or HCFC refrigerants, which are harmful to the environment. Second, absorption systems can utilize a variety of heat sources, including natural gas, steam, solar-heated hot water, and waste heat from a turbine or industrial process. If the source of energy is from waste heat, absorption systems may provide the lowest cost alternative for providing chilled water or refrigeration applications. Third, absorption systems do not require any mechanical compression of the refrigerant, which eliminates the need for a lubricant in the refrigerant. Lubricants can decrease heat transfer in evaporators and condensers.

Two different absorption systems are currently in use. These include (1) a water–lithium bromide system where water is the refrigerant and lithium bromide is the absorbent and (2) a water–ammonia system where the ammonia is the refrigerant and the water is the absorbent.

Evaporator temperatures ranging from $-60°$ to $10°C$ are achievable with absorption systems.[1] For water chilling service, absorption systems generally use water as the refrigerant and lithium bromide as the absorbent solution. For process applications requiring chilled fluid below $7°C$, the ammonia–water pair is used with ammonia serving as the refrigerant.

5.1 Water–Lithium Bromide Absorption Chillers

Water–lithium bromide absorption machines can be classified by the method of heat input. *Indirect-fired* chillers use steam or hot liquids as a heat source. *Direct-fired* chillers use the heat from the firing of fossil fuels. *Heat-recovery* chillers use waste gases as the heat source.

A typical arrangement for a single-stage water–lithium bromide absorption system is shown schematically in Fig. 6. The absorbent, lithium bromide, may be thought of as a carrier fluid bringing spent refrigerant from the low-pressure side of the cycle (the absorber) to the high-pressure side (the generator). There, the waste heat, steam, or hot water that drives the system separates the water from the absorbent by a distillation process. The regenerated absorbent returns to the absorber where it is cooled so it will absorb the refrigerant (water) vapor produced in the evaporator and, thereby, establish the low-pressure level, which controls the evaporator temperature. Thermal energy released during the absorption process is transferred to the cooling water flowing through tubes in the absorber shell.

The external heat exchanger shown in Fig. 6 saves energy by heating the strong liquid flowing to the generator as it cools the hot absorbent flowing from the generator to the absorber. If the weak solution that passes through the regenerator to the absorber does not contain enough refrigerant and is cooled too much, crystallization can occur. Leaks or process upsets that cause the generator to overconcentrate the solution are indicated when this occurs. The slushy mixture formed does not harm the machine, but it interferes with continued operation. External heat and added water may be required to redissolve the mixture.

Single-stage absorption systems are most common when generator heat input temperatures are less than 95°C. The coefficient of performance (COP) of a system is the cooling achieved in the evaporator divided by the heat input to the generator. The COP of a single-

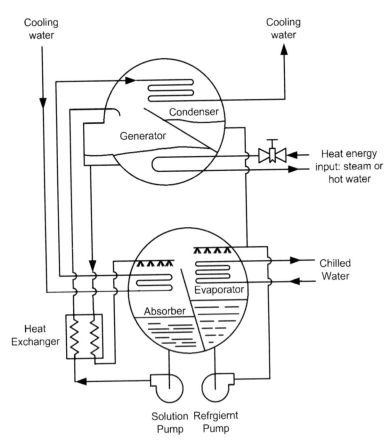

Figure 6 Single-stage water–lithium bromide absorption system.

stage lithium bromide machine generally is 0.65–0.70 for water-chilling duty. The heat rejected by the cooling tower from both the condenser and the absorber is the sum of the waste heat supplied plus the cooling produced, requiring larger cooling towers and cooling water flows than for vapor compression systems.

Absorption machines can be built with a two-stage generator (Fig. 7) with heat input temperatures greater than 150°C. Such machines are called dual-effect machines. The operation of the dual-effect machine is the same as the single-effect machine except that an

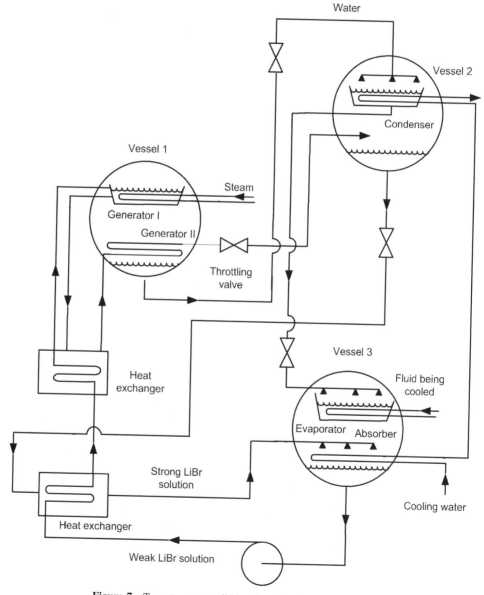

Figure 7 Two-stage water–lithium bromide absorption system.[17]

additional generator, condenser, and heat exchanger are used. Energy from an external heat source is used to boil the dilute lithium bromide (absorbent) solution. The vapor from the primary generator flows in tubes to the second-effect generator. It is hot enough to boil and concentrate absorbent, which creates more refrigerant vapor without any extra energy input. Dual-effect machines typically use steam or hot liquids as input. Coefficients of performance above 1.0 can be obtained with these machines.

5.2   Ammonia–Water Absorption Systems

Ammonia–water absorption technology is used primarily in smaller chillers and small refrigerators found in recreational vehicles.[1] Refrigerators use a variation of the ammonia absorption cycle with ammonia, water, and hydrogen as the working fluids. They can be fired with both gas and electric heat. The units are hermetically sealed. A complete description of this technology can be found in Ref. 1.

Ammonia–water chillers have three major differences from water–lithium bromide systems. First, because the water is volatile, the regeneration of the weak absorbent to strong absorbent requires a distillation process. In a water–lithium bromide system, the generator is able to provide adequate distillation because the absorbent material (lithium bromide) is nonvolatile. In ammonia absorption systems, the absorbent (water) is volatile and tends to carry over into the evaporator where it interferes with vaporization. This problem is overcome by adding a rectifier to purify the ammonia vapor flowing from the generator to the condenser.

A second difference between ammonia–water and water–lithium bromide systems is the operating pressures. In a water–lithium bromide system, evaporating pressures as low as 4–8 kPa are not unusual for the production of chilled water at 5–7°C. In contrast, an ammonia absorption system would run evaporator pressures of between 400 and 500 kPa.

A third difference focuses on the type of heat-transfer medium used in the condenser and absorber. Most lithium bromide systems utilize water cooling in the condenser and absorber, while commercial ammonia systems use air cooling.

6   INDIRECT REFRIGERATION

For indirect refrigeration, the process or refrigeration load is cooled by an intermediate (secondary) liquid, which is itself cooled by refrigerant typically in a conventional vapor-compression cycle (Fig. 8). The secondary liquid can be water, brine, alcohol, or refrigerant. The heat exchanger used to cool the process load may need to be capable of handling corrosive products, high pressures, or high viscosities, and is usually not well suited as a refrigerant evaporator. Other problems preventing direct use of a vapor-compression refrigeration cycle may be remote location, lack of sufficient pressures for the refrigerant liquid feed, difficulties with oil return, or inability to provide traps in the suction line to hold liquid refrigerant. Use of indirect refrigeration simplifies the piping system because it becomes a conventional single- phase liquid-system design.

The indirect or secondary coolant (brine) is cooled in the refrigeration evaporator and then is pumped to the process load. The brine system may include a tank maintained at atmospheric pressure, or may be a closed system pressurized by an inert, dry gas.

Secondary coolants can be separated into four categories:

1. *Coolants with a salt base.* These are water solutions of various concentrations and include the most common brines, that is, calcium chloride and sodium chloride.

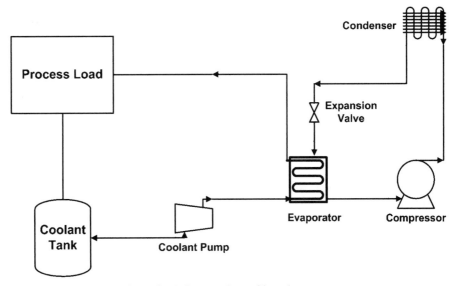

Figure 8 Indirect coolant refrigeration system.

2. *Coolants with a glycol base.* These are water solutions of various concentrations, most commonly ethylene glycol or propylene glycol.

3. *Coolants with an alcohol base.* Where low temperatures are not required, the alcohols are occasionally used in alcohol–water solutions.

4. *Coolants for low-temperature heat transfer.* These usually are pure substances such as methylene chloride, trichloroethylene, R-11, acetone, and methanol.

Coolants containing a mixture of calcium and sodium chloride are the most common refrigeration brines. These are applied primarily in industrial refrigeration and ice skating rinks. Glycols are used to lower the freezing point of water and used extensively as heat-transfer media in cooling systems. Low-temperature coolants include some common refrigerants (R-11, R-30, and R-1120). Because R-11 is a CFC, it cannot be used in any new systems; however, it may still be found in some existing systems. Alcohols and other secondary refrigerants, such as *d*-limonene ($C_{10}H_{16}$), are primarily used by the chemical processing and pharmaceutical industries.

A coolant needs to be compatible with other materials in the system where it is applied. It should have a minimum freezing point approximately 8°C below the lowest temperature to which it is exposed.[1] Table 5 shows a performance comparison of different types of coolants. Some coolants, such as the salts, glycols, and alcohols, are mixed with water to lower the freezing point of water. Different concentrations than listed in Table 5 will result in different freezing temperatures. The flow rate divided by capacity gives a way to compare the amount of flow (L/s) that will be needed to produce a kilowatt of cooling. The low-temperature coolants have the highest flow requirements of the four types of coolants. The heat-transfer factor is a value normalized to propylene glycol. It is based on calculations inside a smooth tube. The salt mixtures and R-30 provide the highest heat-transfer factors of the fluids listed. The energy factor is a measure of the pumping requirements that will be needed for each of the coolants. The low-temperature fluids require the largest pumping requirements.

Table 5 Secondary Coolant Performance Comparisons[a]

Secondary Coolant	Concentration (by weight) %	Freezing Point (°C)	Flow Rate/ Capacity [L/(s-kW)][b]	Heat Transfer Factor[c]	Energy Factor[d]
Salts					
Calcium chloride	22	−22.1	0.0500	2.761	1.447
Sodium chloride	23	−20.6	0.0459	2.722	1.295
Glycols					
Propylene glycol	39	−20.6	0.0459	1.000	1.142
Ethylene glycol	38	−21.6	0.0495	1.981	1.250
Alcohols					
Methanol	26	−20.7	0.0468	2.307	1.078
Low-temperature fluids					
Methylene chloride (R-30)	100	−96.7	0.1146	2.854	3.735
Trichlorethylene (R-1120)	100	−86.1	0.1334	2.107	4.787
Trichlorofluoromethane (R-11)	100	−111.1	0.1364	2.088	5.022
d-Limonene	100	−96.7	0.1160	1.566	2.406

[a]Ref. 1, reprinted by permission from 2002 *ASHRAE Handbook of Refrigeration*. ©American Society of Heating, Refrigerating and Air-Conditioning Engineers, Inc., *www.ashrae.org*.

[b]Based on inlet secondary coolant temperature at the pump of −3.9°C.

[c]Based on a curve fit of the Sieder & Tate heat-transfer equation values using a 27-mm i.d. tube 4.9 m long and a film temperature of 2.8°C lower than the average bulk temperature with a 2.134-m/s velocity. The actual i.d. and length vary according to the specific loading and refrigerant applied with each secondary coolant, tube material, and surface augmentation.

[d]Based on the same pump head, refrigeration load, −6.7°C average temperature, 6 K range, and the freezing point (for water-based secondary coolants) 11 to 13 K below the lowest secondary coolant temperature.

Table 6 shows the general areas of application for the commonly used brines. Criteria for selection are discussed in the following paragraphs. The order of importance depends on the specific application.

Corrosion problems with sodium chloride and calcium chloride brines limit their use. When properly maintained in a neutral condition and protected with inhibitors, they will give 20–30 years of service without corrosive destruction of a closed system. Preventing corrosions requires proper selection of materials, inhibitors, maintaining a clean system, and regular testing for the pH of the system.[1] Glycol solutions and alcohol–water solutions are generally less corrosive than salt brines, but they require inhibitors to suit the specific application for maximum corrosion protection. Methylene chloride, trichloroethylene, and trichlorofluoromethane do not show general corrosive tendencies unless they become contaminated with impurities such as moisture. However, methylene chloride and trichloroethylene must not be used with aluminum or zinc; they also attack most rubber compounds and plastics. Alcohol in high concentrations will attack aluminum. Reaction with aluminum is of concern because, in the event of leakage into the refrigeration compressor system, aluminum compressor parts will be attacked.

Toxicity is an important consideration in connection with exposure to some products and to operating personnel. Where brine liquid, droplets, or vapor may contact food products, as in an open spray-type system, sodium chloride and propylene glycol solutions are acceptable because of low toxicity. All other secondary coolants are toxic to some extent or produce odors, which requires that they be used only inside of pipe coils or a similar pressure-tight barrier.

Table 6 Application Information for Common Secondary Coolants[1,3]

Secondary Coolant	Toxic	Explosive	Corrosive
Salts			
Calcium chloride	No	No	Yes
Sodium chloride	No	No	Yes
Glycols			
Propylene glycol	No	No	Some
Ethylene glycol	Yes	No	Some
Alcohols			
Methanol	Yes	Yes	Some
Ethanol	Yes	Yes	Some
Low-temperature fluids			
Methylene chloride (R-30)	No	No	No
Trichloroethylene (R-1120)	No	No	No
Trichlorofluoromethane (R-11)	No	No	No
d-Limonene	Yes	Yes	Yes

Flash-point and explosive-mixture properties of some coolants require precautions against fire or explosion. Acetone, methanol, and ethanol are in this category but are less dangerous when used in closed systems.

Specific heat of a coolant determines the mass rate of flow that must be pumped to handle the cooling load for a given temperature rise. The low-temperature coolants, such as trichloroethylene, methylene chloride, and trichlorofluoromethane, have specific heats approximately one-third to one-fourth those of the water soluble brines. Consequently, a significantly greater mass of the low-temperature brines must be pumped to achieve the same temperature change.

Stability at high temperatures is important where a brine may be heated as well as cooled. Above 60°C, methylene chloride may break down to form acid products. Trichloroethylene can reach 120°C before breakdown begins.

Viscosities of brines vary greatly. The viscosity of propylene gycol solutions, for example, makes them impractical for use below −7°C because of the high pumping costs and the low heat-transfer coefficient at the concentration required to prevent freezing. Mixtures of ethanol and water can become highly viscous at temperatures near their freezing points, but 190-proof ethyl alcohol has a low viscosity at all temperatures down to near the freezing point. Similarly, methylene chloride and R-11 have low viscosities down to −73°C. In this region, the viscosity of acetone is even more favorable.

Since a secondary coolant cannot be used below its freezing point, certain ones are not applicable at the lower temperatures. Sodium chloride's eutectic freezing point of −20°C limits its use to approximately −12°C. The eutectic freezing point of calcium chloride is −53°C, but achieving this limit requires such an accuracy of mixture that −40°C is a practical low limit of usage.

Water solubility in any open or semi-open system can be important. The dilution of a salt or glycol brine, or of alcohol by entering moisture, merely necessitates strengthening of the brine. But for a brine that is not water-soluble, such as trichloroethylene or methylene chloride, precautions must be taken to prevent free water from freezing on the surfaces of the heat exchanger. This may require provision for dehydration or periodic mechanical removal of ice, perhaps accompanied by replacement with fresh brine.

Vapor pressure is an important consideration for coolants that will be used in open systems, especially where it may be allowed to warm to room temperature between periods of operation. It may be necessary to pressurize such systems during periods of moderate temperature operation. For example, at 0°C the vapor pressure of R-11 is 39.9 kPa (299 mm Hg); that of a 22% solution of calcium chloride is only 0.49 kPa (3.7 mm Hg). The cost of vapor losses, the toxicity of the escaping vapors, and their flammability should be carefully considered in the design of the semi-closed or open system.

Environmental effects are important in the consideration of trichlorofluoromethane (R-11) and other chlorofluorocarbons. This is a refrigerant with a high ozone-depletion potential and halocarbon global-warming potential. The environmental effect of each of the coolants should be reviewed before seriously considering the use of it in a system.

Energy requirements of brine systems may be greater because of the power required to circulate the brine and because of the extra heat-transfer process, which necessitates the maintenance of a lower evaporator temperature.

7   SYSTEM COMPONENTS

There are four major components in any refrigeration system: compressor, condenser, evaporator, and expansion device. Each is discussed below.

7.1   Compressors

Both positive displacement and centrifugal compressors are used in refrigeration applications. With positive displacement compressors, the pressure of the vapor entering the compressor is increased by decreasing the volume of the compression chamber. Reciprocating, rotary, scroll, and screw compressors are examples of positive displacement compressors. Centrifugal compressors utilize centrifugal forces to increase the pressure of the refrigerant vapor. Refrigeration compressors can be used alone or in parallel and series combinations. Features of several of the compressors are described later in this section.

Compressors usually have a variety of protection devices for handling adverse conditions. These include high-pressure controls, high-temperature controls, low-pressure protection, time delay, low voltage and phase loss, and suction line strainer.[18] High-pressure controls are required by Underwriters Laboratories. These can include a high-pressure cutoff or a relief valve. High-temperature devices are designed to protect against overheating and lubrication breakdown. Low-pressure protection is provided to protect the compressor against extremely low pressures, which may cause insufficient lubricant return, freeze-up, or too high a pressure ratio. Time delays are required to prevent damage to the compressor motor from rapid startup after a shutdown. A suction line strainer is used to remove dirt and other particles that may be in the refrigerant line. The specific protection devices will depend on the application and size of the compressor.

Reciprocating Compressors

High-speed, single-stage reciprocating compressors with displacements up to 0.283–0.472 M^3/sec (600–1000 cfm) generally are limited to a pressure ratio of about 9. The reciprocating compressor is basically a constant-volume variable-head machine. It handles various discharge pressures with relatively small changes in inlet-volume flow rate as shown by the heavy line in Fig. 9.

Reciprocating compressors can also be found in an integral two-stage configuration.[18] These can use R-22 or ammonia and can achieve low temperatures from −29 to −62°C. These compressors will consist of multiple cylinders, with the cylinders divided so that the

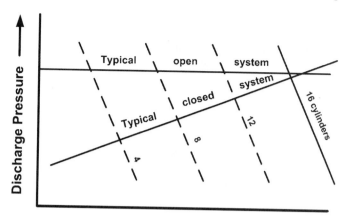

Inlet volume flow rate

Figure 9 Volume–pressure relationships for a reciprocating compressor.

volumetric flow and pressure ratios are balanced. Capacity can be controlled with cylinder unloading.

Open systems and many processes require nearly fixed compressor suction and discharge pressure levels. This load characteristic is represented by the horizontal typical open-system line in Fig. 9. In contrast, condenser operation in many closed systems is related to ambient conditions, for example, through cooling towers, so that on cooler days the condenser pressure can be reduced. When the refrigeration load is lower, less refrigerant circulation is required. The resulting load characteristic is represented by the typical closed-system line in Fig. 9.

The compressor must be capable of matching the pressure and flow requirements imposed upon it by the system in which it operates. The reciprocating compressor matches the imposed discharge pressure at any level up to its limiting pressure ratio. Varying capacity requirements can be met by providing devices that unload individual or multiple cylinders. This unloading is accomplished by blocking the suction or discharge valves that open either manually or automatically. Capacity can also be controlled through the use of variable speed or multispeed motors. When capacity control is implemented on a compressor, other factors at part-load conditions need to considered, such as (1) effect on compressor vibration and sound when unloaders are used, (2) the need for good oil return because of lower refrigerant velocities, and (3) proper functioning of expansion devices at the lower capacities.

Reciprocating compressors employ a lubricant. Oil is pumped into the refrigeration system during operation. Systems must be designed carefully to return oil to the compressor crankcase to provide for continuous lubrication and also to avoid contaminating heat-exchanger surfaces. At very low temperatures (~ −50°C or lower, depending on refrigerant used) oil becomes too viscous to return, and provision must be made for periodic plant shutdown and warm-up to allow manual transfer of the oil.

Reciprocating compressors usually are arranged to start with the cylinders unloaded so that normal torque motors are adequate for starting. When gas engines are used for reciprocating compressor drives, careful torsional analysis is essential.

Rotary Compressors

Rotary compressors include both rolling piston and rotary vane compressors. Rotary vane compressors are primarily used in transportation air-conditioning applications, while rolling piston compressors are usually found in household refrigerators and small air conditioners up to inputs of 2 kW. For a fixed-vane, rolling piston rotary compressor, the shaft is located in the center of the housing while the roller is mounted on an eccentric.[8] Suction gas enters directly into the suction port. As the roller rotates, the refrigerant vapor is compressed and is discharged into the compressor housing through the discharge valve.

One difference between a rotary and a reciprocating compressor is that the rotary is able to obtain a better vacuum during suction.[18] It has low reexpansion losses because there is no high-pressure discharge vapor present during suction as with a reciprocating compressor.

Because rotary vane compressors have a light weight for their displacement, they are ideal for transportation applications. Rotary vane compressors can be used in applications where temperatures drop down to −40 to −51°C, depending whether it is in a single- or two-stage system. Refrigerants R-22, R-404a, and R-717 are currently used with rotary vane compressors.[18]

Scroll Compressors

The principle of the scroll compressor was first patented in 1905.[19] However, the first commercial units were not built until the early 1980s.[20] Scroll compressors are used in building air-conditioning, heat pump, refrigeration, and automotive air-conditioning applications. They range in capacity from 3 to 50 kW.[18] Scroll compressors have two spiral-shaped scroll members that are assembled 180° out of phase (Fig. 10). One scroll is fixed while the other "orbits" the first. Vapor is compressed by sealing vapor off at the edge of the scrolls and reducing the volume of the gas as it moves inward toward the discharge port. Figure 10*a* shows the two scrolls at the instant that vapor enters the compressor and compression begins. The orbiting motion of the second scroll forces the pocket of vapor toward the discharge port while decreasing its volume (Fig. 10*b*–10*h*). In Fig. 10*c* and *f*, the two scrolls open at the ends and allow new pockets of vapor to be admitted into the scrolls for compression. Compression is a nearly continuous process in a scroll compressor.

Scroll compressors offer several advantages over reciprocating compressors. First, relatively large suction and discharge ports can be used to reduce pressure losses. Second, the separation of the suction and discharge processes reduces the heat transfer between the discharge and suction processes. Third, with no valves and reexpansion losses, they have higher volumetric efficiencies. Capacities of systems with scroll compressors can be varied by using variable speed motors or by use of multiple suction ports at different locations within the two spiral members. Fourth, with a smaller number of moving parts, they have the potential to be more reliable and quieter than reciprocating compressors.

Screw Compressors

Screw compressors were first introduced in 1958.[2] These are positive displacement machines available in the capacity range from 15 to 1100 kW, overlapping reciprocating compressors for lower capacities and centrifugal compressors for higher capacities. Both twin-screw and single-screw compressors are used for refrigeration applications.

Fixed suction and discharge ports, used instead of valves in reciprocating compressors, set the "built-in volume ratio" of the screw compressor. This is the ratio of the volume of fluid space in the meshing rotors at the beginning of the compression process to the volume in the rotors as the discharge port is first exposed. Associated with the built-in volume ratio is a pressure ratio that depends on the properties of the refrigerant being compressed. Peak efficiency is obtained if the discharge pressure imposed by the system matches the pressure developed by the rotors when the discharge port is exposed. If the interlobe pressure is greater or less than discharge pressure, energy losses occur but no harm is done to the compressor.

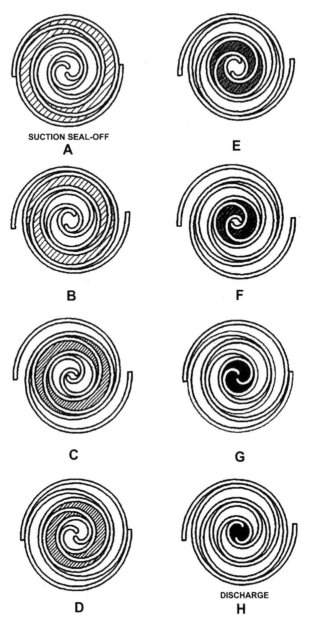

Figure 10 Operation of the fixed and orbiting scrolls in a scroll compressor. (Reprinted by permission from 2000 *ASHRAE Handbook of HVAC Systems and Equipment.* ©American Society of Heating, Refrigerating, and Air-Conditioning Engineers, Inc., *www.ashrae.org.*)

Capacity modulation is accomplished by slide valves that are used to provide a variable suction bypass or delayed suction port closing, reducing the volume of refrigerant actually compressed. Continuously variable capacity control is most common, but stepped capacity control is offered in some manufacturers' machines. Variable discharge porting is available on a few machines to allow control of the built-in volume ratio during operation.

Oil is used in screw compressors to seal the extensive clearance spaces between the rotors, to cool the machines, to provide lubrication, and to serve as hydraulic fluid for the capacity controls. An oil separator is required for the compressor discharge flow to remove the oil from the high-pressure refrigerant so that performance of system heat exchangers will not be penalized and the oil can be returned for reinjection in the compressor.

Screw compressors can be direct driven at two-pole motor speeds (50 or 60 Hz). Their rotary motion makes these machines smooth running and quiet. Reliability is high when the machines are applied properly. Screw compressors are compact so they can be changed out readily for replacement or maintenance. Today, the efficiency of the best screw compressors matches that of reciprocating compressors at full load. Figure 11 shows the efficiency of a single-screw compressor as a function of pressure ratio and volume ratio (Vi). High isentropic and volumetric efficiencies can be achieved with screw compressors because there are no suction or discharge valves and small clearance volumes. Screw compressors have been used with a wide variety of refrigerants, including halocarbons, ammonia, and hydrocarbons.

Centrifugal Compressors

The centrifugal compressor is preferred whenever the gas volume is high enough to allow its use, because it offers better control, simpler hookup, minimum lubrication problems, and lower maintenance. Single-impeller designs are directly connected to high-speed drives or

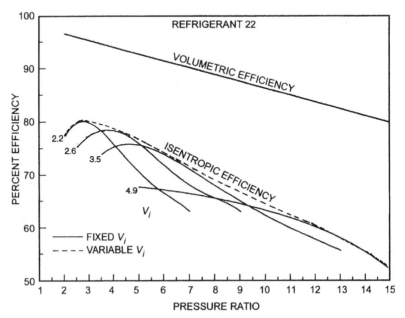

Figure 11 Typical performance of a single-screw compressor. (Reprinted by permission from 2000 *ASHRAE Handbook of HVAC Systems and Equipment.* ©American Society of Heating, Refrigerating, and Air-Conditioning Engineers, Inc., *www.ashrae.org.*)

driven through an internal speed increaser. These machines are ideally suited for clean, noncorrosive gases in moderate-pressure process or refrigeration cycles in the range of 0.236–1.89 m^3/sec (5 cfm). Multistage centrifugal compressors are built for direct connection to high-speed drives or for use with an external speed increaser. Designs available from suppliers generally provide for two to eight impellers per casing covering the range of 0.236–11.8 m^3/sec (500–25,000 cfm), depending on the operating speed. A wide choice of materials and shaft seals to suit any gas composition, including dirty or corrosive process streams, is available.

The centrifugal compressor has a more complex head-volume characteristic than reciprocating machines. Changing discharge pressure may cause relatively large changes in inlet volume, as shown by the heavy line in Fig. 12a. Adjustment of variable inlet vanes or of a diffuser ring allows the compressor to operate anywhere below the heavy line to conditions imposed by the system. A variable-speed controller offers an alternative way to match the compressor's characteristics to the system load, as shown in the lower half of Fig. 12b. The maximum head capability is fixed by the operating speed of the compressor. Both methods have advantages: generally, variable inlet vanes or diffuser rings provide a wider range of capacity reduction; variable speed usually is more efficient. Maximum efficiency and control can be obtained by combining both methods of control.

The centrifugal compressor has a surge point—that is, a minimum-volume flow below which stable operation cannot be maintained. The percentage of load at which the surge point occurs depends on the number of impellers, design–pressure ratio, operating speed, and variable inlet-vane setting. The system design and controls must keep the inlet volume above this point by artificial loading, if necessary. This is accomplished with a bypass–valve and gas recirculation. Combined with a variable inlet-vane setting, variable diffuser ring, or variable speed control, the gas bypass allows stable operation down to zero load.

Compressor Operation

Provision for minimum-load operation is strongly recommended for all installations, because there will be fluctuations in plant load. For chemical plants, this permits the refrigeration system to be started up and thoroughly checked out independently of the chemical process.

Contrasts between the operating characteristics of the positive displacement compressor and the centrifugal compressor are important considerations in plant design to achieve satisfactory performance. Unlike positive displacement compressors, the centrifugal compressor will not rebalance abnormally high system heads. The drive arrangement for the centrifugal compressor must be selected with sufficient speed to meet the maximum head anticipated. The relatively flat head characteristics of the centrifugal compressor necessitates different control approaches than for positive displacement machines, particularly when parallel compressors are utilized. These differences, which account for most of the troubles experienced in centrifugal-compressor systems, cannot be overlooked in the design of a refrigeration system.

A system that uses centrifugal compressors designed for high-pressure ratios and that requires the compressors to start with high suction density existing during standby will have high starting torque. If the driver does not have sufficient starting torque, the system must have provisions to reduce the suction pressure at startup. This problem is particularly important when using single-shaft gas turbine engines or reduced-voltage starters on electric drives. Split-shaft gas turbines are preferred for this reason.

Drive ratings that are affected by ambient temperatures, altitudes, etc., must be evaluated at the actual operating conditions. Refrigeration installations normally require maximum output at high ambient temperatures, a factor that must be considered when using drives such as gas turbines and gas engines.

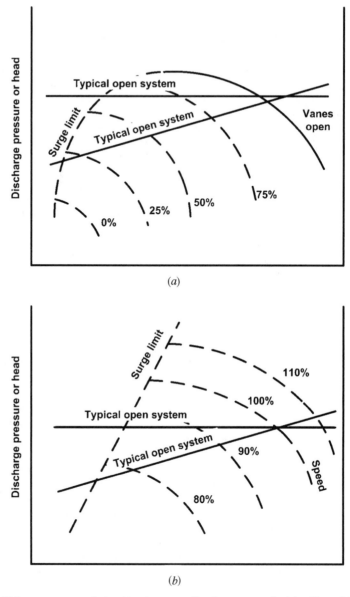

Figure 12 Volume–pressure relationships in a centrifugal compressor[3]: (*a*) with variable inlet-vane control at constant rotational speed; (*b*) with variable speed control at a constant inlet-vane opening.

7.2 Condensers

The refrigerant condenser is used to reject the heat energy added to the refrigerant during compression and the heat energy absorbed in the evaporator. This heat energy is typically rejected to either water or air.

The amount of heat energy added to the refrigerant during compression depends on the compressor power and can become a significant part of the condenser load for low-temperature systems. Common types of water-cooled condensers include shell-and-tube, shell-and-coil, tube-in-tube, and brazed-plate.[18] Shell-and-coil condensers are smaller in size (3.5–50 kW) and circulate the cooling water through coiled tubes inside an external shell. The refrigerant condenses on the outside of the coiled tubes. Tube-in-tube condensers can be found in sizes up to 175 kW and consist of tubes within larger tubes. The refrigerant is condensed either in the annular space between the tubes or inside the inner tube. Brazed-plate condensers can be found in sizes up to 350 kW and are constructed of plates brazed together to form separate channels.[18]

Shell-and-tube condensers can be found in sizes from 3.5 to 35,000 kW. These condensers with finned tubes and fixed tube sheets provide an economical exchanger design for refrigerant use. Figure 13 shows an example of a shell-and-tube condenser. Commercially available condensers conforming to ASME Boiler and Pressure Vessel Code[21] construction adequately meet both construction and safety requirements for this application.

Cooling towers and spray ponds are frequently used for water-cooling systems. These generally are sized to provide 29°C supply water at design load conditions. Circulation rates typically are specified so that design cooling loads are handled with a 5.6°C cooling-water temperature rise. Pump power, tower fans, makeup water (about 3% of the flow rate), and water treatment should be taken into account in operating-cost studies. Water temperatures, which control condensing pressure, may have to be maintained above a minimum value to ensure proper refrigerant liquid feeding to all parts of the system.

River or well water, when available, provides an economical cooling medium. Quantities circulated will depend on initial supply temperatures and pumping cost, but are generally selected to handle the cooling load with 8.3–16.6°C water-temperature range. Water treatment and special exchanger materials frequently are necessary because of the corrosive and scale-forming characteristics of the water. Well water, in particular, must be analyzed for corrosive properties, with special attention given to the presence of dissolved gases, for example, H_2S

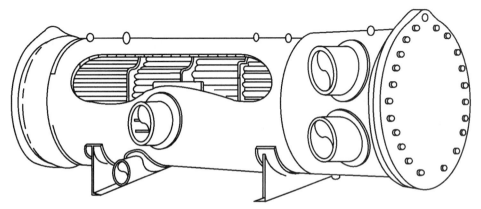

Figure 13 Typical shell-in-tube refrigerant condenser.[3]

and CO_2. These are extremely corrosive to many exchanger materials, yet difficult to detect in sampling. Pump power, water treatment, and special condenser material should be evaluated when considering costs.

Allowances must be made in heat-transfer calculations for fouling or scaling of exchanger surfaces during operation. This ensures sufficient surface to maintain rated performance over a reasonable interval of time between cleanings. Scale-factor allowances are expressed in $m^2 \cdot K/kW$ as additional thermal resistance.

Commercial practice normally includes a scale-factor allowance of 0.088 $m^2 \cdot K/kW$. The long hours of operation usually associated with chemical-plant service and the type of cooling water frequently encountered generally justify a greater allowance to minimize the frequency of downtime for cleaning. Depending on these conditions, an allowance of 0.18 or 0.35 $m^2 \cdot K/kW$ is recommended for chemical-plant service. Scale allowance can be reflected in system designs in two ways: as more heat-exchanger surface or as higher design condensing temperatures with attendant increase in compressor power. Generally, a compromise between these two approaches is most economical. For extremely bad water, parallel condensers, each with 60–100% capacity, may provide a more economical selection and permit cleaning one exchanger while the system is operating.

Air-cooled condensing equipment can also be used in refrigeration systems. With tighter restrictions on the use of water, air-cooled equipment is used even on larger centrifugal-type refrigeration plants, although it requires more physical space than cooling towers. Larger condensers include an array of propeller fans located at the top of the condenser that pull air over the condensing coil. Circulating fans and exchanger surface are usually selected to provide design condensing temperatures of 49–60°C when design ambient dry bulb temperatures range between 35 and 38°C.

The design dry bulb temperature should be carefully considered since most weather data reflect an average or mean maximum temperature. If full load operation must be maintained at all times, care should be taken to provide sufficient condenser capacity for the maximum recorded temperature. This is particularly important when the compressor is centrifugal because of its flat head characteristics and the need for adequate speed. Multiple-circuit or parallel air-cooled condensers must be provided with traps to prevent liquid backup into the idle circuit at light load. Pressure drop through the condenser coil must also be considered in establishing the compressor discharge pressure.

The condensing temperature and pressure must be controlled for the refrigeration system to function optimally. Too high a condensing temperature results in increased power and reduced capacity. Too low a condensing temperature can result in poor performance of the expansion device. Air-cooled condensers often employ fan cycling, modulating dampers, or fan speed control to maintain proper refrigerant condensing temperature and pressure.[18] On condensers with multiple fans, one or more of the fans can each be cycled on and off to maintain refrigerant conditions. When modulating dampers are used, the airflow through the condenser can be controlled from 0 to 100%. Variable-speed drives can also be used to control fan speed and airflow through the condenser.

In comparing water-cooled and air-cooled condensers, the compression power at design conditions is usually higher with air-cooled condensing, because of the larger temperature differential required in air-cooled condensers. However, ambient air temperatures are considerably below the design temperature most of the time, and operating costs frequently compare favorably over a full year. In addition, air-cooled condensers usually require less maintenance, although dirty or dusty atmospheres may affect performance.

7.3 Evaporators

There are special requirements for evaporators in refrigeration service that are not always present in other types of heat-exchanger design. These include problems of oil return, flash-gas distribution, gas–liquid separation, and submergence effects.

Oil Return

When the evaporator is used with reciprocating-compression equipment, it is necessary to ensure adequate oil return from the evaporator. If oil will not return in the refrigerant flow, it is necessary to provide an oil reservoir for the compression equipment and to remove oil mechanically from the low side of the system on a regular basis. Evaporators used with centrifugal compressors do not normally require oil return from the evaporator, since centrifugal compressors pump very little oil into the system. However, even with centrifugal equipment, low-temperature evaporators eventually may become contaminated with oil, which must be reclaimed.

Two-Phase Refrigeration Distribution

As a general rule, the refrigerant is introduced into the evaporator by expanding it from a high-pressure liquid through the expansion device. In the expansion process, a significant portion of the refrigerant flashes off into vapor, producing a two-phase mixture of liquid and vapor that must be introduced properly into the evaporator for satisfactory performance. Improper distribution of this mixture can result in liquid carryover to the compressor and in damage to the exchanger tubes from erosion or from vibration.

Vapor–Liquid Separation

To avoid compressor damage, the refrigerant leaving the evaporator must not contain any liquid. The design should provide adequate separation space or include mist eliminators. Liquid carryover is one of the most common sources of trouble with refrigeration systems.

Submergence Effect

In flooded evaporators, the evaporating pressure and temperature at the bottom of the exchanger surface is higher than at the top of the exchanger surface, owing to the liquid head. This static head, or submergence effect, significantly affects the performance of refrigeration evaporators operating at extremely low temperatures and low-suction pressures.

Beyond these basic refrigeration-design requirements, the chemical industry imposes many special conditions. Exchangers frequently are applied to cool highly corrosive process streams. Consequently, special materials for evaporator tubes and channels of particularly heavy wall thickness are dictated. Corrosion allowances in the form of added material thicknesses in the evaporator may be necessary in chemical service.

High-pressure and high-temperature designs on the process side of refrigerant evaporators are frequently encountered in chemical-plant service. Process-side construction may have to be suitable for pressures seldom encountered in commercial service. Differences between process inlet and outlet temperatures greater than 55°C are not uncommon. In such cases, special consideration must be given to thermal stresses within the refrigerant evaporator. U-tube construction or floating-tube-sheet construction may be necessary. Minor process-side modifications may permit use of less expensive standard commercial fixed-tube-sheet designs. However, coordination between the equipment supplier and chemical-plant designer is necessary to tailor the evaporator to the intended duty. Relief devices and safety precautions common to the refrigeration field normally meet chemical-plant needs, but should be reviewed against individual plant standards. It must be the mutual responsibility of the refrigeration equipment supplier and the chemical-plant designer to evaluate what special features, if any, must be applied to modify commercial equipment for chemical-plant service.

Refrigeration evaporators are usually designed to meet the ASME Boiler and Pressure Vessel Code,[21] which provides for a safe, reliable exchanger at economical cost. In refrigeration systems, these exchangers generally operate with relatively small temperature differentials for which fixed-tube-sheet construction is preferred. Operating pressures in refrigerant evaporators also decrease as operating temperatures are reduced. This relationship results in extremely high factors of safety on pressure stresses, eliminating the need for expensive nickel steels from -59 to $-29°C$. Most designs are readily modified to provide suitable materials for corrosion problems on the process side.

The basic shell-and-tube exchanger with fixed-tube sheets (Fig. 14) is most widely used for refrigeration evaporators. Most designs are suitable for fluids up to 2170 kPa (300 psig) and for operation with up to 38°C temperature differences. Above these limits, specialized heat exchangers generally are used to suit individual requirements.

With the fluid on the tube side, the shell side is flooded with refrigerant for efficient wetting of the tubes (see Fig. 15). Designs must provide for distribution of flash gas and liquid refrigerant entering the shell, and for separation of liquid from the gas leaving the shell before it reaches the compressor.

In low-temperature applications and large evaporators, the exchanger surface may be sprayed rather than flooded. This eliminates the submergence effect or static-head penalty, which can be significant in large exchangers, particularly at low temperatures. The spray cooler (Fig. 16) is recommended for some large coolers to offset the cost of refrigerant inventory or charge that would be necessary for flooding.

Where the Reynolds number in the process fluid is low, as for a viscous or heavy brine, it may be desirable to handle the fluid on the shell side to obtain better heat transfer. In these cases, the refrigerant must be evaporated in the tubes. On small exchangers, commonly referred to as direct-expansion coolers, refrigerant feeding is generally handled with a thermal-expansion valve.

On large exchangers, this can best be handled by a small circulating pump to ensure adequate wetting of all tubes (Fig. 17). An oversized channel box on one end provides space for a liquid reservoir and for effective gas–liquid separation.

7.4 Expansion Devices

The primary purpose of an expansion device is to control the amount of refrigerant entering the evaporator. In the process, the refrigerant entering the valve expands from a relatively

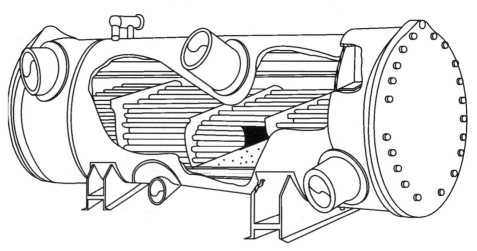

Figure 14 Typical fixed-tube-sheet evaporator.[3]

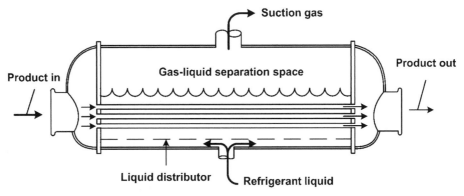

Figure 15 Typical flooded shell-and-tube evaporator.[3]

high-pressure, subcooled liquid to a saturated low-pressure mixture. Other types of flow control devices, such as pressure regulators and float valves, can also be found in some refrigeration systems. Discussion of these can be found in Ref. 1. Five types of expansion devices can be found in refrigeration systems: (1) thermostatic expansion valves, (2) electronic expansion valves, (3) constant-pressure expansion valves, (4) capillary tubes, and (5) short tube restrictors. Each is discussed briefly.

Thermostatic Expansion Valve
The thermostatic expansion valve (TXV) senses the superheat of the gas leaving the evaporator to control the refrigerant flow into the evaporator. Its primary function is to provide superheated vapor to the suction of the compressor. A TXV is mounted near the entrance to the evaporator and has a capillary tube extending from its top that is connected to a small bulb (Fig. 18). The bulb is mounted on the refrigerant tubing near the evaporator outlet. The capillary tube and bulb is filled with a substance called the thermostatic charge.[1] This charge

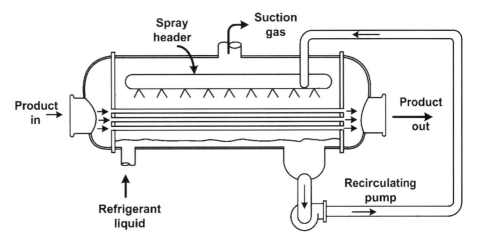

Figure 16 Typical spray-type evaporator.[3]

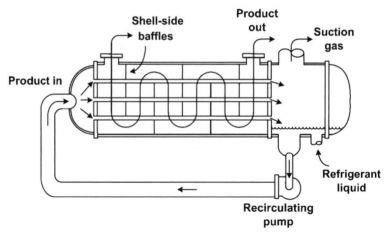

Figure 17 Typical baffled-shell evaporator.[3]

often consists of a vapor or liquid that is the same substance as the refrigerant used in the system. The response of the TXV and the superheat setting can be adjusted by varying the type of charge in the capillary tube and bulb.

The operation of a TXV is straightforward. Liquid enters the TXV and expands to a mixture of liquid and vapor at pressure P_2. The refrigerant evaporates as it travels through the evaporator and reaches the outlet where it is superheated. If the load on the evaporator

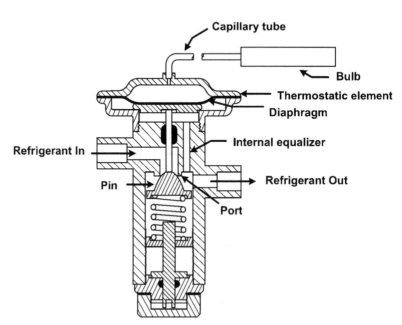

Figure 18 Cross section of a thermal expansion valve. (Reprinted by permission from 2002 *ASHRAE Handbook of Refrigeration.* ©American Society of Heating, Refrigerating, and Air-Conditioning Engineers, Inc., *www.ashrae.org.*)

is increased, the superheat leaving the evaporator will increase. This increase in flow will increase the temperature and pressure (P_1) of the charge within the bulb and capillary tube. Within the top of the TXV is a diaphragm. With an increase in pressure of the thermostatic charge, a greater force is exerted on the diaphragm, which forces the valve port to open and allow more refrigerant into the evaporator. The larger refrigerant flow reduces the evaporator superheat back to the desired level.

The capacity of TXVs is determined on the basis of opening superheat values from 2 to 4°C. TXV capacities are published for a range in evaporator temperatures and valve pressure drops. TXV ratings are based on liquid only entering the valve. The presence of flash gas will reduce the capacity substantially.

Electronic Expansion Valve

The electronic expansion valve (EEV) has become popular in recent years on larger or more expensive systems where its cost can be justified. EEVs can be heat-motor activated, magnetically modulated, pulse-width modulated, and step-motor driven.[1] EEVs can be used with digital control systems to provide control of the refrigeration system based on input variables from throughout the system.

Constant-Pressure Expansion Valve

A constant-pressure expansion valve controls the mass flow of the refrigerant entering the evaporator by maintaining a constant pressure in the evaporator. Its primary use is for applications where the refrigerant load is relatively constant. It is usually not applied where the refrigeration load may vary widely. Under these conditions, this expansion valve will provide too little flow to the evaporator at high loads and too much flow at low loads.

Capillary Tube

Capillary tubes are used extensively in household refrigerators, freezers, and small air conditioners. The capillary tube consists of one or more small-diameter tubes, which connect the high-pressure liquid line from the condenser to the inlet of the evaporator. Capillary tubes range in length from 1 to 6 m and diameters from 0.5 to 2 mm.[17]

After entering a capillary tube, the refrigerant remains a liquid for some length of the tube (Fig. 19). While a liquid, the pressure drops, but the temperature remains relatively constant (from point 1 to 2 in Fig. 19). At point 2, the refrigerant enters into the saturation region where a portion of the refrigerant begins to flash to vapor. The phase change accelerates the refrigerant and the pressure drops more rapidly. Because the mixture is saturated, its temperature drops with the pressure from 2 to 3. In many applications, the flow through a capillary tube is choked, which means that the mass flow through the tube is independent of downstream pressure.[17]

Because there are no moving parts to a capillary tube, it is not capable of making direct adjustments to variations in suction pressure or load. Thus, the capillary tube does not provide as good as performance as TXVs when applied in systems that will experience a wide range in loads.

Even though the capillary tube is insensitive to changes in evaporator pressure, its flow rate will adjust to changes in the amount of refrigerant subcooling and condenser pressure. If the load in the condenser suddenly changes so that subcooled conditions are no longer maintained at the capillary-tube inlet, the flow rate through the capillary tube will decrease. The decreased flow will produce an increase in condenser pressure and subcooling.

The size of the compressor, evaporator, and condenser as well as the application (refrigerator or air conditioner) must all be considered when specifying the length and diameter of capillary tubes. Systems using capillary tubes tend to be much more sensitive to the

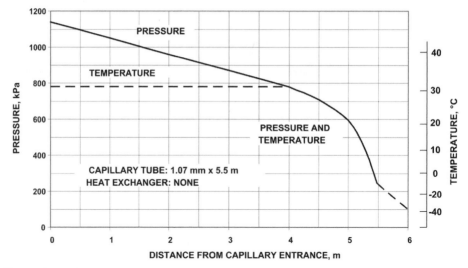

Figure 19 Typical temperature and pressure distribution in a capillary tube. (Reprinted by permission from 2002 *ASHRAE Handbook of Refrigeration.* ©American Society of Heating, Refrigerating, and Air-Conditioning Engineers, Inc., *www.ashrae.org*.)

amount of refrigerant charge than systems using TXVs or EEVs. Design charts for capillary tubes can be found in Ref. 1 for R-12 and R-22.

Short-Tube Restrictor

Short-tube restrictors are applied in many systems that formerly used capillary tubes. Figure 20 illustrates a short-tube restrictor and its housing. The restrictors are inexpensive, reliable, and easy to replace. In addition, for systems such as heat pumps, which reverse cycle, short-tube restrictors eliminate the need for a check valve. Short-tubes vary in length from 10 to

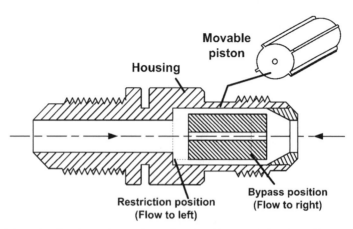

Figure 20 Schematic of a short-tube restrictor. (Reprinted by permission from 2002 *ASHRAE Handbook of Refrigeration.* ©American Society of Heating, Refrigerating, and Air-Conditioning Engineers, Inc., *www.ashrae.org*.)

13 mm, with a length-to-diameter ratio from 3 to 20.[1] Current applications for short-tube restrictors are primarily in air conditioners and heat pumps.

Like a capillary tube, short-tube restrictors operate with choked or near-choked flow in most applications.[22] The mass flow through the orifice is nearly independent of conditions downstream of the orifice. The flow rate does vary with changes in the condenser subcooling and pressure.

In applying short-tube restrictors, there are many similarities to capillary tubes. The size of the system components and type of system must be considered when sizing this expansion device. Sizing charts for the application of short-tube restrictors with R-22 can be found in Ref. 23.

8 DEFROST METHODS

When refrigeration systems operate below 0°C for extended periods of time, frost can form on the heat-transfer surfaces of the evaporator. As frost grows, it begins to block the airflow passages and insulates the cold refrigerant from the warm/moist air that is being cooled by the refrigeration system. With increasing blockage of the airflow passages, the evaporator fan(s) are unable to maintain the design airflow through the evaporator. As airflow drops, the capacity of the system decreases, and eventually reaches a point where the frost must be removed. This is accomplished with a defrost cycle.

Several defrost methods are used with refrigeration systems: hot refrigerant gas, air, and water. Each method can be used individually or in combination with the other.

8.1 Hot Refrigerant Gas Defrost

This method is the most common technique for defrosting commercial and industrial refrigeration systems. When the evaporator needs defrosting, hot gas from the discharge of the compressor is diverted from the condenser to the evaporator by closing control valve 2 and opening control valve 1 in Fig. 21. The hot gas increases the temperature of the evaporator and melts the frost. Some of the hot vapor condenses to liquid during the process. A special tank, such as an accumulator, can be used to protect the compressor from any liquid returning to the compressor.

During defrost operation, the evaporator fans are turned off, which increases the evaporator coil temperature faster. With the fans off, the liquid condensate drains from the coil faster and helps minimize the thermal load to the refrigerated space during defrost. In some instances, electrical heaters are added to increase the speed of the defrost.

Defrost initiation is usually accomplished via demand defrost or time-initiated defrost. Demand-defrost systems utilize a variable, such as pressure drop across the air side of the evaporator or a set of temperature inputs, to determine if frost has built up enough on the coil to require a defrost. Time-initiated defrost relies on a preset number of defrosts per day. The number of defrosts and length of time of each defrost can be adjusted. Ideally, the demand–defrost system provides the most efficient defrost controls on a system, because a defrost is only initiated if the evaporator needs it. Termination of the defrost cycle is accomplished by either a fixed-time interval or a set temperature of the coil.

8.2 Air, Electric, and Water Defrost

If the refrigerated space operates above 0°C, then the air in the space can be used directly to defrost the evaporator. Defrost is accomplished by continuing to operate the evaporator

Evaporator

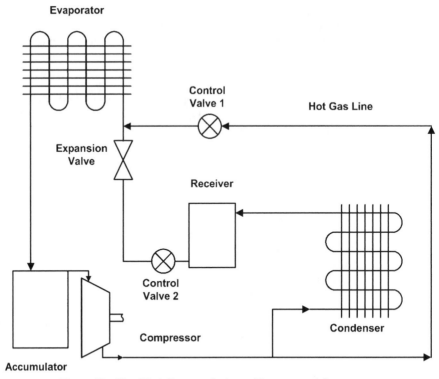

Figure 21 Simplified diagram of a hot-refrigerant-gas defrost system.

blower while the compressor is off. As the frost melts, some of it is evaporated into the airstream while the rest drains away from the coil as liquid water. The evaporated moisture adds an additional load to the evaporator when the compressor starts again.

Electric heat can be applied to the evaporator via a heating element. This technique is usually slower than hot-gas defrost. The heating element can be in direct contact with the evaporator or be located between the fan and evaporator.[1]

Water can also be used to defrost the evaporator. The compressor and fans are shut off while water is sprayed over the evaporator. If the velocity of the water is kept high, it washes or melts the frost off the coil.

9 SYSTEM DESIGN CONSIDERATIONS

Associated with continuous operation are refrigeration startup and shutdown conditions that invariably differ, sometimes widely, from those of the process itself. These conditions, although they occupy very little time in the life of the installation, must be properly accommodated in the design of the refrigeration system. Consideration must be given to the amount of time required to achieve design operating conditions, the need for standby equipment, etc.

In batch processing, operating conditions are expected to change with time, usually in a repetitive pattern. The refrigeration system must be designed for all extremes. Use of brine storage or ice banks can reduce equipment sizes for batch processes.

Closed-cycle operation involves both liquid and gas phases. System designs must take into account liquid-flow problems in addition to gas-flow requirements and must provide for effective separation of the liquid and gas phases in different parts of the system. These factors require careful design of all components and influence the arrangement or elevation of certain components in the cycle.

Liquid pressures must be high enough to feed liquid to the evaporators at all times, especially when evaporators are elevated or remotely located. In some cases, a pump must be used to suit the process requirements. The possibility of operation with reduced pressures caused by colder condensing temperatures than the specified design conditions must also be considered. Depending on the types of liquid valves and relative elevation of various parts of the system, it may be necessary to maintain condensing pressures above some minimum level, even if doing so increases the compression power.

Provision must be made to handle any refrigerant liquid that can drain to low spots in the system upon loss of operating pressure during shutdown. It must not be allowed to return as liquid to the compressor upon startup.

The operating charge in various system components fluctuates depending on the load. For example, the operating charge in an air-cooled condenser is quite high at full load, but is low, that is, essentially dry, at light load. A storage volume such as a liquid receiver must be provided at some point in the system to accommodate this variation. If the liquid controls permit the evaporator to act as the variable storage, the level may become too high, resulting in liquid carry over to the compressor.

Abnormally high process temperatures may occur either during startup or process upsets. Provision must be made for this possibility, for it can cause damaging thermal stresses on refrigeration components and excessive boiling rates in evaporators, forcing liquid to carry over and damage the compressor.

Factory-designed-and-built packages, which provide cooling as a service or utility, can require several thousand kilowatts of power to operate, but in most cases, they require no more installation than connection of power, utilities, and process lines. As a result, there is a single source of responsibility for all aspects of the refrigeration cycle involving the transfer and handling of both saturated liquids and saturated vapors throughout the cycle, oil return, and other design requirements. These packages are custom engineered, including selection of components, piping, controls, base designs, torsional and critical speed analysis, and individual chemical process requirements. Large packages are designed in sections for shipment, but are readily interconnected in the field.

As a general rule, field-erected refrigeration systems should be close-coupled to minimize problems of oil return and refrigerant condensation in suction lines. Where process loads are remotely located, pumped recirculation or brine systems are recommended. Piping and controls should be reviewed with suppliers to assure satisfactory operation under all conditions.

10   REFRIGERATION SYSTEM SPECIFICATIONS

To minimize costly and time-consuming alterations owing to unexpected requirements, the refrigeration specialist who is to do the final design must have as much information as possible before the design is started. Usually, it is best to provide more information than thought necessary , and it is always wise to note where information may be sketchy, missing, or uncertain. Carefully spelling out the allowable margins in the most critical process variables and pointing out portions of the refrigeration cycle that are of least concern is always helpful to the designer. A checklist of minimum information (Table 7) needed by a refrigeration specialist to design a cooling system for a particular application may be helpful.

Table 7   Information Needed for the Design of a Refrigeration System

Process flow sheets and thermal specifications
  Type of process
    Batch
    Continuous
  Normal heat balances
  Normal material balances
  Normal material composition
  Design operating pressure and temperatures
  Design refrigeration loads
  Energy recovery possibilities
  Manner of supplying refrigeration (primary or secondary)
Basic specifications
  Mechanical system details
    Construction standards
      Industry
      Company
      Local plant
    Insulation requirements
    Special corrosion prevention requirements
    Special sealing requirements
      Process streams to the environment
      Process stream to refrigerant
  Operating environment
    Indoor or outdoor location
    Extremes
    Special requirements
  Special safety considerations
    Known hazards of process
    Toxicity and flammability constraints
    Maintenance limitations
  Reliability requirements
    Effect of loss of cooling on process safety
    Maintenance intervals and types that may be performed
  Redundancy requirement
  Acceptance test requirements
Instrumentation and control requirements
  Safety interlocks
  Process interlocks
  Special control requirements
    At equipment
    Central control room
  Special or plant standard instruments
  Degree of automation: interface requirements
  Industry and company control standards
Off-design operation
  Process startup sequence
    Degree of automation
    Refrigeration loads vs. time
    Time needed to bring process on-stream
    Frequency of startup
    Process pressure, temperature, and composition changes during startup
    Special safety requirements
  Minimum load
  Need for standby capability

Table 7   (*Continued*)

Peak-load pressures and temperatures
Composition extremes
Process shutdown sequence
  Degree of automation
  Refrigeration load vs. time
  Shutdown time span
  Process pressure, temperature, and composition changes
  Special safety requirements

Process Flow Sheets. For chemical process designs, seeing the process flow sheets is the best overall means for the refrigeration engineer to become familiar with the chemical process for which the refrigeration equipment is to be designed. In addition to providing all of the information shown in Table 7, they give the engineer a feeling for how the chemical plant will operate as a system and how the refrigeration equipment fits into the process.

Basic Specifications. This portion of Table 7 fills in the detailed mechanical information that tells the refrigeration engineer how the equipment should be built, where it will be located, and specific safety requirements. This determines which standard equipment can be used and what special modifications need to be made.

Instrumentation and Control Requirements. These tell the refrigeration engineer how the system will be controlled by the plant operators. Particular controller types, as well as control sequencing and operation, must be spelled out to avoid misunderstandings and costly redesign. The refrigeration engineer needs to be aware of the degree of control required for the refrigeration system—for example, the process may require remote starting and stopping of the refrigeration system from the central control room. This could influence the way in which the refrigeration safeties and interlocks are designed.

Off-design Operation. It is likely that the most severe operation of the refrigeration system will occur during startup and shutdown. The rapidly changing pressures, temperatures, and loads experienced by the refrigeration equipment can cause motor overloads, compressor surging, or loss of control if they are not anticipated during design.

REFERENCES

1. *ASHRAE Handbook of Refrigeration,* American Society of Heating, Refrigerating and Air-Conditioning Engineers, Atlanta, 2002.
2. R. Thevenot, *A History of Refrigeration Throughout the World,* International Institute of Refrigeration, Paris, France, 1979, pp. 39–46.
3. K. W. Cooper and K. E. Hickman, "Refrigeration" in *Encyclopedia of Chemical Technology,* Vol. 20, 3rd ed., Wiley, New York, 1984, pp. 78–107.
4. C. E. Salas and M. Salas, *Guide to Refrigeration CFC's,* Fairmont Press, Liburn, GA, 1992.
5. U.S. Environmental Protection Agency, "The Accelerated Phaseout of Ozone-Depleting Substances,"*Federal Register,* **58**(236), 65018–65082 (December 10, 1993).
6. U.S. Environmental Protection Agency, "Class I Nonessential Products Ban, Section 610 of the Clean Air Act Amendments of 1990," *Federal Register* **58**(10), 4768–4799 (January 15, 1993).
7. United Nations Environmental Program (UNEP), *Montreal Protocol on Substances That Deplete the Ozone Layer—Final Act,* 1987.

8. G. King, *Basic Refrigeration,* Business News, Troy, MI, 1986.

9. ANSI/ASHRAE 34-2001, *Number Designation and Safety Classification of Refrigerants,* American Society of Heating, Refrigerating and Air-Conditioning Engineers, Atlanta, 2001.

10. G. G. Haselden, *Mech. Eng., 44* (March 1981).

11. ANSI/ASHRAE 15-1994, *Safety Code for Mechanical Refrigeration,* American Society of Heating, Refrigerating and Air-Conditioning Engineers, Atlanta, 1994.

12. *ASHRAE Handbook of Fundamentals,* American Society of Heating, Refrigerating and Air-Conditioning Engineers, Atlanta (Chapter 19), 2001.

13. M. J. Molina and F. S. Rowland, "Stratospheric Sink for Chlorofluoromethanes: Chlorine Atoms Catalyzed Destruction of Ozone," *Nature,* **249,** 810–812 (1974).

14. C. D. MacCracken, "The Greenhouse Effect on ASHRAE," *ASHRAE Journal,* 31(6), 52–55 (1996).

15. U.S. EPA, Ozone Depletion Website, *http://www.epa.gov/ozone/index.html,* United States Environmental Protection Agency, 2004.

16. Houghton et al., *Climate Change 2001: The Scientific Basis,* Cambridge University Press, Cambridge, UK, 2001.

17. W. F. Stoecker and J. W. Jones, *Refrigeration and Air Conditioning,* 2nd ed., McGraw-Hill, New York, 1982.

18. *ASHRAE Handbook of HVAC Systems and Equipment,* American Society of Heating, Refrigerating and Air-Conditioning Engineers, Atlanta, 2000, Chapters 34 and 35.

19. K. Matsubara, K. Suefuji, and H. Kuno, "The Latest Compressor Technologies for Heat Pumps in Japan," in *Heat Pumps,* K. Zimmerman and R. H. Powell, Jr. (eds.), Lewis, Chelsea, MI, 1987.

20. T. Senshu, A. Araik, K. Oguni, and F. Harada, "Annual Energy-Saving Effect of Capacity-Modulated Air Conditioner Equipped with Inverter-Driven Scroll Compressor," *ASHRAE Transactions,* **91**(2) (1985).

21. *ASME Boiler and Pressure Vessel Code,* Sect. VIII, Div. 1, The American Society of Mechanical Engineers, New York, 1980.

22. Y. Kim and D. L. O'Neal, "A Comparison of Critical Flow Models for Estimating Two-Phase Flow of HCFC 22 and HFC 134a Through Short Tube Orifices," *International Journal of Refrigeration,* **18**(6) (1995).

23. Y. Kim and D. L. O'Neal, "Two-Phase Flow of Refrigerant-22 through Short-Tube Orifices," *ASHRAE Transactions,* **100**(1) (1994).

CHAPTER 13
CRYOGENIC SYSTEMS

Leonard A. Wenzel
Lehigh University
Bethlehem, Pennsylvania

1 CRYOGENICS AND CRYOFLUID PROPERTIES

The science and technology of deep refrigeration processing occurring at temperatures lower than about 150 K is the field of cryogenics (from the Greek *kryos,* icy cold). This area has developed as a special discipline because it is characterized by special techniques, requirements imposed by physical limitations, and economic needs, and unique phenomena associated with low-thermal-energy levels.

Compounds that are processed within the cryogenic temperature region are sometimes called cryogens. There are only a few of these materials; they are generally small, relatively simple molecules, and they seldom react chemically within the cryogenic region. Table 1 lists the major cryogens along with their major properties, and with a reference giving more complete thermodynamic data.

All of the cryogens except hydrogen and helium have conventional thermodynamic and transport properties. If specific data are unavailable, the reduced properties correlation can be used with all the cryogens and their mixtures with at least as much confidence as the correlations generally allow. Qualitatively *T–S* and *P–H* diagrams such as those of Figs. 1

Table 1 Properties of Principal Cryogens

Name	T (K)	Normal Boiling Point Liquid Density (kg/m³)	Normal Boiling Point Latent Heat (J/kg·mole)	Critical Point T (K)	Critical Point P (kPa)	Triple Point T (K)	Triple Point P (kPa)	Reference
Helium	4.22	123.9	91,860	5.28	227			1
Hydrogen	20.39	70.40	902,300	33.28	1296	14.00	7.20	2, 3
Deuterium	23.56	170.0	1,253,000	38.28	1648	18.72	17.10	4
Neon	27.22	1188.7	1,737,000	44.44	2723	26.28	43.23	5
Nitrogen	77.33	800.9	5,579,000	126.17	3385	63.22	12.55	6
Air	78.78	867.7	5,929,000					7, 8
Carbon monoxide	82.11	783.5	6,024,000	132.9	3502	68.11	15.38	9
Fluorine	85.06	1490.6	6,530,000	144.2	5571			10
Argon	87.28	1390.5	6,504,000	151.2	4861	83.78		11, 12, 13
Oxygen	90.22	1131.5	6,801,000	154.8	5081	54.39	0.14	6
Methane	111.72	421.1	8,163,000	190.61	4619	90.67	11.65	14
Krypton	119.83	2145.4	9,009,000	209.4	5488	116.00	73.22	15
Nitric oxide	121.50	1260.2	13,809,000	179.2	6516	108.94		
Nitrogen trifluoride	144.72	1525.6	11,561,000	233.9	4530			
Refrigerant-14	145.11	1945.1	11,969,000	227.7	3737	89.17	0.12	16
Ozone	161.28	1617.8	14,321,000	261.1	5454			
Xenon	164.83	3035.3	12,609,000	289.8	5840	161.39	81.50	17
Ethylene	169.39	559.4	13,514,000	282.7	5068	104.00	0.12	18

and 2 differ among cryogens only by the location of the critical point and freezing point relative to ambient conditions.

Air, ammonia synthesis gas, and some inert atmospheres are considered as single materials although they are actually gas mixtures. The composition of air is shown in Table 12. If a thermodynamic diagram for air has the lines drawn between liquid and vapor boundaries where the pressures are equal for the two phases, these lines will not be at constant temperature, as would be the case for a pure component. Moreover, these liquid and vapor states are not at equilibrium, for the equilibrium states have equal Ts and Ps, but differ in composition. That being so, one or both of these equilibrium mixtures is not air. Except for this difference the properties of air are also conventional.

Hydrogen and helium differ in that their molecular mass is small in relation to zero-point-energy levels. Thus quantum differences are large enough to produce measurable changes in gross thermodynamic properties.

Hydrogen and its isotopes behave abnormally because the small molecular weight allows quantum differences stemming from different molecular configurations to affect total thermodynamic properties. The hydrogen molecule consists of two atoms, each containing a single proton and a single electron. The electrons rotate in opposite directions as required by molecular theory. The protons, however, may rotate in opposed or parallel directions. Figure 3 shows a sketch of the two possibilities, the parallel rotating nuclei identifying ortho-hydrogen and the opposite rotating nuclei identifying the para-hydrogen. The quantum mechanics exhibited by these two molecule forms are different, and produce different thermodynamic properties. Ortho- and para-hydrogen each have conventional thermodynamic properties. However, ortho- and para-hydrogen are interconvertible with the equilibrium fraction of pure H_2 existing in para form dependent on temperature, as shown in Table 2. The natural ortho- and para-hydrogen reaction is a relatively slow one and of second order[19]:

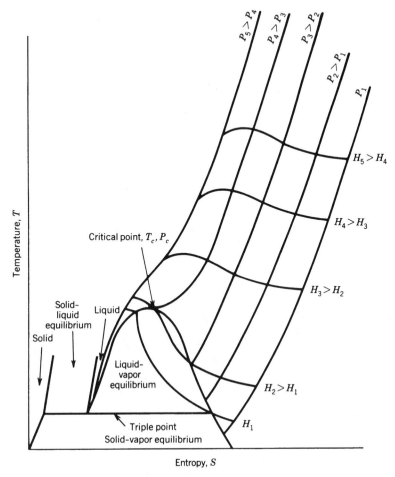

Figure 1 Skeletal T–S diagram.

$$\frac{dx}{d\theta} = 0.0114x^2 \quad \text{at} \quad 20 \text{ K} \tag{1}$$

where θ is time in hours and x is the mole fraction of ortho-hydrogen. The reaction rate can be greatly accelerated by a catalyst that interrupts the molecular magnetic field and possesses high surface area. Catalysts such as NiO_2/SiO_2 have been able to yield some of the highest heterogeneous reaction rates measured.[20]

Normally hydrogen exists as a 25 mole % p-H_2, 75 mole % o-H_2 mix. Upon liquefaction the hydrogen liquid changes to nearly 100% p-H_2. If this is done as the liquid stands in an insulated flask, the heat of conversion will suffice to evaporate the liquid, even if the insulation is perfect. For this reason the hydrogen is usually converted to para form during refrigeration by the catalyzed reaction, with the energy released added to the refrigeration load.

Conversely, liquid para-hydrogen has an enhanced refrigeration capacity if it is converted to the equilibrium state as it is vaporized and warmed to atmospheric condition. In certain applications recovery of this refrigeration is economically justifiable.

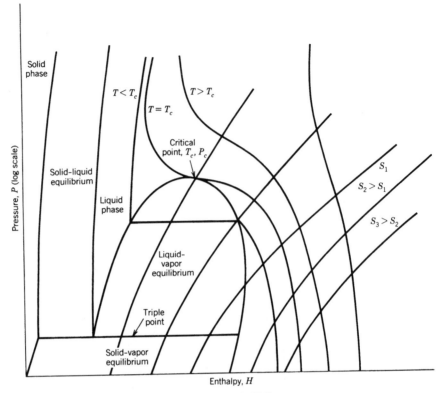

Figure 2 Skeletal *P–H* diagram.

Helium, though twice the molecular weight of hydrogen, also shows the effects of flow molecular weight upon gross properties. The helium molecule is single-atomed and thus free from ortho–para-type complexities. Helium was liquefied conventionally first in 1908 by Onnes of Leiden, and the liquid phase showed conventional behavior at atmospheric pressure.

As temperature is lowered, however, a second-order phase change occurs at 2.18 K (0.05 atm) to produce a liquid called HeII. At no point does solidification occur just by evacuating the liquid. This results from the fact that the relationship between molecular volume, thermal energy (especially zero-point energy), and van der Waals attractive forces is such that the atoms cannot be trapped into a close-knit array by temperature reduction alone. Eventually, it was found that helium could be solidified if an adequate pressure is applied, but that the normal liquid helium (HeI)–HeII phase transition occurs at all pressures

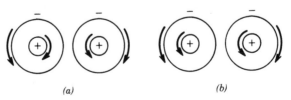

Figure 3 Molecular configurations of (*a*) para- and (*b*) ortho-hydrogen.

Table 2   Equilibrium Para-Hydrogen Concentration as a
Function of T (K)

T (K)	Equilibrium Percentage of Para-Hydrogen
20	99.82
30	96.98
40	88.61
60	65.39
80	48.39
100	38.51
150	28.54
273	25.13
500	25.00

up to that of solidification. The phase diagram for helium is shown in Fig. 4. The HeI–HeII phase change has been called the lambda curve from the shape of the heat capacity curve for saturated liquid He, as shown in Fig. 5. The peculiar shape of the heat capacity curve produces a break in the curve for enthalpy of saturated liquid He as shown in Fig. 6.

HeII is a unique liquid exhibiting properties that were not well explained until after 1945. As liquid helium is evacuated to increasingly lower pressures, the temperature also drops along the vapor-pressure curve. If this is done in a glass vacuum-insulated flask, heat leaks into the liquid He causing boiling and bubble formation. As the temperature approaches 2.18 K, boiling gets more violent, but then suddenly stops. The liquid He is completely quiescent. This has been found to occur because the thermal conductivity of HeII is extremely large. Thus the temperature is basically constant and all boiling occurs from the surface where the hydrostatic head is least, producing the lowest boiling point.

Not only does HeII have very large thermal conductivity, but it also has near zero viscosity. This can be seen by holding liquid He in a glass vessel with a fine porous bottom such that normal He does not flow through. If the temperature is lowered into the HeII

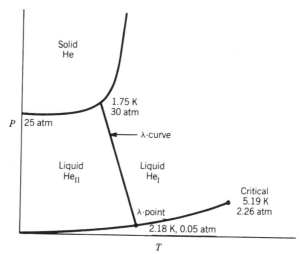

Figure 4   Phase diagram for helium.

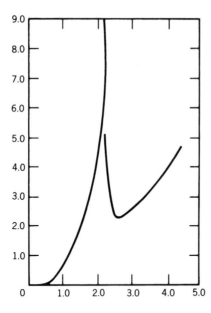

Figure 5   Heat capacity of saturated liquid ⁴He.

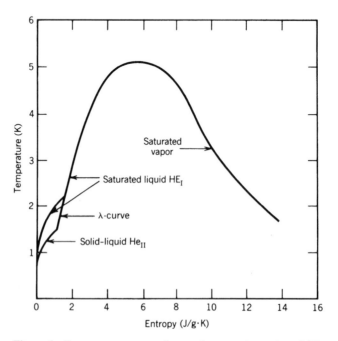

Figure 6   Temperature–entropy diagram for saturation region of ⁴He.

region, the helium will flow rapidly through the porous bottom. Flow does not seem to be enhanced or hindered by the size of the frit. Conversely, a propeller operated in liquid HeII will produce a secondary movement in a parallel propeller separated from the first by a layer of liquid HeII. Thus HeII has properties of finite and of infinitesimal viscosity.

These peculiar flow properties are also shown by the so-called thermal-gravimetric effect. There are two common demonstrations. If a tube with a finely fritted bottom is put into liquid HeII and the helium in the tube is heated, liquid flows from the main vessel into the fritted tube until the liquid level in the tube is much higher than that in the main vessel. A second, related, experiment uses a U-tube, larger on one leg than on the other with the two sections separated by a fine frit. If this tube is immersed, except for the end of the narrow leg, into liquid HeII and a strong light is focused on the liquid He above the frit, liquid He will flow through the frit and out the small tube opening producing a fountain of liquid He several feet high.

These and other experiments[21] can be explained through the quantum mechanics of HeII. The pertinent relationships, the Bose–Einstein equations, indicate that HeII has a dual nature: it is both a "superfluid" which has zero viscosity and infinite thermal conductivity among other special properties, and a fluid of normal properties. The further the temperature drops below the lambda point the greater the apparent fraction of superfluid in the liquid phase. However, very little superfluid is required. In the flow through the porous frit the superfluid flows, the normal fluid is retained. However, if the temperature does not rise, some of the apparently normal fluid will apparently become superfluid. Although the superfluid flows through the frit, there is no depletion of superfluid in the liquid He left behind. In the thermogravimetric experiments the superfluid flows through the frit but is then changed to normal He. Thus there is no tendency for reverse flow.

At this point applications have not developed for HeII. Still, the peculiar phase relationships and energy effects may influence the design of helium processes, and do affect the shape of thermodynamic diagrams for helium.

2 CRYOGENIC REFRIGERATION AND LIQUEFACTION CYCLES

One characteristic aspect of cryogenic processing has been its early and continued emphasis on process efficiency, that is, on energy conservation. This has been forced on the field by the very high cost of deep refrigeration. For any process the minimum work required to produce the process goal is

$$W_{min} = T_0 \, \Delta S - \Delta H \tag{2}$$

where W_{min} is the minimum work required to produce the process goal, ΔS and ΔH are the difference between product and feed entropy and enthalpy, respectively, and T_0 is the ambient temperature. Table 3 lists the minimum work required to liquefy 1 kg-mole of several common cryogens. Obviously, the lower the temperature level the greater the cost for unit result. The evident conflict in H_2 and He arises from their different molecular weights and properties. However, the temperature differences from ambient to liquid H_2 temperature and from ambient to liquid He temperatures are similar.

A refrigeration cycle that would approach the minimum work calculated as above would include ideal process steps as, for instance, in a Carnot refrigeration cycle. The cryogenic engineer aims for this goal while satisfying practical processing and capital cost limitations.

2.1 Cascade Refrigeration

The cascade refrigeration cycle was the first means used to liquefy air in the United States.[22] It uses conveniently chosen refrigeration cycles, each using the evaporator of the previous

Table 3 Minimum Work Required to Liquefy Some Common Cryogens

Gas	Normal Boiling Point (K)	Minimum Work of Liquefaction (J/mole)
Helium	4.22	26,700
Hydrogen	20.39	23,270
Neon	27.11	26,190
Nitrogen	77.33	20,900
Air	78.8	20,740
Oxygen	90.22	19,700
Methane	111.67	16,840
Ethane	184.50	9,935
Ammonia	239.78	3,961

fluid cycle as condenser, which will produce the desired temperature. Figures 7 and 8 show a schematic *T–S* diagram of such a cycle and the required arrangement of equipment.

Obviously, this cycle is mechanically complex. After its early use it was largely replaced by other cryogenic cycles because of its mechanical unreliability, seal leaks, and poor mechanical efficiency. However, the improved reliability and efficiency of modern compressors has fostered a revival in the cascade cycle. Cascade cycles are used today in some base-load natural gas liquefaction (LNG) plants[23] and in the some peak-shaving LNG plants. They are also used in a variety of intermediate refrigeration processes. The cascade cycle is potentially the most efficient of cryogenic processes because the major heat-transfer steps are liquefaction–vaporization exchanges with each stream at a constant temperature. Thus, heat-transfer coefficients are high and ΔTs can be kept very small.

2.2 The Linde or Joule–Thomson Cycle

The Linde cycle was used in the earliest European efforts at gas liquefaction and is conceptually the simplest of cryogenic cycles. A simple flow sheet is shown in Fig. 9. Representation of the cycle as a *P–H* diagram is shown in Fig. 10. Here the gas to be liquefied or used as refrigerant is compressed through several stages each with its aftercooler. It then enters the main countercurrent heat exchanger where it is cooled by returning low-pressure gas. The gas is then expanded through a valve where it is cooled by the Joule–Thomson effect and partially liquefied. The liquid fraction can then be withdrawn, as shown, or used as a refrigeration source.

Making a material and energy balance around a control volume including the main exchanger, JT valve, and liquid receiver for the process shown gives

$$X = \frac{(H_7 - H_2) - Q_L}{H_7 - H_5} \tag{3}$$

where X is the fraction of the compressed gas to be liquefied. Thus process efficiency and even operability depend entirely on the Joule–Thomson effect at the warm end of the main heat exchanger and on the effectiveness of that heat exchanger. Also, if Q_L becomes large due to inadequate insulation, X quickly goes to zero.

Because of its dependence on Joule–Thomson effect at the warm end of the main exchanger, the Joule–Thomson liquefier is not usable for H_2 and He refrigeration without

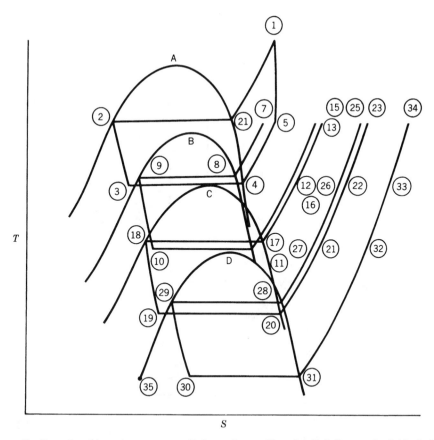

Figure 7   Cascade refrigeration system on *T–S* coordinates. Note that *T–S* diagram for fluids A, B, C, and D are here superimposed. Numbers here refer to Fig. 8 flow points.

precooling. However, if H_2 is cooled to liquid N_2 temperature before it enters the JT cycle main heat exchanger, or if He is cooled to liquid H_2 temperature before entering the JT cycle main heat exchanger, further cooling to liquefaction can be done with this cycle. Even with fluids such as N_2 and CH_4 it is often advantageous to precool the gas before it enters the JT heat exchanger in order to take advantage of the greater Joule–Thomson effect at the lower temperature.

2.3   The Claude or Expander Cycle

Expander cycles have become workhorses of the cryogenic engineer. A simplified flow sheet is shown in Fig. 11. Here part of the compressed gas is removed from the main exchanger before being fully cooled, and is cooled in an expansion engine in which mechanical work is done. Otherwise, the system is the same as the Joule–Thomson cycle. Figure 12 shows a *T–S* diagram for this process. The numbers on the diagram refer to those on the process flow sheet.

If, as before, energy and material balances are made around a control volume including the main exchanger, expansion valve, liquid receiver, and the expander, one obtains

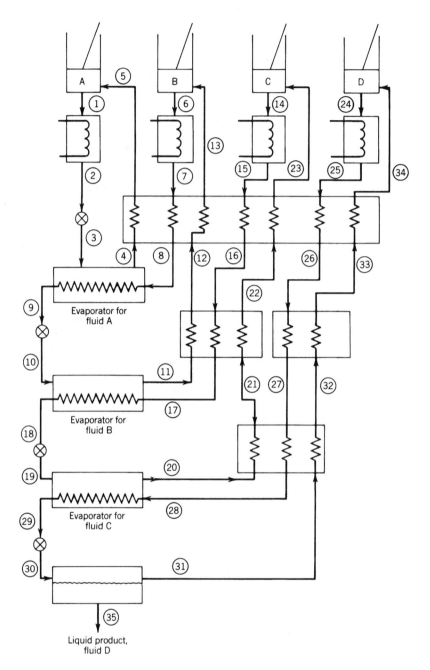

Figure 8 Cascade liquefaction cycle—simplified flow diagram.

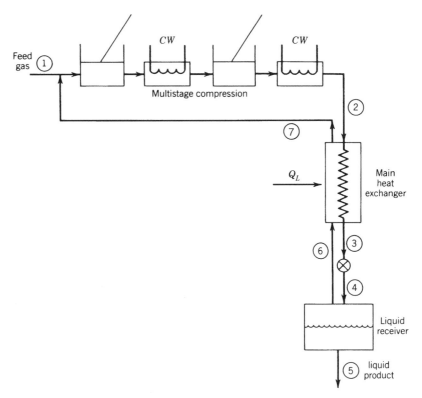

Figure 9 Simplified Joule–Thomson liquefaction cycle flow diagram.

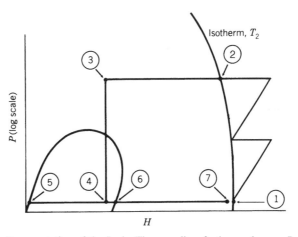

Figure 10 Representation of the Joule–Thomson liquefaction cycle on a *P–H* diagram.

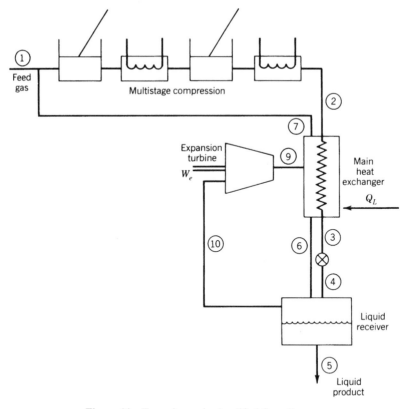

Figure 11 Expander cycle simplified flow diagram.

$$X = \frac{(H_7 - H_2) + Y(H_9 - H_{10}) - Q_L}{H_7 - H_5} \qquad (4)$$

where Y is the fraction of the high-pressure stream that is diverted to the expander.

Here the liquid yield is not so dependent on the shape of the warm isotherm or the effectiveness of heat exchange since the expander contributes the major part of the refrigeration. Also, the limitations applicable to a JT liquefier do not pertain here. The expander cycle will operate independent of the Joule–Thomson effect of the system gas.

The expansion step, line 9–10 on the T–S diagram, is ideally a constant entropy path. However, practical expanders operate at 60–90% efficiency and hence the path is one of modestly increasing entropy. In Fig. 12 the expander discharges a two-phase mixture. The process may be designed to discharge a saturated or a superheated vapor. Most expanders will tolerate a small amount of liquid in the discharge stream. However, this should be checked carefully with the manufacturer, for liquid can rapidly erode some expanders and can markedly reduce the efficiency of others.

Any cryogenic process design requires careful consideration of conditions in the main heat exchanger. The cooling curve plotted in Fig. 13 shows the temperature of the process stream being considered, T_i, as a function of the enthalpy difference $(H_o - H_i)$, where H_o is the enthalpy for the process stream as it enters or leaves the warm end of the exchanger, and H_i is the enthalpy of that same stream at any point within the main exchanger. The

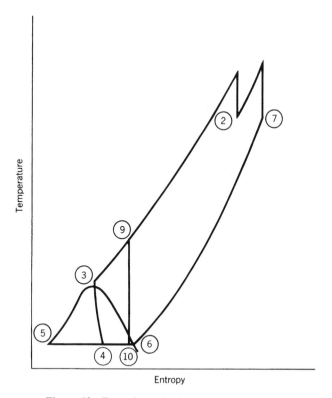

Figure 12 Expander cycle shown on a *T–S* diagram.

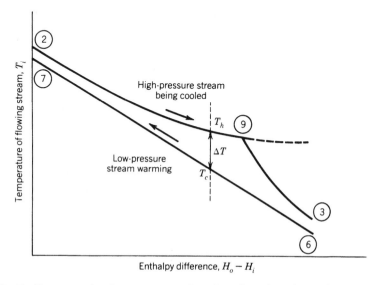

Figure 13 Cooling curves showing temperatures throughout the main exchanger for the expander cycle.

enthalpy difference is the product of the ΔH obtainable from a thermodynamic diagram and the mass flow rate of the process stream. If the mass flow rate changes, as it does at point 9 in the high-pressure stream, the slope will change. $H_o - H_i$, below such a point would be obtained from $H_o - H_i = (H_o - H_i) \cdot (1 - y)$ if the calculation is made on the basis of unit mass of high-pressure gas.

It is conventional practice to design cryogenic heat exchangers so that the temperature of a given process stream will be the same in each of the multiple passages of the exchanger at a given exchanger cross section. The temperature difference between the high- and low-pressure streams $(T_h - T_c)$ at that point is the ΔT available for heat transfer. Obviously, the simple ΔT_{lm} approach to calculation of heat-exchanger area will not be satisfactory here, for that method depends on linear cooling curves. The usual approach here is to divide the exchanger into segments of ΔH such that the cooling curves are linear for the section chosen and to calculate the exchanger area for each section. It is especially important to examine cryogenic heat exchangers in this detail because temperature ranges are likely to be large, thus producing heat-transfer coefficients that vary over the length of the exchanger, and because the curvature of the cooling curves well may produce regions of very small ΔT. In extreme cases the designer may even find that ΔT at some point in the exchanger reaches zero, or becomes negative, thus violating the second law. No exchanger designed in ignorance of this situation would operate as designed.

Minimization of cryogenic process power requirements, and hence operating costs, can be done using classical considerations of entropy gain. For any process

$$W = W_{\min} + \Sigma T_0 \Delta S_T \tag{5}$$

where W is the actual work required by the process, $W_{\min}$ is the minimum work [see Eq. (1)], and the last term represents the work lost in each process step. In that term T_0 is the ambient temperature, and ΔS_T is the entropy gain of the universe as a result of each process step.

In a heat exchanger

$$T_0 \Delta S_T = W_L = T_0 \int \frac{T_h - T_c}{T_h T_c} \, dH_i \tag{6}$$

where T_h and T_c represent temperatures of the hot and cold streams and the integration is carried out from end to end of the heat exchanger.

A comparison of the Claude cycle (so named because Georges Claude first developed a practical expander cycle for air liquefaction in 1902) with the Joule–Thomson cycle can thus be made by considering the W_L in the comparable process steps. In the cooling curve diagram, Fig. 13, the dotted line represents the high-pressure stream cooling curve of a Joule–Thomson cycle operating at the same pressure as does the Claude cycle. In comparison, the Claude cycle produces much smaller ΔTs at the cold end of the heat exchanger. If this is translated into lost work as done in Fig. 14, there is considerable reduction. The Claude cycle also reduces lost work by passing only a part of the high-pressure gas through a valve, which is a completely irreversible pressure reduction step. The rest of the high-pressure gas is expanded in a machine where most of the pressure lost produces usable work.

There are other ways to reduce the ΔT, and hence the W_L, in cryogenic heat exchangers. These methods can be used by the engineer as process conditions warrant. Figure 15 shows the effect of (*a*) intermediate refrigeration, (*b*) varying the amount of low-pressure gas in the exchanger, and (*c*) adding a third cold stream to the exchanger.

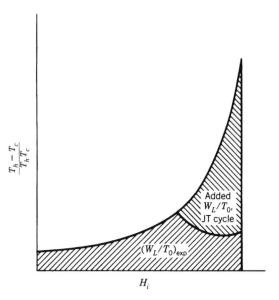

Figure 14 Calculation of W_L in the main heat exchanger using Eq. (6) and showing the comparison between JT and Claude cycles.

2.4 Low-Temperature Engine Cycles

The possibility that Carnot cycle efficiency could be approached by a refrigeration cycle in which all pressure change occurs by compression and expansion has encouraged the development of several cycles that essentially occur within an engine. In general, these have proven useful for small-scale cryogenic refrigeration under unusual conditions as in space vehicles. However, the Stirling cycle, discussed below, has been used for industrial-scale production situations.

The Stirling Cycle

In this cycle a noncondensable gas, usually helium, is compressed, cooled both by heat transfer to cooling water and by heat transfer to a cold solid matrix in a regenerator (see Section 3.3), and expanded to produce minimum temperature. The cold gas is warmed by heat transfer to the fluid or space being refrigerated and to the now-warm matrix in the regenerator and returned to the compressor. Figure 16 shows the process path on a *T–S* diagram. The process efficiency of this idealized cycle is identical to a Carnot cycle efficiency.

In application the Stirling cycle is usually operated in an engine where compression and expansion both occur rapidly with compression being nearly adiabatic. Figure 17 shows such a machine. The compressor piston (1) and a displacer piston (16 with cap 17) operate off the same crankshaft. Their motion is displaced by 90°. The result is that the compressor position is near the top or bottom of its cycle as the displacer is in most rapid vertical movement. Thus the cycle can be traced as follows:

1. With the displacer near the top of its stroke the compressor moves up compressing the gas in space 4.

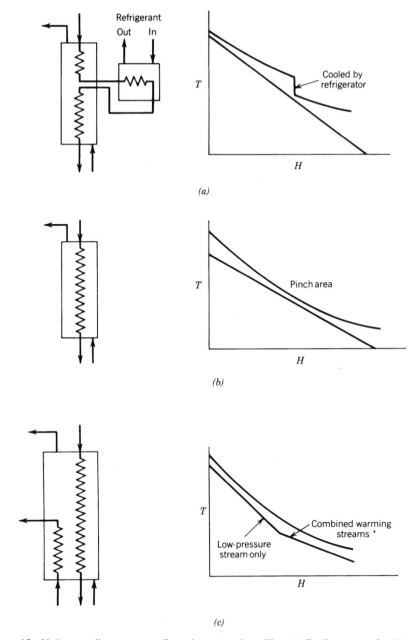

Figure 15   Various cooling curve configurations to reduce W_L: (a) Cooling curve for intermediate refrigerator case. (b) Use of reduced warming stream to control ΔTs. (c) Use of an additional warming stream.

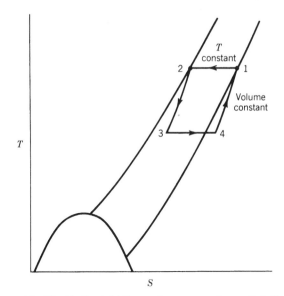

Figure 16 The idealized Stirling cycle represented on a *T–S* diagram.

2. The displacer moves down, and the gas moves through the annular water-cooled heat exchanger (13) and the annular regenerator (14) reaching the upper space (5) in a cooled, compressed state. The regenerator packing, fine copper wool, is warmed by this flow.

3. Displacer and compressor pistons move down together. Thus the gas in (5) is expanded and cooled further. This cold gas receives heat from the chamber walls (18) and interior fins (15) thus refrigerating these solid parts and their external finning.

4. The displacer moves up, thus moving the gas from space (5) to space (4). In flowing through the annular passages the gas recools the regenerator packing.

The device shown in Fig. 17 is arranged for air liquefaction. Room air enters at (23), passes through the finned structure where water and then CO_2 freeze out, and is then liquified as it is further cooled as it flows over the finned surface (18) of the cylinder. The working fluid, usually He, is supplied as needed from the tank (27).

Other Engine Cycles

The Stirling cycle can be operated as a heat engine instead of as a refrigerator and, in fact, that was the original intent. In 1918 Vuilleumier patented a device that combines these two forms of Stirling cycle to produce a refrigerator that operated on a high-temperature heat source rather than on a source of work. This process has received recent attention[24] and is useful in situations where a heat source is readily available but where power is inaccessible or costly.

The Gifford–McMahon cycles[25] have proven useful for operations requiring a light-weight, compact refrigeration source. Two cycles exist: one with a displacer piston that produces little or no work; the other with a work-producing expander piston. Figure 18 shows the two cycles.

In both these cycles the compressor operates continuously maintaining the high-pressure surge volume of P_1, T_1. The sequence of steps for the system with work-producing piston are

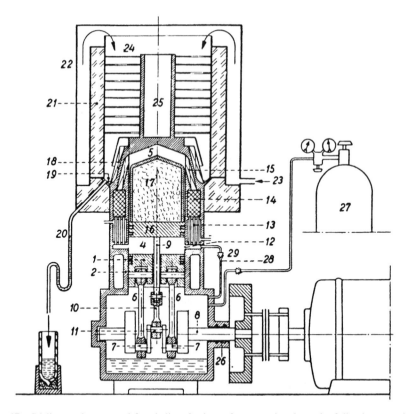

Figure 17 Stirling cycle arranged for air liquefaction reference points have the following meanings: 1, compressor; 2, compression cylinder; 4, working fluid in space between compressor and displacer; 5, working fluid in the cold head region of the machine; 6, two parallel connecting rods with cranks, 7, of the main piston; 8, crankshaft; 9, displacer rod, linked to connecting rod, 10, and crank, 11, of the displacer; 12, ports; 13, cooler; 14, regenerator; 15, freezer; 16, displacer piston, and 17, cap; 18, condenser for the air to be liquefied, with annular channel, 19, tapping pipe (gooseneck) 20, insulating screening cover, 21, and mantel 22; 23, aperture for entry of air; 24, plates of the ice separator, joined by the tubular structure, 25, to the freezer (15); 26, gas-tight shaft seal; 27, gas cylinder supplying refrigerant; 28, supply pipe with one-way valve, 29. (Courtesy U.S. Philips Corp.)

1. Inlet valve opens filling system to P_2.
2. Gas enters the cold space below the piston as the piston moves up doing work and thus cooling the gas. The piston continues, reducing the gas pressure to P_1.
3. The piston moves down pushing the gas through the heat load area and the regenerator to the storage vessel at P_1.

The sequence of steps for the system with the displacer is similar except that gas initially enters the warm end of the cylinder, is cooled by the heat exchanger, and then is displaced by the piston so that it moves through the regenerator for further cooling before entering the cold space. Final cooling is done by "blowing down" this gas so that it enters the low-pressure surge volume at P_1.

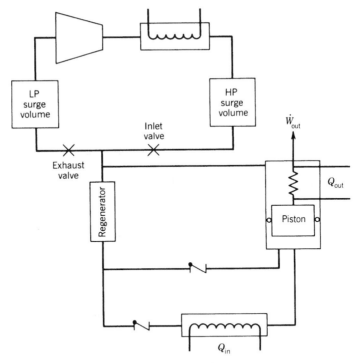

Figure 18 Gifford–McMahon refrigerator. The dashed line and the cooler are present only when the piston is to be used as a displacer with negligible work production.

If the working fluid is assumed to be an ideal gas, all process steps are ideal, and compression is isothermal, the COPs for the two cycles are:

$$\text{COP (work producing)} = 1 \left/ \frac{RT_1 \ln P_2/P_1}{C_{P\,0}T_3[1 - (P_4/P_3)^{(k-1)/k}] - 1} \right.$$

$$\text{COP (displacer)} = \frac{P_3 - P_1}{P_3(T_1/T_3 - P_1/P_3)\ln P_2/P_1}$$

In these equations states 1 and 2 are those immediately before and after the compressor. State 3 is after the cooling step but before expansion, and state 4 is after the expansion at the lowest temperature.

3 CRYOGENIC HEAT-TRANSFER METHODS

In dealing with heat-transfer requirements the cryogenic engineer must effect large quantities of heat transfer over small ΔTs through wide temperature ranges. Commonly heat capacities and/or mass flows change along the length of the heat-transfer path, and often condensation or evaporation takes place. To minimize heat leak these complexities must be handled using exchangers with as large a heat-transfer surface area per exchanger volume as possible.

Compact heat-exchanger designs of many sorts have been used, but only the most common types will be discussed here.

3.1   Coiled-Tube-in-Shell Exchangers

The traditional heat exchanger for cryogenic service is the Hampson or coiled-tube-in-shell exchanger as shown in Fig. 19. The exchanger is built by turning a mandrel on a specially built lathe, and wrapping it with successive layers of tubing and spacer wires. Since longitudinal heat transfer down the mandrel is not desired, the mandrel is usually made of a poorly conducting material such as stainless steel, and its interior is packed with an insulating material to prevent gas circulation. Copper or aluminum tubing is generally used. To prevent uneven flow distribution from tube to tube, tube winding is planned so that the total length of each tube is constant independent of the layer on which the tube is wound. This results in a constant winding angle, as shown in Fig. 20. For example, the tube layer next to the mandrel might have five parallel tubes, whereas the layer next to the shell might have 20 parallel tubes. Spacer wires may be laid longitudinally on each layer of tubes, or they may be wound counter to the tube winding direction, or omitted. Their presence and size depends on the flow requirements for fluid in the exchanger shell. Successive tube layers may be wound in opposite or in the same direction.

After the tubes are wound on the mandrel they are fed into manifolds at each end of the tube bundle. The mandrel itself may be used for this purpose, or hook-shaped manifolds of large diameter tubing can be looped around the mandrel and connected to each tube in the bundle. Finally, the exchanger is closed by wrapping a shell, usually thin-walled stainless steel, over the bundle and welding on the required heads and nozzles.

In application the low-pressure fluid flows through the exchanger shell, and high-pressure fluids flow through the tubes. This exchanger is easily adapted for use by three or more fluids by putting in a pair of manifolds for each tube-side fluid to be carried. However, tube arrangement must be carefully engineered so that the temperatures of all the cooling streams (or all the warming streams) will be identical at any given exchanger cross section. The exchanger is typically mounted vertically so that condensation and gravity effects will not result in uneven flow distribution. Most often the cold end is located at the top so that any liquids not carried by the process stream will move toward warmer temperatures and be evaporated.

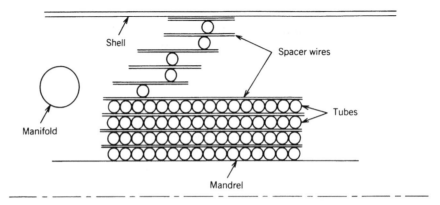

Figure 19   Section of a coiled-tube-in-shell heat exchanger.

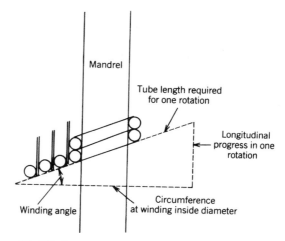

Figure 20 Winding relationships for a coiled-tube-in-shell exchanger.

Heat-transfer coefficients in these exchangers will usually vary from end to end of the exchangers because of the wide temperature range experienced. For this reason, and because of the nonlinear ΔT variations, the exchanger area must be determined by sections, the section lengths chosen so that linear ΔTs can be used and so that temperature ranges are not excessive. For inside tube heat-transfer coefficients with single-phase flow the Dittus–Boelter equation is used altered to account for the spiral flow:

$$\frac{hD}{k} = 0.023 N_{\mathrm{Re}}^{0.8} N_{\mathrm{Pn}}^{0.32} \left(1 + 3.5 \frac{d}{D}\right) \tag{7}$$

where D is the diameter of the helix and d the inside diameter.

For outside heat-transfer coefficients the standard design methods for heat transfer for flow across tube banks with in-line tubes are used. Usually the metal wall resistance is negligible. In some cases adjacent tubes are brazed or soldered together to promote heat transfer from one to the other. Even here wall resistance is usually a very small part of the total heat-transfer resistance.

Pressure drop calculations are made using equivalent design tools. Usually the low-pressure-side ΔP is critical in designing a usable exchanger.

The coiled-tube-in-shell exchanger is expensive, requiring a large amount of hand labor. Its advantages are that it can be operated at any pressure the tube can withstand, and that it can be built over very wide size ranges and with great flexibility of design. Currently these exchangers are little used in standard industrial cryogenic applications. However, in very large sizes (14 ft diameter $\times$ 120 ft length) they are used in base-load natural gas liquefaction plants, and in very small size (finger sized) they are used in cooling sensors for space and military applications.

3.2 Plate-Fin Heat Exchangers

The plate-fin exchanger has become the most common type used for cryogenic service. This results from its relatively low cost and high concentration of surface area per cubic foot of exchanger volume. It is made by clamping together a stack of alternate flat plates and corrugated sheets of aluminum coated with brazing flux. This assembly is then immersed in molten salt where the aluminum brazes together at points of contact. After removal from the bath the salt and flux are washed from the exchanger paths, and the assembly is enclosed in end plates and nozzles designed to give the desired flow arrangement. Usually the exchanger is roughly cubic, and is limited in size by the size of the available salt bath and the ability to make good braze seals in the center of the core. The core can be arranged for countercurrent flow or for cross flow. Figure 21 shows the construction of a typical plate-fin exchanger.

Procedures for calculating heat-transfer and pressure loss characteristics for plate-fin exchangers have been developed and published by the exchanger manufacturers. Table 4 and Fig. 22 present one set of these.

3.3   Regenerators

A regenerator is essentially a storage vessel filled with particulate solids through which hot and cold fluid flow alternately in opposite directions. The solids absorb energy as the hot fluid flows through, and then transfer this energy to the cold fluid. Thus this solid acts as a short-term energy-storage medium. It should have high heat capacity and a large surface area, but should be designed as to avoid excessive flow pressure drop.

In cryogenic service regenerators have been used in two very different applications. In engine liquefiers very small regenerators packed with, for example, fine copper wire have been used. In these situations the alternating flow direction has been produced by the intake and exhaust strokes of the engine. In air separation plants very large regenerators in the form of tanks filled with pebbles have been used. In this application the regenerators have been used in pairs with one regenerator receiving hot fluid as cold fluid enters the other. Switch valves and check valves are used to alternate flow to the regenerator bodies, as shown in Fig. 23.

The regenerator operates in cyclical, unsteady-state conditions. Partial differential equations can be written to express temperatures of gas and of solid phase as a function of time and bed position under given conditions of flow rates, properties of gaseous and solid phases, and switch time. Usually these equations are solved assuming constant heat capacities, thermal conductivities, heat-transfer coefficients, and flow rates. It is generally assumed that flow is uniform throughout the bed cross section, that the bed has infinite conductivity in the radial direction but zero in the longitudinal direction, and that there is no condensation or vaporization occurring. Thermal gradients through the solid particles are usually ignored. These equations can then be solved by computer approximation. The results are often expressed graphically.[26]

An alternative approach compares the regenerator with a steady-state heat exchanger and uses exchanger design methods for calculating regenerator size.[27] Figure 24 shows the temperature–time relationship at several points in a regenerator body. In the central part of the regenerator ΔTs are nearly constant throughout the cycle. Folding the figure at the switch points superimposes the temperature data for this central section as shown in Fig. 25. It is clear that the solid plays only a time-delaying function as energy flows from the hot stream to the cold one. Temperature levels are set by the thermodynamics of the cooling curve such as Fig. 15 presents. Thus, the $q = UA \, \Delta T$ equation can be used for small sections of the regenerator if a proper U can be determined.

During any half cycle the resistance to heat transfer from the gas to the solid packing will be just the gas-phase film coefficient. It can be calculated from empirical correlations for the packing material in use. For pebbles, the correlations for heat transfer to spheres in

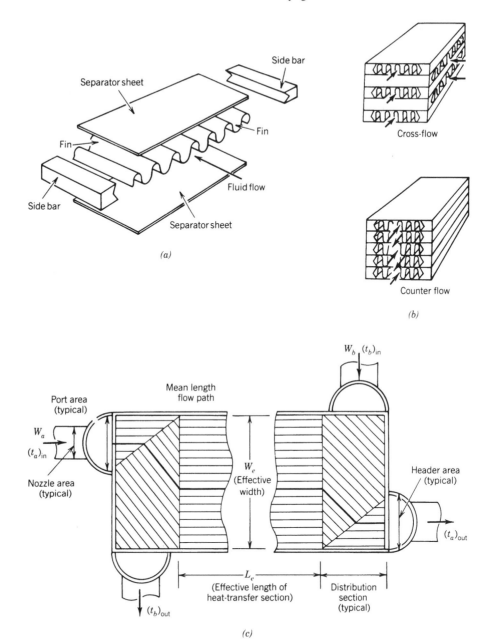

Figure 21 Construction features of a plate-fin heat exchanger. (*a*) Detail of plate and fin. (*b*) Flow arrangements. (*c*) Total assembly arrangement.

Table 4   Computation of Fin Surface Geometrics[a]

Fin Height (in.)	Type of Surface	Fin Spacing (FPI)	Fin Thickness (in.)	A_c'	A_{ht}''	B	r_h	A_r/A_{ht}
0.200	Plain or perforated	14	0.008	0.001185	0.596	437	0.001986	0.751
0.200	Plain or perforated	14	0.012	0.001086	0.577	415	0.001884	0.760
0.250	Plain or perforated	10	0.025	0.001172	0.500	288	0.00234	0.750
0.375	Plain or perforated	8	0.025	0.001944	0.600	230	0.003240	0.778
0.375	Plain or perforated	15	0.008	0.00224	1.064	409	0.00211	0.862
0.250	⅛ lanced	15	0.012	0.001355	0.732	420	0.001855	0.813
0.250	⅛ lanced	14	0.020	0.001150	0.655	378	0.001751	0.817
0.375	⅛ lanced	15	0.008	0.00224	1.064	409	0.002108	0.862
0.455	Ruffled	16	0.005	0.002811	1.437	465	0.001956	0.893

[a] Definition and use of terms:
FPI = fins per inch
A_c' = free stream area factor, ft^2/passage/in. of effective passage width
A_{ht}'' = heat-transfer area factor, ft^2/passage/in./ft of effective length
B = heat-transfer area per unit volume between plates, ft^2/ft^3
r_h = hydraulic radius = cross section area/wetted perimeter, ft
A_r = effective heat-transfer area = $A_{ht} \cdot \eta_0$
A_{ht} = total heat-transfer area
η_0 = weighted surface effectiveness factor
   $= 1 - (A_r/A_{ht})(1 - \eta_f)$
A_f = fin heat-transfer area
η_f = fin efficiency factor = [tanh (mL)]/ml
ml = fin geometry and material factor = $(b/s)\sqrt{2h/k}$
b = fin height, ft
h = film coefficient for heat transfer, Btu/hr·ft^2·°F
k = thermal conductivity of the fin material, Btu/hr·ft·°F
s = fin thickness, ft
U = overal heat transfer coefficient = $1/(A/h_a A_a + A/h_b A_b)$
a,b = subscripts indicating the two fluids between which heat is being transferred
Courtesy Stewart-Warner Corp.

a packed bed[28] is normally used to obtain the film coefficient for heat transfer from gas to solid:

$$h_{gs} = 1.31(G/d)^{0.93} \tag{8}$$

where h_{gs} = heat transfer from gas to regenerator packing or reverse, J/hr·m^2·K
   G = mass flow of gas, kg/hr·m^2
   d = particle diameter, m

The heat that flows to the packing surface diffuses into the packing by a conductive mode. Usually this transfer is fast relative to the transfer from the gas phase, but it may be necessary to calculate solid surface temperatures as a function of heat-transfer rate and adjust the overall ΔT accordingly. The heat-transfer mechanisms are typically symmetrical and hence the design equation becomes

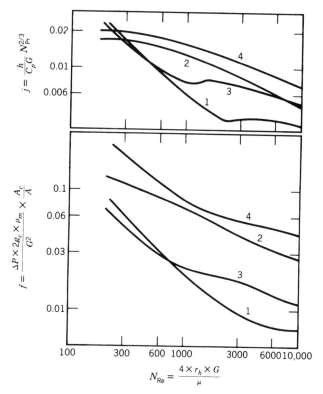

Figure 22   Heat-transfer and flow friction factors in plate and fin heat exchangers. Curves 1: plain fin (0.200 in. height, 14 fins/in.—0.008 in. thick). Curves 2: ruffled fin (0.445 in. height, 16 fins/in.—0.005 in. thick). Curves 3: perforated fin (0.375 in. height, 8 fins/in.—0.025 in. thick). (Courtesy Stewart-Warner Corp.)

$$A = \frac{q}{U\Delta T} = \frac{q}{h/2 \times \Delta T/2} \doteq \frac{4q}{h_{gs}\Delta T}$$

This calculation can be done for each section of the cooling curve until the entire regenerator area is calculated. However, at the ends of the regenerator temperatures are not symmetrical nor is the ΔT constant throughout the cycle. Figure 26 gives a correction factor that must be used to adjust the calculated area for these end effects. Usually a 10–20% increase in area results.

The cyclical nature of regenerator operation allows their use as trapping media for contaminants simultaneously with their heat-transfer function. If the contaminant is condensable, it will condense and solidify on the solid surfaces as the cooling phase flows through the regenerator. During the warming phase flow, this deposited condensed phase will evaporate flowing out with the return media.

Consider an air-separation process in which crude air at a moderate pressure is cooled by flow through a regenerator pair. The warmed regenerator is then used to warm up the returning nitrogen at low pressure. The water and CO_2 in the air deposit on the regenerator

High-pressure gas
feed

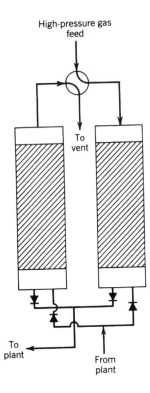

To
vent

To
plant

From
plant

Figure 23 Regenerator pair configuration.

surfaces and then reevaporate into the nitrogen. If deposition occurs at thermodynamic equilibrium, and assuming Raoult's law,

$$y_{H_2O} \text{ or } y_{CO_2} = \frac{P^{\circ}_{H_2O \text{ or } CO_2}}{P} \tag{9}$$

where y = mole fraction of H_2O or CO_2 in the gas phase
P° = saturation vapor pressure of H_2O or CO_2
P = total pressure of flowing stream

This equation can be applied to both the depositing, incoming situation and the reevaporating,
outgoing situation. If the contaminant is completely removed in the regenerator, and the
return gas is pure as it enters the regenerator, the moles of incoming gas times the mole
fraction of contaminant must equal that same product for the outgoing stream if the contaminant does not accumulate in the regenerator. Since the vapor pressure is a function of
temperature, and the returning stream pressure is lower than the incoming stream pressure,
these relations can be combined to give the maximum stream-to-stream ΔT that may exist
at any location in the regenerator. Figure 27 shows the results for one regenerator design
condition. Also plotted on Fig. 27 is a cooling curve for these same design conditions. At
the conditions given H_2O will be removed down to very low concentrations, but CO_2 solids
may accumulate in the bottom of the regenerator. To prevent this it would be necessary to
remove some of the air stream in the middle of the regenerator for further purification and
cooling elsewhere.

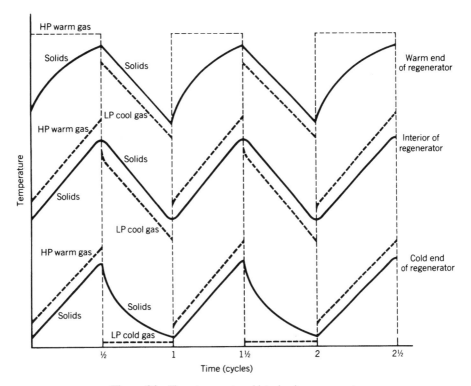

Figure 24 Time–temperature histories in a regenerator.

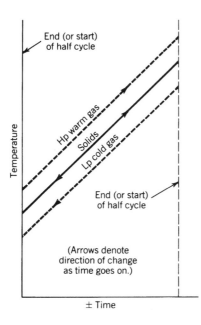

Figure 25 Time–temperature history for a central slice through a regenerator.

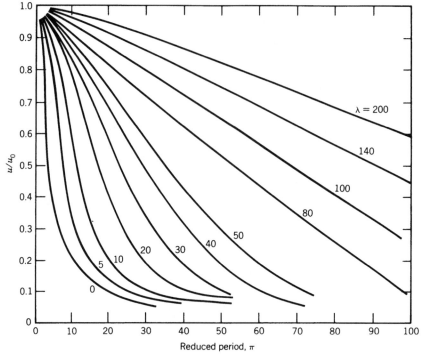

Figure 26   End correction for regenerator heat-transfer calculation using symmetrical cycle theory[27] (Courtesy Plenum Press):

$$\lambda = \frac{4H_0 S(T_c + T_w)}{C_c T_c + C_w T_w} = \text{reduced length}$$

$$\pi = \frac{12 H_0 (T_c + T_w)}{c\rho_s d} = \text{reduced period}$$

$$U_0 = \frac{1}{4}\left(\frac{1}{h} + \frac{0.1d}{k}\right)$$

where T_w, T_c = switching times of warm and cold streams, respectively, hr
  S = regenerator surface area, m²
  U_0 = overall heat transfer coefficient uncorrected for hysteresis, kcal/m²·hr·°C
  U = overall heat transfer coefficient
  C_w, C_c = heat capacity of warm and cold stream, respectively, kcal/hr·°C
  c = specific heat of packing, kcal/kg·°C
  d = particle diameter, m
  ρ_s = density of solid, kg/m³

Cryogenic heat exchangers often are called on to condense or evaporate and two-phase heat-transfer commonly occurs, sometimes on both sides of a given heat exchanger. Heat-transfer coefficients and flow pressure losses are calculated using correlations taken from high-temperature data.[29] The distribution of multiphase processing streams into parallel channels is, however, a common and severe problem in cryogenic processing. In heat exchangers thousands of parallel paths may exist. Thus the designer must ensure that all possible paths

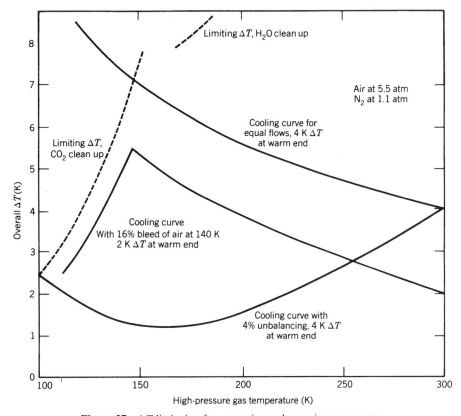

Figure 27 ΔT limitation for contaminant cleanup in a regenerator.

offer the same flow resistance and that the two phases are well distributed in the flow stream approaching the distribution point. Streams that cool during passage through an exchanger are likely to be modestly self-compensating in that the viscosity of a cold gas is lower than that of a warmer gas. Thus, a stream that is relatively high in temperature (as would be the case if that passage received more than its share of fluid) will have a greater flow resistance than a cooler system, so flow will be reduced. The opposite effect occurs for streams being warmed, so that these streams must be carefully balanced at the exchanger entrance.

4 INSULATION SYSTEMS

Successful cryogenic processing requires high-efficiency insulation. Sometimes this is a processing necessity, as in the Joule–Thomson liquefier, and sometimes it is primarily an economic requirement, as in the storage and transportation of cryogens. For large-scale cryogenic processes, especially those operating at liquid nitrogen temperatures and above, thick blankets of fiber or powder insulation, air or N_2 filled, have generally been used. For lower temperatures and for smaller units, vacuum insulation has been enhanced by adding one or many radiation shields, sometimes in the form of fibers or pellets, but often as reflective metal barriers. The use of many radiation barriers in the form of metal-coated plastic sheets wrapped around the processing vessel within the vacuum space has been used for most applications at temperatures approaching absolute zero.

4.1   Vacuum Insulation

Heat transfer occurs by convection, conduction, and radiation mechanisms. A vacuum space ideally eliminates convective and conductive heat transfer but does not interrupt radiative transfer. Thus heat transfer through a vacuum space can be calculated from the classic equation:

$$q = \sigma A F_{12}(T_1^4 - T_2^4) \tag{10}$$

where q = rate of heat transfer, J/sec
σ = Stefan-Boltzmann constant, 5.73×10^{-8} J/sec·m²·K
F_{12} = combined emissivity and geometry factor
T_1, T_2 = temperature (K) of radiating and receiving body, respectively

In this formulation of the Stefan–Boltzmann equation it is assumed that both radiator and receiver are gray bodies, that is, emissivity ϵ and absorptivity are equal and independent of temperature. It is also assumed that the radiating body loses energy to a totally uniform surroundings and receives energy from this same environment.

The form of the Stefan–Boltzmann equation shows that the rate of radiant energy transfer is controlled by the temperature of the hot surface. If the vacuum space is interrupted by a shielding surface, the temperature of that surface will become T_s, so that

$$q/A = F_{1s} (T_1^4 - T_s^4) = F_{s2} (T_s^4 - T_2^4) \tag{11}$$

Since q/A will be the same through each region of this vacuum space, and assuming $F_{1s} = F_{s2} = F_{12}$

$$T_s = \sqrt[4]{\frac{T_1^4 + T_2^4}{2}} \tag{12}$$

For two infinite parallel plates or concentric cylinders or spheres with diffuse radiation transfer from one to the other,

$$F_{12} = 1 \left/ \frac{1}{\epsilon_1} + \frac{A_1}{A_2} \left(\frac{1}{\epsilon_2} - 1 \right) \right. \tag{13}$$

If A_1 is a small body in a large enclosure, $F_{12} = \epsilon_1$. If radiator or receiver has an emissivity that varies with temperature, or if radiation is spectral, F_{12} must be found from a detailed statistical analysis of the various possible radiant beams.[30]

Table 5 lists emissivities for several surfaces of low emissivity that are useful in vacuum insulation.[31]

It is often desirable to control the temperature of the shield. This may be done by arranging for heat transfer between escaping vapors and the shield, or by using a double-walled shield in which is contained a boiling cryogen.

It is possible to use more than one radiation shield in an evacuated space. The temperature of intermediate streams can be determined as noted above, although the algebra becomes clumsy. However, mechanical complexities usually outweigh the insulating advantages.

Table 5   Emissivities of Materials Used for Cryogenic
Radiation Shields

Material	Emissivity at 300 K	Emissivity at 77.8 K	Emissivity at 4.33 K
Aluminum plate	0.08	0.03	
Aluminum foil (bright finish)	0.03	0.018	0.011
Copper (commercial polish)	0.03	0.019	0.015
Monel	0.17	0.11	
304 stainless steel	0.15	0.061	
Silver	0.022		
Titanium	0.1		

4.2   Superinsulation

The advantages of radiation shields in an evacuated space have been extended to their logical conclusion in superinsulation, where a very large number of radiation shields are used. A thin, low emissivity material is wrapped around the cold surface so that the radiation train is interrupted often. The material is usually aluminum foil or aluminum-coated Mylar. Since the conductivity path must also be blocked, the individual layers must be separated. This may be done with glass fibers, perlite bits, or even with wrinkles in the insulating material; 25 surfaces/in. of thickness is quite common. Usually the wrapping does not fill in the insulating space. Table 6 gives properties of some available superinsulations.

Superinsulation has enormous advantages over other available insulation systems as can be seen from Table 6. In this table insulation performance is given in terms of effective thermal conductivity

$$k_e = \frac{q/A}{T/L} \tag{14}$$

where k_e = effective, or apparent, thermal conductivity
  L = thickness of the insulation
  $T = T_1 - T_2$

Table 6   Properties of Various Multilayer Insulations (Warm Wall at 300 K)

Sample Thickness (cm)	Shields per Centimeter	Density (g/cm³)	Cold Wall T (K)	Conductivity (μW/cm·K)	Material[a]
3.7	26	0.12	76	0.7	1
3.7	26	0.12	20	0.5	1
2.5	24	0.09	76	2.3	2
1.5	76	0.76	76	5.2	3
4.5	6	0.03	76	3.9	4
2.2	6	0.03	76	3.0	5
3.2	24	0.045	76	0.85	5
1.3	47	0.09	76	1.8	5

[a] 1, Al foil with glass fiber mat separator; 2, Al foil with nylon net spacer; 3, Al foil with glass fabric spacer; 4, Al foil with glass fiber, unbonded spacer; 5, aluminized Mylar, no spacer.

This insulating advantage translates into thin insulation space for a given rate of heat transfer, and into low weight. Hence designers have favored the use of superinsulation for most cryogen containers built for transport, especially where liquid H_2 or liquid He is involved, and for extraterrestrial space applications.

On the other hand, superinsulation must usually be installed in the field, and hence uniformity is difficult to achieve. Connections, tees in lines, and bends are especially difficult to wrap effectively. Present practice requires that layers of insulation be overlapped at a joint to ensure continuous coverage. Some configurations are shown in Fig. 28. Also, it has been found that the effectiveness of superinsulation drops rapidly as the pressure increases. Pressures must be kept below 10^{-3} torr; evacuation is slow; a getter is required in the evacuated space; and all joints must be absolutely vacuum tight. Thus the total system cost is high.

4.3   Insulating Powders and Fibers

Fibers and powders have been used as insulating materials since the earliest of insulation needs. They retain the enormous advantage of ease of installation, especially when used in air, and low cost. Table 7 lists common insulating powders and fibers along with values of effective thermal conductivity.[32] Since the actual thermal conductivity is a function of temperature, these values may only be used for the temperature ranges shown.

For cryogenic processes of modest size and at temperatures down to liquid nitrogen temperature, it is usual practice to immerse the process equipment to be insulated in a cold

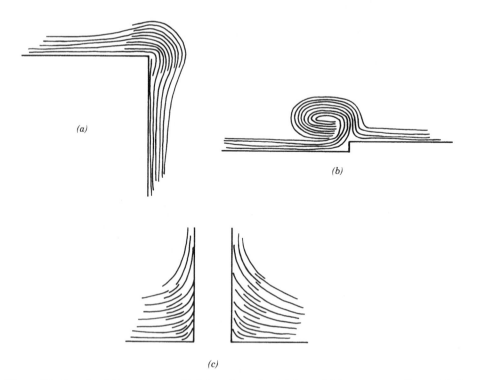

Figure 28 Superinsulation coverage at joints and nozzles: (*a*) Lapped joint at corner. Also usable for nozzle or for pipe bend. (*b*) Rolled joint used at surface discontinuity, diameter change, or for jointure of insulation sections. (*c*) Multilayer insulation at a nozzle.

Table 7 Effective Thermal Conductivity of Various Common Cryogenic Insulating Materials (300 to 76 K)

Material	Gas Pressure (mm Hg)	P (g/cm^2)	K (W/cm·K)
Silica aerogel (250A)	<10^{-4}	0.096	20.8×10^{-6}
	N$_2$ at 628	0.096	195.5×10^{-6}
Perlite (+30 mesh)	<10^{-5}	0.096	18.2×10^{-6}
	N$_2$ at 628	0.096	334×10^{-6}
Polystyrene foam	Air, 1 atm	0.046	259×10^{-6}
Polyurethane foam	Air, 1 atm	0.128	328×10^{-6}
Foamglas	Air, 1 atm	0.144	346×10^{-6}

box, a box filled with powder or fiber insulation. Insulation thickness must be large, and the coldest units must have the thickest insulation layer. This determines the placing of the process units within the cold box. Such a cold box may be assembled in the plant and shipped as a unit, or it can be constructed in the field. It is important to prevent moisture from migrating into the insulation and forming ice layers. Hence the box is usually operated at a positive gauge pressure using a dry gas, such as dry nitrogen. If rock wool or another such fiber is used, repairs can be made by tunneling through the insulation to the process unit. If an equivalent insulating powder, perlite, is used, the insulation will flow from the box through an opening into a retaining bag. After repairs are made, the insulation may be poured back into the box.

Polymer foams have also been used as cryogenic insulators. Foam-in-place insulations have proven difficult to use because as the foaming takes place cavities are likely to develop behind process units. However, where the shape is simple and assembly can be done in the shop, good insulating characteristics can be obtained.

In some applications powders or fibers have been used in evacuated spaces. The absence of gas in the insulation pores reduces heat transfer by convection and conduction. Figure 29 shows the effect on a powder insulation of reducing pressure in the insulating space. Note that the pressures may be somewhat greater than that needed in a superinsulation system.

5 MATERIALS FOR CRYOGENIC SERVICE

Materials to be used in cryogenic service must operate satisfactorily in both ambient and cryogenic temperatures. The repeated temperature cycling that comes from starting up, operating, and shutting down this equipment is particularly destructive because of expansion and contraction that occur at every boundary and jointure.

5.1 Materials of Construction

Metals
Many of the normal metals used in equipment construction become brittle at low temperatures and fail with none of the prewarning of strain and deformation usually expected. Sometimes failure occurs at very low stress levels. The mechanism of brittle failure is still a topic for research. However, those metals that exhibit face-centered-cubic crystal lattice structure do not usually become brittle. The austenitic stainless steels, aluminum, copper, and nickel alloys are materials of this type. On the other hand, materials with body-centered-cubic

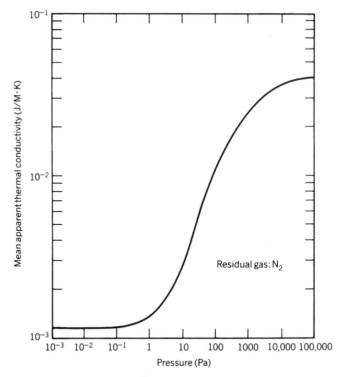

Figure 29 Effect of residual gas pressure on the effective thermal conductivity of a powder insulation—perlite, 30–80 mesh, 300 to 78 K.

crystal lattice forms or close-packed-hexagonal lattices are usually subject to a brittle transformation as the temperature is lowered. Such materials include the low-carbon steels and certain titanium and magensium alloys. Figure 30 shows these crystal forms and gives examples of notch toughness at room temperature and at liquid N_2 temperature for several example metals. In general carbon acts to raise the brittle transition temperature, and nickel lowers it. Additional lowering can be obtained by fully killing steels by deoxidation with silicon and aluminum and by effecting a fine grain structure through normalizing by addition of selected elements.

In selecting a material for cryogenic service, several significant properties should be considered. The toughness or ductility is of prime importance. Actually, these are distinctively different properties. A material that is ductile, as measured by elongation, may have poor toughness as measured by a notch impact test, particularly at cryogenic temperatures. Thus both these properties should be examined. Figures 31 and 32 show the effect of nickel content and heat treatment on Charpy impact values for steels. Figure 33 shows the tensile elongation before rupture of several materials used in cryogenic service.

Tensile and yield strength generally increase as temperature decreases. However, this is not always true, and the behavior of the particular material of interest should be examined. Obviously if the material becomes brittle, it is unusable regardless of tensile strength. Figure 34 shows the tensile and yield strength for several stainless steels.

Fatigue strength is especially important where temperature cycles from ambient to cryogenic are frequent, especially if stresses also vary. In cryogenic vessels maximum stress

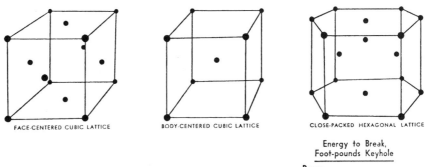

Metal	Crystal Lattice	Energy to Break, Foot-pounds Keyhole	
		Room Temperature	−320°F
Austenitic Stainless Steel	Face-centered Cubic	43	50
Aluminum	Face-centered Cubic	19	27
Copper	Face-centered Cubic	43	50
Nickel	Face-centered Cubic	89	99
Iron	Body-centered Cubic	78	1.5
Titanium	Close-packed Hexagonal	14.5	6.6
Magnesium	Close-packed Hexagonal	4	(3 at −105° F)

(Courtesy—American Society for Metals)

Figure 30   Effect of crystal structure on brittle impact strengths of some metals. (Courtesy American Society for Metals.)

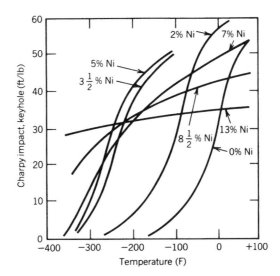

Figure 31   Effect of nickel content in steels on Charpy impact values. (Courtesy American Iron and Steel Institute.)

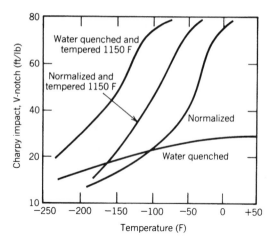

Figure 32   Effect of heat treatment on Charpy impact values of steel. (Courtesy American Iron and Steel Institute.)

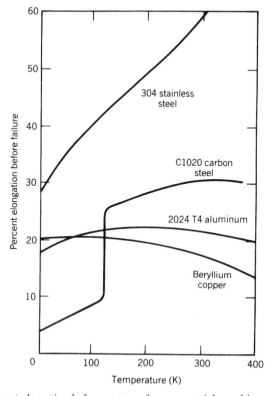

Figure 33   Percent elongation before rupture of some materials used in cryogenic service.[33]

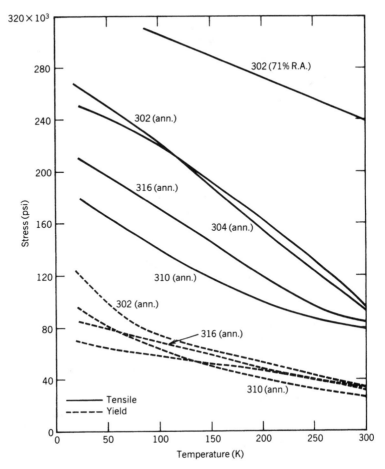

Figure 34 Yield and tensile strength of several AISI 300 series stainless steels.[33] (Courtesy American Iron and Steel Institute.)

cycles for design are about 10,000–20,000 rather than the millions of cycles used for higher-temperature machinery design. Because fatigue strength data for low-temperature applications are scarce, steels used in cryogenic rotating equipment are commonly designed using standard room-temperature fatigue values. This allows a factor of safety because fatigue strength usually increases as temperature decreases.

Coefficient of expansion information is critical because of the stress that can be set up as temperatures are reduced to cryogenic or raised to ambient. This is particularly important where dissimilar materials are joined. For example, a 36-ft-long piece of 18-8 stainless will contract more than an inch in cooling from ambient to the boiling point of liquid H_2. And stainless steel has a coefficient of linear expansion much lower than that of copper or aluminum. This is seen in Fig. 35.

Thermal conductivity is an important property because of the economic impact of heat leaks into a cryogenic space. Figure 36 shows the thermal conductivity of some metals in the cryogenic temperature range. Note that pure copper shows a maximum at very low temperatures, but most alloys show only modest effect of temperature on thermal conduc-

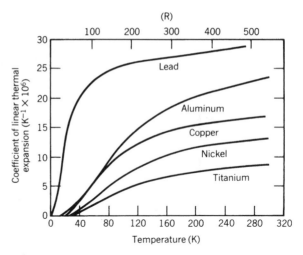

Figure 35 Coefficient of linear thermal expansion of several metals as a function of temperature. (Courtesy American Institute of Chemical Engineers.)

tivity. One measure of the suitability of a material for cryogenic service is the ratio of tensile strength to thermal conductivity. On this basis stainless steel looks very attractive and copper much less so.

The most common materials used in cryogenic service have been the austenitic stainless steels, aluminum alloys, copper alloys, and aluminum-alloyed steels. Fine grained carbon-manganese steel and aluminum-killed steel and the 2.5% Ni steels can be used to temperatures as low as −50°C. A 3.5% Ni steel may be used roughly to −100°C; 5% Ni steels have been developed especially for applications in liquified natural gas processing, that is, for temperatures down to about −170°C. Austenitic stainless steels with about 9% Ni such as the common 304 and 316 types are usable well into the liquid H_2 range (−252°C). Aluminum and copper alloys have been used throughout the cryogenic temperature range. However, in selecting a particular alloy for a given application the engineer should consider carefully all of the properties of the material as they apply to that application.

Stainless steel may be joined by welding. However, the welding rod chosen and the joint design must both be selected for the material being welded and the expected service. For example, 9% nickel steel can be welded using nickel-based electrodes and a 60–80° single V joint design. Inert gas welding using Inconel-type electrodes is also acceptable. Where stress levels will not be high types 309 and 310 austenitic-stainless-steel electrodes can be used despite large differences in thermal expansion between the weld and the base metal.

Dissimilar metals can be joined for cryogenic service by soft soldering, silver brazing, or welding. For copper-to-copper joints a 50% tin/50% lead solder can be used. However, these joints have little ductility and so cannot stand high stress levels. Soft solder should not be used with aluminum, silicon-bronze, or stainless steel. Silver soldering is preferred for aluminum and silicon bronze and may also be used with copper and stainless steel.

Polymers
Polymers are frequently used as structural materials in research apparatus, as windows into cryogenic spaces, and for gaskets, O-rings, and other seals. Their suitability for the intended

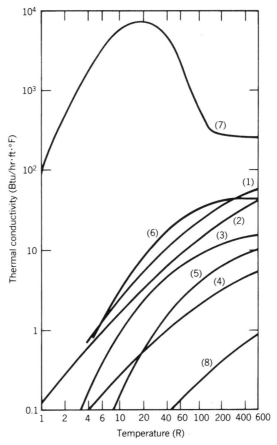

Figure 36 Thermal conductivity of materials useful in low-temperature service. (1) 2024TA aluminum; (2) beryllium copper; (3) K-Monel; (4) titanium; (5) 304 stainless steel; (6) C1020 carbon steel; (7) pure copper; (8) Teflon.[35]

service should be as carefully considered as metals. At this point there is little accumulated, correlated data on polymer properties because of the wide variation in these materials from source to source. Hence properties should be obtained from the manufacturer and suitability for cryogenic service determined case by case.

Tables 8 and 9 list properties of some common polymeric materials. These are not all the available suitable polymers, but have been chosen especially for their compatibility with liquid O_2. For this service chemical inertness and resistance to flammability are particularly important. In addition to these, nylon is often used in cryogenic service because of its machinability and relative strength. Teflon and similar materials have the peculiar property of losing some of their dimensional stability at low temperatures; thus they should be used in confined spaces or at low stress levels.

Glass
Glasses, especially Pyrex and quartz, have proven satisfactory for cryogenic service because of their amorphous structure and very small coefficient of thermal expansion. They are

Table 8 Properties of Polymers Used in Cryogenic Service

Elastomer Type	Silicone Rubber	Vinylidene Fluoride Hexafluoropropylene	Fluorosilicone	Polytrifluorochloroethylene
Trade Name	Silastic[a] Silicone Rubber[e,f]	Viton[b] Fluore[c]	Silastic LS-53[a]	Kel-F[c,d]
Physical and Mechanical Properties				
Durometer range (shore A)	45–60	55–90	50–60	55–90
Specific gravity (base elastomer)	1.17–1.46	1.4–1.85	1.41–1.46	1.4–1.85
Density, lb/in.³ (base elastomer)	0.045	0.051–0.067	0.051	0.051–0.067
Tensile strength; psi:				
Pure gum	Under 400	>2000	1000	350–600
Reinforced	600–1500	—	—	—
Elongation, percent:				
Pure gum	Under 200	>350	200	500–800
Reinforced	200–800	—	—	—
Thermal conductivity, g, Btu/hr/ft²/(°F/ft)	0.13	—	0.13	—
Coefficient of thermal expansion, cubical, in.³/in.³/°F	45×10^{-5}	27×10^{-5}	45×10^{-5}	—
Electrical insulation	Excellent	Excellent	Good	Excellent
Rebound				
Cold	Very good	Good	Very good	—
Hot	Very good	Excellent	Very good	—
Compression set	Good to excellent	Good to excellent	Good	Good to excellent
Resistance Properties				
Temperature:				
Tensile strength at 250°F, psi	850	300–800	—	300–800
Tensile strength at 400°F, psi	400	150–300	—	150–300
Elongation at 250°F, percent	350	100–350	—	100–350
Elongation at 400°F, percent	200	50–160	—	50–160
Low temperature brittle point, °F	−90 to −200	10 to −60	−90	10 to −60
Low temperature range of rapid stiffening, °F	−60 to −120	20 to −30	—	−20 to −30
Drift, room temperature	Poor to excellent	Good	—	Good
Drift, elevated temperature (158° to 212°F)	Excellent	Good to excellent	Excellent	Good to excellent

	Excellent 480	Excellent 450	Excellent 500	Excellent 400
Heat aging (212°F)	Excellent	Excellent	Excellent	Excellent
Maximum recommended continuous service temperature, °F	480	450	500	400
Minimum recommended service temperature, °F	−178	−50	−90	−60
Mechanical:				
Tear resistance	Poor	Poor to good	—	Poor to good
Abrasion resistance	Poor	Good	Poor	Good
Impact resistance (fatigue)	Poor	Poor to good	Poor	Poor to good
Chemical:				
Sunlight aging	Excellent	Excellent	Excellent	Excellent
Weather resistance	Excellent	Excellent	Excellent	→
Oxidation	Excellent	Excellent	Excellent	
Acids:				
Dilute	Very good	Excellent	Excellent	
Concentrated	Good	Good	Very good	
Alkali	Fair to excellent	Poor to fair	Fair to excellent	Poor to fair
Alcohol	Good	Excellent	Poor to good	Excellent
Petroleum products, resistance	Poor to fair	Good to excellent	Excellent	Good to excellent
Coal tar derivatives, resistance	Poor	Excellent	—	Excellent
Chlorinated solvents, resistance	Good	Good	—	Good
Hydraulic oils:				
Silicates	Poor	Good	—	Good
Phosphates	Good	Poor	—	Poor
Water swell resistance	Good	Good	Excellent	Good
Permeability to gases	Good	Excellent	Fair	Excellent

[a] Dow Corning Corp.

[b] E. I. duPont de Nemours.

[c] Minnesota Mining and Manufacturing Co.

[d] CTFE compounded with vinylidine fluoride.

[e] General Electric.

[f] Union Carbon and Carbide.

Table 9 Properties of Polymers Used in Cryogenic Service

Common Name	Fluorinated Ethylene Propylene	Polychlorotrifluoroethylene	Polyvinylidene Fluoride	Polytetrafluoroethylene	Polyimide
Trade Name	Teflon FEP[a]	Kel-F[b]	Kynar[c]	Fluorosint[d,e] Teflon TFE[a] Halon TFE[f]	Kapton H[a] Kapton F[a] Vespel[a] Polymer SP-1[a]
Physical and Mechanical Properties					
Specific gravity	2.14–2.17	2.1–2.2	1.76–1.77	2.13–2.22	1.42
Tensile strength, psi	2700–3100	4500–6000	7000	2000–4500	25,000[g]; 10,500
Elongation, percent	250–330	30–250	100–300	200–400	70[g]; 6–8
Tensile modulus, psi	0.5×10^5	$1.5 \times 10^5 - 3 \times 10^5$	1.2×10^5	0.58×10^5	4.3×10^5
Compressive strength, psi	2200	32,000–80,000	10,000	1700	24,400
Flexural strength, psi	—	7400–9300	—	—	14,000
Impact strength, ft-lb/in. of notch	No break	0.8–5.0	3.5	3.0	0.9
Rockwell hardness	R25	R110–R115	D80 (Shore)	D50–D65 (Shore)	H85–H95
Thermal conductivity, Btu/hr/ft²/(°F/in.)	1.75	0.9	0.9	1.75	2.2
Specific heat, Btu/lbm/°F	0.28	0.22	0.33	0.25	0.27
Coefficient of linear expansion, in./in./°F $\times 10^{-5}$	4.7×10^{-5} to 5.8×10^{-5}	5×10^{-5} to 15×10^{-5}	6.7×10^{-5}	5.5×10^{-5}	28×10^{-5} to 35×10^{-5}
Volume resistivity, ohm-cm	$>2 \times 10^{18}$	1.2×10^{18}	2×10^{14}	$>10^{18}$	10^{18}
Clarity	Transparent to translucent	Transparent to translucent	Transparent to translucent	Opaque	Opaque
Processing Properties					
Molding qualities	Excellent	Excellent	Excellent	—	—
Injection molding temperature, °F	625–760	440–600	450–550	—	—
Mold shrinkage, in./in.	0.03–0.06	0.005–0.010	0.030	—	—
Machining qualities	Excellent	Excellent	Excellent	Excellent	—

Resistance Properties

Mechanical abrasion and wear Tabor CS 17 wheel mg, loss/1000 cycles	—	0.01	17.6	—	—
Temperature:					
Flammability	None	None	Self-extinguishing	None	None
Low temperature brittle point, °F	−420	−400	−80	−420	−420
Resistance to heat, °F (continuous)	400	350–390	300	550	500
Deflection temperature under load, °F	—	258 (66 psi)	300 (66 psi), 195 (264 psi)	250 (66 psi)	—
Chemical:					
Effect of sunlight	None	None	Slight bleaching on long exposure	None	Degrades after prolonged exposure
Effect of weak acids	⟶	⟶	None	⟶	None
Effect of strong acids			Attacked by fuming sulfuric		
Effect of weak alkalies			None		
Effect of strong alkalies			None		Attacked
Effect of organic solvents	⟶	Halogenated compounds cause slight swelling	Resists most solvents		Resistant to most organic solvents

[a] E. I. duPont de Nemours.

[b] Minnesota Mining and Manufacturing Co.

[c] Pennsalt Chemicals Corp.

[d] Polymer Corp. of Pennsylvania.

[e] Polypenco, Inc.

[f] Allied Chemical Corp.

[g] Film.

commonly used in laboratory equipment, even down to the lowest cryogenic temperatures. They have also successfully been used as windows into devices such as hydrogen bubble chambers that are built primarily of metal.

5.2   Seals and Gaskets

In addition to careful selection of materials, seals must be specially designed for cryogenic service. Gaskets and O-rings are particularly subject to failure during thermal cycling. Thus they are best if confined and/or constructed of a metal–polymer combination. Such seals would be in the form of metal rings with C or wedge cross sections coated with a sealant such as Kel-F, Teflon, or soft metal. Various designs are available with complex cross sections for varying degrees of deflection. The surfaces against which these seal should be ground to specified finish. Elastomers such as neoprene and Viton-A have proven to be excellent sealants if captured in a space where they are subjected to 80% linear compression. This is true despite the fact that they are both extremely brittle at cryogenic temperatures without this stress.

Adhesive use at low temperatures is strictly done on an empirical basis. Still, adhesives have been used successfully to join insulating and vapor barrier blankets to metal surfaces. In every case the criteria are that the adhesive must not become crystalline at the operating temperature, must be resistant to aging, and must have a coefficient of contraction close to that of the base surface. Polyurethane, silicone, and various epoxy compounds have been used successfully in various cryogenic applications.

5.3   Lubricants

The lubrication of cryogenic machinery such as valves, pumps, and expanders is a problem that has generally been solved by avoidance. Valves usually have a long extension between the seat and the packing gland. This extension is gas filled so that the packing gland temperature stays close to ambient. For low-speed bearings babbitting is usually acceptable, as is graphite and molybdenum sulfide. For high-speed bearings, such as those in turboexpanders, gas bearings are generally used. In these devices some of the gas is leaked into the rotating bearing and forms a cushion for rotation. If out-leakage of the contained gas is undesirable, N_2 can be fed to the bearing and controlled so that leakage of N_2 goes to the room and not into the cryogenic system. Bearings of this sort have been operated at speeds up to 100,000 rpm.

6   SPECIAL PROBLEMS IN LOW-TEMPERATURE INSTRUMENTATION

Cryogenic systems usually are relatively clean and free flowing, and they often exist at a phase boundary where the degrees of freedom are reduced by one. Although these factors ease measurement problems, the fact that the system is immersed in insulation and therefore not easily accessible, the desire to limit thermal leaks to the system, and the likelihood that vaporization or condensation will occur in instrument lines all add difficulties.

Despite these differences all of the standard measurement techniques are used with low-temperature systems, often with ingenious changes to adapt the device to low-temperature use.

6.1 Temperature Measurement

Temperature may be measured using liquid-in-glass thermometers down to about −40°C, using thermocouples down to about liquid H_2 temperature, and using resistance thermometers and thermistors down to about 1 K. Although these are the usual devices of engineering measurement laboratory measurements have been done at all temperatures using gas thermometers and vapor pressure thermometers.

Table 10 lists the defining fixed points of the International Practical Temperature Scale of 1968. This scale does not define fixed points below the triple point of equilibrium He.[36] Below that range the NBS has defined a temperature scale to 1 K using gas thermometry.[37] At still lower temperatures measurement must be based on the fundamental theories of solids such as paramagnetic and superconducting phenomena.[38]

The usefulness of vapor pressure thermometry is limited by the properties of available fluids. This is evident from Table 11. For example, in the temperature range from 20.4 to 24.5 K there is no usable material. Despite this, vapor pressure thermometers are accurate and convenient. The major problem in their use is that the hydraulic head represented by the vapor line between point of measurement and the readout point must be taken into account. Also, the measurement point must be the coldest point experienced by the device. If not, pockets of liquid will form in the line between the point of measurement and the readout point greatly affecting the reading accuracy.

Standard thermocouples may be used through most of the cryogenic range, but, as shown in Fig. 37 for copper–constantan, the sensitivity with which they measure temperature drops as the temperature decreases. At low temperatures heat transfer down the thermocouple wire may markedly affect the junction temperature. This is especially dangerous with copper wires, as can be seen from Fig. 36. Also, some thermocouple materials, for example, iron, become brittle as temperature decreases. To overcome these difficulties special thermocouple pairs have been used. These usually involve alloys of the noble metals. Figure 37 shows the thermoelectric power, and hence sensitivity of three of these thermocouple pairs.

Resistance thermometers are also very commonly used for cryogenic temperature measurement. Metal resistors, especially platinum, can be used from ambient to liquid He temperatures. They are extremely stable and can be read to high accuracy. However, expensive instrumentation is required because resistance differences are small requiring precise bridge circuitry. Resistance as a function of temperature for platinum is well known.[36]

Table 10 Defining Fixed Points of the International Practical Temperature Scale, 1968

Equilibrium Point	T (K)
Triple point of equilibrium H_2	13.81
Boiling point of equilibrium H_2 ($P = 33330.6$ N/m^2)	17.042
Boiling point of equilibrium H_2 ($P = 1$ atm)	20.28
Boiling point of neon ($P = 1$ atm)	27.102
Triple point of O_2	54.361
Boiling point of O_2 ($P = 1$ atm)	90.188
Triple point of H_2O ($P = 1$ atm)	273.16
Freezing point of Zn ($P = 1$ atm)	692.73
Freezing point of Ag ($P = 1$ atm)	1235.08
Freezing point of Au ($P = 1$ atm)	1337.58

Table 11 Properties of Cryogens Useful in Vapor Pressure Thermometers

Substance	Triple Point (K)	Boiling Point (K)	Critical Point (K)	dP/dT (mm/K)	Hydraulic Heat at Boiling Point (K/cm^2)
^{3}He	—	3.19	3.32	790	0.000054
^{4}He	—	4.215	5.20	715	0.00013
p-H$_2$ (20.4 K equilibrium)	13.80	20.27	32.98	224	0.00023
Ne	24.54	27.09	44.40	230	0.0039
N$_2$	63.15	77.36	126.26	89	0.0067
Ar	83.81	87.30	150.70	80	0.013
O$_2$	54.35	90.18	154.80	79	0.011

At temperatures below 60 K, carbon resistors have been found to be convenient and sensitive temperature sensors. Since the change in resistance per given temperature difference is large (580 ohms/K would be typical at 4 K) the instrument range is small, and the resistor must be selected and calibrated for use in the narrow temperature range required.

Germanium resistors that are single crystals of germanium doped with minute quantities of impurities are also used throughout the cryogenic range. Their resistance varies approximately logarithmically with temperature, but the shape of this relation depends on the amount and type of dopant. Again, the germanium semiconductor must be selected and calibrated for the desired service.

Thermistors, that is, mixed, multicrystal semiconductors, like carbon and germanium resistors, give exponential resistance calibrations. They may be selected for order-of-magnitude resistance changes over very short temperature ranges or for service over wide temperature ranges. Calibration is necessary and may change with successive temperature cycling. For this reason they should be temperature-cycled several times before use. These sensors are cheap, extremely sensitive, easily read, and available in many forms. Thus they are excellent indicators of modest accuracy but of high sensitivity, such as sensors for control action. They do not, however, have the stability required for high accuracy.

6.2 Flow Measurement

Measurement of flow in cryogenic systems is often made difficult because of the need to deal with a liquid at its boiling point. Thus any significant pressure drop causes vaporization,

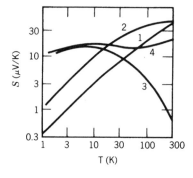

Figure 37 Thermoelectric power of some thermocouples useful for cryogenic temperature measurement (courtesy Plenum Press): 1, copper versus constantan; 2, Au + 2 at % Co versus silver normal (Ag + 0.37 at % Au); 3, Au + 0.03 at % Fe versus silver normal; 4, Au + 0.03 at % Fe versus Chromel.

which disrupts the measurement. This may be avoided by subcooling the liquid before measurement. Where this is possible, most measurement problems disappear, for cryogenic fluids are clean, low-viscosity liquids. Where subcooling is not possible, flow is most often measured using turbine flow meters or momentum meters.

A turbine meter has a rotor mounted axially in the flow stream and moved by the passing fluid. The rate of rotation, which is directly proportional to the volumetric flow rate, is sensed by an electronic counter that senses the passage of each rotor blade. There are two problems in the use of turbine meters in cryogenic fluids. First, these fluids are nonlubricating. Hence the meter rotor must be self-lubricated. Second, during cool-down or warm-up slugs of vapor are likely to flow past the rotor. These can flow rapidly enough to overspeed and damage the rotor. This can be avoided by locating a bypass around the turbine meter shutting off the meter during unsteady operation.

Momentum meters have a bob located in the flow stream to the support of which a strain gage is attached. The strain gage measures the force on the bob, which can be related through drag calculations or correlation to the rate of fluid flow past the bob. These meters are flexible and can be wide of range. They are sensitive to cavitation problems and to overstrain during upsets. Generally, each instrument must be calibrated.

6.3 Tank Inventory Measurement

The measurement of liquid level in a tank is made difficult by the cryogenic insulation requirements. This is true of stationary tanks, but even more so when the tank is in motion, as on a truck or spaceship, and the liquid is sloshing.

The simplest inventory measurement is by weight, either with conventional scales or by a strain gage applied to a support structure.

The sensing of level itself can be done using a succession of sensors that read differently when in liquid than they do in vapor. For instance, thermistors can be heated by a small electric current. Such devices cool quickly in liquid, and a resistance meter can "count" the number of thermistors in its circuit that are submerged.

A similar device that gives a continuous reading of liquid depth would be a vertical resistance wire, gently heated, while the total wire resistance is measured. The cold, submerged, fraction of the wire can be easily determined.

Other continuous reading devices include pressure gages, either with or without a vapor bleed, that read hydrostatic head, capacitance probes that indicate the fraction of their length that is submerged, ultrasonic systems that sense the time required for a wave to return from its reflectance off the liquid level, and light-reflecting devices.

7 EXAMPLES OF CRYOGENIC PROCESSING

Here three common, but greatly different, cryogenic technologies are described so that the interaction of the cryogenic techniques discussed above can be shown.

7.1 Air Separation

Among the products from air separation, nitrogen, oxygen, and argon are primary and are each major items of commerce. In 1994 nitrogen was second to sulfuric acid in production volume of industrial inorganic chemicals, with 932 billion standard cubic feet produced. Oxygen was third at 600 billion standard cubic feet produced. These materials are so widely used that their demand reflects the general trend in national industrial activity. Demand generally increases by 3 to 5%/year. Nitrogen is widely used for inert atmosphere generation in the metals, electronics and semiconductor, and chemical industries, and as a source of deep refrigeration, especially for food freezing and transporation. Oxygen is used in the steel industry for blast furnace air enrichment, for welding and scarfing, and for alloying operation. It is also used in the chemical industry in oxidation steps, for wastewater treatment, for welding and cutting, and for breathing. Argon, mainly used in welding, in stainless steel making, and in the production of specialized inert atmospheres, has a demand of only about 2% of that of oxygen. However, this represents about 25% of the value of oxygen shipments, and the argon demand is growing faster than that of oxygen or nitrogen.

Since all of the industrial gases are expensive to ship long distances, the industry was developed by locating a large number of plants close to markets and sized to meet nearby market demand. Maximum oxygen plant size has now grown into the 3000 ton/day range, but these plants are also located close to the consumer with the product delivered by pipe line. Use contracts are often long-term take-or-pay rental arrangements.

Air is a mixture of about the composition shown in Table 12. In an air separation plant O_2 is typically removed and distilled from liquified air. N_2 may also be recovered. In large plants argon may be recovered in a supplemental distillation operation. In such a plant the minor constituents (H_2–Xe) would have to be removed in bleed streams, but they are rarely collected. When this is done the Ne, Kr, Xe are usually adsorbed onto activated carbon at low temperature and separated by laboratory distillation.

Figure 38 is a simplified flow sheet of a typical small merchant oxygen plant meeting a variety of O_2 needs. Argon is not separated, and no use is made of the effluent N_2. Inlet air is filtered and compressed in the first of four compression stages. It is then sent to an air purifier where the CO_2 is removed by reaction with a recycling NaOH solution in a countercurrent packed tower. Usually the caustic solution inventory is changed daily. The CO_2-free gas is returned to the compressor for the final three stages after each of which the gas is cooled and water is separated from it. The compressed gas then goes to an adsorbent drier where the remaining water is removed onto silica gel or alumina. Driers are usually switched each shift and regenerated by using a slip stream of dry, hot N_2 and cooled to operating temperature with unheated N_2 flow.

The compressed, purified air is then cooled in the main exchanger (here a coiled tube type, but more usually of the plate-fin type) by transferring heat to both the returning N_2 and O_2. The process is basically a variation of that invented by Georges Claude where part of the high-pressure stream is withdrawn to the expansion engine (or turbine). The remainder of the air is further cooled in the main exchanger and expanded through a valve.

Table 12 Approximate Composition of Dry Air

Component	Composition (mole %)
N_2	78.03
O_2	20.99
Ar	0.93
CO_2	0.03
H_2	0.01
Ne	0.0015
He	0.0005
Kr	0.00011
Xe	0.000008

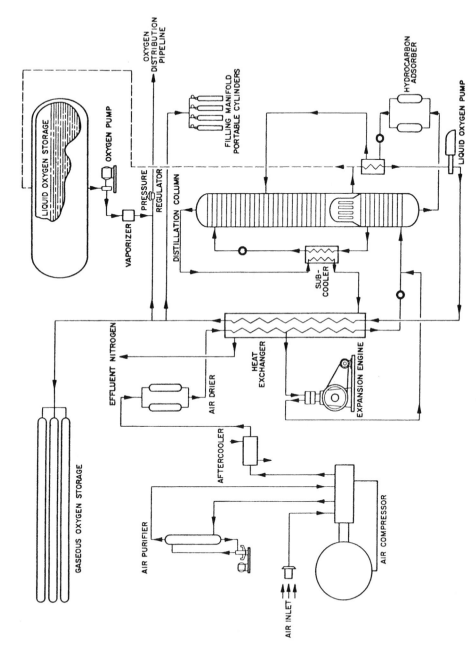

Figure 38 Flow sheet of a merchant oxygen plant. (Courtesy Air Product and Chemicals, Inc.)

513

The combined air stream, nearly saturated or partly liquefied, enters the bottom of the high-pressure column. This distillation column condenses nearly pure N_2 at its top using boiling O_2 in the low-pressure column as heat sink. If the low-pressure column operates at about $140 \ kN/m^2$ (20 psia), the high-pressure column must operate at about $690 \ kN/m^2$ (100 psia). The bottom product, called crude O_2, is about 65 mole % N_2. The top product from the high-pressure column, nearly pure N_2, is used as N_2 reflux in the low-pressure column.

The crude O_2 is fed to an activated carbon bed where hydrocarbons are removed, is expanded to low-pressure column pressure, goes through a subcooler in which it supplies refrigeration to the liquid O_2 product, and is fed to the low-pressure column. The hydrocarbons removed in the adsorber may come in as impurities in the feed or may be generated by decomposition of the compressor oil. If they are not fully removed, they are likely to precipitate in the liquid O_2 at the bottom of the low-pressure column. They accumulate there and can form an explosive mixture with oxygen whenever the plant is warmed up. Acetylene is especially dangerous in this regard because it is so little soluble in liquid oxygen.

The separation of O_2 and N_2 is completed in the low-pressure column. In the column, argon accumulates below the crude O_2 feed and may be withdrawn at about 10 mole % for further distillation. If it is not so removed, it leaves as impurity in the N_2 product. Light contaminants (H_2 and He) must be removed periodically from the top of the condenser/reboiler. Heavy contaminants are likely to leave as part of the O_2 product.

This plant produces O_2 in three forms: liquid, high-pressure O_2 for cylinder filling, and lower-pressure O_2 gas for pipe line distribution. The liquid O_2 goes directly from the low-pressure column to the storage tank. The rest of the liquid O_2 product is pumped to high pressure in a plunger pump after it is subcooled so as to avoid cavitation. This high-pressure liquid is vaporized and heated to ambient in the main heat exchanger. An alternate approach would be to warm the O_2 to ambient at high-pressure column pressure and then compress it as a gas. Cylinder pressure is usually too great for a plate-and-fin exchanger, so if the option shown in this flow sheet is used, the main exchanger must be of the coiled tube sort.

The nitrogen product, after supplying some refrigeration to the N_2 reflux, is warmed to ambient in the shell of the main exchanger. Here the N_2 product is shown as being vented to atmospheric. However, some of it would be required to regenerate the adsorbers and to pressurize the cold box in which the distillation columns, condenser/reboiler, main exchanger, hydrocarbon adsorber, subcoolers, throttling valves, and the liquid end of the liquid oxygen pump are probably contained.

This process is self-cooling. At startup refrigeration needed to cool the unit to operating temperatures is supplied by the expansion engine and the three throttling valves. During that time the unit is probably run at maximum pressure. During routine operation that pressure may be reduced. The lower the liquid O_2 demand, the less refrigeration is required and the lower the operating pressure may be.

7.2 Liquefaction of Natural Gas

Natural gas liquefaction has been commercially done in two very different situations. Companies that distribute and market natural gas have to meet a demand curve with a sharp maximum in midwinter. It has been found to be much more economic to maintain a local supply of natural gas liquid that can be vaporized and distributed at peak demand time than to build the gas pipe line big enough to meet this demand and to contract with the supplier for this quantity of gas. Thus, the gas company liquefies part of its supply all year. The liquid is stored locally until demand rises high enough to require augmenting the incoming gas. Then the stored liquid is vaporized and added to the network. These "peak-shaving" plants consist of a small liquefier, an immense storage capacity, and a large capacity vaporizer. They can be found in most large metropolitan areas where winters are cold, especially in the northern United States, Canada, and Europe.

The second situation is that of the oil/gas field itself. These fields are likely to be at long distances from the market. Oil can be readily transported, since it is in a relatively concentrated form. Gas is not. This concentration is done by liquefaction prior to shipment, thus reducing the volume about 600-fold. Subsequently, revaporization occurs at the port near the market. These "base-load" LNG systems consist of a large liquefaction plant, relatively modest storage facilities near the source field, a train of ships moving the liquid from the field to the port near the market, another storage facility near the market, and a large capacity vaporizer. Such a system is a very large project. Because of the large required investment, world political and economic instability, and safety and environmental concerns in some developed nations, especially the United States, only a few such systems are now in operation or actively in progress. See Table 13 for data on world LNG trade.

Peak-Shaving Plants

The liquefaction process in a peak-shaving installation is relatively small capacity, since it will be operating over the bulk of the year to produce the gas required in excess of normal capacity for two to six weeks of the year. It usually operates in a region of high energy cost but also of readily available mechanical service and spare parts, and it liquefies relatively pure methane. Finally, operating reliability is not usually critical because the plant has capacity to liquefy the required gas in less time than in the maximum available.

For these reasons efficiency is more important than system reliability and simplicity. Cascade and various expander cycles are generally used, although a wide variety of processes have been used including the Stirling cycle.

Figure 39 shows a process in which an N_2 expander cycle is used for low-temperature refrigeration, whereas the methane itself is expanded to supply intermediate refrigeration. This is done because of the higher efficiency of N_2 expanders at low temperature and the reduced need for methane purification. The feed natural gas is purified and filtered and then

Table 13 Data on World LNG Trade

World's LNG Plants, 1994			World's LNG Imports, 1994		World's LNG Trade	
Location	Capacity, Million Metric Tons/yr	Parallel Liq. Trains	Country	Quantity, Million Metric Tons/yr	Year	Amount, Million Metric Tons/yr
Kenai, Alaska	2.9	2	Japan	38.9	1980	22
Skikda, Algeria	6.2	8	S. Korea	4.4	1990	65
Arzew, Algeria	16.4	12	Taiwan	1.7	2000	90–95 (est)
Camel, Algeria	1.3	1	France	6.6	2010	130–160 (est)
Mersa, Libya	3.2	4	Other Europe	7.8		
Das Is., Abu Dhabi	4.3	2	U.S.A.	1.7		
Arun, Indonesia	9.0	5				
Bontang, Indonesia	13.2	7				
Lamut, Brunei	5.3	5				
Bintulu, Malaysia	7.5	3				
Barrup, Australia	6.0	3				

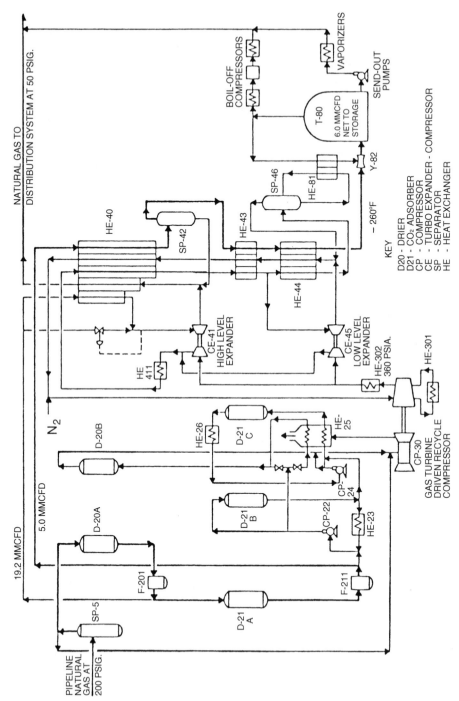

Figure 39 Flow sheet of an LNG process using N_2 refrigeration.

NATURAL GAS TO DISTRIBUTION SYSTEM AT 50 PSIG.

BOIL-OFF COMPRESSORS

VAPORIZERS

T-80
6.0 MMCFD NET TO STORAGE

SEND-OUT PUMPS

Y-82

~ -260°F

SP-46

HE-81

HE-40

SP-42

HE-43

HE-44

CE-41 HIGH LEVEL EXPANDER

CE-45 LOW LEVEL EXPANDER

HE 411

HE-302 360 PSIA.

HE-301

N_2

D-20B

HE-26

D-21 C

HE-25

CP-30

GAS TURBINE DRIVEN RECYCLE COMPRESSOR

CP-24

D-21 B

CP-22

HE-23

19.2 MMCFD

5.0 MMCFD

D-20A

F-201

D-21 A

F-211

SP-5

PIPELINE NATURAL GAS AT 200 PSIG.

KEY

D20 - DRIER
D21 - CO_2 ADSORBER
CP - COMPRESSOR
CE - TURBO EXPANDER - COMPRESSOR
SP - SEPARATOR
HE - HEAT EXCHANGER

split into two streams. The larger is cooled in part of the main exchanger, expanded in a turboexpander, and rewarmed to supply much of the warm end refrigeration, after which it is sent to the distribution system. The smaller fraction is cooled both by methane and by N_2 refrigeration until it is largely liquid, whereupon it goes to storage. Heavier liquids are removed by phase separation along the cooling path. Low-temperature refrigeration is supplied by a two-stage Claude cycle using N_2 as working fluid.

The LNG is stored in very large, insulated storage tanks. Typically such a tank might be 300 ft in diameter and 300 ft high. The height is made possible by the low density of LNG compared to other hydrocarbon liquids. LNG tanks have been built in ground as well as aboveground and of concrete as well as steel. However, the vast majority are aboveground steel tanks.

In designing and building LNG tanks the structural and thermal requirements added to the large size lead to many special design features. A strong foundation is necessary, and so the tank is often set on a concrete pad placed on piles. At the same time the earth underneath must be kept from freezing and later thawing and heaving. Thus electric cables or steam pipes are buried in the concrete to keep the soil above freezing. Over this pad a structurally sound layer of insulation, such as foam glass, is put to reduce heat leak to the LNG. The vertical tank walls are erected onto the concrete pad. The inner one is of stainless steel, the outer one is usually of carbon steel, and the interwall distance would be about 4 ft. The walls are field erected with welders carried in a tram attached to the top of the wall and lifted as the wall proceeds. The wall thickness is, of course, greater at the bottom than it is higher up.

The floor of the tank is steel laid over the foam glass and attached to the inner wall with a flexible joint. This is necessary because the tank walls will shrink upon cooling and expand when reheated. The dish roof is usually built within the walls over the floor. When the walls are completed, a flexible insulating blanket is put on the inside wall and the rest of the interwall space is filled with perlite. The blanket is necessary to counter the wall movement and prevent settling and crushing the perlite. At the end of construction the roof is lifted into position with slight air pressure. Usually this roof has hanging from it an insulated subroof that also rises and protects the LNG from heat leak to the roof. When this structure is in place, it is welded in and cover plates are put over the insulated wall spaces.

For safety considerations these tanks are usually surrounded by a berm designed to confine any LNG that escapes. LNG fire studies have shown such a fire to be less dangerous than a fire in an equivalent volume of gasoline. Still, the mass of LNG is so large that opportunities for disaster are seen as equally large. The fire danger will be reduced if the spill is more closely confined, and hence these berms tend to be high rather than large in diameter. In fact, a concrete tank berm built by the Philadelphia Gas Works is integral with the outside tank wall. That berm is of prestressed concrete thick enough to withstand the impact of a major commercial airliner crash.

Revaporization of LNG is done in large heat exchangers using air or water as heat sink. Shell and tube exchangers, radiators with fan-driven air for warming, and cascading liquid exchangers have all been used successfully, although the air-blown radiators tend to be noisy and subject to icing.

Base-Load LNG Plant

Table 13 lists the base-load LNG plants in operation in 1994. Products from these plants produce much of the natural gas used in Europe and in Japan, but United States use has been low, primarily because of the availability of large domestic gas fields.

In contrast to peak-shaving plants, liquefiers for these projects are large, primarily limited by the size of compressors and heat exchangers available in international trade. Also, these plants are located in remote areas where energy is cheap but repair facilities expensive or nonexistent. Thus, only two types of processes have been used: the classic cascade cycle and the mixed refrigerant cascade. Of these the mixed refrigerant cascade has gradually become dominant because of its mechanical simplicity and reliability.

Figure 40 shows a simplified process flow sheet of a mixed refrigerant cascade liquefier for natural gas. Here the natural gas passes through a succession of heat exchangers, or of bundles in a single heat exchanger, until liquified. The necessary refrigeration is supplied by a multicomponent refrigeration loop, which is essentially a Joule-Thomson cycle with successive phase separators to remove liquids as they are formed. These liquid streams are subcooled, expanded to low pressure, and used to supply the refrigeration required both by the natural gas and by the refrigerant mixture.

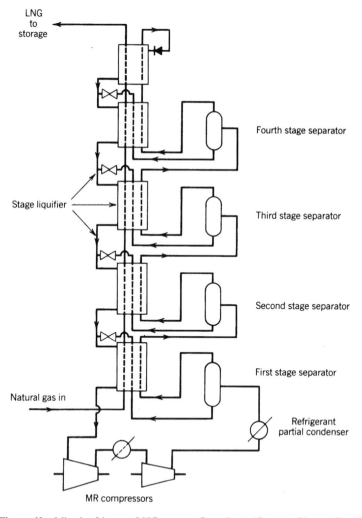

Figure 40   Mixed refrigerant LNG process flow sheet. (Courtesy Plenum Press.)

The success of this process depends on a selection of refrigerant composition that gives a cooling curve with shape closely matching the shape of the natural gas cooling curve. Thus all heat transfer will be across small ΔTs. This is shown in Fig. 41, a cooling curve for a mixed refrigerant cycle. The need to deal with a mixed refrigerant and to control the composition of the refrigerant mixture are the major difficulties with these processes. They complicate design, control, and general operation. For instance, a second process plant, nearly as large as the LNG plant, must be at hand to separate refrigerant components and supply makeup as needed by the liquefier.

Also not shown in this flow sheet is the initial cleanup of the feed natural gas. This stream must be filtered, dried, purified of CO_2 before it enters the process shown here.

As noted above, both compressors and heat exchangers will be at the commercial maximum. The heat exchanger is of the coiled-tube-in-shell sort. Typically it would have ¾-in. aluminum tubes wrapped on a 2–3 ft diameter mandrel to a maximum 14-ft diameter. The exchanger is probably in two sections totaling about 120 ft in length. Shipping these exchanger bundles across the world challenges rail and ship capacities.

Ships used to transport LNG from the terminal by the plant to the receiving site are essentially supertankers with insulated storage tanks. These tanks are usually built to fit the ship hull. There may be four or five of them along the ship's length. Usually they are constructed at the shipyard, but in one design they are built in a separate facility, shipped by barge to the shipyard, and hoisted into position. Boiloff from these tanks is used as fuel for the ship. On long ocean hauls 6–10% of the LNG will be so consumed. In port the

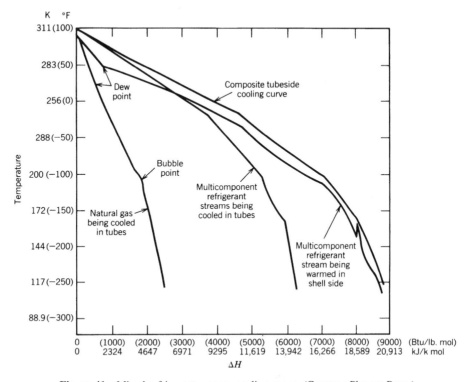

Figure 41 Mixed refrigerant process cooling curve. (Courtesy Plenum Press.)

evaporated LNG must be reliquefied, for which purpose a small liquefier circuit is available onboard.

7.3   Helium Recovery and Liquefaction

Helium exists in minute concentrations in air (see Table 12). However, this concentration is well below the 0.3 vol % that is considered to be the minimum for economic recovery. It exists at higher concentrations in a few natural gas deposits in the United States, as shown in Table 14, and in like concentrations in some deposits in Russia, Poland, and Venezuela. This fossil material is apparently the total world supply.

The vital role that helium plays in welding, superconductivity applications, space program operations, medicine, in certain heat transfer and inert atmosphere needs, and in a wide variety of research requirements lead to the demand that helium be conserved. This was undertaken by the Bureau of Mines after World War II. A series of helium-separation plants was built in the Southwest. Generally these produced an 80% helium stream from high He-content steams of natural gas that would otherwise have gone directly to the municipal markets. The processes used a modified Joule-Thompson cooling system that depended on the methane accompanying the He. This crude He was stored in the Cliffside Field, a depleted gas reservoir, from which it could be withdrawn and purified. Most of these plants shut down during the 1970s because of shifting government policies and budgetary limitations. In 1995 the last of these plants was closed down, as was the Bureau of Mines itself. The fate of the stored crude helium is being debated now (1996).

There are now about 30 billion standard cubic feet of crude He stored in the Cliffside reservoir, more than enough to supply the U.S. government needs estimated, at 10 Bcf

Table 14   Helium in Natural Gases in the United States

A. Composition of Some He-Rich Natural Gases in the United States

Location	Typical Composition (vol %)					
	CH_4	C_2H_6	N_2	CO_2	O_2	He
Colorado (Las Animas Co.)	0		77.6	14.7	0.3	7.4
Kansas (Waubaunsee, Elk, McPherson Cos.)	30	30	66.4	0.2	0	3.4
Michigan (Isabella Co.)	57.9	25.5	14.3	0	0.3	2.0
Montana (Musselshell)			54	30		16
Utah (Grand)	17		1.0	3.5		7.1

B. Estimated Helium Reserves (1994)

Location	Estimated Reserve (SCF)
Rocky Mountain area (Arizona, Colorado, Montana, New Mexico, Utah, Wyoming)	25×10^9
Midcontinent area (Kansas, Oklahoma, Texas)	169×10^9
He stored in the cliffside structure	30×10^9
Total	224×10^9

through 2015. Total demand for U.S. helium is nearly constant at about 3 Bcf/yr (in 1994). Private industry supplies about 89% of this market, the rest coming from the stored government supply. The estimated He resources in helium-rich natural gas in the United States is about 240 Bcf as of 1994. With the stored He, this makes a total supply of about 270 Bcf, probably enough to supply the demand until the middle of the 21st century. Eventually technology will be needed to economically recover He from more dilute sources.

The liquefaction of He, or the production of refrigeration at temperatures in the liquid He range, requires special techniques. He, and also H_2, have negative Joule–Thomson coefficients at room temperature. Thus cooling must first be done with a modified Claude process to a temperature level of 30 K or less. Often expanders are used in series to obtain temperatures close to the final temperature desired. An expansion valve may then be used to effect the actual liquefaction. Such a process is shown in Fig. 42. The goal of this process is the maintenance of a temperature low enough to sustain superconductivity (see below) using a conventional low-temperature superconductor. Since such processes are usually small, and since entropy gains at very low temperature are especially damaging to process efficiency, these processes must use very small ΔT's for heat transfer, require high-efficiency expanders, and must be insulated nearly perfectly. Note that in heat exchanger X4 the ΔT at the cold end is 0.55 K.

8   SUPERCONDUCTIVITY AND ITS APPLICATIONS

For normal electrical conductors the resistance decreases sharply as temperature decreases, as shown in Fig. 43. For pure materials this decrease tends to level off at very low temperatures. This results from the fact that the resistance to electron flow results from two factors: the collision of electrons with crystal lattice imperfections and electron collisions with the lattice atoms themselves. The former effect is not temperature dependent, but the latter is. This relationship has, itself, proven of interest to engineers, and much thought and development has gone toward the building of power transmission lines operating at cryogenic temperatures and taking advantage of the reduced resistance.

8.1   Superconductivity

In 1911 Dr. Onnes of Leiden was investigating the electrical properties of metals at very low temperatures, helium having just been discovered and liquefied. He was measuring the resistance of frozen mercury as the temperature was reduced into the liquid He range. Suddenly the sample showed zero resistance. At first a short circuit was suspected. However, very careful experiments showed that the electrical conductivity of the sample had dropped discontinuously to a very low value. The phenomenon of superconductivity has since been found to occur in a wide range of metals and alloys. The resistance of a superconductor has been found to be smaller than can be measured by the best instrumentation available. Possibly it is zero. Early on this was demonstrated by initiating a current in a superconducting ring which could then be maintained, undiminished, for months.

The phenomenon of superconductivity has been studied ever since in attempts to learn the extent of the phenomena, to develop a theory that will explain the basic mechanism and predict superconductive properties, and to use superconductivity in practical ways.

On an empirical basis it has been found that superconductors are diamagnetic, that is, they exclude a magnetic field, and that they exist within a region bounded by temperature and magnetic field strength. This is shown in Fig. 44. In becoming superconductive a material also changes in specific heat and in thermal conductivity.

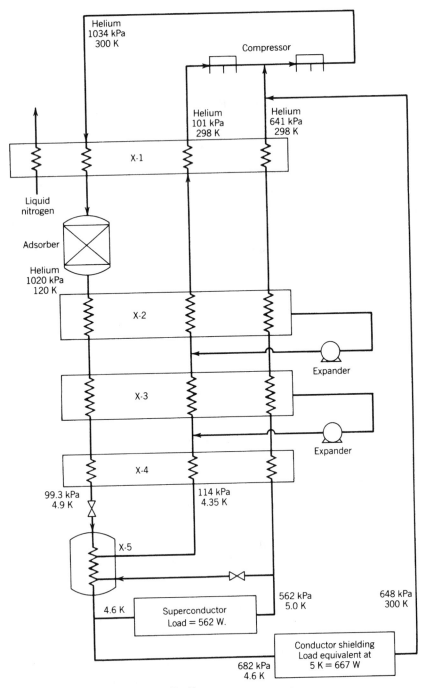

Figure 42 Helium liquefier flow sheet.

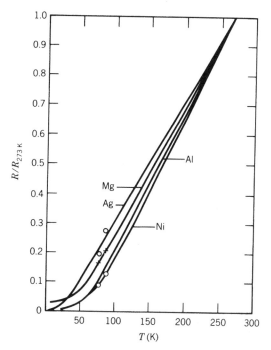

Figure 43 Variation of resistance of metals with temperature.

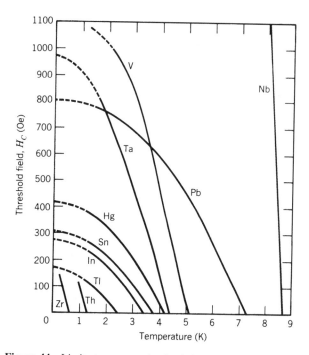

Figure 44 Limits to superconductive behavior of some elements.

The theory of supercoductivity developed after the discovery in 1933 by Meissner of the magnetic field exclusion. This led to a qualitative "two-fluid" model analogous to the theory underlying HeII. Since then this theory has been recast in quantum mechanical theory terms, most completely and successfully by J. Bardeen, L. N. Cooper, and R. Schrieffer of the University of Illinois in 1957 (BCS theory).

The BCS theory accounts for the Meissner effect and for other physical behavior phenomena of superconductors. It does not yet allow the prediction of superconductive transition points for new materials. The theory predicts an energy gap between normal and superconductive states existing simultaneously and visualizes the flow of paired electrons through the crystal lattice and the quantization of the magnetic flux. The quantized flux lines are fundamental to the explanation of the difference between type I superconductors that exhibit perfect Meissner effects, and which have relatively low transition temperatures and field tolerance, and type II superconductors that have imperfect Meissner effects, higher transition temperatures, and greater tolerance of magnetic fields. For example, Nb_3Sn, which is a type II superconductor, can be used for generation of magnetic fields of 100,000 gauss. These materials allow the penetration of magnetic field above a lower critical field strength, H_{c1}, but remain superconductive to a much greater field strength, H_{c2}. At fields above H_{c1}, flux enters the material in the form of quantized bundles that are pinned by dislocations so that the flux does not move easily and lead to normalization of the material.

Thus, both HeII and superconductors may be considered examples of superfluids. Each exhibits a nondissipative process for mass transfer. In HeII the mass transferred is the fluid itself in inviscid flow; with a superconductor it is the electrons flowing without encountering resistance. In both cases a flow velocity greater than a critical value restores normality. In superconductors circulating currents are set up in the penetration layer at the surface to cancel the applied magnetic field. When this field is increased to a critical value, the critical current is reached and the material reverts to the normal state.

8.2 Applications of Superconductivity

Development of applications of superconductivity has proceeded very rapidly over the past decade. However, by far the widest use of superconductivity has been in magnet applications. Superconductive magnets were constructed and tested soon after such practical superconductors as Nb_3Sn and rhodium–zirconium were discovered about 1960. Field strengths as high as 7 T were predicted then. Since then magnets in a wide range of sizes and shapes have been made, tested, and used with field strengths approaching 20 T.

The first three Nb alloys listed in Table 15 are ductile to the point that they are readily fabricated by conventional wire-drawing techniques. These form the cheapest finished product, but cannot be used at high field strengths. The Nb_3Sn, which is the most widely used superconductive material, and V_3Ga are formed into tape by chemical vapor deposition. The tape is clad with copper for stability and stainless steel for strength. Materials for multifilament conductor formation are produced by the bronze process. In this process filaments of Nb or V are drawn down in a matrix of Sn–Cu or Sn–Ga alloy, the bronze. Heat treatment then produces the Nb_3Sn or V_3Ga. The residual matrix is too resistive for satisfactory stabilization. Hence copper filaments are incorporated in the final conductors.

Multifilament conductors are then made by assembling superconductive filaments in a stabilizing matrix. For example, in one such conductor groups of 241 Ni–Ti filaments are sheathed in copper and cupronickel and packed in a copper matrix to make a 13,255-filament conductor. Such a conductor can be wound into an electromagnet or other large-scale electrical device.

Table 15 Some Commercially Available Superconductive Materials[a]

Material	T_c (K)	H_c 2 (T) at 4.2 K	J_c (10^5 A/cm² at 4.2 K)				Fabrication
			2.5 T	5 T	10 T	15 T	
Nb–25 wt% Zr	11	7.0	1.1	0.8	0	0	Fairly ductile
Nb–33 wt% Zr	11.5	8.0	0.9	0.8	0	0	Fairly ductile
Nb–48 wt% Ti	9.5	12.0	2.5	1.5	0.3	0	Ductile
Nb₃Sn	18.0	22.0	17.0	10.0	4.0	0.5	CVD diffusion bronze
V₃Ga	15.0	23.0	5.0	2.5	1.4	0.9	Diffusion bronze

[a] Courtesy Plenum Press.

Superconductive magnets have been used, or are planned to be used for particle acceleration in linear accelerators, for producing the magnetic fields in the plasma step of magnetohydrodynamics, for hydrogen bubble chambers, for producing magnetic "bottles" for nuclear fusion reactors such as the Tokomak, for both levitation and propulsion of ultra high speed trains, for research in solid-state physics, for field windings in motors, and for a host of small uses usually centered on research studies. In fact superconductive magnetics with field strength approaching 10 T are an item of commerce. They are usable where liquid helium temperatures are available and produce magnetic fields more conveniently and cheaply than can be done with a conventional electromagnet. Table 16 lists the superconductive magnets in use for various energy related applications.

Perhaps the most interesting of these applications is in high-speed railroads. Studies in Japan, Germany, Canada, and the United States are aimed at developing passenger trains that will operate at 300 mph and above. The trains would be levitated over the track by superconductive magnets, sinking to track level only at start and stop. Propulsion systems vary but are generally motors often with superconductive field windings. Such railroads are proposed for travel from Osaka to Tokyo and from San Diego to Los Angeles. Design criteria for the Japanese train are given in Table 17.

Superconductive electrical power transmission has been seriously considered for areas of high density use. Superconductors make it possible to bring the capacity of a single line up to 10,000–30,000 MW at a current density two orders of magnitude greater than conventional practice. The resulting small size and reduced energy losses reduce operating costs of transmission substantially.

The economic attraction of a superconductive transmission line depends on the cost of construction and the demand for power, but also on the cost of refrigeration. Thus a shield is built in and kept at liquid N_2 temperature to conserve on helium. Also, superinsulation is used around the liquid N_2 shield.

Other applications of superconductivity have been found in the microelectronics field. Superconductive switches have been proposed as high-speed, high-density memory devices and switches for computers and other electronic circuits. The ability of the superconductor to revert to normal and again to superconductive in the presence or absence of a magnetic field makes an electric gate or a record of the presence of an electric current. However, these devices have been at least temporarily overshadowed by the rapid development of the electronic chip. Ultimately, of course, these chips will be immersed in a cryogen to reduce resistance and dissipate resistive heat.

Table 16 General Characteristics of Superconductive Magnets for Energy Conversion and Storage Systems[40,a]

Application	Magnet Type	Typical Stored Energy in the Winding (MJ)	Operated			Largest Prototype So Far
			dc	Pulsed	Transients	
MHD generators	Dipole magnet with warm aperature, possibly tapered	500–5000	Yes	No	Yes, from the MHD fluid	60-MJ magnet
Homopolar machines	Solenoid	10–100	Yes	No	No	3-MW generator
Synchronous machines	Rotating dipole or quadrupole winding	Power plant machines, 50–100; airborne systems, 0.5–1	Yes	No	Yes, in case of unbalanced load	5-MVA generator
Fusion magnets Tokamak or similar low-β confinement	Toroidal field coils	$\geq 10^5$	Yes	Yes, in case of fast voltage control	Yes, pulsed field harmonic components of poloidal field	—
	Poloidal field coils	$\geq 10^3$	No		Yes, dc field components from the toroidal field	
Mirror confinement	Baseball coils	$\geq 10^5$	Yes	No	No	Baseball coils with 9 MJ
Energy storage; operation of pulsed fusion magnets Theta pinch	No optimal shape defined yet	≥ 100 per unit	No	Yes, transfer time about 30 msec	—	300 kJ
Tokamak	No optimal shape defined yet	$\geq 10^4$	No	Yes, transfer time about seconds	—	
Load leveling in the grid	No optimal shape defined yet	$\geq 10^8$	No	Yes, transfer time about hours	—	—

[a]Courtesy Plenum Press.

Table 17 Design Criteria for Japanese High-Speed Trains[41,a]

Maximum number of coaches/train	16
Maximum operation speed	550 km/hr
Maximum acceleration and deceleration:	
Acceleration	3 km/hr/sec
Deceleration, normal brake	5 km/hr/sec
Deceleration, emergency brake	10 km/hr/sec
Starting speed of levitation	100 km/hr
Effective levitation height (between coil centers)	250 mm
Accuracy of the track	± 10 mm/10 m
Hours of operation	From 6 AM to 12 PM at 15-min intervals
Period of operation without maintenance service	18 hr
Number of superconduction magnets	
Levitation	4 × 2 rows/coach
Guiding and drive	4 × 2 rows/coach
Carriage weight	30 tons
dimensions	25 m × 3.4 m × 3.4 m
Propulsion	Linear synchronous motor

[a]Courtesy Plenum Press.

9 CRYOBIOLOGY AND CRYOSURGERY

Cryogenics has found applications in medicine, food storage and transportation, and agriculture. In these areas the low temperature can be used to produce rapid tissue freezing and to maintain biological materials free of decay over long periods.

The freezing of food with liquid N_2 has become commonplace. Typically the loose, prepared food material is fed through an insulated chamber on a conveyor belt. Liquid N_2 is sprayed onto the food, and the evaporated N_2 flows countercurrent to the food movement to escape the chamber at the end in which the food enters. The required time of exposure depends on the size of individual food pieces and the characteristics of the food itself. For example, hamburger patties freeze relatively quickly because there is little resistance to nitrogen penetration. Conversely, whole fish may freeze rapidly on the surface, but the enclosing membranes prevent nitrogen penetration, so internal freezing occurs by conductive transfer of heat through the flesh. Usually a refrigerated holding period is required after the liquid N_2 spray chambers to complete the freezing process.

The advantages of liquid N_2 food freezing relative to more conventional refrigeration lie in the speed of freezing that produces less tissue damage and less chance for spoilage, and the inert nature of nitrogen, which causes no health hazard for the freezer plant worker or the consumer.

Liquid N_2 freezing and storage has also been used with parts of living beings such as red blood cells, bull semen, bones, and various other cells. Here the concern is for the survival of the cells upon thawing, for in the freezing process ice crystals form which may rupture cell walls upon freezing and thawing. The rate of survival has been found to depend on the rate of cooling and heating, with each class of material showing individual optima. Figure 45 shows the survival fractions of several cell types as a function of cooling velocity. Better than half the red blood cells survive at cooling rates of about 3000 K/min. Such a cooling rate would kill all of the yeast cells.

The mechanism of cell death is not clearly understood, and may result from any of several effects. The cell-wall rupture by crystals is the most obvious possibility. Another is

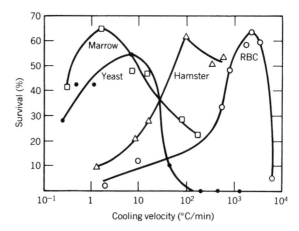

Figure 45 Survival rate for various cells frozen to liquid N_2 temperature.[42] (Courtesy Plenum Press.)

the dehydration of the cell by water migration during the freezing process. In any case the use of additives such as glycerol, dimethyl sulfoxide, pyridine n-oxide, and methyl and dimethyl acetamide has greatly reduced cell mortality in various specific cases. The amount and type of additive that is most effective depends upon the specific cell being treated.

Controlled freezing has proven useful in several surgical procedures. In each of these the destruction of carefully selected cells and/or their removal has been the goal of the operation.

In treating Parkinson disease destruction of some cells in the thalmus can lead to sharp reduction in tremors and muscular rigidity. The operation is done under local anesthetic using a very fine probe consisting of three concentric tubes. Liquid N_2 flows in through the center tube, returning as vapor through the central annulus. The outer annulus is evacuated and insulates all but the probe tip. The surgeon inserts the probe using X-ray pictures for guidance. He or she gently cools that probe tip using temperatures just below freezing. If the patient's tremors subside without other side effects, the right location has been found. Freezing of a quarter inch sphere around the probe tip can proceed.

In ophthalmic surgery cryogenic probes are used to lift cataracts from the lens of the eye. Here the cataract is frozen to the cryo-probe tip and carefully separated from the eye. Liquid N_2 is not needed and Freons or Joule–Thomson cooling is sufficient.

Malignant or surface tumors can also be removed cryogenically. The freezing of such a cell mass helps to prevent the escape of some of the cells into the blood stream or the body cavity.

REFERENCES

1. R. D. McCarty, "Thermodynamic Properties of Helium-4 from 2 to 1500 K at Pressures to 10^8 Pa," *J. Chem. Phys. Ref. Data,* **2**(4), 923 (1973); D. B. Mann, "Thermodynamic Properties of Helium from 3 to 300°K Between 0.5 and 100 Atmospheres," NBS Tech. Note 154 (Note 154A for British Units), Jan. 1962.

2. H. M. Roder and R. D. McCarty, "A Modified Benedict–Webb-Rubin Equation of State for Para-hydrogen-2," NBS Report NBS1R 75-814, June 1975.

3. J. G. Hurst and R. B. Stewart, "A Compilation of the Property Differences of Ortho- and Para-Hydrogen or Mixtures of Ortho- and Para-Hydrogen," NBS Report 8812, May 1965.

4. R. Prydz, K. D. Timmerhaus, and R. B. Stewart, "The Thermodynamic Properties of Deuterium," *Adv. Cryo. Eng.* **13,** 384 (1968).

5. R. D. McCarty and R. B. Stewart, "Thermodynamic Properties of Neon from 25 to 300 K between 0.1 and 200 Atmospheres," Third Symposium on Thermophysical Properties, ASME, 1965, p. 84.

6. R. T. Jacobsen, R. B. Stewart, and A. F. Myers, "An Equation of State of Oxygen and Nitrogen," *Adv. Cryo. Eng.* **18,** 248 (1972).

7. A. Michels, T. Wassenaar, and G. Wolkers, *Appl. Sci. Res.* **A5,** 121 (1955).

8. T. R. Strobridge, "The Thermodynamic Properties of Nitrogen from 64 to 300 K between 0.1 and 200 Atmospheres," NBS Tech. Note 129 (Note 129A for British Units), Jan. 1962 and Feb. 1963.

9. J. G. Hust and R. B. Stewart, "Thermodynamic Properties Valves for Gaseous and Liquid Carbon Monoxide from 70 to 300 K with Pressures to 300 Atmospheres," NBS Tech. Note 202, Nov. 30, 1963.

10. R. Prydz, G. C. Straty, and K. D. Timmerhaus, "The Thermodynamic Properties of Fluorine," *Adv. Cryo. Eng.* **16,** 64 (1971).

11. E. Bendu, "Equations of State Exactly Represent the Phase Behavior of Pure Substances," *Proceedings of the 5th Symposium on Thermosphysical Properties,* ASME, 1970, p. 227.

12. A. L. Gosman, R. D. McCarty, and J. G. Hust, "Thermodynamic Properties of Argon from the Triple Point to 300 K at Pressures to 1000 Atmospheres," NBS Reference Data Series (NSRDS-NSB 27), Mar. 1969.

13. L. A. Weber, "Thermodynamic and Related Properties of Oxygen from the Triple Point to 300 K at Pressures to 330 Atmospheres," NBS Rpt. 9710 (Rpt. 9710A for British Units), June and Aug. 1968; L. A. Weber, *NSB J. Res.* **74A**(1), 93 (1970).

14. R. D. McCarty, "A Modified Benedict–Webb-Rubin Equation of State for Methane Using Recent Experimental Data," *Cryogenics* **14,** 276 (1974).

15. W. T. Ziegle, J. C. Mullins, B. S. Kirk, D. W. Yarborough, and A. R. Berquist, "Calculation of the Vapor Pressure and Heats of Vaporization and Sublimation of Liquids and Solids, Especially Below One Atmosphere Pressure: VI, Krypton," Tech. Rpt. No. 1, Proj. A-764, Georgia Inst. of Tech. Engrg. Expt. Sta., Atlanta, 1964.

16. R. C. Downing, "Refrigerant Equations," ASHRAE Paper 2313, *Trans. ASHRAE,* **80,** Part III, 1974, p. 158.

17. See Ref. 15, "VIII, Xenon," Tech. Rept. No. 3, Projs. A-764 and E-115, 1966.

18. E. Bendu, "Equations of State for Ethylene and Propylene," *Cryogenics* **15,** 667 (1975).

19. R. B. Scott, F. G. Brickwedde, H. C. Urey, and M. H. Wahl, *J. Chem. Phys.* **2,** 454 (1934).

20. A. H. Singleton, A. Lapin, and L. A. Wenzel, "Rate Model for Ortho-Para Hydrogen Reaction on a Highly Active Catalyst," *Adv. Cry. Eng.* **13,** 409–427 (1967).

21. C. T. Lane, *Superfluid Physics,* McGraw-Hill, New York, 1962, pp. 161–177.

22. L. S. Twomey (personal communication).

23. O. M. Bourguet, "Cryogenic Technology and Scaleup Problems of Very Large LNG Plants," *Adv. Crys. Prg.* **18,** K. Timmerhaus (ed.), Plenum Press, New York (1972), pp. 9–26.

24. T. T. Rule and E. B. Quale, "Steady State Operation of the Idealized Vuillurmier Refrigerator," *Adv. Cryo. Eng.* **14,** 1968.

25. W. E. Gifford, "The Gifford–McMahon Cycle," *Adv. Cryo. Eng.* **11,** 1965.

26. H. Hausen, "Warmeubutragung in Gegenstrom, Gluchstrom, und Kiezstrom," Springer-Verlag, Berlin, 1950.

27. D. E. Ward, "Some Aspects of the Design and Operation of Low Temperature Regenerator," *Adv. Cryo. Eng.* **6,** 525 (1960).

28. G. O. G. Lof and R. W. Hawley, *Ind. Ing. Clem.* **40,** 1061 (1948).

29. G. E. O'Connor and T. W. Russell, "Heat Transfer in Tubular Fluid-Fluid Systems," *Advances in Chemical Engineering.* T. B. Drew et al. (eds.), Academic Press, New York, 1978, Vol. 10, pp. 1–56.

30. M. Jacob, *Heat Transfer,* Wiley, New York, 1957, Vol. 2, pp. 1–199.

31. W. T. Ziegh and H. Cheung, *Adv. Cryo. Eng.* **2,** 100 (1960).

32. R. H. Kropschot, "Cryogenic Insulation," *ASHRAE Journal* (1958).

33. T. F. Darham, R. M. McClintock, and R. P. Reed, "Cryogenic Materials Data Book," Office of Tech. Services, Washington, DC, 1962.

34. R. J. Coruccini, *Chem. Eng. Prog.* 342 (July 1957).

35. R. B. Stewart and V. J. Johnson (eds.), "A Compedium of Materials at Low Temperatures, Phases I and II," WADD Tech. Rept. 60-56, NBS, Boulder, CO, 1961.

36. C. R. Barber et al., "The International Practical Temperature Scale," *Metrologia* **5,** 35 (1969).

37. F. G. Brickwedde, H. van Diyk, M. Durieux, J. M. Clement, and J. K. Logan, *J. Res. Natl. Bur. Stds.* **64A,** 1 (1960).

38. R. P. Reis and D. E. Mapother, *Temperature, Its Measurement in Science and Industry,* H. H. Plumb (ed.), **4,** 885–895 (1972).

39. L. L. Sperikr, R. L. Powell, and W. J. Hall, "Progress in Cryogenic Thermocouples," *Adv. Cryo. Eng.* **14,** 316 (1968).

40. P. Komacek, "Applications of Superconductive Magnets to Energy with Particular Emphasis on Fusion Power," *Adv. Cryo. Eng.* **21,** 115 (1975).

41. K. Oshima and Y. Kyotani, "High Speed Transportation Levitated by Superconducting Magnet," *Adv. Cryo. Eng.* **19,** 154 (1974).

42. E. G. Cravalho, "The Application of Cryogenics to the Reversible Storage of Biomaterials," *Adv. Cryo. Eng.* **21,** 399 (1975).

43. K. D. Timmerhaus, "Cryogenics and Its Applications: Recent Developments and Outlook," *Bulletin of the Int. Inst. of Refrigeration* **66**(5), 3 (1994).

CHAPTER 14

INDOOR ENVIRONMENTAL CONTROL

Jelena Srebric
The Pennsylvania State University
University Park, Pennsylvania

The ability to control indoor environment parameters is vital to our everyday lives because we spend most of our time indoors. The engineering systems that enable control of environmental parameters are called heating, ventilating, and air-conditioning (HVAC) systems. Air-conditioning and refrigeration systems are one of the engineering achievements that transformed our lives in the past century (Constable and Somerville 2003). The first applications of HVAC systems were industrial, primarily for ice production and food preservation in the 19th century. The first installations of air-conditioning systems for building environments dates to the early 20th century and includes buildings such as the New York Stock Exchange (1905), the Central Park theater in Chicago (1917), and the Senate (1929). Residential air conditioning became affordable and popular after World War II. Today, air-conditioning systems are present everywhere in residential and commercial buildings and different kinds of vehicles, such as automobiles, space shuttles, and submarines. Based on the U.S. Energy Information Administration survey, 47% of all U.S. households use air conditioning (RECS, 1997). Furthermore, in commercial buildings, more than half of the yearly energy consumption is for HVAC systems (CBECS, 1999), including space heating, cooling, and ventilation. Largely, HVAC systems are major engineering systems that significantly affect the country's overall energy consumption and occupants' well being through control of indoor air parameters.

1 INDOOR ENVIRONMENT PARAMETERS

The indoor environment parameters to be controlled by HVAC depend on the type of the building. In general, residential HVAC systems typically control indoor air temperature by

531

heating in winter and cooling in summer. The humidity control is present in the form of humidification and dehumidification, but not always. Recently, air cleaners became cheaper and, as a result, are installed more widely for their potential to remove dust and other potential allergens. Overall, residential HVAC primarily controls the indoor air temperature. For commercial buildings, HVAC systems are more sophisticated and may control many parameters, such as pressure, temperature, humidity, carbon dioxide levels, and other gaseous or particulate contaminant concentrations.

1.1 Moist Air Parameters

HVAC systems supply treated outdoor air to building spaces. The outdoor air conditions depend on location, elevation, and time of day. When designing an HVAC system, engineers use outdoor air parameters for U.S. standard atmosphere (NASA, 1976). The standard atmosphere contains the following components specified by volume fraction: 78.084% nitrogen (N_2), 20.948% oxygen (O_2), 0.934% argon (A), 0.031% carbon dioxide (CO_2), 0.003% other minor gases. These components form a gas mixture called dry air. The atmospheric air, in addition to the dry air, includes water vapor and different gaseous and particulate contaminants. The standard design of HVAC systems accounts for the atmospheric air as a binary mixture of dry air and water vapor called moist air. Both of these moist air components are considered to obey the ideal gas law:

$$\text{For dry air:} \qquad p_a v_a = R_a T \qquad (1)$$

$$\text{For water vapor:} \quad p_v v_v = R_v T \qquad (2)$$

where p is the gas partial pressure, v is the specific gas volume, T is the gas absolute temperature, and R is the specific gas constant ($R_a = 287$ J/(kg·K); $R_v = 462$ J/(kg·K)).

The ideal gas law proves to be an excellent approximation for the real gas behavior of both moist air components. The following moist air parameters are commonly used in the design of HVAC systems: pressure p, dry bulb temperature T_{DB}, wet bulb temperature T_{WB}, dew point temperature T_{dp}, humidity ratio W, relative humidity ϕ, enthalpy i, and specific volume v.

The atmospheric pressure is the first consideration when designing an HVAC system. The standard barometric pressure at sea level is 101.325 kPa, which linearly decreases with the elevation based on the following equation (ASHRAE, 2001):

$$p = 101.325(1 - 2.25577 \times 10^{-5} H)^{5.2559} \quad \text{[kPa]} \qquad (3)$$

where p is the barometric pressure in [kPa] and H is the elevation in [m].

The total barometric pressure is a sum of the partial pressures of the dry air p_a and water vapor p_v based on the Gibbs-Dalton's law for ideal gases:

$$p = p_a + p_v \qquad (4)$$

The standard atmosphere also has defined standard air temperatures for different elevations, but when designing an HVAC system a more detailed temperature distribution is taken into account. The design temperature varies with geographic location and it is tabulated for most of the United States, Canada, and other world locations (ASHRAE, Chapter 27, 2001). In fact, two different temperatures, dry bulb temperature and wet bulb temperature, are specified for each location. To distinguish these two temperatures, it is necessary to introduce a condition called the saturation of moist air. Saturation is a condition of moist air that occurs because the moist air can contain only a limited amount of water vapor. When saturation occurs, the excess water vapor condenses on nearby surfaces. By definition, the saturation of moist air is the condition where moist air may coexist in a neutral equilibrium with its condensed water on a flat surface.

The dry bulb temperature represents the temperature of the moist air measured by a standard thermometer, while the wet bulb temperature represents temperature of the same moist air under adiabatic saturation. Adiabatic saturation could be achieved in an adiabatically insulated enclosure by evaporation of water at wet bulb temperature until the moist air is fully saturated. The wet bulb temperature was named by the experiment that used a wet cloth wrapped around a regular thermometer rotated at a specific speed to achieve a condition close to that of adiabatic saturation. The wet bulb temperature reflects the moisture content in the air: the lower the wet bulb temperature for the same dry bulb temperature, the lower the content of moisture. The wet bulb temperature can only be equal to or lower than the dry bulb temperature. Saturation occurs when these two temperatures are the same. The temperature for which the air is fully saturated is called the adiabatic saturation temperature, or wet bulb temperature.

Saturation occurs due to the moist air's limited capacity to contain water vapor. It is possible to reach a fully saturated condition in the moist air by to adding moisture or by decreasing the dry bulb temperature under constant pressure. This process is different from adiabatic saturation because the moisture content stays constant. An example could be condensation on a window in a cold climate during the winter season. By decreasing the temperature of a surface such as the window, a first droplet of water appears as soon as the surface reaches the dewpoint temperature. This process is very useful for the dehumidification of moist air by condensation in HVAC system cooling coils.

The humidity ratio W represents the ratio of the mass of water vapor m_v to the mass of dry air m_a in the moist air binary mixture:

$$W = \frac{m_v}{m_a} = \frac{p_v R_a}{p_a R_v} = 0.622 \frac{p_v}{p_a} = 0.622 \frac{p_v}{p - p_v} \quad \left[\frac{kg_{\text{water vapor}}}{kg_{\text{dry air}}} \right] \tag{5}$$

The relative humidity ϕ is the ratio of the water vapor partial pressure p_v to the water vapor partial pressure in a saturated mixture under the same temperature p_s:

$$\phi = \frac{p_v}{p_s} \times 100 \quad [\%] \tag{6}$$

When compared to the wet bulb temperature, both the humidity ratio and the relative humidity more explicitly define the moisture content in the moist air. In principle, the humidity ratio represents a nondimensional weight of the moisture, while the relative humidity represents a nondimensional degree of saturation. The humidity ratio ranges from zero or a few grams of water vapor per kilogram of dry air to 30 or more grams of water vapor per kilogram of dry air. The relative humidity ranges from zero for completely dry air to 100% for saturated moist air.

The enthalpy i of ideal gas mixtures is equal to the sum of the mixture component enthalpies. Therefore, the enthalpy of the moist air is a sum of the dry air enthalpy and the water vapor enthalpy:

$$i = c_{p,a}T + W(i_g + c_{p,v}T) = 1.01\,T + W(2501.3 + 1.86T) \quad \left[\frac{kJ}{kg_{\text{dry air}}} \right] \tag{7}$$

where $c_{p,a}$ is the specific heat of dry air, $c_{p,v}$ is the specific heat of water vapor, and i_g is the enthalpy of saturated water vapor at 0°C.

The specific volume v of the moist air is a ratio of total volume to the mass of dry air, which also could be defined in terms of temperature and pressure based on the ideal gas law:

$$v = \frac{V}{m_{\text{dry air}}} = \frac{R_a T}{p_a} \quad [m^3/kg_{\text{dry air}}] \tag{8}$$

The above moist air parameters are expressed in terms of different units per kilogram of dry air, which is a convention adopted due to the small and volatile mass of the water vapor in the moist air mixture.

1.2 The Psychrometric Chart

All of the parameters defined in the previous section are put together in a chart called the psychrometric chart. The chart enables quick access to values of different moist air parameters without requiring the direct use of equations. This graphical approach to solving HVAC engineering problems is very popular in industry. Modern versions of the psychrometric chart are typically supplied in a software format to enable interactive reading of the parameters. The chart is more traditionally used in a paper-based version as presented in Fig. 1 (ASHRAE, 2001). To simplify the information presented in the psychrometric chart, Fig. 2 presents psychrometric parameters in six separate charts that are merged to form the chart presented in Figure 1. The six charts present the pressure (elevation-dependent), temperatures, humidity ratio, relative humidity, and specific volume.

The most commonly used chart is the sea level chart for the standard barometric pressure of 101.325 kPa, as shown in Fig. 1. Nevertheless, HVAC design for higher elevations should use ASHRAE charts for 750, 1500, and 2250 m or give an exact elevation to psychrometric software. The horizontal axis in the chart represents the dry bulb temperature and the vertical axis represents the humidity ratio, see Fig. 2. The curved boundary on the left side of the chart is the saturation line representative of 100% relative humidity. The dry bulb temperatures are straight, slightly inclined to the left, and not exactly parallel lines. The humidity ratio lines are straight, horizontal, and parallel. At the points where humidity ratio lines intersect the saturation line, a single dew point temperature T_{dp} can be obtained for all of the moist air states along that same humidity ratio line. The relative humidity lines are the only curves in the chart and are evenly spaced. The enthalpy and wet bulb temperature lines are oblique and straight. It is important to notice that enthalpy and wet bulb temperature lines are close to each other, but do not coincide. The enthalpy lines are parallel to each other, while the wet bulb temperature lines diverge from each other. Finally, the specific volume lines are oblique and not parallel.

The dry bulb temperature in the standard psychrometric chart ranges from 0 to 50°C. For other temperature ranges, ASHRAE (2001) provides low-temperature (−40 to 10°C), high-temperature (10 to 120°C) and very high-temperature (100 to 200°C) psychrometric charts.

The protractor shown in the upper left corner of the psychrometric chart (Fig. 1) gives two additional parameters. The inner scale of the protractor represents a sensible heat factor (SHF) and the outer scale gives a ratio of the enthalpy to humidity ratio change. Both parameters are important for air handling processes that take place in HVAC systems.

2 AIR-HANDLING PROCESSES

HVAC systems have different component devices to process and supply air to conditioned spaces. The part of any HVAC system that processes the moist air is called the air-handling unit (AHU). The psychrometric chart is used to study the energy and/or moisture transport that occurs in AHU processes.

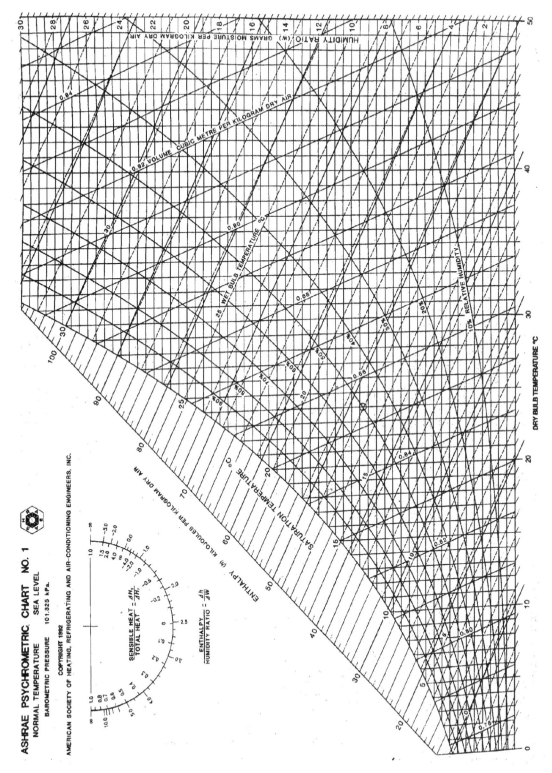

Figure 1 ASHRAE psychrometric chart (ASHRAE, 2001).

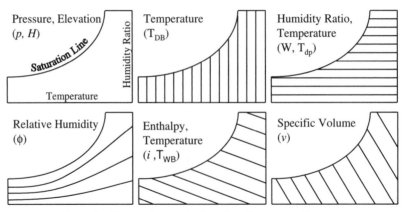

Figure 2   The psychrometric chart lines in separate diagrams.

2.1   Typical HVAC Processes

Several typical processes are present in most HVAC systems, although actual systems might differ significantly in appearance and arrangement. Depending on the role of an HVAC system, a certain process might be present or absent. Typical HVAC processes are

- Sensible heating/cooling
- Cooling and dehumidification
- Heating and humidification
- Adiabatic humidification processes
- Adiabatic mixing of air

Each of the above processes is performed by an HVAC device and corresponds to a particular HVAC device such as heating/cooling coils, humidifiers, or mixing boxes.

The analysis of these different processes uses the first law of thermodynamics and the conservation of mass principle. The first law is applied under the assumptions of steady-state processes and a negligible change in kinetic and potential energy, which are valid assumptions for an HVAC device. The first law for an HVAC device, which is an open thermodynamic system, becomes

$$\dot{Q}_{CV} = \dot{W}_{CV} = \sum_{out} \dot{m}_a i - \sum_{in} \dot{m}_a i \tag{9}$$

where $\dot{Q}_{CV}$ is the heat added to the control volume, $\dot{W}_{CV}$ is the shaft work done by the system (typically zero), $\dot{m}$ is the mass flow rate, and i is the enthalpy of the moist air flowing through the control volume of a device. By convention, heat added to the control volume and work done by the device are positive.

Sensible heating/cooling occurs in a heat exchanger that adds or removes heat without adding or removing the moisture content of the air stream. Figure 3 schematically shows the boundaries of the control volume and the corresponding process line in the psychrometric chart. The air stream changes its condition from state 1 to state 2 by flowing through the heat exchanger. The process is represented as a horizontal line in the psychrometric chart ($W = $ const.). The walls of the heat exchanger and the air ducts are assumed to be adiabatically insulated. The first law for this control volume is

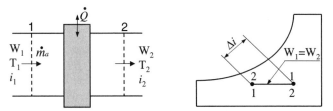

Figure 3 Sensible heating/cooling in a heating or cooling coil.

$$\dot{Q} = \dot{m}_a(i_1 - i_2) = \dot{m}_a(c_{p,a} + Wc_{p,v})(T_1 - T_2) = \dot{m}_a c_{p,a}(T_1 - T_2) \qquad (10)$$

where $Wc_{p,v}$ is much smaller than $c_{p,a}$ and can be neglected for practical applications.

If the surface temperature of the cooling coil is lower than the dew point temperature of the moist air, condensation occurs, resulting in a cooling and dehumidification process. This process is very important and often used to dehumidify air in hot and humid climates. Fig. 4 shows the dehumidification process that changes an air stream from state 1 to state 2. This process also results in water condensation. The first law for the cooling and dehumidification process is

$$\dot{Q} = \dot{m}_a(i_1 - i_2) - \dot{m}_a(W_1 - W_2)i_w \approx \dot{m}_a(i_1 - i_2) \qquad (11)$$

where i_w is the enthalpy of the condensate (water) at the exit air temperature T_2. The first term of the equation is much greater than the second one and can be neglected. Based on mass conservation, the flow rate for the condensate is

$$\dot{m}_w = \dot{m}_a(W_1 - W_2) \qquad (12)$$

The process in the psychrometric chart could be represented by a straight line that connects the two air states, which is the dashed line in Fig. 4. Another way to represent this process is to draw a horizontal line to the saturation line and then a curve that follows the saturation line until point 2. Both sensible and latent heat are exchanged. The sensible heat is proportional to the change in dry bulb temperature (ΔT), while the change in latent heat is proportional to the change in humidity ratio (ΔW). The total heat exchanged during the process is equal to the sum of the sensible and latent heat:

$$\dot{Q} = \dot{Q}_{sensible} + \dot{Q}_{latent} = \dot{m}_a(i_A - i_2) + \dot{m}_a(i_1 - i_A) = \dot{m}_a(i_1 - i_2) \qquad (13)$$

Based on the definition of the sensible and latent heat portions, a new parameter called the sensible heat factor (SHF) is introduced as

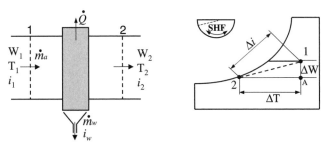

Figure 4 Cooling and dehumidification for hot and humid climates.

$$\text{SHF} = \frac{\dot{Q}_{\text{sensible}}}{\dot{Q}} = \frac{i_A - i_2}{i_1 - i_2} \tag{14}$$

SHF defines the slope for the process line (dashed line). The inner scale of the protractor in the upper left corner of the psychrometric chart gives SHF values. The line connecting the center of the protractor and SHF value is parallel to the process line in the chart.

Another typical process is heating and humidification for cold and dry climates. This process needs a heating coil and a humidifier as presented in Fig. 5. The heat exchanger heats up the air sensibly, while the humidifier adds moisture in either vapor or liquid form without heat exchange with the surroundings. The first law and the conservation of mass equations for this process are

$$\dot{m}_a i_1 + \dot{Q} + \dot{m}_w i_w = \dot{m}_a i_2 \tag{15}$$

$$\dot{m}_w = \dot{m}_a (W_2 - W_1) \tag{16}$$

When the moisture added is liquid at the wet bulb temperature of the incoming air stream 1, the humidification process coincides with the constant enthalpy line ($i_A = i_2$).

Adiabatic humidification of the incoming air stream could be achieved with steam or liquid at an arbitrary temperature. Figure 6 schematically shows the adiabatic humidification without heating process. The outgoing air stream could have different states represented as states 2, 3, or 4 in the example in Fig. 6. The constant enthalpy process 1-2, is achieved with a spray of water at the wet bulb temperature ($T_{\text{WB},1}$). The air state 3 can be reached by adding saturated steam at the dry bulb temperature of the incoming air (T_1). In general, the leaving air could be humidified and cooled if the water enthalpy is between the enthalpy for $T_{\text{WB},1}$ and the saturated liquid at T_1, which would result in a process somewhere in between process 1-2 and process 1-3 on the psychrometric chart. Cooling happens when the water droplets fully evaporate by taking energy from the incoming air stream under the assumption of adiabatic process. The leaving air could be humidified and heated, such as in process 1-4, if the added steam has enthalpy greater than the saturated steam at T_1. Except for processes 1-2 and 1-3, the exact slope of an arbitrary adiabatic humidification process could be determined from the following equation, which is derived from Eqs. (15) and (16) ($\dot{Q} = 0$):

$$\frac{\Delta i}{\Delta W} = \frac{i_2 - i_1}{W_2 - W_1} = i_{\text{water or steam}} \tag{17}$$

The value of the ratio $\Delta i / \Delta W$ can be calculated and used in the protractor to determine the slope of the humidification process, in the same way as SHF factor was used. This ratio is plotted on the outer scale of the protractor.

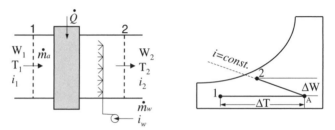

Figure 5   Heating and adiabatic humidification for cold and dry climates.

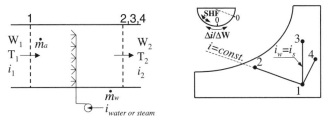

Figure 6 Adiabatic humidification processes.

Adiabatic mixing of two moist air streams is a process often used in AHUs because it enables major energy savings for HVAC systems. For example, instead of taking the entire amount of circulated air from outside in a cold winter day, a certain portion can be recirculated at the room air temperature. A certain percentage of recirculated air intake versus fresh air intake is recommended based on indoor air quality requirements to be discussed in more details in a subsequent chapter. A typical recirculation rate for U.S. commercial buildings is around 80%, which means about 20% fresh air is supplied to the building for dilution of contaminants and occupants' breathing. Figure 7 shows an example of adiabatic mixing for an air stream at state 1 (fresh air) and an air steam at state 2 (recirculated air). The resulting mixture is at state 3 on the mixing process line that connects states 1 and 2. The exact position of the state 3 is proportional to the ratio of the flow rates for the incoming air streams.

2.2 A Simple Air-Handling Unit

A simple AHU incorporates all of the above-mentioned processes for climates that require heating and cooling in different seasons. Nevertheless, individual AHU components may be active only in certain seasons, such as a humidifier in winter and a cooling coil in summer. The simplest AHU would serve a single zone, such as a single space (room), in a building and would be controlled by a thermostat placed in that zone. This kind of HVAC system is typically used for computer rooms, small department stores, or other small individual spaces with heating/cooling loads that are uniform and relatively stable throughout the zone. For nonuniform loads or larger spaces, multiple units could be installed with or without a duct system. Figure 8 presents an example of the single-zone system. Except for the cooling coil, filter, or supply fan, all other devices are optional, depending on the type of space and purpose of the system. The separation of the heating coil into preheat and reheat coils has

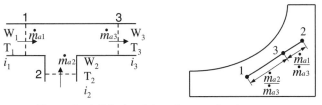

Figure 7 Adiabatic mixing of two moist air streams.

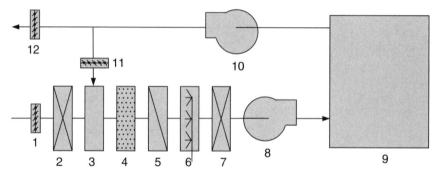

Figure 8   A simple single-zone air-handling system and conditioned space (1, fresh air louver; 2, preheat coil; 3, mixing box; 4, filter; 5, cooling coil; 6, humidifier; 7, reheat coil; 8, supply fan; 9, conditioned space; 10, return fan; 11, recirculation louver; 12, exhaust louver).

two basic functions. First, the reheat coil is used in summer for control of the humidity level in the supply air by subcooling the air to a desired humidity ratio and then reheating the air to the desired supply temperature. Second, the preheat coil prevents freezing problems in the mixing box during cold winter days.

3   THERMAL COMFORT

The primary purpose of HVAC systems is to maintain thermal comfort for building occupants. ANSI/ASHRAE Standard 55-1992 defines thermal comfort as the mind state that expresses satisfaction with thermal environment by a subjective evaluation. Due to the subjectivity of the thermal comfort, HVAC systems are currently designed to satisfy thermal comfort for 80% or more of the building occupants.

3.1   First Law Applied to the Human Body

The human body behaves similar to a heat engine, obeying the first law of thermodynamics. The chemical energy contained in food is converted into thermal energy through the process of metabolism. This thermal energy is used partially to perform work, while the other part has to be released to the surroundings to enable the normal functioning of the human body. The first law of thermodynamics for the human body has the following form (Fanger, 1970):

$$M - W = \dot{Q}_{\text{skin}} + \dot{Q}_{\text{respiration}} = (C_{\text{sk}} + R_{\text{sk}} + E_{\text{sk}}) + (C_{\text{res}} + E_{\text{res}}) \tag{18}$$

where M is the rate of metabolic heat production (W/m$^2_{\text{body surface area}}$), W is the rate of mechanical work, $\dot{Q}$ represents the different heat losses, C the convective heat losses, R the radiative heat losses, and E the evaporative heat losses. Figure 9 schematically presents the energy balance components for the control volume presented by a dashed line.

The metabolic heat production, losses and work are measured in W/m$^2_{\text{body surface area}}$ or in metabolic "met" units (1 met = 58.2 W/m^2). Table 1 gives metabolic heat production rates for typical tasks (ANSI/ASHRAE Standard 55-1992). The table shows that the human body produces a total heat equivalent to one or several light bulbs, depending on the activity level.

The main heat-transfer mechanisms are convection, radiation, and evaporation; conduction is negligible due to the small surface area and high insulation of shoe soles. Each of

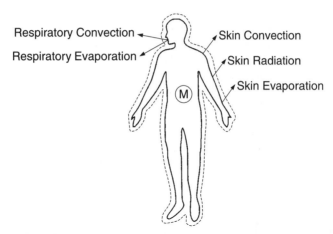

Figure 9 The energy balance for human body (M, metabolism).

the terms in Eq. (18) can be calculated (ASHRAE, 2001). The equations for each term are based on the fundamentals of heat transfer, but also include empirical equations and coefficients. In general, the total heat transfer from the human body depends on environmental and personal factors (Fanger, 1970). The environmental factors are air temperature (affects C), relative humidity (affects E), air velocity near the human body (affects C), and surface temperature of the enclosure and surrounding objects (affects R). The personal factors are activity rate and clothing (body insulation).

3.2 Thermal Comfort Indices

Several thermal comfort indices correlate the perception of thermal comfort with measured environmental parameters, i.e., dry bulb temperature, mean radiant temperature, and humidity levels. The mean radiant temperature (T_{MR}) is a uniform temperature of an imaginary black enclosure that would exchange the same amount of radiative heat with an occupant as the actual nonuniform environment. Further, T_{MR} is used to define an operative temperature T_o, which includes combined radiative and convective heat transfer in the actual nonuniform environment. T_o is approximately equal to the average value of T_{MR} and dry bulb temperature. Finally, the definition of effective temperature (ET*) includes the relative humidity. ET* is the operative temperature combined with 50% relative humidity that would cause the same

Table 1 Metabolic Heat Production Rates for Typical Tasks

Activity	met	W/m²
Reclining	0.8	46.6
Seated and quiet	1.0	58.2
Sedentary activity (office, dwelling, lab, school)	1.2	69.8
Standing, relaxed	1.2	69.8
Light activity, standing (shopping, lab, light industry	1.6	93.1
Medium activity, standing (shop assistant, domestic work, machine work)	2.0	114.4
High activity (heavy machine work, garage work, if sustained)	3.0	174.6

latent and sensible heat transfer for an occupant as would the actual environment (ANSI/ASHRAE Standard 55-1992).

A rationally derived index, the operative temperature (T_o), and an empirical index, the effective temperature (ET*), are used to plot an ASHRAE comfort zone for winter and summer conditions (see Figure 10). The "comfort zone" represents combinations of air temperature and relative humidity that most often produce thermal comfort for a seated North American adult in typical summer or winter clothing. An assumed level of dissatisfaction is 10% of all occupants. The slender lines bordering the comfort zone represent ET*: for winter ET* = 20–23.5°C and for summer ET* = 23–26°C. The ASHRAE comfort zone can be adjusted for different clothing levels, air velocities, activity levels, and human adaptation. Overall, HVAC systems are designed to produce an indoor air state within the ASHRAE comfort zone.

4   INDOOR AIR QUALITY (IAQ)

In addition to the primary function of maintaining thermal comfort, HVAC systems have to maintain good indoor air quality. According to ANSI/ASHRAE Standard 62-2001, acceptable indoor air quality is achieved with air that contains known contaminants at harmful

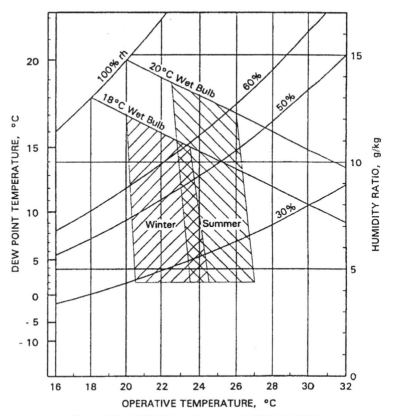

Figure 10   ASHRAE comfort zone (ASHRAE, 2001).

concentrations and the majority of occupants (more than 80%) express satisfaction. Indoor air quality plays a significant role in occupants' well-being, satisfaction, and health, due to the prolonged time periods spent indoors in a modern society.

4.1   Health Issues and Requirements

Buildings with poor air quality have produced many occupant health complaints/illnesses/conditions that have been recorded and labeled "sick building syndrome" (SBS). SBS is discomfort/illness caused by indoor air. Symptoms are often comparable to a cold or influenza, such as headaches, drowsiness, eye irritation, and nose and throat infection. The main difference between SBS and a cold is that the SBS symptoms last much longer than cold symptoms and disappear after the occupant leaves the building. Approximately 30% of new or remodeled buildings produce health complaints as reported by Environmental Protection Agency (EPA). Indoor air contaminant levels can be several magnitudes greater than the outdoor air contaminant levels due to the contaminant releases from building materials, occupants, or different building processes. Table 2 shows a few contaminant sources, permitted levels, and possible health effects according to EPA standards. Regardless of permitted levels, the actual response to a contaminant is based on the individual.

As listed in Table 2, indoor contaminants could be particles or gases and vapors. Particles could be allergens, molds, bacteria, dust, fumes, smoke, or mist and fog. Contaminant gases could be inorganic or organic. Harmful inorganic gases, such as radon (a radioactive gas), have the potential to cause lung cancer due to the particle deposition in lungs. Harmful organic vapors may cause major respiratory irritation and may be perceived as unpleasant odors. In the worst-case scenario, harmful vapors could be completely odorless. Many paints waxes, varnishes, and cleaning products are rich in organic solvents that evaporate even without use. Table 3 shows human irritation responses to total volatile organic compound (TVOC) mixtures based on a human subject study in a controlled laboratory environment (Molhave et al., 1991).

4.2   Problem Mitigation

Numerous indoor air-quality investigations over the last decade by the National Institute for Occupational Safety & Health (NIOSH) have found that primary sources of indoor air-quality problems are

Table 2   Contaminant Sources, Permitted Levels, and Health Effects (EPA)

Contaminants	Sources	Permitted Levels	Health Effects
CO_2	Human, combustion	1000 ppm	Stuffing
CO	Combustion, ETS	15 ppm	Body chemistry
SO_x	Combustion		Irritation, asthma
NO_x	Combustion	100 $\mu g/m^3$	Not very clear
Ra	Soil	4 picocuries/L	Lung cancer
VOCs (formaldehyde)	Combustion, pesticides, building materials, etc.	0.1 ppm	Eyes and mucous membrane irritation
Particulate (0.01 micro-insects)	Outdoor air, activities, ETS, furnishings, pets, etc		Lung diseases Cancer (ETS)

Table 3 Human Irritation Responses to TVOC Mixtures

Concentration (mg/m³)	Health Effect Response	Exposure Definition
<0.20	No irritation	Comfort
0.20–3.0	Irritation & discomfort	Multifactorial exposure range
3.0–25	Exposure effect and headache	Discomfort
>25	Additional neurotoxic effects	Toxic

- Inadequate ventilation, 52%
- Contaminant from inside the building, 16%
- Contaminant from outside the building, 10%
- Microbial contamination, 5%
- Contamination from building fabric, 4%
- Unknown sources, 13%

Appropriate design and maintenance of HVAC system can eliminate the majority of indoor air-quality problems. ANSI/ASHRAE Standard 62-2001 has a prescriptive ventilation rate procedure that defines adequate ventilation. The standard uses CO_2 as an indicator of adequate IAQ because CO_2 is a "marker" for human contaminants. The standard assumes that if CO_2 is kept at a required level, then other contaminants are also successfully diluted. Typically required fresh air quantities are 8–10 L/(s person). The rest of the air may be recirculated in order to save energy used by the HVAC system.

To solve IAQ problems, building owners or managers could use one of the following three strategies: (1) eliminate or modify the contaminant source, (2) dilute the contaminant with outdoor air and air distribution, or (3) use air cleaning (filtering). Even though these strategies sound straightforward, they may not be easy to implement. For example, for mold and fungi prevention, it is important to properly design HVAC systems for humidity control and it is just as crucial to have a properly constructed building envelope. Effective use of air distribution requires proper maintenance and balancing of the building air-distribution systems. Finally, the use of different types of filters and regular replacement of filters can significantly reduce IAQ problems. Table 4 indicates the type of contaminants that certain filters are capable removing; typically, two or more types of filters are combined to achieve proper removal of contaminants.

Recently, in addition to health concerns, HVAC systems have been charged with protecting buildings and occupants from internal or external releases of chemical, biological, or radiological contaminants. These new requirements for the removal of extremely harmful,

Table 4 Use of Different Filters for Different Types of Contaminants

	Media Filtration	HEPA Filters	Activated Carbon Adsorption	UV Photocatalytic Oxidation
Odors			X	X
VOCs			X	X
Bio-aerosols		X		X
Dust	X	X		

high concentration contaminants by HVAC systems are expected to transform the current design procedures.

5 BUILDING THERMAL LOADS

The building thermal loads represent the thermal energy to be removed (cooling loads) or added (heating loads) to a building to keep the occupants thermally comfortable, which is the primary function of an HVAC system. Building thermal loads determine the capacity of the equipment and air-distribution ducts used to condition a building. The procedures for the calculation of cooling and heating thermal loads are slightly different. Nevertheless, both procedures use concepts of indoor/outdoor design conditions and heat transfer through the multilayered building envelope assemblies to calculate the building thermal loads.

Indoor design conditions are based on thermal comfort for occupants (ANSI/ASHRAE Standard 55-1992) and refer to all of the moist air parameters within the thermal comfort zone in winter or summer as shown in Fig. 10. These conditions are maintained by a properly sized HVAC system.

Outdoor design conditions refer to the recommended weather data used for the design of HVAC systems. These data are available from ASHRAE (2001) and include the United States, Canada, and world location data. The recommended data do not reflect the coldest or the hottest temperatures ever measured for a location because designing for ultimate extremes would cause the heating or cooling systems to be critically oversized. Instead, ASHRAE presents HVAC design weather data with statistically determined probability for the "worst" weather conditions to occur. For example, the heating design conditions are published for 99.6 and 99% probability levels, while the cooling design conditions are published for 0.4, 1.0, and 2.0% probability levels. These probability levels refer to a number of hours during a year (12 months = 8760 hours) when the summer dry bulb temperature is higher than the published design value or the winter dry bulb temperature is lower than the published design value. A typical HVAC design would use 99% for heating and 1.0% for cooling probability levels. Therefore, depending on the year, the installed equipment might not have sufficient capacity to properly heat or cool the building during 88 hours in summer and 88 hours in winter, which is acceptable considering the initial and operating costs for an HVAC system.

The outdoor design conditions present the dry bulb temperatures accompanied by other important coincident weather data. The heating design conditions include the coincident extreme and mean wind speed used for estimating infiltration airflow rates. Nevertheless, the outdoor moisture content is not presented because the winter humidity ratio is low and would not significantly influence the capacity of humidifiers. For the cooling conditions, coincident mean wet bulb temperature, dew point temperature, and humidity ratio are used for sizing the cooling and dehumidification equipment.

Once indoor and outdoor design conditions for a particular building site are selected, heat transfer through the multilayered building envelope assemblies can be calculated. Heat transfer through building envelope, comprised of walls, windows, and roofs, includes all three heat-transfer mechanisms: conduction, convection, and radiation. Heat transfer is dynamic due to the continuous change in weather data as well as the change in building occupancy rates, and uses of building equipment. Nevertheless, the simplest HVAC system design procedures assume one-dimensional steady-state heat transfer, a good approximation for practical purposes due to the complexity of the real building heat-transfer phenomena.

Heat-transfer calculations for a building wall assembly use the thermal transmittance, U factor, for envelope assembly components such as windows, walls, floors, or roofs. U

factors are based on thermal resistances, R values, of building components obtained from standardized laboratory experiments that approximate building components as homogenous materials. Thermal transmittance and resistance values are available in published building material tables for standard building components (ASHRAE, 2001), or directly from the component manufacturers. The overall thermal transmittance U_O of the building envelope is

$$U_O = \frac{\sum U_i A_i}{A_O} \tag{19}$$

where A_O is the overall surface area of the envelope, U_i is the thermal transmittance, and A_i is the surface area of each envelope component. Based on the inverse relationship for thermal transmittance and thermal resistance for a building envelope component,

$$U_i = \frac{1}{R_i}; R_i = R_{in} + \sum R_{envelope} + R_{out} \tag{20}$$

where R_{in} and R_{out} are the film resistances for the inner and outer envelope surfaces and $R_{envelope}$ is the thermal resistance of the envelope assembly. The film resistances lump together the effects of thermal radiation and convection at the envelope surfaces, while the component resistance takes into account the conduction through the envelope. The envelope resistance is available in tables (ASHRAE, 2001) for standard components or could be calculated based on the resistance circuits representing the envelope component. The standard thermodynamic calculations for the R-value circuitry apply to building components. For example, a side wall would have siding and sheathing connected is in series ($R = R_{siding} + R_{sheathing}$), while the wood studs and insulation would be thermally connected in parallel:

$$\frac{1}{R} = \frac{1}{R_{wood\ studs}} + \frac{1}{R_{insulation}}$$

The simplest thermal load calculations use U factors for envelope assemblies and the indoor and outdoor design conditions as described above. In general, heating and cooling loads are calculated differently for residential and nonresidential buildings. The main features of residential buildings are 24-hour conditioned small internal loads, single zone, small capacity, dehumidification for cooling only, and thermostats for temperature control (ASHRAE, 2001). All other buildings are considered nonresidential. The following section discusses the calculations of residential and nonresidential building heating and cooling loads.

5.1 Heating Loads

The heating-load calculation procedure for residential buildings is relatively simple. The two heating-load components are the heat losses through the building envelope and the heat required for the heating of outdoor air. Building envelope heat losses occur through structural components such as the roof, windows, floors, walls, and walls below grade, while the outdoor air heat losses are introduced by infiltration through cracks around doors and windows, porous materials, and open doors and windows. The heating load through structural components and windows is

$$\dot{Q} = UA(T_o - T_i) \tag{21}$$

where U is the building envelope component transmittance, A is the component surface area, and T_o and T_i are the indoor and outdoor design temperatures.

The heating load through floors and walls below grade is

$$\dot{Q} = UA(T_{earth} - T_i) \tag{22}$$

where U is the transmittance, A is the area for the walls below grade, T_i is the indoor design temperature, and T_{earth} is design temperature for the ground.

The heating load by infiltration is

$$\dot{Q} = \rho \dot{V}_{infiltration}(T_o - T_i) \tag{23}$$

where ρ is the air density and $\dot{V}_{infiltration}$ is the volume flow rate of the infiltration. The infiltration flow rate can be estimated simply by assuming a certain flow rate for a building structure. This method is called the air exchange method and is suitable for experienced designers. Other methods, such as the crack length method and basic LBNL method, are also available for more detailed calculations (ASHRAE, 2001).

Nonresidential heating-load calculations are similar to the residential building thermal-load calculations because they do not include thermal storage effects or solar radiation heat gains. However, in nonresidential buildings, designers have to account for the fresh air heating required for occupant breathing and dilution of contaminants by ANSI/ASHRAE Standard 62-2001. In addition, nonresidential buildings should comply with ASHRAE/IES Standard 90.1-2001. This standard has been developed since the 1970s when the first energy crisis sparked energy-conscious building design. This building energy performance standard already has been adopted by more than a dozen states in the United States, with the idea that within a decade all of the states will have an energy code in compliance with this standard. The heating of nonresidential buildings is usually accompanied with the simultaneous requirement for cooling of the core building zones without external walls or windows.

5.2 Cooling Loads

Cooling loads occur through structural components, through windows, by infiltration, and due to occupants, appliances, lighting, and other equipment. For cooling-load calculations, the thermal storage factor cannot be neglected. In fact, the building structure, whether light, medium, or heavy, plays a significant role in the building cooling-load distribution. The cooling load is the heat-transfer rate at which energy has to be removed from a space to maintain indoor design conditions within thermal comfort. As an opposite, a heat gain is the rate at which the thermal energy is transferred to or generated within a space. An instantaneous heat gain such as solar radiation is first absorbed by the building structure and later transferred to the indoor air by convection, when it becomes the cooling load. For example, the peak cooling load in a building typically occurs in the afternoon or early evening, which is much later than the time when the actual heat gain has a peak value, creating a thermal lag. The heavier the building structure, the longer the thermal lag.

Residential cooling loads account for the heat gain from the structural components, windows, ventilation (fresh air requirement), infiltration, and occupants. The calculation for structural components uses the cooling-load temperature differences (CLTDs). This CLTD is equivalent to the actual indoor/outdoor temperature difference that would result in the same heating loads. Therefore, the building cooling load based on CLTD is

$$\dot{Q} = UA(\text{CLTD}) \tag{24}$$

where U is the building envelope component transmittance for summer, A is the component surface area, and CLTD is available in tables for single-family and multifamily residences

(ASHARE, 2001). CLTD depends on the building component type, orientation, outdoor design temperature, and daily temperature range.

To accurately account for the thermal radiation effects delayed by thermal storage, the cooling load through windows uses a glass load factor (GLF):

$$\dot{Q} = A(\text{GLF}) \tag{25}$$

where GLF is available in tables for single-family and multifamily residences (ASHRAE, 2001). GLF also accounts for the air-to-air heat conduction and shading.

Another important cooling load is ventilation, which accounts for the air handling of the fresh air requirements for the building occupants (ANSI/ASHRAE Standard 62-2001). Infiltration in summer tends to be much lower than infiltration in winter because the wind velocities and the indoor/outdoor temperature difference are typically much smaller in summer than in winter. The actual infiltration exchange rate depends on whether the building is tight, medium, or loose, and on the indoor/outdoor temperature difference. Suggested infiltration flow rates are tabulated (ASHRAE, 2001). Finally, each occupant gives off a certain amount of heat (~70 W/person) that also needs to be included in the cooling load. The total cooling load is the sum of all defined sensible loads multiplied by the latent load multiplier (LF), which is based on the outdoor design humidity ratio and building air tightness.

In nonresidential buildings, it is important to consider not only the already listed heat gains, but also the effects of multiple zones present in such buildings. An air-handling unit serves multiple zones in nonresidential buildings in contrast to a residential unit that conditions a single zone. The zones are distinguished not only by their heating or cooling requirements, but also by the schedule of these requirements. Therefore, the thermal-load analysis for design of a nonresidential HVAC system should consider the simultaneous effects of different zones, diversification of heat gains for internal loads such as nonuniform occupancy rates, and other unique circumstances (ASHRAE, 2001). Specifically, the calculation of nonresidential building thermal loads require detailed information on building characteristics (materials, size, and shape), configuration (location, orientation, and shading), outdoor design conditions, indoor design conditions, operating schedules (lighting, occupancy, and equipment), date and time, and additional considerations such as the type of air-conditioning system, fan energy, fan location, duct heat loss and gain, duct leakage, and the type and position of the air-return system. In general, thermal-load calculations for nonresidential buildings require use of computer programs no matter which calculation method is used.

6   COMPUTER PROGRAMS

The use of computer programs in building design is now a common practice. From the perspective of HVAC systems, building simulation programs can be classified as either building energy simulation programs or building airflow simulation programs. Building simulation programs are used for yearly HVAC system energy consumption calculations, while building airflow simulation programs are used for infiltration, thermal comfort, indoor air quality, and contaminant distribution calculations. Both simulation technologies enable the design of better buildings.

6.1   Energy Calculation Programs

Computer programs are often employed to perform nonresidential cooling/heating loads or yearly energy consumption analyses due to the complexity of heat transfer through building structures and variable weather data. The original manual methods for the calculation of yearly HVAC system energy consumption, degree day and bin methods, can provide only a general estimate based on the averaged weather data. Their use is limited to first-order energy consumption estimates. The introduction of computer methods in the early 1960s enabled the development of several different simulation methodologies and over a dozen of energy simulation programs.

Energy simulation programs use two different methods for the calculation of heating/cooling loads: the weighting factor method and the energy balance method. The weighting factor method is older and simpler than the energy balance method. Based on building materials, weighting factors are precalculated for different building components. The weighting factors express the convective gains over the total heat gain incoming on a building element in a time sequence. The data based off of the weighting factor for different elements are used to calculate the cooling loads. The energy simulation programs DOE-1 and DOE-2 are the most popular representatives of this simulation technology.

The energy balance method is currently used by the most popular energy simulation programs such as EnergyPlus and ESP-r. The energy balance method uses a heat balance for each zone, typically a single space, in a building. Each building zone surface has conductive, convective, and radiative heat fluxes in balance. All of the energy balance equations are put together in a matrix to resolve all of the zone fluxes simultaneously. The energy balance equation for the zone air is

$$\sum_{i=1}^{N} \dot{q}_{i,c} A_i + \dot{Q}_{\text{lights}} + \dot{Q}_{\text{people}} + \dot{Q}_{\text{appliances}} + \dot{Q}_{\text{infiltration}} - \dot{Q}_{\text{heat\_extraction}} = \frac{\rho \dot{V}_{\text{room}} c_p \Delta T}{\Delta t} \quad (26)$$

where $\sum_{i=1}^{N} \dot{q}_{i,c} A_i$ is the convective heat transfer from enclosure surfaces to room air, N is the number of enclosure surfaces, A_i is the area of surface i, $\dot{Q}_{\text{lights}}$, $\dot{Q}_{\text{people}}$, $\dot{Q}_{\text{appliances}}$, and $\dot{Q}_{\text{infiltration}}$ are the cooling loads of lights, people, appliances, and infiltration, respectively, $\dot{Q}_{\text{heat\_extraction}}$ is the heat extraction via HVAC device (equal to cooling loads), $\rho \dot{V}_{\text{room}} c_p \Delta T / \Delta t$ is the room air energy change, ρ is the air density, $\dot{V}_{\text{room}}$ is the room volume flow rate, c_p is the air specific heat, ΔT is the temperature change of room air, and Δt is the sampling time interval.

The energy balance method results not only in the energy requirement data to heat and cool a building hour by hour, but also in the surface temperature distribution and an average air temperature needed to evaluating the thermal comfort in a building zone.

6.2 Airflow Simulation Programs

The first generation of airflow simulation programs for buildings were multizone models used to estimate the building infiltration rates for use in energy simulation programs. Multizone models define a zone in a building that is connected to other zones by flow paths both among zones and to the outside environment. In each zone (room), perfect mixing is assumed with uniform density, temperature, velocities, and species concentration. Mass balance and energy conservation are applied to each zone

$$\text{Mass balance:} \qquad \frac{dM_i}{dt} = \sum_{j=1}^{n} m_{ij} + m_{\text{source}} + m_{\text{sin } k} \qquad (27)$$

$$\text{Energy conservation:} \qquad \frac{dQ_i}{dt} = \sum_{j=1}^{n} q_{ij} + q_{\text{source}} + q_{\text{sin } k} \qquad (28)$$

where M_i is the mass in zone i; Q_i is the thermal energy in zone i; $\sum_{j=1}^{n} m_{ij}$ is the sum of mass transported to zone i from other zones, $i \neq j$; $\sum_{j=1}^{n} q_{ij}$ is the sum of thermal energy transferred to zone i from other zones, $i \neq j$; m_{source} is the mass (rate) generated by the source in zone i; q_{source} is the thermal energy generated by the sources in zone i; m_{sink} is the mass removed from the zone, and q_{sink} is the thermal energy removed from zone i.

Flow paths (elements) connect zones and allow airflow in one or both directions. The flow between different zones is usually driven by a pressure and/or temperature difference, so the wind data and outdoor temperatures can be used to estimate infiltration flow rates.

Multizone models have benefits and drawbacks. A major benefit of multizone models is that they can quickly calculate the airflow, heat flux, or contaminant transportation between different zones (rooms) in an entire building. Furthermore, specifying building zones is relatively simple because boundary conditions such as walls, windows, and supply/exhaust do not require detailed specifications. The simplicity of multizone models, on the other hand, does not allow them to provide detailed information on the airflow pattern, velocity, temperature, and concentration distributions in individual rooms (zones). However, more complicated programs based on computational fluid dynamics (CFD) are capable of predicting all those distributions with accuracy.

CFD programs numerically solve a set of partial differential equations for the conservation of mass, momentum (Navier-Stokes equations), energy, and species concentration. There are three groups of numerical methods for solving the equations: control volume, finite element, and spectral methods. The control volume method is widely used in CFD for building environment simulations. This CFD model divides a single room into fine control volumes, order(s) of magnitude smaller than the building zones. Typically, a control volume would be anywhere between 1 and 10 cm³. In each control volume, a uniform density, temperature, velocity, and species concentration is assumed. Also, each control volume has to satisfy mass (in the form of the continuity equation), momentum, energy, and species conservation. The major difference between the multizone and CFD models is the additional momentum equations used in the CFD model:

$$\frac{\partial \rho U_i}{\partial t} + \frac{\partial \rho U_i U_j}{\partial x_j} = -\frac{\partial P}{\partial x_i} + \frac{\partial}{\partial x_j}\left[\mu_{\text{eff}}\left(\frac{\partial U_i}{\partial x_j} + \frac{\partial U_j}{\partial x_i}\right)\right] + \rho\beta(T_o - T)g_i \tag{29}$$

where U_j is the mean velocity component in the x_j direction, P is the mean pressure, μ_{eff} is the effective viscosity, β is the thermal expansion coefficient of air, T_0 is the temperature of the reference point, T is the mean temperature, and g_i is the gravity acceleration in the i direction.

The major benefit of CFD models is their capability to provide detailed results for three-dimensional distributions of air velocity, temperature, and species concentration in a room. Figure 11 shows a distribution of carbon dioxide from two occupants in an office with displacement ventilation, which has a supply diffuser at the floor level. However, CFD requires much more computing time and much more detailed boundary conditions compared to the multizone model. For example, the computing time for a single room would be several hours with the CFD model, but only a few seconds with the multizone model. Furthermore, the boundary conditions, such as walls, supply/exhaust, sources, and sinks, must be defined not only by their properties, but also by their detailed locations. As a result, and due to the extensive computation time, using CFD for practical engineering design is challenging, especially when an entire building is being analyzed.

6.3   Coupled Simulation Tools

Each of the building simulation programs has a function in improving building environment design. Due to the rapid increase in computation capacity, a new trend in building simulation technology is under way. Each building simulation program has advantages and disadvan-

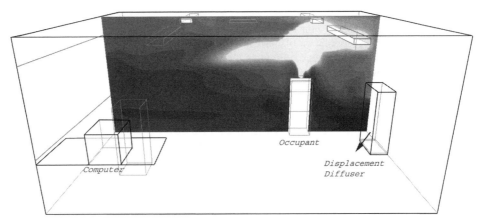

Figure 11 CFD result for CO_2 distribution in an office with displacement ventilation (blue, fresh air with CO_2 around 350 ppm; red, room air with CO_2 over 1000 ppm).

tages, so coupling of simulation tools is the way to mitigate respective weaknesses and create a new generation of building environment simulation programs. The coupling of energy simulation and CFD programs is not a new idea. Chen (1988) was probably first to show how coupled energy simulations and CFD can provide more accurate predictions of cooling loads for different cooling and heating systems. Furthermore, recent coupling of multizone and CFD models (Yuan and Srebric, 2002) has shown how CFD models can help increase the accuracy of multizone models while the multizone model eliminates the requirement of applying CFD to the entire building. Future generations of building simulation software will integrate all the components: energy, CFD, and multizonal modeling. Such a sophisticated tool will allow building HVAC designers to accurately calculate the heating/cooling load, thermal comfort, indoor air quality, and contaminant concentrations while exploring different innovative HVAC solutions.

7 EQUIPMENT FOR ENVIRONMENTAL CONTROL

The capacity of the HVAC equipment based on indoor and outdoor design conditions and U factors for envelope assemblies or load simulation programs gives the HVAC equipment capable of handling the peak design conditions. It is important to notice that besides the performance requirements (removal of contaminants) and capacity requirements (building loads), an HVAC system also needs to satisfy spatial, initial cost, operating cost, reliability, flexibility, maintainability, and other special requirements. HVAC systems are typically coupled with heating/cooling production systems such as boilers for production of hot water or steam and chillers for production of cold refrigerant. Based on the fluid used to condition building spaces, an HVAC system can be all-air, air–water, or all-water systems. In the United States, all-air systems are the most widely used for environmental control. These systems include air-handling, air-distribution, and control systems.

7.1 Air Handling and Distribution Systems

Air handling and distribution systems have a capacity calculated based on the design thermal loads. Heaters, coolers, and humidifiers are sized for the air-handling system, while supply/

exhaust diffusers, ducts and supply/return fans are sized for the air-distribution systems. The air-handling unit calculations use the equations derived from the first law of thermodynamics and conservation of mass as outlined in Section 2 on basic air-handling processes. Once the air-handling unit is selected, the air-distribution system is designed based on the airflow supply rate required by the air-handling unit. The procedure for diffuser design includes the selection of supply and return diffuser types, numbers, and locations. The diffuser selection is based on the diffuser manufacturer catalogues. From each of the diffusers, a duct branch has to be laid out. The duct and diffuser layout enables calculations of the total pressure drop in the distribution system based on the equal friction or static regain method. Finally, based on the total pressure drop, supply/return fans are selected. In this way, HVAC air handling and distribution systems are designed for the peak loads.

7.2   Control Systems

The HVAC systems only occasionally operate under peak loads. Most of the time HVAC systems work under partial or so-called off-design loads. For off-design conditions, an HVAC system needs a control system to regulate its capacity; otherwise the building spaces would be too hot in winter or too cold in summer. Based on the first law of thermodynamics and under uniform indoor design conditions, the regulation could occur by changing the flow rate of the supply air stream while keeping all the other parameters constant. This system of control by variable air volume (VAV) flow rates is the most widely used system in the United States. Another way of controlling an all-air system is by modulating the supply air dry bulb temperature. This system is called constant air volume (CAV) control system. CAV offers fine regulation of indoor air parameters, but it is not often installed due to the high penalties in energy prices.

The essential components of a control system are controlled variable, sensor, controller, and controlled device. The controlled variable is a characteristic or parameter of HVAC system to be regulated. In particular, a "set point" is the desired value, while a "control point" is the actual value of the regulated parameter. For example, the set point is the temperature set on the dial of a thermostat, while the control point is the temperature measured by a thermocouple in the same zone with the thermostat. The "error" or "offset" is the difference between the set and control points. The sensor measures the actual value of the controlled variable, such as the thermostat in the above mentioned example. The controller modifies the action of the controlled device in response to error, and the controlled device acts to modify controlled variable as directed by the controller. The controller could be a valve controller connected the thermocouple and thermostat, while the controlled device could be the steam valve regulating the flow rate of the steam for space heating.

The control system action control could be a two-position (on–off) control or a modulating control. The two-position system is commonly installed with HVAC systems due to its low cost, but it is relatively imprecise. Modulating control systems produce continuously variable output over a range. This is a finer control system than the two-position system, and is typically found in large HVAC systems. The modulating control system could be proportional, proportional plus integral (PI), or proportional plus integral plus derivative (PID). The simplest control system resulting in the control of only required parameters is typically selected. Overly complicated systems prove inefficient as they tend to be difficult and expensive to operate and maintain. Initial cost considerations also play an important role in the selection of a control system.

BIBLIOGRAPHY

ANSI/ASHRAE, Standard 55-1992. "Thermal Environmental Conditions for Human Occupancy," American Society of Heating, Ventilating and Air-Conditioning Engineers, Inc., USA, 1992.

ANSI/ASHRAE, Standard 62-2001. "Ventilation for Acceptable Indoor Air Quality," American Society of Heating, Ventilating and Air-Conditioning Engineers, Inc., USA, 2001.

ASHRAE, *ASHARE Handbook of Fundamentals,* American Society of Heating, Ventilating and Air-Conditioning Engineers, Inc., USA, 2001.

ASHRAE/IES, Standard 90.1-2001. "Energy Standard for Buildings Except Low-Rise Residential Buildings," American Society of Heating, Ventilating and Air-Conditioning Engineers, Inc., USA, 2001.

CBECS, "The Commercial Buildings Energy Consumption Survey," U.S. Energy Information Administration, 1999.

Chen, Q., *Indoor Airflow, Air Quality and Energy Consumption of Buildings,* Thesis, Delft University, The Netherlands, 1988.

Constable, G., and B. Somerville, *A Century of Innovation: Twenty Engineering Achievements That Transformed Our Lives,* Joseph Henry Press, 2003.

Fanger, P. O. *Thermal Comfort Analysis and Applications in Environmental Engineering,* McGraw-Hill, New York, 1970.

Molhave, L., J. G. Jensen, and S. Larsen, "Subjective Reactions to Volatile Organic Compounds as Air Pollutants," *Atmospheric Environment,* **25A**(7), 1283–1293 (1991).

NASA, "U.S Standard Atmosphere," National Oceanic and Atmospheric Administration, National Aeronautics and Space Administration, 1976.

RECS, "The Residential Energy Consumption Survey," U.S. Energy Information Administration, 1997.

Yuan, J., and J. Srebric, "Improved Prediction of Indoor Contaminant Distribution for Entire Buildings," American Society of Mechanical Engineers (ASME), Fluids Engineering Division (Publication) FED, **258**, 111–118 (2002).

CHAPTER 15
THERMAL SYSTEMS OPTIMIZATION

Reinhard Radermacher
University of Maryland
College Park, Maryland

1 INTRODUCTION

Thermal systems include all functional groups of equipment and working fluids that are designed to manage temperature and humidity conditions inside various spaces or materials. Thermal management systems provide comfort, establish and maintain conditions necessary for the functionality of other equipment, or utilize the change of thermo-physical properties of materials for energy conversion. Applications run from thermal management of electronic systems (electronic cooling) to space conditioning and power generation.

Optimization is the systematic procedure that guides system designers in their choice of processes and components such that all requirements for the system are balanced in the best fashion possible. In most applications the designer has to balance several contradicting demands, such as high efficiency and reliability versus low costs and emissions. To keep the design time and associated costs as low as possible, it is essential to take all requirements of the thermal system into account at the earliest possible design stage. Often a great amount of time and costs can be saved if the design engineer has means to evaluate the approximate costs of a design in the early stages of the development.

The optimization of thermal systems usually includes a mixture of technology decisions and the optimization of specific properties of selected components. One example is the decision between tube-fin and microchannel technology for an air-refrigerant heat exchanger in the air-conditioning system of a commercial building and the subsequent optimization of tube diameter/channel geometry and fin spacing. The system designer should find the least expensive designs for each technology that provides the required performance, in this example heat load, while minimizing the fan power consumption. An informed decision can be reached only if the best options of all feasible technologies are compared. Additionally, other factors have to be considered: The best microchannel heat exchanger may be more expensive than the best fin-tube heat exchanger, but may require a smaller fan and thus lead to savings at other system components. This shows that the optimization of thermal systems requires the evaluation of entire system performance and costs. The system designer must conduct the component optimization and selection with the system perspective in mind.

2 OPTIMIZATION TOOLBOX

Optimizing thermal systems requires evaluating, comparing, and modifying large numbers of design options. The system evaluation usually includes engineering factors such as efficiency and reliability, as well as accounting factors such as first and operating costs. To evaluate a large number of design options it is helpful to employ a computer-simulation tool (or a collection of tools) that is capable of predicting the system performance and system costs with sufficient accuracy. Section 2.1 describes a formulation of the simulation of general energy-conversion systems.

The system designer must develop a basis of comparison for the various design options. This can be a single parameter of the system performance or costs, but usually is a combination of many parameters. If the relative importance of the significant parameters is known in advance, the designer can formulate a weighted penalty function, which assigns a characteristic value to each design option. If the relative importance is not known beforehand, the design task becomes a multiobjective optimization problem. In this case the optimization procedure should determine a Pareto-optimum set of solutions (see Section 2.2).

The optimization driver derives new design options based on the comparison of evaluated options. Section 2.2 illustrates a number of optimization drivers with their advantages and disadvantages. The selection of the appropriate driver for the optimization problem is essential for the success of the design process.

2.1 System Evaluation

In most design optimization problems the designer has to take into consideration engineering-level parameters of the thermal system as well as accounting level parameters. The engineering-level evaluation includes parameters such as system efficiency, reliability, noise, vibration, and emissions. This level of evaluation requires a physics-based simulation of the system, which is sufficiently detailed to reflect the effects of relevant component variations on the system performance. The accounting-level evaluation requires a cost model for the system and its components that is sufficiently detailed to reflect the effect of component variations on the overall costs.

Engineering-Level System Simulation

Thermal systems can generally be described as networks of components and their interaction with the environment. System components are connected by junctions, through which they exchange flow rates as a result of driving forces imposed by the states of the junctions. A large number of component models typically encountered in thermal systems are presented throughout this volume of the handbook. The component models should be physics-based descriptions of the components with a level of detail that allows for evaluating the effect of changes in optimization parameters on the component and system performance. For instance, if the heat-transfer area of a heat exchanger is an optimization parameter, the model of the heat exchanger must reflect the effect of a change in heat-transfer area on the performance (as opposed to using a constant heat exchanger effectiveness, for instance).

An appropriate mathematical description of the system results in a system of residual equations, which has to be solved numerically. The residuals can be formulated by the differences of flow rates (mass flow rates, heat flow rates etc.) entering and leaving a junction. Figure 1 illustrates this on the example of a vapor compression system at steady-state conditions. Table 1 lists the residual equations. Flow rates leaving a component are associated with a negative sign; flow rates entering a component have a positive sign. Table 2 gives examples for the forms of the component equations for steady-state conditions.

The residual equations can be passed to an equation solver. A list of solvers is given in the following section. If the network representation of the system is not very complex, in particular, if there are no splits or mergers of streams, the residual equations can be simplified as illustrated in Table 3. Alternative residual and component equations can be formulated

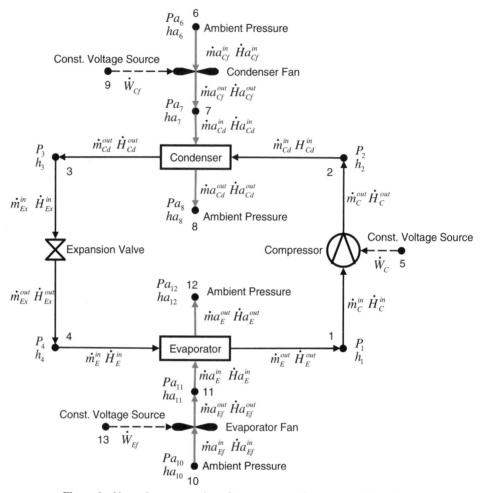

Figure 1 Network representation of vapor compression system with air fans.

and the choice of the appropriate formulation can have a significant influence on the convergence of the simulation.

The convergence of numerical equation solvers typically depends on the quality of the initial guess values. Special care must be given to provide good guess values to the equation solver. Other sources of divergence are invalid inputs to thermophysical property functions and invalid inputs to component models. Both instances can occur during the course of the

Table 1 Residual Equations for Vapor Compression System

$R_1 = \dot{m}_C^{out} + \dot{m}_{Cd}^{in}$	$R_4 = -M_{System} + M_C + M_{Cd} + M_{Ex} + M_E$
$R_2 = \dot{m}_{Cd}^{out} + \dot{m}_{Ex}^{in}$	$R_5 = \dot{m}_{Cf}^{out} + \dot{m}_{Cd}^{in}$
$R_3 = \dot{m}_{Ex}^{out} + \dot{m}_E^{in}$	$R_6 = \dot{m}_{Ef}^{out} + \dot{m}_E^{in}$
$R_4 = \dot{m}_E^{out} + \dot{m}_C^{in}$	

Table 2 Examples for Forms of Component Equations at Steady State

$|\dot{m}_C^{\text{in}}| = |\dot{m}_C^{\text{out}}| = \dot{m}_C$
$\dot{m}_C,\ h_2 = f_C(P_1, h_1, P_2,\ \text{compressor parameter})$

$|\dot{m}_{Cd}^{\text{in}}| = |\dot{m}_{Cd}^{\text{out}}| = \dot{m}_{Cd}$
$|\dot{m}a_{Cd}^{\text{in}}| = |\dot{m}a_{Cd}^{\text{out}}| = \dot{m}a_{Cd}$
$\dot{m}_{Cd},\ h_3,\ \dot{m}a_{Cd},\ ha_8 = f_{Cd}(P_1, h_1, P_2, Pa_7, ha_7, Pa_8,\ \text{condenser parameter})$

$|\dot{m}a_{Cf}^{\text{in}}| = |\dot{m}a_{Cf}^{\text{out}}| = \dot{m}a_{Cf}$
$\dot{m}a_{Cf},\ h_7 = f_{Cf}(Pa_6, ha_6, Pa_7,\ \text{condenser fan parameter})$

....

iteration and may cause the premature termination of the simulation, which would have been successful if the functions handled the instance appropriately. For instance, instead of terminating the simulation if a function call attempts to compute the density of a fluid at negative absolute pressure, the function could return the lower limit of the density and report a warning to the user. The low density may produce a high value of the residual equation and may cause the iteration to step into a more appropriate direction. The success of a simulation can often be greatly enhanced when all functions and subroutines are as robust as possible and return appropriate outputs for all input values.

Other system parameters, such as reliability or noise and vibration, can be associated with the system design through specific correlations provided by component manufacturers, though test data which are accessible to the simulation through data bases or other means available to the system designer. The general approach illustrated by the example of a vapor-compression system can be applied to all thermal systems by substituting the junction properties, flow rates, and component models with the appropriate variables and models.

Equation Solvers
Simultaneous equation solving is frequently an integral part of any mathematical model. The available equation solvers can be classified based on the types of equations they solve, as follows:

1. *Simultaneous linear equations.* These solve problems of the form $\mathbf{AX} = \mathbf{B}$, where $\mathbf{A}$, $\mathbf{X}$, and $\mathbf{B}$ are matrices of order $\mathbf{m} \times \mathbf{n}$, $\mathbf{m}$, and $\mathbf{n}$, respectively. Many routines are available for these types of problems, such as LU decomposition and QR decomposition (Press et al., 2002).

2. *Simultaneous nonlinear equations.* These methods are discussed here.

One-dimensional equations (problems with one variable and one unknown) are seldom encountered in thermal systems simulation and optimization. This section discusses gradient-

Table 3 Simplified Residual Equations for System Without Splits and Mergers

$0 = \dot{m}_C^{\text{out}} + \dot{m}_{Cd}^{\text{in}}$	$R_2 = -M_{\text{System}} + M_C + M_{Cd} + M_{Ex} + M_E$
$0 = \dot{m}_{Cd}^{\text{out}} + \dot{m}_{Ex}^{\text{in}}$	$R_3 = \dot{m}_{Cf}^{\text{out}} + \dot{m}a_{Cd}^{\text{in}}$
$0 = \dot{m}_{Ex}^{\text{out}} + \dot{m}_E^{\text{in}}$	$R_4 = \dot{m}_{Ef}^{\text{out}} + \dot{m}a_E^{\text{in}}$
$R_1 = \dot{m}_E^{\text{out}} + \dot{m}_C^{\text{in}}$	

Table 4 Schemes for Calculation of Iteration Step

Scheme Name	Step Calculation	Comments
Steepest descent	$\mathbf{s}_{sd} \equiv \mathbf{s}_i = -\mathbf{Gf}_i$	Search is in the direction in which $\mathbf{f}$ decreases rapidly.
Newton-Raphson	$\mathbf{s}_n \equiv \mathbf{s}_i = -\mathbf{H}_i\mathbf{f}_i$ $\alpha = 1$	The most basic method of all. Very efficient, with quadratic convergence close to the solution. Involves the calculation of $\mathbf{H}$ at every iteration.
Broyden's method (unmodified)	$\mathbf{s}_i = -\mathbf{H}_i\mathbf{f}_i$ α: $\|\mathbf{f}_{i+1}\| < \|\mathbf{f}_i\|$ $\mathbf{H}_{i+1} = \mathbf{H}_i - (\mathbf{H}_i\mathbf{y}_i - \mathbf{s}_i\alpha_i)\mathbf{s}_i^T\mathbf{H}_i / \mathbf{s}_i^T\mathbf{H}_i\mathbf{y}_i$	The inverse of the Jacobian is required only at the first step of iterations. In subsequent steps, it is updated using the strategy shown.
Levenberg Marquardt method	$\mathbf{s}_{lm} \equiv$ Obtained by solving $(\mathbf{J}^T\mathbf{J} + \mu\mathbf{I})\mathbf{s}_{lm} = -\mathbf{Gf}$ μ is damping parameter, updated at each step.	Method used for nonlinear least-squares problem and nonlinear equation solving. Can be combined with secant updates to avoid repeated Jacobian calculations.
Powell's dogleg method	Alternates between the Newton step and the steepest descent step. Radius of trust region updated at every iteration (Dennis & Schnabel, 1996).	Based on the concept of trust region. Can be combined with secant updates to avoid repeated Jacobian calculations.

based algorithms for solving multiple simultaneous equations. Other methods for one-dimensional root finding are available but not discussed here. The reader is referred to Press et al. (2002).

The general algorithm for a gradient-based nonlinear equation solver can be summarized as follows:

1. Get an initial guess $\mathbf{x}_0$
2. Get an initial value for Jacobian matrix $\mathbf{J}$. $\mathbf{J}$ is the transpose of the matrix of the gradients
3. $\mathbf{x}_i = \mathbf{x}_0$
4. Compute new direction and step $\mathbf{s}$
5. Compute new $\mathbf{x}$ and the length α

$$\mathbf{x}_{i+1} = \mathbf{x}_i + \alpha\mathbf{s}_i$$

Table 5 Different Stopping Criteria Used in Nonlinear Equation Solving

Stopping Criteria	Representation	Comments
Function residuals	$\|\mathbf{f}\| <= \varepsilon$ ε is some tolerance value	Most used criteria
Change in solution in successive steps	$\text{rel}(\mathbf{x}_i)_j <= \varepsilon$ $\text{rel}(x_i)_j = \dfrac{\|(x_i)_j - (x_{i-1})_j\|}{\max\{\|(x_i)_j\|, \text{typ}(x_i)_j\}}$ $(x_i)_j = j^{\text{th}}$ component of x in the ith iteration.	Indication that the algorithm has stalled
Maximum value of the components of $\mathbf{f}$	$\|\mathbf{f}\|_\infty <= \varepsilon$	

6. Test convergence

7. Compute $\mathbf{J}_{i+1}$

8. Repeat from step 4.

Several schemes are available for calculating the new direction $\mathbf{s}$, the step length α, and the convergence criteria and for computing the Jacobian for subsequent time steps. A summary of all of these schemes is provided in the following text. This list of equation solving schemes is not meant to be complete.

Let

$\mathbf{G}$ denotes the gradient matrix of $\mathbf{f}$.

$\mathbf{H}$ denotes the inverse of $\mathbf{J}$, i.e., $\mathbf{J}^{-1}$

$\mathbf{f}$ denotes the vector of function values and $\|\cdot\|$ denotes the Euclidian norm

$\mathbf{y}_i = \mathbf{f}_{i+1} - \mathbf{f}_i$

Other methods based on the second derivatives are also available in the literature.

Cost Estimation

In contrast to engineering-level system simulation, cost estimation is less challenging on a mathematical level. The challenge in accurately estimating the cost of a thermal system arises from the availability and uncertainty of cost data. Cost data vary with the source that they are obtained from, they vary in time, and they may even be subject to the negotiation skills of the system designer. While most companies can accurately predict the cost of in-house manufacturing, equipment cost estimates from suppliers are less accurate. Fuel and other operating costs probably represent the most volatile cost category. Construction and instal-lation costs can contribute significantly to the total costs, depending on size and type of thermal system.

A large number of costs for components in thermal systems are provided in cost-estimating guides such as *Building Construction Cost Data* (Waier, 2004), *Chemical Engineering, Vol. 6: Chemical Engineering Design* (Sinnott, 2003), and *Thermal Design and Optimization* (Bejan et al., 1996) along with methods for estimating fuel, operating, con-struction, and installation costs. These references generally provide component cost data associated with characteristic parameters for the equipment, such as heat-transfer area as a parameter for the costs of the heat exchanger. Whenever possible it is recommended to obtain cost data directly from the equipment supplier.

Cost data typically present a discontinuous variable in the system simulation, since they often do not scale continuously with equipment size. The optimization procedure either can interpolate over the discontinuities and round the optimization result to the closest available equipment size or can treat the cost as a discontinuous function. In this case, it is necessary to use an optimization driver that is capable of handling discontinuous (discrete) variables, such as genetic algorithms.

2.2 Optimization Drivers

An engineering optimization process in its simplest form can be visualized as shown in Fig. 2. There are two basic components, the optimization driver and the simulation tool. The optimization driver is a numerical implementation of some optimization algorithm. This optimization driver needs a function that can, with a given set of inputs (the design variables over which we optimize), provide the output (the objective value that we wish to optimize) and any other relevant simulation information, such as constraints.

Figure 2 Engineering optimization process.

From an implementation point of view, the above approach also demonstrates the power of component-based computer code development. Here we assume that eventually all of the optimization algorithms and the system and component models are used in the form of computer code. Ideally, we would want to have a library of optimization routines or drives from which we can choose an optimization routine or adopt a hybrid approach wherein the intermediate results of one routine are fed to another more effective routine. Similarly, we would want to have a library of component models, for example, a shell-and-tube heat exchanger model, an air-cooled heat exchanger model, a turbine model, a compressor model, etc. These models can then be put together to assemble a thermal system or can be used individually to evaluate component performance. This component-based approach allows us to achieve system-level objectives with the freedom of changing the lowest level (individual component) variables.

The simulation tool is the numerical model of the thermal system or the component, for example, the model of a vapor compression cycle or the model of an air-cooled heat exchanger.

Choosing an Optimization Driver

After having a library of optimization routines and the required component models to design and optimize a system, the next logical question is which optimization routine to choose. This section elaborates on several criteria that come into play when making such a decision.

Objectives. Objectives are performance measures that the designer wants to optimize. A vapor compression cycle system coefficient of performance (COP) is an example of objective. Another example is the cost of the system. The objective function can be linear or nonlinear, continuous or discontinuous in its domain. The designer can have an analytical expression for the objective function or use a result of model simulation.

Independent Variables. Independent variables are the variables over which the problem is optimized. These are the variables that are changed/varied during an optimization process to arrive at an optimal solution. The continuity of the independent variables needs to be considered. For example, the tube length of a heat exchanger is a continuous variable, whereas the fan model number in an air-handling system is a discrete variable. Whether an objective function value exists for all possible values of the independent variables or not, this needs to be gracefully conveyed to the optimization driver so that it can proceed in alternative search directions.

Derivative/Gradient Information. The gradient or the derivative information is required to improve the estimates of a solution in a nonlinear equation solving or an optimization process. Gradient computation can be simple if an analytical expression for the gradients is available. Alternatively the designer can use a finite difference technique. Using a finite difference technique will involve additional function calls of the objective function, which might not be feasible when the objective function is computationally expensive.

Number of Objectives. Many real-world problems are multiobjective. Technically multiobjective optimization is very different from single-objective optimization. In single-objective optimization the driver tries to find a solution that is usually the global minimum or maximum. In a multiobjective optimization problem, there may not exist one solution that is superior to other solutions with respect to all the objectives. As a result the designer has to make a trade-off between such solutions. Many a times, in a multiobjective optimization problem, there exists a set of solutions that is better than the remaining solutions in the design space with respect to all the objectives, but within this set, the solutions are not better than each other with respect to all the objectives. Such a set is called the Pareto set or the trade-off set and the solutions are called Pareto solutions, trade-off solutions, or nondominated solutions. The choice of different optimization routines is covered by A. Ravindran and G.V. Reklaitis in Chapter 24 of the Materials and Mechanical Design volume of this handbook.

The case study discussed in Section 3.1 will provide a practical example of the above decisions. In conclusion of this section, a relatively new search and optimization technique termed genetic algorithms is introduced, which is also used in the case study discussed in Section 3.1.

Genetic Algorithms

Genetic algorithms (GA), first put forth by John Holland (1996) are part of a broader class of evolutionary computation methods. Based on the principal of natural evolution, they mimic on a mathematical level the biological principals of population: generations, inheritance, and selection of individuals based on the survival of the fittest.

Genetic algorithms maintain a population of candidate solutions. Each of these candidates is evaluated for their fitness in terms of the corresponding objective function value. Then the candidates with the best fitness are transformed, based on "crossover" and "mutation" into new candidate solutions. Thus, a new population is created and normally the fitness of the best individual increases from generation to generation.

Some of the advantages of genetic algorithms are explained below:

1. They need only one scalar value, which is the fitness or the objective value of the function that is being optimized. Thus, the objective function is a black box object as far as the genetic algorith is concerned.

2. They do not need any gradient, i.e., first- or second-derivative information. This can result in significant computation savings, if the objective function is computationally expensive.

3. Genetic algorithms maintain a population of candidate solutions, i.e., they simultaneously search in multiple directions as opposed to deterministic search methods that search in one direction at a time.

4. They can handle discrete and continuous variables at the same time.

Representation and Genetic Operators. In a genetic algorithm (binary coded) a given candidate solution is known as a chromosome, and is represented as a series of binary digits (viz. 0's and 1's). These chromosomes are manipulated via genetic operators of selection, mutation, and crossover to form new chromosomes or candidate solutions. More information on genetic algorithms can be found in D. E. Goldberg's book listed in the Bibliography.

Case 2 in the next section demonstrates the use of genetic optimization algorithms for multiobjective optimization problems.

3 METHODOLOGY

The optimization of thermal systems requires the repeated evaluation of physics-based system simulations and cost functions. The number of system evaluations ranges from a few dozen for single-objective low-dimensional optimization tasks with deterministic algorithms to tens of thousands of evaluations for multiobjective, high-dimensional tasks with nondeterministic schemes. Especially for optimization tasks with large numbers of system evaluations, it is critical to reduce the computational time for the system evaluation by implementing fast simulation methods or by increasing the computer capacity, possibly by parallel computing.

Whenever possible it is recommended to break out separate optimization tasks from the entire system simulation. This method is especially helpful for complex thermal systems. At a stage when the major technology decisions for a system are identified and the optimization focuses on the design of individual component parameters it is often possible to optimize functional groups of the thermal system under the assumption of constant boundary condition between the functional group and the rest of the system. In the design of the vapor compression system from Section 2.1 the optimization task may be to find an evaporator–fan combination that provides a certain cooling capacity at minimum cost. Instead of evaluating the performance of the entire system for each evaporator-fan combination during the optimization, the designer can optimize the functional group evaporator, fan, and fan motor. By assuming reasonable values for the evaporator refrigerant inlet pressure and mass flow rate and minimizing the refrigerant pressure drop and cost of the functional group, the designer can find a design close to the optimum. The refrigerant inlet pressure and mass flow rate can be updated repeatedly by evaluating the system performance with the design found in the previous optimization of the functional group. This method can reduce the number of entire system evaluations. However, it may result in a higher number of component evaluations of the functional group and the designer must base her/his decision for the optimization strategy on experience and sound judgment.

The following case studies illustrate optimization strategies for various thermal systems. The case studies are presented as examples to guide the designer in the development of successful optimization applications. While the general scheme for most optimization tasks is identical—find a system with maximum performance and minimum costs—each optimization task has its individual particularities that make a generalized approach very challenging. In the end the design engineer must use his/her sound understanding of the underlying physics, his/her experience, and creativity.

3.1 Case Studies

Case 1—Constrained Single-Objective Optimization of an Automotive Air-Conditioning System

The objective of this example is to maximize the efficiency measured as coefficient of performance (COP) of an automotive air-conditioning system. The constraint is the cooling capacity and the overall volume of the heat exchangers. Space is a valuable commodity in automotive applications and the space available for the heat exchangers of the air-conditioning system usually cannot be changed once the chassis and engine designs are completed.

The optimization follows the approach of optimizing functional groups separated from the rest of the system as laid out in the introduction of Section 3. While the overall goal of the optimization is maximum efficiency of the system, the heat exchangers are optimized separately from the rest of the system. Three performance parameters of the heat exchangers affect the system efficiency: Heat load, refrigerant pressure drop, and air-side pressure drop. The heat load is a constraint determined by the capacity of the system and thus not an optimization objective. The refrigerant and air-side pressure drops are combined in a fitness function f of the form

$$f = \alpha \, \Delta P_{\text{Refrigerant}} + \beta \, \Delta P_{\text{Air}}$$

where α and β are weight functions, which must be determined by the designer based on experience or other parameters. The component with the smallest fitness value will perform best in the air-conditioning system.

The optimization is subject to the constraints of a given heat load and a maximum overall volume of the heat exchangers. The heat load for the evaporator is given by the required cooling capacity. The heat load for the condenser is given by the capacity and an estimated energy efficiency of the system:

$$\dot{Q}_{\text{Condenser}} = \dot{Q}_{\text{Evaporator}} + \dot{W}_{\text{Compressor}}$$

$$\dot{W}_{\text{Compressor}} = \dot{Q}_{\text{Evaporator}} / \text{COP}$$

While the COP of the system depends on the performance of the heat exchangers and is thus not known beforehand, an experienced designer can start with a good guess for the efficiency, perform the optimization, and update the guess to find a solution closer to the global optimum. For this case study only the first step of this iteration is shown.

The following geometric parameters are variables in this optimization (microchannel heat exchangers are used for the condenser and evaporator):

- Fin spacing
- Fin thickness
- Number of tubes per pass
- Number of ports per tube
- Port height

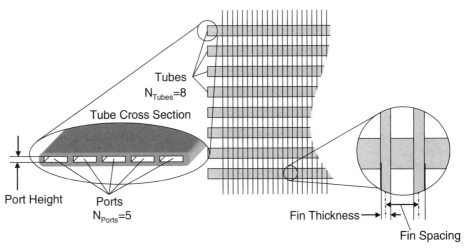

Figure 3 Geometric parameters of microchannel heat exchanger optimization.

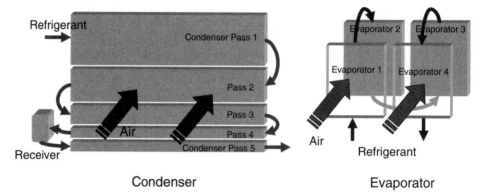

Figure 4 Condenser and evaporator geometry.

Figure 3 illustrates the geometric parameters of the microchannel heat exchangers. The heat exchangers are simulated with a detailed, segmented model. The heat exchanger geometry is coded as a binary string and passed to a genetic algorithm. Figure 4 illustrates the condenser and evaporator geometry. Figure 5 shows the network representation of the refrigerant circuit. The starting point for the optimization is the design of commercially available automotive condensers and evaporators.

Figure 6 shows the pressure drop and fitness value of the evaporator over the generation number. Note that the goal of the optimization was to minimize the value of the fitness function.

Table 6 lists the geometric parameters of the original system and the result of the optimization. Figure 7 shows the efficiency and capacity of the original and optimized systems at various temperature conditions and engine speeds. Figure 8 shows the relative performance improvement of the automotive air-conditioning system due to the optimized heat exchangers.

Case 2—Constrained Multiobjective Optimization of a Condensing Unit

This study performs a multiobjective optimization of a fan-coil unit with respect to minimizing cost and maximizing heat rejection capacity. The condensing unit in consideration is a traditional tube–fin heat exchanger. The individual components of the condensing unit include tubes, fins, fans, cabinet, etc. There is also a manufacturing cost associated with these components.

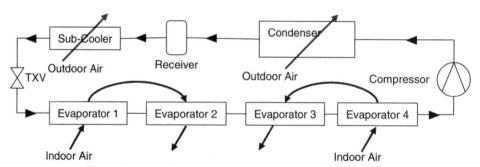

Figure 5 Network representation of automotive refrigerant circuit.

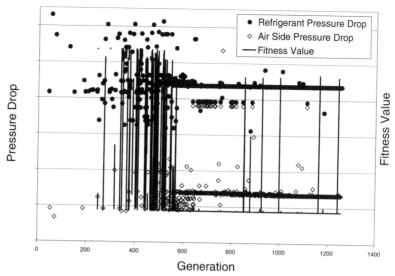

Figure 6 Performance of evaporator over generation number.

The independent variables are

Tube diameter (OD). Off-the-shelf four different tube sizes are available and are being used in this study. The tube diameter sizes are not continuous; as a result there are four discrete choices.

Fin spacing (FPI). Expressed in terms of fins per inch. The study uses 11 different values ranging from 6 to 16 fins per inch.

Tube length. Tubes are cut from a coil of tubes; as a result any tube length is possible. This is a continuous variable. From a heat exchanger point of view, the heat rejection capacity increases with tube length, but so does the cost.

Fan models (fan ID). For the particular coil, 20 different fan models are available, along with the required performance data, such as static pressure drop, noise, power consumption, and frame width.

Number of fans (NFan). The baseline coil length is fairly long, and hence multiple fans are required to drive the air flow across the coil.

Number of parallel circuits. The coil comprises of several parallel refrigerant circuits, with each circuit having a fixed number of tubes. The number of parallel circuits affects the coil height.

Table 6 Geometric Properties of Original and Optimized Heat Exchangers

	Fin Spacing (mm)	Fin Thickness (mm)	Port Height (mm)	No. of Tubes per Pass	No. of Ports per Tube
Original condenser	1.05	0.15	1	18/8/7/4/4	1
Optimized condenser	0.82	0.11	2.48	7/11/5/3/3	4
Original evaporator	0.81	0.15	2.9	9	2
Optimized evaporator	1.31	0.104	2.2	8	4

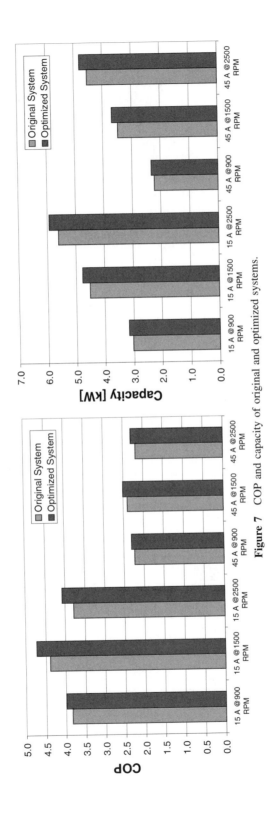

Figure 7 COP and capacity of original and optimized systems.

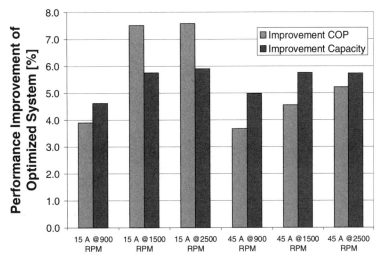

Figure 8 Relative performance improvement of optimized systems.

The constraints in this problem include manufacturing as well as performance-based constraints.

Constraint 1. The total combined width of the fans should be less than the specified cabinet width, so that the fans can fit into the cabinet.

Constraint 2. The fan chosen should be able to provide the required static pressure head for the coil, i.e., the fan pressure head should be equal to the air pressure drop through the coil.

Constraint 3. The coil height must be less than a specified maximum height.

Constraint 4. The refrigerant side pressure drop must be within acceptable lower and upper limits.

For this problem a multiobjective genetic algorithm is used. Figure 9 shows the representation of a single condensing unit as seen by the genetic algorithm.

As seen from the independent variables and the constraints the problem has two objectives, four constraints, one continuous variable, and four discrete variables. This is an example of constrained multiobjective optimization with mixed variables. A multiobjective genetic algorithm is used for this problem.

In Fig. 10, the inputs are generated by the optimization algorithms and are supplied to the condenser model. The condenser model is a very detailed simulation based on a segmented approach. After the condenser model is executed the outputs are transferred to the multiobjective genetic algorithm. The condenser model is coupled with the optimization algorithm, wherein the model is evaluate repeatedly with different input variables until the termination criteria of the genetic algorithm is satisfied. Figure 11 shows the pseudo code for the multiobjective genetic algorithm.

Nt	OD	Fan ID			FPI		NFan	Tube Length						
1	0	0	1	0	1	1	0	0	0	1	1	0	0	0	0	1	1	1	1	0	0	0	0	0

Figure 9 Binary representation of a condensing unit in a genetic algorithm.

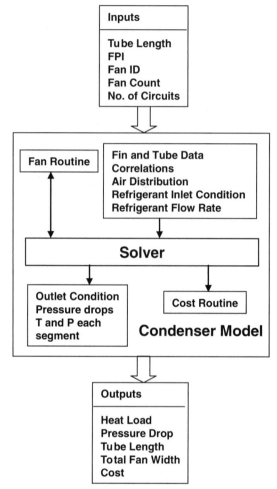

Figure 10 Flow chart of an air-cooled heat exchanger model.

Considering all the choices, the total solutions space consists of over 125,000 different condensing units. The genetic algorithm evaluated only 5500 of them, which is 4.4% of the total solution space.

The results of the optimization runs are shown in Figs. 12–14. Figure 12 shows how the number of Pareto solutions increases with respect to algorithm generations. This is consistent with the principal of natural evolution where successive generations are better or fitter than the previous. Figure 13 shows the Pareto curve for the solution. It can be seen that there are several different choices for the designer to pick from and the decision will have to be based on other factors, such as fan power requirements and pressure drops. The baseline case is also shown in Fig. 13. For the same heat rejection capacity it is possible to reduce the coil cost by 16% or, for the same coil cost, the heat rejection capacity can be increased by 12%. Figure 14 shows other condensing units evaluated by the genetic algorithm, which were found infeasible because they violated one or more of the constraints.

Start
Initialize Condenser Model
Initialize GA
While (GA not done) {
    g = g + 1 "Update generation count"
    Evaluate {
        Decode Variables
        Run Condenser Model
        Calculate constraint violation
        Return (Cost, Heat load) and constraint violation
    }
    Perform Non-Dominated Sorting
    Assign Rank
    Assign Fitness
    Perform Selection, Crossover & Mutation
}
Print Results
End

Figure 11 Pseudo code for multiobjective genetic algorithm used in case study 1.

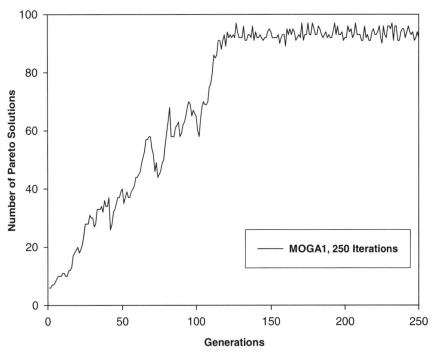

Figure 12 Number of Pareto solutions versus genetic algorithm generations.

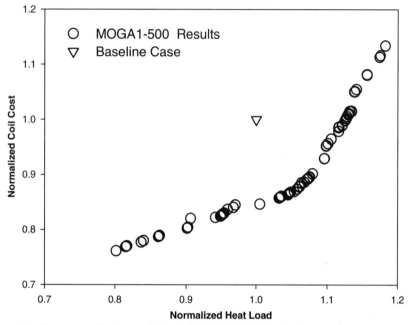

Figure 13 Pareto curve for the results from the multiobjective optimization of a condensing unit.

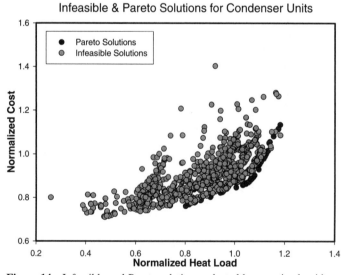

Figure 14 Infeasible and Pareto solution evaluated by genetic algorithm.

For all the given solutions the average reduction in cost is found to be 10%, while the average increase in the heat rejection capacity is found to be 7%. It can be concluded that genetic algorithms are very powerful tools for mixed-variable optimization.

SYMBOLS

COP	coefficient of performance
f	function
G	gradient of function vector
H	inverse of Jacobian **J**
h, ha	Specific enthalpy, specific enthalpy of air
I	Identity matrix
J	Jacobian matrix
$M, \dot{m}$	mass, mass flow rate
$P, Pa, \Delta P$	pressure, air pressure, pressure difference
$\dot{Q}$	heat flow rate
R	residual
W	work
$\dot{W}, P$	Power
x	vector
α, β	Step length, weight function
μ	Damping parameter
ε	Tolerance value
$\|\cdot\|, \|\cdot\|_\infty$	Euclidian norm, infinity norm

BIBLIOGRAPHY

Bejan, A., M. J. Moran, and G. Tsatsaronis, *Thermal Design and Optimization,* Wiley, New York, 1996.

Chaudhari, P. D., U. M. Diwekar, and J. S. Logsdon, "An Automated Approach for the Optimal Design of Heat Exchangers," *Industrial and Engineering Chemistry Research,* **36,** 3685–3693 (1997).

Deb, K., *Multi-Objective Optimization Using Evolutionary Algorithms,* Wiley Interscience Series in Systems and Optimization, Wiley, Chichester, UK, 2001.

Dennis, J. E., and R. B. Schnabel, *Numerical Methods for Unconstrained Optimization and Nonlinear Equations, Classics in Applied Mathematics,* Society of Industrial and Applied Mathematics, 1996.

Fax, D. H., and R. R. Mills, Jr., Generalized Optimal Heat-Exchanger Design, *ASME Gas Turbine and Power Division Semi-annual Meeting,* Paper No 56-SA-19, June 1956.

Fonseca, C. M., and P. J. Fleming, "Genetic Algorithms for Multiobjective Optimization: Formulation, Discussion and Generalization," *Proceedings of the Fifth International Conference on Genetic Algorithms, San Mateo, CA, USA, 1993,* Morgan Kaufmann, 1995, pp. 416–423.

Fonseca, C. M., and P. J. Fleming, "Multiobjective Genetic Algorithms Made Easy: Selection Sharing and Mating Restriction," *First International Conference on 'Genetic Algorithms in Engineering Systems: Innovations and Applications' GALESIA,* Publ. No. 414, IEEE, London, UK, 1995, pp. 45–52.

Fonseca, C. M., and P. J. Fleming, "Multiobjective Optimization and Multiple Constraint Handling with Evolutionary algorithms—Part I: A Unified Formulation, Part II: Application Example," *IEEE Transactions on Systems, Man and Cybernetics—Part A: Systems and Humans,* **28**(1) (1998).

Fonseca, C. M., and P. J. Fleming, "An Overview of Evolutionary Algorithms in Multiobjective Optimization," *Evolutionary Computation,* **3**(1):1–16 (Spring 1995).

Goldberg, D. E., *Genetic Algorithms in Search Optimization and Machine Learning,* Addison-Wesley, 1989.

Hedderich C. P., M. D. Kelleher, and G. N. Vanderplaats, Design and Optimization of Air-Cooled Heat Exchangers, *ASME Journal of Heat Transfer,* **104,** 683–690 (1982).

Holland, J. H., *Adaptation in Natural and Artificial Systems: An Introductory Analysis with Applications to Biology, Control, and Artificial Intelligence,* MIT Press, Cambridge, MA, 1996.

Jiang, H., V. Aute, and R. Radermacher, "A User-friendly Simulation and Optimization Tool for Design of Coils," *Ninth International Refrigeration and Air Conditioning Conference at Purdue,* July 2002.

Narayanan, S., and S. Azarm, "On Improving Multiobjective Genetic Algorithms for Design Optimization, *Structural Optimization,* **18,** 146–155, (1999).

Press, W. H., S. A. Teukolsky, W. T. Vetterling, and B. P. Flannery, *Numerical Recipes in C++: The Art of Scientific Computing,* 2d ed., Cambridge University Press, 2002.

Reklaitis, G. V., A. Ravindran A, and K. M. Ragsdell, *Engineering Optimization—Methods And Applications,* Wiley Interscience Publication, Wiley, New York, 1983.

Sinnott, R. S., *Chemical Engineering,* Butterworth-Heinemann, Oxford, UK, 2003.

Srinivas, N., and K. Deb, "Multiobjective Optimization Using Nondominated Sorting in Genetic Algorithms," *Journal of Evolutionary Computation,* **2**(3) 221–248 (1995).

Tayal, M. C., Y. Fu, and U. M. Diwekar, "Optimal Design of Heat Exchangers: A Genetic Algorithm Framework," *Industrial and Engineering Chemistry Research,* **38,** 456–467 (1999).

Van den Bulck, E., "Optimal Design of Crossflow Heat Exchangers," *ASME Journal of Heat Transfer,* **113,** 341–347 (1991).

Waier, P. R., *Building Construction Cost Data,* 62nd annual edition, R. S. Means, Kingston, MA, 2004.

Wu, J., and S. Azarm, "Metrics for Quality Assessment of a Multiobjective Design Optimization Solution Set," *Journal of Mechanical Design, Transactions of the ASME,* **123,** 18–25 (2001).

PART 2
POWER

CHAPTER **16**
COMBUSTION

Eric G. Eddings
Department of Chemical Engineering
University of Utah
Salt Lake City, Utah

1 FUNDAMENTALS OF COMBUSTION

1.1 Air–Fuel Ratios

Combustion is rapid oxidation, usually for the purpose of changing chemical energy into thermal energy—heat. This energy usually comes from oxidation of carbon, hydrogen, sulfur, or compounds containing C, H, and/or S. The oxidant is usually O_2—molecular oxygen from the air.

One can perform basic chemical reaction balancing to permit determination of the air required to burn a fuel. For example, consider the following reaction:

$$CH_4 + 2O_2 \rightarrow CO_2 + 2H_2O$$

where the units are moles; therefore, 1 mole of methane (CH_4) produces 1 mole of CO_2; or 1000 moles of CH_4 requires 2000 moles of O_2 and produces 2000 moles of H_2O. Knowing that the atomic weight of C is 12, H is 1, N is 14, O is 16, and S is 32, one can determine molecular weights for each of the species involved in the reaction, and thus predict weight flow rates: 16 lb/hr CH_4 requires 64 lb/hr O_2 to burn to 44 lb/hr CO_2 and 36 lb/hr H_2O. If the oxygen for combustion comes from air, it is necessary to know that dry air is 20.99% O_2 by volume and 23.20% O_2 by weight, most of the remainder being nitrogen (N_2).

It is convenient to remember the following ratios:

$$Air/O_2 = 100/20.99 = 4.76 \text{ by volume}$$

$$N_2/O_2 = 3.76 \text{ by volume}$$

$$\text{Air/O}_2 = 100/23.20 = 4.31 \text{ by weight}$$

$$\text{N}_2\text{/O}_2 = 3.31 \text{ by weight}$$

If there is significant humidity in the combustion air, or if some combustion products have been recirculated into the combustion air stream, the moisture or other products will dilute the oxygen concentration. Therefore, the diluted oxygen concentration should be determined and used in the previous ratios, as well as in the following calculations.

Rewriting the previous formula for combustion of methane gives

$$\text{CH}_4 + 2\text{O}_2 + 2(3.76)\text{N}_2 \rightarrow \text{CO}_2 + 2\text{H}_2\text{O} + 2(3.76)\text{N}_2$$

or

$$\text{CH}_4 + 2(4.76)\text{air} \rightarrow \text{CO}_2 + 2\text{H}_2\text{O} + 2(3.76)\text{N}_2$$

Table 1 lists the amounts of air or oxygen required for stoichiometric (quantitatively and chemically correct) combustion of a number of pure fuels, calculated by the above method.

The stoichiometrically correct (perfect, ideal) air/fuel ratio from the above formula is therefore $2 + 2(3.76) = 9.52$ volumes of air per volume of the fuel gas. More than that is called a "lean" ratio, and includes excess air and produces an oxidizing atmosphere. For example, if the actual air/fuel ratio were 10:1, the % excess air would be

$$\frac{10 - 9.52}{9.52} \times 100 = 5.04\%$$

Some fuels (e.g., coal or fuel oil) do not have a simple molecular formula, and cannot be analyzed by the method shown above. These fuels are typically described by an elemental analysis, weight percentages of C, H, O, N, and S, as well as percentages of moisture and

Table 1   Proper Combining Proportions for Stoichiometric Combustion[a]

Fuel	vol O_2 / vol fuel	vol air / vol fuel	wt O_2 / wt fuel	wt air / wt fuel	ft³O_2 / lb fuel	ft³ air / lb fuel	m³O_2 / kg fuel	m³ air / kg fuel
Acetylene, C_2H_2	2.50	11.9	3.08	13.3	36.5	174	2.28	10.8
Benzene, C_6H_6	7.50	35.7	3.08	13.3	36.5	174	2.28	10.8
Butane, C_4H_{10}	6.50	31.0	3.59	15.5	42.5	203	2.65	12.6
Carbon, C	—	—	2.67	11.5	31.6	150	1.97	9.39
Carbon monoxide, CO	0.50	2.38	0.571	2.46	6.76	32.2	0.422	2.01
Ethane, C_2H_6	3.50	16.7	3.73	16.1	44.2	210	2.76	13.1
Hydrogen, H_2	0.50	2.38	8.00	34.5	94.7	451	5.92	28.2
Hydrogen sulfide, H_2S	1.50	7.15	1.41	6.08	16.7	79.5	1.04	4.97
Methane, CH_4	2.00	9.53	4.00	17.2	47.4	226	2.96	14.1
Naphthalene, $C_{10}H_8$	—	—	3.00	12.9	35.5	169	2.22	10.6
Octane, C_8H_{18}	—	—	3.51	15.1	41.6	198	2.60	12.4
Propane, C_3H_8	5.00	23.8	3.64	15.7	43.1	205	2.69	12.8
Propylene, C_3H_6	4.50	21.4	3.43	14.8	40.6	193	2.54	12.1
Sulfur, S	—	—	1.00	4.31	11.8	56.4	0.74	3.52
Coal, bituminous (avg.)	—	—	2.27	9.80	26.9	128	1.68	7.96

[a]Reproduced with permission from *Combustion Handbook.*[1]

noncombustible material (ash). In such a case, the required oxygen can be determined by calculating the number of moles for the elements of interest for a particular mass of fuel (say, 1 pound). Then, the required amount of oxygen can be determined for the moles of each element using the following balanced reactions:

$$C + O_2 \rightarrow CO_2$$

$$H + \tfrac{1}{4}O_2 \rightarrow \tfrac{1}{2}H_2O$$

$$O \rightarrow \tfrac{1}{2}O_2$$

$$S + O_2 \rightarrow SO_2$$

$$N \rightarrow \tfrac{1}{2}N_2$$

The oxygen produced due to the presence of inherent oxygen in the fuel will result in an oxygen "credit" and should be subtracted from the stoichiometric oxygen requirement for complete combustion of C, H, and S. For some fuels like Fuel Oil No. 6, this oxygen will be negligible; however, for coal and wood this can be quite significant and needs to be taken in consideration when determining the stoichiometric air requirement. This type of calculation will determine the moles of oxygen required per pound of fuel, and this value can be converted to moles of air required using the ratios given above. The moles of air can be subsequently converted to mass using a molecular weight of 29.

Although a small amount of the nitrogen in the fuel can be oxidized, it is typically neglected in these calculations since it is generally a minor "sink" for oxygen. The majority of the nitrogen will combine with the N_2 from the combustion air and will pass through the system as a largely inert gas that dilutes the combustion products. This large amount of nitrogen results in a significant inefficiency in combustion systems, because much of the energy released in the combustion reactions is used to heat up this inert gas. Significant efforts have gone into the study of enriched oxygen combustion systems as a means to reduce the energy penalties associated with nitrogen in the air.

Communication problems sometimes occur because some individuals think in terms of air/fuel ratios, others in fuel/air ratios; some in weight ratios, others in volume ratios; and some in mixed metric units (such as normal cubic meters of air per metric tonne of coal), others in mixed American units (such as ft^3 air/gal of oil). To avoid such confusion, the following method is recommended.

It is more convenient to specify air/fuel ratio in unitless terms such as % excess air, equivalence ratio, or stoichiometric ratio. Those experienced in this field prefer to converse in terms of % excess air. The scientific community favors equivalence ratio or stoichiometric ratio. The stoichiometric ratio is the most intuitive to use and explain to newcomers to the field. A stoichiometric ratio (SR) of 1.0 is the correct (stoichiometric) amount; a stoichiometric ratio or SR = 2.0 air is twice as much as necessary, or 100% excess air, SR = 1.2 represents 20% excess air, and so on. Equivalence ratio (ER), widely used in combustion research, is the actual fuel/air ratio divided by the theoretical or stoichiometric fuel/air ratio. It is also the inverse of the stoichiometric ratio, or ER = 1/SR. The Greek letter phi, ϕ, is typically used for ER: $\phi < 1.0$ represents fuel lean or excess air conditions; $\phi > 1.0$ is fuel rich or air deficient conditions; and $\phi = 1.0$ is "on-ratio" or the stoichiometric point. Table 2 lists a number of equivalent terms for convenience in converting values from one method of categorization to another.

Significant levels of excess air are undesirable, because, like the residual N_2, excess air passes through the combustion process without significant chemical reaction; yet it absorbs

Table 2 Equivalent Ways to Express Fuel-to-Air or Air-to-Fuel Ratios[1]

	ϕ^a	SR^b	$\%XS^c$
Fuel rich	2.50	0.40	
(air lean)	1.67	0.60	
	1.25	0.80	
	1.11	0.90	
	1.05	0.95	
Stoichiometric	1.00	1.00	0
Fuel lean	0.95	1.05	5
(air rich)	0.91	1.10	10
	0.83	1.20	20
	0.77	1.30	30
	0.71	1.40	40
	0.63	1.60	60
	0.56	1.80	80
	0.50	2.00	100
	0.40	2.50	150
	0.33	3.00	200
	0.25	4.00	300
	0.20	5.00	400
	0.167	6.00	500
	0.091	11.00	1000
	0.048	21.00	2000

[a] Equivalence ratio.
[b] Stoichiometric ratio.
[c] % excess air.

heat, which it carries out the flue. The percent available heat (best possible fuel efficiency) is highest with approximately zero excess air (see Fig. 1); however, thermodynamic and mixing considerations dictate some level of excess air (typically 10–20%) to achieve a high level of combustion efficiency, or near-complete conversion of fuel to oxidized products (CO_2 and H_2O).

Excess fuel is even more undesirable because it means there is a deficiency of air and some of the fuel cannot be burned. This results in the emission of soot and smoke and carbon monoxide (CO). The accumulation of unburned fuel or partially burned fuel can represent an explosion hazard, as well as a pollutant emission problem.

Enriching the oxygen content of the combustion "air" above the normal 20.9% reduces the nitrogen and thereby reduces the loss due to heat carried up the stack. This also raises the flame temperature, improving heat transfer, especially that by radiation.

Vitiated air (containing less than the normal 20.9% oxygen) results in less fuel efficiency, and may result in flame instability. Vitiated air is sometimes encountered in incineration of fume streams or in staged combustion, or with flue gas recirculation.

1.2 Fuels

Fuels used in practical industrial combustion processes have such a major effect on the combustion that they must be studied simultaneously with combustion. Fuels are covered in

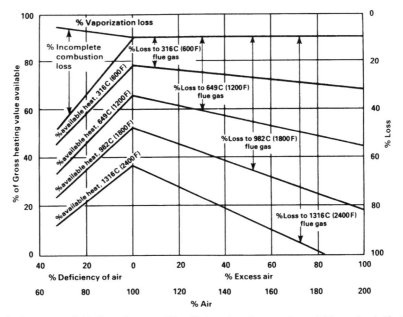

Figure 1 Percent available heat (best possible efficiency) peaks near the stoichiometric air/fuel ratio.[1]

detail in later chapters, so the treatment here is brief, relating only to the aspects having direct bearing on the combustion process.

Gaseous fuels are generally easier to burn, handle, and control than liquid or solid fuels. Molecular mixing of a gaseous fuel with oxygen need not wait for vaporization or mass transport within a solid. Burning rates are limited only by mixing rates and the kinetics of the combustion reactions; therefore, combustion can be compact and intense. Reaction times as short as 0.001 sec and combustion volumes from 10^4 to 10^7 Btu/hr·ft^3 are possible at atmospheric pressure.[2] Gases of low calorific value may be so dilute after mixing with air that their combustion rates will be limited by the mixing time.

Combustion stability means that a flame lights easily and then burns steadily and reliably after the pilot (or direct spark) is programmed off. Combustion stability depends on burner geometry, plus air and fuel flow controls that maintain the point(s) of flame initiation (1) above the fuel's minimum ignition temperature, (2) within the fuel's flammability limits, and (3) with feed speed equal to flame speed—throughout the burner's full range of firing rates and conditions. (Fuel properties are discussed and tabulated in Chapters 17–19.)

Liquid fuels are usually not as easily burned, handled, or controlled as gaseous fuels. Mixing with oxygen can occur only after the liquid fuel is evaporated; therefore, burning rates are limited by vaporization rates. In practice, combustion intensities are usually less with liquid fuels than with high calorific gaseous fuels such as natural gas.

Because vaporization is such an integral part of most liquid fuel burning processes, much of the emphasis in evaluating liquid fuel properties is on factors that relate to vaporization. One of the most critical properties is viscosity, which hinders good atomization, since atomization or the creation of small droplets is the primary method for enhancing vaporization. Much concern is also devoted to properties that affect storage and handling because, unlike gaseous fuels that usually come through public utility main pipelines, liquid fuels must be stored and distributed by the user.

The stability properties (ignition temperature, flammability limits, and flame velocity) are not readily available for liquid fuels, but flame stability is often less critical with liquid fuels.

Solid fuels are frequently more difficult to burn, handle, and control than liquid or gaseous fuels. After initial devolatilization or release of volatile matter, the combustion reaction rate depends on diffusion of oxygen into the remaining char particle, and the diffusion of carbon monoxide back to its surface, where it burns as a gas. Reaction rates are usually low and required combustion volumes high, even with pulverized solid fuels burned in suspension. Some cyclone combustors have been reported to reach the intensities of gas and oil flames.[2]

Waste or by-product fuels and gasified solids are being used more as fuel costs rise. Operations that produce such materials should attempt to consume them as energy sources or to sell them as a fuel to others. Problems with handling the lack of a steady supply, and pollution problems often complicate such fuel usage.

For the precise temperature control and uniformity required in many industrial heating processes, the burning of solids, especially the variable quality solids found in wastes, presents a critical problem. Such fuels are often left to very large combustion chambers, particularly boilers and cement kilns. When solids and wastes must be used as heat sources in small and accurate heating processes, a better approach is to gasify them to produce a synthesis gas, which can be cleaned and then controlled more precisely.

2 THERMAL ASPECTS OF COMBUSTION

The purpose of combustion in industrial applications, for the most part, is to transform chemical energy, available in various types of fuels, to thermal energy or heat to be of use in the processing of gas or liquid streams, or solid objects. Typical examples involve the heating of air, water, and steam for use in heating of other processes or equipment, the heating of metals and nonmetallic minerals during production and processing, the heating of organic streams for use in refining and processing, as well as heating of air for space comfort conditioning. For all of these, it is necessary to have a workable method for evaluating the heat available from a combustion process.

Available heat is the heat accessible for the load (useful output) and to balance all losses other than stack losses (see Fig. 2.). The available heat per unit of fuel is

$$AH = HHV - \text{Total stack gas loss} = LHV - \text{Dry stack gas loss}$$

$$\% \text{ available heat} = 100(AH/HHV)$$

where AH = available heat, HHV = higher heating value, and LHV = lower heating value, as defined in Chapter 18. Figure 3 shows values of % available heat for a typical natural gas; Fig. 4 for a typical heavy fuel oil.

Example Calculation A process furnace is to raise the heat content of 10,000 lb/hr of a load from 0 to 470 Btu/lb in a continuous furnace (no wall storage) with a flue gas exit temperature of 1400°F. The sum of wall loss and opening loss is 70,000 Btu/hr. There is no conveyor loss. Estimate the fuel consumption using 1000 Btu/ft³ natural gas with 10% excess air.

Solution: From Fig. 3, the % available heat = 58.5%. In other words, the flue losses are 100% − 58.5% = 41.5%. The sum of other losses and useful output = 70,000 + (10,000

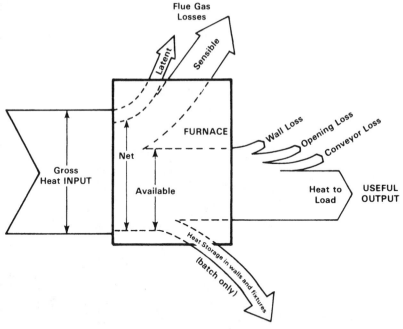

Figure 2 Sankey diagram for a furnace, oven, kiln, incinerator, boiler, or heater—a qualitative and roughly quantitative method for analyzing efficiency of fuel-fired heating equipment.

lb/hr)(470 Btu/lb) = 4,770,000 Btu/hr. This constitutes the "available heat" required. The required gross input is therefore 4,770,000/0.585 = 8,154,000 Btu/hr, or 8154 ft³/hr of natural gas (and about 81,540 ft³/hr of air).

The use of the above precalculated % available heats has proved to be a practical way to avoid long iterative methods for evaluating stack losses and what is therefore left for useful heat output and to balance other losses. For low exit gas temperatures such as encountered in boilers, ovens, and industrial dryers, the dry stack gas loss can be estimated by assuming the total exit gas stream has the specific heat of nitrogen, which is usually a major component of the "poc" (products of combustion):

$$\frac{\text{Dry stack loss}}{\text{Unit of fuel}} = \left(\frac{\text{lb dry proc}}{\text{Unit of fuel}}\right)\left(\frac{0.253 \text{ Btu}}{\text{lb poc (°F)}}\right)(T_{\text{exit}} - T_{\text{in}})$$

or

$$\left(\frac{\text{scf dry poc}}{\text{Unit of fuel}}\right)\left(\frac{0.0187 \text{ Btu}}{\text{scf poc (°F)}}\right)(T_{\text{exit}} - T_{\text{in}})$$

For a gaseous fuel, the "unit of fuel" is usually scf (standard cubic feet), where "standard" is at 29.92 in. Hg and 60°F or nm³ (normal cubic meter), where "normal" is at 1.013 bar and 15°C.

Heat transferred from combustion takes two forms: radiation and convection. Both phenomena involve transfer to a surface. Flame radiation comes from both particle and gas radiation. The visible yellow-orange light normally associated with a flame is actually from

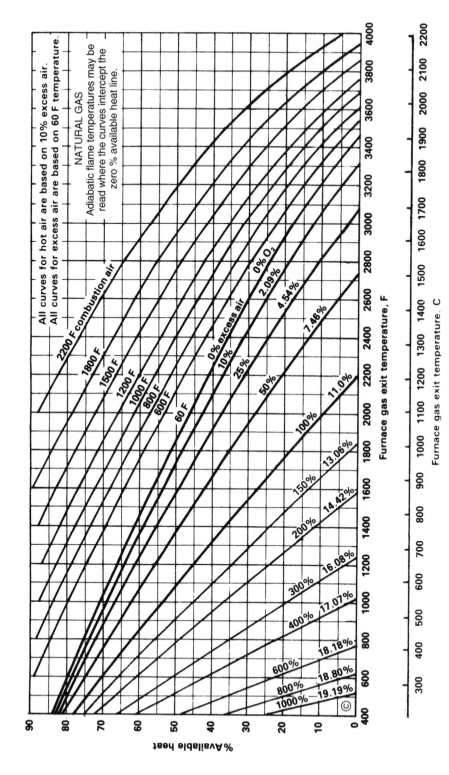

Figure 3 Available heat for 1000 Btu/ft³ natural gas. Examples: In a furnace with 1600°F flue temperature, 60°F air, and 10 % excess air, read that 54% of the gross heat input is available for heating the load and balancing the losses other than stack losses; and, at the x intercept, read that the adiabatic flame temperature will be 3310°F. If the combustion air were 1200°F instead of 60°F, read that the available heat would be 77% and that the adiabatic flame temperature would be 3760°F. It is enlightening to compare this graph with Figure 24 for oxy-fuel firing and oxygen enrichment.

582

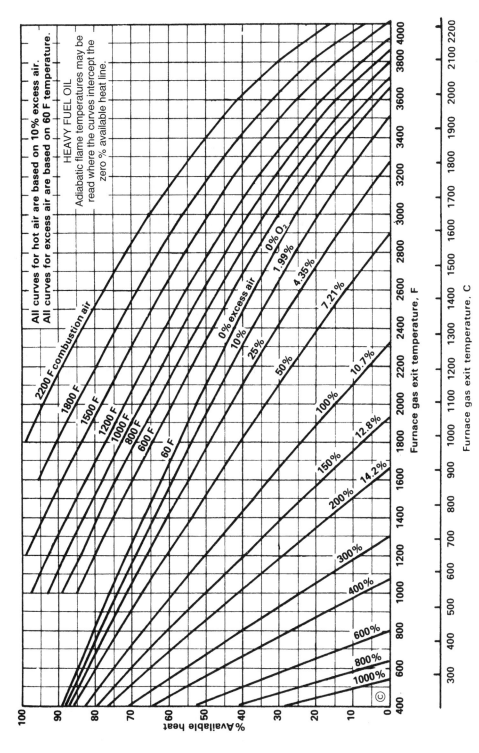

Figure 4 Available heat for 153,120 gross Btu/gal residual fuel oil (heavy, No. 6). With 2200°F gases leaving a furnace, 1000°F air entering the burners, and 10% excess air, 62% of the 153,120 is available; 100% − 62% = 38 % stack loss.

583

solid soot or char particles in the flame, and the "working" portion of this form of heat transfer is in the infrared wavelength range. Because oils have higher C/H ratios than gaseous fuels, oil flames are usually more yellow or luminous than gas flames (although oil flames can be made blue). Gas flames can be made more luminous, by a delayed-mixing burner design, for the purpose of increasing their radiating capability.

Particulate radiation follows the Stefan-Boltzmann law for solids, but depends on the concentration of particles within the flame. Estimating or measuring the particle temperature and concentration in the flame can be difficult.

Gas radiation and blue flame radiation contain more ultraviolet radiation and tend to be less intense. Triatomic gases (CO_2, H_2O, and SO_2) emit radiation that is largely infrared. Gases beyond the tips of both luminous and nonluminous flames continue to emit this gas radiation. As a very broad generalization, blue or nonluminous flames tend to be hotter, smaller, and less intense radiators than luminous flames. Gas radiation depends on the concentrations (or partial pressures) of the triatomic molecules and the beam thickness of their "cloud." Their temperatures are very transient in most combustion applications.

Convection from combustion products beyond the flame tip follows conventional convection formulas—largely a function of velocity. This is the reason for recent emphasis on high-velocity burners in some applications. Flame convection by actual flame impingement is more difficult to evaluate because (1) flame temperatures change so rapidly and are thus difficult to measure or predict, and (2) it involves extrapolating many convection formulas into ranges where good data are lacking.

Refractory radiation is a second stage of heat transfer. The refractory must first be heated by flame radiation and/or convection. A gas mantle, so-called "infrared" burners, and "radiation burners" use flame convection to heat some solid (refractory or metal) to incandescence so that they become good radiators.

3   FLAME AERODYNAMICS

The functions of a properly designed burner can be listed as follows:

1. To introduce desired quantities of fuel and air into a furnace to release a given amount of heat
2. To release heat in a specified heat release pattern
3. To provide stable ignition and safe operation
4. To induce proper mixing of fuel and air
5. To allow for sufficient turndown range
6. To allow for fuel and oxidant composition variations
7. To be simple in operation
8. To have low pollutant omission and minimum noise
9. To maximize efficiency

In short, the burner must be designed so that the flame is retained in a fixed position with respect to the burner, with a predetermined flame pattern. It should have the ability to retain the flame in position over a range of firing rate. We will delineate how burner aerodynamics is often the most important factor influencing essentially all of the above factors.

3.1 Premixed Flames

Premixed Laminar Flame Speed

The premixed laminar flame speed is the rate at which a flame front propagates through a mixture of fuel and air. It is a fundamental physico-chemical property of an oxidant fuel mixture. It is thus not influenced by burner aerodynamics. Very few industrial burners produce laminar flames and so the connection between laminar flame speed and burner throat velocity is indirect, if at all, and not direct. Figure 5 shows premixed laminar flame speeds for a variety of air/fuel mixture. These speeds are controlled by molecular diffusion of species and heat upstream, as well as the overall rate of reaction, and are thus, with the notable exception of H_2, quite low, of the order of 1–2 ft/sec. Under adiabatic conditions, which are extremely rare in industry, the flame will flash back or blow off if the flame speed exceeds or is less than the incoming flow velocity determined by the operator. Under practical laminar flow conditions, the flame front, which is very thin, say about 1 mm, adjusts its position, so that the normal velocity to it is equal to the flame speed.

The simplest theory[3] predicting laminar flame speed, v_f, uses an enthalpy balance to predict:

$$v_f \text{ proportioned to } \left(\frac{k}{\rho C_p} \right)^{1/2} (\text{RR})^{1/2} \tag{1}$$

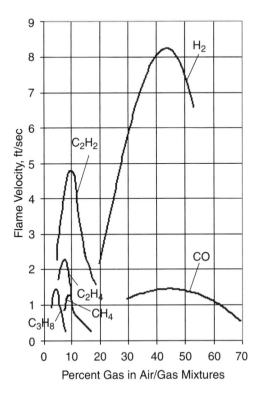

Figure 5 Laminar flame speeds.

where k = thermal conductivity of mixture
ρ = density of mixture
C_p = specific heat of mixture
RR = overall reaction rate

Thus, the flame speed depends exponentially on the flame temperature because the reaction rate has this type of dependence.

Premixed Laminar Flame Stabilization on Tubes

In practice, the tube diameter plays an important role in laminar flame stabilization. This is because the tube walls act as a heat sink, quenching the reaction. When the flow velocity exceeds the flame speed, the flame begins to lift off the end of the tube, whereupon its heat loss to the tube is decreased, its temperature increased, and the flame speed thus also increased, allowing stabilization at a new position above the tube exit. Further increase in flow velocity will blow the flame off and extinguish it, due to increased dilution by the surroundings. At low flow velocities the flame may not necessarily flash back down the tube if the tube diameter is less than half the quenching distance, the latter defined by[4]

$$d_{\text{quenching}} = \frac{k/\rho C_p}{v_f} \tag{2}$$

The quenching distance, $d_{\text{quenching}}$, is an important property, otherwise the flame would always propagate down the low-velocity regions of flow next to the tube wall. Thus, laminar premixed flames, stabilized on a tube, do allow a range in turndown, even though the adiabatic flame speed does not change. The home gas range is an example of this.

Turbulent Premixed Flames (No Recirculation)

In practice, under turbulent flow conditions, the mean incoming velocity capable of supporting a premixed flame is significantly higher than that of the laminar flame speed. This is because turbulence produces eddies that grossly distort or wrinkle the laminar flame front. Thus, while the *mean* velocity is higher than the laminar flame speed, the local velocity normal to the (fluctuating) wrinkled flame front is equal to v_f. At high Reynolds numbers, the mean turbulent flame speed is typically a factor of four or five times the laminar flame speed, or about 10 ft/s, for methane/air mixtures.

From the foregoing it is clear that without recycle of heat via radiation or convective recirculation of hot products back into the ignition zone, it is impossible to stabilize premixed flames at high burner velocities. This introduces the subject of reverse flows. For the purpose of discussion we can define premixed flames as those in which the mixing of fuel and air is rapid and does *not* control flame length.

Aerodynamics of Premixed Flames

We now consider in detail how hot product gas recirculation can be accomplished through burner adjustments and flows. It is useful to consider (1) free jets, (2) confined jets with no swirl, (3) confined jets with swirl, and (4) flow past bluff bodies.

Free Jets. Consider a turbulent jet flowing into a quiescent medium as shown in Figure 6. If the velocity along the center axis is denoted by u_m, and the mean velocity leaving the jet is u_o, the following relationship holds for distances beyond five diameters from the exit:

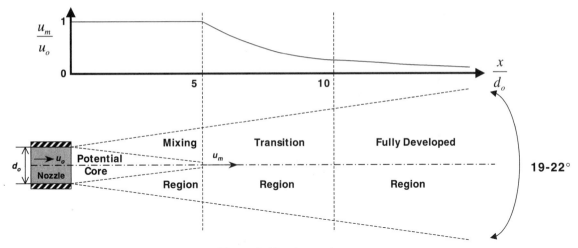

Figure 6 Free jet entrainment.

$$\frac{u_m}{u_o} = A\frac{x}{d_o}^{-1} \tag{3}$$

Thus, the velocity decay is inversely proportional to the distance downstream. The mean radial velocity has a Gaussian distribution

$$\frac{u(r)}{u_m} = \exp\left[-k\left(\frac{r}{x}\right)^2\right] \tag{4}$$

The axial velocity decrease is due to entrainment from the surrounding fluid. If this fluid is cold, the ensuing dilution effect will rapidly cool down a hot jet. The entrained fluid influences the fluid at the center line of the jet at a distance equal to five times the jet diameter. The angle of spread is approximately 19–22° and is independent of u_o. According to Ricou and Spalding[5] the rate of entrainment, m_e, is given by

$$\frac{\dot{m}_e}{\dot{m}_o} = 0.32\left(\frac{\rho_e}{\rho_o}\right)^{1/2}\frac{x}{d_o} - 1 \tag{5}$$

where ρ_e = density of entrained fluid
ρ_o = nozzle density
$\dot{m}_e$ = mass flow entrained
$\dot{m}_o$ = mass flow from nozzle
d_o = nozzle diameter

Confined Jets: Creation of Secondary Recirculation. When the jet is confined, it still entrains surrounding fluid and spreads out. However, the fluid entrained must now consist of product gases, which are recirculated back along the walls. The walls also serve to prevent the jet from spreading out too rapidly. If the walls are hot, this secondary recirculation serves to recycle hot gases back into the ignition zone, which is good for flame stability, If the walls are cold, then heat is transferred by convection and the secondary recirculation acts like a cold diluent, thus providing for poor flame stability.

Confined jets have been analyzed by Thring and Newby[6] (see Fig. 7). They assumed that a jet

1. Expands as a free jet with an angle of 19.4 degrees and thus impinges on a wall at a distance x_p.

2. Entrains as a free jet up to a point C, at position x_c (i.e., entrainment ceases and disentrainment begins) midway between the point where entrainment is zero, x_N, and the point where the jet hits the wall (see Fig. 8).

These assumptions lead to the following predictions for x_N, x_c, and x_p:

$$\frac{x_N}{d_o} = \frac{1}{0.32}\left(\frac{\rho_o}{\rho_e}\right)^{1/2} \qquad \text{(see Eq. (5))} \tag{6}$$

$$x_c = \frac{1}{2}\left[x_p + \frac{d_o}{0.32}\left(\frac{\rho_o}{\rho_e}\right)^{1/2}\right] \tag{7}$$

$$x_p = 5.85L \qquad \text{(Free jet, } a = 19.4°) \tag{8}$$

The mass entrained between N and C is the mass of secondary recirculation and is given by

$$\frac{m_e}{m_o} = \frac{m_{\text{recirc}}}{m_o} = 0.32\left(\frac{\rho_e}{\rho_o}\right)^{1/2}\frac{x_c}{d_o} - 1 \qquad \text{(see Eq. (5))} \tag{9}$$

Defining

$$\theta = \frac{d_o}{2L}\left(\frac{\rho_o}{\rho_e}\right)^{1/2} \tag{10}$$

yields

$$\frac{m_{\text{recirc}}}{m_o} = \frac{0.47}{\theta} - 0.5 \tag{11}$$

which agrees with experimental data to within about 10%.[7] The parameter θ is a similitude parameter and can be used for nozzle/chamber scale-up purposes. We can conclude that both useful analytical predictions and scale-up criteria can be used for this simple premixed

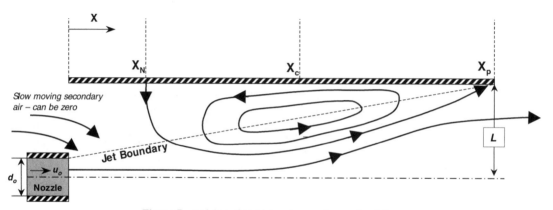

Figure 7 Axial confined jet and secondary recirculation.

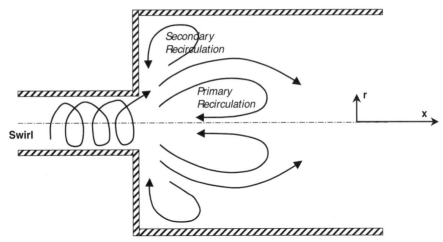

Figure 8 Creation of primary recirculation resulting from swirl.

turbulent enclosed jet. The importance of secondary recirculation for flame stabilization should be again emphasized.

Confined Jets with Swirl: Recirculation. Some recirculation is essential for flame stability. In many cases, secondary recirculation originating from confined jets is insufficient to maintain a constant supply of hot gases and/or active species to maintain ignition. In this case recirculation of hot gas must occur by some other means.

Swirl, or a tangential velocity at the burner exit, will cause a radial pressure gradient, given by

$$\frac{dP}{dr} = \rho \, \frac{v_\theta^2}{r} \qquad \text{(centrifugal force)} \tag{12}$$

if radial velocity and viscosity are neglected, but will in all cases be a positive number. This means that the pressure will be lowest in the center of the vortex. As the swirl dissipates in the combustor the pressure along the center line increases. This adverse pressure gradient prompts a reverse flow down the center, creating primary recirculation of hot gases as shown in Figure 8.

The swirl number, S, is defined as

$$S = \text{axial flow of angular momentum/axial flow of axial momentum}$$

and is given by

$$S = \frac{\left[\int_0^R \rho v_x r v_\theta r \, dr \right]}{\left[\int_0^R \rho v_x v_x r \, dr \right] R} \tag{13}$$

Unfortunately, v_θ is a strong function of r at the burner exit, and knowledge of this dependence is essential to calculate swirl numbers in practical instances. This functionality depends on how swirl is generated. The swirl number, however, is an important parameter describing the flow of swirling flames and can be used for scale-up.

For $S > 0.3$ a reverse flow can occur and increasing swirl will always shorten the flame jet and increase its spread. For weakly swirling flows ($S < 0.3$) the angle of spread is given by

$$a = 4.8 + 14S \tag{14}$$

Flame Stabilization by Bluff Bodies. When a bluff body is placed in the path of a fluid with velocity U, a low-pressure zone will form behind it and viscous forces in the boundary layer will result in flow separation. The result will be a recirculation zone or trailing wake (Fig. 9).

The blow-off velocity can be determined by equating the ignition time τ_i with the contact time in the recirculating wake, τ_c.

Thus,

$$\tau_c = \frac{D}{U_{bo}} = \tau_i \tag{15}$$

yielding

$$U_{bo} = \frac{D}{\tau_i} \tag{16}$$

$$(= 100D \text{ for } \tau_i = 10^{-2} \text{ sec})$$

Thus, for reasonable sized obstacle diameters, D, the blow-off velocity can be many times the maximum turbulent flame speed described above.

The conclusions from this section on premixed turbulent flames are

1. Recirculation of hot products or of heat is essential to maintain stable ignition at high burner velocities.

2. This can be achieved by secondary recirculation in a hot enclosure, where it can be modeled and predicted for scale-up purposes for axial jets only.

3. Swirl is always an effective technique to stabilize a flame but will also shorten it. Swirling flow cannot be easily modeled.

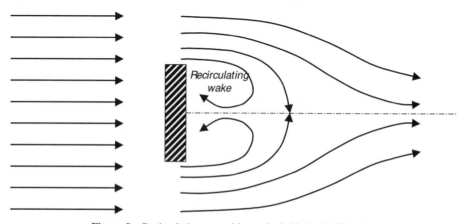

Figure 9 Recirculation caused by wake behind a bluff body.

4. Bluff body stabilization is an inexpensive method to induce recirculating flows, and allows high incoming blow-off velocities.

3.2 Diffusion-Mixed Turbulent Flames

An alternative to the premix burner is the diffusion-mixed type, also called the nozzle-mixed type. The majority of burners are of this type. Raw gas, atomized oil droplets, or pulverized coal is mixed with the air supply by a turbulent diffusion process occurring within the plume of flame. Figure 10 is a schematic representation of a diffusion flame gas burner with this mixing process taking place. Air is introduced through a register that can control the amount passing through, upon the pressure available; often the register is in the form of vanes, which will give the combustion air a swirling motion. Gas jets are introduced by a nozzle or a distributor ring placed in the air entry channel, with the jets disposed for mixing into the air. However, it is only some distance downstream from the gas jets where the velocity has slowed and where enough mixture within the combustible limits has been formed to support a flame. It is only within restricted limits of an air to fuel ratio that the burning will persist.

The actual burning process is an extremely complicated one, and many chemical and physical processes are occurring simultaneously. In the diffusion flame, the rate controlling process is that of the mixing between air and fuel, which involves a transport of both mass and momentum between regions of high velocity and regions of low velocity. Only in a relatively thin interface is there a combustible mixture, and this governs the formation of a sheet of flame. The sheet is very much warped and wrinkled as eddies of fuel are swept out into the air section or as eddies of oxygen-containing air move into the fuel. Pockets of flame will be carried into unignited regions, which will burst into new flames; or they will

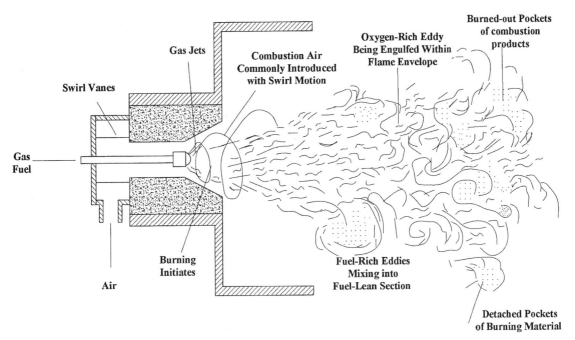

Figure 10 Schematic of a diffusion mixed burner.

be projected into zones that have already been burned out, whereupon the flame is diluted and quenched. Since the flame front cannot move at a very high velocity, it remains always detached from the gas nozzle. A flame front will sweep through a stagnant mixture of cold gases at relatively low velocity, of the order of 2 to 30 ft/sec. At too high a velocity the flame will actually blow out and away, and the flame will become extinguished; special provisions must therefore be made to hold the flame in place.

Some nozzle-mix burners distribute the gas supply through a series of jets of from a pipe ring with many openings into the throat section of air inlet. In this way the air and fuel mixing are well under way before entering the combustion space. Such a burner is not clearly premix or a discrete diffusion-mixed type.

The size and shape of the flame is largely determined by the means used for mixing gas and fuel, and for preheating it to ignition temperature. The resulting flames patterns differ widely from one burner to the next. This is the topic of a later section. The comments above pertain also to oil flame, with the added complications of atomization and liquid phase pyrolysis.

As with the gas flame, many patterns of flame are possible with the oil burner. These depend partly on the spray pattern of the atomizing system, and also on the manner of introducing the combustion air, and its axial and tangential or swirl momentum.

3.3   Turbulent Diffusion Flame Types

Effect of Swirl

Consider the effect of swirl momentum and total (air and fuel) axial momentum. Recall that high swirl attempts to create a primary recirculation, backflow, up the center of the burner axis. However, this reverse flow may be countered by the axial momentum of the fuel jet, and may or may not be sufficient to overcome this forward axial momentum. We consider four cases that are shown schematically in Fig. 11 (nomenclature follows naming conventions outlined by Beer and Chigier[8]).

Type 1A: Zero Swirl, Moderate Axial Momentum. The type 1 flame is long, poorly mixed. and highly luminous. The Type 1A designation refers to an attached flame, and attachment with zero swirl is normally possible only when moderate axial momentum levels are used in conjunction with hot secondary recirculation. Burner throat velocities are typically on the order of 30 m/s (100 ft/sec). This flame is utilized in cement kiln burners and other metal and product heating applications, and in tangentially fired combustors.

Type 1B: Zero Swirl, Very High Axial Momentum. The type 1B designation refers to a high-axial-momentum, zero-swirl flame where ignition is delayed and the high velocity almost blows the flame off. If secondary recirculation causes cold gases (cold walls) to quench the ignition, the flame will go out. This "lifted" flame is not inherently stable.

Type 2: Moderate Swirl ($S < 0.6$), Moderate Axial Momentum. As swirl increases, primary recirculation zones attempt to form, but the axial fuel jet momentum is still sufficient to break through the reverse flow region. This behavior causes a toroidal recirculation zone to form, with primary jet penetration. This configuration is called a type 2 flame, and is stable over a wide range of turndown ratios. This flame is typical of many wall fired burner configurations. Increased swirl will broaden and shorten the flame. The direction of the fuel jet also clearly plays an important role in flame shape.

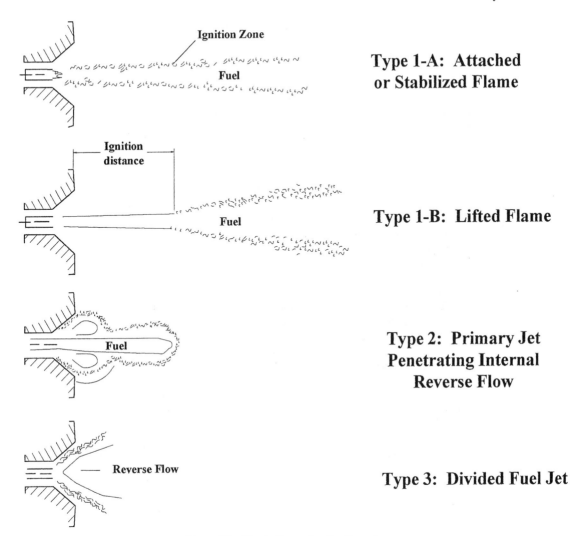

Figure 11 Simple flame classification scheme.

Type 3: High Swirl, Low Axial Momentum (S > 0.6). As swirl is increased, or axial momentum of the fuel jet is decreased, the primary recirculation overcomes the forward thrust of fuel jet and divides it. This forms a type 3 flame, with recirculation down the center, As the swirl number increases, the flame shape can vary from a ball of flame to a conical flame to a flat flame along the walls. Thus, many type 3 flames may look quite different from one another. This flame is stable under most conditions, unless the fuel is injected too rapidly and there is sufficient momentum to penetrate the recirculation zone.

Other Methods of Stabilization
Holding a flame under a variety of operating and firing rates depends on maintaining a localized zone within or adjacent to the normal flame envelope, which will always lie within

the combustion limits for the fuel and oxygen mixture, and in holding the zone where the fluid velocities are not sufficiently high to sweep the flame away. The following are examples of ways that are used to bring about improved stability.

Burner Tile (*Fig. 12a*). A conically shaped ceramix tile piece surrounds the burner opening. The diverging flow area slows the air velocity. The tile is heated by exposure to the flame to a glowing red temperature, and it reradiates and conducts to the incoming mixture. In addition, its conical or cup shape combined with a tangential swirl to the incoming air or fume stream may cause a recirculation zone in the flame center directly opposite the fuel nozzle, which brings hot combustion products and radicals back into contact with the fresh fuel and air mixture for ignition.

Target Rod and Plate (*Fig. 12b and f*). The flame is made to impinge on a cylindrical rod or a flat plate (made of ceramic or temperature-resistant alloy). Eddies form on the downstream side of this target, and these contain pockets of fuel-rich and air-rich mixture. Portions of these are swept off into the stream, forming multiple ignition sources; portions are retained and regenerated on the back of the target. The retained flame serves as a continuous pilot.

Fuel Rich, Low-velocity Pocket (*Fig. 12c*). The pocket is located downstream from the main fuel addition, and has its own fuel supply. The low velocity in the pocket allows a flame to be retained, and to act as a continuous pilot for the main stream. John Zink Company uses a ledge in the wall of the throat of one of its gas burners in this manner.

Graded Air Entry (*Fig. 12d and e*). The gas enters by a channel having a diverging cross section; air is introduced by holes that produce cross-stream jets of air. The diverging channel slows the mixture; meanwhile, some of the air jets are too fuel-rich, some too fuel lean; at some point in the channel they are near optimum. Wake eddies are formed that retain the flame. Many jet engines use variants of this design.

Deflector Plate on Fuel Gun (*Fig. 13*). A vaned conical plate attached to the end of the fuel gun serves as a baffle and swirl-inducing device, holding a low-velocity turbulent wake zone immediately opposite the burner tip. Although the mixture in this region is generally fuel-rich, pockets of burnable mixture recirculate in the zone to hold the flame in place. Additional air sweeps past the plate undisturbed, and mixes with the central wake zone further out from the burner. This arrangement is commonly used with packaged pressure-atomized burner units.

Effects of Inner and Outer Swirl (*Fig. 14*). As noted before, type 3 flames can vary widely in shape. Some examples are shown in Fig. 14 (*top*). If a long, stable flame with primary recirculation is desired, then swirl in the center may be feasible (Fig. 14, *bottom*). This is a form of a delayed mixing flame, with type 1 attribute of luminosity, length, etc., but type 3 properties of stability. It is still essentially a type 3 flame.

4 FIRING SYSTEMS

The following sections describe the most commonly encountered firing systems for gaseous, liquid, and solid fuels. A firing system provides for mixing of fuel with an oxidizer (typically air), in a manner that promotes stable combustion and desired heat release rates.

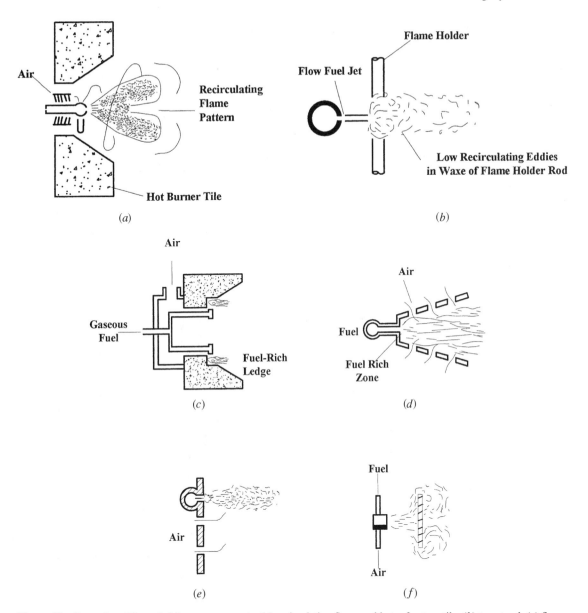

Figure 12 Examples of flame holding arrangements: (*a*) recirculating flame and hot refractory tile; (*b*) target rod; (*c*) flame-holder ledge (after John Zink); (*d*) grated air entry; (*e*) graded air entry, grid plate version; and (*f*) target plate.

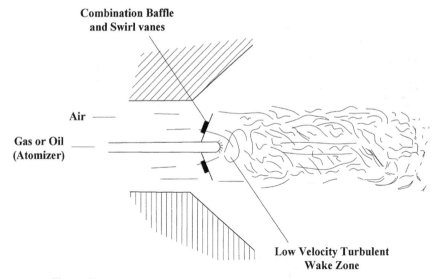

Figure 13   Flow pattern, axial jet burner, and conical oil fuel spray nozzle.

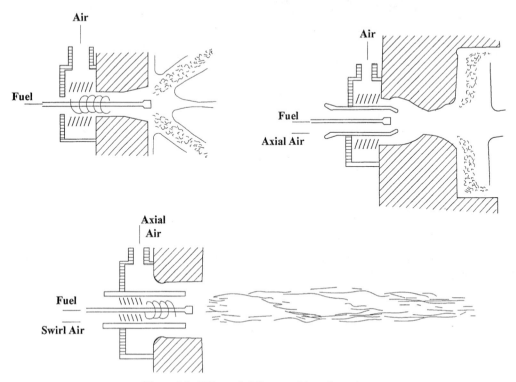

Figure 14   Effects of different swirl configurations.

4.1 Gaseous Fuels

Gaseous fuels are typically the simplest to use as they are typically supplied at pressure and do not require special pumping, heating, or atomizing (as is the case for liquid fuels), or crushing, grinding, or pulverizing (in the case of solid fuels). In general, it is easier to maintain a steady feed rate, and thus a more stable flame, with a gaseous fuel (unless the fuel has a very low heating value—e.g., blast furnace gas).

The previous section on Flame Aerodynamics discussed the basic principles underlying the shaping and stabilizing of both premixed and diffusion flames. Either of these flame types can be applied to gaseous fuels using the common burner types described below. Although presented under Gaseous Fuels, many of these burner types are also applicable for liquid and solid fuels.

Basic Burner Types

Open and natural draft-type burners rely on a negative pressure in the combustion chamber to pull in the air required for combustion, usually through adjustable shutters around the fuel nozzles. The suction in the chamber may be natural draft (chimney effect) or induced by draft fans. A crude "burner" may be nothing more than a gas gun and/or atomizer inserted through a hole in the furnace wall. Fuel–air mixing may be poor, and fuel–air ratio control may be nonexistent. Retrofitting for addition of preheated combustion air is difficult (see Fig. 15).

Sealed-in and power burners have no intentional "free" air inlets around the burner, nor are there air inlets in the form of louvers in the combustion chamber wall. All air inflow is controlled, usually by a forced draft blower or fan pushing the air through pipes or a windbox. These burners usually have a higher air pressure drop at the burner, so air velocities are higher, enabling more through mixing and better control of flame geometry. Air flow can be measured, so automatic air–fuel ratio control is easy (see Fig. 16).

Windbox burners often consist of little more than a long atomizer and a gas gun or gas ring. These are popular for small industrial and large utility boilers where economic reasons have dictated that the required large volumes of air be supplied at very low pressure (2–10 in. we) (in. we = inches of water column). Precautions are necessary to avoid fuel flowback into the windbox (see Fig. 17).

Packaged burners usually consist of bolt-on arrangements with an integral fan and perhaps integral controls. These are widely used for new and retrofit installations from very small up to about 50×10^6 Btu/hr (see Fig. 18).

Recommended Applications

Premix burner systems may be found in any of the above configurations. Most domestic appliances incorporate premixing, using some form of gas injector or inspirator (gas pressure

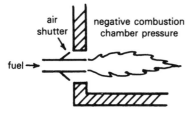

Figure 15   Open, natural draft-type burner.

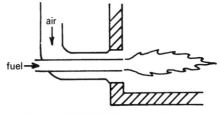

Figure 16   Sealed-in, power burner.

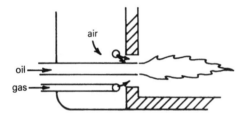

Figure 17 Windbox burner.

inducing air through a venturi). Small industrial multiport burners of this type facilitate spreading a small amount of heat over a large area, as for heating kettles, vats, rolls, small boilers, moving webs, and low-temperature processing of conveyorized products. Large single-port premix burners have been replaced by nozzle-mix burners. Better fuel–air ratio control is possible by use of aspirator mixers. (Air injection provides the energy to draw in the proper proportion of gas.) Many small units have undersized blowers, relying on furnace draft to provide secondary air. As fuel costs rise, the unwarranted excess air involved in such arrangements makes them uneconomical. Mixture manifold inside diameters larger than 4 in. (100 mm) are usually considered too great an explosion risk. For this reason, mixing in a fan inlet is rarely used.

Diffusion flame or nozzle-mix burner systems constitute the most common industrial gas burner arrangement today. They permit a broad range of fuel–air ratios, a wide variety of flame shapes, and multifuel firing. A very wide range of operating conditions are now possible with stable flames, using nozzle-mix burners. For processes requiring special atmospheres, they can even operate with very rich (50% excess fuel) or lean (1500% excess air). They can be built to allow very high velocities (420,000 scfh/in.^{2} of refractory nozzle opening) for emphasizing convection heat transfer.

Others use swirl to elicit centrifugal and coanda effects to cause the flame to scrub an adjacent refractory wall contour, thus enhancing wall radiation (see Fig. 14). By engineering the mixing configuration, nozzle-mix burner designers are able to provide a wide range of mixing rates, from a fast, intense ball of flame ($L/D = 1$) to conventional feather-shaped flame ($L/D = 5$–10) to long flames ($L/D = 20$–50). Changeable flame patterns are also possible.

Aerodynamically staged or delayed-mixing burners are a special form of diffusion flame burners, in which mixing is intentionally slow. (A raw gas torch is an unintentional form of delayed mixing.) Ignition of a fuel with a shortage of air results in polymerization or thermal cracking that forms soot particles only a few microns in diameter. These solids in the flame absorb heat and glow immediately, causing a delayed mix flame to be yellow or orange. The added luminosity enhances flame radiation heat transfer, which is one of the reasons for using delayed-mix flames. The other reason is that delayed mixing permits stretching the heat release over a great distance for uniform heating down the length of a radiant tube or

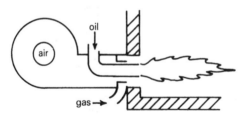

Figure 18 Integral fan burner.

a long kiln or furnace that can only be fired from one end. These types of long flames can also be beneficial for reducing NOx emissions, as peak temperatures are reduced with the staged heat release.

Gaseous fuels can sometimes provide a unique aerodynamic advantage over liquid or solid fuels. Most industrial process burners have traditionally used energy from the air stream to maintain flame stability and flame shape, as discussed previously. Now that higher-pressure fuel supplies are readily accessible in most locations, there is an opportunity to use the energy in the fuel stream for controlling flame stability and shape, thereby permitting the use of lower pressure air sources.

Figure 19 shows a fuel-directed burner for gas and preheated air. Multiple supply passages and outlet port positions permit changing the flame pattern during operation for optimum heat transfer during the course of a furnace cycle. Oil burners or dual-fuel combination burners can be constructed in a similar manner using two-fluid atomizers with compressed air or steam as the atomizing medium.

In summary, desired flame shape, stability, and flexibility can be assured by the selection of an appropriate burner design. If maximum flexibility is needed, nozzle-mix or diffusion flame burners with *adjustable* swirl for secondary air are recommended. Together with flexibility in injector angle, this design allows just about any flame shape to be achieved and to be modified as desired heat flux distributions are varied. Such burners also allow for variation in fuels.

4.2 Liquid Fuels

Much of what has been said above for gas burners applies as well for oil or other liquid fuel burning. Liquids do not burn; therefore, they must first be vaporized and thus additional considerations must be addressed beyond those described above for gaseous fuels.

The liquid fuel can be prevaporized outside of the furnace to produce a hot vapor stream that is directly substitutable for gas in premix burners. Unless there are many burners or they are very small, it is generally more practical (less maintenance) to convert to combination (dual-fuel) burners of the nozzle-mix type. For nozzle-mix burners, the fuel is atomized inside the furnace to enhance vaporization rates as noted below.

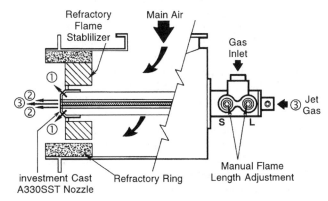

Figure 19 Low NOx fuel-directed gas burner for use with preheated air. (1) Increasing tangential gas flow (adjustment screw S) shortens flame. (2) Increasing forward gas flow (adjustment screw L) lengthens flame. (3) Jet gas—to maintain flame definition as input is reduced. (Courtesy of North American Mfg. Company.)

Atomization of Liquid Fuel

Almost all industrial liquid fuel burners use atomization to aid vaporization by exposing the large surface area (relative to volume) of millions of droplets in the size range of 100–400 μm. Evaporation then occurs at a rapid rate even if the droplets are not exposed to furnace radiation or hot air due to enhanced mass transfer rates.

Pressure atomization (as with a garden hose) uses the pressure energy in the liquid steam to cause the kinetic energy to overcome viscous and surface tension forces. If input is turned down by reducing fuel pressure, however, atomizing quality suffers; therefore, this method of atomization is limited to on–off units or cases where more than 250 psi fuel pressure is available.

Two-fluid atomization is the method most commonly used in industrial burners. Viscous friction by a high-velocity second fluid surrounding the liquid fuel stream literally tears it into droplets. The second fluid may be low-pressure air (<2 psi, or <13.8 kPa), compressed air, gaseous fuel, or steam. Many patented atomizer designs exist, for a variety of spray angles, sizes, turndown ranges, and droplet sizes. Emulsion mixing usually gives superior atomization (uniformly small drops with relatively small consumption of atomizing medium) but control is complicated by interaction of the pressures and flows of the two streams. External mixing is just the opposite. Tip-emulsion atomization provides a compromise between these two mixing limits.

Rotary-cup atomization delivers the liquid fuel to the center of a fast-spinning cup surrounded by an air stream. Rotational speed and air pressure determine the spray angle. This is still used in some large boilers, but the moving parts near the furnace heat have proved to be too much of a maintenance problem in higher-temperature process furnaces and on smaller installations where a strict preventive maintenance program could not be effected.

Sonic and ultrasonic atomization systems create very fine drops, but impart very little motion to them. For this reason they do not work well with conventional burner configurations, but require an all new design.

Liquid Fuel Conditioning

A variety of additives can be used to reduce fuel degeneration in storage, minimize slagging, lessen surface tension, reduce pollution, and lower the dew point. Regular tank draining and cleaning as well as the use of filters are recommended.

Residual oils must be heated to reduce their viscosity for pumping, usually to 500 SSU (100 cSt). For effective atomization, burner manufacturers specify viscosities in the range of 100–150 SSU (22–32 cSt). In all but tropical climates, blended oils (Nos. 4 and 5) also require heating. In Arctic situations, distillate oils need heating. Figure 20 enables one to predict the oil temperature necessary for a specified viscosity. It is best, however, to install extra heating capacity because delivered oil quality may change.

Oil heaters can be steam or electric. If oil flow stops, the oil may vaporize or char. Either situation reduces heat transfer from the heater surfaces, which can lead to catastrophic failure in electric heaters. Oil must be circulated through heaters, and the system must be fitted with protective limit controls.

Hot oil lines must be insulated and traced with steam, induction, or resistance heating. The purpose of tracing is to balance heat loss to the environment. Rarely will a tracing system have enough capacity to heat up an oil line from ambient conditions. When systems are shut down, arrangements must be made to purge the heavy oil from the lines with steam, air, or (preferably) distillate oil. Oil standby systems should be operated regularly, whether needed or not. In cold climates, they should be started before the onset of cold weather and kept circulating all winter.

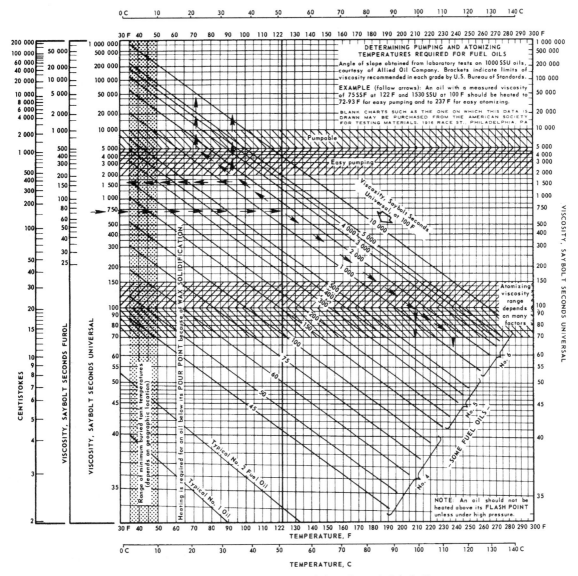

Figure 20 Viscosity–temperature relations for typical fuel oils.

4.3 Solid Fuels

Solid fuel burns by a series of steps: (1) initial particle heatup and drying (most solid fuels have some degree of moisture); (2) evolution of volatile matter from the solid particle; (3) combustion of volatile gases, and (4) residual carbon or char oxidation. The design of a suitable firing system for a given solid fuel must take into account the relative importance of each of these steps, which will in turn depend on fuel characteristics and particle size. For gaseous fuels, only step 3 is required and thus firing system design is focused on appropriate fuel–air mixing to achieve desired heat release and emissions levels. For liquid fuels, steps 1–3 are required and thus firing systems must allow for droplet heatup and evaporation in addition to suitable fuel–air mixing.

Firing systems for solid fuels must provide for these steps as well as ensuring adequate burnout of the residual carbon material (which can be as high as 95% of the fuel in some cases). The timescales for these processes can vary greatly for solid fuels. For example, at typical combustion conditions, volatile release and combustion is on the order of 100s of milliseconds (represented by the luminous flame), whereas char oxidation can range from 1–2 sec for pulverized char particles to several minutes for large chunks of fuel (the "burning embers" or flying sparklers visible in solid fuel furnaces). As a result, firing systems for solid fuels vary greatly, depending on the fuel characteristics and particle size.

Pulverized Fuels

The mostly commonly used pulverized fuel is coal, although a growing trend in the industrial and utility boiler industries is the co-firing of finely ground biomass fuels (e.g., wood, agricultural residues). Burners used with pulverized fuel tend to be either the windbox type (typical for boilers), or the sealed-in type (for smaller-scale applications). Particle sizes typically range from tens of microns to a few hundred microns, with a mean particle size occurring near 60 μm.

The most common type of burner for pulverized coal is a *circular burner,* which consists of a fuel feed pipe in the center and one, two or even three annular air registers, as shown in Fig. 21. The figure illustrates a typical configuration for a modern low NO_x burner. The pulverized coal is introduced in the core, either with or without some sort of bluff body, and then air is introduced in the annular registers. By varying the degree of swirl and the relative velocities, a stable attached flame can be achieved. An example of a commercial circular burner is shown in Fig. 22, where the dual air registers and the vanes used to generate swirl are clearly evident in the cutaway view.

In addition, hardware modifications are often made to the fuel nozzle, including stabilizer rings, ledges, tabs, teeth, bluff bodies to assist with flame stabilization. Also, burner quarls or tiles can provide additional stability by providing a hot surface for the heating of recirculating gases by convection and radiation. Some coal nozzles are designed to segregate fine and coarse particles prior to the nozzle exit, and the very fine coal particles can heat up

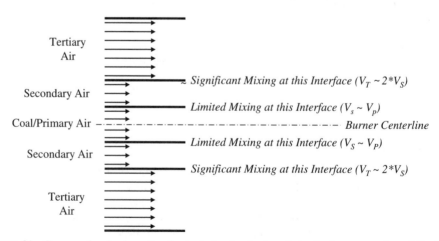

Figure 21   Cross-sectional schematic of a typical pulverized coal circular burner, configured for low NOx firing.

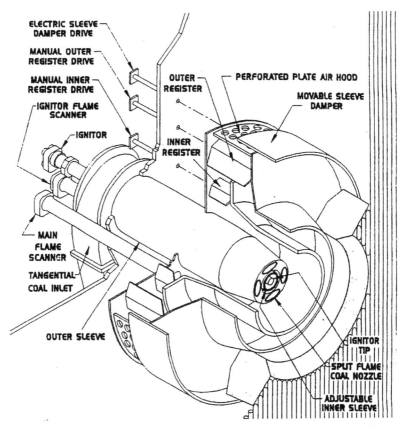

ELECTRIC SLEEVE
DAMPER DRIVE

MANUAL OUTER
REGISTER DRIVE

MANUAL INNER
REGISTER DRIVE

IGNITOR FLAME
SCANNER

IGNITOR

OUTER
REGISTER

PERFORATED PLATE AIR HOOD

MOVABLE SLEEVE
DAMPER

INNER
REGISTER

MAIN
FLAME
SCANNER

TANGENTIAL
COAL INLET

OUTER SLEEVE

IGNITOR
TIP

SPLIT FLAME
COAL NOZZLE

ADJUSTABLE
INNER SLEEVE

Figure 22 Typical low NOx pulverized coal circular burner. (Courtesy of Foster Wheeler Inc.)

rapidly and release volatiles more quickly to assist with flame stability and attachment. An example of such a nozzle is shown in Fig. 22. Combinations of all these hardware modifications have led to the apparent wide variety of burners commercially available.

Circular burners are commonly used in arrays along the front and rear walls in single- and opposed-wall-fired boilers, and along the roof in arch-fired systems.

Tangentially-fired burners, differ in that the burners are generally located in the corners of the boiler and create helical flames in the center. The coal nozzles and air-injection ports are stacked horizontally, as shown in Fig. 23, and can tilt their injection angle up or down to provide control over furnace outlet temperature (and thus steam reheat temperature) as boiler load changes. Tangentially fired systems provide delayed fuel–air mixing that results in long winding flames in the core of the boiler, and such flames have a significant benefit in terms of NO_x reduction. Circular burners can also be configured to provide long delayed-mixing flames for low NO_x operation.

Granular and Larger Fuels

Solid fuels are typically less expensive than liquid or gaseous fuels, but the cost savings are mitigated by handling and fuel preparation costs. The power required for pulverizing coal

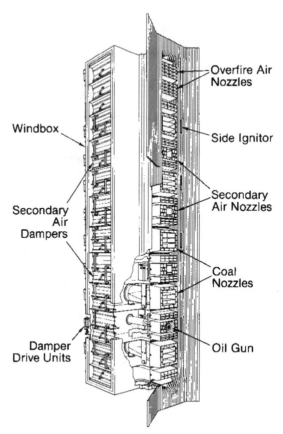

Figure 23 Typical tangentially fired pulverized coal burner arrangement. (Reproduced with permission of Alstom Power Inc., Windsor, CT, from Combustion Fossil Power, copyright 1991).

can be a significant portion of the cost of operating a boiler; therefore, smaller-scale units often utilize firing systems that require less fuel preparation.

Grate-fired systems are generally the simplest to operate, and can handle a wide range of fuel sizes and fuel types. Particles ranging from fractions of inches to several inches in approximate diameter are reliably feed. The fuel is "spread" onto a perforated metal grate and air is blown through the grate to provide oxygen for combustion and to assist in cooling the grate. In some designs, the grate is also water-cooled. Grate-fired systems are among the oldest of solid fuel firing systems, and were widely used in steamships, locomotives, and domestic heating until the advent of convenient means for distributing liquid and gaseous fuels. They are still commonly used for industrial boiler application (process steam generation and small-scale power production).

There are several grate designs, including (1) a traveling grate, in which the grate is moving at slow but constant velocity and the fuel slowly burns out before the residual ash is dumped off the end of the grate into a collecting hopper; (2) a reciprocating grate, which is typically a series of steps that are driven by reciprocating pistons that slowly move the fuel down the steps as it burns out; and (3) a vibrating grate, which is set at a slight angle and is operated by an electric motor with an eccentric cam. The vibrating grate is periodically cycled (every few minutes) to provide some stirring or mixing of the bed and to move fuel toward the end of the grate to dump into the ash hopper.

Typically, fuel size in grate-fired systems is dictated by feeding limitations and desired combustion efficiency targets. Larger fuel size requires greater residence time for acceptable consumption of residual carbon. In many cases, highly irregularly shaped particles are used (such as chipped or shredded wood products or fibrous biomass fuels), and the greatest challenge is often not combustion of the fuel, but maintaining a reliable feed rate due to bridging and channeling in feed chutes, screws and conveyors.

Fluidized bed systems provide the best environment for gas/solid contacting, and utilize fuel that has been typically sized to millimeter-sized particles. A bed of (typically) inert material is fluidized by high-velocity air that is introduced through a distributor plate at the base of the furnace. The bubbling motion of the bed provides for a high degree of mixing and fuel is injected into the hot bubbling bed for combustion. If low-quality fuels are used (high ash content or low-reactivity carbon), increased residence times can be achieved by operating in *circulating* mode. Higher gas velocities are used and some of the bed material and fuel is entrained in the furnace and is separated in a cyclone and reinjected into the furnace.

Combustion temperatures are typically lower than in other combustion systems, and this provides some benefits in terms of NO_x emissions. Also, in many systems the bed material utilized is limestone, which can react with SO_2 generated from the combustion of sulfur in the fuel and form a benign calcium sulfate (gypsum) product.

A wide variety of fuels can be combusted in a fluidized bed, including coal, biomass, and sludges. A significant operational constraint is the melting temperature for the mineral matter present in the fuel. If the ash from the fuel becomes molten or even partially sticky, it may cause the bed material to agglomerate and create particles that can longer be fluidized by the current gas velocity. Such behavior can result in "defluidization" of the bed if a sufficient fraction of the bed becomes sticky and grows to form particles that cannot be entrained and simply fall to the bottom of the furnace.

Cyclone-fired systems utilize similar particle sizes as the fluidized bed systems, but introduce the fuel in a cyclonic fashion into a horizontal barrel. Combustion air is also introduced tangentially to preserve the cyclonic motion. These cyclone barrels typically operate near adiabatic conditions (heat loss is typically <5%) and thus have very high temperatures to promote slagging of the mineral (ash) matter in the fuel. The molten slag runs down the walls of the horizontal barrel and exits through a tap at the bottom near the gas exit. Fuel that is thrown tangentially into the cyclone barrel is trapped in the molten slag layer and burns out as the slag runs down the walls. The hot exhaust gases exit the cyclone barrel and enter a water-wall boiler for subsequent heat extraction and steam production.

Cyclone-fired systems provide a high degree of fuel flexibility due to the slagging operation. Variations in ash composition when varying different fuel mixtures often creates significant operational problems for other types of solid fuel firing systems due to the potential for a transition from dry ash to molten, deposit-forming ash.

Ash Handling

Solid fuels typically contain some fraction of noncombustible material (mineral matter or ash) that must be considered when selecting or designing a solid fuel firing system. Many wood or biomass materials have less than 1% ash; however, a typical coal will range from 6 to 30% ash, with some low rank coals as high as 50 to 60% ash. The disposition of this ash is one of the greatest operational concerns for an operator of a solid-fuel firing system.

In general, the finer fractions of the ash (called "flyash") will be entrained with the exhaust gases and need to be removed with some sort of particulate collection device (see

Chapter 30). The larger ash particles ("bottom ash") will be collected in some sort of hopper at the base of the furnace. Some ash particles will deposit on furnace walls, and if the wall is a heat transfer surface, the deposits must be periodically cleaned by an air or steam jet ("soot blowing") to maintain furnace efficiency.

If ash particles become molten, the deposit will be difficult to remove and can grow to form an obstruction to flow as well as heat transfer. Thus, a knowledge of the sticking temperature of the fuel or fuel mixtures being considered for firing is important to prevent deposition problems. It is for this reason that cyclone-fired systems are more forgiving of fuel switching than other firing systems.

In general, pulverized fuel systems will see approximately 80% of the ash end up as fly ash, with 20% as bottom ash. Grate-fired systems are the reverse, with 20% flyash and 80% bottom ash. Cyclone-fired systems have very little flyash carryover (approximately 5%), as most of the ash is trapped in the slagging barrel.

The carbon content of the ash (often referred to as loss-on-ignition or LOI for the test performed to quantify it) should be monitored periodically. LOI provides an indication of combustion efficiency, since greater carbon-in-ash content implies less fuel is being completely burned out before exiting the furnace.

In addition, the carbon content often dictates the potential value of the flyash as a product. Flyash is often sold as a filler for use in concrete mixes (among other uses); however, carbon content greater than roughly 5% will often render it unusable for this purpose. Thus, flyash with low carbon-in-ash means good combustion efficiency and a ready revenue stream. Flyash with high carbon-in-ash results in poor combustion efficiency and a process waste stream that requires disposal; a financial liability on two counts. Many low NO_x firing system modifications result in an increase in LOI, as most of these modifications delay fuel–air mixing and make it more difficult to completely oxidize the residual char before exiting the furnace.

5   POLLUTANT EMISSIONS

Combustion processes produce gaseous and particulate emissions that can be harmful to human health and the environment. The pollutants of primary concern with gaseous fuels are carbon monoxide (CO) and various nitrogen oxides (NO_x), since generally most sulfur-bearing species have been removed. For liquid fuels, the same pollutants come into play; however, many oils have significant sulfur contents and thus also produce sulfur oxides (SO_x). Solid fuels can also produce CO, NO_x, and SO_x emissions, but also produce particulate emissions. Products of incomplete combustion (PICs) are possible with any of these fuels if the combustion process is poorly designed.

5.1   Control of Nitrogen Oxides (NO_x)

There are various sources of NOx in a combustion system. All fuels have the potential to create *thermal NO_x*, which results from the oxidation of N_2 in the atmosphere at high temperatures. The air used for combustion is typically 79% N_2, and at temperatures above roughly 2600 F, this N_2 can be readily oxidized to NO. In gas-fired systems and many liquid fuel-fired systems, this is the principle source of NO_x emissions. Control technologies focus on reducing peak temperatures to minimize significant NO formation. One common approach is the use of flue gas recirculation (FGR), either external recirculation with subsequent injection into the burner to dilute the combustion process, or internal recirculation using eductive mixing.

Another source of NO_x stems from the presence of nitrogen in the fuel and is referred to as *fuel NO_x*. Fuel NO_x is the primary source of NO for coal combustion, and represents approximately 80% of uncontrolled NO emissions, with thermal NO_x providing the balance. The primary method for controlling fuel NO_x is promoting fuel-rich or oxygen-starved combustion in the main burner zone, which allows the fuel-bound nitrogen (typically in the form of HCN or NH_3) to decay to N_2. Once sufficient time has been given for nitrogen decay, then additional air can be introduced (*staging air*) to complete the combustion process. As more and more of the fuel NO is controlled through air staging, the NO emissions decrease and the percentage of the NO emissions represented by thermal NO_x increases.

The staging of air can be performed through burner aerodynamics, using multiple air registers, as shown in Fig. 21. The innermost air register (secondary air) is operated with a small fraction of the total air, and its velocity is matched to the primary air/coal jet to minimize shear or mixing at the interface between the two streams. Sufficient air is provided in the secondary register to ensure ignition and attachment of the flame, however.

The tertiary stream, which contains the bulk of the combustion air, is operated at a velocity that is significantly higher than the secondary stream and thus promotes shearing or mixing between the secondary and tertiary air streams. The secondary air stream serves as a buffer between the fuel jet and the tertiary air stream and delays the mixing of most of the air with the fuel. The burner can be designed to optimize NO_x reduction and burner stability for a given fuel.

A small amount of NO is formed very early in the flame by reaction of fuel fragments with NO and is referred to as *prompt NO*. Prompt NO is difficult to control and is typically of concern only for gas burners designed to operate at very low NO emission levels (< 20 ppm).

5.2 Control of Other Gaseous Emissions

Carbon monoxide (CO) emissions as well as other hydrocarbon emissions are generally controlled through good combustion practice. Providing adequate air/fuel mixing with sufficient residence time will generally minimize CO emissions. Many of the techniques for reducing NO emissions, however, can result in an increase in CO emissions due to a delay in air/fuel mixing. Thus, any emission strategy needs to be optimized for all pollutants of concern.

Sulfur oxides are formed through oxidation of the sulfur present in the fuel. Essentially all of the sulfur is oxidized to SO_2, and there are no simple methods to convert fuel-bound sulfur to benign products like there is for converting fuel-bound nitrogen to N_2. In practice, fuel-bound sulfur is oxdized to SO_2 in the furnace and is then captured downstream through reaction with sorbents or caustic scrubber liquid. The notable exception is in-bed capture of SO_2 in fluidized beds using limestone.

Another option is eliminating the sulfur prior to combustion. Sulfur is generally removed from most gaseous and liquid fuels (with the exception of heavier fuel oils) through various "sweetening" processes at the well head or the refinery. Sulfur present in coal can be removed to some extent using beneficiation techniques; however, these processes may not be as economical as purchasing a lower-sulfur coal from a more distant source.

5.3 Control of Particulate Emissions

Flyash formed during combustion processes typical has a bimodal size distribution. Larger flyash particles ($1-10$ μm) result from fragmentation and attrition processes, or from molten ash particles that coalesce on the surface of a burning particle and are eventually liberated once the carbonaceous material is mostly consumed. Smaller flyash particles (less than 1 μm) are typically the result of vaporization/condensation processes. Burning coal particles create a locally reducing environment, which promotes reduction of metal oxides to more volatile suboxide or base metal species. These volatile species vaporize, then oxidize as they diffuse away from the particle and then condense as a submicron fume. Some of the more volatile elements will condense further downstream on the surface of this submicron fume, creating fine particulate that can be enriched in trace metals that are often toxic.

Particulate emissions are typically controlled using standard particulate collection devices such as cyclones, fabric filter baghouses, or electrostatic precipitators (ESPs). Cyclones are typically suitable for larger particles only (greater than 1 micron), whereas baghouses and ESPs will provide some measure of capture for submicron particulate.

6 SAFETY CONSIDERATIONS

Operations involving combustion must be concerned about all the usual safety hazards of industrial machinery, with additional consideration for the potential for explosions, fires, burns from hot surfaces, and asphyxiation. Less immediately severe, but long-range health problems related to combustion result from overexposure to noise and pollutants.

Preventing explosions should be the primary operating and design concern of every person in any way associated with combustion operations, because an explosion can be so devastating as to eliminate all other goals of anyone involved. The requirements for an explosion include the first five requirements for combustion (Table 3); therefore, striving for efficient combustion also sets the stage for a potential explosion. The statistical probability of meeting all seven explosion requirements at the same time and place is so small that people often become careless, and therein lies the problem. Continual training and retraining is the key to maintaining awareness.

The lower and upper limits of flammability are the same as the lower and upper explosive limits for any combustible gas or vapor. A summary of the most commonly encountered gases is given in Table 4, and a more extensive listing is provided in Table 3 in Chapter 17

Table 3 Requirements for Combustion and Explosion[a]

Requirements for Combustion	Requirements for Explosion
1. Fuel	1. Fuel
2. Oxygen (air)	2. Oxygen (air)
3. Proper proportion (within flammability limits)	3. Proper proportion (within explosive limits)
4. Mixing	4. Mixing
5. Ignition	5. Ignition
	6. Accumulation
	7. Confinement

[a]There have been incidents of disastrous explosions of unconfined fast-burning gases, but most of the damage from industrial explosions comes from the fragments of the containing furnace that are propelled like shrapnel. Lightup explosions are often only "puffs" if large doors are kept open during startup.

Table 4 Flammability Limits for Common Gaseous Fuels[13]

Fuel	Lean or Lower Limit		Rich or Upper Limit	
	ϕ^a	SR^b	ϕ	SR
Acetylene, C_2H_2	0.19	5.26	∞	~0
Carbon monoxide, CO	0.34	2.94	6.76	0.15
Ethane, C_2H_6	0.5	2	2.72	0.37
Ethylene, C_2H_4	0.41	2.44	>6.1	<0.16
Hydrogen, H_2	0.14	7.14	2.54	0.39
Methane, CH_4	0.46	2.17	1.64	0.61
Propane, C_3H_8	0.51	1.96	2.83	0.35

[a] Equivalence ratio.
[b] Stoichiometric ratio.

on Gaseous Fuels. Table 5 lists flammability data for some common liquids. Dusts generated from solid fuel handling represent an explosion hazard as well; some typical values for carbonaceous fuels and other biomass materials are listed in Table 6. References 9–12 provide a more extensive source of explosion-related data for many industrial solvents, off-gases and fugitive dusts.

Electronic safety control programs for most industrial combustion systems are generally designed to (1) prevent accumulation of unburned fuel when any source of ignition is present or (2) immediately remove any source of ignition when something goes wrong that may result in fuel accumulation.

Of course, the removal of an ignition source is very difficult in a furnace operating above 1400°F. If a burner in such a furnace should be extinguished because it suddenly became too fuel rich, requirement number 3 is negated and there can be no explosion until someone (untrained) opens a port or shuts off the fuel. The only safe procedure is to gradually flood the chamber with steam or inert gas (gradually, so as to not change furnace pressure and thereby cause more air in-flow).

For (1), the best way to prevent unburned fuel accumulation is to have a reliable automatic fuel–air ratio control system coordinated with automatic furnace pressure control. Such a system should also have input control so that the input cannot range beyond the capabilities of either automatic system. The emergency backup system consists of a trip valve that stops fuel flow in the event of flame failure or any of many other interlocks such as low air flow, or high or low fuel flow.

For (2), removal of ignition sources is implemented by automatic shutoff of other burner flames, pilot flames, spark igniters, and glow plugs. In systems where a single flame sensor monitors either a main flame or a pilot flame, the pilot flame must be programmed out when the main flame is proven. If this is not done, such a "constant" or "standing" pilot can "fool" the flame sensor and cause an explosion.

Most operational codes and insuring authorities insist on the use of flame monitoring devices for combustion chambers that operate at temperatures below 1400°F. Some of these authorities point out that even high-temperature furnaces must go through this low-temperature range on their way to and from their normal operating temperature. Another situation where safety regulations and economic reality have not yet come to agreement involves combustion chambers with dozens or even hundreds of burners, such as refinery heaters, ceramic kilns, and heat-transfer furnaces.

Avoiding fuel-fed fires first requires preventing explosions, which often start such fires (see previous discussion). Every building containing a fuel-fired boiler, oven, kiln, furnace,

Table 5 Flammability Data for Liquid Fuels

Liquid Fuel	Flash Point, °F (Closed Cup Method)	(°C)	Flammability Limits (%) Volume in Air Lower	Upper	Autoignition Temperature, °F	(°C)	Vapor density, G(air = 1)	Boiling Temperature, °F	(°C)
Butane, -n	−76	(−60)	1.9	8.5	761	(405)	2.06	31	(−1)
Butane, -iso	−117	(−83)	1.8	8.4	864	(462)	2.06	11	(−12)
Ethyl alcohol (ethanol)	55	(13)	3.5	19	737	(392)	1.59	173	(78)
Ethyl alcohol, 30% in water	85	(29)	3.6	10	—	—	—	203	(95)
Fuel oil, #1	114–185	(46–85)	0.6	5.6	445–560	(229–293)	—	340–555	(171–291)
Fuel oil (diesel), #1-D	>100	(>38)	1.3	6.0	350–625	(177–329)	—	<590	<310
Fuel oil, #2	126–230	(52–110)	—	—	500–705	(260–374)	—	340–640	(171–338)
Fuel oil (diesel), #2-D	>100	>38	1.3	6.0	490–545	(254–285)	—	380–650	(193–343)
Fuel oil, #4	154–240	(68–116)	1	5	505	(263)	—	425–760	(218–404)
Fuel oil, #5	130–310	(54–154)	1	5	—	—	—	—	—
Fuel oil, #6	150–430	(66–221)	1	5	765	(407)	—	—	—
Gasoline, automotive	−50±	(−46±)	1.3–1.4	6.0–7.6	700	(371)	3–4	91–403	(33–206)
Gasoline, aviation	−50±	(−46±)	1	6.0–7.6	800–880	(427–471)	3–4	107–319	(42–159)
Jet fuel, JP-4	−2	(−19)	0.8	6.2	468	(242)	—	140–490	(60–254)
Jet fuel, JP-5	105	(41)	0.6	4.6	400	(204)	—	370–530	(188–277)
Jet fuel, JP-6	127	(53)	—	—	500	(260)	—	250–500	(121–260)
Jet fuel, JP-8	100	(38)	0.7	4.7	410	(210)	5.7	330–510	(165–265)
Jet fuel, JP-10	106	(41)	0.9[a]	6.0[a]	552[a]	(289)	4.8	379	(193)
Kerosene	110–130	(43–54)	0.6	5.6	440–560	(227–293)	4.5	350–550	(177–288)
Methyl alcohol (methanol)	54	(12)	5.5	36.5	878	(470)	1.11	147	64
Methyl alcohol, 30% in water	75	(24)	—	—	—	—	—	167	75
Naphtha, dryclean	100–110	(38–43)	0.8	5.0	440–500	(227–260)	—	300–400	(149–204)
Naphtha, 76%, vm&p	20–45	(−7–+7)	0.9	6.0	450–500	(232–260)	3.75	200–300	(93–149)
Nonane, -n	88	(31)	0.74	2.9	403	(206)	4.41	303	(151)
Octane, -iso	10	(−12)	1.0	6.0	784	(418)	3.93	190–250	(88–121)
Propane	−156	(−104)	2.2	9.6	871	(466)	1.56	−44	(−42)
Propylene	−162	(−108)	2.0	11.1	927–952	(497–511)	1.49	−54	(−48)

[a] Estimated.

610

Table 6 Explosion Characteristics of Various Dusts[12]

| Type of Dust | Ignition Temperature | | Minimum Cloud Explosion Concentration | Limiting Oxygen Percentage[a] |
	Cloud (°C)	Layer (°C)	(g/m³)	(Spark ignition)
Cellulose, alpha	410	300	45	—
Cornstarch, commercial product	400	—	45	—
Lycopodium	480	310	25	C13
Wheat starch, edible	430	—	45	C12
Wood flour, white pine	470	260	35	—
Charcoal, hardwood mixture	530	180	140	—
Coal, Pennsylvania, Pittsburgh (Experimental Mine Coal)	610	170	55	—
Sulfur	190	220	35	C12
Aluminum flake, A 422 extra fine lining, polished	610	326	45	—
Magnesium, milled, Grade B	560	430	30	—
Acrylamide polymer	410	240	40	—
Methyl methacrylate polymer	480	—	30	C11
Cellulose acetate	420	—	40	C14
Polycarbonate	710	—	25	C15
Phenol formaldehyde molding compound, wood flour filler	500	—	30	C14

[a]C = CO_2, numbers indicate percentages; e.g., C13 refers to dilution to 13% oxygen with carbon dioxide as the diluent.

heater, or incinerator should have a spring-operated manual reset fuel shutoff valve outside the building with panic buttons at the guardhouse or at exits to allow shutting off fuel as one leaves the burning building.

Gas-fuel lines should be overhead where crane and truck operators cannot rupture them, or underground. If underground, keep and use records of their locations to avoid digging accidents. Overhead fuel lines must have well-marked manual shutoff valves where they can be reached without a ladder.

Liquid-fuel lines should be underground; otherwise, a rupture will pour or spray fuel on a fire. The greatest contributor to fuel-fed fires is a fuel shutoff valve that will not work. All fuel shutoff valves, manual or automatic, must be tested on a regular maintenance schedule. Such testing may cause nuisance shutdowns of related equipment; therefore, a practical procedure is to have the maintenance crew do the day's end (or week's end) shutdown about once each month. The same test can check for leaking. If it is a fully automatic or manual-reset automatic valve, shutdown should be accomplished by simulating flame failure and, in succession, each of the interlocks.

Because so much depends on automatic fuel shutoff valves, it makes sense (1) to have a backup valve (blocking valve), and (2) to replace it before it hangs up—at least every 5 years, or more often in adverse environments. Maintenance and management people must be ever alert for open side panels and covers on safety switches and fuel shutoff valves, which may be evidence that safety systems have been bypassed or tampered with in some way.

Storage of LP gas, oils, or solid fuels requires careful attention to applicable codes. If the point of use, an open line, or a large leak is below the oil storage elevation, large quantities may siphon out and flow into an area where there is a water heater or other source of ignition. Steam heaters in heavy oil tanks need regular inspections, or leaks can emulsify the oil, causing an overflow. It is advisable to make provision for withdrawing the heater for repair without having to drain the whole tank.

LP gas is heavier than air. Workers have been suffocated by this invisible gas when it leaked into access pits below equipment. Thus, the potential for accidental accumulation should be considered when designing and laying out equipment, to avoid possible explosion or asphyxiation hazards.

Codes and regulations are proliferating, many by local authorities or insuring groups. Most refer, as a base, to publications of the NFPA, the National Fire Protection Association, Batterymarch Park, Quincy, MA 02269. Their publications usually represent the consensus of technically competent volunteer committees from industries involved with the topic.

7 OXY-FUEL FIRING

Commercially "pure" oxygen (90–99% oxygen) is sometimes substituted for air (20.9% oxygen) in various mixture ratios up to pure oxygen to (1) achieve higher flame temperature, (2) get higher % available heat (best possible efficiency), or (3) to try to eliminate nitrogen from the furnace atmosphere, and thereby reduce the probability of NO_x formation.

Figure 24 shows percent available heat and adiabatic flame temperatures (x-intercepts) for various amounts of oxygen enrichment of an existing air supply, and for "oxy-fuel firing" or 100% oxygen.

Oxy-fuel burners have been water-cooled in the past, but their propensity to spring leaks and do terrible damage has led to use of better materials to avoid water cooling. Oxygen

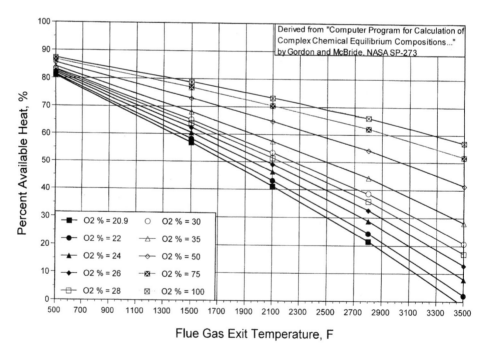

Figure 24 Percents available heat for $\phi = .95$ or 5% excess air with average natural gas, HHV = 1025 Btu/cu ft, with various degrees of oxygen enrichment and for oxy-fuel firing (top curve).

burner nozzles and tiles are subject to much higher temperatures and more oxidizing atmospheres than are air burner nozzles and tiles. Control valves, regulators, and piping for oxygen require special cleaning and material selection.

Acknowledgments

The author would like to acknowledge Jost O.L. Wendt of the University of Arizona for his assistance with the Flame Aerodynamics section, Richard J. Reed of North American Manufacturing Company for contributions from previous editions of this chapter, and the many contributions of the International Flame Research Foundation (IFRF) for their vast body of work on flame stability, shaping and aerodynamics.

REFERENCES

1. R. J. Reed (ed.), *Combustion Handbook,* Vol. I, 3rd ed., North American Mfg. Co., Cleveland, OH, 1986.
2. R. H. Essenhigh, "An Introduction to Stirred Reactor Theory Applied to Design of Combustion Chambers," in *Combustion Technology,* H. B. Palmer and J. M. Beer (eds.), Academic, New York, 1974, pp. 389–391.
3. E. Mallard and H. L. Le Chatelier, *Ann. Mines,* **4,** 379 (1883).
4. B. Lewis and G. von Elbe, *Combustion, Flames and Explosion of Gases,* 2nd ed., Academic Press, New York, 1961, Chapter V.
5. F. P. Ricou and D. B. Spalding, *J. Fluid Mech.,* **11,** 21–32 (1961).
6. M. W. Thring and M. P. Newby, *Fourth International Symposium on Combustion,* 1953, The Combustion Institute, Pittsburgh, PA, p. 789–796.
7. M. A. Field, D. W. Gill, B. B. Morgan, and P. G. W. Hawksley, *Combustion of Pulverized Coal,* The British Coal Utilization Research Association, Letherhead, Surrey, UK, 1967, p. 58.
8. J. M. Beer and N. A. Chigier, *Combustion Aerodynamics,* Krieger, Malabar, FL, 1983.
9. National Fire Protection Association, NFPA No. 86, *Standard for Ovens and Furnaces,* National Fire Protection Association, Quincy, MA, 1995.
10. National Fire Protection Association, *Flash Point Index of Trade Name Ligands,* National Fire Protection Association, Quincy, MA, 1978.
11. National Fire Protection Association, NFPA No. 325M, *Fire Hazard Properties of Flammable Liquids, Gases, Volatile Solids,* National Fire Protection Association, Quincy, MA, 1994.
12. National Fire Protection Association, *Fire Protection Handbook,* 18th ed., Quincy, MA, 1997.
13. S. R. Turns, *An Introduction to Combustion,* McGraw Hill, New York, 1996.
14. Factory Mutual Engineering Corp., *Handbook of Industrial Loss Prevention,* McGraw-Hill, New York, 1967.

CHAPTER 17
GASEOUS FUELS

Richard J. Reed
North American Manufacturing Company
Cleveland, Ohio

1 INTRODUCTION

Gaseous fuels are generally easier to handle and burn than are liquid or solid fuels. Gaseous fossil fuels include natural gas (primarily methane and ethane) and liquefied petroleum gases (LPG; primarily propane and butane). Gaseous man-made or artificial fuels are mostly derived from liquid or solid fossil fuels. Liquid fossil fuels have evolved from animal remains through eons of deep underground reaction under temperature and pressure, while solid fuel evolved from vegetable remains. Figure 1, adapted from Ref. 1, shows the ranges of hydrogen/carbon ratios for most fuels.

2 NATURAL GAS

2.1 Uses and Distribution

Although primarily used for heating, natural gas is also frequently used for power generation (via steam turbines, gas turbines, diesel engines, and Otto cycle engines) and as feedstock for making chemicals, fertilizers, carbon-black, and plastics. It is distributed through intra- and intercontinental pipe lines in a high-pressure gaseous state and via special cryogenic cargo ships in a low-temperature, high-pressure liquid phase (LNG).

Final street-main distribution for domestic space heating, cooking, water heating, and steam generation is at regulated pressures on the order of a few inches of water column to a few pounds per square inch, gage, depending on local facilities and codes. Delivery to commercial establishments and institutions for the same purposes, plus industrial process heating, power generation, and feedstock, may be at pressures as high as 100 or 200 psig (800 or 1500 kPa absolute). A mercaptan odorant is usually added so that people will be aware of leaks.

614

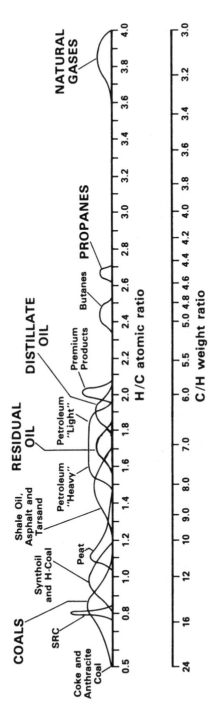

Figure 1 Hydrogen/carbon ratios of fossil and synthetic fuels. (Adapted from Ref. 1.)

Before the construction of cross-country natural gas pipe lines, artificial gases were distributed through city pipe networks, but gas generators are now usually located adjacent to the point of use.

2.2 Environmental Impact

The environmental impact of natural gas combustion is generally less than that of liquid or solid fuels. Pollutants from natural gas may be (1) particulates, if burners are poorly adjusted or controlled (too rich, poor mixing, quenching), or (2) nitrogen oxides, in some cases with intense combustion, preheated air, or oxygen enrichment.

2.3 Sources, Supply, and Storage

Natural gas is found with oil deposits (animal fossils) and coal deposits (plant fossils). As-yet untapped supplies are known to exist (1) near the coast of the Gulf of Mexico in very deep geopressured/geothermal aquifers and (2) in difficult-to-separate Appalachian shale formations.

Except for these hard-to-extract supplies, U.S. natural gas supplies have been variously predicted to last 10–20 years, but such predictions are questionable because of the effects of economic and regulatory variations on consumption, production, and exploration. Except for transoceanic LNG vessels, distribution is by pipe line, using a small fraction of the fuel in compressors to provide pumping power.

Storage facilities are maintained by many local gas utilities as a cushion for changing demand. These may be low-pressure gas holders with floating bell-covers, old wells or mines (medium pressure), or cryogenic vessels for high-pressure liquefied gas.

2.4 Types and Composition

Natural gases are classified as "sweet" or "sour," depending on their content of sulfur compounds. Most such compounds are removed before distribution. Odorants added (so that leaks can be detected) are usually sulfur compounds, but the amount is so minute that it has no effect on performance or pollution.

Various geographic sources yield natural gases that may be described as "high methane," "high Btu," or "high inert."

2.5 Properties

Properties that concern most users of natural gases relate to the heat available from their combustion, flow characteristics, and burnability in a variety of burner types. Strangely, few people pay attention to the properties of their gas until they are forced to substitute another fuel for it. Some properties are listed in Table 1.[2]

2.6 Calorific Value or Heating Value

The gross or higher heating value (HHV) is usually measured in a steady-state calorimeter, which is a small fire-tube heat exchanger with a water-cooled surface area so large that it cools the products of combustion to the temperature at which they entered as fuel and air (usually 60°F). HHV can be calculated from a volumetric analysis and the calorific values of the pure compounds in the gas (Table 2). For example, for a natural gas having the analysis shown in column 2 below, the tabulation shows how a weighted average method can be used to determine the calorific value of the mixture:

Table 1a   Analyses of Typical Gaseous Fuels[2]

Type of Gas	Analysis in % by Volume								
	CH_4	C_2H_6	C_3H_8	C_4H_{10}	CO	H_2	CO_2	O_2	N_2
Acetylene, commercial	(97.1% C_2H_2, 2.5% C_3H_6O)							0.084	0.28
Blast furnace	—	—	—	—	27.5	1.0	11.5	—	60.0
Blue (water), bituminous	4.6	—	—	0.7	28.2	32.5	5.5	0.9	27.6
Butane, commercial, natural gas	—	—	6.0	70.7 n-, 23.3 iso-	—	—	—	—	—
Butane, commercial, refinery gas	—	—	5.0	50.1 n-, 16.5 iso-	(28.3% C_4H_6)				
Carbureted blue, low gravity	10.9	2.5	—	6.1	21.9	49.6	3.6	0.4	5.0
Carbureted blue, heavy oil	13.5	—	—	8.2	26.8	32.2	6.0	0.9	12.4
Coke oven, by-product	32.3	—	—	3.2	5.5	51.9	2.0	0.3	4.8
Mapp	—	—	15.0	10.0	(66.0% C_3H_4, 9.0% C_3H_6)				
Natural, Alaska	99.6	—	—	—	—	—	—	—	0.4
Natural, Algerian LNG, Canvey	87.20	8.61	2.74	1.07	—	—	—	—	0.36
Natural, Gaz de Lacq	97.38	2.17	0.10	0.05	—	—	—	—	0.30
Natural, Groningen, Netherlands	81.20	2.90	0.36	0.14	—	—	0.87	—	14.40
Natural, Libyan LNG	70.0	15.0	10.0	3.5	—	—	—	—	0.90
Natural, North Sea, Bacton	93.63	3.25	0.69	0.27	—	—	0.13	—	1.78
Natural, Birmingham, AL	90.0	5.0	—	—	—	—	—	—	5.0
Natural, Cleveland, OH	82.9	11.9		0.3	—	—	0.2	0.3	4.4
Natural, Kansas City, MO	84.1	6.7	—	—	—	—	0.8	—	8.4
Natural, Pittsburgh, PA	83.4	15.8	—	—	—	—	—	—	0.8
Producer, Koppers–Totzek[a]	0.09	—	—	—	55.1	33.7	9.8	—	1.3
Producer, Lurgi[b]	5.0	—	—	—	16.0	25.0	14.0	—	40.0
Producer, W-G, bituminous[b]	2.7	—	—	—	28.6	15.0	3.4	0	50.3
Producer, Winkler[b]	1	—	—	—	10	12	22	—	55
Propane, commercial, natural gas	—	2.2	97.3	0.5	—	—	—	—	—
Propane, commercial, refinery gas	—	2.0	72.9	0.8	(24.3% C_3H_6)				
Sasol, South Africa	28.0	—	—	—	22.0	48.9	1.0		
Sewage, Decatur	68.0	—	—	—	—	2.0	22.0	—	6.0
SNG, no methanation	79.9	—	—	—	1.2	19.0	0.5	—	—

[a] O_2-blown.

[b] Air-blown.

Col. 1, Constituent	Col. 2, % Volume	Col. 3, HHV from Table 2 (Btu/ft³)	Col. 4 = (Col. 3 × Col. 2)/100
Methane, CH_4	90	1013	912
Ethane, C_2H_6	6	1763	106
Nitrogen, N_2	4	0	0
Total	100%		1018 Btu/ft³

Table 1b Properties of Typical Gaseous Fuels[2]

Type of Gas	Gas Gravity	Calorific Value				Gross Btu/ft³ of Standard Air	Gross kcal/m³ of Standard Air
		Btu/ft³		kcal/m³			
		Gross	Net	Gross	Net		
Acetylene, commercial	0.94	1410	1360	12548	12105	115.4	1027
Blast furnace	1.02	92	91	819	819	135.3	1204
Blue (water), bituminous	0.70	260	239	2314	2127	126.2	1121
Butane, commercial, natural gas	2.04	3210	2961	28566	26350	104.9	932.6
Butane, commercial, refinery gas	2.00	3184	2935	28334	26119	106.1	944.2
Carbureted blue, low gravity	0.54	536	461	4770	4102	106.1	944.2
Carbureted blue, heavy oil	0.66	530	451	4716	4013	101.7	905.0
Coke oven, by-product	0.40	569	509	5064	4530	105.0	934
Mapp	1.48	2406	2282	21411	20308	113.7	1011.86
Natural, Alaska	0.55	998	906	8879	8063	104.8	932.6
Natural, Algerian LNG, Canvey	0.64	1122	1014	9985	9024	104.3	928.2
Natural, Gaz de Lacq	0.57	1011	911	8997	8107	104.1	927.3
Natural, Groningen, Netherlands	0.64	875	789	7787	7021	104.4	927.3
Natural, Libyan LNG	0.79	1345	1223	11969	10883	106.1	928.2
Natural, North Sea, Bacton	0.59	1023	922	9104	8205	105.0	934.4
Natural, Birmingham, AL	0.60	1002	904	8917	8045	106.1	945.1
Natural, Cleveland, OH	0.635	1059	959	9424	8534	106.2	942.4
Natural, Kansas City, MO	0.63	974	879	8668	7822	106.3	946.0
Natural, Pittsburgh, PA	0.61	1129	1021	10047	9086	106.3	945.1
Producer, Koppers–Totzek[a]	0.78	288	271	2563	2412	135.2	1203
Producer, Lurgi[b]	0.80	183	167	1629	1486	125.3	1115
Producer, W-G, bituminous[b]	0.84	168	158	1495	1406	129.2	1150
Producer, Winkler[b]	0.98	117	111	1041	988	188.7	1679
Propane, commercial, natural gas	1.55	2558	2358	22764	20984	107.5	956.6
Propane, commercial, refinery gas	1.77	2504	2316	22283	20610	108.0	961.1
Sasol, South Africa	0.55	500	448	4450	3986	114.9	1022
Sewage, Decatur	0.79	690	621	6140	5526	105.3	936.2
SNG, no methanation	0.47	853	765	7591	6808	105.8	943.3

[a] O₂-blown.
[b] Air-blown.

It is a convenient coincidence that most solid fossil fuels release about 96–99 gross Btu/ft³ of standard *air;* liquid fossil fuels release about 101–104 Btu/ft³; gaseous fossil fuels about 104–108 Btu/ft³.

This would say that the natural gas in the example above should require about 1017 Btu/ft³ gas divided by 106 Btu/ft³ air = 9.6 ft³ air/ft³ gas. Precise stoichiometric calculations would say 0.909(9.53) + 0.06(16.7) = 9.58 ft³ air/ft³ gas.

Table 1c Combustion Characteristics of Typical Gaseous Fuels[2]

Type of Gas	Wobbe Index	Vol. Air. Req'd per Vol. Fuel	Stoichiometric Products of Combustion				Flame Temperature (°F)[b]
			% CO_2 Dry[a]	% H_2O Wet	% N_2 Wet	Total Vol. / Vol. Fuel	
Acetylene, commercial	1559	12.14	17.4	8.3	75.8	12.66	3966
Blast furnace	91.0	0.68	25.5	0.7	74.0	1.54	2559
Blue (water), bituminous	310.8	2.06	17.7	16.3	68.9	2.77	3399
Butane, commercial, natural gas	2287	30.6	14.0	14.9	73.2	33.10	3543
Butane, commercial, refinery gas	2261	30.0	14.3	14.4	73.4	32.34	3565
Carbureted blue, low gravity	729.4	5.05	14.0	18.9	69.8	5.79	3258
Carbureted blue, heavy oil	430.6	5.21	15.7	16.6	70.3	6.03	3116
Coke oven, by-product	961.2	5.44	10.8	21.4	70.1	6.20	3525
Mapp	1947	21.25	15.6	11.9	74.4	22.59	3722
Natural, Alaska	1352	9.52	11.7	18.9	71.6	10.52	3472
Natural, Algeria LNG, Canvey	1423	10.76	12.1	18.3	71.9	11.85	3483
Natural, Gaz de Lacq	1365	9.71	11.7	18.8	71.6	10.72	3474
Natural, Groningen, Netherlands	1107	8.38	11.7	18.4	72.0	9.40	3446
Natural, Kuwait, Burgan	1364	10.33	12.2	18.3	71.7	10.40	3476
Natural, Libya LNG	1520	12.68	12.5	17.4	72.2	13.90	3497
Natural, North Sea, Bacton	1345	9.74	11.8	18.7	71.7	10.77	3473

Table 1c (*Continued*)

Type of Gas	Wobbe Index	Vol. Air. Req'd per Vol. Fuel	Stoichiometric Products of Combustion				Flame Temperature (°F)[b]
			% CO_2 Dry[a]	% H_2O Wet	% N_2 Wet	Total Vol. / Vol. Fuel	
Natural, Birmingham, AL	1291	9.44	11.7	18.6	71.8	10.47	3468
Natural, East Ohio	1336	9.70	11.9	18.7	71.7	10.72	3472
Natural, Kansas City, MO	1222	9.16	11.8	18.5	71.9	10.19	3461
Natural, Pittsburgh, PA	1446	10.62	12.0	18.3	71.9	11.70	3474
Producer, BCR, W. Kentucky	444	3.23	23.3	14.7	66.0	3.88	3514
Producer, IGT, Lignite	562	4.43	18.7	17.5	67.0	5.24	3406
Producer, Koppers–Totzek	326.1	2.13	27.7	12.6	63.2	2.69	3615
Producer, Lurgi	204.6	1.46	18.4	15.5	68.9	2.25	3074
Producer, Lurgi, subbituminous	465	2.49	23.4	19.6	61.5	3.20	3347
Producer, W-G, bituminous	183.6	1.30	18.5	9.8	73.5	2.08	3167
Producer, Winkler	118.2	0.62	24.1	9.3	68.9	1.51	3016
Propane, commercial, natural gas	2029	23.8	13.7	15.5	73.0	25.77	3532
Propane, commercial, refinery gas	2008	23.2	14.0	14.9	73.2	25.10	3560
Sasol, South Africa	794.4	4.30	12.8	21.0	68.8	4.94	3584
Sewage, Decatur	791.5	6.55	14.7	18.4	69.7	7.52	3368
SNG, no methanation	1264	8.06	11.3	19.8	71.1	8.96	3485

[a] Ultimate.

[b] Theoretical (calculated) flame temperatures, dissociation considered, with stoichiometrically correct air/fuel ratio. Although these temperatures are lower than those reported in the literature, they are all computed on the same basis; so they offer a comparison of the relative flame temperatures of various fuels.

Table 2 Calorific Properties of Some Compounds Found in Gaesous Fuels

Compound	Wobbe Index	Gross Heating Value[d] (Btu/ft³)	Net Heating Value (Btu/ft³)	Pounds, Dry poc[a] per std ft³ of Fuel	Pounds H₂O per std ft³ of Fuel	Air Volume per Fuel Volume
Methane, CH_4	1360	1013	921	0.672	0.0950	9.56
Ethane, C_2H_6	1729	1763	1625	1.204	0.1425	16.7
Propane, C_3H_8	2034	2512	2328	1.437	0.1900	23.9
Butane, C_4H_{10}	2302	3264	3034	2.267	0.2375	31.1
Carbon Monoxide, CO	328	323	323	0.255	0	2.39
Hydrogen, H_2	1228	325	279	0	0.0474	2.39
Hydrogen Sulfide, H_2S	588	640	594	0.5855	0.0474	7.17
N_2, O_2, H_2O, CO_2, SO_2	0	0	0	[b]	[c]	0

[a] poc = products of combustion.
[b] Weight of N_2, O_2, CO_2, and SO_2 in fuel.
[c] Weight of H_2O in fuel.
[d] Higher heating value (HHV).

2.7 Net Heating Value

Because a calorimeter cools the exit gases below their dew point, it retrieves the latent heat of condensation of any water vapor therein. But that latent heat is not recapturable in most practical heating equipment because of concern about corrosion; therefore, it is more realistic to subtract the latent heat from HHV, yielding a net or lower heating value, LHV. This is approximately

$$\frac{LHV}{unit\ of\ fuel} = \frac{HHV}{unit\ of\ fuel} - \left(\frac{970\ Btu}{lb\ H_2O} \times \frac{lb\ H_2O}{unit\ of\ fuel}\right)$$

Values for the latter term are listed in Table 2. (Note that available heat was discussed in Chapter 6.)

2.8 Flame Stability

Flame stability is influenced by burner and combustion chamber configuration (aerodynamic and heat transfer characteristics) and by the fuel properties tabulated in Table 3.

2.9 Gas Gravity

Gas gravity, G (Table 1), is the ratio of the actual gas density relative to the density of dry air at standard temperature and pressure (0.0765 lb/ft³). This should not be confused with "specific gravity," which is the ratio of actual density relative to that of water. Gas gravity for natural gases typically ranges from 0.58 to 0.64, and is used in determination of flow rates and pressure drops through pipe lines, orifices, burners, and regulators:

$$Flow = Flow\ coefficient \times Area\ (ft^2) \times \sqrt{2g(psf\ pressure\ drop)/\rho}$$

Table 3 Fuel Properties That Influence Flame Stability[2,a]

Fuel	Minimum Ignition Temperature, °F(°C)	Calculated Flame Temperature, °F(°C)[b]		Flammability Limits, % Fuel Gas by Volume[c]		Laminar Flame Velocity, fps(m/sec)		Percent Theoretical Air for Maximum Flame Velocity
		In Air	In O_2	Lower	Upper	In Air	In O_2	
Acetylene, C_2H_2	581(305)	4770(2632)	5630(3110)	2.5	81.0	8.75(2.67)	—	83
Blast furnace gas	—	2650(1454)	—	35.0	73.5	—	—	—
Butane, commercial	896(480)	3583(1973)	—	1.86	8.41	2.85(0.87)	—	—
Butane, n-C_4H_{10}	761(405)	3583(1973)	—	1.86	8.41	1.3(0.40)	—	97
Carbon monoxide, CO	1128(609)	3542(1950)	—	12.5	74.2	1.7(0.52)	—	55
Carbureted water gas	—	3700(2038)	5050(2788)	6.4	37.7	2.15(0.66)	—	90
Coke oven gas	—	3610(1988)	—	4.4	34.0	2.30(0.70)	—	90
Ethane, C_2H_6	882(472)	3540(1949)	—	3.0	12.5	1.56(0.48)	—	98
Gasoline	536(280)	—	—	1.4	7.6	—	—	57
Hydrogen, H_2	1062(572)	4010(2045)	5385(2974)	4.0	74.2	9.3(2.83)	—	—
Hydrogen sulfide, H_2S	558(292)	—	—	4.3	45.5	—	—	—
Mapp gas, C_3H_4	850(455)	—	5301(2927)	3.4	10.8	—	15.4(4.69)	—
Methane, CH_4	1170(632)	3484(1918)	—	5.0	15.0	1.48(0.45)	14.76(4.50)	90
Methanol, CH_3OH	725(385)	3460(1904)	—	6.7	36.0	—	1.6(0.49)	—
Natural gas	—	3525(1941)	4790(2643)	4.3	15.0	1.00(0.30)	15.2(4.63)	100
Producer gas	—	3010(1654)	—	17.0	73.7	0.85(0.26)	—	90
Propane, C_3H_8	871(466)	3573(1967)	5130(2832)	2.1	10.1	1.52(0.46)	12.2(3.72)	94
Propane, commercial	932(500)	3573(1967)	—	2.37	9.50	2.78(0.85)	—	—
Propylene, C_3H_6	—	—	5240(2893)	—	—	—	—	—
Town gas (Br. coal)	700(370)	3710(2045)	—	4.8	31.0	—	—	—

[a] For combustion with air at standard temperature and pressure.

[b] Flame temperatures are theoretical—calculated for stoichiometric ratio, dissociation considered.

[c] In a fuel–air mix. Example for methane: the lower flammability limit or lower explosive limit, LEL = 5% or 95 volumes air/5 volumes gas = 19.1 air/gas ratio. From Table 2, stoichiometric ratio is 9.56:1. Therefore excess air is $19 - 9.56 = 9.44$ ft³ air/ft³ gas or $9.44/9.56 \times 100 = 99.4\%$ excess air.

where $g = 32.2$ ft/sec^2 and ρ = gas gravity $\times 0.0765$. Unless otherwise emphasized, gas gravity is measured and specified at standard temperature and pressure (60°F and 29.92 in Hg).

2.10   Wobbe Index

Wobbe index or Wobbe number (Table 2) is a convenient indicator of heat input considering the flow resistance of a gas-handling system. Wobbe index is equal to gross heating value divided by the square root of gas gravity; $W = \text{HHV}/\sqrt{G}$.

If air can be mixed with a substitute gas to give it the same Wobbe index as the previous gas, the existing burner system will pass the same gross Btu/hr input. This is often invoked when propane–air mixtures are used as standby fuels during natural gas curtailments. To be precise, the amount of air mixed with the propane should then be subtracted from the air supplied through the burner.

The Wobbe index is also used to maintain a steady input despite changing calorific value and gas gravity. Because most process-heating systems have automatic input control (temperature control), maintaining steady input may not be as much of a problem as maintaining a constant furnace atmosphere (oxygen or combustibles).

2.11   Flame Temperature

Flame temperature depends on burner mixing aerodynamics, fuel/air ratio, and heat loss to surroundings. It is very difficult to measure with repeatability. Calculated adiabatic flame temperatures, corrected for dissociation of CO_2 and H_2O, are listed in Tables 1 and 3 for 60°F air; in Chapter 15 it is listed for elevated air temperatures. Obviously, higher flame temperatures produce better heat-transfer rates from flame to load.

2.12   Minimum Ignition Temperature

Minimum ignition temperature, Table 3, relates to safety in handling, ease of light-up, and ease of continuous self-sustained ignition (without pilot or igniter, which is preferred). In mixtures of gaseous compounds, such as natural gas, the minimum ignition temperature of the mixture is that of the compound with the lowest ignition temperature.

2.13   Flammability Limits

Flammability limits (Table 3, formerly termed "limits of inflammability") spell out the range of air-to-fuel proportions that will burn with continuous self-sustained ignition. "Lower" and "upper" flammability limits [also termed lower explosive limit (LEL) and upper explosive limit (UEL)] are designated in % gas in a gas–air mixture. For example, the flammability limits of a natural gas are 4.3% and 15%. The 4.3% gas in a gas–air mixture means 95.7% must be air; therefore, the "lean limit" or "lower limit" air/fuel ratio is $95.7/4.3 = 22.3{:}1$, which means that more than $22.3{:}1$ (volume ratio) will be too lean to burn. Similarly, less than $(100 - 15)/15 = 5.67{:}1$ is too rich to burn.

For the flammability limits of fuel mixtures other than those listed in Table 3, the Le Chatelier equation[3] and U.S. Bureau of Mines data[4] can be used.

Table 4a   Physical Properties[a] of LP Gases[b,5]

	Propane	Isobutane	Butane
Molecular weight	44.09	58.12	58.12
Boiling point, °F	−43.7	+10.9	+31.1
Boiling point, °C	−42.1	−11.7	−0.5
Freezing point, °F	−305.8	−255.0	−216.9
Density of liquid			
Specific gravity, 60°F/60°F	0.508	0.563	0.584
Degrees, API	147.2	119.8	110.6
Lb/gal	4.23	4.69	4.87
Density of vapor (ideal gas)			
Specific gravity (air = 1)	1.522	2.006	2.006
Ft^3 gas/lb	8.607	6.53	6.53
Ft^3 gas/gal of liquid	36.45	30.65	31.8
Lb gas/1000 ft^3	116.2	153.1	153.1
Total heating value (after vaporization)			
Btu/ft^3	2,563	3,369	3,390
Btu/lb	21,663	21,258	21,308
Btu/gal of liquid	91,740	99,790	103,830
Critical constants			
Pressure, psia	617.4	537.0	550.1
Temperature, °F	206.2	272.7	306.0
Specific heat, Btu/lb, °F			
c_p, vapor	0.388	0.387	0.397
c_v, vapor	0.343	0.348	0.361
c_p/c_v	1.13	1.11	1.10
c_p, liquid 60°F	0.58	0.56	0.55
Latent heat of vaporization at boiling point, Btu/lb	183.3	157.5	165.6
Vapor pressure, psia			
0°F	37.8	11.5	7.3
70°F	124.3	45.0	31.3
100°F	188.7	71.8	51.6
100°F (ASTM), psig max	210		70
130°F	274.5	109.5	80.8

[a] Properties are for commercial products and vary with composition.

[b] All values at 60°F and 14.696 psia unless otherwise stated.

3   LIQUEFIED PETROLEUM GASES

LP gases (LPG) are by-products of natural gas production and of refineries. They consist mainly of propane (C_3H_8), with some butane, propylene, and butylene. They are stored and shipped in liquefied form under high pressure; therefore, their flow rates are usually measured in gallons per hour or pounds per hour. When expanded and evaporated, LPG are heavier than air. Workmen have been asphixiated by LPG in pits beneath leaking LPG equipment.

The rate of LPG consumption is much less than that of natural gas or fuel oils. Practical economics usually limit use to (a) small installations inaccessible to pipe lines, (b) transportation, or (c) standby for industrial processes where oil burning is difficult or impossible.

Table 4b   Physical Properties[a] of LP Gases[b,5]

	Propane	Isobutane	Butane
Flash temperature, °F (calculated)	−156	−117	−101
Ignition temperature, °F	932	950	896
Maximum flame temperature in air, °F			
Observed	3497	3452	3443
Calculated	3573	3583	3583
Flammability limits, % gas in air			
Lower	2.37	1.80	1.86
Higher	9.50	8.44	8.41
Maximum rate flame propagation in 1 in. tube			
Inches per second	32	33	33
Percentage gas in air	4.6–4.8	3.6–3.8	3.6–3.8
Required for complete combustion (ideal gas)			
Air, ft³ per ft³ gas	23.9	31.1	31.1
lb per lb gas	15.7	15.5	15.5
Oxygen, ft³ per ft³ gas	5.0	6.5	6.5
lb per lb gas	3.63	3.58	3.58
Products of combustion (ideal gas)			
Carbon dioxide, ft³ per ft³ gas	3.0	4.0	4.0
lb per lb gas	2.99	3.03	3.03
Water vapor, ft³ per ft³ gas	4.0	5.0	5.0
lb per lb gas	1.63	1.55	1.55
Nitrogen, ft³ per ft³ gas	18.9	24.6	24.6
lb per lb gas	12.0	11.8	11.8

[a] Properties are for commercial products and vary with composition.

[b] All values at 60°F and 14.696 psia unless otherwise stated.

LPG can usually be burned in existing natural gas burners, provided the air/gas ratio is properly readjusted. On large multiple burner installations an automatic propane–air mixing station is usually installed to facilitate quick changeover without changing air/gas ratios. (See the discussion of Wobbe index, Section 2.10.) Some fuel must be consumed to produce steam or hot water to operate a vaporizer for most industrial installations.

Table 4 lists some properties of commercial LPG, but it is suggested that more specific information be obtained from the local supplier.

REFERENCES

1. M. G. Fryback, "Synthetic Fuels—Promises and Problems," *Chemical Engineering Progress* (May 1981).
2. R. J. Reed (ed.), *Combustion Handbook,* Vol. I, North American Mfg. Co., Cleveland, OH, 1986, pp. 12, 36–38.
3. F. E. Vandeveer and C. G. Segeler, "Combustion," in *Gas Engineers Handbook,* C. G. Segeler (ed.), Industrial Press, New York, 1965, pp. 2/75–2/76.
4. H. F. Coward and G. W. Jones, *Limits of Flammability of Gases and Vapors* (U.S. Bureau of Mines Bulletin 503), U.S. Government Printing Office, Washington, DC, 1952, pp. 20–81.
5. E. W. Evans and R. W. Miller, "Testing and Properties of LP-Gases," in *Gas Engineers Handbook,* C. G. Segeler (ed.), Industrial Press, New York, 1965, p. 5/11.
6. W. Trinks, M. H. Mawhinney, R. A. Shannon, R. J. Reed, and J. R. Garvey, *Industrial Furnaces,* 6th ed., Wiley, Hoboken, NJ, 2003.

CHAPTER 18

LIQUID FOSSIL FUELS FROM PETROLEUM

Richard J. Reed
North American Manufacturing Company
Cleveland, Ohio

1 INTRODUCTION

The major source of liquid fuels is crude petroleum; other sources are shale and tar sands. Synthetic hydrocarbon fuels—gasoline and methanol—can be made from coal and natural gas. Ethanol, some of which is used as an automotive fuel, is derived from vegetable matter.

Crude petroleum and refined products are a mix of a wide variety of hydrocarbons—aliphatics (straight- or branched-chained paraffins and olefins), aromatics (closed rings, six carbons per ring with alternate double bonds joining the ring carbons, with or without aliphatic side chains), and naphthenic or cycloparaffins (closed single-bonded carbon rings, five to six carbons).

Very little crude petroleum is used in its natural state. Refining is required to yield marketable products that are separated by distillation into fractions including a specific boiling range. Further processing (such as cracking, reforming, and alkylation) alters molecular structure of some of the hydrocarbons and enhances the yield and properties of the refined products.

Crude petroleum is the major source of liquid fuels in the United States now and for the immediate future. Although the oil embargo of 1973–1974 intensified development of facilities for extraction of oil from shale and of hydrocarbon liquids from coal, the economics do not favor early commercialization of these processes. Their development has been slowed by an apparently adequate supply of crude oil. Tar sands are being processed in small amounts in Canada, but no commercial facility exists in the United States. (See Table 1.)

Except for commercial propane and butane, fuels for heating and power generation are generally heavier and less volatile than fuels used in transportation. The higher the "flash point," the less hazardous is handling of the fuel. (Flash point is the minimum temperature at which the fuel oil will catch fire if exposed to naked flame. Minimum flash points are stipulated by law for safe storage and handling of various grades of oils.) (See Table 5 of Chapter 6.)

For most of the information in this chapter, the author is deeply indebted to John W. Thomas, retired Chief Mechanical Engineer of the Standard Oil Company (Ohio).

Table 1 Principal Uses of Liquid Fuels

Heat and Power

Fuel oil	Space heating (residential, commercial, industrial)
	Steam generation for electric power
	Industrial process heating
	Refinery and chemical feedstock
Kerosene	Supplemental space heating
Turbine fuel	Stationary power generation
Diesel fuel	Stationary power generation
Liquid propane[a]	Isolated residential space heating
	Standby industrial process heating

Transportation

Jet fuel	Aviation turbines
Diesel fuel	Automotive engines
	Marine engines
	Truck engines
Gasoline	Automotive
	Aviation
Liquid propane and butane[a]	Limited automotive use

[a] See Chapter 8 on gaseous fossil fuels.

Properties of fuels reflect the characteristics of the crude. Paraffinic crudes have a high concentration of straight-chain hydrocarbons, which may leave a wax residue with distillation. Aromatic and naphthenic crudes have concentrations of ring hydrocarbons. Asphaltic crudes have a preponderance of heavier ring hydrocarbons and leave a residue after distillation. (See Table 2.)

Table 2 Ultimate Chemical Analyses of Various Crudes[a,6]

Crude Petroleum Source	% wt of					Specific Gravity (at temperature, °F)	Base
	C	H	N	O	S		
Baku, USSR	86.5	12.0		1.5		0.897	
California	86.4	11.7	1.14		0.60	0.951 (at 59°F)	Naphthene
Colombia, South America	85.62	11.91	0.54				
Kansas	85.6	12.4			0.37	0.912	Mixed
Mexico	83.0	11.0	___1.7___		4.30	0.97 (at 59°F)	Naphthene
Oklahoma	85.0	12.9			0.76		Mixed
Pennsylvania	85.5	14.2				0.862 (at 59°F)	Paraffin
Texas	85.7	11.0	___2.61___		0.70	0.91	Naphthene
West Virginia	83.6	12.9		3.6		0.897 (at 32°F)	Paraffin

[a] See, also, Table 7.

2 FUEL OILS

Liquid fuels in common use are broadly classified as follows:

1. Distillate fuel oils derived directly or indirectly from crude petroleum
2. Residual fuel oils that result after crude petroleum is topped; or viscous residua from refining operations
3. Blended fuel oils, mixtures of the above

The distillate fuels have lower specific gravity and are less viscous than residual fuel oils. Petroleum refiners burn a varying mix of crude residue and distilled oils in their process heaters. The changing gravity and viscosity require maximum oil preheat for atomization good enough to assure complete combustion. Tables 3–8 describe oils in current use. Some terms used in those tables are defined below.

Aniline point is the lowest Fahrenheit temperature at which an oil is completely miscible with an equal volume of freshly distilled aniline.

API gravity is a scale of specific gravity for hydrocarbon mixtures referred to in "degrees API" (for American Petroleum Institute). The relationships between API gravity, specific gravity, and density are:

$$\text{sp gr } 60/60°F = \frac{141.5}{°API + 131.5}$$

where °API is measured at 60°F (15.6°C).

$$\text{sp gr } 60/60°F = \frac{\text{lb/ft}^3}{62.3}$$

where lb/ft^3 is measured at 60°F (15.6°C).

Table 3 Some Properties of Liquid Fuels[2]

Property	Gasoline	Kerosene	Diesel Fuel	Light Fuel Oil	Heavy Fuel Oil	Coal Tar Fuel	Bituminous Coal (for Comparison)
Analysis, % wt							
C	85.5	86.3	86.3	86.2	86.2	90.0	80.0
H	14.4	13.6	12.7	12.3	11.8	6.0	5.5
N						1.2	1.5
O						2.5	7
S	0.1	0.1	1.0	1.5	2.0	0.4	1
Boiling range, °F	104–365	284–536	356 up	392 up	482 up	392 up	
Flash point, °F	−40	102	167	176	230	149	
Gravity specific at 59°F	0.73	0.79	0.87	0.89	0.95	1.1	1.25
Heat value, net							
cal/g	10,450	10,400	10,300	10,100	9,900	9,000	7,750
Btu/lb	18,810	18,720	18,540	18,180	17,820	16,200	13,950
Btu/US gal	114,929	131,108	129,800	131,215	141,325		
Residue, % wt at 662°F			15	50	60	60	
Viscosity, kinematic							
Centistokes at 59°F	0.75	1.6	5.0	50	1,200	1,500	
Centistokes at 212°F		0.6	1.2	3.5	20	18	

Table 4 Gravities and Related Properties of Liquid Petroleum Products

°API	Specific Gravity 60°F/60°F (15.6°C/15.6°C)	lb/ gal	kg/ m³	Gross Btu/ gal[a]	Gross kcal/ liter[a]	% H, wt[a]	Net Btu/ gal[a]	Net kcal/ liter[a]	Specific Heat @ 40°F	Specific Heat @ 300°F	Temperature Correction °API/°F[a]	ft³ 60°F air/ gal	Ultimate % CO₂
0	1.076	8.969	1075	160,426	10,681	8.359	153,664	10,231	0.391	0.504	0.045	1581	—
2	1.060	8.834	1059	159,038	10,589	8.601	152,183	10,133	0.394	0.508	—	—	—
4	1.044	8.704	1043	157,692	10,499	8.836	150,752	10,037	0.397	0.512	—	—	18.0
6	1.029	8.577	1028	156,384	10,412	9.064	149,368	9,945	0.400	0.516	0.048	1529	17.6
8	1.014	8.454	1013	155,115	10,328	9.285	148,028	9,856	0.403	0.519	0.050	1513	17.1
10[b]	1.000[b]	8.335[b]	1000[b]	153,881	10,246	10.00	146,351	9,744	0.406	0.523	0.051	1509	16.7
12	0.986	8.219	985.0	152,681	10,166	10.21	145,100	9,661	0.409	0.527	0.052	1494	16.4
14	0.973	8.106	971.5	151,515	10,088	10.41	143,888	9,580	0.412	0.530	0.054	1478	16.1
16	0.959	7.996	958.3	150,380	10,013	10.61	147,712	9,502	0.415	0.534	0.056	1463	15.8
18	0.946	7.889	945.5	149,275	9,939	10.80	141,572	9,426	0.417	0.538	0.058	1448	15.5
20	0.934	7.785	933.0	148,200	9,867	10.99	140,466	9,353	0.420	0.541	0.060	1433	15.2
22	0.922	7.683	920.9	147,153	9,798	11.37	139,251	9,272	0.423	0.545	0.061	1423	14.9
24	0.910	7.585	909.0	146,132	9,730	11.55	138,210	9,202	0.426	0.548	0.063	1409	14.7
26	0.898	7.488	897.5	145,138	9,664	11.72	137,198	9,135	0.428	0.552	0.065	1395	14.5
28	0.887	7.394	886.2	144,168	9,599	11.89	136,214	9,069	0.431	0.555	0.067	1381	14.3
30	0.876	7.303	875.2	143,223	9,536	12.06	135,258	9,006	0.434	0.559	0.069	1368	14.0
32	0.865	7.213	864.5	142,300	9,475	12.47	134,163	8,933	0.436	0.562	0.072	1360	13.8
34	0.855	7.126	854.1	141,400	9,415	12.63	133,259	8,873	0.439	0.566	0.074	1347	13.6
36	0.845	7.041	843.9	140,521	9,356	12.78	132,380	8,814	0.442	0.569	0.076	1334	13.4
38	0.835	6.958	833.9	139,664	9,299	12.93	131,524	8,757	0.444	0.572	0.079	1321	13.3
40	0.825	6.887	824.1	138,826	9,243	13.07	130,689	8,702	0.447	0.576	0.082	1309	13.1
42	0.816	6.798	814.7	138,007	9,189	—	—	—	0.450	0.579	0.085	—	13.0
44	0.806	6.720	805.4	137,207	9,136	—	—	—	0.452	0.582	0.088	—	12.8

Typical Ranges for —
Diesel Fuels: 1D (48), 2D (48), JET A (47)
Aviation Turbine Fuels: JET A (47), JP5 (48), JP4 (48), (56)
Fuel Oils: #6, #5, #4, #2, #1

[a] For gravity measured at 60°F (15.6°C) only.
[b] Same as H₂O.

629

Table 5 Heating Requirements for Products Derived from Petroleum[3]

Commercial Fuels	Specific Gravity at 60°F/60°F (15.6°C)	Distillation Range, °F (°C)	Vapor Pressure,[a] psia (mm Hg)	Latent Btu/gal[b] to Vaporize	Btu/gal[b] to Heat from 32°F (0°C) to		
					Pumping Temperature	Atomizing Temperature	Vapor
No. 6 oil	0.965	600–1000 (300–500)	0.054 (2.8)	764	371	996	3619[c]
No. 5 oil	0.945	600–1000 (300–500)	0.004 (0.2)	749	133	635	3559[c]
No. 4 oil	0.902	325–1000 (150–500)	0.232 (12)	737	—	313	2725[c]
No. 2 oil	0.849	325– 750 (150–400)	0.019 (1)	743	—	—	2704[c]
Kerosene	0.780	256– 481 (160–285)	0.039 (2)	750	—	—	1303[c]
Gasoline	0.733	35– 300 (37–185)	0.135 (7)	772	—	—	1215[c]
Methanol	0.796	148 (64)	4.62 (239)	3140	—	—	3400[d]
Butane	0.582	31 (0)	31 (1604)	808	—	—	976[d]
Propane	0.509	−44 (−42)	124 (6415)	785	—	—	963[d]

[a] At the atomizing temperature or 60°F, whichever is lower. Based on a sample with the lowest boiling point from column 3.

[b] To convert Btu/US gallon to kcal/liter, multiply by 0.666. To convert Btu/US gallon to Btu/lb, divide by 8.335 × sp gr, from column 2. To convert Btu/US gallon to kcal/kg, divide by 15.00 × sp gr, from column 2.

[c] Calculated for boiling at midpoint of distillation range, from column 3.

[d] Includes latent heat plus sensible heat of the vapor heated from boiling point to 60°F (15.6°C).

Table 6 Analyses and Characteristics of Selected Fuel Oils[3]

| Source | Ultimate Analysis (% Weight) | | | | | | ppm if > 50 | % wt Asphaltine | % wt C Residue | °API at 60°F | Flash Point, °F | HV, Btu/lb | | Pour Point, °F | Viscosity, SSU | |
	C	H	N	S	Ash	O[a]						Gross	Net		At 140°F	At 210°F
Alaska	86.99	12.07	0.007	0.31	<0.001	0.62	—	—	—	33.1	—	—	—	—	33.0	29.5
California	86.8	12.52	0.053	0.27	<0.001	0.36	—	—	—	32.6	—	19,330	—	—	30.8	29.5
West Texas	88.09	9.76	0.026	1.88	<0.001	0.24	—	—	—	18.3	—	—	—	—	32.0	28.8
Alaska	86.04	11.18	0.51	1.63	0.034	0.61	50 Ni 67 V	5.6	12.9	15.6	215	18,470	17,580	38	1071	194
California	86.66	10.44	0.86	0.99	0.20	0.85		8.62	15.2	12.6	180	18,230	17,280	42	720	200
DFM (shale)	86.18	13.00	0.24	0.51	0.003	1.07	[b]	0.036	4.1	33.1	182	19,430	18,240	40	36.1	30.7
Gulf of Mexico	84.62	10.77	0.36	2.44	0.027	1.78	—	7.02	14.8	13.2	155	18,240	17,260	40	835	181
Indo/Malaysia	86.53	11.93	0.24	0.22	0.036	1.04	101 V	0.74	3.98	21.8	210	19,070	17,980	61	199	65
Middle East[c]	86.78	11.95	0.18	0.67	0.012	0.41	—	3.24	6.0	19.8	350	19,070	17,980	48	490	131.8
Pennsylvania[d]	84.82	11.21	0.34	2.26	0.067	1.3	65 Na 82 V	4.04	12.4	15.4	275	18,520	17,500	66	1049	240
Venezuela	85.24	10.96	0.40	2.22	0.081	1.10	52 Ni 226 V	8.4	6.8	14.1	210	18,400	17,400	58	742	196.7
Venezuela desulfurized	85.92	12.05	0.24	0.93	0.033	0.83	101 V	2.59	5.1	23.3	176	18,400	17,300	48	113.2	50.5

[a] By difference.
[b] 91 Ca, 77 Fe, 88 Ni, 66 V.
[c] Exxon.
[d] Amerada Hess.

Table 7 ASTM Fuel Oil Specifications[8]

Grade of Fuel Oil[a]	Flash Point, °C (°F) Min	Pour Point, °C (°F) Max	Water and Sediment, Vol % Max	Carbon Residue on 10% Bottoms, % Max	Ash, Weight % Max	Distillation Temperatures, °C (°F) 10% Point Max	90% Point Min	90% Point Max	Saybolt Viscosity, s[d] Universal at 38°C (100°F) Min	Max	Furol at 50°C (122°F) Min	Max	Kinematic Viscosity, cSt[d] At 38°C (100°F) Min	Max	At 40°C (104°F) Min	Max	At 50°C (122°F) Min	Max	Specific Gravity, 60/60°F (deg API) Max	Copper Strip Corrosion Max	Sulfur, % Max
No. 1 A distillate oil intended for vaporizing pot-type burners and other burners requiring this grade of fuel	38 (100)	−18[c] (0)	0.05	0.15	—	215 (420)	—	288 (550)	—	—	—	—	1.4	2.2	1.3	2.1	—	—	0.8499 (35 min)	No. 3	0.5
No. 2 A distillate oil for general purpose heating for use in burners not requiring No. 1 fuel oil	38 (100)	−6[c] (20)	0.05	0.35	—	—	282[c] (540)	338 (640)	(32.6)	(37.9)	—	—	2.0[c]	3.6	1.9[c]	3.4	—	—	0.8762 (30 min)	No. 3	0.5[b]
No. 4 (Light) Preheating not usually required for handling or burning	38 (100)	−6[c] (20)	0.50	—	0.05	—	—	—	(32.6)	(45)	—	—	2.0	5.8	—	—	—	—	0.876[g] (30 max)	—	—
No. 4 Preheating not usually required for handling or burning	55 (130)	−6[c] (20)	0.50	—	0.10	—	—	—	(45)	(125)	—	—	5.8	26.4[h]	5.5	24.0[f]	—	—	—	—	—

632

Grade	Flash Point °C (°F) min	Pour Point	Water and Sediment % vol	Carbon Residue	Ash % mass	Distillation	Kinematic Viscosity				Saybolt Viscosity	
No. 5 (Light) Preheating may be required depending on climate and equipment	55 (130)	—	1.00	—	0.10	—	(>125) (300)	—	—	>26.4 65^f	>24.0 58^f	—	—
No. 5 (Heavy) Preheating may be required for burning and, in cold climates, may be required for handling	55 (130)	—	1.00	—	0.10	—	(>300) (900)	(23) (40)	>65 194^f	58 168^f	(42) (81)	—	—
No. 6 Preheating required for burning and handling	60 (140)	^{g}	2.00^e	—	—	—	(>900) (9000)	(>45) (300)	—	—	—	>92 638^f	—

^{a} It is the intent of these classifications that failure to meet any requirement of a given grade does not automatically place an oil in the next lower grade unless in fact it meets all requirements of the lower grade.

^{b} In countries outside the United States other sulfur limits may apply.

^{c} Lower or higher pour points may be specified whenever required by conditions of storage or use. When pour point less than −18°C (0°F) is specified, the minimum viscosity for grade No. 2 shall be 1.7 cSt (31.5 SUS) and the minimum 90% point shall be waived.

^{d} Viscosity values in parentheses are for information only and not necessarily limiting.

^{e} The amount of water by distillation plus the sediment by extraction shall not exceed 2.00%. The amount of sediment by extraction shall not exceed 0.50%. A deduction in quantity shall be made for all water and sediment in excess of 1.0%.

^{f} Where low-sulfur fuel oil is required, fuel oil falling in the viscosity range of a lower numbered grade down to and including No. 4 may be supplied by agreement between purchaser and supplier. The viscosity range of the initial shipment shall be identified and advance notice shall be required when changing from one viscosity range to another. This notice shall be in sufficient time to permit the user to make the necessary adjustments.

^{g} This limit guarantees a minimum heating value and also prevents misrepresentation and misapplication of this product as Grade No. 2.

^{h} Where low-sulfur fuel oil is required, Grade 6 fuel oil will be classified as low pour +15°C (60°F) max or high pour (no max). Low-pour fuel oil should be used unless all tanks and lines are heated.

Table 8 Application of ASTM Fuel Oil Grades, as Described by One Burner Manufacturer

Fuel Oil	Description
No. 1	Distillate oil for vaporizing-type burners
No. 2	Distillate oil for general purpose use, and for burners not requiring No. 1 fuel oil
No. 4	Blended oil intended for use without preheating
No. 5	Blended residual oil for use with preheating; usual preheat temperature is 120–220°F
No. 6	Residual oil for use with preheaters permitting a high-viscosity fuel; usual preheat temperature is 180–260°F
Bunker C	Heavy residual oil, originally intended for oceangoing ships

SSU (or SUS) is seconds, Saybolt Universal, a measure of kinematic viscosity determined by measuring the time required for a specified quantity of the sample oil to flow by gravity through a specified orifice at a specified temperature. For heavier, more viscous oils, a larger (Furol) orifice is used, and the results are reported as SSF (seconds, Saybolt Furol).

$$\text{kin visc in centistokes} = 0.226 \times SSU - 195/SSU, \text{ for SSU } 32\text{--}100$$

$$\text{kin visc in centistokes} = 0.220 \times SSU - 135/SSU, \text{ for SSU} > 100$$

$$\text{kin visc in centistokes} = 2.24 \times SSF - 184/SSF, \text{ for SSF } 25\text{--}40$$

$$\text{kin visc in centistokes} = 2.16 \times SSF - 60/SSF, \text{ for SSF} > 40$$

$$1 \text{ centistoke (cSt)} = 0.000001 \text{ m}^2/\text{sec}$$

Unlike distillates, residual oils contain noticeable amounts of inorganic matter, ash content ranging from 0.01% to 0.1%. Ash often contains vanadium, which causes serious corrosion in boilers and heaters. (A common specification for refinery process heaters requires 50% nickel–50% chromium alloy for tube supports and hangers when the vanadium exceeds 150 ppm.) V_2O_5 also lowers the eutectic of many refractories, causing rapid disintegration. Crudes that often contain high vanadium are

Venezuela, Bachaqoro	350 ppm
Iran	350–440 ppm
Alaska, North Slope	80 ppm

2.1 Kerosene

Kerosene is a refined petroleum distillate consisting of a homogeneous mixture of hydrocarbons. It is used mainly in wick-fed illuminating lamps and kerosene burners. Oil for illumination and for domestic stoves must be high in paraffins to give low smoke. The presence of naphthenic and especially aromatic hydrocarbons increases the smoking tendency. A "smoke point" specification is a measure of flame height at which the tip becomes smoky. The "smoke point" is about 73 mm for paraffins, 34 mm for naphthalenes, and 7.5 mm for aromatics and mixtures.

Low sulfur content is necessary in kerosenes because:

1. Sulfur forms a bloom on glass lamp chimneys and promotes carbon formation on wicks.

2. Sulfur forms oxides in heating stoves. These swell, are corrosive and toxic, creating a health hazard, particularly in nonvented stoves.

Kerosene grades[9] (see Table 9) in the United States are

No. 1 K: A special low-sulfur grade kerosene suitable for critical kerosene burner applications

No. 2 K: A regular-grade kerosene suitable for use in flue-connected burner applications and for use in wick-fed illuminating lamps

2.2 Aviation Turbine Fuels

The most important requirements of aircraft jet fuel relate to freezing point, distillation range, and level of aromatics. Fluidity at low temperature is important to ensure atomization. A typical upper viscosity limit is 7–10 cSt at 0°F, with the freezing point as low as −60°F.

Aromatics are objectionable because (1) coking deposits from the flame are most pronounced with aromatics of high C/H ratio and less pronounced with short-chain compounds, and (2) they must be controlled to keep the combustor liner at an acceptable temperature.

Jet fuels for civil aviation are identified as jet A and A1 (high-flash-point, kerosene-type distillates), and jet B (a relatively wide boiling range, volatile distillate).

Jet fuels for military aviation are identified as JP4 and JP5. The JP4 has a low flash point and a wide boiling range. The JP5 has a high flash point and a narrow boiling range. (See Table 10.)

Gas turbine fuel oils for other than use in aircraft must be free of inorganic acid and low in solid or fibrous materials. (See Tables 11 and 12.) All such oils must be homogeneous mixtures that do not separate by gravity into light and heavy components.

2.3 Diesel Fuels

Diesel engines, developed by Rudolf Diesel, rely on the heat of compression to achieve ignition of the fuel. Fuel is injected into the combustion chamber in an atomized spray at the end of the compression stroke, after air has been compressed to 450–650 psi and has reached a temperature, due to compression, of at least 932°F (500°C). This temperature

Table 9 ASTM Chemical and Physical Requirements for Kerosene[9]

Property	Limit
Distillation temperature	
10% recovered	401°F (205°C)
Final boiling point	572°F (300°C)
Flash point	100°F (38°C)
Freezing point	−22°F (−30°C)
Sulfur, % weight	
No. 1 K	0.04 maximum
No. 2 K	0.30 maximum
Viscosity, kinematic at 104°F (40°C), centistokes	1.0 min / 1.9 max

Table 10 ASTM Specifications[10] and Typical Properties[7] of Aviation Turbine Fuels

| Property | Specifications | | | Typical, 1979 | | |
	Jet A	Jet A1	Jet B	26 Samples JP4	7 Samples JP5	60 Samples Jet A
Aromatics, % vol	20	20	20	13.0	16.4	17.9
Boiling point, final, °F	572	572	—	—	—	—
Distillation, max temperature, °F						
For 10% recovered	400	400	—	208	387	375
For 20% recovered	—	—	290	—	—	—
For 50% recovered	—	—	370	293	423	416
For 90% recovered	—	—	470	388	470	473
Flash point, min, °F	100	100	—	—	—	—
Freezing point, max, °F	−40	−53	−58	−110	−71	−56
Gravity, API, max	51	51	57	53.5	41.2	42.7
Gravity, API, min	37	37	45	—	—	—
Gravity, specific 60°F min	0.7753	0.7753	0.7507	0.765	0.819	0.812
Gravity, specific 60°F max	0.8398	0.8398	0.8017	—	—	—
Heating value, gross Btu/lb	—	—	—	18,700	18,530	18,598
Heating value, gross Btu/lb min	18,400	18,400	18,400	—	—	—
Mercaptan, % wt	0.003	0.003	0.003	0.0004	0.0003	0.0008
Sulfur, max % wt	0.3	0.3	0.3	0.030	0.044	0.050
Vapor pressure, Reid, psi	—	—	3	2.5	—	0.2
Viscosity, max SSU						
At −4°F	52	—	—	—	—	—
At −30°F	—	—	—	34–37	60.5	54.8

ignites the fuel and initiates the piston's power stroke. The fuel is injected at about 2000 psi to ensure good mixing.

Diesels are extensively used in truck transport, rail trains, and marine engines. They are being used more in automobiles. In addition, they are employed in industrial and commercial stationary power plants.

Fuels for diesels vary from kerosene to medium residual oils. The choice is dictated by engine characteristics, namely, cylinder diameter, engine speed, and combustion wall temperature. High-speed small engines require lighter fuels and are more sensitive to fuel quality

Table 11 Nonaviation Gas Turbine Fuel Grades per ASTM[11]

Grade	Description
No. 0-GT	A naphtha or low-flash-point hydrocarbon liquid
No. 1-GT	A distillate for gas turbines requiring cleaner burning than No. 2-GT
No. 2-GT	A distillate fuel of low ash suitable for gas turbines not requiring No. 1-GT
No. 3-GT	A low ash fuel that may contain residual components
No. 4-GT	A fuel containing residual components and having higher vanadium content than No. 3-GT

Table 12 ASTM Specifications[11] for Nonaviation Gas Turbine Fuels

Property	Specifications				
	0-GT	1-GT	2-GT	3-GT	4-GT
Ash, max % wt	0.01	0.01	0.01	0.03	—
Carbon residue, max % wt	0.15	0.15	0.35	—	—
Distillation, 90% point, max °F	—	(550)[a]	(640)	—	—
Distillation, 90% point, min °F	—	—	(540)	—	—
Flash point, min °F	—	(100)	(100)	(130)	(150)
Gravity, API min	—	(35)	(30)	—	—
Gravity, spec 60°F max	—	0.850	0.876	—	—
Pour point, max °F	—	(0)	(20)	—	—
Viscosity, kinematic					
Min SSU at 100°F	—	—	(32.6)	(45)	(45)
Max SSU at 100°F	—	(34.4)	(40.2)	—	—
Max SSF at 122°F	—	—	—	(300)	(300)
Water and sediment, max % vol	0.05	0.05	0.05	1.0	1.0

[a] Values in parentheses are approximate.

variations. Slow-speed, larger industrial and marine engines use heavier grades of diesel fuel oil.

Ignition qualities and viscosity are important characteristics that determine performance. The ignition qualities of diesel fuels may be assessed in terms of their cetane numbers or diesel indices. Although the diesel index is a useful indication of ignition quality, it is not as reliable as the cetane number, which is based on an engine test:

$$\text{Diesel index} = (\text{Aniline point, °F}) \times (\text{API gravity}/100)$$

The diesel index is an arbitrary figure having a significance similar to cetane number, but having a value 1–5 numbers higher.

The cetane number is the percentage by volume of cetane in a mixture of cetane with an ethylnaphthalene that has the same ignition characteristics as the fuel. The comparison is made in a diesel engine equipped either with means for measuring the delay period between injection and ignition or with a surge chamber, separated from the engine intake port by a throttle in which the critical measure below which ignition does not occur can be measured. Secondary reference fuels with specific cetane numbers are available. Cetane number is a measure of ignition quality and influences combustion roughness.

The use of a fuel with too low a cetane number results in accumulation of fuel in the cylinder before combustion, causing "diesel knock." Too high a cetane number will cause rapid ignition and high fuel consumption.

The higher the engine speed, the higher the required fuel cetane number. Suggested rpm values for various fuel cetane numbers are shown in Table 13.[5] Engine size and operating conditions are important factors in establishing approximate ignition qualities of a fuel.

Too viscous an oil will cause large spray droplets and incomplete combustion. Too low a viscosity may cause fuel leakage from high-pressure pumps and injection needle valves. Preheating permits use of higher viscosity oils.

To minimize injection system wear, fuels are filtered to remove grit. Fine-gauge filters are considered adequate for engines up to 8 Hz, but high-speed engines usually have fabric

Table 13 ASTM Fuel Cetane Numbers for Various Engine Speeds[5]

Engine Speed (rpm)	Cetane Number
Above 1500	50–60
500–1500	45–55
400–800	35–50
200–400	30–45
100–200	15–40
Below 200	15–30

or felt filters. It is possible for wax to crystallize from diesel fuels in cold weather, therefore, preheating before filtering is essential.

To minimize engine corrosion from combustion products, control of fuel sulfur level is required. (See Tables 14 and 15.)

2.4 Summary

Aviation jet fuels, gas turbine fuels, kerosenes, and diesel fuels are very similar. The following note from Table 1 of Ref. 11 highlights this:

> No. 0-GT includes naphtha, Jet B fuel, and other volatile hydrocarbon liquids. No. 1-GT corresponds in general to Spec D396 Grade No. 1 fuel and Classification D975 Grade No. 1-D Diesel fuel in physical properties. No. 2-GT corresponds in general to Spec D396 Grade No. 2 fuel and Classification D975 Grade No. 2 Diesel fuel in physical properties. No. 3-GT and No. 4-GT viscosity range brackets Spec D396 and Grade No. 4, No. 5 (light), No. 5 (heavy), No. 6, and Classification D975 Grade No. 4-D Diesel fuel in physical properties.

3 SHALE OILS

As this is written, there are no plants in the United States producing commercial shale oil. Predictions are that the output products will be close in characteristics and performance to those made from petroleum crudes.

Table 16 lists properties of a residual fuel oil (DMF) from one shale pilot operation and of a shale crude oil.[13] Table 17 lists ultimate analyses of oils derived from shales from a

Table 14 ASTM Diesel Fuel Descriptions[12]

Grade	Description
No. 1D	A volatile distillate fuel oil for engines in service requiring frequent speed and load changes
No. 2D	Distillate fuel oil of lower volatility for engines in industrial and heavy mobile service
No. 4D	A fuel oil for low and medium speed diesel engines
Type CB	For buses, essentially 1D
Type TT	For trucks, essentially 2D
Type RR	For railroads, essentially 2D
Type SM	For stationary and marine use, essentially 2D or heavier

Table 15a ASTM Detailed Requirements for Diesel Fuel Oils[a,h,12]

Grade of Diesel Fuel Oil	Flash Point, °C (°F) Min	Cloud Point, °C (°F) Max	Water and Sediment, Vol% Max	Carbon Residue on 10% Residuum,% Max	Ash, Weight % Max	Distillation Temperatures, °C (°F), 90% Point		Viscosity				Sulfur,[d] Weight % Max	Copper Strip Corrosion Max	Cetane Number[e] Min
								Kinematic, cSt[g] at 40°C		Saybolt, SUS at 100°F				
						Min	Max	Min	Max	Min	Max			
No. 1-D A volatile distillate fuel oil for engines in service requiring frequent speed and load changes	38 (100)	b	0.05	0.15	0.01	—	288 (550)	1.3	2.4	—	34.4	0.50	No.3	40[f]
No. 2-D A distillate fuel oil of lower volatility for engines in industrial and heavy mobile service	52 (125)	b	0.05	0.35	0.01	282[c] (540)	338 (640)	1.9	4.1	32.6	40.1	0.50	No.3	40[f]
No. 4-D A fuel oil for low and medium speed engines	55 (130)	b	0.50	—	0.10	—	—	5.5	24.0	45.0	125.0	2.0	—	30[f]

[a] To meet special operating conditions, modifications of individual limiting requirements may be agreed upon between purchaser, seller, and manufacturer.

[b] It is unrealistic to specify low-temperature properties that will ensure satisfactory operation on a broad basis. Satisfactory operation should be achieved in most cases if the cloud point (or wax appearance point) is specified at 6°C above the tenth percentile minimum ambient temperature for the area in which the fuel will be used. This guidance is of a general nature; some equipment designs, using improved additives, fuel properties, and/or operations, may allow higher or require lower cloud point fuels. Appropriate low-temperature operability properties should be agreed on between the fuel supplier and purchaser for the intended use and expected ambient temperatures.

[c] When cloud point less than −12°C (10°F) is specified, the minimum viscosity shall be 1.7 cSt (or mm²/sec) and the 90% point shall be waived.

[d] In countries outside the United States, other sulfur limits may apply.

[e] Where cetane number by Method D613 is not available, ASTM Method D976, Calculated Cetane Index of Distillate Fuels may be used as an approximation. Where there is disagreement, Method D613 shall be the referee method.

[f] Low-atmospheric temperatures as well as engine operation at high altitudes may require use of fuels with higher cetane ratings.

[g] cSt = 1 mm²/sec.

[h] The values stated in SI units are to be regarded as the standard. The values in U.S. customary units are for information only.

Table 15b ASTM Typical Properties of Diesel Fuels[7]

	All United States, 1981						Eastern United States, 1981																	
	48 Samples			112 Samples			24 Samples			44 Samples			13 Samples			4 Samples		
	No. 1D			No. 2D			Type CB			Type TT			Type RR			Type SM		
Property	Min	Avg	Max	Min	Avg	Max	Min	Avg	Max	Min	Avg	Max	Min	Avg	Max	Min	Avg	Max
Ash, % wt	0.000	0.001	0.005	0.000	0.002	0.020	—	0.001	0.005	—	0.002	0.015	—	0.000	0.001	—	0.001	0.001
Carbon residue, % wt	0.000	0.059	0.067	0.000	0.101	0.300	—	—	0.21	0.101	—	0.25	—	0.121	0.23	—	0.148	0.21
Cetane number	36	46.7	53.0	29.0	45.6	52.4	—	49.8	—	—	45.6	—	—	44.8	—	—	—	—
Distillation, 90% point, °F	445	448	560	493	587	640	451	512	640	451	571	640	540	590	640	482	577	640
Flash point, °F	104	138	176	132	166	240	120	140	240	120	162	240	156	164	192	136	162	180
Gravity, API spec, 60/60°F	37.8	42.4	47.9	22.8	34.9	43.1	—	41.5	—	—	36.3	—	—	33.8	—	—	35.3	—
	0.836	0.814	0.789	0.917	0.850	0.810	—	0.818	—	—	0.843	—	—	0.856	—	—	0.848	—
Sulfur, % wt	0.000	0.070	0.25	0.010	0.283	0.950	—	0.086	0.24	—	0.198	0.46	—	0.283	0.580	—	0.155	0.28
Viscosity, SSU at 100°F	32.6	33.3	35.7	33.8	36.0	40.3	32.9	34.3	40.2	32.9	35.7	40.2	34.2	36.0	37.8	36.0	—	37.8

Table 16 Properties of Shale Oils[13]

Property	DMF Residual	Crude
Ultimate analysis		
Carbon, % wt	86.18	84.6
Hydrogen, % wt	13.00	11.3
Nitrogen, % wt	0.24	2.08
Sulfur, % wt	0.51	0.63
Ash, % wt	0.003	0.026
Oxygen, % wt by difference	1.07	1.36
Conradson carbon residue, %	4.1	2.9
Asphaltene, %	0.036	1.33
Calcium, ppm	0.13	1.5
Iron, ppm	6.3	47.9
Manganese, ppm	0.06	0.17
Magnesium, ppm	—	5.40
Nickel, ppm	0.43	5.00
Sodium, ppm	0.09	11.71
Vanadium, ppm	0.1	0.3
Flash point, °F	182	250
Pour point, °F	40	80
API gravity at 60°F	33.1	20.3
Viscosity, SSU at 140°F	36.1	97
SSU at 210°F	30.7	44.1
Gross heating value, Btu/lb	19,430	18,290
Net heating value, Btu/lb	18,240	17,260

Table 17 Elemental Content of Shale Oils, % wt[14]

Source	Carbon, C Min	Avg	Max	Hydrogen, H Min	Avg	Max	Nitrogen, N Min	Avg	Max	Sulfur, S Min	Avg	Max	Oxygen, O Min	Avg	Max
Colorado	83.5	84.2	84.9	10.9	11.3	11.7	1.6	1.8	1.9	0.7	1.2	1.7	1.3	1.7	2.1
Utah	84.1	84.7	85.2	10.9	11.5	12.0	1.6	1.8	2.0	0.5	0.7	0.8	1.2	1.6	2.0
Wyoming	81.3	83.1	84.4	11.2	11.7	12.2	1.4	1.8	2.2	0.4	1.0	1.5	1.7	2.0	2.3
Kentucky	83.6	84.4	85.2	9.6	10.2	10.7	1.0	1.3	1.6	1.4	1.9	2.4	1.8	2.3	2.7
Queensland Australia (four locations)	80.0	82.2	85.5	10.0	11.1	12.8	1.0	1.2	1.6	0.3	1.9	6.0	1.1	3.0	6.6
Brazil		85.3			11.2			0.9			1.1			1.5	
Karak, Jordan	77.6	78.3	79.0	9.4	9.7	9.9	0.5	0.7	0.8	9.3	10.0	10.6	0.9	1.4	1.9
Timahdit Morocco	79.5	80.0	80.4	9.7	9.8	9.9	1.2	1.4	1.6	6.7	7.1	7.4	1.8	2.0	2.2
Sweden	86.5	86.5	86.5	9.0	9.4	9.8	0.6	0.7	0.7	1.7	1.9	2.1	1.4	1.6	1.7

Table 18 Chemical and Physical Properties of Several Tar Sand Bitumens[15]

	Uinta Basin, Utah	Southeast Utah	Athabasca, Alberta	Trapper Canyon, WY[a]	South TX	Santa Rosa, NM[a]	Big Clifty, KY	Bellamy, MO
Carbon, % wt	85.3	84.3	82.5	82.4	—	85.6	82.4	86.7
Hydrogen, % wt	11.2	10.2	10.6	10.3	—	10.1	10.8	10.3
Nitrogen, % wt	0.96	0.51	0.44	0.54	0.36	0.22	0.64	0.10
Sulfur, % wt	0.49	4.46	4.86	5.52	~10	2.30	1.55	0.75
H/C ratio	1.56	1.44	1.53	1.49	1.34	1.41	1.56	1.42
Vanadium, ppm	23	151	196	91	85	25	198	—
Nickel, ppm	96	62	82	53	24	23	80	—
Carbon residue, % wt	10.9	19.6	13.7	14.8	24.5	22.1	16.7	—
Pour point, °F	125	95	75	125	180	—	85	—
API gravity	11.6	9.2	9.5	5.4	−2.0	5	8.7	10

Viscosities range from 50,000 to 600,000 SSF (100,000 to 1,300,000 cSt).

[a] Outcrop samples.

number of locations.[14] Properties will vary with the process used for extraction from the shale. The objective of all such processes is only to provide feedstock for refineries. In turn, the refineries' subsequent processing will also affect the properties.

If petroleum shortages occur, they will probably provide the economic impetus for completion of developments already begun for the mining, processing, and refining of oils from shale.

4 OILS FROM TAR SANDS

At the time that this is written, the only commercially practical operation for extracting oil from tar sands is at Athabaska, Alberta, Canada, using surface mining techniques. When petroleum supplies become short, economic impetus therefrom will push completion of de-

Table 19 Elemental Composition of Bitumen and Oils Recovered from Tar Sands by Methods C and S[a,15]

	Bitumen	Light Oil C[b]	Heavy Oil C 1–4 Mo.	Heavy Oil C 5–6 Mo.	Product Oil C	Product Oil S[c]
Carbon, % wt	86.0	86.7	86.1	86.7	86.6	85.9
Hydrogen, % wt	11.2	12.2	11.8	11.3	11.6	11.3
Nitrogen, % wt	0.93	0.16	0.82	0.66	0.82	1.17
Sulfur, % wt	0.45	0.30	0.39	0.33	0.43	0.42
Oxygen, % wt	1.42	0.64	0.89	1.01	0.55	1.21

[a] These percentages are site- and project-specific.

[b] C = reverse-forward combustion.

[c] S = steamflood.

[d] By difference.

Table 20 Orimulsion Fuel Characteristics

Density	63 lb/ft³	
Apparent Viscosity	41F/20 sec-1—700 mPa	
	86F/20 sec-1—450 mPa	
	158F/100 sec-1—105 mPa	
Flash point	266°F	
Pour point	32°F	
Higher heating value	12,683 Btu/lb	
Lower heating value	11,694 Btu/lb	
Weight analysis	Carbon	60%
	Hydrogen	7.5%
	Sulfur	2.7%
	Nitrogen	0.5%
	Oxygen	0.2%
	Ash	0.25%
	Water	30%
	Vanadium	300 ppm
	Sodium	70 ppm
	Magnesium	350 ppm

velopments already well under way for mining, processing, and refining of oils from tar sands.

Table 18 lists chemical and physical properties of several tar sand bitumens.[15] Further refining will be necessary because of the high density, viscosity, and sulfur content of these oils.

Extensive deposits of tar sands are to be found around the globe, but most will have to be recovered by some *in situ* technique, firefiooding, or steam flooding. Yields tend to be small and properties vary with the recovery method, as illustrated in Table 19.[15]

5 OIL–WATER EMULSIONS

Emulsions of oil have offered some promise of low fuel cost and alternate fuel supply for some time. The following excerpts from Ref. 16 provide introductory information on a water emulsion with an oil from the vicinity of the Orinoco River in Venezuela. It is being marketed as "Orimulsion" by Petroleos de Veneauels SA and Bitor America Corp of Boca Raton, Florida. It is a natural bitumen, like a liquid coal that has been emulsified with water to make it possible to extract it from the earth and to transport it.

Table 20 shows some of its properties and contents. Although its original sulfur content is high, the ash is low. A low C/H ratio promises less CO_2 emission. Because of handleability concerns, it will probably find use mostly in large steam generators.

REFERENCES

1. "Journal Forecast Supply & Demand," *Oil and Gas Journal,* 131 (Jan. 25, 1982).
2. J. D. Gilchrist, *Fuels and Refractories,* Macmillan, New York, 1963.
3. R. J. Reed, *Combustion Handbook,* 3rd ed., Vol. 1. North American Manufacturing Co., Cleveland, OH, 1986.

4. Braine and King, *Fuels—Solid, Liquid, Gaseous,* St. Martin's Press, New York, 1967.

5. *Kempe's, Engineering Yearbook,* Morgan Grompium, London.

6. W. L. Nelson, *Petroleum Refinery Engineering,* McGraw-Hill, New York, 1968.

7a. E. M. Shelton, *Diesel Oils, DOE/BETC/PPS—81/5,* U.S. Department of Energy, Washington, DC, 1981.

7b. E. M. Shelton, *Heating Oils, DOE/BETC/PPS—80/4,* U.S. Department of Energy, Washington, DC, 1980.

7c. E. M. Shelton, *Aviation Turbine Fuel, DOE/BETC/PPS—80/2,* Department of Energy, Washington, DC, 1979.

8. ANSI/ASTM D396, *Standard Specification for Fuel Oils,* American Society for Testing and Materials, Philadelphia, PA, 1996.

9. ANSI/ASTM D3699, *Standard Specification for Kerosene,* American Society for Testing and Materials, Philadelphia, PA, 1996.

10. ANSI/ASTM D1655, *Standard Specification for Aviation Turbine Fuels,* American Society for Testing and Materials, Philadelphia, PA, 1996.

11. ANSI/ASTM D2880, *Standard Specification for Gas Turbine Fuel Oils,* American Society for Testing and Materials, Philadelphia, PA, 1996.

12. ANSI/ASTM D975, *Standard Specification for Diesel Fuel Oils,* American Society for Testing and Materials, Philadelphia, PA, 1996.

13. M. Heap et al., *The Influence of Fuel Characteristics on Nitrogen Oxide Formation—Bench Scale Studies,* Energy and Environmental Research Corp., Irvine, CA, 1979.

14. H. Tokairin and S. Morita, "Properties and Characterizations of Fischer-Assay-Retorted Oils from Major World Deposits," in *Synthetic Fuels from Oil Shale and Tar Sands,* Institute of Gas Technology, Chicago, IL, 1983.

15. K. P. Thomas et al., "Chemical and Physical Properties of Tar Sand Bitumens and Thermally Recovered Oils," in *Synthetic Fuels from Oil Shale and Tar Sands,* Institute of Gas Technology, Chicago, IL, 1983.

16. J. Makansi, "New Fuel Could Find Niche Between Oil, Coal," *POWER* (Dec. 1991).

17. W. Trinks, M. H. Mawhinney, R. A. Shannon, R. J. Reed, and J. R. Garvey, *Industrial Furnaces,* 6th ed., Wiley, Hoboken, NJ, 2003.

CHAPTER 19

COALS, LIGNITE, PEAT

James G. Keppeler
Progress Materials, Inc.
St. Petersburg, Florida

1 INTRODUCTION

1.1 Nature

Coal is a dark brown to black sedimentary rock derived primarily from the unoxidized remains of carbon-bearing plant tissues. It is a complex, combustible mixture of organic, chemical, and mineral materials found in strata, or "seams," in the earth, consisting of a wide variety of physical and chemical properties.

The principal types of coal, in order of metamorphic development, are lignite, subbituminous, bituminous, and anthracite. While not generally considered a coal, peat is the first development stage in the "coalification" process, in which there is a gradual increase in the carbon content of the fossil organic material, and a concomitant reduction in oxygen.

Coal substance is composed primarily of carbon, hydrogen, and oxygen, with minor amounts of nitrogen and sulfur, and varying amounts of moisture and mineral impurities.

1.2 Reserves—Worldwide and United States

According to the World Coal Study (see Ref. 3), the total geological resources of the world in "millions of tons of coal equivalent" (mtce) is 10,750,212, of which 662,932, or 6%, is submitted as "Technically and Economically Recoverable Resources." Millions of tons of coal equivalent is based on the metric ton (2205 lb) with a heat content of 12,600 Btu/lb (7000 kcal/kg).

A summary of the percentage of technically and economically recoverable reserves and the percentage of total recoverable by country is shown in Table 1. As indicated in Table 1,

Table 1

	Percentage of Recoverable[a] of Geological Resources	Percentage of Total Recoverable Reserves
Australia	5.5	4.9
Canada	1.3	0.6
Peoples Republic of China	6.8	14.9
Federal Republic of Germany	13.9	5.2
India	15.3	1.9
Poland	42.6	9.0
Republic of South Africa	59.7	6.5
United Kingdom	23.7	6.8
United States	6.5	25.2
Soviet Union	2.2	16.6
Other Countries	24.3	8.4
		100.0

[a]Technically and economically recoverable reserves. Percentage indicated is based on total geological resources reported by country.

Source: World Coal Study, *Coal—Bridge to the Future,* 1980.

the United States possesses over a quarter of the total recoverable reserves despite the low percentage of recovery compared to other countries. The interpretation of "technical and economic" recovery is subject to considerable variation and also to modification, as technical development and changing economic conditions dictate. Also, there are significant differences in density and heating values in various coals, and, therefore, the mtce definition should be kept in perspective.

In 1977, the world coal production was approximately 2450 mtce,[3] or about $\frac{1}{270}$th of the recoverable reserves. According to the U.S. Geological Survey, the remaining U.S. Coal Reserves total almost 4000 billion tons,[4] with overburden to 6000 ft in seams of 14 in. or more for bituminous and anthracite and in seams of 2½ ft or more for subbituminous coal and lignite. The U.S. Bureau of Mines and U.S. Geological Survey have further defined "Reserve Base" to provide a better indication of the technically and economically minable reserves, where a higher degree of identification and engineering evaluation is available.

A summary of the reserve base of U.S. coal is provided in Table 2.[5]

1.3   Classifications

Coals are classified by "rank," according to their degree of metamorphism, or progressive alteration, in the natural series from lignite to anthracite. Perhaps the most widely accepted standard for classification of coals is ASTM D388, which ranks coals according to fixed carbon and calorific value (expressed in Btu/lb) calculated to the mineral-matter-free basis. Higher-rank coals are classified according to fixed carbon on the dry basis; the lower-rank coals are classed according to calorific value on the moist basis. Agglomerating character is used to differentiate between certain adjacent groups. Table 3 shows the classification requirements.

Table 2 Demonstrated Reserve Base[a] of Coal in the United States in January 1980, by Rank (Millions of Short Tons)

State[b]	Anthracite	Bituminous	Subbituminous	Lignite	Total[c]
Alabama[d]	—	3,916.8	—	1,083.0	4,999.8
Alaska	—	697.5	5,443.0	14.0	6,154.5
Arizona	—	410.0	—	—	410.0
Arkansas	96.4	288.7	—	25.7	410.7
Colorado[d]	25.5	9,086.1	3,979.9	4,189.9	17,281.3
Georgia	—	3.6	—	—	3.6
Idaho	—	4.4	—	—	4.4
Illinois[d]	—	67,606.0	—	—	67,606.0
Indiana	—	10,586.1	—	—	10,586.1
Iowa	—	2,197.1	—	—	2,197.7
Kansas	—	993.8	—	—	993.8
Kentucky					
Eastern[d]	—	12,927.5	—	—	12,927.5
Western	—	21,074.4	—	—	21,074.4
Maryland	—	822.4	—	—	822.4
Michigan[d]	—	127.7	—	—	127.7
Missouri	—	6,069.1	—	—	6,069.1
Montana	—	1,385.4	103,277.4	15,765.2	120,428.0
New Mexico[d]	2.3	1,835.7	2,683.4	—	4,521.4
North Carolina	—	10.7	—	—	10.7
North Dakota	—	—	—	9,952.3	9,952.3
Ohio[d]	—	19,056.1	—	—	19,056.1
Oklahoma	—	1,637.8	—	—	1,637.8
Oregon	—	—	17.5	—	17.5
Pennsylvania	7,092.0	23,188.8	—	—	30,280.8
South Dakota	—	—	—	366.1	366.1
Tennessee[d]	—	983.7	—	—	983.7
Texas[d]	—	—	—	12,659.7	12,659.7
Utah[d]	—	6,476.5	1.1	—	6,477.6
Virginia	125.5	3,345.9	—	—	3,471.4
Washington[d]	—	303.7	1,169.4	8.1	1,481.3
West Virginia	—	39,776.2	—	—	39,776.2
Wyoming[d]	—	4,460.5	65,463.5	—	69,924.0
Total[c]	7,341.7	239,272.9	182,035.0	44,063.9	472,713.6

[a] Includes measured and indicated resource categories defined by USBM and USGS and represents 100% of the coal in place.

[b] Some coal-bearing states where data are not sufficiently detailed or where reserves are not currently economically recoverable.

[c] Data may not add to totals due to rounding.

[d] Data not completely reconciled with demonstrated reserve base data.

Agglomerating character is determined by examination of the residue left after the volatile determination. If the residue supports a 500-g weight without pulverizing or shows a swelling or cell structure, it is said to be "agglomerating."

The mineral-matter-free basis is used for ASTM rankings, and formulas to convert Btu, fixed carbon, and volatile matter from "as-received" bases are provided. Parr formulas, Eqs.

Table 3 ASTM (D388) Classification of Coals by Rank[a]

Class	Group	Fixed Carbon Limits, Percent (Dry, Mineral-Matter-Free Basis)		Volatile Matter Limits, Percent (Dry, Mineral-Matter-Free Basis)		Calorific Value Limits, Btu/lb Moist,[b] Mineral-Matter-Free Basis)		Agglomerating Character
		Equal to or Greater Than	Less Than	Greater Than	Equal to or Less Than	Equal to or Greater Than	Less Than	
I Anthracitic	1. Metaanthracite	98	—	—	2	—	—	
	2. Anthracite	92	98	2	8	—	—	Nonagglomerating
	3. Semianthracite[c]	86	92	8	14	—	—	
II Bituminous	1. Low-volatile bituminous	78	86	14	22	—	—	
	2. Medium-volatile bituminous	69	78	22	31	—	—	
	3. High-volatile A bituminous	—	69	31	—	14,000[d]	—	
	4. High-volatile B bituminous	—	—	—	—	13,000[d]	14,000	Commonly agglomerating[e]
	5. High-volatile C bituminous	—	—	—	—	11,500	13,000	
		—	—	—	—	10,500	11,500	Agglomerating
III Subbituminous	1. Subbituminous A	—	—	—	—	10,500	11,500	
	2. Subbituminous B	—	—	—	—	9,500	10,500	
	3. Subbituminous C	—	—	—	—	8,300	9,500	
IV Lignitic	1. Lignite A	—	—	—	—	6,300	8,300	
	2. Lignite B	—	—	—	—	—	6,300	

[a] This classification does not include a few coals, principally nonbanded varieties, that have unusual physical and chemical properties and that come within the limits of fixed carbon or calorific value of the high-volatile bituminous and subbituminous ranks. All of these coals either contain less than 48% dry, mineral-matter-free fixed carbon or have more than 15,500 moist, mineral-matter-free British thermal units per pound.

[b] Moist refers to coal containing its natural inherent moisture but not including visible water on the surface of the coal.

[c] If agglomerating, classify in the low-volatile group of the bituminous class.

[d] Coals having 69% or more fixed carbon on the dry, mineral-matter-free basis shall be classified according to fixed carbon, regardless of calorific value.

[e] It is recognized that there may be nonagglomerating varieties in these groups of the bituminous class and that there are notable exceptions in the high-volatile C bituminous group.

(1)–(3), are appropriate in case of litigation. Approximation formulas, Eqs. (4)–(6), are otherwise acceptable.

Parr formulas

$$\text{Dry, MM-Free } FC = \frac{FC - 0.15S}{100 - (M + 1.08A + 0.55S)} \times 100 \tag{1}$$

$$\text{Dry, MM-Free } VM = 100 - \text{Dry, MM-Free } FC \tag{2}$$

$$\text{Moist, MM-Free Btu} = \frac{\text{Btu} - 50S}{100 - (1.08A + 0.55S)} \times 100 \tag{3}$$

Approximation formulas

$$\text{Dry, MM-Free } FC = \frac{FC}{100 - (M + 1.1A + 0.1S)} \times 100 \tag{4}$$

$$\text{Dry, MM-Free } VM = 100 - \text{Dry, MM-Free } FC \tag{5}$$

$$\text{Moist, MM-Free Btu} = \frac{\text{Btu}}{100 - (1.1A + 0.1S)} \times 100 \tag{6}$$

where MM = mineral matter
   Btu = British thermal unit
   FC = percentage of fixed carbon
   VM = percentage of volatile matter
   A = percentage of ash
   S = percentage of sulfur

Other classifications of coal include the International Classification of Hard Coals, the International Classification of Brown Coals, the "Lord" value based on heating value with ash, sulfur, and moisture removed, and the Perch and Russell Ratio, based on the ratio of Moist, MM-Free Btu to Dry, MM-Free VM.

2   CURRENT USES—HEAT, POWER, STEELMAKING, OTHER

According to statistics compiled for the 1996 Keystone Coal Industry Manual, the primary use of coals produced in the United States in recent years has been for Electric Utilities; comprising almost 90% of the 926 million tons consumed in the United States in 1993. Industry accounted for about 8% of the consumption during that year in a variety of Standard Industrial Classification (SIC) Codes, replacing the manufacturing of coke (now about 3%) as the second largest coal market from the recent past. Industrial users typically consume coal for making process steam as well as in open-fired applications, such as in kilns and process heaters.

It should be noted that the demand for coal for coking purposes was greater than the demand for coal for utility use in the 1950s, and has steadily declined owing to more efficient steelmaking, greater use of scrap metal, increased use of substitute fuels in blast furnaces, and other factors. The production of coke from coal is accomplished by heating certain coals in the absence of air to drive off volatile matter and moisture. To provide a suitable by-product coke, the parent coal must possess quality parameters of low ash content, low sulfur content, low coking pressure, and high coke strength. By-product coking ovens, the most predominant, are so named for their ability to recapture otherwise wasted by-products driven off by heating the coal, such as coke oven gas, coal-tar, ammonia, oil, and useful chemicals. Beehive ovens, named for their shape and configuration, are also used, albeit much less extensively, in the production of coke.

3   TYPES

Anthracite is the least abundant of U.S. coal forms. Sometimes referred to as "hard" coal, it is shiny black or dark silver-gray and relatively compact. Inasmuch as it is the most advanced form in the coalification process, it is sometimes found deeper in the earth than bituminous. As indicated earlier, the ASTM definition puts upper and lower bounds of dry, mineral-matter-free fixed carbon percent at 98% and 86%, respectively, which limits volatile matter to not more than 14%. Combustion in turn is characterized by higher ignition temperatures and longer burnout times than bituminous coals.

Excepting some semianthracites that have a granular appearance, they have a consolidated appearance, unlike the layers seen in many bituminous coals. Typical Hardgrove Grindability Index ranges from 20 to 60 and specific gravity typically ranges 1.55 ± 0.10.

Anthracite coals can be found in Arkansas, Colorado, Pennsylvania, Massachusetts, New Mexico, Rhode Island, Virginia, and Washington, although by far the most abundant reserves are found in Pennsylvania.

Bituminous coal is by far the most plentiful and utilized coal form, and within the ASTM definitions includes low-, medium-, and high-volatile subgroups. Sometimes referred to as "soft" coal, it is named after the word bitumen, based on its general tendency to form a sticky mass on heating.

At a lower stage of development in the coalification process, carbon content is less than the anthracites, from a maximum of 86% to less than 69% on a dry, mineral-matter-free basis. Volatile matter, at a minimum of 14% on this basis, is greater than the anthracites, and, as a result, combustion in pulverized form is somewhat easier for bituminous coals. Production of gas is also enhanced by their higher volatility.

The tendency of bituminous coals to produce a cohesive mass on heating lends them to coke applications. Dry, mineral-matter-free oxygen content generally ranges from 5% to 10%, compared to a value as low as 1% for anthracite. They are commonly banded with layers differing in luster.

The low-volatile bituminous coals are grainier and more subject to size reduction in handling. The medium-volatile bituminous coals are sometimes distinctly layered, and sometimes only faintly layered and appearing homogeneous. Handling may or may not have a significant impact on size reduction. The high-volatile coals (A, B, and C) are relatively hard and less sensitive to size reduction from handling than low- or medium-volatile bituminous.

Subbituminous coals, like anthracite and lignite, are generally noncaking. "Caking" refers to fusion of coal particles after heating in a furnace, as opposed to "coking," which refers to the ability of a coal to make a good coke, suitable for metallurgical purposes.

Oxygen content, on a dry, mineral-matter-free basis, is typically 10–20%.

Brownish black to black in color, this type coal is typically smooth in appearance with an absence of layers.

Although high in inherent moisture, these fuels are often dusty in handling and appear much like drying mud as they disintegrate on sufficiently long exposure to air.

The Healy coal bed in Wyoming has the thickest seam of coal in the United States at 220 ft. It is subbituminous, with an average heating value of 7884 Btu/lb, 28.5% moisture,

30% volatile matter, 33.9% fixed carbon, and 0.6% sulfur. Reported strippable reserves of this seam are approximately 11 billion tons.[4]

Lignites, sometimes referred to as "brown coal," often retain a woodlike or laminar structure in which wood fiber remnants may be visible. Like subbituminous coals, they are high in seam moisture, up to 50% or more, and also disintegrate on sufficiently long exposure to air.

Both subbituminous coals and lignites are more susceptible than higher-rank coals to storage, shipping, and handling problems, owing to their tendency for slacking (disintegration) and spontaneous ignition. During the slacking, a higher rate of moisture loss at the surface than at the interior may cause higher rates and stresses at the outside of the particles, and cracks may occur with an audible noise.

Peat is decaying vegetable matter formed in wetlands; it is the first stage of metamorphosis in the coalification process. Development can be generally described as anaerobic, often in poorly drained flatlands or former lake beds. In the seam, peat moisture may be 90% or higher, and, therefore, the peat is typically "mined" and stacked for drainage or otherwise dewatered prior to consideration as a fuel. Because of its low bulk density at about 15 lb/ft^3 and low heating value at about 6000 Btu/lb (both values at 35% moisture), transportation distances must be short to make peat an attractive energy option.

In addition, it can be a very difficult material to handle, as it can arch in bins, forming internal friction angles in excess of 70°.

Chemically, peat is very reactive and ignites easily. It may be easily ground, and unconsolidated peat may create dusting problems.

4   PHYSICAL AND CHEMICAL PROPERTIES—DESCRIPTION AND TABLES OF SELECTED VALUES

There are a number of tests, qualitative and quantitative, used to provide information on coals; these tests will be of help to the user and/or equipment designer. Among the more common tests are the following, with reference to the applicable ASTM test procedure.

A. *"Proximate"* analysis (D3172) includes moisture, "volatile matter," "fixed carbon," and ash as its components.

Percent moisture (D3173) is determined by measuring the weight loss of a prepared sample (D2013) when heated to between 219°F (104°C) and 230°F (110°C) under rigidly controlled conditions. The results of this test can be used to calculate other analytical results to a dry basis. The moisture is referred to as "residual," and must be added to moisture losses incurred in sample preparation, called "air-dry losses" in order to calculate other analytical results to an "as-received" basis. The method which combines both residual and air dry moisture is D3302.

Percent volatile matter (D3175) is determined by establishing the weight loss of a prepared sample (D2013) resulting from heating to 1740°F (950°C) in the absence of air under controlled conditions. This weight loss is corrected for residual moisture, and is used for an indication of burning properties, coke yield, and classification by rank.

Percent ash (D3174) is determined by weighing the residue remaining after burning a prepared sample under rigidly controlled conditions. Combustion is in an oxidizing atmosphere and is completed for coal samples at 1290–1380°F (700–750°C).

Fixed carbon is a calculated value making up the fourth and final component of a proximate analysis. It is determined by subtracting the volatile, moisture, and ash percentages from 100.

Also generally included with a proximate analysis are calorific value and sulfur determinations.

B. *Calorific value,* Btu/lb (J/g, cal/g), is most commonly determined (D2015) in an "adiabatic bomb calorimeter," but is also covered by another method (D3286), which uses an "isothermal jacket bomb calorimeter." The values determined by this method are called gross or high heating values and include the latent heat of water vapor in the products of combustion.

C. *Sulfur* is determined by one of three methods provided by ASTM, all covered by D3177: the Eschka method, the bomb washing method, and a high-temperature combustion method.

The Eschka method requires that a sample be ignited with an "Eschka mixture" and sulfur be precipitated from the resulting solution as $BaSO_4$ and filtered, ashed, and weighed.

The bomb wash method requires use of the oxygen-bomb calorimeter residue, sulfur is precipitated as $BaSO_4$ and processed as in the Eschka method.

The high-temperature combustion method produces sulfur oxides from burning of a sample at 2460°F (1350°C), which are absorbed in a hydrogen peroxide solution for analysis. This is the most rapid of the three types of analysis.

D. *Sulfur forms* include sulfate, organic, and pyritic, and rarely, elemental sulfur. A method used to quantify sulfate, pyritic sulfur, and organic sulfur is D2492. The resulting data are sometimes used to provide a first indication of the maximum amount of sulfur potentially removable by mechanical cleaning.

E. *Ultimate analysis* (D3176) includes total carbon, hydrogen, nitrogen, oxygen, sulfur, and ash. These data are commonly used to perform combustion calculations to estimate combustion air requirements, products of combustion, and heat losses such as incurred by formation of water vapor by hydrogen in the coal.

Chlorine (D2361) and phosphorus (D2795) are sometimes requested with ultimate analyses, but are not technically a part of D3176.

F. *Ash mineral analysis* (D2795) includes the oxides of silica (SiO_2), alumina (Al_2O_3), iron (Fe_2O_3), titanium (TiO_2), phosphorus (P_2O_5), calcium (CaO), magnesium (MgO), sodium (Na_2O), and potassium (K_2O).

These data are used to provide several indications concerning ash slagging or fouling tendencies, abrasion potential, electrostatic precipitator operation, and sulfur absorption potention.

See Section 8 for further details.

G. *Grindability* (D409) is determined most commonly by the Hardgrove method to provide an indication of the relative ease of pulverization or grindability, compared to "standard" coals having grindability indexes of 40, 60, 80, and 110. As the index increases, pulverization becomes easier, that is, an index of 40 indicates a relatively hard coal; an index of 100 indicates a relatively soft coal.

Standard coals may be obtained from the U.S. Bureau of Mines.

A word of caution is given: grindability may change with ash content, moisture content, temperature, and other properties.

H. *Free swelling index* (D720) also referred to as a "coke-button" test, provides a relative index (1–9) of the swelling properties of a coal. A sample is burned in a covered crucible, and the resulting index increases as the swelling increases, determined by comparison of the button formed with standard profiles.

I. *Ash fusion temperatures* (D1857) are determined from triangular-core-shaped ash samples, in a reducing atmosphere and/or in an oxidizing atmosphere. Visual observations are recorded of temperatures at which the core begins to deform, called "initial deformation";

where height equals width, called "softening"; where height equals one-half width, called "hemispherical"; and where the ash is fluid.

The hemispherical temperature is often referred to as the "ash fusion temperature."

While not definitive, these tests provide a rough indication of the slagging tendency of coal ash.

Analysis of petrographic constituents in coals has been used to some extent in qualitative and semiquantitative analysis of some coals, most importantly in the coking coal industry. It is the application of macroscopic and microscopic techniques to identify maceral components related to the plant origins of the coal. The macerals of interest are vitrinite, exinite, resinite, micrinite, semifusinite, and fusinite. A technique to measure reflectance of a prepared sample of coal and calculate the volume percentages of macerals is included in ASTM Standard D2799.

Table 4 shows selected analyses of coal seams for reference.

5 BURNING CHARACTERISTICS

The ultimate analysis, described in the previous section, provides the data required to conduct fundamental studies of the air required for stoichiometric combustion, the volumetric and weight amounts of combustion gases produced, and the theoretical boiler efficiencies. These data assist the designer in such matters as furnace and auxiliary equipment sizing. Among the items of concern are draft equipment for supplying combustion air requirements, drying and transporting coal to the burners and exhausting the products of combustion, mass flow and velocity in convection passes for heat transfer and erosion considerations, and pollution control equipment sizing.

The addition of excess air must be considered for complete combustion and perhaps minimization of ash slagging in some cases. It is not uncommon to apply 25% excess air or more to allow operational flexibility.

As rank decreases, there is generally an increase in oxygen content in the fuel, which will provide a significant portion of the combustion air requirements.

The theoretical weight, in pounds, of combustion air required per pound of fuel for a stoichiometric condition is given by

$$11.53C + 34.34(H_2 - \tfrac{1}{8}O_2) + 4.29S \tag{7}$$

where C, H_2, O_2, and S are percentage weight constituents in the ultimate analysis.

The resulting products of combustion, again at a stoichiometric condition and complete combustion, are

$$CO_2 = 3.66C \tag{8}$$

$$H_2O = 8.94H_2 + H_2O \text{ (wt\% } H_2O \text{ in fuel)} \tag{9}$$

$$SO_2 = 2.00S \tag{10}$$

$$N_2 = 8.86C + 26.41(H_2 - \tfrac{1}{8}O_2) + 3.29S + N_2 \text{ (wt\% nitrogen in fuel)} \tag{11}$$

The combustion characteristics of various ranks of coal can be seen in Fig. 1, showing "burning profiles" obtained by thermal gravimetric analysis. As is apparent from this figure, ignition of lower rank coals occurs at a lower temperature and combustion proceeds at a more rapid rate than higher rank coals. This information is, of course, highly useful to the design engineer in determination of the size and configuration of combustion equipment.

Table 4 Selected Values—Coal and Peat Quality

Parameter	East Kentucky, Skyline Seam (Washed)	Pennsylvania, Pittsburgh #8 Seam (Washed)	Illinois, Harrisburg 5 (Washed)	Wyoming, Powder River Basin (Raw)	Florida Peat, Sumter County In situ	Florida Peat, Sumter County Dry
Moisture % (total)	8.00	6.5	13.2	25.92	86.70	—
Ash %	6.48	6.5	7.1	6.00	0.54	4.08
Sulfur %	0.82	1.62	1.28	0.25	0.10	0.77
Volatile %	36.69	34.40	30.6	31.27	8.74	65.73
Grindability (HGI)	45	55	54	57	36	69[a]
Calorific value (Btu/lb, as received)	12,500	13,100	11,700	8,500	1,503	11,297
Fixed carbon	48.83	52.60	49.1	37.23	4.02	30.19
Ash minerals						
SiO_2	50.87	50.10	48.90	32.02	58.29	
Al_2O_2	33.10	24.60	25.50	15.88	19.50	
TiO_2	2.56	1.20	1.10	1.13	1.05	
CaO	2.57	2.2	2.90	23.80	1.95	
K_2O	1.60	1.59	3.13	0.45	1.11	
MgO	0.80	0.70	1.60	5.73	0.94	
Na_2O	0.53	0.35	1.02	1.27	0.40	
P_2O_5	0.53	0.38	0.67	1.41	0.09	
Fe_2O_3	5.18	16.20	12.20	5.84	14.32	
SO_3	1.42	1.31	1.96	11.35	2.19	
Undetermined	0.84	1.37	1.02	1.12	0.16	
Ash	7.04	7.0	8.23	7.53	4.08	
Hydrogen	5.31	5.03	4.95	4.80	4.59	
Carbon	75.38	78.40	76.57	69.11	69.26	
Nitrogen	1.38	1.39	1.35	0.97	1.67	
Sulfur	0.89	1.73	1.47	0.34	0.77	
Oxygen (by difference)	9.95	6.35	7.03	17.24	19.33	
Chlorine	0.05	0.10	0.40	0.01	0.30	
Ash Fusion Temperatures (°F)						
Initial deformation (reducing)	2800+	2350	2240	2204	1950	
Softening ($H = W$) (reducing)	2800+	2460	2450	2226	2010	
Hemispherical ($H = \frac{1}{2}W$) (reducing)	2800+	2520	2500	2250	2060	
Fluid (reducing)	2800+	2580	2700+	2302	2100	

[a] At 9% H_2O.

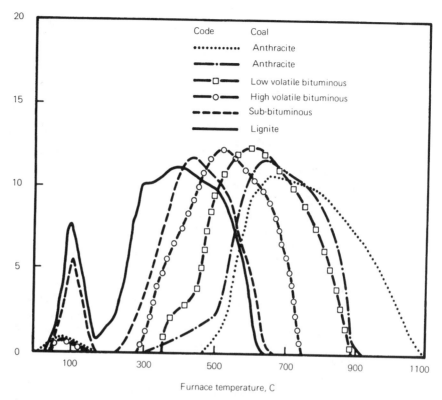

Rate of weight loss,
Mg/minute

Furnace temperature, C

Figure 1 Comparison of burning profiles for coals of different rank (courtesy of The Babcock and Wilcox Company).

The predominant firing technique for combustion of coal is in a pulverized form. To enhance ignition, promote complete combustion, and, in some cases, mitigate the effects of large particles on slagging and particulate capture, guidelines are generally given by the boiler manufacturer for pulverizer output (burner input).

Typical guidelines are as follows:

Coal Class/Group	Percentage Passing a 200-mesh Sieve	Percentage Retained on a 50-mesh Sieve	Allowable Coal/Air Temperature (°F)
Anthracite	80	2.0	200
Low-volatile bituminous	70–75	2.0	180
High-volatile bituminous A	70–75	2.0	170
High-volatile bituminous C	65–72	2.0	150–160
Lignite	60–70	2.0	110–140

These guidelines may vary for different manufacturers, ash contents, and equipment applications and, of course, the manufacturer should be consulted for fineness and temperature recommendations.

The sieve designations of 200 and 50 refer to U.S. Standard sieves. The 200-mesh sieve has 200 openings per linear inch, or 40,000 per square inch. The ASTM designations for these sieves are 75 and 300 μm, respectively.

Finally, agglomerating character may also have an influence on the fineness requirements, since this property might inhibit complete combustion.

6   ASH CHARACTERISTICS

Ash is an inert residue remaining after the combustion of coal and can result in significant challenges for designers and operators of the combustion, ash handling, and pollution control equipment. The quantity of ash in the coal varies widely from as little as 6% or less to more than 30% by weight. Additionally, diverse physical and chemical properties of ashes can pose substantial problems, with slagging, abrasion, and fouling of boilers. Electrostatic precipitators, used for pollution control, can experience material changes in collection efficiency depending on the mineral constituents of the ash.

"Slagging" is a term that generally refers to the formation of high-temperature fused ash deposits on furnace walls and other surfaces primarily exposed to radiant heat.

"Fouling" generally refers to high-temperature bonded ash deposits forming on convection tube banks, particularly superheat and reheat tubes.

Indication of ash-slagging tendencies can be measured by tests such as viscosity–temperature tests or by ash-softening tests. In addition, there are many empirical equations that are used to provide information as to the likelihood of slagging and fouling problems.

ASTM Standard number D1857 is the most common test used for slagging indication. In this test, ash samples are prepared as triangular cones and then are heated at a specified rate. Observations are then made and recorded of temperatures at prescribed stages of ash deformation, called initial deformation, softening temperature, hemispherical temperature, and fluid temperature. These tests are conducted in reducing and/or oxidizing atmospheres.

Another method used, although far more costly, involves measurement of the torque required to rotate a platinum bob suspended in molten slag. A viscosity–temperature relationship is established as a result of this test, which is also conducted in reducing and/or oxidizing atmospheres. A slag is generally considered liquid when its viscosity is below 250 poise, although tapping from a boiler may require a viscosity of 50–100 poise. It is plastic when its viscosity is between 250 and 10,000 poise. It is in this region where removal of the slag is most troublesome.

Ash mineral analyses are used to calculate empirical indicators of slagging problems. In these analyses are included metals reported as equivalent oxide weight percentages of silica, alumina, iron, calcium, magnesium, sodium, potassium, titania, phosphorous, and sulfur, as follows:

$$SiO_2 + Al_2O_3 + Fe_2O_3 + CaO + MgO + Na_2O + K_2O + TiO_2 + P_2O_5 + SO_3 = 100\%$$

Some ratios calculated using these data are:

Base: Acid Ratio, B/A

$$\frac{B}{A} = \frac{Base}{Acid} = \frac{Fe_2O_3 + CaO + MgO + Na_2O + K_2O}{SiO_2 + Al_2O_3 + TiO_2}$$

It has been reported[1] that a base/acid ratio in the range of 0.4 to 0.7 results typically in low ash fusibility temperatures and, hence, more slagging problems.

Slagging Factor, R_s

$$R_s = B/A \times \% \text{ sulfur, dry coal basis}$$

It has been reported[15] that coals with bituminous-type ashes exhibit a high slagging potential with a slagging factor above 2 and severe slagging potential with a slagging factor of more than 2.6. Bituminous-type ash refers to those ashes where iron oxide percentage is greater than calcium *plus* magnesium oxide.

Silica/Alumina Ratio

$$\frac{\text{Silica}}{\text{Alumina}} = \frac{SiO_2}{Al_2O_3}$$

It has been reported[1] that the silica in ash is more likely to form lower-melting-point compounds than is alumina and for two coals having the same base/acid ratio, the coal with a higher silica/alumina ratio should result in lower fusibility temperatures. However, it has also been reported[2] that for low base/acid ratios the opposite is true.

Iron/Calcium Ratio

$$\frac{\text{Iron}}{\text{Dolomite}} = \frac{Fe_2O_3}{CaO + MgO}$$

This ratio and its use are essentially the same as the iron/calcium ratio.

Silica Percentage (SP)

$$SP = \frac{SiO_2 \times 100}{SiO_2 + Fe_2O_3 + CaO + MgO}$$

This parameter has been correlated with ash viscosity. As silica ratio increases, the viscosity of slag increases. Graphical methods[2] are used in conjunction with this parameter to estimate the T_{250} temperature—the temperature where the ash would have a viscosity of 250 poise. Where the acidic content is less than 60% and the ash is lignitic, the *dolomite percentage* (*DP*) is used in preference to the silica percentage, along with graphs to estimate the T_{250}:

$$DP = \frac{(CaO + MgO) \times 100}{Fe_2O_3 + CaO + MgO + Na_2O + K_2O}$$

where the sum of the basic and acidic components are adjusted, if necessary, to equal 100%. For bituminous ash or lignitic-type ash having acidic content above 60%, the base/acid ratio is used in conjunction with yet another graph.

Fouling Factor (R_F)

$$R_F = \text{Acid base} \times \% \text{ Na}_2 \text{ (bituminous ash)}$$

or

$$R_F = \% \text{ Na}_2O \text{ (lignitic ash)}$$

For bituminous ash, the fouling factor[17] is "low" for values less than 0.1, "medium" for values between 0.1 and 0.25, "high" for values between 0.25 and 0.7, and "severe" values above 0.7. For lignitic-type ash, the percentage of sodium is used, and low, medium, and high values are <3.0, 3.0–6.0, and >6.0, respectively.

The basis for these factors is that sodium is the most important single factor in ash fouling, volatilizing in the furnace and subsequently condensing and sintering ash deposits in cooler sections.

Chlorine has also been used as an indicator of fouling tendency of eastern-type coals. If chlorine, from the ultimate analysis, is less than 0.15%, the fouling potential is low; if between 0.15 and 0.3, it has a medium fouling potential; and if above 0.3, its fouling potential is high.[1]

Ash resistivity can be predicted from ash mineral and coal ultimate analyses, according to a method described by Bickelhaupt.[14] Electrostatic precipitator sizing and/or performance can be estimated using the calculated resistivity. For further information, the reader is referred to Ref. 14.

7 SAMPLING

Coals are by nature heterogeneous and, as a result, obtaining a representative sample can be a formidable task. Its quality parameters such as ash, moisture, and calorific value can vary considerably from seam to seam and even within the same seam. Acquisition of accurate data to define the nature and ranges of these values adequately is further compounded by the effects of size gradation, sample preparation, and analysis accuracy.

Inasmuch as these data are used for such purposes as pricing, control of the operations in mines, and preparation plants, determination of power plant efficiency, estimation of material handling and storage requirements for the coal and its by-products, and in some cases for determination of compliance with environmental limitations, it is important that samples be taken, prepared, and analyzed in accordance with good practice.

To attempt to minimize significant errors in sampling, ASTM D2234 was developed as a standard method for the collection of a gross sample of coal and D2013 for preparation of the samples collected for analysis. It applies to lot sizes up to 10,000 tons (9080 mg) per gross sample and is intended to provide an accuracy of $\pm \frac{1}{10}$th of the average ash content in 95 of 100 determinations.

The number and weight of increments of sample comprising the gross sample to represent the lot, or consignment, is specified for nominal top sizes of $\frac{5}{8}$ in. (16 mm), 2 in. (50 mm), and 6 in. (150 mm), for raw coal or mechanically cleaned coal. Conditions of collection include samples taken from a stopped conveyor belt (the most desirable), full and partial stream cuts from moving coal consignments, and stationary samples.

One recommendation made in this procedure and worth special emphasis is that the samples be collected under the direct supervision of a person qualified by training and experience for this responsibility.

This method does not apply to the sampling of reserves in the ground, which is done by core drilling methods or channel sampling along outcrops as recommended by a geologist or mining engineer. It also does not apply to the sampling of coal slurries.

A special method (D197) was developed for the collection of samples of pulverized coals to measure size consist or fineness, which is controlled to maintain proper combustion efficiency.

Sieve analyses using this method may be conducted on No. 16 (1.18 mm) through No. 200 (750 mm) sieves, although the sieves most often referred to for pulverizer and classifier performance are the No. 50, the No. 100, and the No. 200 sieves. The number refers to the quantity of openings per linear inch.

Results of these tests should plot as a straight line on a sieve distribution chart. A typical fineness objective, depending primarily on the combustion characteristics might be for 70%

of the fines to pass through a 200-mesh sieve and not more than 2% to be retained on the 50-mesh sieve.

D2013 covers the preparation of coal samples for analysis, or more specifically, the reduction and division of samples collected in accordance with D2234. A "referee" and "nonreferee" method are delineated, although the "nonreferee" method is the most commonly used. "Referee" method is used to evaluate equipment and the nonreferee method.

Depending on the amount of moisture and, therefore, the ability of the coal to pass freely through reduction equipment, samples are either predried in air drying ovens or on drying floors or are processed directly by reduction, air drying, and division. Weight losses are computed for each stage of air drying to provide data for determination of total moisture, in combination with D3173 or 3302 (see Section 6).

Samples must ultimately pass the No. 60 (250 mm) sieve (D3173) or the No. 8 (2.36 mm) sieve (D3302) prior to total moisture determination.

Care must be taken to adhere to this procedure, including the avoidance of moisture losses while awaiting preparation, excessive time in air drying, proper use of riffling or mechanical division equipment, and verification and maintenance of crushing equipment size consist.

8 COAL CLEANING

Partial removal of impurities in coal such as ash and pyritic sulfur has been conducted since before 1900, although application and development has intensified during recent years owing to a number of factors, including the tightening of emissions standards, increasing use of lower quality seams, and increasing use of continuous mining machinery. Blending of two or more fuels to meet tight emissions standards, or other reasons, often requires that each of the fuels is of a consistent grade, which in turn may indicate some degree of coal cleaning.

Coal cleaning may be accomplished by physical or chemical means, although physical coal cleaning is by far the most predominant.

Primarily, physical processes rely on differences between the specific gravity of the coal and its impurities. Ash, clay, and pyritic sulfur have a higher specific gravity than that of coal. For example, bituminous coal typically has a specific gravity in the range 1.12–1.35, while pyrite's specific gravity is between 4.8 and 5.2.

One physical process that does not benefit from specific gravity differences is froth flotation. It is only used for cleaning coal size fractions smaller than 28 mesh ($\frac{1}{2}$ mm). Basically the process requires coal fines to be agitated in a chamber with set amounts of air, water, and chemical reagents, which creates a froth. The coal particles are selectively attached to the froth bubbles, which rise to the surface and are skimmed off, dewatered, and added to other clean coal fractions.

A second stage of froth flotation has been tested successfully at the pilot scale for some U.S. coals. This process returns the froth concentrate from the first stage to a bank of cells where a depressant is used to sink the coal and a xanthate flotation collector is used to selectively float pyrite.

The predominant commercial methods of coal cleaning use gravity separation by static and/or dynamic means. The extent and cost of cleaning naturally depends on the degree of end product quality desired, the controlling factors of which are primarily sulfur, heating value, and ash content.

Although dry means may be used for gravity separation, wet means are by far the more accepted and used techniques.

The first step in designing a preparation plant involves a careful study of the washability of the coal. "Float and sink" tests are run in a laboratory to provide data to be used for judging application and performance of cleaning equipment. In these tests the weight percentages and composition of materials are determined after subjecting the test coal to liquid baths of different specific gravities.

Pyritic sulfur and/or total sulfur percent, ash percent, and heating value are typically determined for both the float (called "yield") and sink (called "reject") fractions.

Commonly, the tests are conducted on three or more size meshes, such as 1½ in. × 0 mesh, ⅜ in. × 100 mesh, and minus 14 mesh, and at three or more gravities of such as 1.30, 1.40, and 1.60. Percentage recovery of weight and heating value are reported along with other data on a cumulative basis for the float fractions. An example of this, taken from a Bureau of Mines study[11] of 455 coals is shown in Table 5.

Many coals have pyrite particles less than one micron in size (0.00004 in.), which cannot be removed practically by mechanical means. Moreover, the cost of coal cleaning increases as the particle size decreases, as a general rule, and drying and handling problems become more difficult. Generally, coal is cleaned using particle sizes as large as practical to meet quality requirements.

It is interesting to note that a 50-μm pyrite particle (0.002 in.) inside a 14-mesh coal particle (0.06 in.) does not materially affect the specific gravity of a pure coal particle of the same size.

Washability data are usually organized and plotted as a series of curves, including:

1. Cumulative float–ash, sulfur %

2. Cumulative sink ash %

Table 5   Cumulative Washability Data[a]

Product	Recovery Weight	% Btu	Btu/lb	Ash %	Sulfur (%) Pyritic	Sulfur (%) Total	lb SO$_2$/ MBtu
Sample Crushed to Pass 1½ in.							
Float—1.30	55.8	58.8	14,447	3.3	0.26	1.09	1.5
Float—1.40	90.3	93.8	14,239	4.7	0.46	1.33	1.9
Float—1.60	94.3	97.3	14,134	5.4	0.53	1.41	2.0
Total	100.0	100.0	13,703	8.3	0.80	1.67	2.4
Sample Crushed to Pass ⅜ in.							
Float—1.30	58.9	63.0	14,492	3.0	0.20	1.06	1.5
Float—1.40	88.6	93.2	14,253	4.6	0.37	1.26	1.8
Float—1.60	92.8	96.8	14,134	5.4	0.47	1.36	1.9
Total	100.0	100.0	13,554	9.3	0.77	1.64	2.4
Sample Crushed to Pass 14 Mesh							
Float—1.30	60.9	65.1	14,566	2.5	0.16	0.99	1.4
Float—1.40	88.7	93.1	14,298	4.3	0.24	1.19	1.7
Float—1.60	93.7	97.1	14,134	5.4	0.40	1.29	1.8
Total	100.0	100.0	13,628	8.8	0.83	1.72	2.5

[a] State: Pennsylvania (bituminous); coal bed: Pittsburgh; county: Washington; raw coal moisture: 2.0%.

3. Elementary ash % (not cumulative)

4. % recovery (weight) versus specific gravity

Types of gravity separation equipment include jigs, concentrating tables, water-only cyclones, dense-media vessels, and dense-media cyclones.

In jigs, the coal enters the vessel in sizes to 8 in. and larger and stratification of the coal and heavier particles occurs in a pulsating fluid. The bottom layer, primarily rock, ash, and pyrite, is stripped from the mixture and rejected. Coal, the top layer, is saved. A middle layer may also be collected and saved or rejected depending on quality.

Concentrating tables typically handle coals in the $\frac{3}{8}$ in. $\times$ 0 or $\frac{1}{4}$ in. $\times$ 0 range and use water cascading over a vibrating table tilted such that heavier particles travel to one end of the table while lighter particles, traveling more rapidly with the water, fall over the adjacent edge.

In "water-only" and dense-media cyclones, centrifugal force is used to separate the heavier particles from the lighter particles.

In heavy-media vessels, the specific gravity of the media is controlled typically in the range of 1.45 to 1.65. Particles floating are saved as clean coal, while those sinking are reject. Specific gravity of the media is generally maintained by the amount of finely ground magnetite suspended in the water, as in heavy-media cyclones. Magnetite is recaptured in the preparation plant circuitry by means of magnetic separators.

Drying of cleaned coals depends on size. The larger sizes, $\frac{1}{4}$ in. or $\frac{3}{8}$ in. and larger, typically require little drying and might only be passed over vibrating screens prior to stockpiling. Smaller sizes, down to, say, 28 mesh, are commonly dried on stationary screens followed by centrifugal driers. Minus 28-mesh particles, the most difficult to dry, are processed in vibrating centrifuges, high-speed centrifugal driers, high-speed screens, or vacuum filters.

REFERENCES

1. J. G. Singer, *Combustion—Fossil Power Systems,* Combustion Engineering, Inc., Windsor, CT, 1981.

2. *Stream—Its Generation and Use,* Babcock & Wilcox, New York, 1992.

3. C. L. Wilson, *Coal—Bridge to the Future,* Vol. 1, Report of the World Coal Study, Ballinger, Cambridge, MA, 1980.

4. *Keystone Coal Industry Manual,* McGraw-Hill, New York, 1996.

5. *1979/1980 Coal Data,* National Coal Association and U.S. DOE, Washington, DC, 1982.

6. R. A. Meyers, *Coal Handbook,* Dekker, New York, 1981.

7. ASTM Standards, Part 26: *Gaseous Fuels; Coal and Coke, Atmospheric Analysis,* Philadelphia, PA 1982.

8. F. M. Kennedy, J. G. Patterson, and T. W. Tarkington, *Evaluation of Physical and Chemical Coal Cleaning and Flue Gas Desulfurization,* EPA 600/7-79-250, U.S. EPA, Washington, DC, 1979.

9. Gibbs and Hill, Inc., *Coal Preparation for Combustion and Conversion,* Electric Power Research Institute, Palo Alto, CA, 1978.

10. *Coal Data Book,* President's Commission on Coal, U.S. Government Printing Office, Washington, DC, 1980.

11. J. A. Cavallaro and A. W. Deubrouch, *Sulfur Reduction Potential of the Coals of the United States,* RI8118, U.S. Dept. of Interior, Washington, DC, 1976.

12. R. A. Schmidt, *Coal in America,* 1979.

13. Auth and Johnson, *Fuels and Combustion Handbook,* McGraw-Hill, New York, 1951.

14. R. E. Bickelhaupt, *A Technique for Predicting Fly Ash Resistivity,* EPA 600/7-79-204, Southern Research Institute, Birmingham, AL, August, 1979.

15. R. C. Attig and A. F. Duzy, *Coal Ash Deposition Studies and Application to Boiler Design,* Bobcock & Wilcox, Atlanta, GA, 1969.

16. A. F. Duzy and C. L. Wagoner, *Burning Profiles for Solid Fuels,* Babcock & Wilcox, Atlanta, GA, 1975.

17. E. C. Winegartner, *Coal Fouling and Slagging Parameters,* Exxon Research and Engineering, Baytown, TX, 1974.

18. J. B. McIlroy and W. L. Sage, Relationship of Coal-Ash Viscosity to Chemical Composition, *The Journal of Engineering for Power,* Babcock & Wilcox, New York, 1960.

19. R. P. Hensel, *The Effects of Agglomerating Characteristics of Coals on Combustion in Pulverized Fuel Boilers,* Combustion Engineering, Windsor, CT, 1975.

20. W. T. Reed, *External Corrosion and Deposits in Boilers and Gas Turbines,* Elsevier, New York, 1971.

21. O. W. Durrant, *Pulverized Coal—New Requirements and Challenges,* Babcock & Wilcox, Atlanta, GA, 1975.

CHAPTER 20

SOLAR ENERGY APPLICATIONS

Jan F. Kreider
Kreider and Associates, LLC
and Joint Center for Energy Management
University of Colorado
Boulder, Colorado

1 SOLAR ENERGY AVAILABILITY

Solar energy is defined as that radiant energy transmitted by the sun and intercepted by earth. It is transmitted through space to earth by electromagnetic radiation with wavelengths ranging between 0.20 and 15 μm. The availability of solar flux for terrestrial applications varies with season, time of day, location, and collecting surface orientation. In this chapter we shall treat these matters analytically.

1.1 Solar Geometry

Two motions of the earth relative to the sun are important in determining the intensity of solar flux at any time—the earth's rotation about its axis and the annual motion of the earth and its axis about the sun. The earth rotates about its axis once each day. A solar day is defined as the time that elapses between two successive crossings of the local meridian by the sun. The local meridian at any point is the plane formed by projecting a north–south longitude line through the point out into space from the center of the earth. The length of a solar day on the average is slightly less than 24 hr, owing to the forward motion of the earth in its solar orbit. Any given day will also differ from the average day owing to orbital eccentricity, axis precession, and other secondary effects embodied in the equation of time described below.

Declination and Hour Angle
The earth's orbit about the sun is elliptical with eccentricity of 0.0167. This results in variation of solar flux on the outer atmosphere of about 7% over the course of a year. Of more

663

importance is the variation of solar intensity caused by the inclination of the earth's axis relative to the ecliptic plane of the earth's orbit. The angle between the ecliptic plane and the earth's equatorial plane is 23.45°. Figure 1 shows this inclination schematically.

The earth's motion is quantified by two angles varying with season and time of day. The angle varying on a seasonal basis that is used to characterize the earth's location in its orbit is called the solar "declination." It is the angle between the earth–sun line and the equatorial plane as shown in Fig. 2. The declination δ_s is taken to be positive when the earth–sun line is north of the equator and negative otherwise. The declination varies between +23.45° on the summer solstice (June 21 or 22) and −23.45° on the winter solstice (December 21 or 22). The declination is given by

$$\sin \delta_s = 0.398 \cos[0.986(N - 173)] \tag{1}$$

in which N is the day number, counted from January 1.

The second angle used to locate the sun is the solar-hour angle. Its value is based on the nominal 360° rotation of the earth occurring in 24 hr. Therefore, 1 hr is equivalent to an angle of 15°. The hour angle is measured from zero at solar noon. It is denoted by h_s and

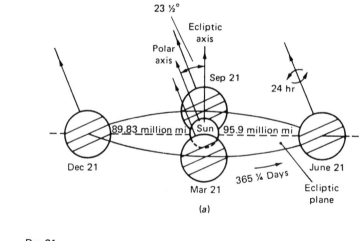

(a)

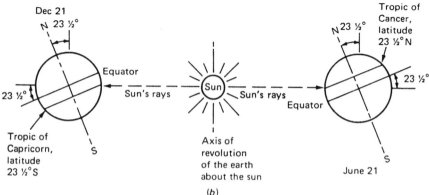

(b)

Figure 1 (a) Motion of the earth about the sun. (b) Location of tropics. Note that the sun is so far from the earth that all the rays of the sun may be considered as parallel to one another when they reach the earth.

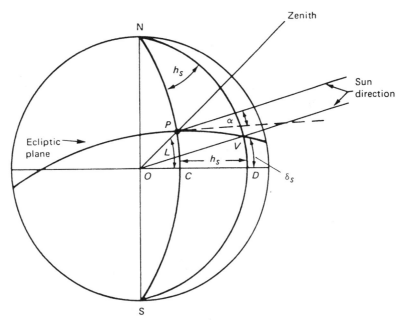

Figure 2 Definition of solar-hour angle h_s (*CND*), solar declination δ_s (*VOD*), and latitude L (*POC*): P, site of interest. (Modified from J. F. Kreider and F. Kreith, *Solar Heating and Cooling*, revised 1st ed., Hemisphere, Washington, DC, 1977.)

is positive before solar noon and negative after noon in accordance with the right-hand rule. For example 2:00 PM corresponds to $h_s = -30°$ and 7:00 AM corresponds to $h_s = +75°$.

Solar time, as determined by the position of the sun, and clock time differ for two reasons. First, the length of a day varies because of the ellipticity of the earth's orbit; and second, standard time is determined by the standard meridian passing through the approximate center of each time zone. Any position away from the standard meridian has a difference between solar and clock time given by [(local longitude − standard meridian longitude)/ 15) in units of hours. Therefore, solar time and local standard time (LST) are related by

$$\text{Solar time} = \text{LST} - \text{EoT} - (\text{Local longitude} - \text{Standard meridian longitude})/15 \quad (2)$$

in units of hours. EoT is the equation of time which accounts for difference in day length through a year and is given by

$$\text{EoT} = 12 + 0.1236 \sin x - 0.0043 \cos x + 0.1538 \sin 2x + 0.0608 \cos 2x \quad (3)$$

in units of hours. The parameter x is

$$x = \frac{360(N - 1)}{365.24} \quad (4)$$

where N is the day number.

Solar Position

The sun is imagined to move on the celestial sphere, an imaginary surface centered at the earth's center and having a large but unspecified radius. Of course, it is the earth that moves, not the sun, but the analysis is simplified if one uses this Ptolemaic approach. No error is introduced by the moving sun assumption, since the relative motion is the only motion of interest. Since the sun moves on a spherical surface, two angles are sufficient to locate the sun at any instant. The two most commonly used angles are the solar-altitude and azimuth angles (see Fig. 3) denoted by α and a_s, respectively. Occasionally, the solar-zenith angle, defined as the complement of the altitude angle, is used instead of the altitude angle.

The solar-altitude angle is related to the previously defined declination and hour angles by

$$\sin \alpha = \cos L \cos \delta_s \cos h_s + \sin L \sin \delta_s \tag{5}$$

in which L is the latitude, taken positive for sites north of the equator and negative for sites south of the equator. The altitude angle is found by taking the inverse sine function of Eq. (5).

The solar-azimuth angle is given by

$$\sin a_s = \frac{\cos \delta_s \sin h_s}{\cos \alpha} \tag{6}$$

To find the value of a_s, the location of the sun relative to the east–west line through the site must be known. This is accounted for by the following two expressions for the azimuth angle:

$$a_s = \sin^{-1}\left(\frac{\cos \delta_s \sin h_s}{\cos \alpha}\right), \qquad \cos h_s > \frac{\tan \delta_s}{\tan L} \tag{7}$$

$$a_s = 180° - \sin^{-1}\left(\frac{\cos \delta_s \sin h_s}{\cos \alpha}\right), \qquad \cos h_s < \frac{\tan \delta_s}{\tan L} \tag{8}$$

Table 1 lists typical values of altitude and azimuth angles for latitude $L = 40°$. Complete tables are contained in Refs. 1 and 2.

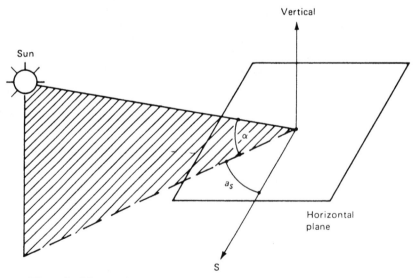

Figure 3 Diagram showing solar-altitude angle α and solar-azimuth angle a_s.

Table 1 Solar Position for 40°N Latitude

Date	Solar Time AM	Solar Time PM	Solar Position Altitude	Solar Position Azimuth	Date	Solar Time AM	Solar Time PM	Solar Position Altitude	Solar Position Azimuth
January 21	8	4	8.1	55.3	July 21	5	7	2.3	115.2
	9	3	16.8	44.0		6	6	13.1	106.1
	10	2	23.8	30.9		7	5	24.3	97.2
	11	1	28.4	16.0		8	4	35.8	87.8
	12		30.0	0.0		9	3	47.2	76.7
						10	2	57.9	61.7
February 21	7	5	4.8	72.7		11	1	66.7	37.9
	8	4	15.4	62.2		12		70.6	0.0
	9	3	25.0	50.2					
	10	2	32.8	35.9	August 21	6	6	7.9	99.5
	11	1	38.1	18.9		7	5	19.3	90.9
	12		40.0	0.0		8	4	30.7	79.9
						9	3	41.8	67.9
March 21	7	5	11.4	80.2		10	2	51.7	52.1
	8	4	22.5	69.6		11	1	59.3	29.7
	9	3	32.8	57.3		12		62.3	0.0
	10	2	41.6	41.9					
	11	1	47.7	22.6	September 21	7	5	11.4	80.2
	12		50.0	0.0		8	4	22.5	69.6
						9	3	32.8	57.3
April 21	6	6	7.4	98.9		10	2	41.6	41.9
	7	5	18.9	89.5		11	1	47.7	22.6
	8	4	30.3	79.3		12		50.0	0.0
	9	3	41.3	67.2					
	10	2	51.2	51.4	October 21	7	5	4.5	72.3
	11	1	58.7	29.2		8	4	15.0	61.9
	12		61.6	0.0		9	3	24.5	49.8
						10	2	32.4	35.6
May 21	5	7	1.9	114.7		11	1	37.6	18.7
	6	6	12.7	105.6		12		39.5	0.0
	7	5	24.0	96.6					
	8	4	35.4	87.2	November 21	8	4	8.2	55.4
	9	3	46.8	76.0		9	3	17.0	44.1
	10	2	57.5	60.9		10	2	24.0	31.0
	11	1	66.2	37.1		11	1	28.6	16.1
	12		70.0	0.0		12		30.2	0.0
June 21	5	7	4.2	117.3	December 21	8	4	5.5	53.0
	6	6	14.8	108.4		9	3	14.0	41.9
	7	5	26.0	99.7		10	2	20.0	29.4
	8	4	37.4	90.7		11	1	25.0	15.2
	9	3	48.8	80.2		12		26.6	0.0
	10	2	59.8	65.8					
	11	1	69.2	41.9					
	12		73.5	0.0					

1.2   Sunrise and Sunset

Sunrise and sunset occur when the altitude angle $\alpha = 0$. As indicated in Fig. 4, this occurs when the center of the sun intersects the horizon plane. The hour angle for sunrise and sunset can be found from Eq. (5) by equating α to zero. If this is done, the hour angles for sunrise and sunset are found to be

$$h_{sr} = \cos^{-1}(-\tan L \tan \delta_s) = -h_{ss} \qquad (9)$$

in which h_{sr} is the sunrise hour angle and h_{ss} is the sunset hour angle.

Figure 4 shows the path of the sun for the solstices and the equinoxes (length of day and night are both 12 hr on the equinoxes). This drawing indicates the very different azimuth and altitude angles that occur at different times of year at identical clock times. The sunrise and sunset hour angles can be read from the figures where the sun paths intersect the horizon plane.

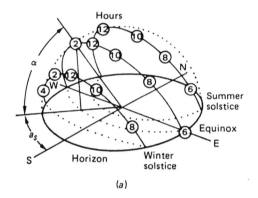

(a)

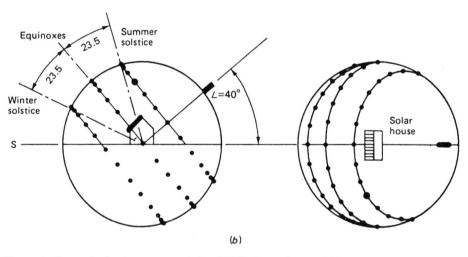

(b)

Figure 4   Sun paths for the summer solstice (6/21), the equinoxes (3/21 and 9/21), and the winter solstice (12/21) for a site at 40°N; (a) isometric view; (b) elevation and plan views.

Solar Incidence Angle

For a number of reasons, many solar collection surfaces do not directly face the sun continuously. The angle between the sun–earth line and the normal to any surface is called the incidence angle.

The intensity of off-normal solar radiation is proportional to the cosine of the incidence angle. For example, Fig. 5 shows a fixed planar surface with solar radiation intersecting the plane at the incidence angle i measured relative to the surface normal. The intensity of flux at the surface is $I_b \times \cos i$, where I_b is the beam radiation along the sun–earth line; I_b is called the direct, normal radiation. For a fixed surface such as that in Fig. 5 facing the equator, the incidence angle is given by

$$\cos i = \sin \delta_1(\sin L \cos \beta - \cos L \sin \beta \cos a_w)$$
$$+ \cos \delta_s \cos h_s(\cos L \cos \beta + \sin L \sin \beta \cos a_w) \qquad (10)$$
$$+ \cos \delta_s \sin \beta \sin a_w \sin h_s$$

in which a_w is the "wall" azimuth angle and β is the surface tilt angle relative to the horizontal plane, both as shown in Fig. 5.

For fixed surfaces that face due south, the incidence angle expression simplifies to

$$\cos i = \sin(L - \beta)\sin \delta_s + \cos(L - \beta)\cos \delta_s \cos h_s \qquad (11)$$

A large class of solar collectors move in some fashion to track the sun's diurnal motion, thereby improving the capture of solar energy. This is accomplished by reduced incidence

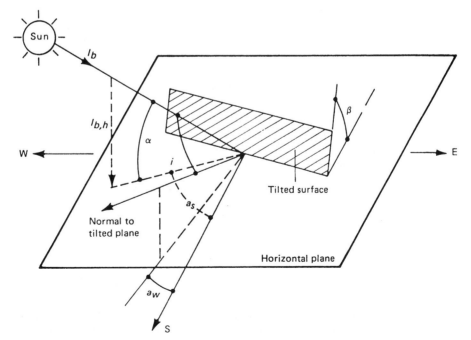

Figure 5 Definition of incidence angle i, surface tilt angle β, solar-altitude angle α, wall-azimuth angle a_w, and solar-azimuth angle a_s for a non-south-facing tilted surface. Also shown is the beam component of solar radiation I_b and the component of beam radiation $I_{b,h}$ on a horizontal plane.

angles for properly tracking surfaces vis-á-vis a fixed surface for which large incidence angles occur in the early morning and late afternoon (for generally equator-facing surfaces). Table 2 lists incidence angle expressions for nine different types of tracking surfaces. The term "polar axis" in this table refers to an axis of rotation directed at the north or south pole. This axis of rotation is tilted up from the horizontal at an angle equal to the local latitude. It is seen that normal incidence can be achieved (i.e., cos i = 1) for any tracking scheme for which two axes of rotation are present. The polar case has relatively small incidence angles as well, limited by the declination to $\pm 23.45°$. The mean value of cos i for polar tracking is 0.95 over a year, nearly as good as the two-axis case for which the annual mean value is unity.

1.3   Quantitative Solar Flux Availability

The previous section has indicated how variations in solar flux produced by seasonal and diurnal effects can be quantified. However, the effect of weather on solar energy availability cannot be analyzed theoretically; it is necessary to rely on historical weather reports and empirical correlations for calculations of actual solar flux. In this section this subject is described along with the availability of solar energy at the edge of the atmosphere—a useful correlating parameter, as seen shortly.

Extraterrestrial Solar Flux

The flux intensity at the edge of the atmosphere can be calculated strictly from geometric considerations if the direct-normal intensity is known. Solar flux incident on a terrestrial surface, which has traveled from sun to earth with negligible change in direction, is called beam radiation and is denoted by $I_{b,0}$. The extraterrestrial value of I_b averaged over a year is called the solar constant, denoted by I_{sc}. Its value is 429 Btu/hr·ft^2 or 1353 W/m^2. Owing to the eccentricity of the earth's orbit, however, the extraterrestrial beam radiation intensity varies from this mean solar constant value. The variation of $I_{b,0}$ over the year is given by

$$I_{b,0}(N) = \left[1 + 0.034 \cos \left(\frac{360N}{265} \right) \right] \times I_{sc} \tag{12}$$

in which N is the day number as before.

In subsequent sections the total daily, extraterrestrial flux will be particularly useful as a nondimensionalizing parameter for terrestrial solar flux data. The instantaneous solar flux on a horizontal, extraterrestrial surface is given by

$$I_{b,h0} = I_{b,0}(N) \sin \alpha \tag{13}$$

as shown in Fig. 5. The daily total, horizontal radiation is denoted by I_0 and is given by

$$I_0(N) = \int_{t_{sr}}^{t_{ss}} I_{b,0}(N) \sin \alpha \, dt \tag{14}$$

$$I_0(N) = \frac{24}{\pi} I_{sc} \left[1 + 0.034 \cos \left(\frac{360N}{265} \right) \right] \times (\cos L \cos \delta_s \sin h_{sr} + h_{sr} \sin L \sin \delta_s) \tag{15}$$

in which I_{sc} is the solar constant. The extraterrestrial flux varies with time of year via the variations of δ_s and h_{sr} with time of year. Table 3 lists the values of extraterrestrial, horizontal flux for various latitudes averaged over each month. The monthly averaged, horizontal, extraterrestrial solar flux is denoted by $\overline{H}_0$.

Table 2 Solar Incidence Angle Equations for Tracking Collectors

Description	Axis (Axes)	Cosine of Incidence Angle ($\cos i$)
Movements in altitude and azimuth	Horizontal axis and vertical axis	1
Rotation about a polar axis and adjustment in declination	Polar axis and declination axis	1
Uniform rotation about a polar axis	Polar axis	$\cos \delta_s$
East–west horizontal	Horizontal, east–west axis	$\sqrt{1 - \cos^2 a \, \sin^2 a_s}$
North–south horizontal	Horizontal, north–south axis	$\sqrt{1 - \cos^2 \alpha \, \cos^2 a_s}$
Rotation about a vertical axis of a surface tilted upward L (latitude) degrees	Vertical axis	$\sin (\alpha + L)$
Rotation of a horizontal collector about a vertical axis	Vertical axis	$\sin \alpha$
Rotation of a vertical surface about a vertical axis	Vertical axis	$\cos \alpha$
Fixed "tubular" collector	North–south tilted up at angle β	$\sqrt{1 - [\sin (\beta - L) \cos \delta_s \cos h_s + \cos (\beta - L) \sin \delta_s]^2}$

Table 3 Average Extraterrestrial Radiation on a Horizontal Surface $\overline{H}_0$ in SI Units and in English Units Based on a Solar Constant of 429 Btu/hr·ft² or 1.353 kW/m²

Latitude, Degrees	January	February	March	April	May	June	July	August	September	October	November	December
SI units, W·hr/m²·day												
20	7415	8397	9552	10,422	10,801	10,868	10,794	10,499	9791	8686	7598	7076
25	6656	7769	9153	10,312	10,936	11,119	10,988	10,484	9494	8129	6871	6284
30	5861	7087	8686	10,127	11,001	11,303	11,114	10,395	9125	7513	6103	5463
35	5039	6359	8153	9869	10,995	11,422	11,172	10,233	8687	6845	5304	4621
40	4200	5591	7559	9540	10,922	11,478	11,165	10,002	8184	6129	4483	3771
45	3355	4791	6909	9145	10,786	11,477	11,099	9705	7620	5373	3648	2925
50	2519	3967	6207	8686	10,594	11,430	10,981	9347	6998	4583	2815	2100
55	1711	3132	5460	8171	10,358	11,352	10,825	8935	6325	3770	1999	1320
60	963	2299	4673	7608	10,097	11,276	10,657	8480	5605	2942	1227	623
65	334	1491	3855	7008	9852	11,279	10,531	8001	4846	2116	544	97
English units, Btu/ft²·day												
20	2346	2656	3021	3297	3417	3438	3414	3321	3097	2748	2404	2238
25	2105	2458	2896	3262	3460	3517	3476	3316	3003	2571	2173	1988
30	1854	2242	2748	3204	3480	3576	3516	3288	2887	2377	1931	1728
35	1594	2012	2579	3122	3478	3613	3534	3237	2748	2165	1678	1462
40	1329	1769	2391	3018	3455	3631	3532	3164	2589	1939	1418	1193
45	1061	1515	2185	2893	3412	3631	3511	3070	2410	1700	1154	925
50	797	1255	1963	2748	3351	3616	3474	2957	2214	1450	890	664
55	541	991	1727	2585	3277	3591	3424	2826	2001	1192	632	417
60	305	727	1478	2407	3194	3567	3371	2683	1773	931	388	197
65	106	472	1219	2217	3116	3568	3331	2531	1533	670	172	31

Terrestrial Solar Flux

Values of instantaneous or average terrestrial solar flux cannot be predicted accurately owing to the complexity of atmospheric processes that alter solar flux magnitudes and directions relative to their extraterrestrial values. Air pollution, clouds of many types, precipitation, and humidity all affect the values of solar flux incident on earth. Rather than attempting to predict solar availability accounting for these complex effects, one uses long-term historical records of terrestrial solar flux for design purposes.

The U.S. National Weather Service (NWS) records solar flux data at a network of stations in the United States. The pyranometer instrument, as shown in Fig. 6, is used to measure the intensity of horizontal flux. Various data sets are available from the National Climatic Center (NCC) of the NWS. Prior to 1975, the solar network was not well maintained; therefore, the pre-1975 data were rehabilitated in the late 1970s and are now available from the NCC on magnetic media. Also, for the period 1950–1975, synthetic solar data have been generated for approximately 250 U.S. sites where solar flux data were not recorded. The predictive scheme used is based on other widely available meteorological data. Finally, from 1977 to the mid-1990s the NWS recorded hourly solar flux data at a 38-station network with improved instrument maintenance. In addition to horizontal flux, direct-normal data were recorded and archived at the NCC. Figure 7 is a contour map of annual, horizontal flux for the United States based on recent data.

The principal difficulty with using NWS solar data is that they are available for horizontal surfaces only. Solar-collecting surfaces normally face the general direction of the sun and are, therefore, rarely horizontal. It is necessary to convert measured horizontal radiation to radiation on arbitrarily oriented collection surfaces. This is done using empirical approaches to be described.

Hourly Solar Flux Conversions

Measured, horizontal solar flux consists of both beam and diffuse radiation components. Diffuse radiation is that scattered by atmospheric processes; it intersects surfaces from the entire sky dome, not just from the direction of the sun. Separating the beam and diffuse components of measured, horizontal radiation is the key difficulty in using NWS measurements.

The recommended method for finding the beam component of total (i.e., beam plus diffuse) radiation is described in Ref. 1. It makes use of the parameter k_T called the clearness index and defined as the ratio of terrestrial to extraterrestrial hourly flux on a horizontal surface. In equation form k_T is

$$k_T \equiv \frac{I_h}{I_{b,h0}} = \frac{I_h}{I_{b,0}(N)\,\sin\,\alpha} \tag{16}$$

in which I_h is the measured, total horizontal flux. The beam component of the terrestrial flux is then given by the empirical equation

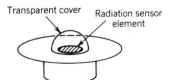

Transparent cover   Radiation sensor
element

Figure 6 Schematic drawing of a pyranometer used for measuring the intensity of total (direct plus diffuse) solar radiation.

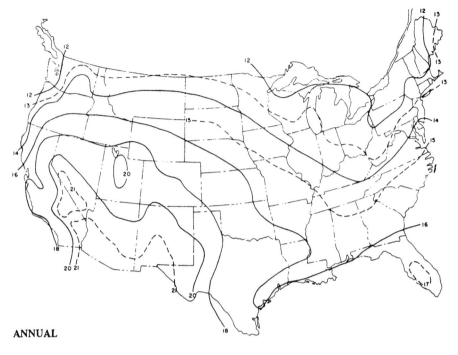

ANNUAL

*1 mJ/m² = 88.1 Btu/ft².

Figure 7 Mean daily solar radiation on a horizontal surface in megajoules per square meter for the continental United States.

$$I_b = (ak_T + b)I_{b,0}(N) \tag{17}$$

in which the empirical constants a and b are given in Table 4. Having found the beam radiation, the horizontal diffuse component $I_{d,h}$ is found by the simple difference

$$I_{d,h} = I_h - I_b \sin \alpha \tag{18}$$

Table 4 Empirical Coefficients for Eq. (17)

Interval for K_T	a	b
0.00, 0.05	0.04	0.00
0.05, 0.15	0.01	0.002
0.15, 0.25	0.06	−0.006
0.25, 0.35	0.32	−0.071
0.35, 0.45	0.82	−0.246
0.45, 0.55	1.56	−0.579
0.55, 0.65	1.69	−0.651
0.65, 0.75	1.49	−0.521
0.75, 0.85	0.27	0.395

The separate values of horizontal beam and diffuse radiation can be used to find radiation on any surface by applying appropriate geometric "tilt factors" to each component and forming the sum accounting for any radiation reflected from the foreground. The beam radiation incident on any surface is simply $I_b \cos i$. If one assumes that the diffuse component is isotropically distributed over the sky dome, the amount intercepted by any surface tilted at an angle β is $I_{d,h} \cos^2(\beta/2)$. The total beam and diffuse radiation intercepted by a surface I_c is then

$$I_c = I_b \cos i + I_{d,h} \cos^2(\beta/2) + \rho I_h \sin^2(\beta/2) \tag{19}$$

The third term in this expression accounts for flux reflected from the foreground with reflectance ρ.

Monthly Averaged, Daily Solar Flux Conversions

Most performance prediction methods make use of monthly averaged solar flux values. Horizontal flux data are readily available (see the appendix), but monthly values on arbitrarily positioned surfaces must be calculated using a method similar to that previously described for hourly tilted surface calculations. The monthly averaged flux on a tilted surface $\bar{I}_c$ is given by

$$\bar{I}_c = \bar{R}\bar{H}_h \tag{20}$$

in which $\bar{H}_h$ is the monthly averaged, daily total of horizontal solar flux and $\bar{R}$ is the overall tilt factor given by Eq. (21) for a fixed, equator-facing surface:

$$\bar{R} = \left(1 - \frac{\bar{D}_h}{\bar{H}_h}\right)\bar{R}_b + \frac{\bar{D}_h}{\bar{H}_h}\cos^2\frac{\beta}{2} + \rho\sin^2\frac{\beta}{2} \tag{21}$$

The ratio of monthly averaged diffuse to total flux, $\bar{D}_h/\bar{H}_h$ is given by

$$\frac{\bar{D}_h}{\bar{H}_h} = 0.775 + 0.347\left(h_{sr} - \frac{\pi}{2}\right) - \left[0.505 + 0.261\left(h_{sr} - \frac{\pi}{2}\right)\right]\cos\left[(\bar{K}_T - 0.9)\frac{360}{\pi}\right] \tag{22}$$

in which $\bar{K}_T$ is the monthly averaged clearness index analogous to the hourly clearness index. $\bar{K}_T$ is given by

$$\bar{K}_T \equiv \bar{H}_h/\bar{H}_0$$

where H_0 is the monthly averaged, extraterrestrial radiation on a horizontal surface at the same latitude at which the terrestrial radiation $\bar{H}_h$ was recorded. The monthly averaged beam radiation tilt factor $\bar{R}_b$ is

$$\bar{R}_b = \frac{\cos(L - \beta)\cos\delta_s \sin h'_{sr} + h'_{sr}\sin(L - \beta)\sin\delta_s}{\cos L \cos\delta_s \sin h_{sr} + h_{sr}\sin L \sin\delta_s} \tag{23}$$

The sunrise hour angle is found from Eq. (9) and the value of h'_{sr} is the smaller of (1) the sunrise hour angle h_{sr} and (2) the collection surface sunrise hour angle found by setting $i = 90°$ in Eq. (11). That is, h'_{sr} is given by

$$h'_{sr} = \min\{\cos^{-1}[-\tan L \tan\delta_s], \cos^{-1}[-\tan(L - \beta)\tan\delta_s]\} \tag{24}$$

Expressions for solar flux on a tracking surface on a monthly averaged basis are of the form

$$\bar{I}_c = \left[r_T - r_d\left(\frac{\bar{D}_h}{\bar{H}_h}\right)\right]\bar{H}_h \tag{25}$$

in which the tilt factors r_T and r_d are given in Table 5. Equation (22) is to be used for the diffuse to total flux ratio $\bar{D}_h/\bar{H}_h$.

Table 5 Concentrator Tilt Factors[2] (monthly averaged)

Collector Type	$r_T^{a,b,c,d}$	r_d^e
Fixed aperture concentrators that do not view the foreground	$[\cos(L - \beta)/(d \cos L)]\{-ah_{coll} \cos h_{sr}(i = 90°)$ $+ [a - b \cos h_{sr}(i = 90°)] \sin h_{coll}$ $+ (b/2)(\sin h_{coll} \cos h_{coll} + h_{coll})\}$	$(\sin h_{coll}/d)\{[\cos(L + \beta)/\cos L] - [1/(CR)]\} + (h_{coll}/d)\{[\cos h_{sr}/(CR)]$ $- [\cos(L - \beta)/\cos L] \cos h_{sr} \ (i = 90°)\}$
East–west axis tracking[f]	$(1/d) \int_0^{h_{coll}} \{[(a + b \cos x)/\cos L] \times \sqrt{\cos^2 x + \tan^2 \delta_s}\} \, dx$	$(1/d) \int_0^{h_{coll}} \{(1/\cos L)\sqrt{\cos^2 x + \tan^2 \delta_s} - [1/(CR)][\cos x - \cos h_{sr}]\} \, dx$
Polar tracking	$(ah_{coll} + b \sin h_{coll})/(d \cos L)$	$(h_{coll}/d)\{(1/\cos L) + [\cos h_{sr}/(CR)]\} - \sin h_{coll}/[d(CR)]$
Two-axis tracking	$(ah_{coll} + b \sin h_{coll})/(d \cos \delta_s \cos L)$	$(h_{coll}/d)[1/\cos \delta_s \cos L] +[\cos h_{sr}/(CR)] - h_{coll}/[d(CR)]$

[a] The collection hour angle value h_{coll} not used as the argument of trigonometric functions is expressed in radians; note that the total collection interval, $2h_{coll}$, is assumed to be centered about solar noon.

[b] $a = 0.409 + 0.5016 \sin(h_{sr} - 60°)$.

[c] $c = 0.6609 - 0.4767 \sin(h_{sr} - 60°)$.

[d] $d = \sin h_{sr} - h_{sr} \cos h_{sr}$; $\cos h_{sr} \ (i = 90°) = -\tan \delta_s \tan(L - \beta)$.

[e] CR is the collector concentration ratio.

[f] Use elliptic integral tables to evaluate terms of the form of $\int_0^h \sqrt{\cos^2 x + \tan^2 \delta_s} \, dx$ contained in r_T and r_d.

2   SOLAR THERMAL COLLECTORS

The principal use of solar energy is in the production of heat at a wide range of temperatures matched to a specific task to be performed. The temperature at which heat can be produced from solar radiation is limited to about 6000°F by thermodynamic, optical, and manufacturing constraints. Between temperatures near ambient and this upper limit very many thermal collector designs are employed to produce heat at a specified temperature. This section describes the common thermal collectors.

2.1   Flat-Plate Collectors

From a production volume standpoint, the majority of installed solar collectors are of the flate-plate design; these collectors are capable of producing heat at temperatures up to 100°C. Flat-plate collectors are so named since all components are planar. Figure 8a is a partial isometric sketch of a liquid-cooled flat-plate collector. From the top down it contains a glazing system—normally one pane of glass, a dark colored metal absorbing plate, insulation to the rear of the absorber, and, finally, a metal or plastic weatherproof housing. The glazing system is sealed to the housing to prohibit the ingress of water, moisture, and dust. The piping shown is thermally bonded to the absorber plate and contains the working fluid by which the heat produced is transferred to its end use. The pipes shown are manifolded together so that one inlet and one outlet connection, only, are present. Figure 8b shows a number of other collector designs in common use.

The energy produced by flat-plate collectors is the difference between the solar flux absorbed by the absorber plate and that lost from it by convection and radiation from the upper (or "front") surface and that lost by conduction from the lower (or "back") surface. The solar flux absorbed is the incident flux I_c multiplied by the glazing system transmittance τ and by the absorber plate absorptance α. The heat lost from the absorber in steady state is given by an overall thermal conductance U_c multiplied by the difference in temperature between the collector absorber temperature T_c and the surrounding, ambient temperature T_a. In equation form the net heat produced q_u is then

$$q_u = (\tau\alpha)I_c - U_c(T_c - T_a)   (\text{W}/\text{m}^2)   \tag{26}$$

The rate of heat production depends on two classes of parameters. The first—T_c, T_a, and I_c—having to do with the operational environment and the condition of the collector. The second—U_c and $\tau\alpha$—are characteristics of the collector independent of where or how it is used. The optical properties τ and α depend on the incidence angle, both dropping rapidly in value for $i > 50$–$55°$. The heat loss conductance can be calculated,[1] but formal tests, as subsequently described, are preferred for the determination of both $\tau\alpha$ and U_c.

Collector efficiency is defined as the ratio of heat produced q_u to incident flux I_c, that is,

$$\eta_c \equiv q_u/I_c   \tag{27}$$

Using this definition with Eq. (26) gives the efficiency as

$$\eta_c = \tau\alpha - U_c\left(\frac{T_c - T_a}{I_c}\right)   \tag{28}$$

The collector plate temperature is difficult to measure in practice, but the fluid inlet temperature $T_{f,i}$ is relatively easy to measure. Furthermore, $T_{f,i}$ is often known from characteristics of the process to which the collector is connected. It is common practice to express the efficiency in terms of $T_{f,i}$ instead of T_c for this reason. The efficiency is

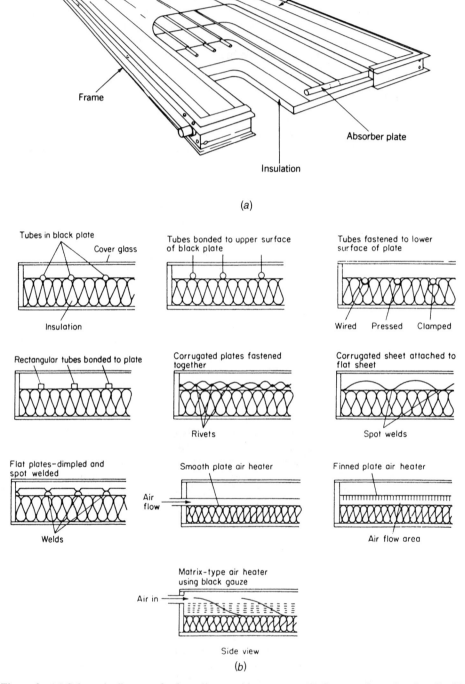

Figure 8 (*a*) Schematic diagram of solar collector with one cover. (*b*) Cross sections of various liquid- and air-based flat-plate collectors in common use.

$$\eta_c = F_R \left[\tau\alpha - U_c \left(\frac{T_{f,i} - T_a}{I_c} \right) \right] \qquad (29)$$

in which the heat removal factor F_R is introduced to account for the use of $T_{f,i}$ for the efficiency basis. F_R depends on the absorber plate thermal characteristics and heat loss conductance.[2]

Equation (29) can be plotted with the group of operational characteristics $(T_{f,i} - T_a)/I_c$ as the independent variable as shown in Fig. 9. The efficiency decreases linearly with the abscissa value. The intercept of the efficiency curve is the optical efficiency $\tau\alpha$ and the slope is $-F_R U_c$. Since the glazing transmittance and absorber absorptance decrease with solar incidence angle, the efficiency curve migrates toward the origin with increasing incidence angle, as shown in the figure. Data points from a collector test are also shown on the plot. The best-fit efficiency curve at normal incidence ($i = 0$) is determined numerically by a curve-fit method. The slope and intercept of the experimental curve, so determined, are the preferred values of the collector parameters as opposed to those calculated theoretically.

Selective Surfaces
One method of improving efficiency is to reduce radiative heat loss from the absorber surface. This is commonly done by using a low emittance (in the infrared region) surface having high absorptance for solar flux. Such surfaces are called (wavelength) selective surface and

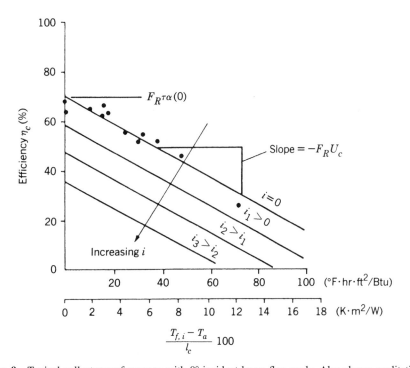

Figure 9 Typical collector performance with 0° incident beam flux angle. Also shown qualitatively is the effect of incidence angle i, which may be quantified by $\overline{\tau\alpha}(i)/\overline{\tau\alpha}(0) = 1.0 + b_0(1/\cos i - 1.0)$, where b_0 is the incidence angle modifier determined experimentally (ASHRAE 93-77) or from the Stokes and Fresnel equations.

are used on very many flat-plate collectors to improve efficiency at elevated temperature. Table 6 lists emittance and absorptance values for a number of common selective surfaces. Black chrome is very reliable and cost effective.

2.2 Concentrating Collectors

Another method of improving the efficiency of solar collectors is to reduce the parasitic heat loss embodied in the second term of Eq. (29).[3] This can be done by reducing the size of the absorber relative to the aperture area. Relatively speaking, the area from which heat is lost is smaller than the heat collection area and efficiency increases. Collectors that focus sunlight onto a relatively small absorber can achieve excellent efficiency at temperatures above which flat-plate collectors produce no net heat output. In this section a number of concentrators are described.

Trough Collectors

Figure 10 shows cross sections of five concentrators used for producing heat at temperatures up to 650°F at good efficiency. Figure 10a shows the parabolic "trough" collector representing the most common concentrator design available commercially. Sunlight is focused onto a circular pipe absorber located along the focal line. The trough rotates about the absorber centerline in order to maintain a sharp focus of incident beam radiation on the absorber. Selective surfaces and glass enclosures are used to minimize heat losses from the absorber tube.

Figures 10c and 10d show Fresnel-type concentrators in which the large reflector surface is subdivided into several smaller, more easily fabricated and shipped segments. The smaller reflector elements are easier to track and offer less wind resistance at windy sites; futhermore, the smaller reflectors are less costly. Figure 10e shows a Fresnel lens concentrator. No reflection is used with this approach; reflection is replaced by refraction to achieve the focusing effect. This device has the advantage that optical precision requirements can be relaxed somewhat relative to reflective methods.

Figure 10b shows schematically a concentrating method in which the mirror is fixed, thereby avoiding all problems associated with moving large mirrors to track the sun as in the case of concentrators described above. Only the absorber pipe is required to move to maintain a focus on the focal line.

The useful heat produced Q_u by any concentrator is given by

Table 6 Selective Surface Properties

Material	Absorptance[a] α	Emittance ϵ	Comments
Black chrome	0.87–0.93	0.1	
Black zinc	0.9	0.1	
Copper oxide over aluminum	0.93	0.11	
Black copper over copper	0.85–0.90	0.08–0.12	Patinates with moisture
Black chrome over nickel	0.92–0.94	0.07–0.12	Stable at high temperatures
Black nickel over nickel	0.93	0.06	May be influenced by moisture
Black iron over steel	0.90	0.10	

[a]Dependent on thickness.

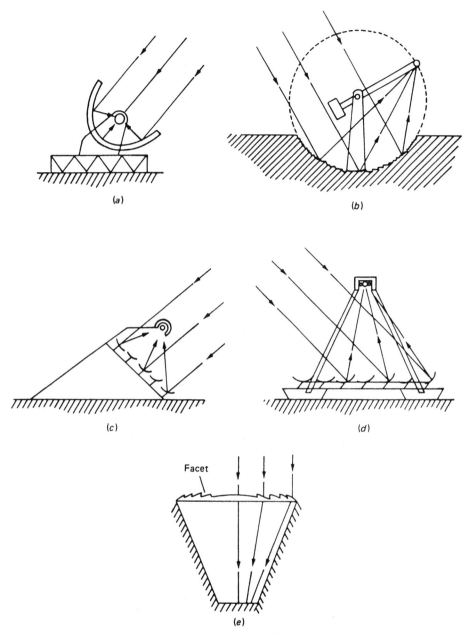

Figure 10 Single-curvature solar concentrators: (*a*) parabolic trough; (*b*) fixed circular trough with tracking absorber; (*c*) and (*d*) Fresnel mirror designs; and (*e*) Fresnel lens.

$$Q_u = A_a \, \eta_0 I_c - A_r U_c'(T_c - T_a) \tag{30}$$

in which the concentrator optical efficiency (analogous to $\tau\alpha$ for flat-plate collectors) is η_0, the aperture area is A_a, the receiver or absorber area is A_r, and the absorber heat loss conductance is U_c'. Collector efficiency can be found from Eq. (27) and is given by

$$\eta_c = \eta_0 - \frac{A_r}{A_a} U_c' \left(\frac{T_c - T_a}{I_c} \right) \tag{31a}$$

The aperture area-receiver area ratio $A_a/A_r > 1$ is called the geometric concentration ratio CR. It is the factor by which absorber heat losses are reduced relative to the aperture area:

$$\eta_c = \eta_0 - \frac{U_c'}{\text{CR}} \left(\frac{T_c - T_a}{I_c} \right) \tag{31b}$$

As with flat-plate collectors, efficiency is most often based on collector fluid inlet temperature $T_{f,i}$. On this basis, efficiency is expressed as

$$\eta_c = F_R \left[\eta_0 - U_c \left(\frac{T_{f,i} - T_a}{I_c} \right) \right] \tag{32}$$

in which the heat loss conductance U_c on an aperture area basis is used ($U_c = U_c'/\text{CR}$).

The optical efficiency of concentrators must account for a number of factors not present in flat-plate collectors including mirror reflectance, shading of aperture by receiver and its supports, spillage of flux beyond receiver tube ends at off-normal incidence conditions, and random surface, tracking, and construction errors that affect the precision of focus. In equation form the general optical efficiency is given by

$$\eta_0 = \rho_m \tau_c a_r f_t \delta F(i) \tag{33}$$

where ρ_m is the mirror reflectance (0.8–0.9), τ_c is the receiver cover transmittance (0.85–0.92), α_r is the receiver surface absorptance (0.9–0.92), f_t is the fraction of aperture area not shaded by receiver and its supports (0.95–0.97), δ is the intercept factor accounting for mirror surface and tracking errors (0.90–0.95), and $F(i)$ is the fraction of reflected solar flux intercepted by the receiver for perfect optics and perfect tracking. Values for these parameters are given in Refs. 2 and 4.

Compound Curvature Concentrators

Further increases in concentration and concomitant reductions in heat loss are achievable if "dishtype" concentrators are used. This family of concentrators is exemplified by the paraboloidal dish concentrator, which focuses solar flux at a point instead of along a line as with trough collectors. As a result the achievable concentration ratios are approximately the square of what can be realized with single curvature, trough collectors. Figure 11 shows a paraboloidal dish concentrator assembly. These devices are of most interest for power production and some elevated industrial process heat applications.

For very large aperture areas it is impractical to construct paraboloidal dishes consisting of a single reflector. Instead the mirror is segmented as shown in Fig. 12. This collector system called the central receiver has been used in several solar thermal power plants in the 1- to 15-MW range. This power production method is discussed in the next section.

The efficiency of compound curvature dish collectors is given by Eq. (32), where the parameters involved are defined in the context of compound curvature optics.[4] The heat loss term at high temperatures achieved by dish concentrators is dominated by radiation; therefore, the second term of the efficiency equation is represented as

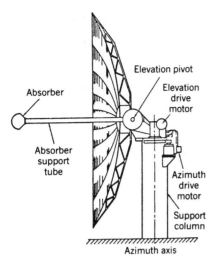

Figure 11 Segmented mirror approximation to paraboloidal dish. Average CR is 118, while maximum local CR is 350.

$$\eta_c = \eta_0 - \frac{\epsilon_r \sigma (T_c^4 - T_a'^4)}{\text{CR}} \tag{34}$$

where ϵ_r the infrared emittance of the receiver, σ is the Stefan-Boltzmann constant, and T_a' is the equivalent ambient temperature for radiation depending on ambient humidity and cloud cover. For clear, dry conditions T_a' is about 15–20°F below the ambient dry bulb temperature. As humidity decreases, T_a approaches the dry bulb temperature.

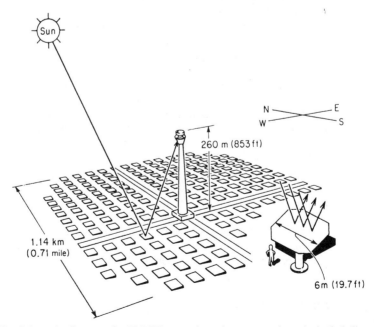

Figure 12 Schematic diagram of a 50-MWe central receiver power plant. A single heliostat is shown in the inset to indicate its human scale. (From Electric Power Research Institute (EPRI).)

The optical efficiency for the central receiver is expressed in somewhat different terms than those used in Eq. (33). It is referenced to solar flux on a horizontal surface and therefore includes the geometric tilt factor. For the central receiver, the optical efficiency is given by

$$\eta_0 = \phi \wp \rho_m \alpha_r f_t \delta \tag{35}$$

in which the last four parameters are defined as in Eq. (33). The ratio of redirected flux to horizontal flux is $\wp$ and is given approximately by

$$\wp = 0.78 + 1.5(1 - \alpha/90)^2 \tag{36}$$

from Ref. 4. The ratio of mirror area to ground area ϕ depends on the size and economic factors applicable to a specific installation. Values for ϕ have been in the range 0.4–0.5 for installations made through 1985.

2.3 Collector Testing

To determine the optical efficiency and heat loss characteristics of flat-plate and concentrating collectors (other than the central receiver, which is difficult to test because of its size), testing under controlled conditions is preferred to theoretical calculations. Such test data are required if comparisons among collectors are to be made objectively. As of the mid-1980s very few consensus standards had been adopted by the U.S. solar industry. The ASHRAE Standard Number 93-77 applies to flat-plate collectors that contain either a liquid or a gaseous working fluid.[5] Collectors in which a phase change occurs are not included. In addition, the standards do not apply well to concentrators, since additional procedures are needed to find the optical efficiency and aging effects. Testing of concentrators uses sections of the above standard where applicable plus additional procedures as needed; however, no industry standard exists. (The ASTM has promulgated standard E905 as the first proposed standard for concentrator tests.) ASHRAE Standard Number 96-80 applies to very-low-temperature collectors manufactured without any glazing system.

Figure 13 shows the test loop used for liquid-cooled flat-plate collectors. Tests are conducted with solar flux at near-normal incidence to find the normal incidence optical efficiency $(\tau\alpha)_n$ along with the heat loss conductance U_c. Off-normal optical efficiency is determined in a separate test by orienting the collector such that several substantially off-normal values of $\tau\alpha$ or η_0 can be measured. The fluid used in the test is preferably that to be used in the installed application, although this is not always possible. If operational and test fluids differ, an analytical correction in the heat removal factor F_R is to be made.[2] An additional test is made after a period of time (nominal one month) to determine the effect of aging, if any, on the collector parameters listed above. A similar test loop and procedure apply to air-cooled collectors.[5]

The development of full system tests has only begun. Of course, it is the entire solar system (see next section) not just the collector that ultimately must be rated in order to compare solar and other energy-conversion systems. Testing of full-size solar systems is very difficult owing to their large size and cost. Hence, it is unlikely that full system tests will ever be practical except for the smallest systems such as residential water heating systems. For this one group of systems a standard test procedure (ASHRAE 95-81) exists. Larger-system performance is often predicted, based on component tests, rather than measured.

3 SOLAR THERMAL APPLICATIONS

One of the unique features of solar heat is that it can be produced over a very broad range of temperatures—the specific temperature being selected to match the thermal task to be

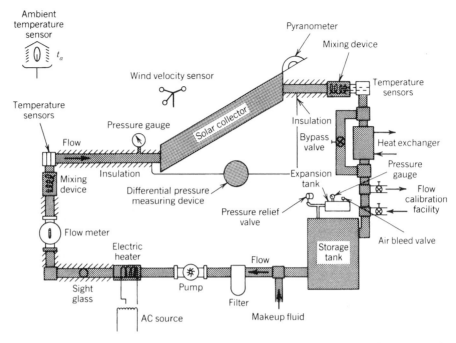

Figure 13 Closed-loop testing configuration for the solar collector when the transfer fluid is a liquid.

performed. In this section the most common thermal applications will be described in summary form. These include low-temperature uses such as water and space heating (30–100°C), intermediate temperature industrial processes (100–300°C), and high-temperature thermal power applications (500–850°C and above). Methods for predicting performance, where available, will also be summarized. Nonthermal solar applications are described in the next section.

3.1 Solar Water Heating

The most often used solar thermal application is for the heating of water for either domestic or industrial purposes.[6] Relatively simple systems are used, and the load exists relatively uniformly through a year resulting in a good system load factor. Figure 14*a* shows a single-tank water heater schematically. The key components are the collector (0.5–1.0 ft²/gal day load), the storage tank (1.0–2.0 gal/ft² of collector), a circulating pump, and controller. The check valve is essential to prevent backflow of collector fluid, which can occur at night when the pump is off if the collectors are located some distance above the storage tank. The controller actuates the pump whenever the collector is 15–30°F warmer than storage. Operation continues until the collector is only 1.5–5°F warmer than the tank, at which point it is no longer worthwhile to operate the pump to collect the relatively small amounts of solar heat available.

The water-heating system shown in Fig. 14*a* uses an electrical coil located near the top of the tank to ensure a hot water supply during periods of solar outage. This approach is only useful in small residential systems and where nonsolar energy resources other than electricity are not available. Most commercial systems are arranged as shown in Fig. 14*b*, where a separate preheat tank, heated only by solar heat, is connected upstream of the

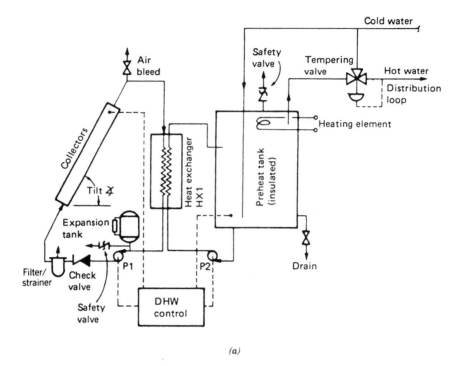

(a)

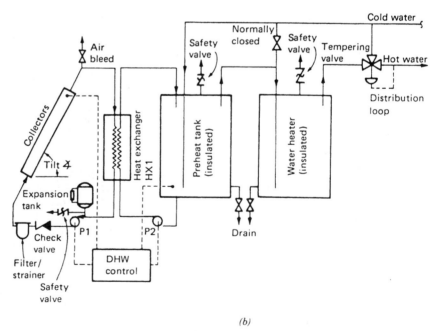

(b)

Figure 14   (*a*) Single-tank indirect solar water-heating system. (*b*) Double-tank indirect solar water-heating system. Instrumentation and miscellaneous fittings are not shown.

nonsolar, auxiliary water heater tank or plant steam-to-water heat exchanger. This approach is more versatile in that any source of backup energy whatever can be used when solar heat is not available. Additional parasitic heat loss is encountered, since total tank surface area is larger than for the single tank design.

The water-heating systems shown in Fig. 14 are of the indirect type, that is, a separate fluid is heated in the collector and heat thus collected is transferred to the end use via a heat exchanger. This approach is needed in locations where freezing occurs in winter and anti-freeze solutions are required. The heat exchanger can be eliminated, thereby reducing cost and eliminating the unavoidable fluid temperature decrement between collector and storage fluid streams, if freezing will never occur at the application site. The exchanger can also be eliminated if the "drain-back" approach is used. In this system design the collectors are filled with water only when the circulating pump is on, that is, only when the collectors are warm. If the pump is not operating, the collectors and associated piping all drain back into the storage tank. This approach has the further advantage that heated water otherwise left to cool overnight in the collectors is returned to storage for useful purposes.

The earliest water heaters did not use circulating pumps, but used the density difference between cold collector inlet water and warmer collector outlet water to produce the flow. This approach is called a "thermosiphon" and is shown in Fig. 15. These systems are among the most efficient, since no parasitic use of electric pump power is required. The principal difficulty is the requirement that the large storage tank be located above the collector array, often resulting in structural and architectural difficulties. Few industrial solar water-heating systems have used this approach, owing to difficulties in balancing buoyancy-induced flows in large piping networks.

3.2 Mechanical Solar Space Heating Systems

Solar space heating is accomplished using systems similar to those for solar water heating. The collectors, storage tank, pumps, heat exchangers, and other components are larger in proportion to the larger space heat loads to be met by these systems in building applications. Figure 16 shows the arrangement of components in one common space heating system. All

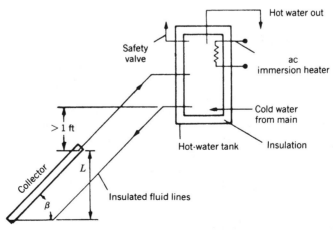

Figure 15 Passive thermosiphon single-tank direct system for solar water heating. Collector is positioned below the tank to avoid reverse circulation.

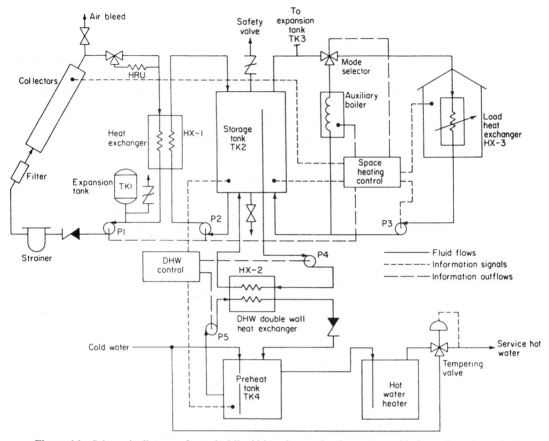

Figure 16 Schematic diagram of a typical liquid-based space heating system with domestic water preheat.

components except the solar collector and controller have been in use for many years in building systems and are not of special design for the solar application.

The control system is somewhat more complex than that used in nonsolar building heating systems, since two heat sources—solar and nonsolar auxiliary—are to be used under different conditions. Controls using simple microprocessors are available for precise and reliable control of solar space heating systems.

Air-based systems are also widely used for space heating. They are similar to the liquid system shown in Fig. 16 except that no heat exchanger is used and rock piles, not tanks of fluid, are the storage media. Rock storage is essential to efficient air-system operation since gravel (usually 1–2 in. in diameter) has a large surface-to-volume ratio necessary to offset the poor heat transfer characteristics of the air working fluid. Slightly different control systems are used for air-based solar heaters.

3.3 Passive Solar Space Heating Systems

A very effective way of heating residences and small commercial buildings with solar energy and without significant nonsolar operating energy is the "passive" heating approach.[1] Solar flux is admitted into the space to be heated by large sun-facing apertures. In order that overheating not occur during sunny periods, large amounts of thermal storage are used, often also serving a structural purpose. A number of classes of passive heating systems have been identified and are described in this section.

Figure 17 shows the simplest type of passive system known as "direct gain." Solar flux enters a large aperture and is converted to heat by absorption on dark colored floors or walls. Heat produced at these wall surfaces is partly conducted into the wall or floor serving as stored heat for later periods without sun. The remaining heat produced at wall or floor surfaces is convected away from the surface thereby heating the space bounded by the surface. Direct-gain systems also admit significant daylight during the day; properly used, this can reduce artificial lighting energy use. In cold climates significant heat loss can occur through the solar aperture during long, cold winter nights. Hence, a necessary component of efficient direct-gain systems is some type of insulation system put in place at night over the passive aperture. This is indicated by the dashed lines in the figure.

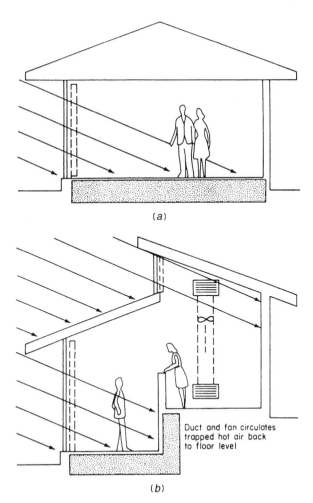

(a)

Duct and fan circulates
trapped hot air back
to floor level

(b)

Figure 17 Direct-gain passive heating systems: (a) adjacent space heating; (b) clerestory for north zone heating.

The second type of passive system commonly used is variously called the thermal storage wall (TSW) or collector storage wall. This system, shown in Fig. 18, uses a storage mass interposed between the aperture and space to be heated. The reason for this positioning is to better illuminate storage for a significant part of the heating season and also to obviate the need for a separate insulation system; selective surfaces applied to the outer storage wall surface are able to control heat loss well in cold climates, while having little effect on solar absorption. As shown in the figure, a thermocirculation loop is used to transport heat from the warm, outer surface of the storage wall to the space interior to the wall. This air flow convects heat into the space during the day, while conduction through the wall heats the space after sunset. Typical storage media include masonry, water, and selected eutectic mixtures of organic and inorganic materials. The storage wall eliminates glare problems associated with direct-gain systems, also.

The third type of passive system in use is the attached greenhouse or "sunspace" as shown in Fig. 19. This system combines certain features of both direct-gain and storage wall systems. Night insulation may or may not be used, depending on the temperature control required during nighttime.

The key parameters determining the effectiveness of passive systems are the optical efficiency of the glazing system, the amount of directly illuminated storage and its thermal characteristics, the available solar flux in winter, and the thermal characteristics of the building of which the passive system is a part. In a later section, these parameters will be quantified and will be used to predict the energy saved by the system for a given building in a given location.

3.4 Solar Ponds

A "solar pond" is a body of water no deeper than a few meters configured in such a way that usual convection currents induced by solar absorption are suppressed.[5] The oldest method for convection suppression is the use of high concentrations of soluble salts in layers near the bottom of the pond with progressively smaller concentrations near the surface. The surface layer itself is usually fresh water. Incident solar flux is absorbed by three mechanisms. Within a few millimeters of the surface the infrared component (about one-third of the total

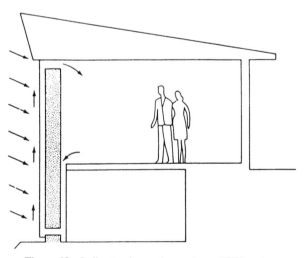

Figure 18 Indirect-gain passive system—TSW system.

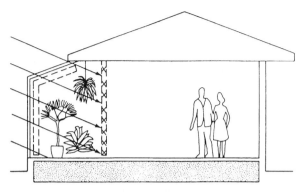

Figure 19   Greenhouse or attached sun-space passive heating system using a combination of direct gain into the greenhouse and indirect gain through the thermal storage wall, shown by cross-hatching, between the greenhouse and the living space.

solar flux energy content) is completely absorbed. Another third is absorbed as the visible and ultraviolet components traverse a pond of nominal 2-m depth. The remaining one-third is absorbed at the bottom of the pond. It is this component that would induce convection currents in a freshwater pond thereby causing warm water to rise to the top where convection and evaporation would cause substantial heat loss. With proper concentration gradient, convection can be completely suppressed and significant heat collection at the bottom layer is possible. Salt gradient ponds are hydrodynamically stable if the following criterion is satisfied:

$$\frac{d\rho}{dz} = \frac{\partial\rho}{\partial s}\frac{ds}{dz} + \frac{\partial\rho}{\partial T}\frac{dT}{dz} > 0 \tag{37}$$

where s is the salt concentration, ρ is the density, T is the temperature, and z is the vertical coordinate measured positive downward from the pond surface. The inequality requires that the density must decrease upward.

Useful heat produced is stored in and removed from the lowest layer as shown in Fig. 20. This can be done by removing the bottom layer of fluid, passing it through a heat exchanger, and returning the cooled fluid to another point in the bottom layer. Alternatively,

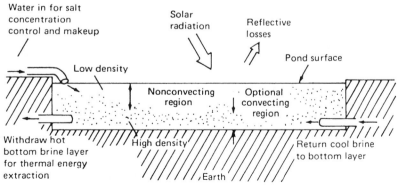

Figure 20   Schematic diagram of a nonconvecting solar pond showing conduits for heat withdrawal, surface washing, and an optional convecting zone near the bottom.

a network of heat-removal pipes can be placed on the bottom of the bond and the working fluid passed through for heat collection. Depending on the design, solar ponds also may contain substantial heat storage capability if the lower convective zone is relatively thick. This approach is used when uniform heat supply is necessary over a 24-hr period but solar flux is available for only a fraction of the period. Other convection-suppression techniques and heat-removal methods have been proposed but not used in more than one installation at most.

The requirements for an effective solar pond installation include the following. Large amounts of nearly free water and salt must be available. The subsoil must be stable in order to minimize changes in pond shape that could fracture the waterproof liner. Adequate solar flux is required year around; therefore, pond usage is confined to latitudes within 40° of the equator. Freshwater aquifers used for potable water should not be nearby in the event of a major leak of saline water into the groundwater. Other factors include low winds to avoid surface waves and windblown dust collection within the pond (at the neutral buoyancy point), low soil conductivity (i.e., low water content) to minimize conduction heat loss, and durable liner materials capable of remaining leakproof for many years.

The principal user of solar ponds has been the country of Israel. Ponds tens of acres in size have been built and operated successfully. Heat collected has been used for power production with an organic Rankine cycle, for space heating, and for industrial uses. A thorough review of solar pond technology is contained in Ref. 5. A method for predicting the performance of a solar pond is presented in the next section.

3.5 Solar Thermal Power Production

Solar energy has very high thermodynamic availability owing to the high effective temperature of the source. Therefore, production of shaft power and electric power therefrom is thermodynamically possible. Two fundamentally different types of systems can be used for power production: (1) a large array of concentrating collectors of several tens of meters in area connected by a fluid or electrical network and (2) a single, central receiver using mirrors distributed over a large area but producing heat and power only at one location. The determination of which approach is preferred depends on required plant capacity. For systems smaller than 10 MW the distributed approach appears more economical with existing steam turbines. For systems greater than 10 MW, the central receiver appears more economical.[7] However, if highly efficient Brayton or Stirling engines were available in the 10–20 kW range, the distributed approach would have lowest cost for any plant size. Such systems will be available by the year 2000.

The first U.S. central receiver began operating in the fall of 1982. Located in the Mojave Desert, this 10-MW plant (called "Solar One") is connected to the southern California electrical grid. The collection system consists of 1818 heliostats totaling 782,000 ft^2 in area. Each 430 ft^2 mirror is computer controlled to focus reflected solar flux onto the receiver located 300 ft above the desert floor. The receiver is a 23-ft diameter cyclinder whose outer surface is the solar absorber. The absorbing surface is coated with a special black paint selected for its reliability at the nominal 600°C operating temperature. Thermal storage consisting of a mixture of an industrial heat transfer oil for heat transport and of rock and sand has a nominal operating temperature of 300°C. Storage is used to extend the plant operating time beyond sunset (albeit at lower turbine efficiency) and to maintain the turbine, condenser, and piping at operating temperatures overnight as well as to provide startup steam the following morning. The plant was modernized in 1996 and operated for several more years.

Solar-produced power is not generally cost effective currently. The principal purpose of the Solar One experiment and other projects in Europe and Japan is to acquire operating experience with the solar plant itself as well as with the interaction of solar and nonsolar power plants connected in a large utility grid. Extensive data collection and analysis will answer questions regarding long-term net efficiency of solar plants, capacity displacement capability, and reliability of the new components of the system—mirror field, receiver, and computer controls.

3.6 Other Thermal Applications

The previous sections have discussed the principal thermal applications of solar energy that have been reduced to practice in at least five different installations and that show significant promise for economic displacement of fossil or fissile energies. In this section two other solar-conversion technologies are summarized.

Solar-powered cooling has been demonstrated in many installations in the United States, Europe, and Japan. Chemical absorption, organic Rankine cycle, and desiccant dehumidifaction processes have all been shown to be functional. Most systems have used flat-plate collectors, but higher coefficients of performance are achievable with mildly concentrating collectors. Reference 6 describes solar-cooling technologies. To date, economic viability has not been generally demonstrated, but further research resulting in reduced cost and improved efficiency is expected to continue.

Thermal energy stored in the surface layers of the tropical oceans has been used to produce electrical power on a small scale. A heat engine is operated between the warmest layer at the surface and colder layers several thousand feet beneath. The available temperature difference is of the order of 20°C, therefore, the cycle efficiency is very low—only a few percent. However, this type of power plant does not require collectors or storage. Only a turbine capable of operating efficiently at low temperature is needed. Some cycle designs also require very large heat exchangers, but new cycle concepts without heat exchangers and their unavoidable thermodynamic penalties show promise.

3.7 Performance Prediction for Solar Thermal Processes

In a rational economy the single imperative for use of solar heat for any of the myriad applications outlined heretofore must be cost competitiveness with other energy sources— fossil and fissile. The amount of useful solar energy produced by a solar-conversion system must therefore be known along with the cost of the system. In this section the methods usable for predicting the performance of widely deployed solar systems are summarized. Special systems such as the central receiver, the ocean thermal power plant, and solar cooling are not included. The methods described here require a minimum of computational effort, yet embody all important parameters determining performance.

Solar systems are connected to end uses characterized by an energy requirement or "load" L and by operating temperature that must be achievable by the solar-heat-producing system. The amount of solar-produced heat delivered to the end use is the useful energy Q_u. This is the net heat delivery accounting for parasitic losses in the solar subsystem. The ratio of useful heat delivered to the requirement L is called the "solar fraction" denoted by f_s. In equation form the solar fraction is

$$f_s \equiv \frac{Q_u}{L} \tag{38}$$

Empirical equations have been developed relating the solar fraction to other dimensionless groups characterizing a given solar process. These are summarized shortly.

A fundamental concept used in many predictive methods is the solar "utilizability" defined as that portion of solar flux absorbed by a collector that is capable of providing heat to the specified end use. The key characteristic of the end use is its temperature. The collector must produce at least enough heat to offset losses when the collector is at the minimum temperature T_{min} usable by the given process. Figure 21 illustrates this idea schematically. The curve represents the flux absorbed over a day by a hypothetical collector. The horizontal line intersecting this curve represents the threshold flux that must be exceeded for a net energy collection to take place. In the context of the efficiency equation [Eq. (32)], this critical flux I_{cr} is that which results in a collector efficiency of exactly zero when the collector is at the minimum usable process temperature T_{min}. Any greater flux will result in net heat production. From Eq. (32) the critical intensity is

$$I_{cr} = \frac{U_c(T_{min} - T_a)}{\tau\alpha} \tag{39}$$

The solar utilizability is the ratio of the useful daily flux (area above I_{cr} line in Fig. 21) to the total absorbed flux (area $A_1 + A_2$) beneath the curve. The utilizability denoted by ϕ is

$$\phi = \frac{A_1}{A_1 + A_2} \tag{40}$$

This quantity is a solar radiation statistic depending on I_{cr}, characteristics of the incident solar flux and characteristics of the collection system. It is a very useful parameter in predicting the performance of solar thermal systems.

Table 7 summarizes empirical equations used for predicting the performance of the most common solar-thermal systems. These expressions are given in terms of the solar fraction defined above and dimensionless parameters containing all important system characteristics. The symbols used in this table are defined in Table 8. In the brief space available in this chapter, all details of these prediction methodologies cannot be included. The reader is referred to Refs. 1, 3, 7, and 8 for details.

4   PHOTOVOLTAIC SOLAR ENERGY APPLICATIONS

In this section the principal nonthermal solar conversion technology is described. Photovoltaic cells are capable of converting solar flux directly into electric power. This process, first demonstrated in the 1950s, holds considerable promise for significant use in the future. Major

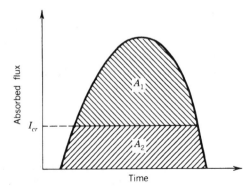

Figure 21   Daily absorbed solar flux ($A_1 + A_2$) and useful solar flux (A_1) at intensities above I_{cr}.

Table 7 Empirical Solar Fraction Equations[a]

System Type	f_s Expression	Time Scale
Water heating and liquid-based space heating	$f_s = 1.029P_s\text{s} - 0.065P_L - 0.245P_s^2 + 0.0018P_L^2$ $+ 0.00215P_s^3$	Monthly
Space heating—air-based systems	$f_s = 1.040P_s - 0.065P_L - 0.159P_s^2 + 0.00187P_L^2$ $+ 0.0095P_s^3$	Monthly
Passive direct gain	$f_s = PX + (1 - P)(3.082 - 3.142\overline{\phi})(1 - e^{-0.329x})$	Monthly
Passive storage wall	$f_s = Pf_\infty + 0.88(1 - P)(1 - e^{-1.26f_\infty})$	Monthly
Concentrating collector systems	$f_s = F_R\overline{\eta}_0\overline{I}_cA_cN\overline{\phi}'/L$	Monthly
Solar ponds (pond radius R to provide annual pond temperature $\overline{T}_p$)	$R = \dfrac{2.2\,\overline{\Delta T} + [4.84(\Delta T)^2 + \overline{L}(0.3181\,\overline{I}_p - 0.1592\Delta T)]^{1/2}}{\overline{I}_p - 0.5\Delta T}$	Annual

[a] See Table 8 for symbol definitions.

cost reductions have been accomplished. In this section the important features of solar cells are described.

Photovoltaic conversion of sunlight to electricity occurs in a thin layer of semiconductor material exposed to solar flux. Photons free electric charges, which flow through an external circuit to produce useful work. The semiconductor materials used for solar cells are tailored to be able to convert the majority of terrestrial solar flux; however, low-energy photons in the infrared region are usually not usable. Figure 22 shows the maximum theoretical conversion efficiency of seven common materials used in the application. Each material has its own threshold band-gap energy, which is a weak function of temperature. The energy contained in a photon is $E = h\nu$. If E is greater than the band-gap energy shown in this figure, conversion can occur.

Figure 22 also shows the very strong effect of temperature on efficiency. For practical systems it is essential that the cell be maintained as near to ambient temperature as possible.

Solar cells produce current proportional to the solar flux intensity with wavelengths below the band-gap threshold. Figure 23 shows the equivalent circuit of a solar cell. Both internal shunt and series resistances must be included. These result in unavoidable parasitic loss of part of the power produced by the equivalent circuit current source of strength I_s. Solving the equivalent circuit for the power P produced and using an expression from Ref. 1 for the junction leakage I_J results in

$$P = [I_s - I_0(e^{e_0\nu/kT} - 1)]V \tag{41}$$

in which e_0 is the electron charge, k is the Boltzmann constant, and T is the temperature. The current source I_s is given by

$$I_s = \eta_0(1 - \rho_c)\alpha e_0 n_p \tag{42}$$

in which η_0 is the collector carrier efficiency, ρ_c is the cell surface reflectance, α is the absorptance of photons, and n_p is the flux density of sufficiently energetic photons.

In addition to the solar cell, complete photovoltaic systems also must contain electrical storage and a control system. The cost of storage presents another substantial cost problem

Table 8   Definition of Symbols in Table 7

Parameters		Definition	Units[a]
P_L		$P_s = \dfrac{F_{hx}F_R U_c(T_r - \overline{T}_a)\Delta t}{L}$	None
	F_{hx}	$F_{hx} = \left\{\left[1 + \dfrac{F_R U_c A_c}{(\dot{m}C_p)_c}\right]\left[\dfrac{(\dot{m}C_p)_c}{(\dot{m}C_p)_{min}\epsilon} - 1\right]\right\}^{-1},$ collector heat exchanger penalty factor	None
	$F_R U_c$	Collector heat-loss conductance	But/hr·ft²·°F
	A_c	Collector area	ft²
	$(\dot{m}C_p)_c$	Collector fluid capacitance rate	Btu/hr·°F
	$(\dot{m}C_p)_{min}$	Minimum capacitance rate in collector heat exchanger	Btu/hr·°F
	ϵ	Collector heat-exchanger effectiveness	None
	T_r	Reference temperature, 212°F	°F
	$\overline{T}_a$	Monthly averaged ambient temperature	°F
	Δt	Number of hours per month	hr/month
	L	Monthly load	Btu/month
P_s		$P_s = \dfrac{F_{hx}F_R\overline{\tau\alpha}\overline{I}_c N}{L}$	None
	$F_R\overline{\tau\alpha}$	Monthly averaged collector optical efficiency	None
	$\overline{I}_c$	Monthly averaged, daily incident solar flux	Btu/day·ft²
	N	Number of days per month	day/month
(P'_L – to be used for water heating only)		$P'_L = P_L \dfrac{(1.18T_{wo} + 3.86T_{wi} - 2.32\overline{T}_a - 66.2)}{212 - \overline{T}_a}$	None
	T_{wo}	Water output temperature	°F
	T_{wi}	Water supply temperature	°F
	P_L	(See above)	None
P		$P = (1 - e^{-0.294Y})^{0.652}$	None
	Y	Storage-vent ratio, $Y = \dfrac{C\Delta T}{\overline{\phi}\overline{I}_c\overline{\tau\alpha}A_c}$	None
	C	Passive storage capacity	Btu/°F
	ΔT	Allowable diurnal temperature saving in heated space	°F
$\overline{\phi}$		Monthly averaged utilizability (see below)	None
X		Solar-load ratio, $X = \dfrac{\overline{I}_c\overline{\tau\alpha}A_c N}{L}$	None
	L	Monthly space heat load	Btu/month
f_∞		Solar fraction with hypothetically infinite storage, $f_\infty = \dfrac{\overline{Q}_i + L_w}{L}$	None
	$\overline{Q}_i$	Net monthly heat flow through storage wall from outer surface to heated space	Btu/month
	L_w	Heat *loss* through storage wall	Btu/month
$F_R\overline{\eta}_0$		Monthly averaged concentrator optical efficiency	None
$\overline{\phi}'$		Monthly average utilizability for concentrators	None
R		Pond radius to provide diurnal average pond temperature $\overline{T}_p$	m

Table 8 (*Continued*)

Parameters		Definition	Units[a]
ΔT		$\overline{\Delta T} = \overline{T}_p - \overline{\overline{T}}_a$	°C
	$\overline{T}_p$	Annually averaged pond temperature	°C
	$\overline{\overline{T}}_a$	Annually averaged ambient temperature	°C
$\overline{L}$		Annual averaged load at $\overline{T}_p$	W
$\overline{I}_p$		Annual averaged insolation absorbd at pond bottom	W/m²
$\overline{\phi}$		Monthly flat-plate utilizability (equator facing collectors), $\overline{\phi} = \exp\{[A + B(\overline{R}_N/\overline{R})](\overline{X}_c + C\overline{X}_c^2)\}$	None
	A	$A = 7.476 - 20.0\overline{K}_T + 11.188\overline{K}_T^2$	None
	B	$B = -8.562 + 18.679\overline{K}_T - 9.948\overline{K}_T^2$	None
	C	$C = -0.722 + 2.426\overline{K}_T + 0.439\overline{K}_T^2$	None
	$\overline{R}$	Tilt factor, see Eq. (21)	None
	$\overline{R}_N$	Monthly averaged tilt factor for hour centered about noon (see Ref. 9)	None
	$\overline{X}_c$	Critical intensity ratio, $\overline{X}_c = \dfrac{I_{cr}}{r_{T,N}\overline{R}_N\overline{H}_h}$	None
	$r_{T,N}$	Fraction of daily total radiation contained in hour about noon, $r_{T,N} = r_{d,n}[1.07 + 0.025 \sin(h_{sr} - 60)]$	day/hr
		$r_{d,n} = \dfrac{\pi}{24} \dfrac{1 - \cos h_{sr}}{h_{sr} - h_{sr}\cos h_{sr}}$	day/hr
	I_{cr}	Critical intensity [see Eq. (39)]	Btu/hr·ft²
$\overline{\phi}'$		Monthly concentrator utilizability, $\overline{\phi}' = 1.0 - (0.049 + 1.49\overline{K}_T)\overline{X} + 0.341\overline{K}_T\overline{X}^2$ $0.0 < \overline{K}_T < 0.75, 0 < \overline{X} < 1.2)$ $\overline{\phi}' = 1.0 - \overline{X} \; (\overline{K}_T > 0.75, 0 < \overline{X} < 1.0)$	None
	$\overline{X}$	Concentrator critical intensity ratio, $\overline{X} = $ $\dfrac{U_c(T_{f,i} - \overline{T}_a)\Delta t_c}{\overline{\eta}_0 \overline{I}_c}$	None
	$T_{f,i}$	Collector fluid inlet temperature—assumed constant	°F
	Δt_c	Monthly averaged solar system operating time	hr/day

[a] USCS unit shown except for solar ponds; SI units may also be used for all parameters shown in USCS units.

in the widespread application of photovoltaic power production. The costs of the entire conversion system must be reduced by an order of magnitude in order to be competitive with other power sources. Vigorous research in the United States, Europe, and Japan has made significant gains in the past decade.

Figure 24 shows the effects of illumination intensity and cell temperature. Temperature affects the performance in a way that the voltage and thus the power output decrease with increasing temperature.

Efficiency is given by the ratio of useful output to insolation input:

$$n = E/(A_c I_c) \tag{43}$$

E is the area of the maximum rectangle inscribable within the IV curve shown in the figure, part *a*. A_c is the collector area and I_c is the incoming collector plane insolation. The char-

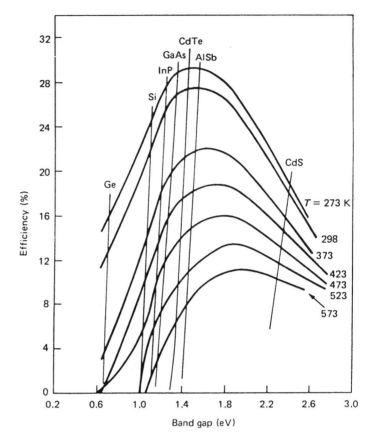

Figure 22 Maximum theoretical efficiency of photovoltaic converters as a function of band-gap energy for several materials.

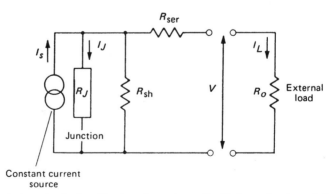

Figure 23 Equivalent circuit of an illuminated *p-n* photocell with internal series and shunt resistances and nonlinear junction impedance R_J.

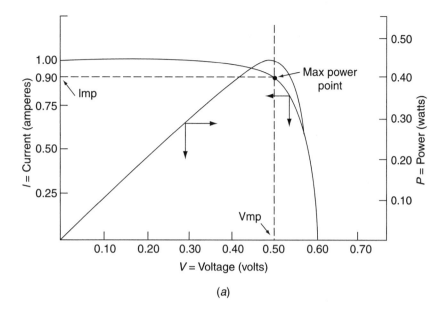

(a)

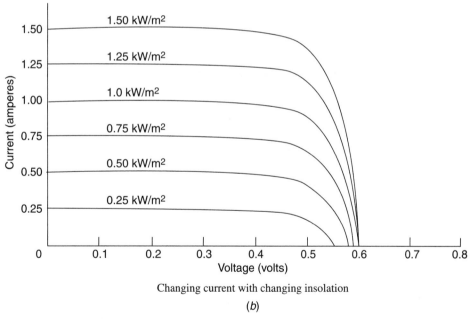

Changing current with changing insolation

(b)

Figure 24 (*a*) Current-voltage (IV) characteristic of a typical silicon cell depicting nominal power output as the area of the maximal rectangle that can be inscribed within the IV curve; (*b*) typical IV curves showing the effects of illumination level and (*c*) the effects of cell temperature.

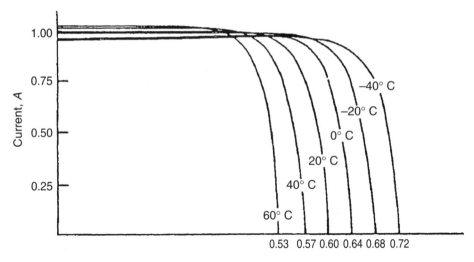

Changing voltage with changing temperature

(c)

Figure 24 (*Continued*)

acteristics noted in the IV curves below show that efficiency is essentially independent of insolation and varies inversely with cell temperature. A useful efficiency equation is

$$n = n_R (1 - \beta[T_C - T_R]) \tag{44}$$

where n_R is the reference efficiency (mfr's data)

β is the temperature coefficient (0.004/°C)

T_C is the PV cell temperature (°C)

T_R is the cell reference temperature at which the efficiency is n_R (°C)

This expression can be used with weather data to find the electrical output E from this expression, which follows from the definition of efficiency

$$E = n A_c I_c = \{n_R (1 - \beta[T_C - T_R])\} A_c I_c$$

Typical efficiencies for crystalline solar cells are 15–20% at rated conditions.

REFERENCES

1. Y. Goswami, F. Kreith, and J. F. Kreider, *Principles of Solar Engineering,* 2nd ed., Taylor & Francis, New York, 2000.
2. M. Collares-Pereira and A. Rabl, "The Average Distribution of Solar Energy," *Solar Energy* **22,** 155–164 (1979).
3. J. F. Kreider, *Medium and High Temperature Solar Processes,* Academic Press, New York, 1979.
4. ASHRAE Standard 93-77, *Methods of Testing to Determine the Thermal Performance of Solar Collectors,* ASHRAE, Atlanta, GA, 1977.

5. H. Tabor, "Solar Ponds—Review Article," *Solar Energy* **27,** 181–194 (1981).

6. J. F. Kreider and F. Kreith, *Solar Heating and Cooling,* Hemisphere/McGraw-Hill, New York, 1982.

7. M. Collares-Pereira and A. Rabl, "Simple Procedure for Predicting Long Term Average Performance of Nonconcentrating and of Concentrating Solar Collectors," *Solar Energy* **23,** 235–253 (1979).

8. J. F. Kreider, *The Solar Heating Design Process,* McGraw-Hill, New York, 1982.

CHAPTER 21

GEOTHERMAL RESOURCES AND TECHNOLOGY: AN INTRODUCTION

Peter D. Blair*
National Academy of Sciences
Washington, DC

1 INTRODUCTION

Geothermal energy is heat from the Earth's interior. Nearly all of geothermal energy refers to heat derived from the Earth's molten core. Some of what is often referred to as geothermal heat derives from solar heating of the surface of the Earth, although it amounts to a very small fraction of the energy derived from the Earth's core. For centuries, geothermal energy was apparent only through anomalies in the Earth's crust that permit the heat from the Earth's molten core to venture close to the surface. Volcanoes, geysers, fumaroles, and hot springs are the most visible surface manifestations of these anomalies.

The Earth's core temperature is estimated by most geologists to be around 5000 to 7000°C. For reference, that is nearly as hot as the surface of the sun (although, substantially cooler than the sun's interior). And although the Earth's core is cooling, it is doing so very slowly in a geologic sense, since the thermal conductivity of rock is very low and, further, the heat being radiated from the Earth is being substantially offset by radioactive decay and solar radiation. Some scientists estimate that over the past three billion years, the Earth may have cooled several hundred degrees.

Geothermal energy has been used for centuries, where it is accessible, for aquaculture, greenhouses, industrial process heat, and space heating. It was first used for production of electricity in 1904 in Lardarello, Tuscany, Italy with the first commercial geothermal power plant (250 kWe) developed there in 1913. Since then geothermal energy has been used for electric power production all over the world, but most significantly in the United States, the Philippines, Mexico, Italy, Japan, Indonesia, and New Zealand. Table 1 lists the current levels of geothermal electric power generation installed worldwide.

---

*Peter D. Blair, PhD, is Executive Director of the Division on Engineering and Physical Sciences of the National Academy of Sciences (NAS) in Washington, DC. The views expressed in the chapter, however, are his own and not necessarily those of the NAS.

Table 1 Worldwide Geothermal Power Generation (2002)

Country	Installed Capacity (mWe)	Electricity Generation (millions kWh/yr)
United States	2,850	15,900
Philippines	1,848	8,260
Mexico	743	5,730
Italy	742	5,470
Japan	530	3,350
Indonesia	528	3,980
New Zealand	364	2,940
El Salvador	105	550
Nicaragua	70	276
Costa Rica	65	470
Iceland	51	346
Kenya	45	390
China	32	100
Turkey	21	90
Russia	11	30
Azores	8	42
Guadalupe	4	21
Taiwan	3	—
Argentina	0.7	6
Australia	0.4	3
Thailand	0.3	2
TOTALS	7,953	47,967

Source: U.S. Department of Energy.[1]

2 GEOTHERMAL RESOURCES

Geothermal resources are traditionally divided into the following three basic categories or types that are defined and described later in more detail:

1. *Hydrothermal convection systems,* which include both vapor-dominated and liquid-dominated systems
2. *Hot igneous resources,* which include hot dry rock and magma systems
3. *Conduction-dominated resources,* which include geopressured and radiogenic resources

These basic resource types are distinguished by basic geologic characteristics and the manner in which heat is transferred to the Earth's surface, as noted in Table 2. At present only hydrothermal resources are exploited commercially, but research and development activities around the world are developing the potential of the other categories, especially hot dry rock. The following discussion includes a description of and focuses on the general characterization of the features and location of each of these resource categories in the United States.

2.1 The United States Geothermal Resource Base

The U.S. Geological Survey (USGS) compiled an assessment of geothermal resources in the United States in 1975[2] and updated it in 1978[3] which characterizes a "geothermal resource

Table 2 Geothermal Resource Classification

Resource Type	Temperature Characteristics
Hydrothermal convection resources (heat carried upward from depth by convection of water or steam)	
a. Vapor dominated	~240°C
b. Liquid (hot-water) dominated	
1. High temperature	150–350°C
2. Intermediate temperature	90–150°C
3. Low temperature	<90°C
Hot igneous resources (rock intruded in molten form from depth)	
a. Molten material present—magma systems	>659°C
b. No molten material—hot dry rock systems	90–650°C
Conduction-dominated resources (heat carried upward by conduction through rock)	
a. Radiogenic (heat generated by radioactive decay)	30–150°C
b. Sedimentary basins (hot fluid in sedimenary rock)	30–150°C
c. Geopressured (hot fluid under high pressure)	150–200°C

base" for the United States based on geological estimates of all stored heat in the earth above 15°C within six miles of the surface. The defined base ignores the practical "recoverability" of the resource but provides to first order a sense of the scale, scope, and location of the geothermal resource base in the United States.

The U.S. geothermal resource base includes a set of 108 known geothermal resource areas (KGRAs) encompassing over three million acres in the 11 western states. The USGS resource base captured in these defined KGRAs does not include the lower-grade resource base applicable in direct uses (space heating, greenhouses, etc.), which essentially blankets the entire geography of the nation, although once again ignoring the issues of practical recoverability. Since the 1970s, many of these USGS-defined KGRAs have been explored extensively and some developed commercially for electric power production. For more details see Blair et al.[4]

2.2 Hydrothermal Resources

Hydrothermal convection systems are formed when underground reservoirs carry the Earth's heat toward the surface by convective circulation of steam in the case of *vapor-dominated resources* or water in the case of *liquid-dominated resources*. Vapor-dominated resources are extremely rare on Earth. Three are located in the United States: The Geysers and Mount Lassen in California and the Mud Volcano system in Yellowstone National Park.* All remaining KGRAs in the United States are liquid-dominated resources (located in Figure 1).

Vapor-Dominated Resources
In vapor-dominated hydrothermal systems, boiling of deep subsurface water produces water vapor, which is also often superheated by hot surrounding rock. Many geologists speculate that as the vapor moves toward the surface, a level of cooler near-surface rock induces

---

*Other known vapor-dominated resources are located at Larderello and Monte Amiata, Italy, and at Matsukawa, Japan.

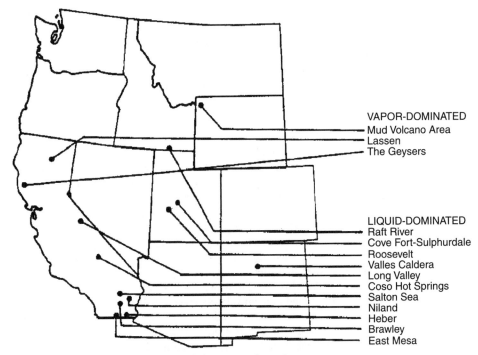

Figure 1 U.S. known geothermal resource areas.

condensation, which along with the cooler groundwater from the margins of the reservoir, serves to recharge the reservoir. Since fluid convection takes place constantly, the temperature in the vapor-filled area of the reservoir is relatively uniform and a well drilled into this region will yield high-quality* superheated steam, which can be circulated directly in a steam turbine generator to produce electricity.

The most commercially developed geothermal resource in the world today in known as The Geysers in northern California, which is a very high-quality, vapor-dominated hydrothermal convection system. At The Geysers steam is delivered from the reservoir from a depth of 5000–10000 feet and piped directly to turbine generators to produce electricity. Power production at The Geysers began in 1960, growing to a peak generating capacity in 1987 of over 2000 mWe. Since then it has declined to around 1200 mWe, but still accounts for over 90% of the total U.S. geothermal electric generating capacity.

Commercially produced vapor-dominated systems at The Geysers, Lardarello (Italy), and Matsukawa (Japan) all are characterized by reservoir temperatures in excess of 230°C.†

---

*High-quality steam is often referred to as *dry* steam since it contains no entrained liquid water spray. Most steam boilers are designed to produce high-quality steam. Steam with entrained liquid has significantly lower heat content than dry steam. Superheated steam is steam generated at a higher temperature than its equivalent pressure, created either by further heating of the steam (known as *superheating*), usually in a separate device or section of a boiler known as a *superheater*, or by dropping the pressure of the steam abruptly, which allows the steam drop to a lower pressure before the extra heat can dissipate.

† The temperature of dry steam is 150°C, but steam plants are most cost-effective when the resource temperature is above about 175°C.

Accompanying the water vapor in these resources are very small concentrations (less than 5%) of noncondensable gases (mostly carbon dioxide, hydrogen sulfide, and ammonia). The Mount Amiata Field in Italy is actually different type of vapor-dominated resource, characterized by somewhat lower temperatures than The Geysers-type resource and by much higher concentrations of noncondensable gases. The geology of Mount Amiata-type resources is less well understood than The Geysers-type vapor-dominated resources, but may turn out to be more common because its existence is more difficult to detect.

Liquid-Dominated Resources

Hot-water or wet steam hydrothermal resources are much more commonly found around the globe than dry steam deposits. Hot-water systems are often associated with a hot spring that discharges at the surface. When wet steam deposits occur at considerable depths (also relatively common), the resource temperature is often well above the normal boiling point of water at atmospheric pressures. These temperatures are know to range from 100 to 700°C at pressures of 50–150 psig. When the water from such resources emerges at the surface, either through wells or through natural geologic anomalies (e.g., geysers), it flashes to wet steam. As noted later, converting such resources to useful energy forms requires more complex technology than that used to obtain energy from vapor-dominated resources.

One of the reasons dealing with wet steam resources is more complex is that the types of impurities found in them vary considerably. Commonly found dissolved salts and minerals include sodium, potassium, lithium, chlorides, sulfates, borates, bicarbonates, and silica. Salinity concentrations can vary from thousands to hundreds of thousands of parts per million. The Wairekei Fields in New Zealand and the Cerro Prieto Fields in Mexico are examples of currently well-developed liquid-dominated resources and in the United States many such resources are in development or under consideration for development.

2.3   Hot Dry Rock and Magma Resources

In some areas of the western United States, geologic anomalies such as tectonic plate movement and volcanic activity have created pockets of impermeable rock covering a magma chamber within six or so miles of the surface. The temperature in these pockets increases with depth and the proximity to the magma chamber, but, because of the impermeability of the rock, they lack a water aquifer. Hence, they are often referred to as *hot dry rock* (HDR) deposits.

A number of schemes for useful energy production from HDR resources have been proposed, but all of them involve creation of an artificial aquifer that is used to bring the heat to the surface. The basic idea is to introduce artificial fractures that connect a production and injection well. Cold water is injected from the surface into the artificial reservoir where the water is heated then returned to the surface through a production well for use in direct-use or geothermal power applications. The concept is being tested by the U.S. Department of Energy at Fenton Hill near Los Alamos, New Mexico.

A typical HDR resource extraction system design is shown in Fig. 2. The critical parameters affecting the ultimate commercial feasibility of HDR resources are the geothermal gradient throughout the artificial reservoir and the achievable well flow rate from the production well.

Perhaps even more challenging than HDR resource extraction is the notion of extracting thermal energy directly from shallow (several kilometers in depth) magma intrusions beneath volcanic regions. Little has been done to date to develop this kind of resource.

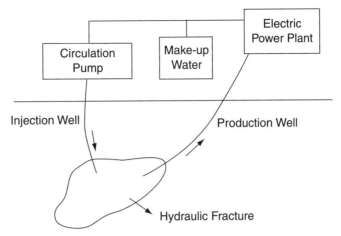

Figure 2 Hot dry rock geothermal resource conversion.

2.4 Geopressured Resources

Near the Gulf Coast of the United States are a number of deep sedimentary basins that are geologically very young, less than 60 million years. In such regions, fluid located in subsurface rock formations carry a part of the overburden load, thereby increasing the pressure within the formation. If the water in such a formation is also confined in an insulating clay bed, the normal heat flow of the Earth can raise the temperature of the water considerably. The water in such formations is typically of somewhat lower salinity as well, compared to adjacent aquifers, and, in many cases, is saturated with large amounts of recoverable methane. Such formations are referred to as *geopressured* and are considered by some geologists to be promising sources of energy in the coming decades.

The promise of geopressured geothermal resources lies in the fact that they may be able to deliver energy in three forms: (1) mechanical energy, since the gas and liquids are resident in the formations under high hydraulic pressure, (2) the geothermal energy stored in the liquids, and (3) chemical energy, since, as noted above, many geopressured resources are accompanied by high concentrations of methane or natural gas.

Geopressured basins exist in several areas within the United States, but those considered the most promising are located in the Texas–Louisiana coast. They are of particular interest because they are very large in terms of both areal extent and thickness and because the geopressured liquids (mostly high-salinity brine) are suspected to include high concentrations of methane.

In past evaluations of the Gulf Coast region, a number of so called "geopressured fairways" were identified, which are thick sandstone bodies expected to contain geopressured fluids of at least 150°C. Detailed studies of the fairways of the Frio Formation in East Texas were completed in 1979, although only one, Brazoria, met the requirements for further well testing and remains the subject of interest by researchers.

3 GEOTHERMAL ENERGY CONVERSION

For modern society, geothermal energy has a number of important advantages. While not immediately renewable like solar and wind resources, the energy within the earth is vast and essentially inexhaustible, i.e., with a lifetime of billions of years; environmental impacts associated with geothermal energy conversion are generally modest and local compared with other alternatives; and energy production is generally very reliable and available day and night. In addition, geothermal energy is not generally affected by weather, although there may be seasonal differences in plant efficiency. Finally, geothermal plants take little space and can be made unobtrusive even in areas of high scenic value, where many geothermal resources are located.

Most economical applications of geothermal energy, at least at this point in the development of the necessary technology, hinge on the availability and quality of the resource. On one hand, geothermal resources are far less pervasive than solar or wind resources, but, on the other hand, as technology continues to develop, the use of lower quality but much more common geothermal resources may increase their development substantially.

Since use of geothermal energy involves interaction with a geologic system, the characteristics and quality of the resource involves some natural variability (far less than with solar or wind), but, more importantly, the utilization of the geothermal resource can be affected profoundly by the way in which the resource is tapped. In particular, drawing steam or hot water from a geothermal aquifer at a rate higher than the rate at which the aquifer is refreshed will reduce the temperature and pressure of the resource available for use locally and can precipitate geologic subsidence evident even at the surface. The consequences of resource utilization for the quality of the resource are especially important since geothermal energy is used immediately and not stored, in contrast to the case of oil and gas resources, and would undermine the availability, stability, and reliability of the commercially produced energy from the resource and diminish its value.

Reinjecting geothermal fluids that remain after the water (or steam) has been utilized in a turbine (or other technology that extracts the useful heat from the fluid) helps preserve the fluid volume of the reservoir and is now a common practice for environmental reasons and to mitigate subsidence.* Nonetheless, even with reinjection, the heat content of a well-developed geothermal reservoir will gradually decline, as typified by the history at The Geysers.

A variety of technologies are in current use to convert geothermal energy to useful forms. These can very generally be grouped into three basic categories: (1) direct use, (2) electric power generation, and (3) geothermal heat pumps. Each category utilizes the geothermal resource in a very different way.

3.1   Direct Uses of Geothermal Energy

The heat from geothermal resources is frequently used directly without a heat pump or to produce electric power. Such applications generally use lower-temperature geothermal resources for space heating (commercial buildings, homes, greenhouses, etc.), industrial processes requiring low-grade heat (drying, curing, food processing, etc.), or aquaculture.

Generally, these applications use heat exchangers to extract the heat from geothermal fluids delivered from geothermal wells. Then, as noted earlier, the spent fluids are then injected back into the aquifer through reinjection wells. The heat exchangers transfer the

---

*Reinjection of water is common in oil and gas field maintenance to preserve the volume and pressure of the resource in those fields and the basic concept is applicable in geothermal fields as well to both mitigate the environmental impacts of otherwise disposing of spent geothermal liquids and to maintain the volume, temperature, and pressure of the geothermal aquifer.

heat from the geothermal fluid usually to fresh water that is circulated in pipes and heating equipment for the direct use applications.

Such applications can be very efficient in small end-use applications such as greenhouses, but it is generally necessary for the applications to be located close to the geothermal heat source. Perhaps the most spectacular and famous example of direct use of geothermal energy is the city of Reykjavik, Iceland, which is heated almost entirely with geothermal energy.

3.2 Electric Power Generation

Geothermal electric power generation generally uses higher-temperature geothermal resources (above 110°C). The appropriate technology used in power conversion depends on the nature of the resources.

As noted earlier, for vapor-dominated resources, it is possible to use direct steam conversion. For higher-quality liquid-dominated hydrothermal resources, i.e., with temperatures greater than 180°C, power plants can be used to separate steam (flashed) from the geothermal fluid and then feed the steam into a turbine that turns a generator. For lower-quality resources so-called *binary* power plants can increase the efficiency of electric power production from liquid-dominated resources.

In a manner similar to direct uses of geothermal energy, binary power plants use a secondary working fluid that is heated by the geothermal fluid in a heat exchanger. In binary power plants, however, the secondary working fluid is usually a substance such as isobutane, which is easily liquified under pressure but immediately vaporizes when the pressure is released at lower temperatures than that of water. Hence, the working-fluid vapor turns the turbine and is condensed prior to reheating in a heat exchanger to form a closed-loop working cycle.

In all versions of geothermal electric power generation, the spent geothermal fluids are ultimately injected back into the reservoir. Geothermal power plants vary in capacity from several hundred kWe to hundreds of mWe. In the United States, at the end of 2002, there were 43 geothermal power plants, mostly located in California and Nevada. In addition, Utah has two operating plants and Hawaii has one. The power-generating capacity at The Geysers remains the largest concentration of geothermal electric power production in the world, producing almost as much electricity as all the other U.S. geothermal sites combined.

Muffler[3] estimates that identified hydrothermal resources in the United States could provide as much as 23,000 mWe of electric generating capacity for 30 years, and undiscovered hydrothermal resources in the nation could provide as much as five times that amount. Kutscher[5] observes that if hot dry rock resources become economically recoverable in the United States, they would be "sufficient to provide our current electric demand for tens of thousands of years," although currently economically tapping hot dry rock resources remains largely elusive and speculative. To explore that potential, a variety of research and development program activities are underway sponsored by the U.S. Government.[1]

The following sections explore more specifically the technologies of direct steam, flash, and binary geothermal energy conversion along with the strategy of combining geothermal energy with fossil (oil, coal, or natural gas) in power generation.

Direct Steam Conversion

Electric power generation using the geothermal resources at The Geysers in California and in central Italy, which were referred to earlier as The Geysers-type vapor-dominated resources, is a very straightforward process relative to the processes associated with other kinds of geothermal resources. A simplified flow diagram of direct steam conversion is shown in Fig. 3. The key components of such a system include the steam turbine–generator, condenser, cooling towers, and some smaller facilities for degassing and removal of entrained solids and for pollution control of some of the noncondensable gases.

The process begins when the naturally pressurized steam is piped from production wells to a power plant, where it is routed through a turbine generator to produce electricity. The geothermal steam is supplied to the turbine directly, except for the relatively simple removal of entrained solids in gravity separators or the removal of noncondensable gases in degassing vessels. Such gases include carbon dioxide, hydrogen sulfide, methane, nitrogen, oxygen, hydrogen, and ammonia. In modern geothermal plants additional equipment is added to control, in particular, the hydrogen sulfide and methane emissions from the degassing stage. Release of hydrogen sulfide is generally recognized as the most important environmental issue associated with direct steam conversion plants at The Geysers' generating facilities. The most commonly applied control technology for abatement of toxic gases such as hydrogen sulfide in geothermal power plants is known as the Stratford process.

As the "filtered" steam from the gravity separators and degassing units expands in the turbine it begins to condense. It is then exhausted to a condenser, where it cools and condenses completely to its liquid state and is subsequently pumped from the plant. The condensate is then almost always reinjected into the subterranean aquifer at a location somewhat removed from the production well. Cooling in the condenser is provided by a piping loop between the condenser and the cooling towers. The hot water carrying the heat extracted from the condensing steam line from the turbine is routed to the cooling tower where the heat is rejected to the atmosphere. The coolant fluid, freshly cooled in the cooling tower, is then routed back to the condenser, forming the complete cooling loop (as shown in Fig. 3).

Reinjection of geothermal fluids in modern geothermal systems is almost always employed to help preserve reservoir volume and to help mitigate air and water pollutant emissions on the surface. However, as the geothermal well field is developed and the resource produced, effective reservoir maintenance becomes an increasingly important issue. For example, in The Geysers, noted earlier as a highly developed geothermal resource, as the geothermal fluids are withdrawn and reinjected, the removal of the heat used in power

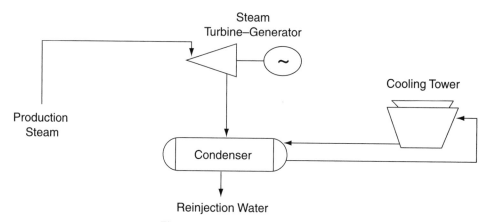

Figure 3   Direct steam conversion.

generation causes the reservoir temperature to decline. The cooling reservoir then contracts*
and this is observed at the surface as subsidence. Geophysicists Mossop and Segall[6] observe
that subsidence at The Geysers has been on the order of 0.05 m per year since the early
1970s.

Because of the quality of the resource and the simplicity of the necessary equipment,
direct steam conversion is the most efficient type of geothermal electric power generation.
A typical measure of plant efficiency is the amount of electric energy produced per pound
of steam at a standard temperature (usually around 175°C). For example, the power plants
at The Geysers produce 50–55 Whr of electricity per pound of 176°C steam used, which is
a very high-quality geothermal resource. Another common measure of efficiency is known
as the geothermal resource utilization efficiency (GRUE), defined as the ratio of the net
power output of a plant to the difference in the thermodynamic availability of the geothermal
fluid entering the plant and that of that fluid at ambient conditions. Power plants at The
Geysers operate at a GRUE of 50–56%.

Flashed Steam Conversion

Most geothermal resources do not produce dry steam, but rather a pressurized two-phase
mixture of steam and water often referred to as wet steam. When the temperature of the
geothermal fluid in this kind of resource regime is greater than about 180°C, plants can use
the flashed steam energy conversion process. Figure 4 is a simplified schematic that illustrates

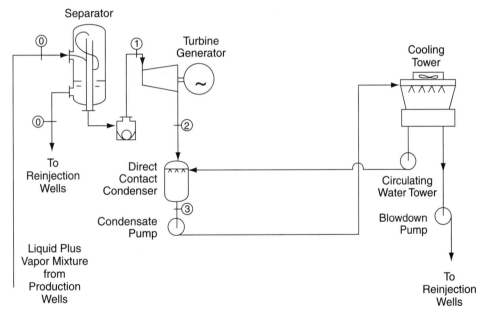

Figure 4 Flashed steam conversion.

---

*Most researchers conclude that the extraction, reinjection, and associated temperature decline causes
strain due to a combination of thermoelastic and poroelastic deformations, which results in surface
subsidence, e.g., Mossop and Segall.[6]

the flashed steam power generation process used in such plants. In addition to the key components used in direct steam conversion plants (i.e., turbine, condenser, and cooling towers), flashed steam plants include a component called a separator or flash vessel.

The flash conversion process begins with the geothermal fluid flows from the production well(s) flows under its own pressure into the separator, where saturated steam is flashed from the liquid brine. That is, as the pressure of the fluid emerging from the resource decreases in the separator, the water boils or "flashes" to steam and the water and steam are separated. The steam is diverted into the power production facility and the spent steam and remaining water are then reinjected into the aquifer.

Many geothermal power plants use multiple stages of flash vessels to improve the plant efficiency and raise power generation output. Figure 5 is a simplified schematic illustrating a two-stage or "dual-flash" system. Such systems are designed to extract additional energy from geothermal resource by capturing energy from both high and lower temperature steam.

In the two-stage process, the unflashed fluid leaving the initial flash vessel enters a second flash vessel that operates at a lower pressure, causing additional steam to be flashed. This lower-pressure steam is supplied to the low-pressure section of the steam turbine, recovering energy that would have been lost if a single-stage flash process had been used. The two-stage process can result in a 37% or better improvement in plant performance compared with a single-stage process. Additional stages can be included as well, resulting in successively diminishing levels of additional efficiency improvement. For example, addition of a third stage can add an additional 6% in plant performance.

Binary Cycle Conversion

For lower-quality geothermal resource temperatures, i.e., usually below about 175°C, flash power conversion is not efficient enough to be cost effective. In such situations, it becomes more efficient to employ a binary cycle. In the binary cycle, heat is transferred from the geothermal fluid to a volatile working fluid (usually a hydrocarbon such as iobutane or

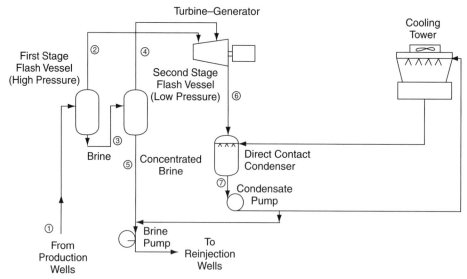

Figure 5 Two-stage flash conversion.

isopentane) that vaporizes and is passed through a turbine. Such plants are called binary since the secondary fluid is used in a Rankine power production cycle, and the primary geothermal fluid is used to heat the working fluid. These power plants generally have higher equipment costs than flash plants because the system is more complex.

Figure 6 is a simplified schematic illustrating the key components in the binary cycle conversion process. Geothermal brine from the production well(s) passes through a heat exchanger, where it transfers heat to the secondary working fluid. The cooled brine is then reinjected into the aquifer. The secondary working fluid is vaporized and superheated in the heat exchanger and expanded through a turbine, which drives an electric generator. The turbine exhaust is condensed in a surface condenser, and the condensate is pressurized and returned to the heat exchanger to complete the cycle. A cooling tower and a circulating water system reject the heat of condensation to the atmosphere.

A number of variations of the binary cycle have been designed for geothermal electric power generation. For example, a regenerator may be added between the turbine and condenser to recover energy from the turbine exhaust for condensate heating and to improve plant efficiency. The surface-type heat exchanger, which passes heat from the brine to the working fluid, may be replaced with a direct contact or fluidized-bed type exchanger to reduce plant cost. Hybrid plants combining the flashed steam and binary processes have also been evaluated in many geothermal power generation applications.

The binary process is proving to be an attractive alternative to the flashed steam process at geothermal resource locations that produce high-salinity brine. First, since the brine can remain in a pressurized liquid state throughout the process and does not pass through the turbine, problems associated with salt precipitation and scaling as well as corrosion and erosion can be greatly reduced. In addition, binary cycles offer the additional advantage that a working fluid can be selected that has superior thermodynamic characteristics to steam, resulting in a more efficient conversion cycle. Finally, because all the geothermal brine is reinjected into the aquifer, binary cycle plants do not require mitigation of gaseous emissions,

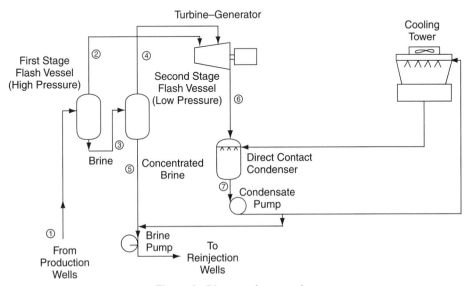

Figure 6 Binary cycle conversion.

and reservoir fluid volume is maintained. Larger binary plants are typically constructed as a series of smaller units or modules, so maintenance can be completed on individual modules without shutting down the entire plant, thereby minimizing the impact on total plant output.

Some Additional System Selection Considerations

The overall efficiency of energy conversion processes for liquid-dominated resources is dependent primarily on the resource temperature and to a lesser degree on brine salinity and the concentration of noncondensable gases. System efficiency can generally be improved by system modifications, but such modifications usually involve additional cost and complexity. Figure 7 shows an empirical family of curves relating power production per unit of geothermal brine consumed for both two-stage flash and binary conversion systems.

The level of hydrogen sulfide emissions is an important consideration in geothermal power plant design. Emissions of hydrogen sulfide at liquid-dominated geothermal power plants are generally lower than for direct steam processes. For example, steam plants emit 30–50% less hydrogen sulfide than direct steam plants. Binary plants would generally not emit significant amounts of hydrogen sulfide because the brine remains contained and pressurized throughout the entire process.

Finally, the possibility of land surface subsidence caused by withdrawal of the brine from the geothermal resource can be an important design consideration. Reinjection of brine is the principal remedy for avoiding subsidence by maintaining reservoir volume and has the added environmental benefit of minimizing other pollution emissions to the atmosphere. However, faulty reinjection can contaminate local fresh groundwater. Also in some plant designs if all brine is reinjected, an external source of water is required for plant cooling water makeup.

Hybrid Geothermal/Fossil Energy Conversion

Hybrid fossil/geothermal power plants use both fossil energy and geothermal heat to produce electric power. A number of alternative designs exist. First, a geothermal preheat system involves using geothermal brine to preheat the feedwater in an otherwise conventional fossil-

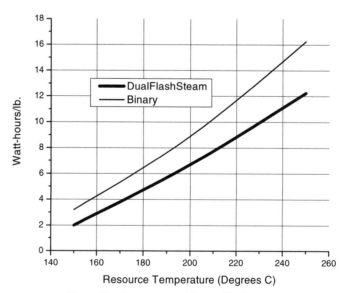

Figure 7 Net geothermal brine effectiveness.

fired power plant. Another variation is fossil superheat concept that incorporates a fossil-fired heater to superheat geothermal steam prior to expansion in a turbine.

3.3 Geothermal Heat Pumps

Geothermal heat pumps (GHP), sometimes also referred to as groundwater heat pumps, use the earth's typical diffuse low-grade heat found in the very shallow subsurface (usually between 30 and 300 ft in depth) usually in space heating applications. In most geographic areas in the United States, GHP can deliver 3–4 times more energy than it consumes in the electricity needed to operate and can be used over a wide range of earth temperatures.

The GHP energy-conversion process works much like a refrigerator, except that it is reversible, i.e., the GHP can move heat either into the earth for cooling or out of the earth for heating, depending on whether it is summer or winter. GHP can be used instead of or in addition to direct uses of geothermal energy for space or industrial process heating (or cooling), but the shallow resource used by GHP is available essentially anywhere, constrained principally by land use and economics, especially initial installation costs.

The key components of the GHP system include a ground refrigerant-to-water heat exchanger, refrigerant piping and control valves, a compressor, an air coil (used to heat in winter and to cool and dehumidify in summer), a fan, and control equipment. This system is illustrated in Fig. 8.

The GHP energy-conversion process begins with the ground heat exchanger, which is usually a system of pipes configured either as a closed- or open-loop system. The most common configuration is the closed loop, in which high-density polyethylene pipe is buried horizontally at a depth of at least 4–6 ft deep or vertically at a depth of 100–400 ft. The pipes are typically filled with a refrigerant solution of antifreeze and water, which acts as a heat exchanger. That is, in winter, the fluid in the pipes extracts heat from the earth and carries it into the building. In the summer, the system reverses the process and takes heat from the building and transfers it to the cooler ground.

GHP systems deliver heated or cooled air to residential or commercial space through ductwork just like conventional heating, ventilating, and air conditioning (HVAC) systems. An indoor coil and fan called an air handler also contains a large blower and a filter just like conventional air conditioners.

Ground-Loop GHP Systems

There are four basic types of ground-loop GHP systems: (1) horizontal, (2) vertical, (3) pond/lake, and (4) open-loop configurations; the first three of these are all closed-loop sys-

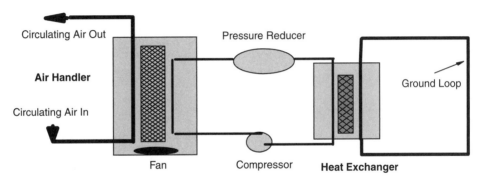

Figure 8 Geothermal heat pump system configuration.

tems. Selection of one of these system types depends on climate, soil conditions, available land, and local installation costs. The following briefly describes each of the approaches to GHP ground-loop systems.

- *Horizontal.* Considered generally most cost-effective for residential installations, especially for new construction where sufficient land is available, the installation entails two pipes buried in trenches that form a loop.

- *Vertical.* Vertically oriented systems are often used for large commercial buildings and schools where the land area required for horizontal loops would be prohibitive or where the surrounding soil is too shallow for trenching in a horizontal system or when a goal is to minimize the disturbance to existing landscaping. In such systems holes are drilled about 20 ft apart and 100–400 ft deep and pipes are installed and connected at the bottom to form the loop.

- *Pond/Lake.* If the site has a suitable water body accessible, a supply line pipe can be run from the building to the water and coiled under the surface of the water body to prevent freezing in winter.

- *Open-Loop.* An open-loop system uses water from well(s) or a surface body of water as the heat exchange fluid that circulates directly through the GHP system. Once the water has circulated through the heat exchanger, the water returns to the ground through another well or by surface discharge. This option can be used only where there is an adequate supply of relatively clean water and where its use is permitted under local environmental codes and regulations.

REFERENCES

1. U.S. Department of Energy, Office of Energy Efficiency and Renewable Energy, Strategic Plan, DOE/GO-102002-1649, October 2002.
2. White, D. E., and D. L. Williams (eds.), "Assessment of Geothermal Resources of the United States—1975," U.S. Geological Survey, Arlington, VA, Circular 726, 1975.
3. Muffler, L. J. P. (ed.), "Assessment of Geological Resources of the United States—1978," U.S. Geological Survey, Arlington, VA, Circular 790, 1979.
4. Blair, P. D., T. A. V. Cassel, and R. H. Edelstein, *Geothermal Energy: Investment Decisions and Commercial Development,* Wiley, New York, 1982.
5. Kutscher, C. F., "The Status and Future of Geothermal Electric Power," NREL/CP-550-28204, National Renewable Energy Laboratory, Golden, CO, August 2000.
6. Mossop, A., and P. Segall, "Subsidence at the Geysers Geothermal Field," *Geophysical Research Letters,* **24**(14), 1839–1842 (1997).

CHAPTER 22

PUMPS, FANS, BLOWERS, AND COMPRESSORS

Keith Marchildon
Keith Marchildon Chemical Process Design, Inc.
Kingston, Ontario

David Mody
Fluor Canada
Kingston, Ontario

1 INTRODUCTION TO FLUID MOVERS

Pumps, fans, blowers, and compressors are all devices that move fluids across an adverse pressure difference, i.e., from a region of lower pressure to a region of higher pressure. The fluid may require the higher pressure to overcome frictional losses in subsequent piping, to participate in a high-pressure operation such as a chemical reactor, or to serve as the drive medium in a hydraulic or pneumatic system. Or the objective may lie on the inlet side of the device where it is desired to maintain vacuum in some region, in which case the pressure on the outlet side may simply be atmospheric. In any of these cases there may or may not be a change in net velocity.

The two broad categories of fluid to be moved are liquids and gases. Liquids are moved by pumps; gases are moved by fans, blowers, and compressors. The main differences between the moving of liquids and the moving of gases is that gases undergo significant changes in volume and temperature if the rise in pressure is appreciable. Along with liquids and gases, there are more complex fluid media, such as liquid–gas mixtures, liquids that partially vaporize, gases that condense, slurries that consist of a liquid containing solid particles, and gas–particulate mixtures, all of which may require special handling or special equipment.

Fluid movers fall into two general types: kinetically driven and positive displacement. Kinetically driven devices impart internal velocity to the fluid, then convert this momentum to pressure at the exit. Positive-displacement devices trap incremental volumes of lower-pressure fluid and transport it forcibly into the region of higher pressure.

Another distinction among fluid movers is whether they operate in a rotary or nonrotary manner. This distinction applies across both the kinetically driven and positive-displacement devices.

In Table 1, the most common of the fluid movers are listed according to the above categorization. This scheme constitutes the organization of the chapter.

2 LIQUID MOVERS—PUMPS

A pump is any device that transfers a liquid from a region of lower pressure into one of higher pressure. This movement may or may not be accompanied by a change of velocity but that effect is incidental. Generally it is possible to ignore changes of density except for great changes in pressure or for liquids containing a gaseous phase. In the present treatment density is assumed constant. Some considerations in choosing a pump are

- Pressure rise to be effected
- Liquid flow rate
- Range of flow rates
- Required accuracy of flow rate
- Liquid viscosity
- Suction-side pressure

A common concern with most pumps is the possibility of vaporization in the pump inlet. Not only does the appearance of a gaseous phase reduce the capacity of the pump but the possible subsequent collapse of bubbles as pressure recovers ("cavitation") may severely erode surfaces in the device. Because this consideration is common to almost all pumps (but not jet pumps), a simple example is given in Fig. 1 of the calculation of pressure at pump inlet. This pressure has the specific designation of *net positive suction head—available* or NPSHa and is defined as the difference between the pressure at the pump inlet and the vapor pressure of the liquid at the temperature at the pump inlet, both pressures expressed as *head (meters) of liquid*. In the example the available NPSH (m) is equal to

$$[P_S - \Delta P - P_{\text{VAP}}(T_2)]/(\rho g) - Z$$

Table 1   Classification of Fluid Movers

	Liquid Movers		Gas and Vapor Movers	
	Rotary	Nonrotary	Rotary	Nonrotary

Kinetically driven	Centrifugal pumps
	Axial pumps
	Mixed-flow
	  pumps
		Jet eductors	Centrifugal fans and
			  blowers
			Axial fans
			Mixed-flow fans and
			  blowers
				Ejectors or
				  recompressors

Positive
  displacement
	Gear pumps
	Screw pumps
	Vane pumps
	Progressive-cavity
	  pumps
	Peristaltic pumps
		Reciprocating pumps:
		Diaphragm, piston,
		plunger
			Screw compressors
			Vane compressors and
			  vacuum pumps
			Liquid ring vacuum pumps
			Lobe compressors and
			  vacuum pumps
				Reciprocating
				  compressors

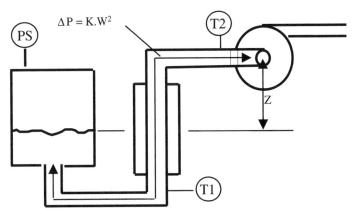

Figure 1 Net positive head availability for liquid pump.

where ΔP is the frictional pressure drop (n/m^2)
$P_{\text{VAP}}(T_2)$ is the vapor pressure of the liquid calculated at temperature T_2 (n/m^2)
  ρ is the density of the liquid, averaged over the elevation change (kg/m^3)
  g is terrestrial acceleration due to gravity, i.e., $9.8\ \text{m/s}^2$
  K is the coefficient for frictional pressure drop, obtained from standard formulas or graphs and incorporating the contributions of pipe lengths, elbows, and other fittings ($\text{n/m}^2/(\text{kg/s})^2$)

The unit conversion, kilogram·meter/(second² newton) = 1, is applied where required. As seen, the frictional pressure drop increases with flow rate of liquid, W (kg/s), causing NPSHa to decrease. The question of how high a value of NPSHa is required for a given pump is considered subsequently.

In practice the great majority of pumps are centrifugal. They are relatively inexpensive and better able to handle liquids containing inhomogeneities such as particulate matter. They are available at much larger volumetric throughputs than other types of pumps. Single-stage centrifugal pumps can generate pressures in excess of 10 atm and multistage pumps can exceed 100 atm. They do not operate well with volatilizing liquids or with liquids containing a significant fraction of gas, and in this case the jet pump is more appropriate. Centrifugal pumps lose efficiency with liquids of higher viscosity and such liquids are better handled with certain types of positive-displacement pumps. Also, when the required pressure rise is high, positive displacement is sometimes the better choice. These considerations are quantified in the following discussions.

2.1 Kinetically Driven, Rotary

Kinetically driven devices are governed by Bernouilli's mechanical energy balance and make use of the interconvertibility of pressure energy and kinetic energy. In the case of centrifugal and axial pumps, a velocity is imparted internally to the fluid and is then converted to pressure by deceleration as the exit of the pump is approached. In the case of jet pumps, a *motive fluid* is accelerated to create an internal region of low pressure, into which the fluid to be pumped is drawn, and the mixture of fluids is then decelerated back to a higher pressure.

Bernouilli's equation is derivable from Newton's second law, which states that a body accelerates at a rate proportional to the ratio of force to mass. This law leads to the definition of *momentum,* the product of mass and velocity, which is changed linearly by the application of external force. In Fig. 2, fluid flow through the cross-hatched region is at steady state so there is no time-wise change of momentum in this region. Momentum flows into the region at the rate WU and out at the rate $W(U + \Delta U)$. The force applied by pressure is PA on the left-hand face and $(P + \Delta P)A$ on the right-hand face. The momentum balance is

$$0 = \delta(\text{Momentum})/\delta(\text{Time}) = WU - W(U + \Delta U) + PA - (P + \Delta P)A$$

$$= - W \, \Delta U - A \, \Delta P = - UA\rho \, \Delta U - A \, \Delta P \qquad (1)$$

where W is weight flow (kg/s)
      U is velocity (m/s)
      P is pressure (n/m^2, i.e., pascals)
      A is area normal to the direction of flow (m^2)
      ρ is liquid density (kg/m^3)

Canceling A and letting the region become infinitesimal gives

$$0 = U\rho \, d(\text{U}) + d(P) \qquad (2)$$

which is the differential form of Bernouilli's equation. Integrating (using an average value for density) gives

$$U^2/2 + P/\rho = \text{constant} \qquad (3)$$

which is the familiar integral form, without the elevation and work terms. The unit conversion (kg-m/s^2/n $= 1$) must be applied to P to convert it from units of n/m^2 to kg-m/s.

In most rotary pumps the velocity imparted to the fluid is *centrifugal,* with the fast-moving fluid ending up in a circular motion around the periphery of the device and slowing down as it enters the tangential outlet nozzle. The arrangement is shown schematically in Fig. 3. Less common are *axial-flow* pumps in which a propeller type of rotor imparts velocity in the direction parallel to the axis of rotation, this velocity being then converted into pressure. In both cases the rotor is known as the *impeller,* which comprises a hub to which are attached vanes or blades.

Euler's Equation for Centrifugal Machines
Using Fig. 4, the Euler equation for idealized centrifugal pump operation can be derived. The basis is the *angular momentum balance.* At any distance from the center of rotation, a constant weight of liquid flows outward, namely W (kg/s), the pumping rate. At any specific distance r, this fluid has a tangential velocity c_t due to the motion of the impeller. The rate of transfer of angular momentum, $c_t r$, across the imaginary surface at r is

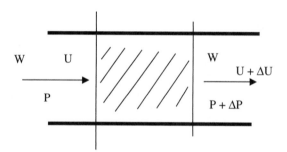

Figure 2   Continuous-flow momentum balance.

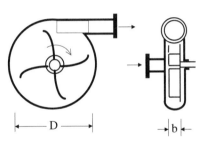

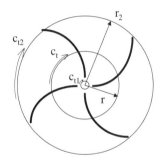

Figure 3   Centrifugal pump schematic.   **Figure 4**   Torque balance in centrifugal pump.

$$MMR_r = Wc_r r \quad (\text{kg m}^2/\text{s}^2) \tag{4}$$

If the entire interior volume of the pump casing is considered, the net increase in angular momentum is

$$\Delta MMR = W(c_{t2} r_2 - c_{t1} r_1) \tag{5}$$

To impart this increase, the impeller must exert a torque **T** (N m) equal to ΔMMR, and hence must be supplied with power (N m/s or watts) of

$$Pw_S = \mathbf{T} \cdot \boldsymbol{\omega} = W (c_{t2} r_2 - c_{t1} r_1) \, \omega = W(c_{t2} v_2 - c_{t1} v_1) \tag{6}$$

where ω is impeller rotational speed in radians per second, equal to $2\pi N$
  N is the impeller speed in revolutions per second
  v is the speed (m/s) of the impeller at distance r, i.e., $2\pi N r$

The subscripts 1 and 2 refer to the inner and outer extremes of the impeller.
At the same time the power received by the fluid as it passes through the pump is given by

$$Pw_R = Q \, \Delta P = (W/\rho) \, \Delta P = WHg \tag{7}$$

where Q is the volumetric flow rate of liquid
  ΔP is the net pressure increase
  H is the pressure increase expressed in head of liquid (m)
  g is terrestrial acceleration due to gravity (m/s²)

Ideally, Pw_R equals Pw_S, which leads to Euler's equation for turbo machines (rotating fluid devices):

$$H = (c_{t2} v_2 - c_{t1} v_1)/g \tag{8}$$

In many cases the velocities c_{t1} and v_1 at the innermost point of the impeller arms or vanes are much less than at the outermost point, so that approximately

$$H = c_{t2} v_2 / g \tag{9}$$

Idealized Pump Characteristics

The estimation of the liquid tangential velocity c_{t2} is explained with reference to Fig. 5. If the impeller vanes had no curvature, then c_{t2} would simply equal v_2, which is shown by vector AD in Fig. 5. However, for stable operation, the vanes of centrifugal pumps are usually designed with a rearward curvature. Assuming that the liquid follows the curvature, its velocity at the tip of the vane, relative to the vane, is denoted by vector AC in Fig. 5. But the

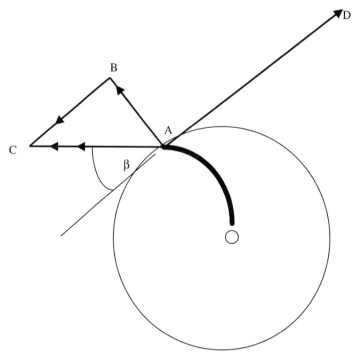

Figure 5   Fluid velocity vectors in centrifugal pump.

radial component of velocity AB is equal to the volumetric flow rate of liquid divided by the peripheral area:

$$Q/(\pi Db)$$

where b is the width (m) of the impeller vanes and D is the diameter (m) of the impeller (see Fig. 3).

The tangential component of the velocity relative to vane tip is denoted by vector BC, which equals AB times $\cot \beta$. Consequently,

$$c_{t2} = v_2 - [Q/(\pi Db)] \cot \beta \tag{10}$$

and, from Eq. (9),

$$gH = c_{t2} v_2 = v_2^2 - v_2 [Q/(\pi Db)] \cot \beta$$

$$= (\pi DN)^2 - (NQ/b) \cot \beta \tag{11}$$

It can be seen that the relationship is independent of liquid density, which makes it convenient to use head H and volumetric throughput Q as the operating variables rather than, say, pressure rise and weight flow, which would lead to a density-specific relation:

$$\Delta P = (\pi DN)^2 \rho - (NW/b) \cot \beta \tag{12}$$

Equation (11) suggests that the change in head across the pump is a maximum at zero flow and decreases linearly as volumetric flow increases, as shown in Fig. 6. This idealized *operating line* or *characteristic curve* is never, in fact, reached, because turbulence, flow sep-

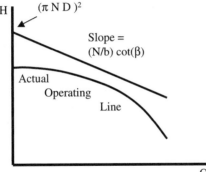

Figure 6 shows H plotted against Q, with $(\pi N D)^2$ marked at top, Slope = $(N/b)\cot(\beta)$, and an "Actual Operating Line" below.

Figure 6 Centrifugal pump head behavior.

aration, flow rotation, and other nonidealities in the pump enclosure cause a loss of efficiency. Equation (11) becomes the inequality of Eq. (13):

$$gH < (\pi DN)^2 - (NQ/b)\cot\beta \tag{13}$$

The ratio of left-hand side to right-hand side is defined as η_H, the hydraulic efficiency. η_H is one of three efficiency factors, the other two being η_M, the mechanical efficiency, accounting for frictional losses between drive and impeller, and η_V, the volumetric efficiency, accounting mainly for internal leakage around the impeller. Typically, $1 - \eta_H - \eta_V$ is 0.25–0.75 the magnitude of $1 - \eta_M$.

Pump Performance Description
Figure 7 shows typical operating data for a pump, all at one rotational speed, typically 1750 or 3500 rpm. Such data are obtained experimentally. Plotted against volumetric flow are four quantities:

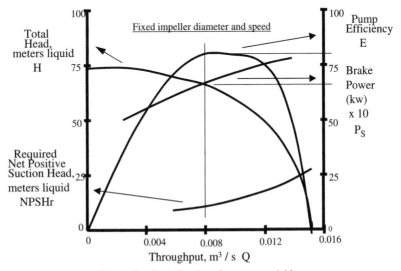

Figure 7 Centrifugal performance variables.

- The increase in liquid head (m) across the pump
- The efficiency of the pump
- The power (kW) that must be provided by the shaft (*brake power*)
- The required net positive suction head, **NPSHr**

The head-versus-throughput curve is typical of many centrifugal pumps. In this case, at a throughput of about 0.015 m³/s the head falls to zero and the pump reaches its pumping capacity. The efficiency is defined as the ratio of received power Pw_r to brake power Pw_s. Received power is given as

$$Pw_r = HQ\rho g \tag{14}$$

If the fluid is water, density ρ is 1000 kg/m³. The value of g is 9.8 m/s². For example, at Q of 0.008 m³/s, H is 68 m. From Eq. (14) received power is calculated as 5.3 kW. At this volumetric throughput, the brake power is seen to be 6.6 kW. The efficiency is calculated as 5.3/6.6 or 80%, which is what is indicated on the plot.

From Eq. (14), the received power is obviously zero when Q is zero but brake power is still required to rotate the rotor. Consequently, efficiency is zero at this end of the graph. Brake power increases as Q increases but efficiency increases faster. Efficiency passes through a maximum at some throughput. That is the ideal condition under which to operate. Fortunately, such maxima are generally broad but in practice a centrifugal pump cannot (or should not) be operated through the whole range of throughputs from zero up to the point of zero head.

The required net positive suction head NPSHr is seen to rise as volumetric throughput rises. This is expected because the inlet frictional pressure losses, which are the source of the NPSH requirement, increase as approximately the second power of flow rate. Considering that the available head NPSHa may decrease with flow rate, it is important to recalculate the difference between available and required heads whenever an increase in throughput is planned. The required suction head is determined by the pump manufacturer and is the value below which the head increase across the pump falls by some amount, usually 3%. It is recommended that the pump supply system be designed to maintain the available head NPSHa at least 0.5 m greater than the required head NPSHr.

Affinity Rules and Specific Speed

Equation (11) indicates that, to a first approximation, the achievable head

$$H \text{ is proportional to } N^2 \tag{15}$$

and

$$H \text{ is proportional to } D^2 \tag{16}$$

Although these ratios are taken from the zero-Q form of the equation, they are good approximations under other conditions, including the preferred condition of operation, i.e., the H-versus-Q combination that yields maximum efficiency. These relationships are known as *affinity rules*.

Examining Fig. 5 and assuming that ideal operation implies maintaining the same ratio of liquid radial velocity to impeller tip velocity, i.e., same ratio of vector *AB* to *AD*, we find the ratio

$$\frac{Q/(\pi Db)}{v_2} = \frac{Q/(\pi Db)}{\pi DN} = \text{a constant} \tag{17}$$

For geometrically similar pumps the width b is proportional to diameter D, so Eq. (17) leads to two more affinity rules:

$$Q \text{ is proportional to } N \tag{18}$$

$$Q \text{ is proportional to } D^3 \tag{19}$$

Since received power, Pw_r is proportional to the product of head H and volumetric throughput Q, then

$$Pw_r \text{ is proportional to } N^3 \tag{20}$$

$$Pw_r \text{ is proportional to } D^5 \tag{21}$$

On the basis of these proportionalities, three coefficients are defined:

$$\text{Head coefficient} = H/(ND)^2 \tag{22}$$

$$\text{Flow coefficient} = Q/(ND^3) \tag{23}$$

$$\text{Power coefficient} = Pw/(N^3 D^5). \tag{24}$$

These coefficients are functions of the pump style and proportions.

Another quantity of great usefulness is obtained by dividing the 0.5 power of the flow coefficient by the 0.75 power of the head coefficient. The result is the elimination of the diameter D and the derivation of the specific speed,

$$N_S = N Q^{1/2}/H^{3/4} \tag{25}$$

The values of Q and H are taken at the operating condition of maximum efficiency for the speed N.

Specific speed is commonly quoted in units of (revolutions/minute) (gallons/minute)$^{1/2}$/ (feet)$^{3/4}$ and it usually ranges from 500 to 15,000. In units of (revolutions/s) (m^3/s)$^{1/2}$/(m)$^{3/4}$ the range is 0.15 to 5, the conversion factor being 0.0003225. A low value of N_S indicates a pump designed primarily for head development, typically a pump with narrow elongated vanes, shrouded on each side to minimize leakage. A high value indicates a pump designed for flow, typically an axial-flow device.

As an example consider a pump operating at its preferred, maximum-efficiency condition with N of 3500 rpm, Q of 225 gpm, and H of 420 ft. The specific speed is therefore

$$(3500 \times 225^{1/2})/420^{3/4} \quad \text{or} \quad 566$$

Assume that the speed drops to 2625 rpm. From Eq. (18) the flow drops to $(225) \times (2625)/(3500)$ or 169 gpm, and from Eq. (15) the head drops to $(420) \times (2625/3500)^2$ or 236 ft. These are the conditions at the new point of optimum operation. Recalculating specific speed gives

$$(2625 \times 169^{1/2})/236^{3/4} = 566$$

as before.

Generalized Pump Performance Plot

Figure 7 applies to the case of a fixed impeller diameter. In the considerations to this point it has been implicit that the impeller diameter is the largest that can be accommodated by the pump casing. One of the degrees of freedom available to the user of pumps is to operate with various diameters of impellers. Changing the diameter of the impeller in a pump casing of fixed diameter has the same effect as changing pump speed. To predict the effect, the affinity rules for pump speed [Eqs. (15), (18), and (20)] are used. If a 7-in. impeller is replaced by a 5-in. impeller, the ideal head decreases by the factor $(5/7)^2$, the ideal throughput decreases by $5/7$ and the corresponding power decreases by $(5/7)^3$. The reason for using an impeller of diameter less than the maximum possible diameter is that the pump may be oversized for the application. Reducing the impeller diameter can bring the ideal and the actual throughput closer together.

Figure 8 presents a more complete picture of pump performance, taking into account the possibility of using different impeller sizes. Because the curves for power and for required NPSH depend very heavily on the diameter of the impeller, we do not show each curve separately; instead, lines of constant NPSHr and Pw$_s$ are presented.

Pumping Systems and Flow Control

The pump is now considered as part of an overall system of piping and vessels. The inlet side has already been considered under the NPSH discussion. A typical downstream configuration, as sketched in Fig. 9, is now examined. The pump is supplying liquid to a vessel operating at a constant pressure of 345 kPa gauge, so even with zero flow the pump must supply this pressure. If the liquid has density of 1000 kg/m^3 this pressure requires 345,000/9.8/1000 or 35 m of liquid head. Assuming that the pump is fed with liquid at atmospheric pressure, then it must develop this head. The *system curve* shows the head that the pump must deliver at different throughputs: it intersects the zero-flow axis at 35 m. When there is flow the required head increases because of friction loss in the piping. The pump characteristic curves are all for the same impeller diameter but for different pump speeds. The point

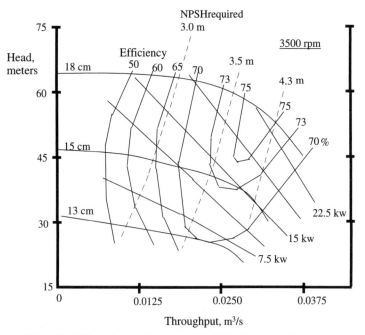

Figure 8   Effect of rotor diameter on centrifugal pump performance.

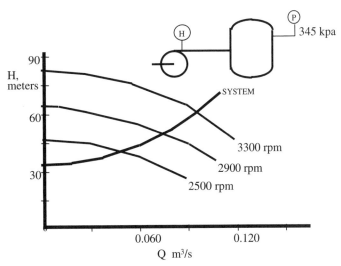

Figure 9   Pumping system.

of intersection of each curve with the system curve defines the flow that can be achieved. If this is not the desired flow then some options for shifting it are to

- Change the pump speed
- Change the impeller diameter
- Allow some recycle flow

These options are discussed below.

If the liquid changed to one with density 900 kg/m³ but the vessel stayed at 345 kPa, then the required no-flow head would increase to 35 × 1000/900 or 39 m. For constant weight flow the frictional head loss will also increase, approximately as the inverse of the square of liquid density according to standard pipe-flow calculations. These changes would shift the system curve and the points of intersection.

Of the three options for controlling flow rate, changing the pump speed appears to be the simplest, but it requires installation of a variable-speed drive, which costs more than drives with standard speeds such as 1750 and 3500 rpm. Changing the impeller diameter is a good option if the pump is oversized and if changes in desired flow rate are infrequent.

The third option, using recycle, is illustrated in Fig. 10. In this approach the pump is deliberately sized to provide flow at a rate above (10–20%) the highest rate that is ever expected. The pump operates at or near this rate all the time, with the excess volume going back to a feed tank. The process line is equipped with a flow meter and a valve, the valve being intermittently or continually manipulated to control the flow rate. The recycle line, to the feed tank, is fitted with an orifice to restrict flow but it always provides an open path for pressure relief in case the control valve ever shuts completely. The heads and flows are shown in Fig. 11. There are three *system curves:* one for the recycle loop *AC*, one for the main stream *AB*, and one for the combined flow. The *AC* curve remains fixed unless the piping or orifice is changed. The *AB* curve becomes steeper or more shallow depending on whether the valve is closing or opening. When this happens the flows and pressures adjust automatically so that the new combined system curve maintains its operating point at its intersection with the pump curve.

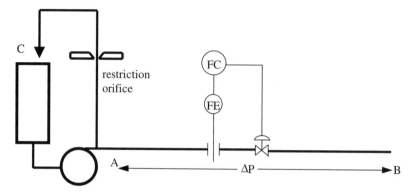

Figure 10   Flow control with a centrifugal pump.

Axial and Mixed-Flow Pumps

Axial-flow kinetic devices (sometimes called *turbine pumps*) have impellers that propel the liquid in the direction of net flow. Their specific speed, N_s, is in the high range, e.g., between 10,000 and 15,000, so they are used in applications where high head does not need to be developed. Figure 12 shows a schematic view of a vertically mounted axial-flow pump. The rotor and blades are driven by a top-mounted motor. This configuration is useful where liquid must be withdrawn from a low point. The vertical axial-flow pump might be used to raise a low-lying liquid and to feed it to a second pump capable of developing a higher head. In some cases the drive and pump are an integral unit, with the drive sealed against moisture infiltration. These units are *submersible pumps* and are commonly situated near the bottom of wells.

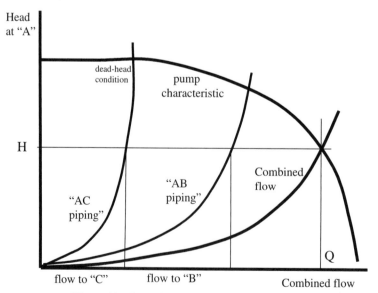

Figure 11   Operating points with recycle stream.

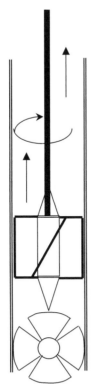

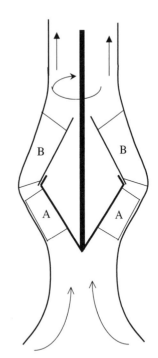

Figure 12 Axial flow pump.   **Figure 13** Mixed centrifugal and axial flow.

A mixed-flow pump combines axial and radial motion, as shown in Fig. 13. The angled blades *A* are attached to a top-driven rotor. The rotor is kept at the center of the pipe by a fixed hub, which is attached to the wall by struts *B*.

Although these types of pumps are not used for high heads, their characteristic curve of *H*-versus-*Q* is steeper than for pumps of lower specific speed. As a result, if their volumetric flow is throttled to levels much lower than design, the head and the power consumption rise sharply. For this reason they should not be "dead-headed," i.e., have their flow cut to zero.

2.2 Kinetically Driven, Jet Pumps

The term *jet pump* encompasses a collection of devices for the entraining and pumping of gases, gas–liquid mixtures, and liquids. Only liquid pumping is considered here. The driving force is a stream of liquid, the *motive* or *primary liquid,* which accelerates through a nozzle into the interior of the pump and creates a zone of low pressure in acccordance with Bernouilli's equation, Eq. (3). Into this zone is drawn the liquid to be pumped, which may or may not be the same substance as the motive stream. The streams combine into a single stream, which then recovers pressure as it passes through an expanding diffuser. Figure 14 shows such a device.

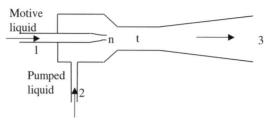

Figure 14   Liquid flow jet pump.

The hydrodynamics of this device are described by Bernouilli's equation and by the mass balance. For a pump of specified size and proportions the performance can be calculated as the relation between

$$N = (P_3 - P_2)/(P_1 - P_3) \tag{26}$$

and

$$M = Q_2/Q_1 \tag{27}$$

where P_1 is the pressure of the motive liquid before acceleration
  P_2 is the pressure of the pumped liquid before it enters the pump
  P_3 is the pressure at the exit of the pump,
  Q_1 is the volumetric flow of the motive liquid, and
  Q_2 is the volumetric flow of the pumped liquid.

With some simplifying but reasonable assumptions, the following relationship is found:

$$N = \text{Num}/(1 + K_1 - \text{Num}) \tag{28a}$$

where

$$\text{Num} = (2 A_n/A_t) \{1 + (A_n/A_t) [(\rho_2/\rho_1) M^2/(1 - A_n/A_t)]$$
$$- (1 + K_{t3}) (1 + M^2)\} - (M^2/c^2) (1 + K_2) \tag{28b}$$

where A_n is the cross-sectional area of the nozzle
  A_t is the cross-sectional area of the throat
  ρ_1 is the density of the motive liquid
  ρ_2 is the density of the pumped liquid
  K_1 is the number of velocity heads frictional pressure loss in the nozzle
  K_2 is the number of velocity heads frictional loss in the pumped liquid entry
  K_{t3} is the number of velocity heads frictional pressure loss in the throat and diffuser (calculable from standard pipe friction equations)
  c is the ratio $(A_t - A_n)/A_n$

The efficiency of a jet pump is the flow work done on the pumped liquid divided by the flow work extracted from the motive liquid:

$$\eta = Q_2 (P_3 - P_2)/[Q_1 (P_1 - P_3) \tag{29}$$

As an example, consider the operation of a pump under the following conditions

• Same motive and pumped liquid, i.e., $\rho_2 = \rho_1$
• A_n/A_t equal to 0.2
• $K_1 = K_2 = K_{t3} = 0.1$

Applying Eqs. (26)–(29) yields predictions in Table 2. The data are plotted in Fig. 15. Analogously with many other types of pumps

- The best pressure performance occurs at zero flow
- There is a condition of maximum efficiency

The efficiency is low compared to mechanically driven pumps but this type of pump is very forgiving of the condition of fluid with which it is presented. It can handle liquids, gases, and mixtures. It is recommended that, in liquid service, the chosen point of operation be somewhat to the left of the point of maximum efficiency to avoid the possibility of cavitation. A point about two-thirds of the way between zero flow and maximum-efficiency flow is suitable.

2.3 Positive Displacement, Rotary

The positive-displacement family of pumps is so named because there is a direct connection between pump action and liquid motion, with no reliance on an uncertain conversion between kinetic energy and pressure. Kinetic energy plays, at most, a subsidiary role in the action of these devices. Some of them are primarily used for moving highly viscous liquids, where it would be difficult to generate kinetic energy in the first place. Some are used for developing high pressure, which would require extensive staging in a kinetically driven device. Some are used to achieve high accuracy of liquid delivery rate with no need for a flow meter to monitor the rate. As with the kinetically driven pumps, the broad division is between rotary and nonrotary units. Four types of rotary pump are discussed and all are suited for viscous liquids.

Gear Pumps

Gear pumps are primarily used for high-viscosity liquids. Two or more gears trap liquid in the space between the gear teeth and the casing wall and convey it from inlet to outlet. Obviously, it is essential to minimize paths through which liquid could flow backward, i.e., between the intermeshing gears, over the tips of the gears, and over the top and bottom faces of the gears. This is especially important if the pump is raising the pressure significantly. The pumping rate can be described by an equation of the form

Table 2 Performance of Liquid–Liquid Jet Pump

$M = Q_2/Q_1$	$N = (P_3 - P_2)/(P_1 - P_3)$	$\%\eta = Q_2(P_3 - P_2)/Q_1(P_1 - P_3)$ $= MN100$
0	0.4785	0
0.1	0.4610	4.61
0.2	0.4434	8.87
0.3	0.4257	12.77
0.5	0.3903	19.52
0.7	0.3550	24.85
1.0	0.3022	30.22
1.5	0.2160	32.41
2.0	0.1331	26.62
2.2	0.1010	22.23
2.5	0.0543	13.57
2.7	0.0240	6.49
2.8	0.0092	2.58

Performance of Liquid-Liquid Jet Pump

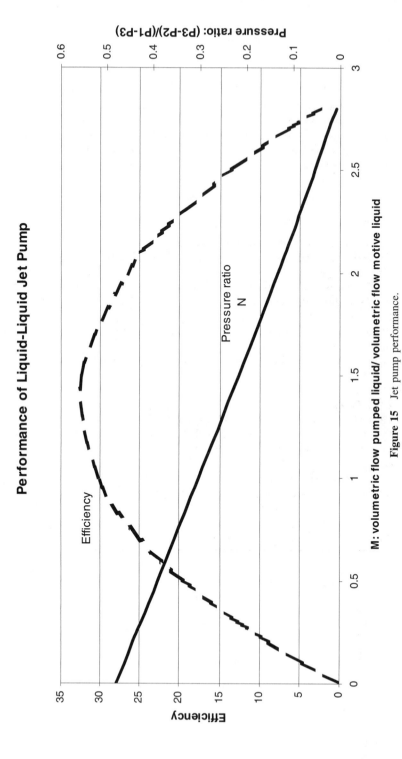

Figure 15   Jet pump performance.

$$Q = \alpha \times \text{rpm} - \beta \, \Delta P / \mu \qquad (30)$$

where

Q is volumetric throughput

μ is liquid viscosity

ΔP is pressure rise generated by the pump

α and β are constants that depend on the style, size, and clearances of the pump

If clearances are too high the pump loses much or all of its pumping efficiency. Gear pumps are widely used in the polymer industries, where viscosities of thousands of pascal-seconds are encountered and where pressures of tens of megapascals are required to force these liquids through pipes and vessels. The concept of the gear pump is illustrated in Fig. 16.

Screw Pumps

Screw pumps are related to the gear pump in that they act by pushing liquid along the inner surface of the casing, in this case the screw barrel. The most common embodiment is a single screw in a single barrel but other models make use of two (or more) screws in parallel intersecting barrels, where the screws may corotate or counterrotate. Screw pumps with a single screw and those with corotating twin screws are not true positive-displacement pumps because liquid is able to flow back along the screw channels. However, this backflow is quantitatively predictable and the action of the pump is well described by the *screw equation*, which is similar to that of the gear pump:

$$Q = \alpha \times \text{rpm} - \beta \times \Delta P / (\mu L) \qquad (31)$$

where L is the filled length of the screw.

The constant β depends on both the cross-sectional area of the screw channels and on the clearances between screw and barrel. Screw pumps can still pump against significant

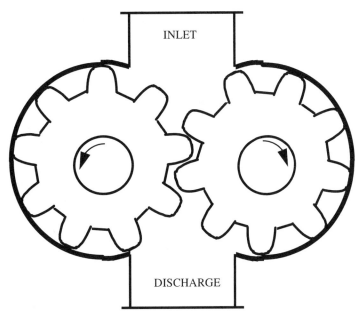

Figure 16 Gear pump.

pressures in spite of the backflow term. As viscosity increases the backflow term decreases, which is why single-screw and corotating twin-screw pumps are widely used for viscous liquids. The two types of screw pump are illustrated in Figs. 17 and 18.

Figure 19 shows a counterrotating twin-screw pump. In this diagram the screws are *fully intermeshing* and as such make the pump truly a positive displacement one. There is no reverse flow along the screw channels; rather, the screws form discrete pockets of liquid, which are carried down the barrel to the high-pressure exit. The only backflow is through clearances between screws and between screws and wall. Counterrotating screws can also be made less than fully intermeshing, in which case there is backflow in the channels.

Progressive Cavity Pumps

The principle of the *progressive cavity* was formulated in 1929 in France by Rene Moineau; hence, the well-known Moyno pump. As shown in Fig. 20, the device consists of

- A barrel
- An elastomeric lining moulded with the shape of a two-start helix, bonded into the barrel
- A metal one-start helical rotor turning within the barrel and lining

The rotating screw forms pockets with the lining and these pockets move progressively toward one or the other end of the barrel, depending on which way the screw is turning. The pump is able to handle high loadings of even abrasive solids and can pump liquids with viscosity up to 1000 Pa-s. Displacement is positive although there is backleakage (calculable) if pumping against a pressure. The Moyno and other brands of progressive cavity pumps have other attractive features, although they are often used to pump unattractive materials like sewage sludge.

Peristaltic Pumps

Peristalsis is the mechanism by which muscular contractions move materials through various passages in the body. The peristaltic pump mimics this process by trapping and moving liquids through a flexible tube. The advantage is that there is no contact between pump mechanism and the liquid. This type of pump is restricted to low-pressure applications. Figure 21 shows the principle.

2.4 Positive Displacement, Reciprocating

Reciprocating pumps comprise a family of devices in which pistons or plungers work back and forth in a cylinder to force fluids into a region of high pressure. The fluid may be either

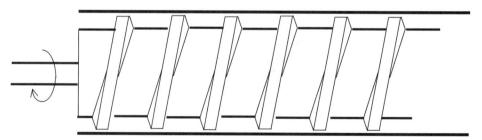

Figure 17   Single-screw screw pump.

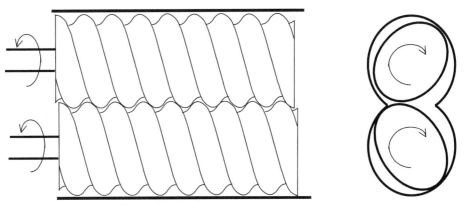

Figure 18   Twin-screw co-rotating screw pump.

liquid or gas. In the latter case the device acts as a *compressor,* which is described more fully in Section 3.4. A variation is the *diaphragm pump,* which works on the same principle but avoids metal-to-fluid contact and also avoids leakage of fluid around the driver. Such a pump is shown schematically in Fig. 22. The fluid enters and leaves a reciprocating pump through check valves: each valve opens when the pressure upstream of it is greater than the pressure downstream of it. This action allows the pump chamber to fill with low-pressure fluid and to expel fluid into higher pressure. The necessity for the valves to open quickly and close quickly limits the viscosity that can be handled.

The pumping rate of reciprocating pumps can be varied in two ways: by varying the speed of reciprocation or by varying the stroke length. A set of pumps may be "ganged" together on a common drive, in situations where a group of liquids are being supplied to a single destination at a variable overall rate but always in the same proportion to one another. The proportions are adjusted from time to time as desired by adjusting the stroke lengths of the individual pump heads.

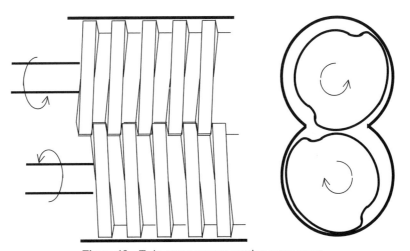

Figure 19   Twin-screw counter-rotating screw pump.

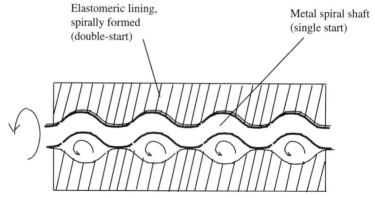

Figure 20 Progressive cavity pump.

Reciprocating pumps deliver a flow that pulsates at the frequency of the reciprocation. This characteristic may or may not be acceptable. If it is not, then it can be partially redressed by installing a pulsation dampener downstream of the pump. In specifying the strength of the piping downstream of the pump it must be remembered that the instantaneous flow of liquid can reach values up to π times the average flow. Experience is that reciprocating pumps must work against an adverse (rising) pressure gradient for the valves to open and close properly.

3 GAS MOVERS

Four types of device are used for moving gases. Depending on the pressure range and pressure change a gas may be moved by a

- Vacuum pump
- Fan
- Blower
- Compressor

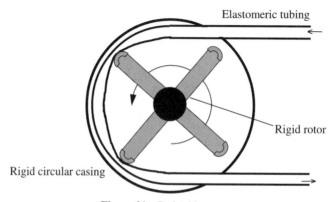

Figure 21 Peristaltic pump.

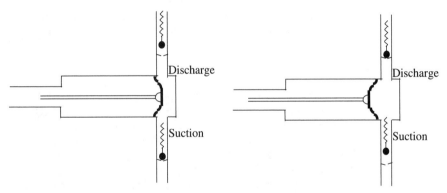

Figure 22 Diaphragm reciprocating pump.

By contrast, all liquid movers have the same designation: *pumps*. Figure 23 shows qualitatively the ranges of application of the gas movers. The pressure on both axes at the origin of the plot is one atmosphere absolute.

Vacuum pumps create an internal zone of low pressure into which the gas in a region of subatmospheric pressure is induced to flow. A single stage of vacuum pumping discharges to atmospheric pressure. With multiple stages, the discharge of intermediate stages is to subatmospheric pressure.

Fans generally add only a small amount of pressure to a gas, generally no more than 20 kPa (60 in. of water). Fluid compressibility can be ignored in the calculations. Typically, fans pull vapors from a slightly subatmospheric region and discharge to atmosphere or they pull from the atmosphere and discharge to a space that is slightly above atmospheric pressure. For example, in the former case, they may be removing unwanted vapors; in the latter, they may be supplying fresh air.

Blowers and compressors impart significant positive pressure to gases. Such devices may have several stages, where the suction pressure of the first is atmospheric and that of subsequent stages is higher. In these devices it is necessary to take account of change in density with pressure and also the heat evolved by work ($P\,dV$) done on the gas.

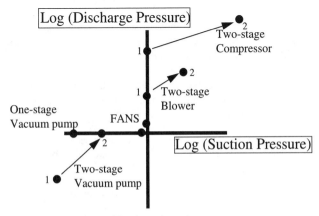

Figure 23 Overview of gas movers.

The distinction among these devices is not always as sharp as implied by the diagram. A particular gas mover may be called a fan by one person and a blower by another.

3.1   Kinetically Driven, Rotary

The devices are similar to the corresponding liquid movers, but there are important differences due to low density, density change, and temperature effects.

This category comprises fans, centrifugal blowers, and centrifugal compressors. These units all operate, at least ideally, according to Euler's equation and are described by Eq. (13), which was derived for centrifugal pumps. As with liquids, these movers are said to generate a *head* of fluid rather than a pressure. For a fluid centrifugal mover of given size and operating at a given speed, the volumetric throughput is independent of fluid density and the pressure generated is proportional to the fluid density. There are some differences in application:

- In the case of liquids the impeller vanes are always curved backward (in Fig. 5, β less than 90 degrees) but this is not always true for gas movers.
- Adjustable inlet guide vanes are sometimes provided to reduce entrance losses or to reduce the density of the gas as a method of flow control.
- Net positive suction head is not a consideration.

Fans

There are five styles of fan. They are listed in Table 3 along with their characteristics. Figure 24 shows them schematically and also the general shape of their performance curves. In four of the five cases, the flow enters at the center of the axis of rotation and exits tangentially at the outer periphery of the rotating blades; these fans are classified as "centrifugal." In the axial case, the flow enters and leaves in a direction parallel to the axis of rotation. In the other four cases, the internal exit passage of the fan body can be arranged to permit axial (or "in-line") exit if desired. In the axial case flow straighteners may be provided downstream to remove swirl from the stream.

The fan characteristics for a typical radial-blade fan are shown in Fig. 25. As is almost universally true for fluid movers, as throughput increases the pressure rise decreases. When pressure rise falls to zero, the capacity limit of the fan has been reached, except if the fan is being used to regulate the flow in a falling-pressure situation. Unlike some of the other four styles of fan, there is no region of the pressure–throughput curve where pressure briefly *increases* with flow; in those cases, the anomaly causes unstable operation unless the fan is operated to the right of it, i.e., in the falling-pressure zone. Brake power increases steadily with increasing flow, so a motor should be provided that can handle maximum flow if there is a chance that it may occur.

The power theoretically required by the fan is given by the product of the volumetric throughput and the pressure rise. For commonly used units of expression, the units' conversion is as follows

$$
\begin{array}{ccccccc}
\text{Power} = & \text{Volumetric flow} & \times & \text{Pressure rise} & \times & 0.0361 & \times & 144 & \times & 1/33{,}000 \\
\text{hp} & \text{ft}^3/\text{min} & & \text{in. water} & & \text{lb/in.}^2/\text{in. water} & & \text{in.}^2/\text{ft}^2 & & \text{hp/ft-lb/min}
\end{array}
$$

or

$$
\begin{array}{ccc}
\text{Power} = & \text{Volumetric flow} & \times & \text{Pressure rise} \\
\text{W} & \text{m}^3/\text{s} & & \text{n/m}^2
\end{array}
$$

Table 3 Fan Types

Configuration	Features
Backward inclined blades	Best mechanical efficiency
	Quietest
	Nonoverloading power characteristic
Axial	Used for domestic fans
	Can be most compact
	Relatively low pressure rise
	Maximum power required at zero flow
	Can be provided with vanes in the housing to allow more pressure generation
Forward curved blades	Runs more slowly and quietly
	Used for medium volumes at low pressure rise
	Power increases with flow
	Theoretically pressure increases with flow but practically these fans are designed and used with a falling pressure curve
Radial-tip blades	A hybrid between backward inclined and radial
	More efficient than simple radial
	Can handle some particulate contamination
	Power increases with flow
Radial blades	The most common fan found in industry
	Wide range of applicability
	Can handle significant amounts of particulates
	Relatively low rpm
	Stable throughout its range
	Power increases with flow
	Least efficient but still acceptable

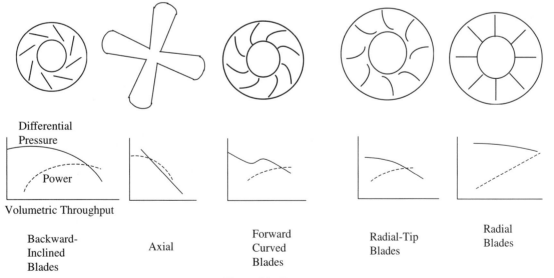

Figure 24 Fan types.

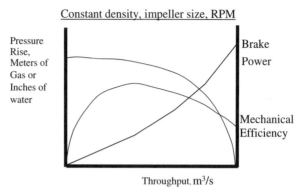

Constant density, impeller size, RPM

Pressure Rise, Meters of Gas or Inches of water

Brake Power

Mechanical Efficiency

Throughput, m³/s

Figure 25 Fan performance variables.

The operation of a fan must be considered in the context of the overall system of which it is a part. The flow of gas or vapor encounters frictional resistance either upstream or downstream of the fan and this resistance tends to be proportional to the square of the flow rate. The operating point of the fan is determined by the intersection of the curve for pressure versus throughput for the fan with the curve for the frictional resistance versus throughput, as shown in the Fig. 26. If the frictional resistance curve changes due, say, to a change in the design of ducting or to the adjustment of a damper, then the point of intersection changes. In the figure, shifting from the DP-lo resistance curve to the DP-hi curve causes the operating point to change from *A* to *B*. If it is important to maintain throughput, then the fan rpm has to be increased, i.e., to point *C*. If, for some reason, pressure must be kept constant, then the rpm must be decreased.

The effect of fan speed on throughput, pressure rise, and power consumption must be considered. The *fan affinity rules* are similar to those for liquid centrifugal pumps:

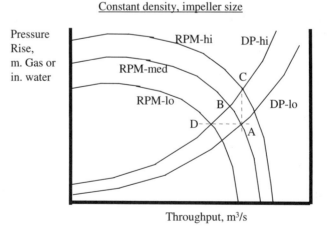

Constant density, impeller size

Pressure Rise, m. Gas or in. water

RPM-hi   DP-hi

RPM-med

C

RPM-lo   B   DP-lo

D   A

Throughput, m³/s

Figure 26 Fan system.

- Throughput is proportional to the first power of rpm.
- Pressure rise is proportional to the second power of rpm and also to gas density.
- Power consumption is proportional to the third power of rpm and to gas density.

When a fan manufacturer provides characteristic operating curves for a particular fan, the curves have been determined experimentally in a correctly designed setting, i.e., where there are no obstructions and sharp bends near the inlet and outlet of the fan. In the purchaser's setting, some of these obstacles may of necessity be present. The best approach is to remove the obstacles. If this is not possible, then the effect of each of them can be predicted by consulting a standard published by the Air Movement and Control Association (see Bibliography). The effect is characterized by a downward correction in the curve of pressure rise versus throughput. In specifying a fan and knowing the unavoidable deficiencies in the system, the engineer can take account of these effects in making sure to order a fan that can overcome them and still provide the required performance.

Centrifugal Compressors
The heart of a centrifugal compressor is its impeller (sometimes called the wheel). The principle of operation is similar for centrifugal compressors, blowers, and fans and the impeller design is similar to that of a fan. Most machines use a backward curved impeller for its high efficiency and wider range of operation as compared to radial or forward curved. Since the gas leaving the impeller has significant velocity the casing design employs a diffuser (static vanes) to reduce velocity and gain static pressure, as shown in Fig. 27. (This technique is also sometimes used with centrifugal pumps.)

Centrifugal compressors are extremely popular because most are close to being oil-free. Although oil that is used in the compressor can create an aerosol, the special sealing systems used in most centrifugals reduce oil contamination to very low levels. Centrifugals are also popular because of the very large capacities ($>$100,000 ACFM) that are possible with a single compressor, combined with fairly high pressures (1500 to 5000 psig with multiple stages). Centrifugals are economically attractive when flows are high, and they are the only choice in many high-flow/pressure situations.

If other compressors are available that will meet the same pressure/flow requirements, centrifugals are preferred when the process requirement allows for a fixed pressure ratio and requires oil-free gas. If centrifugal and reciprocating compressors have similar costs, the centrifugal is selected for its reliability advantages and lower life-cycle costs. Centrifugal units can be built to run without any installed spares (99+ reliability) if API (American

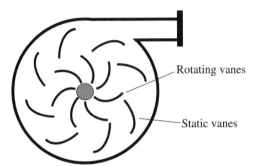

Figure 27 Centrifugal compressor with diffuser vanes.

Petroleum Institute) 617 standards are followed, whereas reciprocating units usually require an installed spare.

It should be kept in mind that centrifugal units often have higher installed costs than reciprocal and screw compressors for the same pressure/flow range, and they are inflexible to changes in pressure ratios, i.e., their capacities drop significantly as the discharge pressure rises.

Centrifugal compressors, like centrifugal pumps, at a given speed and throughput generate a constant head rather than constant pressure. This means that when curves for a centrifugal device are shown in psi or kPa, they are for a particular gas molecular weight (usually air). When the same compressor is operating with a different gas, the compressor will produce the same head but not the same differential pressure. For this reason, low molecular weight gases (less than 10) are not practically handled by centrifugal devices because to achieve a reasonable pressure, the required head is too large for any industrially available centrifugal compressor; a reciprocating compressor is normally used instead.

Centrifugal compressors, due to the nature of their impeller design, exhibit a rising head-versus-flow curve at low flow. This can cause the compressor to be unstable at flows less than some rate, called the *surge point*. Compressor manufacturers provide the controls to prevent the compressor from reaching this potentially catastrophic state of flow. A single stage of compression can operate over a range of about 50–100% of maximum throughput, but a multistage compressor (typically 8 stages) is restricted to a range of about 75–100%.

Axial Compressors
Axial compressors, as shown schematically in Fig. 28, operate with a rotor fitted with successive rows of blades, which move the gas forward. Between the rows of rotating blades are rows of static blades, which remove swirl and keep the flow axial. The space between rotor and barrel becomes progressively smaller, causing the gas to speed up and acquire kinetic energy. The blades are aerodynamically shaped to achieve maximum thrust and minimum drag. Axial compressors are typically 8–10% more efficient than centrifugal compressors.

3.2   Kinetically Driven, Ejectors

The gas-driven *ejector* is the nonrotary member of the kinetically driven family. Ejectors continue to be a common means of creating and maintaining vacuum, chiefly because of

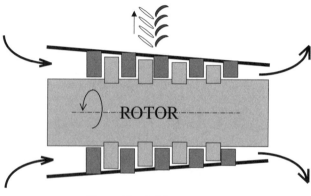

Figure 28   Axial compressor.

their low capital cost and the absence of moving parts. They are used not only for vacuum but in other applications where it is desired to combine two streams of different pressures into a single stream of intermediate pressure (the devices are sometimes call thermocompressors). The principle is to create a zone of very low pressure by raising a high-pressure *motive* stream (e.g., steam) to supersonic velocity. Suction gas or vapor is drawn into this zone where it combines with the motive fluid. The mixed stream, also at supersonic velocity, is slowed down and its pressure recovers to a level intermediate between that of the motive and suction streams. The arrangement and process path is shown schematically in Fig. 29. A two-stage ejector system is shown in Fig. 30, in this case using a precondenser and an intercondenser to reduce the vapor load, as can be done for a condensable motive gas like steam. The method of control is shown, whereby a "bleed" stream of external or higher-pressure gas is allowed into the suction stream. Ejectors are sometimes configured up to six in series.

The design graph is given in Fig. 31. The left-hand curves show the achievable combinations of suction pressure/motive pressure and suction-weight flow/motive-weight flow for various ratios of discharge-throat area to motive-nozzle area. The right-hand curves show the achievable combinations, again of suction-to-motive pressure and discharge-to-suction pressure.

Consider an example: Design an ejector to support an absolute pressure of 60 mm Hg using 100 psig steam as the motive fluid and discharging to atmosphere.

$$\text{Motive pressure} = P_m = (100 + 14.7) \times 6.8948 = 791 \text{ kPa}$$

$$\text{Suction pressure} = P_s = 60 \times 0.13332 = 8.0 \text{ kPa}$$

$$\text{Discharge pressure} = P_d = 101.3 \text{ kPa}$$

From the right-hand curves, for P_d/P_s of 12.7 and P_s/P_m of 0.01, the required area ratio (A_2/A_1) is about 15. From the left-hand curves, the ratio of suction flow to motive fluid flow is around 0.04. To size the system, an estimate must be available of the required suction flow, W_s, based either on the speed we wish to "pump down" the system or on estimates of

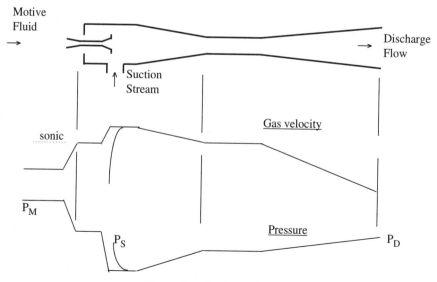

Figure 29 Gas-driven ejector.

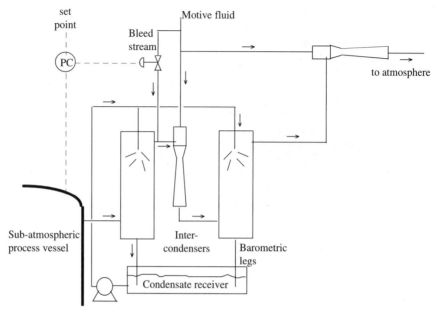

Figure 30 Ejector system.

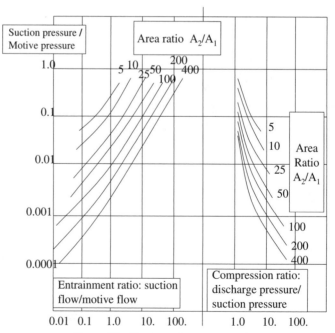

Figure 31 Ejector performance curves.

in-leakage into the vacuum. Then the motive flow, W_m, is 1/0.04 times the suction flow and the ejector can be sized by calculating A_1 from the formula

$$W_m = 146,000 P_m A_1 / \sqrt{T_m} \qquad \text{if air is the motive fluid} \qquad (32)$$

$$W_m = 113,000 P_m A_1 / \sqrt{T_m} \qquad \text{if steam is the motive fluid} \qquad (33)$$

where W_m is in kilograms/hour, P_m in kilopascals, A_1 in square meters, and T_m in degrees Kelvin. The predictive curves can be considered accurate to about $\pm 20\%$.

3.3   Positive Displacement, Rotary

Positive displacement gas movers are used both for vacuum generation and for compression to above-atmospheric pressures. Some comparisons may be made in these two applications among the various available devices, including the kinetically driven ones already discussed.

In the case of vacuum, there are four principal methods (see Table 4):

- Jet ejectors
- Liquid ring
- Oil-sealed (usually rotary vane)
- Dry (usually rotary lobe)

Ejectors were discussed in Section 3.2. The other three vacuum producers are discussed here. Compressors are compared in Fig. 32 and Table 5. Four devices are discussed, all of which can effect a significant compression, with compression ratios ranging from about 2 to 20.

Sliding-Vane Compressor

The principle of the rotary sliding vane compressor is illustrated in Fig. 33. The in-and-out vane action gathers and traps the incoming gas, then compresses it as it reaches the discharge port. These compressors are used both for the generation of pressure and for the production of vacuum. In the latter case the effectiveness is enhanced if oil is used to lubricate and seal the friction surfaces.

Liquid Ring Pump

This unit is sometimes also called a *liquid-piston pump*. Its main use is for vacuum generation. Liquid-ring pumps have been superseding ejectors, based mainly on their better thermal efficiency, relatively low cost (compared with other mechanical devices), and robustness to upsets and to the presence of liquid or solids in the vapor stream. The liquid ring is shown schematically in Fig. 34: water or other fluid is swirled by vanes around the periphery of an

Table 4   Capabilities of Vacuum Producers

Type	Suction Pressure, mm Hg abs (Pa)	Capacity, cfm (m³/s)
Steam ejectors		
One stage	75 (10,000)	10–1,000,000 (0.005–500)
Six stages	0.003 (0.4)	
Liquid ring	10–75 (1300–10,000)	3–10,000 (0.0015–5)
Oil-sealed, rotary vane	0.001–1 (0.13–130)	3–50, 50–800 (0.0015–0.4)
Rotary blower	60–300 (8000–40,000)	30–30,000 (0.015–15)

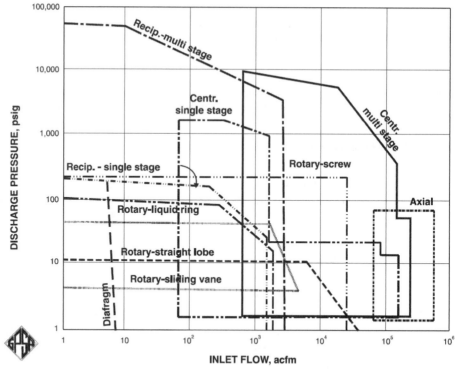

Figure 32 Compressor range comparison. (Courtesy of Gas Processors Suppliers Association.)

Table 5 A Comparison of Common Gas Compressors

Type of Compressor	Oil-lubricated Screw	Oil-free Screw	Centrifugal	Oil-lubricated Reciprocating	Oil-free Reciprocating
Typical capacities, ACFM	150–10,000	250–50,000	200–100,000+	0–6000	0–6000
Maximum inlet pressure P_1, psig	Generally 100, but can go to 700	Generally 150, but can go to 225	Close to P_2	Close to P_2	Close to P_2
Maximum discharge pressure P_2, psig	865	400	1500–5000	6000	
Maximum discharge temperature T_2, °F	250[a]	400–500	350–500	300–400	<300
Maximum compression ratio per stage	23:1	5:1	1.5–3:1[b]	5:1	
Adiabatic efficiency at design pt.	70–85%, 50% at 15:1 compression ratio	70–85%	70–88%	80–92%	>75%
Reliability/availability	98–99.5%	99–99.5%	98–99.5%	92–95%	92–95%

[a] Most lubricants break down at 280°F.
[b] Centrifugal compression ratio is for a single stage.

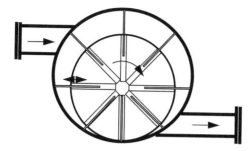

Figure 33   Rotary sliding vane compressor.

eccentric chamber. Non-liquid-filled pockets open up, expand, compress, and collapse twice (in the design shown) around the circuit. Suction gas or vapor is pulled in at the points marked I and discharged at the points marked D. The liquid may in fact be a process fluid and may be the liquid form of the vapor being handled.

Lobe Compressor

The cycle of operation of this type of device is shown in Fig. 35, where the darkness of the ellipses indicates the magnitude of the pressure. Lobe compressors are at the low end of the compression scale and are often classed as *blowers* instead.

Screw Compressor

Screw compressors operate using two dissimilar helical rotors, one meshing into the other. The rotors mesh together to create and convey cavities that draw gas from the inlet and deliver it to the discharge. In Fig. 36 the darkness of the ellipses indicates the magnitude of the pressure: flow is from top to bottom in the diagram. Viewed from the drive end, the right-hand rotor turns in the counterclockwise direction and the left-hand rotor in the clockwise direction. The rotors have very tight clearances and are prevented from touching each other using timing gears. Oil-lubricated compressors inject oil to lubricate and seal the rotors and to draw away heat of compression (thus improving the isentropic n value). Although the oil is in contact with the process gas and the two contaminate each other, recent advances in filtration can ensure oil-free gas to about 1 part per billion. Screw compressors are reported to be used in low molecular weight applications (i.e., hydrogen).

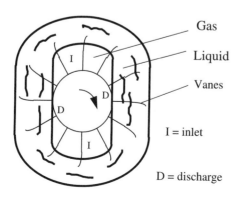

Gas

Liquid

Vanes

I = inlet

D = discharge

Figure 34   Liquid-ring compressor.

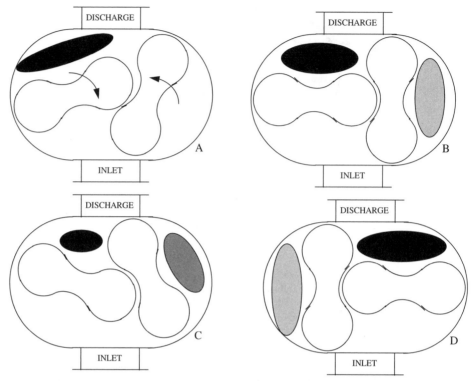

Figure 35 Lobe blower/compressor.

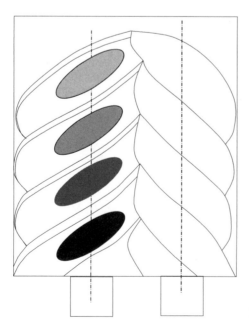

Figure 36 Screw compressor.

3.4 Positive Displacement, Reciprocating

The principle of the reciprocating compressor is shown in Fig. 37. The inlet and discharge of the cylinder are each fitted with a one-way valve, of which the spring-loaded valves in the diagram are one type. On the suction stroke (piston moving left), the inlet valve is pulled open and the discharge valve is pulled shut. On the discharge or compression stroke the reverse is true.

In the double-stroke design, there is a compression chamber on both sides of the piston. However, in any piston compressor, there is always some leakage between the piston and cylinder or between the drive rod and its cylinder, which could either contaminate the material being pumped or alternatively cause release of the material being pumped. The solution to this issue is to use a diaphragm reciprocating compressor. In such a compressor, the piston is replaced by a sealed flexible diaphragm. The diaphragm is pushed forward and pulled back either by an attached rod or by hydraulic oil behind it. Diaphragm compressors provide the highest pressures of any compressor (about 15,000 psig, 100,000 kpa), but are limited to no more than 100 ACFM (0.05 m³/s).

3.5 Work, Temperature Rise, and Efficiency of Compression

Given the gas weight rate and given the desired or intended compression ratio, P_2/P_1, it is necessary to be able to estimate the power required, and the temperature rise per stage of compression. Compressive work results in an increase in gas enthalpy and this means temperature rise. Heat may have to be extracted from the compressed gas to avoid equipment damage and to reduce the volume of the gas. Isothermal compression would be ideal but difficult to achieve. Adiabatic operation is closer to reality in most cases. A first analysis of compression may assume this condition and is described here.

Adiabatic Treatment

The pressure–volume relationship during an adiabatic change is given by

$$PV^\gamma = C \quad \text{a constant} \tag{34}$$

where γ is the ratio of specific heat at constant pressure to specific heat at constant volume.

The work of compression (negative because it is being done on the gas) is

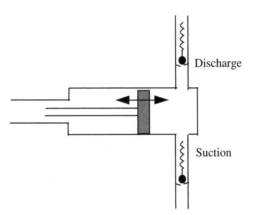

Figure 37 Reciprocating compressor (piston-type).

$$-W_f = P_1 V_1 + \int P \, d(V) - P_2 V_2 = \int V \, d(P) \tag{35}$$

Substituting from (34) for V,

$$-W_f = [(\gamma P_1 V_1)/(\gamma - 1)][(P_2/P_1)^{(\gamma-1)/\gamma} - 1] \tag{36}$$

The product $P_1 V_1$ may be replaced by ZRT_1/MW

where Z is the compressibility factor
      R is the gas constant (8314 joules per kilogram-mole per kelvin)
      T_1 is the initial gas temperature (kelvins)
    MW   is the molecular weight of the gas

In practice, the required work is always greater, so an *adiabatic efficiency* factor η_{ad} is inserted into the denominator of the right-hand side of Eq. (36).
If the gas follows the ideal gas law then

$$T = PV/\text{a constant} = P \, (P/C)^{1/\gamma}/\text{a constant} \tag{37}$$

So

$$T_2/T_1 = (P_2/P_1)^{(\gamma-1)/\gamma} \tag{38}$$

For example, if P_2/P_1 is $5/1$ and if γ is 1.4, then T_2/T_1 is 1.58. If the initial temperature is 50°C, i.e., 323 K, then the final temperature is 512 K or 239°C.
In practice, the final temperature is always somewhat greater than this prediction. If the temperature T_2 predicted by Eq. (38) is called T_{ad}, then the adiabatic efficiency is

$$\eta_{ad} = (T_{ad} - T_1)/(T_2 - T_1) \tag{39}$$

where T_2 is the actual exit temperature. Then

$$T_2 = T_1 + \{[(P_2/P_1)^{(\gamma-1)/\gamma}] * T_1 - T_1\}/\eta_{ad} \tag{40}$$

Polytropic Treatment

A different approach, favored by compressor manufacturers, is to recognize that adiabatic behavior is not exactly followed and to do the above calculation based on a revised *polytropic* version of Eq. (34):

$$PV^n = C \quad \text{a constant} \tag{41}$$

where n is experimentally observed by tests and is generally greater than γ. For instance, if n is 1.45, then the modified form of Eq. (38)

$$T_2/T_1 = (P_2/P_1)^{(n-1)/n} \tag{42}$$

leads to a prediction of T_2 equal to 259°C.
Use of the polytropic approach does not completely eliminate the need for an efficiency factor but the residual inefficiency is smaller and more constant. So the work of compression is calculated by a modification of Eq. (36) as

$$-W_f = [(\gamma P_1 V_1)/(\gamma - 1)][(P_2/P_1)^{(n-1)/n} - 1]/\eta_p \tag{43}$$

where η_p is the polytropic efficiency factor. The value of n and an estimate of η_p are supplied by the compressor manufacturer.
W_f has units of joules per kilogram or of foot-pounds-force per pound. Because of the latter units, $-W_f$ is called the *adiabatic head* in units of feet.
The power required by the compressor is given by

$$Pw_r = -W_f \ (J/kg) \times \text{weight rate of gas (kg/s) W} \tag{44}$$

or

$$Pw_r = -W_f \ (ft) \times \text{weight rate of gas (lb/min)}/(60 \times 550) \ \text{hp} \tag{45}$$

Use of Thermodynamic Charts

If a chart of *entropy*-versus-*enthalpy* is available for the gas then it may be used directly to make the adiabatic estimate and then to correct it for efficiency. Gases such as hydrocarbons deviate from the ideal gas laws significantly, and therefore thermodynamic charts or equations are the preferred method of calculating horsepower and discharge temperatures. In the example steam is compressed from an initial state of 30 psia and 320°F (point 1 in Fig. 38) to a pressure of 50 psia following a line of constant entropy (isentropic). It can be seen that the initial enthalpy is approximately 1200 Btu/lb and the enthalpy at point 2 is 1245 Btu/lb. The difference in enthalpy is the ideal adiabatic horsepower or 45 Btu/lb (0.018 hp/lb/hr). If the compressor actually has an adiabatic efficiency of 50% then the exit conditions of the gas would be at point 3, 50 psia, 514°F, and the total horsepower consumed would be 90 Btu/lb (0.035 hp/lb/hr). Compressor required head can be calculated from the equation:

$$\text{Head} = \text{Enthalpy}/g \quad (\text{where } g \text{ is the gravitational constant}) \tag{46}$$

Pressure/enthalpy charts for hydrocarbon gases are readily available from sources such as the gas processors suppliers association.

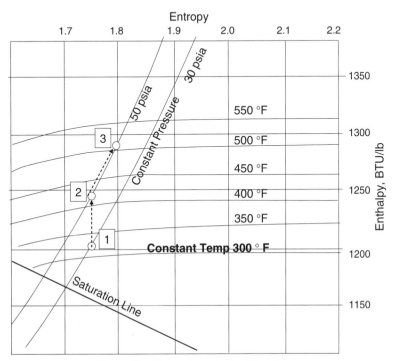

Figure 38 Thermodynamic chart for steam.

Total Compressor Power

The power calculated above was the gas power (GP). To obtain the total power requirements, the losses due to friction in bearings, seals, and speed-increasing gears must be added. An approximate equation for this is

$$\text{Brake power BP} = \text{GP} + \text{Mechanical losses} \tag{47a}$$

where

$$\text{Mechanical losses} = \text{GP}^{0.4} \tag{47b}$$

Field Determination of Compression Horsepower

The complexity of using these calculations to determine energy use or motor horsepower for a multistage compressor, in combination with having to know the adiabatic and mechanical efficiencies (which may not be available), has led to a simplified empirical approach. Using a current meter to determine the electrical power usage of the compressor and an air flow meter to determine the gas flow through the compressor, a value for the air flow SCFM $(\text{ft}^3/\text{min})/\text{kW}$ of electricity can be determined. Typical values for compressing air are 6 SCFM/kW for 100 psig air, and 4.2 SCFM/kW for 175 psig air.

Required Number of Stages

For most available compressors, compression ratios $(P_{\text{disch}}/P_{\text{inlet}})$ usually vary from 1.05 to 7 per stage, but a ratio of 3.5 to 4 per stage is reasonable for most compressors. Sealing systems are usually limited to about 300°F (148°C) and this in turn limits the stage compression ratio. Some gases, such as oxygen, chlorine, and acetylene, require that the temperatures be maintained below 200°F (93°C). The number of compression and cooling stages can be estimated based on meeting this requirement. Theoretically, the minimum power used occurs when all stages have the same compression ratio, and thus

$$\text{Compression ratio per stage} = (\text{Total compression ratio})^{(1/\# \text{ stages})} \tag{48}$$

BIBLIOGRAPHY

Air Movement and Control Association International, Inc., Publication 201-02, *Fans and Systems*, 1990.

American Petroleum Institute Standard 610, *Centrifugal Pumps for Petroleum, Petrochemical and Natural Gas Industries*, 10th ed., 2004.

Bloch, H. P., *A Practical Guide to Compressor Technology*, McGraw-Hill, New York, 1995.

Crane Technical Paper 410, *Flow of Fluids through Valves, Fittings, and Pipe*, New York, 2000.

Gas Processors Suppliers Association Data Book, 12th ed., 2004.

Hydraulic Institute Engineering Data Book, 2nd ed. Hydraulic Institute, Parsippany NJ, 1990.

International Standards Organization, Document BS EN ISO 5198, *Centrifugal, Mixed-Flow, and Axial Pumps—Code for Hydraulic Performance Tests—Precision Class*, 1999.

Karassik, I. J., J. P. Messina, P. Cooper, and C. C. Heald, *Pump Handbook*, 3rd ed., McGraw-Hill, New York, 1998.

Munson, B. R., D. F. Young, and T. H. Okiishi, *Fundamentals of Fluid Mechanics*, Wiley, New York, 1998.

CHAPTER 23

NUCLEAR POWER

William Kerr
Department of Nuclear Engineering
University of Michigan
Ann Arbor, Michigan

William Updegrove
Consultant
Tucson, Arizona

1 HISTORICAL PERSPECTIVE

1.1 The Birth of Nuclear Energy

The first large-scale application of nuclear energy was in a weapon. The second use was in submarine propulsion systems. Subsequent development of fission reactors for electric power production has been profoundly influenced by these early military associations, both technically and politically. It appears likely that the military connection, tenuous though it may be, will continue to have a strong political influence on applications of nuclear energy.

Fusion, looked on by many as a supplement to, or possibly as an alternative to fission for producing electric power, was also applied first as a weapon. Most of the fusion systems now being investigated for civilian applications are far removed from weapons technology. A very few are related closely enough that further civilian development could be inhibited by this association.

1.2 Military Propulsion Units

The possibilities inherent in an extremely compact source of fuel, the consumption of which requires no oxygen, and produces a small volume of waste products, was recognized almost immediately after World War II by those responsible for the improvement of submarine propulsion units. Significant resources were soon committed to the development of a compact, easily controlled, quiet, and highly reliable propulsion reactor. As a result, a unit was produced which revolutionized submarine capabilities.

The decisions that led to a compact, light-water-cooled and -moderated submarine reactor unit, using enriched uranium for fuel, were undoubtedly valid for this application. They have been adopted by other countries as well. However, the technological background and experience gained by U.S. manufacturers in submarine reactor development was a principal factor in the eventual decision to build commercial reactors that were cooled with light water and that used enriched uranium in oxide form as fuel. Whether this was the best approach for commercial reactors is still uncertain.

1.3 Early Enthusiasm for Nuclear Power

Until the passage, in 1954, of an amendment to the Atomic Energy Act of 1946, almost all of the technology that was to be used in developing commercial nuclear power was classified. The 1954 Amendment made it possible for U.S. industry to gain access to much of the available technology, and to own and operate nuclear power plants. Under the amendment the Atomic Energy Commission (AEC), originally set up for the purpose of placing nuclear weapons under civilian control, was given responsibility for licensing and for regulating the operation of these plants.

In December 1953 President Eisenhower, in a speech before the General Assembly of the United Nations, extolled the virtues of peaceful uses of nuclear energy and promised the assistance of the United States in making this potential new source of energy available to the rest of the world. Enthusiasm over what was then viewed as a potentially inexpensive and almost inexhaustible new source of energy was a strong force which led, along with the hope that a system of international inspection and control could inhibit proliferation of nuclear weapons, to formation of the International Atomic Energy Agency (IAEA) as an arm of the United Nations. The IAEA, with headquarters in Vienna, continues to play a dual role of assisting in the development of peaceful uses of nuclear energy, and in the development of a system of inspections and controls aimed at making it possible to detect any diversion of special nuclear materials, being used in or produced by civilian power reactors, to military purposes.

2 CURRENT POWER REACTORS AND FUTURE PROJECTIONS

Although a large number of reactor types have been studied for possible use in power production, the number now receiving serious consideration is rather small.

2.1 Light-Water-Moderated Enriched-Uranium-Fueled Reactor

The only commercially viable power reactor systems operating in the United States today use LWRs. This is likely to be the case for the next decade or so. France has embarked on a construction program that will eventually lead to productions of about 90% of its electric power by LWR units. Great Britain has under consideration the construction of a number of LWRs. The Federal Republic of Germany has a number of LWRs in operation with additional units under construction. Russia and a number of other Eastern European countries are operating LWRs, and are constructing additional plants. Russia is also building a number of smaller, specially designed LWRs near several population centers. It is planned to use these units to generate steam for district heating. The first one of these reactors is scheduled to go into operation soon near Gorki.

2.2 Gas-Cooled Reactor

Several designs exist for gas-cooled reactors. In the United States the one that has been most seriously considered uses helium for cooling. Fuel elements are large graphite blocks containing a number of vertical channels. Some of the channels are filled with enriched uranium fuel. Some, left open, provide a passage for the cooling gas. One small power reactor of this type is in operation in the United States. Carbon dioxide is used for cooling in some European designs. Both metal fuels and graphite-coated fuels are used. A few gas-cooled reactors are being used for electric power production both in England and in France.

2.3 Heavy-Water-Moderated Natural-Uranium-Fueled Reactor

The goal of developing a reactor system that does not require enriched uranium led Canada to a natural-uranium-fueled, heavy-water-moderated, light-water-cooled reactor design dubbed Candu. A number of these are operating successfully in Canada. Argentina and India each uses a reactor power plant of this type, purchased from Canada, for electric power production.

2.4 Liquid-Metal-Cooled Fast Breeder Reactor

France, England, Russia, and the United States all have prototype liquid-metal-cooled fast breeder reactors (LMFBRs) in operation. Experience and analysis provide evidence that the plutonium-fueled LMFBR is the most likely, of the various breeding cycles investigated, to provide a commercially viable breeder. The breeder is attractive because it permits as much as 80% of the available energy in natural uranium to be converted to useful energy. The LWR system, by contrast, converts at most 3%–4%.

Because plutonium is an important constituent of nuclear weapons, there has been concern that development of breeder reactors will produce nuclear weapons proliferation. This is a legitimate concern, and must be dealt with in the design of the fuel cycle facilities that make up the breeder fuel cycle.

2.5 Fusion

It may be possible to use the fusion reaction, already successfully harnessed to produce a powerful explosive, for power production. Considerable effort in the United States and in a number of other countries is being devoted to development of a system that would use a controlled fusion reaction to produce useful energy. At the present stage of development the fusion of tritium and deuterium nuclei appears to be the most promising reaction of those that have been investigated. Problems in the design, construction, and operation of a reactor system that will produce useful amounts of economical power appear formidable. However, potential fuel resources are enormous, and are readily available to any country that can develop the technology.

3   CATALOG AND PERFORMANCE OF OPERATING REACTORS WORLDWIDE

Worldwide, the operation of nuclear power plants in 1982 produced more than 10% of all the electrical energy used. Table 1 contains a listing of reactors in operation in the United States and in the rest of the world.

4   U.S. COMMERCIAL REACTORS

As indicated earlier, the approach to fuel type and core design used in LWRs in the United States comes from the reactors developed for marine propulsion by the military.

4.1   Pressurized-Water Reactors

Of the two types developed in the United States, the pressurized water reactor (PWR) and the boiling water reactor (BWR), the PWR is a more direct adaptation of marine propulsion reactors. PWRs are operated at pressures in the pressure vessel (typically about 2250 psi) and temperatures (primary inlet coolant temperature is about 564°F with an outlet temperature about 64°F higher) such that bulk boiling does not occur in the core during normal operation. Water in the primary system flows through the core as a liquid, and proceeds through one side of a heat exchanger. Steam is generated on the other side at a temperature slightly less than that of the water that emerges from the reactor vessel outlet. Figure 1 shows a typical PWR vessel and core arrangement. Figure 2 shows a steam generator.

The reactor pressure vessel is an especially crucial component. Current U.S. design and operational philosophy assumes that systems provided to ensure maintenance of the reactor core integrity under both normal and emergency conditions will be able to deliver cooling water to a pressure vessel whose integrity is virtually intact after even the most serious accident considered in the safety analysis of hypothesized accidents required by U.S. licensing. A special section of the ASME Pressure Vessel Code, Section III, has been developed to specify acceptable vessel design, construction, and operating practices. Section XI of the code specifies acceptable inspection practices.

Practical considerations in pressure vessel construction and operation determine an upper limit to the primary operating pressure. This in turn prescribes a maximum temperature for water in the primary. The resulting steam temperature in the secondary is considerably lower than that typical of modern fossil-fueled plants. (Typical steam temperatures and pressures are about 1100 psi and 556°F at the steam generator outlet.) This lower steam temperature has required development of massive steam turbines to handle the enormous steam flow of the low-temperature steam produced by the large PWRs of current design.

Table 1 Operating Power Reactors (2004)

Country	Reactor Type[a]	Number in Operation	Net MWe
Argentina	PHWR	2	935
Armenia	PWR	1	378
Belgium	PWR	7	5728
Brazil	PWR	2	1901
Bulgaria	PWR	4	2722
Canada	PHWR	17	12080
China	PWR	15	11471
Czech Republic	PWR	6	3472
Finland	PWR	2	890
	BWR	2	1420
France	PWR	54	57140
	HTGR	5	7000
Germany	PWR	14	15822
	BWR	7	6989
Hungary	PWR	4	1755
India	BWR	2	300
	PHWR	8	1395
Japan	PWR	22	17298
	BWR	26	22050
Korea (South)	PWR	9	7541
	PHWR	10	10000
Lithuania	LGR	2	2760
Mexico	BWR	2	1308
Netherlands	PWR	1	452
	BWR	1	55
Pakistan	PHWR	1	125
Romania	PHWR	1	655
Russia	LGR	11	10175
	PWR	13	9064
	LMFBR	1	560
Slovenia	PWR	1	676
Slovokia	PWR	6	2472
South Africa	PWR	2	1840
Spain	BWR	2	1389
	PWR	7	5712
Sweden	BWR	9	7370
	PWR	3	2705
Switzerland	BWR	2	1385
	PWR	3	1665
Taiwan	BWR	4	3104
	PWR	2	1780
UK	GCR	20	3360
	AGR	14	8180
	PWR	1	1188

Table 1 (*Continued*)

Country	Reactor Type[a]	Number in Operation	Net MWe
Ukraine	LGR	2	1850
	PWR	12	10245
United States	BWR	37	32215
	PWR	72	67458

[a]PWR = pressurized water reactor; BWR = boiling water reactor; AGR = advanced gas-cooled reactor; GCR = gas-cooled reactor; HTGR = high-temperature gas-cooled reactor; LMFBR = liquid-metal fast-breeder reactor; LGR = light-water-cooled graphite-moderated reactor; HWLWR = heavy-water-moderated light-water-cooled reactor; PHWR = pressurized heavy-water-moderated-and-cooled reactor; GCHWR = gas-cooled heavy-water-moderated reactor.

4.2 Boiling-Water Reactors

As the name implies, steam is generated in the BWR by boiling, which takes place in the reactor core. Early concerns about nuclear and hydraulic instabilities led to a decision to operate military propulsion reactors under conditions such that the moderator–coolant in the core remains liquid. In the course of developing the BWR system for commercial use, solutions have been found for the instability problems.

Although some early BWRs used a design that separates the core coolant from the steam which flows to the turbine, all modern BWRs send steam generated in the core directly to the turbine. This arrangement eliminates the need for a separate steam generator. It does, however, provide direct communication between the reactor core and the steam turbine and condenser, which are located outside the containment. This leads to some problems not found in PWRs. For example, the turbine–condenser system must be designed to deal with radioactive nitrogen-16 generated by an (n, p) reaction of fast neutrons in the reactor core with oxygen-16 in the cooling water. Decay of the short-lived nitrogen-16 (half-life 7.1 sec) produces high-energy (6.13-MeV) highly penetrating gamma rays. As a result, the radiation level around an operating BWR turbine requires special precautions not needed for the PWR turbine. The direct pathway from core to turbine provided by the steam pipes also affords a possible avenue of escape and direct release outside of containment for fission products that might be released from the fuel in a core-damaging accident. Rapid-closing valves in the steam lines are provided to block this path in case of such an accident.

The selection of pressure and temperature for the steam entering the turbine that are not markedly different from those typical of PWRs leads to an operating pressure for the BWR pressure vessel that is typically less than half that for PWRs. (Typical operating pressure at vessel outlet is about 1050 psi with a corresponding steam temperature of about 551°F.)

Because it is necessary to provide for two-phase flow through the core, the core volume is larger than that of a PWR of the same power. The core power density is correspondingly smaller. Figure 3 is a cutaway of a BWR vessel and core arrangement. The in-vessel steam separator for removing moisture from the steam is located above the core assembly. Figure 4 is a BWR fuel assembly. The assembly is contained in a channel box, which directs the two-phase flow. Fuel pins and fuel pellets are not very different in either size or shape from those for PWRs, although the cladding thickness for the BWR pin is somewhat larger than that of PWRs.

4.3 High-Temperature Gas-Cooled Reactors

Experience with the high-temperature gas-cooled reactor (HTGR) in the United States is limited. A 40-MWe plant was operated from 1967 to 1974. A 330-MWe plant has been in

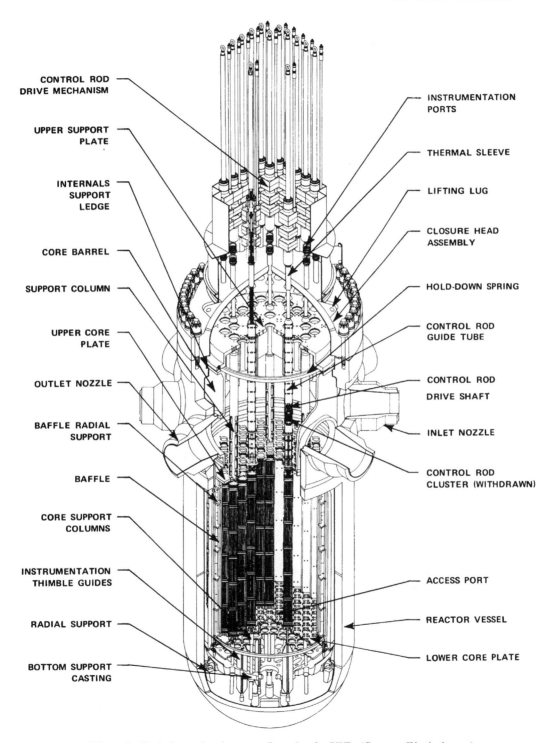

CONTROL ROD
DRIVE MECHANISM

UPPER SUPPORT
PLATE

INTERNALS
SUPPORT
LEDGE

CORE BARREL

SUPPORT COLUMN

UPPER CORE
PLATE

OUTLET NOZZLE

BAFFLE RADIAL
SUPPORT

BAFFLE

CORE SUPPORT
COLUMNS

INSTRUMENTATION
THIMBLE GUIDES

RADIAL SUPPORT

BOTTOM SUPPORT
CASTING

INSTRUMENTATION
PORTS

THERMAL SLEEVE

LIFTING LUG

CLOSURE HEAD
ASSEMBLY

HOLD-DOWN SPRING

CONTROL ROD
GUIDE TUBE

CONTROL ROD
DRIVE SHAFT

INLET NOZZLE

CONTROL ROD
CLUSTER (WITHDRAWN)

ACCESS PORT

REACTOR VESSEL

LOWER CORE PLATE

Figure 1 Typical vessel and core configuration for PWR. (Courtesy Westinghouse.)

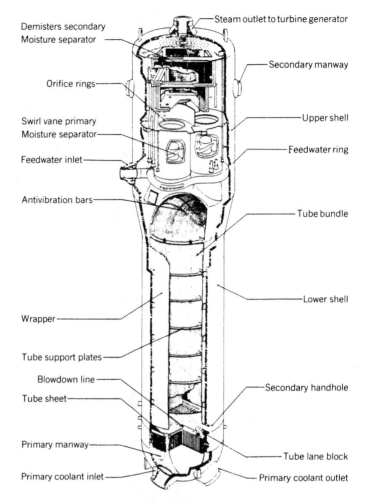

Demisters secondary Moisture separator

Orifice rings

Swirl vane primary Moisture separator

Feedwater inlet

Antivibration bars

Wrapper

Tube support plates

Blowdown line

Tube sheet

Primary manway

Primary coolant inlet

Steam outlet to turbine generator

Secondary manway

Upper shell

Feedwater ring

Tube bundle

Lower shell

Secondary handhole

Tube lane block

Primary coolant outlet

Figure 2   Typical PWR steam generator.

operation since 1976. A detailed design was developed for a 1000-MWe plant, but plans for its construction were abandoned.

Fuel elements for the plant in operation are hexagonal prisms of graphite about 31 in. tall and 5.5 in. across flats. Vertical holes in these blocks allow for passage of the helium coolant. Fuel elements for the larger proposed plant were similar. Figure 5 shows core and vessel arrangement. Typical helium-coolant outlet temperature for the reactor now in operation is about 1300°F. Typical steam temperature is 1000°F. The large plant was also designed to produce 1000°F steam.

The fuel cycle for the HTGR was originally designed to use fuel that combined highly enriched uranium with thorium. This cycle would convert thorium to uranium-233, which is also a fissile material, thereby extending fuel lifetime significantly. This mode of operation also produces uranium-233, which can be chemically separated from the spent fuel for further

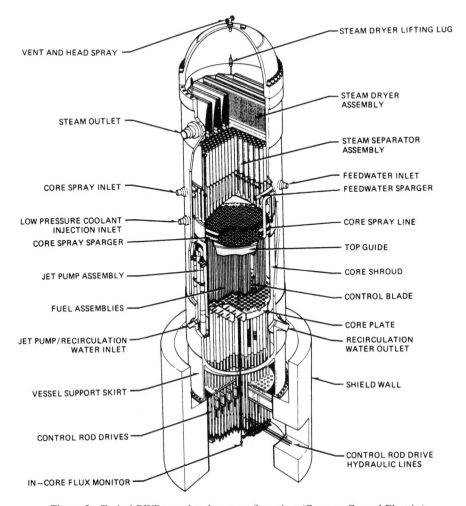

STEAM DRYER LIFTING LUG

VENT AND HEAD SPRAY

STEAM DRYER ASSEMBLY

STEAM OUTLET

STEAM SEPARATOR ASSEMBLY

FEEDWATER INLET
FEEDWATER SPARGER

CORE SPRAY INLET

CORE SPRAY LINE

LOW PRESSURE COOLANT INJECTION INLET

CORE SPRAY SPARGER

TOP GUIDE

JET PUMP ASSEMBLY

CORE SHROUD

FUEL ASSEMBLIES

CONTROL BLADE

CORE PLATE

JET PUMP/RECIRCULATION WATER INLET

RECIRCULATION WATER OUTLET

VESSEL SUPPORT SKIRT

SHIELD WALL

CONTROL ROD DRIVES

CONTROL ROD DRIVE HYDRAULIC LINES

IN−CORE FLUX MONITOR

Figure 3 Typical BWR vessel and core configuration. (Courtesy General Electric.)

use. Recent work has resulted in the development of a fuel using low-enriched uranium in a once-through cycle similar to that used in LWRs.

The use of graphite as a moderator and helium as coolant allows operation at temperatures significantly higher than those typical of LWRs, resulting in higher thermal efficiencies. The large thermal capacity of the graphite core and the large negative temperature coefficient of reactivity make the HTGR insensitive to inadvertent reactivity insertions and to loss-of-coolant accidents. Operating experience to date gives some indication that the HTGR has advantages in increased safety and in lower radiation exposure to operating personnel. These possible advantages plus the higher thermal efficiency that can be achieved make further development attractive. However, the high cost of developing a large commercial unit, plus the uncertainties that exist because of the limited operating experience with this type reactor have so far outweighed the perceived advantages.

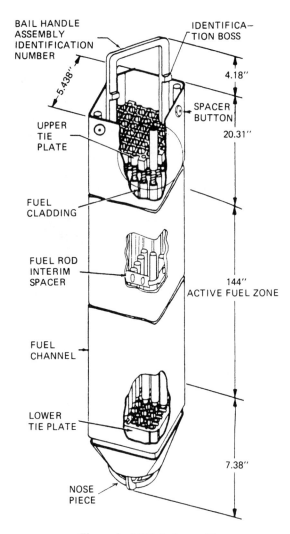

BAIL HANDLE
ASSEMBLY
IDENTIFICATION
NUMBER

IDENTIFICA-
TION BOSS

5.438″

4.18″

SPACER
BUTTON

20.31″

UPPER
TIE
PLATE

FUEL
CLADDING

FUEL ROD
INTERIM
SPACER

144″
ACTIVE FUEL ZONE

FUEL
CHANNEL

LOWER
TIE PLATE

7.38″

NOSE
PIECE

Figure 4 BWR fuel assembly.

As the data in Table 1 indicate, there is significant successful operating experience with several types of gas-cooled reactors in a number of European countries.

5 BASIC ENERGY PRODUCTION PROCESSES

Energy can be produced by nuclear reactions that involve either fission (the splitting of a nucleus) or fusion (the fusing of two light nuclei to produce a heavier one). If energy is to result from fission, the resultant nuclei must have a smaller mass per nucleon (which means they are more tightly bound) than the original nucleus. If the fusion process is to produce energy, the fused nucleus must have a smaller mass per nucleon (i.e., be more tightly bound)

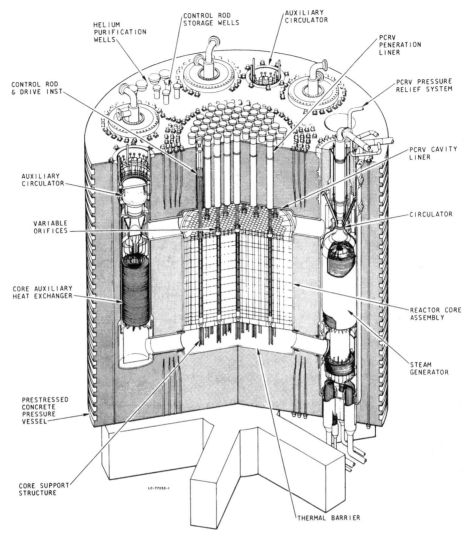

Figure 5 HTGR pressure vessel and core arrangement. (Used by permission of Marcel Dekker, Inc., New York.)

than the original nuclei. Figure 6 is a curve of nuclear binding energies. Observe that only the heavy nuclei are expected to produce energy on fission, and that only the light nuclei yield energy in fusion. The differences in mass per nucleon before and after fission or fusion are available as energy.

5.1 Fission

In the fission process this energy is available primarily as kinetic energy of the fission fragments. Gamma rays are also produced as well as a few free neutrons, carrying a small

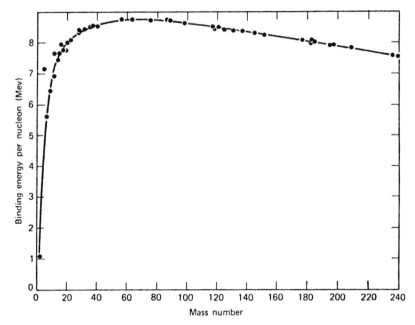

Figure 6 Binding energy per nucleon versus mass number.

amount of kinetic energy. The radioactive fission products decay (in most cases there is a succession of decays) to a stable nucleus. Gamma and beta rays are produced in the decay process. Most of the energy of these radiations is also recoverable as fission energy. Table 2 lists typical energy production due to fission of uranium by thermal neutrons, and indicates the form in which the energy appears. The quantity of energy available is of course related to the nuclear mass change by

Table 2 Emitted and Recoverable Energies from Fission of ^{235}U

Form	Emitted Energy (MeV)	Recoverable Energy (MeV)
Fission fragments	168	168
Fission product decay		
β rays	8	8
γ rays	7	7
Neutrinos	12	—
Prompt γ rays	7	7
Fission neutrons (kinetic energy)	5	5
Capture γ rays	—	3–12
Total	207	198–207

$$\Delta E = \Delta mc^2$$

Fission in reactors is produced by the absorption of a neutron in the nucleus of a fissionable atom. In order to produce significant quantities of power, fission must occur as part of a sustained chain reaction, that is, enough neutrons must be produced in the average fission event to cause at least one new fission event to occur when absorbed in fuel material. The number of nuclei that are available and that have the required characteristics to sustain a chain reaction is limited to uranium-235, plutonium-239, and uranium-233. Only uranium-235 occurs in nature in quantities sufficient to be useful. (And it occurs as only 0.71% of natural uranium.) The other two can be manufactured in reactors. The reactions are indicated below:

$$^{238}\text{U} + n \rightarrow {}^{239}\text{U} \rightarrow {}^{239}\text{Np} \rightarrow {}^{239}\text{Pu}$$

Uranium-239 has a half-life of 23.5 min. It decays to produce neptunium-239, which has a half-life of 2.35 days. The neptunium-239 decays to plutonium, which has a half-life of about 24,400 years.

$$^{232}\text{Th} + n \rightarrow {}^{233}\text{Th} \rightarrow {}^{233}\text{Pa} \rightarrow {}^{233}\text{U}$$

Thorium-233 has a half-life of 22.1 min. It decays to protactinium-233, which has a half-life of 27.4 days. The protactinium decays to produce uranium-233 with a half-life of about 160,000 years.

5.2 Fusion

Fusion requires that two colliding nuclei have enough kinetic energy to overcome the Coulomb repulsion of the positively charged nuclei. If the fusion rate is to be useful in a power-producing system, there must also be a significant probability that fusion-producing collisions occur. These conditions can be satisfied for several combinations of nuclei if a collection of atoms can be heated to a temperature typically in the neighborhood of hundreds of millions of degrees and held together for a time long enough for an appreciable number of fusions to occur. At the required temperature the atoms are completely ionized. This collection of hot, highly ionized particles is called a plasma. Since average collision rate can be related to the product of the density of nuclei, n, and the average containment time, τ, the $n\,\tau$ product for the contained plasma is an important parameter in describing the likelihood that a working system with these plasma characteristics will produce a useful quantity of energy.

Examination of the fusion probability, or the cross section for fusion, as a function of the temperature of the hot plasma shows that the fusion of deuterium (^{2}H) and tritium (^{3}H) is significant at temperatures lower than that for other candidates. Figure 7 shows fusion cross section as a function of plasma temperature (measured in electron volts) for several combinations of fusing nuclei. Table 3 lists several fusion reactions that might be used, together with the fusion products and the energy produced per fusion.

One of the problems with using the D–T reaction is the large quantity of fast neutrons that results, and the fact that a large fraction of the energy produced appears as kinetic energy of these neutrons. Some of the neutrons are absorbed in and activate the plasma-containment-system walls, making it highly radioactive. They also produce significant damage in most of the candidate materials for the containment walls. For these reasons there are some who advocate that work with the D–T reaction be abandoned in favor of the development of a system that depends on a set of reactions that is neutron-free.

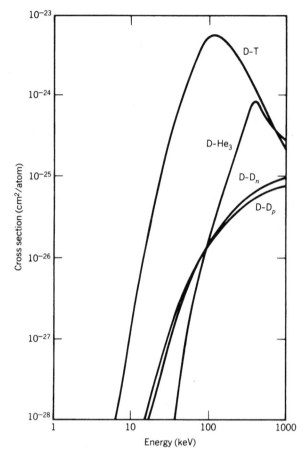

Figure 7 Fusion cross section versus plasma temperature.

Table 3 Fusion Reactions

$^3H + {}^2H \rightarrow n + {}^4He$	$+17.6$ MeV
$^3He + {}^2H \rightarrow p + {}^4He$	$+18.4$ MeV
$^2H + {}^2H \rightarrow p + {}^3H$	$+4.0$ MeV
$^2H + {}^2H \rightarrow n + {}^3He$	$+3.3$ MeV
$^6Li + p \rightarrow {}^3He + {}^4He$	$+4.0$ MeV
$^6Li + {}^3He \rightarrow {}^4He + p + {}^4He$	$+16.9$ MeV
$^6Li + {}^2H \rightarrow p + {}^7Li$	$+5.0$ MeV
$^6Li + {}^2H \rightarrow {}^3H + p + {}^4He$	$+2.6$ MeV
$^6Li + {}^2H \rightarrow {}^4He + {}^4He$	$+22.4$ MeV
$^6Li + {}^2H \rightarrow n + {}^7Be$	$+3.4$ MeV
$^6Li + {}^2H \rightarrow n + {}^3He + {}^4He$	$+1.8$ MeV

Another problem with using the D–T reaction is that tritium does not occur in nature in sufficient quantity to be used for fuel. It must be manufactured. Typical systems propose to produce tritium by the absorption in lithium of neutrons resulting from the fusion process. Natural lithium consists of ^{6}Li (7.5%) and ^{7}Li (92.5%). The reactions are

$$^6\text{Li} + n \rightarrow {}^4\text{He} + {}^3\text{H} + 4.8 \text{ MeV (thermal neutrons)}$$

and

$$^7\text{Li} + n \rightarrow {}^4\text{He} + {}^3\text{H} + n + 2.47 \text{ MeV (threshold reaction)}$$

Considerations of neutron economy dictate that most of the neutrons produced in the fusion process be absorbed in lithium in order to breed the needed quantities of tritium. The reactions shown produce not only tritium, but also additional energy. The (^{6}Li,n) reaction, for example, produces 4.7 MeV per reaction. If this energy can be recovered, it effectively increases the average available energy per fusion by about 27%.

6   CHARACTERISTICS OF THE RADIATION PRODUCED BY NUCLEAR SYSTEMS

An important by-product of the processes used to generate nuclear power is a variety of radiations in the form of either particles or electromagnetic photons. These radiations can produce damage in the materials that make up the systems and structures of the power reactors. High-energy neutrons, for example, absorbed in the vessel wall make the steel in the pressure vessel walls less ductile.

Radiation also causes damage to biological systems, including humans. Thus, most of the radiations must be contained within areas from which people are excluded. Since the ecosystem to which humans are normally exposed contains radiation as a usual constituent, it is assumed that some additional exposure can be permitted without producing undue risk. However, since the best scientific judgment concludes that there is likely to be some risk of increasing the incidence of cancer and of other undesirable consequences with any additional exposure, the amount of additional exposure permitted is small and is carefully controlled, and an effort is made to balance the permitted exposure against perceived benefits.

6.1   Types of Radiation

The principal types of radiation encountered in connection with the operation of fission and fusion systems are listed in Table 4. Characteristics of the radiation, including its charge and energy spectrum, are also given.

Table 4   Radiation Encountered in Nuclear Power Systems

Name	Description	Charge (in units of electron charge)	Energy Spectrum (MeV)
Alpha	Helium nucleus	+2	0 to about 5
Beta	Electron	+1, −1	0 to several
Gamma	Electromagnetic radiation	0	0 to about 10
Neutron		0	0 to about 20

Alpha particles are produced by radioactive decay of all of the fuels used in fission reactors. They are, however, absorbed by a few millimeters of any solid material and produce no damage in typical fuel material. They are also a product of some fusion systems and may produce damage to the first wall that provides a plasma boundary. They may produce damage to human lungs during the mining of uranium when radioactive radon gas may be inhaled and decay in the lungs. In case of a catastrophic fission reactor accident, severe enough to generate aerosols from melted fuels, the alpha-emitting materials in the fuel might, if released from containment, be ingested by those in the vicinity of the accident, thus entering both the lungs and the digestive system.

Beta particles are produced by radioactive decay of many of the radioactive substances produced during fission reactor operation. The major source is fission products. Although more penetrating than alphas, betas produced by fission products can typically be absorbed by at most a few centimeters of most solids. They are thus not likely to be harmful to humans except in case of accidental release and ingestion of significant quantities of radioactive material. A serious reactor accident might also release radioactive materials to a region in the plant containing organic materials such as electrical insulation. A sufficient exposure to high-energy betas can produce damage to these materials. Reactor systems needed for accident amelioration must be designed to withstand such beta irradiations.

Gamma rays are electromagnetic photons produced by radioactive decay or by the fission process. Photons identical in characteristics (but not in name) are produced by decelerating electrons or betas. When produced in this way, the electromagnetic radiation is usually called X rays. High-energy (above several hundred keV) gammas are quite penetrating, and protection of both equipment and people requires extensive (perhaps several meters of concrete) shielding to prevent penetration of significant quantities into the ecosystem or into reactor components or systems that may be subject to damage from gamma absorption.

Neutrons are particles having about the same mass as that of the hydrogen nucleus or proton, but with no charge. They are produced in large quantities by fission and by some fusion interactions including the D-T fusion referred to earlier. High-energy (several MeV) neutrons are highly penetrating. They can produce significant biological damage. Absorption of fast neutrons can induce a decrease in the ductility of steel structures such as the pressure vessel in fission reactors or the inner wall of fusion reactors. Fast-neutron absorption also produces swelling in certain steel alloys.

7 BIOLOGICAL EFFECTS OF RADIATION

Observations have indicated that the radiations previously discussed can cause biological damage to a variety of living organisms, including humans. The damage that can be done to human organisms includes death within minutes or weeks if the exposure is sufficiently large, and if it occurs during an interval of minutes or at most a few hours.

Radiation exposure has also been found to increase the probability that cancer will develop. It is considered prudent to assume that the increase in probability is directly proportional to exposure. However, there is evidence to suggest that at very low levels of exposure, say an exposure comparable to that produced by natural background, the linear hypothesis is not a good representation. Radiation exposure has also been found to induce mutations in a number of biological organisms. Studies of the survivors of the two nuclear weapons exploded in Japan have provided the largest body of data available for examining the question of whether harmful mutations are produced in humans by exposure of their forebears to radiation. Analyses of these data have led those responsible for the studies to conclude that the existence of an increase in harmful mutations has not been demonstrated unequivocally. However, current regulations of radiation exposure, in order to be conservative, assume that increased exposure will produce an increase in harmful mutations. There is also evidence to suggest that radiation exposure produces life shortening.

The Nuclear Regulatory Commission has the responsibility for regulating exposure due to radiation produced by reactors and by radioactive material produced by reactors. The standards used in the regulatory process are designed to restrict exposures to a level such that the added risk is not greater than that from other risks in the workplace or in the normal environment. In addition, effort is made to see that radiation exposure is maintained as "low as reasonably achievable."

8 THE CHAIN REACTION

Setting up and controlling a chain reaction is fundamental to achieving and controlling a significant energy release in a fission system. The chain reaction can be produced and controlled if a fission event, produced by the absorption of a neutron, produces more than one additional neutron. If the system is arranged such that one of these fission-produced neutrons produces, on the average, another fission, there exists a steady-state chain reaction. Competing with fission for the available neutrons are leakage out of the fuel region and absorptions that do not produce fission.

We observe that if only one of these fission-neutrons produces another fission, the average fission rate will be constant. If more than one produces fission, the average fission rate will increase at a rate that depends on the average number of new fissions produced for each preceding fission and the average time between fissions.

Suppose, for example, each fission produced two new fissions. One gram of uranium-235 contains 2.56×10^{21} nuclei. It would therefore require about 71 generations ($2^{71} \sim 2.4 \times 10^{21}$) to fission 1 g of uranium-235. Since fission of each nucleus produces about 200 MeV, this would result in an energy release of about 5.12×10^{23} MeV or 5.12×10^{10} J. The time interval during which this release takes place depends on the average generation time. Note, however, that in this hypothesized situation only about the last 10 generations contribute any significant fraction of the total energy. Thus, for example, if a generation could be made as short as 10^{-8} sec, the energy production rate could be nearly 5.12×10^{17} J/sec/g.

In power reactors the generation time is typically much larger than 10^{-8} sec by perhaps four or five orders of magnitude. Furthermore, the maximum number of new fissions produced per old fission is much less than two. Power reactors (in contrast to explosive devices) cannot achieve the rapid energy release hypothesized in the above example, for the very good reasons that the generation time and the multiplication inherent in these machines make it impossible.

8.1 Reactor Behavior

As indicated, it is neutron absorption in the nuclei of fissile material in the reactor core that produces fission. Furthermore, the fission process produces neutrons that can generate new fissions. This process sustains a chain reaction at a fixed level, if the relationship between neutrons produced by fission and neutrons absorbed in fission-producing material can be maintained at an appropriate level.

One can define neutron multiplication k as

$$k = \frac{\text{Neutrons produced in a generation}}{\text{Neutrons produced in the preceding generation}}$$

A reactor is said to be critical when k is 1. We examine the process in more detail by following neutron histories. The probability of interaction of neutrons with the nuclei of some designated material can be described in terms of a mean free path for interaction. The inverse, which is the interaction probability per unit path length, is also called macroscopic cross section. It has dimensions of inverse length.

We designate a cross section for absorption, Σ_a, a cross section for fission, Σ_f, and a cross section for scattering, Σ_s. If, then, we know the number of path lengths per unit time, per unit volume, traversed by neutrons in the reactor (for monoenergetic neutrons this will be $n\nu$, where n is neutron density and ν is neutron speed), usually called the neutron flux, we can calculate the various interaction rates associated with these cross sections and with a prescribed neutron flux, as a product of the flux and the cross section.

A diagrammatic representation of neutron history, with the various possibilities that are open to the neutrons produced in the fission process, is shown below:

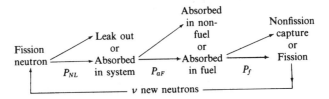

where P_{NL} = probability that neutron will not leak out of system before being absorbed

$\quad P_{aF}$ = probability that a neutron absorbed is absorbed in fuel

$\quad P_f$ = probability that a neutron absorbed in fuel produces a fission

In terms of the cross sections for absorption in fuel, Σ_a^F and for absorption, Σ_a

$$P_{aF} = f = \Sigma_a^F / \Sigma_a$$

where f is called the utilization factor. We can describe P_f as

$$P_f = \Sigma_f^F / \Sigma_a^F$$

Making use of the average number of neutrons produced per fission ν, we calculate the quantity η, the average number of neutrons produced per neutron absorbed in fuel, as

$$\eta = \nu \, \Sigma_f^F / \Sigma_a^F$$

With these definitions, and guided by the preceding diagram, we conclude that the number of offspring neutrons produced by a designated fission neutron can be calculated as

$$N = \eta f P_{NL}$$

We conclude that the multiplication factor k is thus equal to $N/1$ and write

$$k = \eta f P_{NL}$$

Alternatively, making use of the earlier definitions we write

$$k = (\nu \Sigma_f^F / \Sigma_a^F)(\Sigma_a^F / \Sigma_a)$$

and if we describe Σ_a as

$$\Sigma_a = \Sigma_a^F + \Sigma_a^{nF}$$

where Σ_a^{nF} is absorption in the nonfuel constituents of the core, we have

$$k = \nu\Sigma_f^F P_{NL}/(\Sigma_a^F + \Sigma_a^{nF})$$

Observe that from this discussion one can also define a neutron generation time l as

$$l = N(t)/L(t)$$

where $N(t)$ and $L(t)$ represent, respectively, the neutron population and the rate of neutron loss (through absorption and leakage) at a time t.

For large reactors, the size of those now in commercial power production, the nonleakage probability is high, typically about 97%. For many purposes it can be neglected. For example, small changes in multiplication, produced by small changes in concentration of fissile or nonfissile material in the core, can be assumed to have no significant effect on the nonleakage probability, P_{NL}. Under these circumstances, and assuming that appropriate cross-sectional averaging can be done, the following relationships can be shown to hold. If we rewrite an earlier equation for k as

$$k = \eta f \left(n_f \sigma_f \Big/ \sum_i n_i \sigma_i \right)$$

where η_f, η_i represent, respectively, the concentration of fissile and nonfissile materials, and

$$\eta_f \sigma_f = \Sigma_f$$
$$n_i \sigma_i = \Sigma_{a_i}$$

where the last equation is the macroscopic cross section of the ith nonfissile isotope.

Variation of k with the variation in concentration of the fissile material (i.e., n_f) is given by

$$\frac{\delta k}{k} = \frac{\delta n_f}{n_f} \left(\frac{\sum_i n_i \sigma_i}{\eta_f \sigma_f + \sum_i \eta_i \sigma_i} \right)$$

This says that the fractional change in multiplication is equal to the fractional change in concentration of the fissile isotope times the ratio of the neutrons absorbed in all of the nonfissile isotopes to the total neutrons absorbed in the core.

Variation of k with variation in concentration of the jth nonfissile isotope is given by

$$\frac{\delta k}{k} = -\frac{\delta m_j}{m_j} \left(\frac{n_j \sigma_j}{n_f \sigma_f + \sum_i n_i \sigma_i} \right), \qquad i \neq j$$

This says that the fractional increase in multiplication is equal to the fractional decrease in concentration of the jth nonfissile isotope times the ratio of neutrons absorbed in that isotope to total neutrons absorbed. Although these are approximate expressions, they provide useful guidance in estimating effects of small changes in the material in the core on neutron multiplication.

8.2 Time Behavior of Reactor Power Level

We assume that the fission rate, and hence the reactor power, is proportional to neutron population. We express the rate of change of neutron population $N(t)$ as

$$l\left(\frac{dN}{dt}\right) = (k - 1)N$$

that is, in one generation the change in neutron population should be just the excess over the previous generation, times the multiplication k. The preceding equation has as a solution

$$N = N_0 \exp\left[\left(\frac{k - 1}{l}\right)t\right]$$

where N_0 is the neutron population at time zero. One observes an exponential increase or decrease, depending on whether k is larger or smaller than unity. The associated time constant or e-folding time is

$$\tau = l/(k - 1)$$

For a $k - 1$ of 0.001 and l of 10^{-4} sec, the e-folding time is 0.10 sec. Thus, in 1 sec the power level increases by e^{10} or about 10^4.

8.3   Effect of Delayed Neutrons on Reactor Behavior

Dynamic behavior as rapid as that described by the previous equations would make a reactor almost impossible to control. Fortunately there is a mode of operation in which the time constant is significantly greater than that predicted by these oversimplified equations. A small fraction of neutrons produced by fission, typically about 0.7%, come from radioactive decay of fission products. Six such fission products are identified for uranium fission. The mean time for decay varies from about 0.3 to about 79 sec. For an approximate representation, it is reasonable to assume a weighted mean time to decay for the six of about 17 sec. Thus, about 99.3% of the neutrons (prompt neutrons) may have a generation time of, say, 10^{-4} sec, while 0.7% have an effective generation time of 17 sec plus that of the prompt neutrons. An effective mean lifetime can be estimated as

$$\bar{l} = (0.993)l + 0.007(l + \bar{\lambda}^{-1})$$

For an l of 10^{-4} sec and a $\bar{\lambda}^{-1}$ of 17 sec we calculate

$$l \simeq 0.993 \times 10^{-4} + 0.007(10^{-4} + 17) \simeq 0.12 \text{ sec}$$

This suggests, given a value for $k - 1$ of 10^{-3}, an e-folding time of 120 sec. Observe that with this model the delayed neutrons are a dominant factor in determining time behavior of the reactor power level. A more detailed examination of the situation reveals that for a reactor slightly subcritical on prompt neutrons alone, but supercritical when delayed neutrons are considered (such a reactor is said to be "delayed supercritical"), the delayed neutrons almost alone determine time behavior. If, however, the multiplication is increased to the point that the reactor is critical on prompt neutrons alone (i.e., "prompt critical"), the time behavior is determined by the prompt neutrons, and changes in power level may be too rapid to be controlled by any external control system. Reactors that are meant to be controlled are designed to be operated in a delayed critical mode. Fortunately, if the reactor should inadvertently be put in a prompt critical mode, there are inherent physical phenomena that decrease the multiplication to a controllable level when a power increase occurs.

9   POWER PRODUCTION BY REACTORS

Most of the nuclear-reactor-produced electric power in the United States, and in the rest of the world, comes from light-water-moderated reactors (LWRs). Nuclear power reactors produce heat that is converted, in a thermodynamic cycle, to electrical energy. The two types now in use, the pressurized water reactor (PWR) and the boiling water reactor (BWR), use fuel that is very similar, and produce steam having about the same temperature and pressure. In both types water serves both as a coolant and a moderator. We will examine some of the salient features of each system and identify some of the differences.

9.1 The Pressurized-Water Reactor

The arrangement of fuel in the reactor core and of the core in the pressure vessel is shown in Fig. 1. As indicated earlier, bulk boiling is avoided by operation at high pressures. Liquid water is circulated through the core by large electric-motor-driven pumps located outside the pressure vessel in the cold leg of the piping that connects the vessel to a steam generator. Current designs use from two to four separate loops, each containing a steam generator. Each loop contains at least one pump. One current design uses two pumps in the cold leg of each loop. A schematic of the arrangement is shown in Fig. 8. Reactor pressure vessel and primary coolant loops, including the steam generator, are located inside a large containment vessel. A typical containment structure is shown in Fig. 9.

The containment, typically a massive 3- to 4-ft-thick structure of reinforced concrete, with a steel inner-liner, has two principal functions: protection of pressure vessel and primary loop from external damage (e.g., tornadoes, aircraft crashes) and containment of fission products that might be released outside the primary pressure boundary in case of serious damage to the reactor in an accident.

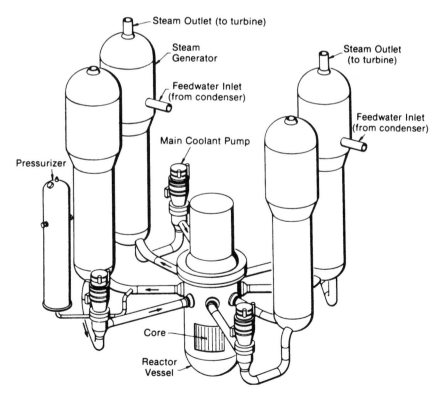

Figure 8 Typical arrangement of PWR primary system.

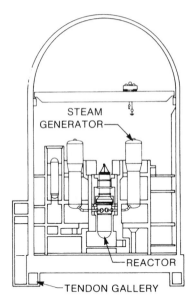

Figure 9 Typical large dry PWR containment.

The steam generator is markedly different from the boiler in a fossil-fueled plant. It is essentially a heat exchanger containing several thousand metal tubes that carry the hot water coming from the reactor vessel outlet. Water surrounding the outside of these tubes is converted to steam. The rest of the energy-conversion cycle is similar in principle to that found in a fossil-fueled plant.

Experience with reactor operation has indicated that very careful control of water chemistry is necessary to preclude erosion and corrosion of the steam generator tubes (SGT). An important contributor to SGT damage has been leakage in the main condenser, which introduces impurities into the secondary water system. A number of early PWRs have retubed or otherwise modified their original condensers to reduce contamination caused by in-leakage of condenser cooling water.

The performance of steam generator tubes is of crucial importance because these tubes are part of the primary pressure boundary. SGT rupture can initiate a loss of coolant accident. Also, leakage or rupture of SGTs usually leads to opening of the steam system safety valves because of the high primary system pressure. Since these valves are located outside containment, this accident sequence can provide an uncontrolled path for release of any radioactive material in the primary system directly to the atmosphere outside containment.

The reactor control system controls power level by a combination of solid control rods, containing neutron-absorbing materials that can be moved into and out of the core region, and by changing the concentration of a neutron-absorbing boron compound (typically boric acid) in the primary coolant. Control rod motion is typically used to achieve rapid changes in power. Slower changes, as well as compensation for burnup of uranium-235 in the core, are accomplished by boron-concentration changes. In the PWR the control rods are inserted from the top of the core. In operation enough of the absorber rods are held out of the core to produce rapid shutdown, or scram, when inserted. In an emergency, if rod drive power should be lost, the rods automatically drop into the core, driven by gravity.

The PWR is to some extent load following. Thus, for example, an increase in turbine steam flow caused by an increase in load produces a decrease in reactor coolant–moderator temperature. In the usual mode of operation a decrease in moderator temperature produces an increase in multiplication (the size of the effect depends on the boron concentration in the coolant-moderator), leading to an increase in power. The increase continues (accompanied by a corresponding increase in moderator-coolant temperature) until the resulting decrease in reactor multiplication leads to a return to criticality at an increased power level. Since the size of the effect changes significantly during the operating cycle (as fuel burnup increases the boron concentration is decreased), the inherent load-following characteristic must be supplemented by externally controlled changes in reactor multiplication.

A number of auxiliaries are associated with the primary. These include a water purification and makeup system, which also permits varying the boron concentration for control purposes, and an emergency cooling system to supply water for decay heat removal from the core in case of an accident that causes loss of the primary coolant.

Pressure in the primary is controlled by a pressurizer, which is a vertical cylindrical vessel connected to the hot leg of the primary system. In normal operation the bottom 60% or so of the pressurizer tank contains liquid water. The top 40% contains a steam bubble. System pressure can be decreased by water sprays located in the top of the tank. A pressure increase can be achieved by turning on electric heaters in the bottom of the tank.

9.2   The Boiling-Water Reactor

Fuel and core arrangement in the pressure vessel is shown in Fig. 3. Boiling in the core produces a two-phase mixture of steam and water, which flows out of the top of the core. Steam separators above the vessel water level remove moisture from the steam, which goes directly to the turbine outside of containment. Typically about one-seventh of the water flowing through the core is converted to steam during each pass. Feedwater to replace the water converted to steam is distributed around the inside near the top of the vessel from a spray ring. Water is driven through the core by jet pumps located in the annulus between the vessel wall and the cylindrical core barrel that surrounds the core and defines the upward flow path for coolant.

Because there is direct communication between the reactor core and the turbine, any radioactive material resulting, for example, from leakage of fission products out of damaged fuel pins, from neutron activation of materials carried along with the flow of water through the core, or from the nitrogen-16 referred to earlier, has direct access to turbine and condenser. Systems must be provided for removal from the coolant and for dealing with these materials as radioactive waste.

Unlike the PWR, the BWR is not load following. In fact, normal behavior in the reactor core produces an increase in the core void volume with an increase in steam flow to the turbine. This increased core voiding will increase neutron leakage, thereby decreasing reactor multiplication, and leading to a decrease in reactor power in case of increased demand. To counter this natural tendency of the reactor, a control system senses an increase in turbine steam flow and increases coolant flow through the core. The accompanying increase in core pressure decreases steam voiding, increasing multiplication and producing an increase in reactor power.

Pressure regulation in the BWR is achieved primarily by adjustment of turbine throttle setting to achieve constant pressure. An increase in load demand is sensed by the reactor control system and produces an increase in reactor power. Turbine valve position is adjusted to maintain constant steam pressure at the throttle. In rapid transients, which involve decreases in load demand, a bypass valve can be opened to send steam directly to the condenser, thus helping to maintain constant pressure.

BWRs in the United States make use of a pressure-suppression containment system. The hot water and steam released during a loss-of-coolant accident are forced to pass through a pool of water, condensing the steam. Pressure buildup is markedly less than if the two-phase mixture is released directly into containment. Figure 10 shows a Mark III containment structure of the type being used with the latest BWRs. Passing the fission products and the hot water and steam from the primary containment through water also results in significant removal of some of the fission products. The designers of this containment claim decontamination factors of 10,000 for some of the fission products that are usually considered important in producing radiation exposure following an accident.

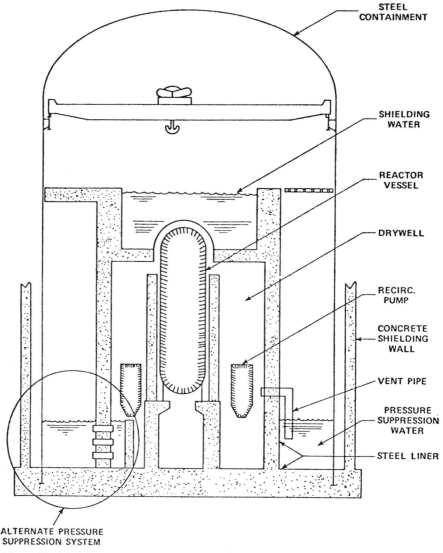

Figure 10   Mark III containment for BWR. (Used by permission of Pergamon Press, Inc., New York.)

As previously indicated, the control system handles normal load changes by adjusting coolant flow in the core. For rapid shutdown and for compensating for core burnup, the movable control rods are used. In a normal operating cycle several groups of control rods will be in the core at the beginning of core life, but will be completely out of the core at the end of the cycle. At any stage in core life some absorber rods are outside the core. These can be inserted to produce rapid shutdown.

Because of the steam-separator structure above the core, control rods in the BWR core must be inserted from the bottom. Insertion is thus not gravity assisted. Control rod drive is hydraulic. Compressed gas cylinders provide for emergency insertion if needed.

No neutron absorber is dissolved in the coolant, hence, control of absorber concentration is not required. However, cleanup of coolant containments, both solid and gaseous, is continuous. Maintaining a low oxygen concentration is especially important for the inhibition of stress-assisted corrosion cracking that has occurred in the primary system piping of a number of BWRs.

The current review process includes a detailed analysis of reactor system behavior under both normal and accident conditions. The existing approach involves the postulating of a set of design basis accidents (DBAs) and carrying out a deterministic analysis, which must demonstrate that the consequences of the hypothesized accidents are within a defined acceptable region. A number of the accident scenarios used for this purpose are of sufficiently low probability that they have not been observed in operating reactors. It is not practical to simulate the accidents using full-scale models or existing reactors. Analysis of reactor system behavior under the hypothesized situations must depend on analytical modeling. A number of large and complicated computer codes have been developed for this purpose.

Although the existing approach to licensing involves analysis of DBAs that can cause significant damage to the reactor power plant, none of the DBAs produces any calculable damage to personnel. Indeed, core damage severe enough to involve melting of the core is not included in any of the sequences that are considered. However, in the design of the plant allowance is made, on a nonmechanistic basis, for consequences beyond those calculated for the DBA. This part of the design is not based on the results of an analytical description of a specific serious accident, but rather on nonmechanistic assumptions meant to encompass a bounding event.

This method of analysis, developed over a period of about two decades, has been used in the licensing and in the regulation of the reactors now in operation. It is likely to be a principal component of the licensing and regulatory processes for at least the next decade. However, the accident at Three Mile Island in 1979 convinced most of those responsible for reactor analysis, reactor operation, and reactor licensing that a spectrum of accidents broader than that under the umbrella of the DBA should be considered.

In the early 1970s, under the auspices of the Atomic Energy Commission, an alternative approach to dealing with the analysis of severe accidents was developed. The result of an application of the method to two operating reactor power plants was published in 1975 in an AEC report designated as WASH-1400 or the Reactor Safety Study. This method postulates accident sequences that may lead to undesirable consequences such as melting of the reactor core, breach of the reactor containment, or exposure of members of the public to significant radiation doses. Since a properly designed and operated reactor will not experience these sequences unless multiple failures of equipment, serious operator error, or unexpected natural calamities occur, an effort is made to predict the probability of the required multiplicity of failures, errors, and calamities, and to calculate the consequences should such a sequence be experienced. The risk associated with the probability and the consequences can then be calculated.

A principal difficulty associated with this method is that the only consequences that are of serious concern in connection with significant risk to public health and safety are the result of very-low-probability accident sequences. Thus, data needed to establish probabilities are either sparse or nonexistent. Thus, application of the method must depend on some appropriate synthesis of related experience in other areas to predict the behavior of reactor systems. Because of the uncertainty introduced in this approach, the results must be interpreted with great care. The method, usually referred to as probabilitistic risk analysis (PRA), is still in a developmental stage, but shows signs of some improvement. It appears likely that for some time to come PRA will continue to provide useful information, but will be used, along with other forms of information, only as one part of the decision process used to judge the safety of power reactors.

BIBLIOGRAPHY

Dolan, T., *Fusion Research, Vol III* (*Technology*), Pergamon Press, New York, 1980.

Duderstadt, J. J., and L. J. Hamilton, *Nuclear Reactor Analysis,* Wiley, New York, 1975.

El-Wakil, M. M., *Nuclear Heat Transport,* American Nuclear Society, La Grange Park, IL, 1978.

Foster, A. L., and R. L. Wright, *Basic Nuclear Engineering,* Allyn & Bacon, Boston, MA, 1973.

Graves, H. W., Jr., *Nuclear Fuel Management,* Wiley, New York, 1979.

Lamarsh, J. R., *Introduction to Nuclear Engineering,* 2nd ed., Addison-Wesley, Reading, MA, 1983.

Rahn, F. J., et al., *A Guide to Nuclear Power Technology,* Wiley, New York, 1984.

CHAPTER 24

GAS TURBINES

Harold E. Miller
GE Energy
Schenectady, New York

Todd S. Nemec
GE Energy
Schenectady, New York

1 INTRODUCTION

1.1 Basic Operating Principles

Gas turbines are heat engines based on the Brayton thermodynamic cycle, which is one of the four cycles that account for most heat engines in use. The other three cycles are the Otto, Diesel, and Rankine. The Otto and Diesel cycles are cyclic in regard to energy content, while the Brayton (gas turbine) and Rankine (steam turbine) cycles are steady-flow, continuous energy transfer cycles. The Rankine cycle involves condensing and boiling of the working fluid (steam) and utilizes a boiler to transfer heat to the working fluid. The working fluid in the Otto, Diesel, and Brayton cycles is generally air, or air plus combustion products; these cycles are usually internal combustion cycles wherein the fuel is burned in the working fluid. In summary, the Brayton cycle is differentiated from the Otto and Diesel cycles in that it is continuous, and from the Rankine in that it relies on internal combustion and does not involve a phase change in the working fluid.

In all cycles, the working fluid experiences induction, compression, heating, expansion, and exhaust. In a nonsteady cycle, these processes are performed in sequence in the same closed space—one formed by a piston and cylinder—and operate on the working fluid one mass at a time. In contrast, the working fluid flows without interruption through a steam turbine power plant or gas turbine engine, passing continuously from one single purpose device to the next.

Gas turbines are used to power aircraft and land vehicles, to drive generators (alternators) to produce electric power, and to drive other devices such as pumps and compressors. Gas turbines in production range in output from below 50 kW to over 200 MW. Design philosophies and engine configurations vary significantly across the industry. For example, aircraft engines are optimized for high power-to-weight ratios, while heavy-duty, industrial and utility gas turbines are heavier, since they are designed for low cost and long life in severe environments.

Figure 1 shows the arrangement of a simple gas turbine engine. The rotating compressor acts to raise the pressure of the working fluid and force it into the combustor. The turbine is rotated by fluid expanding from a high pressure at the combustor discharge to ambient air pressure at the turbine exhaust. The resulting mechanical work drives the mechanically connected compressor and output load device.

The nomenclature of the gas turbine is not standardized. In this chapter the following descriptions of terms apply:

- *Blading* refers to all rotating and stationary airfoils in the gas path.
- *Buckets* are turbine (expander) section rotating blades; the term is derived from steam turbine practice.
- *Nozzles* are turbine section stationary blades.
- *Combustors* are the combustion components in contact with the working fluid; major combustor components are *fuel nozzles* and *combustion liners*. The combustor configurations are *annular, can-annular,* and *silo.* Can-annular and silo type combustors have transition pieces, which conduct hot gas from the combustion liners to the first stage nozzles.
- A *compressor stage* consists of a row of rotor blades, all at one axial position in the gas turbine, and the stationary blade row downstream of it.
- A *turbine stage* consists of a set of nozzles occupying one axial location and the set of buckets immediately downstream.
- *Discs* and *wheels* are interchangeable terms; rotating blading is attached either to a monolithic rotor structure or to individual discs or wheels designed to support the blading against centrifugal force and the aerodynamic loads of the working fluid.

Gas turbine performance is established by three basic parameters: mass flow, pressure ratio, and firing temperature. Compressor, combustor, and turbine efficiency have significant, but secondary, effects on performance, as do inlet and exhaust systems, turbine gas path and rotor cooling and heat loss through turbine and combustor casings. In gas turbine catalogues and other descriptive literature, mass flow is usually quoted as compressor inlet flow, although turbine exit flow is sometimes quoted. Output is proportional to mass flow.

Pressure ratio is quoted as the compressor pressure ratio. Aircraft engine practice defines the ratio as the total pressure at the exit of the compressor blading *divided by* the total pressure at the inlet of the compressor blading. Industrial/utility turbine manufacturers generally refer to the static pressure in the plenum downstream of the compressor discharge diffuser (upstream of the combustor) *divided by* the total pressure downstream of the inlet filter and upstream of the inlet of the gas turbine. Similarly, there are various possibilities for defining turbine pressure ratio. All definitions yield values within one or two percent of one another. Pressure ratio is the primary determiner of simple cycle gas turbine efficiency. High pressure results in high simple cycle efficiency.

Firing temperature is defined differently by each manufacturer, and the differences are significant. Heavy-duty gas turbine manufacturers use three definitions:

(a)

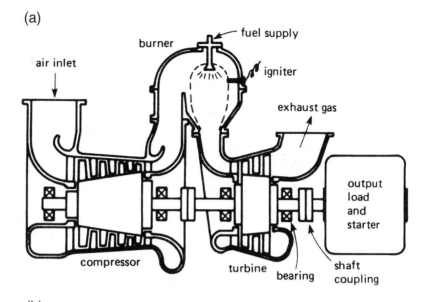

(b)

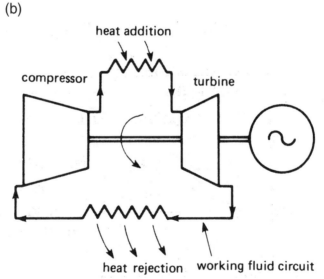

Figure 1 Simple engine types: (*a*) open cycle and (*b*) closed cycle.

1. There is an ISO definition of firing temperature, which is a calculated temperature. The compressor discharge temperature is increased by a calculated enthalpy rise based on the compressor inlet air flow and the fuel flow. This definition is valuable in that it can be used to compare gas turbines or to monitor changes in performance on through calculations made on the basis of field measurements. Knowledge of the secondary flows within the gas turbine is not required to determine ISO firing temperature.

2. A widely used definition of firing temperature is the average total temperature in the exit plane of the first stage nozzle. This definition is used by General Electric for its industrial engines.

3. Westinghouse (now part of Siemens-Westinghouse), and several other manufacturers refer to "turbine inlet temperature," the temperature of the gas entering the first stage nozzle. Turbine inlet temperature is approximately 100°C above nozzle exit firing temperature, which is in turn approximately 100°C above ISO firing temperature.

Since firing temperature is commonly used to compare the technology level of competing gas turbines, for comparison purposes it is important to use one definition of this parameter.

Aircraft engines and aircraft-derivative industrial gas turbines have other definitions. One nomenclature establishes numerical stations (in which station 3.9 is combustor exit, and station 4.0 is first-stage nozzle exit). Thus, $T_{3.9}$ is very close to turbine inlet temperature and $T_{4.0}$ is approximately equal to GE's firing temperature. There are some subtle differences that relate to the treatment of the leakage flows near the first stage nozzle.

Firing temperature is a primary determiner of power density (specific work) and combined cycle (Brayton-Rankine) efficiency. High firing temperature increases the power produced by a gas turbine of a given physical size and mass flow. The pursuit of higher firing temperatures by all manufacturers of the large, heavy-duty gas turbines used for electrical power generation is driven by the economics of high combined cycle efficiency.

Pressures and temperatures used in the following descriptions of gas turbine performance will be total pressures and temperatures. Absolute, stagnation, or total values are those measured by instruments that face into the approaching flow to give an indication of the energy in the fluid at any point. The work done in compression or expansion is proportional to the change of stagnation temperature in the working fluid, in the form of heating during a compression process or cooling during an expansion process. The temperature ratio, between the temperatures before and after the process, is related to the pressure ratio across the process by the expression $T_b/T_a = (P_b/P_a)^{(\gamma-1)/\gamma}$, where γ is the ratio of working fluid specific heats at constant pressure and volume. The temperature and pressure are stagnation values. It is the interaction between the temperature change and ratio, at different starting temperature levels, which permits the engine to generate a useful work output.

This relationship between temperature and pressure can be demonstrated by a simple numerical example using the Kelvin scale for temperature. For a starting temperature of 300 K (27°C), a temperature ratio of 1.5 yields a final temperature of 450 K and a change of 150°C. Starting instead at 400 K, the same ratio would yield a change of 200°C and a final temperature of 600 K. The equivalent pressure ratio would ideally be 4.13, as calculated from solving the preceding equation for P_b/P_a; $P_b/P_a = T_b/T_a^{\gamma/\gamma-1} = 1.5^{1.4/0.4} = 4.13$. These numbers show that, working over the same temperature ratio, the temperature change and, therefore, the work involved in the process vary in proportion to the starting temperature level.[1]

This conclusion can be depicted graphically. If the temperature changes are drawn as vertical lines *ab* and *cd*, and are separated horizontally to avoid overlap, the resultant is Fig. 2*a*. Assuming the starting and finishing pressures to be the same for the two processes, the thin lines through *ad* and *bc* depict two of a family of lines of constant pressure, which diverge as shown. In this ideal case, expansion processes could be represented by the same diagram, simply by proceeding down the lines *ba* and *cd*. Alternatively, if *ab* is taken as a compression process, *bc* as heat addition, *cd* as an expansion process, and *da* as a heat rejection process, then the figure *abcda* represents the ideal cycle to which the working fluid of the engine is subjected.

Over the small temperature range of this example, the assumption of constant gas properties is justified. In practice, the 327°C (600 K) level at point *d* is too low a temperature

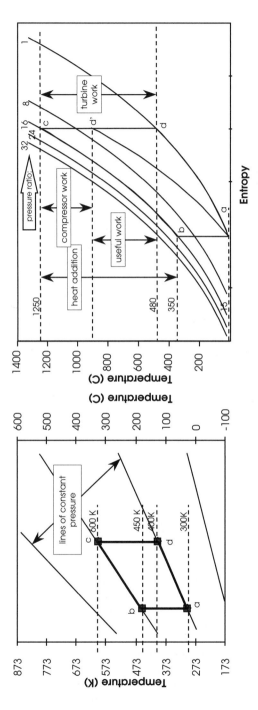

Figure 2 Temperature changes (*a*) and temperature–entropy diagram (*b*) for ideal simple gas turbine cycles.

from which to start the expansion. Fig. 2*b* is more realistic. Here, the lines of constant pressure have been constructed for ideal gas–air properties, which are dependent on temperature. Expansion begins from a temperature of 1250°C. With a pressure ratio of 16 : 1, the end point of the expansion is approximately 480°C. Now, *ab* represents the work input required by the compressor. Of the expansion work capacity *cd*, only the fraction *cd'* is required to drive the compressor. An optical illusion makes it appear otherwise, but line *ad'* is displaced vertically from line *bc* by the same distance everywhere. The remaining 435°C, line *d'd*, is energy that can be used to perform useful external work, by further expansion through the turbine or by blowing through a nozzle to provide jet thrust.

Now consider line *bc*. The length of its vertical projection is proportional to the heat added. The ability of the engine to generate a useful output arises from its use of the energy in the input fuel flow, but not all of the fuel energy can be recovered usefully. In this example, the heat input proportional to 1250 − 350 = 900°C compares with the excess output proportional to 435°C (line *d'd*) to represent an efficiency of (435/900), or 48%. If more fuel could be used, raising the maximum temperature level at the same pressure, then more useful work could be obtained at nearly the same efficiency.

The line *da* represents heat rejection. This could involve passing the exhaust gas through a cooler, before returning it to the compressor, and this would be a closed cycle. However, almost universally, *da* reflects discharge to the ambient conditions and intake of ambient air (Fig. 1*b*). Figure 1*a* shows an open-cycle engine, which takes air from the atmosphere and exhausts back to the atmosphere. In this case, line *da* still represents heat rejection, but the path from *d* to *a* involves the whole atmosphere and very little of the gas finds its way immediately from *e* to *a*. It is fundamental to this cycle that the remaining 465°C, the vertical projection of line *da*, is wasted heat because point *d* is at atmospheric pressure. The gas is therefore unable to expand further, so can do no more work.

Designers of simple cycle gas turbines—including aircraft engines—have pursued a course of reducing exhaust temperature through increasing cycle pressure ratio, which improves the overall efficiency. Figure 3 is identical to Fig. 2*b* except for the pressure ratio that has been increased from 16 : 1 to 24 : 1. The efficiency is calculated in the same manner. The total turbine work is proportional to the temperature difference across the turbine, 1250 − 410 = 840°C. The compressor work, proportional to 430 − 15 = 415°C, is subtracted from the turbine temperature drop 840 − 415 = 425°C. The heat added to the cycle is proportional to 1250 − 430 = 820°C. The ratio of the net work to the heat added is 425/820 = 52%. The approximately 8% improvement in efficiency is accompanied by a 70°C drop in exhaust temperature. When no use is made of the exhaust heat, the 8% efficiency may justify the mechanical complexity associated with higher pressure ratios. Where there is value to the exhaust heat, as there is in combined Brayton-Rankine cycle power plants, the lower pressure ratio may be superior. Manufacturers forecast their customer requirements and understand the costs associated with cycle changes and endeavor to produce gas turbines featuring the most economical thermodynamic designs.

The efficiency levels calculated in the preceding example are very high because many factors have been ignored for the sake of simplicity. Inefficiency of the compressor increases the compressor work demand while turbine inefficiency reduces turbine work output, thereby reducing the useful work output and efficiency. The effect of inefficiency is that, for a given temperature change, the compressor generates less than the ideal pressure level while the turbine expands to a higher temperature for the same pressure ratio. There are also pressure losses in the heat addition and heat rejection processes. There may be variations in the fluid mass flow rate and its specific heat (energy input divided by consequent temperature rise) around the cycle. These factors can easily combine to reduce the overall efficiency.

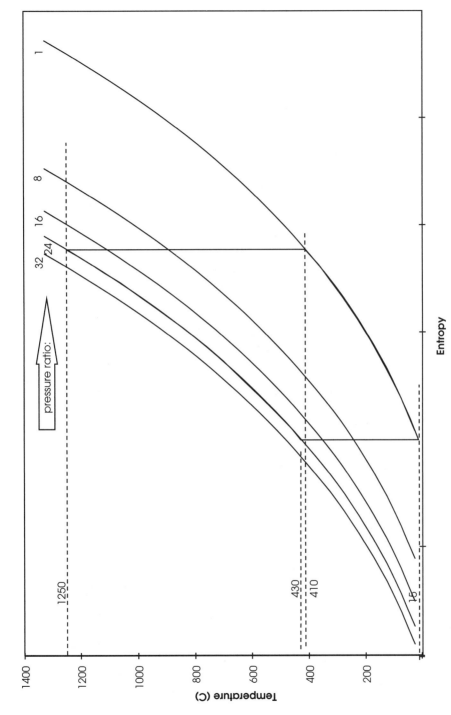

Figure 3 Simple cycle gas turbine temperature–entropy diagram for high pressure ratio (24 : 1) and 1250°C firing temperature.

1.2 A Brief History of Gas Turbine Development and Use

It was not until the year 1791 that John Barber patented the forerunner of the gas turbine, proposing the use of a reciprocating compressor, a combustion system, and an impulse turbine. Even then, he foresaw the need to cool the turbine blades, for which he proposed water injection.

The year 1808 saw the introduction of the first explosion type of gas turbine, which in later forms used valves at entry and exit from the combustion chamber to provide intermittent combustion in a closed space. The pressure thus generated blew the gas through a nozzle to drive an impulse turbine. These operated successfully but inefficiently for Karavodine and Holzwarth from 1906 onward, and the type died out after a Brown, Boveri model was designed in 1939.[1]

Developments of the continuous flow machine suffered from lack of knowledge, as different configurations were tried. In 1872 Stolze designed an engine with a seven-stage axial flow compressor, heat addition through a heat exchanger by external combustion, and a ten-stage reaction turbine. It was tested from 1900 to 1904 but did not work because of its very inefficient compressor. Parsons was equally unsuccessful in 1884, when he tried to run a reaction turbine in reverse as a compressor. These failures resulted from the lack of understanding of aerodynamics prior to the advent of aircraft. As a comparison, in typical modern practice, a single-stage turbine drives about six or seven stages of axial compressor with the same mass flow.

The first successful dynamic compressor was Rateau's centrifugal type in 1905. Three assemblies of these (with a total of 25 impellers in series giving an overall pressure ratio of 4) were made by Brown, Boveri and used in the first working gas turbine engine, which was built by Armengaud and Lemale in the same year. The exhaust gas heated a boiler behind the turbine to generate low-pressure steam, which was directed through turbines to cool the blades and augment the power. Low component efficiencies and flame temperature (828 K) resulted in low work output and an overall efficiency of 3%. By 1939, the use of industrial gas turbines had become well established and experience with the Velox boiler led Brown, Boveri into diverging applications; a Hungarian engine (Jendrassik) with axial flow compressor and turbine used regeneration to achieve an efficiency of 0.21; and the Sun Oil Co. (USA) was using a gas turbine engine to improve a chemical process.[1]

The history of gas turbine engines for aircraft propulsion dates from 1930, when Frank Whittle saw that its exhaust gas conditions ideally matched the requirements for jet propulsion and took out a patent.[1] His first model was built by British Thomson-Houston and ran as the Power Jets Type U in 1937, with a double-sided centrifugal compressor, a long combustion chamber that was curled round the outside of the turbine, and an exhaust nozzle just behind the turbine. Problems of low compressor and turbine efficiency were matched by hardware problems and the struggle to control the combustion in a very small space. In 1938 reverse flow, can-annular combustors were introduced with the aim still being to keep the compressor and turbine as close together as possible to avoid shaft whirl problems (Fig. 4). Whittle's first flying engine was the W1, with 850-lb thrust, in 1941. It was made by Rover, whose gas turbine establishment was taken over by Rolls-Royce in 1943. A General Electric version of the W1 flew in 1941. A parallel effort at General Electric led to the development of a successful axial-flow compressor. This was incorporated in the first turboprop engine, the TG100, later designated the T31. This engine was first tested in May 1943 and produced 1200 hp from an engine weighing under 400 kg. Flight testing followed in 1949. An axial-compressor turbojet version was also constructed. Designated the J35, it flew in 1946. The compressor of this engine evolved to the compressor of the GE MS3002 industrial engine that was introduced in 1950 and is still commercially available.[2]

Figure 4 Photo of an early Whittle-type jet engine with portions cut away to show double-sided compressor and reverse-flow combustion chambers. (Photo by Mark Doane.)

A Heinkel experimental engine flew in Germany in 1939. Several jet engines were operational by the end of the Second World War, but the first commercial engine did not enter service until 1953—the Rolls-Royce Dart turboprop in the Viscount, followed by the turbojet de Havilland Ghost in the Comet of 1954. The subsequent growth of the use of jet engines has been visible to most of the world, and has forced the growth of design and manufacturing technology. [1] By 1970 a range of standard configurations for different tasks had become established, and some aircraft engines were established in industrial applications and in ships.

Gas turbines entered the surface transportation fields also during their early stages of development. The first railway locomotive application was in Switzerland in 1941, with a 2200-hp, Brown, Boveri engine driving an electric generator and electric motors driving the wheels. The engine efficiency approached 19%, using regeneration. The next decade saw several similar applications of gas turbines by some 43 different manufacturers. A successful application of gas turbines to transportation was the 4500 draw-bar horsepower engine based on the J35 compressor. Twenty-five locomotives so equipped were delivered to the Union Pacific railroad between 1952 and 1954. The most powerful locomotive gas turbine was the 8500-hp unit offered by General Electric to the Union Pacific railroad for long distance freight service.[3] This became the basis of the MS5001 gas turbine, which is the most common heavy-duty gas turbine in use today. Railroad applications continue today, but they rely on a significantly different system. Japan Railway operates large, stationary gas turbines to generate power transmitted by overhead lines to their locomotives.

Automobile and road vehicle use started with a Rover car of 1950, followed by Chrysler and other companies, but commercial use has been limited to trucks, particularly by Ford. Automotive gas turbine development has been largely independent of other types and has forced the pace of development of regenerators. Of course, no history would be complete without mention of the Pratt & Whitney-engined race car campaigned by Andy Granatelli and driven by Parnelli Jones at the 1967 Indianapolis 500, which led most of the race before being sidelined by a bearing failure.

1.3   Subsystem Characteristics and Capabilities

The three subsystems that comprise the gas turbine proper are the compressor, combustor, and turbine. Technologies developed for these subsystems enable the operation and contribute to the value of the gas turbine.

Compressors

Compressors used in gas turbines are of the dynamic type, wherein air is continuously ingested and raised to the required pressure level—usually, but not necessarily, between 8 and 40 atm. Larger gas turbines use axial types; smaller ones use radial outflow, centrifugal compressors. Some smaller gas turbines use both—an axial flow compressor upstream of a centrifugal stage.

Axial compressors feature an annular flow path, larger in cross-section area at the inlet than at the discharge. Multiple stages of blades alternately accelerate the flow of air and allow it to expand, recovering the dynamic component and increasing pressure. Both rotating and stationary stages consist of cascades of airfoils as can be seen in Fig. 5. Physical characteristics of the compressor determine many aspects of the gas turbine's performance. Inlet annulus area establishes the mass flow of the gas turbine. Rotor speed and mean blade diameter are interrelated since optimum blade velocities exist. A wide range of pressure ratios can be provided, but today's machines feature compressions from below 8:1 to as high as 40:1. The higher pressure ratios are achieved using two compressors operating in series at different rotational speeds. The number of stages required is partially dependent on the pressure ratio required, but also on the sophistication of the blade aerodynamic design that is applied. Generally, the length of the compressor is a function of mass flow and pressure ratio, regardless of the number of stages. Older designs have stage pressure ratios of 1.15:1. Low aspect ratio blading designed with three-dimensional analytical techniques has stage pressure ratios of 1.3:1. There is a trend toward fewer stages of blades of more complicated configuration. Modern manufacturing techniques make more complicated forms more practical to produce, and minimizing parts count usually reduces cost.

Centrifugal compressors are usually chosen for machines of below 2 or 3 MW in output, where their inherent simplicity and ruggedness can largely offset their lower compression

a   b   c

Figure 5   Diagram and photos of centrifugal compressor rotor and axial compressor during assembly. (Courtesy of General Electric Company.)

efficiency. Such compressors feature a monolithic rotor with shaped passages leading from the inlet circle or annulus to a volute at the outer radius where the compressed air is collected and directed to the combustor. The stator contains no blades and only interconnecting passages, and simply provides a boundary to the flow path (three sides of which are machined or cast into the rotor). Two or more rotors can be used in series to achieve the desired pressure ratio within the mechanical factors that limit rotor diameter at a given rotational speed.[4]

Two efficiency definitions are used to describe compressor performance. Polytropic efficiency characterizes the aerodynamic efficiency of low pressure-ratio individual stages of the compressor. Isentropic, or adiabatic, efficiency describes the efficiency of the cycle's first thermodynamic process shown in Fig. 6 (the path from *a* to *b*). From the temperatures shown for the compression process on this figure, the isentropic efficiency can be calculated. The isentropic temperature rise is for the line *ab*: 335°C. The actual rise is shown by line *ab'*, and this rise is 372°C. The compressor efficiency, η_c, is 90%.

Successful compressor designs achieve high component efficiency while avoiding compressor surge or stall—the same phenomenon experienced when airplane wings are forced to operate at too high an angle of attack at too low a velocity. Furthermore, blade and rotor structures must be designed to avoid vibration problems. These problems occur when natural frequencies of components and assemblies are coincident with mechanical and aerodynamic stimuli, such as those encountered as blades pass through wakes of upstream blades. The stall phenomenon may occur locally in the compressor or even generally, whereupon normal flow through the machine is disrupted. A compressor must have good stall characteristics to operate at all ambient pressures and temperatures, and to operate though the start, acceler-

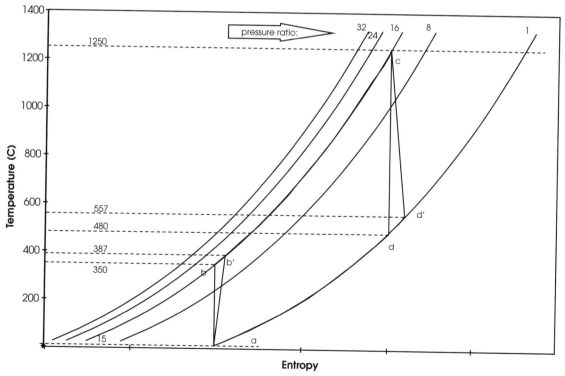

Figure 6   Temperature–entropy diagram showing the effect of compressor and turbine efficiency.

ation, load, load-change, unload, and shutdown phases of turbine operation. Compressors are designed with features and mechanisms for avoiding stall. These include air bleed at various points, variable-angle stator (as opposed to rotor) blades, and multiple spools.

Recent developments in the field of computational fluid dynamics (CFD) provide analytical tools that allow designers to substantially reduce aerodynamic losses due to shock waves in the supersonic flow regions. Using this technique, stages that have high tip Mach numbers can attain efficiencies comparable to those of completely subsonic designs. With these tools, compressors can be designed with higher tip diameters, hence higher flows. The same tools permit the design of low aspect ratio, high stage pressure ratio blades for reducing the number of blade rows. Both capabilities contribute to lower cost gas turbine designs with no sacrifice in performance.

Gas Turbine Combustors

The gas turbine combustor is a device for mixing large quantities of fuel and air and burning the resulting mixture. A flame burns hottest when there is just enough fuel to react with the available oxygen (which is called a stoichiometric condition). Here combustion produces the fastest chemical reaction and the highest flame temperatures, compared with excess air (fuel lean) and excess fuel (fuel rich) conditions, where reaction rates and temperatures are lower. The term "equivalence ratio" is used to describe the ratio of fuel to air relative to the stoichiometric condition. An equivalence ratio of 1.0 corresponds to the stoichiometric condition. At fuel-lean conditions, the ratio is less than 1; and, conversely, when fuel rich, it is greater than 1. The European practice is to use the reciprocal, which is the lambda value (λ). In a gas turbine—since air is extracted from the compressor for cooling the combustor, buckets, nozzles, and other components and to dilute the flame (as well as support combustion)—the overall equivalence ratio is far less than the value in the flame zone; ranging from 0.4 to 0.5 ($\lambda = 2.5$ to 2).[5]

Historically, the design of combustors required providing for the near-stoichiometric mixture of fuel and air locally. The combustion in this near-stoichiometric situation results in a diffusion flame of high temperature. Near-stoichiometric conditions produce a stable combustion front without requiring designers to provide significant flame-stabilizing features. Since the temperatures generated by the burning of a stoichiometric mixture greatly exceed those at which materials are structurally sound, combustors have to be cooled, and also the gas heated by the diffusion flame must be cooled by dilution before it becomes the working fluid of the turbine. Emissions implications are discussed below.

Gas turbine operation involves a start-up cycle that features ignition of fuel at 20% of rated operating speed, where air flow is proportionally lower. Loading, unloading, and part-load operation, however, require low fuel flow at full compressor speed, which means full air flow. Thermodynamic cycles are such that the lowest fuel flow per unit mass flow of air through the turbine exists at full speed and no-load. The fuel flow here is about 1/6 of the full-load fuel flow. Hence, the combustion system must be designed to operate over a 6:1 range of fuel flows with full rated air flow.

Manufacturers have differed on gas turbine combustor construction in significant ways. Three basic configurations have been used: annular, can-annular, and silo combustors. All have been used successfully in machines with firing temperatures up to 1100°C. Annular and can-annular combustors feature a combustion zone uniformly arranged about the centerline of the engine. All aircraft engines and most industrial gas turbines feature this type of design. A significant number of units equipped with silo combustor have been built as well. Here, one or two large combustion vessels are constructed on top or beside the gas turbine. All manufacturers of large machines have now abandoned silo combustors in their state-of-the-art large gas turbine products. The can-annular, multiple combustion chamber assembly consists of an arrangement of cylindrical combustors, each with a fuel-injection system, and a transition piece that provides a flow path for the hot gas from the combustor to the inlet of the turbine. Annular combustors have fuel nozzles at their upstream end and an inner and outer liner surface extending from the fuel nozzles to the entrance of the first stage stationary blading. No transition piece is needed.

The current challenge to combustion designers is to provide the cycle with a sufficiently high firing temperature while simultaneously limiting the production of oxides of nitrogen, NO_x, which refers to NO and NO_2. Very low levels of NO_x have been achieved in special low-emission combustors. NO_x is formed from the nitrogen and oxygen in the air when it is heated. The nitrogen and oxygen combine at a significant rate at temperatures above 1500°C, and the formation rate increases exponentially as temperature increases. Even with the high gas velocities in gas turbines, NO_x emissions will reach 200 parts per million by volume, dry (ppmvd), in gas turbines with conventional combustors and no NO_x abatement features. Emissions standards throughout the world vary, but many parts of the world require gas turbines to be equipped to control NO_x to below 25 ppmvd at base load.

Emissions

Combustion of common fuels necessarily results in the emission of water vapor and carbon dioxide. Combustion of near-stoichiometric mixtures results in very high temperatures. Oxides of nitrogen are formed as the oxygen and nitrogen in the air combine, and this happens at gas turbine combustion temperatures. Carbon monoxide forms when the combustion process is incomplete. Unburned hydrocarbons (UHC) are discharged as well when combustion is incomplete. Other pollutants are attributed to fuel; principal among these is sulfur. Gas turbines neither add nor remove sulfur; hence, what sulfur enters the gas turbine in the fuel exits as SO_2 in the exhaust.

Carbon dioxide emissions are no longer considered totally benign due to global concern for the "greenhouse effect" and global warming. Carbon in fuel is converted to CO_2 in the combustion process, so for any given fuel, increasing cycle efficiency is the only means of reducing exhaust of the CO_2 associated with each MW generated. Figure 7 gives the relationship between efficiency and exhaust CO_2 for a gaseous and a solid fuel. Several studies have been made for further reducing the release of CO_2 to the environment, and these involve either removing the carbon from the fuel or removing the CO_2 from the exhaust. In either case, the scheme requires an enabling technology for sequestering the carbon or carbon dioxide. Some schemes involve gas turbine design considerations, as they require some integration with air separation processes or working fluids with high water content, hence high thermal transport properties.

Much of the gas turbine combustion research and development since the 1980s focused on lowering NO_x production in mechanically reliable combustors, while maintaining low CO and UHC emissions. Early methods of reducing NO_x emissions included removing it from the exhaust by selective catalytic reduction (SCR), and by diluent injection, which is the injection of water or steam into the combustor. These methods continue to be employed. The lean-premix combustors now in general use are products of ongoing research.

Thermal NO_x is generally regarded as being generated by a chemical reaction sequence called the Zeldovich mechanism,[6] and the rate of NO_x formation is proportional to temperature as shown in Fig. 8. In practical terms, a conventional gas turbine emits approximately 200 ppmvd when its combustors are not designed to control NO_x. This is because a significant portion of the combustion zone has stoichiometric or near-stoichiometric conditions, and temperatures are high. Additional oxygen and, of course, nitrogen on the boundary of the flame is heated to sufficiently high temperatures and held at these temperatures for sufficient time to produce NO_x.

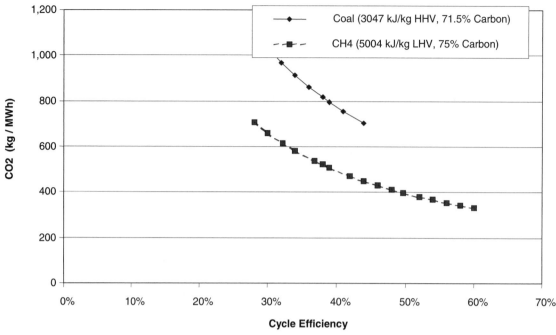

Figure 7   Coal versus natural gas: kg CO_2 per MWh. (Coal and CH_4 quoted at different heating values; published efficiencies are consistent with each definition.)

Water- and steam-injected combustors achieve low flame temperatures by placing diluent in the neighborhood of the reacting fuel and air. Among low-NO_x combustion systems operating today, water and steam injection is the most common means of flame temperature reduction. Several hundred large industrial turbines operating with steam or water injection have accumulated over 2.5 million hours of service. Water is not the only diluent used for NO_x control. In the case of integrated gasification combined cycle (IGCC) plants, nitrogen and CO_2 are available, which can be introduced into the combustion region.[7]

Water or steam injection can achieve levels that satisfy many current standards, but water consumption is sometimes not acceptable to the operator because of cost, availability, or the impact on efficiency. Steam injection sufficient to reduce NO_x emissions to 25 ppmvd can increase fuel consumption in combined cycle power plants by over 3%. Water injection increases fuel use by over 4% for the same emissions level. In base-load power plants, fuel cost is so significant that it has prompted the development of systems that do not require water.[8]

In all combustion processes, when a molecule of methane combines with two molecules of oxygen, a known and fixed amount of heat is released. When only these three molecules are present, a minimum amount of mass is present to absorb the energy not radiated and the maximum temperature is realized. Add to the neighborhood of the reaction the nitrogen as found in air (four times the volume of oxygen involved in the reaction) and the equilibrium temperature is lower. By adding even more air to the combustion region, more mass is available to absorb the energy, and the resulting observable temperature is lower still. The same can be achieved through the use of excess fuel. Thus, by moving away from the stoichiometric mixture, observable flame temperature is lowered, and the production of NO_x

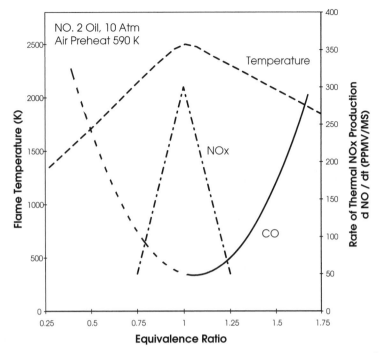

Figure 8 NO$_x$ formation rate driven by temperature. (Drawn from figure in Davis.[5]).

is also reduced. On a microscopic level, lean-burning, low-NO$_x$ combustors are designed to force the chemical reaction to take place in such a way that the energy released is in the neighborhood of as much mass not taking part in the reaction as possible. By transferring heat to neighboring material immediately, the time-at-temperature is reduced. On a larger scale, a high measurable temperature will never be reached in a well-mixed lean system. Thus, NO$_x$ generation is minimized. Both rich mixture and lean mixture systems have led to low NO$_x$ schemes. Those featuring rich flames followed by lean burning zones are sometimes suggested for situations where there is nitrogen in the fuel; most of today's systems are based on lean burning.

Early lean, premix, dry, low-NO$_x$ combustors were operated in GE gas turbines at the Houston Light and Power Wharton Station (USA) in 1980, in Mitsubishi units in Japan in 1983, and were introduced in Europe in 1986 by Siemens KWU. These combustors control the formation of NO$_x$ by premixing fuel with air prior to its ignition while conventional combustors mix essentially at the instant of ignition. Dry low-NO$_x$ combustors, as the name implies, achieve NO$_x$ control without consuming water, and without imposing efficiency penalties on combined-cycle plants.

Figures 9 and 10 show dry low-NO$_x$ combustors developed for large gas turbines. In the GE system, several premixing chambers are located at the head end of the combustor. A fuel nozzle assembly is located in the center of each chamber. By manipulating valves external to the gas turbine, fuel can be directed to several combinations of chambers and to various parts of the fuel nozzles. This is to permit the initial ignition of the fuel, and to maintain a relatively constant local fuel–air ratio at all load levels. There is one flame zone, immediately downstream of the premixing chambers. The Westinghouse combustor illus-

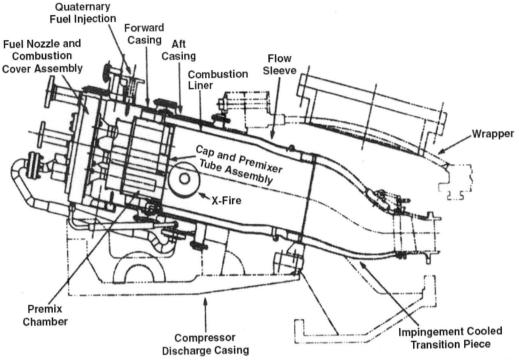

Figure 9   GE DLN-2 lean-premix combustor designed for low emissions at high firing temperatures. (Courtesy of General Electric Company.)

trated in Fig. 10 has three concentric premixing chambers. The two nearest the centerline of the combustor are designed to swirl the air passing through them in opposite directions and discharge into the primary combustion zone. The third, which has a longer passage, is directed to the secondary zone. Modulating fuel flow to the various mixing passages and combustion zones ensures low NO_x production over a wide range of operating temperatures. Both the combustors shown are designed for state-of-the-art, high firing temperature gas turbines.

Low-NO_x combustors feature multiple premixing features and a more complex control system than more conventional combustors, to achieve stable operation over the required range of operating conditions. The reason for this complexity can be explained with the aid of Fig. 11. Conventional combustors operate with stability over a wide range of fuel–air mixtures; between the rich and lean flammability limits. A sufficiently wide range of fuel flows could be burned in a combustor with a fixed air flow, to match the range of load requirements from no-load to full-load. In a low-NO_x combustor, the fuel–air mixture feeding the flame must be regulated between the point of flame loss and the point where the NO_x limit is exceeded. When low gas turbine output is required, the air premixed with the fuel must be reduced to match the fuel flow corresponding to the low power output. The two combustors shown previously hold nearly constant fuel–air ratios over the load range by having multiple premixing chambers, each one flowing a constant fraction of the compressor discharge flow. By directing fuel to only some of these passages at low load, the design achieves both part load and optimum local fuel–air ratio. Three, four, or more sets of fuel

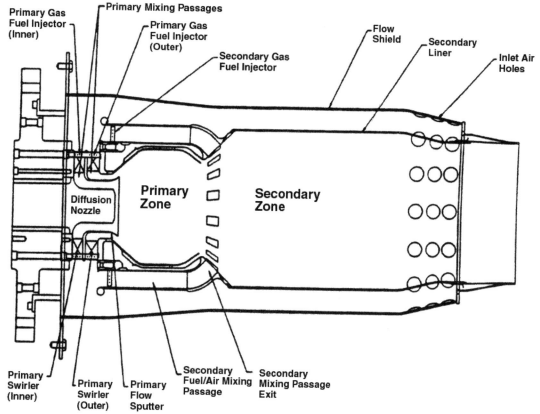

Figure 10 Westinghouse dry low-NO$_x$ combustor for advanced gas turbines. (Courtesy of Westinghouse Corporation.)

passages are not uncommon, and premixed combustion is maintained to approximately 50% of the rated load of the machine.[5,9]

Catalytic combustion systems are under investigation for gas turbines. These systems have demonstrated stable combustion at lower fuel–air ratios than those using chamber, or nozzle, shapes to stabilize flames. They offer the promise of simpler fuel regulation and greater turn-down capability than low-NO$_x$ combustors now in use. In catalytic combustors, the fuel and air react in the presence of a catalytic material, which is deposited on a structure having multiple parallel passages or mesh. Extremely low NO$_x$ levels have been observed in laboratories with catalytic combustion systems.

Turbine

Figure 12 shows an axial flow turbine. Radial in-flow turbines similar in appearance to centrifugal compressors are also produced for some smaller gas turbines. By the time the extremely hot gas leaves the combustor and enters the turbine, it has been mixed with compressor discharge air to cool it to temperatures that can be tolerated by the first-stage blading in the turbine: temperatures ranging from 950°C in first-generation gas turbines to over 1500°C in turbines currently being developed and in state-of-the-art aircraft engines. Less dilution flow is required as firing temperatures approach 1500°C.

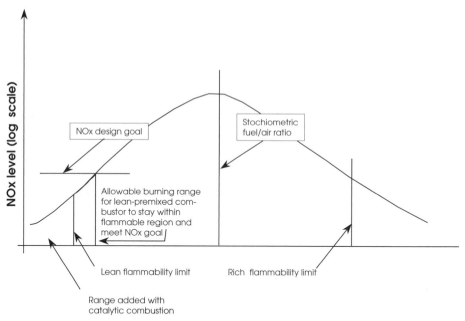

Figure 11 Fuel–air mixture ranges for conventional and premixed combustors. (Courtesy of Westinghouse Corporation.)

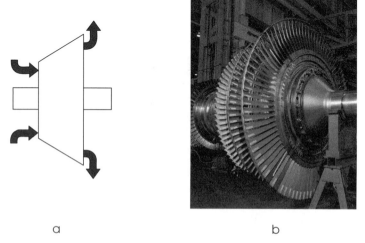

a   b

Figure 12 Turbine diagram and photo of an axial flow turbine during assembly. (Courtesy of General Electric Company.)

The first-stage stationary blades, or nozzles, are located at the discharge of the combustor. Their function is to accelerate the hot working fluid and turn it so as to enter the following rotor stage at the proper angle. These first-stage nozzles are subjected to the highest gas velocity in the engine. The gas entering the first-stage nozzle can regularly be above the melting temperature of the structural metal. These conditions produce high heat transfer to the nozzles, so that cooling is necessary.

Nozzles are subjected to stresses imposed by aerodynamic flow of the working fluid, pressure loading of the cooling air, and thermal stresses caused by uneven temperatures over the nozzle structure (Fig. 13). First-stage nozzles can be supported at both ends, by the inner and outer sidewalls. But later-stage nozzles, because of their location in the engine, can be supported only at the outer end intensifying the effect of aerodynamic loading.

The rotating blades of the turbine, or buckets, convert the kinetic energy of the hot gas exiting the nozzles to shaft power used to drive the compressor and load devices (Fig. 14). The blade consists of an airfoil section in the gas path, a dovetail or other type of joint connecting the blade to the turbine disk, and often a shank between the airfoil and dovetail allowing the dovetail to run at lower temperature than the root of the airfoil. Some bucket designs employ tip shrouds to limit deflection at the outer ends of the buckets, raise natural vibratory frequencies, and provide aerodynamic benefits. Exceptions from this configuration are radial inflow turbines like those common to automotive turbochargers and axial turbines, wherein the buckets and wheels are made of one piece of metal or ceramic.

The total temperature of the gas relative to the bucket is lower than that relative to the preceding nozzles. This is because the tangential velocity of the rotor-mounted airfoil is in a direction away from relative to the gas stream and thus reduces the dynamic component of total temperature. Also, the gas temperature is reduced by the cooling air provided to the upstream nozzle and the various upstream leakages.

Buckets and the disks on which they are mounted are subject to centrifugal stresses. The centrifugal force acting on a unit mass at the blades' midspan is 10,000–100,000 times that of gravity. Midspan airfoil centrifugal stresses range from 7 kg/mm^2 (10,000 psi) to over 28 kg/mm^2 (40,000 psi) at the airfoil root in the last stage (longest buckets).

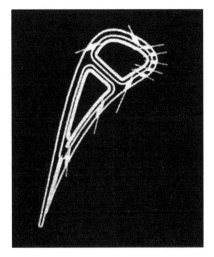

Figure 13 Gas turbine nozzle. Sketch shows cooling system of one airfoil. (Courtesy of General Electric Company.)

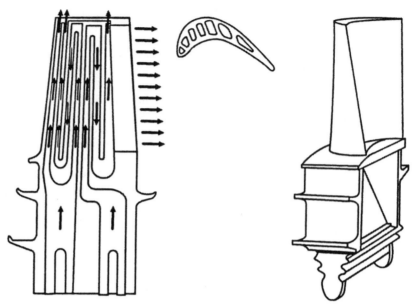

Figure 14   Gas turbine first-stage air-cooled bucket. Cutaway view exposes serpentine cooling passages. (Courtesy of General Electric Company.)

Turbine efficiency is calculated similarly to compressor efficiency. Figure 6 also shows the effect of turbine efficiency. Line *cd* represents the isentropic expansion process and *cd'* the actual. Turbine efficiency, η_t, is the ratio of the vertical projections of the lines. Thus, $(1250 - 557)/(1250 - 480) = 90\%$. It is possible at this point to compute the effect of a 90% efficient compressor and a 90% efficient turbine upon the simple-cycle efficiency of the gas turbine represented in the figure. The turbine work is proportional to 693°C, and the compressor work to 372°C. The heat added by combustion is proportional to 887°C, the temperature rise from *b'* to *c*. The ratio of the useful work to the heat addition is thus 36.2%. It was shown previously that the efficiency with ideal components is approximately 48.3%.

The needs of gas turbine blading have been responsible for the rapid development of a special class of alloys. To tolerate higher metal temperatures without decrease in component life, materials scientists and engineers have developed, and continue to advance, families of temperature-resistant alloys, processes and coatings. The "superalloys" were invented and continue to be developed primarily in response to turbine needs. These are usually based on group VIIIA elements: cobalt, iron, and nickel. Bucket alloys are austenitic with gamma/ gamma-prime, face-centered cubic structure (Ni_3 Al). The elements titanium and columbium are present and partially take the place of aluminum with beneficial hot corrosion effect. Carbides are present for grain boundary strength, along with some chromium to further enhance corrosion resistance. The turbine industry has also developed processes to produce single-crystal and directionally solidified components, which have even better high-temperature performance. Coatings are now in universal use, which enhance the corrosion and erosion performance of hot gas path components.[10]

Cooling

Metal temperature control is addressed primarily through airfoil cooling, with cooling air being extracted from the gas turbine flow ahead of the combustor. Since this air is not heated by the combustion process, and may even bypass some turbine stages, the cycle is less efficient than it would be without cooling. Further, as coolant reenters the gas path, it produces quenching and mixing losses. Hence, for efficiency, the use of cooling air should be minimized. Turbine designers must make trade-offs among cycle efficiency (firing temperature), parts lives (metal temperature), and component efficiency (cooling flow).

In early first-generation gas turbines, buckets were solid metal, operating at the temperature of the combustion gases. In second-generation machines, cooling air was conducted through simple, radial passages, to keep metal temperatures below those of the surrounding gas. In today's advanced-technology gas turbines, most manufacturers utilize serpentine air passages within the first-stage buckets, with cooling air flowing out the tip, leading, and trailing edges. Leading edge flow is used to provide a cooling film over the outer bucket surface. Nozzles are often fitted with perforated metal inserts attached to the inside of hollow airfoils. The cooling air is introduced inside of the inserts. It then flows through the perforations, impinging on the inner surface of the hollow airfoil. The cooling thus provided is called impingement cooling. The cooling air then turns and flows within the passage between the insert and the inner surface of the airfoil, cooling it by convection until it exits the airfoil in either leading edge film holes or trailing edge bleed holes.

The effectiveness of cooling, η, is defined as the ratio of the difference between gas and metal temperatures to the difference between the gas temperature and the coolant temperature:

$$\eta = (T_g - T_m)/(T_g - T_c)$$

Figure 15 portrays the relationship between this parameter and a function of the cooling air flow. It can be seen that, while increased cooling flows have improved cooling effectiveness, there are diminishing returns with increased cooling air flow.

Cooling can be improved by precooling the air extracted from the compressor. This is done by passing the extracted air through a heat exchanger prior to using it for bucket or nozzle cooling. This does increase cooling, but presents several challenges such as increasing temperature gradients and the cost and reliability of the cooling equipment. Recent advanced gas turbine products have been designed with both cooled and uncooled cooling air.

Other cooling media have been investigated. In the late 1970s, the U.S. Department of Energy sponsored the study and preliminary design of high-temperature turbines cooled by water and steam. Nozzles of the water-cooled turbine were cooled by water contained in closed passages and kept in the liquid state by pressurization; no water from the nozzle circuits entered the gas path. Buckets were cooled by two-phase flow; heat was absorbed as the coolant was vaporized and heated. Actual nozzles were successfully rig tested. Simulated buckets were tested in heated, rotating rigs. Recent advanced land-based gas turbines have been configured with both buckets and nozzles cooled with a closed steam circuit. Steam, being a more effective cooling medium than air, permits high firing temperatures and, since it does not enter the gas path, eliminates the losses associated with cooling air mixing with the working fluid. The coolant, after being heated in the buckets and nozzles, returns to the steam cycle of a combined cycle plant. The heat carried away by the steam is recovered in a steam turbine.

Steam cooling of gas path airfoils—both stationary and rotating—has been reduced to practice in gas turbines, and such gas turbines are available commercially. The first so-called "H" technology, 50-Hz power plant operated at Baglan Bay, in the United Kingdom, has been observed to produce more than 530 MW when operated at an ambient temperature of 4.5°C. The level of power was produced by a single gas turbine, single steam turbine combined cycle plant.

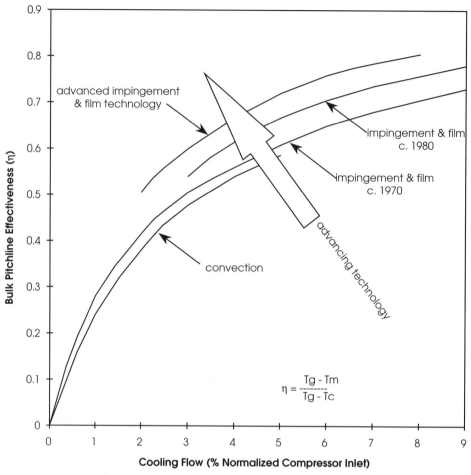

Figure 15 Evolution of turbine airfoil cooling technology.

1.4 Controls and Accessories

Controls

The control system is the interface between the operator and the gas turbine. More correctly, the control system of modern industrial and utility gas turbines interfaces between the operator, and the entire power plant, including the gas turbine, generator, and all related accessories. In combined-cycle power plants where in addition a steam turbine, heat recovery steam generator, condensing system, and all related accessories are present, the control system interfaces with these as well.

Functions provided are described in the Gas Turbine Operation section (1.5) plus protection of the turbine from faults such as overspeed, overheating, combustion anomalies, cooling system failures, and high vibrations. Also, controls facilitate condition monitoring, problem identification and diagnosis, and monitoring of thermodynamic and emissions performance. Sensors placed on the gas turbine include speed pickups, thermocouples at the inlet, exhaust, compressor exit, wheelspaces, bearings, oil supplies, and drains. Vibration monitors are placed on each bearing. Pressures are also monitored at the compressor exit. Multiple thermocouples in the exhaust can detect combustor malfunction by noting abnormal differences in exhaust temperature from one location to another. Multiple sensors elsewhere allow the more sophisticated control systems to self-diagnose, to determine if a problem reading is an indication of a dangerous condition or the result of a sensor malfunction.

Control system development over the past two decades has contributed greatly to the improved reliability of power generation gas turbines. The control systems are now all computer-based. Operator input is via keyboard and cursor movement. Information is displayed to the operator via color graphic displays and tabular and text data on color monitors.

Inlet Systems

Inlet systems filter and direct incoming air, and provide attenuation of compressor noise. They also can include heating and cooling devices to modify the temperature of the air drawn into the gas turbine. Since fixed-wing aircraft engines operate most of the time at high altitudes where air is devoid of heavier and more damaging particles these engines are not fitted with inlet air treatment systems performing more than an aerodynamic function. The premium placed on engine weight makes this so. Inertial separators have been applied to helicopter engines to reduce their ingestion of particulates.

Air near the surface of the earth contains dust and dirt of various chemical compositions. Because of the high volume of air taken into a gas turbine, this dirt can cause erosion of compressor blades, corrosion of turbine and compressor blades, plugging of passages in the gas path as well as cooling circuits. The roughening of compressor blade surfaces can be due to particles sticking to airfoil surfaces, erosion, or because of corrosion caused by their chemical composition. This fouling of the compressor can, over time, reduce mass flow and lower compressor efficiency. Both effects will reduce the output and efficiency of the gas turbine. "Self-cleaning" filters collect airborne dirt. When the pressure drop increases to a preset value, a pulse of air is used to reverse the flow briefly across the filter medium, cleaning the filter. More conventional, multistage filters also find application.

Under low ambient temperature, high-humidity conditions, it is possible to form frost or ice in the gas turbine inlet. Filters can be used to remove humidity by causing frost to form on the filter element; this frost is removed by the self-cleaning feature. Otherwise, a heating element can be installed in the inlet compartment. These elements use higher-temperature air extracted from the compressor. This air is mixed with the ambient air raising its temperature. Compressors of most robust gas turbines are designed so that these systems are required only at part load or under unusual operating conditions.

Inlet chillers have been applied on gas turbines installed in high ambient temperature, low-humidity regions of the world. The incoming air is cooled by the evaporation of water. Cooling the inlet air increases its density and increases the output of the gas turbine.

Exhaust Systems

The exhaust systems of industrial gas turbines perform three basic functions. Personnel must be protected from the high-temperature gas, and from the ducts that carry it. The exhaust gas must be conducted to an exhaust stack or to where the remaining heat from the gas turbine cycle can be effectively used. The exhaust system also contains baffles and other features employed to reduce the noise generated by the gas turbine.

Enclosures and Lagging

Gas turbines are enclosed for four reasons: noise, heat, fire protection, and visual aesthetics. Gas turbines are sometimes provided for outdoor installation, where the supplier includes a sheet metal enclosure, which may be part of the factory-shipped package. Other times, gas turbines are installed in a building. Even in a building, the gas turbine is enclosed for the benefit of maintenance crews or other occupants. Some gas turbines are designed to accommodate an insulating wrapping that attaches to the casings of the gas turbine. This prevents maintenance crews from coming into contact with the hot casings when the turbine is operating and reduces some of the noise generated by the gas turbine. Proponents cite the benefit of lowering the heat transferred from the gas turbine to the environment. Theoretically, more heat is carried to the exhaust, which can be used for other energy needs. Others contend that the larger internal clearances resulting from hotter casings would offset this gain by lower component efficiencies.

Where insulation is not attached to the casings—and sometimes when it is—a small building-like structure is provided. This structure is attached to either the turbine base or the concrete foundation. Such a structure provides crew protection and noise control, and assists in fire protection. If a fire is detected on the turbine, within the enclosure, its relative small volume makes it possible to quickly flood the area with CO_2 or other fire-fighting chemical. The fire is thereby contained in a small volume and more quickly extinguished. Even in a building, the noise control provided by an enclosure is beneficial, especially in buildings containing additional gas turbines, or other equipment. By lowering the noise 1 m from the enclosure to below 85 or 90 dba it is possible to safely perform maintenance on this other equipment, yet continue to operate the gas turbine. Where no turbine enclosure is provided within a building, the building becomes part of the fire-protection and acoustic system.

Fuel Systems

The minimum function required of a gas turbine fuel system is to deliver fuel from a tank or pipeline to the gas turbine combustor fuel nozzles at the required pressure and flow rate. The pressure required is somewhat above the compressor discharge pressure, and the flow rate is that called for by the controls. On annular and can-annular combustors, the same fuel flow must be distributed to each nozzle to ensure minimum variation in the temperature to which gas path components are exposed. Other fuel system requirements are related to the required chemistry and quality of the fuel.

Aircraft engine fuel quality and chemistry are closely regulated, so extensive on-board fuel-conditioning systems are not required. Such is not the case in many industrial applications. Even the better grades of distillate oil may be delivered by ocean-going tanker and run the risk of sodium contamination from the saltwater sometimes used for ballast. Natural gas now contains more of the heavier liquified petroleum gases. Gas turbines are also fueled with crude oil, heavy oils, and various blends. Some applications require the use of nonlubricating fuels such as naphtha. Most fuels today require some degree of on-site treatment.

Complete liquid fuel treatment includes washing, to remove soluble trace metals such as sodium, potassium, and certain calcium compounds. Filtering the fuel removes solid oxides and silicates. Inhibiting the vanadium in the fuel with magnesium compounds in a ratio of three parts of magnesium by weight to one part of vanadium, limits the corrosive action of vanadium on the alloys used in high-temperature gas path parts.

Gas fuel is primarily methane, but it contains varying levels of propane, butane, and other heavier hydrocarbons. When levels of these heavier gases increase, the position of the flame in the combustor may change, resulting in local hot spots, which could damage first-stage turbine stator blades. Also, sudden increases could cause problems for dry low-NO_x premixed combustors. These combustors depend on being able to mix fuel and air in a combustible mixture before the mixture is ignited. Under some conditions, heavier hydrocarbons can self-ignite in these mixtures at compressor exit temperatures, thus causing flame to exist in the premixing portion of the combustor. The flame in the premixing area would have to be extinguished and reestablished in the proper location. This process interferes with normal operation of the machine.

Lubricating Systems

Oil must be provided to the bearings of the gas turbine and its driven equipment. The lubricating system must maintain the oil at sufficiently low temperature to prevent deterioration of its properties. Contaminants must be filtered out. Sufficient volume of oil must be in the system so that any foam has time to settle out. Also, vapors must be dealt with; preferably they are recovered and the oil returned to the plenum. The oil tank for large industrial turbines is generally the base of the lubricating system package. Large utility machines are provided with tanks that hold over 12,000 L of oil. The oil is generally replaced after approximately 20,000 h of operation. More oil is required in applications where the load device is connected to the gas turbine by a gearbox.

The lubrication system package also contains filters and coolers. The turbine is fitted with mist-elimination devices connected to the bearing air vents. Bearings may be vented to the turbine exhaust, but this practice is disappearing for environmental reasons.

Cooling Water and Cooling Air Systems

Several industrial gas turbine applications require the cooling of some accessories. The accessories requiring cooling include the starting means, lubrication system, atomizing air, load equipment (generator/alternator), and turbine support structure. Water is circulated in the component requiring cooling, then conducted to where the heat can be removed from the coolant. The cooling system can be integrated into the industrial or power plant hosting the gas turbine, or can be dedicated to the gas turbine. In this case the system usually contains a water-to-air heat exchanger with fans to provide the flow of air past finned water tubes.

Water Wash Systems

Compressor fouling related to deposition of particles that are not removed by the air filter can be dealt with by water washing the compressor. A significant benefit in gas turbine efficiency over time can be realized by periodic cleaning of the compressor blades. This cleaning is most conveniently done when the gas turbine is fitted with an automatic water-wash system. Washing is initiated by the operator. The water is preheated and detergent is added. The gas turbine rotor is rotated at a low speed and the water is sprayed into the compressor. Drains are provided to remove wastewater.

1.5 Gas Turbine Operation

Like other internal combustion engines, the gas turbine requires an outside source of starting power. This is provided by an electrical motor or diesel engine connected through a gear box to the shaft of the gas turbine (the high-pressure shaft in a multishaft configuration.) Other devices can be used (including the generator of large, electric utility gas turbines) by using a variable-frequency power supply. Power is normally required to rotate the rotor past the gas turbine's ignition speed of 10–15%, on to 40–80% of rated speed where the gas turbine is self-sustaining, meaning the turbine produces sufficient work to power the compressor and overcome bearing friction, drag, etc. Below self-sustaining speed, the component efficiencies of the compressor and turbine are too low to reach or exceed this equilibrium.

When the operator initiates the starting sequence of a gas turbine, the control system acts by starting auxiliaries such as those that provide lubrication and the monitoring of sensors provided to ensure a successful start. Then the control system calls for application of torque to the shaft by the starting means. In many industrial and utility applications, the rotor must be rotated for a period of time to purge the flow path of unburned fuel, which may have collected there. This is a safety precaution. Thereafter, the light-off speed is achieved; ignition takes place and is confirmed by sensors. Ignition is provided by either a spark-plug type device or by a liquified petroleum gas torch built into the combustor. Fuel flow is then increased to increase the rotor speed. In large gas turbines, a warm-up period of one minute or so is required at approximately 20% speed. The starting means remains engaged, since the gas turbine has not reached its self-sustaining speed. This reduces the thermal gradients experienced by some of the turbine components and extends their low cycle fatigue life.

The fuel flow is again increased to bring the rotor to self-sustaining speed. For aircraft engines, this is approximately the idle speed. For power-generation applications, the rotor continues to be accelerated to full speed. In the case of these alternator-driving gas turbines, this is set by the speed at which the alternator is synchronized with the power grid to which it is to be connected.

The speed and thrust of aricraft engines are interrelated. The fuel flow is increased and decreased to generate the required thrust. The rotor speed is a principally a function of this fuel flow, but also depends on any variable compressor or exhaust nozzle geometry changes programmed into the control algorithms. Thrust is set by the pilot to match the current requirements of the aircraft—through takeoff, climb, cruise, maneuvering, landing, and braking.

At full speed, the power-generation gas turbine and its generator (alternator) must be synchronized with the power grid in both speed (frequency) and phase. This process is computer-controlled, and involves making small changes in turbine speed until synchronization is achieved. At this point, the generator is connected with the power grid. The load of a power-generation gas turbine is set by a combination of generator (alternator) excitement and fuel flow. As the excitation is increased, the mechanical work absorbed by the generator increases. To maintain a constant speed (frequency) the fuel flow is increased to match that required by the generator. The operator normally sets the desired electrical output and the turbine's electronic control increases both excitation and fuel flow until the desired operating conditions are reached.

Normal shutdown of a power-generation gas turbine is initiated by the operator, and begins with the reduction of load, reversing the loading process described immediately previous. At a point near zero load, the breaker connecting the generator to the power grid is opened. Fuel flow is decreased, and the turbine is allowed to decelerate to a point below 40% speed, whereupon the fuel is shut off and the rotor is allowed to a stop. The rotors of large turbines should be turned periodically to prevent temporary bowing from uneven cooldown, which will cause vibration on subsequent start-ups. Turning of the rotor for cooldown is accomplished by a ratcheting mechanism on smaller gas turbines, or by operation of a motor associated with shaft-driven accessories, or even the starting mechanism on others. Aircraft engine rotors do not tend to exhibit the bowing just described. Bowing is a phenomenon observed in massive rotors left stationary surrounded by cooling, still air, which, due to free convection, is cooler at the 6:00 position than at the 12:00 position. The large rotor assumes a similar gradient, and because of proportional thermal expansion assumes a bowed shape. Because of the massiveness of the rotor, this shape persists for several hours, and could remain present at the time when the operator wishes to restart the turbine.

2 GAS TURBINE PERFORMANCE

2.1 Gas Turbine Configurations and Cycle Characteristics

There are several possible mechanical configurations for the basic simple cycle, or open cycle, gas turbine. There are also some important variants on the basic cycle—intercooled, regenerative, and reheat cycles.

The simplest configuration is shown in Fig. 16. Here the compressor and turbine rotors are connected directly to one another, and to shafts by which turbine work in excess of that required to drive the compressor can be applied to other work-absorbing devices. Such devices are the propellers and gear boxes of turboprop engines, electrical generators, ships' propellers, pumps, gas compressors, vehicle gear boxes and driving wheels, and the like. A variation is shown in Fig. 17, where a jet nozzle is added to generate thrust. Through aerodynamic design, the pressure drop between the turbine inlet and ambient air is divided so that part of the drop occurs across the turbine and the remainder across the jet nozzle. The pressure at the turbine exit is set so that there is only enough work extracted from the working fluid by the turbine to drive the compressor (and mechanical accessories). The remaining energy accelerates the exhaust flow through the nozzle to provide jet thrust.

The simplest of multishaft arrangements appears in Fig. 18. For decades, such arrangements have been used in heavy-duty turbines applied to various petrochemical and gas pipeline uses. Here, the turbine consists of a high-pressure and a low-pressure section. There is no mechanical connection between the rotors of the two turbines. The high-pressure (h.p.) turbine drives the compressor and the low-pressure (l.p.) turbine drives the load—usually a gas compressor for a process, gas well or pipeline. Often, there is a variable nozzle between the two turbine rotors, which can be used to vary the work split between the two turbines. This offers the user an advantage. When it is necessary to lower the load applied to the driven equipment—for example, when it is necessary to reduce the flow from a gas pumping station—fuel flow would be reduced. With no variable geometry between the turbines, both would drop in speed until a new equilibrium between low- and high-pressure speeds occurs. By changing the nozzle area between the rotors, the pressure drop split is changed, and it is possible to keep the high-pressure rotor at a high, constant speed and have all the speed drop occur in the low-pressure rotor. By doing this, the compressor of the gas turbine continues to operate at or near its maximum efficiency, contributing to the overall efficiency of the gas turbine, and providing high part-load efficiency. This two-shaft arrangement is one of those applied to aircraft engines in industrial applications. Here, the high-pressure section is essentially identical to the aircraft turbojet engine, or the core of a fan-jet engine. This high-pressure section then becomes the "gas generator," and the free-turbine becomes what is referred to as the "power turbine." The modern turbofan engine is somewhat similar in

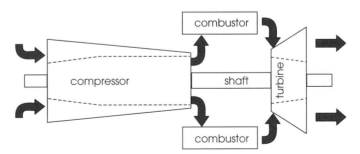

Figure 16 Simple-cycle, single-shaft gas turbine schematic.

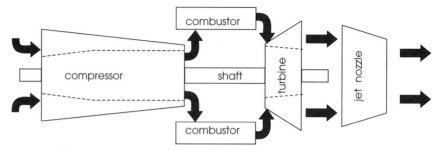

Figure 17   Single-shaft, simple-cycle gas turbine with jet nozzle; simple turbojet engine schematic.

that a low-pressure turbine drives a fan, which forces a concentric flow of air outboard of the gas generator aft, adding to the thrust provided by the engine. In the case of modern turbofans, the fan is upstream of the compressor and is driven by a concentric shaft inside of the hollow shaft connecting the high-pressure compressor and high-pressure turbine.

Figure 19 shows a multishaft arrangement common to today's high-pressure turbojet and turbofan engines. The high-pressure compressor is connected to the high-pressure turbine, and the low-pressure compressor to the low-pressure turbine, by concentric shafts. There is no mechanical connection between the two rotors (high-pressure and low-pressure) except via bearings and the associated supporting structure and the shafts operate at speeds mechanically independent of one another. The need for this apparently complex structure arises from the aerodynamic design constraints encountered in very high-pressure ratio compressors. By having the higher-pressure stages of a compressor rotating at a higher speed than the early stages, it is possible to avoid the low annulus height flow paths that contribute to poor compressor efficiency. The relationship between the speeds of the two shafts is determined by the aerodynamics of the turbines and compressors, the load on the loaded shaft and the fuel flow. The speed of the high-pressure rotor is allowed to float, but is generally monitored. Fuel flow and adjustable compressor blade angles are used to control the low-pressure rotor speed. Turbojet engines, and at least one industrial aero-derivative engine, have been configured just as shown in Fig. 19.

Additional industrial aero-derivative engines have gas generators configured as shown, and have power turbines as shown in Fig. 18.

The next three configurations reflect deviations from the basic Brayton gas turbine cycle. To describe them, reference must be made back to the temperature–entropy diagram.

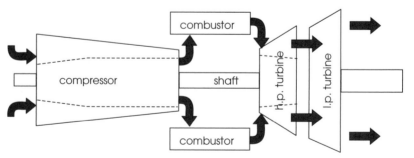

Figure 18   Industrial two-shaft gas turbine schematic showing high-pressure (h.p.) gas generator rotor and separate, free-turbine low-pressure (l.p.) rotor.

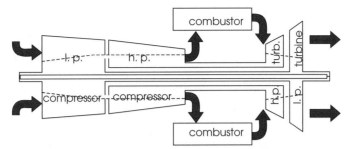

Figure 19 Schematic of multishaft gas turbine arrangement typical of those used in modern, high pressure ratio aircraft engines. Either a jet nozzle, for jet propulsion, or a free power turbine, for mechanical drive, can be added aft of the low-pressure (l.p.) turbine.

Intercooling is the cooling of the working fluid at one or more points during the compression process. Figure 20 shows a low-pressure compression, from points a to b. At point b, heat is removed at constant pressure, moving to point c. At point c, the remaining compression takes place (line cd), after which heat is added by combustion (line de). Following combustion, expansion takes place (line ef) and finally the cycle is closed by discharge of air to the environment (line fa), closing the cycle. Intercooling lowers the amount of work required for compression, because work is proportional to the sum of line ab and line cd, and this is less than that of line ad', which would be the compression process without the intercooler. Lines of constant pressure are closer together at lower temperatures, due to the

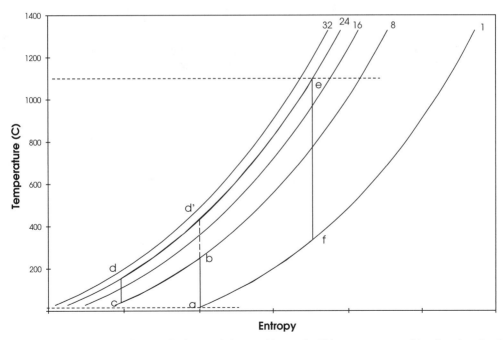

Figure 20 Temperature–entropy diagram for intercooled gas turbine cycle. Firing temperature arbitrarily selected at 1100°C and pressure ratio at 24:1.

same phenomenon that explains higher turbine work than compressor work over the same pressure ratio. Although the compression process is more efficient with intercooling, more fuel is required by this cycle. Note the line *de* as compared with the line *d'e*. It is clear that the added vertical length of line *de* versus *d'e* is greater than the reduced vertical distance achieved in the compression cycle. For this reason, when the heat in the partially compressed air is rejected, the efficiency of an intercooled cycle is generally lower than a similar simple cycle. Some benefit may be observed when comparing real machinery with intercooling applied early in the compression process, but it is, arguably, small. Furthermore, attempts to utilize low-quality heat in a cost-effective manner are usually not successful.

The useful work, which is proportional to *ef* less the sum of *ab* and *cd*, is greater than the useful work of the simple *ad'efa* cycle. Hence, for the same turbomachinery, more work is produced by the intercooled cycle—an increase in power density. This benefit is somewhat offset by the fact that relatively large heat-transfer devices are required to accomplish the intercooling. The intercoolers are roughly the size and volume of the turbomachinery and its accessories.

The preceding comments compare intercooled with simple cycles at fixed pressure ratio and firing temperature. The comparison ignores a potential benefit. Supercharging existing compressors and adding intercooling means pressure ratio can be increased without the mechanical implications of high compressor exit temperature. Higher cycle efficiency will follow from a higher pressure ratio.

An intercooled gas turbine is shown schematically in Fig. 21. A single-shaft arrangement is shown to demonstrate the principal, but a multishaft configuration could also be used. The compressor is divided at some point where air can be taken off-board, cooled, and brought back to the compressor for the remainder of the compression process.

The compressor discharge temperature of the intercooled cycle (point *d*) is lower than that of the simple cycle (point *d'*). Often, cooling air used to cool turbine and combustor components is taken from, or from near, the compressor discharge. An advantage often cited for intercooled cycles is the lower volume of compressor air that has to be extracted. Critics of intercooling point out that the cooling of the cooling air only, rather than the full flow of the machine, would offer the same benefit with smaller heat exchangers. Only upon assessment of the details of the individual application can the point be settled.

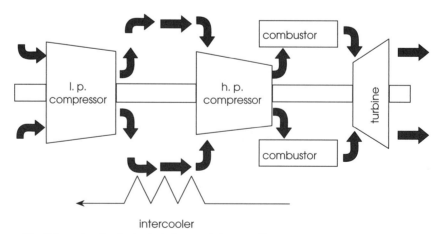

Figure 21 Schematic of a single-shaft, intercooled gas turbine. In this arrangement, both compressor groups are fixed to the same shaft. Concentric, multishaft, and series arrangements are also possible.

The temperature–entropy diagram for a reheat or refired gas turbine is shown in Fig. 22. The cycle begins with the compression process shown by line *ab*. The first combustion process is shown by line *bc*. At point *c*, a turbine expands the fluid (line *cd*) to a temperature associated with an intermediate pressure ratio. At point *d*, another combustion process takes place, returning the fluid to a high temperature (line *de*). At point *e*, the second expansion takes place, returning the fluid to ambient pressure (line *ef*); thereafter, the cycle is closed by discharge of the working fluid back to the atmosphere.

An estimate of the cycle efficiency can be made from the temperatures corresponding to the process end points of the cycle in Fig. 22. By dividing the turbine temperature drops less the compressor temperature rise by the sum of the combustor temperature rises, one calculates an efficiency of approximately 49%. This, of course, reflects perfect compressor, combustor, and turbine efficiency and pure air as the working fluid. Actual efficiencies and properties and consideration of turbine cooling produce less optimistic values.

A simple cycle with the same firing temperature and exhaust temperature would be described by the cycle *ab'efa*. The efficiency calculated for this cycle is approximately 38%, significantly lower than for the reheat cycle. This is really not a fair comparison, since the simple cycle has a pressure of only 8:1, whereas the refired cycle operates at 32:1. The ALSTOM GT26 shown in Fig. 23, and the 60-Hz version, the GT24, are current examples of a refired gas turbine.

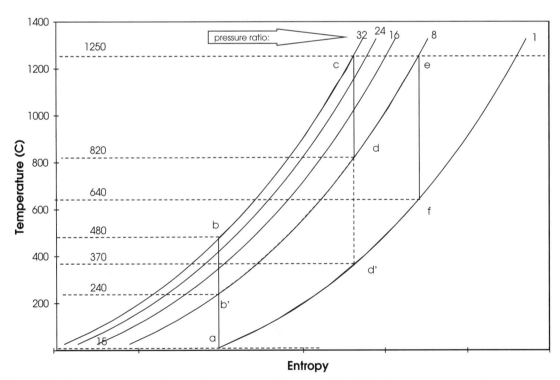

Figure 22 Temperature–entropy diagram for a reheat, or refired, gas turbine. Firing temperatures were arbitrarily chosen to be equal, and to be 1250°C. The intermediate pressure ratio was chosen to be 8:1, and the overall pressure ratio was chosen to be 32:1. Dashed lines are used to illustrate comparable simple gas turbine cycles.

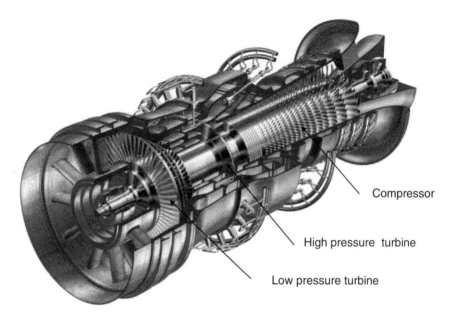

Figure 23 ALSTOM GT26 gas turbine. (Courtesy of ALSTOM.)

A simple-cycle gas turbine with the same pressure ratio and firing temperature would be described by the cycle *abcd'a*. Computing the efficiency, one obtains a value of approximately 54%, more efficient than the comparable reheat cycle. However, there is another factor to be considered. The exhaust temperature of the reheat cycle is 270°C higher than for the simple cycle gas turbine. When applied in combined cycle power plants (these are discussed later) this difference is sufficient to allow optimized reheat cycle-based plants more efficient than simple cycle-based plants of similar overall pressure ratio and firing temperature. Figure 24 shows the arrangement of a single-shaft, reheat gas turbine.

Regenerators, or recouperators, are devices used to transfer the heat in a gas turbine exhaust to the working fluid, after it exits the compressor but before it is heated in the

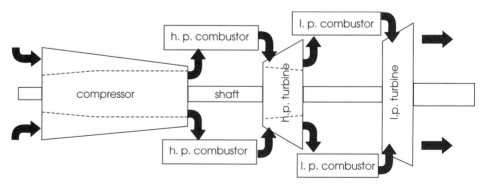

Figure 24 Schematic of a reheat, or refired, gas turbine. This arrangement shows both turbines connected by a shaft. Variations include multiple-shaft arrangements and independent components or component groups arranged in series.

combustor. Figure 25 shows the schematic arrangement of a gas turbine with regenerator. Such gas turbines have been used extensively for compressor drives on natural gas pipelines, and have been tested in wheeled vehicle propulsion applications. Regeneration offers the benefit of high efficiency from a simple, low-pressure gas turbine without resort to combining the gas turbine with a steam turbine and a boiler to make use of exhaust heat. Regenerative gas turbines with modest firing temperature and pressure ratio have comparable efficiency to advanced, aircraft-derived simple-cycle gas turbines.

The temperature–entropy diagram for an ideal, regenerative, gas turbine appears in Fig. 26. Without regeneration, the 8:1 pressure ratio, 1000°C firing temperature gas turbine has an efficiency of $[(1000 - 480) - (240 - 15)]/(1000 - 240) = 38.8\%$ by the method used repeatedly above. Regeneration, if perfectly effective, would raise the compressor discharge temperature to the turbine exhaust temperature, 480°C. This would reduce the heat required from the combustor, reducing the denominator of this last equation from 760 to 520°C, and thereby increasing the efficiency to 56.7%. Such efficiency levels are not realized in practice because of real component efficiencies and heat transfer effectiveness in real regenerators. The relative increase in efficiency between simple and regenerative cycles is as indicated in this example.

Figure 26 has shown the benefit of regeneration in low-pressure ratio gas turbines. As the pressure ratio is increased, the exhaust temperature decreases, and the compressor discharge temperature increases. The dashed line $ab'cd'a$ shows the effect of increasing the pressure to 24:1. Note that the exhaust temperature, d', is lower than the compressor discharge temperature, b'. Here regeneration is impossible. As the pressure ratio (at constant firing temperature) is increased from 8:1 to nearly 24:1, the benefit of regeneration decreases and eventually vanishes. There is, of course, the possibility of intercooling the high pressure

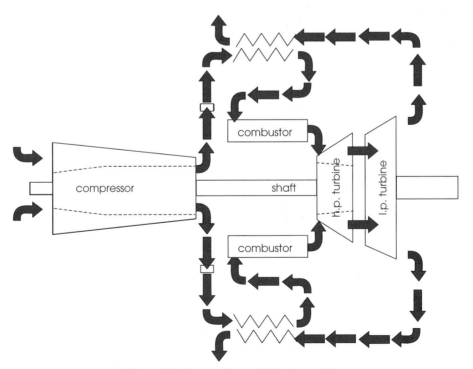

Figure 25 Regenerative, multishaft gas turbine.

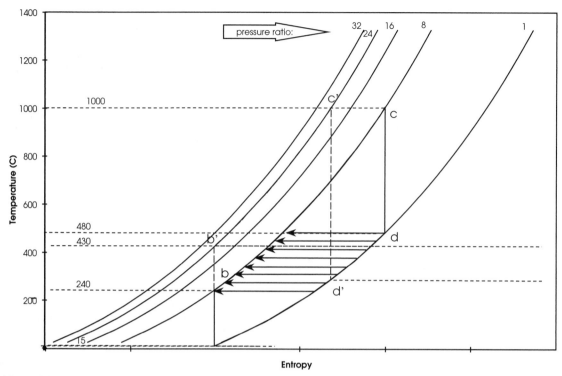

Figure 26   Temperature–entropy diagram comparing an 8:1 pressure ratio, ideal, regenerative cycle with a 24:1 pressure ratio simple cycle, both at a firing temperature of 1000°C.

ratio compressor, reducing its discharge temperature to where regeneration is again possible. Economic analysis and detailed analyses of the thermodynamic cycle with real component efficiencies is required to evaluate the benefits of the added costs of the heat-transfer and air-handling equipment.

2.2   Trends in Gas Turbine Design and Performance

Output or Size
Higher power needs can be met by increasing either the number or the size of gas turbines. Where power needs are high, economics generally favor large equipment. The specific cost (cost per unit power) of gas turbines decreases as size increases, as can be shown in Fig. 27. Note that the cost decreases, but at a decreasing rate; the slope remains negative at the maximum current output for a single gas turbine, around 300 MW. Output increases are accomplished by increased mass flow and increased firing temperature. Mass flow is roughly proportional to the inlet annulus area of the compressor. There are four ways of increasing this:

1. *Lower rotor speed while scaling root and tip diameter proportionally.* This results in geometric similarity and low risk, but is not possible in the case of synchronous gas turbines where the shaft of the gas turbine must rotate at either 3600 or 3000 rpm to generate 60 or 50 Hz (respectively) alternating current.

SC Plants

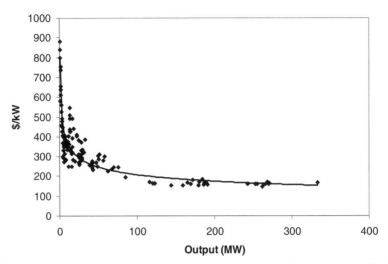

Figure 27 Cost of simple-cycle, generator-drive electric power generation equipment. (Plotted from data published by *Gas Turbine World Magazine,* 2003 [Ref. 11].)

2. *Increase tip diameter.* Designers have been moving the tip velocity into the transsonic region. Modern airfoil design techniques have made this possible while maintaining good aerodynamic efficiency.

3. *Decrease hub diameter.* This involves increasing the solidity near the root, since the cross section of blade roots must be large enough to support the outer portion of the blade against centrifugal force. The increased solidity interferes with aerodynamic efficiency. Also, where a drive shaft is designed into the front of the compressor (cold end drive) and where there is a large bearing at the outboard end of the compressor, there are mechanical limits to reducing the inlet inner diameter.

4. *Reduce the thickness of the blades themselves.*

Firing Temperature

Firing temperature increases provide higher output per unit mass flow and higher combined-cycle efficiency. Efficiency is improved by increased firing temperature wherever exhaust heat is put to use. Such uses include regeneration/recoupation, district heating, supplying heat to chemical and industrial processes, Rankine bottoming cycles, and adding a power turbine to drive a fan in an aircraft engine. The effect of firing temperature on the evolution of combined Brayton-Rankine cycles for power generation is illustrated in Fig. 28.

Firing temperature increases when the fuel flow to the engine's combustion system is increased. The challenge faced by designers is to increase firing temperature without decreasing the reliability of the engine. A metal temperature increase of 15°C will reduce bucket creep life by 50%. Material advances and more increasingly more aggressive cooling techniques must be employed to allow even small increases in firing temperature. These technologies have been discussed previously.

Maintenance practices represent a third means of keeping reliability high while increasing temperature. Sophisticated life-prediction methods and experience on identical or similar turbines are used to set inspection, repair, and replacement intervals.

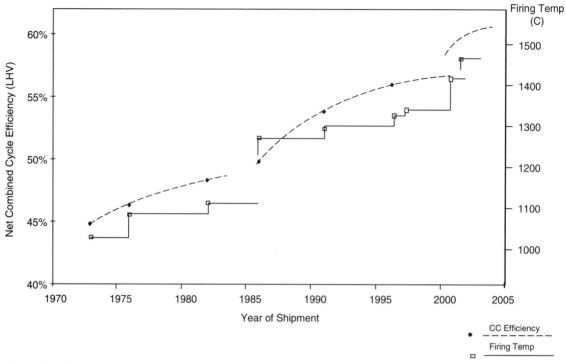

Figure 28   History of power-generation, combined-cycle efficiency and firing temperature, illustrating the trends to higher firing temperature and its effect on efficiency.

Coupled with design features that reduce the time required to perform maintenance, both planned and unplanned downtime can be reduced to offset shorter parts lives, with no impact on reliability. Increased firing temperature usually increases the cost of the buckets and nozzles (developments involve exotic materials or complicated cooling configurations). Although these parts are expensive, they represent a small fraction of the cost of an entire power plant. The increased output permitted by the use of advanced buckets and nozzles is generally much higher, proportionally, than the increase in power plant cost and maintenance cost, hence increased firing temperature tends to lower specific power plant cost.

Pressure Ratio

Two factors drive the choice of pressure ratio. First is the primary dependence of simple cycle efficiency on pressure ratio. Gas turbines intended for simple cycle application, such as those used in aircraft propulsion, emergency power, or power where space or weight is a primary consideration, benefit from higher pressure ratios.

Combined-cycle power plants do not necessarily benefit from high pressure ratios. At a given firing temperature, an increase in pressure ratio lowers the exhaust temperature. Lower exhaust temperature means less power from the bottoming cycle, and a lower efficiency bottoming cycle. So, as pressure ratio is increased, the gas turbine becomes more efficient and the bottoming cycle becomes less efficient. There is an optimum pressure ratio for each firing temperature, all other design rules held constant. Figure 29 shows how specific output

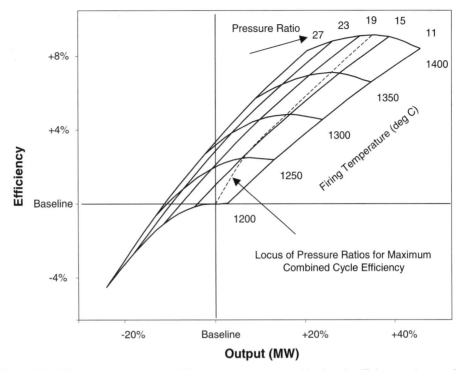

Figure 29   Effect of pressure ratio and firing temperature on combined-cycle efficiency and output for fixed air flow and representative gas and steam turbine designs.

and combined cycle efficiency are affected by gas turbine firing temperature and pressure ratio for a given type of gas turbine and steam cycle. At each firing temperature there is a pressure ratio for which the combined-cycle efficiency is highest. Furthermore, as firing temperature is increased, this optimum pressure ratio is higher as well. This fact means that, as firing temperature is increased in pursuit of higher combined-cycle efficiency, pressure ratio must also be increased.

Reducing the flow area through the first-stage nozzle of the turbine increases pressure ratio. This increases the pressure ratio per stage of the compressor. There is a point at which increased pressure ratio causes the compressor airfoils to stall. Stall is avoided by either adding stages (reducing the pressure ratio per stage) or increasing the chord length, and applying advanced aerodynamic design techniques. For a very high pressure ratio a simple, single-shaft rotor with fixed stationary airfoils cannot deliver the necessary combination of pressure ratio, stall margin, and operating flexibility. Features required to meet all design objectives simultaneously include variable-angle stationary blades in one or more stages; extraction features, which can be used to bleed air from the compressor during low-speed operation; and multiple rotors that can be operated at different speeds.

Larger size, higher firing temperature, and higher pressure ratio are pursued by manufacturers to lower cost and increase efficiency. Materials and design features evolve to accomplish these advances with only positive impact on reliability.

3   APPLICATIONS

3.1   Use of Exhaust Heat in Industrial Gas Turbines

By adding equipment for converting exhaust energy to useful work, the thermal efficiency of a gas turbine-based power plant can be increased by 10% to over 30%. Of these numerous schemes, the most significant is the fitting of a heat recovery steam generator (HRSG) to the exhaust of the gas turbine and delivering the steam produced to a steam turbine. Both the steam turbine and gas turbine drive one or more electrical generators.

Figure 30 displays the combining of the Brayton and Rankine cycles. The Brayton cycle *abcda* has been described already. It is important to point out that the line *da* now represents heat transferred in the HRSG. In actual plants, the turbine work is reduced slightly by the backpressure associated with the HRSG. Point *d* would be above the 1:1 pressure curve, and the temperature drop proportionately reduced.

The Rankine cycle begins with the pumping of water into the HRSG, line *mn*. This process is analogous to the compression in the gas turbine, but rather than absorbing 50% of the turbine work, consumes only about 5%, since the work required to pump a liquid is less than that required to compress a gas. The water is heated (line *no*) and evaporated (*op*). The energy for this is supplied in the HRSG by the exhaust gas of the gas turbine. More energy is extracted to superheat the steam as indicated by line *pr*. At this point, superheated steam is delivered to a steam turbine and expanded (*rs*) to convert the energy therein to mechanical work.

The addition of the HRSG reduces the output of the gas turbine only slightly. The power required by the mechanical devices (like the feedwater pump) in the steam plant is also small. Therefore, most of the steam turbine work can be added to the net gas turbine work

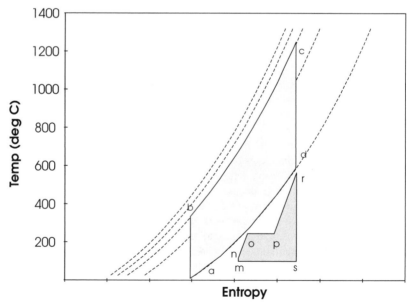

Figure 30   Temperature-entropy diagram illustrating the combining of a gas turbine (*abcda*) and steam turbine cycle (*mnoprsm*). The heat wasted in process *da* in simple cycle turbines supplies the heat required by processes *no, op,* and *pr*.

with almost no increase in fuel flow. For combined-cycle plants based on industrial gas turbines where exhaust temperature is in the 600°C class, the output of the steam turbine is about half that of the gas turbine. Their combined-cycle efficiency is approximately 50% higher than simple-cycle efficiency. For high pressure ratio gas turbines with exhaust temperature near 450°C the associated steam turbine output is close to 25% of the gas turbine output, and efficiency is increased by approximately 25%. The thermodynamic cycles of the more recent large, industrial gas turbines have been optimized for high combined-cycle efficiency. They have moderate to high simple-cycle efficiency and relatively high exhaust temperatures. Figure 30 has shown that net combined-cycle efficiency (lower heating value) of approximately 55% has been realized as of this writing, and levels of 60% and beyond are under development.

Figure 31 shows a simple combined-cycle arrangement, where the HRSG delivers steam at one pressure level. All the steam is supplied to a steam turbine. Here, there is neither steam reheat nor additional heat supplied to the HRSG. There are many alternatives.

Fired HRSGs have been constructed where heat is supplied both by the gas turbine and by a burner in the exhaust duct. This practice lowers overall efficiency, but accommodates the economics of some situations of variable load requirements and fuel availability. In other applications steam from the HRSG is supplied to nearby industries or used for district heating, lowering the power-generation efficiency, but contributing to the overall economics in specific applications.

Efficiency of electric power generation benefits form more complicated steam cycles. Multiple-pressure, nonreheat cycles improve efficiency as a result of additional heat-transfer surface in the HRSG. Multiple-pressure, reheat cycles, such as shown in Fig. 32, match the performance of higher exhaust temperature gas turbines (600°C). Such systems are the most

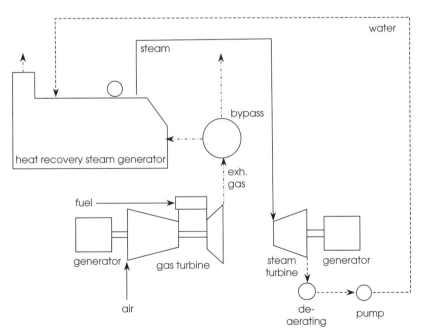

Figure 31 Schematic of simple combined cycle power plant. A single-pressure, nonreheat cycle is shown.

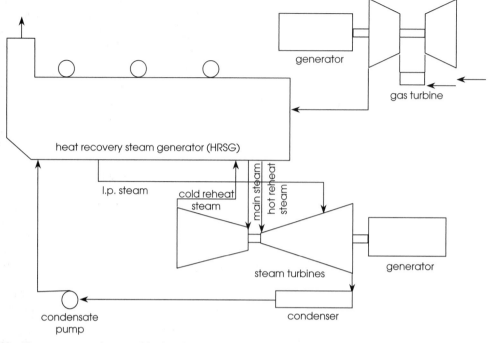

Figure 32 Three-pressure, reheat combined-cycle arrangement. The highest power generation efficiency is currently achieved by such plants.

efficient currently available, but are also the most costly. The relative performance for several combined-cycle arrangements are shown in Table 1.[12] The comparison was made for plants using a gas turbine in the 1250°C firing temperature class.

Plant costs for simple-cycle gas turbine generators is lower than that for steam turbines and most other types of power plant. Since combined-cycle plants generate two-thirds of their power with the gas turbine, their cost is between that of simple-cycle gas turbine plants and steam turbine plants. Their efficiency is higher than either. The high efficiency and low cost make combined-cycle plants extremely popular. Very large commitments to this type of plant have been made around the world. Table 2 shows some of the more recent to be put into service.

Table 1 Comparison of Performance for Combined-Cycle Arrangements Based on Third-Generation (1250°C firing temperature) Industrial Gas Turbines

Steam Cycle	Relative Net Plant Output (%)	Relative Net Plant Efficiency (%)
Three pressure, reheat	Base	Base
Two pressure, reheat	−1.1	−1.1
Three pressure, nonreheat	−1.2	−1.2
Two pressure, nonreheat	−2.0	−2.0

Table 2 Recent, Large, Multiunit Gas Turbine-Based Combined-Cycle Power Plants

Station	Country	Number of GTs	Rating (MW)
Seo-Inchon 1-4	Korea	16	3950
Chiba 2-3	Japan	8	2850
Hsinta 1-5	Taiwan	15	2450
Gila River 1-4	USA	8	2350
El Dorado 1-4	USA	8	2330
Ratchaburi	Thailand	6	2310
Lumut	Malaysia	9	2230
Phu My 1-3	Vietnam	7	2190
Black Point	Hong Kong	6	2080

There are other uses for gas turbine exhaust energy. Regeneration, or recoupment, uses the exhaust heat to raise the temperature of the compressor discharge air before the combustion process. Various steam injection arrangements have been used as well. Here, an HRSG is used as in the combined cycle arrangements shown in Fig. 32, but instead of expanding the steam in a steam turbine, it is introduced into the gas turbine, as illustrated in Fig. 33. It may be injected into the combustor, where it lowers the generation of NO_x by cooling the combustion flame. This steam increases the mass flow of the turbine and its heat is converted to useful work as it expands through the turbine section of the gas turbine.

Steam can also be injected downstream of the combustor at various locations in the turbine, where it adds to the mass flow of the working fluid. Many gas turbines can tolerate steam injection levels of 5% of the mass flow of the air entering the compressor; others can accommodate 16% or more, if distributed appropriately along the gas path of the gas turbine. Gas turbines specifically designed for massive steam injection have been proposed and studied. These proposals arise from the fact that the injection of steam into gas turbines of existing designs has significant reliability implications. There is a limit to the level of steam injection into combustors without flame stability problems and loss of flame. Adding steam to the gas flowing through the first-stage nozzle increases the pressure ratio of the machine and reduces the stall margin of the compressor. Addition of steam to the working fluid expanding in the turbine increases the heat-transfer coefficient on the outside surfaces of the

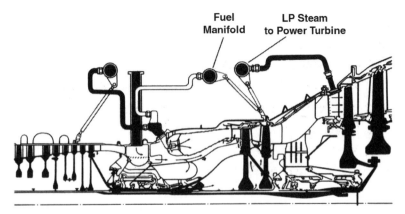

Figure 33 LM500 aero-derivative gas turbine with steam injection. (Courtesy of General Electric Company.)

blading, raising the temperature of these components. The higher work increases the aerodynamic loading on the blading, which may be an issue on latter-stage nozzles, and increases the torque applied to the shafts and rotor flanges. Design changes can be made to address the effects of steam in the gas path.[13]

Benefits of steam injection are an increase in both efficiency and output over those of the simple-cycle gas turbine. The improvements are less than those of the steam turbine and gas turbine combined cycles, since the pressure ratio of the steam expansion is much higher in a steam turbine. Steam turbine pressures may be greater than 100 atm gas turbines no higher than 40. Steam injection cycles are less costly to produce since there is no steam turbine. There is, of course, higher water consumption with steam injection, since the expanded steam exits the plant in the gas turbine exhaust.

3.2 Integrated Gasification Combined Cycle

In many parts of the world, coal is the most abundant and inexpensive fuel. Coal-fired boilers raising steam that is expanded in steam turbine generators is the conventional means of converting this fuel to electricity. Pulverized coal plants with flue gas desulfurization operate at over 40% overall efficiency and have demonstrated the ability to control sulfur emissions from conventional boiler systems. Gas turbine combined-cycle plants are operating with minimal environmental impact on natural gas at 55% efficiency, and 60% is expected with new technologies. A similar combined-cycle plant that could operate on solid fuel would be an attractive option.

Competing means of utilizing coal with gas turbines have included direct combustion, indirect-firing, and gasification. Direct combustion in conventional, on-engine combustors has resulted in rapid, ash-caused erosion of bucket airfoils. Off-base combustion schemes—such as pressurized fluidized bed combustors—have not simultaneously demonstrated the high exit temperature needed for efficiency and low emissions. Indirect firing raises compressor discharge temperature by passing it through a heat exchanger. Metal heat exchangers are not compatible with the high turbine inlet temperature required for competitive efficiency. Ceramic heat exchangers have promise, but their use will necessitate the same types of emission controls required on conventional coal-fired plants. Power plants with the gasification process, desulfurization, and the combined-cycle machinery integrated have been successfully demonstrated, with high efficiency, low emissions, and competitive first cost. Significant numbers of integrated gasification combined-cycle (IGCC) plants are operating or under construction.

Fuels suitable for gasification include several types of coal and other low-cost fuels. Those studied include

- Bituminous coal
- Sub-bituminous coal
- Lignite
- Petroleum coke
- Heavy oil
- Orimulsion
- Biomass

Fuel feed systems of several kinds have been used to supply fuel into the gasifier at the required pressure. Fuel type, moisture content size, and the particular gasification process need to be considered in selecting a feed system.

Several types of gasifiers have been designed to produce fuel with either air or oxygen provided. The system shown in Fig. 34 features a generic oxygen-blown gasifier, and a system for extracting some of the air from the compressor discharge, and dividing it into oxygen and nitrogen. An oxygen-blown gasifier produces a fuel about one-third of the heating value of natural gas. The fuel produced by the gasifier (after sulfur removal) is about 40% CO, 30% H_2. Most of the remaining 30% is H_2O and CO_2 which are inert, and act as diluents in the gas turbine combustor, reducing NO_x formation. A typical lower heating value is 1950 K-Cal/m^3. The fuel exits the gasifier at a temperature higher than that at which it can be cleaned. The gas is cooled by either quench or heat-exchange and cleaned. Cleaning is done by water spray scrubber or dry filtration to remove solids that are harmful to the turbine and potentially harmful to the environment. This is followed by a solvent process that absorbs H_2S.

Some gas turbine models can operate on coal gas without modification. The implications for the gas turbine relate to the volume of fuel, which is three times or more higher than that of natural gas. When the volume flow through the first-stage nozzle of the gas turbine increases, the backpressure on the compressor increases. This increases the pressure ratio of the cycle and decreases the stall margin of the compressor. Gas turbines with robust stall margins need no modification. Others can be adapted by reducing inlet flow by inlet heating or by closing off a portion of the inlet. (Variable inlet stator vanes can be rotated toward a closed position.) The volume flow through the turbine increases as well. This increases the heat transfer to the buckets and nozzles. To preserve parts lives, depending on the robustness of the original design, the firing temperature may have to be reduced. The increased flow

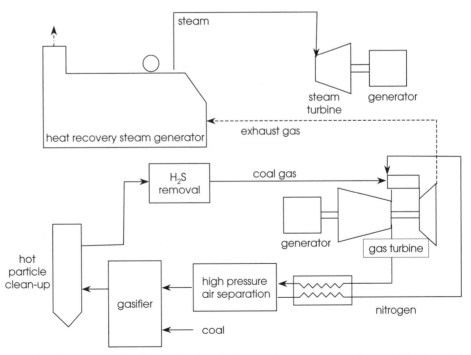

Figure 34 Integrated gasification combined-cycle diagram. Air compressed in the gas turbine is cooled, then separated into oxygen and nitrogen. Oxygen is fed to the gasifier while the nitrogen is sent back to the gas turbine for NO_x control. Coal is partially burned in the gasifier. The gas produced is cleaned and flows to the gas turbine as fuel.

and decreased firing temperature, if required, result in higher gas turbine output than developed by the same gas turbine fired on natural gas.

3.3   Applications in Electricity Generation

Before 1965, the generating capacity of gas turbines shipped per year was below 2 GW. In 2003, in sharp contrast, about 75 GW of electrical generation power plants were commissioned. Approximately 16% of the entire world's current generating capacity is now produced by gas turbines, either in simple cycle or combined cycle. In 2003, over 12% of new power plants featured simple-cycle gas turbines, and over 50% were combined-cycle plants, which derive two-thirds of their capacity from gas turbines. Thus, 45% of new power plant capacity is provided by gas turbine generators. This compares to 40% for steam turbine generators, either alone or in combined cycle. Hydroelectric plants, nuclear and other means, provide the remaining additions.

The 30-fold increase in the rate of gas turbine installations between 1965 and 2003 was due to several factors. In the late 1960s and early 1970s, there was an increasing need for peaking power, particularly in the United States. Gas turbines, because of their low cost, low operating crew size, and fast installation time, were the engine of choice. Because of the seasonal and daily variations in the demand for electric power, generating companies could minimize their investment in plant and equipment by installing a mixture of expensive but efficient base load plants (steam and nuclear) run over 8000 hours per year, and far less expensive (but less efficient) plants, which would operate only a few hundred hours per year. However, rapid progress in efficiency, reliability, availability and environmental impact would soon follow.

Existing industrial gas turbines and newly designed larger units, whose operating speed was chosen to match the requirements of a directly coupled alternator, met the demand for peaking power. The experience on these early units resulted in improvements in efficiency, reliability, and cost-effectiveness. Much of the technology needed to improve the value of industrial gas turbines came from aircraft engine developments, as it still does. Beginning in the 1970s, with the rapid rise in oil prices and associated natural gas prices, electric utilities focused on ways of improving the efficiency of generating plants. Combined-cycle plants are the most thermally efficient fuel-burning plants. Furthermore, their first cost is lower than all other types of plants except for simple-cycle gas turbine plants.

The only drawback to gas turbine plants was the requirement for more noble fuels; natural gas and light distillates are usually chosen to minimize maintenance requirements. Coal is abundant in many parts of the world and costs significantly less that oil or gas per unit energy. Experiments in the direct firing of gas turbines on coal have been conducted without favorable results. Other schemes for using coal in gas turbines include indirect firing, integrating with a fluidized bed combustor, and integrated gasification. The last of these offers the highest efficiency due to its ability to deliver the highest temperature gas to the turbine blading. Furthermore, integrated gasification is the most environmentally benign means of converting coal to electricity. The technology has been demonstrated in several plants, including early technology demonstration plants and commercial power-generating facilities.

3.4   Engines in Aircraft

Like electricity generation, gas turbine design for aircraft engines begins with an understanding of customer requirements. Utility metrics, such as cost of electricity, internal rate of return, and project net present value, are equivalent to cost per seat-mile for a commercial turbofan, or specific mission performance objectives for a military engine.

In addition to optimizing for best economics or performance, electric power and aircraft engine designs are usually subject to a number of constraints. From the utility perspective, these could include kilowatts or fuel consumption on a hot or cold day, or power response during grid events. For an aircraft engine customer, mission objectives such as airspeed, range, payload, maneuverability, runway length, or engine-out rate of climb may limit the available solutions. For both design environments, engineers will optimize the cycle selection variables (such as overall pressure ratio, firing temperature, and bypass ratio) in conjunction with technology (materials, cooling, airfoil aerodynamics) to produce a solution that optimizes economics for an acceptable risk.

Three of the key propulsion parameters to an aircraft are thrust, thrust-specific fuel consumption (SFC, or lbm/h fuel flow/lbf thrust), and weight. Gas turbines currently provide the best combination of these three parameters for flight Mach numbers ranging from about 0.3 through supersonic speeds. Normally, aspirated piston engines (both Otto and Diesel cycles) are favored for aircraft operating at low speeds and altitudes; turbocharged versions are used at higher speeds and at higher altitudes, up to about 400–500 km/hr. Pulse jet, scramjets, and rocket propulsion take over from gas turbines at high supersonic and hypersonic speeds. The two most commonly used aircraft gas turbine configurations are the turboprop and turbofan. Gas turbines are particularly appropriate for aircraft applications because of their efficiency at high airspeeds, high power to weight, and good high-power efficiency.

Equations for thrust and TSFC are shown below:

Net Thrust: $\qquad F_n = W_1(V_j - V_o) + A_e(P_j - P_o)$ $\qquad\qquad$ for turbojet

$\qquad\qquad\qquad F_n = W_c(V_j - V_o) + W_d(V_d - V_o) + A_e(P_j - P_o)$ $\qquad$ for turbofan

Specific Thrust: $\qquad \text{ST} = F_n/W_1$

Specific Fuel Consumption: $\qquad \text{SFC} = W_f/F_n$ $\qquad$ general definition

$\qquad\qquad\qquad\qquad\qquad \text{SFC}_{\text{tp}} = W_f/\text{hp}/\text{h}$ $\qquad$ sometimes applied to turboprop and shaft engines

where W_1 = total inlet mass flow rate, W_c = portion of W_1 passing through the core engine, W_d = portion of W_1 passing through bypass duct, V_o = flight velocity, V_j = exhaust jet velocity, A_e = exhaust area, P_j = exhaust jet static pressure, P_o = ambient pressure, and W_f = fuel flow rate.

The gas turbine flight speed range mentioned above is not met with a single gas turbine configuration, however. The equations above show that to produce thrust at higher flight velocity, higher exhaust velocity is required. Each configuration—turboprop, turbofan, and turbojet—has its own range of potential and optimum exhaust velocities, which, in addition to thrust, are set as a function of SFC and weight objectives.

While the prior power-generation thermal efficiency conclusions still hold true for aircraft turbines, higher firing temperature and overall pressure ratio are selected in conjunction with cooling component efficiency technology to select the right balance between specific power and thermal efficiency. Two new efficiency definitions are required, however, to quantify the aircraft fuel efficiency metric (SFC): overall and propulsive efficiency. SFC is proportional to the flight velocity times the inverse of overall efficiency, while overall efficiency is the product of thermal and propulsive efficiencies:

$$\text{SFC} \sim = V_o / \eta_{\text{overall}}$$

where $\eta_{\text{overall}} = \eta_{\text{thermal}} \times \eta_{\text{ps}}$
$\eta_{\text{ps}} = \text{const} \times (F_n \, V_o)/W_f$ or $\eta_{\text{ps}} = \text{const} \times V_o/\text{SFC}$

This more extensive efficiency definition adds computational complexity and new design variables to mission design and optimization problem: bypass ratio and fan pressure ratio.

Because of its high exhaust velocity, the twin-spool turbojet shown in Fig. 19 will have a high specific thrust and low frontal area, which are ideal for high flight speeds. To improve the fuel consumption at lower speeds, the exhaust velocity could be lowered to the ideal level for a turbojet (two times flight velocity) by extending the low-pressure compressor blades and adding a duct around this extension, creating a turbofan. The duct airflow divided by core airflow is defined as the bypass ratio. The pressure ratio of the low-pressure compressor (now called the fan) is now available to optimize efficiency, but is also available to meet other mission objectives, such as specific power. For unmixed core and exhaust streams, fan pressure ratio impacts the turbine specific power and relative exhaust velocities between the two streams. In a mixed-flow exhaust, common to military turbofans, fan pressure ratio communicates to the core exhaust through a common static pressure during mixing of the two streams. This communication can be controlled by area sizing before these flows mix, and can be used to modify the cycle performance match at both design and off-design flight conditions.

In this example, thrust has increased because core flow was unchanged. If bypass ratio were increased at constant thrust, the total flow will increase faster than the core flow decreases, driving up weight. As technology has improved, however (with better materials, aerodynamics, manufacturing, and heat transfer technologies), core specific power has gone up, leading to increased bypass ratio and lower weight at a given thrust.

Trends in specific thrust and SFC with flight Mach number are shown in Figs. 35 and 36. The figures show that at lower Mach numbers, the turbofan engines have relatively high propulsion efficiency (low SFC). The need for improved efficiency in the high-subsonic speed regime has produced a focus on turbofan engines rather than turbojets. At lower speeds, turboprop engines are preferred.

Installation Effects

Specific thrust curves shown in Fig. 35 and most other gas turbine-only comparisons are done on an uninstalled basis. Here, thrust means net thrust as it is used in the equations above. The curves are particularly convenient for showing generic effects of technology or of changing design parameters. For actual cycle design, optimization, control scheduling, operability studies, and aircraft performance calculations, however, installation effects must be considered to develop the optimum propulsion system. The effect of installation parameters and features on performance is more significant in aircraft applications than in power generation. Aircraft engine installation effects include items such as inlet and nozzle drag, inlet ram recovery, horsepower and bleed extraction, and inlet pressure distortion.

Inlets and Nozzles

For aircraft and propulsion performance analysis, thrust is converted from gross levels (produced by the exhaust) to net, which subtracts inlet ram drag. Installed corrections also account for real losses, and are used to align the bookkeeping between aircraft and propulsion manufacturers. These corrections include inlet spillage drag, inlet ram recovery (pressure losses), and nozzle drag. The design of inlets and nozzles, and scheduling of inlet airflow, is a compromise between performance and aeromechanical considerations. Supersonic inlets can become more complex in an effort to reduce shock losses.

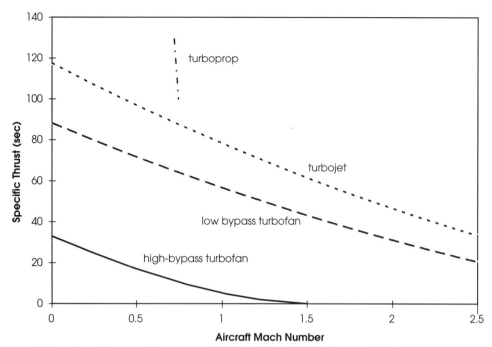

Figure 35 Low values of specific thrust give higher propulsive efficiency at low Mach numbers. (Redrawn from a figure in Ref. 14.)

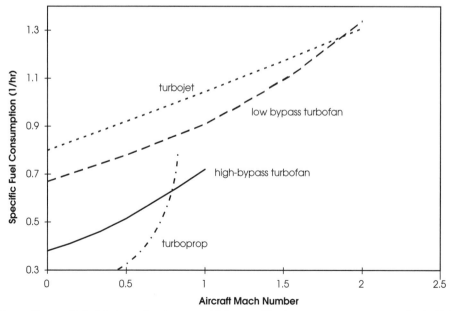

Figure 36 As flight Mach number increases, higher specific thrusts are necessary to maintain high propulsive efficiency and reduce SFC. (Redrawn from a figure in Ref. 14.)

As design speeds increase, inlets increase in complexity from a normal shock design, used for subsonic, transonic, and low supersonic designs, to external compression, to mixed internal/external compression designs. Military aircraft such as the Boeing F-15 use variable-geometry inlet ramps to reduce recovery losses, which improves performance and operability across its flight envelope, while paying a penalty in weight and cost.

Exhaust nozzle configurations are driven by the throat area, efficiency, and cost requirements of the propulsion system. Fixed-area convergent nozzles are satisfactory for subsonic, nonafterburning designs. Variable-area convergent nozzles maintain high efficiency up to low supersonic speeds with afterburning engines. Variable area convergent–divergent nozzles are required for good performance for afterburning engines operating at higher supersonic speeds.

Figures of Merit and Cycle Design Variables

Business, Commercial, and Transport Aircraft. Turbofans use two or three concentric spools, and bypass the fan exit air from the outer spool, usually through an independent duct nozzle. Commercial aircraft are characterized by requirements for relatively low thrust-to-weight, high efficiency during cruise, low noise, and low emissions. The low thrust-to-weight requirement means gas turbine size, or scale factor, is set by the take off thrust specification. Cruise conditions are generally at a high fraction of full power, which is ideal for high gas turbine thermal efficiency.

Military Turbofans. Most modern military aircraft use a twin-spool turbofan configuration similar to commercial aircraft, but will have additional complexity driven by requirements across a larger flight envelope. Military aircraft designed for supersonic operation and high thrust to weight require low drag and low weight. Emerging military aircraft designs include the engines as an integral part of the fuselage, and associated engine designs with higher specific power as opposed to the pylon-mounted large fans engines used in commercial and transport aircraft. Military engine cruise thrust is usually a lower fraction of peak thrust, and cruise operation is frequently interrupted by maximum power operation during the mission, decreasing its significance and adding to the frequency and severity of fatigue loading cycles.

Afterburners are designed to be a cost-effective means of achieving high thrust for short duration. During afterburner operation (referred to as augmentation) fuel is introduced and burned downstream of the last turbine stage and diffuser to increase thrust through higher exhaust exit velocity. The temperature increase causes the exiting gas to expand. Since mass flow is fixed and set by the compressor and fan, exit velocity must increase to satisfy continuity. Afterburner performance is measured by percentage increase in thrust due to augmentation. Turbojets, low-bypass ratio turbofans, and gas turbines with high exit fuel–air from the gas-generator core have reduced potential for augmentation percentage because of stoichiometric fuel/air ratio and exhaust temperature limits.

Military turbofans have historically led the development of gas turbine technologies. Absolute performance metrics have driven materials development, advanced turbine cooling schemes, variable cycle technology, and controls software.

Figure 37 compares the engines selected for, or competing for, recent applications. All of the larger commercial transport applications and newer military applications are met by turbofan engines.[16]

There are multiple ratings for several basic engine designs. The range of ratings is due to the practice of fine-tuning engine performance for particular applications, incremental performance gains over time, and optional features. This comparison is a snapshot of performance over a particular time. Relative ratings change often as manufacturers continue to apply new technologies and improve designs. One of the newest and most powerful turbofan

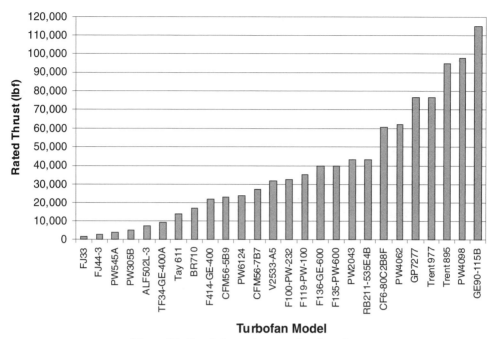

Figure 37 Rated thrust of current aircraft engines.

engines is shown in Fig. 38. It is a two-rotor engine. The one-stage fan and three-stage, low-pressure compressor are joined on one shaft connected to a six-stage, low-pressure turbine. The ten-stage, high-pressure compressor is driven by the two-stage, high-pressure turbine, both joined on another shaft that can rotate at a higher speed. The ratio of the air mass flow through the duct to the air flowing through the compressor, combustor, and turbines is 9:1.

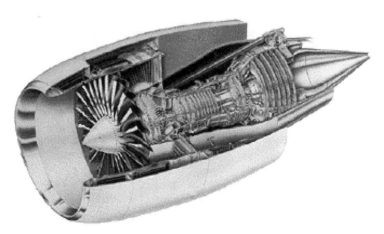

Figure 38 Sectional view of the GE 90, a high bypass ratio, two-shaft, turbofan engine rated at over 80,000 lbf thrust. (Courtesy of General Electric Company.)

The overall pressure ratio is 40:1, and the rated thrust is in the 85,000-lb class. The engine is over 3 m in diameter. A new feature for aircraft engines is the double-domed, lean-premixed, fuel-staged dry low-NO_x combustor. The GE 90, PW4084, and 800-series RB.211 Trent high bypass ratio turbofan engines have been built for use on the Boeing 777 aircraft.

Control Scheduling. Aircraft engine control scheduling has also led the field in gas turbine development. Fast thrust transients, wide ranges of inlet conditions, airframe bleed and power demands, mission demands such as in-flight refueling, and high reliability have necessitated more reliable and smarter controls. To create such controls, designers make the best use of a wide range of control effectors, including primary combustor and augmentor fuel flow, fan and compressor vane angles, bleed valves, exhaust nozzle area and area ratio, and other variable geometry. The advent of digital electronic controls in the early 1980s led to significant improvements in this regard. Controls technology has produced increased monitoring and diagnostic capability, which allows real-time sensor fault identification and corrective action.

Auxiliary Power Units. Another aircraft engine type is the auxiliary power unit (APU). It is a small turboshaft engine that provides air-conditioning and electric and hydraulic power while the main engines are stopped, but its main function is to start the main engines without external assistance. The APU is usually started from batteries by an electric motor. When it reaches its operating speed, a substantial flow of air is bled from its compressor outlet and is ducted to drive the air turbine starters on the main engines. The fuel flow is increased when the turbine air supply is reduced by the air bleed, to provide the energy required for compression. These engines are also found on ground carts, which may be temporarily connected to an aircraft to service it. They may also have uses in industrial plants requiring air pressure at 3 or 4 bar.

3.5 Engines for Surface Transportation

This category includes engines for rail, road, off-road, and over-water transport. The low weight and high power density of gas turbines are assets in all cases, but direct substitution for Diesel or Otto cycle engines is unusual. When the economics of an application favor high power density or high driven-device speed, or when some heat recovery is possible, gas turbines become the engines of choice. Surface vehicle engines include the array of turboshaft and turboprop derivatives, free-turbine aero derivative and industrial gas turbines, and purpose-built gas turbines. Applications exist for engines ranging from around 100 hp to nearly 40,000 hp.

Truck, bus, and automobile gas turbine engines are, for the most part, in the development stage. Current U.S. Department of Energy initiatives are supporting development of gas turbine automobile engines of superior efficiency and low emissions. Production cost similar to current power plants is also a program goal. Additional requirements must be met, such as fast throttle response and low fuel consumption at idle. The balancing of efficiency, first cost, size, and weight have led to different cycle and configuration choices than for aircraft or power generation applications. Regenerative cycles with low pressure ratios have been selected. Parts count and component costs are addressed through the use of centrifugal compressors, integral blade-disk axial turbines and radial inflow turbines. Low pressure ratio designs support the low stage count. It is possible to achieve the necessary pressure ratio with one centrifugal compression stage, and in one turbine stage, or one each high pressure and power turbine. The small size of parts and the selection of radial inflow or integral blade-disk turbines make ceramic materials an option. Single-can combustors are also employed to control cost. Prototypes have been built and operated in the United States, Europe, and Japan.[17]

The most successful automotive application of gas turbines is the power plant for the M1 Abrams Main Battle Tank. The engine uses a two-spool, multistage, all-axial flow, gas generator plus power turbine. The cycle is regenerative. Output and cost appear too high for highway vehicle application.

Ship propulsion by gas turbine is more commonplace. One recent report summarizing orders and installations over an 18-month period listed 10 orders for a total of 64 gas turbines—75 MW in all. Military applications accounted for 55% of the total. The remaining 45% were applied to fast ferries and similar craft being built in Europe, Australia, and Hong Kong.[11] Gas turbine outputs in the 3- to 5-MW range, around 10 MW, and in the 20- to 30-MW range account for all the applications. Small industrial engines were selected in the 3- to 5-MW range and aero derivative, free-turbine engines accounted for the remainder.

Successful application of gas turbines aboard ship requires protection from the effects of saltwater, and in the case of military vessels, maneuver and sudden seismic loads. In addition to the common problems with saltwater-induced corrosion of unprotected metal, airborne sodium (in combination with sulfur usually found in the fuel or air) presents a problem for buckets, nozzles, and combustors. Hot corrosion—also called sulfidation—has led to the development of alloys that combine the creep strength of typical aircraft engine bucket and nozzle alloys, with superior corrosion resistance. Inconel 738 was the first such alloy, and this set of alloys is used in marine propulsion engines. Special corrosion-resistant coatings are applied to further improve the corrosion resistance of nickel-based superalloy components. The level of sodium ingested by the engine can also be controlled with proper inlet design and filtration.[10]

Although there was a period when gas turbines were being applied as prime movers on railroad locomotives, the above report contained only one small railroad application.

4 EVALUATION AND SELECTION

4.1 Maintenance Intervals, Availability, and Reliability

Service requirements of aircraft and industrial gas turbines differ from other power plants principally in the fact that several key components operate at very high temperatures, and thus have limited lives and have to be repaired or replaced periodically to avoid failures during operation. Components that must be so maintained include combustion chambers, buckets, and nozzles. Occasionally, other components, such as wheels or casings, may require inspection or retirement.

Wear-out mechanisms in hot gas path components include creep, low cycle fatigue, corrosion, and oxidation. All combustors, buckets, and nozzles have a design life, and if operated for significantly longer than this design life, will fail in one these modes. Repair or replacement is required to avoid failure. Most of the failure mechanisms give some warning prior to loss of component integrity. Corrosion and oxidation are observable by visual inspection, and affected parts can be repaired or replaced. The creep deflection of nozzles can be detected by measuring changes in clearances. Low-cycle fatigue cracks can occur in nozzles, buckets, and combustors without causing immediate failure of these components. These can be detected visually or by more sophisticated nondestructive inspection techniques. Tolerance of cracks depends on the particular component design, service conditions, and what other forces or temperatures are imposed on the component at the location of the crack.

Inspection intervals are set by manufacturers (based on analysis, laboratory data, and field experience) so that components with some degree of distress can be removed from service or repaired prior to component failure.

Bucket creep often gives no advance warning. There are several factors that make this so. First, the ability of bucket alloys to withstand alternating stress and the rate of creep progression are both affected by the existence of creep void formation. Local creep void formation is difficult to observe even in individual buckets subjected to radiographic and other nondestructive inspections. Destructive inspection of samples taken from a turbine are not useful in predicting the conditions of a particular bucket in a stage suffering the most advanced creep damage. This is due to the statistical distribution of creep conditions in a sample set. Such a large number of samples would be required to accurately predict the condition of the worst part in a set that the cost of such an inspection would be higher than the set of replacement components. Because of this, creep failure can be avoided by the retirement of sets of buckets as the risk of the failure of the weakest bucket in the set increases to a preselected level.

Some of the wear-out mechanisms are time-related while others are start-related. Thus, the actual service profile is significant to determining when to inspect or retire gas path components. Manufacturers differ in the philosophy applied. Some aircraft engine maintenance recommendations have been based on a particular number of mission hours of operation. Each mission contains a number of hours at takeoff conditions, a number at cruise, a number of rapid accelerations, thrust reversals, etc. Components' lives have been calculated and expressed in terms of a number of mission cycles. Thus, the life of any component can be expressed in hours, even if the mechanism of failure expected is low cycle fatigue, related to the number of thermal excursions to which the component is exposed. Inspection and component retirement intervals (based on mission-hours) can be set to detect distress and remove or repair components before the actual failure is likely to occur. Actual starts and actual hours are becoming more commonplace measures for aircraft engines.

Industrial gas turbine manufacturers have historically designed individual products to be suitable for both continuous duty and frequent starts and stops. A particular turbine model may be applied to missions ranging from twice-daily starts to continuous operation for over 8000 hr per year and virtually no start cycles. To deal with this, manufacturers of industrial gas turbines have developed two ways of expressing component life and inspection intervals. One is to set two criteria for inspection—one based on hours and one based on starts. The other is to develop a formula for "equivalent hours," which counts each start as a number of additional hours of operation. These two methods are illustrated in Fig. 39. The figure is a simplification in that it considers only normal starts and base-load hours. Both criteria evaluate hours of operation at elevated firing temperature, fast starts, and emergency shutdown events as more severe than normal operating hours and starts. Industry practice is to establish maintenance factors that can be used to account for effects that shorten the intervals between inspections. Table 3 gives typical values. The hours-to-inspection, or starts-to-inspection in Fig. 39 would be divided by the factor in Table 3.

The values shown here are similar to those used by manufacturers, but are only approximate, and recommendations are modified and updated periodically. Also, the number, extent and types of inspections vary across the industry. To compare the frequency of inspection recommended for competing gas turbines, the evaluator must forecast the number of starts and hours expected during the evaluation period and, using the manufacturers' recommendation and other experience, determine the inspection frequency for the particular application.

Reliability and availability have specific definitions where applied to power-generation equipment[18]:

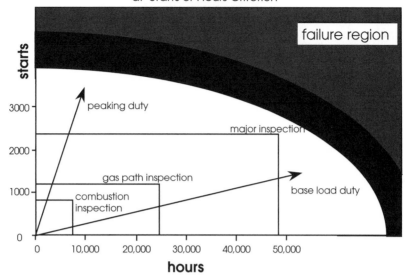

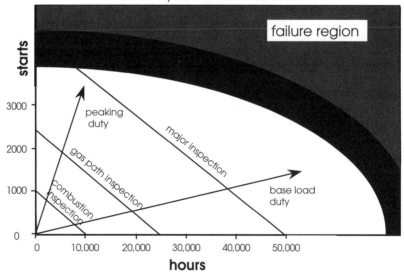

Figure 39 Inspection interval criteria compared. Starts-or-hours criterion shown requires major inspection after 48,000 hr or 2400 starts. Equivalent-hours criterion shown reflects each start being equivalent to 10 hr, and major inspection required after 50,000 equivalent hours.

Table 3 Maintenance Factors—Industrial Gas Turbine
Nozzles and Buckets

		Hours Factors
Fuel	Natural gas	1
	Distillate	1–1.5
	Residual	3–4
Peak load	Elevated firing temperature	5–10
Diluent injection	Water or steam	1–2
		Starts Factors
Trip from full load		6–10+
Fast load		2–4
Emergency start		10–20

Reliability = (1 − (FOH/PH), expressed as a percentage.
FOH = total forced outage hours
PH = period hours (8760 hr per year).

Reliability is used to reflect the percentage of time for which the equipment is not on forced outage. It is the probability of not being forced out of service when the unit is needed, and includes forced outage hours while in service, while on reserve shutdown, and while attempting to start.

Availability = (1 − (UH/PH), expressed as a percentage.
UH = total unavailable hours (forced outage, failure to start, unscheduled maintenance hours, maintenance hours)
PH = period hours

Availability reflects the probability of being available, independent of whether or not the unit is needed, and includes all unavailable hours normalized by period hours.

There are some minor differences in the definitions across the industry, which reflect the way different databases treat particular types of events, but the equations given above reasonably represent industry norms. Availability and reliability figures used in power-generation industry literature reflect the performance of not only the turbomachinery, but of the generator, control system, and accessories. Historically, less than half of the unavailability and forced outage hours are due to the turbomachinery.

Availability is affected by the frequency of inspections, duration of inspections, as well as the duration of forced outages. Improvements in analytical capability, understanding of material behavior, operating practices, and design sophistication have led to improvements in both availability and reliability over the past decades. The availability of industrial gas turbines has grown from 80% in the early 1970s to better than 95% in the mid-1990s.

4.2 Selection of Engine and System

In the transportation field, gas turbines are the engine of choice in large, and increasingly in small, aircraft where the number of hours per year flown is sufficiently high that the higher speed and lower fuel and service costs attributable to gas turbines justifies the higher first cost. Private automobiles, which operate nominally 400 hours per year, and where operating characteristics favor the Otto and Diesel cycles, are not likely to be candidates for gas turbine power, since exhaust-driven superchargers are a more acceptable application of turbomachinery technology to this market. Long-haul trucks, buses, and military applications may be served by gas turbines, if the economics that made them commonplace on aircraft can be applied.

Gas turbine technology finds application in mechanical drive and electric power generation. In mechanical drive application, the turbine rotor shaft typically drives a pump, compressor, or vehicle drive system. Mechanical drive applications usually employ "two-shaft" gas turbines in which the output shaft is controllable in speed to match the varying load/speed characteristic of the application. In electric power generation, the shaft drives an electrical generator at a constant synchronous speed. Mechanical drive applications typically find application for gas turbines in the 5- to 25-MW range and over the last 4 years this market was over 3000 MW per year. Power generation applications are typically in the larger size ranges, from 25 to 250 MW, and have recently averaged over 60,000 MW per year.

Gas turbine technology competes with other technologies in both power generation and mechanical drive applications. In both applications, the process for selecting which thermodynamic cycle or engine type to apply is similar. Table 4 summarizes the four key choices in electric power generation.

Steam turbine technology utilizes an externally fired boiler to produce steam and drive a Rankine cycle. This technology has been used in power generation for nearly a century. Because the boiler is fired external to the working fluid (steam), any type of fuel may be used, including coal, distillate oil, residual oil, natural gas, refuse, and biomass. The thermal efficiencies are typically in the 30% range for small (20- to 40-MW) industrial and refuse plants to 35% for large (400-MW) power-generation units, to 40% for large, ultraefficient, ultrasupercritical plants. These plants are largely assembled and erected at the plant site and have relatively high investment cost per kilowatt of output. Local labor costs and labor productivity influence the plant cost. Thus, the investment cost can vary considerably throughout the world.

Diesel technology uses the Diesel cycle in a reciprocating engine. The diesels for power generation are typically medium speed (800 rpm). The diesel engine has efficiencies from 40 to 45% on distillate oil. If natural gas is the fuel, the Diesel cycle is not applicable, but a spark ignition system based on the Otto cycle can be employed. The Otto cycle leads to three percentage points lower efficiency than the diesel. Diesel engines are available in

Table 4 Fossil Fuel Technologies for Mechanical Drive and Electric Power Generation

Technology	Power Cycle	Performance Level	Primary Advantages	Primary Disadvantages
Steam turbine	Rankine cycle	30–40%	Custom size Solid fuels Dirty fuels	Low efficiency Rel. high $/kW Slow load change
Gas turbine	Brayton cycle	30–40%	Packaged power plant Low $/kW Med fast starts Fast load delta	Clean fuels Ambient dependence
Combined cycle	Brayton topping/ Rankine bottoming	45–60%	Highest efficiency Med. $/KW Limited fast load delta	Clean fuels Ambient dependence Med. start times
Diesel	Diesel cycle	40–50%	Rel. high efficiency Packaged power plant Fast start Fast construction	High maint. Small size (5 MW)

smaller unit sizes than the gas turbines that account for most of the power generated for mechanical drive and power generation (1–10 MW). The investment cost of medium-speed diesels is relatively high per kilowatt of output, when compared with large gas turbines, but is lower than that of gas turbines in this size range. Maintenance cost of diesels per kilowatt of output is typically higher than gas turbine technology.

The *life-cycle cost* of power-generation technology projects is the key factor in their application. The life-cycle cost includes the investment cost charges and the present worth of annual fuel and operating expenses. The investment cost charges are the present worth costs of financing, depreciation, and taxes. The fuel and operating expenses include fuel consumption cost, maintenance expenses, operational material costs (lubricants, additives, etc.), and plant operation and maintenance labor costs. For a combined-cycle technology plant, investment charges can contribute 20%, fuel 70%, and operation and maintenance costs 10%. The magnitude and composition of costs is very dependent on technology and geographic location.

One way to evaluate the application of technology is to utilize a *screening curve* as shown in Fig. 40. This chart represents one particular power output and set of economic conditions, and is used here to illustrate a principal, not to make a general statement on the relative merits of various power-generation means. The screening curve plots the total $/kW/year annual life-cycle cost of a power plant versus the number of hours per year of operation. At zero hours of operation (typically of a standby plant used only in the event of loss of power form other sources), the only life-cycle cost component is from investment financing charges and any operating expense associated with providing manpower to be at the site. As the operating hours increase toward 8000 hr per year, the cost of fuel, maintenance, labor, and direct materials are added into the annual life-cycle cost.

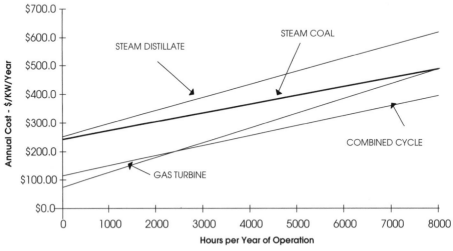

Figure 40 Hypothetical screening curve for selecting power generation technology from among various thermodynamic cycles and fuel alternatives. This curve would indicate the most economic choice for few operating hours per year is the simple cycle gas turbine, and the combined cycle for base-load applications. Position of the lines depends greatly on anticipated relative fuel costs. [*http://www. asme.org/igti/resources/articles/hybrid6.html*]

If the application has only a few hours per year of operation (less than 2000) the simple-cycle gas turbine technology has typically the lowest annual life-cycle cost and is therefore chosen. The simple-cycle gas turbine has the lowest annual life-cycle cost in this region in view of its low investment cost. If the application has more than 2000 hr per year of operation, then the combined-cycle technology provides the lowest annual life-cycle cost and is selected for application.

Other technology choices are the higher investment cost alternatives of the coal-fired steam turbine technology and the IGCC technology. In the example of Fig. 40 these technologies do not have the lowest annual life-cycle cost in any region. Consequently, they would not find application. However, the screening curve of Fig. 40 is based on a specific set of fuel prices and investment costs. In other regions of the world, coal prices may be less or natural gas prices may be higher. In this case, the coal technologies may be the lowest annual life-cycle cost in the 6000- to 8000-hr range. These technologies would then be selected for application.

In summary, there is a range of fuel prices and investment costs for power-generation technology. This range influences the applicability of the power-generation technology. In some countries with large low-priced coal resources, coal steam turbine technology is the most widely used. Where natural gas is available and modestly priced, gas turbine and combined-cycle technology is frequently selected.

REFERENCES

1. R. Harman, "Gas Turbines," in *Mechanical Engineers' Handbook,* M. Kutz (d.), Wiley, 1986, pp. 1984–2013.
2. D. E. Brandt, "The History of the Gas Turbine with an Emphasis on Schenectady General Electric Developments." GE Company, Schenectady, NY, 1994.
3. A. N. Smith and J. D. Alrich, "Gas Turbine Electric Locomotives in Operation in the USA," *Combustion Engine Progress,* c. 1957.
4. H. E. Miller and E. Benvenuti, "State of the Art and Trends in Gas-Turbine Design and Technology," European Powerplant Congress, Liége, 1993.
5. L. B. Davis, "Dry Low NO_x Combustion Systems for GE Heavy-Duty Gas Turbines." 38th GE Turbine State-of-the-Art Technology Seminar, GE Company, Schenectady, NY, 1994.
6. J. Zeldovich, "The Oxidation of Nitrogen in Combustion and Explosions," *Acta Physicochimica USSR,* **21**(4), 577–628 (1946).
7. D. M. Todd and H. E. Miller, "Advanced Combined Cycles as Shaped by Environmental Externality," Yokohama International Gas Turbine Congress, Yokohama, Japan, 1991.
8. G. Leonard and S. Correa, "NO_x Formation in Premixed High-Pressure Lean Methane Flames," ASME Fossil Fuel Combustion Symposium, New Orleans, ASME/PD Vol. 30, S. N. Singh (ed.), 1990, pp. 69–74.
9. R. J. Antos, "Westinghouse Combustion Development 1996 Technology Update," Westinghouse Electric Corporation Power Generation Business Unit, Orlando, FL, 1996.
10. C. T. Sims, N. S. Stoloff, and W. C. Hagel (eds.), *Superalloys II,* Wiley, New York, 1987.
11. R. Farmer, ed., *Gas Turbine World 2003 Handbook,* Vol. 23, Pequot, Fairfield, CT, 2003.
12. D. L. Chase et al., "GE Combined Cycle Product Line and Performance," Publication GER-3574E, GE Company, Schenectady, NY, 1994.
13. M. W. Horner, "GE Aeroderivative Gas Turbines—Design and Operating Features," 38th GE Turbine State-of-the-Art Technology Seminar, GE Company, Schenectady, NY, 1994.
14. T. W. Fowler (ed.), *Jet Engines and Propulsion Systems for Engineers,* University of Cincinnati, Cincinnati, OH, 1989.
15. P. G. Hill and C. R. Peterson, *Mechanics and Thermodynamics of Propulsion,* Addison-Wesley, Reading, MA, 1965.

16. A. L. Velocci Jr. (ed.), *2004 Aviation Week & Space Technology Aerospace Source Book,* McGraw-Hill, New York, 2004, pp. 126–140.

17. D. G. Wilson, "Automotive Gas Turbines: Government Funding and the Way Ahead," *Global Gas Turbine News,* **35**(4), ASME IGTI, Atlanta, 1995.

18. R. F. Hoeft, "Heavy-Duty Gas Turbine Operating and Maintenance Considerations," 38th GE Turbine State-of-the-Art Technology Seminar, GE Company, Schenectedy, NY, August 1994.

19. H. Roxbee Cox, "British Aircraft Gas Turbines," *Journal of the Aeronautical Sciences,* **13**(2) (1946).

20. R. Gusso and H. E. Miller, "Dry Low NO$_x$ Combustion Systems for GE/Nuovo Pignone Heavy Duty Gas Turbines," Flowers '92 Gas Turbine Congress, Florence, Italy, 1992.

21. C. Houllier, "The Limitation of Pollutant Emissions into the Air from Gas Turbines—Draft Final Report," CITEPA No. N 6611-90-007872, Paris, 1991.

CHAPTER 25

WIND TURBINES

Todd S. Nemec
GE Energy
Schenectady, New York

1 MARKET AND ECONOMICS

Wind power has long been used for grain-milling and water-pumping applications. Significant technical progress since the 1980s, however, driven by advances in aerodynamics, materials, design, controls, and computing power, has led to economically competitive electrical energy production from wind turbines. Technology development, favorable economic incentives (due to its early development status and environmental benefits), and increasing costs of power from traditional fossil sources have led to significant worldwide sales growth since the early 1980s. Production has progressed at an even faster pace, beginning in the late 1990s. Figure 1 shows the new U.S. installations since 1980.

The spike in U.S. wind turbine installations from 1982 to 1985 was due to generous tax incentives (up to 50% in California[1]), access to excellent wind resources, and high fossil fuel prices. Today Germany, the United States, Spain, and Denmark lead in installed MW, although significant growth is occurring worldwide.[2] From an energy-share standpoint, the northern German state, Shleswig-Holstein, produces approximately 30% of its electric energy from wind power, while Denmark produces about 20%.[3]

Like other power-producing technologies, wind turbines are measured on their ability to provide a low cost of electricity (COE) to customers, and high project net present value (NPV) to the plant owners. Unlike fossil plants, however, fuel (wind energy) is free. This causes COE to be dominated by the ratio of costs per unit energy, rather than a combination of capital costs, fuel cost, and thermal efficiency. For customers purchasing based on highest NPV, high power sale prices and energy production credits can drive turbine optimization to a larger size (and/or energy capture per rated MW), and higher COE, than would be expected from a typical optimization for lowest COE. Operation and maintenance costs (in cents per kWh) for wind turbines trend higher than those for fossil plants, primarily due to their lower power density. The ability to predict and trade life cycle costs versus energy improvement, from new technologies, is a key contributor to efficient technology development and market success.

2 CONFIGURATIONS

The most popular configuration for power generating wind turbines is the upwind 3-bladed horizontal axis wind turbine (HAWT) shown in Fig. 2. Upwind refers to the position of the blades relative to the tower.

837

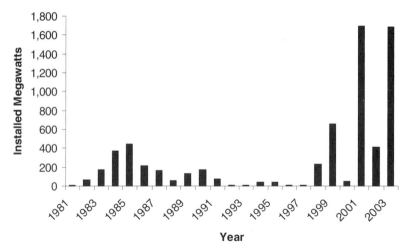

Figure 1 U.S. wind turbine installations (MW) by year. (*Source:* AWEA Annual Global Market Reports.)

Figure 2 Wind farm in Montfort, Wisconsin. (Courtesy GE Energy, © 2004, General Electric Company.)

Wind turbine configurations can be traced back to vertical-axis drag-type machines used for milling grain, which had the theoretical potential to achieve an 8% power coefficient, or percent energy extracted from the wind.[4] Modern vertical axis wind turbines (VAWT), like HAWTs, use the much more effective "lift" principle to produce power. VAWTs have been built in both Darrieus (curved blades connected at one or both ends) and "H" (separate vertical blades; also called giromill) configurations, although neither has been put into widespread use. VAWT aerodynamics are somewhat more complex, with a constantly changing angle of attack, and analyses has generally concluded that their power coefficient entitlement is lower than HAWTs.[5] Figure 3 shows the nacelle cutaway view of a horizontal-axis turbine.

The rotor, made up of the blades and hub, rotates a drivetrain through the low-speed shaft connected to a gearbox, high-speed shaft, and generator (or from the low speed shaft to a direct-drive generator). The nacelle consists of the base frame and enclosure; it houses the drivetrain, various systems, and electronics required for turbine operation. Towers are made of steel or steel-reinforced concrete. Steel towers use either a tubular or lattice type construction. Today's turbine configuration has evolved from both scaling up and adding features to small wind turbine designs, and from private and government-sponsored development of large machines.

3 POWER PRODUCTION AND ENERGY YIELD

Turbines extract energy from the wind according to following formula, derived from the first law of thermodynamics:

$$P = C_p \; \tfrac{1}{2} \; \rho \; A \; U^3$$

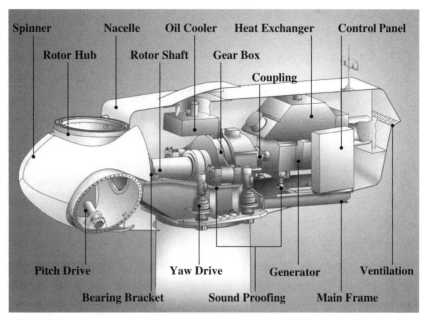

Figure 3 GE 1.5s nacelle, cutaway view. (Courtesy GE Energy, © 2004, General Electric Company.)

where P = power
$\quad C_p$ = power coefficient
$\quad \rho$ = air density
$\quad A$ = rotor swept area
$\quad U$ = air velocity at hub height

This equation shows power to be a function of air density and swept area, while varying by the cube of wind speed. These functions are not exact in real calculations, however, as aerodynamic and drivetrain characteristics restrict power coefficient over much of the operating range. The maximum theoretical power coefficient with zero airfoil drag and other simplifying assumptions is 59.3%, while modern turbines deliver peak coefficients in the mid-40% range.

The peak efficiency corresponds to a rotor exit air velocity of one-third the initial wind speed. This wake effect, along with site geographic, turbulence, and wind rose data, is significant when planning turbine spacing and arrangement on a multiturbine wind farm. Turbulence acts to reduce the velocity reduction immediately behind the turbine by reenergizing the wake, while it also spreads the energy loss over larger area.[4] Crosswind spacing, depending on wind characteristics, can usually be much closer than downwind spacing—crosswind tower spacing is on the order of 3–5 rotor diameters.

The power equation also provides insight into the basic power and mass scaling relationships. Power increases as a function of area, a function of diameter squared, while mass is a function of volume, or diameter cubed. This is true for aerodynamically load-limited components. Most electrical capacities and costs scale with rated power, while some part sizes are independent of turbine size. The cubed-squared relationship between component mass and power is the same found in most power generation cycles, such as gas turbines. Larger wind turbines are made economically possible by both reducing this 3/2 exponent, by improving technology and design strategy, such as using more advanced materials, and from increased leverage over fixed costs.

Wind turbine performance is characterized by its power curve, Figure 4, which shows the gross power produced as a function of wind speed. This curve assumes clean airfoils, standard control schedules, a given wind turbulence intensity, and a sea-level air density. Three items to note on the power curve are the cut-in speed, cut-out speed, and generator rating. Cut-in speed is determined by the wind speed where the aerodynamic torque is enough to overcome losses. Cut-out speed set to balance the power production in high winds with design loads and costs. Both speeds have dead-band regions around them to minimize the number of start–stop transients during small changes in wind speed.

Like cut-out speed, selecting the rotor diameter relative to generator rating relative requires balancing higher energy production at high wind speeds while minimizing costs.

Economic and design analysis has proven that turbines designed for high wind-speed operation should have a larger generator relative to rotor size, while those designed for low wind speed will have a larger rotor for a given generator size.[6] Advances in design and controls technology have not only helped turbines scale economically to larger sizes, but have allowed turbines to run larger rotor diameters at a given rating.[3]

Representative power curves are shown for two methods of limiting power in high winds. The pitch-controlled blade curve shows a constant generator output above rated power, while the curve for a stall-controlled (fixed-pitch) turbine delivers a peaked profile. Gross annual energy production (in kilowatt-hours, kWh) is calculated by multiplying the wind probability (in annual hours) at each wind speed by the power curve kW at that same wind speed, then adding up the total. Because of the higher probability for low wind speeds and turbine design/economic tradeoffs, wind turbines operate at 25 to 45% net capacity factor, depending on the turbine and site. Capacity factor is the fraction of energy produced relative

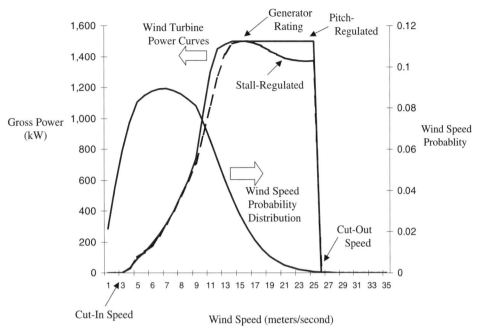

Figure 4 Wind speed probability distribution and notional power curves. (Courtesy GE Energy, © 2004, General Electric Company.)

to the rated capacity, shown below. This is much lower than the 95%+ levels achieved by dispatchable fossil-fuel power plants.

Annual energy yield = 5,000,000 kWh (assumed: measured from a kWh meter, or calculated based on turbine and wind conditions)

Rated capacity = 1500 kW (rating) × 8760 hr/year

Capacity factor = 5,000,000/(1500 × 8760) = 38%

Some of the gross-to-net loss is bookkept as availability losses (1% to 4% of energy produced), which are caused by both forced and scheduled outages. Other losses can total less than 15% and include array interference effects, electrical collection losses, blade soiling, and control losses.

Predictive performance analysis generally assumes that wind speed probability follows a Rayleigh distribution (Weibull distribution with shape factor equal to 2) along with an average wind speed at the hub height. In-depth and site analyses will use modifications to this statistical model, or will use data unique to a given site. While the previous example considers the effects of vertical wind shear on average wind speed at hub height, detailed energy yield analysis and loads calculations need to consider the effects of vertical velocity distribution. Taller towers allow turbines to see a higher average wind speed due to reduced friction with the ground and other objects at lower heights.

4 ROTOR AND DRIVETRAIN DESIGN

Rotor and drivetrain are ultimately optimized to yield the best economics for the turbine's mission. This is part of a multidisciplinary process involving aerodynamics, weight, materials, aeroelasticity, life, first cost, operating cost, frequency response, controls strategy, configuration options/technology availability, noise, site characteristics, supply chain, and a customer value equation. One modern mission requirement is quieter operation for land-based turbines—various noise sources correlate with tip speed raised to powers as high as the five, among other variables.[4] Aerodynamic characteristics that are selected include number of blades, tip speed ratio, blade radius, solidity, blade twist, chord length, airfoil section, and so on in greater detail. As an example of this process, consider that at a given rated power, higher tip speed ratio (higher rotor speed at a given wind speed):

- Reduces main shaft torque requirements and component sizes (costs)
- Increases noise
- Decreases rotor solidity (to maintain or increase the power coefficient), which reduces blade chord length and thickness (for a given number of blades)
- Makes it more difficult to fabricate blades and achieve strength objectives

Prior to generating power, wind turbines were generally configured as direct-drive units to pump water or mill grain. These applications place a high emphasis on torque coefficient at zero rotor speed, defining their ability to start under load. A high power coefficient was sacrificed by using high blade solidity (blade area divided by rotor disk area) and low tip speed ratios (tip speed divided by wind speed) to achieve high torque.[7] When transferring power through an electrical connection, such as a generator, rotor design generally favors higher power coefficients via low solidity and high tip speed ratio, resulting in modern high aspect-ratio blade shapes. Three blades have generally been favored over two because the power coefficient is higher at lower tip speed ratios, and for several structural dynamic considerations: out-of-plane bending loads are higher on 2-blade designs due to wind shear, tower shadow effect, effects of an upward-tiled shaft (to improve tower clearance), and from yaw-induced moments due to a changing moment of inertia.[8] Several of these loads can be eliminated in 2-blade rotors with the use of a teetering hub.

As described above, a wind turbine will generally be designed with little excess weight, or structural design, margin to optimize life-cycle economics. Design misses, such as higher weight blades, have subsequent effect in weight, cost, and/or life of drivetrain, tower, and foundation components. Turbine fatigue and ultimate loads are driven by four categories: aerodynamic, gravity, dynamic interactions, and control.[9]

Drivetrains absorb the rotor loads and distribute them to the bedplate for transmission to the tower and foundation. They also serve to convert torque into electrical power via the nacelle-mounted generator. Direct-drive generators, used by some manufacturers, turn at the rotor rpm, use a higher number of generator poles, and use power electronics to convert this rotor rpm into 50 or 60 Hz ac current. Geared drivetrains use a gearbox to drive a high-speed shaft connected to a smaller generator (with a fewer number of poles). Most manufacturers are employing variable-speed and pitch control using power electronics, pitch controllers, gearboxes, and induction (asynchronous) generators to optimize cost, energy yield, and improve grid power quality.

Design advances are evident in the lighter and more compact drivetrains. ENERCON GmbH has used direct-drive generators since the early 1990s.[10] These, and mixed solutions that use a single-stage gearbox to step up to a smaller low-speed generator, have been receiving more attention by other manufacturers as power electronic costs have come down.

REFERENCES

1. M. Gielecki et al, "Renewable Energy 2000, Issues and Trends: Incentives, Mandates, and Government Programs for Promoting Renewable Energy," U.S. DOE Energy Information Association (EIA), 2001.

2. Joint EWEA/AWEA press release, "Global Wind Power Growth Continues to Strengthen," March 10, 2004.

3. European Wind Energy Association (EWEA), "Wind Energy—The Facts: An Analysis of Wind Energy in the EU-25," 2003, retrieved from EWEA web site www.ewea.org on October 15, 2004.

4. D. Spera (ed.), *Wind Turbine Technology*, ASME Press, New York, 1995.

5. R. Harrison et al, *Large Wind Turbines: Design and Economics,* Wiley, New York, 2000.

6. D. J. Malcolm and A. C. Hansen, "WindPACT Turbine Rotor Design, Specific Rating Study," National Renewable Energy Laboratory (NREL), 2003.

7. J. A. C. Kentfield, *The Fundamentals of Wind-Driven Water Pumps,* Gordon & Breach Science, Amsterdam, 1996.

8. T. Burton et al. *Wind Energy Handbook,* Wiley, New York, 2001.

9. J. F. Manwell et al. *Wind Energy Explained: Theory, Design and Application*, Wiley, West Sussex, UK, 2003.

10. ENERCON History, retrieved October 21, 2004 from Enercon website: www.enercon.de.

CHAPTER 26

STEAM TURBINES

William G. Steltz
Turboflow International, Inc.
Palm City, Florida

1 HISTORICAL BACKGROUND

The process of generating power depends on several energy-conversion processes, starting with the chemical energy in fossil fuels or the nuclear energy within the atom. This energy is converted to thermal energy, which is then transferred to the working fluid, in our case, steam. This thermal energy is converted to mechanical energy with the help of a high-speed turbine rotor and a final conversion to electrical energy is made by means of an electrical generator in the electrical power-generation application. The presentation in this section focuses on the electrical power application, but is also relevant to other applications, such as ship propulsion.

Throughout the world, the power-generation industry relies primarily on the steam turbine for the production of electrical energy. In the United States, approximately 77% of installed power-generating capacity is steam turbine-driven. Of the remaining 23%, hydroelectric installations contribute 13%, gas turbines account for 9%, and the remaining 1% is split among geothermal, diesel, and solar power sources. In effect, over 99% of electric power generated in the United States is developed by turbomachinery of one design or another, with steam turbines carrying by far the greatest share of the burden.

Part of this chapter was previously published in J. A. Schetz and A. E. Fuhs (eds.), *Handbook of Fluid Dynamics and Fluid Machinery,* Vol. 3, *Applications of Fluid Dynamics,* New York, Wiley, 1996, Chapter 27.

Steam turbines have had a long and eventful life since their initial practical development in the late 19th century due primarily to efforts led by C. A. Parsons and G. deLaval. Significant developments came quite rapidly in those early days in the fields of ship propulsion and later in the power-generation industry. Steam conditions at the throttle progressively climbed, contributing to increases in power production and thermal efficiency. The recent advent of nuclear energy as a heat source for power production had an opposite effect in the late 1950s. Steam conditions tumbled to accommodate reactor designs, and unit heat rates underwent a step change increase. By this time, fossil unit throttle steam conditions had essentially settled out at 2400 psi and 1000°F with single reheat to 1000°F. Further advances in steam powerplants were achieved by the use of once-through boilers delivering supercritical pressure steam at 3500–4500 psi. A unique steam plant utilizing advanced steam conditions is Eddystone No. 1, designed to deliver steam at 5000 psi and 1200°F to the throttle, with reheat to 1050°F and second reheat also to 1050°F.

Unit sizes increased rapidly in the period from 1950 to 1970; the maximum unit size increased from 200 to 1200 mW (a sixfold increase) in this span of 20 years. In the 1970s, unit sizes stabilized, with new units generally rated at substantially less than the maximum size. At the present time, however, the expected size of new units is considerably less, appearing to be in the range of 350–500 mW.

In terms of heat rate (or thermal efficiency), the changes have not been so dramatic. A general trend showing the reduction of power station heat rate over an 80-year period is presented in Fig. 1. The advent of regenerative feedwater heating in the 1920s brought about a step change reduction in heat rate. A further reduction was brought about by the introduction of steam reheating. Gradual improvements continued in steam systems and were recently supplemented by the technology of the combined cycle, the gas turbine/steam turbine system (see Fig. 2). In the same period of time that unit sizes changed by a factor of six (1950 to 1970), heat rate diminished by less than 20%, a change that includes the combined cycle. In reality, the improvement is even less, as environmental regulations and the energy required to satisfy them can consume up to 6% or so of a unit's generated power.

The rate of improvement of turbine cycle heat rate is obviously decreasing. Powerplant and machinery designers are working hard to achieve small improvements both in new designs and in retrofit and repowering programs tailored to existing units. Considering the worth of energy, what, then, are our options leading to thermal performance improvements and the management of our energy and financial resources? Exotic energy-conversion processes are a possibility: MHD, solar power, the breeder reactor, and fusion are some of the longer-range possibilities. A more near-term possibility is through the improvement (increase) of steam conditions. The effect of improved steam conditions on turbine cycle heat rate is shown in Fig. 3, where heat rate is plotted as a function of throttle pressure with parameters of steam temperature level. The plus mark indicates the placement of the Eddystone unit previously mentioned.

2   THE HEAT ENGINE AND ENERGY CONVERSION PROCESSES

The mechanism for conversion of thermal energy is the *heat engine,* a thermodynamic concept, defined and sketched out by Carnot and applied by many, the power generation industry in particular. The heat engine is a device that accepts thermal energy (heat) as input and converts this energy to useful work. In the process, it rejects a portion of this supplied heat as unusable by the work production process. The efficiency of the ideal conversion process is known as the *Carnot efficiency.* It serves as a guide to the practitioner and as a limit for

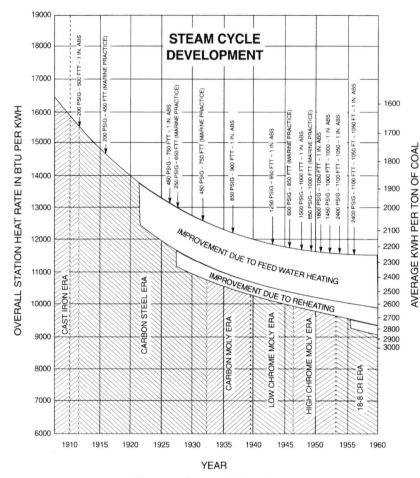

Figure 1 Steam cycle development.

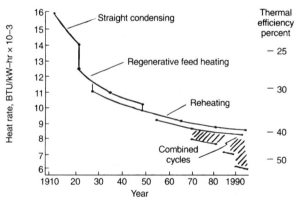

Figure 2 Fossil-fueled unit heat rate as a function of time.

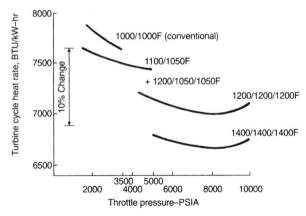

Figure 3 Comparison of turbine cycle heat rate as a function of steam conditions.

which no practical process can exceed. The Carnot efficiency is defined in terms of the absolute temperatures of the heat source T_{hot} and the heat sink T_{cold} as follows:

$$\text{Carnot efficiency} = \frac{T_{hot} - T_{cold}}{T_{hot}} \tag{1}$$

Consider Fig. 4, which depicts a heat engine in fundamental terms consisting of a quantity of heat supplied, heat added, a quantity of heat rejected, heat rejected. and an amount of useful work done, work done. The *thermal efficiency* of this basic engine can be defined as

$$\text{Efficiency} = \frac{\text{Work done}}{\text{Heat added}} \tag{2}$$

This thermal efficiency is fundamental to any heat engine and is, in effect, a measure of the heat rate of any turbine-generator unit of interest. Figure 5 is the same basic heat engine redefined in terms of turbine cycle terminology, that is, heat added is the heat input to the steam generator, heat rejected is the heat removed by the condenser, and the difference

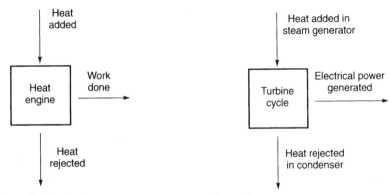

Figure 4 The basic heat engine.

Figure 5 The basic heat engine described in today's terms.

is the work done (power) produced by the turbine cycle. Figure 6 is a depiction of a simple turbine cycle showing the same parameters, but described in conventional terms. *Heat rate* is now defined as the quantity of heat input required to generate a unit of electrical power (kW).

$$\text{Heat rate} = \frac{\text{Heat added}}{\text{Work done}} \tag{3}$$

The units of heat rate are usually in terms of Btu/kW·hr.

Further definition of the turbine cycle is presented in Fig. 7, which shows the simple turbine cycle with pumps and a feedwater heater included (of the open type). In this instance, two types of heat rate are identified: (1) a *gross heat rate,* in which the turbine-generator set's natural output (i.e., gross electrical power) is the denominator of the heat rate expression, and (2) a *net heat rate,* in which the gross power output has been debited by the power requirement of the boiler feed pump, resulting in a larger numeric value of heat rate. This procedure is conventional in the power-generation industry, as it accounts for the inner requirements of the cycle needed to make it operate. In other, more complex cycles, the boiler feed pump power might be supplied by a steam turbine-driven feed pump. These effects are then included in the heat balance describing the unit's performance.

The same accounting procedures are true for all cycles, regardless of their complexity. A typical 450-mW fossil unit turbine cycle heat balance is presented in Fig. 8. Steam conditions are 2415 psia/1000°F/1000°F/2.5 in. Hga, and the cycle features seven feedwater heaters and a motor-driven boiler feed pump. Only pertinent flow and steam property parameters have been shown, to avoid confusion and to support the conceptual simplicity of heat rate. As shown in the two heat rate expressions, only two flow rates, four enthalpies, and two kW values are required to determine the gross and net heat rates of 8044 and 8272 Btu/kWh, respectively.

To supplement the fossil unit of Fig. 8, Fig. 9 presents a typical nuclear unit of 1000 mW capability. Again, only the pertinent parameters are included in this sketch for simplicity. Steam conditions at the throttle are 690 psi with ¼% moisture, and the condenser pressure is 3.0 in. Hga. The cycle features six feedwater heaters, a steam turbine-driven feed pump, and a moisture separator reheater (MSR). The reheater portion of the MSR takes throttle

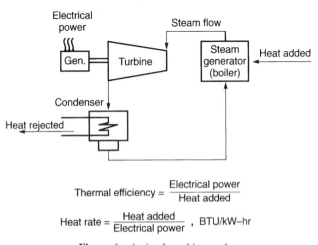

Figure 6   A simple turbine cycle.

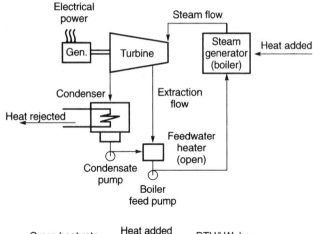

$$\text{Gross heat rate} = \frac{\text{Heat added}}{\text{Electrical power}} \text{ , BTU/kW–hr}$$

$$\text{Net heat rate} = \frac{\text{Heat added}}{\text{Electrical power–BFP power}} \text{ , BTU/kW–hr}$$

Figure 7   A simple turbine cycle with an open heater and a boiler feed pump.

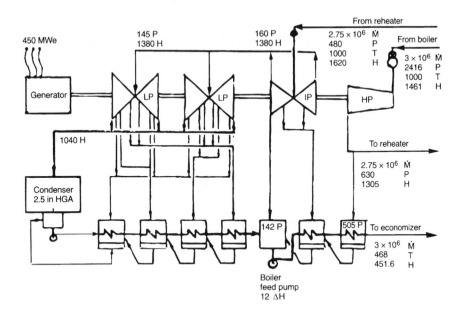

$$\text{Gross heat rate} = \frac{3,000,000 \, (1461 - 451.6) + 2,760,000 \, (1520 - 1305)}{450,000} = 8044 \text{ BTU/kW–hr}$$

$$\text{Net heat rate} = \frac{3,000,000 \, (1461 - 451.6) + 2,760,000 \, (1520 - 1305)}{450,000 - 12,400} = 8272 \text{ BTU/kW–hr}$$

Figure 8   Typical fossil unit turbine cycle heat balance.

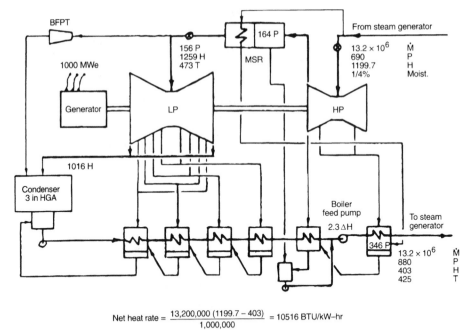

$$\text{Net heat rate} = \frac{13,200,000 \, (1199.7 - 403)}{1,000,000} = 10516 \text{ BTU/kW–hr}$$

Figure 9 Typical nuclear unit turbine cycle heat balance.

steam to heat the low-pressure (LP) flow to 473°F from 369°F (saturation at 164 psia). In this cycle, the feed pump is turbine-driven by steam taken from the MSR exit; hence, only one heat rate is shown, the net heat rate, 10,516 Btu/kWh. This heat rate comprises only four numbers, the throttle mass flow rate, the throttle enthalpy, the final feedwater enthalpy, and the net power output of the cycle.

For comparative purposes, the expansion lines of the fossil and nuclear units of Figs. 8 and 9 have been superimposed on the *Mollier diagram* of Fig. 10. It is easy to see the great difference in steam conditions encompassed by the two designs and to relate the ratio of cold to hot temperatures to their Carnot efficiencies. In the terms of Carnot, the maximum fossil unit thermal efficiency would be 61% and the maximum nuclear unit thermal efficiency would be 40%. The ratio of these two Carnot efficiencies (1.53) compares somewhat favorably with the ratio of their net heat rates (1.27).

To this point, emphasis has been placed on the *conventional* steam turbine cycle, where conventional implies the central station power-generating unit whose energy source is either a fossil fuel (coal, oil, gas) or a fissionable nuclear fuel. Figure 2 has shown a significant improvement in heat rate attributable to *combined-cycle* technology, that is, the marriage of the gas turbine used as a *topping unit* and the steam turbine used as a *bottoming unit*. The cycle efficiency benefits come from the high firing temperature level of the gas turbine, current units in service operating at 2300°F, and the utilization of its waste heat to generate steam in a heat-recovery steam generator (HRSG). Figure 11 is a heat balance diagram of a simplified combined cycle showing a two-pressure-level HRSG. The purpose of the two-pressure-level (or even three-pressure-level) HRSG is the minimization of the temperature differences existing between the gas turbine exhaust and the evaporating water/steam mixture. Second law analyses (commonly termed *availability* or *exergy analyses*) result in improved cycle thermal efficiency when integrated average values of the various heat-exchanger

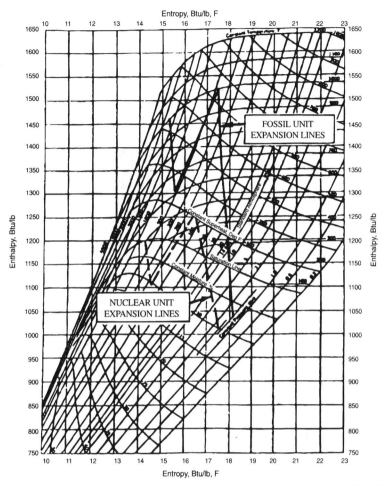

Entropy, Btu/lb, F

FOSSIL UNIT
EXPANSION LINES

NUCLEAR UNIT
EXPANSION LINES

Entropy, Btu/lb, F

Figure 10 Fossil and nuclear unit turbine expansion lines superimposed on the Mollier diagram.

temperature differences are small. The smaller, the better, from an efficiency viewpoint; however, the smaller the temperature difference, the larger the required physical heat transfer area. These second law results are then reflected by the cycle heat balance, which is basically a consequence of the first law of thermodynamics (conservation of energy) and the conservation of mass. As implied by Fig. 11, a typical combined-cycle schematic, the heat rate is about 6300 Btu/kW·hr, and the corresponding cycle thermal efficiency is about 54%, about ten points better than a conventional standalone fossil steam turbine cycle.

A major concept of the Federal Energy Policy of 1992 is the attainment of an Advanced Turbine System (ATS) thermal efficiency of 60% by the year 2000. Needless to say, significant innovative approaches will be required in order to achieve this ambitious level. The several approaches to this end include the increase of gas turbine inlet temperature and probably pressure ratio, reduction of cooling flow requirements, and generic reduction of blade path aerodynamic losses. On the steam turbine side, reduction of blade path aerodynamic losses and most likely increased inlet steam temperatures to be compatible with the gas turbine exhaust temperature are required.

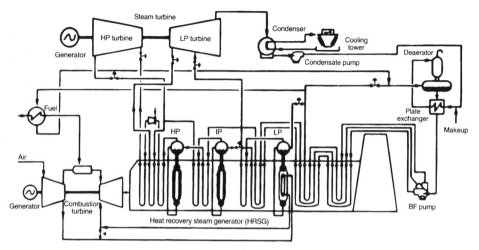

Figure 11 A typical combined-cycle plant schematic.

A possibility that is undergoing active development is the use of an ammonia/water mixture as the working fluid of the gas turbine's bottoming cycle in place of pure water. This concept known as the *Kalina cycle*[1] promises a significant improvement to cycle thermal efficiency primarily by means of the reduction of losses in system *availability*. Physically, a practical ammonia/water system requires a number of heat exchangers, pumps and piping, and a turbine that is smaller than its steam counterpart due to the higher pressure levels that are a consequence of the ammonia/water working fluid.

3 SELECTED STEAM THERMODYNAMIC PROPERTIES

Steam has had a long history of research applied to the determination of its thermodynamic and transport properties. The currently accepted description of steam's thermodynamic properties is the ASME Properties of Steam publication.[2] The Mollier diagram, the plot of enthalpy versus entropy, is the single most significant and useful steam property relationship applicable to the steam turbine machinery and cycle designer/analyst (see Fig. 12).

There are, however, several other parameters that are just as important and that require special attention. Although not a perfect gas, steam may be treated as such, provided the appropriate perfect gas parameters are used for the conditions of interest. The cornerstone of perfect gas analysis is the requirement that $pv = RT$. For nonperfect gases, a factor Z may be defined such that $pv = RZT$ where the product RZ in effect replaces the particular gas constant R. For steam, this relationship is described in Fig. 13, where RZ has been divided by J, Joule's constant.

A second parameter pertaining to perfect gas analysis is the isentropic expansion exponent given in Fig. 14. (The definition of the exponent is given in the caption on the figure.) Note that the value of γ well represents the properties of steam for a *short* isentropic expansion. It is the author's experience that accurate results are achievable at least over a 2:1 pressure ratio using an average value of the exponent.

The first of the derived quantities relates the critical flow rate of steam[3] to the flow system's inlet pressure and enthalpy, as in Fig. 15. The critical (maximum) mass flow rate M, assuming an isentropic expansion process and equilibrium steam properties, is obtained

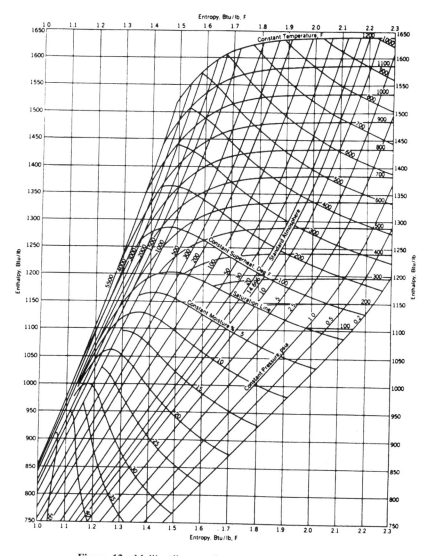

Figure 12 Mollier diagram ($h \cdot s$) for steam. (From Ref. 4.)

by multiplying the ordinate value K by the inlet pressure p_1 in psia and the passage throat area A in square inches:

$$\dot{M}_{\text{critical}} = p_1 K A \qquad (4)$$

The actual steam flow rate can then be determined as a function of actual operating conditions and geometry.

The corresponding choking velocity (acoustic velocity in the superheated steam region) is shown in Figs. 16 and 17 for superheated steam and wet steam, respectively. The range of Mach numbers experienced in steam turbines can be put in terms of the *wheel speed*

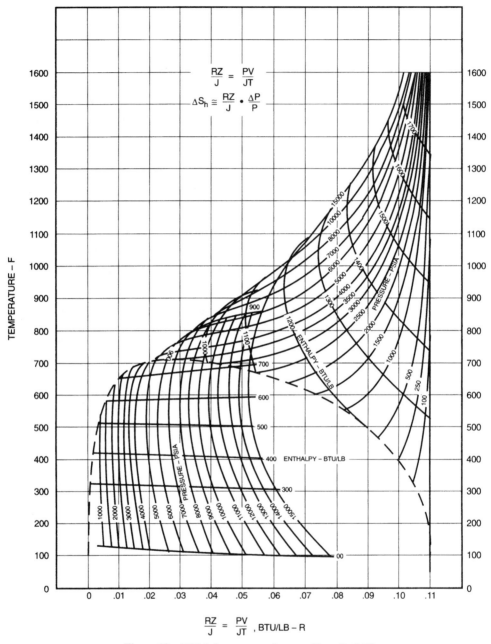

Figure 13 (RZ/J) for steam and water. (From Ref. 5.)

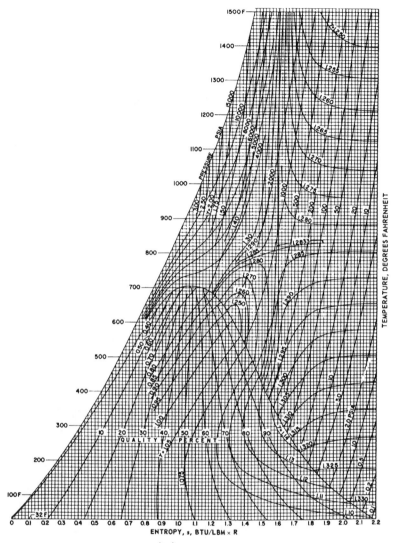

Figure 14 Isentropic exponent, $\gamma = -\dfrac{v}{p}\left(\dfrac{\partial p}{\partial v}\right)_s$, pv^γ = constant for a short expansion. (From Ref. 2.)

Mach number, that is, the rotor tangential velocity divided by the local acoustic velocity. In the HP turbine, wheel speed is on the order of 600 ft/sec, while the acoustic velocity at 2000 psia and 975°F is about 2140 ft/sec; hence, the wheel speed Mach number is 0.28. For the last rotating blade of the LP turbine, its tip wheel speed could be as high as 2050 ft/sec. At a pressure level of 1.0 psia and an enthalpy of 1050 Btu/lb, the choking velocity is 1275 ft/sec; hence, the wheel speed Mach number is 1.60. As Mach numbers relative to either the stationary or rotating blading are approximately comparable, the steam turbine designer must negotiate flow regimes from incompressible flow, low subsonic Mach number of 0.3, to supersonic Mach numbers on the order of 1.6.

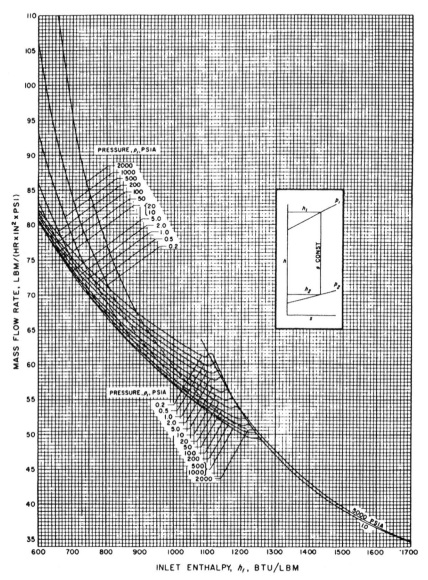

Figure 15 Critical (choking) mass flow rate for isentropic process and equilibrium conditions. (From Ref. 2.)

Another quite useful characteristic of steam is the product of pressure and specific volume plotted versus enthalpy in Figs. 18 and 19 for low-temperature/wet steam and superheated steam, respectively. If the fluid were a perfect gas, this plot would be a straight line. In reality, it is a series of nearly straight lines, with pressure as a parameter. A significant change occurs in the wet steam region, where the pressure parameters spread out at a slope different from that of the superheated region. These plots are quite accurate for determining specific volume and for computing the often used *flow number*

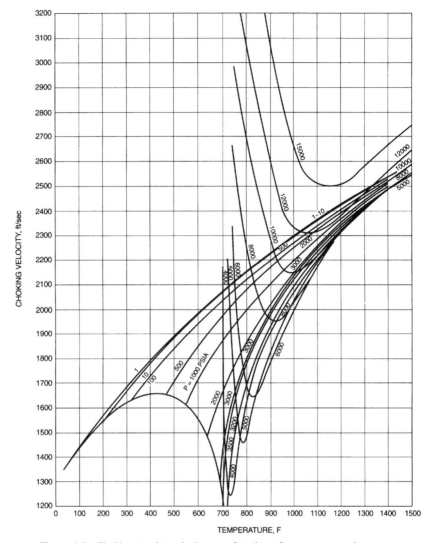

Figure 16   Choking (sonic) velocity as a function of temperature and pressure.

$$\frac{\dot{M}\sqrt{pv}}{p} \tag{5}$$

A direct application of the above-mentioned approximations is the treatment by perfect gas analysis techniques of applications where the working fluid is a mixture of air and a significant amount of steam (*significant* implies greater than 2–4%). Not to limit the application to air and steam, the working fluid could be the products of combustion and steam, or other arbitrary gases and steam.

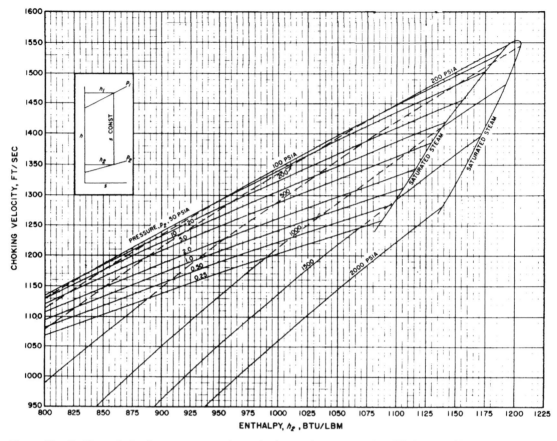

Figure 17   Choking velocity for water–steam mixture for isentropic process and equilibrium conditions. (From Ref. 2.)

4   BLADE PATH DESIGN

The accomplishment of the thermal to mechanical energy-conversion process in a steam turbine is, in general, achieved by successive expansion processes in alternate stationary and rotating blade rows. The turbine is a heat engine, working between the thermodynamic boundaries of maximum and minimum temperature levels, and as such is subject to the laws of thermodynamics prohibiting the achievement of engine efficiencies greater than that of a Carnot cycle. The turbine is also a dynamic machine in that the thermal to mechanical energy-conversion process depends on blading forces, traveling at rotor velocities, developed by the change of momentum of the fluid passing through a blade passage. The laws of nature as expressed by Carnot and Newton govern the turbine designer's efforts and provide common boundaries for his achievements.

The purpose of this section is to present the considerations involved in the design of steam turbine blade paths and to indicate means by which these concerns have been resolved. The means of design resolution are not unique. An infinite number of possibilities exist to achieve the specific goals, such as efficiency, reliability, cost, and time. The real problem is the achievement of all these goals simultaneously in an optimum manner, and even then, there are many approaches and many very similar solutions.

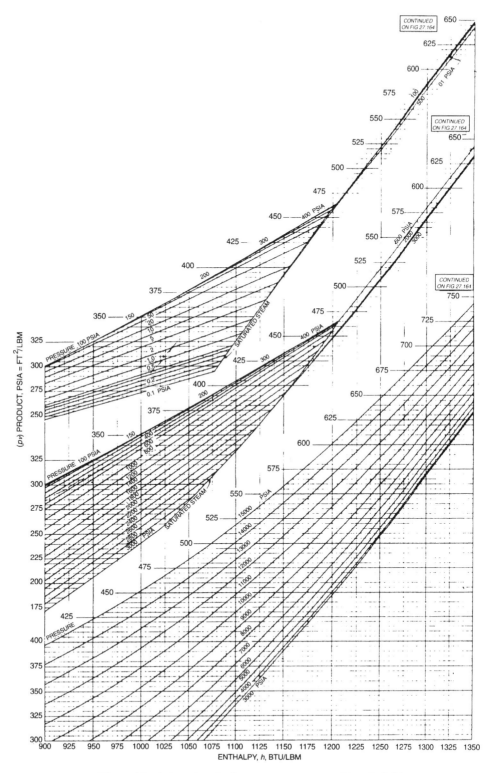

Figure 18 (ρv) product for low-temperature steam. (From Ref. 2.)

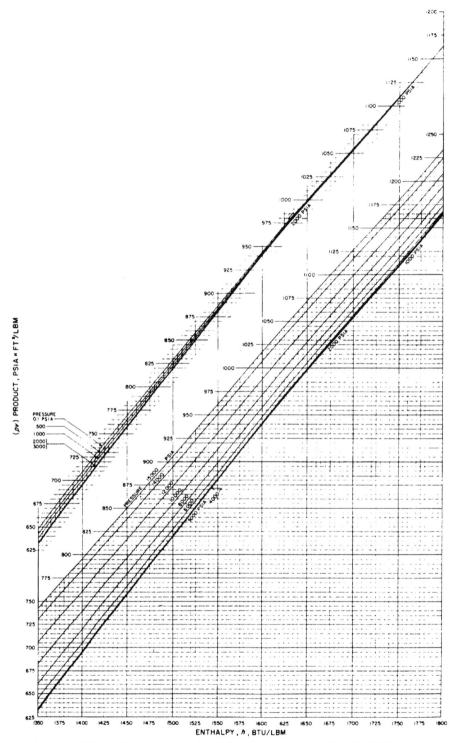

Figure 19   (ρv) product for high-temperature steam. (From Ref. 2.)

4.1   Thermal to Mechanical Energy Conversion

The purpose of turbomachinery blading is the implementation of the conversion of thermal energy to mechanical energy. The process of conversion is by means of curved airfoil sections that accept incoming steam flow and redirect it in order to develop blading forces and resultant mechanical work. The force developed on the blading is equal and opposite to the force imposed on the steam flow, that is, the force is proportional to the change of fluid momentum passing through the turbine blade row. The magnitude of the force developed is determined by application of Newton's laws to the quantity of fluid occupying the flow passage between adjacent blades, a space that is in effect, a *streamtube*. The assumptions are made that the flow process is steady and that flow conditions are uniform upstream and downstream of the flow passage.

Figure 20 presents an isometric view of a turbine blade row defining the components of steam velocity at the inlet and exit of the blade passage streamtube. The tangential force impressed on the fluid is equal to the change of fluid momentum in the tangential direction by direct application of Newton's laws.

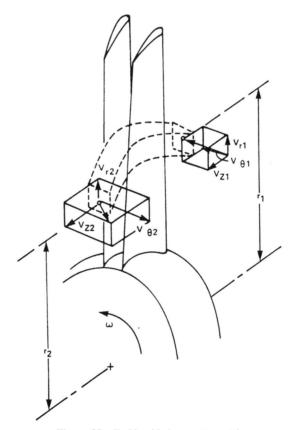

Figure 20   Turbine blade row streamtube.

$$F_{\theta \text{on fluid}} = \int_{A_2} V_{\theta} d\dot{M} - \int_{A_1} V_{\theta}\, d\dot{M} \tag{6}$$

$$F_{\theta \text{on fluid}} = \int_{A_2} \rho_2 V_{Z2} V_{\theta_2}\, dA - \int_{A_1} \rho_1 V_{Z1} V_{\theta_1}\, dA \tag{6a}$$

Since the flow is assumed steady and the velocity components are assumed constant within the streamtube upstream and downstream of the blade row, the force on the fluid is

$$F_{\theta \text{on fluid}} = \frac{\dot{M}}{g}(V_{\theta_2} - V_{\theta_1}) \tag{6b}$$

If the entering and leaving streamtube radii are identical, the power P developed by the moving blade row is

$$P = \frac{\dot{M}}{g} r\omega(V_{\theta 1} - V_{\theta 2}) \tag{7}$$

where the algebraic sign is changed to determine the fluid force on the blade. With unequal radii, implying unequal tangential blade velocities, the change in angular fluid momentum requires the power relation to be

$$P = \frac{\dot{M}}{g} \omega(r_1 V_{\theta 1} - r_2 V_{\theta 2}) \tag{7a}$$

Consideration of the general energy relationship (the first law) expressed as

$$\delta Q - \delta W = du + d(pv) + \frac{VdV}{g} + dz \tag{8}$$

(Q is heat, W is work, and u is internal energy) and applied to the inlet and exit of the streamtube yields

$$\delta W = dh + \frac{VdV}{g} = dh_0 \tag{9}$$

where h is static enthalpy and h_0 is stagnation enthalpy. The change in total energy content of the fluid must equal that amount absorbed by the moving rotor blade in the form of mechanical energy

$$h_{01} - h_{02} = \frac{\omega}{g}(r_1 V_{\theta 1} - r_2 V_{\theta 2}) \tag{10}$$

or

$$h_1 + \frac{V_1^2}{2g} - h_2 - \frac{V_2^2}{2g} = \frac{\omega}{g}(r_1 V_{\theta 1} - r_2 V_{\theta 2}) \tag{10a}$$

Both of these may be recognized as alternate forms of *Euler's turbine equation.*

In the event the radii r_1 and r_2 are the same, Euler's equation may be expressed in terms of the velocity components relative to the rotor blade as

$$h_{01} - h_{02} = \frac{U}{g}(W_{\theta 1} - W_{\theta 2}) \tag{10b}$$

or

$$h_1 + \frac{V_1^2}{2g} - h_2 - \frac{V_2^2}{2g} = \frac{U}{g}(W_{\theta 1} - W_{\theta 2}) \qquad (10c)$$

4.2   Turbine Stage Designs

The means of achieving the change in tangential momentum, and hence blade force and work, are many and result in varying turbine stage designs. Stage design and construction falls generally into two broad categories, *impulse* and *reaction*. The former implies that the stage pressure drop is taken entirely in the stationary blade passage, and that the relative entering and leaving rotor blade steam velocities are equal in magnitude. Work is achieved by the redirection of flow through the blade without incurring additional pressure decrease. At the other end of the scale, one could infer that for the reaction concept, all the stage pressure drop is taken across the rotor blade, while the stator blade merely redefines steam direction. The modern connotation of reaction is that of 50% *reaction,* wherein half the pressure drop is accommodated in both stator and rotor. Reaction can be defined in two ways, on an enthalpy-drop or a pressure-drop basis. Both definitions ratio the change occurring in the rotor blade to the stage change. The pressure definition is more commonly encountered in practice.

Figure 21 presents a schematic representation of impulse and reaction stage designs. The impulse design requires substantial stationary diaphragms to withstand the stage pressure

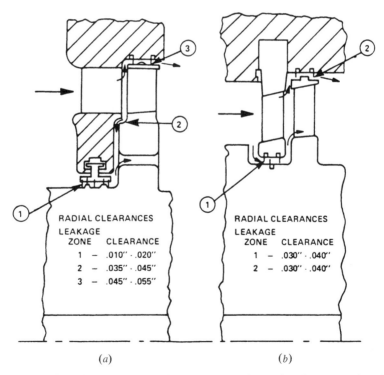

RADIAL CLEARANCES
LEAKAGE
ZONE   CLEARANCE
1 — .010″ · .020″
2 — .035″ · .045″
3 — .045″ · .055″

RADIAL CLEARANCES
LEAKAGE
ZONE   CLEARANCE
1 — .030″ · .040″
2 — .030″ · .040″

(*a*)   (*b*)

Figure 21   Comparison of impulse and reaction stage geometries: (*a*) impulse construction; (*b*) reaction construction.

drop and tight sealing at the inner diameter to minimize leakage. The reaction design (50%) can accept somewhat greater seal clearances under the stator and over the rotor blade and still achieve the same stage efficiency.

The work per stage potential of the impulse design is substantially greater than that of the reaction design. A comparison is presented in Fig. 22 of various types of impulse designs with the basic 50% reaction design. A typical impulse stage (*Rateau stage*) can develop 1.67 times the work output of the reaction stage; for a given energy availability, this results in 40% fewer stages.

Performance characteristics of these designs are fundamentally described by the variation of stage aerodynamic efficiency as a function of velocity ratio. The *velocity ratio* can be defined many ways dependent on the reference used to define the steam speed *V*. The velocity ratio in general is

$$v = \frac{U}{V} \tag{11}$$

where *U* is the blade tangential velocity.

The relative level of stage aerodynamic efficiency is indicated in Fig. 22 for these various geometries ranging from the symmetrical reaction design (50% reaction) to a three-rotating-row impulse design. As the work level per design increases, the aerodynamic efficiency decreases. Some qualifying practicalities prevent this from occurring precisely this way in practice. That is, the impulse concept can be utilized in the high-pressure region of the turbine, where densities are high and blade heights are short. When blade heights necessarily increase to pass the required mass flow, significant radial changes occur in the flow conditions, resulting in radial variations in pressure and, hence, reaction. All impulse designs have some amount of reaction dependent on the ratio of blade hub to tip diameters. Figure 23 presents the variation of reaction in a typical impulse stage as a function of the blade diameter/base diameter ratio.

Further complications arise when comparing the impulse to the reaction stage design. The rotor blade turning is necessarily much greater in the impulse design, the work done can be some 67% greater, and the attendant blade section aerodynamic losses tend to be

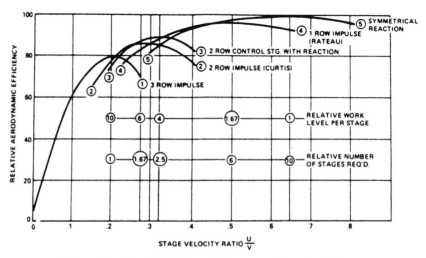

Figure 22   The effect of stage design on aerodynamic efficiency.

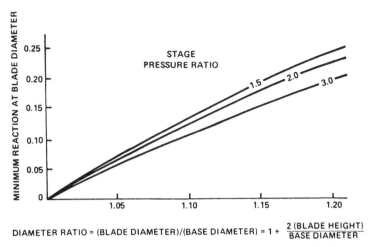

Figure 23 Blade reaction required to prevent negative base reaction.

greater. As an extreme case, Fig. 24 presents blade losses as a function of turning angle, indicating much greater losses incurred by the impulse rotor blade design.

Test data are, of course, the best source of information defining these performance relationships. Overall turbine test data provide the stage characteristic, which includes all losses incurred. The fundamental blade section losses are compounded by leakages around blade ends, three-dimensional losses due to finite length blades and their end walls, moisture losses if applicable, disk and shroud friction, and blading incidence losses induced by the variation in the velocity ratio itself. A detailed description of losses has been presented by Craig and Cox[6] and more recently by others including Kacker and Okapuu.[7]

4.3 Stage Performance Characteristics

Referring again to Fig. 20, the velocities relative to the rotor blade are related to the stator velocities and to the wheel speed by means of the appropriate velocity triangles, as in Fig.

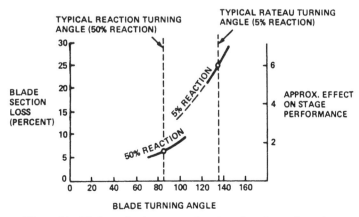

Figure 24 Blade section losses as a function of turning and reaction.

25. Absolute velocities are denoted by V and relative velocities by W. The blade sections schematically shown in this figure are representative of reaction blading. The concepts and relationships are representative of all blading.

The transposition of these velocities and accompanying steam conditions are presented on a Mollier diagram for superheated steam in Fig. 26. The stage work will be compatible with Euler's turbine equation shown in Eqs. (10). The *stage work* is the change in total enthalpy ($\Delta h_w = h_{00} - h_{02}$) of the fluid passing through the stage. The total pressures and temperatures before and after the stage, the local thermodynamic conditions, and total pressure and total temperature relative to the rotor blade (p_{0_r} and T_{0_r}) are all indicated in Fig. 26.

This ideal description of the steam expansion process defines the local conditions of a particular streamtube located at a certain blade height. For analysis calculations and the

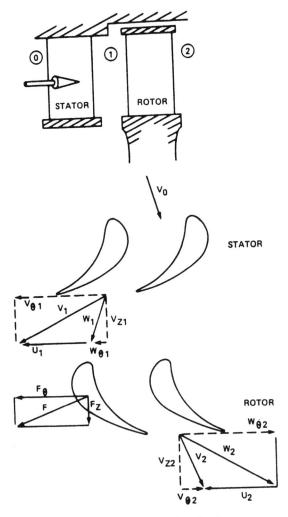

Figure 25 Stage velocity triangles.

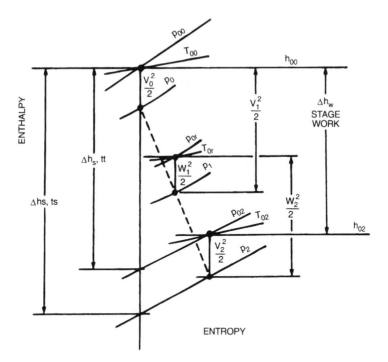

Figure 26 Stage thermodynamic conditions on a Mollier diagram.

prediction of turbine performance, the mean diameter is usually chosen as representative of the stage performance. This procedure is, in effect, a one-dimensional analysis, and it represents the turbine performance well if appropriate corrections are made for three-dimensional effects, leakage, and moisture. In other words, the blade row, or stage efficiency, must be known or characterized as a function of operating parameters. The velocity ratio, for example, is one parameter of great significance. Dimensional analysis of the individual variables bearing on turbomachinery performance has resulted in several commonly used dimensionless quantities:

$$v = \text{velocity ratio } (U/V) \tag{12}$$

$$\rho = \text{pressure ratio } (p_0/p) \tag{13}$$

$$\frac{\dot{M}\sqrt{\theta}}{\delta} = \text{referred flow rate} \tag{14}$$

$$\frac{N}{\sqrt{\theta}} = \text{referred speed} \tag{15}$$

$$\eta = \text{efficiency} \tag{16}$$

where $\theta = T_0/T_{\text{ref}}$ and $\delta = p_0/p_{\text{ref}}$.

The *referred flow rate* is a function of pressure ratio. In gas dynamics, the referred flow rate through a converging nozzle is primarily a function of the pressure ratio across the nozzle (Fig. 27). When the pressure ratio p_0/p is approximately 2 (depending on the ratio of specific heats), the referred mass flow maximizes, that is,

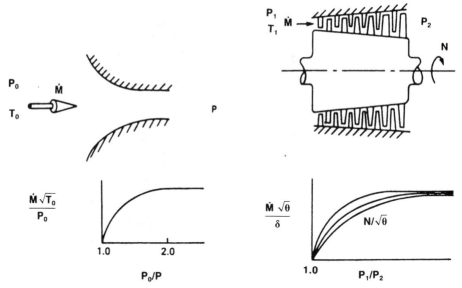

Figure 27 Gas dynamic flow relationships. **Figure 28** Turbine flow relationships.

$$\frac{\dot{M}\sqrt{T_0}}{p_0} = \text{constant} \tag{17}$$

This referred flow rate is also termed a *flow number* and takes the following form, which is more appropriate for steam turbine usage.

$$\frac{\dot{M}\sqrt{p_0 v_0}}{p_0} = \text{flow number} \tag{18}$$

The turbine behaves in a manner similar to a nozzle, but it is also influenced by the wheel-speed Mach number, which is represented by $N/\sqrt{T_0}$. In effect, a similar flow rate–pressure ratio relationship is experienced, as shown schematically in Fig. 28.

A combined plot describing turbine performance as a function of all these dimensionless quantities is shown schematically in Fig. 29. Power-producing steam turbines, however, run at constant speed and at essentially constant flow number, pressure ratio, and velocity ratio, and in effect operate at nearly a fixed point on this performance map. Exceptions to this are the control stage, if the unit has one, and the last few stages of the machine. The control stage normally experiences a wide range of pressure ratios in the situation where throttle pressure is maintained at a constant value and a series of governing valves admits steam to the control stage blading through succeeding active nozzle arcs.

The last stage of the low pressure end exhausts to an essentially constant pressure zone maintained by the condenser. Some variation occurs as the condenser heat load changes with unit flow rate and load. At a point several stages upstream where the pressure ratio to the condenser is sufficiently high, the flow number is maintained at a constant value. As unit load reduces, the pressure level and mass flow rate decrease simultaneously and the pressure ratio across the last few stages reduces in value. This change in pressure ratio changes the stage velocity ratio, and the performance level of these last few stages change. Figure 30 indicates the trend of operation of these stages as a function of load change superimposed

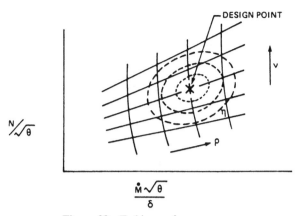

Figure 29 Turbine performance map.

on the dimensionless turbine performance map. It is these last few stages, say two or three, that confront the steam turbine designer with the greatest challenge.

4.4 Low-Pressure Turbine Design

The low-pressure element of the steam turbine is normally a self-contained design, in the sense that it comprises a rotor carrying several stages, a cylinder with piping connections, and bearings supporting the rotor. Complete power-generating units may use from one to three low-pressure elements in combination with high-pressure and intermediate-pressure elements depending on the particular application.

Symmetric double-flow designs utilize from 5 to 10 stages on each end of the low-pressure element. The initial stage accepts flow from a centrally located plenum fed by large-diameter, low-pressure steam piping. In general, the upstream stages feature constant cross-section blading, while the latter stages require blades of varying cross section. Varying design requirements and criteria result in these latter twisted and tapered low pressure blades comprising from 40 to 60% of the stages in the elements.

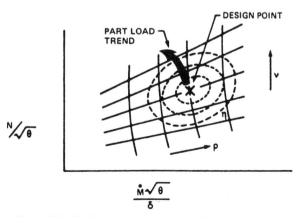

Figure 30 Turbine performance map with part load trend.

The particular design criterion used significantly affects the end product. For example, the low-pressure end (one-half of the element) of a nuclear unit might contain 10 stages and rotate at 1800 rpm and generate 100 mW, while the fossil machine would spin at 3600 rpm, carry five or six stages, and develop 50 mW. The physical size could double and the weight increase by a factor of 8.

Further differences in design are immediately implied by the steam conditions to be accepted by the element. The nuclear unit inlet temperature would be approximately 550°F (288°C), as compared to the fossil condition of 750°F (399°C). The nuclear unit must also be designed to effectively remove blade path moisture in order to achieve high thermal performance levels (heat rate). Moisture-removal devices in both the high-pressure and low-pressure elements are normally incorporated for this purpose. Blading physical erosion is also an important consideration. Material removal by the high speed impingement of water droplets can be minimized by proper design. Moore and Sieverding[8] describe the phenomena controlling wet steam flow in turbines.

Flow Field Considerations

Substantial advances have been achieved in the aerothermodynamic processes of low-pressure turbines. A significant factor is the availability of high-speed digital computers in the solution of the system of turbomachinery aerothermodynamic equations. The coupling of the solution of this complex equation system with the necessary verification as derived from experimental testing programs is the subject of this section.

The development of the system of aerothermodynamic equations is traced from basic considerations of the several conservation equations to working equations, which are then generally solved by numerical techniques. Pioneering work in the development of this type of equation systems was performed by Wu.[9] Subsequent refinement and adaptation of his fundamental approach has been made by several workers in recent years.

The general three-dimensional representation of the conservation equations of mass, momentum, and energy can be written for this adiabatic system

$$\frac{\partial \rho}{\partial t} + \nabla \cdot \rho \overline{V} = 0 \tag{19}$$

$$\frac{\partial \overline{V}}{\partial t} + \overline{V} \cdot \nabla \overline{V} = -\frac{1}{\rho} \nabla p \tag{20}$$

$$\nabla h = T \nabla s + \frac{1}{\rho} \nabla p \tag{21}$$

Assuming the flow to be steady and axisymmetric, the conservation equations can be expanded in cylindrical coordinates as

$$\frac{\partial}{\partial r}(\rho r b V_r) + \frac{\partial}{\partial z}(\rho r b V_z) = 0 \tag{22}$$

$$V_r \frac{\partial}{\partial r}(r V_\theta) + V_z \frac{\partial}{\partial z}(r V_\theta) = 0 \tag{23}$$

$$V_r \frac{\partial V_z}{\partial r} + V_z \frac{\partial V_z}{\partial z} = -\frac{1}{\rho}\frac{\partial p}{\partial z} \tag{24}$$

$$V_r \frac{\partial V_r}{\partial r} + V_z \frac{\partial V_r}{\partial z} - \frac{V_\theta^2}{r} = -\frac{1}{\rho}\frac{\partial p}{\partial r} \tag{25}$$

$$\frac{1}{\rho} dp = dh_0 - Tds - \frac{1}{2} d(V_\theta^2 + V_z^2 + V_r^2) \tag{26}$$

where

$$h_0 = h + \tfrac{1}{2}(V_\theta^2 + V_z^2 + V_r^2) \tag{27}$$

and b has been introduced to account for *blade blockage.* The properties of steam may be described by equations of state, such that $p = p(p, T)$ and $h = h(p, T)$. These functions are available in the form of the steam tables.[2]

As the increase in system entropy is a function of the several internal loss mechanisms inherent in the turbomachine, it is necessary that definitive known (or assumed) relationships be employed for its evaluation. In the design process, this need can usually be met, while the performance analysis of a given geometry usually introduces loss considerations more difficult to evaluate completely. The matter of loss relationship has been discussed with regard to stage design and will be further reviewed in a later section. In general, however, the entropy increase can be determined as a function of aerodynamic design parameters, that is,

$$\Delta s = f(V_i, W_i, \text{Mach number}) \tag{28}$$

Equations (22) through (25) form a system of nine equations in nine unknowns, V_r, V_θ, V_z, ρ, p, T, h, h_0, and s, the solution of which defines the low pressure turbine flow field.

The *meridional plane,* defined as that plane passing through the turbomachine axis and containing the radial and axial coordinates, can be used to describe an additional representation of the flow process. The velocity V_m (see Fig. 31) represents the streamtube meridional plane velocity with direction proportional to the velocity components V_r and V_z. If the changes in entropy and total enthalpy along the streamlines are known, or specified, it can be shown that Eqs. (24) and (25) are equivalent and that Eq. (23) is also satisfied. Application of Eq. (22) to the rotating blade in effect described Euler's turbine equation when equated to the blade force producing useful work. In this situation, it is most convenient to choose

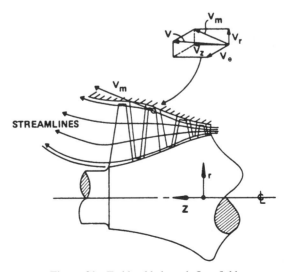

Figure 31 Turbine blade path flow field.

Eq. (25), the commonly known *radial equilibrium equation,* as the relationship for continued evaluation.

4.5 Flow Field Solution Techniques

Two commonly encountered techniques have been employed to solve this set of fundamental differential equations and relationships. They are usually referred to as the *streamline curvature* and *matrix solution* techniques. The streamline curvature technique is structured to evaluate meridional velocities and to trace streamlines in the flow field, allowing the calculation of the streamline curvature itself. The streamline curvature technique has been developed to a high degree of sophistication and has been applied to axial flow compressor design as well as to axial flow turbine problems.

The matrix approach utilizes a stream function satisfying the equation of continuity. The calculation procedure then determines the stream function throughout the flow field. This technique has also been developed satisfactorily with particular application to low-pressure steam turbines. The matrix approach asserts the existence of a stream function that is defined such that

$$V_r = -\frac{1}{\rho r b} \frac{\partial \psi}{\partial z} \tag{29}$$

and

$$V_z = \frac{1}{\rho r b} \frac{\partial \psi}{\partial r} \tag{30}$$

With this definition, ψ, V_r, and V_z identically satisfy Eq. (19). The combination of the equations of momentum, energy, and continuity, Eqs. (25), (26), (29), (30), and the definition of total enthalpy, Eq. (27), results in

$$\frac{\partial}{\partial r}\left(\frac{1}{\rho r b}\frac{\partial \psi}{\partial r}\right) + \frac{\partial}{\partial z}\left(\frac{1}{\rho r b}\frac{\partial \psi}{\partial z}\right) = \frac{1}{V_z}\left[\frac{\partial h_o}{\partial r} - \frac{V_\theta}{r}\frac{\partial (rV_\theta)}{\partial r} - T\frac{\partial s}{\partial r}\right] \tag{31}$$

This is the basic flow field equation, which can then be solved by numerical techniques. This equation is an elliptic differential equation, provided the meridional Mach number is less than unity. This is probably the case in all large steam turbines. The absolute Mach number can of course exceed unity and usually does in the last few blade rows of the low-pressure turbine.

The streamline curvature approach satisfies the same governing equations but solves for the meridional velocity V_m rather than the stream function ψ, as in the matrix approach. Equations (25) and (26) can be combined and expressed in terms of directions along the blade row leading or trailing edges. An equation used to describe the variation of meridional velocity is

$$\frac{dV_m^2}{dl} + A(l)V_m^2 = B(l) \tag{32}$$

where l is the coordinate along the blade edge and the coefficients A and B depend primarily on the slopes of the streamlines in the meridional plane. An iterative process is then employed to satisfy the governing equations.

4.6   Field Test Verification of Flow Field Design

The most conclusive verification of a design concept is the in-service evaluation of the product with respect to its design parameters. Application of a matrix-type flow field design program has been made to the design of the last three stages of a 3600-rpm low-pressure end. Figure 32 is indicative of the general layout of this high-speed fossil turbine.

The aerodynamic design process is to a great degree an iterative process, that is, the specification of the design parameters (for example, work per stage and radial work distribution) are continuously adjusted in order to optimize the flow field design. Just as important, feedback from mechanical analyses must be accommodated in order to achieve reliable long low-pressure blades.

A key design criterion is the requirement that low-pressure blades must be vibrationally tuned in that the lowest several natural modes of vibration must be sufficiently removed from harmonics of running speed during operation. This tuning process is best represented by the *Campbell diagram* of Fig. 33. The first four modes of the longest blade are represented as a function of running speed in rpm. At the design speed, the intersection of the mode lines and their bandwidths indicate nonresonant operation. A comparison is also shown between full-size laboratory test results and shop test data. The laboratory results[10] were obtained by means of strain gage signals transmitted by radio telemetry techniques. In the shop tests, the blades were excited once per revolution by a specially designed steam jet. The strain gage signals were delivered to recording equipment by means of slip rings.

If it turns out that the mode lines intersect a harmonic line at running speed, a resonant condition exists that could lead to a fatigue failure of the blade. The iterative process occurs prior to laboratory and shop verification. Analytic studies and guidance from experience gained from previous blade design programs enable the mechanical designer to determine blade shape changes that will eliminate the resonance problem. This information is then

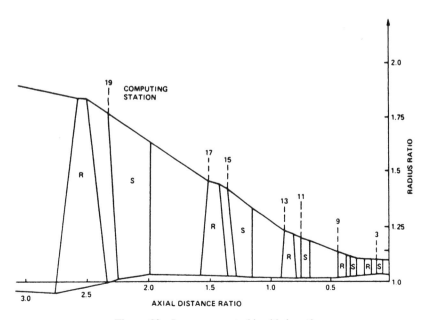

Figure 32 Low-pressure turbine blade path.

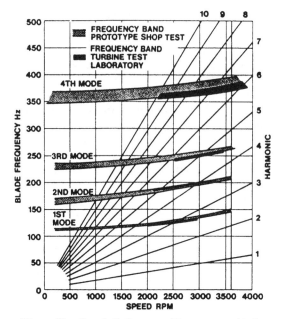

Figure 33 Campbell diagram of low-pressure blade.

incorporated by the aerodynamicist into the flow field design. As this process continues, manufacturing considerations are incorporated to ensure the design can be produced satisfactorily.

Construction and testing then followed: test data was obtained from several sources, a field test and an in-house test program.[10] Figure 34 displays low-pressure turbine efficiency versus exhaust volumetric flow for the advanced design turbine, designed by means of the matrix-type flow field design process, and original design turbine. Although the same range

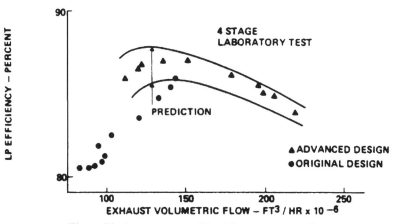

Figure 34 Low-pressure turbine laboratory test results.

of exhaust volumetric flow was not achievable in both test series, the improvement is apparent and is further indicated by the extended performance line of the original design as determined by predictive techniques.

Field test results were obtained in which the original design and the advanced design were evaluated by procedures defined by the ASME Performance Test Code. Figure 35 presents these results in the form of low-pressure turbine efficiency, again as a function of exhaust volumetric flow. The improvement in turbine efficiency is equivalent to 70 Btu/kWh in turbine cycle heat rate (18 kcal/kWh).

Aerodynamic losses in the blade path are an integral part of performance determination, as they directly affect the overall efficiency of the turbine. The losses at part load, or off-design operation, are additive to those existing at the design point and may contribute significantly to a deterioration of performance. Many factors influence these incremental losses, but they are due primarily to blading flow conditions that are different from those at the optimum operating condition. For example, lower (or higher) mass flow rates through the machine have been shown to change the pressure levels, temperatures, velocities, flow angles, moisture content, and so on within the blading. These changes induce flow conditions relative to the blading, which create additional losses. As different radial locations on the same blade are affected to a different degree, the overall effect must be considered as a summation of all individual effects for that blade. In fact, different blade rows, either stationary or rotating, are affected in a like manner, and the complete effect would then be the summation of the individual effects of all the blade rows.

Detailed verification of the design process has been obtained from the analysis and comparison of these internal flow characteristics to expected conditions, as determined by off-design calculations employing the same principles as incorporated in the design procedure. From measurements of total and static pressure and a knowledge of the enthalpy level at the point of interest, it is possible to determine the steam velocity, Mach number, and local mass flow rate. The flow angle is measured simultaneously. Traverse data for the advanced design turbine are presented in Figs. 36 and 37 for the last stationary blade inlet. Flow incidence is presented as a function of blade height for high and low values of specific mass flow. Good agreement is indicated between prediction and the traverse data. As flow

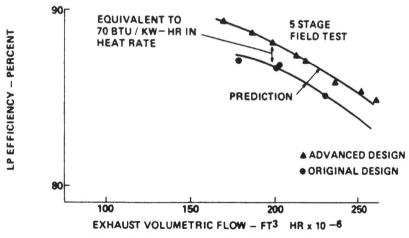

Figure 35 Low-pressure turbine field test results.

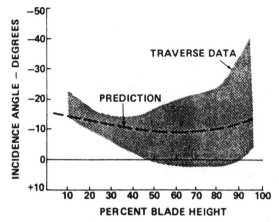

Figure 36 Last stationary blade incidence angle at high-end loading—12,000 lb/hr-ft².

rate (and load) decrease, the correspondence between test and calculation becomes less convincing, indicating that loss mechanisms existing under part-load operation are not as well defined as at high load near the design point.

By calculation, utilizing the rotor blade downstream traverse data, it is possible to determine the operating conditions relative to the rotor blade. That is, the conditions as seen by a traveler on the rotor blade can be determined. Figure 38 presents the Mach number leaving the last rotor blade (i.e., relative to it) as a function of blade height compared to the expected variation. Very high Mach numbers are experienced in this design and are a natural consequence of the high rotor blade tip speed, which approximates 2010 ft/sec, which converts to a wheel speed Mach number of 1.61.

In a manner similar to that applied at the inlet of the last stationary blade, the characteristics at the exit of this blade can be determined. Results of this traverse are shown in Fig. 39, where the effects of the stator wakes can be identified. The high Mach numbers experienced at the stationary blade hub make this particular measurement extremely difficult.

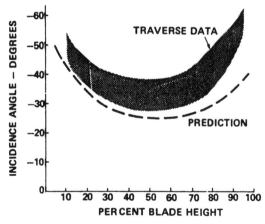

Figure 37 Last stationary blade incidence angle at low end loading—6,000 lb/hr-ft².

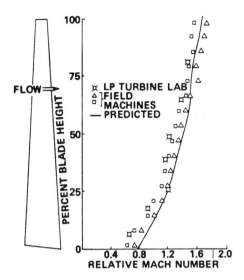

Figure 38 Mach number relative to last rotating blade.

The comparison shows good agreement, with expectations for this difficult measurement location.

The combination of these stationary blade exit traverses with velocity triangle calculations enables the flow conditions at the inlet of the rotor blade to be determined. Flow incidence as a function of blade height (diameter) is presented in Fig. 40 for the high-end load condition of Fig. 36. Significant flow losses can be incurred by off-design operation at high incidence levels. The hub and tip are particularly sensitive, and care must be taken in the design process to avoid this situation.

Further confirmation of the design process can be achieved by evaluation of the kinetic energy leaving the last stage. This energy is lost to the turbine and represents a significant

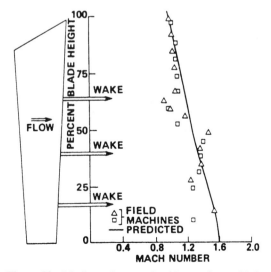

Figure 39 Mach number at exit of last stationary blade.

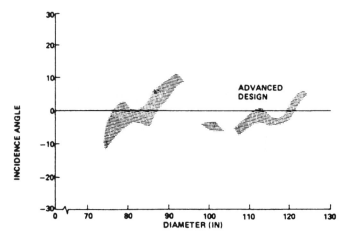

Figure 40 Last rotating blade incidence angle at high-end loading.

power output if it were possible to convert it to useful work. This kinetic energy, or *leaving loss*, is defined as

$$\overline{V^2} = \int_{\text{Hub}}^{\text{Tip}} (\rho V) \frac{V^2 \, dA}{\dot{M}} \tag{33}$$

and plotted in Fig. 41 as a function of exhaust volumetric flow. Excellent agreement has been achieved between the test data and prediction.

The foregoing represents conventional practice in the design of low-pressure blading as well as the design and analysis of upstream blading located in the HP and IP cylinders. The capability of solving the Navier–Stokes equations is now available. Drawbacks to this approach are the complexities of constructing a three-dimensional model of the subject of interest and the computational time required. The current philosophy regarding the use of Navier–Stokes solvers is that their practical application is to new design concept developments in order to determine valid and optimum resolutions of these concepts. In this process, design guidelines and cause-and-effect relationships are developed for subsequent application to production machinery. As of this writing, their day-to-day use is prohibited by manpower

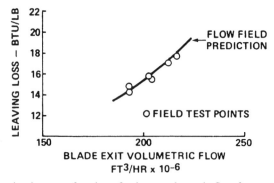

Figure 41 Leaving loss as a function of exhaust volumetric flow for an advanced design.

and computer time requirements. Nevertheless, significant improvements in, and changes to, the conventional design process have been made by the Navier–Stokes solvers. For the steam turbine aerodynamicist, the most influential approaches are those developed by Denton,[11] a program *not* a Navier–Stokes solver but having three-dimensional inviscid capability, and that by Dawes,[12] a truly three-dimensional Navier–Stokes solver.

4.7 Blade-to-Blade Flow Analysis

The blade-to-blade flow field problem confronts the turbine designer with a variety of situations, some design, and some analysis of existing designs. The flow regime can vary from low subsonic (incompressible) to transonic with exit Mach numbers approaching 1.8. The questions to be resolved in blade design are what section will satisfy the flow field design, and how efficiently it will operate.

From the mechanical viewpoint, will the blade section, or combination of sections, be strong enough to withstand the steady forces required of it, centrifugal and steam loading, and will it be able to withstand the unknown unsteady forces impressed upon it? The structural strength of the blade can be readily evaluated as a function of its geometry, material characteristics, and steady stresses induced by steady loads. The response of the blade to unsteady forces, generally of an unknown nature, is a much more formidable problem. Tuning of the long, low pressure blade is a common occurrence, but as blades become shorter, in the high and intermediate pressure elements, and in the initial stages of the low-pressure element, tuning becomes a difficult task. Blades must be designed with sufficient strength, or margin, to withstand these unsteady loads and resonance conditions at high harmonics of running speed.

The aerodynamic design of the blade depends on its duty requirements, that is, what work is the blade expected to produce? The fluid turning is defined by the flow field design, which determines the radial distribution of inlet and exit angles. Inherent in the flow field design is the expectation of certain levels of efficiency (or losses) that have defined the angles themselves. The detailed blade design must then be accomplished in such a manner as to satisfy, or better, these design expectations. Losses then, are a fundamental concern to the blade designer.

4.8 Blade Aerodynamic Considerations

Losses in blade sections, and blade rows, are a prime concern to the manufacturer and are usually considered proprietary information, as these control and establish its product's performance level in the market place. These data and correlative information provide the basic guidelines affecting the design and application of turbine–generator units.

Techniques involved in identifying and analyzing blade sections are more fundamentally scientific and have been the subject of development and refinement for many years. Current processes utilize the digital computer and depend on numerical techniques for their solution.

Early developments in the analysis of blade-to-blade flow fields[13] utilized analogies existing between fluid flow fields and electrical fields. A specific application can be found in Ref. 11.

Transonic Blade Flow

The most difficult problem has been the solution of the transonic flow field in the passageway between two blades. The governing differential equations change from the elliptic type in the subsonic flow regime to the hyperbolic type in the supersonic regime, making a uniform approach to the problem solution mathematically quite difficult. In lieu of, and also in support of, analytic procedures to define the flow conditions within a transonic passage, experimental data have been heavily relied upon. These test programs also define the blade section losses, which is invaluable information in itself. Testing and evaluating blade section performance is an art (and a science) unto itself. Of the several approaches to this type of testing, the air cascade is by far the most common and manageable. Pressure distributions (and hence Mach number) can be determined as a function of operating conditions such as inlet flow direction (incidence), Reynolds number, and overall pressure ratio. Losses also are determinable. Optical techniques are also useful in the evaluation of flow in transonic passages. Constant density parameters from interferometric photos can be translated into pressure and Mach number distributions.

A less common testing technique is the use of a free surface water table wherein local water depths are measured and converted to local Mach number.[14] The analogy between the water table and gas flow is valid for a gas with a ratio of specific heats equal to 2, an approximation, to be sure. A direct comparison of test results for the same blade cascade is presented in Fig. 42. The air test data were taken from interferometric photos and converted to parameters of constant Mach number shown in heavy solid lines. Superimposed on this figure are test data from a free-surface water table in lighter lines. A very good comparison is noted for these two sets of data from two completely different blade-testing concepts. The air test is more useful in that the section losses can be determined at the same time the optical studies are being done. The water table, however, produces reasonable results, comparable to the air cascade as far as blade surface distributions are concerned, and is an inexpensive and rapid means of obtaining good qualitative information and, with sufficient care in the experiment, good quantitative data.

Analytical Techniques

Various analytic means have been employed for the determination of the velocity and pressure distributions in transonic blade passages. A particular technique, termed *time marching,* solves the unsteady compressible flow equations by means of successive calculations in time utilizing a flow field comprised of finite area, or volume, elements. Figure 43 presents the general calculation region describing the flow passage between two blades. A result of the time marching method is shown in Fig. 44, which presents a plot of local Mach number along both the suction and pressure sides of a transonic blade. Parameters of exit isentropic

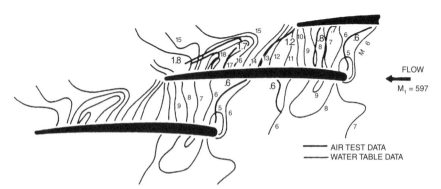

Figure 42 Comparison of water table test results with interferometry air test data for Mach number distribution in transonic blade.

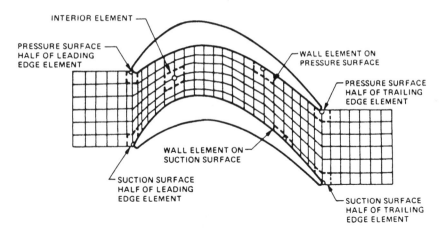

INTERIOR ELEMENT

PRESSURE SURFACE
HALF OF LEADING
EDGE ELEMENT

WALL ELEMENT ON
PRESSURE SURFACE

PRESSURE SURFACE
HALF OF TRAILING
EDGE ELEMENT

WALL ELEMENT ON
SUCTION SURFACE

SUCTION SURFACE
HALF OF LEADING
EDGE ELEMENT

SUCTION SURFACE
HALF OF TRAILING
EDGE ELEMENT

Figure 43 Blade-to-blade calculational region.

Mach number show a significant variation in local surface conditions as the blade's overall total to static pressure ratio is varied. These data were obtained in air from a transonic cascade facility. In summary, sophisticated numerical processes are available and in common use for the determination of blade surface pressure and velocity distributions. These predictions have been verified by experimental programs.

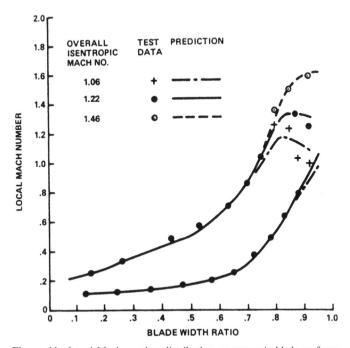

Figure 44 Local Mach number distribution on transonic blade surfaces.

5 RECENT DEVELOPMENTS AND HAPPENINGS

5.1 The Industry Trend

Since the foregoing material has been written (last published in 1998, Ref. 15), about a dozen years have passed. These recent years might be termed "The Combined Cycle Era," wherein the gas turbine has been superposed atop the steam turbine cycle, yielding significant improvements in unit thermal efficiency. This concept has been known and considered for more than 50 years but not implemented seriously until recently due to the prior lack of maturation of power-producing land gas turbines. Gas turbine reliability and availability and the combined cycle's need for a high temperature gas turbine exhaust have all now achieved satisfactory levels, which made the combined cycle economically and technically feasible.

Exact thermal efficiency values of these operating combined cycles are somewhat elusive, as all combined cycles are not equal, nor are all steam turbine cycles. Suffice it to say the most efficient combined cycle thermal efficiency currently approaches 60%, this compared to a nominal steam cycle thermal efficiency of about 42%. The difference, some 18 percentage points, is an enormous improvement in power-producing cycle thermal efficiency. In the writer's experience, the achievement of one or even one-half of one percent in steam turbine cycle efficiency was a significant gain.

Steam turbine operating conditions have shown signs of gradual increases in inlet steam temperature in an attempt to keep pace with ever increasing gas turbine exhaust temperatures. The norm of 1000°F steam turbine inlet temperature is now in some instances on the order of 1050 to 1100°F.

As the combined cycle continued development, the gas turbine and the steam turbine became more and more interdependent. In some instances, steam has been used to cool certain hot components of the gas turbine, stator vanes and their support structures, rotating blades and disks, and combustor ducting, for example. As the specific heat capacity of steam is about twice that of air, less mass flow is required to achieve the same cooling duty. Furthermore, the heat absorbed by the steam is utilized in the steam turbine providing a positive increment of steam turbine power output.

5.2 Repowering

In concert with the combined cycle era, many older coal-fired units have become candidates for "repowering." Repowering implies the use of these existing steam turbine-generator units as the bottoming cycle upon which new gas turbine units provide the topping cycle. Along with a new steam generator (the heat recovery steam generator, HRSG), and possible modifications to the steam turbine, a combined cycle emerges. These steam turbine units are generally those of some 50- to 300-MW capacity and perhaps 25–40 years old. Careful analysis of the repowered system is required to match the new equipment with the older steam cycle equipment.

The general effect on the steam turbine cycle is the elimination of most feedwater heaters to maximize the steam turbine mass flow rate and its power output. Due to the HRSG operating characteristics, additional steam is developed in excess of that delivered to the steam turbine throttle. This excess is introduced to the steam turbine midway through its expansion process, at the liquified petroleum (LP) inlet, for example. The effect is a trend toward increasing the mass flow rate through the final LP stages. To avoid exceeding their maximum flow rates, one or more low-pressure feedwater heaters may be retained in service to extract some 5–15% of the LP mass flow rate to avoid overloading the final LP stages.

Needless to say, the entire steam turbine flow path requires careful scrutiny to ensure reliable and safe operation in this combined cycle, repowered, configuration.

5.3 The Technology Trend

The greatest impact on recent design technology has been the availability of bigger, better, and faster digital computers and their user systems. In the flow and thermodynamics areas, the fundamental equations of motion are solvable, the Navier-Stokes equations, in particular. A generalization of these calculational procedures is termed computational fluid dynamics (CFD). But, as per usual, not all CFD procedures are the same in content, capability, or applicability. In solid mechanics, the fundamental relationships of stress and strain are combined with elements of plasticity and fatigue to allow life-expectancy evaluations of critical components. However fast the computations are, the results are only as good as the model defining the problem and the integrity and technical content of the computational procedure used. It has been the writer's experience that very few engineers in a given engineering organization are cognizant of, and competent with, the actual technical processes used within these important design and analysis tools.

The end results of recent design efforts are many. The most impressive deal with the steam turbine's low-pressure end, the region where small improvements translate to significant effects on unit output and thermal efficiency. Longer last-stage rotating blades head this list, the thermal efficiency attraction being a reduction in absolute kinetic energy passing from the last rotor blade to the condenser, which is then lost as far as power generation is concerned. Long steel and titanium last rotating row blades are currently in service. One consequence of longer (larger tip diameter) last-row blades is an increase of Mach number level within the stage. To more effectively utilize the velocity generated at high Mach numbers (perhaps 1.8 or more relative to the last rotor blade tip), converging-diverging (C-D) sections are designed into the rotor blade's outer diameter sections. The effect of these C-D sections is to generate a velocity in the direction of the blade section orientation to a value greater than that velocity which would occur in a purely converging blade section. In other words, the C-D sections generate a greater tangential velocity than conventional converging sections and hence generate more useful work.

A further application of these CFD procedures is to certain stationary (nonrotating), components. The low-pressure turbine exhaust system is a significant case wherein the development of a diffusing exhaust hood (system) leads to a lower pressure at the last rotating blade exit. The effect on the blade path is an increase of blading available energy and the generation of greater power output. Steam turbine inlets and extraction locations are all candidates for the application of CFD analyses. As mentioned previously, experimental verification of analytic exercises cannot be excluded from consideration. Mechanically, the last rotating blade row, as well as most, if not all blade rows, is generally designed as a meshed full circumferential structure utilizing integral shrouds. This is accomplished at the blade row tip by means of mating shroud surfaces on adjacent blades that come into contact as a result of the blade's slight rotation due to the effects of centrifugal force when at running speed. This technique appears to be a generic means for the minimization of the rotating blade's vibratory response to unsteady forces.

Also, as a result of past developmental efforts, both analytic and experimental, many blade rows (both stationary and rotating), are constructed as bowed, leaned, or swept structures, the primary positive effect being the reduction of secondary flow losses occurring in the vicinity of the juncture of blade and casing (be it either at the blade hub or tip). A secondary effect (perhaps primary to the mechanical designer) is that the impingement of upstream trailing edge wakes on rotating blade rows varies radially as the rotating structure passes the stationary structure. That is, the unsteady force effect does not occur simultaneously on all sections of the downstream rotating blade row. The effect occurs at different instants of time at the different blade heights, thereby smearing out the unsteady force effects induced by blade wakes.

The bowed, leaned, or swept concept has been made possible in one manufacturing sense by the use of 5-axis milling machines (machining centers), capable of automatically manufacturing these complex shapes from stock material. By contrast, the gas turbine also takes advantage of these concepts, usually by means of cast blades.

One significant casualty during recent times is the essential disappearance of the Kalina cycle, referred to in Section 2. The exact reason is unknown to the writer, but it is suspected that the characteristics of the working fluid (a mixture of ammonia and water) were not compatible with turbine materials at the higher temperatures.

6 CONCLUSION

As to the future of electric power generation: what comes next? The large coal-fired steam units of yesteryear have in part been supplanted by new units based on the combined gas turbine/steam turbine cycle. King Coal was being replaced by Prince Gas, the gas turbine fuel of choice. Nuclear power still remains in the doldrums in the United States, and is diminishing generally throughout the world, with only a few exceptions. However, as change is inevitable, external forces are again coming into play. Fuel prices, in particular for oil and gas, have recently relegated the efficient gas-fired combined cycle to a part-time operation status. A potential resolution of this plight apparently lies in the procedures of coal gasification. The "hydrogen economy," which to a significant extent will rely on coal and/or nuclear power in order to come to fruition, has recently been the subject of much attention and speculation. In these regards, King Coal appears to be on the rebound.

What actually happens will depend on economics and politics: technology will adapt to the challenge.

REFERENCES

1. A. I. Kalina, "Recent Improvements in Kalina Cycles: Rationale and Methodology," *American Power Conf.* **55** (1993).
2. ASME, *The 1967 ASME Steam Tables,* ASME, New York.
3. W. G. Steltz, "The Critical and Two Phase Flow of Steam," *J. Eng. Power* **83**, Series A, No. 2 (1961).
4. Babcock and Wilcox, *Steam, Its Generation and Use,* Babcock and Wilcox, 1992.
5. Westinghouse, *Westinghouse Steam Charts,* Westinghouse Electric Co., 1969.
6. H. R. M. Craig and H. J. A. Cox, "Performance Estimation of Axial Flow Turbines," *Proc. Inst. of Mech. Engs.* **185**(32/71) (1970–1971).
7. S. C. Kacker and U. Okapuu, "A Mean Line Prediction Method for Axial Flow Turbine Efficiency," *J. Eng. Power* **104**, 111–119 (1982).
8. M. J. Moore and C. H. Sieverding, *Two Phase Steam Flow in Turbines and Separators,* McGraw-Hill, New York, 1976.
9. C. H. Wu, "A General Theory of Three-Dimensional Flow in Subsonic and Supersonic Turbomachines of Axial-, Radial-, and Mixed-Flow Types," NASA TN 2604, 1952.

10. W. G. Steltz, D. D. Rosard, P. H. Maedel, Jr., and R. L. Bannister, "Large Scale Testing for Improved Reliability," American Power Conference, 1977.
11. J. D. Denton, "The Calculation of Three-Dimensional Viscous Flow Through Multistage Turbines," *J. Turbomachinery* **114** (1992).
12. W. N. Dawes, "Toward Improved Throughflow Capability: The Use of Three-Dimensional Viscous Flow Solvers in a Multistage Environment," *J. Turbomachinery* **114** (1992).
13. R. P. Benedict and C. A. Meyer, "Electrolytic Tank Analog for Studying Fluid Flow Fields within Turbomachinery," ASME Paper 57-A-120, 1957.
14. R. P. Benedict, "Analog Simulation," *Electro-Technology* (December 1963).
15. M. Kutz (ed.), *Mechanical Engineers' Handbook,* 2d ed., Wiley, New York, 1998.

CHAPTER 27

INTERNAL COMBUSTION ENGINES

Ronald Douglas Matthews
Department of Mechanical Engineering
The University of Texas at Austin
Austin, Texas

An internal combustion engine is a device that operates on an open thermodynamic cycle and is used to convert the chemical energy of a fuel to rotational mechanical energy. This rotational mechanical energy is most often used directly to provide motive power through an appropriate drive train, such as for an automotive application. The rotational mechanical energy may also be used directly to drive a propeller for marine or aircraft applications. Alternatively, the internal combustion engine may be coupled to a generator to provide electric power or may be coupled to hydraulic pump or a gas compressor. It may be noted that the favorable power-to-weight ratio of the internal combustion engine makes it ideally suited to mobile applications and therefore most internal combustion engines are manufactured for the motor vehicle, rail, marine, and aircraft industries. The high power-to-weight ratio of the internal combustion engine is also responsible for its use in other applications where a lightweight power source is needed, such as for chain saws and lawn mowers.

This chapter is devoted to discussion of the internal combustion engine, including types, principles of operation, fuels, theory, performance, efficiency, and emissions.

1 TYPES AND PRINCIPLES OF OPERATION

This chapter discusses internal combustion engines that have an intermittent combustion process. Gas turbines, which are internal combustion engines that incorporate a continuous combustion system, are discussed in a separate chapter.

Internal combustion (IC) engines may be most generally classified by the method used to initiate combustion as either spark ignition (SI) or compression ignition (CI or diesel) engines. Another general classification scheme involves whether the rotational mechanical energy is obtained via reciprocating piston motion, as is more common, or directly via the rotational motion of a rotor in a rotary (Wankel) engine (see Fig. 1). The physical principles of a rotary engine are equivalent to those of a piston engine if the geometric considerations are properly accounted for, so that the following discussion will focus on the piston engine and the rotary engine will be discussed only briefly. All of these IC engines include five general processes:

1. An intake process, during which air or a fuel–air mixture is inducted into the combustion chamber
2. A compression process, during which the air or fuel–air mixture is compressed to higher temperature, pressure, and density
3. A combustion process, during which the chemical energy of the fuel is converted to thermal energy of the products of combustion
4. An expansion process, during which a portion of the thermal energy of the working fluid is converted to mechanical energy
5. An exhaust process, during which most of the products of combustion are expelled from the combustion chamber

The mechanics of how these five general processes are incorporated in an engine may be used to more specifically classify different types of internal combustion engines.

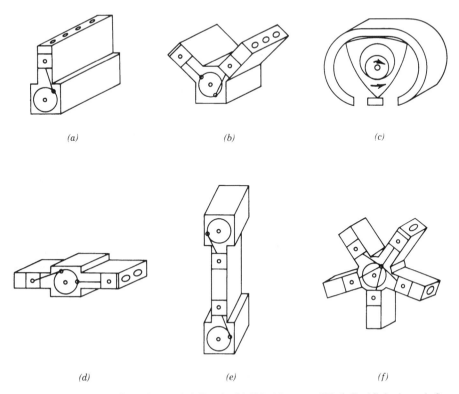

Figure 1 IC engine configurations: (*a*) inline 4; (*b*) V6; (*c*) rotary (Wankel); (*d*) horizontal, flat, or opposed cylinder; (*e*) opposed piston; (*f*) radial.

1.1   Spark Ignition Engines

In SI engines, the combustion process is initiated by a precisely timed discharge of a spark across an electrode gap in the combustion chamber. Before ignition, the combustible mixture may be either homogeneous (i.e., the fuel–air mixture ratio may be approximately uniform throughout the combustion chamber) or stratified (i.e., the fuel–air mixture ratio may be more fuel-lean in some regions of the combustion chamber than in other portions). In all SI engines, except the direct injection stratified charge (DISC) SI engine, the power output is controlled by controlling the air flow rate (and thus the volumetric efficiency) through the engine and the fuel–air ratio is approximately constant (and approximately stoichiometric) for almost all operating conditions. The power output of the DISC engine is controlled by varying the fuel flow rate, and thus the fuel–air ratio is variable while the volumetric efficiency is approximately constant. The fuel and air are premixed before entering the combustion chamber in all SI engines except the direct injection SI engine. These various categories of SI engines are discussed below.

Homogeneous Charge SI Engines

In the homogeneous charge SI engine, a mixture of fuel and air is inducted during the intake process. Traditionally, the fuel was mixed with the air in the venturi section of a carburetor. More recently, as more precise control of the fuel–air ratio became desirable, throttle body fuel injection took the place of carburetors for most automotive applications. Even more recently, intake port fuel injection has almost entirely replaced throttle body injection. The five processes mentioned above may be combined in the homogeneous charge SI engine to produce an engine that operates on either a 4-stroke cycle or on a 2-stroke cycle.

4-Stroke Homogeneous Charge SI Engines. In the more common 4-stroke cycle (see Fig. 2), the first stroke is the movement of the piston from top dead center (TDC—the closest approach of the piston to the cylinder head, yielding the minimum combustion chamber volume) to bottom dead center (BDC—when the piston is farthest from the cylinder head, yielding the maximum combustion chamber volume), during which the intake valve is open and the fresh fuel–air charge is inducted into the combustion chamber. The second stroke is the compression process, during which the intake and exhaust valves are both in the closed position and the piston moves from BDC back to TDC. The compression process is followed by combustion of the fuel–air mixture. Combustion is a rapid hydrocarbon oxidation process (not an explosion) of finite duration. Because the combustion process requires a finite, though very short, period of time, the spark is timed to initiate combustion slightly before the piston reaches TDC to allow the maximum pressure to occur slightly after TDC (peak pressure should, optimally, occur after TDC to provide a torque arm for the force caused by the high cylinder pressure). The combustion process is essentially complete shortly after the piston has receded away from TDC. However, for the purposes of a simple analysis and because combustion is very rapid, to aid explanation it may be approximated as being instantaneous and occurring while the piston is motionless at TDC. The third stroke is the expansion process or power stroke, during which the piston returns to BDC. The fourth stroke is the exhaust process, during which the exhaust valve is open and the piston proceeds from BDC to TDC and expels the products of combustion. The exhaust process for a 4-stroke engine is actually composed of two parts, the first of which is blowdown. When the exhaust valve opens, the cylinder pressure is much higher than the pressure in the exhaust manifold and this large pressure difference forces much of the exhaust out during what is called "blowdown" while the piston is almost motionless. Most of the remaining products of combustion are forced out during the exhaust stroke, but an "exhaust residual" is always left in the

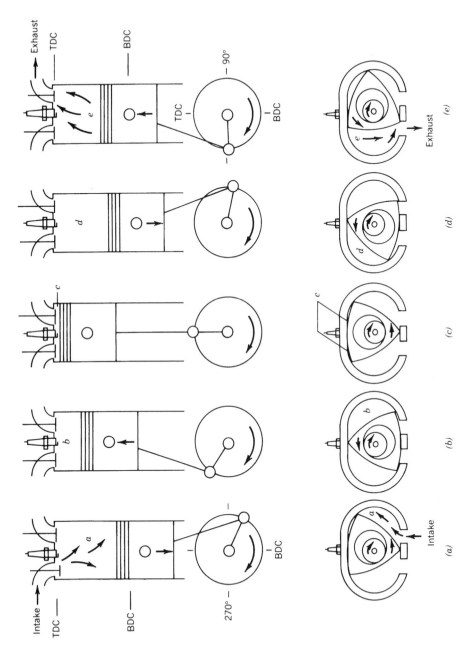

Figure 2 Schematic of processes for 4-stroke SI piston and rotary engines (for 4-stroke CI, replace spark plug with fuel injector): (*a*) intake, (*b*) compression, (*c*) spark ignition and combustion (for CI, fuel injection and combustion (for CI, fuel injection, and autoignition), (*d*) expansion or power stroke, (*e*) exhaust.

combustion chamber and mixes with the fresh charge that is inducted during the subsequent intake stroke. Once the piston reaches TDC, the intake valve opens and the exhaust valve closes and the cycle repeats, starting with a new intake stroke.

This explanation of the 4-stroke SI engine processes implied that the valves open or close instantaneously when the piston is either at TDC or BDC, when in fact the valves open and close relatively slowly. To afford the maximum open area at the appropriate time in each process, the exhaust valve opens before BDC during expansion, the intake valve closes after BDC during the compression stroke, and both the intake and exhaust valves are open during the valve overlap period since the intake valve opens before TDC during the exhaust stroke while the exhaust valve closes after TDC during the intake stroke. Considerations of valve timing are not necessary for this simple explanation of the 4-stroke cycle but do have significant effects on performance and efficiency. Similarly, spark timing will not be discussed in detail but does have significant effects on performance, fuel economy, and emissions.

The rotary (Wankel) engine is sometimes perceived to operate on the 2-stroke cycle because it shares several features with 2-stroke SI engines: a complete thermodynamic cycle within a single revolution of the output shaft (which is called an eccentric shaft rather than a crank shaft) and lack of intake and exhaust valves and associated valve train. However, unlike a 2-stroke, the rotary has a true exhaust "stroke" and a true intake "stroke" and operates quite well without boosting the pressure of the fresh charge above that of the exhaust manifold. That is, the rotary operates on the 4-stroke cycle.

2-Stroke Homogeneous Charge SI Engines. Alternatively, these five processes may be incorporated into a homogeneous charge SI engine that requires only two strokes per cycle (see Fig. 3). All commercially available 2-stroke SI engines are of the homogeneous charge type. That is, any nonuniformity of the fuel–air ratio within the combustion chamber is unintentional in current 2-stroke SI engines. The 2-stroke SI engine does not have valves, but rather has intake "transfer" and exhaust ports that are normally located across from each other near the position of the crown of the piston when the piston is at BDC. When the piston moves toward TDC, it covers the ports and the compression process begins. As previously discussed, for the ideal SI cycle, combustion may be perceived to occur instantaneously while the piston is motionless at TDC. The expansion process then occurs as the high pressure resulting from combustion pushes the piston back toward BDC. As the piston approaches BDC, the exhaust port is generally uncovered first, followed shortly thereafter by uncovering of the intake transfer port. The high pressure in the combustion chamber relative to that of the exhaust manifold results in "blowdown" of much of the exhaust before the intake transfer port is uncovered. However, as soon as the intake transfer port is uncovered, the exhaust and intake processes can occur simultaneously. However, if the chamber pressure is high with respect to the pressure in the transfer passage, the combustion products can flow into the transfer passage. To prevent this, a reed valve can be located within the intake transfer passage, as illustrated in Fig. 3. Alternatively, a disc valve that is attached to the crankshaft can be used to control timing of the intake transfer process. Independent of when and how the intake transfer process is initiated, the momentum of the exhaust flowing out the exhaust port will entrain some fresh charge, resulting in short-circuiting of fuel out the exhaust. This results in relatively high emissions of unburned hydrocarbons and a fuel economy penalty. This problem is minimized but not eliminated by designing the port shapes and/or piston crown to direct the intake flow toward the top of the combustion chamber so that the fresh charge must travel a longer path before reaching the exhaust port. After the piston reaches BDC and moves back up to cover the exhaust port again, the exhaust process is over. Thus, one of the strokes that is required for the 4-stroke cycle has been eliminated

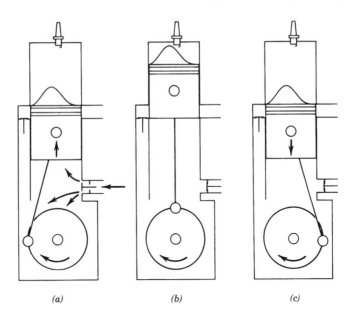

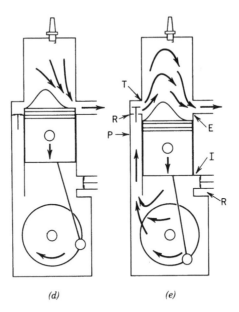

Figure 3 Processes for 2-stroke crankcase compression SI engine (for CI engine, replace spark plug with fuel injector): (*a*) compression of trapped working fluid and simultaneous intake to crankcase, (*b*) spark ignition and combustion (for CI, fuel injection, and autoignition), (*c*) expansion or power, (*d*) beginning of exhaust, (*e*) intake and "loop" scavenging. E, exhaust port; I, intake port; P, transfer passage; R, read valve; T, transfer port.

by not having an exhaust stroke. The penalty is that the 2-stroke has a relatively high exhaust residual fraction (the mass fraction of the remaining combustion products relative to the total mass trapped upon port closing).

As the piston proceeds from TDC to BDC on the expansion stroke, it compresses the fuel–air mixture which is routed through the crankcase on many modern 2-stroke SI engines. To prevent backflow of the fuel–air mixture back out of the crankcase through the carburetor, a reed valve may be located between the carburetor exit and the crankcase, as illustrated in Fig. 3. This crankcase compression process of the fuel–air mixture results in the fuel–air mixture being at relatively high pressure when the intake transfer port is uncovered. When the pressure in the combustion chamber becomes less than the pressure of the fuel–air mixture in the crankcase, the reed valve in the transfer passage opens and the intake charge flows into the combustion chamber. Thus, the 4-stroke's intake stroke is eliminated in the 2-stroke design by having both sides of the piston do work.

Because it is important to fill the combustion chamber as completely as possible with fresh fuel–air charge and thus important to purge the combustion chamber as completely as possible of combustion products, 2-stroke SI engines are designed to promote scavenging of the exhaust products via fluid dynamics (see Figs. 3 and 6). Scavenging results in the flow of some unburned fuel through the exhaust port during the period when the transfer passage reed valve and the exhaust port are both open. This results in poor combustion efficiency, a fuel economy penalty, and high emissions of hydrocarbons. However, since the 2-stroke SI engine has one power stroke per crankshaft revolution, it develops as much as 80% more power per unit weight than a comparable 4-stroke SI engine, which has only one power stroke per every two crankshaft revolutions. Therefore, the 2-stroke SI engine is best suited for applications for which a very high power per unit weight is needed and fuel economy and pollutant emissions are not significant considerations.

Stratified Charge SI Engines

All commercially available stratified charge SI engines in the United States operate on the 4-stroke cycle, although there has been a significant effort to develop a direct injection (stratified) 2-stroke SI engine. They may be subclassified as being either divided chamber or direct injection SI engines.

Divided Chamber. The divided chamber SI engine, as shown in Fig. 4, generally has two intake systems: one providing a stoichiometric or slightly fuel-rich mixture to a small pre-chamber and the other providing a fuel-lean mixture to the main combustion chamber. A spark plug initiates combustion in the prechamber. A jet of hot reactive species then flows through the orifice separating the two chambers and ignites the fuel-lean mixture in the main chamber. In this manner, the stoichiometric or fuel-rich combustion process stabilizes the fuel-lean combustion process that would otherwise be prone to misfire. This same stratified charge concept can be attained solely via fluid mechanics, thereby eliminating the complexity of the prechamber, but the motivation is the same as for the divided chamber engine. This overall fuel-lean system is desired since it can result in decreased emissions of the regulated pollutants in comparison to the usual, approximately stoichiometric, combustion process. Furthermore, lean operation produces a thermal efficiency benefit. For these reasons, there have been many attempts to develop a lean-burn homogeneous charge SI engine. However, the emissions of the oxides of nitrogen (NO_x) peak for a slightly lean mixture before decreasing to very low values when the mixture is extremely lean. Unfortunately, most lean-burn homogeneous charge SI engines cannot operate sufficiently lean—before encountering ignition problems—that they produce a significant NO_x benefit. The overall lean-burn strat-

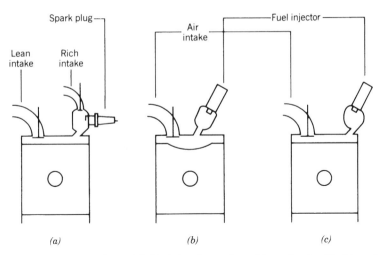

Figure 4 Schematic cross sections of divided chamber engines: (*a*) prechamber SI, (*b*) prechamber IDI diesel, (*c*) swirl chamber IDI diesel.

ified charge SI engine avoids these ignition limits by producing an ignitable mixture in the vicinity of the spark plug but a very lean mixture far from the spark plug. Unfortunately, the flame zone itself is nearly stoichiometric, resulting in much higher emissions of NO_x than would be expected from the overall extremely lean fuel–air ratio. For this reason, divided chamber SI engines are becoming rare. However, the direct injection process offers promise of overcoming this obstacle, as discussed below.

Direct Injection. In the direct injection engine, only air is inducted during the intake stroke. The direct injection engine can be divided into two categories: early and late injection.

The first 40 years of development of the direct injection SI engine focussed upon late injection. This version is commonly known as the *direct injection stratified charge* (DISC) engine. As shown in Fig. 5, fuel is injected late in the compression stroke near the center of the combustion chamber and ignited by a spark plug. The DISC engine has three primary advantages:

1. A wide fuel tolerance, that is, the ability to burn fuels with a relatively low octane rating without knock (see Section 2).
2. This decreased tendency to knock allows use of a higher compression ratio, which in turn results in higher power per unit displacement and higher efficiency (see Section 3).
3. Since the power output is controlled by the amount of fuel injected instead of the amount of air inducted, the DISC engine is not throttled (except at idle), resulting in higher volumetric efficiency and higher power per unit displacement for part load conditions (see Section 3).

Unfortunately, the DISC engine is also prone to high emissions of unburned hydrocarbons. However, more recent developments in the DISC engine aim the fuel spray at the top of the piston to avoid wetting the cylinder liner with liquid fuel to minimize emissions of unburned hydrocarbons. The shape of the piston, together with the air motion and ignition location,

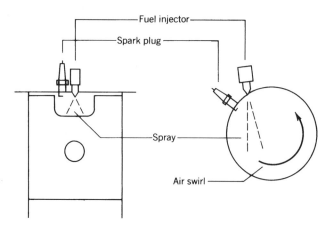

Figure 5 Schematic of DISC SI engine combustion chambers.

ensure that there is still an ignitable mixture in the vicinity of the spark plug even though the overall mixture is extremely lean. However, the extremely lean operation results in a low power capability. Thus, at high loads this version of the direct injection engine uses early injection timing, as discussed below.

Early injection results in sufficient time available for the mixture to become essentially completely mixed before ignition, given sufficient turbulence to aid the mixing process. A stoichiometric or slightly rich mixture is used to provide maximum power output and also ensures that ignition is not a difficulty.

1.2 Compression Ignition (Diesel) Engines

CI engines induct only air during the intake process. Late in the compression process, fuel is injected directly into the combustion chamber and mixes with the air that has been compressed to a relatively high temperature. The high temperature of the air serves to ignite the fuel. Like the DISC SI engine, the power output of the diesel is controlled by controlling the fuel flow rate while the volumetric efficiency is approximately constant. Although the fuel–air ratio is variable, the diesel always operates overall fuel-lean, with a maximum allowable fuel–air ratio limited by the production of unacceptable levels of smoke (also called soot or particulates). Diesel engines are inherently stratified because of the nature of the fuel-injection process. The fuel–air mixture is fuel-rich near the center of the fuel-injection cone and fuel-lean in areas of the combustion chamber that are farther from the fuel injection cone. Unlike the combustion process in the SI engine, which occurs at almost constant volume, the combustion process in the diesel engine ideally occurs at constant pressure. That is, the combustion process in the CI engine is relatively slow, and the fuel–air mixture continues to burn during a significant portion of the expansion stroke (fuel continues to be injected during this portion of the expansion stroke) and the high pressure that would normally result from combustion is relieved as the piston recedes. After the combustion process is completed, the expansion process continues until the piston reaches BDC. The diesel may complete the five general engine processes through either a 2-stroke cycle or a 4-stroke cycle. Furthermore, the diesel may be subclassified as either an indirect injection diesel or a direct injection diesel.

Indirect injection (IDI) or divided chamber diesels are geometrically similar to divided chamber stratified charge SI engines. All IDI diesels operate on a 4-stroke cycle. Fuel is injected into the prechamber and combustion is initiated by autoignition. A glow plug is also located in the prechamber, but is only used to alleviate cold start difficulties. As shown in Fig. 4, the IDI may be designed so that the jet of hot gases issuing into the main chamber promotes swirl of the reactants in the main chamber. This configuration is called the *swirl chamber IDI diesel*. If the system is not designed to promote swirl, it is called the *prechamber IDI diesel*. The divided chamber design allows a relatively inexpensive pintle-type fuel injector to be used on the IDI diesel.

Direct injection (DI) or "open" chamber diesels are similar to DISC SI engines. There is no prechamber, and fuel is injected directly into the main chamber. Therefore, the characteristics of the fuel-injection cone have to be tailored carefully for proper combustion, avoidance of knock, and minimum smoke emissions. This requires the use of a high-pressure, close-tolerance, fuel-injection system that is relatively expensive. The DI diesel may operate on either a 4-stroke or a 2-stroke cycle. Unlike the 2-stroke SI engine, the 2-stroke diesel often uses a mechanically driven blower for supercharging rather than crankcase compression and also may use multiple inlet ports in each cylinder, as shown in Fig. 6. Also, one or more exhaust valves in the top of the cylinder may be used instead of exhaust ports near the bottom of the cylinder, resulting in "through" or "uniflow" scavenging rather than "loop" or "cross" scavenging.

2 FUELS AND KNOCK

Knock is the primary factor that limits the design of most IC engines. Knock is the result of engine design characteristics, engine operating conditions, and fuel properties. The causes of knock are discussed in this section. Fuel characteristics, especially those that affect either knock or performance, are also discussed in this section.

2.1 Knock in Spark Ignition Engines

Knock occurs in the SI engine if the fuel–air mixture autoignites too easily. At the end of the compression stroke, the fuel–air mixture exists at a relatively high temperature and pressure, the specific values of which depend primarily on the compression ratio and the intake manifold pressure (which is a function of the load). The spark plug then ignites a flame that travels toward the periphery of the combustion chamber. The increase in temper-

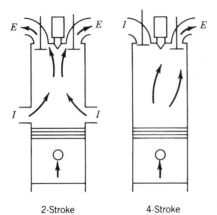

2-Stroke   4-Stroke

Figure 6 Schematic of 2-stroke and 4-stroke DI diesels. The 2-stroke incorporates "uniflow" scavenging.

ature and number of moles of the burned gases behind the flame front causes the pressure to rise throughout the combustion chamber. The "end gases" located in the peripheral regions of the combustion chamber (in the "unburned zone") are compressed to even higher temperatures by this increase in pressure. The high temperature of the end gases can lead to a sequence of chemical reactions that are called *autoignition*. If the autoignition reactions have sufficient time at a sufficiently high temperature, the reaction sequence can produce strongly exothermic reactions such that the temperature in the unburned zone may increase at a rate of several million K/sec, which results in knock. That is, if the reactive end gases remain at a high temperature for a sufficient period of time (i.e., longer than the "ignition delay time"), then the autoignition reactions will produce knock. Normal combustion occurs if the flame front passes through the end gases before the autoignition reactions reach a strongly exothermic stage.

For most fuels, autoignition is characterized by three stages that are dictated by the unburned mixture (or end gas) temperature. Here, it is important to note that the temperature varies with crank angle due to compression by the piston motion and, after ignition, due to compression by the expanding flame front, and the entire temperature history shifts up or down due to the effects of load, ambient air temperature, etc. At "low" temperatures, the reactivity of the end gases increases with increasing temperature. As the temperature increases further, the rate of increase of the reactivity either slows markedly or even decreases (the so-called "negative temperature coefficient" regime). When the temperature increases to even higher values (typically, above ~900 K), the reactivity begins to increase extremely strongly, the autoignition reactions reach an energy liberating stage, enough energy may be released during this stage to initiate a "high"-temperature (>1000 K) chemical mechanism,[1,2] and a runaway reaction occurs. If the rate of energy release is greater than the rate of expansion, then a strong pressure gradient will result. The steep pressure wave thus established will travel throughout the combustion chamber, reflect off the walls, and oscillate at the natural frequency characteristic of the combustion chamber geometry. This acoustic vibration results in an audible sound called *knock*. It should be noted that the flame speeds associated with knock are generally considered to be lower than the flame speeds associated with detonation (or explosion).[3,4] Nevertheless, the terms *knock* and *detonation* are often used interchangeably in reference to end gas autoignition.

The tendency of the SI engine to knock will be affected by any factors that affect the temperature and pressure of the end gases, the ignition delay time, the end gas residence time (before the normal flame passes through the end gases), and the reactivity of the mixture. The flame speed is a function of the turbulence intensity in the combustion chamber, and the turbulence intensity increases with increasing engine speed. Thus, the end gases will have a shorter residence time at high engine speed and there will be a decreased tendency to knock. As the load on the engine increases, the throttle plate is opened wider and the pressure in the intake manifold increases, thereby increasing the end gas pressure (and, thereby, temperature), resulting in a greater tendency to knock. Thus, knock is most likely to be observed for SI engines used in motor vehicles at conditions of high load and low engine speed, such as acceleration from a standing start.

Other factors that increase the knock tendency of an SI engine[1-5] include increased compression ratio, increased inlet air temperature, increased distance between the spark plug and the end gases, location of the hot exhaust valve near the region of the end gases that is farthest from the spark plug, and increased intake manifold temperature and pressure due to pressure boosting (supercharging or turbocharging). Factors that decrease the knock tendency of an SI engine[1-5] include retarding the spark timing, operation with either rich or lean mixtures (and thus the ability to operate the DISC SI engine at a higher compression ratio, since the end gases for this engine are extremely lean and therefore not very reactive), and increased inert levels in the mixture (via exhaust gas recirculation, water injection, etc.). The fuel characteristics that affect knock are quantified using octane rating tests, which are discussed in more detail in Section 2.3. A fuel with higher octane number has a decreased tendency to knock.

2.2 Knock in the Diesel Engine

Knock occurs in the diesel engine if the fuel–air mixture does not autoignite easily enough. Knock occurs at the beginning of the combustion process in a diesel engine, whereas it occurs near the end of the combustion process in an SI engine. After the fuel injection process begins, there is an ignition delay time before the combustion process is initiated. This ignition delay time is not caused solely by the chemical delay that is critical to autoignition in the SI engine, but is also due to a physical delay. The physical delay results from the need to vaporize and mix the fuel with the air to form a combustible mixture. If the overall ignition delay time is high, then too much fuel may be injected prior to autoignition. This oversupply of fuel will result in an energy release rate that is too high immediately after ignition occurs. In turn, this will result in an unacceptably high rate of pressure rise and cause the audible sound called *knock*.

The factors that will increase the knock tendency of a diesel engine[1,3,5] are those that decrease the rates of atomization, vaporization, mixing, and reaction, and those that increase the rate of fuel injection. The diesel engine is most prone to knock under cold start conditions because

1. The fuel, air, and combustion chamber walls are initially cold, resulting in high fuel viscosity (poor mixing and therefore a longer physical delay), poor vaporization (longer physical delay), and low initial reaction rates (longer chemical delay).
2. The low engine speed results in low turbulence intensity (poor mixing, yielding a longer physical delay) and may result in low fuel-injection pressures (poor atomization and longer physical delay).
3. The low starting load will lead to low combustion temperatures and thus low reaction rates (longer chemical delay).

After a diesel engine has attained normal operating temperatures, knock will be most liable to occur at high speed and low load (exactly the opposite of the SI engine). The low load results in low combustion temperatures and thus low reaction rates and a longer chemical delay. Since most diesel engines have a gear-driven fuel-injection pump, the increased rate of injection at high speed will more than offset the improved atomization and mixing (shorter physical delay).

Because the diesel knocks for essentially the opposite reasons than the SI engine, the factors that increase the knock tendency of an SI engine will decrease the knock tendency of a diesel engine: increased compression ratio, increased inlet air temperature, increased intake manifold temperature and pressure due to supercharging or turbocharging, and decreased concentrations of inert species. The knock tendency of the diesel engine will be increased if the injection timing is advanced or retarded from the optimum value and if the fuel has a low volatility, a high viscosity, and/or a low "cetane number." The cetane rating test and other fuel characteristics are discussed in more detail in the following section.

2.3 Characteristics of Fuels

Several properties are of interest for both SI engine fuels and diesel fuels. Many of these properties are presented in Table 1 for the primary reference fuels, for various types of gasolines and diesel fuels, and for the alternative fuels that are of current interest.

Table 1 Properties for Various Fuels

Name	Formula	MW	AF_s	LHV_p	$h_f{}^a$	$h_v{}^*$	sg^b	RON	MON	Ref.
Primary Reference Fuels										
Iso-octane	C_8H_{18}	114	15.1	44.6	-224.3^c	35.1^c	0.69	100	100	5
Normal heptane	C_7H_{16}	100	15.1	44.9	-187.9^c	36.6^c	0.68	0	0	5
Normal hexadecane	$C_{16}H_{34}$	226	14.9	44.1	-418.3^d	50.9^e	0.77	—	0^f	5
Alternative Fuels										
Average CNG	$CH_{3.88}$	17.4	16.3	47.9	-79.6	—	0.60^b	> 120	> 120	26, 27
LPG as propane[1]	C_3H_8	44	15.6	46.3	-103.9	15.1	0.50	112	97	5, 22
Methanol	CH_3OH	32	6.4	21.2	-201.3^g	37.5	0.79	112	91	5, 21
Ethanol	C_2H_5OH	46	9.0	27.8	-235.5^g	42.4	0.78	111	92	5, 21
*Gasolines***										
1988 U.S. avg.	$C_8H_{14.53}$	111	14.5	42.6	-189.9^h	NA^j	0.75	92.0	82.6	28
Certification[2]	$C_8H_{14.69}$	111	14.5	42.6	-202.4^h	NA	0.74	96.7	87.5	28
Cal. Phase 2 RFG[3]	$C_8H_{16.28}O_{0.24}$	116	14.2	41.6	-273.3^h	NA	0.74	NA	NA	29
Aviation	C_8H_{17}	113	14.9	41.9–43.1	-390.3^i	NA	0.72	NA	NA	3
Diesel Fuels										
Automotive	$C_{12}H_{23.7}$	168	14.7	40.6–44.4	-445.0^i	90.1–131.7	0.81–0.85	—	—	3
No. 1D	$C_{12}H_{26}$	170	15.0	42.4	-596.0	45.4	0.88	—	—	1
No. 2D	$C_{13}H_{28}$	184	15.0	41.8	-747.6	44.9	0.92	—	—	1
No. 4D	$C_{14}H_{30}$	198	15.0	41.3	-894.6	46.0	0.96	—	—	1

^{a} Of vapor phase fuel at 298 K in MJ/kmole, except when noted otherwise.

^{b} sg is ρ_F at 20°C/ρ_w at 4°C (1000 kg/m³), except values from Ref. 1 (reference temp. is 15°C) and CNG (referenced to air).

^{c} From Ref. 23.

^{d} Calculated.

^{e} At 1 atm and boiling temperature.

^{f} Estimate from Ref. 1, p. 147.

^{g} Reference 22.

^{h} Enthalpy of formation is for the liquid fuel (rather than the gaseous fuel), as calculated from fuel properties.

^{i} Enthalpy of formation is for the liquid fuel as calculated from the average heating value.

^{j} Typically 35–40.

* At 298 K and corresponding saturation pressure, except when noted otherwise.

** As C8.

[1] Liquefied petroleum gas has a variable composition, normally dominated by propane.

[2] The properties of emissions certification gasoline vary somewhat.

[3] Properties of California Phase 2 Reformulated Gasoline from a sample of Arco EC-X.

The stoichiometry, or relative amount of air and fuel, in the combustion chamber is usually specified by the air–fuel mass ratio (AF), the fuel–air mass ratio (FA = 1/AF), the equivalence ratio (ϕ), or the excess air ratio (λ). Measuring instruments may be used to determine the mass flow rates of air and fuel into an engine so that AF and FA may be easily determined. Alternatively, AF and FA may be calculated if the exhaust product composition is known, using any of several available techniques.[5] The equivalence ratio normalizes the actual fuel–air ratio by the stoichiometric fuel–air ratio (FA$_s$), where "stoichiometric" refers to the chemically correct mixture with no excess air and no excess fuel. Recognizing that the stoichiometric mixture contains 100% "theoretical air" allows the equivalence ratio to be related to the actual percentage of theoretical air (TA, percentage by volume or mole):

$$\phi = FA/FA_s = AF_s/AF = 100/TA = 1/\lambda \tag{1}$$

The equivalence ratio is a convenient parameter because $\phi < 1$ refers to a fuel-lean mixture, $\phi > 1$ to a fuel-rich mixture, and $\phi = 1$ to a stoichiometric mixture.

The stoichiometric fuel–air and air–fuel ratios can be easily calculated from a reaction balance by assuming "complete combustion" [only water vapor (H_2O) and carbon dioxide (CO_2) are formed during the combustion process], even though the actual combustion process will almost never be complete. The reaction balance for the complete combustion of a stoichiometric mixture of air with a fuel of the atomic composition C_xH_y is

$$C_xH_y + (x + 0.25y)O_2 + 3.764(x + 0.25y)N_2 = xCO_2 + 0.5yH_2O + 3.764(x + 0.25y)N_2 \tag{2}$$

where air is taken to be 79% by volume "effective nitrogen" (N_2 plus the minor components in air) and 21% by volume oxygen (O_2) and thus the nitrogen-to-oxygen ratio of air is 0.79/0.21 = 3.764. Given that the molecular weight (MW) of air is 28.967, the MW of carbon (C) is 12.011, and the MW of hydrogen (H) is 1.008, then AF$_s$ and FA$_s$ for any hydrocarbon fuel may be calculated from

$$AF_s = 1/FA_s = (x + 0.25y) \times 4.764 \times 28.967/(12.011x + 1.008y) \tag{3}$$

The stoichiometric air–fuel ratios for a number of fuels of interest are presented in Table 1.

The energy content of the fuel is most often specified using the constant-pressure lower heating value (LHV$_p$). The lower heating value is the maximum energy that can be released during combustion of the fuel if (1) the water in the products remains in the vapor phase, (2) the products are returned to the initial reference temperature of the reactants (298 K), and (3) the combustion process is carried out such that essentially complete combustion is attained. If the water in the products is condensed, then the higher heating value (HHV) is obtained. If the combustion system is a flow calorimeter, then the constant-pressure heating value is measured (and, most usually, this is HHV$_p$). If the combustion system is a bomb calorimeter, then the constant-volume heating value is measured (usually HHV$_v$). The constant-pressure heating value is the negative of the standard enthalpy of reaction (ΔH_R^{298}, also known as the heat of combustion) and ΔH_R^{298} is a function of the standard enthalpies of formation h_f^{298} of the reactant and product species. For a fuel of composition C_xH_y, Eq. (4) may be used to calculate the constant-pressure heating value (HV$_p$), given the enthalpy of formation of the fuel, or may be used to calculate h_f^{298} of the fuel, given HV$_p$:

$$HV_p = -\Delta H_R^{298} = \frac{(h_{f,C_xH_y}^{298} - \beta h_{v,C_xH_y}) + 393.418x + 0.5y(241.763 + 43.998\alpha)}{12.011x + 1.008y} \tag{4}$$

In Eq. (4): (1) $\alpha = 0$ if the water in the products is not condensed (yielding LHV$_p$) and $\alpha = 1$ if the water is condensed (yielding HHV$_p$); (2) $\beta = 0$ if the fuel is initially a vapor and $\beta = 1$ if the fuel is initially a liquid; (3) h_{v,C_xH_y} is the enthalpy of vaporization per kmole* of fuel at 298 K; (4) the standard enthalpies of formation are CO_2: -393.418 MJ/kmole, H_2O: -241.763 MJ/kmole, O_2: 0 MJ/kmole, N_2: 0 MJ/kmole; (5) the enthalpy of vaporization of H_2O at 298 K is 43.998 MJ/kmole; and (6) the denominator is simply the molecular weight of the fuel yielding the heating value in MJ/kg of fuel. Also, the relationship between the constant volume heating value (HV_v) and HV_p for a fuel C_xH_y is

$$HV_v = HV_p + \frac{2.478(0.25y - 1)}{12.011x + 1.008y} \tag{5}$$

Of the several heating values that may be defined, LHV_p is preferred for engine calculations since condensation of water in the combustion chamber is definitely to be avoided and since an engine is essentially a steady-flow device and thus the enthalpy is the relevant thermodynamic property (rather than the internal energy). For diesel fuels, HHV_v may be estimated from nomographs, given the density and the "mid-boiling point temperature" or given the "aniline point," the density, and the sulfur content of the fuel.[6]

Values for LHV_p, h_f^{298}, and h_v^{298} for various fuels of interest are presented in Table 1.

The specific gravity of a liquid fuel (sg_F) is the ratio of its density (ρ_F, usually at either 20°C or 60°F) to the density of water (ρ_w, usually at 4°C):

$$sg_F = \rho_F/\rho_w \qquad \rho_F = sg_F \cdot \rho_w \tag{6}$$

For gaseous fuels, such as natural gas, the specific gravity is referenced to air at standard conditions rather than to water. The specific gravity of a liquid fuel can be easily calculated from a simple measurement of the American Petroleum Institute gravity (API):

$$sg_F = 141.5/(API + 131.5) \tag{7}$$

Tables are available (SAE Standard J1082 SEP80) to correct for the effects of temperature if the fuel is not at the prescribed temperature when the measurement is performed. Values of sg_F for various fuels are presented in Table 1.

The knock tendency of SI engine fuels is rated using an octane number (ON) scale. A higher octane number indicates a higher resistance to knock. Two different octane-rating tests are currently used. Both use a single-cylinder variable-compression-ratio SI engine for which all operating conditions are specified (see Table 2). The fuel to be tested is run in the engine and the compression ratio is increased until knock of a specified intensity (standard knock) is obtained. Blends of two primary reference fuels are then tested at the same compression ratio until the mixture is found that produces standard knock. The two primary reference fuels are 2,2,4-trimethyl pentane (also called iso-octane), which is arbitrarily assigned an ON of 100, and n-heptane, which is arbitrarily assigned an ON of 0. The ON of the test fuel is then simply equal to the percentage of iso-octane in the blend that produced the same knock intensity at the same compression ratio. However, if the test fuel has an ON above 100, then iso-octane is blended with tetraethyl lead instead of n-heptane. After the knock tests are completed, the ON is then computed from

$$ON = 100 + \frac{28.28T}{1 + 0.736T + (1 + 1.472T - 0.035216T^2)^{1/2}} \tag{8}$$

where T is the number of milliliters of tetraethyl lead per U.S. gallon of iso-octane. The two different octane rating tests are called the Motor method (American Society of Testing and

---

*A kmole is a mole based on a kg, also referred to as a kg-mole.

Table 2 Test Specifications That Differ for Research and Motor Method Octane Tests—ASTM D2699-82 and D2700-82

Operating Condition	RON	MON
Engine speed (rpm)	600	900
Inlet air temperature (°C)	[a]	38°C
FA mixture temperature (°C)	[b]	149°C
Spark advance	13°BTDC	[c]

[a] Varies with barometric pressure.

[b] No control of fuel–air mixture temperature.

[c] Varies with compression ratio.

Materials, ASTM Standard D2700-82) and the Research method (ASTM D2600-82), and thus a given fuel (except these two primary reference fuels) will have two different octane numbers: a Motor octane number (MON) and a Research octane number (RON). The Motor method produces the lowest octane numbers, primarily because of the high intake manifold temperature for this technique, and thus the Motor method is said to be a more severe test for knock. The "sensitivity" of a fuel is defined as the RON minus the MON of that fuel. The "antiknock index" is the octane rating posted on gasoline pumps at service stations in the United States and is simply the average of RON and MON. Octane numbers for various fuels of interest are presented in Table 1.

The standard rating test for the knock tendency of diesel fuels (ASTM D613-82) produces the cetane number (CN). Because SI and diesel engines knock for essentially opposite reasons, a fuel with a high ON will have a low CN and therefore would be a poor diesel fuel. A single-cylinder variable-compression-ratio CI engine is used to measure the CN, and all engine operating conditions are specified. The compression ratio is increased until the test fuel exhibits an ignition delay of 13°. Here, it should be noted that ignition delay rather than knock intensity is measured for the CN technique. A blend of two primary reference fuels (n-hexadecane, which is also called n-cetane: CN = 100; and heptamethyl nonane, or isocetane: CN = 15) are then run in the engine and the compression ratio is varied until a 13° ignition delay is obtained. The CN of this blend is given by

$$\text{CN} = \%\ n\text{-cetane} + 0.15 \times (\%\ \text{heptamethyl nonane}) \tag{9}$$

Various blends are tried until compression ratios are found that bracket the compression ratio of the test fuel. The CN is then obtained from a standard chart. General specifications for diesel fuels are presented in Table 3 along with characteristics of "average" diesel fuels for light duty vehicles.

Many other thermochemical properties of fuels may be of interest, such as vapor pressure, volatility, viscosity, cloud point, aniline point, mid-boiling-point temperature, and additives. Discussion of these characteristics is beyond the scope of this chapter but is available in the literature.[1,4–7]

3 PERFORMANCE AND EFFICIENCY

The performance of an engine is generally specified through the brake power (bp), the torque (τ), or the brake mean effective pressure (bmep), while the efficiency of an engine is usually

Table 3 Diesel Fuel Oil Specifications—ASTM D975-81

Property	Units	Fuel Type		
		1D[b]	2D[c]	4D[d]
Minimum flash point	°C	38	52	55
Maximum H$_2$O and sediment	Vol %	0.05	0.05	0.50
Maximum carbon residue	%	0.15	0.35	—
Maximum ash	Wt %	0.01	0.01	0.10
90% distillation temperature, min/max	°C	—/288	282/338	—/—
Kinematic viscosity,^{a} min/max	mm^2/sec	1.3/2.4	1.9/4.1	5.5/24.0
Maximum sulfur	Wt %	0.5	0.5	2.0
Maximum Cu strip corrosion	—	No. 3	No. 3	—
Minimum cetane number	—	40	40	30

^{a} At 40°C.

^{b} Preferred for high-speed diesels, especially for winter use, rarely available. 1976 U.S. average properties[17,24]: API, 42.2; 220°C midboiling point; 0.081 wt % sulfur; and cetane index of 49.5. The cetane index is an approximation of the CN, calculated from ASTM D976-80 given the API and the midpoint temperature, and accurate within ± 2 CN for $30 \leq CN \leq 60$ for 75% of distillate fuels tested.

^{c} For high-speed diesels (passenger cars and trucks) 1972 U.S. average properties[17,24]: API = 35.7; MBP = 261°C; 0.253 wt % sulfur; cetane index^e = 48.4.

^{d} Low- and medium-speed diesels.

specified through the brake specific fuel consumption (bsfc) or the overall efficiency (η_e). Experimental and theoretical determination of important engine parameters is discussed in the following sections.

3.1 Experimental Measurements

Engine dynamometer (dyno) measurements can be used to obtain the various engine parameters using the relationships[5,8,9]

$$bp = LRN/9549.3 = LN/K \tag{10}$$

$$\tau = LR = 9549.3 \, bp/N \tag{11}$$

$$bmep = 60,000 \, bp \, X/DN \tag{12}$$

$$bsfc = \dot{m}_F/bp \tag{13}$$

$$\eta_e = \frac{3600 \, bp}{\dot{m}_F LHV_p} = \frac{3600}{bsfc \, LHV_p} \tag{14}$$

Definitions and standard units* for the variables in the above equations are presented in the symbols list. The constants in the above equations are simply unit conversion factors. The brake power is the useful power measured at the engine output shaft. Some power is used to overcome frictional losses in the engine and this power (the friction power, fp) is not

---

* Standard units are not in strict compliance with the International System of units, in order to produce numbers of convenient magnitude.

available at the output shaft. The total rate of energy production within the engine is called the *indicated power* (ip)

$$ip = bp + fp \tag{15}$$

where the friction power can be determined from dyno measurements using:

$$fp = FRN/9549.3 = FN/K \tag{16}$$

The efficiency of overcoming frictional losses in the engine is called the *mechanical efficiency* (η_M), which is defined as

$$\eta_M = bp/ip = 1 - fp/ip \tag{17}$$

The definitions of ip and η_M allow determination of the indicated mean effective pressure (imep) and the indicated specific fuel consumption (isfc):

$$imep = bmep/\eta_M = 60{,}000 \text{ ip } X/DN \tag{18}$$

$$isfc = \eta_M bsfc = \dot{m}_F/ip \tag{19}$$

Three additional efficiencies of interest are the volumetric efficiency (η_v), the combustion efficiency (η_c), and the indicated thermal efficiency (η_{ti}).

The volumetric efficiency is the effectiveness of inducting air into the engine[5,10–13] and is defined as the actual mass flow rate of air ($\dot{m}_A$) divided by the theoretical maximum air mass flow rate ($\rho_A DN/X$):

$$\eta_v = \frac{\dot{m}_A}{\rho_A DN/X} \tag{20}$$

The combustion efficiency is the efficiency of converting the chemical energy of the fuel to thermal energy (enthalpy) of the products of combustion.[12–14] Thus,

$$\eta_c = -\Delta H^{298}_{R,\text{act}}/\text{LHV}_p \tag{21}$$

The actual enthalpy of reaction ($\Delta H^{298}_{R,\text{act}}$) may be determined by measuring the mole fractions in the exhaust of CO_2, CO, O_2, and unburned hydrocarbons (expressed as "equivalent propane" in the following) and calculating the mole fractions of H_2O and H_2 from atom balances. For a fuel of composition $C_x H_y$,

$$-\Delta H^{298}_{R,\text{act}} = \frac{h^{298}_{f,C_x H_y} + x(Y_{CO_2}393.418 + Y_{H_2O}241.763 + Y_{CO}110.600 + Y_{C_3H_8}103.900)}{(Y_{CO_2} + Y_{CO} + 3Y_{C_3H_8})(12.011x + 1.008y)} \tag{22}$$

where Y_i is the mole fraction of species i in the "wet" exhaust. In Eq. (22), a carbon balance was used to convert moles of species i per mole of product mixture to moles of species i per mole of fuel burned and the molecular weight of the fuel appears in the denominator to produce the enthalpy of reaction in units of MJ per kg of fuel burned. If a significant amount of soot is present in the exhaust (e.g., a diesel under high load), then the carbon balance becomes inaccurate and an oxygen balance would have to be substituted.

The indicated thermal efficiency is the efficiency of the actual thermodynamic cycle. This parameter is difficult to measure directly, but may be calculated from

$$\eta_{ti} = \frac{3600 \text{ ip}}{\eta_c \dot{m}_F \text{LHV}_p} = \frac{3600}{\text{isfc } \eta_c \text{LHV}_p} \tag{23}$$

Because the engine performance depends on the air flow rate through it, the ambient temperature, barometric pressure, and relative humidity can affect the performance parameters and efficiencies. It is often desirable to correct the measured values to standard atmospheric conditions. The use of correction factors is discussed in the literature,[5,8,9] but is beyond the scope of this chapter.

The four fundamental efficiencies η_{ti}, η_c, η_v, and η_M are related to the global performance and efficiency parameters in the following section. Methods for modeling these efficiencies are also discussed in the following section.

3.2 Theoretical Considerations and Modeling

A set of *exact* equations relating the fundamental efficiencies to the global engine parameters is[12,13]

$$\text{bp} = \eta_{ti}\eta_c\eta_v\eta_M \, \rho_A \, D \, N \, \text{LHV}_p \text{FA}/(60X) \tag{24}$$

$$\text{ip} = \eta_{ti}\eta_c\eta_v \, \rho_A \, D \, N \, \text{LHV}_p \text{FA}/(60X) \tag{25}$$

$$\tau = 1000 \, \eta_{ti}\eta_c\eta_v\eta_M \, \rho_A \, D \, \text{LHV}_p \text{FA}/(2\pi X) \tag{26}$$

$$\text{bmep} = 1000 \, \eta_{ti}\eta_c\eta_v\eta_M \, \rho_A \, \text{LHV}_p \text{FA} \tag{27}$$

$$\text{imep} = 1000 \, \eta_{ti}\eta_c\eta_v \, \rho_A \, \text{LHV}_p \text{FA} \tag{28}$$

$$\text{bsfc} = 3600/(\eta_{ti}\eta_c\eta_M \, \text{LHV}_p) \tag{29}$$

$$\text{isfc} = 3600/(\eta_{ti}\eta_c \, \text{LHV}_p) \tag{30}$$

$$\eta_e = \eta_{ti}\eta_c\eta_M \tag{31}$$

where, again, the constants are simply units conversion factors. Equations (24)–(31) are of interest because they (1) can be derived solely from physical and thermodynamic considerations,[12] (2) can be used to explain observed engine characteristics,[12] and (3) can be used as a base for modeling engine performance.[13] For example, Eqs. (27) and (28) demonstrate that the mean effective pressure is useful for comparing different engines because it is a measure of performance that is essentially independent of displacement (D), engine speed (N), and whether the engine is a 2-stroke ($X = 1$) or a 4-stroke ($X = 2$). Similarly, Eqs. (24), (27), and (29) show that a diesel should have less power, lower bmep, and better bsfc than a comparable SI engine because the diesel generally has about the same LHV_p and η_c, higher η_{ti} and η_v, but lower η_M and much lower FA.

The performance of an engine may be theoretically predicted by modeling each of the fundamental efficiencies (η_{ti}, η_c, η_v, and η_M) and then combining these models using Eqs. (2)–(31) to yield the performance parameters. Simplified models for each of these fundamental efficiencies are discussed below. More detailed engine models are available with varying degrees of sophistication and accuracy,[2,4,11,13,15,16,25] but, because of their length and complexity, are beyond the scope of this chapter.

The combustion efficiency may be most simply modeled by assuming complete combustion. It can be shown[13] that for complete combustion

$$\eta_c = 1.0 \qquad \phi \le 1 \tag{32a}$$

$$\eta_c = 1.0/\phi \qquad \phi \ge 1 \tag{32b}$$

Equation (32) implies that η_c is only dependent on FA. It has been shown that for homogeneous charge 4-stroke SI engines using fuels with a carbon-to-hydrogen ratio similar to that of gasoline, η_c is approximately independent of compression ratio, engine speed, ignition timing, and load. Although no data are available, it is expected that this is also true of 4-stroke stratified charge SI engines and diesel engines (at least up to the point of production of appreciable smoke). Such relationships will be less accurate for 2-stroke SI engines due to fuel short-circuiting. As shown in Fig. 7, Eq. (32) is accurate within 5–10% for fuel-lean combustion with accuracy decreasing to about 20% for the very fuel-rich equivalence ratio of 1.5 for the 4-stroke SI engine. Also shown in Fig. 7 is a quasiequilibrium equation that is slightly more accurate for fuel-lean systems and much more accurate for fuel-rich combustion:

$$\eta_c = 0.959 + 0.129\phi - 0.121\phi^2 \qquad 0.5 \le \phi \le 1.0 \qquad (33a)$$

$$\eta_c = 2.594 - 2.173\phi + 0.546\phi^2 \qquad 1.0 \le \phi \le 1.5 \qquad (33b)$$

The indicated thermal efficiency may be most easily modeled using air standard cycles. The values of η_{ti} predicted in this manner will be too high by a factor of about 2 or more, but the trends predicted will be qualitatively correct.

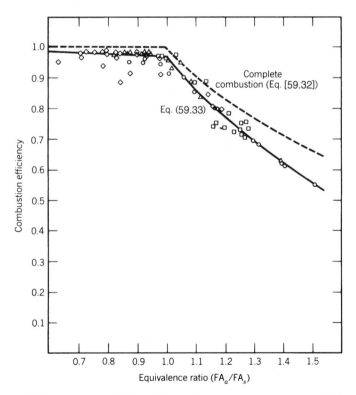

Figure 7 Effect of equivalence ratio (normalized fuel–air ratio) on combustion efficiency for several different 4-stroke SI engines operating on indolene or iso-octane.[13] Model predictions of Eqs. (32) and (33) also shown.

The SI engine ideally operates on the air standard Otto cycle, for which[2,4,5,10,11,14]

$$\eta_{ti} = 1 - (1/\mathrm{CR})^{k-1} \tag{34}$$

where CR is the compression ratio and k is the ratio of specific heats of the working fluid. The assumptions of the air standard Otto cycle are (1) no intake or exhaust processes and thus no exhaust residual and no pumping loss, (2) isentropic compression and expansion, (3) constant-volume heat addition and therefore instantaneous combustion, (4) constant-volume heat rejection replacing the exhaust blowdown process, and (5) air is the sole working fluid and is assumed to have a constant value of k. The errors in this model primarily result from failure to account for (1) a working fluid with variable composition and variable specific heats, (2) the finite duration of combustion, (3) heat losses, (4) fluid mechanics (especially as affecting flow past the intake and exhaust valves and the effects of these on the exhaust residual fraction), and (5) pumping losses. The air standard Otto cycle P–v and T–s diagrams are shown in Fig. 8. Equation (34) indicates that η_{ti} for the SI engine is only a function of CR. While this is not strictly correct, it has been shown[13] that, for the homogeneous charge 4-stroke SI engine, η_{ti} is not as strongly dependent on equivalence ratio, speed, and load as on CR. The predicted effect of CR on η_{ti} is compared with engine data in Fig. 9, showing that the theoretical trend is qualitatively correct. Thus, dividing the result of Eq. (34) by 2 will yield a reasonable estimate of the indicated thermal efficiency but will not reflect the effects of speed, load, spark timing, valve timing, or other engine design and operating conditions on η_{ti}.

The traditional simplified model for η_{ti} of the CI engine is the air standard Diesel cycle, but the air standard dual cycle is more representative of most modern diesel engines. Figure 8 compares the P–v and T–s diagrams for the air standard diesel, dual, and Otto cycles. For the air standard diesel cycle, it can be shown that[2,4,5,11,14]

$$\eta_{ti} = 1 - (1/\mathrm{CR})^{k-1} \frac{r_T^k - 1}{k(r_T - 1)} \tag{35}$$

where r_T is the ratio of the temperature at the end of combustion to the temperature at the beginning of combustion, and is thus a measure of the load. Assumptions for this cycle are (1) no intake or exhaust processes, (2) isentropic compression and expansion, (3) constant-pressure heat addition (combustion), (4) constant-volume heat rejection, and (5) air is the sole working fluid and has constant k. For the air standard dual cycle, it can be shown that[2,4,5,10,14]

$$\eta_{ti} = 1 - (1/\mathrm{CR})^{k-1} \frac{r_P r_v^k - 1}{(r_P - 1) + k r_P (r_v - 1)} \tag{36}$$

where r_P is the ratio of the maximum pressure to the pressure at the beginning of the combustion process and r_v is the ratio of the volume at the end of the combustion process to the volume at the beginning of the combustion process (TDC). The assumptions are the same as those for the air standard diesel cycle except that combustion is assumed to occur initially at constant volume with the remainder of the combustion process occurring at constant pressure. Equations (35) and (36) indicate that the η_{ti} of the diesel is a function of both compression ratio and load, predictions that are qualitatively correct.

The volumetric efficiency is the sole efficiency for which a simplified model cannot be developed from thermophysical principles without the need for iterative calculations. Factors affecting η_v include heat transfer, fluid mechanics (including intake and/or exhaust tuning and valve timing), and exhaust residual fraction. In fact, a 4-stroke engine with tuned intake (exhaust tuning is more important for the 2-stroke) and/or pressure boosting may have

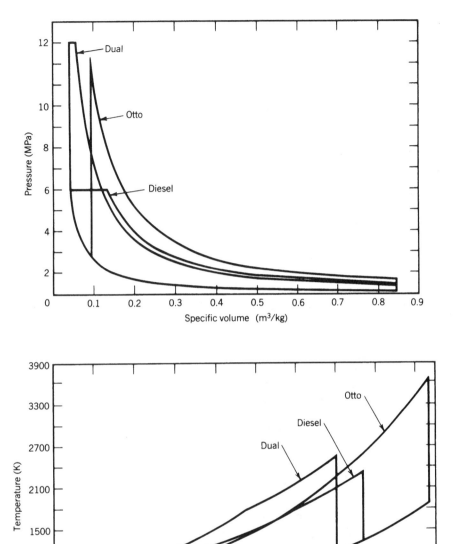

Figure 8 Comparison of air standard Otto, diesel, and dual-cycle $P–v$ and $T–s$ diagrams with $k = 1.3$. Otto with CR = 9:1, $\phi = 1.0$, C_8H_{18}. Diesel with CR = 20:1, $\phi = 0.7$, $C_{12}H_{26}$. Dual at same conditions as diesel, but with 50% of heat added at constant volume.

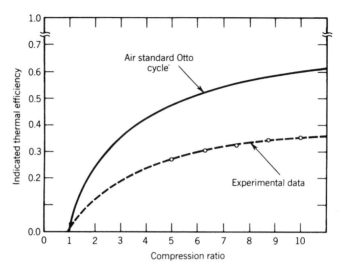

Figure 9 Effect of compression ratio on indicated thermal efficiency of SI engine. Dashed line—experimental data for single cylinder 4-stroke SI engine.[13] Solid line—prediction of air standard Otto cycle [Eq. 34].

greater than 100% volumetric efficiency since Eq. (20) is referenced to the air inlet density rather than to the density of the air in the intake manifold. However, for unboosted engines, observed engine characteristics allow values to be estimated for η_v (for subsequent use in Eqs. (24)–(31)). For the unthrottled DISC SI engine, the diesel, and the SI engine at wide open throttle (full load), η_v is approximately independent of operating conditions other than engine speed (which is important due to valve timing and tuning effects). A peak value for η_v of 0.7–1.0 may be assumed for 4-stroke engines with untuned intake systems, recognizing that there is no justification for choosing any particular number unless engine data for that specific engine are available. For 4-stroke engines with tuned intake systems, a peak value of ~1.15–1.2 might be assumed. Because most SI engines control power output by varying η_v, the effect of load on η_v must be taken into account for this type of engine. Fortunately, other engine operating conditions have much less effect on part load η_v than does the load,[13] so that only load need be considered for this simplified approach. As shown in Fig. 10 for the 4-stroke SI engine, η_v is linearly related to load (imep). Because choked flow is attained at no load, a value for η_v at zero load at roughly one-half that assumed at full load may be used and a linear relationship between η_v and load (or intake manifold pressure) may then be used (the intake manifold pressure may drop well below that for choked flow during idle and deceleration, but the flow is still choked).

The mechanical efficiency may be most simply modeled by first determining η_{ti}, η_c, and η_v and then calculating the indicated power using Eq. (25). The friction power may then be calculated from an empirical relationship,[4] which has been shown to be reasonably accurate for a variety of production multicylinder SI engines (and it may apply reasonably well to diesels):

$$\text{fp} = 1.975 \times 10^{-9} \, D \, S \, CR^{1/2} N^2 \tag{37}$$

where S is the stroke in mm. It should be noted that oil viscosity (and therefore oil and coolant temperatures) can significantly affect fp, and that Eq. (37) applies for normal oper-

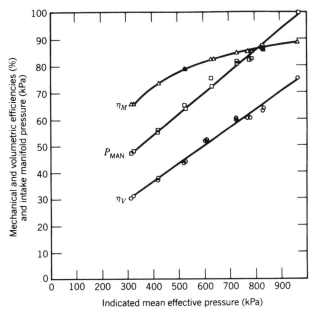

Figure 10 Effect of load (imep) on volumetric efficiency, mechanical efficiency, and intake manifold absolute pressure for 4-stroke V6 SI engine at 2000 rpm.[13]

ating temperatures. Given fp and ip, η_M may be calculated using Eq. (17). Since load strongly affects ip and speed strongly affects fp, then η_M is more strongly dependent on speed and load than on other operating conditions. Figures 10 and 11 demonstrate the effects of speed and load on η_M for a 4-stroke SI engine. Similar trends would be observed for other types of engines.

The accuracy of the performance predictions discussed above depends on the accuracies of the prediction of η_{ti}, η_c, η_v, and fp. Of these, the models for η_c and fp are generally acceptable. Thus, the primary sources for error are the models of η_{ti} and η_v. As stated earlier, highly accurate models are available if more than qualitative performance predictions are needed.

3.3 Engine Comparisons

Table 4 presents comparative data for various types of engines. One of the primary advantages of the SI engine is the wide speed range attainable owing to the short ignition delay, high flame speed, and low rotational mass. This results in high specific weight (bp/W) and high power per unit displacement (bp/D). Additionally, the energy content (LHV$_p$) of gasoline is about 3% greater than that of diesel fuel (on a mass basis; diesel fuel has ~13% more energy per gallon than gasoline). This aids both bmep and bp. The low CR results in relatively high η_M at full load and low engine weight. The low weight, combined with the relative mechanical simplicity (especially of the fueling system), results in low initial cost. The high full load η_M and the high FA (relative to the diesel) result in high bmep and bp. However, the low CR results in low η_{ti} (especially at part load), which in turn causes low η_e, high part load bsfc, and poor low speed τ. Because η_{ti} increases with load, reasonable bsfc is attained near full load. Also, since the product of η_{ti} and N initially increases faster

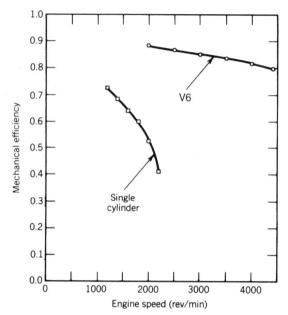

Figure 11 Effect of engine speed on mechanical efficiency of 4-stroke SI V6 engine[13] and of 4-stroke SI single-cylinder research engine.

Table 4 Comparative Data for Motor Vehicle Engines[g]

Engine Type	CR	Peak N (rpm)	bmep (kPa)	bp/D (kW/liter)	bp/W (kW/kg)	bsfc (g/kWhr)	η_e (%)
Spark Ignition							
Motorcycles							
2-stroke	6.5–11	4500–7000[e]	400–550	20–50	0.17–0.40	600–400	14–18
4-stroke	6–10	5000–7500[e]	700–1000	30–60	0.18–0.40	340–270	25–31
Passenger cars							
2-stroke	6–8	4500–5400	450–600	30–45	0.18–0.40	480–340	17–18
4-stroke	7–11[c]	4000–7500[f]	700–1000	20–50	0.25–0.50	380–300	20–28
Rotary	8–9	6000–8000	950–1050	35–45	0.62–1.1	380–300	22–27
Trucks (4-stroke)	7–8	3600–5000	650–700	25–30	0.15–0.40	400–300	16–27
Diesel							
Passenger cars[a]	12–23[d]	4000–5000	500–750	18–22	0.20–0.40	340–240	23–28
Trucks—NA[b]	16–22	2100–4000	600–900	15–22	0.14–0.25	245–220	23–33
Trucks—TC[b]	15–22	2100–3200	1200–1800	18–26	0.14–0.29	245–220	—

[a] Exclusively IDI in United States. 2-stroke and 4-stroke DI may be used in other countries.

[b] NA: naturally aspirated, TC: turbocharged. Trucks are primarily DI.

[c] U.S. average about 9.

[d] U.S. average about 22.

[e] Many modern motorcycle engines have peak speeds exceeding 10,000 rpm.

[f] Production engines capable of exceeding 6000 rpm are rare.

[g] Adopted with permission from Adler and Bazlen.[3] Design trends to be read left to right.

with engine speed than η_M decreases, then good medium speed torque is attained and can be significantly augmented by intake tuning such that η_v peaks at medium engine speed. Another disadvantage of the SI engine is that the near stoichiometric FA used results in high engine-out emissions of the gaseous regulated exhaust pollutants: NO_x (oxides of nitrogen), CO (carbon monoxide), and HCs (unburned hydrocarbons). On the other hand, the nature of the premixed combustion process results in particulate emission levels that are almost too low to measure.

One of the primary advantages of the diesel is that the high CR results in high η_{ti} and thus good full load bsfc and, because of the relationship between η_{ti} and load for the diesel, much higher part load bsfc than the SI engine. The higher η_{ti} and the high η_v produce high η_e and good low speed τ. The low volatility of diesel fuel results in lower evaporative emissions and a lower risk of accidental fire. The high CR also results in relatively low η_M and high engine weight. The high W and the mechanical sophistication (especially of the fuel-injection system) lead to high initial cost. The low η_M, coupled with the somewhat lower LHV_p and the much lower FA in comparison to the SI engine, yield lower bmep. The low bmep and limited range of engine speeds yield low bp and low bp/D. The low bp and high W produce low bp/W. The diesel is ideally suited to supercharging, since supercharging inhibits knock. However, the lower exhaust temperatures of the diesel (500–600°C)[3] means that there is less energy available in diesel exhaust and thus it is somewhat more difficult to turbocharge the diesel. The diffusion-flame nature of the combustion process produces high particulate and NO_x levels. The overall fuel-lean nature of the combustion process produces relatively low levels of HCs and very low levels of CO. Diesel engines are much noisier than SI engines.

The IDI diesel is used in all diesel-powered passenger cars sold in the United States, which are attractive because of the high bsfc of the diesel in comparison to the SI engine. The IDI diesel passenger car has approximately 35% better fuel economy on a kilometers per liter of fuel basis (19% better on a per unit energy basis, since the density of diesel fuel is about 16% higher than that of gasoline) than the comparable SI-engine-powered passenger car.[17] The IDI diesel is currently used in passenger cars rather than the DI diesel because of the more limited engine speed range of the DI diesel and because the necessarily more sophisticated fuel-injection system leads to higher initial cost. Also, the IDI diesel is less fuel sensitive (less prone to knock), is generally smaller, runs more quietly, and has fewer odorous emissions than the DI diesel.[17]

The engine speed range of DI diesels is generally more limited than that of the IDI diesel. The DI diesel is easier to start and rejects less heat to the coolant. The somewhat higher exhaust temperature makes the DI diesel more suitable for turbocharging. At full load, the DI emits more smoke and is noisier than the IDI diesel. In comparison to the IDI diesel, the DI diesel has 15–20% better fuel economy.[17] This is a result of higher η_{ti} (which yields better bsfc and η_e) due to less heat loss because of the lower surface-to-volume ratio of the combustion chamber in absence of the IDI's prechamber.

The primary advantage of the 2-stroke engine is that it has one power stroke per revolution ($X = 1$) rather than one per every two revolutions ($X = 2$). This results in high τ, bp, and bp/W. However, the scavenging process results in very high HC emissions, and thus low η_c, bmep, bsfc, and η_e. When crankcase supercharging is used to improve scavenging, η_v is low,[3] being in the range of 0.3–0.7. Blowers may be used to improve scavenging and η_v, but result in decreased η_M. The mechanical simplicity of the valveless crankcase compression design results in high η_M, low weight, and low initial cost. Air-cooled designs have an even lower weight and initial cost, but have limited CR because of engine-cooling considerations. In fact, both air-cooled and water-cooled engines have high thermal loads because of the lack of no-load strokes. For the 2-stroke SI engine, the CR is also limited by the need to inhibit knock.

The primary advantages of the rotary SI engine are the resulting low vibration, high engine speed, and relatively high η_v. The high η_v produces high bmep and good low-speed τ. The high bmep and high attainable engine speed produce high bp. The valveless design results in decreased W. The high bp and low W result in very high bp/W. Because η_e and bsfc are independent of η_v, these parameters are approximately equal for the rotary SI and the 4-stroke piston SI engines.

4 EMISSIONS AND FUEL ECONOMY REGULATIONS*

Light-duty vehicles sold in the United States are subject to federal and state regulations regarding exhaust emissions, evaporative emissions, and fuel economy. Heavy-duty vehicles do not have to comply with any fuel economy standards, but are required to comply with standards for exhaust and evaporative emissions. Similar regulations are in effect in many foreign countries.

4.1 Light-Duty Vehicles

Light-duty-vehicle (LDV) emissions regulations are divided into those applicable to passenger cars and those applicable to light-duty trucks [defined as trucks of less than 6000 lb gross vehicle weight (GVW) rating prior to 1979 and less than 8500 lb after, except those with more than 45 ft^2 frontal area]. In turn, the light-duty trucks (LDTs) are divided into four categories: LDT1s are "light light-duty trucks" (GVW < 6000 lb) that have a loaded vehicle weight (LVW) of <3750 lb and must meet the same emissions standards as passenger cars; LDT2s are also light light-duty trucks and have 3751 < LVW < 5750 lb; LDT3s are "heavy light-duty trucks" (GVW > 6000 lb) that have an adjusted loaded vehicle weight (ALVW) of 3751 < ALVW < 5750 lb; LDT4s are heavy light-duty trucks with 5751 < LVW < 8550 lb. Standards for the light-duty trucks in the LDT2–LDT4 categories are not presented or discussed for brevity, but the test procedure described below is used for these vehicles as well as for passenger cars.

U.S. federal and state of California exhaust emissions regulations for passenger cars are presented in Table 5. Emissions levels are for operation over the federal test procedure (FTP) transient driving cycle. The regulated emissions levels must be met at the end of the "useful life" of the vehicle. For passenger cars, the useful life is currently defined as five years or 50,000 miles, but the Clean Air Act Amendments (CAAA) of 1990 redefined the useful life of passenger cars and LDT1s as 10 years/100,000 miles and 120,000 miles for LDT2–LDT4 light duty trucks. The longer useful life requirement is being phased-in and all vehicles must meet the longer useful life emissions standards in 2003. The first two phases of the FTP are illustrated in Fig. 12. Following these two phases (together, these are the LA-4 cycle), the engine is turned off, the vehicle is allowed to hot soak for 10 minutes, the engine is restarted, and the first 505 seconds are repeated as phase 3. The LA-4 cycle is intended to represent average urban driving habits and therefore has an average vehicle speed of only 19.5 mi/hr (31.46 km/hr) and includes 2.29 stops per mile (1.42 per kilometer). The LA-4 test length is 7.5 miles (11.98 km) and the duration is 1372 seconds with 19% of this time spent with

---

*The regulations are specified in mixed English and metric units and these specified units are used in this section to avoid confusion.

Table 5 Federal and California Emissions Standards for Passenger Cars

	Useful Life (Miles)	THC (g/mi)	NMHC[a] (g/mi)	NMOG (g/mi)	CO (g/mi)	NO$_x$ (g/mi)	Particulates (g/mi)	HCHO (mg/mi)
Precontrol avg.		10.6			84.0	4.0		
Fed. 1975–77	50K	1.5			15.0	3.1		
Cal. 1975–77	50K	0.9			9.0	2.0		
Fed. 1977–79	50K	1.5			15.0	2.0		
Cal. 1977–79	50K	0.41			9.0	1.5		
Fed. 1980	50K	0.41			7.0	2.0		
Cal. 1980	50K	—	0.39		9.0	1.0		
Fed. 1981	50K	0.41			3.4	1.0		
Cal. 1981	50K		0.39		7.0	0.7		
Fed. 1982–84	50K	0.41			3.4	1.0	0.6	
Cal. 1982	50K		0.39		7.0	0.7	0.6	
Cal. 1983–84	50K		0.39		7.0	0.4	0.6	
Fed. 1985–90	50K	0.41			3.4	1.0	0.2	
Cal. 1985–92	50K		0.39		7.0	0.4	0.2	
Fed. Tier 0 (1991–94)	50K[b]	0.41	—	—	3.4	1.0		—
	100K[c]	0.80	—	—	10.0	1.2		—
Cal. Tier 1	50K		0.25	—	3.4	0.4		15[d]
	100K		0.31	—	4.2	0.6		15
Fed. Tier 1[e] (1994–)	50K		0.25	—	3.4	0.4		—
	100K		0.31	—	4.2	0.6		—
Fed. Tier 2 (2003)	100K		0.125	—	1.7	0.2		—
TLEV—alt. fuel	50K		—	0.125	3.4	0.4		15
	100K		—	0.156	4.2	0.6		18
TLEV—bi-fuel[f]	50K		—	0.250	3.4	0.4		15
	100K		—	0.310	4.2	0.6		18
LEV—alt. fuel	50K		—	0.075	3.4	0.2		15
	100K		—	0.090	4.2	0.3		18
LEV—bi-fuel[f]	50K		—	0.125	3.4	0.2		15
	100K		—	0.156	4.2	0.3		18
ULEV—alt. fuel	50K		—	0.040	1.7	0.2		8
	100K		—	0.055	2.1	0.3		11
ULEV—bi-fuel[f]	50K		—	0.075	1.7	0.2		8
	100K		—	0.090	2.1	0.3		11

[a] Organic material nonmethane hydrocarbon equivalent (OMNHCE) for methanol vehicles.

[b] Passenger cars can elect Tier 0 with the lower useful life, but cannot elect Tier 0 at the full useful life.

[c] LDTs that elect Tier 0 must use the full useful life.

[d] This formaldehyde standard must be complied with at 50,000 miles rather than at the end of the full useful life.

[e] Federal Tier 1 standards also require that all light-duty vehicles meet a 0.5% idle CO standard and a cold (20°F FTP) CO standard of 10 gm/mi.

[f] Flexible fuel and bi-fuel vehicles must certify both to the dedicated (alt. fuel) standard (while using the alternative fuel) *and* the bi-fuel standard when using the conventional fuel (e.g., certification gasoline).

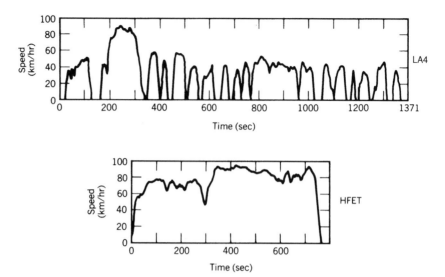

Figure 12 FTP and HFET transient driving cycles.

the vehicle idling. The motor vehicle manufacturer will commit at least two vehicles of each type to the certification procedure. The first vehicle will be used to accumulate 50,000 miles (or 100,000 or 120,000, depending on the vehicle type and the useful life to which the vehicle is certified) so that a deterioration factor (DF) can be determined for each of the regulated species. The DF is determined by periodic FTP testing over the useful life and is intended to reveal how the emissions levels change with mileage accumulation. The second vehicle is subjected to the FTP test after 4000 miles of accumulation. The emission level of each regulated species is then multiplied by the DF for that species and the product is used to determine compliance with the regulatory standard. All data are submitted to the U.S. Environmental Protection Agency (EPA), which may perform random confirmatory testing, especially of the 4000-mile vehicles.

The 1981 standards represented a 96% reduction in HC and CO and 76% reduction in NO$_x$ from precontrol levels. Beginning in 1980, the California HC standard was specified as either 0.41 g of total hydrocarbons (THCs) per mile or 0.39 g of nonmethane hydrocarbons (NMHCs) per mile, in recognition that methane is not considered to be photochemically reactive and therefore does not contribute to smog formation.[17,18] The federal CAAA of 1990 also recognized that methane does not contribute to smog formation and thus instituted a NMHC standard, rather than a THC standard, as part of the new emissions standards phased in from 1994 to 1996. Similarly, because the goal of the hydrocarbon standards is control of ambient ozone, the current California rules regulate nonmethane organic gases (NMOG) rather than NMHCs. Furthermore, in recognition that each of the organic gases in vehicle exhaust has a different efficiency in ozone-formation chemistry, the California rulemaking does not regulate measured NMOG, but rather "reactivity-adjusted" NMOG. The measured NMOG is multiplied by a reactivity adjustment factor (RAF) and the resulting product is then judged against the NMOG standard. The RAF is both fuel-specific and emissions level-specific. The California emissions levels are TLEV (transitional low-emission vehicle), LEV (low-emission vehicle), ULEV (ultra-low-emission vehicle), and ZEV (zero-emission vehicle, or electric vehicle). For LEV vehicles, CARB has established the following RAFs: 1.0 for gasoline, 0.94 for California Phase 2 reformulated gasoline, 0.43 for CNG, 0.50 for LPG, and 0.41 for M85. The particulate emissions standards shown in Table 5 apply only to light-duty diesel (LDD) vehicles; light-duty gasoline (LDG) vehicles emit essentially immeasurable quantities of particulates, and therefore are not required to demonstrate compliance with the particulate standards.

LDVs are also required to meet evaporative emissions standards, as determined using a SHED (sealed housing for evaporative determination) test. The precontrol average for evaporative emissions was 50.6 g per test, which is approximately equivalent to 4.3 g HC per mile.[19] The federal regulations are 6 g/test for 1978–80 and 2 g/test thereafter. The California standard is also 2 g/test. Evaporative emissions from the LDD are essentially immeasurable and thus are not required to demonstrate compliance with this regulation. California allows LDDs a 0.16-g/mi credit on exhaust HCs in recognition of the negligible evaporative HC emissions of these vehicles. In the early 1990s, although retaining the standard of 2 g/test, California instituted new rules requiring use of a new SHED test design, requiring that the test period increase from one day to four days, requiring that the maximum temperature increase to 104°F from 84°F and compliance for 120,000 miles/10 years instead of 50,000 miles/5 years. A similar federal SHED test will be implemented in the near future.

Beginning in 1978, the motor vehicle manufacturers were required to meet the corporate average fuel economy (CAFE) standards shown in Table 6 (a different, and lower, set of standards is in effect for "low-volume" manufacturers). The CAFE standard is the mandated average fuel economy, as determined by averaging over all vehicles sold by a given manufacturer. The CAFE requirement is based upon the "composite" fuel economy (mpg_c, in mi/gal), which is a weighted calculation of the fuel economy measured over the FTP cycle (mpg_u, "urban" fuel economy) plus the fuel economy measured over the highway fuel economy test (HFET, as shown in Fig. 12) transient driving cycle (mpg_h):

$$mpg_c = \frac{1.0}{0.55/mpg_u + 0.45/mpg_h} \tag{38}$$

The manufacturer is assessed a fine of $5 per vehicle produced for every 0.1 mpg below the CAFE standard. Beginning in 1985, an adjustment factor was added to the measured CAFE to account for the effects of changes in the test procedure. Additionally, the motor vehicle manufacturer is subjected to a "gas guzzler" tax for each vehicle sold that is below a minimum composite fuel economy, with the minimum level being: 1983, 19.0 mpg; 1984, 19.5 mpg; 1985, 21.0 mpg; 1986 on, 22.5 mpg. In 1986, the tax ranged from $500 for each

Table 6 CAFE Standards for Passenger Cars Required by the Energy Policy and Conservation Act of 1975

Model Year	Average mpg
1978	18.0
1979	19.0
1980	20.0
1981	22.0[a]
1982	24.0[a]
1983	26.0[a]
1984	27.0[a]
1985	27.5[a]

[a] Set by the Secretary of the U.S. Department of Transportation.

vehicle with 21.5–22.4 mpg up to the maximum of $3850 for each vehicle with less than 12.5 mpg.

The federal and state regulations are intended to require compliance based on the best available technology. Therefore, future standards may be subjected to postponement, modification, or waiver, depending on available technology. For more detailed information regarding the standards and measurement techniques, refer to the U.S. Code of Federal Regulations, Title 40, Parts 86 and 88, and to the Society of Automotive Engineers (SAE) Recommended Practices.[20]

4.2   Heavy-Duty Vehicles

The test procedure for heavy-duty vehicles (HDVs) is different from that for the LDV. Because many manufacturers of heavy-duty vehicles offer the customer a choice of engines from various vendors, the engines rather than the vehicles must be certified. Thus, the emissions standards are based on grams of pollutant emitted per unit of energy produced (g/bhp-hr, mixed metric and English units, with bhp meaning brake horsepower) during engine operation over a prescribed transient test cycle. The federal standards for 1991 and newer heavy-duty engines are divided into two gross vehicle weight categories for spark ignition engines (less than and greater than 14,000 lb), but a single set of standards is used for diesel engines. The standards also depend on the fuel used (diesel, gasoline, natural gas, etc.), as presented in Table 7. Beginning in 1985, the heavy-duty gasoline (HDG) vehicle was required to meet an idle CO standard of 0.47% and an evaporative emissions SHED Test standard of 3.0 g/test for vehicles with GVW less than 14,000 lb and 4.0 g/test for heavier vehicles. Prior to 1988, diesels were required only to meet smoke opacity standards (rather than particulate mass standards) of 20% opacity during acceleration, 15% opacity during lugging, and 50% peak opacity. Particulate mass standards for heavy duty diesels came into effect in 1988.

Table 7   Emissions Standards[a] for 1991–2000 Model Year Heavy-Duty Engines (in g/bhp-hr)

GVW Category	Engine	Fuel	THCs[b]	NMHCs	CO	NO$_x$[c]	PM
≤ 14,000 lb	SI	Gasoline	1.1	NA	14.4	5.0	NA
≤ 14,000 lb	SI	Methanol	1.1	NA	14.4	5.0	NA
≤ 14,000 lb	SI	CNG	NA	0.9	14.4	5.0	NA
≤ 14,000 lb	SI	LPG	1.1	NA	14.4	5.0	NA
> 14,000 lb	SI	Gasoline	1.9	NA	37.1	5.0	NA
> 14,000 lb	SI	Methanol	1.9	NA	37.1	5.0	NA
> 14,000 lb	SI	CNG	NA	1.7	37.1	5.0	NA
> 14,000 lb	SI	LPG	1.9	NA	37.1	5.0	NA
Any (> 8501 lb)	CI	Diesel	1.3	NA	15.5	5.0	0.25[d]
Any (> 8501 lb)	CI	Methanol	1.3	NA	15.5	5.0	0.25[d]
Any (> 8501 lb)	CI	CNG	NA	1.2	15.5	5.0	0.10
Any (> 8501 lb)	CI	LPG	1.3	NA	15.5	5.0	0.10

[a] Other standards apply to clean fuel vehicles.

[b] Organic material hydrocarbon equivalent (OMHCE) for methanol.

[c] NO$_x$ standard decreases to 4.0 g/bhp-hr beginning in 1998.

[d] Particulate standard decreased to 0.10 g/bhp-hr beginning in 1994.

4.3   Nonhighway Heavy-Duty Standards

Stationary engines may be subjected to federal, state, and local regulations, and these regulations vary throughout the United States.[19] Diesel engines used in off-highway vehicles and in railroad applications may be subjected to state and local visible smoke limits, which are typically about 20% opacity.[19]

SYMBOLS

AF	air–fuel mass ratio [-]*
AF_s	stoichiometric air–fuel mass ratio [-]
ALVW	adjusted loaded vehicle weight = (curb weight + GVW)/2 [lb]
API	American Petroleum Institute gravity [°]
ASTM	American Society of Testing and Materials
BDC	bottom dead center
bmep	break mean effective pressure [kPa]
bp	brake power [kW]
bsfc	brake specific fuel consumption [g/kW-hr]
CARB	California Air Resources Board
CI	compression ignition (diesel) engine
CN	cetane number [-]
CO	carbon monoxide
CO_2	carbon dioxide
CR	compression ratio (by volume) [-]
C_xH_y	average fuel molecule, with x atoms of carbon and y atoms of hydrogen
D	engine displacement [liters]
DI	direct injection diesel engine
DISC	direct injection stratified charge SI engine
DF	emissions deteriorator factor [-]
F	force on dyno torque arm with engine being motored [N]
FA	fuel–air mass ratio [-]
FA_s	stoichiometric fuel–air mass ratio [-]
fp	friction power [kW]
FTP	federal test procedure transient driving cycle
GVW	gross vehicle weight rating
HCs	hydrocarbons
$h_{f,i}^{298}$	standard enthalpy of formation of gaseous species i at 298 K [MJ/kmole of species i]†

---

*Standard units, which do not strictly conform to metric practice, in order to produce numbers of convenient magnitude.

†A kmole is a mole based upon a kg; also called a kg-mole.

$h_{v,i}^{298}$	enthalpy of vaporization of species i at 298 K [MJ/kmole of species i]
ΔH_{RX}^{298}	enthalpy of reaction (heat of combustion) at 298 K [MJ/kg of fuel]
HDD	heavy-duty diesel vehicle
HDG	heavy-duty gasoline vehicle
HDV	heavy-duty vehicle
HFET	highway fuel economy test transient driving cycle
HV	heating value [MJ/kg of fuel]
HHV_p	constant-pressure higher heating value [MJ/kg of fuel]
HHV_v	constant-volume higher heating value [MJ/kg of fuel]
HV_p	constant-pressure heating value [MJ/kg of fuel]
HV_v	constant-volume heating value [MJ/kg of fuel]
IDI	indirect injection diesel engine
ILEV	inherently low emission vehicle
imep	indicated mean effective pressure [kPa]
ip	indicated power [kW]
isfc	indicated specific fuel consumption [g/kW-hr]
k	ratio of specific heats
K	dyno constant [N/kW-min]
L	force on dyno torque arm with engine running [N]
LDD	light-duty diesel vehicle
LDG	light-duty gasoline vehicle
LDT	light-duty truck
LDV	light-duty vehicle
LEV	low-emission vehicle
LHV_p	constant-pressure lower heating value [MJ/kg of fuel]
LHV_v	constant-volume lower heating value [MJ/kg of fuel]
LVW	loaded vehicle weight (curb weight plus 300 lb)
$\dot{m}_A$	air mass flow rate into engine [g/hr]
$\dot{m}_F$	fuel mass flow rate into engine [g/hr]
MON	octane number measured using Motor method [-]
mpg_c	composite fuel economy [mi/gal]
mpg_h	highway (HFET) fuel economy [mi/gal]
mpg_u	urban (FTP) fuel economy [mi/gal]
MW	molecular weight [kg/kmole]
N	engine rotational speed [rpm]
N_2	molecular nitrogen
NMHC	nonmethane hydrocarbons
NMOG	nonmethane organic gases (NMHCs plus carbonyls and alcohols)
NO_x	oxides of nitrogen
O_2	molecular oxygen
ON	octane number of fuel [-]
P	pressure [kPa]
R	dyno torque arm length [m]

RAF	reactivity adjustment factor
RON	octane number measured using Research method [-]
r_P	pressure ratio (end of combustion/end of compression) [-]
r_T	temperature ratio (end of combustion/end of compression) [-]
r_v	volume ratio (end of combustion/end of compression) [-]
s	entropy [kJ/kg-K]
S	piston stroke [nm]
SAE	Society of Automotive Engineers
sg_F	specific gravity of fuel
SHED	sealed housing for evaporative determination
SI	spark ignition engine
T	milliliters of tetraethyl lead per gallon of iso-octane [ml/gal]
T	temperature [K]
TA	percent theoretical air
TDC	top dead center
THCs	total hydrocarbons
TLEV	transitional low emission vehicle
ULEV	ultra low emission vehicle
v	specific volume [m^3/kg]
W	engine mass [kg]
X	crankshaft revolutions per power stroke
x	atoms of carbon per molecule of fuel
y	atoms of hydrogen per molecule of fuel
Y_i	mole fraction of species i in exhaust products [-]
ZEV	zero emission vehicle
α	1.0 for HHV, 0.0 for LHV [-]
β	1.0 for liquid fuel, 0.0 for gaseous fuel [-]
η_c	combustion efficiency [-]
η_e	overall engine efficiency [-]
η_M	mechanical efficiency [-]
η_{ti}	indicated thermal efficiency [-]
η_v	volumetric efficiency [-]
ϕ	equivalence ratio, FA/FA$_s$ [-]
ρ_A	density of air at engine inlet [kg/m^3]
ρ_F	density of fuel [kg/m^3]
ρ_w	density of water [kg/m^3]
λ	excess air ratio, AF/AF$_s$ [-]
τ	engine output torque [N-m]

REFERENCES

1. C. F. Taylor, *The Internal Combustion Engine in Theory and Practice,* Vol. 2, MIT Press, Cambridge, MA, 1979.

2. R. S. Benson and N. D. Whitehouse, *Internal Combustion Engines,* Pergamon Press, New York, 1979.

3. U. Adler and W. Baslen (eds.), *Automotive Handbook,* Bosch, Stuttgart, 1978.

4. L. C. Lichty, *Combustion Engine Processes,* McGraw-Hill, New York, 1967.

5. E. F. Obert, *Internal Combustion Engines and Air Pollution,* Harper & Row, New York, 1973.

6. Fuels and Lubricants Committee, "Diesel Fuels—SAE J312 APR82," in *SAE Handbook,* Vol. 3, pp. 23.34–23.39, 1983.

7. Fuels and Lubricants Committee, "Automotive Gasolines—SAE J312 JUN82," in *SAE Handbook,* Vol. 3, pp. 23.25–23.34, 1983.

8. Engine Committee, "Engine Power Test Code—Spark Ignition and Diesel—SAE J1349 DEC80," in *SAE Handbook,* Vol. 3, pp. 24.09–24.10, 1983.

9. Engine Committee, "Small Spark Ignition Engine Test Code—SAE J607a," in *SAE Handbook,* Vol. 3, pp. 24.14–24.15, 1983.

10. C. F. Taylor, *The Internal Combustion Engine in Theory and Practice,* Vol. 1, MIT Press, Cambridge, MA, 1979.

11. A. S. Campbell, *Thermodynamic Analysis of Combustion Engines,* Wiley, New York, 1979.

12. R. D. Matthews, "Relationship of Brake Power to Various Efficiencies and Other Engine Parameters: The Efficiency Rule," *International Journal of Vehicle Design* **4**(5), 491–500 (1983).

13. R. D. Matthews, S. A. Beckel, S. Z. Miao, and J. E. Peters, "A New Technique for Thermodynamic Engine Modeling," *Journal of Energy* **7**(6), 667–675 (1983).

14. B. D. Wood, *Applications of Thermodynamics,* Addison-Wesley, Reading, MA, 1982.

15. P. N. Blumberg, G. A. Lavoie, and R. J. Tabaczynski, "Phenomenological Model for Reciprocating Internal Combustion Engines," *Progress in Energy and Combustion Science* **5**, 123–167 (1979).

16. J. N. Mattavi and C. A. Amann (eds.), *Combustion Modeling in Reciprocating Engines,* Plenum, New York, 1980.

17. Diesel Impacts Study Committee, *Diesel Technology,* National Academy Press, Washington, DC, 1982.

18. R. D. Matthews, "Emission of Unregulated Pollutants from Light Duty Vehicles," *International Journal of Vehicle Design* **5**(4), 475–489 (1983).

19. Environmental Activities Staff, *Pocket Reference,* General Motors, Warren, MI, 1983.

20. Automotive Emissions Committee, "Emissions," in *SAE Handbook,* Vol. 3, Section 25, 1983.

21. Fuels and Lubricants Committee, "Alternative Automotive Fuels—SAE J1297 APR82," in *SAE Handbook,* Vol. 3, pp. 23.39–23.41, 1983.

22. R. C. Weast (ed.), *CRC Handbook of Chemistry and Physics,* 55th ed., CRC Press, Cleveland, OH, 1974.

23. ASTM Committee D-2, *Physical Constants of Hydrocarbons C_1 to C_{10},* American Society for Testing and Materials, Philadelphia, PA, 1963.

24. E. M. Shelton, *Diesel Fuel Oils, 1976,* U.S. Energy Research and Development Administration, Publication No. BERC/PPS-76/5, 1976.

25. C. M. Wu, C. E. Roberts, R. D. Matthews, and M. J. Hall, "Effects of Engine Speed on Combustion in SI Engines: Comparisons of Predictions of a Fractal Burning Model with Experimental Data," SAE Paper 932714, also in SAE Transactions: *Journal of Engines* **102**(3), 2277–2291, 1994.

26. R. D. Matthews, J. Chiu, and D. Hilden, "CNG Compositions in Texas and the Effects of Composition on Emissions, Fuel Economy, and Driveability of NGVs," SAE Paper 962097, 1996.

27. W. E. Liss and W. H. Thrasher, *Variability of Natural Gas Composition in Select Major Metropolitan Areas of the United States,* American Gas Association Laboratories Report to the Gas Research Institute, GRI Report No. GRI-92/0123, 1991; also see SAE Paper 9123464, 1991.

28. R. H. Pahl and M. J. McNally, Fuel Blending and Analysis for the Auto/Oil Air Quality Improvement Research Program, SAE Paper 902098, 1990.

29. S. C. Mayotte, V. Rao, C. E. Lindhjem, and M. S. Sklar, *Reformulated Gasoline Effects on Exhaust Emissions: Phase II: Continued Investigation of Fuel Oxygenate Content, Oxygenate Type, Volatility, Sulfur, Olefins, and Distillation Parameters,* SAE Paper 941974, 1994.

30. J. Kubesh, S. R. King, and W. E. Liss, *Effect of Gas Composition on Octane Number of Natural Gas Fuels,* SAE Paper 922359, 1992.

31. K. Owen and T. Coley, *Automotive Fuels Handbook,* Society of Automotive Engineers, Warrendale, PA, 1990.

BIBLIOGRAPHY

Combustion and Flame, The Combustion Institute, Pittsburgh, PA, published monthly.

Combustion Science and Technology, Gordon & Breach, New York, published monthly.

International Journal of Vehicle Design, Interscience, Jersey, C.I., UK, published quarterly.

Proceedings of the Institute of Mechanical Engineers (Automobile Division), I Mech E, London.

Progress in Energy and Combustion Science, Pergamon, Oxford, UK, published quarterly.

SAE Technical Paper Series, Society of Automotive Engineers, Warrendale, PA.

Symposia (International) on Combustion, The Combustion Institute, Pittsburgh, PA, published biennially.

Transactions of the Society of Automotive Engineers, SAE, Warrendale, PA, published annually.

CHAPTER 28

FUEL CELLS

Matthew M. Mench
Department of Mechanical and Nuclear Engineering
The Pennsylvania State University
University Park, Pennsylvania

1 INTRODUCTION

In 1839, Sir William Grove conducted the first known demonstration of the fuel cell. It operated with separate platinum electrodes in oxygen and hydrogen submerged in a dilute sulfuric acid electrolyte solution, essentially reversing a water electrolysis reaction. Early development of fuel cells had a focus on use of coal to power fuel cells, but poisons formed by the gasification of the coal limited the fuel-cell usefulness and lifetime.[1] High-temperature solid oxide fuel cells (SOFCs) began with Nernst's 1899 demonstration of the still-used yttria-stabilized zirconia solid-state ionic conductor, but significant practical application was not realized.[2] The molten carbonate fuel cell (MCFC) utilizes a mixture of alkali metal carbonates retained in a solid ceramic porous matrix that become ionically conductive at elevated (>600°C) temperatures and was first studied for application as a direct coal fuel cell in the 1930s.[3] In 1933, Sir Francis Bacon began development of an alkaline-based oxygen–hydrogen fuel cell that achieved a short-term power density of 0.66 W/cm², high even for today's standards. However, little additional practical development of fuel cells occurred until the late 1950s, when the space race between the United States and the Soviet Union catalyzed development of fuel cells for auxiliary power applications. Low-temperature polymer electrolyte fuel cells (PEFCs) were first invented by William Grubb at General Electric in 1955 and generated power for NASA's Gemini space program. However, short operational lifetime and high catalyst loading contributed to a shift to alkaline fuel cells (AFCs) for the NASA Apollo program, and AFCs still serve as auxiliary power units (APUs) for the Space Shuttle orbiter.

After the early space-related application development of fuel cells went into relative abeyance until the 1980s. The phosphoric acid fuel cell (PAFC) became the first fuel cell system to reach commercialization in 1991. Although only produced in small quantities (twenty to forty 200-kW units per year) by United Technologies Company (UTC), UTC has installed and operated over two hundred forty-five 200-kW units similar to that shown in Fig. 1 in 19 countries worldwide for applications such as reserve power for the First National Bank of Omaha. As of 2002, these units have successfully logged over five million hours of operation with 95% fleet availability.[1]

Led by researchers at Los Alamos National Laboratory in the mid-1980s, resurgent interest in PEFCs was spawned through the development of an electrode assembly technique that enabled an order-of-magnitude reduction in noble-metal catalyst loading. This major breakthrough and ongoing environmental concerns, combined with availability of a non-hydrocarbon-based electrolyte with substantially greater longevity than those used in the Gemini program, has resurrected research and development of PEFCs for stationary, automotive, and portable power applications.

The area of research and development toward commercialization of high-temperature fuel cells, including MCFC and SOFC systems, has also grown considerably in the past two decades, with a bevy of demonstration units in operation and commercial sales of MCFC systems.

The science and technology of fuel cell engines are both fascinating and constantly evolving. This point is emphasized by Fig. 2, which shows the registered fuel-cell-related patents in the United States, Canada, and United Kingdom since 1975. An acceleration of the patents granted in Japan and South Korea is also well underway, led by automotive

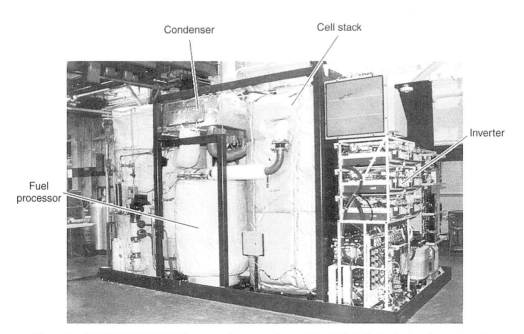

Figure 1   A 200-kW PC25 PAFC power plant manufactured by United Technologies Corporation.[49]

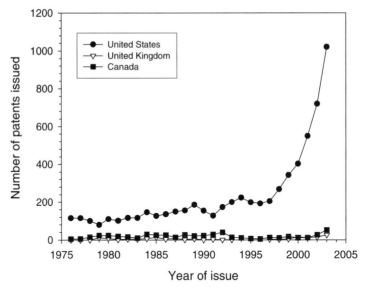

Figure 2 Timeline of worldwide patents in fuel cells (source: www.uspto.gov, www.patent.gov.uk, www.patent.gov.uk).

manufacturer development. The nearly exponential growth in patents granted in this field is obvious and is not likely to wane in the near future.

Potential applications of fuel cells can be grouped into four main categories: (1) transportation, (2) portable power, (3) stationary power, and (4) niche applications. Although automotive fuel-cell applications have a great potential, they are also probably the least likely to be implemented on a large scale in the near future. The existing combustion engine technology market dominance will be difficult to usurp, considering its low comparative cost (~$30/kW), high durability, high power density, suitability for rapid cold start, and high existing degree of optimization. Additionally, the recent success of high-efficiency hybrid electric/combustion engine technology adds another rapidly evolving target fuel cells must match to compete.

Perhaps where fuel cells show the most promise for ubiquitous near-term implementation is in portable power applications, such as cell phones and laptop computers. Toshiba has recently developed a hand-held direct methanol fuel cell for portable power that is planned for sale in 2005. The 8.5-g direct methanol fuel cell (DMFC) is rated at 100 mW continuous power (up to 20 hr) and measures 22 mm × 56 mm × 4.5 mm with a maximum of 9.1 mm for the concentrated methanol fuel tank.[4] Passive portable fuel cells can potentially compete favorably with advanced Li ion batteries in terms of gravimetric energy density of ~120–160 Wh/kg and volumetric energy density of ~230–270 Wh/L. Additionally, the cost of existing premium power battery systems is already on the same order as contemporary fuel cells, with additional development anticipated. With replaceable fuel cartridges, portable fuel-cell systems have the additional advantage of instant and remote rechargeability that can never be matched with secondary battery systems.

Stationary and distributed power applications include power units for homes or auxiliary and backup power generation units. Stationary applications (1–500 kW) are designed for nearly continuous use and therefore must have far greater lifetime than automotive units.

Distributed power plants are designed for megawatt-level capacity, and some have been demonstrated to date. In particular, a 2-MW MCFC was recently demonstrated by Fuel cell Energy in California.[5] A plot showing the estimated number of demonstration and commercial units in the stationary power category from 1986 to 2002 is given in Fig. 3. Earlier growth corresponded mostly to PAFC units, although recently most additional units have been PEFCs. Not surprisingly, the exponential growth in the number of online units follows a similar qualitative trend to the available patents granted for various fuel-cell technologies shown in Fig. 2. The early rise in stationary units in 1997 was primarily PAFC systems sold by United Technologies Center Fuel Cells. Data are estimated from the best available compilation available online at Ref. 6, and some manufacturers do not advertise prototype demonstrations, so that numbers are not exact. However, the trend is clear.

The fundamental advantages common to all fuel cell systems include the following:

1. A potential for a relatively high operating efficiency, scalable to all size power plants.
2. If hydrogen is used as fuel, greenhouse gas emissions are strictly a result of the production process of the fuel stock used.
3. No moving parts, with the significant exception of pumps, compressors, and blowers to drive fuel and oxidizer.
4. Multiple choices of potential fuel feedstocks, from existing petroleum, natural gas, or coal reserves to renewable ethanol or biomass hydrogen production.
5. A nearly instantaneous and remote recharge capability compared to batteries.

The following technical limitations common to all fuel-cell systems must be overcome before successful implementation can occur:

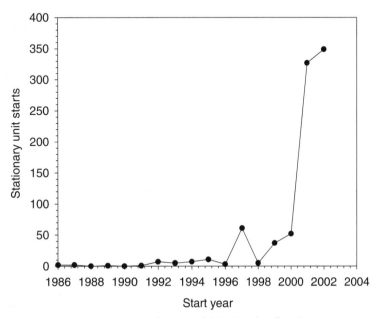

Figure 3 Estimated trend in the number of projects initiated to install stationary power sources since 1986. (Based on data from Ref. 50.)

1. Alternative materials and construction methods must be developed to reduce fuel-cell system cost to be competitive with the automotive combustion engine (~$30/kW) and stationary power systems (~$1000/kW). The cost of the catalyst no longer dominates the price of most fuel-cell systems, although it is still significant. Manufacturing and mass production technology is now a key component to the commercial viability of fuel-cell systems.

2. Suitable reliability and durability must be achieved. The performance of every fuel cell gradually degrades with time, due to a variety of phenomena. The automotive fuel cell must withstand load cycling and freeze–thaw environmental swings with an acceptable level of degradation from the beginning-of-lifetime (BOL) performance over a lifetime of 5000 hr (equivalent to 150,000 miles at 30 mph). A stationary fuel cell must withstand over 40,000 hr of steady operation under vastly changing external temperature conditions.

3. Suitable system power density and specific power must be achieved. The U.S. Department of Energy year 2010 targets for system power density and specific power are 650 W/kg and 650 W/L for automotive (50-kW) applications, 150 W/kg and 170 W/L for auxiliary (5–10-kW-peak) applications, and 100 W/kg and 100 W/L for portable (megawatt to 50-W) power systems.[7] Current systems fall well short of these targets.

4. Fuel storage and delivery technology must be advanced if pure hydrogen is to be used. The issue of hydrogen infrastructure (i.e., production, storage, and delivery), is not addressed herein but is nevertheless daunting in scope. This topic is addressed in detail in Ref. 8.

5. Fuel reformation technology must be advanced if a hydrocarbon fuel is to be used for hydrogen production.

6. Desired performance and longevity of system ancillary components must be achieved. New hardware (e.g., efficient transformers and high-volume blowers) will need to be developed to suit the needs of fuel-cell power systems.

The particular limitations and advantages of several different fuel-cell systems will be discussed in this chapter in greater detail.

2 BASIC OPERATING PRINCIPLES, EFFICIENCY, AND PERFORMANCE

Figure 4 shows a schematic of a generic fuel cell. Electrochemical reactions for the anode and cathode are shown for the most common fuel-cell types. Table 1 presents the various types of fuel cells, operating temperature, electrolyte material, and likely applications. The operating principle of a fuel cell is similar to a common battery, except that a fuel (hydrogen, methanol, or other) and oxidizer (commonly air or pure oxygen) are brought separately into the electrochemical reactor from an external source, whereas a battery has stored reactants. Referring to Fig. 4, separate liquid- or gas-phase fuel and oxidizer streams enter through flow channels separated by the electrolyte/electrode assembly. Reactants are transported by diffusion and/or convection to the catalyzed electrode surfaces, where electrochemical reactions take place. Some fuel cells (alkaline and polymer electrolyte) have a porous (typical porosity ~0.5–0.8) contact layer between the electrode and current-collecting reactant flow channels that functions to transport electrons and species to and from the electrode surface. In PEFCs, an electrically conductive carbon paper or cloth diffusion media (DM) layer (also

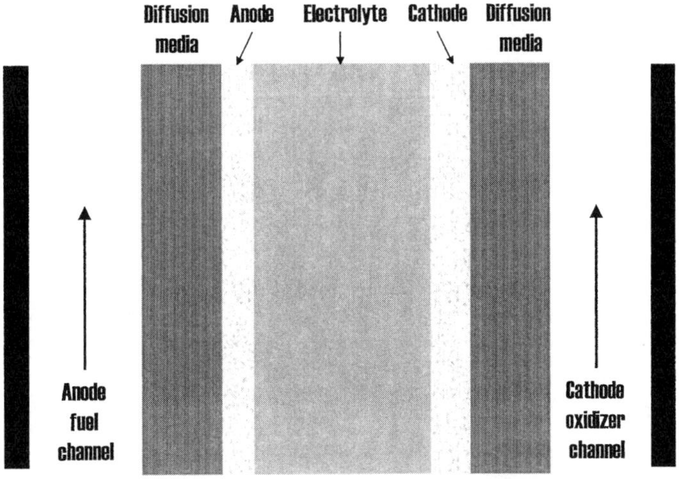

Anode reaction:

H_2 PEFC: $H_2 \rightarrow 2e^- + 2H^+$
DMFC: $CH_3OH + H_2O \rightarrow 6e^- + 6H^+ + CO_2$
SOFC: $H_2 + O^{2-} \rightarrow H_2O + 2e^-$
AFC: $2H_2 + 4OH^- \rightarrow 4H_2O + 4e^-$
MCFC: $2H_2 + 2CO_3^{2-} \rightarrow 2H_2O + 2CO_2 + 4e^-$
PAFC: $H_2 \rightarrow 2e^- + 2H^+$

Cathode reaction:

H_2 PEFC: $1/2O_2 + 2e^- + 2H^+ \rightarrow H_2O$
DMFC: $1/2O_2 + 2e^- + 2H^+ \rightarrow H_2O$
SOFC: $O_2 \rightarrow 2O^{2-} + 4e^-$
AFC: $O_2 + 4e^- + 2H_2O \rightarrow 4OH^-$
MCFC: $O_2 + 2CO_2 + 4e^- \rightarrow 2CO_3^{2-}$
PAFC: $1/2O_2 + 2e^- + 2H^+ \rightarrow H_2O$

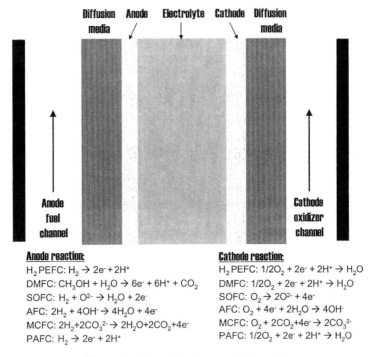

Figure 4 Schematic of a generic fuel cell.

called gas diffusion layer, or GDL) serves this purpose and covers the anode and cathode catalyst layer.

At the anode electrode, the electrochemical oxidation of the fuel produces electrons that flow through the bipolar plate (also called cell interconnect) to the external circuit, while the ions migrate through the electrolyte. The electrons in the external circuit drive the load and return to the cathode catalyst where they recombine with the oxidizer in the cathodic oxidizer reduction reaction (ORR). The outputs of the fuel cell are thus threefold: (1) chemical products, (2) waste heat, and (3) electrical power.

A number of fuel-cell varieties have been developed to differing degrees, and the most basic nomenclature of fuel cells is related to the electrolyte utilized. For instance, a SOFC has a solid ceramic oxide electrolyte, and a PEFC has a flexible polymer electrolyte. Additional subclassification of fuel cells beyond the basic nomenclature can be assigned in terms of fuel used (e.g., hydrogen PEFC or direct methane SOFC) or the operating temperature range.

Each fuel-cell variant has particular advantages that engender use for particular applications. In general, low-temperature fuel cells (e.g., PEFCs, AFCs) have advantages in start-up time and potential efficiency, while high-temperature fuel cells (e.g., SOFCs, MCFCs) have an advantage in raw materials (catalyst) cost and quality and ease of rejection of waste heat. Medium-temperature fuel cells (e.g., PAFCs) have some of the advantages of both high- and low-temperature classification. Ironically, a current trend in SOFC development is to enable lower temperature (<600°C) operation, while a focus of current PEFC research is

Table 1 Fuel-Cell Types, Descriptions, and Basic Data

Fuel-Cell Type	Electrolyte Material	Operating Temperature	Major Poison	Advantages	Disadvantages	Most Promising Applications
AFC	Solution of potassium hydroxide in water	60–250°C (modern AFCs <100°C)	CO_2	High efficiency, low oxygen reduction reaction losses	Must run on pure oxygen without CO_2 contaminant	Space applications with pure O_2/H_2 available
PAFC	Solution of phosphoric acid in porous silicon carbide matrix	160–220°C	Sulfur, high levels of CO	1–2% CO tolerant, good-quality waste heat, demonstrated, durability	Low-power density, expensive; platinum catalyst used; slow start-up; loss of electrolyte	Premium stationary power
SOFC	Yttria (Y_2O_3) stabilized zirconia (ZrO_2)	600–1000°C	Sulfur	CO tolerant, fuel flexible, high-quality waste heat, inexpensive catalyst	Long start-up time, durability under thermal cycling, inactivity of electrolyte below ~600°C	Stationary power with cogeneration, continuous-power applications
MCFC	Molten alkali metal (Li/K or Li/Na) carbonates, in porous matrix	600–800°C	Sulfur	CO tolerant, fuel flexible, high-quality waste heat, inexpensive catalyst	Electrolyte dissolves cathode catalyst, extremely long start-up time, carbon dioxide must be injected to cathode, electrolyte maintenance	Stationary power with cogeneration, continuous-power applications
PEFC[a]	Flexible solid perfluorosulfonic acid polymer	30–100°C	CO, Sulfur, metal ions, peroxide	Low-temperature operation, high efficiency, high H_2 power density, relatively rapid start-up	Expensive catalyst, durability of components not yet sufficient, poor-quality waste heat, intolerance to CO, thermal and water management	Portable, automotive, and stationary applications

[a] Includes DMFC and direct alcohol fuel cells (DAFCs).

to operate at higher temperature (>120°C). Although the alkaline and phosphoric acid fuel cell had much research and development in the past and the MCFC is still under development, fuel-cell technologies under the most aggressive development are the PEFC and SOFC.

2.1 Description of a Fuel-Cell Stack

A single cell can be made to achieve whatever current and power are required simply by increasing the size of the active electrode area and reactant flow rates. However, the output voltage of a single fuel cell is always less than 1 V for realistic operating conditions, limited by the fundamental electrochemical potential of the reacting species involved. Therefore, for most applications and for compact design, a fuel-cell stack of several individual cells connected in series is utilized. Figure 5 is a schematic of a generic planar fuel-cell stack assembly and shows the flow of current through the system. For a stack in series, the total current is proportional to the active electrode area of the cells in the stack and is the same through all cells in the stack, and the total stack voltage is the sum of the individual cell voltages. For applications that benefit from higher voltage output, such as automotive stacks, it is typical to have over 200 fuel cells in a single stack.

Other components necessary for fuel-cell system operation include subsystems for oxidizer delivery, electronic control including voltage regulation, fuel storage and delivery, fuel recirculation/consumption, stack temperature control, and systems sensing of control parameters. For the PEFC, separate humidification systems are also needed to ensure optimal performance and stability. A battery is often used to start reactant pumps/blowers during start-up. In many fuel cells operating at high temperature, such as a SOFC or MCFC, a

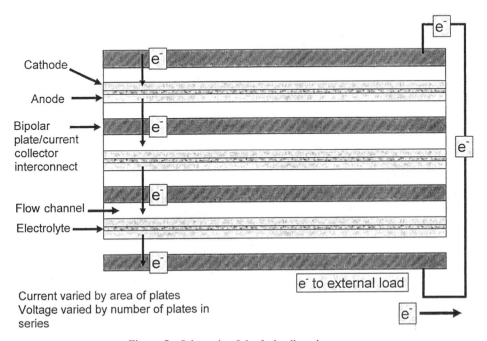

Figure 5 Schematic of the fuel-cell stack concept.

preheating system is needed to raise cell temperatures during start-up. This is typically accomplished with a combustion chamber that burns fuel and oxidizer gases. In all commercial fuel cells, provision must be made for effluent recovery. Fuel utilization efficiency is not 100% due to concentration polarization limitation on performance, so that unused fuel must be actively recycled, utilized, or converted prior to exhaust to the environment. Potential effluent management schemes include the use of recycling pumps, condensers (for liquid fuel), secondary burners, or catalytic converters.

2.2   Performance and Efficiency Characterization

The single-cell combination shown in Fig. 4 provides a voltage dependent on operating conditions such as temperature, pressure, applied load, and fuel/oxidant flow rates. The thermal fuel-cell voltage corresponds to a 100% efficient fuel cell and is shown as

$$E_{\text{th}} = \frac{\Delta H}{nF} \tag{1}$$

This is the total thermal voltage potential available if all chemical energy was converted to electrical potential. This is not possible, however, due to entropy change during reaction.

Consider a generalized global fuel cell reaction:

$$\nu_A A + \nu_B B \rightleftharpoons \nu_C C + \nu_D D \tag{2}$$

The maximum possible electrochemical potential can be calculated from the Nernst equation as

$$E^{\circ} = \frac{-\Delta G^0\,(T)}{nF} + \frac{RT}{nF} \ln\!\left(\frac{a_A^{\nu_A} a_B^{\nu_B}}{a_C^{\nu_C} a_D^{\nu_D}}\right) \tag{3}$$

where $\Delta G^0\,(T)$ is evaluated at the fuel-cell temperature and standard pressure and the pressure dependency is accounted for in the second term on the right. The activity of gas-phase reactants, besides water vapor, can be calculated from

$$a_i = \frac{y_i P}{P^0} \tag{4}$$

where P^0 is a standard pressure of 1 atm. Under less than fully saturated local conditions, the activity of water vapor can be shown to be the relative humidity:

$$a_{\text{H}_2\text{O}} = \frac{y_{\text{H}_2\text{O}} P}{P_{\text{sat}}\,(T)} = \text{RH} \tag{5}$$

where P_{sat} is the saturation pressure of water vapor at the fuel-cell temperature. For values of saturation pressure >1 atm ($\sim 100°C$), a standard pressure of 1 atm should be used for the water vapor activity.

Because a fuel cell directly converts chemical energy into electrical energy, the maximum theoretical efficiency is not bound by the Carnot cycle but is still not 100% due to entropy change via reaction. The maximum thermodynamic efficiency is simply the ratio of the maximum achievable electrochemical potential to the maximum thermal potential:

$$\eta_{\text{th}} = \frac{E^\circ}{E_{\text{th}}} = \frac{\Delta G}{\Delta H} = \frac{\Delta H - T\,\Delta S}{\Delta H} = 1 - \frac{T\,\Delta S}{\Delta H} \qquad (6)$$

The actual operating thermodynamic (voltaic) efficiency of the fuel cell is the actual fuel cell voltage E_{fc} divided by E_{th}. The operating efficiency is really a ratio of the useful electrical ouput and heat output. If water is generated by the reaction, as is the case with most fuel-cell systems, efficiency and voltage values can be based on the low heating value (LHV, all water generated is in the gas phase) or the high heating value (HHV, all water generated is in the liquid phase). Representation with respect to the LHV accounts for the latent heat required to vaporize liquid water product, which is the natural state of the fuel-cell system.

Table 2 shows a selection of fuel-cell reactions and the calculated maximum theoretical efficiency at 298 K. Typical values for maximum efficiency at the open circuit calculated from Eq. (6) range from 60 to 90% and vary with temperature according to sign of the net entropy change. That is, according to Eq. (6), the efficiency will decrease with temperature if the net entropy change is negative and increase with temperature if ΔS is positive (since ΔH is negative for an exothermic reaction). Based on the Le Chatelier principle, we can predict the qualitative trend in the functional relationship between maximum efficiency and temperature. If the global fuel-cell reaction ($\nu_A + \nu_B = \nu_C + \nu_D$) has more moles of gas-phase products than reactants (e.g., $\nu_C + \nu_D > \nu_A + \nu_B$), ΔS will be positive and η_{th} will *increase* with temperature. Liquid- or solid-phase species have such low relative entropy compared to gas-phase species that they have negligible impact on ΔS. These fuels yield a theoretical maximum efficiency greater than 100%! Physically, this means the reaction would absorb heat from the environment and convert the energy into a voltage potential. An example of this is a direct carbon fuel cell. Although use of ambient heat to generate power with an efficiency greater than 100% seems like an amazing possibility, it is of course not realistic in practice due to various losses. If the number of gas-phase moles is the same between products and reactants, η_{th} is basically invariant with temperature and near 100%, as in the methane-powered fuel cell with gas-phase water produced. Most fuel cells, including those with hydrogen fuel, have less product gas-phase moles compared to the reactants, and ΔS is negative. Thus, for these fuel cells η_{th} will *decrease* with operating temperature.

Figure 6 shows the calculated maximum thermodynamic efficiency of a hydrogen–air fuel cell with temperature compared to that of a heat engine. Since the maximum heat engine efficiency is the Carnot efficiency, it is an increasing function of temperature. Note that at a certain temperature above 600°C, the theoretical efficiency of the hydrogen–air fuel cell actually becomes *less* than that of a heat engine. In fact, high-efficiency combined-cycle gas

Table 2 Some Common Fuel-Cell Reactions and Maximum Theoretical Efficiency at 298 K

Fuel	Global Reaction	n (electrons per mole fuel)	Maximum η_{th} (HHV)
Hydrogen	$H_2 + \tfrac{1}{2}O_2 \rightarrow H_2O_l$	2	83
Methanol	$CH_3OH + \tfrac{3}{2}O_2 \rightarrow CO_2 + 2H_2O_l$	6	97
Methane	$CH_4 + 2O_2 \rightarrow CO_2 + 2H_2O_l$	8	92
Formic acid	$HCOOH + \tfrac{1}{2}O_2 \rightarrow CO_2 + H_2O_l$	2	106
Carbon monoxide	$CO + \tfrac{1}{2}O_2 \rightarrow CO_2$	2	91
Carbon	$C_s + \tfrac{1}{2}O_2 \rightarrow CO$	2	124

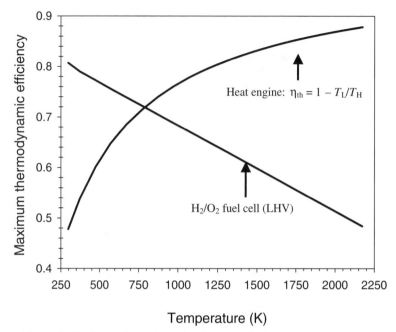

Figure 6 Maximum thermodynamic efficiency of a fuel cell and heat engine.

turbines can now achieve power conversion efficiencies that rival high-temperature SOFCs. It is important to realize that fuel-cell systems are not inherently more efficient than heat engine alternatives. In practice, a 100-kW system operated by Dutch and Danish utilities has demonstrated an operating efficiency of 46% (LHV) over more than 3700 hr of operation.[9] Combined fuel cell/bottoming cycle and cogeneration plants can achieve operational efficiencies as high as 80% with very low pollution. Another major advantage of fuel cells compared to heat engines is that efficiency is not a major function of device size, so that high-efficiency power for portable electronics can be realized, whereas small-scale heat engines have very low efficiencies due to heat transfer from high surface area–volume ratio. In terms of automotive applications, fuel cell hybrids and stand-alone systems operating with a variety of (but not all) fuel feedstocks have the potential for greater than double the equivalent mileage as conventional vehicles.[10]

Thermodynamic efficiency is not the entire picture, however, as the overall fuel-cell efficiency must consider the utilization of fuel and oxidizer. The appropriate mass flow rate of reactants into the fuel cell is determined by several factors related to the minimum requirement for electrochemical reaction, thermal management, and issues related to the particular type of fuel cell. However, the *minimum*-flow requirement is prescribed by the electrochemical reaction. An expression for the molar flow rate of species required for the electrochemical reaction can be shown from Faraday's law as[11]

$$\dot{n}_k = \frac{iA}{n_k F} \qquad (7)$$

where i and A represent the current density and total electrode area, respectively, and n_k represents the electrons generated in the global electrode reaction per mole of reactant k. For fuel cells, the stoichiometric ratio or stoichiometry for an electrode reaction is defined as the ratio of reactant delivered to that required for the electrochemical reaction.

$$\text{Anode:}\quad \xi_a = \frac{\dot{n}_{\text{fuel,actual}}}{\dot{n}_{\text{fuel,required}}} = \frac{\dot{n}_{\text{fuel,actual}}}{iA/(n_{\text{fuel}}F)} \qquad \text{Cathode:}\quad \xi_c = \frac{\dot{n}_{\text{ox,actual}}}{\dot{n}_{\text{ox,required}}} = \frac{\dot{n}_{\text{ox,actual}}}{iA/(n_{\text{ox}}F)} \qquad (8)$$

The stoichiometry can be different for each electrode and must be greater than unity. This is due to the fact that zero concentration near the fuel-cell exit will result in zero voltage from Eq. (3). Since the current collectors are electrically conductive, a large potential difference cannot exist and cell performance will decrease to zero if reactant concentration goes to zero. As a result of this requirement, fuel and oxidizer utilization efficiency is never 100%, and some system to recycle or consume the effluent fuel from the anode is typically required to avoid releasing unused fuel to the environment. Thus, the overall operating efficiency of a fuel cell can be written as the product of the voltaic and Faradaic (fuel utilization) efficiencies:

$$\eta_{\text{fc}} = \eta_{\text{th}} \frac{1}{\xi_a} \frac{1}{\xi_c} \qquad (9)$$

2.3 Polarization Curve

Figure 7 is an illustration of a typical polarization curve for a fuel cell with negative ΔS, such as the hydrogen–air fuel cell, showing five regions labeled I–V. The polarization curve, which represents the cell voltage–current relationship, is the standard figure of merit for evaluation of fuel-cell performance. Also shown in Fig. 7 are the regions of electrical and heat generation. Since the thermodynamically available power not converted to electrical power is converted to heat, the relationship between current and efficiency can be clearly seen by comparing the relative magnitude of the voltage potential converted to waste heat and to electrical power. Region V is the departure from the maximum thermal voltage, caused by entropy generation. In practice, the open-circuit voltage (OCV) achieved is somewhat less than that calculated from the Nernst equation. Region IV represents this departure from the calculated maximum open-circuit voltage. This loss can be very significant and for PEFCs is due to undesired species crossover through the thin-film electrolyte and resulting mixed potential at the electrodes. For other fuel cells, there can be some loss generated by internal currents from electron leakage through the electrolyte. This is especially a challenge in SOFCs. Beyond the departure from the theoretical open-circuit potential, there are three major classifications of losses that result in a drop of the fuel-cell voltage potential, shown in Fig. 7: (1) activation (kinetic) polarization (region I), (2) ohmic polarization (region II), and (3) concentration polarization (region III). It should be noted that voltage loss, polarization, and overpotential are all interchangeable and refer to a voltage loss. The operating voltage of a fuel cell can be represented as the departure from ideal voltage caused by these polarizations:

$$E_{\text{fc}} = E^\circ - \eta_{a,a} - |\eta_{a,c}| - \eta_r - \eta_{m,a} - |\eta_{m,c}| \qquad (10)$$

where E° is the theoretical Nernst open-circuit potential of the cell [Eq. (3)] and the activation overpotential at the anode and cathode are represented by $\eta_{a,a}$ and $\eta_{a,c}$ respectively. The

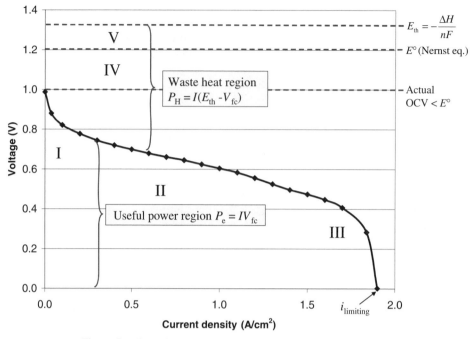

Figure 7 Illustration of a typical polarization curve for a fuel cell.

ohmic (resistive) polarization is shown as η_r. The concentration overpotential at the anode and cathode are represented as $\eta_{m,a}$ and $\eta_{m,c}$, respectively. Cathode polarization losses are negative relative to the standard hydrogen electrode, so the absolute value is taken. Additional losses in region IV of Fig. 7 attributed to species crossover or internal currents can also be added to Eq. (10) as needed. Activation and concentration polarizations occur at both anode and cathode, while the resistive polarization represents ohmic losses throughout the fuel cell and thus includes ionic, electronic, and contact resistances between all fuel-cell components carrying current. It is important to note that the regions on the polarization curve of dominance kinetic, ohmic, or mass-transfer polarizations are not discrete. That is, all modes of loss contribute throughout the entire current density range and there is no discrete "ohmic loss" region where other polarizations are not also contributing to total deviation below the Nernst potential. Although the activation overpotential dominates in the low-current region, it still contributes to the cell losses at higher current densities where ohmic or concentration polarization dominates. Thus, each region is not discrete, and all types of losses contribute throughout the operating current regime.

Activation polarization, which dominates losses at low current density, is the voltage overpotential required to overcome the activation energy of the electrochemical reaction on the catalytic surface and is commonly represented by a Butler–Volmer equation at each electrode[11]:

$$i_{fc} = i_o \left[\exp\left(\frac{\alpha_a Fn}{RT}\eta\right) - \exp\left(\frac{\alpha_c Fn}{RT}\eta\right) \right] \tag{11}$$

where α_a and α_c are the charge-transfer coefficients for the anode and cathode, respectively. The fraction of the electrical overpotential (η) resulting in a change of the rate of reaction for the cathodic branch of this electrode is shown as α_c. Obviously, $\alpha_a + \alpha_c = 1$. Here, n is the number of exchange electrons involved in the elementary electrode reaction, which is typically different from the n used in Eqs. (3) and (7). The exchange current density i_o represents the activity of the electrode for a particular reaction. In hydrogen PEFCs, the anode i_o for hydrogen oxidation is so high relative to the cathode i_o for oxygen reduction that the anode contribution to this polarization is often neglected for pure hydrogen fuel. On the contrary, if neat hydrogen is not used, significant activation polarization losses at both electrodes are typical (e.g., the DMFC). It appears from Eq. (11) that activation polarization should increase with temperature. However, i_o is a highly nonlinear function of the kinetic rate constant of reaction and the local reactant concentration and can be modeled with an Arrhenius form as

$$i_o = i_o^o \, \exp^{E_A(RT)} \left(\frac{C_{\mathrm{ox}}}{C_{\mathrm{ref}}} \right)^{\gamma} \left(\frac{C_f}{C_{\mathrm{ref}}} \right)^{\upsilon} \tag{12}$$

Thus, i_o is an exponentially increasing function of temperature, and the net effect of increasing temperature is to decrease activation polarization. For this reason, high-temperature fuel cells such as SOFCs or MCFCs typically have very low activation polarization and can use less exotic catalyst materials. Accordingly, the effect of an increase in electrode temperature is to decrease the voltage drop within the activation polarization region shown in Fig. 7. For various fuel-cell systems, however, the operating temperature range is dictated by the electrolyte and materials properties, so that temperature cannot be arbitrarily increased to reduce activation losses.

At increased current densities, a primarily linear region is evident on the polarization curve. In this region, reduction in voltage is dominated by internal ohmic losses (η_r) through the fuel cell that can be represented as

$$\eta_r = iA \left(\sum_{k=1}^{n} r_k \right) \tag{13}$$

where each r_k value is the area-specific resistance of individual cell components, including the ionic resistance of the electrolyte, and the electric resistance of bipolar plates, cell interconnects, contact resistance between mating parts, and any other cell components. With proper design and assembly, ohmic polarization is typically dominated by electrolyte conductivity for all fuel-cell types.

At very high current densities, mass-transport limitation of fuel or oxidizer to the corresponding electrode causes a sharp decline in the output voltage. This is referred to as concentration polarization. This region of the polarization curve is really a combined mass-transport/kinetic related phenomenon, as the surface reactant concentration is functionally related to the exchange current density, as shown in Eq. (12). In general, reactant transport to the electrode is limited to some value depending on, for example, operating conditions, porosity and tortuosity of the porous media, and input stoichiometry. If this limiting mass-transport rate is approached by the consumption rate of reactant [Eq. (7)], the surface concentration of reactant will approach zero, and, from Eq. (3), the fuel-cell voltage will also approach zero. The Damköler number (Da) is a dimensionless parameter that is the ratio of the characteristic electrochemical reaction rate to the rate of mass transport to the reaction surface. In the limiting case of infinite kinetics (high Damköler number), one can derive an expression for η_m as

$$\eta_m = -\frac{RT}{nF}\ln\left(1 - \frac{i}{i_l}\right) = -B\ln\left(1 - \frac{i}{i_l}\right) \tag{14}$$

where i_l is the limiting current density and represents the maximum current produced when the surface concentration of reactant is reduced to zero at the reaction site. The limiting current density (i_l) can be determined by equating the reactant consumption rate to the mass transport rate to the surface, which in itself can be a complex calculation. The strict application of Eq. (14) results in a predicted voltage drop-off that is much more abrupt than actually observed. To accommodate a more gradual slope, an empirical coefficient B is often used to fit the model to the experiment.[12] Concentration polarization can also be incorporated into the exchange current density and kinetic losses, as in Eq. (12).

2.4 Heat Management

While PEFC systems can achieve a high relative operating efficiency, the inefficiencies manifest as dissipative thermal losses. At the cell level, if waste heat is not properly managed, accelerated performance degradation or catastrophic failure can occur. The total waste heat rate can be calculated as

$$P_{\text{waste}} = kI(E_{\text{th}} - E_{\text{fc}}) \tag{15}$$

where k is the number of cells in series in the stack.

This heat generation can be broken down into components and shown as[13]

$$q'' = -i_{\text{fc}}\left(\frac{\Delta H}{nF} - \frac{\Delta G}{nF} - \eta_{a,a} - |\eta_{a,c}| - i_{\text{fc}}\sum_{k=1}^{n} r_k\right)$$
$$+ i_{\text{fc}}^2 \sum_{k=1}^{n} r_k - i_{\text{fc}}\left(-\eta_{a,a} - |\eta_{a,c}| + \frac{T\,\Delta S}{nF}\right) \tag{16}$$

The first term on the right-hand side of Eq. (16) is Joule heating and is thus an $i^2 r$ relationship. The second and third terms of Eq. (16) represent the heat flux generated by activation polarization in the anode and cathode catalyst layers. This assumes the concentration dependence on exchange current density is included in the activation polarization terms. The third term in Eq. (16) is a linearly varying function of current density and represents the total Peltier heat generated via entropy change by reaction. The functional relationship derived in Eq. (16) is shown in Fig. 8, a plot of the heat generation via Peltier, Joule, and kinetic heating as a function of current density for a typical PEFC.[14] The ionic conductivity for the electrolyte was chosen to be 0.1 S/cm, based on the assumption of a fully humidified membrane in contact with vapor-phase water,[15] and other parameters were chosen as typical values. This plot should be viewed only as a guide to the qualitative behavior of the heat generation with current density, as each fuel cell and different operating conditions will have much different distributions. For example, the SOFC typically has quite low activation polarization generated heat, due to the high operating temperatures. Note that an assumption of ohmic heating dominance is not always accurate, and entropic heat generation can be quite significant and cannot be ignored. For higher temperature fuel cells with low activation overpotential, the heat generation is dominated by ohmic and entropic terms.

2.5 Degradation

The lifetime of a fuel cell is expected to compete with existing power systems it would replace. As a result, the automotive fuel cell must withstand load cycling and freeze–thaw

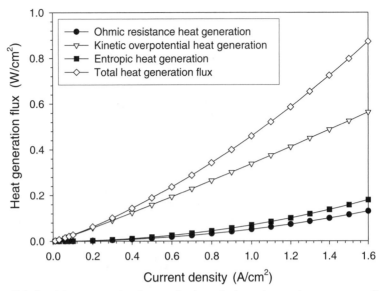

Figure 8 Calculated heat generation from activation, ohmic, and entropic sources as a function of current for a typical PEFC.

environmental swings with minimal degradation over a lifetime of 5000 hr. A stationary fuel cell, meanwhile, must withstand over 40,000 hr of steady operation with minimal downtime. The fuel-cell environment is especially conducive to degradation, since a voltage potential difference exists which can promote undesired reaction, and some fuel cells operate at high temperature or have corrosive electrolytes. Transient load cycling between high- and low-power points has also been shown to accelerate degradation, so that steady-state degradation rates may not be truly representative for transient systems. Many different modes of physicochemical degradation are known to exist, including the following:

1. *Catalyst or Electrolyte Poisoning and Degradation.* Since the catalyst and electrolyte control the reaction and ohmic polarization, any poisoning or other degradation of these components will adversely affect cell performance. Some minor species in air and re-formed gas product, such as carbon monoxide, will foul a platinum catalyst operating at low temperatures (<150°C). As a result, PEFCs are extremely susceptible to CO poisoning. Some surface absorption of species can be reversible, including adsorbed CO. Sulfur is also a major contaminant that will greatly reduce performance in most fuel cells in extremely low (<ppb) concentrations. Many other low-level impurities can greatly harm fuel-cell performance; for example, CO_2 will degrade the electrolyte in AFCs. Given that the electrolyte is an ion conductor, when unintended ions are present in the fuel-cell system by corrosion or other impurities, the electrolyte will absorb these impurities, which can alter the ionic conductivity of the media. In SOFCs, carbonaceous residue from internal performance can foul the anode catalyst.

2. *Electrolyte Loss.* In some cases, electrolyte material is lost through a variety of physicochemical mechanisms. For polymer electrolyte fuel cells, the polymer itself can degrade physically and chemically, particularly from peroxide radical attack.[16] This

results in loss of mass and conductivity in the electrolyte and possible catastrophic pinhole formation. For liquid electrolyte systems such as the AFC, PAFC, and MCFC, the finite vapor pressure of the liquid phase results in a steady but predictable loss of electrolyte through the reactant flow streams, which must be replenished with regularity or performance will suffer.

3. *Morphology Changes or Loss in Catalyst Layer or Other Components.* For all fuel cells, the catalyst layer electrochemical active surface area (ECSA) is a determining factor in overall power density, and nanosize catalysts and supports are present in a complex three-dimensional electrode structure designed to simultaneously optimize electron, ion, and mass transfer. As a result, any morpholological changes can result in reduced performance. Commonly observed phenomena include catalyst sintering, dissolution and migration, catalyst oxidation, supporting material oxidation (e.g., carbon corrosion for carbon-supported catalysts), and Oswald ripening.[17] These effects are most often irreversible. Other components can also be chemically or physically altered, such as the porosity distribution or hydrophobicity of the GDL in a PEFC.

4. *Corrosion of Other Components.* Oxidation of other components such as the current collector can become a major loss in fuel cells over time. This is especially relevant in high-temperature MCFCs and SOFCs, where the corrosion process is accelerated by the high temperature. Chromium used in SOFCs in stainless steel interconnects is believed to cause cathode degradation. Low-temperature PEFCs can also suffer losses from current collector corrosion, and proper coatings with high electronic conductivity must be used.[18]

2.6   Hydrogen PEFC

The hydrogen (H_2) PEFC is seen by many as the most viable alternative to heat engines and battery replacement for automotive, stationary, and portable power applications. The H_2 PEFC is fueled either by pure hydrogen or from a diluted hydrogen mixture generated from a hydrocarbon re-formation process. A H_2 PEFC fuel-cell stack power density of greater than 1.0 kW/L is typical.

2.7 H_2 PEFC Performance

Hydrogen PEFCs operate at 60–100°C. The anode and cathode catalyst is commonly ~2 nm platinum or platinum–ruthenium powder supported on significantly larger size carbon particles with a total (anode and cathode) platinum loading of ~0.4 mg/cm². This represents a major breakthrough in required catalyst loading from the 28 mg/cm² of the original 1960s H_2 PEFC. As a result, the catalyst is no longer the dominating factor in fuel-cell cost, although it is still higher than needed to reach long-term goals of another 20-fold reduction in loading or elimination of precious metals.[19] The state-of-the-art H_2 PEFC can reach nearly 0.7 V at 1 A/cm² under pressurized conditions at 80°C. There is always a desire to operate at high voltages because of increased efficiency and reduced flow requirements. However, power density typically peaks below 0.6 V, and heat generation can cause accelerated degradation, so there is a size trade-off for high-voltage operation. Typical anode and cathode stoichiometry requirements are low, with typical values of 2 or less for the cathode and even lower values for the anode.

2.8 Technical Issues in H₂ PEFC

The H_2 PEFC has many technical issues that complicate performance and control. Besides issues of manufacturing, ancillary system components, cost, and market acceptance, the main remaining technical challenges for the fuel cell include (1) water and heat management, (2) durability, and (3) freeze–thaw cycling capability.

Water and Heat Management

For PEFCs, waste heat affects the water distribution by increasing temperature and thus the local equilibrium saturation pressure of the gases. At a typical PEFC operating temperature of 80°C and atmospheric pressure, each 1°C change in temperature results in an approximately 5% change in equilibrium saturation pressure.[20] Thus, the thermal and water management and control are inexorably coupled at the individual cell and stack level, and even small variations in temperature can dramatically affect optimal humidity, locations of condensation/vaporization, membrane longevity, and a host of other phenomena. Due to heat generation, relatively high temperature gradients up to 10°C between the electrolyte and current collector can occur.[21] Analysis has shown that the through-plane thermal conductivities of carbon cloth gas diffusion media and Nafion electrolyte material are approximately 0.15 and 0.1 W/mK, respectively.[22,23] The main barrier to the heat transport is through the GDL, which acts as a thermal insulator to limit conductive heat transfer. Once through the GDL, the majority of heat transfer is typically through the landings and not to the reactant in the flow channels.[14]

For high-power (>kW) fuel-cell stacks, waste heat must be properly managed with cooling channels, which take up space and require parasitic pumping losses. The choice of coolant is based on the necessary properties of high specific heat, nonconductive, noncorrosive, sufficient boiling/freezing points for operation in all environments, and low viscosity. Laboratory systems typically use deionized water, although practical systems exposed to the environment must use a lower freezing point nonconductive solution.

Water management and humidification are major issues in H_2 PEFC performance. The most common electrolyte used in PEFCs is a perfluorosulfonic acid–polytetrafluoroethylene (PTFE) copolymer in acid (H^+) form, known by the commercial name Nafion (E. I. du Pont de Nemours). Nafion electrolyte conductivity is primarily a function of water content and temperature, shown as[15]

$$\sigma_e = 100 \exp\left[1268\left(\frac{1}{303} - \frac{1}{T(\text{K})}\right)\right](0.005139\lambda - 0.00326) \qquad (\mho \cdot \text{m}^{-1}) \qquad (17)$$

where

$$\lambda = 0.043 + 17.18a - 39.85a^2 + 36.0a^3 \text{ for } 0 < a \leq 1$$
$$\lambda = 14 + 14(a - 1)$$
$$a = \text{water activity} = y_{H_2O}P/P_{sat}(T) = \text{RH}$$

A plot of Nafion conductivity as a function of humidity, based on Eq. (17), is given in Fig. 9. It is obvious that ionic conductivity is severely depressed without sufficient water. Alternatively, excessive water at the cathode can cause flooding, that is, liquid water accumulation at the cathode surface that prevents oxygen access to the reaction sites. Flooding is most likely near the cathode exit under high-current-density, high-humidification, low-temperature, and low-flow-rate conditions. However, the term "flooding" has been rather nebulously applied in the literature to date, representing a general performance loss resulting

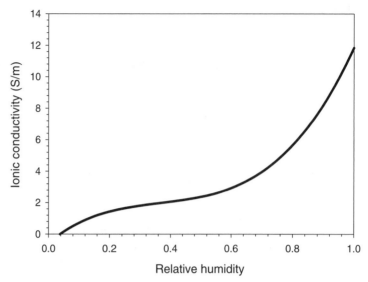

Figure 9 Electrolyte conductivity of Nafion as a function of relative humidity at 80°C.

from liquid water accumulation blocking reactant transport to the electrode. There are actually six discrete regions that can suffer flooding losses in the PEFC: the anode and cathode catalyst layers, the anode and cathode gas diffusion layers, and the anode and cathode flow channels. Figure 10 is a radiograph image of a PEFC under severely flooded conditions with significant diffusion media and channel-level flooding. In this image, the water accumulation was determined to be primarily in the anode flow channels. This phenomenon can occur for low-flow-rate, low-power, and high-fuel-utilization conditions, which indicates that channel level flooding is not solely a cathode issue.

Depending on operating conditions, flow field design, and material properties, a membrane can have a highly nonhomogeneous water (and therefore ionic conductivity) distribution. The membrane beneath a long channel may be dried by hot inlet flow ideally saturated near the middle of the cell and experiencing flooding near the exit. It is difficult in practice to maintain an ideal water distribution throughout the length of the cell. Thus water transport is an especially important issue in PEFC design. Even a slight reduction in ohmic losses through advanced materials, thinner electrolytes, or optimal temperature/water distribution can significantly improve fuel-cell performance and power density.

Figure 11 shows a schematic of the water-transport and generation modes in the PEFC. At the cathode surface, the oxygen reduction reaction results in water production proportional to the current density via Faraday's law. Water transport through the electrolyte occurs by diffusion, electroosmotic drag, and hydraulic permeation from a pressure difference across the anode and cathode. Diffusion through the electrolyte can be represented using Fick's law, and appropriate expressions relating to diffusion coefficients for Nafion can be found.[24] An electroosmotic drag coefficient (λ_{drag}) of 1–5 H_2O/H^+ of Nafion membranes has been shown for a fully hydrated Nafion 117 membrane.[25] The drag coefficient was shown to be a nearly linearly increasing function of temperature from 20 to 120°C. Thus, the water delivered to the cathode by this transport mode can be 2–10 times greater than that generated by reaction. Hydraulic permeation of water through the membrane under gas-phase pressure

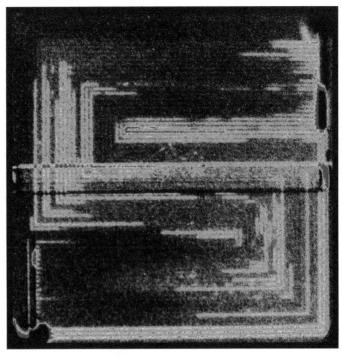

Figure 10 Neutron radiograph of 50-cm² active area fuel cell showing severe channel-level flooding at low current density for a H₂ PEFC.[51]

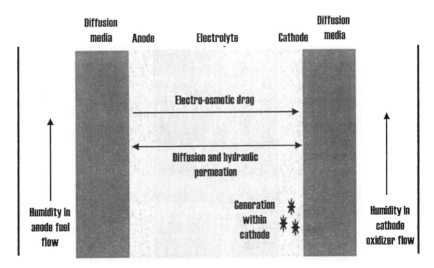

Figure 11 Schematic of water transport and generation modes in the PEFC.

difference between the anode and cathode is usually small for H_2 PEFCs. However, capillary pressure differences can result in a net flux of water via this mode of transport.[26] Combining the different forms of water transport through the membrane, the bulk molar water transport and creation at the cathode can be shown as

$$\dot{n}_{H_2O} = -DA\frac{\Delta C_{c,a}}{\Delta x} + \frac{iA}{F}(\lambda_{\text{drag}} + 0.5) - \dot{n}_{\text{pressure driven},a,c} \tag{18}$$

where the last term on the right represents the flow of water by imbalanced pressure forces, which can be further represented by capillary- and gas-phase pressure terms. Since the reduction reaction and electroosmotic drag result in water transport to and generation at the cathode surface, the flux of water by diffusion can be either to or from the anode surface, depending on local flow and humidity conditions. The flux by pressure difference is a function of tailorable material and operating properties and can flow toward either electrode, depending on the desire of the designer.

To achieve a proper water balance, many technologies have been employed, including tailored temperature or pressure gradients to absorb the net water generated. Flow field design also has a profound effect on the local liquid water distribution and is too detailed a topic to discuss in this chapter. The ambient relative humidity also should affect the system water balance; however, this effect is typically quite small due to the vast difference between typical ambient temperature and operating temperature $P_{g,\text{sat}}$ values. For example, even fully saturated inlet flow drawn from an ambient air source at 20°C ($P_{g,\text{sat}} = 2.338$ kPa) contains only 5% of the water required for saturation in a fuel cell operating at 80°C ($P_{g,\text{sat}} = 47.39$ kPa).

Durability and Freeze–Thaw Cycling

For PEFCs, durability and freeze–thaw cycling capability are major issues. Durability issues in PEFCs have been discussed for fuel cells in general. For PEFCs, carbon support corrosion, morphological changes, susceptibility to chemical poisoning (especially carbon monoxide) catalyst loss, and electrolyte degradation all contribute to an operational degradation rate that is on the order of microvolts per hour under ideal steady-state conditions, which is still too high.[27] Load cycling is known to initiate even greater degradation rate for a variety of reasons, so that steady-state laboratory testing typically underestimates true performance loss. Under nonoptimal higher temperature or low-humidity conditions, longevity is even less. Additionally, further system size and weight reductions are needed for automotive packaging requirements.

For automotive and stationary applications, PEFCs must withstand vastly changing environments as low as −40°C. Because the electrolyte contains water, freezing results in some ice formation and volume change, which can easily result in damage to the electrode structure and interfacial contact between the various layer structures. To compete with existing combustion-based technology, fuel-cell stacks must achieve over 100 cold starts at −20°C with drivable power in under 5 sec.

2.9   Direct Methanol Fuel Cell

The liquid-fed DMFC is seen as the most viable alternative to lithium ion batteries in portable applications because DMFC systems require less ancillary equipment and can therefore be more simplified compared to an H_2 PEFC. While both H_2 PEFCs and DMFCs are strictly PEFCs (same electrolyte), the DMFC feeds a liquid solution of methanol and water to the anode as fuel. The additional complexities of the low-temperature internal re-formation prevent the DMFC from obtaining the same level of fuel-cell power density as the H_2 PEFC. For the DMFC, both anode and cathode activation polarizations are significant and are the same order of magnitude. However, reduced performance compared to the H_2 PEFC is tolerable in light of other advantages of the DMFC:

1. Because anode flow is mostly liquid (gaseous CO_2 is a product of methanol oxidation), there is no need for a separate cooling or humidification subsystem.

2. Liquid fuel used in the anode results in lower parasitic pumping requirements compared to gas flow. In fact, an emerging class of passive DMFC designs operate without any external parasitic losses, instead relying on natural forces such as capillary action, buoyancy, and diffusion to deliver reactants.

3. The highly dense liquid fuel stored at ambient pressure eliminates problems with fuel storage volume. With highly concentrated methanol as fuel (>10 M), passive DMFC system power densities can compare favorably to advanced Li ion batteries.

2.10 Technical Issues of the DMFC

Four main technical issues affecting performance remain: (1) water management, (2) methanol crossover, (3) managing two-phase transport in the anode, and (4) high activation polarization losses and catalyst loading. While significant progress has been made by various groups to determine alternative catalysts, total catalyst loading is still on the order of 10 mg/cm². Typically a platinum–ruthenium catalyst is utilized on the anode to reduce polarization losses from CO intermediate poisoning, and a Pt catalyst is used on the cathode.[28]

External humidification is not needed in the DMFC, due to the liquid anode solution, but prevention of cathode flooding is critical to ensure adequate performance. Flooding is more of a concern for DMFCs than H_2 PEFCs because of constant diffusion of liquid water to the cathode. To prevent flooding, cathode airflow must be adequate to remove water at the rate that it arrives and is produced at the cathode surface. Assuming thermodynamic equilibrium, the minimum stoichiometry required to prevent liquid water accumulation in the limit of zero water diffusion through the membrane can be shown by equating Eqs. (1) and (2) as[29]

$$\xi_{c,\min} = \frac{2.94 / P_{g,\text{sat}}}{P_t - P_{g,\text{sat}}} + 1 \qquad (19)$$

The factor of 1 in Eq. (19) is a result of the consumption of oxygen in the cathode by an electrochemical oxygen reduction reaction. For most cases, the minimum cathode stoichiometry for a DMFC is determined by flooding avoidance rather than electrochemical requirements. Therefore, optimal cathode stoichiometries are significantly greater than unity. It should be noted that Eq. (19) is purely gas phase and therefore does not allow for water removal in the liquid phase, as droplets, a capillary stream out of the cathode or to the anode, or even entrained as a mist in the gas flow. Some recirculation of the water is needed, or, in a totally passive system design, the hydrophobicity of the GDL and catalyst can be tailored to pump water back to the anode via capillary action.

Another critical issue in the DMFC is methanol crossover from the anode to the cathode. This is a result of diffusion, electroosmotic drag, and permeation from pressure gradients. Therefore, an expression for the methanol crossover through the membrane can be written similar to Eq. (18) with different transport properties. When crossover occurs, the mixed potential caused by the anodic reaction on the cathode electrode reduces cell output and the true stoichiometry of the cathode flow.

Of the three modes of methanol crossover, diffusion (estimated as $10^{-5.4163-999.778/T}$ m^2/s)[30] is dominant under normal conditions, especially at higher temperatures. Since the driving potential for oxidation is so high at the cathode, the methanol that crosses over is almost completely oxidized to CO_2, which sets up a sustained maximum activity gradient in methanol concentration across the electrolyte. The electroosmotic drag coefficient of methanol (estimated as 0.16 CH_3OH/H^+,[31] or $2.5y$, where y is the mole fraction of CH_3OH in solution[32]) is relatively weak owing to the nonpolar nature of the molecule. To prevent crossover so that more concentrated (and thus compact) solutions of methanol can be utilized as fuel, various diffusion barriers have been developed. That is, a porous filter in the GDL or separating the fuel from the fuel channel can be used to separate concentrated methanol solution from the membrane electrode assembly (MEA), greatly reducing crossover through the electrolyte.[33] An earlier solution was use of a thicker electrolyte to reduce methanol crossover, but the concomitant loss in performance via increased ohmic losses through the electrolyte is unsatisfactory.

Several other transport-related issues are important to DMFC performance. The anode side is a two-phase system primarily consisting of methanol solution and product CO_2. The methanol must diffuse to the catalyst, while the reaction-generated CO_2 must diffuse outward from the catalyst. At high current densities, CO_2 can become a large volume fraction (>90%) in the anode flow field. Carbon monoxide removal from the catalyst sites is critical to ensure adequate methanol oxidation. Other disadvantages of the DMFC are related to use of methanol. Methanol is toxic, can spread rapidly into groundwater, has a colorless flame, and is more corrosive than gasoline.

3   SOLID OXIDE FUEL CELL

The SOFC and MCFC represent high-temperature fuel-cell systems. The current operating temperature of most SOFC systems is around 800–1000°C, although new technology has demonstrated 600°C operation, where vastly simplified system sealing and materials solutions are feasible. High electrolyte temperature is required to ensure adequate ionic conductivity (of O^{2-}) in the solid-phase ceramic electrolyte and reduces activation polarization so much that cell losses are typically dominated by internal cell ohmic resistance through the electrolyte. Typical SOFC open-circuit cell voltages are around 1 V, very close to the theoretical maximum, and operating current densities vary greatly depending on design. While the theoretical maximum efficiency of the SOFC is less than the H_2 PEFC because of increased temperature, activation polarization is extremely low, and operating efficiencies as high as 60% have been attained for a 220-kW cogeneration system.[34]

There has been much recent development in the United States on SOFC systems, incubated by the Department of Energy Solid State Energy Conversion Alliance (SECA) program. The 10-year goal of the SECA program is to develop kilowatt-size SOFC APU units at \$400/kW with rated performance achievable over the lifetime of the application with less than 0.1% loss per 500 hr operation by 2021.

The solid-state, high-temperature (600–1000°C) SOFC system eliminates many of the technical challenges of the PEFC while suffering unique limitations. The SOFC power density varies greatly depending on cell design but can achieve above 400 mW/cm^2 for some designs. In general, a SOFC system is well suited for applications where a high operating temperature and a longer start-up transient are not limitations, or where conventional fuel feedstocks are desired.

The main advantages of the SOFC system include the following:

1. High operating temperature greatly reduces activation polarization and eliminates the need for expensive catalysts. This also provides a tolerance to a variety of fuel stocks and enables internal re-formation of complex fuels.
2. High-quality waste heat, enabling a potential for high overall system efficiencies (~80%) utilizing a bottoming or cogeneration cycle.[35]
3. Tolerance to CO, which is a major poison to Pt-based low-temperature PEFCs.

3.1 Technical Issues of SOFC

Besides manufacturing and economic issues, which are beyond the scope of this chapter, the main technical limitations of the SOFC include operating temperature, long start-up time, durability, and cell-sealing problems resulting from mismatched thermal expansion of materials. For additional details, an excellent text for SOFC was written by Minh and Takahashi.[36]

The high operating temperature of the SOFC requires long start-up time to avoid damage due to nonmatched thermal expansion properties of materials. Another temperature-related limitation is that no current generation is possible until a critical temperature is reached in the solid-state electrolyte, where oxygen ionic conductivity of the electrolyte becomes non-negligible. Commonly used electrolyte conductivity is nearly zero until around 650°C,[12] although low-temperature SOFC operation at 500°C using doped ceria (CeO_2) ceramic electrolytes has shown feasibility.[37] In many SOFC designs, a combustor is utilized to burn fuel and oxidizer effluent to preheat the cell to light-off temperature and hasten start-up and provide a source of heat for cogeneration. In addition, the combustor effectively eliminates unwanted hydrogen or CO, which is especially high during start-up when fuel-cell performance is low. Additionally, electrolyte, electrode, and current collector materials must have matched thermal expansion properties to avoid internal stress concentrations and damage during both manufacture and operation.

The desire for lower temperature operation of the SOFC is ironically opposite to the PEFC, where higher operating temperature is desired to simplify water management and CO poisoning issues. Lower temperature (~400–500°C) operation would enable rapid start-up, use of common metallic compounds for cell interconnects, reduce thermal stresses, reduce the rate of some modes of degradation, and increase reliability and reduced manufacturing costs. Despite the technical challenges, the SOFC system is a good potential match for many applications, including stationary cogeneration plants and auxiliary power.

Durability of SOFCs is not solely related to thermal mismatch issues. The electrodes suffer a strong poisoning effect from sulfur (in the ppb range), requiring the use of sulfur-free fuels. Additionally, anode oxidation of nickel catalysts can decrease performance and will do so rapidly if the SOFC is operated below 0.5 V. As SOFC reduced operating temperature targets are achieved and use of inexpensive metallic interconnects become feasible, accelerated degradation from metallic interaction is possible. Metals utilizing chromium (e.g., stainless steels) have shown limited lifetimes in SOFCs and can degrade the cathode catalyst through various physicochemical pathways.[38]

3.2 Performance and Materials

In the SOFC system, yttria- (Y_2O_3-) stabilized zirconia (ZrO_2) is most often used as the electrolyte. In contrast to PEFCs, O^{2-} ions are passed *from the cathode to the anode* via oxygen vacancies in the electrolyte instead of H^+ ions *from the anode to the cathode*. Other cell components such as interconnects and bipolar plates are typically doped ceramic, cermet, or metallic compounds.

There are four different basic designs for the SOFC system: planar, sealless tubular, monolithic, and segmented cell-in-series designs. Two of the designs, planar and tubular, are the most promising for continued development. The other designs have been comparatively limited in development to date. The planar configuration looks geometrically similar to the generic fuel-cell shown in Fig. 4. The three layer anode–electrolyte–cathode structure can be an anode-, cathode-, or electrolyte-supported design, meaning the structural support is provided by a thicker layer of one of the structures (anode, cathode, or electrolyte supported). Since excessive ionic and concentration losses result from electrolyte- and cathode-supported structures, respectively, many designs utilize an anode-supported structure, although ribbed supports or cathode-supported designs are utilized in some cases.

For the planar design, the flow channel material structure is used as support for the electrolyte, and a stacking arrangement is employed. Although this design is simple to manufacture, one of the major limitations is difficulty in sealing the flow fields at the edges of the fuel cell. Sealing is a key issue in planar SOFC design because it is difficult to maintain system integrity over the large thermal variation and reducing/oxidizing environment over many start-up and shut-down load cycles. Compressive, glass, cermet, glass–ceramic, and hybrid seals have been used with varied success for this purpose.

The second major design is the sealless tubular concept pioneered by Westinghouse (now Siemens-Westinghouse) in 1980. A schematic of the general design concept is shown in Fig. 12. Air is injected axially down the center of the fuel cell, which provides preheating of the air to operation temperatures before exposure to the cathode. The oxidizer is provided at adequate flow rates to ensure negligible concentration polarization at the cathode exit, to maintain desired cell temperature, and to provide adequate oxidizer for effluent combustion with unused fuel. The major advantage of the tubular configuration is that the difficult high-temperature seals needed for the planar SOFC design are eliminated. Tubular designs have been tested in 100-kW atmospheric pressure and 250-kW pressurized demonstration systems with little performance degradation with time (less than 0.1% per 1000 hr) and efficiencies of 46 and 57% (LHV), respectively.[35]

One drawback of this type of tubular design is the more complex and limited range of cell fabrication methods. Another drawback is high internal ohmic losses relative to the planar design due to the relatively long in-plane path that electrons must travel along the electrodes to and from the cell interconnect. Some of these additional ionic transport losses have been reduced by use of a flattened tubular SOFC design with internal ribs for current flow, called the high-power-density (HPD) design by Siemens-Westinghouse and shown schematically in Fig. 13.[39] This design can also experience significant losses due to limited oxygen transport through the porous (~35% porosity) structural support tube used to provide rigidity to the assembly. The internal tube can also be used as the anode, reducing these losses through the higher diffusivity of hydrogen.

The monolithic and segmented cell-in-series designs are less developed, although demonstration units have been constructed and operated. A schematic of the monolithic cell design is shown in Fig. 14. In the early 1980s, the corrugated monolithic design was developed based on the advantage of HPD compared to other designs. The HPD of the monolithic design is a result of the high active area exposed per volume and the short ionic paths through the electrolyte, electrodes, and interconnects. The primary disadvantage of the monolithic SOFC design, preventing its continued development, is the complex manufacturing process required to build the corrugated system.

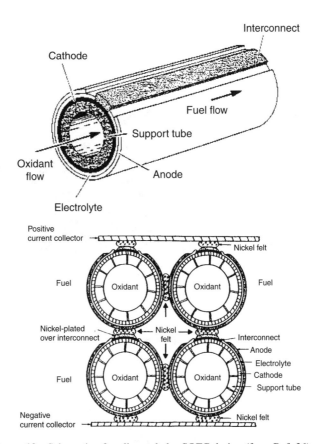

Figure 12 Schematic of sealless tubular SOFC design (from Ref. 36).

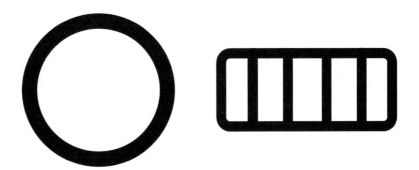

Standard tubular design concept   Flattened tubular design concept

Figure 13 End-view schematic of conventional tubular and flattened high-power-density SOFC concepts for reduced ionic transport losses.

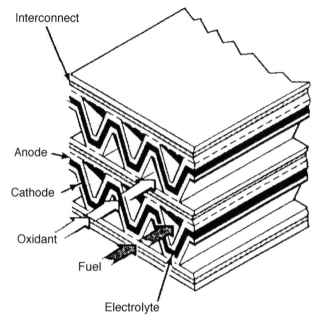

Interconnect

Anode

Cathode

Oxidant

Fuel

Electrolyte

Figure 14 Schematic of the monolithic SOFC design (from Ref. 36).

The segmented cell-in-series design has been successfully built and demonstrated in two configurations: the bell-and-spigot and banded configurations shown schematically in Fig. 15. The bell-and-spigot configuration uses stacked segments with increased electrolyte thickness for support. Ohmic losses are high because electron motion is along the plane of the electrodes in both designs, requiring short individual segment lengths (~1–2 cm). The banded configuration avoids some of the high ohmic losses of the bell-and-spigot configuration with a thinner electrolyte but suffers from the increased mass-transport losses associated with the porous support structure used. The main advantage of the segmented cell design is a higher operating efficiency than larger area single-electrode configurations. The primary disadvantages limiting development of the segmented cell designs include the necessity for many high-temperature gas-tight seals, relatively high internal ohmic losses, and requirement for manufacture of many segments for adequate power output. Other cell designs, such as radial configurations and more recently microtubular designs, have been developed and demonstrated to date.

4 OTHER FUEL CELLS

Many other fuel-cell varieties and configurations, too numerous to enumerate here, have been developed to some degree. The most developed to date have been the phosphoric acid, alkaline, and molten carbonate fuel cells, all three of which have had some applied and commercial success.

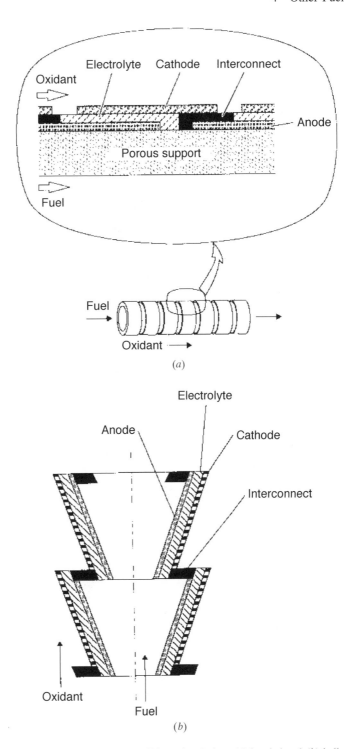

Figure 15 Schematic of the segmented cell-in-series design: (*a*) banded and (*b*) bell-and-spigot configuration (from Ref. 36).

4.1 Alkaline Fuel Cells

Alkaline fuel cells utilize a solution of potassium hydroxide in water as an alkaline, mobile (liquid) electrolyte. Alkaline fuel cells were originally developed as an APU for space applications by the Soviet Union and the United States in the 1950s and served in the Apollo program as well as on the current Space Shuttle orbiter. Figure 16 shows AFCs for installation on Apollo mission service modules in 1964.[40] Alkaline fuel cells were chosen for space applications for their high efficiency and robust operation. Both circulated and static electrolyte designs have been utilized. The AFC operates at around 60–250°C with greatly varied electrode design and operating pressure. More modern designs tend to operate at the lower range of temperature and pressure. The primary advantages of the AFC are the cheaper cost of materials and electrolyte and high operating efficiency (60% demonstrated for space applications) due to use of an alkaline electrolyte. For alkaline electrolytes, ORR kinetics are much more efficient than acid-based electrolytes (e.g., PEFCs, PAFCs) enabling high relative operating efficiencies. Since space applications typically utilize pure oxygen and

Figure 16 Pratt & Whitney technicians assemble alkali fuel cells for Apollo service modules, 1964.[40]

hydrogen for chemical propulsion, the AFC was well suited. However, the electrolyte suffers an intolerance to even small fractions of CO_2 found in air, which reacts to form potassium carbonate (K_2CO_3) in the electrolyte, gravely reducing performance over time. For terrestrial applications, CO_2 poisoning has limited the lifetime of AFCs systems to well below that required for commercial application, and filtration of CO_2 is too expensive for practical use. Due to this limitation, relatively little commercial development of the AFC beyond space applications has been realized.

4.2 Molten Carbonate Fuel Cells

Currently MCFCs are commercially available from several companies, including a 250-kW unit from Fuel Cell Energy in the United States and several other companies in Japan. Some megawatt-size demonstration units are installed worldwide, based on natural gas or coal-based fuel sources which can be internally re-formed within the anode of the MCFC. The MCFCs operate at high temperature (600–700°C) with a molten mixture of alkali metal carbonates (e.g., lithium and potassium) or lithium and sodium carbonates retained in a porous ceramic matrix. In the MCFC, CO_3^{2-} ions generated at the cathode migrate to the anode oxidation reaction. The MCFC design is similar to a PAFC, in that both have liquid electrolytes maintained at precise levels within a porous ceramic matrix and electrode structure by a delicate balance of gas-phase and capillary pressure forces. A major advantage of the MCFC compared to the PAFC is the lack of precious-metal catalysts, which greatly reduces the system raw material costs. Original development on the MCFC was mainly funded by the U.S. Army in the 1950s and 1960s, and significant advances of this liquid electrolyte high-temperature fuel-cell alternative to the SOFC were made.[41] The U.S. Army desired operation of power sources from logistic fuel, thus requiring high temperatures with internal fuel re-formation that can be provided by the MCFC. Development waned somewhat after this early development, but advances have continued and initial commercialization has been achieved. The steadily increasing performance of MCFCs throughout years of development is illustrated in Fig. 17. The main advantages of MCFCs include the following:

1. The MCFC can consume CO as a fuel and generates water at the anode, thus making it ideal for internal re-formation of complex fuels.
2. As with the SOFC, high-quality waste heat is produced for bottom-cycle or cogeneration applications.
3. Non-noble-metal catalysts are used, typically nickel–chromium or nickel–aluminum on the anode and lithiated nickel oxide on the cathode.

The main disadvantages of MCFCs include the following:

1. Versus the SOFC (*the other high-temperature fuel cell*), the MCFC has a highly corrosive electrolyte that, coupled with the high operating temperature, accelerates corrosion and limits longevity of cell components, especially the cathode catalyst. The cathode catalyst (nickel oxide) has a significant dissolution rate into molten carbonate electrolyte.
2. Extremely long start-up time (the MCFC is generally suitable only for continuous power operation) and nonconductive electrolyte at low temperatures.
3. Carbon dioxide must be injected into the cathode to maintain ionic conductivity. This can be accomplished with recycling from the anode effluent or injection of combustion product but complicates the system design.

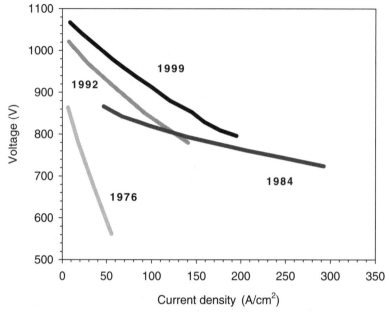

Figure 17 Relative performance of MCFCs at 1 atm pressure. (Adapted from Ref. 12.)

4. Electrolyte maintenance is an engineering technical challenge. The liquid electrolyte interface between the electrodes is maintained by a complex force balance involving gas-phase and electrolyte-liquid capillary pressure between the anode and cathode. Significant spillage of the electrolyte into the cathode can lead to catalyst dissolution but is difficult to eliminate.

5. Vapor pressure of the electrolyte is nonnegligible and leads to loss of electrolyte through reactant flows.

4.3 Phosphoric Acid Fuel Cells

The PAFC was originally developed for commercial application in the 1960s. The PAFC has an acidic, mobile (liquid) electrolyte of phosphoric acid contained by a porous silicon carbide ceramic matrix and operates at around 160–220°C. Like the MCFC, the electrolyte is bound by capillary and gas pressure forces between porous electrode structures. The PAFC is in many ways similar to the PEFC, except the acid-based electrolyte is in liquid form and the operating temperature is slightly higher. Over two hundred 200-kW commercial PAFC units were developed and sold by International Fuel Cells (now United Technologies Research Fuel Cells) and many are still in operation. However, ubiquitous commercial application has not been achieved, primarily due to the high cost of approximately $4500/kW, about five times greater than cost targets for conventional stationary applications. The main advantages of the PAFC include the following:

1. The high operating temperature provides better waste heat than the PEFC and allows operation with 1–2% CO in the fuel stream, which is much better for use of re-formed fuel compared to the PEFC, which cannot tolerate more than 10 ppm of CO without significant performance loss.[42]

2. The acid electrolyte does not need water for conductivity, making water management very simple compared to the PEFC. Only the product water from the cathode reduction reaction needs to be removed, a relatively simple task at elevated temperatures.

3. The demonstrated long life and commercial success for premium stationary power of the PAFC.

The main disadvantages of the PAFC include the following:

1. The PAFC is a bulky, heavy system compared to the PEFC. Area specific power is less than for the PEFC (0.2–0.3 W/cm²).[43]

2. The use of platinum catalyst with nearly the same loading as PEFCs.

3. The liquid electrolyte has finite vapor pressure, resulting in continual loss of electrolyte in the vapor phase and a continual need for replenishment or recirculation. Modern PAFC design includes cooling and condensation zones to mitigate this loss.

4. The relatively long warm-up time until the electrolyte is conductive at ∼160°C (although much less than for the MCFC or SOFC).

4.4 Other Alternatives

Many other fuel-cell systems exist, and new versions are constantly being developed. Most are simply existing fuel-cell systems with a new fuel. For example, PEFCs based on a direct alcohol solution offer alternatives to DMFCs for portable power application and include those based on formic acid,[44] dimethyl ether,[45] ethylene glycol, dimethyl oxalate, and others.[46] A completely different concept is the biological or microbial fuel cell (MFC). In the MFC, electricity is generated by anerobic oxidation of organic material by bacteria. The catalytic activity and transport of protons is accomplished using biological enzymes or exogenous mediators.[47,48] Although relative performance is very low, on the order of 1–100 mW/m², the potential for generating some power or simply power-neutral decomposition and treatment of organic waste matter such as sewage water is potentially quite significant to society.

Based on the continued growth and expansion of fuel-cell science, it is evident that, despite lingering technical challenges, continued growth and development of a variety of fuel cell systems will evolve toward implementation in many, but certainly not all, potential applications. In some cases, development of existing or new power sources or existing technical barriers will ultimately doom ubiquitous application of fuel cells, while some applications are likely to enjoy commercial success.

NOMENCLATURE

a activity coefficient of species, unitless
A area term, cm²
C molar concentration, mol/cm³
D diffusion coefficient, cm²/s
E voltage, V
E_a activation energy of electrochemical reaction, J/mol
F Faraday constant, charge on 1 mol of electrons, 96,487 As/mol electron

G Gibbs free energy, kJ/kg
H enthalpy, kJ/kg
I current, A
i current density, A/cm^2
i_o exchange current density, A/cm^2
i_l mass-limited current density, A/cm^2
M   solution molarity, mol/L
n electrons per mole oxidized or reduced, e$^-$/mol
$\dot{n}$ molar flow rate, mol/s
P pressure, Pa, and power, W
q'' heat flux, W/cm^2
r area specific resistance, Ω/cm^2
R universal gas constant, 8.314 J/mol·K
RH   relative humidity, unitless
S entropy, kJ/kg·K
T temperature, K
y mole fraction, unitless

Greek Letters

α charge transfer coefficient
$\Delta(x)$ change of parameter x
γ reaction order for elementary oxidizer reaction
η polarization, V
λ electroosmotic drag coefficient, mol/mol H$^+$, and water saturation in electrolyte
σ conductivity, $\mho \cdot m^{-1}$
ν stoichiometric coefficient of balanced equation; reaction order for elementary fuel reaction
ξ stoichiometric flow ratio

Superscripts

0   standard conditions

Subscripts

a activation or anode
c cathode
e electrolyte
fc   fuel cell
f fuel
H$_2$O   water
i species i
k individual cell components; number of cells in a stack
m mass transport
ox   oxidizer
r resistive

ref   reference
sat   at saturation conditions
th   thermal

REFERENCES

1. M. L. Perry and T. F. Fuller, "A Historical Perspective of Fuel Cell Technology in the 20th Century," *J. Electrochem. Soc.,* **149**(7), pp. S59–S67 (2002).

2. W. Nernst, "Über die elektrolytische Leitung fester Körper bei sehr hohen Temperaturen," *Z. Elektrochem.,* No. 6, 41–43 (1899).

3. E. Baur and J. Tobler, *Z. Elektrochem. Angew. Phys. Chem.,* **39,** 180 (1933).

4. Toshiba Press Release of June 24, 2004: http://www.toshiba.com/taec/press/dmfc_04_222.shtml.

5. Y. Mugikura, "Stack Materials and Stack Design," in *Handbook of Fuel Cells—Fundamentals, Technology and Applications,* Vol. 4, W. Vielstich, A. Lamm, and H. A. Gasteiger (eds.), Wiley, Chichester, UK, 2003, pp. 907–920.

6. http://www.fuelcells.org.

7. United States Department of Energy 2003 Multi-Year Research, Development and Demonstration Plan for Fuel Cells: http://www.eere.energy.gov/hydrogenandfuelcells/mypp/pdfs/3.4_fuelcells.pdf.

8. D. Sperling, and J. S. Cannon, *The Hydrogen Energy Transition: Moving Toward the Post Petroleum Age in Transportation,* New York, Elsevier, 2004.

9. O. Yamamoto, "Solid Oxide Fuel Cells: Fundamental Aspects and Prospects," *Electrochem. Acta,* **45,** 2423–2435 (2000).

10. M. Wang, "Fuel Choices for Fuel Cell Vehicles: Well-to-Wheels Energy and Emission Impacts," *J. Power Sources,* **112,** 307–312 (2002).

11. A. J. Bard and L. R. Faulkner, *Electrochemical Methods,* Wiley, New York, 1980.

12. J. Larmine and A. Dicks, *Fuel Cell Systems Explained,* 2nd ed., Wiley, Chichester, UK, 2003.

13. A. B. LaConti, M. Hamdan, and R. C. McDonald, "Mechanisms of Membrane Degradation," in *Handbook of Fuel Cells—Fundamentals, Technology and Applications,* Vol. 3, W. Vielstich, A. Lamm, and H. A. Gasteiger (eds.), 3, Wiley, Chichester, UK, 2003, pp. 647–662.

14. D. J. Burford, *Real-Time Electrolyte Temperature Measurement in an Operating Polymer Electrolyte Membrane Fuel Cell,* Thesis, Penn State University, University Park, PA, 2004.

15. T. E. Springer, T. A. Zawodzinski, and S. Gottesfeld, *J. Electrochem. Soc.,* **136,** 2334 (1991).

16. J. Divisek, "Low Temperature Fuel Cells," in *Handbook of Fuel Cells—Fundamentals, Technology and Applications,* Vol. 1, W. Vielstich, A. Lamm, and H. A. Gasteiger (eds.), Wiley, Chichester, UK, 2003, pp. 99–114.

17. D. P. Wilkinson and J. St-Pierre, "Durability," in *Handbook of Fuel Cells—Fundamentals, Technology and Applications,* Vol. 3, W. Vielstich, A. Lamm, and H. A. Gasteiger (eds.), Wiley, Chichester, UK, 2003, pp. 611–626.

18. D. A. Shores and G. A. DeLuga, "Basic Materials and Corrosion Issues," in *Handbook of Fuel Cells—Fundamentals, Technology and Applications,* Vol. 2, W. Vielstich, A. Lamm, and H. A. Gasteiger (eds.), Wiley, Chichester, UK, 2003, pp. 273–285.

19. http://www.eere.energy.gov/hydrogenandfuelcells/fuelcells/fc_parts.html#catalyst.

20. M. J. Moran and H. N. Shapiro, *Fundamentals of Engineering Thermodynamics,* 3rd ed., Wiley, New York, 1995.

21. D. Burford, T. Davis, and M. M. Mench, "*In situ* Temperature Distribution Measurement in an Operating Polymer Electrolyte Fuel Cell," *Proceedings of 2003 ASME International Mechanical Engineering Congress and Exposition,* November 15–21, 2003.

22. D. Burford, S. He, and M. M. Mench, "Heat Transport and Temperature Distribution in PEFCs," *Proceedings of 2004 ASME International Mechanical Engineering Congress and Exposition,* November 13–19, 2004.

23. P. J. S. Vie and S. Kjelstrup, "Thermal Conductivities from Temperature Profiles in the Polymer Electrolyte Fuel Cell," *Electrochim. Acta,* **49,** 1069–1077 (2004).

24. S. Motupally, A. J. Becker, and J. W. Weidner, "Diffusion of Water in Nafion 115 Membranes," *J. Electrochem. Soc.,* **147**(9), 3171–3177 (2000).

25. X. Ren and S. Gottesfeld, "Electro-Osmotic Drag of Water in Poly(perfluorosulfonic acid) Membranes," *J. Electrochem. Soc.,* **148**(1), A87–A93 (2001).

26. U. Pasaogullari and C. Y. Wang, "Two-Phase Transport and the Role of Micro-Porous Layer in Polymer Electrolyte Fuel Cells," *Electrochim. Acta,* **49,** 4359–4369 (2004).

27. S. Cleghorn, J. Kolde, and W. Liu, "Catalyst Coated Composite Membranes," in *Handbook of Fuel Cells—Fundamentals, Technology and Applications,* Vol. 3, W. Vielstich, A. Lamm, and H. A. Gasteiger (eds.), Wiley, Chichester, UK, 2003, pp. 566–575.

28. J. Müller, G. Frank, K. Colbow, and D. Wilkinson, "Transport/Kinetic Limitations and Efficiency Losses," in *Handbook of Fuel Cells—Fundamentals, Technology and Applications,* Vol. 4, W. Vielstich, A. Lamm, and H. A. Gasteiger (eds.), Wiley, Chichester, UK, 2003, pp. 847–855.

29. M. M. Mench and C. Y. Wang, "An *in Situ* Method for Determination of Current Distribution in PEM Fuel Cells Applied to a Direct Methanol Fuel Cell," *J. Electrochem. Soc.,* **150**(1), A79–A85 (2003).

30. C. L. Yaws, *Handbook of Transport Property Data: Viscosity, Thermal Conductivity, and Diffusion Coefficients of Liquids and Gases,* Gulf, Houston, TX, 1995.

31. J. Cruickshank and K. Scott, "The Degree and Effect of Methanol Crossover in the Direct Methanol Fuel Cell," *J. Power Sources,* **70,** 40–47 (1998).

32. X. Ren, T. E. Springer, T. A. Zawodzinski, and S. Gottesfeld, J. *Electrochem. Soc.,* **147,** 466 (2000).

33. Y. Pan, *Integrated Modeling and Experimentation of Electrochemical Power Systems,* Thesis, Penn State University, University Park, PA, 2004.

34. R. F. Service, "New Tigers in the Fuel Cell Tank," *Science,* **288,** 1955–1957 (2000).

35. S. C. Singhal, "Science and Technology of Solid-Oxide Fuel Cells," *MRS Bulletin,* March 2000, pp. 16–21.

36. N. Minh and T. Takahashi, *Science and Technology of Ceramic Fuel Cells,* Elsevier, New York, 1995.

37. R. Doshi, V. L. Richards, J. D. Carter, X. Wang, and M. Krumpelt, "Development of Solid Oxide Fuel Cells that Operate at 500°C," *J. Electrochem. Soc.,* **146,** 1273–1278 (1999).

38. Y. Matsuzaki and I. Yasuda, "Dependence of SOFC Cathode Degradation by Chromium-Containing Alloy on Compositions of Electrodes," *J. Electrochem. Soc.,* **148**(2), A126–A131 (2001).

39. D. Stöver, H. P. Buchkremer, and J. P. P. Huijsmans, "MEA/Cell Preparation Methods: Europe/USA," in *Handbook of Fuel Cells—Fundamentals, Technology and Applications,* Vol. 4, W. Vielstich, A. Lamm, and H. A. Gasteiger (eds.), Wiley, Chichester, UK, 2003, pp. 1015–1031.

40. Image 059-016 from the Science Service Historical Image Collection, The Smithsonian Institution: http://americanhistory.si.edu/csr/fuelcells/alk/alk3.htm.

41. B. S. Baker (ed.), *Hydrocarbon Fuel Cell Technology,* Academic, New York, 1965.

42. R. D. Breault, "Stack Materials and Stack Design," in *Handbook of Fuel Cells—Fundamentals, Technology and Applications,* Vol. 4, W. Vielstich, A. Lamm, and H. A. Gasteiger (eds.), Wiley, Chichester, UK, 2003, pp. 797–810.

43. *Fuel Cell Handbook,* 5th ed., EG&G Services Parsen, Inc., Science Applications International Corporation, 2000.

44. M. Zhao, C. Rice, R. I. Masel, P. Waszczuk, and A. Wieckowski, "Kinetic Study of Electro-oxidation of Formic Acid on Spontaneously-Deposited Pt/Pd Nanoparticles—CO Tolerant Fuel Cell Chemistry," *J. Electrochem. Soc.,* **151**(1), A131–A136 (2004).

45. M. M. Mench, H. M. Chance, and C. Y. Wang "Dimethyl Ether Polymer Electrolyte Fuel Cells for Portable Applications," *J. Electrochem. Soc.,* **151,** A144–A150 (2004).

46. E. Peled, T. Duvdevani, A. Aharon, and A. Melman, "New Fuels as Alternatives to Methanol for Direct Oxidation Fuel Cells," *Electrochem. Solid-State Lett.,* **4**(4), A38–A41 (2001).

47. H. Liu, R. Narayanan, and B. Logan, "Production of Electricity during Wastewater Treatment Using a Single Chamber Microbial Fuel Cell," *Environ. Sci. Technol.* **38,** 2281–2285 (2004).

48. T. Chen, S. Calabrese Barton, G. Binyamin, Z. Gao, Y. Zhang, H.-H. Kim, and A. Heller, "A Miniature Biofuel Cell," *J. Am. Chem. Soc.,* **123,** 8630–8631 (2001).

49. J. M. King and B. McDonald, "Experience with 200 kW PC25 Fuel Cell Power Plant," in *Handbook of Fuel Cells—Fundamentals, Technology and Applications,* Vol. 3, W. Vielstich, A. Lamm, and H. A. Gasteiger (eds.), Wiley, West Chichester, UK, 2003, pp. 832–843.

50. http://www.fuelcells.org.

51. N. Pekula, K. Heller, P. A. Chuang, A. Turhan, M. M. Mench, J. S. Brenizer, and K. Ünlü, "Study of Water Distribution and Transport in a Polymer Electrolyte Fuel Cell Using Neutron Imaging," *Nuclear Instruments and Methods in Physics Research Section: A Accelerators Spectrometers Detectors and Associated Equipment,* Volume 542, Issues 1–3, No. 21, pp. 134–141, 2005.

CHAPTER 29

FLUID POWER SYSTEMS

Andrew Alleyne
University of Illinois, Urbana–Champaign
Urbana, Illinois

1 INTRODUCTION

The use of fluids for power delivery has been a part of human civilization for many centuries. For example, much of the growth in the United States textile industry in the 1820s and 1830s can be credited to the abundant supply of hydraulic power available in the northeastern United States through lakes and man-made canals.[1] Early hydraulic systems utilized gravitational potential to convert the energy stored in the fluid to some type of useful mechanical energy. Currently, fluid power systems are ubiquitous in our everyday society. All facets of our lives are touched by some form of pressurized fluid distributing power. Simple transportation examples would be fuel delivery systems or braking systems in cars and buses. Many manufacturing systems also use fluid power for presses and other types of forming applications. In fact, fluid power is seen as a vital and necessary component for many types of engineering systems. This is particularly true in mobile applications where a power-generation component, such as an internal combustion engine, is coupled to the fluid power system. In these cases, the inherent power density advantages of fluid power make it a very attractive choice over other types of actuation.

One key advantage of fluid power systems is the high power density available. This means a very large force can be generated in a very compact space; something that is useful for applications such as aircraft where weight and volume are at a premium. Another advantage is the ability to hold loads for long periods of time, since a continuous load can be applied without requiring power flow. Any possible heat generated near the point of power application can be drawn away to a remote location for heat rejection. A third advantage is the ability to move large inertias very rapidly. The bandwidth versus power characteristics of modern fluid power systems gives them an advantage over other types of actuation systems at higher loads.

Analysis of a typical fluid power application requires a complete systems-level understanding. Figure 1 illustrates the interconnection of several different components to form an overall system as would be found in a mobile application. Each interconnected component

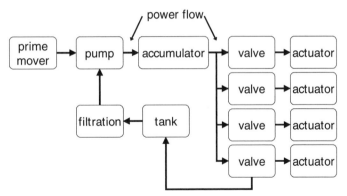

Figure 1 Schematic of a fluid power system.

is itself a complex system made up of various subsystems, as will be described in subsequent sections.

Most fluid power systems have the basic components of *power generation, power distribution, power modulation,* and *power application.* Using Fig. 1 as an example, the power-generation component consists of the prime mover coupled to the pumping device to convert stored chemical energy such as diesel fuel to mechanical power and, subsequently, to high-pressure fluid power. The variables of power in this system include the fluid pressure and flow rate since power = pressure × flow. The power distribution aspect of the system is depicted by the interconnections, which usually consist of hoses and couplings. These conduits of power transport the high-pressure fluid to various locations within the system. At a specific location, the available power is modulated, or metered, often by the use of a valve device. The valve device determines how much power is actually applied to a particular load. The power application component is some type of actuator, usually a piston or motor, which converts the fluid power to mechanical power and applies it to some load. The following sections elaborate in detail on the basic components outlined here.

In addition to the four primary classes of components, several other subsystems are critical to proper overall functionality and are illustrated in Fig. 1. These include the filters for maintaining fluid cleanliness, accumulators for temporarily storing energy, as well as tank storage systems for containing the bulk of the fluid used in the system and providing some centralized place for heat rejection. These subsystems are addressed after the primary components and in less detail. For a more in-depth understanding of these system components, the reader is referred to the excellent and comprehensive book of Yeaple.[2]

2 SYMBOLS AND TERMINOLOGY

As is the case in many other fields, fluid power systems have their own unique symbols and terminology. The following tables illustrate some of the basic hydraulic symbols and terms that are commonly encountered and are referred to later on in this chapter. Section 3 gives more detail on the physical realization of these different system components. Table 1 gives symbols used for representing the hydraulic lines that carry pressurized fluid between different components. Table 2 illustrates symbols for fluid power motors and pumps. These are the components that convert rotary mechanical power to fluid power (pumps) or convert fluid power to rotary mechanical power (motors). Table 3 illustrates symbols for different energy

Table 1 Hydraulic Line Symbols

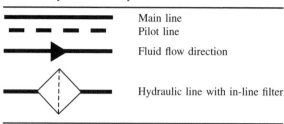

	Main line
	Pilot line
	Fluid flow direction
	Hydraulic line with in-line filter

or fluid storage elements. Table 4 gives they symbols for a wide assortment of valves as well as the types of mechanical or electromechanical devices that move the valves. Table 5 illustrates the symbols used to describe cylinders which, like motors, are also used to transform fluid power to mechanical power. However, unlike fluid power motors, the cylinders in Table 5 convert fluid power into rectilinear mechanical power.

The symbols shown in the tables can be connected to form complicated hydraulic diagrams. A simple example, which is revisited in Section 3 when discussing valves, can be seen in Fig. 2. Here we see a 4-way, 3-position valve with a solenoid actuator and a spring return that is connected to a bidirectional hydraulic motor. The work ports on the valve are labeled as follows: P = pump pressure, T = tank or reservoir, A = work port connected to one side of the motor, B = work port connected to other side of the motor.

As stated before, fluid power systems deal with the flow of power in terms of pressure and volumetric flow. Therefore, in a system dynamics sense they can easily be seen to be

Table 2 Motors and Pumps

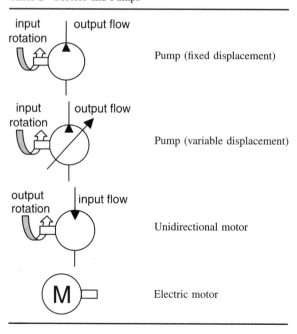

input rotation / output flow	Pump (fixed displacement)
input rotation / output flow	Pump (variable displacement)
output rotation / input flow	Unidirectional motor
M	Electric motor

Table 3 Reservoirs and Energy Storage Elements

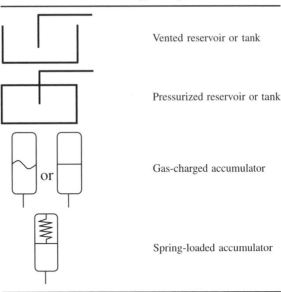

Vented reservoir or tank

Pressurized reservoir or tank

Gas-charged accumulator

Spring-loaded accumulator

quite analogous to electrical systems. The pressure is analogous to voltage and flow is analogous to current. An engineer who is comfortable with creating and analyzing simple electronic circuits could perform similar tasks with a fluid power system once the symbols are clearly understood. Table 6 gives a better comparison between the electrical and fluid power systems.

In addition to the analogous components shown in Table 6 there are several other parallels between electrical and fluid power systems. The hydraulic pump could be connected to a pressure relief valve to give a constant pressure supply, much like a dc battery or other constant voltage source. Moreover, instrumentation devices such as ammeters or voltmeters are analogous to flow meters and pressure sensors, respectively.

3 SYSTEM COMPONENTS

3.1 Hydraulic Oils

The oil used in a hydraulic circuit is designed to withstand high pressures and large pressure drops without deterioration in performance. The majority of common fluids are petroleum-based, often called mineral oil, with other fluids developed for targeted applications. The primary other types of common fluids are synthetic ones, including phosphate ester or silicone based fluids. A primary advantage of these types of synthetic fluids is their fire-resistant nature, making them essential in applications where volatility is catastrophic (e.g., aircraft, submarines). The most fire-resistant type of fluid that can be used is a water–glycol mixture that has the low volatility of water but yet does not have the drawback of freezing in low-temperature applications. In addition to the basic types of hydraulic oil, many additives are used to aid in fluid performance. For example, metal dithiophosphates are often added to inhibit oxidation and to serve as wear-reducing agents. High molecular mass compounds

Table 4 Common Valves and Valve Actuators

	Fixed orifice
	Variable orifice
	Check valve (not spring loaded)
	Check valve (spring loaded)
	Variable relief valve
	2-way, 2 position valve
	3-way, 2 position valve
	4-way, 2 position valve
	4-way, 3 position valve
	Valve return spring
	Manual or lever valve actuation
	Solenoid or electric actuation

Table 5   Cylinders

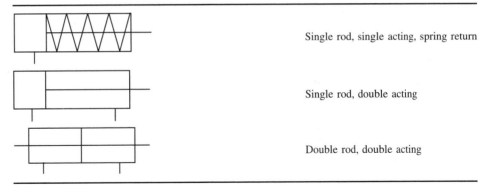

Single rod, single acting, spring return

Single rod, double acting

Double rod, double acting

such as polysiloxanes can act as antifoaming agents to reduce the amount of foam developed in the tank or reservoir. Further information on oil types can be found in Refs. 2 and 3 or from product literature available from major oil manufacturers.

Regardless of their chemical makeup, all of the oils must satisfy basic performance criteria. Critical criteria are (1) viscosity and lubricity (2) stability (thermal, absorption), and (3) incompressibility. Viscosity of a fluid indicates the amount of shear stress necessary for relative motion between two adjacent surfaces with the fluid between them. Figure 3 illustrates a block sliding on a patch of fluid.

A is the area of the underside of the block and t is the thickness of the fluid patch. The force required to move the block is proportional to the velocity of the block and inversely proportional to the fluid thickness:

$$F = \nu A \frac{\dot{x}}{t} \tag{1}$$

The viscosity term ν determines the resistance to motion determined by the internal friction of the fluid. Viscosity is important because of the relatively close tolerances of fluid power components such as valves and pumps. Highly viscous fluids will have a resistance to shear

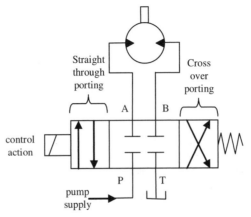

Figure 2   Diagram of a valve-driven motor.

Table 6   Electrical–Hydraulic Analogy

Electrical Components	Electrical Diagram	Fluid Power Diagram	Fluid Power Components
Ground			Reservoir or tank
Current source	i_S		Pump
Voltage regulator	VR		Relief valve
Capacitor			Accumulator
Resistor			Orifice
Diode			Check valve
Electric motor	M		Hydraulic motor

flow and will tend to damp the motion of the various components, thereby making it more difficult to pump fluid up to higher pressure. Simultaneously, low-viscosity fluids tend to leak easily around passages within fluid power components and reduce overall system efficiency. Related to the concept of viscosity is that of lubricity. High-lubricity fluids are able to reduce the wear on the moving mechanical components, thereby increasing the life of the

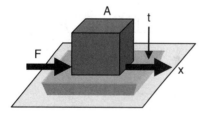

Figure 3   Viscosity schematic.

overall system. Low-lubricity fluids, like water, need to have additives present to prevent excessive metal on metal contact.

Stability is an important aspect of a hydraulic fluid. Essentially, the ideal hydraulic fluid would satisfy the performance requirements and do so irrespective of its operating environment. That is not the case in practice. The viscosity and lubricity of most oils change dramatically with temperature, going from high viscosity at low temperatures to low viscosity at high temperatures. The extreme operating limits for viscosity on most systems range between 1.0 and 5000 centistokes. Anything outside those limits will not be practical. In general, good fluids are less sensitive in their viscosity or lubricity to changes in temperature. Similar aspects apply for insensitivity to pressure. In addition to being stable with respect to physical properties, it is usually desirable for the oil to be environmentally insensitive with respect to its chemical properties. In particular, for reasons to be discussed shortly, the oil should not absorb much air from the surrounding environment if it is exposed. The oil should have a minimal reaction to materials (e.g., rubber, polymers, metals) contained in hoses, seals, filters, valves, or pumps. One example of an undesirable interaction is the potential for swelling of most elastomer seals in contact with typical hydraulic fluids, which can cause unwanted internal stresses on components. In addition to the desired inertness with respect to the components it affects, a good hydraulic fluid should remain chemically stable with little settling of any additives introduced so that its performance will be unaffected by periods of inactivity.

Fluid compressibility is crucial to the overall performance of the system. Thinking of the fluid in a closed chamber (e.g., cylinder) as a spring to store energy, it becomes apparent that a more compressible fluid will act as a softer spring, thereby reducing the natural frequency of the closed chamber. This can be seen in practice as a change in resonant frequency for fluid power systems when the oil properties change. In most cases, it is better to have higher resonant frequencies due to the fluid compressibility rather than lower resonant frequencies.

Unfortunately, the natural tendency of hydraulic fluids is to absorb air from the environment. This air becomes dissolved in the hydraulic fluid, causing it to become more compressible than if there were no air. The compressibility of the fluid is usually defined by the *bulk modulus,* which is defined as the percentage change in volume of a fluid for a given change in pressure. Define a volume, V, under some pressure, P, as shown in Fig. 4. Using Δ to denote the change in a variable, the definition of the fluid's bulk modulus, β, is

$$\frac{1}{\beta_{\text{fluid}}} = -\frac{\Delta V}{V}\frac{1}{\Delta P} \tag{2}$$

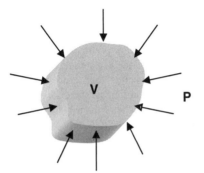

Figure 4   Fluid volume under pressure.

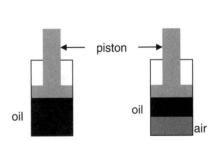

Figure 5   Schematic of a fluid chamber with and without dissolved air.

where the negative sign indicates that an increase in pressure results in a decrease in fluid volume. If the hydraulic oil has air dissolved in it, the air and the oil act as separate springs in series for the purpose of the overall compressibility. An examination of Fig. 5 illustrates this concept. On the right, the air can be seen to be a compressible volume in series with the oil. A force on the piston will act to compress the air more than the oil as would be the case if there were two mechanical springs in series.

Therefore,

$$\frac{1}{\beta_{\text{total}}} = \frac{1}{\beta_{\text{fluid}}} + \frac{V_{\text{air}}}{V_{\text{total}}} \frac{1}{\beta_{\text{air}}} \tag{3}$$

Since the air is so much more compressible than the fluid, a small amount of air ($<1\%$) can have a drastic effect on the overall system stiffness. For example,[4] suppose we have a fluid with 1% entrained air, by volume, that is under 500 psi pressure. If the ratio of specific heat for air is 1.4, then the bulk modulus of the air equals $1.4 \times 500 = 700$ psi. A typical bulk modulus for a petroleum-based fluid is approximately 2×10^5 psi. Therefore,

$$\frac{1}{\beta_{\text{total}}} = \frac{1}{2 \times 10^5} + 0.01 \times \frac{1}{700} = 1.9 \times 10^{-5} \frac{\text{in}^2}{\text{lb}} \tag{4}$$

$$\Rightarrow \beta_{\text{total}} = 5.2 \times 10^4 \text{ psi} \tag{5}$$

Addition of 1% dissolved air reduces the total bulk modulus to only one-fourth of the ideal value of the oil with no air. In practice, it is not uncommon to have much more than 1% air dissolved in a fluid, which will have serious consequences for high-performance hydraulics relying on very low compressibility to achieve high bandwidth and high performance. Therefore, it is important for the designer of the overall system to fully understand the role that oil properties, and the cause for their changes, can have on system performance.

Another oil property that needs to be understood by the system designer is environmental toxicity, which has serious impact on where fluid power systems can be applicable. For example, due to the chemical nature of mineral oil hydraulic fluids, typical fluid power systems are not employed in the food processing industry. Instead, more exotic types of fluid power systems need to be used, including water-based hydraulics, to maintain the necessary system sterility. A final issue of concern is the flammability of the fluid, since many connection or component failures can result in a rapid dispersal of a fine fluid mist, which can be a significant hazard. Further discussion on oil characteristics can be found in Refs. 2–4.

3.2   Hydraulic Hoses

Hoses are a component of fluid power systems that are often overlooked in an overall systems analysis, but they are important to overall system performance. In a systems sense, the hoses are responsible for distributing the power of the pressurized fluid from a central generation location to remote locations where the power is then converted to mechanical power for application to some load. If the hoses are inefficient, then the system will use more energy than necessary. More critically, if the hoses are not properly suited to carry the power being distributed, then failure can result. This would be analogous to sending more electrical current through a wire than it was rated for.

Most hoses in typical hydraulic fluid power systems are relatively complex composite devices constructed from multiple layers of material. Figure 6 illustrates two separate types of hoses with the primary difference being the reinforcing material. The core tube transports the high-pressure fluid and is fabricated from a material that will not react with the fluid being carried. This tube can be made from rubber or some type of thermoplastic polymer.

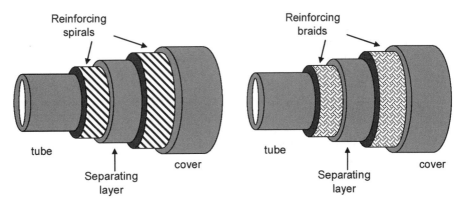

Figure 6 Hydraulic hose construction.

This core is then surrounded by one or more layers of reinforcing material that adds rigidity and strength to the hose. The figure on the left uses reinforcing wire spirals to strengthen the tube, while the one on the right uses reinforcing braids. The braids can be either metal or some synthetic fiber, whereas the spirals are primarily metal. The design of the reinforcing material ensures that the hose can be both strong and flexible. Wire reinforcement gives higher strength, whereas fiber reinforcement gives lower weight and greater flexibility. Flexibility is often desired in hydraulic hose for several reasons. First, it often needs to move with the system that it is attached to; for example, as cylinders extend on an earthmoving vehicle. Secondly, flexibility aids in the design and construction of the overall systems these hoses get integrated into because a flexible hose is easier to route throughout a machine. Finally, a small amount of flexibility with associated viscoelastic behavior can aid in the damping of pressure pulsations occurring in the system, although this usually means reduced efficiency due to hose losses. If maximum efficiency is desired, and flexibility is not needed, then the hoses can be replaced by solid metal tubing.

There are intermediate separating layers between the reinforcing layers and this is particularly true for wire reinforcements. The separation layer ensures that there is no fretting or wear due to relative motion of two wire layers adjacent to each other. If the reinforcing layers are made of a synthetic fiber braid, then the need for a separating layer is lessened. The total system is then housed by a cover layer that is primarily for protection and resistance to abrasion caused by the external environment.

As mentioned, the flexibility of the hoses depicted in Figure 6 can be a system advantage. However, the flexibility also means compliance in the hose. For example, Table 7

Table 7 Volumetric Expansion of Hydraulic Hose Types[5]

| Hose Model | Volumetric Expansion at Maximum Working Pressure | | Equation for Volumetric Expansion |
	psi	cm³/ft	$y = (cm^3/ft)$; $x = (psi)$
77C-06	4000	0.8	$y = 0.0002x$
77C-08	4000	1.2	$y = 0.0003x$
701-12	5000	7.42	$y = 0.0007x + 3.76$
701-8	6000	5.61	$y = 0.0004x + 3.295$
701-6	6500	5.6	$y = 0.0003x + 3.87$

illustrates the volumetric expansion for several particular hoses made by a major U.S. manufacturer.[5] This compliance can introduce an additional dynamic element to the overall system since the fluid inertia contained in the hose, along with the hose inertia itself, can couple with the hose compliance to produce resonant modes within the hose. Therefore, if pressure pulsations from a pumping device happen to occur at one of the resonant frequencies of the hose section, then severe performance degradation, and possible hose/connector failure, can occur. An excellent and detailed analysis on the resonant modes found in hydraulic hoses, termed transmission line dynamics, can be obtained in Ref. 3.

If the pressurized fluid is to be transported through hoses where the length is significantly greater than the inner diameter, then there is likely to be some power loss due to the friction between the fluid and the walls of the hose tube. This power loss manifests itself as a pressure drop or head loss between two points in the hose. A pressure drop or pipe head loss (h_l) correlation can be obtained by the Darcy–Weisbach equation[6]:

$$h_l = f \frac{L}{D} \frac{V^2}{2g} \tag{6}$$

where f = friction factor
  L = pipe length
  D = piper inner diameter
  V = fluid velocity
  g = gravitational constant (9.81 m/s^2)

The empirical relationship given in Eq. (6) is quite general and is valid for pipe flow of any cross section. It applies to both laminar and turbulent flow with the basic assumptions of a straight hose, a constant cross-sectional area, and a constant friction factor. The key term to be empirically determined is the friction factor f. There exist well-established correlations for both laminar and turbulent flow:

$$f = \begin{cases} \dfrac{64}{\mathrm{Re}_d} & \text{Hagen-Poiseuille relation (laminar flow)} \\[2ex] \dfrac{0.316}{(\mathrm{Re}_d)^{1/4}} & \text{Blassius relation (turbulent flow)} \end{cases} \tag{7}$$

The friction factor has been modeled in the literature by various methods and is dependent on factors like pipe roughness and flow condition. In addition, there exist correlations specifically for flow conditions that are in the transition between laminar and turbulent flow. The reader is referred to Ref. 6 for further details on pipe losses. This includes correction factors to account for bends in the piping, which cause additional losses due to momentum changes in the fluid. Although these pipe losses may be small, it is still important to minimize then to achieve maximum overall system efficiency.

3.3 Hydraulic Pumps

The purpose of a pump is to take mechanical power, in the form of a rotational shaft powered by a combustion engine or electric machine, and convert that to fluid power, in the form of high-pressure fluid moving at some flow rate. The majority of pumping devices in use are termed positive-displacement pumps. This means that for a given rotation of a pump's input shaft, there is an amount of fluid expelled out of the pump's discharge port that is primarily related to the size or displacement of the pumping components. The amount of fluid displaced per revolution is much less sensitive to aspects such as the pressure in the line resulting in the terminology of positive displacement. Two basic classes of pumps are available in current fluid power systems: fixed-displacement pumps and variable-displacement pumps.

Fixed-displacement pumps are usually less costly than variable-displacement pumps because they have fewer components. The most basic types of fixed-displacement pumps are gear pumps, vane pumps, lobe pumps, and piston pumps. Figure 7 demonstrates the working of the gear pump. As the input shaft turns, fluid on the suction side of the pump is pulled into the gears. Then small portions of fluid are carried over the gears, much like a water-wheel, to the discharge side of the pump. The space between each gear tooth acts as a separate volume to transport fluid from the low-pressure suction side to the high-pressure discharge side.

As the speed of the input shaft increases, the flow rate at the discharge port also increases. This increase is usually proportional to the speed across the rated range of the pump, and the proportionality constant is fixed for a given pressure at the discharge port. A similar type of action is responsible for the pumping behavior of both vane or lobe pumps. Figure 8 illustrates the constant proportionality between flow and speed for a vane pump. As can be seen, for a fixed discharge pressure the output flow scales linearly with the input shaft rotational speed.

Figure 9 shows a schematic similar to Fig. 7 for a vane pump and Fig. 10 does so for a lobe pump. Similar to the gear pump, the vane pump uses vanes or fins to sweep the fluid from the suction inlet around to the discharge section of the pump. The lobe pump uses the lobes much like the gear pump uses teeth to do the same thing.

The primary differences between these types of fixed-displacement pumps are the applications they are best suited for. For example, the gear pump is better suited for higher-pressure applications than the vane pump because of the ability of the gear teeth to hold higher loads than the vanes. However, the vane pump may be better suited for pumping fluids with low lubricity due to the reduced metal to metal contact. The relatively large volumes between the lobes on the lobe pump allow it to handle small bits of solids or contaminants better than the vane or the gear pumps.

Should high pressures be needed (e.g., 3000 psi or more), then the choice of positive-displacement pumps is the piston pump. In this subsystem, individual pistons move back and forth to pump fluid from the low-pressure side of the pump to the high-pressure side. Typical piston pumps are axial and radial piston pumps. A fixed displacement axial piston pump is usually constructed with the pumping pistons and input shaft offset by an angle. This is termed a bent-axis piston pump and is shown in Fig. 11. As the input shaft rotates, the distance between the piston and the intake section of the valve plate increases. This

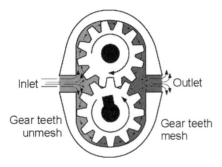

Figure 7   Gear pump schematic.

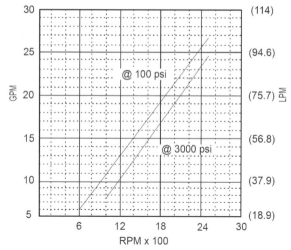

Figure 8   Parker PFVH25 series vane pump flow data (displacement = 40 cc/rev).

draws fluid into the piston chamber. After the piston has reached its maximum displacement it enters the discharge section of the valve plate where its distance to the plate decreases. This decrease forces fluid out of the high-pressure side of the pump.

As illustrated in Fig. 12, a radial piston pump operates similarly to an axial piston pump except that the pistons are now arranged radially. An eccentricity between the pump outer housing, or stator, and the piston housing, or rotor, performs the same actions as the swash-plate in terms of moving the pistons in and out of their pumping chambers. Reciprocating devices like piston pumps are very good for generating higher pressures than the continuous devices such as the vane or gear pumps and they tend to be more efficient. On the other hand, the gear and vane pumps tend to be quieter and less expensive to produce.

In many fluid power systems, there is a need to provide a relatively continuous supply pressure from the pump even while the amount of flow, and hence power, demanded can be varying. This would be analogous to having a constant electrical supply voltage while drawing a varying load. Fixed-displacement devices can solve this problem by running at varying speeds or by running at a fixed speed and incorporating a relief valve device at the pump

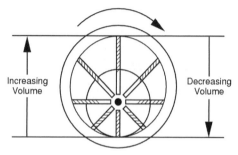

Figure 9   Vane pump schematic.

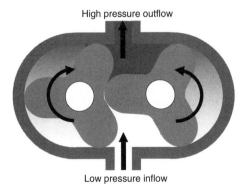

High pressure outflow

Low pressure inflow

Figure 10 Lobe pump schematic.

outlet. The relief valve simply diverts high-pressure flow back to the tank when it is not needed and is usually a cheaper option than the variable-speed device to drive the pump. Unfortunately, generating high-pressure fluid and then dumping it over a relief device is not a very efficient use of energy. Therefore, one option that is common is to have a variable-displacement pumping device. Of the different types of positive-displacement pumping devices given above, the most amenable to variation of displacement is the axial piston pump. A variable-displacement axial piston pump contains a swashplate, or wobble plate, attached to the pump input shaft. This allows variation of the relative angle of the pistons and the input shaft that the bent-axis configuration in Fig. 11 could not achieve. By varying the angle of the swashplate, it is possible to vary the amount of fluid that is drawn into and expelled from the pump during each revolution. Figure 13 shows a variable-displacement pumping device illustrating the movable swashplate.

The swashplate setting can be altered to affect the displacement. A very common usage for this type of variable-displacement device is to incorporate *load sensing* into its behavior. A mechanical feedback path from the pump outlet can be used to drive a spring-loaded pilot

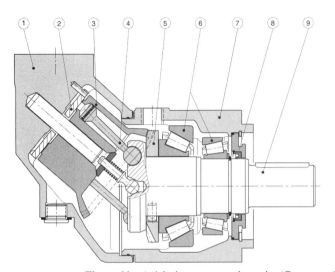

1. Barrel housing
2. Valve plate
3. Cylinder barrel
4. Piston with piston ring
5. Timing gear
6. Tapered roller bearing
7. Bearing housing
8. Shaft seal
9. Output/input shaft

Figure 11 Axial piston pump schematic. (Courtesy of Parker Hannifin, Inc.)

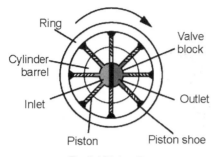

Radial Piston Pump

Figure 12 Radial piston pump schematic.

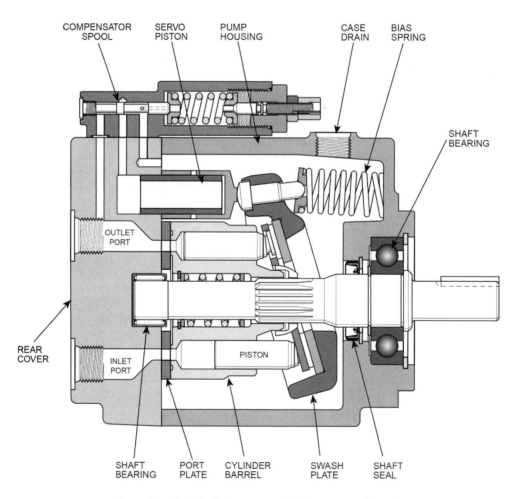

Figure 13 Variable displacement axial piston pump schematic.

valve, which in turn provides a flow to pivot the swashplate. Should the flow out of the pump be too great for the current application's usage, an excessive pressure buildup will occur at the pump discharge outlet. Through the pressure feedback mechanism, this will then cause the pump to destroke or reduce its displacement. In turn, this will lower the flow coming out of the pump per revolution and reduce the output pressure the level set by the spring preload on the pilot valve. A similar sequence of events will regulate the pump displacement to increase should the output pressure drop below some desired value. This mechanical feedback system can therefore provide the correct amount of fluid power for a given application without undersupplying or wasting any fluid that can lead to a large increase in efficiency. A block diagram of a load-sensing variable-displacement pump is shown in Fig. 14.

In addition to the mechanical feedback device, it is possible to have an electronically controlled pilot valve to drive the pump displacement. This electrohydraulic pump arrangement has an increased cost and complexity but also provides increased flexibility. It is possible to electronically sense the pump pressure output and feed that information back to the valve driving the swashplate displacement. This would mimic the mechanical load-sensing behavior mentioned earlier. In addition, direct electronic control of the pump displacement can be used to provide user-specified flow to a system such as a hydrostatic transmission.[7] The variable displacement can be used as a variable transmission to drive a hydraulic motor at different speeds.

When considering pump characteristics the end user should consider the efficiency of the pump, since that will have a direct effect on the overall efficiency of the system. Two types of efficiencies are associated with hydraulic pumps: volumetric efficiency and overall efficiency. Theoretically, the output flow of a pump would be equal to the pump rotational speed (Ω) multiplied by the pump displacement (D):

$$Q_{\text{theoretical}} = D\Omega \tag{8}$$

However, there are many internal passages within pumps and, with high pressure at the outlet, leakage can occur within these passages such that not all of the flow coming into the pump actually makes it to the pump exit. The volumetric efficiency of a pump is the ratio of actual output flow of the pump to the theoretical output flow:

$$\eta_{\text{volumetric}} = \frac{Q_{\text{actual}}}{D\Omega} \tag{9}$$

Volumetric efficiency is affected greatly by internal leakages, which are in turn affected by the pressure at the discharge of the pump. As discharge pressures increase, the volumetric efficiency of a pump will decrease. Due to good design and close manufacturing tolerances, modern pumps have volumetric efficiencies that tend to be quite high; usually above 90% for a properly functioning pump.

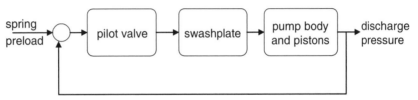

Figure 14 Load sensing axial piston pump diagram.

The overall efficiency of a pump is determined by the ratio between the mechanical power input (torque × speed) to the driving shaft and the fluid power (pressure × flow) that exits the pump discharge:

$$\eta_{\text{total}} = \frac{PQ}{\tau\Omega} \tag{10}$$

The overall efficiency of a pump is related to the volumetric efficiency but they are not the same. Overall efficiency is primarily governed by the energy losses associated with the pump. The principal losses are due to friction. This includes viscous friction, which hinders the movement of pump parts within the fluid, particularly if the pump parts are moving a thin film of hydraulic fluid. Energy losses are also due to friction contained in bearings and seals. The fluid viscosity has a large part to play in both the volumetric and overall efficiency. Low-viscosity fluid means easy pumping and a higher overall efficiency. However, low viscosity also means more internal leakage and therefore lower volumetric efficiency.

Typically, the construction of different pumps will affect their efficiency levels. All pumps have low overall efficiencies at flow rates far below their rated pressures. Typically, gear and vane pumps will have a lower overall efficiency at higher pressures than a piston pump. Figure 15 illustrates a typical performance chart for a variable displacement axial piston pump.

3.4   Hydraulic Valves

In a fluid power system, the hydraulic valve serves to restrict the flow of fluid through the system. In this sense, it is analogous to the electrical resistance as depicted in Table 6. This

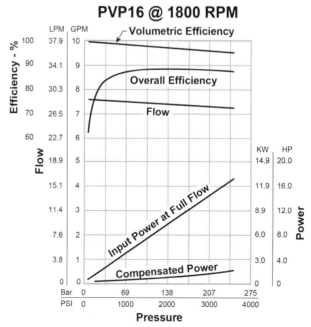

Figure 15   Performance of a pressure compensated variable-displacement axial piston pump.

flow restriction can be used to modulate or meter the amount of fluid going into the chamber of a cylinder or motor. Additionally, this restriction can act to divert the flow of fluid from one path to another. An example of this is the relief valve that diverts flow from the main pressurized line back to the low-pressure tank. Since all valves act as a restriction, it is simplest to consider them as orifices from the point of view of basic fluid mechanics.[6] If a further assumption is made that the fluid is incompressible, this allows for the simple approximation of the valve opening as a sharp-edged orifice. From Bernoulli's principle[6] we can determine the relative velocity downstream of an orifice if we are given flow conditions across the orifice. Considering u to denote upstream of the orifice and d to denote downstream we can calculate

$$v_d^2 - v_u^2 = \frac{2}{\rho}(P_u - P_d) \tag{11}$$

The fluid density is denoted as ρ and the pressures by P. The volumetric fluid flow can be determined from the fluid velocities (v) depicted in Eq. (11). It is essentially the velocity times the area of the pipe or channel carrying the fluid. Then, assuming incompressible flow,

$$A_u v_u = A_d v_d \tag{12}$$

This can be used with (11) to determine the velocity downstream:

$$A_u v_u = A_d v_d = Q = \frac{A_d}{\sqrt{1 - (A_d/A_u)^2}}\sqrt{\frac{2}{\rho}(P_u - P_d)} \tag{13}$$

Equation (13) is an idealization and it is necessary to add an empirical correction factor to account for the loss in energy due to the internal fluid friction. In addition, the downstream area used is the *vena contracta*[4] rather than the area of the actual orifice itself thereby necessitating another correction factor. These two correction factors can be lumped together into a *discharge coefficient, C_d*:

$$Q = C_d A_o \sqrt{\frac{2}{\rho}(P_u - P_d)} \quad \text{or} \quad Q = C_d A_o \sqrt{(P_u - P_d)} \tag{14}$$

Where A_o is the physical orifice area. Either one of the equations in (14) can be used as long as the correct definition of discharge coefficient accompanies it:

$$C_d = \frac{C_v C_c}{\sqrt{1 - C_c^2(A_o/A_u)^2}} \quad \text{or} \quad C_d = \frac{C_v C_c}{\sqrt{1 - C_c^2(A_o/A_u)^2}}\sqrt{\frac{2}{\rho}} \tag{15}$$

C_v ($\sim$1.0) is the correction coefficient due to internal friction losses, and C_c is the contraction coefficient, which is the ratio of the vena contracta to the orifice area A_o. The two empirical coefficients, C_v and C_c, are usually difficult to determine analytically. Oftentimes, the discharge coefficient is simply determined as a lumped parameter from calibrated pressure versus flow data. Reasonable values of C_d for most conditions would range between 0.5 and 0.8.

Many different types of valves are available. Three of the most common ones that will be mentioned briefly here are related to the shape of the metering orifice: the ball valve, poppet valve, and spool valve. The ball valve and poppet valve operate similarly in that there is an opening or orifice on which the ball or poppet is seated. The name of the valve comes from the device that seals the opening. Figure 16 illustrates both a ball valve and poppet valve. As either the ball or poppet lift off of the seat, an annular area is uncovered through which fluid can pass. This annulus is a metering orifice. Usually, flow will come

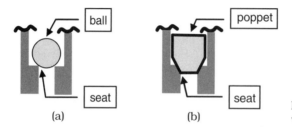

Figure 16 Ball valve (*a*) and poppet valve (*b*).

from the passage below the ball or poppet as drawn in Fig. 16. As will be shown shortly with a relief valve example, the ball or poppet is usually connected to a spring that applies a load attempting to keep the valve in a normally closed when insufficient pressure is present below the seat. The benefit of the poppet style valve over the ball valve is the ability to seal better with less leakage across the valve.

A third class of valve is called a spool valve. This is because the metering orifice is created by the sliding of a spool within a sleeve as illustrated in Fig. 17. The spool is usually cylindrical in geometry. As it moves within the sleeve it will cover or uncover different ports. In Fig. 17 the valve spool covers up the lower port while potentially allowing fluid to flow through the upper port. These ports are connected to different components within the overall fluid power system, thereby allowing this type of a valve to be used to redirect flow from one path to another. This will be further described shortly.

One common function for a ball or poppet type of hydraulic valve is the simple pressure relief valve. The simplest representation of a pressure relief valve is shown schematically in Fig. 18. When the pressure in the pressure port becomes large enough, the force on the ball will overcome the preloaded spring force and open a direct flow connection between the high-pressure line and the return line to the low-pressure tank or reservoir. Therefore, the pressure in the high-pressure line will be maintained at some preset level that is based on the spring preload. The spring preload may be set manually by means of a screw that would be located at the top of the valve. Alternatively, the spring preload could be set by a solenoid, which would enable an electronically adjustable relief pressure setting. This would result in a pressure control valve rather than just a relief valve.

Probably the most common function for a sliding spool valve is the directional control valve. This valve is called directional because is used to direct the flow from a high-pressure source, such as a pump, to a motor or cylinder to perform work on an external environment. A schematic diagram of a four-way directional control valve can be seen in Fig. 19, and its connection to a load was shown earlier in Table 4 and Fig. 2. By shifting the direction of

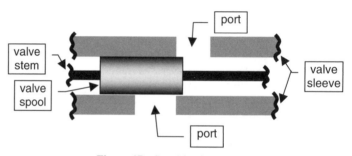

Figure 17 Spool-in-sleeve valve.

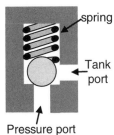

Pressure port

Figure 18 Relief valve schematic.

the valve, different connections between the pump, tank, and work ports (A and B) of the valve can be established. For example, if the valve in Fig. 19 is shifted to the right, there will be a connection between port A and the tank as well as a connection between port B and the high-pressure supply from the pump. Therefore, the fluid will flow into the side of the actuator connected with B causing motion to occur. A motion of the valve to the left will cause a connection between port B and the tank as well as port A and the supply.

Several different terms or characteristics can be used to describe valve construction and their function. As depicted in Table 4 valves can often defined by their *ways;* there are two-way, three-way, and four-way valves. A two-way valve is essentially an on–off valve. The valve is either opened or closed and the flow path cannot be changed. The flow simply moves between the pressure supply port and the exhaust port. The relief valve or ball and poppet valves illustrated in Figs. 16 and 18 are examples of two-way valves. A three-way valve is one that has a pressure port, an exhaust port, and a work port that meters fluid to some actuator in order to enable it to perform work. A four-way valve will have a supply port, an exhaust port, and two outlet ports. Figure 19 is an illustration of a four-way valve because there are 4 basic ports: 2 work ports plus a supply port (P) and an exhaust port (T). Here, P and T are commonly used to represent *pump* and *tank*.

Another valve characteristic associated with directional valves is the type of center configuration: open centered or closed centered. An open-centered valve has a direct path between the tank and the pump when the valve spool is centered. This is useful when the pump is fixed displacement because it minimizes the power loss from the pump during zero valve position. Also, the open-centered valves can provide a faster response from valve command input to flow output. Oftentimes open-centered valves are also referred to as *underlapped* valves. The closed-centered valves block the flow from pump to tank when the valve is in the centered position. This is inadvisable with a fixed-displacement pump because it ends up dumping high-pressure flow over the relief valve with a large waste of energy. In this case, a pressure-compensated pump is advised. Closed-centered valves are good for

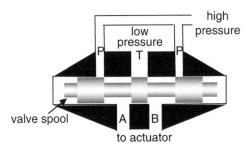

Figure 19 Four-way directional control valve.

holding loads in place, even when the pump is not providing power, because very little leakage can occur across the valve. Closed-centered valves are also called *overlapped* valves. In Fig. 19 the spool dimensions are drawn so as to just cover the pressure and return ports exactly. This is termed *critically lapped*. This means there is flow for a given valve motion away from center and no flow for valve at its center position. Table 8 illustrates the flow characteristics of differently lapped valves. It should be noted that the flow characteristics depicted do not differentiate whether the high-pressure fluid is directed toward port A or B; only the magnitude of the flow is shown. If the spool lands did not sufficiently block the P and T ports, the valve would be considered underlapped. For an underlapped valve, there will always be flow through the valve. This leakage flow can be undesirable in some cases because it represents power that is not utilized for useful work. If the spool lands completely covered the P and T ports with plenty of room to spare, the valve would be considered overlapped. Overlapped valves minimize leakage through the valve, thereby leading to higher overall system efficiency. However, the cost of doing so is to introduce a deadzone or dead-band in the relationship between valve motion and flow. This deadband can have serious consequences for closed-loop control schemes utilizing this valve.

As illustrated in Table 4, there is more than one way to move a valve to generate flow. Manual control of valve motion can be accomplished by connecting the valve, through some type of linkage, to a lever or device that can be operated by a human. Figure 20 illustrates a Parker Hannifin D1VL-series four-way directional control valve operated by a lever. The valve is centered by springs on either side of the valve spool. By shifting the lever in or out, flow can be directed from a high-pressure source to either the A or B work ports. Other manual methods for moving a spool include foot pedals and buttons (for two-way valves).

If electronic control or automation is required from the fluid power system then the input can no longer be manual but must be generated electronically. This leads to electro-

Table 8 Configurations for 4-Way Directional Control Valves

Valve Lap	Valve Schematic	Valve Flow
Underlapped		Leakage flow around null
Overlapped		deadzone
Critically lapped		

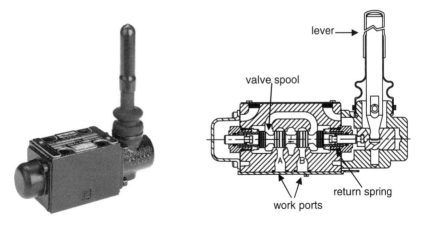

Figure 20 Manual four-way directional control valve. (Courtesy of Parker Hannifin, Inc.)

hydraulic fluid power systems. The simplest and most common electronic actuator to generate valve motion or spring preloading is a solenoid. A solenoid consists of a wire coil surrounding a metal plunger. When current flows through the solenoid, it generates a magnetic field that produces a force on the plunger. Depending on the configuration of the solenoid, its energized state can either retract or extend the plunger. Often, for directional control valves, two separate solenoids are used as shown in Fig. 21 for a Parker Hannifin D1FW-series valve. The left solenoid is used to actuate the valve to the right in the figure and vice versa for the other solenoid. If neither solenoid is engaged, the return springs on either side of the valve will bring it back to a neutral position. The use of dual solenoids is due to the fact that it is simpler to have two single-action solenoids than one electromagnetic device that can retract and extend the plunger attached to the valve spool. For a two-way valve, such as a relief valve, a single solenoid is sufficient for it acts only to provide a preload to the spring that seats the valve.

Although solenoids are rugged and cost-effective, one disadvantage is their nonlinear behavior. As the plunger moves further out of the energized coil, the electromagnetic force decreases in a nonlinear fashion. This is shown schematically in Fig. 22. Additionally, the force generation characteristics associated with the solenoid are greatly affected by temperature, which can change if the solenoid is repeatedly energized or held on for an extended period of time. This temperature dependence is also a nonlinear one. Vaughan and Gamble[8] develop a good model of solenoid characteristics for use in an electrohydraulic system. These

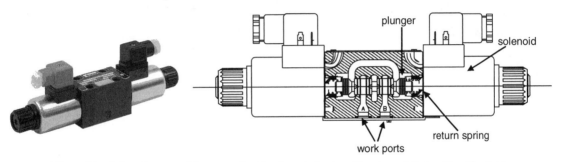

Figure 21 Solenoid operated four-way directional control valve. (Courtesy of Parker Hannifin, Inc.)

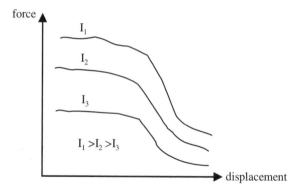

Figure 22 Nonlinear force generation characteristics for a solenoid.

shortcomings make it difficult to provide high-performance precision control from a simple solenoid valve. One method to compensate for this is to use a more expensive direct-drive linear motor in place of a solenoid. A single direct-drive motor can actuate a plunger in two directions. Figure 23*a* shows a cross section of a direct-drive linear motor for electrohydraulic valve actuation. Figure 23*b* shows the single-sided attachment of the linear motor to a valve spool. This type of arrangement gives a much more linear response than the solenoid depicted in Fig. 21.

Another method to compensate for the valve nonlinearity is to close a feedback loop locally about the valve position. Figure 24 shows a schematic of a valve where a lower-precision solenoid actuation is used to move a spool valve. However, the nonlinearity of the solenoid is compensated by the closed loop feedback of the spool position using an integrated position sensor. The actual control electronics depicted in Fig. 24 can be either digital or analog, depending on the sophistication of the valve. Proportional valves with closed-loop feedback are often called servo valves. This term comes from the fact that the valve position

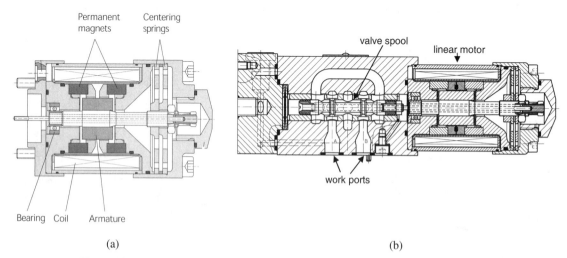

Figure 23 Direct–drive linear motor for electrohydraulic valve. (Courtesy of Moog, Inc.)

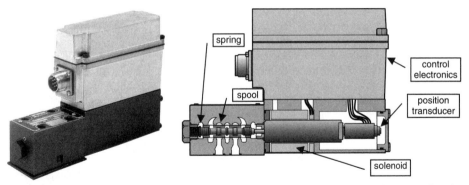

Figure 24 Parker D1FH proportional valve with closed loop spool feedback. (Courtesy of Parker Hannifin, Inc.)

is only a function of the command given. The closed-loop electronics eliminate nearly all of the variability in spool position as a function of commanded reference. Additionally, this effect can be enforced up to very high-frequency motions. Figure 25 illustrates a frequency response of a particular proportional servo valve for different amplitudes of motion. As can be seen, the valve responds reliably to reference command changes up to 100 Hz for small signal amplitudes, which is fast enough for many applications. In addition to being fast, this electronic feedback allows the valve to be programmable. For instance, the valve motion could be made nonlinear with respect to command so that it was very sensitive to commands (high gain) around null and less sensitive (low gain) once the valve was open a certain percentage.

In addition to the electronic position sensing feedback for valve positioning, it is also possible to have an internal mechanical valve feedback. Typically this is obtained by a

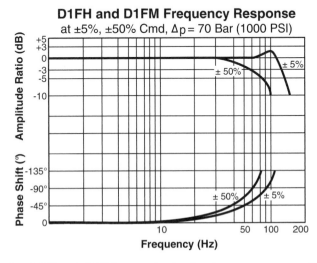

Figure 25 Parker D1FH and D1FM proportional valve frequency response. (Courtesy of Parker Hannifin, Inc.)

mechanical feedback device, such as a spring, between a light and fast first stage and a heavier second stage that contains the actual spool. Figure 26 shows a classical servo valve that operates with a flapper-nozzle arrangement.[9] The fast first stage is actually a torque motor that provides a rotation for a given torque input. This electrical effect produces flapper motion that uses differential pressure to generate a hydraulic amplification, which then affects the spool motion. The motion of the main spool is communicated back to the first-stage torque motor by means of a slim wire attached to the middle of the main spool. The flexure of this wire has a force effect on the spool that is proportional to displacement, thereby providing a proportional mechanical feedback. These types of valves can be designed to have very rapid responses, even beyond those depicted in Figs. 23 and 24. However, they are susceptible to contamination because of the very fine manufacturing tolerances. These manufacturing tolerances also make them generally more expensive than their electronic feedback counterparts. There are other mechanical servo valve designs using jet pipe feedback or other means. The reader is referred to Refs. 4 and 9 for further insight into mechanical servo valve design and construction.

The basic idea behind the mechanical servo valve in Fig. 26 is the idea of a multistage hydraulic valve. This idea is used often used when a fast response is needed in conjunction with a high flow rate. A high flow rate for a directional valve indicates a large spool inertia. For example, the directional valves on large earthmoving equipment can have spools weighing several kilograms. Solenoids would not be sufficiently powerful to move such an inertia or resist the flow forces[4] that would arise. Therefore, it is common in these cases to use a small valve to drive a large valve. This is depicted conceptually in Fig. 27. The small quick valve utilizes the hydraulic amplification of pressure against the second-stage spool end to gain a valve system with high flow and fast response.

3.5   Cylinders and Motors

Hydraulic cylinders and motors are the components that convert fluid power into mechanical power by applying forces and torques to some load environment. Hydraulic cylinders convert

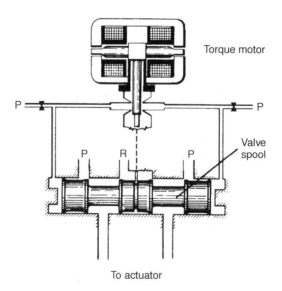

Figure 26   Two-stage flapper-nozzle servo valve. (Courtesy Moog Inc.)

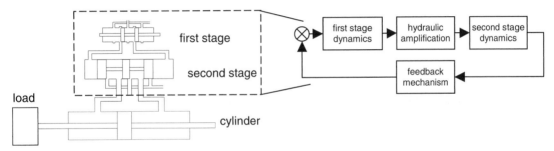

Figure 27 General multistage valving.

fluid power into linear mechanical power and hydraulic motors convert fluid power into rotary mechanical power. The basic hydraulic cylinder consists of a piston and a rod inside of a sleeve or barrel with two end caps. This configuration was shown schematically in Table 5. A more detailed schematic is shown in Fig. 28 for a single-rod cylinder.

The piston can have rods on either end (double acting) or only on one end (single acting) although only one rod is shown in Fig. 28. The price to be paid for a double-acting rod is a more complicated and expensive device that occupies more volume, since allowances have to be made for extensions of rods in two directions. The benefits to be gained are uniformity in behavior during retraction and extension. The hydraulic cylinder is capped at two ends and, as shown in the figure, sometimes reinforced by tie rods that hold the end caps on and provide additional strength. Ports at either end of the cylinder allow fluid to flow into and out of the actuator on either side of the piston. High-performance seals between the piston circumference and the cylinder body ensure that minimal amounts of fluid leak across the piston between the low- and high-pressure chambers. There are also seals on the end cap through which the rod extends to prevent the leakage of fluid to the external environment. Additionally, modern cylinders often have wipers at the seal interface between the rods and the end caps to ensure that dirt and contamination from the external environment do not find their way into the hydraulic fluid.

Figure 28 shows a relatively simple schematic of a hydraulic cylinder. In addition to the aspects shown above, the normal seals on the piston can be replaced by hydrodynamic seals if seal friction is to be reduced for high-performance systems. Additionally, many

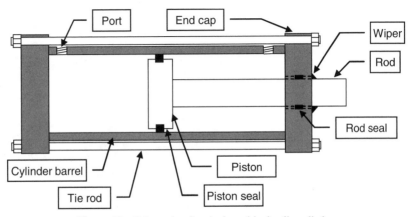

Figure 28 Schematic of a single-rod hydraulic cylinder.

hydraulic cylinders incorporate "snubbers" or hydraulic cushioning devices to minimize shock induced by reaching the end of the cylinder's stroke. Further details may be found in Ref. 2.

Hydraulic motors primarily come in several varieties. In most cases, the motors simply act like pumps running in reverse. For example, consider the gear motor schematic shown in Fig. 29. The high-pressure inlet flow acts on the lower sets of gear teeth and forces them to rotate in the directions of the arrows. The segments of two meshing teeth tend to oppose rotation, making the net torque available be an effective function of one tooth. Pressure on the outer gear teeth that ride along the casing of the motor is effectively neutralized because of the identical teeth. The net torque from this section is zero as the oil is carried around and over to the exit port. The teeth at the top of the motor have only tank low pressure that they face to oppose them, which leads to the pressure difference between the inlet high pressure and the outlet tank pressure to provide torque to move the gears. For the actual motor, one of the gears would be coupled to an output shaft that would rotate and provide the external torque to whatever load was attached. As we can see, the basic principle is exactly the same as the gear pump except the input and output are reversed. For the gear pump, the input is shaft torque to the gears and the output is the high-pressure fluid. For the gear motor, the input is the high-pressure fluid and the output is the shaft torque.

Similar to the gear motor, most other hydraulic motors act with a dual relationship to their hydraulic pumping counterparts. For example, the hydraulic axial piston motor uses high-pressure fluid to force pistons to move. The piston motion causes a swashplate to rotate. Additionally, the pitch of the swashplate can be made variable, just as in a pump, so that the displacement per revolution of the motor can be altered. This would result in a variable-displacement motor that can give a significant control degree of freedom to an overall system. Since the basic behavior of the other types of motors (e.g., axial-piston, radial-piston, vane, etc.) are so similar to their pumping counterparts, the reader is referred to Ref. 2 for further detailed discussion.

3.6   Other Components

There are several components, other than the primary ones described above, that make up an overall fluid power system such as the one depicted in Fig. 1. Without these other components, the behavior of most fluid power systems would be either impossible or very sub-

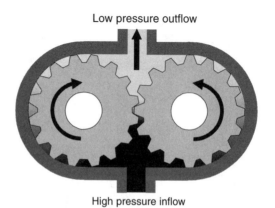

Figure 29   Schematic of a gear motor.

optimal. One of the key components in many fluid power systems is the accumulator. This is effectively the same as an electric capacitor and is used for direct storage of fluid power in the form of highly pressurized fluid. Physically, an accumulator consists of a chamber with a movable interface separating the hydraulic fluid from some other type of gas, often nitrogen. There are two primary types of accumulators, piston and bladder types. Figure 30 shows schematics of both types of systems. For the piston types of accumulators, a linear spring element could be substituted instead of the inert gas to act as an energy storage element.

Accumulators serve many different purposes in fluid power systems. If they are placed immediately downstream of a positive displacement pumping device, such as an axial piston pump, they can act to smooth out pressure ripples that may come from the individual piston strokes. Schematically, this is shown in Fig. 31. In addition to smoothing out pressure pulsations from pumps, accumulators can be placed immediately upstream of valves. In doing so, they will absorb any system shocks that might occur from sudden valve closings. This is often called surge compensation. Accumulators can also be used to store fluid power that can be used to briefly run safety critical systems in the event of a power failure. In this role, the energy stored in the accumulator can be used to safely bring the system or machine to a shutdown condition in a properly controlled fashion. Finally, accumulators can often be used to provide additional flow to a system so that maximum flow available is greater than the instantaneous flow the pump can provide. Therefore, mobile systems, such as aircraft, can be made lighter and more efficient by using smaller pumping and reservoir components.

Another component associated with fluid power systems that should be considered in the overall design is the tank or oil reservoir. The tank serves as more than a repository for oil. Tanks must be appropriately sized so as to provide a suitable heat sink for rejection of the heat developed as the fluid is worked throughout the system. As the fluid goes over several pressure drops it can become heated. As it returns to the tank, it has the opportunity to cool down before going into the system again. It is often sufficient in stationary applications to size the tank appropriately to be an effective source of heat rejection using passive means. However, in mobile applications such as aerospace where space is at a premium, active cooling by means of embedded heat exchangers is often necessary to get by with a smaller tank size.

Several of the components mentioned in this section are given to the reader in an introductory form. That does not mean that they are not crucial to the overall system performance. Rather, given the compact nature of this chapter, there is insufficient space to detail all of the system components in depth. However, there are several excellent texts the reader may turn to for further details on these components and others.[2–4] Some of the other components include filters for keeping contamination out of the oil and couplings for connecting hoses to components.

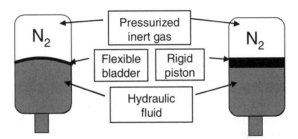

Figure 30 Bladder (*left*) and piston (*right*) types of accumulators.

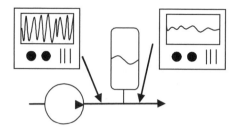

Figure 31   Filtering action of an accumulator placed at a pump exit.

4   SYSTEM DYNAMIC BEHAVIOR

The previous section detailed several of the components that make up a fluid power system. These components are useful only when they are combined and integrated into an overall system. As mentioned at the beginning of this chapter, care must be exercised when connecting each of these subsystems since the interconnected system can exhibit dynamics that were not apparent in the individual subsystem.

Each component has its own dynamic behavior that can be examined separately before interconnection. Here we can consider the dynamics of a variable displacement axial piston pump. The simplest dynamic models are the linear input–output transfer functions.[10] Under many situations, the pump can be treated as a constant gain between input rotational speed and output flow. This is the case when the pump is operating in a steady-state environment:

$$Q_{\text{pump}} = D_{\text{pump}}\omega_{\text{pump}} \tag{16}$$

As mentioned in relation to Fig. 31, slight variations in flow due to individual motion of the pistons are usually smoothed out by the inclusion of an accumulator at the pump output. There are cases where the pump displacement D_{pump} can be varied to change the output flow in a controlled manner. For these systems, the pump displacement is varied by changing the swashplate angle, termed here as α:

$$D_{\text{pump}} = K_{\text{displ}}\alpha \tag{17}$$

In electrohydraulic pumps, the swashplate angle can be changed electronically by means of a solenoid valve providing a pilot force to the swashplate. The simplified swashplate dynamics can usually be modeled as a simple second-order transfer function[11] about some nominal operating position, α_0:

$$\left.\frac{\alpha(s)}{u(s)}\right|_{\alpha_0} = \frac{K_{\text{swashplate}}(\alpha_0)}{s^2 + 2\xi(\alpha_0)\omega_n(\alpha_0)s + \omega(\alpha_0)_n^2} \tag{18}$$

The actual system dynamics are nonlinear, so Eq. (18) is valid for a specific displacement angle since the damping ratio and natural frequency often change as a function of the swashplate displacement angle. The input signal is usually amplified and sent to a control valve, which operates against a spring-loaded hydraulic amplification to affect a net torque on the swashplate. Depending on the internal pump configuration and the operating pressure set by the relief valve, the dynamics can be either underdamped or overdamped. Many manufacturers design the pump to be significantly overdamped and so the dynamics can be further simplified as a first-order system if the faster pole is neglected. Additionally, flow limitations in the internal pump feedback often will limit the pump to a rate-limited dynamic behavior.

For electronically driven single-stage valve components with centering springs the valve usually acts as a simple second-order system. Therefore, the relationship between the valve motion (x_v) and the input signal from some pulse width modulated amplifier is given as

$$\frac{x_v(s)}{u(s)} = \frac{G_v}{m_v s^2 + b_v s + k_v} \tag{19}$$

where G_v = valve gain
$\quad m_v$ = valve mass
$\quad k_v$ = valve centering spring
$\quad b_v$ = valve damping
$\quad u$ = control voltage

A two-stage servo valve, such as the one shown in Fig. 26, possesses different dynamics than the single-stage directional valves. A simplified block diagram is given in Fig. 32,[9] with the related system parameters given subsequently:

$\quad K_t$ = torque motor gain
$\quad K_h$ = hydraulic amplification gain
$\quad A_v$ = spool end area
$\quad k_f$ = net stiffness on armature
$\quad \omega_n$ = 1st-stage (flapper) natural freq.
$\quad \zeta$ = 1st-stage (flapper) damp. ratio
$\quad K_w$ = feedback wire stiffness

This representation ignores the nonlinear effects of the two-stage valve such as flow saturation in the first stage due to flapper rotation limits, but is useful for analyzing small signal valve response. The transfer function from input current, $i(s)$, to output valve position, $x_v(s)$, can be written as

$$\frac{x_v(s)}{i(s)} = \frac{\left(\dfrac{K_t K_h}{K_f A_v}\right)}{s\left(\dfrac{s^2}{\omega_n^2} + \dfrac{2\zeta}{\omega_n} s + 1\right) + \left(\dfrac{K_t K_h}{K_f A_v}\right)} \tag{20}$$

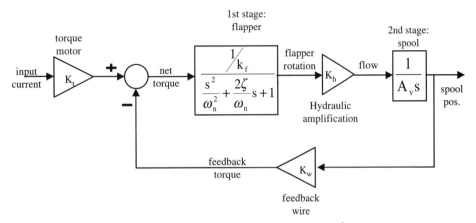

Figure 32 Two-stage servo valve dynamics.[9]

The simplified transfer function for a two-stage servo valve usually has three poles, two complex and one real. The response due to the two complex roots is often much faster than the response due to the real root. Consequently, the model could be simplified to a first-order system for lower-frequency analysis.

Examples of the components given above can be put together to form an overall system to perform a given task. Consider, for example, the electrohydraulic valve-driven, double-ended hydraulic cylinder moving an inertia with both damping and a spring resistance. The overall block diagram for the system's dynamics appears as shown in Fig. 33,

where A = piston area
$\quad\quad m$ = load mass
$\quad\quad b$ = viscous damping on load
$\quad\quad x$ = cylinder position
$\quad\quad F$ = cylinder force generated

In the absence of load forces, the transfer function from valve command input to cylinder position output is given as a fifth-order transfer function:

$$\frac{x(s)}{u(s)} = \frac{1}{m_v s^2 + b_v s + k_v}$$

$$\times \frac{K_q/A}{\left(\dfrac{V_t m}{\beta_e A^2}\right)s^3 + \left(\dfrac{(K_c + C_{tm})m}{A^2} + \dfrac{bV_t}{\beta_e A^2}\right)s^2 + \left(1 + \dfrac{(K_c + C_{tm})b}{A^2} + \dfrac{kV_t}{\beta_e A^2}\right)s + \dfrac{(K_c + C_{tm})k}{A^2}}$$

$$(21)$$

As can be seen by Eq. (21), the dynamics of this relatively simple fluid power system can be quite complex, with dynamic behavior more intricate than the behavior of a single component. If one were to try and close a control loop around this system, one would have to take into account all the dynamics associated with the system. Moreover, if this simple cylinder linear motion system were to be incorporated into an overall system, such as a primary flight control surface on an aircraft, it could become even more complex. A thorough knowledge of system dynamics is an engineer's best tool in this situation.

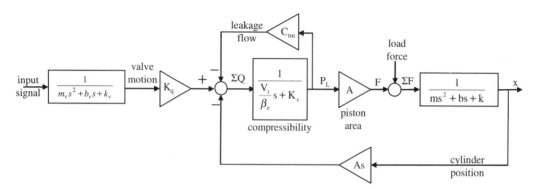

Figure 33 Linearized dynamics of a hydraulic cylinder.

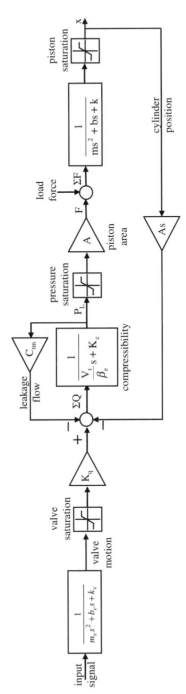

Figure 34 Saturations in hydraulic cylinder model.

5 COMMON NONLINEARITIES

The description of fluid power systems in the previous section was simplified to determine linear, time-invariant, dynamic models. There are some key nonlinearities that are common in fluid power systems and greatly affect their dynamic performance. Here we mention several of the major ones that can affect dynamic system performance.

5.1 Saturations and Deadzones

One of the primary nonlinearities in any hydraulic system is saturation. This can occur at several places. The valve has a finite displacement, which leads to a saturation. Similarly, a cylinder will have a constraint on its motion. Finally, the system pressure is usually limited to be below 3000 psi in most operating systems through the use of a relief valve at the pressure supply as shown in Fig. 5. If these saturations were to be incorporated into the dynamics of a valve-controlled cylinder, the modified block diagram would be as shown in Fig. 34.

Another nonlinearity mentioned in the section on components is the characteristic associated with the valve's center position. As mentioned earlier, the valve can be underlapped, critically lapped, or overlapped. This corresponds to the covering of the ports by the spool lands. An overlapped valve, while minimizing fluid leakage and hence system power, does produce a deadzone in the relationship between valve displacement and fluid flow. An underlapped valve allows flow between the pump and tank even when the valve is in its center position thereby wasting some energy. However, an underlapped valve has a higher flow gain at low valve displacements, thereby making the system response faster. Figure 35 depicts flow gains for these different conditions.

5.2 Hysteresis and Asymmetry

Another nonlinear valve characteristic that will be present is the hysteresis in the valve motion. This hysteresis can be significant for low-performance, solenoid-operated, proportional valves. However, for most high-performance servo valves or electronically controlled directional valves the hysteresis problem is negligible. An additional nonlinearity of fluid

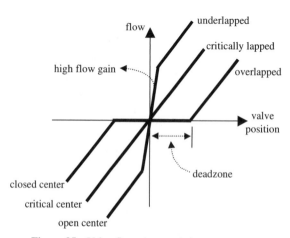

Figure 35 Valve flow characteristics near center.

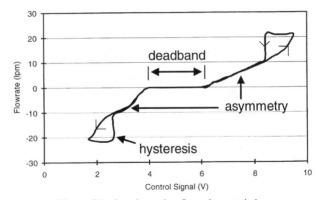

Figure 36 Steering valve flow characteristics.

power systems is their asymmetry. Frequently these systems are asymmetric, due to either intentional design or imperfections in the actual components. The single-ended cylinder of Fig. 28 is intentionally designed to have a different flow-to-velocity characteristic during retraction and extension to minimize the space necessary for the actuator. Quite often, valves will display asymmetrical flow profiles from valve position to flow. This can be done intentionally, to match a single-ended cylinder with an asymmetric valve in order to have an input–output system relationship that is symmetric. The valve flow characteristics can also be asymmetric due to imperfectly manufactured valves, valve tolerances, or component wear. Figure 36 shows the flow rate response of a large electrohydraulic steering valve for an 8000-kg farm tractor.[12] This is a solenoid-operated valve; therefore, it has a large amount of hysteresis. The asymmetry in the flow characteristic is very easy to see, as is the deadband. Figure 36 also illustrates a valve that shifts entirely to one side if the input signal is zero. The spring loading is only on one side, so a nonzero input is required to maintain zero flow.

5.3 Friction

The final nonlinearity that is discussed is friction. This is present to some degree in all fluid power systems, particularly in piston or motor seals as well as the sliding valve spool itself. As with purely electromechanical systems, it can be the key barrier to precise motion control. For all its importance, there are still widely varying models of friction for systems analysis. The simplest is the Coulomb model. However, this does not capture the complicated stick-slip motion of friction. This stick-slip motion necessitates the inclusion of a presliding displacement in the frictional model. A good analytical model that captures the stick slip phenomenon as well as the dynamic behavior of the sliding friction is given by Lischinsky et al.[13]

REFERENCES

1. D. Billington, *The Innovators: The Engineering Pioneers Who Made America Modern,* Wiley, New York, 1996.
2. F. Yeaple, *Fluid Power Design Handbook,* 3rd ed., Marcel Dekker, New York, 1996.
3. J. Watton, *Fluid Power Systems: Modeling, Simulation, Analog and Microcomputer Control,* Prentice-Hall, Englewood Cliffs, NJ, 1989.
4. H. E. Merritt, *Hydraulic Control Systems,* Wiley, New York, 1967.

5. Parker Hannifin Hose Products Tech Alert #42, Parker Hannifin Corp., 2004.

6. F. White, *Fluid Mechanics,* 5th ed., McGraw Hill, New York, 2002.

7. R. Zhang, A. Alleyne, and E. Prasetiawan, "Modeling and H-2/H-infinity MIMO Control of an Earthmoving Vehicle Powertrain," *ASME Journal of Dynamic Systems, Measurement, and Control,* **124,** 625–636 (2002).

8. N. D. Vaughan and J. B. Gamble, "The Modelling and Simulation of a Proportional Solenoid Valve" *ASME Journal of Dynamic Systems, Measurement, and Control,* **118,** 120–125 (1996).

9. W. Thayer, "Transfer Functions for Moog Servovalves," Moog Technical Bulletin #103, 1965.

10. G. Franklin, J. Powell, and A. Emani-Naeini, *Feedback Control of Dynamic Systems,* 4th ed., Prentice-Hall, Upper Saddle River, NJ, 2002.

11. G. Schoenau, R. Burton, and G. Kavanagh, "Dynamic Analysis of a Variable Displacement Pump," *ASME Journal of Dynamic Systems, Measurement, and Control,* **112,** 122–132 (1990).

12. T. Stombaugh, *Automatic Guidance of Agricultural Vehicles at Higher Speeds,* Thesis, University of Illinois, Urbana, 1997.

13. P. Lischinsky, C. Canudas de Wit, and G. Morel, "Friction Compensation for an Industrial Hydraulic Robot." *IEEE Control Systems Magazine,* **19,** 25–32 (1999).

CHAPTER **30**

AIR POLLUTION CONTROL TECHNOLOGIES

C. A. Miller
U.S. Environmental Protection Agency
Research Triangle Park, North Carolina

1 INTRODUCTION

Air pollution and its control have been of interest since the advent of the Industrial Revolution. Although it has been recognized for several hundred years that poor air quality is harmful to health, development of technologies to control pollutant emissions has been an important engineering topic for only the past few decades. Many human activities generate airborne materials that pose risks to health or environmental quality. Particularly in the last 40 years, steps have been taken to reduce the release of these materials into the ambient air by modifying these activities or by installing technologies to reduce the emission of air pollutants. Air pollution sources are generally divided into two major categories, mobile and stationary, although area sources (sources that do not have a distinct point of emissions) are often considered to be a third category. This chapter discusses the principles and applications of technologies for reducing air pollution emissions from stationary point sources, with an emphasis on the control of combustion generated air pollution. Major stationary sources include utility power boilers, industrial boilers and heaters, metal smelting and processing plants, and chemical and other manufacturing plants.

Of primary concern are those pollutants that, in sufficient concentrations in the ambient air, adversely impact human health and/or the quality of the environment. Those pollutants for which health and environmental protection criteria define specific acceptable levels of ambient concentrations are known in the United States as "criteria pollutants." These pollutants are carbon monoxide (CO), nitrogen dioxide (NO_2), ozone, particulate matter (PM) smaller than 10 μm in aerodynamic diameter (PM_{10}) and smaller than 2.5 μm in aerodynamic diameter ($PM_{2.5}$), sulfur dioxide (SO_2), and lead (Pb). Ambient concentrations of NO_2 are usually controlled by limiting emissions of both nitrogen oxide (NO) and NO_2, which combined are referred to as oxides of nitrogen (NO_x). NO_x and SO_2 are important in the formation of acid precipitation and $PM_{2.5}$, and NO_x and volatile organic compounds (VOCs) can react in the lower atmosphere to form ozone, which can cause damage to lungs as well as to property.

Other compounds such as benzene, polycyclic aromatic hydrocarbons (PAHs), other trace organics, and mercury and other metals, are emitted in much smaller quantities, but are more toxic and in some cases accumulate in biological tissue over time. These compounds have been grouped together as hazardous air pollutants (HAPs) or "air toxics," and have been subject to increased regulatory control over the past decade. Of increasing interest are emissions of compounds such as carbon dioxide (CO_2), methane (CH_4), or nitrous oxide (N_2O) that have the potential to affect the global climate by increasing the level of solar radiation trapped in the Earth's atmosphere, and compounds such as chlorofluorocarbons (CFCs), which react with and destroy ozone in the stratosphere, reducing the atmosphere's ability to screen out harmful ultraviolet radiation from the sun.

Air pollution control requirements in the United States can be set at the federal, state, or local levels. At the federal level, the Clean Air Act and subsequent amendments are the most important laws for the majority of pollutants and emission sources. Other requirements are set by the Resource Conservation and Recovery Act and other legislation concerned with hazardous waste. Generally, state requirements may be equally or more stringent than federal requirements.

The U.S. Environmental Protection Agency (EPA) identifies acceptable concentrations of pollutants in ambient air, based on scientific evidence of health or environmental damage caused by exposure to the specific air pollutant. EPA then sets national ambient air quality standards (NAAQS) for the criteria pollutants, and the states and localities then determine the appropriate methods to achieve and maintain those standards. Emission limits for specific source types can therefore vary across states or localities, depending on the need to reduce ambient pollutant levels. EPA also sets minimum pollution performance requirements for new pollution sources, known as new source performance standards or NSPS. For some pollutants (such as HAPs), EPA is required by law to set limits on the annual mass of emissions to reduce the total health risk associated with exposure to these pollutants. Other approaches include the limiting of the total national mass emissions of pollutants such as SO_2; this allows emissions trading to occur between different emission sources and between different regions of the country while ensuring that a limited level of SO_2 is emitted into the atmosphere and thus reducing the formation of acid precipitation and $PM_{2.5}$.

The regulatory structure can significantly impact the type of emission control technology that is appropriate for a given situation. The averaging time over which emissions are reported (hourly, daily, annually, etc.), the number of permitted exceedances of the emission limit, and the penalties for violating the permit can all have significant implications for control technology system design and operation. These factors should be as clearly understood as possible when evaluating control technology options.

2   CONTROL OF SULFUR OXIDES

Sulfur oxide emissions are most often the result of burning a fuel that contains relatively minor concentrations of sulfur. The most common sulfur oxides are SO_2 and sulfur trioxide (SO_3), with SO_2 making up by far the greater total emissions. Of the total 18.8 million tons of SO_2 emitted in 1999, 82% was from stationary point source combustion processes. The remainder is largely split between noncombustion industrial processes and transportation. With the large fraction of emissions being from stationary point sources, flue gas desulfurization (FGD) methods are widely used to reduce SO_2 emissions. Conventional FGD systems use relatively mature technologies capable of SO_2 reductions greater than 90%.

There are several commercially available FGD processes. These can be grouped as either wet or dry, and as once-through or regenerable. Wet FGD systems produce a wet slurry waste or by-product, and the flue gas leaving the absorber is saturated with water. In dry processes, dry waste material or by-product is produced, and flue gas leaving the absorber is not water-saturated. Once-through processes generate a waste or by-product from the sorbent material, while in regenerable systems, the sorbent is regenerated for further use. As the sorbent is regenerated, SO_2 is released from the sorbent and may be further processed to yield sulfuric acid, elemental sulfur, or liquid SO_2. Figure 1 provides an overview of FGD technology types.[1]

To achieve high levels of SO_2 reduction, wet FGD processes must ensure good contact between the scrubber liquid and the flue gas. This can be accomplished using a variety of approaches. A venturi scrubber (see Fig. 2) uses a narrowing of the flue gas flow path to confine the gas path. At the narrowest point, the scrubber liquor is sprayed into the flue gas, allowing the spray to cover as great an area of gas flow as possible.

Perforated plate scrubbers (Fig. 3a) usually are designed with the gas flowing upward and the liquid flowing in the opposite direction. The flow of the gas through the perforations is sufficiently high to retard the counterflow of the liquid, creating a liquid layer on the plate through which the gas must pass. This ensures good contact between the liquid and gas phases. Bubble cap designs also rely on a layer of liquid on the plate, but create the contact of the two phases through the design of the caps. Gas passes up into the cap and back down through narrow openings into the liquid. The liquid level is regulated by overflow weirs, through which the liquid passes to the next lower level. The gas pressure drop in this type of system increases with the height of the liquid and the gas flow rate.

Packed bed scrubbers (Fig. 3b) utilize a bed of inert material, often plastic, to facilitate mixing of the scrubber liquor and the exhaust gas. The gas flows upward through the packing as the liquor flows downward. Both gas and liquor are forced to flow over the packing material, which improves the gas–liquid contact and thus, the rate of reaction between the pollutant and the scrubber liquor. The primary requirements for the packing are to evenly distribute the gas and liquid across the tower cross section, provide adequate surface area for the reactions to occur, and allow the gas to pass through the bed without excessive pressure drop.

While the focus of this discussion is removal of SO_2, other water-soluble acid gases such as hydrogen chloride (HCl) can be controlled using scrubbers of similar fundamental design, although the alkaline liquor may differ from what is used to control SO_2.

Roughly 70% of wet once-through FGD scrubber capacity for utility applications uses limestone (primarily calcium carbonate, $CaCO_3$) in aqueous slurry as the scrubbing liquor, approximately 15% of capacity uses lime (CaO), and the majority of the remaining wet FGD

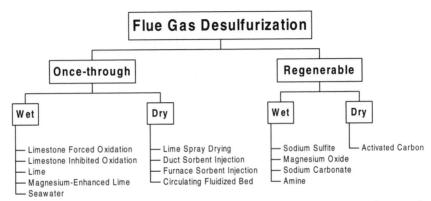

Figure 1 Types of flue gas desulfurization technologies. The most common types in use today are once-through, wet systems using lime or limestone.

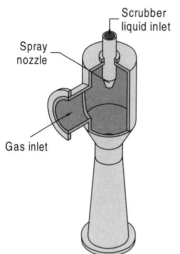

Figure 2 Cutaway view of a venturi scrubber design. Exhaust gases and the spray of scrubber liquid pass through the venturi throat, where the increased velocity improves mixing.

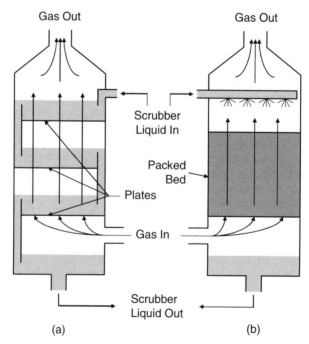

Figure 3 Schematic of plate scrubber (*a*) and packed tower scrubber (*b*) flue gas desulfurization towers.

capacity uses magnesium-enhanced lime (MEL). The reaction of the lime and limestone slurries with SO_2 forms calcium sulfite ($CaSO_3$) or calcium sulfate ($CaSO_4$) in an aqueous solution. Because both these compounds have low water solubility, they may precipitate out of solution and create scale in the system piping and other components. Care must therefore be taken during operation to minimize scale deposition by keeping the concentrations of $CaSO_3$ and $CaSO_4$ below the saturation point during operation. Wet scrubbers typically have high SO_2 removal efficiencies (90–98%) and relatively low flue gas energy (i.e., fan) requirements. In some cases, however, annualized costs may be higher than for dry scrubbers (see below) due to higher fan power requirements or increased maintenance due to excessive scaling. Because the scrubbing liquor is significantly more alkaline in MEL systems compared to lime- and limestone-based processes, less slurry is required and smaller absorber towers are needed, which can result in cost savings in some applications.

Smaller industrial scrubbers typically use a liquor reagent such as sodium carbonate or sodium hydroxide. Alkali compounds other than lime or limestone can also be used. Magnesium oxide (MgO) is used to form a slurry of magnesium hydroxide [$Mg(OH)_2$] to absorb the SO_2 and form magnesium sulfite or sulfate. The solid can be separated from the slurry, allowing the regeneration of the MgO and producing a relatively high concentration (10–15%) stream of SO_2. The SO_2 stream is then often used to produce sulfuric acid.

Ammonia has also been demonstrated as a reagent in once-through FGD systems. Ammonia-based systems are attractive because they produce a by-product that can be marketed as a fertilizer. A full-scale demonstration using ammonia achieved over 90% SO_2 reduction, and analysis of the costs estimated a net profit to the utility due to the sale of the resulting ammonium sulfate.[2] Ammonia storage issues and the potential formation of PM due to excess ammonia in the flue gas (ammonia slip) have inhibited the commercial applications of this approach. Use of diammonium phosphate as the ammonia source may reduce some of these concerns, but such advances have not yet been widely tested.[3]

Dual alkali scrubbing systems are wet regenerative processes that use two chemicals in a dual-loop arrangement. A typical design uses a more expensive sodium oxide or sodium hydroxide scrubbing liquor, which forms principally sodium sulfite (Na_2SO_3) when sprayed into the SO_2-containing flue gases. The spent liquor is then sent to the secondary loop where a less expensive alkaline material such as lime is added. The calcium sulfate or sulfite precipitates out of the liquor, and the sodium-based liquor is regenerated for reuse in the scrubber. The calcium sulfate/sulfite is separated from the liquor and dried, and the solids are usually sent to a landfill for disposal. SO_2 removal efficiencies for such systems are typically 75% and higher, with many systems capable of reductions greater than 90%.

Spray dryer absorbers (SDAs), also referred to as lime spray dryers (LSDs), are the most common dry once-through FGD designs, accounting for about 85% of the capacity of dry FGD systems in use. Like wet FGDs, SDAs also use aqueous lime slurry to capture the SO_2 in the flue gas. However, SDAs create a much finer spray, resulting in rapid evaporation of the water droplets and leaving the lime particles suspended in the flue gas flow. As SO_2 contacts these particles, reactions occur to create $CaSO_4$. The suspended particulate is then captured by a particle removal system, often a fabric filter (see the discussion on particle controls below). An advantage of the dry scrubber is its lower capital and operating cost compared to the wet scrubber, and the production of a dry, rather than wet, waste material for disposal. The flue gases are also not water-saturated as they are for wet systems, which can minimize the impacts of SO_3 formation. Dry systems are typically less efficient than wet scrubbers, providing removal efficiencies of 70–90%.

Dry sorbent injection is the direct injection of a solid calcium-based material, such as hydrated lime, limestone, or dolomite, into the furnace (furnace sorbent injection or FSI) or the flue gas duct upstream of the particle control device (duct sorbent injection) for the purpose of SO_2 capture. Depending on the amount of SO_2 removal required, furnace sorbent injection can be used instead of FGD. SO_2 removal efficiencies of up to 70% have been demonstrated for FSI,[4] although 50% reductions are more typical. The effectiveness of FSI is dependent on the calcium to sulfur (Ca:S) ratio, furnace temperature, and humidity in the flue gas. A Ca:S ratio of 2 is typically used. Furnace sorbent injection effectiveness decreases with increasing furnace temperature, and increases as flue gas humidity levels decrease. Neither type of sorbent injection system is currently in wide commercial application.

Systems that use FSI require adequate capacity in their particulate removal equipment to remove the additional solid material injected into the furnace. In addition, increased soot blowing is also required to maintain clean heat-transfer surfaces and prevent reduced heat-transfer efficiencies when FSI is used.

Fluid bed combustion (FBC) is another technology that employs a dry method of SO_2 reduction, in a similar manner to furnace sorbent injection. In such systems, the fluidized bed contains a calcium-based solid that removes the sulfur as the coal is burned in the bed. FBC is limited to new plant designs, since it is an alternative design significantly different than conventional steam generation systems, and is not suitable for retrofit to existing boilers. FBC systems typically remove 70–90% of the SO_2 generated in the combustion reactions.

2.1   Alternative Control Strategies

Coal cleaning (or fuel desulfurization) is also an option for removing a portion of the sulfur in the as-mined fuel. A significant portion of Eastern and Midwestern bituminous coals are currently cleaned to some degree to remove both sulfur and mineral matter. Cleaning may be done by crushing and screening the coal, or by washing with water or a dense medium consisting of a slurry of water and magnetite. Washing is typically done by taking advantage of the different specific gravities of the different coal constituents. The sulfur in coal is typically in the form of iron pyrite (pyritic sulfur) or organic sulfur contained in the carbon structure of the coal.* The sulfur reduction potential (or washability) of a coal depends on the relative amount and distribution of pyritic sulfur. The washability of U.S. coals varies from region to region, and ranges from less than 10% to greater than 50%. For most eastern U.S. high-sulfur coals, the sulfur reduction potential normally does not exceed 30%, limiting the use of physical coal cleaning for compliance coal production. Cleaning usually results in the generation of a solid or liquid waste that must be either disposed of or recycled.

Fuel switching is a further option for reducing SO_2 emissions. Fuel switching most often involves the change from a high-sulfur fuel to a lower-sulfur fuel of the same type. For coal, the most common change is from a higher-sulfur Eastern coal to lower-sulfur Western coals. In some instances, the change of coals may also result in restrictions to plant operability, usually due to changes in the slagging and fouling characteristics of the coal. However, many plants have found that the costs of compliance with emission limits using a fuel switching approach outweigh the operational challenges. In some instances, fuel switching can also involve a change from high-sulfur coal to natural gas. In this case, not only are SO_2 emissions reduced, but the lack of nitrogen in natural gas also yields a reduction in NO_x emissions. Particulate emissions are also significantly reduced when using natural gas instead of coal, as are emissions of trace metals.

---

* Sulfur in coal may also be in the form of sulfates, particularly in weathered coal. Pyritic and organic sulfur are the two most common forms of sulfur in coal.

2.2 Residue Disposal and Utilization

Flue gas desulfurization results in significant quantities of solid and/or liquid material that must be removed from the plant process. In some cases, the residues can be used as is, or processed to produce higher-quality materials for a number of applications. The cost of residue disposal can account for a significant portion of the total cost of SO_2 removal, particularly where landfilling costs are high. Early waste management approaches focused on landfilling, and as such costs increased, more attention was given to utilization options.

In limestone forced oxidation (LSFO) systems, the oxidation of the spent scrubbing slurry produces $CaSO_4$ from the $CaSO_3$ in the slurry, which can then be processed to form a salable commercial-grade gypsum product. Some impurities can be removed by means of filtration and removal of the smaller particles, followed by the hydration of the $CaSO_4$ to form gypsum ($CaSO_4 : 2H_2O$) and dewatering of the final solids. Depending on the quality of the final product, the resulting solids can be used in building materials, soil stabilization, road base, aggregate products, or agricultural applications. Spray dryer by-products have a higher free lime content, making them less acceptable as a building materials. The most likely end use of these residues is as a road base stabilization material. Residue disposal can play a major role in determining the appropriate technology. For instance, limestone inhibited oxidation (LSIO) is usually preferred over LSFO when landfilling is a viable option, because, unlike LSFO, LSIO does not require the use of air compressors, thereby reducing operating costs.

2.3 Costs of Control

Many factors are involved in the costs of applying SO_2 control technologies, including the amount of sulfur in the coal, the level of control required, and the plant size and configuration (particularly for retrofit applications). There are several ways to evaluate costs: capital cost per unit of plant capacity, total annual cost per unit output, or total cost per unit SO_2 removed. Table 1 presents predictions of total capital required for installation of three FGD technol-

Table 1 Predicted and Reported Capital Costs of Three FGD Technologies for Coal-fired Utility Boilers[1]

LSFO[a]			LSD[a]			MEL[b]	
Plant size (MW)	Predicted cost ($/kW)	Reported cost ($/kW)	Plant size (MW)	Predicted cost ($/kW)	Reported cost ($/kW)	Plant size (MW)	Predicted cost ($/kW)
239	400	317	267	205	231	200	272
316	368	348	362	203	172	316	213
511	213	215	508	222	189	511	175
600	189	180				600	162
750	226	236				750	197
1300	164	200				1300	163
1700	174	195				1700	179

[a] Predicted and reported costs are for comparable coal sulfur contents and system designs.
[b] Predicted costs are based on a 2% sulfur coal.

ogies (LSFO, LSD, and MEL) as a function of plant size. These predictions are compared to actual reported costs where available.

Note that the capital costs can increase as plant size increases. This is due to the modular nature of the absorber towers. The predicted costs assumed absorber towers would be installed in 700-MW increments, resulting in substantial increases of cost as each incremental absorber tower is required.

Control costs can also be estimated based on total annualized costs per kilowatt-hour, which include operating and maintenance costs and amortization costs of capital. Annualized costs can also be used as the basis for calculating costs in terms of dollars per ton of SO_2 removed, by estimating total annual costs and dividing by total SO_2 removed, based on the coal sulfur, plant size, and estimated control efficiency. Table 2 presents annualized operating costs per kilowatt-hour and cost per ton of SO_2 removed for LSFO, LSD, and MEL.

The data in Table 2 illustrate that care must be taken when comparing costs of different technologies. For instance, the annualized costs for FGD on a 500-MW plant burning 1% coal are estimated to be 4.08 mills/kWh for LSD and 4.14 mills/kWh for an MEL system. Even though the annualized costs are lower for the LSD system, the estimated dollar per ton of SO_2 removed is lower for the MEL system ($459/ton vs. $493/ton). This underscores the need to fully understand the cost basis used in technology cost comparisons and the factors that are driving the decision to install an FGD system.

3 CONTROL OF NITROGEN OXIDES

3.1 NO_x Formation Chemistry

Nitrogen oxides (NO_x) formed by the combustion of fuel in air are typically composed of greater than 90% NO in external combustion systems, with NO_2 making up the remainder.

Table 2 Predicted Annualized Total Costs per Unit Plant Output and Cost per Ton of SO_2 Removed for Three FGD Technologies as a Function of Plant Size and Coal Sulfur Content.[1]

Plant size (MW)	LSFO[a] mills/kWh ($/ton SO_2 removed)		LSD[b] mills/kWh ($/ton SO_2 removed)		MEL[c] mills/kWh ($/ton SO_2 removed)	
	1% sulfur coal	3% sulfur coal	1% sulfur coal	3% sulfur coal	1% sulfur coal	3% sulfur coal
100	9.65 (1103)	10.57 (403)	7.30 (881)	9.53 (383)	7.17 (795)	8.34 (308)
300	5.14 (588)	6.03 (230)	4.88 (589)	6.81 (274)	5.03 (557)	6.18 (228)
500	4.24 (484)	5.12 (195)	4.08 (493)	5.96 (240)	4.14 (459)	5.30 (196)
800	3.53 (404)	4.39 (167)	3.74 (451)	5.59 (225)	4.12 (457)	5.28 (195)
1000	3.91 (447)	4.76 (181)	3.60 (435)	5.44 (219)	4.35 (483)	5.51 (203)
1500	3.42 (391)	4.25 (162)	3.42 (413)	5.25 (211)	4.79 (531)	5.95 (220)

[a] Assumed SO_2 reduction of 95%.
[b] Assumed SO_2 reduction of 90%.
[c] Assumed SO_2 reduction of 98%.

Unfortunately, NO is not amenable to flue gas scrubbing processes as is SO_2. An understanding of the chemistry of NO_x formation and destruction is helpful in understanding emission control technologies for NO_x.

There are three major pathways to formation of NO in combustion systems, usually referred to as thermal NO_x, fuel NO_x, and prompt NO_x.[5] Thermal NO_x is created when the oxygen (O_2) and nitrogen (N_2) present in the air are exposed to the high temperatures of a flame, leading to a dissociation of O_2 and N_2 molecules and their recombination into NO. The rate of this reaction is highly temperature dependent; therefore, a reduction in peak flame temperature can significantly reduce the level of NO_x emissions. Thermal NO_x is important in all combustion processes that rely on air as the oxidizer.

Fuel NO_x is due to the presence of nitrogen in the fuel, and is usually the greatest contributor to total NO_x emissions in uncontrolled coal flames. By limiting the presence of O_2 in the region where the nitrogen devolatilizes from the solid fuel, the formation of fuel NO_x can be greatly diminished. NO formation reactions depend on the presence of hydrocarbon radicals and O_2, and since the hydrocarbon–oxygen reactions are much faster than the nitrogen–oxygen reactions, a controlled introduction of air into the devolatilization zone leads to the oxygen preferentially reacting with the hydrocarbon radicals (rather than with the nitrogen) to form water and CO. Finally, the combustion of CO is completed. Since this reaction does not promote NO production, the total rate of NO_x production is reduced in comparison with uncontrolled flames. This staged combustion can be designed to take place within a single burner flame or within the entire furnace, depending on the type of control applied (see below). Fuel NO_x is important primarily in coal combustion systems, although it is also important in systems that use heavy oils, since both fuels contain significant amounts of fuel nitrogen.

Prompt NO_x forms at a rate faster than equilibrium would predict for thermal NO_x formation. Prompt NO_x forms from nonequilibrium levels of oxide (O) and hydroxide (OH) radicals, through reactions initiated by hydrocarbon radicals with molecular nitrogen, reactions of O atoms with N_2 to form N_2O, and finally the reaction of N_2O with O to form NO. Prompt NO_x can account for more than 50% of NO_x formed in fuel rich hydrocarbon flames; however, prompt NO_x does not typically account for a significant portion of the total NO emissions from combustion sources.

3.2 Combustion Modification NO_x Controls

Because the rate of NO_x formation is so highly dependent on temperature as well as local chemistry within the combustion environment, NO_x is ideally suited to control by means of modifying the combustion conditions. There are several methods of applying these combustion modification NO_x controls, ranging from reducing the overall excess air levels in the combustor to burners specifically designed for low NO_x emissions.

Low excess air (LEA) operation is the simplest form of NO_x control, and relies on reducing the amount of combustion air fed into the furnace. LEA can also improve combustion efficiency where excess air levels are much too high. The drawbacks to this method are the relatively low NO_x reduction and the potential for increased emissions of CO and unburned hydrocarbons if excess air levels are dropped too far. NO_x emission reductions using LEA range between 5 and 20%, at relatively minimal cost if the reduction of combustion air does not also lead to incomplete combustion of fuel. Incomplete combustion significantly reduces combustion efficiency, increasing operating costs, and may result in high levels of CO or even carbonaceous soot emissions.[6]

Overfire air (OFA) is a simple method of staged combustion in which the burners are operated with very low excess air or at substoichiometric (fuel-rich) conditions. The remaining combustion air is introduced above the primary flame zone to complete the combustion process and achieve the required overall stoichiometric ratio (see Fig. 4). The LEA or fuel-rich conditions result in lower peak flame temperatures and reduced levels of oxygen in the regions where the fuel-bound nitrogen devolatilizes from the solid fuel. These two effects result in lower NO_x formation in the flame zone, and therefore lower emissions. Recent field studies showed approximately 20% reductions of NO_x emissions using advanced OFA in a coal fired boiler.[7] OFA can be used for coal, oil, and natural gas, and to some degree for solid fuels such as municipal solid waste and biomass when combusted on stoker-grate units.

OFA typically requires special air injection ports above the burners, as well as the associated combustion air ducting to the ports. In some cases, additional fan capability is required to ensure that the OFA is injected with enough momentum to penetrate the flue gases. Emissions of CO are usually not adversely affected by operation with OFA. However, use of OFA can result in higher levels of carbon in fly ash when used in coal-fired applications, but proper design and operation may minimize this disadvantage. Another disadvantage to the application of OFA is the often corrosive nature of the flue gases in the fuel-rich zone. If adequate precautions are not taken, this can lead to increased corrosion of boiler tubes.

Flue gas recirculation is a combustion modification technique used to reduce the peak flame temperature by mixing a relatively small fraction (~5%) of the combustion gases back into the flame zone, as illustrated in Fig. 4. This method is especially effective for fuels with little or no nitrogen, such as natural gas combustion systems. However, in many instances the recirculation system requires a separate fan to compress the hot gases, and the fan capital and operating costs can be substantial. The resulting NO_x reductions can be significant, however, and emission reductions as high as 50% have been achieved.[8]

In the past 15 years, burners for both natural gas and coal have undergone major design improvements intended to incorporate the principles of staging and flue gas recirculation into the flow patterns of the fuel and air injected by the burner. These burners are generically referred to as *low NO_x burners* (LNBs), and are the most widely used NO_x control technology. Staging of fuel and air that is the basis for combustion modification NO_x control is achieved

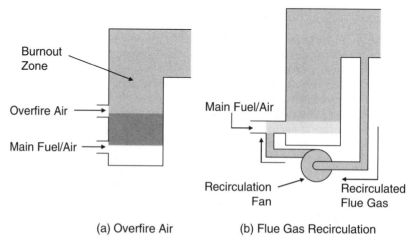

(a) Overfire Air   (b) Flue Gas Recirculation

Figure 4 Schematics of two combustion modification NO_x control technologies applied to industrial and power generation boilers.

in LNBs by creating separate flow paths for the air and fuel. This is in contrast to earlier burner designs, in which the fuel and air flows were designed to mix as quickly and as turbulently as possible. While these highly turbulent flames were very successful in achieving rapid and complete combustion, they also resulted in very high peak flame temperatures and high levels of oxygen in the fuel devolatilization region, with correspondingly high levels of NO_x emissions. The controlled mixing of fuel and air flows typical of LNBs significantly reduced the rates of fuel and air mixing, leading to lower flame temperatures and considerable reductions of oxygen in the devolatilization regions of the flame, thereby reducing the production and emission of NO_x.

Low NO_x burners may further reduce the formation of NO_x by inducing flue gases into the flame zone through recirculation. Careful design of the fluid dynamics of the air and fuel flows acts to recirculate the partially burned fuel and products of combustion back into the flame zone, further reducing the peak flame temperature and thus the rate of NO_x production. In some burner designs, this use of recirculated flue gas is taken a step further by the use of flue gas that has been extracted from the furnace, compressed, and fed back into the burner along with the fuel and air. These burners are typically used in natural gas-fired applications, and are among the "ultralow NO_x burners" that can achieve emission levels as low as 5 ppm.

LNBs are standard on most new facilities. Some difficulties may be encountered during retrofit applications if the furnace dimensions do not allow for the longer flame lengths typical of these burners. The flame lengths can increase considerably due to the more controlled mixing of the fuel and air and, if adequate furnace lengths are not available, impingement of the flame on the opposite wall can lead to rapid cooling of the flame, and therefore increased emissions of CO and organic compounds, as well as reduced heat-transfer efficiency from the flame zone to the heat-transfer fluid.

More precise control of air and fuel flows is often required for LNBs compared to conventional burners due to the reliance of many LNB designs on fluid dynamics to stage the air and fuel flows in particular patterns. Slight changes in the flow patterns can lead to significant drops in burner and boiler efficiencies, higher CO and organic compound emissions, and even damage to the burner from excessive coking of the fuel on the burner. In addition, the more strict operating conditions may impact the burners' ability to properly operate using fuels with different properties, primarily for coal-fired units. Coals with lower volatility or higher fuel nitrogen content may hamper NO_x reduction, and changes in the coals' slagging properties may lead to fouling of the burner ports. Further, improper air distribution within the burner may result in high levels of erosion within the burner, degrading performance and reducing operating life. An example of a typical pulverized coal LNB design is shown in Fig. 5.

A further method of controlling NO_x emissions through modification of the combustion process is *reburning*. Reburning is applied by injecting a portion of the fuel downstream of the primary burner zone, thereby creating a fuel-rich reburn zone in which high levels of hydrocarbon radicals react with the NO formed in the primary combustion zone to create H_2O, CO, and N_2. This is quite different from the other combustion modification techniques, which reduce NO_x emissions by preventing its formation. Reburning can use coal, oil, natural gas, or other hydrocarbon fuel as the reburn fuel, regardless of the fuel used in the main burners. Natural gas is an ideal reburn fuel, as it does not contain any fuel-bound nitrogen. Coals that exhibit rapid devolatilization and char burnout are also suitable for use as reburn fuels. Advanced reburning uses one or more chemical reagents in addition to the reburning fuel to enhance the NO_x reduction performance.

In most applications, between 10 and 20% of the total heat input to the furnace is introduced in the reburn zone in the form of reburn fuel. The main burners are operated at

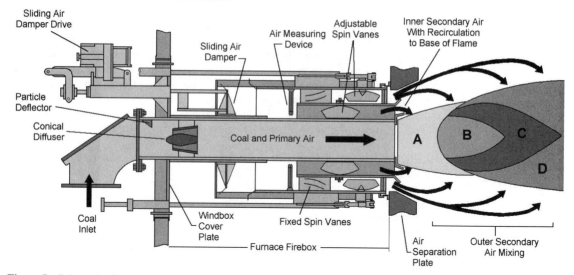

Figure 5   Schematic of a modern low-NO$_x$ burner for boilers burning pulverized coal. (Courtesy of The Babcock and Wilcox Company.)

slightly fuel lean stoichiometries. This usually results in lower NO$_x$ levels leaving the primary zone, since the low excess air and lower flame temperatures produce lower NO$_x$. Above the primary zone, but far enough to allow for the combustion process to be nearly completed, the reburn fuel is introduced, and a reburn zone stoichiometry of 0.8–0.9 is created. Finally, sufficient air is injected downstream to burn out the remaining combustible materials (primarily CO) and reach the desired overall furnace stoichiometry (normally near 1.2). Reburning requires adequate furnace volume to allow the injection of the reburn fuel and the overfire air, as well as time for the combustion reactions to be completed.[9] Advanced reburning systems may utilize the injection of chemical reagents in addition to the reburn fuel to provide additional NO$_x$ reductions or to reduce the amounts of reburn fuel required for a given NO$_x$ reduction level. Reburning applied to full-scale utility boilers has resulted in NO$_x$ emissions ranging from 50 to 65%.

3.3   Postcombustion NO$_x$ Controls

In some cases, either it is not possible to modify the combustion process, or the levels of NO$_x$ reduction are beyond the capabilities of combustion modifications alone. In these instances, postcombustion controls must be used. There are two primary postcombustion NO$_x$ control technologies, *selective noncatalytic reduction* (SNCR) and *selective catalytic reduction* (SCR). Several systems have also been developed for scrubbing NO$_x$; however, because these remove only NO$_2$, they are not in broad commercial operation.

SNCR systems rely on the injection of a nitrogen-based reagent into a relatively high temperature zone of the furnace, and rely on the chemical reaction of the reagent with the NO to produce N$_2$, N$_2$O, and H$_2$O. Removal efficiencies of up to 75% can be achieved with SNCR systems, but lower removal rates are typical. The SNCR reaction is highly temperature dependent and, if not conducted properly, can result in either increased NO$_x$ emissions or considerable emissions of ammonia. Where the reagent is injected in larger amounts than the available NO, or where it is injected into a temperature too low to permit rapid reaction, the ammonia will pass through to the stack in the form of "ammonia slip." The reagents most commonly used are ammonia (NH_3) and urea (NH_2CONH_2), although other chemicals have also been used, including cyanuric acid, ammonium sulfate, ammonium carbamate, and hydrazine hydrate. A number of proprietary reagents are also offered by several vendors, but all rely on similar chemical reaction processes. Proprietary reagents are used to vary the location and width of the temperature window, and to reduce the amount of ammonia slip to acceptable levels (typically 10–20 ppm).

The optimum temperature for SNCR systems will vary depending on the reagent used, but ranges between 870 and 1150°C (1600 and 2100°F). Increased NO_x reductions can be obtained by using increasing amounts of reagent, although excessive use of reagent can lead to emissions of ammonia or, in some cases, conversion of the nitrogen in the ammonia to NO. Reduction efficiencies increase as the base NO level increases and, for systems with a low baseline NO level, removal efficiencies of less than 30% are not unusual. Adequate mixing of the reagent into the flue gases is also important in maximizing the performance of the SNCR process, and can be accomplished by the use of a grid of small nozzles across the gas path, by adjusting the spray atomization to control droplet trajectories, or with the use of an agent such as steam or air to transport the reagent into the flue gas.

Where chlorine is present, a detached visible plume of ammonium chloride (NH_4Cl) may be formed if it is present in high enough levels. As plume temperature drops as it mixes in the atmosphere, the NH_4Cl changes from a liquid to a solid, resulting in a visible white plume. Although these plumes may occur at NO_x or particulate emission levels within permit conditions, they can result in perceptions of uncontrolled or excessive pollutant emissions.

SNCR systems typically have low capital cost, but much higher operating cost compared to low NO_x burners due to the use of reagents. In some applications that have wide variations in load, additional injection locations may be required to ensure that the reagent is being injected into the proper temperature zone. In this case, more complex control and piping arrangements are also required.

SCR systems similarly rely on the use of an injected reagent (usually ammonia or urea) to convert the NO to N_2 and H_2O in the presence of a catalyst, and at lower temperatures (usually around 315–370°C [600–700°F]) than SNCR systems. Catalysts are typically titanium and/or vanadium based, and are installed in the flue gas streams at various locations in the gas path, depending on the available volume, desired temperatures, and potential for solid particle plugging of the catalyst. SCR systems have not usually been installed in pulverized-coal-fired systems due to the difficulties associated with plugging and fouling of the catalyst by the fly ash, poisoning of the catalyst by arsenic, and similar difficulties. However, recent tests have indicated the ability of SCR catalysts to maintain their performance over an extended period in pulverized-coal-fired applications.

Parameters of importance to SCR systems include the space velocity (volumetric gas flow per hour, per volume of catalyst), linear gas flow velocity, operating temperature, and baseline NO_x level. System designs must balance the increasing NO_x reductions with operating considerations such as catalyst cost, pressure drops across the catalyst bed, increased rate of catalyst deactivation, and increased NH_3 requirements. As NO_x reductions increase, the life of the catalyst decreases and the required amount of NH_3 injected increases. NO_x emissions can be reduced by over 90% if adequate catalyst and reagent are present, and injection temperatures are optimized. For such reduction levels, catalysts may require replacement in as little as 2 years. It is possible to increase catalyst life where lower reductions are suitable.

Operational problems such as catalyst plugging and fouling can significantly reduce the effectiveness of SCR systems. Plugging can be a problem where the fuel used (e.g., coal) has a high particle content. Interactions between sulfur oxides and the injected reagent can lead to ammonium sulfate or bisulfate formation, which can result in fouling of the catalyst.

In addition, the catalyst can convert SO_2 into sulfur trioxide (SO_3), which readily forms vapor-phase sulfuric acid (H_2SO_4) in the high moisture environment of the exhaust stack. H_2SO_4 has a high dew point, which can result in the condensation of acidic liquid onto equipment and subsequent corrosion problems or in the formation of visible acid aerosol plumes (see below).

SCR systems are often more expensive to install than other NO_x removal systems due to the relatively high catalyst cost (10,600–14,100 \$/m³ [300–400 \$/ft³]). However, SCR systems can also remove higher levels of NO_x, resulting in costs in terms of \$/ton of NO_x removed that are often competitive with other methods. Where very low NO_x emissions are required, SCR systems may be the only method of achieving the emission standard. SCR capital costs can be significant, particularly if large NO_x reductions are desired. In most cases, the largest portion of the cost is for the catalyst, which must be replaced periodically (approximately every 3–4 years). Costs for NH_3 must also be considered, but these costs are typically lower than for SNCR systems.

Hybrid systems combine SCR and SNCR by injecting a reagent into the furnace sections at the appropriate temperatures to take advantage of the SNCR NO_x reduction reactions, then passing the flue gases through a catalyst section to further reduce NO_x and provide some control of ammonia slip. Emissions of over 80% have been demonstrated on small-scale boilers using the hybrid approach.

Typical NO_x control performance and costs are shown in Fig. 6.[10]

4 CONTROL OF PARTICULATE MATTER

Particulate matter (PM) control technologies employ one or more of several techniques for removing particles from the gas stream in which they are suspended. These techniques are mechanical collection, wet scrubbing, electrostatic precipitation, and filtration.

Mechanical collection systems typically rely on the difference in inertial forces between the particles and the gas to separate the two. Although gravity can also be used as the separation force in settling chambers, the chambers must be very large and are effective only

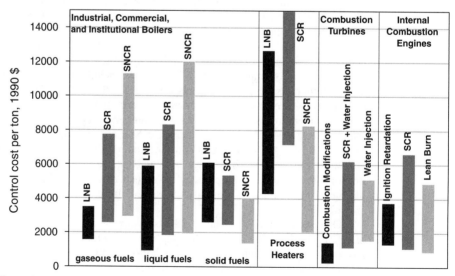

Figure 6 Cost effectiveness of NO_x controls for different source types in U.S. dollars per ton (1990 \$).[10]

for particles that are larger than about 10 μm in aerodynamic diameter.* Most common are cyclones, which induce a spinning motion in the gas, forcing the heavier particles to the outside of the gas stream and against the inner cyclone wall, separating them from the gas. In most designs, particle-laden gas is tangentially introduced into a cylindrical chamber, which creates a vortex in the gas stream. This forces high-inertia particles to strike the wall. These particles then fall to the bottom of the cylinder, and the gas turns inward and upward, exiting at the top of the cylinder. For particles that are larger than 10 μm in aerodynamic diameter, cyclones can remove about 90% of particle mass, with removal efficiencies dropping off very quickly as particle size decreases.

Cyclones in parallel can be used to handle large gas volumes with relatively little pressure drop. When staged in series (a "multiclone"), cyclones can effectively reduce smaller particles, but at an increasingly high pressure drop across the device. To collect smaller particles, the gas velocity must be increased or the cylinder diameter decreased, both of which require significant increases in fan power to force the gas through the cyclones. However, most multiclone applications are designed to achieve higher control efficiencies for larger particles rather than to improve control efficiencies for smaller particles. Cyclones are also used as the initial stage of a multistage gas-cleaning process, where they remove the larger particles before the gas enters a higher efficiency control device.

Cyclones are often used in applications where the particles have been formed by grinding, crushing, or cutting processes, such as stone and aggregate industries, cement production, or woodworking industries. Cyclones are also applied to processes where the solid particles are recovered for further use in the process stream, such as recovery of catalyst particles in petroleum processing. Wet or sticky particles may not be effectively removed, as they can stick to the cyclone walls and degrade the performance of the unit.

Particles can also be removed by *wet scrubbing*. In some cases, wet scrubbing of particulates is done by creating a fine spray of liquid droplets, which enhance the collection of small particles in the gas stream. This is known as diffusion capture since it relies on the Brownian motion of the particles and droplets to lead to capture of the particles. Other methods include direct interception or inertial impaction, similar to mechanical collectors, but in the presence of a liquid (usually water) to assist in removing additional particles from the gas stream. Wet scrubbers can require significant energy inputs to decrease liquid droplet sizes, increase the momentum of the gas stream, or a combination. Greater energy is needed to maintain collection efficiency as particle size drops, but wet scrubbers can effectively control particles as small as 1 μm in diameter. Some particles can be captured by wet flue gas desulfurization systems; however, wet scrubbing does not usually perform as well as ESPs or fabric filters in the capture of small particles. ESP or fabric filter systems are often required in addition to adequately control particulate emissions in applications where high ash content fuels are used. Nevertheless, ESPs, fabric filters, and wet scrubbers can give equivalent particulate removal efficiencies if the systems are properly designed for the specific source and collection efficiency desired.

Wet scrubbers are often used with particles that are already wet, with particle/gas mixtures that may be flammable, or in applications where the gas temperature must be significantly reduced before it is released or used. Another suitable application for wet scrub-

---

*When considering particle diameter, it is appropriate to specify the context in which the diameter is placed. For instance, "aerodynamic diameter" refers to the diameter of a sphere that falls through air at the same rate as the particle being measured. "Optical diameter" refers to the diameter of a sphere that absorbs or refracts the same amount of light as the particle being measured. In other systems, one may refer to the "electrostatic diameter" or other reference that appropriately reflects the forces or measurement methods being applied.

bers is for situations where simultaneous control of gas-phase and particulate matter is required.

Large industrial and utility sources generally use electrostatic precipitators or fabric filters to remove fine particles from high-volume gas streams. Typical PM removal efficiencies are shown in Table 3 for cyclones, venturi scrubbers, fabric filters, and electrostatic precipitators.

Electrostatic precipitators (ESPs) operate by inducing an electrical charge onto the particles and then passing them through an electric field. This exerts a force on the charged particles, forcing them toward an electrode where they are collected. The basic configuration of an ESP consists of one or more high-voltage electrodes, which produce an ion-generating corona, in combination with a grounded collecting electrode. The generated ions charge the particles, and the high voltage between the electrodes results in an electric field that forces the particles toward the collecting electrode. The particles are removed from the collecting electrode by periodic rapping of the electrode, causing the particles to fall into a collection hopper below. There are two main ESP configurations, plate and tube, illustrated in Fig. 7. Large power-generation units typically use a plate configuration, and smaller industrial units often use tube system.

ESP performance can be significantly affected by the resistivity of the incoming particles. This is because the ability of the particles to conduct electricity decreases as resistivity increases, reducing the performance of the unit. Although ESPs can be designed to work in any resistivity range, optimum resistivity levels are in the range of 10^8 to 10^{10} ohms-cm. Chemical additives can be introduced to the flue gas to reduce the effect of the high resistivity in a practice referred to as gas conditioning. A commonly used gas-conditioning reagent in coal-fired power plants is sulfur trioxide (SO_3). Pulsing the electrodes with intense, periodic, high-voltage direct current can also improve performance with high-resistivity particles. Performance can also be improved by using separate charging and collection stages, which allows optimization of each process.

Pressure drops in ESPs are usually less than those of other particulate control devices for the same collection efficiencies, but they require significant amounts of electrical energy input to the electrodes. ESPs can remove 90–99.5% of the total particulate mass in the gas stream. However, particles are generally collected at lower efficiencies for particles between 1.0 and 0.1 μm (see Fig. 8). In some applications where there are significant levels of chlorine present, the combination of temperature and gas composition can lead to formation in the ESP of trace organics, including polychlorinated dibenzodioxins (PCDDs) and polychlorinated dibenzofurans (PCDFs). Proper control of temperatures can significantly reduce or eliminate this problem.

Industrial-scale ESPs may have collecting electrodes 7.6–13.7 m (25–45 ft) high and 1.5–4.6 m (5–15 ft) long, with 60 or more gas flow lanes per section. Large units may have eight consecutive sections, and gas flows of up to 85,000 m³/min (3,000,000 ft³/min). Voltages applied to the corona generating electrode in industrial-scale ESPs may range from 30 to 80 kV.

Most ESP designs are "dry," and should not be used for applications where the particles are sticky or the flue gas stream contains significant levels of entrained droplets. "Wet" ESP designs flush the collected particles with continuous application of water to allow handling of wet or sticky particles. However, the effluent from wet ESPs may require treatment before ultimate disposal of the resulting waste water. Because of high potential for spark generation, explosive gases or particles should not be treated using ESPs.

In addition to large-scale industrial applications, ESPs have also been used for control of particulates in indoor air handling systems, as well as for cleaning air circulating through road and rail tunnels.

Table 3 Cost and Performance Estimates for Particulate Matter Control Technologies

Technology	Annualized Costs per scfm (per sm³/sec)	Cost Effectiveness per Short Ton (per metric ton)	Control Efficiency	Gas Flow	Particle Loading
Cyclone	$1.30–$13.50 ($2800–$29,000)	$0.43–$400 ($0.47–$440)	90%	1060–25,400 scfm (0.5–12 sm³/sec) per cyclone	1–100 g/scf (2.3–230 g/sm³)
Venturi	$5.7–$193 ($12,000–$409,000)	$70–$2400 ($77–$2600)	99%	500–100,000 scfm (0.2–480 sm³/sec)	0.1–50 g/scf (1–110 g/sm³)
Pulse jet fabric filter	$6–$39 ($13,000–$83,000)	$42–$266 ($46–$293)	99–99.9% (new units) 95–99.9% (older units)	100–100,000+ scfm (0.10–50+ sm³/sec)	0.5–10 g/ft³ (1–23 g/m³)
Reverse air fabric filter	$8–$50 ($17,000–$106,000)	$53–$337 ($58–$372)	99–99.9% (new units) 95–99.9% (older units)	100–100,000+ scfm (0.10–50+ sm³/sec)	0.5–10 g/ft³ (1–23 g/m³)
Dry ESP (wire-tube configuration)	$9–$26 ($19,000–$55,000)	$43–$640 ($47–$710)	99–99.9%	1,000–100,000 scfm (0.5 to 50 sm³/sec)	0.5–5 g/scf (1–10 g/m³)
Dry ESP (wire-plate configuration)	$4–$38 ($9100–$81,000)	$35–$236 ($38–$260)	99–99.9%	200,000–1,000,000 scfm (100–500 sm³/sec)	1–50g/scf (2–110 g/m³)

Source: Data from EPA Air Pollution Control Technology Fact Sheets.[11]

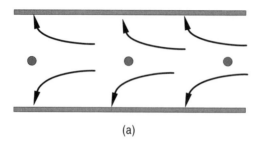

(a)

(b)

Figure 7 Schematic of electrostatic precipitator configurations, viewed from above. (*a*) Plate configuration: particle-laden gas flows in from left, and the electric field forces particles to the side plates (top and bottom of the figure). (*b*) Tube configuration: particle-laden gas flows into the plane of the page, and is forced to the inside surface of the tube.

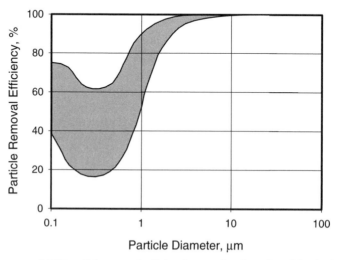

Figure 8 The range of ESP particle removal efficiencies as a function of particle size is shown in the shaded area. The mechanisms of particle removal by ESP result in a minimum collection efficiency between 0.1 and 1.0 μm in diameter. Well-operated modern units will be closer to the upper end of the shaded area.

Fabric filters can also be used for industrial-scale removal of particles. There are several filter configurations, including bags, envelopes, and cartridges. Bags are most common for large-volume gas streams, and are distributed in a baghouse. Industrial-scale baghouses may contain from less than 100 to several thousand individual bags, depending on the gas flow rate, particle concentration, and allowable pressure drop.

Filter bags can be operated with gas flowing from inside to outside the bags, or with the gas flowing from outside to inside, as illustrated in Fig. 9. In the inside-to-outside configuration, the particles are collected on the inner bag surface, and the bags are cleaned periodically by reversing the gas flow intermittently for a short time, mechanical shaking of the bags, or a combination of the two. These are usually referred to as reverse air designs. For outside-to-inside bag configurations, a short pulse of air is injected down the inside of the bag to remove the particles that have collected on the outer bag surface. A rigid internal cage is used to prevent the collapse of the outside-to-inside bags due to the pressure of the flue gas flow. These designs are typically known as pulse jet filters. Inside-to-outside (reverse air) bags are up to 11 m (36 ft) long and 30 cm (12 in.) in diameter, while outside to inside (pulse jet) bags are smaller, with lengths up to 7.3 m (24 ft) and diameters up to 20 cm (8 in.).

Fabric filters require higher fan power to overcome the resulting pressure drop of the flue gases than do ESPs, but little additional power is required to operate the cleaning systems. Filter materials are chosen based on the temperature and gas composition characteristic of the flue gas to be cleaned. Inside-to-outside bags are typically made of woven fibers, while outside-to-inside bags are usually felted fibers. Filter fabrics have been developed with a catalyst coating to improve reduction of VOCs or NO_x, but these systems require additional improvements before they are ready for large-scale commercial use.

Overall mass removal efficiencies of fabric filters range from 95 to 99.9%. However, the performance of fabric filters is usually independent of the chemical composition of the

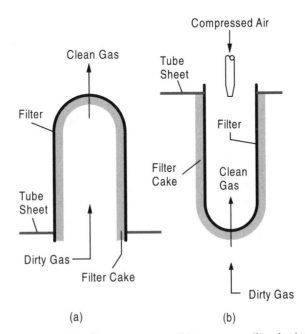

Figure 9 Fabric filter arrangements: (*a*) reverse gas; (*b*) pulse jet.

particulate matter. They are not used when the gas stream generated by the process equipment includes corrosive materials that could chemically attack the filter media, or when there are sticky or wet particles in the gas stream. These materials accumulate on the filter media surface and block gas movement.

Both ESPs and fabric filters can require heaters in the hoppers to maintain proper collection conditions. The energy required for heating to maintain these conditions may be a major part of the total energy requirement for the systems.

Hybrid systems combine features of ESPs and fabric filters to improve particle removal efficiencies per unit energy. Systems such as the electrostatically stimulated fabric filter (ESFF) or the compact hybrid particle collector (COHPAC) are designed largely as retrofit technologies for coal-fired utility applications that can improve particle collection performance at plants using older ESP designs.

Electrostatic stimulation of fabric filtration modifies a fabric filter system to charge incoming dust particles, which are then collected on fabric filters. Filter bags for these designs are specially made to include wires or conductive threads that produce an electric field parallel to the fabric surface. The field alters the deposition pattern such that the pressure drop across the filter fabric is reduced for a given collection efficiency compared to filters without charging. A number of variations exist on the ESFF idea of combining particle charging with fabric filtration.

A third approach is the advanced hybrid particle control (AHPC) system, which places filter bags between ESP plates. The AHPC uses perforated ESP collection plates, which allow a fraction of the particles to reach the filter bags. The plates are designed to capture approximately 90% of the total particle mass before it reaches the filter. As with the other hybrid systems, the AHPC is designed to capture a higher fraction of total particle mass than the ESP alone, and at much lower pressure drop (and therefore fan power) than a stand-alone fabric filter.

In general, hybrid systems have the advantages of improved particle collection efficiency compared to ESPs, and improved collection per unit energy compared to fabric filters. For coal-fired power plants, these systems can improve collection of submicron particles and increase collection of total particle mass in a system that is either within the existing particle control system footprint, or only slightly larger, which is particularly important in retrofit applications. Use of fabric filters in general also improves the feasibility of control strategies that rely on injection of solid sorbent materials, such as activated carbon for mercury control.

5 CONTROL OF GASEOUS ORGANIC COMPOUNDS

Organic compounds are broadly characterized as those that are composed of carbon (C), hydrogen (H), oxygen (O), and often nitrogen (N). Some organic compounds may also contain chlorine (Cl), fluorine (F), or sulfur (S). Generally, organic compounds can be decomposed by reaction with oxygen. For nonhalogenated organic compounds, complete oxidation results in decomposition to carbon dioxide (CO_2) and water (H_2O):

$$C_xH_yO_zN_a + (x + y/4 - z/2)O_2 \rightarrow xCO_2 + y/2H_2O + a/2N_2$$

These reactions are generally exothermic and occur more rapidly at higher temperatures. At high enough organic concentrations, these reactions can be self-sustaining, which is defined as combustion.

At lower organic compound concentrations, an auxiliary or supplemental fuel (usually natural gas) is used to maintain temperature levels adequate to rapidly decompose the pollutant to acceptable levels. Use of catalysts can reduce the temperature at which decomposition occurs, thereby reducing the amount of supplemental fuel needed. In some cases, biological processes can be used to decompose the organic compound.

If the organic compound is not completely oxidized due to low temperatures, inadequate mixing of the oxygen and organic compound, or insufficient residence time at elevated temperatures, some amount of the original pollutant can be emitted, as well as organic compounds that are formed during the complex reactions that occur during thermal oxidation. When other elements are present in the process gas feed, particularly chlorine, the products of incomplete combustion or oxidation that are formed can be more toxic than the original pollutant of concern. Generally, the concentrations of these compounds are significantly lower than the original compound. Even so, these PICs should be considered when evaluating performance of these systems.

Thermal Oxidizers
There is no clear definition of the difference between thermal oxidation and combustion or incineration, which is appropriate since the chemical process is essentially the same. Thermal oxidation of organic compounds (like combustion) is the high-temperature reaction of the hydrocarbon material with air to form primarily CO_2 and H_2O. In many, if not most, cases, the concentration of the organic compound in the process gas is not sufficient to maintain the reactions without addition of a supplemental fuel, usually natural gas. Often, the oxygen content of the untreated gas is sufficient to maintain the reactions without additional air.

In general, thermal oxidation can remove over 98% of the organics in the process gas, and often achieve 99% reduction of the organic compound. This can vary if the system is operated outside its design conditions, or is not adequately maintained. Significant changes in process gas composition or concentration can also result in poor performance if appropriate measures are not taken to account for such changes.

There are several variations of thermal oxidizer design. *Direct fired* systems are the simplest design, and consist of a means to feed process gas, supplemental gas, and air into a (usually) insulated combustion chamber. The gases and air are mixed by the feed nozzles and ignited, resulting in a destruction of the organic compounds in the process gas. The exhaust is then released directly to the atmosphere. *Recuperative* systems recover a portion of the heat generated by the combustion process through conventional air-to-air heat exchangers. Tube-in-shell or similar heat exchangers are used to transfer heat from the treated gases to the untreated gases, or if necessary, to the supporting oxidation air. *Regenerative* systems utilize porous beds of high heat content material to extract heat from the treated gases over a period of time. The air flows are then adjusted to route the flow of the relatively colder untreated gas through the heated bed, allowing the heat to then flow from the bed material into the untreated gas. These systems require dual beds through which the flow alternates between hot, treated gas and cold, untreated gas. These systems are more complex than recuperative systems, since they require the heat exchanger beds, piping, and flow controls necessary to achieve the alternating flows.

Catalytic Oxidation
A catalyst can be used to lower the temperature at which the oxidation reactions occur. Theoretical reactor temperatures required by thermal oxidation systems to achieve a 99.99% reduction of organic compounds such as benzene, toluene, or methyl chloride are in the approximate range of 1300–1600°F. This compares to reaction temperatures of 300–500°F to oxidize 80% of similar compounds using catalytic oxidation. The economic trade-off between thermal and catalytic systems is the reduction in auxiliary fuel cost required to

achieve adequate destruction of the VOCs of concern. Catalysts can be configured as fixed beds near the exhaust of a simple or recuperative thermal oxidation system or can be included in the bed material of regenerative systems.

Catalytic oxidation has until recently been used only for organic compounds containing C, H, and O. However, catalysts using chromia and alumina and oxides of cobalt, copper, and/or manganese have been developed to oxidize chlorinated organics. Platinum-based catalysts can be used to oxidize sulfur-containing VOCs, but these catalysts deactivate rapidly if chlorine is present in the gas stream. Arsenic, lead, and phosphorus are generally considered poisons for most catalysts, and their presence in the treated gases will likely mean that catalytic oxidation cannot be effectively used, unless the concentrations of these elements is sufficiently low.

Particles can also degrade catalyst performance by blinding the pores of fixed-bed systems. Fluid-bed designs are more robust in their ability to operate in the presence of particles, but can lose catalyst material due to the mechanical interactions of the bed material.

Recuperative, regenerative, and catalytic systems are all more expensive to purchase and install than simple thermal oxidation systems, but offset the initial cost with reduced operating cost, particularly in the cost of auxiliary fuel. The highest capital cost is for catalytic systems, but they can also operate at lower temperatures, resulting in reduced heat loss and thus, reduced auxiliary fuel costs.

An example cost comparison developed in EPA's *Cost Control Manual* illustrates the benefits of catalytic oxidation for one case. Calculations were made for a 20,000 scfm air flow containing 1000 ppm each of benzene and methyl chloride. A 98% control efficiency of and a 70% heat recovery were assumed. Although the total capital investment was estimated to be $889,000 for the catalytic system compared to $483,000 for the thermal oxidation system, the total annualized costs favored the catalytic system. Accounting for capital recovery costs, the catalytic system was estimated to cost $316,000 per year compared to $422,000 per year for thermal oxidation. This estimate included replacement of the fixed metal oxide catalyst bed every 2 years and a natural gas cost of $3.13/1000 ft^3.[12]

Concentrators

In some cases, the concentration of the organic compound may not be in the appropriate range for optimum application of a particular technology. Thermal oxidation systems, for instance, can be significantly more efficient at higher organic compound concentrations as the amount of supplemental fuel depends strongly on the amount of pollutant-laden air to be treated. Similarly, high organic levels can result in energy releases that lead to temperatures that are higher than catalytic oxidizers can handle. In the latter case, dilution with ambient air can bring the organic concentration down to appropriate levels.

Carbon Bed Filters

Organic compounds may also be removed and recovered by absorption of the organics on a solid material, usually activated carbon. Low to medium concentrations of organic compounds in air can be effectively captured by passing the gases through a bed of carbon particles, where the organic molecules are held by physical attraction. In such cases, most organic compounds can be removed or desorbed from the carbon bed by passing heated uncontaminated gases, usually steam or in some instances hot combustion gases, through the solid adsorbent. The higher organic concentrations of the gas streams in this regenerating phase of operation can then be more easily removed by condensation and/or distillation into a liquid form that can be recovered for reuse or proper disposal.

Carbon bed filters will become saturated with the organic compounds being collected, and at some point the capture efficiency will drop significantly. This is known as breakthrough, and can occur relatively rapidly once the saturation point is reached. The time to breakthrough is primarily a function of the carbon bed mass and the concentration of the organic compound(s) in the gas stream. Greater bed mass and lower organic concentrations will result in longer breakthrough times. Breakthrough times will also depend on the organic compound being collected, even for equivalent bed mass and concentrations.

In some cases, the organic compound may form chemical bonds with the bed material, leading to chemisorption. Chemisorbed species are much more difficult to remove from the bed, and are often impossible to remove during the desorption process. Chemisorption often occurs at relatively low rates (3–5% of the total absorbed organic mass), leading to eventual degradation of the bed's absorption capacity. Once the capacity has been degraded to such an extent that it is no longer effective, the bed material must be disposed of by landfilling or incineration.

There are several different designs of adsorption systems, the most common of which are regenerable fixed bed and canister systems. Regenerable fixed bed systems can be used to remove organic compounds from gas streams of several hundred to several hundred thousand cfm, with concentrations of several ppb or as high as 2500 ppm. Fixed bed systems can be operated continuously or intermittently. In either case, separate adsorption and desorption phases of operation are required.

For an intermittently operating system, the source routes the polluted gas stream through the bed during normal operation, allowing the bed to remove the organic compound. After a specified time, the source is shut down and the unit's desorption cycle begins and the organic compound is removed from the bed using steam or other heated gas. The desorption cycle is usually composed of three steps: removal of the organic compound by the hot gas (usually steam); drying of the bed using compressed or blown air; and cooling the bed to its operating temperature, usually by using a fan (often with the same fan used for the drying step). Once the desorption cycle has been completed (often within 1–2 hours), the bed is ready to begin the absorption cycle again.

In a continuously operating system, two or more beds are used, as shown in Fig. 10. One of the beds operates in the absorption cycle and the others are either being desorbed or are idle. A continuously operating system does not necessarily require twice as much bed material and supporting structure as an intermittent system. In cases where the desorption cycle is significantly shorter than the absorption cycle, three or more beds can be used in staggered cycles, reducing the amount of carbon bed material to less than twice that of an intermittent system.

Canister systems are similar to fixed bed systems in their absorption capabilities. However, they are not intended for on-site desorption, and are designed to be returned to a central site or the vendor for desorption and organic compound recovery. Originally, canister units were relatively small, such as 55-gallon drums. More recently, vendors have offered larger containers without desorption capability, some treating as much as 30,000 cfm of exhaust gas.[12] Canister units are typically used for lower volume intermittent gas streams, such as those from storage tanks vents.

Biofilters

Rather than relying on chemical reactions (such as FGD or thermal oxidizers) or mechanical separation (such as fabric filters or cyclones), biofilters utilize microorganisms to biologically convert organic pollutants into CO_2 and water. Similar to chemical reactions, biofilters are strongly dependent on the residence time needed for adequate destruction of the pollutant. While many chemical reactions occur very rapidly, biological processes tend to be significantly slower. Even so, they can occur rapidly enough to allow their use in engineering applications.

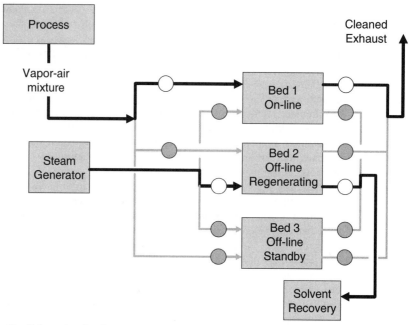

Figure 10   Schematic of a three-bed carbon filter. This configuration allows for continuous operation of the filter system, with additional capacity for upset or maintenance operation.

Figure 11 illustrates an open biofilter system, such as those used to control odors from concentrated animal-feeding or egg-laying operations, or for storage of manure. The basic components (fans, ductwork, compost bed, and wetting system) are similar for many biofilter designs, although industrial applications may use a closed multistage design to control emissions from processes with higher concentrations of VOCs.

In the simplest systems, VOC-laden air is circulated through beds of organic material, usually compost mixed with shredded or chipped wood or other biomass. The organic bed

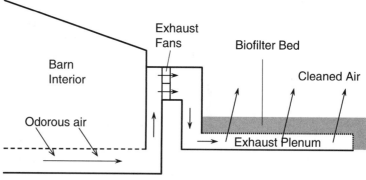

Figure 11   Schematic of an open biofilter system for controlling odors from animal feeding operations.

material removes a significant amount of the odorous organic compounds, such as hydrogen sulfide (H_2S), from the untreated exhaust air stream. These open systems are well suited for applications in which there is adequate open space to accommodate the filter's footprint and in which the escape of modest amounts of off-gases is acceptable, i.e., in noncritical odor-control applications.

The effectiveness of a compost biofilter is a function of the contact time between the odorous air and the bed material. Typical bed depths are 10–18 inches, resulting in a mass requirement of 50–85 square feet of compost material per 1000 cfm of exhaust air flow. Because of the greater flow resistance of the bed, more powerful exhaust fans are required to maintain proper air flow. To ensure the odorous air is exhausted through the biofilter, changes may be required to remove or modify vents and reduce air leaks through doors, windows, and other components.

The effectiveness of the biofilter also depends on the moisture content of the compost, which usually must be maintained at adequate levels (40–60% by weight) using a sprinkler system. The moisture content can change significantly depending on the flow through the filter bed and the relative humidity and organic content of the odorous exhaust air. Maintenance of proper bed moisture can be an iterative process, but there are guidelines that can be used to estimate the amount of water required. Figure 12 shows an estimated relationship between face velocity (V_f) and moisture requirements for various VOC concentrations. The face velocity is calculated by dividing the gas residence time in seconds (t_r) per meter of bed thickness, into 3600:

$$V_f = 3600/t_r$$

The moisture loss in Fig. 12 is given in g m^{-2} h^{-1}, approximately equivalent to 0.001 inches of water loss per each meter of bed thickness. The estimates shown in Fig. 12 are

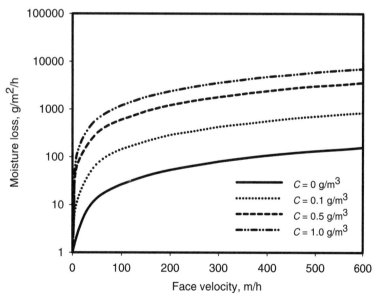

Figure 12   Moisture loss from an open-bed biofilter as a function of exhaust gas face velocity and organic compound concentration.

for an outside wet bulb temperature of 20°C, a gas stream with 95% relative humidity, an upper heating value of the VOC stream of 27.5 MJ/kg (11,800 Btu/lb), and complete VOC removal.[13]

The University of Minnesota Extension Service has estimated construction costs of approximately $100 per 1000 cfm of off-gas flow, with annual operating expenses for the larger fans, bed moisture maintenance, and bed replacement estimated at approximately $3 per year per 1000 cfm. van Lith et al. estimate total installed capital costs of 4–12 $/m^3 of bed material per 1000 cfm of off-gas flow for small (less than 10,000 cfm) systems, and roughly half that for larger (greater than 40,000 cfm) systems.[13] For some larger systems, these costs may be competitive with incineration systems.

Flares

Flares are most often used to destroy hydrocarbons that are released from industrial processes such as organic chemical production and petroleum refining. Flares are also used to incinerate combustible gas emissions from landfills and other processes that generate off-gases from biological processes, or to destroy natural gas that is released during petroleum drilling operations. The flares seen on offshore drilling and production platforms are examples of the latter application. In some processes, flares are used largely to destroy process gases that would otherwise be released during a process upset, and the flame must therefore be maintained by burning natural gas. In most cases, however, the waste gases are fed to the flare continuously.

The most simple flare designs are those that simply force the combustible gases through a nozzle (usually elevated), with the combustion air mixing with the gases through entrainment and diffusion. Simple flares will provide adequate destruction (usually greater than 99%) of the hydrocarbons in the gases when the compounds in the flared gases are relatively low in molecular weight. As the compounds become heavier, however, the mixing of the air and combustible gases may not occur rapidly enough to prevent the flare from smoking. In such cases, assisted flares should be used.

Flare combustion can be assisted by either air or steam. Air-assisted flares are usually composed of a process gas pipe surrounded by a concentric air duct. The air is forced through the duct with blowers, and the relatively high-velocity air generates turbulence, which results in a greater degree of mixing than is the case with unassisted flares. The improved mixing results in higher combustion efficiencies and therefore lower smoke levels, particularly for heavier hydrocarbons that unassisted flares may not fully burn. Steam-assisted flares can prevent smoking from an even wider range of flare gases, including those with the heaviest hydrocarbon compounds. However, high-pressure steam must be available at the flare site, which may limit the applicability of these flare designs. Steam-assisted flares use the high-pressure steam to generated turbulent mixing of the air and fuel, again to improve combustion efficiencies and reduce smoke formation.

Recently, flares that incorporate water injection have been developed for reducing flare smoking and noise. Particularly for high molecular weight flare gases, high-pressure injection of water in appropriate amounts and at the proper locations appears to provide significant reductions in flare combustion noise compared to flares without such injection. In addition to reducing noise and smoke, water injection can also reduce flare radiation. This can be an important consideration for applications such as offshore platforms, where radiation limits are needed to maintain safe working and operating conditions. Reduced radiation can result in reduced flare boom length or an increase in allowable flare gas flow.

One of the key parameters in proper flare operation is flare stability. Flare gas velocities that are too low can result in flares that do not burn steadily and that generate temperatures that are lower than desirable for maintaining combustion. Excessive flare gas velocities can lead to flare liftoff from the flare head that is too high for good radiative heat transfer from the flare head to the base of the flame. Windy conditions and variations in flare gas heat content will also impact the stability, and need to be adequately monitored to ensure that the conditions at the flare are such that stability is maintained. Flare stability can vary within bounds, but the combustion efficiency of the flare will drop off rapidly if stability is allowed to go outside those bounds.

For high-pressure gas flares, several flare tip designs can be used to increase the rate of mixing with the combustion air. Sonic or supersonic tips can be used to increase the entrainment of air into the gas stream, resulting in improved combustion efficiency and reduced flare radiation. A "Coanda" flare tip generates similar increases in air/gas mixing using a different design. In some cases, several small sonic tips can be used in place of a single large tip to generate the high velocity sonic flows necessary to create the increased air entrainment and resulting performance improvements.

Enclosed flares do not use significantly different technology, but rather use conventional flares (simple or steam-, air-, or gas-assisted) situated inside a large vertical duct. The enclosure acts to prevent the light and heat from the flare from being as perceptible to observers. Flare noise is also reduced in enclosed systems, which are often used in landfill gas flares and other flares near areas of concentrated population.

Fugitive Leak Detection and Repair

A considerable amount of VOC and volatile HAP emissions, especially in refineries and chemical plants, are attributed to leaks from piping and piping components such as pumps, valves, and gauges. The means of controlling these emissions typically require tightening connections or replacing seals, which not only reduce or eliminate these fugitive emissions, but also prevent the loss of product. In areas where significant reductions of VOCs are needed to achieve ambient air-quality standards, facilities may be required to have a leak detection and repair (LDAR) program.

By themselves, leak reductions can be quite cost-effective by preventing loss of products or process intermediate compounds. The difficulty and expense is in locating the leaks. The conventional LDAR approach relies on the use of portable "sniffers" that are held close to each component and detect the presence of hydrocarbons. This approach requires that each possible leak site be checked manually, which is very time-consuming and expensive for a large facility. However, an analysis of LDAR program results indicated that only about 2% of components are found to be leaking, and that over 80% of emissions come from less than 10% of the leaking components.[14] Many LDAR programs require that each component be checked regularly, often on a monthly basis. Therefore, while the regular inspections do find leaks, the programs are very time-intensive with relatively low rates of emission reductions.

An alternative approach is a "Smart LDAR" approach, which relies on more frequent inspections using remote sensing systems to identify leaks and repair the largest of them. Man-portable lasers are under development for use in backscatter absorption gas imaging (BAGI) that captures reflected laser light using video cameras to visually identify leaks from scanned components. Other systems that use larger equipment can be used to do a facility-wide scan to identify emissions from large equipment such as storage tanks that are not typically checked during routine LDAR inspections. The development and use of these systems are resulting in more reliable and accurate systems that are acceptable alternatives to conventional detectors.

6 CONTROL OF MERCURY

Emissions of mercury (Hg) into the atmosphere from anthropogenic sources come from several processes—chloralkali plants, manufacturing of Hg-containing products, incineration of hazardous and municipal wastes, and combustion of mercury-containing fuels, primarily coal. Control of these emissions is difficult, both because Hg concentrations in incineration and combustion flue gases tend to be very small (on the order of 1–10 ppm, and even less in coal flue gases), and because Hg is difficult to capture and contain.

Chloralkali plants use Hg in the process to form caustic soda, and most of the Hg is retained in liquid effluents. However, leaks and spills of the process liquids lead to evaporation of Hg vapor, which then escapes into the atmosphere. Reduction of these emissions is achieved by changing to Hg-free production methods and by reducing fugitive emissions of the process liquids.

In waste incineration systems, Hg is released from the thermal degradation of mercury batteries, dental amalgams, switches (including thermostats), manometers, and fluorescent light tubes when they are introduced into the incinerator or combustor. The level of Hg is high enough that activated carbon can be used to significantly control emissions. A more effective emissions control strategy is to prevent the introduction of Hg into the waste stream, which is increasingly effective as manufacturers turn toward Hg-free products and the waste disposal industry continues to improve its ability to segregate Hg-containing wastes for disposal through means other than combustion.

Removal of Hg using activated carbon can be achieved by either injecting the carbon into the flue gas or by passing the gas through an activated carbon bed filter. Alternative sorbent materials to activated carbon are available, and although they usually cost more per unit mass to purchase, they can capture higher levels of Hg per mass of sorbent. Therefore, careful evaluation is required to adequately determine the lowest-cost alternative.

Mercury control in coal-fired power plants is less straightforward due to the low levels of mercury in flue gases and the presence of different forms of Hg. Mercury in coal flue gases can exist in elemental (Hg^0) or ionic (Hg^{2+}) forms, which behave very differently with respect to control methods. Sorbent-based control technologies are effective for Hg^0, but not for Hg^{2+}. On the other hand, Hg^{2+} is water soluble and can be captured in wet flue gas desulfurization systems, while Hg^0 cannot be effectively controlled by FGD units.

For coal-fired units in general, fabric filters show significantly higher Hg collection efficiencies than other types of particle control technologies. Ionic mercury appears to be more easily captured than Hg^0, which means that Hg emissions from bituminous coal (which form higher levels of Hg^{2+}) are more readily controlled than emissions from sub-bituminous coal or lignite. Preliminary studies indicate that Hg concentrations from bituminous coal combustion can be reduced by as much as 98% by conventional control technologies (ESPs, fabric filters, and FGD systems), while units burning sub-bituminous coal or lignite have significantly lower reductions across the air-pollution control systems. Injection of activated carbon shows promise for increasing Hg capture, but this approach has not yet been demonstrated over extended periods of operation.

For all forms of Hg control, a key issue is the disposal of pollution control residues. Whether the residues are in solid or liquid form, it is important to ensure that the captured Hg is not released into the environment via other routes, such as leaching into the soil or groundwater. The presence of Hg in the pollution control system effluents can also have adverse effects on the salability byproducts such as gypsum from FGD residues or cement from coal fly ash.

REFERENCES

1. Srivastava, R. K., "Controlling SO_2 Emissions: A Review of Technologies," EPA/600/R-00/093, Office of Research and Development, Washington, DC 20460, November 2000.

2. Borio, R., P. Rader, and M. Walters, "Ammonia Scrubbing: Creating Value from SO_2 Compliance," EPRI-DOE-EPA Combined Utility Air Pollution Control Symposium: The MegaSymposium: Volume 1: SO_2 Controls, EPRI TR-113187-V1, Atlanta, GA, August 1999, pp. 7-13–7-27.

3. Ellison, W., "Worldwide Progress in Ammonia FGD Application," EPRI-DOE-EPA Combined Utility Air Pollution Control Symposium: The MegaSymposium, Volume 1: SO_2 Controls, EPRI TR-113187-V1, Atlanta, GA, August 1999, pp. 7-1–7-11.

4. Nolan, P. S., T. J. Purdon, M. E. Peruski, M. T. Santucci, M. J. DePero, R. V. Hendriks, and D. G. Lachapelle, "Results of the EPA LIMB Demonstration at Edgewater," Proceedings: 1990 SO_2 Control Symposium, EPA-600/9-91-015a, Office of Research and Development, Research Triangle Park, NC, May 1991.

5. Bowman, C. T., "Control of Combustion-Generated Nitrogen Oxide Emissons: Technology Driven by Regulation," XXIV Symposium (International) on Combustion, The Combustion Institute, Pittsburgh, PA, 1992, pp. 859–876.

6. Lim, K. J., L. R. Waterland, C. Castaldini, Z. Chiba, and E. B. Higginbotham, "Environmental Assessment of Utility Boiler Combustion Modification NO_x Controls, Volume 1: Technical Results," EPA-600/7-80-075a (NTIS PB80-220957), Office of Research and Development, Research Triangle Park, NC, April 1980.

7. Wilson, S. M., J. N. Sorge, L. L. Smith, and L. L. Larsen, "Demonstration of Low NO_x Combustion Control Technologies on a 500 MWe Coal-Fired Utility Boiler," in Proceedings of the 1991 Joint Symposium on Stationary Combustion NO_x Control, EPA-600/R-92-093a, Office of Research and Development, Research Triangle Park, NC, July 1992.

8. Bayard de Volo, N., L. Radak, R. Aichner, and A. Kokkinos, "NO_x Reduction and Operational Performance of Two Full-Scale Utility Gas/Oil Burner Retrofit Installations," in Proceedings of the 1991 Joint Symposium on Stationary Combustion NO_x Control, EPA-600/R-92-093a, Office of Research and Development, Research Triangle Park, NC, July 1992.

9. U.S. Environmental Protection Agency, "Control of NO_x Emissions by Reburning," EPA-625/R-96-001, Office of Research and Development, Washington, DC, February 1996.

10. U.S. Environmental Protection Agency, "Nitrogen Oxides: Why and How They Are Controlled," EPA-456/F-99-006R, Office of Air Quality Planning and Standards, Research Triangle Park, NC, November 1999.

11. U.S. Environmental Protection Agency, "Air Pollution Control Technology Fact Sheets," EPA-452/F-03-003, -005, -017, -025, -026, -027, and -028, Office of Air Quality Planning and Standards, Research Triangle Park, NC, July 2003.

12. U.S. Environmental Protection Agency, "EPA Air Pollution Control Cost Manual, Sixth Edition," EPA 452/B-02-001, Office of Air Quality Planning and Standards, Research Triangle Park, NC, July 2002.

13. van Lith, C., G. Leson, and R. Michelson, "Evaluating Design Options for Biofilters," *Journal of the Air & Waste Management Association,* **47**, 37–48 (1997).

14. Taback, H., "Analysis of Refinery Screening Data," American Petroleum Institute Publication 310, American Petroleum Institute, Washington, DC, November 1997.

CHAPTER **31**

WATER POLLUTION CONTROL TECHNOLOGY

Carl A. Brunner
Retired from U.S. Environmental Protection Agency

J. F. Kreissl
Environmental Consultant
Villa Hills, Kentucky

1 INTRODUCTION

Various degrees of pollution management have been carried out for centuries. Roman sewers are well known. For much of this time, emphasis was on reducing aesthetic problems in cities. When a connection was made between disease and microorganisms harbored in human wastes, the urgency to remove these wastes from human contact increased. Although this connection was suspected earlier, it was not proven until the 19th century. The early pollution-management methods did not include treatment and did not have a significant impact on the environment. Concerns about the growing impacts of wastes created by mankind upon the environment are relatively recent. In the 19th century, people began to notice that waterways were showing serious signs of pollution and a few states began to consider remedial measures.[1] Widespread use of treatment of wastewater did not occur prior to the early 20th century. The most commonly employed method of solid waste disposal well into the 20th century was simply dumping on the land or sometimes in local water bodies. Municipal incinerators began to be used by the end of the 19th century. The more acceptable sanitary landfill did not become a common substitute for the open dump until after World War II. Similarly, significant efforts to curb air pollution were not initiated until after World War II. Strong concern for pollution control, not only for health reasons, but also to improve the environment, originated in the 1960s with the development of very active environmental organizations. The federal government and some state governments responded by passing much stronger legislation than had been in force up to that time. The result of this legislation and strong public interest in the environment has been a greatly increased level of treatment of solid, liquid, and gaseous discharges.

This chapter will familiarize the reader with the most important treatment methods for treating waste materials, primarily liquid wastes. To understand why these methods are used, it is useful to know something about the regulatory framework that has developed, primarily as a result of federal legislation. Wastewater discharges have been regulated under a long series of legislative acts, but the key legislation that made very great changes to the degree of control required was the Federal Water Pollution Control Act Amendments of 1972. This act has been amended a number of times since and the overall legislation is commonly called the *Clean Water Act*. Under this legislation, specific requirements have been made for the direct discharges of both municipal and industrial wastewaters to a receiving water body and for the discharge of industrial wastewater to the public sewer system, or indirect discharge. For municipal wastewater, the minimum requirement is either for a reduction of biochemical oxygen demand (BOD) and suspended solids (SS) of at least 85% or effluent values of both BOD and SS not exceeding 30 mg/L, whichever is more stringent. For discharges to streams where the minimum requirement is viewed as having too great an impact, more stringent standards can be required, including parameters in addition to BOD and SS.

The US Environmental Protection Agency (EPA) has developed effluent guidelines for direct discharge to receiving streams for 51 industrial categories. Thirty-two of those industrial categories must meet EPA-specified effluent quality when discharging to municipal sewer systems.

For specific information in a community, the first point of contact should be local wastewater officials. Above the local level would be the state environmental agency and finally the U.S. Environmental Protection Agency (EPA), which maintains ten Regional Offices across the country. Changes in clean water legislation will continue to be made, possibly with significant impact on discharges.

In recent years, the Safe Drinking Water Act (SDWA) has become a major waste control regulatory mechanism, particularly in areas served by ground water, as sources of drinking water supplies. The SDWA has banned certain industrial wastes from being disposed to soils in areas without surface discharging, central collection and treatment systems, and required source water protection plans can force additional restrictions on discharges within the zone of influence of municipal wells.

Another law that has had an impact on control of wastes is the Resource Conservation and Recovery Act (RCRA). RCRA had its origin in 1976 as amendments to the Solid Waste Act. Regulations arising from RCRA and amendments are the primary method of control for hazardous wastes. For a waste to be hazardous, it must exhibit one or more of the following characteristics: ignitability, reactivity, corrosivity, or failure to pass a defined extraction procedure. The regulations are quite complex. One part deals with the land-disposal ban for listed substances. Hundreds of materials are included along with the best demonstrated available technology (BDAT) for controlling each. Questions regarding whether wastes are hazardous and their control should be referred to the state environmental agency.

Although this chapter deals with waste treatment, it is important to point out that a widespread attitude has emerged that favors prevention of pollution over pollution control. There are many facets to a pollution-prevention program, including changes in raw materials to less hazardous materials and changes in manufacturing methods. From the standpoint of waste treatment, there should eventually be measurable reduction in discharges or at least a reduction in the strengths of wastes being discharged. To some degree, wastewater treatment should change to in-plant reuse and recovery systems with polluting materials and clean water being returned to the manufacturing operation.

2 MUNICIPAL WASTEWATER TREATMENT

Every day, billions of gallons of municipal wastewaters are discharged from plants ranging in size from those that handle the discharge from a single house to more than one billion gallons per day. There are differences in the composition of municipal wastewater from city to city because of differences in the industrial and commercial contributions, and differences in overall contaminant concentration because of varying degrees of stormwater or ground-water entrance into the sanitary sewer system. Even with these differences, however, there is enough similarity that a small group of technologies is able to handle the treatment of most of the sources. Municipal wastewater contains an enormous number of chemical compounds including a wide range of metals and organic materials. The exact composition of all contaminants is never known. Two very common chemical elements, phosphorus and nitrogen, must be emphasized because they act as fertilizer in the receiving water body and can lead to excessive growth of nuisance plants such as algae.

Nitrogen is normally considered the limiting nutrient in marine waters, and its removal is dictated in many facilities discharging to estuaries and coastal waters. In fresh waters phosphorus is the limiting nutrient, and may need to be removed. Both nutrient removal schemes can be quite expensive depending on the degree of removal required by the permit.

In addition, the wastewater contains large numbers of microorganisms, including human pathogens. Ideally, wastewater treatment would reduce all of the contaminants to low levels. In reality, the methods that are used have the primary function of reducing the oxygen demand of those materials that would biologically degrade in the receiving water body. When degradation occurs in the receiving water, oxygen is consumed. Because the desirable stream biota require oxygen, its concentration should not be depressed significantly below the ambient level. As already indicated, minimum requirements have been placed on the BOD in the discharge because this parameter gives a partial measure of the potential for decreasing dissolved oxygen.

For sensitive receiving waters a high removal of BOD alone does not reduce the oxygen demand of the treatment plant effluent sufficiently. A typical secondary effluent also contains, on the average, about 20 mg/L of reduced organic plus ammonium nitrogen. That represents about 92 mg/L of oxygen demand. Therefore, in oxygen-sensitive receiving waters, nitrogen must either be oxidized to nitrate or be removed to minimize oxygen demand.

By settling to the bottom of the receiving water body, SS can also have a direct negative effect on sediment quality and bottom-dwelling aquatic life and habitat.

Nearly all municipal wastewater is treated by biological methods. The more common technologies are well described in the textbook literature.[2] These methods take advantage of the ability of a large group of natural microorganisms to utilize the organic materials in the wastewater for food. These organics include soluble materials and many of the insoluble materials that are solubilized by the complicated biochemical activity taking place. The method used by nearly all large cities and treating the greatest volume of wastewater is the activated sludge process. A second method, the trickling filter, was formerly very common, but has been displaced in many urban communities by activated sludge. There are a number of variations of these two methods, including combinations. Although these two methods account for a large fraction of total treatment capacity, there are a variety of other, usually less complicated, methods used predominantly by small communities.

Figure 1 is a diagram of a typical sequence of treatment operations and processes that are found in many treatment plants. Incoming wastewater passes through preliminary treatment, which might include screens for removal of very large objects, and a short-term settling operation for removal of small to moderate but heavy particles (grit). The wastewater is usually pumped to a high enough level that it can flow through the remainder of the plant by gravity. Pumping may precede preliminary treatment or may follow at least some of the preliminary treatment to reduce wear on the pumps.

Following preliminary treatment is primary settling, which is intended to remove large organic particles not removed in preliminary treatment, and results in about 50–60% SS and

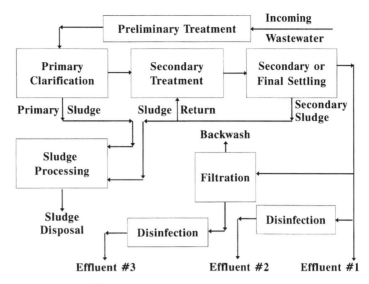

Figure 1 Municipal wastewater treatment.

about 30% BOD removal. This operation is carried out in tanks with circular or rectangular configurations. Residence time is up to two hours. Although the principal action is settling, there is also some biological activity. If the settled sludge is not removed quickly enough, biological activity can result in unpleasant odors and floating sludge. In large plants, there may be many settlers operating in parallel. A sludge-removal mechanism moves the sludge in the bottom of each settler to an opening for discharge to the sludge-processing system. Primary treatment is not required for all plants, but is almost always used in large plants.

The partially treated wastewater or primary effluent flows next to secondary treatment, where the major biological activity occurs. In the commonly used activated sludge process, treatment is carried out in tanks supplied with aeration devices. These may be various kinds of diffusers, mounted in or near the bottom of the tanks, that bubble air through the water or surface-mounted mixers or mechanical aerators. In addition to providing oxygen to the water, the aeration devices keep solids in suspension throughout the volume of the aeration tanks, which may be very large. In the aeration tanks, aerobic degradation of the organic matter in the wastewater takes place. Commonly a level of SS of about 2000 mg/L or higher is maintained. The very large number of microorganisms in these solids, commonly referred to as activated sludge, accomplish the biological degradation. The residence time in the aeration tanks is often about four to six hours, but there can be great variation from high rate treatment to extended aeration. In the latter, the residence time might be 24 hours. Extended aeration plants may not require primary treatment. Water leaving the aeration tanks flows to the secondary or final settling tanks. Although these settling tanks are used with technologies other than activated sludge, they play a unique role with activated sludge. As indicated in Fig. 1, part of the sludge removed from these tanks is returned to the aeration tanks. The amount of sludge returned is adjusted to maintain the desired SS concentration in these tanks.

The frequently used trickling filter form of secondary treatment consists of one or more beds of rock or other packing. The wastewater is dispersed over the upper surface of the filter bed and is collected at the bottom in an underdrain system. Biological slime accumulates on the packing. Oxygen is usually provided by the natural flow of air through the packing, although forced air may be employed. There is a continuous process of sloughing of the biological growths from the trickling filter media that carries by the wastewater flow to the secondary or final clarifier.

Many trickling filters operate with return of some of the water from the underdrain system back to the top of the filter, where it is mixed with incoming feed water. Only rarely is there any return of material from the final settler.

There are a number of variations and combinations of activated sludge and trickling filter or fixed film devices that are being operated. The rotating biological contractor (RBC) combines the two concepts. RBC systems include basins which contain some biologically active solids, but they also are equipped with banks of partially submerged vertical disks that rotate. Biomass accumulates on the disks. Oxygen is supplied to the system primarily through exposure of the rotating disks to the air. Adequate flocculation of the biomass from a trickling filter to provide effective settling in the final settler can be a problem. Adding a short-residence-time aeration tank following the trickling filter along with returning some sludge from the final settling tank has been found to improve flocculating and settling. These systems are referred to as trickling filter-solids contact.

The primary function of the final settling tanks is the removal of suspended matter from the water to a level that meets discharge requirements. For the standards of 30 mg/L BOD and 30 mg/L SS, this degree of treatment should be adequate, assuming there is no upset in the biological system. The settling tanks are of circular and rectangular configuration and typically have an overflow rate of 400 to 600 gallons/day per square foot. The important function of providing sludge to the aeration tanks in the activated sludge process has already been mentioned.

In the 21st century there have been significant trends toward substitution of membrane bioreactors for conventional biological reactors and secondary clarifiers, particularly in smaller facilities. The benefits of fine-bubble diffusion have resulted in greatly increased numbers of treatment plants using this more-efficient alternative. Biological process variations account for much of the nutrient (nitrogen and phosphorus) removal, often with some chemical addition (in the case of phosphorus) to further reduce the effluent concentrations to meet permitted levels. Improved control systems allow the plants to accomplish their goals with reduced energy demands, particularly regarding activated sludge aeration energy.

Figure 1 indicates three possible routes for the secondary effluent leaving the final settlers. Direct discharge to a receiving water is allowed in some locations. Disinfection is frequently required. To meet discharge requirements significantly less than the 30 mg/l requirements would necessitate a minimum of additional SS removal. Additional SS removal may also be a practical approach for meeting a phosphorus requirement.

Disinfection is frequently carried out using either elemental chlorine or a chlorine source, such as sodium hypochlorite. The chlorine is mixed with the water in a chlorine contact chamber, which often has a residence time of about 30 minutes. Because the addition of chlorine to waters containing organic materials has the potential for producing toxic chlorinated species, there is increasing concern about the health and ecological impacts of chlorination. Possible chemical substitutes are chlorine dioxide, which does not result in significant formation of chlorinated organics, and ozone. Use of these chemicals is more costly and at this time is rare. A non-chemical method, ultraviolet radiation (UV), is being used increasingly as a substitute for the traditional chlorination. This is a rapidly evolving technology with further improvements in the radiation sources being likely. In a typical installation, the water is passed through a chamber containing banks of tubular lamps. The lower the SS concentration, the more effective is the disinfection, which is dependent upon sufficient radiation reaching all of the water. There is also the obvious question of the ability of UV (and to a lesser degree with chemical methods) to disinfect within particles constituting the suspended matter. Parker and Darby reviewed the literature on this question[3] and carried out tests using rapid stirring to break up solids particles in secondary effluent that had been UV disinfected with low pressure UV lamps. They found shielding of coliforms as measured by increases in these organisms in the water phase after stirring and stated the need for greater emphasis on pretreatment of the wastewater before disinfection, including particulate removal.

In the past decade, there has been a strong trend toward use of UV disinfection to eliminate the safety problems associated with chlorine disinfectants and nearly universal regulatory requirements for dechlorination before discharge. The most complete coverage of disinfection can be found in Crites and Tchobanoglous.[4]

The usual method in large plants for removal of SS below the concentration in secondary effluent is in-depth filtration. An in-depth filter contains two feet or more of sand, or a combination of sand covered with larger-particle-size-coal (dual media), or sand and coal over a fine, high-density material such as garnet (multimedia). Water is introduced at the top of the filter bed. These filters strain much of the remaining SS from the water by several mechanisms, including incorporation in biological slime. Eventually the solids deposits increase the pressure drop across the filters to the point that they must be backwashed. Backwashing suspends the filter media particles and removes most of the deposited solids. After backwashing, the media particles in the dual-media and multimedia filters naturally grade themselves with larger particles (coal) at the top. The reason for considering a combination of media such as sand and coal of a larger particle size is the increased solids holding capacity and increased run length between backwashes. The multimedia system should result in the highest degree of solids removal. To avoid the need for disposal, the backwash stream must be returned to an appropriate point in the treatment sequence, possibly before secondary treatment or the final settler.

In Fig. 1, the concentrated solids streams are referred to as sludge and are shown leading to sludge processing. To improve public acceptance of the treated sludge for use as a soil amendment, there is a tendency in the profession to refer to adequately stabilized sludge as biosolids. Although the volume of sludge produced is only a few percent of the volume of water treated, its processing constitutes a significant fraction of the overall treatment activity. A variety of methods exist that vary somewhat with plant size and possible modes of disposal for the treated sludge or biosolids. Metcalf and Eddy mention eight categories of sludge-processing operations or processes, under which are included 34 specific methods (Ref. 2, p. 767). In the liquid handling sequence of a sewage treatment plant, much of the capital cost is in concrete. In the sludge treatment sequence, there is greater opportunity for companies to manufacture major parts of the system. Both U.S. and foreign manufacturers are responding. Unless the sludge contains unusually high concentrations of hazardous materials, such as certain metals, it is very useful as a soil amendment. EPA and many environmental groups favor this beneficial use of sludge. Where application to the soil is not practical, incineration or landfilling are alternatives. For any of these methods, substantial processing of the primary and secondary sludge streams is necessary. Because their solids concentrations are only a few percent or less, these streams are usually first thickened. The simplest method is gravity thickening. If the sludge is to be incinerated, to reduce the fuel requirement to the incinerator, it must be further dewatered by mechanical means, such as filtration using belt filters or plate and frame filters, or centrifugation. Solids concentrations of up to 30% are desirable. For other than incineration, a very common method of sludge treatment is digestion. The objective of this biological process is to stabilize or further reduce the biodegradable content of the sludge and to reduce the concentration of pathogens. Digestion is commonly carried out under anaerobic conditions in large plants, but is also frequently conducted aerobically. Usually, the temperature is elevated, if possible, by using the heat generated by biological activity or, in the case of anaerobic digestion, by burning of methane produced by the digestion process. Another method to stabilize sludge is by adding chemicals which raise the pH and form a cement-like residue. In small plants, it is common to stabilize sludge by long-term storage in lagoons. Recent sludge disposal regulations determine to a considerable extent the condition of the sludge for landfilling, beneficial use on land and production of marketable products. For large-scale spreading on agricultural land, digestion may be sufficient. Composting has become common for producing a marketable product. The process has encountered odor problems that necessitate careful control of the operation.

Although very large plants will consist of a system typified in Fig. 1, occasionally plants of 10–20 million gallons per day and many smaller plants may have somewhat simpler systems such as treatment ponds, or lagoons, wetland systems, and land application systems. Most of these systems will include some form of preliminary treatment. They may or may not include primary sedimentation. Biological degradation occurs in the ponds, on the surface of the land, or underground. These methods are often land-intensive and are most appropriate in locations where large expanses of inexpensive land are available.

It should be noted that approximately 25% of the United States is not serviced by sewers, and about 80% of the nation's treatment plants service populations of 10,000 or less. These smaller facilities have been receiving more intense scrutiny from federal and state regulators recently, owing to their locations in sensitive headwaters to which they have historically discharged and their past record of violations of their discharge permits. The unsewered systems overwhelmingly discharge to the soil close to the source. The small sewered systems are being scrutinized to stimulate better management for themselves and for their surrounding unsewered populations. The technologies employed by these small treatment facilities have largely mirrored the large city practice during the last few decades, a practice that has had a disturbing impact on the communities, the receiving waters, and water resources in general. Modern approaches to these smaller populations emphasize centrally managed, distributed, soil-based effluent dispersal as close as possible to the sources. This approach maintains water balances within watersheds, minimizes the potential for catastrophic problems, and maximizes the involvement of the community. However, the emphasis of this chapter is on the other end of the scale, i.e., large urban wastewater systems and industrial wastewater treatment.

In the 1960s, there was much enthusiasm for advanced-waste-treatment systems that would produce very-high-quality water from municipal wastewater. To justify the cost of these systems, the water had to be reused for a high-quality purpose, including potable water. Although a pilot system operated at Denver, Colorado, indicated that, by all practical measures, water of potable quality could be produced, no full-scale plant has been constructed in the United States for the direct potable reuse of treated wastewater. There have been a small number of full-scale plants constructed, however, that do produce very-high-quality water. The systems include, among other technologies, chemical clarification, activated carbon treatment, and in at least one case, reverse osmosis. These technologies are discussed in Section 3.

The interest in reuse has not waned, but it has taken a less "high-tech" turn. There is much more emphasis today on using the natural processes of the soil in lieu of the high capital cost and skilled operation and maintenance demands of these very sophisticated technologies. Now the emphasis is on mixing with natural ground water and storage for long periods before extraction of the mixed water sources. In water-short countries such as Israel, these practices have been used for several decades. In the United States the driving force for reuse has been nonpotable needs such as saltwater intrusion minimization along the coasts and because the treatment requirements for many nonpotable reuse applications in water-short areas are less costly than meeting stringent discharge permit requirements for surface water discharge.

Table 1 summarizes the municipal wastewater treatment technologies included above.

Capital and operating costs have not been included because of the wide variability, depending on site-specific conditions. As expected, there is an economy of scale for both capital and operating costs. At the time of writing, federal funds (primarily loans) are available through each state to aid in payment for construction of municipal sewage-treatment plants.

3 INDUSTRIAL WASTEWATER AND HAZARDOUS WASTE TREATMENT

Although nonhazardous industrial wastewater and hazardous wastes are regulated under different legislation, methods for their control may be similar or even identical. In contrast to municipal wastewater, there is a very great diversity in the types of waste to be treated and an equally great variety in the kinds of treatment that might be used. Because it is impossible to include discussion of all of these methods, a few commonly used methods have been selected. It must be emphasized that biological treatment is used for treatment of many organic industrial wastewaters. Systems described for municipal wastewater treatment are applicable for this purpose.

3.1 Chemical Precipitation and Clarification

Chemical clarification can be very effective for removal of contaminants from aqueous streams where the contaminants are particulate in nature. Figure 2 is a diagram of a typical chemical clarification system. To obtain a high degree of solids removal requires the formation of flocculent particles that will trap other particles and that will settle well. Flocculating chemicals such as aluminum sulfate (usually referred to as *alum*) or various iron salts

Table 1 Municipal Wastewater Treatment Technologies

Technology	Usual Function
Preliminary treatment	Removal of large objects or reduction in size
	Removal of grit
Primary settling	Removal of large organic particles not removed in preliminary treatment
	Results in about 30% BOD reduction
Secondary treatment	Most common use is for further BOD and SS reduction to levels of 30 mg/L or less
	Can be designed to remove phoshorous, nitrify ammonia and denitrify nitrate
Disinfection	Reduction in levels of pathogenic microorganisms
Filtration (sand, dual media, multimedia)	Reduction of SS and BOD where effluent requirements are less than 30 mg/L
	Can be used to improve effectiveness of disinfection
Sludge processing	Preparation of sludge for beneficial use or disposal
	Can involve a wide variety of technologies for thickening, dewatering, biologically or chemically stabilizing and thermally treating, including drying and incineration

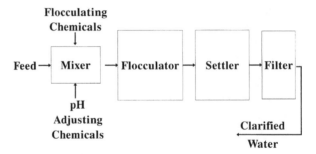

Figure 2   Chemical clarification system.

are rapidly mixed with the feed under pH conditions that will cause the insoluble hydroxides of these metals to form. The water then flows to the flocculator, where it is slowly agitated. Under these conditions, large gelatinous particles form that settle to the bottom of the settler and are removed as sludge. The overflow from the settler may be of satisfactory clarity. If solids removal to very low levels is needed, a filter can be added to the system. In large installations, a commonly used type of filter would be a sand, a dual-media, or a multimedia filter. Similarly to the situation with municipal wastewater treatment, a large clarification system would consist largely of concrete tanks. For small systems, a number of manufacturers provide package systems which may have unique flocculator–settler combinations and filters.

As already indicated, clarification may be the only treatment needed for some industrial wastewaters where the intent is to reduce BOD and SS—for example, discharge to a public sewer. By carrying out precipitation of soluble materials, it may be possible to extend use of clarification to removing some hazardous materials, such as toxic metals. Lime can be used for this purpose where the hydroxide form of the metal is insoluble. In some situations, the precipitate that forms from lime addition settles well without adding other chemicals. If not, inorganic flocculants or a wide range of organic flocculants called polyelectrolytes can be added. Other materials that can be used for metals precipitation are compounds containing sulfide and carbonate, most likely sodium sulfide and sodium carbonate. Sulfides of many metals are very insoluble; some carbonates are insoluble. Because very toxic hydrogen sulfide can be formed from sulfides at low pH, care must be taken to prevent pH reduction after precipitation has occurred. Sulfide dose must be controlled closely to prevent excess sulfide in the water leaving the system, which is a potential source of hydrogen sulfide. Also, sulfide is a reducing agent that exerts an oxygen demand until it is converted to sulfate.

Chemical coagulation and chemical precipitation may function as pretreatment prior to discharge to a public sewer system or prior to other treatment technologies where particulate matter causes problems. Two examples would be granular activated carbon treatment and certain membrane processes.

Chemical coagulation has been used for municipal as well as industrial wastewaters. In municipal applications the advantage is that is also removes phosphorus as well as BOD and SS. The disadvantages are primarily related to the volume and weight of sludge produced and difficulties with sludge processing. In the case of lime precipitation, the sludge produced is even more voluminous.

3.2   Activated Carbon Adsorption

A very commonly used method for removing many organic materials from both gas and liquid streams is activated carbon adsorption. This discussion will involve only liquid (water) treatment, although much of what is said pertains also to treatment of gas streams. Activated carbon is made either from coal or organic materials that are carbonized and activated, usually by a thermal treatment that creates a pore structure within each carbon particle. This pore structure creates a large internal surface area that allows, under some circumstances, percentage amounts of contaminants, based on the mass of carbon, to be adsorbed from solution. The amount adsorbed varies from solute to solute, depending on the attraction between the carbon surface and the adsorbed material. The amount of adsorption for any solute is a function of the concentration of that solute in the aqueous phase. For a simple system with a single solute, adsorption capacity data can be obtained as a function of liquid concentration to produce a relationship called an isotherm. Very often the Freundlich isotherm is used to correlate the data. The form of this equation is as follows:

$$q = K_1 C^{1/n}$$

where q = amount of solute adsorbed, mass/mass carbon
   K_1 = a constant for the specific solute.
   C = the solute concentration in solution, mass/volume solution
   n = a constant for the specific solute

For low concentrations, it can often be assumed that there is a linear relationship between q and C, as follows:

$$q = K_2 C$$

where K_2 = a constant for the specific solute.

Observing either equation, it can be seen that the amount adsorbed decreases with decrease in solution concentration, approaching zero as the solution concentration approaches zero. For multicomponent systems, the isotherm for one component depends upon the overall composition. It is important to remember, however, that in general the amount of adsorption, q, increases or decreases with increase or decrease of that component in solution. To devise contactor systems that will make efficient use of the carbon, this principle must be kept in mind.

Activated carbon is available as a very fine powder or in a granular form with an average particle size of about one millimeter. Aqueous wastes are usually contacted with powdered carbon in a mixing and settling operation, conducted batchwise or as a continuous operation. A single tank can be used for batch operation by first mixing and then settling. Often a flocculating chemical must be added to attain adequate separation of the carbon. Continuous operation requires a mixer and a settler. Filtration of the effluent might be necessary for good removal of the carbon. Ultimately, in batch operation, and clearly in continuous operation, carbon is in contact with water of effluent concentration. Recalling the isotherm discussion, the loading of organic on the carbon will be low. More effective use of the carbon could be made using a multistage operation, but the cost of the system would be substantially increased. Use of powdered carbon does not lend itself easily to carbon reactivation.

Treatment with granular activated carbon is usually carried out by passing water through columns of the adsorbent. Figure 3 is a diagram of a two-column system using downflow operation. Operation can also be carried out in an upflow mode, either at a rate that allows carbon particles to remain as a packed bed or at a high enough rate to fluidize the bed. Conducting carbon treatment as shown in Fig. 3 results in much higher loading of organics on the carbon than is usually possible with powdered carbon treatment. In this case, carbon initially is in contact with water of feed concentration and, in terms of an isotherm, can adsorb an amount equivalent to that concentration. As the water passes downward, the concentration of contaminants in the water decreases, producing an S-shaped concentration curve

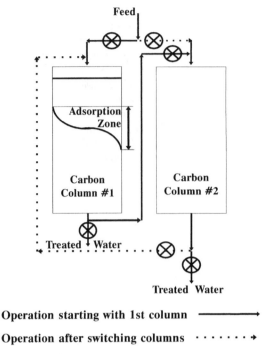

Operation starting with 1st column ⟶

Operation after switching columns ⋯⋯⋯→

Figure 3   Granular activated carbon treatment.

(solution concentration versus column length) designated as the adsorption zone. Assuming the length of the carbon bed is somewhat longer than the adsorption zone, effluent concentrations of adsorbable materials can approach zero. As the carbon becomes loaded, the adsorption zone moves down the column and eventually breakthrough of contaminants occurs. At that point, it becomes necessary to replace the carbon with new or reactivated carbon. If the adsorption zone is very short, a single column may be sufficient. If the adsorption zone is long, multiple columns in series can be used to increase the effectiveness of carbon utilization. With multiple columns, breakthrough is delayed until the lower end of the adsorption zone reaches the bottom of the last column. At that point, only the carbon in the first column needs to be replaced and the flow needs to be changed, as shown by the dotted lines in Fig. 3. With this mode of operation, high loading of most of the carbon in the system can be attained. If treatment cannot be interrupted, a standby column or multiple systems in parallel must be provided. Granular carbon is frequently reactivated because of the high cost of replacement carbon. Also, reactivation eliminates the need for and the cost of disposal of the contaminated carbon. Reactivation is a thermal process that is similar in function to the original activation. Reactivation can be done on-site if the size of the operation is large. Some companies provide a reactivation service if on-site reactivation is not practical. When dealing with carbon loaded with hazardous materials, careful attention must be given during reactivation to assure that the hazardous materials are either destroyed during reactivation or are controlled by an adequate air pollution-control system. Carbon loss during reactivation may be about 10%.

Operation of granular activated carbon systems has a number of problems, two of which can occur in essentially all cases. Activated carbon and ordinary steel form an electrolytic cell that corrodes the steel. To protect the metal, carbon columns are lined with rubber or coated with a nonporous organic coating. If the coating sustains damage, rapid corrosion of the tank wall can occur. Biological activity usually occurs in carbon systems when they have operated for a period of time. Surprisingly, this activity can be found even in systems treating toxic materials. If the activity remains at a low enough level that serious clogging does not occur, the result can be beneficial because biological degradation reduces loading on the carbon. If biological activity does cause clogging, backwash of the carbon columns is necessary. Not only does backwashing create an added operating cost, but it results in a biomass residual that must be further handled. Biological activity can create anaerobic conditions resulting in odor problems.

Activated carbon treatment of normal municipal wastewaters is rarely justified, owing to the costs and demands of operation. However, it is an effective process for industrial wastewaters and certain municipal wastewaters where soluble organic matter must be removed to meet effluent requirements.

3.3   Air Stripping

Air stripping is useful for removal of volatile organic materials from a variety of aqueous streams, including industrial wastewaters, drinking water sources, contaminated groundwater, and contaminated soil itself. The treatment method is practical for dilute solutions with concentrations less than about 100 mg/L[4] and for hydrophobic compounds, such as chlorinated solvents, that exhibit large Henry's law constants. Where the solutes being stripped are hazardous materials, air stripping may result in an exhaust air stream that is in violation of air pollution requirements, necessitating some form of air pollution control.

For an aqueous system with one contaminant at low concentrations, the equilibrium concentration of contaminant in air contacting the aqueous phase bears a linear relationship to the liquid concentration. This line passes through the origin. The slope of this curve is given by the Henry's law constant. For multicomponent systems, the Henry's law constant of any one component will almost certainly be affected by the presence of the other components, but generally, for each component, the concentration in the gas phase in equilibrium with the liquid will increase as aqueous phase concentration increases.

Air stripping can be carried out by simply bubbling air through water in a tank. In a batch operation, assuming one contaminant, the effect on a plot of gas concentration versus liquid concentration would be to move down that curve toward the origin. In a single-tank continuous operation, concentration of the liquid in the air stripping vessel is essentially that of the treated water and, therefore, the maximum concentration in the air must be very low. In either batch or continuous operation with this simple system, the amount of air required can be very large. This mode of operation may be acceptable if the air can be discharged without a need for pollutant removal. Much more efficient operation results from using a countercurrent contacting device. A commonly used example is a packed tower with downflow of the contaminated water countercurrent to the upward flow of air. The packing can be wood slats, rings of ceramic or other materials, or plastic packing of various geometries. The packing provides a large surface for mass transfer of contaminants from the aqueous phase to the air to take place. In a countercurrent contactor, treated water leaving from the bottom contacts clean air just as it would in a simple single-stage device. As the air travels upward, however, it contacts water of increasing contaminant concentrations. The result is a continuing transfer of the contaminant to the air. If the contactor were very long, representing a large number of equilibrium contacting stages, the air concentration would approach the value predicted by Henry's law for the feed, resulting in a greatly reduced air volume compared to the equivalent situation with a single-stage device. In an actual case, a compromise would be made that resulted in minimizing the cost of the overall operation, including air pollution control.

Although the packing voids in air strippers would ordinarily be substantial, there is still the opportunity for clogging by suspended matter in the feed, biological growth, and possibly precipitation of inorganic material as a result of chemical oxidation. Chemical clarification or other methods of solids removal might be necessary.[5] If a significant precipitating oxidation reaction is expected, preoxidation with removal of the precipitate would be necessary.

Air stripping is not was widely used today because it is recognized that it essentially just transfers the pollutants from one medium to another, e.g., water to air. The present emphasis is on either transferring the pollutant to a solid phase from which it can be destroyed or destroying it directly via other treatment means, such as new bio-fermenting processes.

3.4   Steam Stripping

Steam stripping can be used to remove volatile organic materials from aqueous streams at higher concentrations than is practical for air stripping. In addition, the method can remove lower-volatility materials than can reasonably be removed by air stripping. The process is more complicated, however, than air stripping and has higher capital and operating costs.

In its simplest form, steam stripping can be carried out by sparging steam into a vessel of water, either batchwise or as a continuous process, resulting in heating the water to boiling followed by conversion of some of the water to water vapor. The resulting vapor is condensed, producing a more concentrated aqueous solution of the volatile materials. Operating in this manner requires a large amount of energy and produces a condensate that is still quite dilute. The treatment method can be made much more energy efficient by utilizing a countercurrent stripping tower, as described for air stripping. The tower might be packed, but it might also contain multiple trays, typical of distillation columns. Steam stripping is actually a distillation operation. The water to be treated would be fed continuously to the top of the stripping tower typically after heat exchange with the hot treated water leaving the bottom of the tower. Steam would be injected at the bottom of the tower and would flow upward countercurrent to the downward flowing water. Just as in the case of air stripping, the water vapor rising through the tower would contact increasingly concentrated liquid water until it exited at the top of the tower approaching equilibrium with the feed water. The contaminated water vapor would then be condensed. This system not only makes efficient use of energy, but produces the highest possible concentration of contaminants in the condensate. Depending on solubility of the contaminants, it is possible for a separate contaminant phase to be formed that might be reusable in the manufacturing operation. Unlike air stripping, this process produces only a small amount of uncondensable gas to be discharged to the atmosphere. Steam stripping is used only for certain industrial waste applications.

3.5   Membrane Technologies

Natural membranes of various types have been used for centuries to remove solid materials from liquids. With the advent of synthetic polymers, mostly developed since World War II, the usefulness of membranes has been greatly expanded. It is now possible to produce membranes with removal capabilities ranging from what is usually considered ordinary filtration or removal of solid particles through ultrafiltration and microfiltration, which can remove large molecules, to reverse osmosis, which can effectively remove even inorganic ions such as sodium and chloride to very low levels. A very extensive discussion of membrane technologies is given by Kirk and Othmer[6] and Crites and Tchobanoglous.[4] Much of the early development of these membranes was done for the purpose of producing potable water from mineralized sources such as brackish water and seawater. Potable water production continues to provide a significant market. Reverse osmosis, because it has such good rejection, has widespread potential in industrial wastewater treatment and in industrial processing for separating contaminated water into reusable water and a concentrate stream that also may be reusable. An example of the latter is in treatment of plating rinsewaters.

In addition to membrane systems such as reverse osmosis, where water passes through the membrane, leaving contaminants behind, there are systems in which certain contaminants pass through the membrane. One of these that has been in use for many years is electrodialysis and a much newer method is pervaporation. In the latter method, volatile contaminants are vaporized through a membrane for which a vacuum or a stream of a carrier gas is maintained on the downstream side.

Reverse Osmosis

Ultrafiltration, microfiltration, nanofiltration and reverse osmosis have similar system configurations. Reverse osmosis will be described, but much of what is stated applies also to the other operations. Reverse osmosis utilizes membranes that can exclude a large fraction of almost all solutes. The earliest membranes were made of cellulose acetate formulated in such a way that a very thin rejecting layer formed on the surface, with a much more porous layer beneath. Other materials are now available including thin-film composites. During operation, water under pressure is forced through the membrane. This water is called the permeate. To treat at practical flow rates, a large membrane area is needed. This need has been accommodated by development of several high area-to-volume configurations. One of these is the spiral wound module, which contains sandwiches of membrane and spacer material that are wrapped around an inner permeate collection tube and placed in a cylindrical pressure vessel. The permeate-carrying compartment is sealed around the edges. Feed entering one end of the cylinder is forced through the appropriate spaces in the membrane roll. Concentrate not passing through the membrane exits at the opposite end of the pressure vessel. Another high-surface configuration is the hollow fiber module. These modules contain bundles of reverse osmosis membrane in the form of fine hollow fibers. The ends of the fibers are mounted in a material such as epoxy. There are a number of specific configurations. These modules operate by having the water pass from the outside of the fibers into the hollow center, where it flows to one or both ends of the bundle. Two other configurations include tubular modules and plate-and-frame modules. Tubular modules contain much larger-diameter tubes with the reverse osmosis membrane in the interior of the tubes. Plate-and-frame modules contain stacks of circular membranes on supports with spacers between the membranes. The stacks are connected so that feed water flows upward through them and permeate is collected from each support. Neither the tubular nor plate-and-frame configurations have the surface-to-volume ratio that is possible with the other two configurations. The tubular modules have the advantage of being the easiest to clean.

To operate reverse osmosis, the feed water is forced under pressures of up to several hundred pounds per square inch through the modules. The minimum theoretical pressure to force water through the membrane is that which just exceeds the difference in osmotic pressure of the water on the upstream and downstream sides of the membrane. The osmotic pressure increases throughout the system as the contaminants in that stream are concentrated. In actuality, the operating pressure is usually well above the osmotic pressure. The water flows through the system until it emerges as a residual concentrate. The volume of this residual is an important consideration if it contains hazardous materials that interfere with simple methods of disposal. There are limits to which the water can reasonably be concentrated. Usually there is a minimum rate at which the water should flow along the membrane surface to minimize fouling and concentration gradients, which can lead to unexpected precipitation of scaling materials on the membrane surface. As the feed stream is concentrated and reduced in volume by water passing through the membranes, the flow rate will decrease. In practice, a combination of staging of the membrane modules with decreasing membrane area in each stage, and some degree of recycling of the residual flow from a module back to that module to be mixed with new feed, could be used to maintain proper hydraulic conditions. As concentration on the upstream side increases, the concentration in the permeate also increases and may exceed either a discharge requirement or a reuse quality requirement. Finally, however, the concentrated stream reaches the point at which solubility limits are exceeded and precipitates begin to form. High turbulence can minimize precipitation or scaling on the membrane, but cannot totally prevent it. Calcium carbonate is a scale that commonly forms from treatment of natural waters because these waters contain calcium and a pH-dependent mixture of carbonate and bicarbonate alkalinity. To prevent this scale, feed water is acidified to convert the alkalinity to CO_2.

Fouling of the membranes is almost always found during operation of reverse osmosis and is also a problem with other membrane processes. Fouling is sometimes considered the same as scaling, but it differs from scale formation in not being the result of precipitation. The two can be found to occur together. Fouling results primarily from deposition on the membrane of suspended matter from the water. There is also the possibility of slimes forming from biological activity. One way to minimize fouling is to pretreat the feed water for removal of suspended matter. There are enzyme and chemical rinses that have been used to reduce fouling.

Membrane Bioreactor (MBR)

As noted earlier in the chapter, the potential for the MBR is accelerating worldwide. In effect, these systems combine an activated sludge system with a microfiltration or nanofiltration membrane. The concentrate stays in the activated sludge and builds up until it is necessary to remove excess solids. The permeate from these membranes should be free of SS, bacteria, parasites, and most of the biodegradable organics and priority pollutants (toxic organics and inorganics). Thus, the permeate requires only disinfection (UV performance is enhanced by this effluent quality) for a wide spectrum of reuse opportunities, particularly where virus removal is necessary. MBRs are being used at a number of smaller treatment facilities at this time. As yet, these applications are too new to get a complete set of operation and maintenance costs, which are highly dependent on the membrane life to replacement, its cleaning and maintenance requirements, and the frequency of excess residuals removal, which is dependent on the rate of rejection by the membrane. However, several world organizations and a number of U.S. engineers and their clients are optimistic about such a small-footprint system, owing to the absence of need for a secondary clarifier. This MBR system may also have numerous applications for industrial wastewater treatment where SS, BOD, and other pollutants can be removed by this system.

Electrodialysis

This treatment method is used to remove part of the ionic materials from water. It has been used for many years to partially demineralize heavily mineralized groundwater for production of potable water. The equipment consists of one or more stacks of membranes that have the ability to transmit either cations or anions, but not a significant amount of water. In each stack, the cation and anion membranes alternate and are separated by spacers that form a path for water passage. When a direct electric current is imposed on the stack of membranes, cations will tend to move toward the negative electrode and anions to the positive electrode. The ions can only pass through one membrane, however, before being blocked. The result is that the alternating spacer compartments become diluting compartments and concentrating compartments. All of the diluting streams are connected and produce the treated water. All of the concentrating streams are connected and produce a concentrate that must be disposed or possibly reused in the manufacturing operation.

Use of electrodialysis for pollution control is less flexible than reverse osmosis and some other treatment methods because it is not able to reduce the ionic content of the treated water to very low levels that might be required for discharge. Uses for industrial wastewater treatment would most likely occur where useful materials can be recovered from the concentrate stream and where the product water is acceptable for in-plant uses. Electrodialysis is subject to fouling and scale formation and operates most effectively on feed waters with very low concentration of suspended matter.

Pervaporation

As already indicated, this treatment method functions by the passage of volatile materials from an aqueous solution through a membrane to a gas phase consisting of a carrier gas or produced by a vacuum. Without considering the details of what occurs in the membrane, this method can be looked upon as a membrane-assisted evaporation that increases significantly the ratio of volatile materials to water in the gas phase compared to ordinary evaporation without a membrane. The treatment method accomplishes results similar to air stripping. It has the advantage over air stripping when using vacuum operation of capturing the volatiles in a highly concentrated form that may in some situations allow recovery and reuse. This technology is just beginning to reach the stage of commercial use, but could see rapid expansion in its application to process and waste streams.

3.6 Ion Exchange

Ion exchange has been used for many years for the softening of water and for production of deionized water. The technology is also being used for recovery of useful materials from in-plant streams and for treatment of some industrial wastewaters. With increased regulation of metals in wastewaters, increase in use can be expected. Traditionally, ion exchange has been used to remove only inorganic ionic materials by direct exchange with other ions on a solid matrix. Resins are now being produced by ion exchange resin manufacturers that remove some organic materials and selectively remove inorganic materials by more complicated mechanisms than simple ion exchange.

Ion exchange is carried out using granular, usually resinous, materials. The traditional ion exchange resins are cation or acid resins and anion or base resins. Furthermore, there are strong and weak acid resins and strong and weak base resins. The following equations describe two typical ion exchange reactions:

$$R_1H + M^+ \rightarrow R_1M + H^+$$
$$R_2OH + A^- \rightarrow R_2A + OH^-$$

where R_1 = a cation or acid resin
R_2 = an anion or base resin
M^+ = a metal ion
A^- = an anion

Both of these reactions are easily reversible. If the resin in the form R_1M were treated with acid, M^+ would be liberated and the resin would revert to the acid form R_1H. If the base resin were treated with a base, A^- would be liberated and the resin would revert to the base form R_2OH. If water containing the neutral compound MA were passed through a column of R_1H and then a column of R_2OH, the water could be demineralized. R_1H could be a strong acid or a weak acid cation resin. The difference is that the strong acid resin will function over a wide range of pH. Weak acid resins have a stronger affinity for H^+ than the strong acid resins and do not function well at low pH (high H^+ concentration). If there were a large amount of M^+ in solution, the large amount of H^+ potentially entering the water phase might result in only partial M^+ removal. A strong acid resin would be a safer choice. Weak base resins, on the other hand, function well under acid conditions. A weak base resin could be used to complete the demineralization. If the weak base resin were placed before the cation exchange resin, it would not remove A^- significantly from a solution of a neutral salt.

As the ion exchange materials are exposed to more and more water, their ion exchange capacity is finally exhausted. The active sites in this case would have either M or A attached. At this point, the resin columns must be regenerated, the cation exchange resin with an acid such as sulfuric acid and the anion exchange resin with a base such as sodium hydroxide. The primary reason for using the weak acid and weak base resins whenever possible is their greater ease of regeneration. The most effective regeneration, utilizing only a slight excess of regenerant, would be accomplished by passing the regenerant solutions countercurrent to the direction of flow of the water being treated. In many systems, however, the resins are first backwashed before regeneration. Backwashing mixes the resin particles and eliminates the advantage of countercurrent regeneration. In these systems, both treatment and regeneration would most likely be downflow. Backwashing is done to flush from the beds particulate matter that might cause fouling and loss of ion exchange capacity. For most effective operation, the feed to an ion exchange system should have a very low concentration of particulate matter. Both the backwash and the used regenerant streams represent a waste. The volume of this waste would usually be only a small percentage of the volume of water treated. The objective of ion exchange is to concentrate the contaminants in a small volume. For industrial wastes containing hazardous materials, disposal becomes a problem. In these cases, recovery for reuse of the hazardous material has obvious advantages. There is also an obvious advantage to reusing the clean product water.

The above discussion was aimed at giving a general understanding of ion exchange. For industrial waste treatment, deionization would not be a likely objective. Possibly only a cation exchanger would be involved, or, as in the case of one chromium recovery system,[7] a cation exchanger for removing metallic ion impurities followed by an anion exchanger for removal of chromate. In addition, there are some organic removing resins that may use an acid or base for regenerant or may use a solvent.

The widest use for ion exchange has been for water softening in individual homes. It has also been used in some wastewater reuse facilities.

Table 2 summarizes the industrial wastewater treatment technologies included above. Capital and operating costs have not been included because of wide variability likely for different applications. Demineralization technologies can be expected to be somewhat more expensive than the other technologies.

3.7   Other Methods

There are a great many other technologies that could be included. In some cases, the methods are more closely associated with chemical processing than pollution control. In some cases,

Table 2 Industrial Wastewater and Hazardous Waste Treatment Technologies

Technology	Contaminants Removed	Typical Applications
Chemical precipitation and clarification	Particulates BOD associated with particulates Trace metals when appropriate precipitants are used	Treatment prior to discharge to a public sewer Treatment prior to particulate sensitive technologies such as granular activated carbon and membrane processes
Activated carbon adsorption	A very broad range of soluble organic materials	Many industrial wastes containing soluble organic materials
Air stripping	Highly volatile materials	Dilute solutions of volatile materials, including industrial wastewaters, drinking water sources, contaminated groundwater and contaminated soil
Steam stripping	Volatile materials of lower volatility than appropriate for air stripping and at higher concentrations	Dilute to moderately concentrated solutions of a wide range of volatile materials
Reverse osmosis	A high proportion of organic and inorganic contaminants	To produce very-high-quality water
Membrane bioreactor	SS, bacteria, parasites, toxic and biodegradable organics	High-level treatment in small areas
Electrodialysis	Inorganic ions	To remove a significant fraction of minerals from water
Pervaporation	Highly volatile materials	Not yet widely used Uses similar to air stripping
Ion exchange	Primarily inorganic ions	Deionization For recovery of ionic materials from industrial aqueous streams and industrial wastewaters

they have very limited or very specific uses. Some technologies might show promise, but are not far enough along in their stage of development to be considered for full-scale use.

The emphasis in municipal wastewater treatment has traditionally been on biological and physical processes rather than more traditional chemical processes. Many of the unit processes and treatment trains commonly employed in water pollution control facilities are being questioned, particularly as they apply to smaller facilities. Even the traditional wastewater collection systems and concepts are being reevaluated. The idea that all the wastewater should be transferred to a single distant treatment facility for dispersal directly into a receiving stream has numerous shortcomings, particularly if nontraditional values are included in that evaluation. However, for the near future, the continued use of activated sludge treatment is certainly likely in the larger urban areas. Indeed, those facilities that serve more than 100,000 people service more than one-third of the U.S. population and are generally landlocked and have little area to expand. Thus, the foreseeable future will likely see updating of these facilities with enhanced versions of the same processes, but sludge processing will be a fertile ground for new concepts.

Industrial wastewater systems, however, have only to satisfy performance standards, and will thus have significant potential for new ideas and approaches. The private sector provides an opportunity for new ways to meet treatment goals, recover raw materials, and reuse as much of its waste that can be financially justified. Recent emphasis in this area is primarily a result of ever more stringent effluent requirements and the relative costs of recycle and reuse compared to meeting those requirements with no return on the investment.

REFERENCES

1. E. L. Armstrong (ed.), *History of Public Works in the United States 1776–1976,* American Public Works Association, Chicago, IL, 1976, Chapters 12 and 13.
2. Metcalf & Eddy, Inc., revised by G. Tchobanoglous and F. L. Burton, *Wastewater Engineering Treatment, Disposal and Reuse,* 3rd ed., McGraw-Hill, New York, 1991.
3. J. A. Parker and J. L. Darby, "Particle-Associated Coliform in Secondary Effluents: Shielding from Ultraviolet Light Disinfection," *Water Environment Research* **67,** 1065–1075 (1995).
4. R. Crites and G. Tchobanoglous, *Small and Decentralized Wastewater Management Systems,* WCB McGraw-Hill, New York, 1998, Chapter 12.
5. J. V. Boegel, "Air Stripping and Steam Stripping," in *Standard Handbook of Hazardous Waste Treatment and Disposal,* H. M. Freeman (ed.), McGraw-Hill, New York, 1989, p. 6.108.
6. M. Grayson (ed.), *Kirk–Othmer Encyclopedia of Chemical Technology,* 3rd ed., Wiley, New York, 1982.
7. C. J. Brown, "Ion Exchange," in H. M. Freeman (ed.), *Standard Handbook of Hazardous Waste Treatment and Disposal,* McGraw-Hill, New York, 1989, p. 6.66.

INDEX

A

C

D

N

W